A

DICTIONARY OF

TWENTIETH
CENTURY
WORLD
BIOGRAPHY

A
DICTIONARY OF
TWENTIETH
CENTURY
WORLD
BIOGRAPHY

Consultant Editor

ASA BRIGGS

BCA

LONDON · NEW YORK · SYDNEY · TORONTO

This edition published 1992
by BCA by arrangement with
Oxford University Press

Oxford is a trade mark of Oxford University Press

Printed in Great Britain by
The Bath Press Ltd.
Bath, Avon

CN 1697

Foreword

As the twentieth century draws to its close, attempts will be made to identify and interpret its shape and to place it in long-term perspective. It has been a century of great events, many of them surprising, and as it has unfolded history has increasingly become world history. It has also become social history in that much of what has happened has had its origins in social forces working from below.

Historians have responded to change. Yet there have been tendencies in the twentieth century to underplay the role of individual personality in history, to concentrate on collecting statistics, and to search for and, if possible, to identify what have recently been called 'megatrends'. Many writers, including biographers, have also challenged the view that any men—or women—can be 'great'. Schools of literary critics have demanded 'the total absence of biographical consideration'.

Biography is not dead, however, and biographies sell in large numbers. So long as there is curiosity about human beings, including those who made—or appeared to be making—decisions, biography will remain alive even when there are attempts to supersede or to suppress it.

This *Dictionary of Twentieth-Century World Biography* is a work not of investigation or of interpretation but of reference, and, as such, it is conceived on lines appropriate to the century. It includes biographies of individuals from different countries and cultures who have contributed to thought as well as to action. It breaks, therefore, with traditions of 'national biography' and it covers very varied aspects of human achievement. As communication and information have spread and as the world has shrunk, both the scenario and the pace have changed. There has also been increasing interest in the possible shapes of the future. There is a need, therefore, for a new reference system. It must be reliable, however, and this is the main criterion of all the entries in this dictionary, each of which is written in plain, unpretentious prose.

Of course, since the number of entries has to be restricted and the choices are many, the task of selection carries with it an inevitable element of interpretation. Names are bound to be missing for which there can be—and will be—extremely strong claims. There are always people who never figured in *Who's Who*s who figure in historical biographies, and there are many people in *Whos's Who*s who disappear from history altogether. The people who are included in this dictionary are people about whom it is useful at this point late in the twentieth century to be able to have easily accessible biographical information.

How many of them will figure in dictionaries of biography a hundred years hence is a different question. Time changes estimates of people as

well as events. It will be interesting, too, to see how many of the people
who are still children in the late stages of this century will figure prominently in the late twenty-first century. The youngest figures in this
dictionary have already accomplished enough to merit assessment yet their
most significant achievements may well lie far ahead.

The biographical information in this dictionary—about, for example,
national backgrounds, education, and sex as well as age—itself lends to
further compartmentalization. That will only be one element, however, in
the future writing of twentieth-century history. The range of personal
qualities displayed (or sometimes hidden) by the characters in this dictionary
is as wide as it was in any previous period of history, and historians will
have to consider their impact. They will have to consider also the relations,
where they existed, between the different people described in these pages,
including thinkers and 'men of action', for thought and action cannot be
compartmentalized and biographies do not run in parallel. They intersect.
Thomas Carlyle was not alone in describing history as 'the essence of
innumerable biographies'.

Asa Briggs

October 1991

Consultant editor	Asa Briggs
Editors	Alan Isaacs
	Elizabeth Martin
	David Pickering
Contributors	D'Arcy Adrian-Vallance
	Joan Merrick Ashley
	Donald Clarke
	E. S. Dwight
	Rosalind Fergusson
	David Foot
	Derek Gjertsen
	Robert S. Hine
	Jonathan Law
	T. H. Long
	Kirsty Melville
	Stephanie L. Pain
	N. J. Priestnall
	Judith Ravenscroft
	Doreen Sherwood
	Jennifer Speake

Preface

The object of this biographical dictionary is to provide, in a compact and readable form, a brief outline of the lives of the twentieth-century men and women from all parts of the world who have had an impact, either directly or indirectly, on the lives of the rest of us.

In selecting the 1,750 or so subjects of this dictionary we started from the premiss that we would exclude any one who died before the turn of the century or who had not been active for at least part of their life in this century. We were anxious to include those who have made their mark in the fields of sport and popular entertainment, as well as the world's great thinkers, creators, and men and women of action. Sex, colour, creed, or language, were, of course, not in any way criteria for or against entry. In practice, the task of selection became a three-tier operation. At the top of the cake the entrants virtually selected themselves: Chaplin, Churchill, Einstein, Hitler, Picasso, Rachmaninov, and Shaw tower to such heights in achievement or notoriety that their inclusion requires no feat of judgement. Others, a middle tier, are obligatory entrants because of the offices they hold: the Queen, the Pope, the president of the USA. However, the subjects forming the third and largest tier had to be selected individually from the ranks of the famous and infamous. At this level, each field of human endeavour had to be investigated to make the selection. In practice this cannot be done impartially because all of us are to some extent ignorant and prejudiced. The editors can only claim to have been aware of these frailties and to have done their best in spite of them.

In the entries themselves, we have attempted to provide a sketch of the family background and education of the subject, followed by a description of the events that have brought them to fame or infamy, and a summary of their achievements or misdemeanours. Private lives are mentioned if they are relevant, but gossip has been excluded.

If a name is set in small capitals in an entry, it means that the person referred to has an entry in the dictionary. The titles of works of non-English-speaking writers are given in their original language with the date of publication; if an English translation has been published, the title and date of publication of the translated version are also given. Occasionally a translation of a foreign title is provided (in inverted commas) even if the work has not been published in translation.

A.I.
E.A.M.
D.P.
1991

A

Aalto, (Hugo) Alvar (Henrik) (1898–1976) *Finnish architect and designer, whose early neoclassical style changed in the late 1920s to the international modern, in such buildings as the Viipuri library and the MIT hall of residence.*

Born in Kuortane, Aalto studied at the Helsinki Technical University and in 1938 went to the USA, where he taught at MIT and the Cambridge College of Architecture in Massachusetts. After World War II he returned to Finland and carried on an international practice.

His typically Scandinavian buildings include the Viipuri library (1927–35), the Paimio convalescent home (1929–33), the town hall at Säynatsälo (1951), and the Finlandia Concert Hall in Helsinki (1971), his last building. Outside Scandinavia his important buildings include the MIT hall of residence (1947) and the Maison Carré near Paris (1956–58), which he finished with his own bentwood furniture, which he had first designed in 1932.

Abbas, Ferhat (1899–1989) *Algerian nationalist leader who became an active member of the National Liberation Front (FLN), although he sought to achieve Algerian independence by parliamentary means.*

Born in Taher, the son of a Muslim civil servant, Abbas attended French schools in Philippeville and Constantine before studying at the University of Algiers. He briefly served in the French army before working as a chemist in Sétif. Abbas first became actively involved in politics in 1938, when he helped to organize the Union Populaire Algérienne (Algerian People's Union), an organization advocating equal rights for the Algerians and French under French colonial rule together with the maintenance of the Algerian language and culture. In 1943 he published the 'Manifesto of the Algerian People', calling for an autonomous Algerian state; in the following year he founded the Amis du Manifeste et de la Liberté (Friends of the Manifesto and Liberty) and in 1946 established and led the Union Democratique du Manifeste Algérien (Democratic Union of the Algerian Manifesto). For the next nine years he attempted to cooperate

with the French in setting up an Algerian state, being elected to both the French constituent assembly and the Algerian assembly. In 1956, despairing of any progress, he joined the Front de Libération Nationale (FLN) in the hope of securing Algerian independence by revolution. He was elected president of the provisional government of the Algerian republic (based in Tunisia) in 1958, and in 1962, when Algeria finally gained independence after a bitter struggle with France, he became president of the constituent assembly. He was expelled from the FLN in 1963 for opposing its proposed constitution, drawn up without the participation of the constituent assembly, and was held in detention by the newly elected president, BEN BELLA.

Abbas wrote several books, including *Le Jeune Algérien* (1931), *La Nuit coloniale* (1963), and *Autopsie d'une guerre* (1980).

Abdul Rahman, Tunku (1903–90) *First prime minister of Malaya (1957–63) and of Malaysia (1963–70). A skilled negotiator, he was one of the architects of modern Malaysia.*

Born in Alor Star, the son of a former sultan of Kedah, Abdul Rahman was educated at schools in Malaya and Thailand before attending Cambridge University, where he graduated in 1925. He studied law at the Inner Temple in London but failed the bar exams. On his return to Malaya in 1931 he became a district officer in the civil service. After World War II he returned to London and qualified as a barrister (1949).

Abdul Rahman began his political career in 1945, when he became one of the founders of the United Malays National Organization (UMNO); in 1951 he was elected president of the organization. In this role he formed a coalition with the Malayan-Chinese Association (1952) and the Malayan Indian Congress (1955), which became the Alliance Party. As leader of the Alliance Party he was involved in negotiations for the independence of Malaya (1957) and became the first Malayan prime minister. In order to resist the claims of SUKARNO of Indonesia, Abdul Rahman also founded the Association of Southeast Asia (1961), which later became the Association of Southeast Asian Nations (ASEAN). In 1963 he played a major role in the creation of the federation of Malaysia (comprising Malaya, Singapore, North Borneo, and Sarawak), becoming its prime minister with a resounding victory at the polls. Despite Singapore's withdrawal from the federation in 1965 and domestic agitation over his policy concerning the Indian and Chinese minorities, he remained in power until 1970, when he retired. In 1988 he came out of retirement to make public his opposition to the government's repressive policy towards its opponents.

Abdul Rahman was also a writer of some note and had several of his works published. They include the play *Mahsuri* (1941; filmed 1958), and *Looking Back* (1977), an autobiography and history of Malaysia.

Abercrombie, Sir (Leslie) Patrick (1879–1957) *British town planner and architect, who is known for his postwar* The Greater London Plan. *He was knighted in 1945.*

Born in Ashton-upon-Mersey, he was educated at Liverpool University, where he became professor of civic design (1915–35). He then moved to University College, London, where he was professor of town planning from 1935 to 1946. He won the 1913 competition for replanning Dublin and wrote the standard prewar textbook *Town and Country Planning* (1933). His first London plan, *The County of London Plan* (with J. H. Forshaw; 1943), was extended in 1944, with the help of a team of specialists, to *The Greater London Plan*, which was influential in planning the transport, population distribution, industry, green belt, and other amenities of Greater London. He prepared plans for other UK towns and regions including Edinburgh, Plymouth, Hull, Bath, Bristol, Sheffield, Bournemouth, and the West Midlands.

Achebe, Chinua (1930–) *Nigerian novelist and poet of Ibo descent. He was awarded the Nobel Prize for Literature in 1989.*

Born in Ogidi, in eastern Nigeria, the son of a mission teacher, Achebe was educated at the University College of Ibadan. In 1954 he joined the Nigerian Broadcasting Corporation and in 1961 became director of external broadcasting. During this period he wrote the four novels that won him international acclaim: *Things Fall Apart* (1958), *No Longer At Ease* (1960), *Arrow of God* (1964), and *A Man of the People* (1966). All four focus on the moral and practical dilemmas of Africans caught up in the disorientating clash between western values and traditional lifestyles, and have been translated into many languages.

Leaving broadcasting, Achebe taught at the University of Nigeria at Nsukka (1967–72), and followed this by a spell in the USA before returning to Nsukka as professor of English. He published a book of poems, *Beware Soul Brother*, in 1972 and won the Commonwealth Poetry Prize the same year. He has also published short stories on traditional African themes. In 1979 he was awarded the Order of the Federal Republic. A further novel, *Anthills of the Savannah*, was published in 1987.

Acheson, Dean Gooderham (1893–1971) *US lawyer and statesman who became secretary of state (1949–53) under President Harry S. Truman.*

Acheson was born into a wealthy patrician family in Middletown, Connecticut, and studied at Yale and at Harvard law school. After a spell with the US navy during World War I, he was selected from his class of law graduates to serve as assistant to Louis Brandeis (1856–1941), judge of the Supreme Court. He entered politics in 1933, as undersecretary of the Treasury in ROOSEVELT's administration, and in 1941 became an assistant secretary in the State Department. In 1945, the year in which Truman became president of the USA, he rose to the office of undersecretary of state.

In 1947 Acheson played an important role in the development of the Truman Doctrine, which pledged support to Greece and Turkey, and of the MARSHALL Plan, which offered economic aid to war-devastated Europe – two moves that marked a significant turning point in US foreign policy. During his first year as secretary of state to President Truman, Acheson was instrumental in the formation of the North Atlantic Treaty Organization (NATO). In his dealings with the Far East, however, he met with mounting resistance from Republican members of Congress, and his staunch defence of the State Department against McCARTHY's accusations of espionage and subversion led to widespread denunciation. The situation was exacerbated by the apparent exclusion of Korea from US protection in Asia and the subsequent outbreak of the Korean War. A resolution was passed in December 1950 calling for Acheson's resignation; with President Truman's support he remained in office but found himself obliged to make concessions to Congress in order to gain support for his foreign policies.

Shortly after the election of President EISENHOWER in 1952, Acheson left the State Department and returned to his law practice. He retained his interest in foreign affairs, however, and served as adviser to subsequent Democratic presidents, notably John F. KENNEDY. In 1969 he published a book of memoirs of his years with the State Department, *Present at the Creation*, which won the 1970 Pulitzer Prize in history.

Adamov, Arthur (1908–70) *Russian-born French dramatist, one of the most prominent exponents of the Theatre of the Absurd.*

Adamov was born into a wealthy Armenian family in Kislovodsk; at the age of four he moved with his family to Germany. Having completed his education in Paris, he settled there in 1924 and became involved with surrealist groups, editing their journal *Discontinuité* and writing poetry. In 1938 he had a nervous breakdown; the neuroses that had plagued him since childhood and that were to form the bizarre inspiration for many of his plays are revealed in his confessional work *L'Aveu* (1946).

Adamov began writing for the theatre in 1947. He sought to express the loneliness and helplessness of man and the futility of any quest for the meaning of life. In *La Parodie*, first performed in the early 1950s, the central characters bombard each other with questions about time against the background of a clock with no hands. *L'Invasion* (1950), *La grande et la petite manoeuvre* (1950), *Tous contre tous* (1953), and *Le Professeur Taranne* (1953) depict in nightmarish images the cruelty of social conventions and pressures and were influenced by Antonin ARTAUD's Theatre of Cruelty.

In the mid-1950s Adamov turned to a more political style of drama, beginning with his best-known play, *Le Ping-Pong* (1955). The central image of the play, a pinball machine in an amusement arcade, is a symbol of the capitalist system to which men willingly submit in an endless futile game of chance. After *Paolo Paoli* (1957) Adamov's plays became increasingly radical: *Le Printemps 71* (1961), about the Paris Commune, *La Politique des restes* (1963), and *Off Limits* (1969) are laced with Marxist propaganda. Adamov committed suicide in 1970.

Adams, Ansel Easton (1902–84) *US photographer, noted for memorable images of his native landscape and his valuable contributions to the development of photographic techniques.*

The son of a San Francisco businessman, Adams developed an abiding love of the natural world as a result of trips to the Sierra Nevada and the Yosemite National Park. He began taking photographs at the age of fourteen and joined the Sierra Club, an organization devoted to conservation and education. Adams's first ambition was to be a concert pianist but in 1927 he published his first portfolio of photographs, *Parmelian Prints of the High Sierras*. In 1930 he met the photographer Paul Strand (1890–1976) and was struck by the sharpness and detail of Strand's work, which contrasted with the soft-focus impressionistic style then fashionable. In 1932 Adams and several fellow photographers, including Edward Weston (1866–1958), formed an informal group known as 'f. 64'. Their work was characterized by the use of large-format cameras and small apertures to produce sharp images with maximum depth of field. In the 1930s, Adams began teaching photographic techniques: the plates in his book *Making A Photograph* (1935) exemplify his superb technical skill. The celebrated photographer Alfred STIEGLITZ arranged Adams's first one-man show in New York in 1936. Adams was instrumental in establishing a department of photography at New York's Museum of Modern Art in 1940, and in 1946 he founded one of the first academic photography departments – at the California School of Fine Arts.

In 1937, Adams moved his home to the Yosemite Valley, to the landscape that is the subject of much of his work. His images testify to his passion for conserving the wilderness and many of his collections have been published by the Sierra Club, including *This is the American Earth* (1960), with text by Nancy Newhall. Although working mainly in the American West, Adams also photographed in Alaska and Hawaii and took still lifes, architectural studies, and portraits.

Adams, Richard George (1920–) *British novelist.*

Adams was educated at Bradford College, Berkshire, and Worcester College, Oxford, from which he graduated with a degree in modern history. Towards the end of a career in the civil service (1948–74) Adams wrote *Watership Down* (1972), which won him the Carnegie Medal and became an international best-seller; it was filmed in 1978. Ostensibly a children's book about a rabbit colony, *Watership Down* contains epic and narrative elements that appeal also to an adult readership. It was followed by *Shardik* (1974) and *The Plague Dogs* (1977; filmed 1982), the latter a disturbing indictment of the use of animals in research laboratories. His other books include the novels *The Girl in a Swing* (1980), *Maia*

(1984), and *Traveller* (1989); he also wrote a collection of short stories, *The Iron Wolf* (1980).

Adenauer, Konrad (1876–1967) *German statesman and first chancellor of the Federal Republic of Germany (1949–63).*

Born in Cologne, the son of a clerk, Adenauer was educated at Freiburg, Munich, and Bonn universities, before practising law in Cologne. Entering municipal politics in 1906, he was elected deputy mayor of Cologne in 1909 and lord mayor in 1917, the same year in which he became a Centre Party member of the Provincial Diet and Prussian State Council (of which he was chairman from 1920 to 1933). In 1933 he was dismissed from all these offices by GOERING because of his opposition to the Nazi regime, for which he was twice imprisoned (in 1934 and 1944).

In 1945, under Allied occupation, Adenauer was reinstated as lord mayor of Cologne. Co-founding the Christian Democratic Union the same year, he became chairman in 1946 and was elected as the first chancellor of the Federal Republic in 1949, being re-elected in 1953, 1957, and 1961. He retired in 1963 at the age of eighty-seven.

Known affectionately as 'der Alte' ('the Old Fellow') by the German public, Adenauer gained respect and popularity as a result of the political and economic transformation West Germany achieved under his leadership following World War II. He was an advocate of strengthening political and economic ties within the western bloc through NATO and the Common Market, but was criticized for his restraint during the construction of the Berlin Wall in 1961. He was made an honorary fellow of the Weizman Institute of Science, Israel, in 1966 and granted the freedom of the cities of Bonn and Cologne on his seventy-fifth birthday.

Adler, Alfred (1870–1937) *Austrian psychiatrist who founded a school of thought based on the psychology of the individual and introduced the concept of the inferiority feeling (later called inferiority complex).*

Adler qualified in medicine from the University of Vienna Medical School in 1895 and practised ophthalmology before taking up psychiatry. Initially he was a prominent member of FREUD's circle of psychoanalysts, but differences in their ideas became evident early in the association. In 1907 Adler first put forward the idea that people attempt to compensate psychologically for physical disabilities and the feelings of inferiority that they produce; an inability to compensate adequately causes neurosis and mental illness. He disagreed with Freud that mental illness was caused by sexual conflicts in infancy and confined the role of sexuality to a small part in the greater striving to overcome feelings of inferiority.

By 1911, his break with Freud was complete and he and his followers formed their own school to develop the ideas of individual psychology, maintaining that an individual's main motive is to seek perfection in order to achieve superiority and overcome feelings of inadequacy. Adler's methods of psychotherapy were supportive and aimed to encourage good human relationships and greater social interest, thus helping patients who were emotionally disabled by their inferiority feelings to become mature and socially useful. In 1921 Adler founded the first child guidance clinic in Vienna and went on to establish many more.

Adler, Larry (Lawrence Adler; 1914–   ) *US harmonica player.*

Born in Baltimore, Adler won a talent competition at the age of thirteen by playing a Beethoven minuet on what had hitherto been regarded as a musical toy. He thereafter elevated the instrument to concert-hall status by means of his astonishing virtuosity and musicality. VAUGHAN WILLIAMS, Malcolm ARNOLD, Darius MILHAUD, and Paul HINDEMITH have all written works for him and his reputation is worldwide. Adler himself has also written film scores, including *Genevieve* (1954), for which he played in the sound track.

With other show-business personalities, he was blacklisted by the American entertainment industry after 1949 for his alleged left-wing sympathies. Since then he has lived for many years in England, where he has created a second career for himself as a journalist and writer.

Adorno, Theodor Wiesengrund (Theodor Wiesengrund; 1903–69) *German philosopher and sociologist, a leading member of the Frankfurt school.*

Born in Frankfurt, the son of a Jewish wine merchant, he took his mother's maiden name, Adorno, during World War I. He was educated at the University of Frankfurt and went on to study musical composition in Vienna under Alban BERG before returning to Frankfurt to teach. Banned from teaching by the Nazis in 1933, Adorno spent three unhappy years in Oxford before moving in 1938 to the USA. He first worked in New York but in 1941, with his Frankfurt colleagues Horkheimer and MARCUSE, moved to California. Although Adorno produced much of his best work during his stay in the USA, he was keen to return to Europe, and when the University of Frankfurt announced plans to reopen the Institute for Social Research in 1949 Adorno willingly agreed to serve with Horkheimer as its joint director.

With Horkheimer he had earlier written *Die Dialektik der Aufklärung* (1947; translated as *Dialectic of the Enlightenment*, 1972), in which they argued that the rationalism of the Enlightenment had led not only to the domination of nature but also of man. The crucial weapon of this enslavement had been the concept of reason itself. There thus seemed to Adorno to be an inevitable oppression in any form of philosophical theorizing. Marxism could not be exempted from this general complaint, nor could science. The only remedy, Adorno proposed in his *Negative Dialektik* (1966; translated as *Negative Dialectics*, 1972), was the systematic and conscious rejection of all theories. Adorno's views were seized upon and adopted by the student revolutionaries of the 1960s.

Adorno also wrote extensively on problems of aesthetics and music.

Adrian, Edgar Douglas, 1st Baron (1889–1977) *British neurophysiologist, whose work on the electrical properties of the nervous system earned him the Nobel Prize for Physiology or Medicine in 1932. He received the OM in 1942 and the title Baron Adrian of Cambridge in 1955.*
Born in London, Adrian graduated in medicine from Trinity College, Cambridge, in 1915. After serving in the Royal Army Medical Corps during World War I, he returned to Cambridge University, becoming professor of physiology (1937), master of Trinity College (1951–65), and ultimately chancellor of the University (1968–75). His early research centred on measuring and recording the electrical impulses in the nervous system. Using very fine electrodes and amplification equipment, he managed to record impulses from single nerve fibres and showed how the frequency of electrical discharges was the basic method of signalling in both sensory and motor nerve cells. In 1934 Adrian turned his attention to the electrical activity of the brain, recording and analysing the various wave patterns and contributing greatly to the newly founded technique of electroencephalography.

Adrian was president of the Royal Society (1950–55). His books include *The Basis of Sensation* (1927), *The Mechanism of Nervous*

Action (1932), and *The Physical Background of Perception* (1947).

Aga Khan III (Sultan Sir Mohammed Shah; 1877–1957) *Imam (leader) of the Nizari branch of the Ismaili sect from 1885, who was a prominent leader of the Muslim communities in India.*
Of Persian descent, he was born in Karachi (then in India) and was educated in three traditions – the Islamic, oriental, and western. In 1906 the Aga Khan led a Muslim deputation to the British Viceroy of India, Lord Minto (1845–1914), a meeting that influenced the provision in the subsequent Morley–Minto reforms (1909) for separate Muslim electorates in India. The Aga Khan was then president of the All-India Muslim League for the first three years of its existence (1906–09). Always committed to friendship with Britain (he became a member of the Privy Council in 1934), in World War I he urged his followers to give their support to the Allies. He later played an important part in the Round Table conferences (1930–32) on the future Indian constitution. He also represented India at the World Disarmament Conference at Geneva (1932) and at the League of Nations (1932–37), becoming president of the League in 1937. During World War II, owing mainly to ill-health, he withdrew from politics. He was a breeder of thoroughbred horses, and his stables produced five Derby winners. He was married four times.

Aiken, Howard Hathaway (1900–73) *US mathematician and pioneer of the modern computer.*
Born in Hoboken, New Jersey, Aiken was educated at the universities of Harvard, Wisconsin, and Chicago. After some time in industry with the Madison Gas and Engineering Company and with Westinghouse, Aiken returned in 1939 to Harvard, where he became professor of applied mathematics and, from 1946 until his retirement in 1961, director of the computation laboratory. In the late 1930s Aiken worked on the design of a fully automatic computer. It was a mechanical device depending for its operation on the use of punched cards. Completed in 1944, with the help of IBM, and known as the ASCC (Automatic Sequence Controlled Calculator), or the Harvard Mark I, it was the world's first automatic digital computer. Its lifespan was extremely short, however. As a mechanical device with moving parts it was too much a machine of the past, despite its speed and power. In the following year, Aiken's ASCC was made obsolete with the appearance of

ENIAC (Electronic Numerical Integrator and Calculator), the world's first electronic computer.

Alain-Fournier (Henri-Alban Fournier; 1886–1914) *French novelist.*
Born in the rural village of La Chapelle-d'Angillon, Cher, where his father was a schoolteacher, Alain-Fournier was educated at the local school and subsequently in Paris. After his military service (1907–09) he became a journalist, writing literary columns. He was killed in action in 1914, at the first Battle of the Marne.

Alain-Fournier completed just one novel in his short lifetime: *Le Grand Meaulnes* (1913; translated as *The Lost Domain*, 1959). Based in part on the author's own experience, it centres on the young hero's search for a beautiful girl he met by chance at a party in a dilapidated country house. The pervading atmosphere is one of nostalgia, powerfully evoked in Alain-Fournier's prose, as Meaulnes yearns for his lost world of enchantment. The setting of the novel, the familiar countryside of Alain-Fournier's childhood, is also realistically and vividly depicted.

At his untimely death in 1914 Alain-Fournier left manuscripts of poetry and short stories, collected in *Miracles* (1924); a second unfinished novel, *Colombe Blanchet*; and letters to his close friend and future brother-in-law, published posthumously as *Correspondance avec Jacques Rivière* (1948).

Albee, Edward Franklin (1928–) *US dramatist whose success in the commercial theatre in the 1960s was followed by more experimental, less accessible, works.*
Albee was born in Washington, DC, and educated at Columbia University. His first play, *Zoo Story*, was staged in Berlin in 1959 with Samuel BECKETT's *Krapp's Last Tape*, an association that led Albee to be mistakenly classed as a Theatre of the Absurd writer. *Who's Afraid of Virginia Woolf?* (1962; filmed 1966), probably his most popular play, clarified the playwright's position as social critic, whose main concern was to emphasize the importance of human relationships in the face of destructive materialism. Subsequent plays, such as *Tiny Alice* (1965) and *A Delicate Balance* (1966), demonstrated his increasing preoccupation with abstract issues and are generally considered less successful. He has also dramatized works of fiction, including Carson McCULLERS's *Ballad of the Sad Café* (1963) and Giles Cooper's *Everything in the Garden* (1968). The obscurity of *Box, Quota-tions from Chairman Mao Tse Tung* (both 1968), *Seascape* (1975), and *Marriage Play* (1986) reflect the playwright's view of life in which reality has lost all meaning. Despite various criticisms, however, and the inaccessibility of his later highly experimental work, Albee's mastery of language and dramatic technique has never been disputed; he is regarded as a major dramatist in the US theatre.

Albers, Joseph (1888–1976) *German-born US painter, designer, and influential teacher of art.*
Before 1920 Albers divided his time between teaching and art studies. He then entered the newly created Bauhaus school, which aimed at the union of all the arts with modern architecture and with industry. When Albers began to teach there three years later, it had become the most important school of design in Germany. At the Bauhaus, Albers rejected the emotional self-expression and representational style of his early work in favour of constructivist art built up by intellectual calculation and the use of simple geometric forms. The glass pictures and windows that he created represented careful investigation into the relationships of line, colour, and shape. He also designed utility objects and furniture, including the first laminated chair for mass production.

When the Nazis closed the Bauhaus in 1933, Albers moved to the USA, where he spread Bauhaus ideas through his teaching at Black Mountain College. From 1950 to 1958 he was chairman of the department of architecture and design at Yale University. Here he began the long series of paintings and lithographs for which he is best known: *Homage to the Square*. Of all geometric forms Albers preferred the square for its non-natural man-made quality. The works in this series consist of superimposed squares of colour and reflect his preoccupation with the interaction of colours. This work and his creation of visual ambiguities and illusions by the juxtaposition of colours anticipated op art.

Alberti, Rafael (1902–) *Spanish poet.*
Born near Cádiz into a formerly well-to-do family, Alberti was educated at a local Jesuit college. His position as a charity student left a bitterness that may have contributed to his later allegiance to Marxism. Although his first interest was painting, in 1923 he started writing verse and his first volume, *Marinero en tierra* (1925), was awarded the National Prize for Literature in 1924 while it was still in manuscript. This was followed by two other volumes (*La amante*, 1926; *Cal y canto*, 1927),

written in traditional Spanish forms of ballad, tercet, etc. In these early books the subject is mainly the idyllic world of his youth; the memory and recovery of an ideal unspoiled world was indeed to remain the dominant theme of much of his more mature work, even in its more political and committed phase. His fifth volume, usually considered his greatest, was *Sobre los ángeles* (1929; translated as *Concerning the Angels*, 1967). The result of an intense personal crisis that affected him both emotionally and physically, the book breaks with the traditional forms of his previous poems and uses a longer flexible line in a frequently surrealistic rendering of the 'angels' (representing internal states, such as jealousy).

In 1931 Alberti became involved in anti-monarchical activities and two years later proclaimed his commitment to revolution and the Communist Party; his subsequent work in the 1930s (for example *Consignas*, 1933) reflects his political engagement. His plays written at this time, such as *Fermín Galán* (1931), were less successful than his poetry. With the overthrow of the Republic in 1937 he went into exile, living mainly in Argentina (until 1963) and then in Rome, returning to Spain in 1977. He campaigned for a seat in the Cortes – reading his poetry rather than making political speeches – and was elected as Communist representative for Cádiz. After a few months, however, he resigned to devote himself entirely to his work. During his long exile Alberti confirmed his position as Spain's most important politically committed poet and produced a substantial body of work that gradually won international recognition. He was awarded the Lenin Peace Prize in 1967. Among his many volumes are *Retornos de lo vivo lejano* (1952), the autobiographical *La arboleda perdida* (2 vols, 1959; translated as *The Lost Grove*, 1978), poems on painting entitled *A la pintura* (illustrated edition, 1968), and *Alberti tal cual* (1978), poems written and recited during his political campaign. *Selected Poems* (1966) was edited and translated by Ben Belitt.

Alcock, Sir John William (1892–1919) *British aviator who, in 1919 with Arthur Brown (1886–1948), achieved the first nonstop transatlantic flight. Both men received knighthoods in recognition of their success.*

Alcock was born in Manchester, the son of a horse dealer, and worked at the Empress motor works in Manchester as an apprentice before going to Brooklands airfield as an aircraft mechanic. He received his aviator's licence in November 1912 and worked for the Sunbeam company as a test pilot until the outbreak of World War I. He served as an instructor in the Royal Naval Air Service before flying on bombing raids in the eastern Mediterranean. In September 1917 he was forced to ditch in the sea while heading for a raid on Constantinople (now Istanbul) and Alcock and his crew were captured and imprisoned for the duration of the war. He left the RAF in March 1919, having been awarded the Distinguished Service Cross for gallantry. The prize of £10,000 offered by the *Daily Mail* before the war for the first nonstop transatlantic flight still stood. Vickers provided Alcock with a modified twin-engined Vimy biplane for his attempt. With Arthur Brown, a fellow aviator who acted as navigator on the flight, he took off from Newfoundland at 1.58 pm local time (4.13 pm GMT) on 14 June 1919. They encountered bad weather conditions, lost radio contact, and, abandoning their original intention of flying to London, landed in a bog at Clifden, Galway, at 8.25 am GMT after a flight lasting 16 hrs 28 mins. Their success was recognized throughout the world as heralding the era of long-distance air travel. But Alcock was not to see it. On 18 December 1919, in spite of bad weather, he set out from Brooklands to deliver a Vickers Viking amphibious plane to Paris. He crash-landed near Rouen and was fatally injured.

Aldington, Richard (Edward Godfree Aldington; 1892–1962) *British writer, the author or editor of over two hundred books in different genres.*

Forced to leave University College, London, for financial reasons, Aldington determined upon a literary career. He quickly established his reputation as a poet, becoming a prominent figure in the literary life of London. Through Ezra POUND he met the US poet Hilda Doolittle (H.D.; 1886–1961), whom he married in 1913. These three originated the imagist movement in poetry, favouring new rhythms, common speech, clear imagery, and freedom of choice of subject-matter; Aldington's first collection, *Images 1910–1915*, was published in 1915.

Aldington suffered the effects of gas and shell shock in World War I. After the war, now separated from H.D., he lived for some time in a cottage lent to him by D. H. LAWRENCE. Nonetheless he continued active as poet, critic, and translator, and became assistant editor of T. S. ELIOT's journal *The Criterion* in 1921.

Increasingly alienated from English life, Aldington went abroad in 1928 and never again lived for long in his native country. His powerful anti-war novel, *Death of a Hero* (1929), was an international best-seller, being particularly acclaimed in the USSR. It was followed by *The Colonel's Daughter* (1931), which satirized English village life, and *All Men are Enemies* (1933).

Between 1935 and 1947 Aldington lived mainly in the USA. Besides doing the writer's obligatory stint in Hollywood (1942–46), he published his autobiography (*Life for Life's Sake*, 1941), a well-received biography of Wellington, and numerous editions and anthologies. In 1947 he returned permanently to France. His most notable – and often violently controversial – books in his last years were biographies: of D. H. Lawrence (1950), Norman Douglas (1954), T. E. LAWRENCE (1955), and Frédéric Mistral (1956).

Aleksandrov, Pavel Sergeevich (1896–1982) *Russian mathematician, who was a leading member of the Moscow school of topology.*

Born in Bodorodska, Aleksandrov graduated from Moscow University in 1917. He joined the faculty in 1921 and in 1929 was appointed to the chair of mathematics. He planned, in collaboration with the Swiss mathematician H. Hopf, to produce a definitive multi-volume topological treatise. Only the first volume, *Topologie I* (1935), was published. It was nonetheless extremely influential and did much to encourage the development of the Moscow school of topology. Aleksandrov's own work, initially in set-theoretic topology, later moved into the field of algebraic topology.

Alexander, Samuel (1859–1938) *Australian-born British philosopher, best known for his* Space, Time and Deity *(1920).*

The son of a saddler, Alexander was educated in Melbourne and, after his arrival in Britain in 1877, at Balliol College, Oxford. In 1882 he was appointed fellow of Lincoln College and thereby became the first professing Jew to hold a fellowship of any Oxford or Cambridge college. In 1893 Alexander accepted the chair of philosophy at Owens College (later to become Manchester University), where he remained until his retirement in 1924.

As a metaphysician in the grand manner, Alexander published a full account of his system in the two large volumes of his *Space, Time and Deity*. His aim, he declared, was to describe and identify 'in concrete experience...the ultimates which science has left over'. He objected to a relational view of space and time, arguing that the terms of any relation must themselves be spatio-temporal. Instead, he attempted to clarify his strange claim that things are themselves a specification of space-time. Within this framework Alexander adopted a realist theory of knowledge while accepting God as an ideal rather than an actual being. Although much admired in his day, Alexander's work is now little read.

Alexander of Tunis, Harold Rupert Leofric George, 1st Earl (1891–1969) *British general and one of the leading military figures of World War II, who commanded Allied forces in the North Africa campaign and the invasion of Italy. He was created a viscount in 1946, an earl in 1952, and awarded the OM in 1959.*

The third son of the Earl of Caledon, Alexander attended the Royal Military College, Sandhurst, and in 1911 was commissioned in the Irish Guards. He was twice wounded while leading his battalion in France during World War I and ended the war as a brigadier-general. Between the wars he served in eastern Europe, Turkey, and India and in 1937 became the youngest general in the British army.

At the outbreak of World War II, Alexander commanded the 1st Division on the western front and fought a rearguard action at Dunkirk to enable the evacuation of the British expeditionary force. After home defence duties in command of the 1st Corps, he was posted to Burma in February 1942, where he managed to extricate British troops northwards from Rangoon into Assam in the face of superior Japanese ground and air forces. The following August, CHURCHILL appointed Alexander commander in chief in the Middle East, replacing Sir Claude Auchinleck (1884–1981). Alexander's army commander was General Bernard MONTGOMERY, who soon achieved first significant victories against Erwin ROMMEL's Afrika Korps. Following the Anglo–US invasion of Algeria in November 1942, Alexander was made deputy to EISENHOWER in coordinating the entire North African campaign, which culminated in the Battle of Tunis on 13 May 1943 and the capitulation of the Axis forces.

Just two months later, US and British forces invaded Sicily. Alexander's objectives were now to knock Italy out of the war and to engage the maximum number of German divisions on the Italian front while preparations for the D-Day landings went ahead. With the surrender of Italy on 3 September 1943, Allied forces invaded the Italian mainland. In October, Alexander was appointed supreme Allied commander on the Mediterranean front. In spite of difficult terrain and a series of strong German defensive positions, Alexander's forces slowly drove the enemy northwards, capturing Rome on 4 June 1944; to mark this, Alexander was promoted to field-marshal. By the end of April 1945, all Italy had been liberated.

After the war, Alexander served as governor-general of Canada (1946–52) and then spent an unhappy period as Churchill's defence minister. He resigned in 1954.

Alexandra (1872–1918) *Consort of NICHOLAS II, the last tsar of Russia. Her influence played a fatal part in the overthrow of the Russian monarchy.*

Alexandra, a grand-daughter of the British Queen Victoria, was the daughter of Louis XIV, Duke of Hesse-Darmstadt in Germany. Her marriage to Nicholas was arranged in 1894. Her dominance over her husband made her intensely disliked at court, and she became almost fanatically involved in the Orthodox religion. After her son was found to be a victim of haemophilia, she turned for help to a 'holy man', Grigori Yefimovich RASPUTIN. His influence on her and, after the tsar's departure for the Russian front in World War I, on the government was notoriously corrupt and Alexandra was, erroneously, believed to be a German agent. Nicholas returned too late to salvage the situation and, after the October Revolution in 1917, the royal family was arrested and shot at Ekaterinburg.

Ali, Muhammad (Cassius Marcellus Clay; 1942–) *US boxer and world heavyweight champion (1964–67; 1974–78; 1978–79). He is generally accepted as the finest heavyweight boxer of all time.*

Shortly after winning an Olympic gold medal in the light-heavyweight class at Rome in 1960, Ali turned professional. He won his first professional fight in his home town of Louisville, Kentucky, and of his next seventeen victories all but three were gained on a knock-out. In a famous nontitle fight at Wembley, England, he was knocked down by Henry Cooper but still won in the fifth round.

He won the world heavyweight title in 1964 by beating Sonny Liston and successfully defended it until 1967, when he had his licence to box withdrawn because he refused, as his pledge to the Black Muslim faith, to serve in the US armed forces. This kept him out of boxing for more than three years. He suffered his first defeat in 1971, to Joe FRAZIER on points, but in 1974 regained the world title by knocking out George Foreman. In 1978, after

losing the title to Leon Spinks, he became the only boxer to become world champion three times when he defeated Spinks in a return match later that year.

Ali announced his retirement in 1981 after two unsuccessful bids for a fourth world title, against Larry Holmes and Trevor Berbick. He was always a colourful and controversial character, in and out of the ring. His habit of nominating the round in which he intended to beat his opponent added to the appeal of this innate showman. The blows he took during his career did permanent damage to his brain and after his retirement he was confirmed to be suffering from Parkinson's disease.

Allen, Woody (Allen Stewart Konigsberg; 1935–) *US actor, director, and writer, who has created the image of the bespectacled and neurotic misfit in a sophisticated world, a portrayal much enhanced by the exploitation of his own unimpressive physical stature.*

Born in Brooklyn, New York, Allen dropped out of college at eighteen and set himself up as a gag writer and comedy writer, contributing to magazines and television shows. He switched to being a stand-up comic in the early 1960s and, after a period performing his own material in nightclubs, *What's New Pussycat?* (1965) gave him his break into films, as screenwriter and actor. This was followed by *Casino Royale* (1967). His two Broadway hits, *Don't Drink the Water* and *Play It Again Sam*, were made into films (1969 and 1972 respectively), the latter being particularly memorable for the portrayal of the classic screen hero Humphrey BOGART, humorously contrasted with the unimposing self-deprecating anti-hero played by Allen. *Everything You Always Wanted to Know About Sex But Were Afraid to Ask* (1972) followed.

Diane Keaton (1946–), who played opposite him in *Play It Again Sam*, starred in several of his other films, including *Sleeper* (1973), *Love and Death* (1975), and the Oscar-winning *Annie Hall* (1977), for which he received an award as director and co-screenwriter. His first break from comedy came with *Interiors* (1978), for which he was nominated for an Academy Award as best director. More recent films include *Manhattan* (1979), *Zelig* (1983), *Broadway Danny Rose* (1984), the Oscar-winning *Hannah and Her Sisters* (1986), and *Crimes and Misdemeanours* (1990).

Allenby, Edmund Henry Hynman, 1st Viscount (1861–1936) *British general who led the Allied expeditionary force that overwhelmed Turkish and German forces in Pales-*

tine during the closing stages of World War I. He received a knighthood in 1915 and was created Viscount Allenby of Megiddo in 1919.

Allenby attended the Royal Military College, Sandhurst, and in 1882 was commissioned into the 6th Inniskilling Dragoons. He served in Bechuanaland (1884–85), Zululand (1888), and the Boer War, and at the outbreak of World War I in 1914 was commander of a cavalry division sent to France as part of the British Expeditionary Force. He commanded the Cavalry Corps and headed the 5th Corps at the Battle of Ypres (May 1915). In October 1915 he was appointed commander of the 3rd Army, positioned north of the River Somme.

Nicknamed 'the Bull' because of his short temper and intransigent nature, Allenby failed to consolidate advances made during the first assault at the Battle of Arras in 1917, enabling a German counter-attack and eventual stalemate, which cost 160 000 Allied and German casualties. In April 1917, Allenby replaced General Dobell in Egypt and quickly reorganized British forces and command structure. On 31 October, Allenby's forces captured Beersheba and drove a wedge through the Turkish lines. The Turks retreated into Palestine, pursued by Allenby, who captured Jerusalem on 9 December. Elements of his army were withdrawn to Europe and he spent the spring and summer of 1918 training raw replacements. On 19 September, using cavalry and RAF air cover, Allenby's forces attacked, breaching the enemy lines at Megiddo. 80 000 enemy troops were killed or captured and Turkey duly surrendered. Thus ended one of the last great cavalry-led campaigns in military history.

Allenby was promoted to field-marshal in 1919 and appointed special high commissioner for Egypt, a very difficult job in the circumstances. He retired in 1925.

Allende (Gossens), Salvador (1908–73) *President of Chile (1970–73). The first Marxist to be elected to power, he was awarded the Lenin Peace Prize in 1972.*

Born in Valparaiso, the son of a lawyer, Allende graduated in medicine from the University of Chile in 1932. As a student he was politically active and was jailed for his participation in the takeover of Santiago University during the uprising against the Ibañez regime. In 1933 he co-founded the Chilean Socialist Party, and for the next four years he worked as a coroner's assistant.

Allende was elected to the chamber of deputies in 1937. He served as minister of health (1939–42) in the liberal leftist coalition gov-

ernment of Cerda and in 1945 was elected to the senate. He first ran for the presidency in 1952 but it was not until 1970 (his fourth attempt) that he was successful. Once in power he immediately began to introduce socialist measures, including the nationalization of several industries, the redistribution of land, and the opening of diplomatic relations with communist countries. He was soon faced with a severe economic crisis exacerbated by industrial unrest and the withdrawal of foreign investment. In 1973 he was overthrown and killed in a military coup led by General Pinochet (1915–   ).

Alpher, Ralph Asher (1921–   ) *US physicist known for his work on the big-bang theory and his prediction of the microwave background it produced.*

Born in Washington, Alpher obtained his PhD from George Washington University in 1948, having worked there under George GAMOW, with whom he collaborated on a number of papers. Their best-known paper, written with Hans BETHE, concerns the origins of the elements and is known as the Alpher–Bethe–Gamow theory (often referred to as the $\alpha\beta\gamma$-theory). It gave a theoretical background to the big-bang theory of Abbé Georges Lemaître (1894–1966), proposing realistic mechanisms by which the elements could have been formed in the original cosmic explosion.

In the same year, Alpher published a paper with Robert C. Herman predicting that the big bang would have produced intense radiation that gradually lost energy as the universe expanded, which would now have a characteristic temperature of about 5 K. Although the radar experts approached by Alpher at that time said that such radiation, if it existed, would be undetectable, it was in fact discovered in 1964 by Arno Penzias (1933–   ) and Robert Wilson (1936–   ). This discovery had a major impact on cosmology and astrophysics.

Amin Dada, Idi (1925–   ) *Ugandan president and chief of the armed forces (1971–79), who became a brutal dictator until he fled his country after defeat by the Tanzanian army. He was chairman of the Organization of African Unity (1975–76).*

Born in the West Nile province, the son of a Muslim-Kakwa family, Amin was educated at the local primary school and joined the army in 1946, serving in the King's African Rifles during the Mau Mau uprising in Kenya. He was heavyweight boxing champion of Uganda from 1951 to 1969. He rose through the ranks to become a major-general and was appointed commander of the armed forces in 1968. In 1970, he was put in charge of the General Service Unit (a security force) by President OBOTE. Rallying a majority of the army in 1971, he took over the Ugandan government in a coup, while Obote was at a conference of the Commonwealth heads of state. Amin declared himself president for life in 1976, instigating a reign of terror in which many of Obote's supporters were murdered, the Asian minority was expelled, and the electorate were wooed by mocking the former colonial powers. However, Amin was overthrown in 1979 during a Tanzanian invasion, provoked by an unsuccessful Ugandan attack, that enabled Obote to return to power in 1980. Fleeing Uganda, Amin was resident in Libya in 1979 and has lived in Jeddah, Saudi Arabia, since 1980.

Amis, Sir Kingsley (1922–   ) *British novelist and poet. He was knighted in 1990.*

Born in London and educated at the City of London School and St John's College, Oxford, Amis combined his writing with a career as university teacher of English, first at Swansea (1949–61), then at Cambridge (1961–63), and later in the USA. His first novel, *Lucky Jim* (1954), was a high-spirited satire on academic and middle-class aspirations in provincial Britain, and was an immediate success. Through this and subsequent novels, such as *That Uncertain Feeling* (1955), *Take a Girl Like You* (1960), and *One Fat Englishman* (1963), Amis became associated with the so-called Angry Young Men who satirized the manners and morals of Britain in the 1950s and early 1960s. Later novels show a more sombre tendency in Amis's fiction, exemplified in *Ending Up* (1974), *Jake's Thing* (1978), *Stanley and the Women* (1984), *Difficulties with Girls* (1988), and *The Folks that Live on the Hill* (1990). His novel *The Old Devils* won the Booker Prize in 1986. In 1991 he published his *Memoirs*. He has also published or edited works on science fiction and edited the *New Oxford Book of Light Verse* (1978), besides publishing his own poems. He was married (1965–83) to the novelist Elizabeth Jane Howard (1923–   ) and his son Martin Amis (1949–   ) is also a novelist, whose books include *The Rachel Papers* (1974), *London Fields* (1989), and *Time's Arrow* (1991).

Amundsen, Roald (1872–1928) *Norwegian explorer who, in 1911, led the first expedition to reach the South Pole. On a previous Arctic expedition he had located the site of the magnetic north pole and was the first to navigate the Northwest Passage.*

Amundsen's ambition was to be an explorer. He gave up his studies at medical school and, after military service and work on merchant ships, was accepted in 1897 as first mate on the *Belgica* as part of the Belgian Antarctic Expedition. In 1900 Amundsen purchased the 47-ton sloop *Gjoa*, and in June 1903 set sail for the Northwest Passage. His expedition reached the magnetic north pole by sledge over the ice but it was not until August 1905 that the *Gjoa* broke through the ice to reach the Beaufort Sea and the Pacific.

In 1909 Amundsen was preparing an expedition to the North Pole when news broke that Robert Peary (1856–1920) had already reached it. Amundsen nevertheless set sail on 7 June 1910 in his ship *Fram*, but turned southwards to Antarctica. The expedition landed at the Bay of Whales on 3 January 1911 in the Antarctic summer. After establishing forward supply bases, Amundsen and four companions set out for the Pole on 20 October 1911 with four sledges and fifty-two dogs. They reached the area on 14 December and two days later established the exact position of the South Pole. They had beaten Robert F. SCOTT and his team who were heading for the same objective. Leaving the Norwegian flag and a note for Scott, Amundsen's group returned safely to base. Amundsen received many awards and honours for his achievement, including the gold medal of the National Geographical Society. He wrote a book describing his exploit, *The South Pole* (1913).

During World War I Amundsen built up a successful shipping business and in 1925 was able to finance an attempt to fly over the North Pole. He was forced down but, the following May, succeeded in the airship *Norge* piloted by the Italian, Umberto Nobile. In 1928, Amundsen was killed in a plane crash while searching for Nobile, who had been forced down while flying in the Polar region.

Anderson, John (1893–1962) *Scottish-born Australian philosopher.*

The son of a radical headmaster, Anderson was educated at the University of Glasgow, where he gained his MA in 1917. He taught at the universities of Cardiff, Glasgow, and Edinburgh before emigrating to Australia in 1927 to take up the appointment of professor of philosophy at the University of Sydney, a post he held until his retirement in 1958.

Anderson began his philosophical career as an absolute idealist but, under the influence of

William James (1842–1910), he developed a more naturalistic and a more realistic approach. Although Anderson's views were too idiosyncratic and too personally expressed to fit into any neat classification, he consistently rejected ultimates of any kind while seeking to defend, on his own terms, traditional formal logic. His views can best be seen in his posthumously published collected papers, *Studies in Empirical Philosophy* (1962).

A man of marked intellectual independence, Anderson was a bitter and courageous foe of authoritarian views of any kind. His critical comments on censorship and Christianity were not always appreciated and there were several attempts to have him dismissed from his Sydney post. An inspired teacher, Anderson attracted disciples and created in Sydney an approach to philosophy and a thriving department that have long survived his death.

Anderson, Sherwood (1876–1941) *US novelist and short-story writer, best known for his fictionalized accounts of small-town life in the American midwest.*

Anderson was born in Camden, Ohio, one of seven children of a travelling sign painter and harness maker. After completing a limited education at the age of fourteen, he had a variety of jobs before finding himself the owner of a successful paint manufacturing company at thirty-six. He then abruptly abandoned both his family and business and moved to Chicago to take up a career in writing.

His first book, *Windy McPherson's Son* (1916), was followed by *Marching Men* (1917) and a book of verse, *Mid American Chants* (1918). The impact on Anderson of Freudian psychology was apparent in a basically pessimistic series of character studies of small-town life in *Winesburg, Ohio* (1919), the novel that established his reputation as a successful writer. After the publication of *Poor White* (1921), Anderson expanded his circle of literary contacts, meeting James JOYCE in Europe and encouraging the career of William FAULKNER, whom he 'discovered' in New Orleans. Of his later works, the collection of short stories *The Triumph of the Egg* (1921) and *Horses and Men* (1923) are considered major achievements in their portrayal of human experience. Although Anderson's literary status has diminished in recent years, he is still acknowledged both as an influence on US literature and as a master storyteller, a title he claimed for himself in his autobiography, *A Story Teller's Story* (1924).

Andrić, Ivo (1892–1975) *Yugoslav short-story writer, novelist, and essayist. He was awarded the Nobel Prize for Literature in 1961.*

Born near Travnik, Bosnia, of middle-class parents, Andrić was educated at Sarajevo and studied Slavic languages at the universities of Zagreb, Cracow, and Vienna. As a student he joined a revolutionary nationalistic group, Young Bosnia, and consequently was imprisoned by the Austrian authorities during World War I. After the war he completed his university studies at Graz and entered the Yugoslav diplomatic corps, eventually serving as ambassador to Berlin (1940). During World War II he remained in German-occupied Belgrade; after the war and the establishment of socialism, he devoted himself solely to his literary activities.

Andrić's earliest work was in verse. After a few poems that appeared in an anthology and translations of Walt Whitman and August Strindberg, he published two works of poetic prose, *Ex Ponto* (1918) and *Nemiri* (1920), before turning to the short story, a form of which he became one of the most distinguished contemporary practitioners. *Put Alije Djerzeleza* (1920) was followed by three story collections (1924, 1931, 1936) based on life in Turkish-ruled Bosnia. The uneasy coexistence of conflicting Christian (Orthodox and Catholic) and Muslim cultures provided a richly dramatic and often violent subject matter that Andrić rendered in faultless prose and with a sweeping historical vision. The stories assume a symbolic significance as instances of the tragic inevitability of historical change and death. Immediately after World War II Andrić published his most famous work, the novel *Na Drini ćuprija* (1945; translated as *The Bridge on the Drina*, 1959), which covers a vast stretch of time; the bridge of the title serves as a symbol connecting past and present as well as different cultures (east and west). *Travnička hronika* (1945; translated as *Bosnian Chronicle*, 1963) concerns the town of Travnik during the Napoleonic wars.

After the war Andrić continued to write stories as well as much nonfiction, including critical essays on Yugoslav writers, travel books, and memoirs. He also published two shorter novels: *Gospodjica* (1945; translated as *The Woman from Sarajevo*, 1966) and *Prokleta avlija* (1954; translated as *Devil's Yard*, 1962).

Andropov, Yuri Vladimirovich (1914–84) *Soviet statesman. He was general secretary of the Soviet Communist Party (1982–84), head* *of the KGB (1967–82), and president of the Soviet Union (1983–84).*

Born in Nagutskaia, Stavropol Krai, the son of a railway worker, Andropov was educated at the Inland Waterways Transport College at Petrozavodsk State University. In the course of his studies he worked as a telegraph worker, an apprentice cinematograph mechanic, and a seaman (1930–32). After his graduation in 1936 Andropov became a Young Communist League (Komsomol) organizer at the Volodanski shipyards in Rybinsk before securing election as secretary and then first secretary of the Yaroslavl Oblast Committee of the All-Union Komsomol (1936–40) and the Karelian Komsomol (1940–44). As Komsomol head of the then Karelo-Finnish Republic, he organized guerrillas behind German lines during World War II.

Andropov, who had an excellent command of English, rare among Soviet leaders, joined the Soviet Communist Party (CPSU) in 1939. He began working for the party in 1944 when he served as first secretary of the Petrozavodsk City Committee, later becoming second secretary of the Karelian Central Committee (1947). In 1951 he moved to the secretariat staff of the CPSU. He was ambassador to Hungary (1953–57) during the uprising there (1956), and then given responsibility for CPSU relations with other communist countries (1958–62), becoming secretary of the CPSU central committee (1962–67). In 1967 he was appointed chairman of the State Security Committee (KGB) of the Soviet Council of Ministers and a candidate member of the Politburo central committee. Six years later he became a full member. He resigned as head of the KGB in 1982 to become a member of the Presidium of the Supreme Soviet and general secretary of the central committee of the CPSU. He became president of the Soviet Union in 1983 and died in office.

Anne, Princess (Anne Elizabeth Alice Louise, the Princess Royal; 1950–) *Daughter of* ELIZABETH II *of the United Kingdom.*

Currently eighth in line to the throne, Princess Anne attended Benenden School in Kent before embarking on her public career. A keen horsewoman, she won a silver medal in the Individual European Three-Day Event in 1971 (when she was also voted BBC Sports Personality of the Year) and represented Great Britain at the Montreal Olympics in 1976. In 1973 she married Captain Mark Phillips (1948–), also an accomplished rider; they have a son Peter

(1977–) and a daughter Zara (1981–). Their separation was announced in 1989.

As a young woman Princess Anne acquired a reputation for brusqueness towards the media, but since her untiring work as president of the Save the Children Fund, on whose behalf she has visited many Third World countries, she has become one of the most popular and respected members of the royal family. In recognition of her involvement on behalf of this and many other charitable organizations, she was granted the title Princess Royal by the Queen in 1987. She is also president of the British Olympic Association and in 1981 replaced her grandmother as Chancellor of the University of London.

Annigoni, Pietro (1910–88) *Italian painter of portraits and religious subjects.*

Born in Milan, Annigoni studied at the Academy of Fine Arts in Florence, where he had his first exhibition in 1932. He helped to found a group of modern realist painters, who exhibited in Milan in 1947 and then in Rome and Florence. In the 1950s he worked in England and exhibited in London, Paris, and New York. His technique was unusual for a twentieth-century artist in that he practised the methods of the old masters, painting mainly in tempera. Annigoni's best-known portraits include members of European royal families – Queen ELIZABETH II (1955 and 1970) and Prince Philip (1957) – as well as other prominent figures, such as President KENNEDY (1961) and Dame Margot FONTEYN. His religious paintings include altarpieces and, from the 1960s, numerous frescos, notably the cycle of the life of Christ in the church of S Michele Arcangelo, near Florence. In 1977 he published his autobiography, *An Artist's Life.*

Anouilh, Jean (1910–87) *French dramatist.*

Anouilh was born in Bordeaux. Influenced by GIRAUDOUX, he wrote his first play at the age of nineteen: *L'Hermine* (1929; translated as *The Ermine*, 1955). In 1931 he became secretary to director Louis Jouvet; *L'Hermine* was first performed the following year. *Le Voyageur sans bagage* (1937; translated as *Traveller Without Luggage*, 1959) established Anouilh as a writer of considerable potential; this was recognized by André Barsacq, who directed many of Anouilh's plays over the next decade.

Anouilh divided his plays into five categories. His *pièces roses* are fantasies, such as *Le Bal des voleurs* (1932; translated as *Thieves' Carnival*, 1952), which was first performed in 1938 under the direction of André Barsacq. The darker, more pessimistic, plays are classified as *pièces noires*; these include *Antigone* (1944; translated in 1946), a reworking of the Greek myth with undertones of the contemporary political situation in occupied France, and *Médée* (1946; translated in 1957). In these and other plays Anouilh reveals his obsession with purity, doomed to destruction by the immoral society in which the hero or heroine is forced to move. Some of his characters accept a compromise; others, such as Antigone and Joan of Arc (in *L'Alouette*), prefer to die rather than sacrifice their ideals.

The *pièces brillantes* are lighter social comedies, such as *L'Invitation au château* (1947; translated as *Ring Round the Moon*, 1950) and *La Répétition ou l'amour puni* (1950), which features a device frequently used by Anouilh, the 'play within a play'. The fourth group, the *pièces grinçantes*, includes *Pauvre Bitos ou le dîner de têtes* (1956; translated as *Poor Bitos*, 1963). Among the more optimistic of Anouilh's plays are his *pièces costumées* – *L'Alouette* (1953; translated as *The Lark*, 1955), which tells the story of Joan of Arc, and *Becket ou l'honneur de Dieu* (1959; translated as *Becket, or The Honour of God*, 1962). His later plays include *La Grotte* (1961), *Cher Antoine ou l'amour raté* (1969), *Ne réveillez pas madame* (1970), and *Le Nombril* (1981; translated as *The Navel*).

A gifted theatrical craftsman, Anouilh enjoyed considerable and lasting popularity both in France and abroad. He also wrote for the cinema; his films include *Monsieur Vincent* (1947) and *Deux sous de violettes* (1951).

Antonioni, Michelangelo (1912–) *Italian film director. His films concentrate upon the study of character and, through the use of complex metaphoric plots and masterful camera-work, illuminate such themes as suicide and the environmental alienation of man.*

Antonioni, who was born in Ferrara and studied economics and commerce at Bologna University, came to films after working in a bank. He wrote film critiques for newspapers and contributed to the journal *Cinema* before briefly attending the Centro Sperimentale, the famous film school in Rome. In 1942 he worked with ROSSELLINI and CARNÉ and made his debut as a director in the documentary *Gente del Po* (*The People of the Po Valley*) in 1943. His first feature film, *Cronaca di un amore* (1950; *Story of a Love Affair*) was hardly noticed. However, his reputation grew with 'Tentato Suicidio' (his contribution to the episodic *Amore in città* (1953; *Love in the City*) and *Le amiche* (1955; *The Girl Friends*),

which won the Golden Lion Award at the Venice Film Festival. International acclaim came with the prize-winning *L'avventura* (1960), starring Monica Vitti (1931–). This, together with his two other prize-winning films, *La notte* (1961) and *L'eclisse* (1962), are now regarded as constituting a trilogy. In 1964 he made his first colour film, *Il deserto rosso* (1964; *The Red Desert*). His films since then have included *Blow-Up* (1967), *Zabriskie Point* (1970), *The Passenger* (1975), *The Oberwald Mystery* (1981), and *Identification of a Woman* (1982). In 1989 he released a further documentary, *Fumbha Mela, Roma*.

Appleton, Sir Edward Victor (1892–1965) *British physicist, who discovered and investigated the properties of the ionosphere. He was knighted in 1941 and awarded the Nobel Prize for Physics in 1947.*

Born in Bradford, he studied physics at Cambridge (1910–13) and spent World War I in the Royal Engineers. Much concerned with the persistent problem of the fading of radio signals during the war, Appleton turned after the war to the study of the propagation of electromagnetic waves. It had been proposed by Oliver HEAVISIDE and Arthur KENNELLY in 1902 that some waves (known as sky waves) were reflected back to earth by an electrified layer in the upper atmosphere. Appleton set out to confirm this suggestion experimentally and to explore the nature of the layer, which eventually became known as the ionosphere. Suspecting that fading was caused by interference, Appleton arranged for the BBC to vary the frequency of their transmitter while he recorded the strength of the signal received some miles away in Cambridge. He found a strengthening of the signal when the ground waves interfered constructively with the sky waves. Appleton calculated the height of the reflecting layer to be about 95 km and went on to show that it had a complex structure. The top layer (F-region) of the ionosphere is often known as the Appleton layer. This work had enormous practical and theoretical implications for radio transmission.

From 1924 to 1936 Appleton was Wheatstone Professor of Experimental Physics at King's College, London. After a spell as Jacksonian Professor of Natural Philosophy at Cambridge, he was appointed secretary to the Department of Scientific and Industrial Research (1939–49). During his period with the DSIR Appleton was one of the key figures in the development of the British nuclear programme, both for military and industrial applications. In 1944 he moved to Edinburgh University, where he was vice chancellor until his death.

Aquino, (Maria) Corazon (1933–   ) *Filipino stateswoman and president (1986–   ).*
The daughter of a prominent landowning family, Corazon Cojuangco was educated in the USA, where she obtained a BA before marrying the political journalist Benigno Aquino in 1955. Her wealth and connections helped to launch Benigno on a political career and assisted him in becoming leader (1967) of the opposition Liberal party and the main challenger to President Ferdinand E. Marcos (1917–89). In 1972 Marcos assumed dictatorial powers and imprisoned political rivals including Benigno Aquino, who later (1977) received a death sentence. Following international protests, the death sentence was commuted in 1980 and Aquino was allowed to travel to the USA, where he stayed for three years. Corazon, who accompanied him, was also beside him when he returned to the Philippines to contest presidential elections in 1983. On arrival at Manila airport, Benigno Aquino was shot dead: a later enquiry pointed to a high-level conspiracy within the country's armed forces. Corazon, who had not previously assumed a public role, became the symbol of moral opposition to the corrupt Marcos regime and addressed large protest rallies. In 1986 she stood against Marcos in presidential elections. Although Marcos was declared victorious, independent observers charged the regime with widespread fraud and Aquino's followers proclaimed her president. Faced with the desertion of his allies and mounting public disorder, Marcos fled to the USA. As president, Aquino moved quickly to end the repression and corruption of the Marcos years but has been accused of doing too little to right economic injustices. She has faced continuing problems from communist insurgents and from sections of the military, who mounted unsuccessful coup attempts in 1986, 1987, and 1990.

Arafat, Yasser (1929–   ) *Palestinian leader and chairman of the Palestinian Liberation Organization (1968–   ). Popularly known as 'Abu Ammar' ('the builder'), Arafat became well known throughout the world in 1974 when, on behalf of the PLO, he addressed the United Nations General Assembly, being the first representative of a nongovernmental organization to appear before it.*
Born in Jerusalem, the son of a provincial merchant, Arafat was first exposed to political activity at his local secondary school, where there was a strong anti-Zionist movement. His father and brother were also active in opposing the growth of Jewish armed groups in Palestine. Arafat studied civil engineering at Cairo University and was elected president of the League of Palestinian Students (1952–56). He briefly served with the Egyptian army during the 1956 Suez crisis, then moved to Kuwait, where he worked as an engineer (1957–65).

Arafat's first active involvement with Palestinian independence was in 1956, when he helped establish Al Fatah, an Arab guerrilla movement largely inspired by NASSER's revolution in Egypt. Nine years later this movement came into prominence as the leading military faction of the Palestinian Liberation Organization (PLO), founded in 1964. Arafat was elected chairman of the PLO in 1968 and in 1971 was appointed general commander of the Palestinian Revolution Forces. He has remained in charge of the PLO since then, although his forced departure from Beirut (1982) and opposition from more radical elements within the movement (1983) have challenged his position. In 1988 he recognized the state of Israel and denounced terrorist methods. Under Arafat's leadership the PLO announced a Palestinian state in 1990 but many countries refused to recognize it. Arafat's support of Saddam Hussein's regime in Iraq during the Gulf War of 1991 damaged his reputation internationally.

Aragon, Louis (1897–1982) *French poet, novelist, journalist, and essayist.*
Aragon was born in Paris, where as a young man he became involved with dadaism and surrealism. With André BRETON and Philippe Soupault he co-founded the surrealist review *Littérature* in 1919. His first collection of poetry, *Feu de joie*, appeared in 1920; this was followed by *Le Mouvement perpétuel* (1925) and *La Grande Gaîté* (1929). In 1926 he produced his first novel, *Le Paysan de Paris*, which celebrated in surrealist terms the everyday wonders of the city. Together with his essays in *Traité du style* (1928), these works established Aragon as a leading surrealist writer.

In 1927 Aragon joined the Communist Party. A visit to the Soviet Union in 1930 so impressed him that he subsequently broke off his association with the surrealists and committed himself to writing for the communist cause, becoming editor of the party newspaper, *Ce Soir*, in 1937. Meanwhile, in 1928, he had met the Russian-born writer Elsa Triolet, who was to become his lifelong companion and the inspiration for such lyric poems as *Les Yeux d'Elsa* (1942), *Elsa* (1959), and *Le Fou d'Elsa* (1963).

During World War II Aragon was an active member of the intellectual Resistance, publishing the intensely patriotic poems of *Le Crève-coeur* (1941) and *La Diane française* (1945). *Le Monde réel* (1933–44), a cycle of four novels, and the six volumes of *Les Communistes* (1949–51) are chronicles of the march of communism, laden with Marxist propaganda. *La Semaine sainte* (1958; translated as *Holy Week*, 1961), one of Aragon's best-known novels, is a Marxist view of the France of 1815. In 1953 Aragon founded the communist weekly *Les Lettres Françaises*, a review of arts and literature.

A versatile and prolific writer in a career spanning more than sixty years, Aragon also produced such works as a poetic autobiography, *Le Roman inachevé* (1956), and a collection of critical essays on classical authors, *La Lumière de Stendhal* (1954).

Arden, John (1930–   ) *British playwright.* Arden studied at Cambridge and Edinburgh and trained as an architect, but from the late 1950s devoted himself to the theatre. His early plays, which included *The Waters of Babylon* (1957), *Live Like Pigs* (1958), and *The Happy Haven* (1960), were grotesque comedies of modern life. With the publication of his dramas *Sergeant Musgrave's Dance* (1959) and *Armstrong's Last Goodnight* (1964) he was acknowledged as one of the leading dramatists of the 1960s. He was particularly acclaimed for his use of colloquial speech to express such deep, dangerous, and often irresolvable issues as pacifism and violence. Since the mid-1960s many of his plays have been written in collaboration with his wife, Margaretta D'Arcy. Major commercial success has eluded Arden, despite the praise of the critics, and in recent years he has forsaken the professional theatre in order to stage his works with fringe theatre groups. More recent plays include *Left-Handed Liberty* (1965), *Don Quixote* (1980), *Garland for a Hoar Head* (1982), and *Whose is the Kingdom?* (1988). He has also published a collection of essays, *To Present the Pretence* (1977), and such novels as *Silence Among the Weapons* (1982) and *Books of Bale* (1988).

Arendt, Hannah (1906–75) *German-born US political philosopher, well known for her work on the nature of totalitarianism, violence, revolution, and other features of modern political life.*

Born in Hanover, she began as a pupil of HUSSERL and Karl Jaspers (1883–1969) at Heidelberg, obtaining her PhD in 1928, but with the rise of Hitler she moved to Paris (1933), where she worked for several years for Zionist organizations. In 1940 she once more fled the threat of Nazism and sought refuge in the USA, where she worked for various publishers while continuing also to help Jewish organizations. In 1963 Arendt accepted her first academic appointment at the University of Chicago; later she taught at Berkeley, Columbia, and Princeton.

Arendt established her reputation with *The Origins of Totalitarianism* (1951). It was one of the first works to argue that the totalitarian regimes of Hitler's Germany and Stalin's Russia had common elements and roots. They were both, she declared, antisemitic, imperialistic, and nationalistic. A later work, *Eichmann in Jerusalem* (1961), with its emphasis on 'the banality of evil' and its suggestion that the Jews played some part in their own genocide, provoked considerable controversy and brought her work before a much wider public. She continued to explore the parameters of contemporary political life in two later works, *On Revolution* (1963) and *On Violence* (1970).

Up to her death Arendt was trying to develop a more systematic and theoretical position. Her work remained incomplete although part of it was published posthumously in *The Life of the Mind* (1978).

Armstrong, Edwin Howard (1890–1954) *US electrical engineer, who invented the superheterodyne radio receiver and the FM radio.*

The son of a publisher, he was educated at Columbia University, New York. He joined the faculty shortly after graduating in 1913 and became professor of electrical engineering in 1934, a post he held until his death. A highly inventive engineer with forty-two patents to his name, he was responsible for two very important innovations that have had a profound impact on the development of radio. The first arose from his work with the US Signals during World War I and involved attempts to detect distant aircraft by the electromagnetic signals they produced. To tune in to such high frequencies Armstrong designed a 'superheterodyne circuit' that allowed the complicated task of tuning to be performed by the simple turning of a knob. This device enabled radio to become available for everyone. Although he received half a million dollars for his invention from Westinghouse, it also involved him in a prolonged, expensive, and bitter legal battle with Lee DE FOREST, the inventor of the triode valve.

Armstrong's second major invention proved no less contentious. It arose from his plan of reducing the static that ruined so much of early broadcasting. Realizing that much of this static arose as a result of amplitude modulation, he devised a means of removing it completely by using frequency modulation (FM). Surprisingly, RCA and other broadcasting companies objected to the proposal in the 1930s and further delays were caused by the outbreak of World War II. Depressed by obstacles put in his way by the US government and the broadcasting companies and impoverished by the constant litigation that accompanied his work, Armstrong committed suicide in 1954.

Armstrong, Louis (1900–71) *Black US jazz trumpet player and singer, also called 'Satchmo'. His superb playing, abrasive voice, and irrepressible personality made him one of the best-loved entertainers of the century.*

Born in New Orleans, he learned to play the cornet while confined in a waifs' home at the age of thirteen for firing a revolver in the street. After his release he did a variety of odd jobs before playing in the jazz bands of King Oliver and Fletcher HENDERSON. In 1925–28 he made about sixty records as the leader of small groups, now known as the Hot Fives and the Hot Sevens. These records made him the first great solo star in the history of popular music. During this period he switched from the cornet to the larger brighter-sounding trumpet. From 1931 to 1947 he led various big bands, then formed a small group again, called the All Stars. As well as complete mastery of his instrument, he had an instinctive understanding of harmony, a flamboyant lyricism, and a unique sense of phrasing, all of which combined to make him one of the main influences in popular music.

As early as 1932 he toured Europe, eventually playing in nearly every country in the world. He had a hit record as late as 1964 ('Hello, Dolly!') and appeared in many films, notably *The Birth of the Blues* (1941), *High Society* (1956), and *Satchmo the Great* (1957).

Armstrong, Neil Alden (1930–) *US astronaut who, in 1969, became the first person to set foot on the surface of the moon.*

Armstrong entered Purdue University in 1947 as a naval air cadet to study aeronautical engineering but his course was interrupted by service in the Korean War, in which he flew seventy-eight combat missions and was shot down on one occasion. At the end of the war he completed his course and received a BS degree in 1955. He joined the National Advisory Committee for Aeronautics (NACA) – the forerunner of NASA – as a civilian test pilot, flying supersonic jets, including the experimental X-15 plane, which was capable of over 4000 mph. In 1962 he began training as an astronaut and was command pilot of *Gemini 8* in which, with David R. Scott (1932–) on 16 March 1966, he performed the first manual space docking with an unmanned rocket. Armstrong and Scott returned safely to earth in spite of problems with a faulty booster rocket. But Armstrong's most memorable moment was at 02.56 GMT on 21 July 1969. As commander of the *Apollo 11* mission, he became the first person to set foot on the moon. He and Edward 'Buzz' Aldrin (1930–) landed their lunar module, *Eagle*, in the Sea of Tranquillity at 20.17 GMT on 20 July, pronouncing the famous words, 'That's one small step for a man, one giant leap for mankind.' On the lunar surface they set up a television camera, took samples of lunar dust, deployed scientific instruments, and made a series of observations, providing invaluable information for lunar scientists. At 17.54 GMT, *Eagle* took off to rejoin the orbiting command module, *Columbia*, piloted by Michael Collins (1930–). After returning to earth and spending eighteen days in quarantine, the crew duly became national and international celebrities, making a world tour of twenty-two nations. Armstrong was later appointed professor of aerospace engineering at the University of Cincinnati (1971–79) and has held directorships in various US business corporations.

Armstrong, Warwick Windridge (1879–1947) *Australian cricket all-rounder who appeared in fifty test matches; as captain of his country in ten matches he was never on the losing side. In all first-class matches he scored 16 158 runs and took 832 wickets.*

Armstrong was born at Kyneton, Victoria, and educated at University College, Armadale. He played for Victoria from 1898 until 1922 and made his test debut in 1902, against England. During his career he was renowned, apart from his ability as an all-rounder, for his outspoken opinions and his enormous size: when he retired at the age of forty-two he weighed 22 stone.

Armstrong toured England on four occasions (1902, 1905, 1909, and 1921). He was in particularly brilliant form in 1905, when he scored more than 2000 runs and took 130 wickets. He was a hard-hitting batsman and bowled leg breaks. His frequent battles on behalf of players' conditions made him more popular with teammates than with the cricket authorities: he had revolted against the Board of Control in 1912 and was one of a number of players who refused to tour England that year. Probably his greatest triumph came in 1920–21, when England lost all five tests to the Australians and he scored three hundreds in the process.

Arnold, Malcolm Henry (1921–   ) *British composer and conductor. He was created a CBE in 1970.*

Arnold was born in Northampton. After winning an open scholarship to the Royal College of Music (1938), where he studied composition with Gordon Jacob (1895–1984) and trumpet with Ernest Hall, he won the Cobbett Prize (1941) and joined the London Philharmonic Orchestra, becoming first trumpet in 1942. After war service (1944–45) he joined the BBC Symphony Orchestra for a season, returning to the London Philharmonic in 1946. In 1948 he won the Mendelssohn scholarship, enabling him to spend a year in Italy; since then he has devoted himself to composition and conducting. Exeter University made him an honorary DMus in 1969, as did Durham University in 1982 and Leicester University in 1984.

Arnold writes for the orchestra with the technical bravura of a professional player, ensuring that the music is readily understandable, the tunes attractive, and the texture clear, using repetition rather than development. He considers Berlioz to have been the greatest influence on his writing, which is prolific and includes nine symphonies, concertos for a variety of instruments, chamber works, including the early *Three Shanties for Wind Quintet* (1952), five ballets, and more than eighty film scores, including *The Bridge on the River Kwai* (1957).

Arp, Jean (or **Hans**) (1887–1966) *German-born French sculptor, painter, and poet.*

Born in Strasbourg, Alsace, when it was still part of Germany, Arp was French at heart though German by birth. During his life he moved frequently, establishing contact with many major avant-garde artists of the day. He studied in Weimar and Paris, met KLEE in Switzerland in 1909 and KANDINSKY in Munich in 1912, where he exhibited with the Blaue Reiter group. In Berlin the following year he exhibited with the expressionists and in 1914 in Paris he associated with avant-garde artists, such as DELAUNAY and MODIGLIANI.

He spent World War I in Switzerland, where he exhibited his first abstract works and helped to found the Zürich dada group. With the artist Sophie Taeuber (1889–1943), whom he married in 1922, he experimented with reliefs, cut-outs, constructions, collages, and random compositions using wood, card, paper, and other media. Unlike many early abstractions, Arp's works consisted mainly of organic curvilinear shapes. After the war he worked on collective compositions with Max ERNST in Cologne. He published *The Isms of Art* with El Lissitzky (1890–1941) in 1925, and after moving to Meudon, near Paris (1926), exhibited with the Paris surrealists. Though still producing painted wood reliefs, by the 1930s he was producing fully developed supple rounded three-dimensional forms of sculpture in marble and bronze. He was a founder member of the Abstraction-Création group in 1931.

Back in Meudon after spending World War II in Grasse, Arp published a collection of poems and in 1948 poetry and essays. In the USA in 1950 he did a sculpture for Harvard University and in 1958 a mural relief for the UNESCO building in Paris. Arp won the International Prize for Sculpture at the Venice Biennale in 1954.

Arrabal, Fernando (1932–   ) *Spanish playwright.*

Born at Melilla, Morocco, Arrabal was educated at a military school and at an academy in Valencia. Among the formative influences of his youth he has mentioned his reading of Dostoevsky, PROUST, KAFKA, CAMUS, and FAULKNER, the films of CHAPLIN and the MARX BROTHERS, *Alice in Wonderland*, IONESCO, and BECKETT. The mysterious fate of his father also had an important bearing on his work. An army officer who sympathized with the Republican cause, his father was arrested in 1936 and condemned to death; the sentence was subsequently changed to thirty years' imprisonment. He escaped from prison in 1942, having apparently become insane in the interval and been committed to a psychiatric ward; nothing further was heard of him. Arrabal's mother and her parents, by whom he was raised, were devout and conservative and suppressed all information about his father. At seventeen Arrabal accidentally found family papers relating to his father, which aroused a lasting suspicion that he had been betrayed by his mother. The shock of this discovery is reflected in the violence of his plays, in which fantasy, farce,

19 Artaud, Antonin

ritual, and erotic elements combine in a world of baroque horror. The effect has been justly compared to the nightmarish qualities of Goya and Kafka.

Arrabal's first two plays were written in Spanish. In 1954 he met his future wife, a French student, whom he married in 1958. He has since written in French and lived in France, apart from one disastrous visit to Spain in 1967, when he was arrested on a petty charge of blasphemy and freed only after appeals by Sartre and others. Arrabal gained an international reputation with a Passion play, *Le Cimetière des voitures* (1957; translated as *The Car Cemetery*, 1962), in which three jazz musicians (clowns modelled on the Marx Brothers), who live amid the automotive wreckage of contemporary civilization, act out a drama of betrayal and death. *L'Architecte et l'empereur d'Assyrie* (1966; translated as *The Architect and the Emperor of Assyria*, 1969), perhaps his best known work, reflects his participation in the 'Panic Group', a movement proclaimed in 1962 in a manifesto by Arrabal, the Mexican film director Jodorowsky, and others (the allusion is to the god Pan, representing inexplicable terror as well as comedy). A highly prolific dramatist and extremely active literary figure, Arrabal has lectured widely, has charge of a theatre review, and has produced several film scripts of his plays. Several volumes of his *Théâtre* have been published. Volume I (1958) contains (among others) *Les Deux Bourreaux* (1956; translated as *The Two Executioners*, 1962), *Fando et Lis* (1955), and *Le Cimetière des voitures*; Volume II (1961) includes *Guernica* (1959) and *Pique-nique en campagne* (1952; translated as *Picnic on the Battlefield*, 1967); Volume III (1965) includes *Le Grand Cérémonial* and *Cérémonie pour un noir assassiné*. Among more recent works are *La Ballade du train fantôme* (1974), ironically set in the deserted mining town of Madrid, New Mexico, and *Baal Babylone*, first performed in 1980.

Arrau, Claudio (1903–91) *Chilean-born US pianist.*

Born in Chilán, Arrau was a child prodigy who first played in public in Santiago at the age of five. After two years' tuition with Paoli, the Chilean government gave him a scholarship to study at Stern's Conservatory in Berlin, where he was a pupil of Martin Krause (1912–18) and won many awards. His first Berlin recital in 1914 was followed by extensive European tours, during which he played under such conductors as Furtwängler and Nikisch. In 1921

he returned to South America, giving concerts in Argentina and Chile. Arrau's first London concert was in 1922, when he shared the platform with Melba and Bronislaw Huberman (1882–1947); he toured the USA in 1923. From 1924 to 1940 he taught at Stern's Conservatory and in 1935 played the entire keyboard works of J. S. Bach in a series of concerts in Berlin; after this he declined to play Bach again in public, feeling that the music was unsuitable on a modern piano. He gave his last public performance in 1989.

Arrau was an unusually unostentatious pianist. His reputation lay chiefly in his interpretation of the works of Beethoven, Brahms, Schumann, Liszt, and Chopin and in the meticulous musicianship and intellectual penetration that accompanied his virtuoso technique. He was awarded the International UNESCO Music Prize in 1983.

Artaud, Antonin (1896–1948) *French actor, director, and drama theoretician, who originated the theory of the Theatre of Cruelty.*

Artaud was born in Marseilles and began his career as an actor. He became interested in surrealism and symbolist drama, contributing to *Revolution Surréaliste* and *Nouvelle Revue Français*, and with fellow-surrealist Roger Vitrac (1899–1952) founded the Théâtre Alfred Jarry (1927), named after an important figure in the avant-garde theatre. He also wrote the script for the surrealist film *La Coquille et le clergyman* (1926) and appeared in a number of other films, most notably as Marat in Gance's *Napoléon* (1927) and in Dreyer's *La Passion de Jeanne d'Arc* (1928).

Influenced by Balinese dancing and oriental drama, Artaud began to develop his theory of the Theatre of Cruelty. His idea was to strip drama of its 'civilizing' dialogue and contrived theatrical concepts and to return it to its symbolic and ritualistic roots. Thus restored, drama could then act as a liberating force, releasing man from his veneer of civilizing conventions and returning him to his primitive self, thereby bringing him nearer the truth of his own nature. *Les Cenci* (1935), derived from Shelley and Stendhal, was his only play based on his theoretical writings. Although it was unsuccessful, his collected essays, published as *Le Théâtre et son double* (1938; translated as *The Theatre and Its Double*, 1958) had a powerful influence on such dramatists and directors as Jean-Louis Barrault, Roger Blin (1907–84), and Peter Brook.

A lifelong sufferer from physical and mental illness, Artaud spent most of the remainder of his life from 1937 in mental institutions.

Asch, Sholem (1880–1957) *Polish-born Jewish novelist and playwright, who wrote in Yiddish.*

Asch was born in Kutno, the fifteenth child of a dealer in cattle and sheep. He was educated in Hebrew schools before attending rabbinical college. His first successful novel, *The Village* (1904), about the life of Jews in rural Poland, staked his claim to be among the foremost of the writers of the Yiddish revival in eastern Europe. His controversial play *The God of Vengeance* (1907), one of several on religious themes, was widely performed.

In 1910 Asch moved to the USA, where he did most of his writing and became a US citizen (1920), but he also spent much time travelling; as a Zionist, his journeys frequently took him to Palestine. His novels fall into two main categories: those dealing with contemporary themes and those centred on major figures in the Judaeo-Christian tradition. Major representatives of the former group are *Three Cities* (1933), about events prior to the Soviet revolution, and *East River* (1948), perhaps his best novel, about Jews and Roman Catholics in the poorer quarters of New York. His novels about New Testament figures won international acclaim: *The Nazarene* (1939), *The Apostle* (1943) about St Paul, and *Mary* (1949). Among his other novels were *Sabbatai Zevi* (1930), *Moses* (1952), and *The Prophet* (1955), about Deutero-Isaiah. In 1953 Asch moved to England, where he died four years later.

Ashcroft, Dame Peggy (Edith Margaret Emily Ashcroft; 1907–91) *British actress. She was made a DBE in 1956.*

Born in Croydon, Ashcroft studied drama at the Central School of Dramatic Art and began her long and distinguished career playing Margaret in *Dear Brutus* (1926) at the Birmingham Repertory Theatre. She made her London debut at the Playroom Six as Bessie in *One Day More* (1927). Naomi in *Jew Süss* (1929), Desdemona to Paul ROBESON's Othello, and Juliet in GIELGUD's 1935 production of *Romeo and Juliet* were just a few of her many notable performances. Memorable, too, were her portrayals of Lady Teazle in Sheridan's *The School for Scandal* (1937–38) and Cecily Cardew in Wilde's *The Importance of Being Earnest* (1939).

Her first appearance in New York was at the Martin Beck Theatre as Lise in *High Tor*

(1939). In Oslo she gave one of her most highly acclaimed performances in the title role of *Hedda Gabler* (1955) and was awarded the King's Gold Medal. *The Chalk Garden* (1956) and *The Lovers of Viorne* (1971) brought Evening Standard awards and in 1962 she was chosen Best Actress at the Paris Theatre Festival while touring in *The Hollow Crown*.

She has also appeared on television, notably in *The Jewel in the Crown* (1983) adapted from Paul SCOTT's *Raj Quartet*, and in films – for her performance in David LEAN's *A Passage to India* (1984) she won an Oscar for best supporting actress. She also won a Venice Film Festival Award for her performance in the television play *She's Been Away* (1989). In 1962 the Ashcroft Theatre in Croydon was named in her honour. Other honours included honorary doctorates and special awards from the British Theatre Association (1982) and BAFTA (1990).

Ashdown, Paddy (Jeremy John Durham Ashdown; 1941–) *British politician, leader of the Social and Liberal Democrats since 1988.*

Educated at Bedford School, Ashdown enlisted in the Royal Marines in 1959, rising to the rank of captain, with command of the 2 Special Boat Squadron. A gifted linguist, he served with the British UN mission to Geneva from 1971 to 1976. In 1983 he captured the safe Tory seat of Yeovil for the Liberals and was almost immediately tipped as a future party leader. He became Liberal spokesman on trade and industry (1983–86) and education and science (1987). In July 1988 he was elected leader of the Social and Liberal Democrats, formed from a merger of the Liberals and the Social Democratic Party earlier that year. As leader he has adopted a high public profile and is widely credited with reversing the decline in his party's fortunes that followed the acrimonious merger negotiations.

Ashkenazy, Vladimir Davidovich (1937–) *Soviet-born pianist and conductor. He lived in England from 1963 to 1968 and thereafter in Iceland, taking Icelandic nationality in 1972.*

A child prodigy, Ashkenazy made his Moscow debut in 1945 and studied for ten years at the Moscow Central School of Music under Anaida Sumbatyan. In 1955 he joined Lev Oborin's class at the Moscow Conservatory, gaining second prize in the fifth Warsaw International Chopin Competition in the same year. The following year he won the gold medal at the Queen Elizabeth International Competition in Brussels and toured the USA while still a student (1958). In 1962 he shared first prize with John Ogdon (1937–89) in the second Moscow Tchaikovsky Competition, making a triumphant London debut at the Festival Hall in 1963.

Ashkenazy specializes in the music of Mozart and is renowned for his interpretations of the music of SCRIABIN, RACHMANINOV, and PROKOFIEV. His playing has exceptional sensitivity in its delicacy of phrasing and tone colour; his repertoire includes the orchestral and chamber works of Beethoven, Schubert, Schumann, Chopin, and Liszt. He frequently conducts piano concertos from the keyboard and is increasingly active as conductor in purely orchestral works. He was appointed music director of the Royal Philharmonic Orchestra in 1987; since 1989 he has also been chief conductor of the Berlin Radio Symphony Orchestra.

Ashton, Sir Frederick William Mallandaine (1904–88) *British choreographer and co-founder of the Royal Ballet. He is regarded as the architect of the lyrical style of classical dancing and a major contributor to the mid-twentieth-century ballet boom. He was knighted in 1962, made a CH in 1970, and received the OM in 1977.*

Born in Ecuador, Ashton became interested in ballet after seeing Anna PAVLOVA dance in Lima in 1917. In 1926 he studied dance in London with Marie RAMBERT, who encouraged him to choreograph his first ballet, *A Tragedy of Fashion* (1926). In 1928 he joined Ida Rubinstein's company in Paris and worked with Léonide MASSINE and Bronislava Nijinska (1891–1972), both of whom influenced his early work. Ashton returned to London in 1935 and became resident choreographer and principal dancer for the Vic-Wells Ballet, founded in 1931 by Ninette DE VALOIS, which became the Royal Ballet in 1956. Although he was never acclaimed as a virtuoso dancer, Ashton's interpretations of such roles as Kostchei in STRAVINSKY's *The Firebird* and the timid stepsister in his own version of PROKOFIEV's *Cinderella* were well received. As a choreographer Ashton developed his witty and elegant style in numerous works for the Royal Ballet, including *Façade* (1931), *Cinderella* (1948), *Ondine* (1958), and his most popular piece, *La Fille mal gardée* (1960), a recreation of a French ballet first staged in 1789. For many years Margot FONTEYN was his favourite interpreter but in later years he also choreographed for younger talents, such as Anthony Dowell,

Lynn Seymour, Antoinette Sibley, and Nadia Nernia.

In 1963 Ashton succeeded de Valois as a director of the company. After his retirement in 1970 he choreographed one major ballet, *A Month in the Country* (1976), and several shorter pieces.

Asplund, (Erik) Gunnar (1885–1940) *Swedish architect, who developed the Swedish idiom in modern architecture. Many of his works are in Stockholm, some are in Gothenburg.*

Born in Stockholm, he was educated at the university there, where he became professor of architecture (1931–40). His principal buildings include the Stockholm city library (1920–28), with its drum-shaped reading room, the extension to the Gothenburg town hall (1934–37), and the Stockholm Crematorium (1935–40). He also built numerous schools, which inspired many similar buildings in other European countries.

Asquith, Herbert Henry, 1st Earl of Oxford and (1852–1928) *British statesman, Liberal leader (1908–26), and prime minister (1908–16). Though his leadership of the nation was unconvincing, he is regarded as having been an outstanding parliamentarian. He accepted a peerage in 1925.*

Asquith was born in Yorkshire, the son of a Congregationalist businessman. He became a barrister and, in 1886, MP for East Fife. In 1894 he married, as his second wife, Margot Tennant (1865–1945), who became famous as a political hostess. Following the landslide Liberal victory in 1905 he was chancellor of the exchequer until succeeding CAMPBELL-BANNERMAN as party leader and prime minister in 1908. His government was responsible for some outstanding legislation. The Parliament Act 1911, giving legislative supremacy to the House of Commons, was a response to the rejection by the Lords of LLOYD GEORGE's 'people's budget' (1909). To secure the Bill's passage Asquith persuaded George V to threaten to create enough Liberal peers to overcome the opposition to reform. 1911 also saw the passage of the National Insurance Act, which introduced insurance against sickness and unemployment.

Asquith's leadership in World War I was less successful. He brought the Conservatives into a coalition government in 1915; however, his failure to consult his colleagues on the matter fatally divided the party. The split deepened when, with the failure of the Dardanelles expedition (1915) and the continuing stalemate on the Western Front, Asquith introduced conscription, a measure regarded by some Liberals as authoritarian. His unpopularity increased with the brutal suppression of the Easter Rising in Dublin (1916), and manoeuvres by Lloyd George succeeded in forcing his resignation. The Liberal split was now irrevocable, and Asquith refused to serve under Lloyd George.

Assad, Hafiz al (1928–   ) *Syrian statesman and soldier; president since 1971.*
Born into a poor family, Assad was educated at military academies in Syria and the Soviet Union, graduating as an air-force pilot. His support for the Ba'ath party led to his dismissal from the armed forces in 1961. After the Ba'athists' seizure of power in 1963, Assad was appointed commander-in-chief of the air force (1965) and, following a second coup, minister of defence (1966). In 1970 a power struggle between the civilian and military wings of the party brought Assad to power: he was elected president the following year.

Domestically, Assad's role has been marked by economic progress (based on oil revenues and increased irrigation) and the repression of political opponents. An uprising of Muslim extremists (1979–82) was put down with great brutality. Assad was re-elected in plebiscites in 1978 and 1985, becoming the longest-serving ruler of independent Syria. In foreign affairs, Assad's role has been characterized by friendliness to the Soviet Union, shifting relations with his Arab neighbours, and (until tentative negotiations were begun in 1991) uncompromising hostility to Israel. In 1976 Syrian troops intervened in the Lebanese civil war and went on to occupy most of the country, clashing with invading Israeli forces in 1982. Allegations that Syria was sponsoring international terrorism led to deteriorating relations with the west in the 1980s. These were repaired in 1991 when Assad's hostility to the Ba'athist regime in Iraq led him to side with the coalition forces in the Gulf War. Subsequently Syria played a key role in the release of Western hostages held by various political factions in Lebanon.

Astaire, Fred (Frederick Austerlitz; 1899–1987) *US dancer and film star, described by both George BALANCHINE and Rudolph NUREYEV as the world's greatest dancer.*
Born in Omaha, Nebraska, he and his sister Adele began to dance in the music halls at an early age. What became a seasoned vaudeville act broke up, however, when Adele married and retired from the theatre in 1932. The thirty-three-year-old Astaire, now on his own, was attracted to Hollywood, where he submitted himself for the customary screen test. The verdict 'Can't act. Slightly bald. Can dance a little...' is now part of film history. Nonetheless he was given a small part in *Dancing Lady* (1933) with Joan CRAWFORD. This moderate success was followed by a series of musicals made with the unknown singer and dancer Ginger ROGERS, which established Astaire as an international star. *Flying Down to Rio* (1933), *The Gay Divorcée* (1934), *Roberta* (1935), *Top Hat* (1935), *Follow the Fleet* (1936), *Swing Time* (1936), *Shall We Dance?* (1937), and *The Story of Vernon and Irene Castle* (1939) capitalized upon Astaire's wit, elegance, and virtuoso dancing, supported by the rags-to-riches success story and the unobtrusive dancing of Rogers as Astaire's foil. After this partnership came to an end in 1939, Astaire made many more successful musical films with other partners, including *Easter Parade* (1948) with Judy GARLAND, *Daddy Long Legs* (1955) with Leslie Caron (1931–   ), *Funny Face* (1957) with Audrey Hepburn (1929–   ), and *Silk Stockings* (1957) with Cyd Charisse (1921–   ). He also played several dramatic roles, including a part in the apocalyptic saga *On the Beach* (1959). As an octogenarian Astaire continued to play small straight parts, but he is best remembered for his inventive choreography and his incomparable grace and skill as a dancer.

Aston, Francis William (1877–1945) *British scientist, who surveyed the periodic table of the elements with his mass spectrograph, which he designed himself and which has since become a standard tool in atomic physics. He was awarded the 1922 Nobel Prize for Chemistry.*

The son of a metal merchant, Aston studied chemistry at Birmingham University, where he pursued his first research interests before accepting an appointment as works chemist in a Wolverhampton brewery in 1900. Aston returned to research in 1903, when he joined J. H. Poynting (1852–1914) at Birmingham University to work on methods of developing efficient X-ray discharge tubes. In 1910 Aston moved to the Cavendish Laboratory at Cambridge, where he remained for the rest of his life, apart from the years of World War I spent at the Royal Aircraft Establishment, Farnborough. In 1919 Aston made accurate measurements of the relative atomic masses of a large number of elements using a mass spectrograph that he had built himself. An improved spectrograph, ready for use in 1927, enabled him to complete a survey of all the known elements by 1935. For this Aston worked alone in a remote corner of the Cavendish.

Aston's work was fundamental to the elucidation of atomic and nuclear structure so successfully carried out at the Cavendish under RUTHERFORD during the inter-war years. He also demonstrated that the existence of the isotopes, first described by Frederick SODDY in 1913, was widespread throughout the periodic table. His measurements achieved a high degree of accuracy and enabled small discrepancies between Prout's integral values of relative atomic masses to be revealed. These differences led Aston very close to the discovery of atomic energy.

Astor, Nancy Witcher Langhorne, Viscountess (1879–1964) *American-born British politician, the first woman to sit in the House of Commons (1919–45).*

She was born in Danville, Virginia, the daughter of a Confederate soldier who became a wealthy tobacco auctioneer. After her first marriage ended in divorce, she visited England, where she met, and in 1906 married, the proprietor of the *Observer* newspaper, Waldorf, later 2nd Viscount, Astor (1879–1952). Lady Astor presided over a fashionable circle of friends at their home at Clivedon until her husband became Conservative MP for Plymouth (1910), when they moved to his constituency. In 1919 he succeeded to the viscountcy and, following a widely publicized and extravagant campaign, his wife was elected in his place as MP for Plymouth. Lady Astor was the first woman to sit in the Commons; although Countess Markiewicz (?1868–1927), a Sinn Feiner, had been elected in 1918, she had not taken her seat. A woman of independence, Lady Astor spoke out for the causes about which she felt deeply, notably temperance and women's rights, rather than dutifully following the party line. In the 1930s she supported CHAMBERLAIN's appeasement policy, but was strongly anti-Nazi. Similarly, though hotly opposed to communism (in 1931 she visited the Soviet Union with George Bernard SHAW), she condemned the anticommunism of Senator McCARTHY in the 1950s. Lady Astor retired from politics in 1945.

Atatürk, Kemal (Mustafa Kemal; 1881–1938) *Turkish statesman and first president of modern Turkey (1923–38). He took the name Atatürk (father of the Turks) in 1934, to encourage the western use of surnames. Al-*though he was a ruthless dictator he modernized many aspects of Turkish life.*

Born in Salonika, the son of a customs official, Atatürk was educated at the military academy in Istanbul. After his graduation in 1902, he advanced through the ranks of the army, serving in the Turkish cavalry against the Italians in Libya (1911) and the Bulgars in the Balkan wars (1912–13). During World War I he served as divisional commander at the Dardanelles, which he successfully defended against British, French, and Anzac troops.

Opposed to the postwar emasculation of Turkey, Atatürk resigned from the army in 1919 and began to work for Turkish sovereignty. He established a provisional government (1920) in opposition to the government under Allied control and successfully led Turkish forces (1921–22) against the Greeks, who were occupying much of Asia Minor. Following the peace treaty signed with Britain in 1923 and the proclamation of the Turkish republic, when the Ottoman sultan was deposed, he became president. He immediately embarked on a programme of sweeping social and political reform, beginning with the abolition of the Caliphate, the traditional religious institution. He remained in power until his death from cirrhosis in 1938.

Atget, (Jean) Eugène (1856–1927) *French photographer, whose studies of Paris and Parisians are now widely acclaimed.*

Atget was born in Bordeaux into a bourgeois family but, orphaned young, was brought up by an uncle – a stationmaster in the Gironde. In his teens, Atget went to sea but he later turned to acting. During this time he met the actress who was to be his lifelong companion. They toured provincial theatres together but Atget's short stature denied him all but minor roles. At the age of forty Atget finally settled in Paris. He first tried painting and then, in 1898, set himself up as 'photographe d'art'. He managed to scrape a living by selling his photographs of the city's historic buildings, fountains, and other architectural features to museums and galleries. However, he also recorded other aspects of Paris: the people both in splendour and poverty, the streets and houses, and the natural beauty of the trees and flowers. Atget's work is even more remarkable in view of the relatively simple equipment he used, requiring long exposure times. He therefore often worked in the early mornings to avoid traffic, giving his pictures a characteristic light and atmosphere. Artists, including BRAQUE and UTRILLO, bought his pictures as aides-mémoire and he was making a modest living. World War I caused a reversal in his fortunes, although he still sold his work sporadically and in 1921 received a commission to photograph the brothels of Paris to illustrate a book. Only towards the end of his life did his work receive due recognition. In 1926, four of his pictures were published in *La Révolution surréaliste*, including his famous study of tailor's dummies in a shop window. In his final year, Atget's work came to the attention of Man RAY's assistant, Berenice Abbott, who was largely responsible for preserving his pictures and presenting them to the world.

Atiyah, Sir Michael Francis (1929–)
British mathematician, who has made significant contributions to topology and algebraic geometry. He was knighted in 1983. Since 1990 he has been president of the Royal Society.

Born in London, Atiyah was educated at Victoria College in Egypt, Manchester Grammar School, and Trinity College, Cambridge. He was Savilian Professor of Geometry at Oxford (1963–69) before accepting the chair of mathematics at the Institute for Advanced Studies at Princeton in 1969. He was Royal Society Research Professor at the Mathematical Institute in Oxford (1973–90) and was appointed Master of Trinity College, Cambridge, in 1990. Atiyah has developed a number of sophisticated modern mathematical techniques in the fields of algebraic geometry and differential equations. Some of them have been deployed in such areas as thermodynamics and particle physics. His collected works were published in 1988.

Attenborough, Sir David Frederick (1926–) *British naturalist and broadcaster, whose much-acclaimed television series have presented to a wide audience the spectacular array of plants and animals living on this planet. He was made a CBE in 1974 and became a fellow of the Royal Society in 1983. He was knighted in 1985.*

The younger brother of the film director Sir Richard ATTENBOROUGH, David Attenborough read zoology and geology at Clare College, Cambridge. After a brief spell in the Royal Navy, he joined a publishing house as an editorial assistant (1949–52) before moving to BBC television as a trainee producer. Attenborough conceived the idea of a TV series based on animals in their natural habitats and in 1954 the first *Zoo Quest* programmes were filmed in Sierra Leone. In the second series, Attenborough took over as presenter

and also wrote the first of several books to accompany the programmes, *Zoo Quest to Guiana* (1956). In 1965, Attenborough was appointed controller of the BBC's newly created second channel, BBC 2. He was responsible for overseeing the production of such notable series as Jacob Bronowski's *The Ascent of Man* and Kenneth Clarke's *Civilization*. In 1968 he was made director of programmes for both BBC TV channels but four years later he resigned to return to filmmaking and writing. *The Tribal Eye* (1976), a series concerned with art in so-called 'primitive' societies, was followed by an extensive series about evolution called *Life on Earth* (1978). It was highly praised and the companion book, *Life on Earth* (1979), became a best-seller. This was followed by the equally successful series, *The Living Planet* (1983), in which Attenborough presented examples of how plants and animals are adapted to their environments. A further series, *The Trials of Life* (1990), examined the processes by which various species survive. Attenborough was a member of the Nature Conservancy Council (1973–82) and is a trustee of the UK branch of WWF (Worldwide Fund for Nature), the British Museum, and the Royal Botanic Gardens at Kew. In 1991 he was elected president of the British Association. He is known for his passionate advocacy of international action to save wildlife and their habitats from destruction by human activities.

Attenborough, Sir Richard Samuel (1923–) *British actor, film producer, and director. He was knighted in 1976.*

Born in Cambridge, Attenborough studied at the Royal Academy of Dramatic Art where he met Sheila Sim (1922–), who became his wife. In 1942 he made his film debut in Noël COWARD's *In Which We Serve*, co-directed by David LEAN, and his West End stage debut as Ralph Berger in *Awake and Sing* at the Arts Theatre.

After war service with the RAF (1943–46), during which he worked with the RAF film unit, he resumed his acting career both on stage and in films. Of his early roles he is perhaps best remembered for his portrayal of the young thug Pinkie in the Boulting Brothers' film of Graham GREENE's *Brighton Rock* (1947), a part he had earlier created on stage at the Garrick in 1943. During the 1950s he made numerous films including *Private's Progress* (1955), *Brothers in Law* (1957), and *Dunkirk* (1958), as well as appearing with his wife in the origi-

25 Auden, W(ystan) H(ugh)

nal cast of Agatha CHRISTIE's record-breaking *Mousetrap*.

At the end of the fifties he set up two production companies with Bryan Forbes (1926–   ) and others. This resulted in such grimly realistic films as *The Angry Silence* (1959), in which he acted and was co-producer with Forbes, and *Seance on a Wet Afternoon* (1963), which he produced and acted in and which won him Best Actor Award at the San Sebastian Film Festival as well as a British Film Academy Award.

With these and other successes behind him he turned to directing. *Oh! What a Lovely War* (1968) won sixteen international awards, including the Hollywood Golden Globe. This was followed by other award-winning films: *Young Winston* (1972) and *A Bridge Too Far* (1976). In 1982 he fulfilled a longstanding ambition by directing and producing the internationally acclaimed *Gandhi*, which won the Hollywood Golden Globe and eight Oscars. A leading spokesman for the British cinema, he enjoyed further critical success in 1987 with the film *Cry Freedom*, about the death of Black activist Steve Biko in South Africa. He served as chairman (1982–87) of the British film company Goldcrest and was deputy chairman (1980–86) of the Channel Four television company. In 1987 he became Goodwill Ambassador for UNICEF.

Attlee, Clement Richard, 1st Earl (1883–1967) *British statesman and Labour prime minister (1945–51), whose government created the 'welfare state'. He was created an earl in 1955.*

Born in London, the son of a lawyer, Attlee became a socialist in 1907, influenced by his reading and by direct contact with poverty in London's East End. He qualified as a barrister and then taught at the London School of Economics from 1913 to 1923. With a distinguished record at Gallipoli and in France during World War I, Major Attlee, as he was known, became mayor of Stepney in 1919 and MP for Limehouse in 1922. Under Ramsay MACDONALD, he held various junior posts in the governments of 1924 and 1929–31 but then refused to serve in the national government. In 1935 he was elected leader of the Labour Party and the opposition, with strong trade union support, advocating 'all practicable support' to the opponents of fascism in Europe while opposing massive rearmament. During World War II, he served in CHURCHILL's coalition as lord privy seal (1940–42), secretary for the dominions (1942–43), and lord president of

the Council (1943–45). From 1942 to 1945 he was deputy prime minister.

With Labour's dramatic victory in 1945, Attlee became the first Labour premier to have an overall majority in the Commons. Quietly but firmly, he carried through in difficult economic circumstances a wide-ranging series of major reforms that laid the foundations of the welfare state. The Bank of England was nationalized, as were the coal industry, transport, gas and electricity, communications, and, later, iron and steel manufacture. A programme of social security, based on W. H. BEVERIDGE's report of 1942, was implemented with the National Insurance Act (1946); the National Health Service was introduced by Aneurin BEVAN. Foreign policy, in the hands of Ernest BEVIN, caused controversy within the party, the left wing opposing reliance on the USA against the Soviet Union. Attlee appointed Lord MOUNTBATTEN to oversee the granting of independence to India and the establishment of the state of Pakistan. Burma and Ceylon also achieved independence under Attlee's administration. But his government failed to deal adequately with Palestine (Britain withdrew in 1948) or with Africa and his last years in office – years of continuing austerity – were dogged by political divisions and misfortunes.

Attlee lost the election of 1951, remaining party leader until succeeded by Hugh GAITSKELL in 1955. He published his memoirs, *As It Happened*, in 1954.

Auden, W(ystan) H(ugh) (1907–73) *British poet, who became a naturalized US citizen in 1946.*

Born in York and brought up in Birmingham, where his father became professor of public health at the university, Auden was educated at Gresham's School and Christ Church, Oxford. While still at Oxford, in the 1920s, Auden established his pre-eminence among the young writers who became the leading left-wing poets of the next decade. *Poems* (1930), *The Orators* (1932), and *Look, Stranger!* (1936) made his reputation as a witty and technically accomplished lyricist. He collaborated with his friend Christopher ISHERWOOD on several plays, including *The Dog Beneath the Skin* (1935), and *The Ascent of F6* (1936). He also edited the *Oxford Book of Light Verse* (1938).

After a spell as an ambulance driver in the Spanish Civil War, Auden realized that the antifascists were powerless to stop the imminent European war; in 1939 he and Isherwood emigrated to the USA. Auden settled in New

York but never entirely severed connections with Britain, especially Oxford, where he was professor of poetry from 1956 to 1961. Auden continued to publish volumes of new poetry, among them *Another Time* (1940), *For the Time Being* (1944), *The Age of Anxiety* (1947, which was awarded a Pulitzer Prize), *Nones* (1951), *The Shield of Achilles* (1956), *Homage to Clio* (1960), and *City Without Walls* (1969). He also collaborated with Chester Kallman on libretti, including STRAVINSKY's *The Rake's Progress* (1951) and Hans Werner HENZE's *Elegy for Young Lovers* (1961). His selections from the writings of Kierkegaard (1952, 1955) are indicative of the way his thought turned away from the Marxist preoccupations of the 1930s to a kind of Christian existentialism. He also edited several anthologies, wrote critical essays, and made a number of translations. In 1966 he published his *Collected Shorter Poems 1927–57* and two years later his *Collected Longer Poems.*

Auric, Georges (1899–1983) *French composer, best known for his ballets and film scores.*

Born at Lodève, he studied first at the Montpellier Conservatory, then at the Paris Conservatoire (1913) and with Vincent D'INDY at the Schola Cantorum (1914–16). He became acquainted with SATIE and COCTEAU, under whose influence he and five other young French composers (including MILHAUD, HONEGGER, and POULENC) formed the group Les Six as a reaction against the influence of Wagner and Debussyan impressionism on French music. During the 1920s Auric wrote several pieces for DIAGHILEV's Ballets Russes, notably *Les Matelots* (1925), and was one of the pianists in the first performance of STRAVINSKY's *Les Noces.* His piano sonata (1930–31), in fact, shows the considerable influence that Stravinsky had on his work. His film score for Cocteau's *Le Sang d'un poète* (1929) marked the beginning of his important contribution to this field, which included the scores for René CLAIR's *À nous la liberté* (1932), Cocteau's *La Belle et la bête* (1946), *The Lavender Hill Mob* (1951), and *Moulin Rouge* (1952).

After World War II Auric turned again to ballet, his later works including *Phèdre* (1950) and *Coup de feu* (1952). He also held a number of administrative posts, among them director of the Paris Opéra and of the Opéra-Comique (1962–68), in which he helped to re-establish French operatic life.

Austin, Herbert, 1st Baron (1866–1941) *British automobile engineer. The founder of Austin motors, he was ennobled in 1936.*

The son of a farmer, Austin was educated at Rotherham Grammar School and, in preparation for a career as an architect, at Brampton Commercial College. More interested in engineering than architecture, Austin joined an uncle in Australia, where he served an apprenticeship with a Melbourne engineering firm. Soon after his return to Britain in 1893 he joined the Wolsey Sheep Shearing Machine Company, whom he persuaded to invest £2000 in plant to manufacture cars under Austin's personal management. The arrangement lasted from 1896 to 1905, when Austin resigned from Wolsey to set up as an independent manufacturer in a disused printing works in Longbridge (near Birmingham) with a capital of £20,000.

By 1914 Austin was successfully producing a thousand cars a year, but the great days of Austin did not begin until the 1920s with the launch of the Austin Seven, popularly known as the baby Austin. Nearly 300 000 models were produced before 1939, which justified Austin's claim that it provided a vehicle for the man who owned a motor cycle and 'yet has the ambition to become a motorist'. Profits in 1939 rose to nearly two million pounds, which made Austin an extremely wealthy man. However, unlike his contemporary and rival, Lord NUFFIELD, he did not create one of the great fortunes of his day. As an engineer he was more interested in designing cars than increasing his personal wealth; at the time of his death he was working on the design of the new cars he planned to build after World War II. His name lives on in many of BL's models, some of which are still made at Longbridge.

Austin, John Langshaw (1911–60) *British philosopher, the central figure of the Oxford ordinary language school of philosophy.*

Born in Lancaster, the son of an architect, Austin was educated at Balliol College, Oxford, where he remained for his entire academic career; he held the position of White's Professor of Moral Philosophy from 1952 until his early death from cancer in 1960. During World War II he served in the Intelligence Corps, rising to the rank of lieutenant colonel and playing an important role in the planning of the Normandy landings.

Austin was in many ways the most original and powerful British philosopher of the early postwar years. He felt that it was time for philosophy to make a genuine advance and that this could only be achieved by philosophers eschewing ambitious programmes, abandoning jargon, and tackling cooperatively a range of simpler and more clearly defined problems. He also began by insisting on the need to attend to the meanings of the terms used in ordinary language before attempting to build complex theories of perception, truth, and knowledge. Such an approach is best seen in his two important papers 'A plea for excuses' (1956) and 'Ifs and cans' (1956) as well as in his posthumously published lectures, *Sense and Sensibilia* (1962).

Austin, however, advanced from merely seeing the study of ordinary language as an indispensable preliminary to philosophical analysis to perceiving the need for a theory of language itself. He distinguished between statements that are constative, i e that can be true or false, and those he called performative, such as 'I promise', which do something other than simply assert something to be or not to be the case. Precisely how utterances come to have what Austin called an 'illocutionary force' was the main theme of his posthumously published lectures, *How to Do Things with Words* (1962).

Despite the powerful influence exercised by Austin over many of his philosophical colleagues he left no disciples. His methods were too idiosyncratic and too demanding to be easily assumed. Consequently, while many have discussed the problems raised by Austin, no one has really attempted to complete his programme.

Avery, Oswald Theodore (1877–1955) *US bacteriologist, who established that deoxyribonucleic acid (DNA) was responsible for causing heritable genetic changes in bacteria, thereby contributing to its identification as the carrier of genetic information in all living organisms.*

Born in Halifax, Nova Scotia, Avery attended Colgate University (graduating AB in 1900) and received his medical degree from Columbia University in 1904. In 1913, after working at the Hoagland Laboratory, New York, researching and lecturing in bacteriology and immunology, he moved to the Rockefeller Institute Hospital. His studies of various types of *Pneumococcus* bacteria responsible for causing lobar pneumonia revealed that differences in immunological specificity, i e antigenic characteristics, were determined by variations in polysaccharide components of the cell surface coat. In 1944, Avery and co-workers published the results of their work on genetic transformation, again using strains of *Pneumo-*

coccus. It was known that an extract from a virulent strain of this bacterium was able to transform a harmless strain into the virulent form. By chemically purifying and analysing this transforming principle, Avery and his colleagues showed that it was not a protein, as had hitherto been thought, but DNA.

Avon, 1st Earl of See EDEN, (ROBERT) ANTHONY, 1ST EARL OF AVON.

Ayckbourn, Alan (1939–) *British playwright and director. He was made a CBE in 1987.*

After several years as a stage manager and actor in repertory, he was appointed director of productions (now artistic director) at the Stephen Joseph Theatre in the Round at Scarborough in 1970. The theatre soon became established as a proving ground for Ayckbourn's prolific output of plays, prior to the transfer of many of them to London's West End. The first of his plays to make this transition was *Mr Whatnot* (1964), although the first to attract critical attention was *Relatively Speaking* (1967). This bitter comedy set the pattern for much of Ayckbourn's early work, playing on the foibles and frustrations of British suburbia while adroitly toeing the line between hilarious and easily accessible farce and the author's bleak view of contemporary middle-class society.

Ayckbourn's plays are particularly notable for their technical inventiveness. *Absurd Person Singular* (1973), for instance, is set in three kitchens on three successive Christmas Eves; *The Norman Conquests* (1974) is a trilogy of full-length plays with the same setting; *Bedroom Farce* (1977), which was presented at the National Theatre, portrays events in three separate bedrooms simultaneously; while *Way Upstream* (1982) requires the use of a real boat on a flooded stage.

Although Ayckbourn's early plays were criticized for their shallowness, his later plays have been characterized by increasing sombreness, with the lines between farce and tragedy stretched ever tauter. Ayckbourn's box-office appeal and his continued success with the critics (marked by his many awards and television adaptations of his plays) has established him as the leading writer for the commercial theatre in the UK. His most recent works include *A Chorus of Disapproval* (1985), *Man of the Moment* (1990), and *Wildest Dreams* (1991), his 42nd play.

Ayer, Sir Alfred Jules (1910–89) *British philosopher, responsible for introducing the* *principles of logical positivism of the Vienna Circle to British philosophers. He was knighted in 1970.*

Born in London, Ayer was educated at Christ Church, Oxford, where he was a pupil of Gilbert RYLE. On Ryle's suggestion Ayer, after graduating in 1932, enrolled at the University of Vienna and attended the regular meetings of the Viennese positivists centred around SCHLICK. Ayer became converted to the exciting new doctrines vigorously argued by the logical positivists and consequently, on his return to Oxford (1933), he began to prepare an account of their work that eventually became *Language, Truth, and Logic* (1936), one of the most successful philosophical works of the century. In it Ayer followed the positivists in formulating the verification principle, with which he could demonstrate the meaninglessness of all metaphysics and theology, and the vacuousness of all ethical propositions. Such views did not impress many of Oxford's more traditional philosophers, who resisted for some time attempts to find a suitable academic appointment for Ayer in Oxford. Eventually he was appointed to the staff of his own college, Christ Church. After service with the Welsh Guards during World War II Ayer returned briefly to Oxford in 1945 before being appointed (in 1946) Grote Professor of the Philosophy of Mind and Logic at University College, London. He became Wykeham Professor of Logic at Oxford in 1959, a post held until his retirement in 1977.

Although in his later work Ayer qualified many and rejected some of the claims made in his first book, he continued to adhere to the spirit if not the letter of the positivist programme laid down in the Vienna of the early 1930s. Working mainly in the field of epistemology, he tried to show in such works as *The Foundations of Empirical Knowledge* (1940) and *The Problem of Knowledge* (1956) how we could justifiably, on empiricist assumptions, come to know truths about such phenomena as the external world, other minds, and the past. These two works were supplemented by four collections of his philosophical essays published in 1954, 1963, 1969, and 1972.

Ayub Khan, Mohammed (1907–74) *President of Pakistan (1958–69). His attempt to establish a new democratic system in Pakistan failed because of lack of popular support.*

Born in Abbottabad of Indo-Iranian origin, the son of an officer in the British army, Ayub Khan was educated at the University of Aligarh and at the Royal Military College,

Sandhurst. He was commissioned in the British army in 1928 and during World War II served as a battalion commander in Burma (1942–45). When Pakistan achieved independence in 1947 he became the first commander-in-chief of the Pakistani army (1951–58). He also served as minister of defence (1954–55) and, when martial law was declared in 1958, was appointed supreme commander of the armed forces. Shortly afterwards he took over the presidency from Iskander Mirza (1899–1969), who was sent into exile. Promoting himself to field marshal in 1959, he ruled Pakistan through the army for the next ten years.

His policy of 'basic democracies', which encouraged political responsibility at a local level, was formalized in a new constitution (promulgated 1962). Martial law was lifted the same year, but all civil liberties were abolished. In foreign policy he softened Pakistan's approach towards China but entered into conflict with India over Kashmir in 1962 and 1965. Opposition to the peace settlement reached between India and Pakistan at Tashkent in 1966, as well as to his increasingly repressive style of government, led to his ultimate downfall. He was forced to resign in 1969 in favour of General Agha Yahya Khan (1917–80) after a series of riots in protest against the imprisonment of political dissidents.

Azikiwe, Nnamdi (1904–) *First president of Nigeria (1963–66), considered by many as the father of modern Nigeria. He has been leader of the Nigerian People's Party since 1979.*

Born in northern Nigeria of Ibo descent, Azikiwe was the son of an army clerk. He was educated in mission schools before going to the USA in 1925 to study political science and anthropology at Lincoln and Pennsylvania universities. Returning to Africa in 1934, he became editor-in-chief of the *African Morning Post* in Ghana before moving to Lagos in 1937 to establish the *West African Pilot*.

Azikiwe became involved in politics in 1944, when he helped to organize a gathering of forty political, labour, and educational groups under the auspices of the National Council of Nigeria and the Cameroons (NCNC). Becoming president of the NCNC in 1946, he led a delegation to London in 1947 to oppose the constitution being implemented by the governor. Elected to the legislative council of Nigeria in 1947, he entered the western house of assembly in 1952 but changed to the eastern house in 1954 because the controlling Yoruba Action group obstructed his election to the federal assembly. He was premier of the eastern region until 1959, when he was elected governor-general and commander-in-chief of the Federation of Nigeria. Azikiwe became president of the newly independent Nigeria in 1963. Following the military coup in January 1966, Azikiwe joined the Biafran secessionist government, travelling abroad to seek international recognition for the new state. Favouring reunification in 1969, he remained out of the country until the war was over, returning in 1972 to become chancellor of Lagos University. In 1979, as leader of the new Nigerian People's Party, he re-entered the political arena in an unsuccessful attempt to contest the presidential election.

Azorín (José Martínez Ruiz; 1874–1967) *Spanish essayist, novelist, and playwright.*

Born at Monóvar, Alicante, Azorín was initially interested in the law, which he studied at Valencia, Granada, and Salamanca. In 1896 he settled in Madrid and embarked on a career in journalism. At first supporting liberal ideals for the reform and modernization of Spain, which would drive it in the direction of other European democracies, he developed in time a more conservative view as he came to appreciate the values unique to the nation in its history, customs, and landscape. His writings, which consist mainly of short impressionistic sketches originally published in journals and newspapers, typify the peculiarly Spanish sensitivity to the tension between past and present of the group of writers known as the 'Generation of 1898'.

Between 1907 and 1919 Azorín played an active role in politics and was elected five times as a representative of the Conservative Party in the Cortes. He also actively backed liberal policies during the Republic (1931–36) but withdrew to Paris during the civil war (1936–39), adopting a neutral position. On returning to Madrid and adjusting to the changed circumstances of FRANCO's rule, he devoted himself to a final series of autobiographical novels and several unsuccessful humorous books; he also continued to publish pieces in newspapers and to compile his memoirs.

Following several early autobiographical impressionistic novels – for example, *Antonio Azorín* (1903), the novel from which he drew his pseudonym, and *Las confesiones de un pequeño filósofo* (1904) – Azorín's work (until the mid-1920s) consisted chiefly of journalistic vignettes. In the 1920s he was influenced by surrealism, as reflected in the experimental stories of *Blanco en azul* (1929; translated as *The Sirens and Other Stories*, 1931). He also became an influential drama critic and was responsible for publicizing avant-garde European playwrights in Spain. He wrote a number of plays at this time, including *Brandy, mucho brandy* (1927) and *Cervantes o la casa encantada* (1931). An edition of his 'complete' works appeared in 1953, but over twenty further volumes have been published since.

B

Baade, Wilhelm Heinrich Walter (1893–1960) *German-born US astronomer, who made valuable contributions to knowledge of stellar and galactic evolution.*

The son of a schoolteacher, Baade was educated at the universities of Munster and Göttingen, where he obtained his PhD in 1919. He began his career at the Hamburg Observatory, but moved to the USA in 1931 in search of bigger telescopes. The rest of his career, until his retirement in 1958, was spent at the Mount Wilson and Palomar Observatories in California.

Baade's most significant work was carried out in the 1940s, when the Los Angeles blackout imposed during World War II provided an unusually dark sky. As an enemy alien he was permitted to use the 100-inch telescope while other astronomers were doing their war work. In these favourable conditions he began a careful survey of the Andromeda galaxy, finding that, for the first time, stars in the central region could be resolved. He went on to distinguish two types of stars: population I stars – young, hot, and blue but found only in the arms of the galaxy; and the older population II stars – found in the central galactic area. Baade was able to use the distinction between the two stellar types to more than double the estimated age of the universe. He is also remembered for his discovery of two minor planets, Hidalgo in 1920 and Icarus in 1949.

Babbitt, Milton Byron (1916–) *US composer whose works show the influence of serialism and include much electronic music.*

Born in Philadelphia and originally a student of mathematics at Pennsylvania University, Babbitt later transferred to the faculty of music at New York University, where he became fascinated with the serial techniques of SCHOENBERG, BERG, and WEBERN. After graduation he studied with Roger SESSIONS. His first twelve-tone work, *Composition for Orchestra*, was completed in 1941 when he was on the staff of Princeton University. In 1946 he produced a paper on *The Function of the Set Structure in the Twelve-Tone System*, the first formal investigation into Schoenberg's compositional method. Babbitt became Conant Professor of Music at Princeton in 1960 and, among other posts, has taught at the New England Conservatory of Music and the Juilliard School.

Babbitt has been obsessed with the exploration of total serialism, in which rhythm, dynamics, timbre, and register, as well as notes, are all equally controlled and inseparable; to quote him: 'the twelve-tone set must absolutely determine every aspect of the piece.... I believe in cerebral music and I never choose a note unless I know why I want it there'. He believes the layman should not be expected to understand advanced contemporary music any more than he expects to understand advanced physics or mathematics. He was the first composer to work with the newly developed RCA synthesizer, which enabled him to realize a previously impossible precision in the execution of his complex works. These include *Composition for Twelve Instruments* (1948), the song cycle *Du* (1951), *Composition for Tenor and Six Instruments* (1960), *Philomel* (1964) for soprano and tape, *Reflections* (1975), *Melismata* (1982), and *Canonic Form* (1983). He won a Pulitzer Prize in 1982.

Babel, Isaak Emmanuilovich (1894–1941) *Russian short-story writer and playwright.*

The son of a tradesman, Babel was born and grew up in the Jewish community of Odessa. Determined to establish himself as a writer outside this closed Yiddish-speaking society, he went to St Petersburg where he met GORKI, who published his first two stories, written in Russian, in his magazine *Khronika* (1916). In World War I Babel at first served with the Tsarist army but in 1917 joined the Bolsheviks. His experience as a war correspondent with the First (Cossack) Cavalry in Poland in 1920, under the command of Semyon Budyonny (1883–1973), formed the basis of his first collection of stories, *Konarmiya* (1926; translated as *Red Cavalry*). The book brought him worldwide recognition, being received not so much as fiction as an eye-witness account of events of historical importance. The viewpoint of the stories, however, was not one of simplistic propaganda. The intellectual narrator observes not only the extreme violence and chaos of the Polish campaign but also the humanity of some of those involved. The stories caused some controversy in Russia and Budyonny accused Babel of neglecting to take into account the revolutionary objective of the campaign. Babel's position was secure, however, through the continued support of Gorki. In his second collection, *Odesskiye rasskazy* (1924; translated in *The Collected Stories*, 1955), Babel turned to the shady and brawling low life of the Odessa Jewish community of his youth. A second series of Odessa tales (written 1925–30) were autobiographical in theme. *Zakat* (1928), his first play, also dealing with the Odessa underworld but from a less comic point of view, was produced by the Moscow Art Theatre, but neither it nor *Mariya* (1935), a play about the October Revolution, met with much success.

Although Babel was sympathetic to the revolution, his writing had a moral complexity ill-suited to the dogma of socialist realism required of the arts under STALIN in the 1930s. He attempted to revise his stories (in 1932 and 1936) to conform to the new line but otherwise published nothing of importance. While Gorki's patronage had protected him from the first Stalinist purges, he was arrested in 1939 and, according to later Soviet reports, died in 1941. He was 'rehabilitated' in 1957 and his collected works, including new material, have been republished in the Soviet Union (1957 and 1966).

Bacon, Francis (1909–) *British painter, whose powerful and original pictures with their lurid colour and distorted figures convey a personal vision of the repulsiveness and horror of the human condition.*

Bacon was born in Dublin, of English parents. With little formal training, he began painting in the 1920s in London, where he worked as an interior designer. In 1934, after failing to attract encouraging critical attention, he virtually ceased painting. However, early in 1945 his *Three Studies for Figures at the Base of a Crucifixion*, painted the previous year, was exhibited. Horrific and awe-inspiring, this triptych, now in the Tate Gallery, London, made him overnight the most controversial painter in Britain. In the 1950s and 1960s his work was shown in many important international exhibitions. Another triptych of studies for a crucifixion, painted in 1962, is now in the Guggenheim Museum, New York. The crucifixion in this work becomes a repulsive butchery enacted in some indefinably sordid arena. Other well-known works are *Figure in a Landscape* (1946) and the portraits that he painted from photographs and film stills, such as the

screaming nurse from EISENSTEIN's film *The Battleship Potemkin* and Pope Innocent X from the painting by Velázquez.

Of his later work John Rothenstein wrote in 1974, 'The undisguised horrific subjects of many of his early paintings … have on the whole been replaced by subjects not intrinsically horrific but imbued with his obsession with human cruelty, vulnerability, loneliness and the pitiful indignity of men, and occasionally women, in solitude, unobserved.'

Baeck, Leo (1873–1956) *German Jewish rabbi and theologian who was one of the leading exponents of liberal Jewish theology in the twentieth century and spiritual leader of Germany's Jews during their struggle against Nazi persecution before and during World War II.*

Ordained in 1897, Baeck served as a rabbi in Oppeln, Silesia (1897–1907), in Düsseldorf (1907–12), and in Berlin (1912–42), where he also lectured in midrash and homiletics at the Hochschule für die Wissenschaft des Judentums. His *Das Wesen des Judentums* (1905; translated as *The Essence of Judaism*, 1948), in which Baeck stressed the dynamic evolutionary nature of religion, established him as an outstanding Jewish scholar. A later (1922) edition expanded Baeck's philosophy of a 'religion of polarity', which he regarded as a dialectic between 'divine mystery' and the 'ethical imperative'. He considered the Jews as exemplars of a supreme Judaistic morality and the sole progenitors of the faith.

In 1933, Baeck was appointed co-leader, with Otto Hirsch, of the National Agency of Jews in Germany and was increasingly engaged in the fight with the Nazis for Jewish rights. Baeck was arrested and imprisoned in Theresienstadt concentration camp in 1943. There he gave philosophy classes to the inmates and served as unofficial pastor. He also wrote *This People Israel: The Meaning of Jewish Existence* (1955), exploring the philosophical implications of the existence of a Jewish state. On 8 May 1945, Theresienstadt was liberated, the day before Baeck's scheduled execution. After the war he settled in England and served as president of the World Union for Progressive Judaism.

Baekeland, Leo Hendrik (1863–1944) *Belgian-born US chemist who discovered Bakelite.*

Born in Ghent and educated at the University of Ghent, Baekeland gained his doctorate in 1884. He was appointed shortly afterwards to the chair of chemistry at Bruges University, but after a honeymoon trip to the USA chose in 1889 to leave Belgium permanently. Baekeland initially worked in the field of photography, inventing Velox, a special photographic paper that permitted pictures to be printed in artificial light. He hoped to sell the discovery to Eastman Kodak for $25,000. However, before he could begin the negotiation George EASTMAN offered him one million dollars for the invention. With this totally unexpected fortune Baekeland retired to Yonkers, New York, to work in his private laboratory.

Here Baekeland dedicated himself to finding a substitute for shellac. As it takes 150 000 lac insects six months to produce one pound of shellac, there was clearly a great fortune to be made for anyone who could produce the substance artificially. Baekeland was aware, as indeed were most chemists, of the sticky resin that forms when phenols and aldehydes are heated together; with no apparent use, it merely clogged up valuable equipment. He found that when subjected to prolonged heating under pressure, the resin turned out to have a surprising number of useful properties. It was hard, insoluble, could be machined, moulded, coloured, dyed and, though light, was remarkably strong. It was in fact the first thermosetting plastic, named Bakelite in 1909. As president of the Bakelite Corporation (1910–39), Baekeland saw that his product gained worldwide use in both industry and the home.

Baez, Joan (1941–) *US folksinger and political activist, who spoke and sang for a generation of young people opposed to the Vietnam War. She is a Chevalier de la Légion d'honneur.*

She began her career by singing in coffee houses in Boston and New York and at the Newport Folk Festival in 1959; her first records were released in 1960. Her voice, singing mostly traditional ballads to her own guitar accompaniment, came as a revelation to many young people after years of neglect of folk music in the USA. Her fourth album, a live concert recording, included 'We Shall Overcome', which became the anthem of the civil rights and anti-war movements. Her tenth release was a set of songs by her friend Bob DYLAN. In 1965 she founded an Institute for the Study of Non-Violence, and her political activity took precedence over her music; in 1968 she married David Harris, who went to prison the next year for refusing to be drafted (they were later divorced). In 1975 she made a return to full-time singing, confessing frankly that

she needed the money. In *Diamonds and Rust* she had an electric folk-rock band and sung contemporary songs; the title song related to her affair with Dylan. She had remained active in politics, opposing the coup in Chile in 1973 and touring Latin America in 1981. Her later recordings include a live *European Tour* (1981) and *Ballad Book* (1984). In 1968 she published an autobiography, *Daybreak*; subsequent publications include *One Bowl of Porridge: Memoirs of Somalia* (1987).

Bailey, David (1938–) *British photographer whose images of London in the 'swinging sixties' have come to epitomize that era.*

The son of a tailor living in London's East End, Bailey left school at fifteen for a succession of dead-end jobs. While on National Service in Singapore in 1956, he bought a cheap copy of a Rolleiflex camera and discovered his vocation. In 1959, after working as a photographer's odd-job boy, he became assistant to one of London's leading fashion photographers, John French. His contract with the fashion magazine *Vogue* in 1960 signalled the take-off for Bailey's career. Using a 35-mm camera and outdoor locations, he brought a new reportage look to fashion features while his work with model Jean Shrimpton made her one of the emblems of the early 1960s. Bailey himself became a prominent figure of the swinging sixties' pop culture, and his idiosyncratic, often jokey, portraits of fellow celebrities remain as enduring images of that decade. Many were included in *Goodbye Baby and Amen* (1969), his personal review of those years. Bailey's work formed part of an exhibition at the National Portrait Gallery in 1971 and two years later he held his first one-man-show.

Beady Minces (1973) showed the diversity of Bailey's subjects, including landscapes, portraits, nudes, and reportage. *Trouble and Strife* (1980) is a collection of nude studies of his third wife, Marie Helvin; subsequent collections include *Nudes 1981–84* (1984) and *Imagine* (1985). At the same time Bailey has covered such other topics as life in a New Guinea tribe, the Vietnamese boat-people, and the townscape of his north London neighbourhood (*David Bailey's NW1*; 1982). He has also ventured into film-making, with a 1966 short entitled *GG Passion* and three television documentaries. His second wife was the French film actress Catherine Deneuve (1943–).

Baird, John Logie (1888–1946) *British electrical engineer, a pioneer in the development of television.*

Born in Helensburgh, Scotland, the son of a clergyman, Baird displayed even as a boy all the instincts and aptitudes of an inventor. He was educated in Glasgow at the Royal Technical College and the university and, pronounced unfit for military service during World War I, became superintendent of an electric power company. After eight years trying to make his fortune in a variety of unlikely ventures, Baird found himself in Hastings recovering from a breakdown and penniless. At this point he began to seek ways to use his engineering skills and in 1923 conducted his first experiments using radio waves to transmit pictures. In the following year he succeeded in transmitting an image of a Maltese cross a few yards. With this primitive demonstration, Baird was able to persuade the wealthy store owner Gordon Selfridge to provide financial backing. In 1926 Baird gave the first public demonstration of his system and several companies were then set up with substantial capital invested by a number of backers. In 1928 he produced both colour and stereoscopic television and, a generation before the satellite age, succeeded in transmitting a picture across the Atlantic. The pictures themselves were of very poor quality.

Despite this poor quality the BBC agreed in 1929 to run a series of experimental broadcasts for the thirty TV sets then in existence. Soon, however, a powerful challenge to Baird was mounted by EMI. When the BBC finally agreed to introduce regular television broadcasts in 1936, they decided to test the competing EMI and Baird systems by inviting them to transmit alternately. The trial period lasted until February 1937, when Baird's mechanical-scanning system was dropped in favour of EMI's all-electronic system. Baird sought other markets for his system by trying to persuade cinema owners to install large-screen television in their theatres for showing live news. Finding little interest in his plans, Baird returned once more to the development of colour television. In some financial difficulties, Baird hoped that when the BBC resumed transmitting in 1946 they would at last invest in some of his newer systems. Before this could happen Baird died suddenly from pneumonia, the same month that the BBC restored its postwar television service.

Baker, Dame Janet (Abbott) (1933–) *British mezzo-soprano. She was created a CBE in 1970 and a DBE in 1976.*

Baker studied with Helene Isepp and later with Meriel St Clair in London, winning second prize in the Kathleen Ferrier Competition in 1956, when she also joined the Glyndebourne chorus. Her operatic debut in 1957 with the Oxford University Opera Club (as Rosa in Smetana's *The Secret*) marked the beginning of her regular appearances in opera and on the concert platform throughout Britain. In 1959 Baker played Eduige in Handel's *Rodelinda* and the heroic roles in his *Ariodante* (1964) and *Orlando* (1966); she also sang in Raymond Leppard's revival of early Italian opera: Diana/Jupiter (1970) in Cavalli's *Calisto* and Penelope (1972) in Monteverdi's *The Return of Ulysses.*

Her debut at Covent Garden was as Hermia in Britten's *A Midsummer Night's Dream* (1966), and BRITTEN wrote for her the part of Kate Julian in his television opera *Owen Wingrave* (1971) and the dramatic cantata *Phaedra* (1975). Her performances as Dido in both Purcell's *Dido and Aeneas* and Berlioz's *The Trojans* won great acclaim, and she excelled as an interpreter of Bach, Schubert, and Mahler. Baker's personal magnetism won her a large new audience in her New York debut of 1966. She retired from opera in her prime, her farewell performance (1982) being Orfeo in Gluck's opera of that name, and has published her reminiscences, *Full Circle* (1982).

Balanchine, George (Georgi Melitonovitch Balantchivadze; 1904–83) *Russian-born US dancer and choreographer.*

Son of a Georgian folk musician, Balanchine studied at the Imperial Ballet School and the Petrograd Conservatory of Music. In his early years he was torn between his love of dance and his desire to become a composer. He choreographed his first piece, *La Nuit*, at the age of sixteen when he began to experiment in various balletic styles and types of music. In 1924, while touring Europe, he was engaged by DIAGHILEV in Berlin and within a year, at the age of twenty-one, was appointed principal choreographer to the nomadic Ballets Russes. During the next four years Balanchine choreographed ten ballets, the most significant of which were *Apollo* (1928), set to the music of STRAVINSKY, and *The Prodigal Son* (1929), set to the music of PROKOFIEV.

Balanchine travelled widely during this period and in 1934 he became a co-founder of the School of American Ballet in New York. In the same year he also choreographed his first American ballet, *Serenade*, which further developed his interest in preserving and transforming classical Russian ballet. Balanchine insisted that the company's high standard of performance was maintained and in 1948 the troupe became the New York City Ballet – recognized as one of the world's most important companies. Balanchine choreographed one hundred ballets and revivals for the company with a wide range of musical scores. The music of Stravinsky remained the primary inspiration throughout his career, however, and was used in works such as *Agon* (1957), *Monumentum pro Gesualdo* (1960), and *Rubies* (1967). Other works included *Concerto barocco* (1941), *Danses concertantes* (1944), *Theme and Variations* (1947), and *Symphony in Three Movements* (1972).

Balcon, Sir Michael (1896–1977) *British film producer and production executive. He was knighted in 1948.*

Born in Birmingham, Balcon began his career as a regional film distributor in 1919 and produced his first film, *Woman to Woman*, in 1923 with Alfred HITCHCOCK as art director, screenwriter, and assistant director. Subsequently Balcon founded Gainsborough Pictures (1928) and became director of production with Gaumont-British (1931) and MGM-British (1936), during which time he was responsible for films such as *Man of Aran* (1933), *The Man Who Knew Too Much* (1934), and *The Thirty-Nine Steps* (1935).

But the studio with which his name became synonymous was Ealing, where he was director and chief of production from 1937 to 1959. Classic comedies that were produced under his guidance included *Kind Hearts and Coronets, Passport to Pimlico,* and *Whisky Galore* (all 1949), *The Man in the White Suit* (1951), *The Lavender Hill Mob* (1952), and the last Ealing comedy, *The Ladykillers* (1955); many of these films featured Alec GUINNESS. *Hue and Cry* (1947), *The Blue Lamp* (1950), *The Cruel Sea* (1953), *The Divided Heart* (1954), and *Dunkirk* (1958) were among his other notable films. In 1964 he also served for a time as chairman of British Lion Films.

Baldwin, James Arthur (1924–87) *US novelist, playwright, and essayist. His literary work and polemical essays are among the most influential expressions of the predicament and aspirations of US blacks in the 1950s and 1960s.*

Baldwin was born in Harlem, the eldest of nine children, and educated at New York City schools. His stepfather was a preacher and at fourteen Baldwin followed in his footsteps, preaching at 'storefront' churches in Harlem.

Although he soon lost his faith, his gifts as a speaker continued to stir audiences during his many appearances later as a spokesman for the civil rights movement. At seventeen he left home to devote himself to writing. In 1948 he moved to Europe, living mainly in Paris. *Go Tell It on the Mountain* (1953), drawing on youthful experiences and covering one day in the lives of the members of a Harlem church, was immediately recognized as a major novel and established Baldwin as a leading black writer. *Giovanni's Room* (1956), set in Paris, dealt with sexual, especially homosexual, and racial relationships, subjects further explored in *Another Country* (1962). Baldwin wrote extensively as a journalist in the 1950s and on returning to America in 1956 committed these talents to the civil-rights struggle. In the 1960s he published increasingly challenging essays on racial questions. His first collection of essays, *Notes of a Native Son* (1955), was followed by two others, *Nobody Knows My Name* (1961) and *No Name in the Street* (1972). His play *Blues for Mr Charley* (1964) concerns a racial murder and was based on a notorious incident of the time. A collection of short stories appeared as *Going to Meet the Man* (1965).

Baldwin's writings on race have had a lasting impact on the national conscience in America, none more so than the two angry letters published as *The Fire Next Time* (1963). Detailing the degrading treatment and alienation of black people, Baldwin finds some hope that the condition of blacks may yet force the country to see the speciousness of its national myths of virtue, equality, etc. A similar view informs *A Rap on Race* (1971), a taped conversation with Margaret MEAD in which Baldwin is optimistic about the future, 'but not the future of this civilization'. Later works included the novels *Just Above My Head* (1979), the story of a gospel singer, and *Evidence of Things not Seen* (1986).

Baldwin, Stanley, 1st Earl (1867–1947) *British statesman and Conservative prime minister (1923–24; 1924–29; 1935–37). He accepted an earldom in 1937.*

Born at Bewdley, into a wealthy steel-manufacturing family, Baldwin managed the family business until he became an MP in 1908. In 1916 he became parliamentary private secretary to the chancellor of the exchequer Bonar LAW, and was then financial secretary to the Treasury (1917–21) and president of the Board of Trade (1921–22). In 1922 he joined the opposition to LLOYD GEORGE and became chancellor of the exchequer in Bonar Law's government (1922–23). Succeeding Bonar Law in 1923, he lost the election in December as a result of his policy of protective tariffs, but returned to power the following year. Under Baldwin, Neville CHAMBERLAIN, as minister of health, introduced notable social reforms, including a contributory pensions Act (1925) and an unemployment insurance Act (1929). Baldwin proclaimed a state of emergency in response to the General Strike of 1926. When it was over, his government passed the severe Trade Disputes Act 1927. This, together with widespread unemployment, led to his defeat in the 1929 election, fought by Baldwin under the slogan 'safety first'.

He served as lord president of the Council (1931–35) in the coalition national government, before succeeding MACDONALD as prime minister. His last government was responsible for the Hoare–Laval pact (1935), an Anglo-French plan to allow Italy to annex Ethiopia. Baldwin attempted to atone for the unpopularity of this measure by adopting an implacable moral stand against Edward VIII's desire to marry Mrs Simpson – a stand that ended in the king's abdication (see WINDSOR, DUKE OF). Baldwin retired in favour of Neville Chamberlain in 1937. He has been blamed for Britain's lack of military strength and for forcing Chamberlain to appease HITLER and MUSSOLINI in order to buy time to rearm.

Balenciaga, Cristóbal (1895–1972) *Spanish couturier who became one of the leading international designers of haute couture in the 1940s and 1950s.*

Balenciaga was born in a Basque village on the Bay of Biscay, where his father was a fisherman. A boyhood interest in sewing, encouraged by his mother, was turned to a source of income following his father's death. His first patron was a local aristocrat, the Marquesa de Casa Torres, who commissioned Balenciaga to make her a dress. After studying tailoring in San Sebastián, he started his own business there, moving to Madrid in 1932. The Spanish civil war of 1937 forced him to leave for Paris, where he opened premises in the prestigious Avenue George V.

Specializing in individually tailored and fitted designs, Balenciaga produced garments noted for their simplicity and boldness of design. His collections of the early 1950s contributed to the move away from the tightly waisted New Look originated by Christian DIOR to a looser semi-fitted style. This culminated in Balenciaga's tunic dress of 1955 and his 1957 chemise, which in the hands of his protégé, Hubert de Givenchy, became known as the 'sack'. Always a retiring man, Balenciaga closed his business in 1968. However, shortly before his death in 1972 he accepted, as his final commission, the design of the wedding dress for FRANCO's grand-daughter, Maria del Carmen Franco.

Balfour, Arthur James, 1st Earl of (1848–1930) *British statesman and Conservative prime minister (1902–05). He is famous for the Balfour Declaration (1917), which stated Britain's support for 'the establishment in Palestine of a national home for the Jewish people'. He was created an earl in 1922.*

Balfour was born in Scotland, a nephew of Robert Cecil, 3rd Marquess of Salisbury. He entered parliament as a Conservative MP in 1874. In 1887 he became chief secretary for Ireland in Lord Salisbury's government. He opposed the scandal of English absentee landlordism – 'killing home rule by kindness' was his phrase – but for his ruthless suppression of insurrection he was known as 'Bloody Balfour'. After succeeding his uncle as prime minister in 1902, he introduced the Education Act 1902, the Irish Land Purchase Act 1903, and the Committee of Imperial Defence (1904) to coordinate defence strategy in the Empire. By the entente cordiale (1904) Britain's supremacy in Egypt and French supremacy in Morocco were mutually recognized. In 1905 his government resigned after the party split over tariffs, but Balfour remained as party leader until 1911. He was first lord of the admiralty (1915–16) and then foreign secretary (1916–19), but the conduct of World War I was largely in LLOYD GEORGE's hands.

The Balfour Declaration, in a letter dated 2 November 1917 to Lord Rothschild (chairman of the British Zionist Federation), expressed support for the establishment of a Jewish nation in Palestine provided that safeguards be secured for 'existing non-Jewish communities'. Balfour played an important role at imperial conferences in the 1920s, helping to draft the Statute of Westminster (1931), which gave legislative independence to the self-governing dominions. Balfour was also a philosopher, the author of *Defence of Philosophical Doubt* (1879) and *Theism and Humanism* (1914).

Balla, Giacomo (1871–1958) *Italian futurist painter, who created some of the earliest non-objective paintings.*

Born in Turin, the son of a photographer, Balla worked as a lithographer and studied art before settling in Rome to work as a painter. In 1900 he visited Paris, where he learned the divisionist technique of the neoimpressionists. He introduced this technique to BOCCIONI and Gino Severini (1883–1966) who used it to convey movement and speed, which they saw as the most important elements of modern life. Through them Balla met other members of the futurist movement and in 1910 he signed the manifesto of futurist painters. His first futurist painting was *Dynamism of a Dog on a Leash*, which was based on multi-exposure photographic studies of movement and conveys the effect of a dog running by depicting each movement of the legs, tail, and leash simultaneously.

He soon moved towards less superficial representations of movement in such works as *Automobile and Noise* (1912) and *Car in a Race* (1914), which are an attempt to portray abstract expressions of speed and noise. He also experimented with sculptural constructs involving movement, colour, and sound but in about 1930 reverted to nonexperimental figurative painting.

Banda, Hastings Kamuzu (1905–   ) *First prime minister (1964–   ) and president (1966–   ) of Malawi. Welcomed as a national hero on his return to Nyasaland in 1958 after forty years abroad, Banda has established an austere one-party regime in the republic he helped to create.*

Born in Kasungu, the son of a peasant, Banda worked as an interpreter in the Rand goldmines in South Africa before travelling to the USA (1923). He studied political science and philosophy at the universities of Indiana and Chicago and in 1937 graduated in medicine from Meharry Medical College, Nashville. Continuing his medical studies in London and Scotland, he practised in the north of England and then in London.

Banda first became involved in African politics when he co-founded the Nyasaland African Congress, while working in London. Opposed to the Federation of Rhodesia and Nyasaland (1953), he left Britain and practised medicine in Ghana before returning to Nyasaland in 1958, as head of the Nyasaland African Congress. In 1959 he was imprisoned during a state of emergency, but was released in 1960 and became head of the new Malawi Congress Party. He then participated in talks with Britain over the constitutional future of Nyasaland and served as minister of natural resources and local government (1961–63). Appointed prime minister in 1963 (the year the Federation was disbanded), he became prime minister of independent Malawi in 1964 and was named president of the Republic of Malawi in 1966. In 1971 he was appointed president for life.

In more recent years Banda has developed a reputation as an autocratic leader and has been criticized in many black African countries for opening a dialogue and trading with white South Africa.

Bandaranaike, Sirimavo Ratwatte Dias (1916–   ) *Prime minister of Ceylon (later Sri Lanka) (1960–65; 1970–77). A fervent nationalist and socialist, she was the first woman in the world to be elected a prime minister.*

Born in Ratnapura, the daughter of a Sinhalese landowner, she was educated at St Bridget's convent, Colombo. In 1940 she married S. W. R. D. Bandaranaike (Ceylonese prime minister 1956–59). She spent the next twenty years supporting her husband's political advancement and promoting social reform for women, as a volunteer at a village institute affiliated to the All Ceylon Women's Association.

Mrs Bandaranaike became leader of the Sri Lankan Freedom Party (SLFP) in 1960, following her husband's assassination (1959), and continued to pursue his policies. Pressing for the adoption of Sinhalese instead of English as the official language, the encouragement of Buddhism as the official religion, and the nationalization of Ceylon's major industries, she led the SLFP to victory in the general election of 1960 and was appointed prime minister. Defeated in 1965, she returned to power at the head of a socialist coalition in 1970 with an increased majority. In 1972 she adopted a new constitution, establishing Ceylon as the Republic of Sri Lanka. Opposition to her socialist programmes, stringent measures taken to alleviate economic problems, and her inability to resolve conflict between various ethnic groups led to a crushing defeat in the 1977 elections. In 1980 she was found guilty on six charges of the misuse of power, deprived of her civil rights for six years, and expelled from parliament. In 1988 she emerged as the leading opposition candidate for presidential office, which was eventually won by Ranasinghe Premadasa (1924–   ).

Bankhead, Tallulah (1903–68) *US actress, whose popularity was due as much to her beauty, wit, and distinctly husky voice as to her acting ability.*

Daughter of a US congressman, Bankhead was born in Huntsville, Alabama. She made her stage and screen debut in 1918 and although she went on to make films at intervals, including HITCHCOCK's *Lifeboat* (1944) and (as Catherine the Great) *A Royal Scandal* (1945), it was in the theatre that she had her greatest success. She appeared on the London stage in the twenties in popular productions of *The Gold Diggers*, *The Garden of Eden*, and *Her Cardboard Lover*. Her most notable performances, however, came in the USA in the more dramatically challenging *The Little Foxes* (1939), followed by Thornton WILDER's Pulitzer prizewinning play *The Skin of Our Teeth* (1942), for which Bankhead received New York Critics Awards. Other successes included COWARD's *Fallen Angels*, in a part offered to her by 'The Master' after Somerset MAUGHAM had turned her down for the part of Sadie Thompson in *Rain*.

Her marriage to the actor John Emery in 1937 ended three years later. She began appearing in cabaret in the fifties and in 1952 published her autobiography, *Tallulah*.

Bannister, Sir Roger Gilbert (1929–)
British middle-distance runner and doctor. He was the first man to run the mile in under four minutes. He was knighted in 1975.

While still a medical student, Bannister won British (1951, 1953–54) and Empire (1954) championships in the mile run. He finally broke the four-minute barrier at Oxford in May 1954, with a time of 3 minutes 59.4 seconds. Later the same year, in his last race, he won the gold medal in the European 1500 metres event. After his retirement he wrote several papers on the physiology of exercise and neurology and a book, *First Four Minutes* (1955). He is now a consultant neurologist. In 1984 he was appointed master of Pembroke College, Oxford.

Banting, Sir Frederick Grant (1891–1941)
Canadian physician who, with the US physiologist Charles Herbert Best (1899–1978), isolated the hormone insulin in a form that could be used to treat diabetes. Banting received the 1923 Nobel Prize for Physiology or Medicine for this work and was later (1934) knighted.

Banting was born in Ontario, the fourth son of a farmer. In 1910 he began to train as a Methodist minister at the Victoria College of the University of Toronto but two years later transferred to the medical school. In 1915, before completing his studies, he joined the Royal Canadian Army Air Corps but was soon sent back to complete his studies, graduating in 1916 and then being awarded a commission. He served in France and England and was awarded the Military Cross in 1918. After World War I he returned to Canada to work at the Hospital for Sick Children, Toronto. In 1920 he began a private practice in London, Ontario, and worked as a part-time demonstrator at the University of Western Ontario to supplement his income.

In 1921, under the guidance of the Scottish physiologist John MacLeod (1876–1936), he began his work on diabetes mellitus. With his assistant Best, who was then a student, he succeeded in extracting insulin – the hormone that regulates carbohydrate metabolism – from the pancreas of dogs. He and MacLeod shared the Nobel Prize for this work but Banting shared his prize money with Best. In the same year Banting became professor and head of the newly created Banting and Best Department of Medical Research at the University of Toronto and in 1930 he was made director of the Banting Institute for Medical Research in Toronto. Banting's later researches concerned the function of the adrenal cortex, cancer, and aviation medicine. He was killed in a plane crash in 1941 in connection with his aviation studies.

Barber, Samuel (1910–81) *US composer whose work is marked by a strong traditional element of melodic structure.*

Born and brought up in West Chester, Pennsylvania, Barber became a student at the Curtis Institute, studying piano, composition, and conducting. His excellent baritone voice led him to consider a career in singing, and in 1935 he gave recitals on the NCB radio and recorded his *Dover Beach* for voice and string quartet (1931). He travelled extensively in Europe on scholarships (including the Prix de Rome) and found his style in a natural romanticism allied to classical forms. The popular *Adagio for Strings* is an orchestral transcription of the second movement of his string quartet (1936). His opera *Vanessa* (1957), with libretto by MENOTTI, was first performed at the Metropolitan Opera and played there for two subsequent seasons (1958/59 and 1964/65); an opera commissioned for the opening of the new Lincoln Center, *Antony and Cleopatra* (1966), designed by ZEFFIRELLI, was less successful, probably due to its cumbersome production.

Barber's harmony is basically that of late-nineteenth-century diatonicism; his music is lyrical and often dramatic and comparable with that of Brahms. Barber was in no way an innovator: 'I write as I feel', he said. Theatre works include three operas and two ballets; there are two symphonies (1936, revised 1943; 1944, revised 1947), concertos for violin, cello, and piano, and the *Capricorn Concerto* (1944) for flute, oboe, trumpet, and strings. His choral works include *Prayers of*

Kierkegaard (1954), and he has also composed instrumental pieces and songs.

Barbirolli, Sir John (1899–1970) *British conductor and cellist. He was knighted in 1949 and created a CH in 1969.*

Born in London, the son of an Italian violinist and his French wife, Barbirolli won scholarships to Trinity College of Music and the Royal Academy of Music. At seventeen, he became the youngest member of the Queen's Hall Orchestra and gave his first solo recital at the Aeolian Hall in 1917. He returned from two years' army service to orchestral and solo playing and to the beginning of his conducting career. He conducted the British National Opera Company on tour and in London (1928) and was guest conductor at Covent Garden (1929–33). He went to Sadler's Wells in 1934 and returned to Covent Garden for the coronation season of 1937. In 1936, as a result of his growing reputation, he was invited to be guest conductor of the New York Philharmonic Symphony Orchestra, and later succeeded Toscanini as conductor.

During World War II Barbirolli returned to Britain and was instrumental in rebuilding the Hallé Orchestra, of which he was permanent conductor from 1943 to 1958; he was appointed conductor-in-chief in 1958, and – in recognition of his devoted service – conductor laureate for life in 1968. He was guest conductor again at Covent Garden for the 1951 to 1954 seasons, principal conductor of the Houston (Texas) Symphony Orchestra (1961–67), and – from 1961 until his death – guest conductor of the Berlin Philharmonic Orchestra.

Among many honours, Barbirolli received the Gold Medal of the Royal Philharmonic Society and the Freedom of the City of Manchester. He was twice married; his second wife was the oboist Evelyn Rothwell. Barbirolli was a renowned interpreter of the music of Elgar, Delius, and of Vaughan Williams, who dedicated his eighth symphony to him; later in his career he gave increasingly profound and moving interpretations of the music of Mahler. His colourful personality – together with his enthusiasm and commitment to the music being performed – fired orchestra and audience alike.

Barbusse, Henri (1873–1935) *French novelist.*

Barbusse was born in Asnières, Seine. His early writings combined realism with symbolism and made little impression; they include the poems *Pleureuses* (1895) and the novels *Les Suppliants* (1903) and *L'Enfer* (1908). In 1910 he became editor of *Je Sais Tout*. Having enlisted with the infantry in World War I, he was invalided out in 1917 with two citations for gallantry.

Barbusse's best-known work is his novel *Le Feu: journal d'une escouade* (1916; translated as *Under Fire*, 1917), for which he was awarded the Prix Goncourt. Based on his experiences in World War I, it sought to portray the grim reality of a soldier's life in the trenches and was a reaction against the idealized depiction of the glory of war. As such it stood out from and outlived other war novels of the time. Barbusse's pacifist leanings led to a sympathy for the communist cause, which was increasingly expounded in his later works, such as *Clarté* (1919), *Paroles d'un combattant* (1921), and *Staline* (1935), to the detriment of their literary merit. He died in Moscow.

Bardeen, John (1908–91) *US physicist, whose invention of the point-contact transistor and the theory to explain superconductivity won him two Nobel Prizes, the first physicist to do so.*

Born in Madison, Wisconsin, the son of an anatomy professor, Bardeen studied electrical engineering at the University of Wisconsin. After working for three years as a geophysicist he entered Princeton as a graduate student, gaining his PhD in mathematical physics in 1936. Bardeen subsequently held brief appointments at Harvard and the University of Minnesota, before joining the Naval Ordnance Laboratory in World War II. In 1945 Bardeen joined the Bell Telephone Laboratories, where a new group had been formed to work on the development of solid-state devices. Here, in collaboration with W. H. Brattain, Bardeen published a paper introducing the transistor (1948). For this discovery, which changed the whole electronics industry, and indeed many aspects of society, Bardeen won the first of his Nobel Prizes (1956) in conjunction with Brattain and W. B. Shockley.

In 1951 Bardeen moved to the University of Illinois as professor of physics and electrical engineering. Here, with his colleagues L. N. Cooper (1930–) and J. R. Schrieffer (1931–), he tackled the problem of superconductivity. Although the phenomenon was first described by K. Onnes in 1911, no theoretical explanation had been accepted for it. The crucial insight of the BCS (Bardeen–Cooper–Schrieffer) theory is that at very low temperatures, under certain conditions, electrons can form bound pairs (Cooper pairs). For his work on superconductivity Bardeen was awarded his second Nobel Prize for Physics (1972), sharing it with Cooper and Schrieffer.

Bardot, Brigitte (Camille Javal; 1934–) *French film star, known as the 'Sex Kitten' of the 1950s.*

Bardot was born into an industrialist family in Paris. Before entering films she was a model, appearing on the cover of *Elle* magazine. She was introduced to films by Marc Allégret, whose assistant director at the time, Roger Vadim, later became her husband (1952–57) and was influential in her career. (Bardot's second and third husbands were the actor Jacques Charrier and the industrialist Gunther Sachs.) She made her debut in Jean Boyer's *Le Trou normand* (1952; *Crazy for Love*) but stardom and international fame came with Vadim's first film as director, *Et Dieu créa la femme* (1956; *And God Created Woman*), the film that launched Bardot as a sex symbol. Subsequent films included *En cas de malheur* (1959; *Love is My Profession*), *La Vérité* (1960; *The Truth*), *Vie privée* (1962; *A Very Private Affair*), *Le Repos du guerrier* (1962; *Love on a Pillow*), *Le Mépris* (1964; *Contempt*), *Viva Maria* (1965), *Shalako* (1968), and *The Novices* (1970).

Bardot's personal life, which has been the subject of much publicity, has often been far from happy. She was the subject of Simone DE BEAUVOIR's *Brigitte Bardot and the Lolita syndrome* (1959), Jacques Rozier's film *I paparazzi* (1963), and *Dear Brigitte* (1965), which was made in the USA and in which she played herself. Since her retirement as an actress Bardot has devoted herself to the cause of animal welfare and lives in seclusion in the south of France.

Barenboim, Daniel (1942–) *Israeli pianist and conductor of Jewish Ukrainian descent. He became an Officier de la Légion d'honneur in 1987.*

Born in Buenos Aires, where he made his piano debut at the age of seven, Barenboim and his family moved to Europe in 1951 and to Israel in 1952. He studied the piano in Salzburg with Edwin Fischer (1886–1960) (playing the Bach D minor concerto at the Mozarteum at the age of nine) and with Igor Markevich (1912–83), and composition in Paris with Nadia BOULANGER. He made his English debut in 1955, his New York debut in 1957, and his conducting debut in Israel in 1962.

Barenboim has made his reputation as a virtuoso pianist, with complete recordings of the Mozart piano concertos and the Beethoven piano sonatas and concertos; as a pianist-conductor, particularly with the English Chamber Orchestra; and as a chamber musician, especially in partnership with the violinists Pinchas Zukerman and Itzak Perlman and the cellist Jacqueline DU PRÉ, whom he married in 1967. He is also known as an accompanist in Lieder, especially with FISCHER-DIESKAU. In 1975 Barenboim became musical director of the Orchestre de Paris, which he left in controversial circumstances in 1988; he subsequently became music director of the Chicago Symphony Orchestra. He has been awarded the Beethoven medal (1958) and the Paderewski medal (1963).

Bar-Hillel, Yehoshua (1915–75) *Austrianborn Israeli logician, philosopher, and linguist.*

Bar-Hillel obtained his doctorate from the Hebrew University, Jerusalem, in 1947. He worked for several years in the USA, first at Chicago, where he was much influenced by CARNAP, and later at the Massachusetts Institute of Technology, where he became interested in the problem of developing an artificial language. After 1961, however, he was based in Jerusalem, where he was professor of logic and the philosophy of science at the Hebrew University.

As a pupil of Carnap, Bar-Hillel worked initially on problems in the fields of semantics and inductive logic. Later work was concerned with the possibility of machine translation and, under the influence of CHOMSKY, on the relationship between grammar, logic, and language. This later work is best seen in two collections of his essays, *Language and Information* (1964) and *Aspects of Language* (1970).

Barnard, Christiaan Neethling (1922–) *South African heart surgeon who carried out the first heart transplant in 1967.*

Barnard graduated from the University of Cape Town in 1946 and became resident surgeon at the Groote Schuur Hospital, where he discovered that intestinal atresia (congenital obstruction in the small intestine) is caused by an inadequate blood supply to the fetus during pregnancy and developed a procedure to correct it. He later took a scholarship at the University of Minnesota, USA (1956–58), where he specialized in cardiothoracic surgery; on obtaining his doctorate, he returned to the Groote Schuur Hospital to concentrate on open-heart surgery. During this period he designed the Barnard artificial heart valve and began experimenting on heart transplants in dogs. In 1967 he led a team of surgeons in successfully replacing the heart of Louis Washkansky with that of an accident victim. However, the patient died eighteen days later from infections contracted as a result of the destruction of his immune system by the drugs used to suppress rejection of the donor heart by his body. Barnard retired as a surgeon in 1983 and later entered politics.

Barnes, Sydney Francis (1873–1967) *England cricketer, often regarded as the finest test bowler in the long history of cricket. He chose to play relatively little first-class cricket: in only 27 test appearances he took a total of 189 wickets.*

Barnes was born in Staffordshire. In 1894 he made his first-class debut for Warwickshire, for whom he played sporadically over three seasons before joining Lancashire, where he remained till 1903. That was the extent of his county cricket. The remainder of his career (he retired in 1935) was spent substantially with Staffordshire and in the North Staffordshire, Lancashire, Bradford, and Central Lancashire Leagues. He had a magnificent record in league cricket.

Between 1901 and 1914 he made his modest number of appearances for England, meeting with great success in Australia and South Africa. Faster than medium pace, he could turn the ball both ways and would vary his deliveries at will according to the conditions and circumstances. He earned praise from all the great cricketers of his day and it was regretted that he apparently preferred to play league cricket for much of the time.

Baroja (y Nessi), Pío (1872–1956) *Spanish novelist.*

Of Basque and Italian descent, Baroja, the son of a mining engineer, was born at San Sebastián. He took a degree in medicine at the University of Madrid but practised only briefly. His first work, written in the 1890s, brough him to the attention of AZORÍN and in 1902 he decided to devote himself solely to his writing. The trilogy *La lucha por la vida* (1904; translated as *The Quest*, 1922, *Weeds*, 1923, and *Red Dawn*, 1924) established his reputation internationally and was notable for portraying the life of the Madrid poor in a realistic unrhetorical style that faithfully rendered colloquial speech. Among his many other novels are *Paradox, Rey* (1906; translated as *Paradox, King*, 1931), *César o nada* (1910; translated as *Caesar or Nothing*, 1919), and *El árbol de la ciencia* (1911; translated as

The Tree of Knowledge, 1928). In 1913 he embarked on a huge series of novels, *Memorias de un hombre de acción* (1913–34), twenty-two volumes forming a panoramic portrait of Spain that are tenuously unified around the central historical figure, Baroja's relative Eugenio de Aviraneta e Ibargoyan (1792–1872).

Baroja was the major novelist of the 'Generation of 1898', a group of writers concerned with Spain's unique character, history, and landscape. Careless of the fine points of plot, structure, or even character development, his work aimed at capturing the variety of Spanish life and realistically treated subjects new to the Spanish novel. Of his enormous output – sixty-six novels, eight volumes of memoirs, nine books of essays, three biographies, nine collections of stories, two plays, and one book of poems – his most characteristic and powerful work was virtually all completed before the civil war.

Barrault, Jean-Louis (1910–) *French actor, director, and theatre manager. In recognition of his contribution to French theatre he was made an Officier de la Légion d'honneur and a Commandeur des Arts et des lettres.*

Barrault began his career with Charles Dullin (1885–1949) at the Atelier (1931) and later studied mime with Étienne Decroux; he made his debut as director with *Autour d'une mère* (1935), adapted with CAMUS from a William FAULKNER novel.

While at the Comédie-Français (1940–47) he met and married Madeleine Renaud and together they founded their own repertory company at the Théâtre Maligny (1947–56), staging modern and classical plays. At the theatre and on tour Barrault worked with the company both as actor and director. He became co-director of the Théâtre du Palais-Royal (1958) and director of the Théâtre National de l'Odéon (1959–68), but was dismissed from the post after the student riots of 1968, l'Odéon having become a focal point during the demonstrations. He also worked at the Théâtre des Nations (1965–67 and 1972–74). In 1974 he created the Théâtre d'Orsay inside a huge tent at the Gare d'Orsay, which moved to the Palais des Glaces, a disused ice-skating rink, in 1980. Barrault has appeared in a number of films – including *Les Beaux Jours* (1935), *Les Enfants du paradis* (1945), and *The Longest Day* (1962) – and in many television programmes. Barrault's theatrical ideas are discussed in his autobiographies, *Réflexions sur le théâtre* (1949; translated as *Reflections on the Theatre, 1951) and Nouvelles Réflexions sur le théâtre* (1959; translated as *The Theatre of Jean-Louis Barrault*, 1961); other publications include *Souvenirs pour demain* (1972) and *Saisir le present* (1984).

Barrie, Sir J(ames) M(atthew) (1860–1937) *Scottish playwright and novelist. The author of* Peter Pan, *he was made a baronet in 1913 and in 1922 was awarded the OM.*

Born in Kirriemuir in Scotland, the son of a weaver, Barrie worked his way from local schools to Edinburgh University, largely with the support and encouragement of his mother. After graduating he moved south to begin a journalistic career, first in Nottingham, then in London. He published sketches, collaborated on plays, and in 1891 achieved notice as a novelist with *The Little Minister*, which he later (1897) sucessfully adapted for the stage. In 1892 the farcical *Walker, London* launched his career as a playwright, a career of almost uninterrupted success. Among his prose writings of the 1890s was *Margaret Ogilvy* (1896), a tribute to his mother; much of his early prose exploited his Scottish background.

In the early 1900s Barrie abandoned fiction for the theatre with the comedies *Quality Street* (1902), *The Admirable Crichton* (1902), and *Little Mary* (1903). Of the three, *The Admirable Crichton*, an astringent comedy of 'natural' class distinction, survives the best. In 1904, despite doubts expressed by Barrie's theatrical mentors, *Peter Pan* was staged at the Duke of York's Theatre in London; based on a fantasy created one summer for the young sons of a friend, it has proved the most lasting of Barrie's successes. *What Every Woman Knows* (1908) explores another characteristic Barrie theme: the effect of worldly success upon personal relationships. Throughout this period he was also writing a number of well-received one-act plays. During World War I Barrie's best plays were *A Kiss for Cinderella* (1916) and *Dear Brutus* (1917). The latter is the archetypal Barrie play with its fantasy, pathos, and regret for lost opportunities, focusing on the characters of the childless and estranged artist and his wife.

In later years Barrie's output slackened. Honorary degrees were showered upon him as an elder statesman of the British literary scene. From 1930 he was chancellor of Edinburgh University.

Barrymore, Lionel (1878–1954) *US actor, best known for his portrayal of irascible but lovable characters in numerous films in the 1930s and 1940s.*

Lionel Barrymore was born in Philadelphia into one of the great theatrical families: his father was the British-born actor Maurice Barrymore and his mother was Georgie Drew, daughter of Irish actor-manager John Drew and American actress Louisa Lane. His brother John Barrymore (1882–1942) and his sister Ethel Barrymore (1879–1959) also became distinguished actors.

Lionel made his stage debut with his grandmother's company in *The Road to Ruin* (1893), after which he appeared in such plays as *The Humming Bird*, with Charles Frohman's company, and J. M. BARRIE's *Pantaloon* (1905) before ill-health interrupted his career. He studied painting in Paris but returned to the stage in *Peter Ibbetson* (1917). Between then and 1925 he achieved outstanding successes in several plays, including *The Copperhead* (1918) and *The Jest* (1919), and was particularly admired for his performance as Macbeth. *Man or Devil* (1925) marked the end of his stage career, after which he devoted himself to films. Barrymore's film career had begun with D. W. GRIFFITH's *Friends* (1909). With the coming of sound he directed the outstanding *Madame X* (1929) and received an Oscar as best actor for his portrayal of the father in *A Free Soul* (1931). Other memorable films included *Grand Hotel* (1932), *Rasputin and the Empress* (1932; with John and Ethel Barrymore), *Camille* (1937), and the *Dr Kildare* series, as Dr Gillespie.

Barrymore also wrote books, including a biography of his family called *We Barrymores*, and composed a symphony.

Barth, Karl (1886–1968) *Swiss theologian and, as founder of the 'neo-orthodox' movement, arguably the most significant and radical innovator in Protestant theology of the twentieth century.*

The son of a Protestant historian and theologian, Barth studied theology in Bern, Berlin, Tübingen, and Marburg. As minister of Safenwil in the Aargau canton, he became a member of the Religious Socialist Movement, publicly supporting his parishioners in their struggle for improved wages and conditions. Barth was shocked by arguments in support of German war aims advanced by prominent German theologians and in 1919 he published his seminal work, *Der Römerbrief* (translated as *The Epistle to the Romans*, 1933). A rebuttal to liberal nineteenth-century Protestant theology, Barth's book established a neo-orthodox or 'theocentric' approach, which emphasized the discontinuity between man and God – the 'otherness of God'. It shook the foundations of the theological, establishment and Barth was offered professorships in theology at Göttingen (1921), Münster (1925), and Bonn (1930). Barth insisted that the teachings of Christ as related in the Bible were the only route to an understanding of God.

The rise of HITLER's National Socialists in Germany precipitated a crisis in the German Protestant Church. Barth wholly repudiated the efforts of the 'German Christians' to reconcile themselves with Nazism and, with NIEMÖLLER, he founded the Confessing Church, which, according to its constitutive Barmen Declaration of 1934, was to 'trust and obey only Jesus Christ and no other'. Barth's refusal to swear an oath of allegiance to Hitler led to his dismissal from Bonn in 1935. He moved to Basel University as professor of theology. After the war, Barth showed a willingness to comprehend communist aims and warned against the cold war. He was opposed to nuclear weapons. Elaboration of his theocentric doctrines is presented in the massive *Kirchliche Dogmatik* (1932; translated as *Church Dogmatics*, 1961).

Barthes, Roland (1915–80) *French critic.*
Barthes was born in Cherbourg. A leading exponent of the *nouvelle critique* of the 1950s and 1960s, his approach to literary criticism was first presented in *Le Degré zéro de l'écriture* (1953; translated as *Writing Degree Zero*, 1972). The ideas he put forward in *Mythologies* (1957; translated in 1972) led Barthes towards the theory of semiology, the science of signs and symbols developed by Ferdinand de Saussure in the early twentieth century; *Eléments de sémiologie* (1964; translated in 1967) defined the theory in more detail. Severely criticized for his attack on traditional methods of literary analysis in *Sur Racine* (1963; translated as *On Racine*, 1964), Barthes replied with *Critique et vérité* (1966); the dispute between the *nouveaux critiques* and those of earlier schools continued to rage for some time. Barthes's later works include an autobiographical novel, *Roland Barthes par Roland Barthes* (1975; translated in 1977), the successful *Fragments d'un discours amoureux* (1977; translated as *A Lover's Discourse: Fragments*, 1978), and an analysis of the photograph, *La Chambre claire* (1980; translated as *Camera Lucida*, 1982).

Bartók, Béla (1881–1945) *Hungarian composer, pianist, and student of folk music.*

Bartók was born in Nagyszentmiklós and his first piano teacher was his widowed mother; he gave his first concert at the age of ten, with a programme including one of his own compositions. In 1894 he began to study with László Erkel and in 1899, instead of taking up a scholarship at the Vienna Conservatory, decided to go to the Budapest Academy of Music to study piano with István Thomán and composition with János Koessler. At this period Brahms, Liszt, and Wagner were important influences. Crucial to his development was a performance of Richard STRAUSS's *Thus Spake Zarathustra* in Budapest in 1902, an overwhelming experience leading him to study Strauss's works and compose his first major orchestral work, the unpublished *Kossuth Symphony* (1903), a symphonic poem in ten sections. His first published work was *Rhapsody for Piano and Orchestra* (1904). In 1905 Bartók formed a lasting friendship with Zoltán KODÁLY, which led to their joint research and publication of Hungarian and other folk songs.

In 1907 Bartók was appointed professor of piano at the Budapest Academy. Although this demonstrated official recognition, the fact that in Budapest there was no conductor or orchestra capable of performing his works – and that he himself was an inadequate conductor – led him to withdraw from concert life and concentrate on folksong research. However, in 1917, the Italian conductor Egisto Tango prepared and performed Bartók's ballet *The Wooden Prince* (1914–16) with great success, and the following year his performance of *Prince Bluebeard's Castle* (1911) established Bartók as a major composer. Tours of Europe, playing his own piano works, consolidated his reputation. In 1923 he divorced his first wife, whom he had married in 1909, and married a pupil, Ditta Pásztory. In 1927 Bartók toured the USA, and when political pressures in Hungary in the late 1930s became intolerable, he and his wife emigrated to the USA in 1940. To survive financially they undertook arduous concert tours and his health began to fail. He was unable to complete commissioned scores and died of leukaemia in straitened circumstances, saying to his doctor, 'The trouble is that I have to go with so much still to say.'

Bartók's musical language is wholly original, although often stimulated by other composers (among them, STRAVINSKY and DEBUSSY). Basically homophonic and harmonically adventurous, it owes much to Hungarian folk music, is often percussive and always dynamic, and is remarkable for sections of 'night music' based on notated bird and insect sounds. Of greatest importance are the six string quartets, spanning his creative life from 1908 to 1939. Other works include three piano concertos, the third written during the final weeks of his life (1926; 1930–31; 1945); the *Cantata Profana* (1930) for soloists, chorus, and orchestra; *Concerto for Orchestra* (1943); *Music for Strings, Percussion and Celesta* (1936); *Sonata for Two Pianos and Percussion* (1937); and *Mikrokosmos* (1926–37), six books of piano pieces ranging from the simplest to the most difficult.

Barton, Sir Edmund (1849–1920) *Australia's first prime minister (1901–03) and a high court judge. He was knighted in 1902.*
The son of a sharebroker and estate agent, Barton was born and educated in Sydney, where he studied law. He entered politics in 1879, when he won the University seat in the New South Wales legislative assembly. Prominent in the federation movement, Barton led the delegation to London that persuaded the British government to adopt the proposed Australian constitution (1900). He became prime minister but resigned in 1903 and took a seat on the bench of the newly formed high court, where he remained for seventeen years.

A learned and able judge who was one of the finest public speakers of his day, Barton was awarded honorary degrees from Oxford, Cambridge, and Edinburgh.

Baruch, Bernard Mannes (1870–1965) *US financier and economic adviser.*
Born in South Carolina, the son of a Jewish doctor who had emigrated from east Prussia, Baruch was educated at the College of the City of New York and began his career as an office junior in a glassware business. He then joined a firm of stockbrokers, becoming a partner at the age of twenty-five and amassing an enormous personal fortune by speculation by the time he was thirty-two. His political influence grew from his friendship with Woodrow WILSON, and subsequently with President F. D. ROOSEVELT and with Winston CHURCHILL, whom he met at Versailles when serving as economic adviser to the American Peace Commission.

Baruch held several public posts, including the chairmanship of the War Industries Board in 1918, and he represented the USA on the United Nations Atomic Energy Commission (1946–51), to which he presented his plan for the international control of atomic energy, named after him. In 1964 Baruch presented Princeton University with his papers to be used in the study of twentieth-century public affairs.

The extent of his powerful connections is obvious from the twelve hundred letters from nine presidents and seven hundred letters from Churchill.

Baruch wrote two autobiographical works – *My Own Story* (1958) and *The Public Years* (1961) – as well as *A Philosophy for our Times* (1953).

Basie, Count (William Basie; 1904–84) *Black US jazz pianist, organist, and bandleader. He led one of the best-known big bands during the swing era, and continued to do so for many years thereafter.*
Born in Red Bank, New Jersey, Basie began his musical career by playing the piano in Harlem dance halls. As a youngster he admired the organ playing of Fats WALLER in a silent cinema in Harlem and arranged to take piano lessons from him. Subsequently he toured theatres and accompanied vaudeville acts; in 1928–29 he played with Walter Page's Blue Devils and then with Bennie Moten (1894–1935), whose band was regarded as the best in the midwest. Basie broadcast from Kansas City in 1936 with a group called the Barons of Rhythm, acquiring the name 'Count' from a radio announcer. Producer John Hammond heard the band and organized a tour for them. The augmented band had several great soloists and an extremely competent rhythm section; success followed an engagement at the Savoy Ballroom in Harlem in early 1938. Basie kept his band together until well into the 1960s, except for a period between 1950 and 1952 when he led an octet. He toured the world with changing personnel, appearing and recording with singers Tony Bennett (1926–), Ella FITZGERALD, and Frank SINATRA.

He was as modest as Duke ELLINGTON about his own stride style of piano playing, reducing it as a leader to a percussive cueing of the band from the keyboard. However, his real worth as a jazz pianist was always evident in his playing with small groups.

Bateson, William (1861–1926) *British geneticist, whose work reaffirmed the fundamental importance of Mendel's genetic principles and who, in 1907, originated the term 'genetics' for the science of heredity.*
Born in Whitby, Yorkshire, Bateson received his BA from Cambridge University in 1883 and two years later was appointed fellow of St John's College. In 1908 he became the first British professor of genetics (at Cambridge) and in 1910 was appointed director of the John Innes Horticultural Institution. From the outset

a zoologist, Bateson was among the first to observe the possible evolutionary relationship between echinoderms and primitive chordates. Most significant, however, was his series of breeding experiments from 1900 onwards that helped re-establish the truth of Mendel's principles of heredity (established by experiments with pea plants). By using the domestic fowl, Bateson demonstrated that these principles applied to animals as well as plants. He also discovered examples of hereditary characters that exhibited incomplete dominance and, most notably, linkage, i e a tendency to be inherited together rather than independently. However, Bateson was opposed to the increasingly held view that chromosomes were the carriers of genes and proposed his own vibratory theory of inheritance. Involving unorthodox ideas about inherent bodily forces, this was consistent with Bateson's antagonism to what he saw as the utilitarianism in many aspects of evolutionary theory and led to his estrangement from the scientific mainstream.

His books include *Mendel's Principles of Heredity* (1909) and *Problems of Genetics* (1913). In 1910 he founded, with his colleague R. C. Punnett, the *Journal of Genetics*.

Batista y Zaldívar, Fulgencio (1901–73) *Cuban president (1940–44; 1952–59). His second, corrupt, dictatorship was terminated by Fidel CASTRO.*

Born in Banes, Cuba, of mixed descent, Batista was educated at an American Quaker School, after which he worked in a variety of trades. In 1921 he enlisted in the Cuban National Army but resigned from his active unit two years later to take up clerical work in the army. By the early 1930s he had risen to the rank of sergeant and had a considerable personal following.

In 1953, during a general strike in opposition to the regime of General Machado, Batista instigated the 'sergeants' revolt' and indirectly took control of the provisional government. Cultivating the support of the army, civil service, and trade unions, he ruled through a succession of presidents until 1940, when he himself was elected president. He was defeated in 1944 and left Cuba for the USA, where he lived for five years. On his return in 1949 he was elected to the senate and in 1952 again secured the presidency. He remained in power until 1959 when, faced with strong opposition from Castro and his rebel forces, he fled to the Dominican Republic. He lived in exile in Europe until his death in 1973.

During his first term as president Batista provided the strong leadership that fostered economic growth. He is more widely known, however, for his second term and the brutal and repressive measures he introduced to maintain power, including imprisonment of political opponents, corruption, and control of the press, universities, and congress.

Bax, Sir Arnold Edward Trevor (1883–1953) *British composer. He was knighted in 1937 and served as Master of the King's Music from 1941 until his death, which occurred shortly after he was made KCVO.*

Born in London, Bax studied composition at the Royal Academy of Music with Frederick Corder (1900–05); although he never played the piano in public, he was a fine pianist and an expert score reader. A visit to Russia in 1910 was influential, but more lasting was his sympathy for and understanding of the Celtic revival, particularly of Irish literature, which fostered in him a lifelong love of Ireland, its folksong, and its scenery. With this powerful focus allied to a natural melodic gift, an expansive inventiveness, and an idiomatic notion of harmony, he produced a number of evocative tone poems, of which *Tintagel* (1917) is most frequently performed.

Between 1921 and 1939 he wrote seven symphonies, each classically conceived but with three instead of the usual four movements. The first four symphonies are cyclic works, in that themes in the first movements recur in later movements. Among other works are two ballets, choral pieces, chamber music, and a large number of songs and piano pieces. A self-confessed romantic, he wrote as he felt, unmoved by modern trends. Many tokens of esteem include the Gold Medal of the Royal Philharmonic Society (1931).

Baylis, Lilian Mary (1874–1937) *British theatrical manager, who founded the Old Vic and Sadler's Wells companies.*

Born in London, the daughter of singers Edward Baylis and Elizabeth Cons, Lilian began her career as a violinist, first in London and then in South Africa, where the family settled in 1890. In 1898 she returned to London to assist her aunt, Emma Cons, run the Royal Victoria Coffee Music Hall in Lambeth, which was a temperance hall housed in the 'Old Vic', the Victoria Theatre, where a mixture of entertainment, lectures, and temperance meetings were put on for the working classes. After the death of her aunt (1912), Lilian renamed the hall 'The People's Opera House'. Eventually drama was added to opera and during World

War I the Shakespeare Company at the Old Vic became firmly established. Between 1914 and 1923 all of the plays of Shakespeare in the First Folio were produced there. She subsequently acquired Sadler's Wells, Islington, which she rebuilt and opened in 1931. Initially intended as a second 'Old Vic', it later became the home of opera and ballet and evolved into the Royal Ballet in 1956.

A religious woman, Lilian Baylis was endowed with a strong will and intense determination. Her achievement in founding the Old Vic and Sadler's Wells companies, often against tremendous odds, justifiably earned her a place among the great names of English theatrical history.

Beaton, Sir Cecil Walter Hardy (1904–80) *British photographer noted for his characteristically theatrical approach to fashion photography and portraits. He received a knighthood in 1972.*

The son of a timber merchant, Beaton showed an early interest in the theatre, painting, and photography that persisted through Harrow and Cambridge, and in the 1920s he began taking portraits. His work soon carried the distinctive Beaton style, with subjects posed amid highly elaborate settings to produce a dramatic or humorous tableau. Beaton staged his first exhibition in 1929. In the same year, a visit to the USA led to a contract with *Vogue* magazine and during the 1930s Beaton became internationally famous for his fashion features and portraits of film stars and celebrities. However, in 1939, antisemitic connotations in one of Beaton's *Vogue* features led to controversy and his resignation, although he later resumed his association with the magazine.

Beaton travelled widely and took full advantage of the new compact cameras to capture people and places on film in such work as *Cecil Beaton's New York* (1939). During World War II, he was commissioned by the Ministry of Information to record aspects of the war in Britain, North Africa, and Indo-China. His books of war images include *Air of Glory* (1941) and *Winged Squadrons* (1942). After the war, Beaton's career diversified into the designing of costumes and sets for theatre, films, and ballet. His design and costumes for the film *My Fair Lady* earned him two Oscars in 1965. To the public, however, Beaton is perhaps best remembered for his many portraits of the British royal family. His other books include *Persona Grata* (with Kenneth Tynan; 1953), *Glass of Fashion* (1954), *Quail*

in Aspic (1962), and *The Best of Beaton* (1968).

Beatty, David, 1st Earl (1871–1936) *British admiral who took part in the Battle of Jutland – the major naval battle of World War I. He was awarded the OM and created Earl Beatty of the North Sea and of Brooksby in 1919.*

The son of a cavalry officer, Beatty was a naval cadet at thirteen and by 1892 had become a lieutenant. In 1896 he was chosen as second-in-command of a small fleet of gunboats sent to assist KITCHENER's army in the Sudan. His display of daring in out-manoeuvring Arab forces and capturing the town of Dongola earned him the DSO. He returned to Sudan the following year, seeing action ashore during the advance on Omdurman, and at the age of twenty-seven found himself promoted to commander. Beatty's meteoric rise continued. Following his gallantry in the 1900 Boxer Rebellion in China, he was promoted to captain. In 1911, by now rear-admiral, he was appointed naval secretary by Winston CHURCHILL, then first lord of the Admiralty.

At the outbreak of World War I, Beatty commanded a battlecruiser squadron that sank two German cruisers in Heligoland Bight on 28 August 1914. A further engagement with enemy vessels on 24 January 1915 at Dogger Bank revealed woeful shortcomings in communication between British ships, although it was judged a British victory. But it was the Battle of Jutland on 31 May 1916 that impressed Britain's naval strength on Germany. With characteristic zeal, Beatty directed his battlecruisers to engage leading vessels of the German fleet before his support squadron was in position. There followed a fierce battle in which the *Indefatigable* and *Queen Mary* were sunk. Beatty's critics have pointed to his failure to keep Admiral JELLICOE, commanding the main body of the fleet, fully informed. However, Beatty was promoted to admiral in 1916. He improved communications between ships during engagements and introduced the convoy system to protect merchant shipping in the North Sea. In November 1919, Beatty was appointed first sea lord and supervised the postwar reorganization of the navy.

Beaverbrook, (William) Max(well) Aitken, 1st Baron (1879–1964) *Canadian-born British financier, politician, and newspaper publisher, who was a minister in both World Wars and proprietor of the* Daily Express *group of papers. He received a knighthood in 1911 and was made a baronet in 1916 and a baron in 1917.*

The son of a Presbyterian minister, Max Aitken progressed from selling insurance to head the Canada Cement Company, born out of a series of controversial mergers in the early 1900s. He moved to England in 1910 and, encouraged by his friend, Andrew Bonar LAW, was elected Conservative MP for Ashton-under-Lyne in the general election of December 1910. Following the outbreak of World War I, Aitken served as Canadian government representative at British GHQ, St Omer, and created the Canadian War Records Office. But by 1916 he was back in Whitehall playing a persuasive role in LLOYD GEORGE's climb to power following ASQUITH's downfall. However, he received no government post until 1918 when, as Lord Beaverbrook, he briefly served as chancellor of the Duchy of Lancaster and minister of information. Meanwhile, in December 1916, Beaverbrook had purchased the *Daily Express*. In collaboration with its long-standing editor, R. D. Blumenfeld, he gave the paper a livelier and more entertaining character to restore profitability. In December 1918 the *Sunday Express* was launched and in 1923 Beaverbrook acquired the *Evening Standard*. Through his newspapers, 'the Beaver' supported Neville CHAMBERLAIN's attempts to appease HITLER in the late 1930s. In 1940, CHURCHILL appointed Beaverbrook minister of aircraft production, a task that he successfully accomplished. He later became minister of supply, but in 1942 he left office. In 1943 he was appointed lord privy seal.

Beaverbrook's memoirs of British politics include *Politicians and the War* (2 vols, 1928–32), *Men and Power 1917–1918* (1956), and *The Decline and Fall of Lloyd George* (1963). He also wrote *The Abdication of King Edward VIII* (1966).

Bechet, Sidney (1897–1959) *Black US jazz clarinettist and saxophonist.*

He began teaching himself as a small boy and was playing in jazz bands in his native New Orleans at the age of twelve. From 1919 to 1921 he lived in Europe, where the Swiss conductor Ernest Ansermet (1883–1969) described him as an 'artist of genius'. In 1925 Bechet worked briefly with Duke ELLINGTON, but returned to Europe (1925–28). He settled permanently in France in 1951.

Like his great contemporary Louis ARMSTRONG, Bechet learned to play in the New Orleans ensemble style, but was too great a soloist with too much imagination to be permanently contained by it. From about 1921 Bechet concentrated on the soprano saxo-phone, on which he became a virtuoso. He never learned to read music and never abandoned the unorthodox fingering he had taught himself as a child, yet he was one of the best-known jazz musicians in the world and a national figure in France.

Becker, Boris (1967–) *German tennis player, who in 1985 became the youngest ever winner of the men's singles championship at Wimbledon.*

Born at Leimen, near Heidelberg, his first major success came in 1983 when he won the West German junior championship. The power of Becker's serve and his speed in crossing the court to return apparently unreachable shots thrilled crowds and made him a formidable opponent, particularly on grass courts. His victory at Wimbledon in 1985, in which he beat Kevin Curren in the final, was the first occasion that the championship had been won by an unseeded player; he was just seventeen years old. He repeated his success at Wimbledon in 1986, again guided by his coach Ion Tiriac, who had directed his career since 1984; he won once more in 1989 and reached the finals again in 1991, only to lose to Michael Stich. He has had only mixed results in other major tournaments, especially when playing on hard courts.

Beckett, Samuel (1906–89) *Irish writer, who lived most of his life in France. He was awarded the Prix Formentor in 1961 and the Nobel Prize for Literature in 1969.*

Born in Dublin into a middle-class family, Beckett attended Trinity College, Dublin. In 1929 he first visited Paris, where he began writing. Returning to Trinity, he taught French for two years (1930–32) but then left Ireland for good. After an unsatisfactory period in London (1933) he moved back to Paris, where he became the friend and secretary of James JOYCE, who profoundly influenced his work. During this period he published poems, a novel (*Murphy*, 1938), and short stories.

In 1940 Beckett joined a Resistance network in Paris but was forced to escape to Free France in 1942, narrowly evading the Gestapo. His most prolific period began in 1946 when, writing in French, he produced a trilogy of novels, *Molloy* (1951; translated 1956), *Malone meurt* (1952; translated as *Malone Dies*, 1956), and *L'Innommable* (1953; translated as *The Unnamable*, 1960). He also wrote the play *En attendant Godot* (1952; translated as *Waiting for Godot*, 1954), in which two tramps, passing the time as they wait in vain for the enigmatic Godot, embody Beckett's vision of the emptiness and futility of the human condition. Only a desperate vaudevillian humour keeps them from suicide. *Godot* was followed by *Fin de partie* (1957; translated as *Endgame*), *Krapp's Last Tape* (1959), and *Happy Days* (1960), all displaying Beckett's obsession with the passage of time, death, and nothingness: Winnie, the main character of *Happy Days*, is literally sinking into her grave, buried first to her waist, then to her neck. During this period Beckett also wrote several radio plays, notably *Cascando* (1964). His later plays are increasingly short and enigmatic – existential jokes more than theatrical events.

Beckmann, Max (1884–1950) *German painter and graphic artist.*
Born in Leipzig, Beckmann studied at the Weimar Academy (1900–03) and then worked in Berlin. His early paintings were influenced by impressionism, and he also produced works with biblical and mythological themes, reflecting his admiration of medieval art. At the start of World War I (1914) he served as a medical orderly but was discharged following a nervous breakdown. These experiences of war profoundly influenced his later work. Settling in Frankfurt in 1915, he began to paint figurative compositions conveying a harsh vision of an evil and malicious contemporary world. Brutality and oppressive apathy permeate these pictures, as in *The Night* (1919), a scene of torture that, like many of his works, reflects a social reality of the time as well as being a powerfully symbolic composition. This combination of realism with allegorical and symbolic significance has been described as transcendental realism. Certain symbolic objects, such as candles and musical instruments, recur in his pictures. His characteristic style was one of simplified forms in a crowded almost two-dimensional space. He also painted a number of self-portraits.

Beckmann was dismissed from his teaching post in Frankfurt by the Nazis in 1933, the same year he painted *Robbery of Europe* and the triptych *Departure*. He went first to Berlin, then to Paris and Amsterdam, and finally (in 1947) to the USA, where he taught and produced work that was lighter and less harsh. In the last two years of his life Beckmann received academic honours and prizes in the USA and at the Venice Biennale.

Beecham, Sir Thomas (1879–1961) *British conductor, who did much to stimulate public interest in new and neglected music. He was knighted in 1916, shortly before succeeding to his baronetcy, and was made a CH in 1957.*

Born in St Helen's, Lancashire, the son of Sir Joseph Beecham, the wealthy pharmaceutical manufacturer, Beecham had no special musical training. After a brief spell at Wadham College, Oxford, he formed first the New Symphony Orchestra (1905) and then the Beecham Symphony Orchestra (1907), specializing in performances of eighteenth-century and contemporary music. It was during this time that he first produced the works of DELIUS, of whom he was a friend and outstanding interpreter, against general opposition. This championship culminated in his arranging and conducting Delius festivals in London (1929 and 1946) and publishing a biography of the composer in 1959. From 1909 Beecham turned his attention to opera, producing, as conductor or impresario, some 120 operas during his career. Notable among these were the works of Richard STRAUSS. In 1911 he was responsible for the first appearance of DIAGHILEV's Ballets Russes in London, which led to an interest in Russian opera, and in 1915 he formed the Beecham Opera Company (which in 1923 became the British National Opera Company, later (1929) absorbed into Covent Garden). In 1932 he founded the London Philharmonic Orchestra and, after war years spent mostly in the USA, the Royal Philharmonic Orchestra (1947).

Beecham's musical tastes were wide, but he excelled in Mozart and Haydn and was responsible for introducing the music of Delius, Strauss, and SIBELIUS to the public. Virtually self-taught in music, he developed his own expressive style of conducting and often conducted from memory. He was equally famous for his sharp wit and quick repartee. Beecham published his autobiography, *A Mingled Chime*, in 1944.

Beerbohm, Sir (Henry) Max(imilian) (1872–1956) *British author and cartoonist. He was knighted in 1939.*

Born in London, the son of an emigré Lithuanian corn merchant, Beerbohm was educated at Charterhouse and Merton College, Oxford. After returning to London he first attracted attention with his talented caricatures in *Strand Magazine* (1892). From 1894 he contributed to *The Yellow Book* and in 1898 succeeded G. B. SHAW as theatre critic of the *Saturday Review*. He also published volumes of caricatures and essays. By now he was a fashionable wit in London literary society and the friend of many artists and writers.

In 1910 Beerbohm married the American actress Florence Kahn and settled in Rapallo,

Italy. In 1911 he published *Zuleika Dobson*, his only novel – a sparkling fantasy of Oxford life. Back in England during World War I, he wrote his humorous imaginary biographies *Seven Men* (1919). His exhibitions of caricatures in 1921 and 1923 aroused such controversy that he withdrew some of the drawings of royalty and returned to Rapallo. His final volume of impeccably written essays, *A Variety of Things*, appeared in 1928. On a visit to London (1935) he was persuaded to give a radio talk. He achieved an extraordinary triumph in the new medium, and gave occasional broadcasts for the rest of his life; some of these were published in *Mainly on the Air* (1946). From 1939 to 1947 the Beerbohms remained in England, but later returned to Rapallo where they both died.

Begin, Menachem (1913–) *Israeli statesman and prime minister (1977–84). He was awarded the Nobel Peace Prize in 1978, jointly with President SADAT of Egypt.*

Born in Brest-Litovsk in Russian Poland, Begin attended the Mizrachi Hebrew school before studying law at the University of Warsaw, from which he graduated in 1935. During the 1930s he took an active part in the Betar Zionist Youth Movement, becoming leader in 1939. Arrested by the Russians and imprisoned in a Siberian concentration camp (1940–41), he joined the Polish army in exile on his release and made his way to Palestine (1942), where he served from 1943 to 1948 as commander-in-chief of the Irgun Zvai Leumi, a militant Zionist group.

Following Israel's independence in 1948, Begin founded the Herut (Freedom) Party, which evolved from the Irgun Zvai Leumi. Elected to the Knesset (parliament) the same year, he served as leader of the opposition until 1967, when he joined the National Unity government. He was appointed minister without portfolio (1967–70) and in 1973 took up the position of joint chairman of the Likud (Unity) coalition. When, in 1977, the Likud Party was successful at the polls, he became prime minister. He retained this position until 1984, when he resigned.

His hard line on Arab-Israeli relations, particularly with regard to the retaining of territories occupied by Israel during the Arab-Israeli War (1967), softened in a series of meetings with the Egyptian president, Anwar SADAT. The result was a peace treaty between Egypt and Israel and both leaders shared the Nobel Peace Prize in 1978. Begin has written several books, including *The Revolt: Personal Mem-*

oirs of the Commander of Irgun Zvai Leumi (1949), *White Nights* (1957), and *In the Underground* (1978).

Behan, Brendan (1923–64) *Irish playwright and poet.*

Behan was born in Dublin, one of three sons of a house-painter. He was an anti-British activist from an early age, and in 1940 he was captured on an IRA sabotage mission in England. He was sentenced to three years in a Borstal institution and subsequently deported. This experience was later recalled in his autobiographical *Borstal Boy* (1958). In 1942 he was involved in a terrorist shooting, for which he received a fourteen-year prison sentence, part of which he served at Mountjoy Prison, Dublin. This became the setting for his play *The Quare Fellow* (1954), which evokes the horror and humour in a prison the night before the execution of an axe-murderer ('the quare fellow'). The play also became a key text in the anti-hanging debate of the time. *The Hostage* (1958) is a play about a British soldier held captive by terrorists; this too is memorable for its black humour in the face of violence and imminent death. Behan was released from prison under amnesty in 1946, but was reimprisoned briefly for minor offences in 1947 and 1948.

Behan wrote for radio and published poems (in Irish). He also wrote short stories and in 1963 published a volume of his contributions to the *Irish Press*, works that show his outstanding skills as a raconteur and observer of the Irish scene. An alcoholic from childhood, Behan died prematurely in a diabetic coma in a Dublin hospital.

Behrens, Peter (1868–1940) *German architect and designer, whose factories were among the first to be treated as architecture. A prolific self-taught architect, he was also an influential teacher.*

Born in Hamburg, Behrens studied art in Hamburg, Karlsruhe, Düsseldorf, and Munich before establishing himself as a successful painter. However, by about 1890 he was influenced by the Arts and Crafts movement of William Morris and over the next ten years designed a variety of different objects as well as typefaces. In 1900 his success came to the notice of the Grand Duke of Hesse, who invited him to join the artists' colony at Darmstadt, of which he was patron. There Behrens designed his first house (for himself), which led him into architecture. By 1907 he had been appointed architect to AEG, and for them he built the turbine factory in Berlin (1909), which had an architectural form of its own unrelated to its functional purpose of housing machinery.

During the next ten years Behrens's style developed in the direction of neoclassicism, influenced to some extent by the expressionist trend in Europe. The German embassy in St Petersburg (1911) and the I. G. Farben office in Höchst (1920) are examples of this trend.

As a teacher, Behrens had a considerable influence on GROPIUS, MIES VAN DER ROHE, and LE CORBUSIER, all of whom worked in his office at AEG. Behrens also ran a masterclass for postgraduate architects at the Vienna Academy (1922–36). During the Nazi regime Behrens was on the staff of the Berlin Academy – a measure of the extent to which he was prepared to compromise.

Beiderbecke, (Leon) Bix (1903–31) *US jazz cornetist and pianist. Although he died at the age of twenty-eight, he was one of a handful of white musicians who profoundly influenced the early development of jazz.*

Born in Davenport, Iowa, he studied the piano as a child, and was playing the cornet professionally while still at school. In 1922 he was expelled from a military academy after less than a year; in 1923–24 he played and recorded with a band called the Wolverines and thereafter with Jean Goldkette, Frankie Trumbauer, and Paul Whiteman. He died of alcoholism.

His recordings are mostly of poor technical quality but his tone was described by fellow musicians as bell-like, as though the notes were struck by a mallet. With this exceptional tone, his perfect intonation, and a deep sense of melody, he influenced a generation of both black and white musicians. He was an indecisive and ineffectual man, who never bothered to learn to read music really well. He did not live long enough to become as well known to the public as he was to other musicians. Nevertheless, largely as a result of his tragic lifestyle and early death, he became a jazz legend.

Belloc, (Joseph) Hilaire (Pierre René) (1870–1953) *British author of half-French parentage.*

Born in St Cloud, France, Belloc was brought by his widowed mother to England in 1872. He was educated at Cardinal Newman's Oratory School near Birmingham, and his Roman Catholic upbringing was a strong influence throughout his life. He travelled in France and the USA before going up to Oxford (1893), where he made many friends and gained a name as a speaker. He failed, however, to obtain a university teaching post. In 1896, the year of his marriage, he published the first of his popular books of light verse, *The Bad Child's Book of Beasts.*

In 1899 the Bellocs moved to London, where Belloc soon became prominent as a Liberal journalist and met his lifelong friend G. K. CHESTERTON. *The Path to Rome* (1902) proved to be Belloc's greatest success in the genre of travel writing, at which he excelled. He also published several novels, collections of essays, and historical works during this period. Another success of these years was *Cautionary Tales for Children* (1908). In addition to his extensive literary activities, Belloc became a Liberal MP.

In 1914 Belloc's wife died. Although this was a shattering blow, he continued writing to support his family. In the 1920s and 1930s he produced numerous historical and biographical studies, including an ambitious *History of England* (1925–31). Throughout this period he wrote increasingly from a Roman Catholic viewpoint. World War II brought personal tragedy to Belloc: he was deeply grieved by the fall of France (1940), his youngest son was killed in action (1941), and he himself suffered a severe stroke (1942) from which he never fully recovered.

Bellow, Saul (1915–) *US writer, regarded by many as the leading figure in mid-twentieth-century US fiction. Among numerous awards, he has received the Nobel Prize for Literature (1977) and the Pulitzer Prize (1976). He was appointed Commander de la Légion d'honneur in 1983.*

Born in Quebec, Canada, the son of Russian Jewish immigrant parents, Bellow moved with his parents at the age of nine to Chicago, where he completed his education at Chicago, Northwestern, and Wisconsin universities. His first novel, *Dangling Man* (1944), was followed by *The Victim* (1947) and *The Adventures of Augie March* (1953), for which he won his first National Book Award (1954). *Henderson the Rain King* (1959) confirmed his virtuosity as a writer and *Herzog* (1964), which has been called his most personal novel incorporating many autobiographical features, was critically acclaimed. Other novels include *Mr Sammler's Planet* (1970), *Humboldt's Gift* (1975), *The Dean's December* (1982), *More Die of Heartbreak* (1987), and *The Bellarosa Connection* (1989). Among other publications is the collection of short stories *Him with His Foot in His Mouth* (1984) and three ventures in drama – *The Wrecker* (1954), *The Last Analysis* (1964), and *A Wen* (1965), generally considered less successful than his novels.

The overall tone of Bellow's style is both ironic and optimistic. The linking theme of all his fiction is the problem of the alienated individual, who must learn to be reconciled with external reality and the human condition. Bellow has been married and divorced three times and has a son by each marriage. He continues to combine a career as writer with that of university professor, having been a visiting lecturer at Princeton and New York universities and associate professor at the University of Minnesota.

Ben Bella, Mohammed Ahmed (1916–   ) *Algerian prime minister (1962–63) and president (1963–65). A leading figure in the Algerian struggle for independence, he was awarded the Lenin Peace Prize in 1964.*

Born in Marnia, the son of a peasant, Ben Bella was conscripted into the French army in 1937. He served in France and Italy during World War II, except for three years (1940–43) when he returned to Algeria to work on the family farm.

In 1947 Ben Bella became leader of the Organisation Speciale, an underground movement devoted to Algerian independence. Imprisoned by the French in 1950, he escaped to Cairo in 1952, where he founded and led the National Liberation Front (FLN), which instigated the Algerian war of independence. He was re-arrested by the French in 1956 and not released until 1962, shortly before the conclusion of the war, when he became prime minister of the provisional government. The following year he was elected president of independent Algeria. Ben Bella initiated several social and economic reforms, particularly related to land and education. He encouraged closer ties with other Arab nations but was reluctant to align his country too closely with the Soviet Union. He was overthrown in a military coup led by Colonel Boumédienne in 1965 and detained until 1979. In 1982 he was appointed chairman of the International Islamic Commission for Human Rights in London. He returned to Algeria in 1990, after nine years in exile, to lead the opposition to the ruling regime.

Benchley, Robert Charles (1889–1945) *US humorist, screenwriter, and actor.*

He graduated from Harvard in 1912 and worked for various New York City newspapers and magazines. He was theatre critic for *Life* magazine (1920–29) before becoming associated with *The New Yorker* (1929–40). Despite successes in Hollywood, where he wrote, produced, and acted in a number of his own short films, and despite some writing for radio, Benchley disliked screen and radio work and considered himself primarily a writer. His theme, the many vexations of an ordinary man, recurs in endless comic variation in some dozen collections of his pieces, including *Of All Things* (1921), *The Treasurer's Report* (1930), *My Ten Years in a Quandary* (1936), *After 1903–What?* (1938), and *Inside Benchley* (1942). The short film *How to Sleep* won an Academy Award in 1935, and six of his film scripts were posthumously gathered in *The Reel Benchley* (1950).

Beneš, Edvard (1884–1948) *Co-founder of Czechoslovakia, who served as minister of foreign affairs (1919–35), prime minister (1921–22), and president (1935–38; 1945–48).*

Born in Kozlany, Bohemia, the son of a peasant, Beneš was educated at the universities of Prague, Dijon, and Paris. He was appointed professor at the Prague Academy of Commerce in 1909, but moved to France in 1915 to work for the Czech independence movement. In 1918 he helped Tomáš MASARYK to found Czechoslovakia, serving as minister of foreign affairs from 1919 to 1935, during which period he established close ties with France and the Soviet Union, negotiating the formation of the Little Entente and championing the League of Nations (he was chairman of the council of the League six times). During this period he was also prime minister (1921–22).

Beneš succeeded Masaryk as president in 1935 but resigned in 1938 over the Munich Agreement, which ceded Sudetenland to Germany. During World War II he came to London as head of the Czechoslovakian provisional government-in-exile (1941–45), returning to Prague in 1945 to re-establish a government with himself as president. He resigned three years later when Czechoslovakia became a communist state. He was the author of several publications on international history, including his memoirs *From Munich to New War and New Victory*, which were published in English, in incomplete form, in 1954.

Ben-Gurion, David (1886–1973) *The first prime minister of Israel (1948–53; 1955–63), who became an elder statesman and a symbol of the aspirations of the Israeli people.*

Born in Plonsk, Poland, the son of an unlicensed lawyer, Ben-Gurion attended local Jewish schools in Poland before emigrating with his father in 1906 to Palestine. There he worked as a farm labourer and became an active Zionist. He studied law at the universities of Salonika and Constantinople, but was forced to leave Turkey because of his Zionist activities (1915). He fought with the British against the Turks in Palestine during World War I.

Ben-Gurion became active in the Zionist movement as secretary-general of the Jewish Labour Federation (1921–33). In 1930 he was elected leader of the Mapai Party, the predominant socialist faction within the Zionist movement. In 1935 he was appointed chairman of the Jewish Agency, the executive body of the World Zionist Organization. When the state of Israel was proclaimed in 1948, he became the first prime minister and minister for defence. He served in these positions until 1953 and again from 1955 to 1963, when he retired. He briefly returned to the political arena (1965–67) to lead the breakaway Labour (Rafi) Party, but resigned his seat in the Knesset (parliament) in 1970.

Benn, Anthony (Neil) Wedgwood (1925–) *British politician, on the left of the Labour Party.*

Benn was Labour MP for Bristol South-East from 1950 to 1960, when he inherited the title Viscount Stansgate and was consequently debarred from sitting in the House of Commons. His subsequent campaign for the right for individuals to disclaim hereditary titles achieved its aim with passage of the Peerage Act 1963. Benn was then re-elected as MP for Bristol South-East. He held office in the WILSON and CALLAGHAN administrations as postmaster-general (1964), minister of technology (1966–70), minister for trade and industry (1974–75), and secretary of state for energy (1975–79). He was chairman of the Labour Party Executive (1971–72) and of the Home Policy Committee. In 1979 he refused a place in the shadow cabinet, and in 1981 made a bid for the deputy leadership, which he lost to Denis HEALEY. He was also an unsuccessful candidate in the party leadership elections in 1976 and 1988. Benn lost his seat in the 1983 election, but was re-elected to parliament in a by-election at Chesterfield in the following year.

Benn's underlying conviction is that the state apparatus (and the parliamentary Labour Party) should be made more accountable to the people; he also envisages an extended role for the trade unions. He is one of the most gifted orators in contemporary British politics; his publications include three volumes of diaries.

Bennett, (Enoch) Arnold (1867–1931) *British novelist, playwright, and critic.*

Bennett was born in Hanley in the Potteries, the son of a solicitor. In 1888 he went to London as a solicitor's clerk, a profession he soon abandoned for the more congenial job of journalist. He became assistant editor (1893), then editor (1896–1900), of the journal *Woman* and also experimented with narrative prose; *The Grand Babylon Hotel* and *Anna of the Five Towns*, novels widely different in their styles, were published almost simultaneously in 1902.

The same year Bennett moved to Paris, where he married a Frenchwoman (1907). Here he wrote his most famous novel, *The Old Wives' Tale* (1908), which established him as a novelist who was at his best when dealing with the beliefs, cultural interests, and industrial surroundings of the people among whom he had passed his early life. He also wrote the first volumes of the *Clayhanger* trilogy: *Clayhanger* (1910) and *Hilda Lessways* (1911). Among his other productions of this prolific period were philosophical articles, later published as collections.

In 1912 Bennett resettled permanently in England. His enormously popular play, *The Great Adventure* (1913), was based on his own novel *Buried Alive* (1908). During World War I he was active as a political propagandist as well as keeping up his other writing. The last *Clayhanger* novel, *These Twain*, appeared in 1916.

In the postwar period Bennett was lionized on both the literary and social scenes; he satirized the latter in *Lord Raingo* (1926). *Riceyman Steps* (1923), a portrait of miserliness, was his last acknowledged masterpiece as a novelist. In his later years his writing retained its brilliance but lacked the depth and power of his earlier work. He died of typhoid.

Bennett, Richard Rodney (1936–) *British composer and pianist. He was created CBE in 1977.*

Born at Broadstairs, Kent, Bennett won a scholarship to the Royal Academy of Music (1953) and studied with Lennox BERKELEY and Howard Ferguson (1908–). He was soon addicted to serialism (see SCHOENBERG, ARNOLD FRANZ WALTER) and in 1957 went to study with BOULEZ in Paris on a French government scholarship; he described this time as a 'period of violent stylistic change'. On return to London he established himself as a composer and pianist of wit and brilliance with an enthusiasm for jazz. Later works tend to a richer texture. In 1970 he was visiting professor in composition at the Peabody Conservatory, Baltimore.

Bennett's compositions include *Nocturnall upon St Lucie's Day*, an early cantata with words by Donne; several operas, including *Mines of Sulphur* (1963) and *Victory* (1970); two symphonies (1965; 1967); concertos, including *Concerto for Orchestra* (1976); music for children, including the opera *All the King's Men* (1969); cinema music in a variety of films, including scores for *Far From the Madding Crowd* (1967), *Murder on the Orient Express* (1974), and *Yanks* (1980); and the ballet *Isadora* (1981).

Bentley, Edmund Clerihew (1875–1956) *British writer, inventor of the form of humorous verse known as the clerihew.*

A Londoner, Bentley was educated at St Paul's School, where he met his great friend G. K. CHESTERTON and conceived the idea of the clerihew:

> Sir Humphry Davy
> Abominated gravy.
> He lived in the odium
> Of having discovered sodium.

Such *jeux d'esprit* were collected into a mock 'Dictionary of Biography' which was eventually published with the original sketches by Chesterton as *Biography for Beginners* (1905).

After a successful but academically undistinguished career at Oxford, Bentley returned to London to practise law. He kept up his output of light verse and reviews and in 1901 abandoned the law for journalism, working first on the *Daily News* and from 1912 on the *Daily Telegraph*. His detective novel *Trent's Last Case* (1913), a brilliant parody of the 'infallible sleuth' tale of the Sherlock Holmes tradition, was an immediate and lasting success. More clerihews appeared in *More Biography* (1929), *Baseless Biography* (1939), and *Clerihews Complete* (1951).

Berdyaev, Nikolai Alexandrovich (1874–1948) *Russian anti-communist religious philosopher.*

Born in Kiev into an aristocratic family, Berdyaev was very much the revolutionary student. For such activity he was exiled from Kiev in 1899 for three years. He supported the 1917 revolution and, as a Marxist, was appointed professor of philosophy at Moscow University in 1920. By 1922 his commitment to revolutionary Marxist materialism had receded and, in disagreement with the authorities, Berdyaev was exiled from the Soviet Union. After spending two years in Berlin he settled in France, founding his own Academy for Religious Philosophy at Clamart, near Paris.

Berdyaev, in such works as *The Meaning of History* (1936) and *The Destiny of Man* (1937), sought to develop an original metaphysics. Although far from easy to understand, it is clear that central to much of his thought is the concept of freedom. Man is free, he declared, not just in the sense that he can choose to act morally or immorally, but that he can also determine for himself just what is and what is not moral and immoral. It is also clear that Berdyaev's thought bears little relation to traditional Christian philosophy and indeed, at some points, seems to approach more closely to an earlier gnosticism with his dismissal of material reality and his claim that truth is best attained not by following rational procedures but by insight. Thus, Berdyaev's views have been too obscurely expressed and too personally held to have had any significant impact on the main development of religious philosophy.

Berenson, Bernard (1865–1959) *Lithuanian-born US art historian and connoisseur whose work set new standards of criticism.*

Berenson was born into a poor family, which emigrated to Boston in 1872. He managed to obtain a place at Harvard University, from which he graduated in 1887 in languages and literature, but transferred to the study of art history after his first visit to Italy in the same year. Inspired by the beauty of Italy, Berenson soon became known as a connoisseur sought by many potential art purchasers. He became artistic adviser to the international art dealer Lord Duveen (1869–1939) and to Isabella Stewart Gardner (1840–1924), for whom he acquired Titian's *Rape of Europa*. Berenson also set about compiling catalogues of Italian Renaissance artists and was soon a world expert on that sphere of art history. His book *Italian Painters of the Renaissance* (1952), in which Berenson also displayed his concise writing style and discriminating eye, was regarded as definitive. Other major works were *The Venetian Painters of the Renaissance* (1894), *Drawings of the Florentine Painters* (1938), and works on the methodology of art history and criticism.

From 1887 Berenson spent most of his life in Italy. His reminiscences of the war years, passed in Tuscany, were published as *Rumour and Reflection, 1941–1944* (1952) and he also produced some excellent diaries, published in *The Passionate Sightseer* (1960). When he died, he bequeathed his Italian villa, with its library and art collection, to Harvard University.

Berg, Alban (1885–1935) *Austrian composer who, with* SCHOENBERG, *was one of the leading exponents of twelve-note composition.*

Against a cultivated musical Viennese family background, Berg early developed a love of music and literature and started composing during childhood. His meeting with Schoenberg in 1904 determined his future course: he studied with him for the next six years (harmony, counterpoint, analysis, orchestration, form, and composition) and achieved a mastery of his craft and, with it, the confidence of the maturing artist. His gratitude is shown in the dedication of three of his finest works to Schoenberg: *Three Orchestral Pieces* (1913–14), *Kammerkonzert* (1923–25), and his second opera, *Lulu* (unfinished). His mastery of the principles of musical form, although he composed in an atonal idiom (i e without the traditional use of key relationships), was combined with the ability to build a dramatic climax. In 1914 he saw a performance of Georg Büchner's drama *Woyzeck* that inspired his opera *Wozzeck* (1915–21), for which he wrote his own libretto, dedicating the opera to Alma Mahler, Gustav Mahler's widow. *Wozzeck* – a deeply disturbing tragedy of the misfit Wozzeck in a world of bullying and deceit – is a masterpiece of musical architecture, using such forms as sonata, fugue, passacaglia, march, and fantasia. It was staged by the Berlin State Opera (1925) after Berg had made a successful concert arrangement of *Three Fragments of Wozzeck*. In the 1920s he turned to chamber music: the *Lyric Suite* (1925–26) for string quartet is a particularly powerful and personal work in which he first uses Schoenberg's serial technique. In 1928 he arranged the three central movements for string orchestra. His next opera, *Lulu*, uses twelve-tone writing throughout; based on Wedekind's double drama, *Earth Spirit* and *Pandora's Box*, and again using Berg's own libretto, it portrays Lulu as the essence of female sexuality, paradoxically destroying and being destroyed by those to whom she gives herself. The story of her encounters, ending with her death at the hands of Jack the Ripper, is transformed by Berg's music to a cosmic and compassionate level. All the musical interludes derive from a single twelve-note series and each of the main characters has his or her own musical form. Act III, unfinished at the composer's death, was suppressed by his wife, Helene, and not orchestrally completed (by Paul Cerha) and performed (by the Paris Opéra) until 1979. A violin concerto (1935) was written for the violinist Louis Krasner and

dedicated to Alma Mahler's daughter, Manon Gropius, who had recently died: 'to the memory of an angel'. Berg's output, although not large, is in the Viennese tradition of formal coherence. He was a man of noble ideals and a writer and conversationalist of charm and intellectual depth.

Bergman, Ingmar (1918–) *Swedish film and theatre director.*

An early interest in drama, which he actively pursued while studying literature and art at the University of Stockholm, led Bergman into becoming a trainee director at a Stockholm theatre.

Bergman began his film career as a scriptwriter in 1941. His complete script *Frenzy* (1944) won considerable praise and led to his debut as a director the following year. He has since directed a long string of highly successful films. His upbringing in a religious household (his father was a Lutheran pastor and chaplain to the Swedish royal family) undoubtedly fostered his preoccupation with such religious themes as God, the Devil, death, and purity, which dominate his highly psychologically and philosophically charged work.

Early films included *Prison* (1949), which was directed by Bergman and was drawn from his own script, and *Thirst* (1949). It was, however, the last of his summer films, *Smiles of a Summer Night* (1955), that won the Cannes Film Festival Award and brought him international fame. His earlier summer films were *Summer Interlude* (1950) and *Summer with Monika* (1952). *The Seventh Seal* (1956) won the Grand Prix and *Wild Strawberries* (1957) established him as a master of the cinema. Important films that followed included *Persona* (1966), *Cries and Whispers* (1972), *Autumn Sonata* (1978), and *Fanny and Alexander* (1983). His stage productions include *Hamlet* (1986) and *Miss Julie* (1987) at the National Theatre in London.

Charges of tax fraud, later withdrawn, resulted in a period of self-imposed exile during the seventies. In 1978 he returned to Sweden and resumed his work as director at Stockholm's Royal Dramatic Theatre. To honour him, the Swedish Film Institute established the Ingmar Bergman Annual Prize for excellence in film-making. His publications include the autobiography *The Magic Lantern* (1988).

Bergman, Ingrid (1915–82) *Swedish-born actress.*

Bergman first made her name on stage and screen in Sweden before embarking on an international career in Hollywood. Gustaf

Molander's *Intermezzo* (1936), made in Sweden, was followed by David O. Selznick's 1939 American remake of the film with Leslie HOWARD (known in the UK as *Escape to Happiness*). After this success came such classics as *Casablanca* (1943) with Humphrey BOGART, *For Whom the Bell Tolls* (1943), and *Gaslight* (1944), for which she received her first Academy Award. Equally memorable were HITCHCOCK's *Spellbound* (1945), *Notorious* (1946), and *Under Capricorn* (1949. Then came the neorealist *Stromboli* (1950), directed without a script by Roberto ROSSELLINI. She had by this time left her husband, Dr Peter Lindstrom, for Rossellini: the adverse publicity that surrounded the long affair led to her exile from Hollywood and a decline in her popularity. Among the other films she made with Rossellini were *Europa '51* (1952) and *Giovanna d'Arco al rogo* (1954; *Joan at the Stake*). The affair became a marriage in 1950, although a Rome court refused to recognize either her divorce from Lindstrom or her marriage to Rossellini. This obstruction to wedlock, however, turned out to be irrelevant in 1958 when they separated; Bergman subsequently (1960) married Swedish theatre producer Lars Schmidt.

Shortly after making Jean RENOIR's *Eléna et les hommes* (1956; *Paris Does Strange Things*), Bergman re-established her career in the USA with *Anastasia* (1956), for which she won her second Oscar. Notable among the films that followed were *The Inn of the Sixth Happiness* (1958), *Murder on the Orient Express* (1974) – her third Oscar-winning film – Ingmar BERGMAN's *Autumn Sonata* (1978), which brought an Oscar nomination, and, shortly before her death, the biographical TV film of Golda MEIR. Throughout her career Bergman also had a number of major stage successes.

Bergson, Henri Louis (1859–1941) *French philosopher, best known for his concept of creative evolution. In 1928 he was awarded the Nobel Prize for Literature.*

Bergson was born in Paris, the son of a musician. He graduated and subsequently taught at the École Normale Supérieure before being appointed in 1900 to the chair of philosophy at the Collège de France, a post he held until his retirement in 1924.

Bergson established his reputation with his first book, *Essai sur les données immédiates de la conscience* (1889; translated as *Time and Free Will*, 1910). He distinguished between time, which was abstract, quantitative, and successive, and duration, which was continuous, experienced, qualitative, and personal. Thus for Bergson the real world of experience was distinct from the world described by science. He rejected the picture of a world of inert matter responding deterministically to causal stimuli for a world containing an *élan vital*, or 'vital spirit'. In his later *Matière et mémoire* (1896; translated as *Matter and Memory*, 1911) Bergson used a comparable distinction between pure memory and motor or habit memory to argue for the independence of mental phenomena from the physicochemical processes described by physiologists.

Such views were developed further in his best known work, *L'Évolution créatrice* (1907; translated as *Creative Evolution*, 1911). As before, Bergson attacked scientific materialism. He rejected the Darwinian view of evolution by natural selection and proposed instead that life possessed an inherent creative drive that led to the production of new forms. This view had little impact on biologists but exercised considerable influence on such writers as G. B. SHAW, who dramatized Bergson's views in his *Back to Methuselah* (1921).

Bergson also became widely known for his work on the philosophy of humour, *Le Rire* (1900; translated as *Laughter*, 1910).

Beria, Laventi Pavlovich (1899–1953) *Soviet politician. As head of the Soviet secret police (1938–53) he was a close associate of Josef STALIN.*

Born in Merkheuli, Georgia, the son of a peasant, Beria was attending the Polytechnical Institute in Baku when he joined the Bolsheviks (1917). After graduating in 1919, he participated in the October Revolution and was appointed director of the secret police in Georgia (1921–31). During the 1930s he rose to national prominence under Stalin's patronage, being elected to the central committee of the Communist Party in 1934.

In 1938 Beria was appointed People's Commissar of Internal Affairs (NKVD). As head of the secret police he was directly involved in the infamous 'purge trials' in which Stalin's opponents were eliminated. He also instigated other terrorist activities, including deportation of people from regions near the Baltic to forced labour camps. During World War II he continued to rise through the party and government ranks, eventually becoming a full member of the Politburo in 1946. When Stalin died in 1953, rumour suggested that he planned to take over the premiership. Feared by other leaders, who believed that he had deliberately hastened

Stalin's death, he was arrested and charged with espionage and various other offences. His fate is not certain but it was officially announced in December 1953 that he had been tried and shot as a traitor. Beria is now widely regarded within the Soviet Union as a reprehensible figure who, as a close associate of Stalin, has had his name deleted from official Soviet publications.

Berio, Luciano (1925–) *Italian avant-garde composer, resident in the USA since 1963.*

Born at Oneglia into a musical family, Berio was first taught by his father and grandfather and later studied with Ghedini at the Milan Conservatory and with Dallapiccola at the Berkshire Music Center, Tanglewood. Like BOULEZ he is committed to a total serialism, in which the placement, duration, timbre, and intensity of each note is predetermined according to the chosen series. Within these limitations, Berio writes music of great emotional intensity and is constantly experimenting with different moving and static groupings of instruments and singers and with the use of prerecorded tape and electronic sound. He was associated with Bruno Maderna (1920–73) in the opening of the Milan Studio di Fonologia. He has been on the teaching staff of many institutions, including the Darmstadt summer courses, Dartington, Tanglewood, the Juilliard School (where he formed the Juilliard Ensemble for the performance of contemporary music), Harvard University, and Boulez's Institut de Recherche et de Coordination Acoustique/Musique in Paris. He was married (1950–64) to the soprano Cathy Berberian (1928–83), chosen interpreter of many of his works. He settled in the USA in 1963, following an invitation to teach at Mills College, Oakland.

His music is often ostensibly orientated to contemporary political or human dilemmas: in *Passaggio* (1963), for example, the female protagonist is brought to degradation by the materialism of society. Berio is a prolific composer: examples of some of his compositional techniques are seen in *Omaggio a Joyce* (1958), in which a female voice reads a passage from James JOYCE's *Ulysses*, which is reconstructed electronically and used as compositional material; *Circles* (1960), for singer, harp, and percussion, the voice merging with and separating from the instruments as the singer moves round the stage; and a series of *Sequenzas* (1958–75) for virtuoso solo instruments (flute, harp, piano, trombone, viola,

oboe, percussion, violin, recorder), which are of particular interest in their extension of instrumental technique. More recent works include *Un re in ascolto* (1982), *Requies* (1983), and *Voci* (1984).

Beriosova, Svetlana (1932–) *Lithuanian-born British ballerina who danced the entire classical repertoire and created leading roles in many modern ballets.*

Beriosova was brought to the USA in 1940 by her father, the dancer Nicholas Beriosoff, and studied ballet there with Anatole Vilzak and Ludmila Shollar. She made her debut with the Ottawa Ballet Company in *Les Sylphides* and *The Nutcracker* in 1947, at the age of fifteen, and then danced with the Grand Ballet de Monte Carlo and the Metropolitan Ballet. During this period she danced in classical works, such as *Swan Lake*, and also created leading roles in new ballets, such as John Taras's *Designs with Strings* and Frank Staff's *Fanciulla delle rose*.

Joining Sadler's Wells Theatre Ballet as leading dancer in 1950, Beriosova created the main role in BALANCHINE's *Trumpet Concerto* as well as playing more roles from the classical repertoire. Two years later she transferred to Sadler's Wells (now Royal) Ballet as soloist and became the company's prima ballerina in 1955. Major successes since that time have included leading roles in CRANKO's *The Shadow* (1953), *The Prince of the Pagodas* (1957), and *Antigone* (1959); MACMILLAN's *Le Baiser de la fée* (1960) and *Diversions* (1961); and ASHTON's *Rinaldo and Armida* (1955) and *Persephone* (1961), in which her role involved both speaking and dancing.

Beriosova has travelled widely with the Royal Ballet and has danced at all the greatest venues, including La Scala, Milan. She also featured in the film of *The Soldier's Tale* (1966) and has made many television appearances.

Berkeley, Busby (William Berkeley Enos; 1895–1976) *US choreographer and film director.*

Born in Los Angeles into a theatrical family, the son of a stage director and an actress, Busby Berkeley had appeared on stage by the time he was five. After World War I, in which he served as a field artillery lieutenant, he returned to the stage and became a leading Broadway dance director.

Sam GOLDWYN introduced him to films, choreographing such Eddie Cantor films as *Whoopee* (1930), but the work for which he is best remembered was done with Warners, in whose films he produced kaleidoscopic patterns of rhythmically moving dancers (known as 'Busby's Babes'), using huge casts on tiered stages, revolving platforms, and even in swimming pools. Symmetry and movement were often enhanced by enormous numbers of pianos, harps, and cascading waterfalls, the whole effect being achieved through the novel use of moving cameras. The *Gold Diggers* series (1933–37), *Babes In Arms* (1939), *Ziegfeld Girl* (1941), and *Lady Be Good* (1941) are a few of the remarkable films he choreographed. He also directed such films as *Strike Up the Band* (1940), *For Me and My Gal* (1942), *The Gang's All Here* (1943; his first colour film), and *Take Me Out To the Ball Game* (1949).

With changing tastes fewer films came his way, *Jumbo* (1962) being his last. Interest in his unique work revived in the 1960s, however, when his films were shown on television.

Berkeley, Sir Lennox Randal Francis (1903–89) *British composer, knighted in 1974. Berkeley was not in the main tradition of British composers: being partly of French ancestry and having studied with Nadia Boulanger in Paris, he had more in common with Fauré and the neoclassicism of Stravinsky.*

Born at Oxford, he read modern languages at the University before taking up music as a career. In 1936 Berkeley met Britten at the ISCM Festival in Barcelona, the start of a long friendship and the impetus for a joint composition: the *Mont Juic Suite*, based on folksongs heard together in the Mont Juic park, outside Barcelona.

Berkeley's music has a natural melodic flow, interesting instrumental textures, and a sobriety and restraint that sometimes suggest a great intensity of feeling. It was tonally based until the early 1960s, when his second symphony (1956–58), a transitional work, showed him to be moving towards a more complex atonality; this eventually led to a serialism that brought with it the same problem that Schoenberg had had to face: that of sustaining length. Berkeley has written some of his best works for particular performers: the *Four Poems of St Teresa of Avila* (1947) for Kathleen Ferrier, *Five Songs* (1946) for Pierre Bernac (1899–1979) and Francis Poulenc, a guitar sonatina (1957) and concerto (1974) for Julian Bream, and *Songs of the Half-Light* (1964) for Peter Pears. His works include four operas, four symphonies, music for ballet, theatre, film, and radio, concertos, and some

deeply-felt sacred liturgical choral music (he became a Roman Catholic in 1928).

Berlin, Irving (Israel Baline; 1888–1989) *US songwriter. He wrote more than fifteen hundred songs, many of which acquired worldwide familiarity. Bing Crosby's recording of his 'White Christmas' became the best-selling record of all time.*

Born in Russia, he emigrated to the USA with his family in 1893. He was a street singer and a singing waiter in 1907, when he began writing lyrics; a printer's error gave him the name 'Berlin'. His first big success was 'Alexander's Ragtime Band' (1911), which became an international hit. He formed his own music publishing company in 1919, and contributed to many shows and revues, including *Annie Get Your Gun* (1946) and *Call Me Madam* (1953), which were filmed in 1950 and 1953 respectively; other musical films included *Top Hat* (1935), *On the Avenue* (1937), *Holiday Inn* (1942), and *Easter Parade* (1948).

He never learned to read music or to play the piano properly, but his songs possessed a quality that made people of all colours and creeds want to sing them.

Bernadotte, Folke, Count (1895–1948) *Swedish diplomat, remembered for his work with the Swedish Red Cross during World War II and for his role as mediator in the Arab/Jewish conflict, for which he was assassinated.*

Born in Stockholm, the nephew of King Gustavus V of Sweden, Bernadotte was educated at the Karlberg and Stromsholm military schools, becoming a cavalry officer in the Royal House Guards. Towards the end of World War I he assisted the Red Cross in arranging the exchange of prisoners. He continued to show an interest in the Red Cross during the 1920s and 1930s and encouraged the growth of the scouting movement in Scandinavia. At the outbreak of World War II, when he was president of the Swedish boy scouts, he used his military skills to train scouts in anti-aircraft work and as medical assistants, integrating the scout organization with the Swedish defence system.

Bernadotte's most significant contribution to World War II was his work as vice-president of the Swedish Red Cross (he became president in 1946). He arranged the exchange of British and German prisoners and the transfer of Danish and Norwegian political prisoners from German concentration camps to a single camp supervised by the Swedish YMCA. In 1945 he acted as an intermediary between Himmler and the Allies, conveying the German offer of capitulation. In May 1948 he was appointed by the UN Security Council to mediate in the Arab-Jewish conflict in Palestine; in September of that year he was assassinated by members of the Stern Gang, a group of extreme Zionists.

A man of integrity, Bernadotte was well respected as a diplomat. His book *The Curtain Falls*, which describes his meetings with Himmler, was published in English in 1945.

Berlin, Sir Isaiah (1909–) *Latvian-born British philosopher and historian of ideas. He was knighted in 1957 and made a member of the OM in 1971.*

Born in Riga, Berlin was educated at Corpus Christi College, Oxford, where he remained for his entire academic career. He was Chichele Professor of Social and Political Theory (1957–67) and subsequently president of Wolfson College until his retirement in 1975.

Berlin spent the 1930s engaged in conventional philosophical research. In the early 1940s, however, he became persuaded that philosophical insight required the command of greater skills in logic than he possessed; consequently he decided to devote the rest of his career to the study of the history of ideas. It was a field in which, with his earlier study, *Karl Marx* (1939), he had already excelled. Much of his later thought was directed to identifying the intellectual roots of such modern growths as nationalism, romanticism, populism, liberalism, and historicism. There were two eighteenth-century thinkers who, Berlin argued, had done most to transform traditional modes of thought. He discussed their contributions in the appropriately named *Vico and Herder* (1976).

Berlin has not been content merely to chronicle the development of such ideas in modern thought. Instead, he has sought, in his *Historical Inevitability* (1954), *Four Essays on Liberty* (1969), and several other works, to expose the central errors and distortions inherent in historical determinism. Berlin's various attempts to develop an alternative pluralism can be traced in the four volumes of his *Collected Papers* (1975–80).

Bernal, John Desmond (1901–71) *British physicist. His pioneering work in the field of X-ray crystallography enabled the structure of many complex molecules to be elucidated.*

Bernal came from an Irish farming family. Brought up as a Catholic, he was educated at Stonyhurst and Cambridge, where he abandoned Catholicism and became (1923) an active member of the Communist Party. After Cambridge, Bernal spent four years at the Royal Institution in London learning the practical details of X-ray crystallography from Sir William BRAGG. When he returned to Cambridge in 1927 he planned a research programme to reveal the complete three-dimensional structure of complex molecules, including those found exclusively in living organisms, by the techniques of X-ray crystallography.

In 1933 Bernal succeeded in obtaining photographs of single-crystal proteins and went on to study the tobacco mosaic virus. It was not, however, Bernal's own achievements in crystallography, as much as those of his pupils and colleagues, such as Dorothy HODGKIN and Max PERUTZ, that brought about the revolution in biochemistry and launched the subject of molecular biology.

In 1937 Bernal was appointed professor of physics at Birkbeck College, London. His attempts to develop the department were interrupted by the outbreak of World War II. Despite his known membership of the Communist Party, and against the advice of the security forces, Bernal spent much of the war as adviser to Earl MOUNTBATTEN. In 1945 he returned to Birkbeck College and in 1963 was appointed to a chair of crystallography. In the same year he suffered a stroke and although he continued to work for some time, a second and more severe stroke in 1965 paralysed him down one side and virtually ended Bernal's scientific life. His books include *The Social Function of Science* (1939), *Science in History* (1954), *World Without War* (1958), and *The Origin of Life* (1967).

Bernanos, Georges (1888–1948) *French novelist and polemical writer.*

Born in Paris of partly Spanish descent, Bernanos was a staunch royalist and edited the royalist weekly *L'Avant-garde* from 1913 until his enlistment in World War I. Vehement and uncompromising, his political pamphlets attacked a range of targets. *La Grande Peur des bien-pensants* (1931) denounced the French middle classes; *Les Grands Cimetières sous la lune* (1938; translated as *A Diary of My Times*, 1938) condemned FRANCO's initiation of the Spanish civil war.

Bernanos made his name as a novelist with *Sous le Soleil de Satan* (1926; translated as *Star of Satan*, 1940), which depicts a priest's battle against Satan; this theme of the struggle between good and evil is further developed in *La Joie* (1929; translated as *Joy*, 1948) and in Bernanos's best-known novel, *Journal d'un curé de campagne* (1936; translated as *The Diary of a Country Priest*, 1937). The hero of the latter is a young and inexperienced priest who is constantly frustrated in his endeavours to bring his form of Christianity to an unreceptive parish.

In 1938, disturbed by political developments in Europe and shocked by the Munich agreement with HITLER, Bernanos left for Brazil with his wife and six children. He continued to express his views in pamphlets, such as *Scandale de la vérité* (1939) and *Lettre aux Anglais* (1942; translated as *Plea for Liberty*, 1944), and in radio broadcasts to his compatriots during World War II. While in Brazil he completed and published the novel *Monsieur Ouine* (1943; translated as *The Open Mind*, 1945), which had a mixed reception on its publication in Paris (1946).

At the end of World War II Bernanos returned to France. *Dialogues des Carmélites*, a film script set at the time of the French Revolution and based on the martyrdom of a group of nuns, was completed just before his death.

Bernstein, Leonard (1918–90) *US conductor and versatile composer, whose works range from symphonies to popular musicals.*

Born at Lawrence, Massachusetts, he studied at Harvard University, the Curtis Institute, and also, during the summers of 1940 and 1941, at Tanglewood under KOUSSEVITSKY, who – impressed by his talent – made him his assistant there in 1942. He quickly attracted notice as a conductor (and pianist) and made his name overnight when he deputized at short notice for Bruno WALTER in 1944. He conducted the New York Philharmonic Orchestra from 1945 to 1948, and again from 1957 to 1969, when he was made conductor laureate for life. He toured extensively in Latin America, Europe, Asia, and the USA.

As a composer, Bernstein wrote works in widely different styles, from the *Chichester Psalms* (1965), a festival commission from Chichester Cathedral, to *West Side Story* (1957), his best-known musical. The style of his large instrumental and choral works is a diffuse virtuosity, juxtaposing a romantic intensity with jazz and Latin American elements. His book, *The Unanswered Question* (1973), reprints a series of lectures given at Harvard University.

Bernstein, Sidney Lewis, Baron (1899–) *British businessman who with his brother, Cecil, built up the Granada Group of companies, which has had an important influence on*

British cinema and television. He was created Baron Bernstein of Leigh in 1969.

Born into a Jewish family living in Ilford, Bernstein left school at fifteen and worked briefly as an engineering apprentice before joining his father, who ran a small chain of cinemas. After his father died in 1921, Bernstein took over. He ran the Edmonton Empire as a music hall for five years (1922–27) and in 1925 was one of the founders of The Film Society, which sought to present the best of the new European cinema to British audiences. In the late 1920s, Bernstein embarked on an ambitious programme of cinema construction and renovation, which produced many extravagant renaissance-style picture palaces. In 1930, Noël COWARD's *Private Lives* opened at Bernstein's Phoenix Theatre in London. By the end of the 1930s, Bernstein owned some thirty cinemas.

During World War II, Bernstein acted as films adviser to the Ministry of Information (1940–45) and, in collaboration with Alfred HITCHCOCK, compiled a documentary film of Nazi atrocities, which was suppressed after the war by the British government. Bernstein continued his association with Hitchcock, in Hollywood, as producer of such films as *Rope* (1948), *Under Capricorn* (1949), and *I Confess* (1952). After this, however, he returned to England, and in 1956 Bernstein's Granada Television broadcast to the north of England as one of the first companies in the newly formed independent television network. Bernstein's business interests have also extended to publishing, television rentals, and motorway services. As a lifelong socialist, Bernstein was a keen supporter of the Labour Party and a close friend of Harold WILSON. Since 1983 he has been chairman of Manchester's Royal Exchange Theatre.

Berryman, John (1914–72) *US poet, writer, and critic, who was extremely influential in America during the 1950s and 1960s.*

Born at McAlester, Oklahoma, Berryman was brought up as a strict Roman Catholic and never escaped from a sense of religious guilt, which was strengthened by his father's suicide when Berryman was twelve years old. At Columbia University he was influenced by the poet Mark Van Doren (1894–1972) and began to publish his own poems in magazines, mostly in the style of W. B. YEATS. Several volumes of poetry followed and he also won awards for his short stories, such as *The Lovers* and *The Imaginary Jew*, which won a major competition.

Homage to Mistress Bradstreet (1956), an innovative monologue, established Berryman as a leading US writer, and an equally daring long poem, *77 Dream Songs* (1964), won a Pulitzer Prize. Berryman's reputation as a poet continued to grow, despite bouts of alcoholism and depression. In 1968 he enlarged his 'dream song' sequence by adding to it *His Toy, His Dream, His Rest*. A more transparently autobiographical note was sounded in *Berryman's Sonnets* (1967; written in the late 1940s) and *Love and Fame* (1970). Apart from writing poetry, Berryman taught at several US universities and produced a number of critical pieces, including *The Freedom of the Poet* (1974). In 1972 his inability to control his alcoholism drove Berryman to suicide by leaping from a bridge over the frozen Mississippi at Minneapolis. A book of poems, *Delusions, etc.* (1972), and a novel, *Recovery* (1973), were published after his death.

Bertolucci, Bernardo (1940–   ) *Italian film director.*

Son of the poet and film critic Attilio Bertolucci, Bernardo was born in Parma. As a child he wrote poetry and his first volume of verse, *In Search of Mystery*, won a national prize while he was still a student at Rome University. He left his studies to work as assistant director on PASOLINI's *Accattone!* (1961) and the following year made his debut as a director with *La commare secca* (1962). Critical acclaim, however, came with his next film, *Prima della rivoluzione* (1964; *Before the Revolution*), which won the Max Ophüls Prize. *La strategia del ragno* (1970; *The Spider's Strategy*) and *Il conformista* (1970; *The Conformist*) are among the notable films that followed. Bertolucci's radical political viewpoint and remarkable use of setting and camera built up his reputation for producing stringently emotional work. The controversial and sexually explicit *Last Tango in Paris* (1972), featuring Marlon BRANDO, which was something of a departure from his earlier work, became a huge box-office success. Later films include the epic *Novecento* (1976; *1900*), which originally lasted for over five hours, *La luna* (1979), and *Tragedy of a Ridiculous Man* (1982). He enjoyed his greatest commercial success to date in 1988 with *The Last Emperor*, about the fall of the imperial dynasty in China, which won nine Oscars. His films since then include *The Sheltering Sky* (1990).

Best, Charles Herbert (1899–1978) See BANTING, SIR FREDERICK GRANT.

Best, George (1946–   ) *British Association footballer, who played for Manchester United and Northern Ireland and was regarded as one of the most talented footballers in the world.*

Born in Belfast, Best signed for Manchester United almost straight from school and his exceptional ability was soon evident. As an attacking winger with remarkable ball control, he went on to win championship medals and a European Cup winners' medal with the First Division club. In 1968–69 he was European and English Footballer of the Year. By 1973, however, his lack of self-discipline and wayward behaviour were becoming apparent. Best played in America, had a spell with Fulham and played briefly for a succession of other clubs of varying status, during which time his domestic and alcohol problems were publicized almost as much as his great footballing skills once were. It is particularly regretted that he never played for Northern Ireland in a World Cup final.

Bethe, Hans Albrecht (1906–   ) *German-born US physicist, who discovered the source of energy in the sun and the stars. He was awarded the 1967 Nobel Prize for Physics.*

The son of a university professor, Bethe was educated at the University of Munich, where he obtained his PhD under the supervision of Arnold SOMMERFELD in 1928. He taught initially at the University of Tübingen but as a Jew was dismissed in 1933. He abandoned Germany and after two years in Britain accepted in 1935 the post of professor of physics at Cornell University, a post he retained until his retirement in 1975.

Bethe established his reputation with a series of papers, known as Bethe's Bible (1936–37), in which he reviewed the current state of nuclear physics. In the following year he went on to explain how the stars were able to produce prodigious amounts of energy for vast periods of time. Bethe proposed a cyclic nuclear reaction in which, starting with carbon-12, four protons fuse to form two helium nuclei, leaving the original carbon-12 to enter into the reaction once more. The small mass loss accounted for the energy released. It was for this work that Bethe was awarded his Nobel Prize. Bethe also collaborated with Ralph ALPHER and George GAMOW on their well-known paper (1948) concerning the origin of the elements of the universe.

During World War II Bethe worked on the military application of nuclear fission. He was appointed head of the theoretical division at Los Alamos in 1943. Later, when his old colleague and director J. R. OPPENHEIMER was under investigation by the security authorities, Bethe gave evidence in his favour. In the 1960s he spoke out for the test-ban treaty and in the 1970s he argued strongly that nuclear power was essentially safe and the only real alternative to fossil fuels.

Betjeman, Sir John (1906–84) *British poet and author. He was knighted in 1969 and appointed poet laureate in 1972. Among many other honours he was awarded the Queen's Medal for Poetry (1960).*

After a somewhat lonely childhood in London and education at Marlborough, Betjeman went up to Oxford. Here he failed to distinguish himself academically but found his métier as a poet and made many friends. His earliest publications, the poems *Mount Zion* (1932) and the architectural essays *Ghastly Good Taste* (1933), reveal his preoccupations with suburban dreariness, ecclesiastical architecture, High-Church Anglicanism, death, and English topography. Other volumes of verse followed: *Continual Dew* (1937), *Old Lights for New Chancels* (1940), *New Bats in Old Belfries* (1945), and *Selected Poems* (1948). *A Few Late Chrysanthemums* (1954) won him an appreciative readership, and his *Collected Poems* (1958) was a best-seller. In 1960 his verse autobiography, *Summoned by Bells*, was an immediate success.

Betjeman's interest in the English landscape was revealed in the several county guides he wrote alone or in collaboration with John PIPER. He also did a great deal to raise public awareness of the merits of Victorian and Edwardian architecture during a period in which this was unfashionable. His poetry continues to be popular among readers who might otherwise read very little twentieth-century poetry but who like his use of traditional verse forms, his gentle satire, his self-deprecation, and his affectionate description of the English scene.

Betti, Ugo (1892–1953) *Italian playwright.* Born at Camerino in the Marches, as a child Betti moved with his family to Parma, where he later studied law, taking his degree in 1914. He volunteered and served with distinction as an artillery officer in World War I, was captured by the Austrian forces in 1917, and interned in Germany. After the war he returned to Parma and his legal studies and was appointed a magistrate in the 1920s. He married in 1930 and was appointed to the Court of Appeals in Rome, where he served until 1943. Although he was criticized for having contin-

ued as a judge under MUSSOLINI, there is no evidence of fascist sympathies in his work. For the last decade of his life he was employed in legal archives in Rome while devoting himself primarily to his literary work.

Although he had published volumes of verse, several short-story collections, and one novel during his lifetime, few of Betti's works, apart from his later poems, won much critical notice. His twenty-five plays constitute his important work, and only after the Paris production of the best of these, *Delitto all' isola delle capre* (1950; translated as *Crime on Goat Island*, 1961), a few months before his death, did his dramatic work find an international audience. This and his other most widely produced plays – which include *Frana allo scalo nord* (1935; translated as *Landslide*, 1964), *Ispezione* (1947; translated as *The Inquiry*, 1966), and *Corruzione al palazzo di giustizia* (1949; translated as *Corruption at the Palace of Justice*) – take the form of harrowing legal examinations that result in a relentless exposure of the real motives, evil, and guilt that lie beneath the social surfaces of the characters.

Bevan, Aneurin (1897–1960) *British Labour politician and leader of the radical wing of the party after World War II.*

'Nye' Bevan was born in Monmouthshire, the son of a miner. As MP for Ebbw Vale for over thirty years (1929–60), he was the spokesman for the miners of South Wales. In 1934 he married Jenny Lee (later Baroness Lee 1904–88), herself a Labour MP (1929–31; 1945–70) and later minister for the arts (1967–70). From 1940 to 1945 he edited the left-wing Labour weekly *Tribune*, attacking CHURCHILL's wartime government, including its Labour members. His oratory, though brilliant, could be abusive, and Churchill called him a 'merchant of discourtesy'. As minister of health (1945–51) under ATTLEE, Bevan established the National Health Service (1948), effectively nationalizing the hospitals and providing free medical treatment and drugs for all. Leader of the Keep Left group (the 'Bevanites') in the party, he resigned from the government in 1951 in protest against the introduction of health-service charges (he was then minister of labour). An unsuccessful challenger for the party leadership in 1955 (won by Hugh GAITSKELL), Bevan was deputy leader from 1959 to 1960. His autobiography, *In Place of Fear*, was published in 1952.

Beveridge, William Henry, 1st Baron (1879–1963) *British economist and social reformer, author of the famous Beveridge Report*

(Report on Social Insurance and Allied Services, 1942), which formed the basis of the welfare state in the UK. He was made a baron in 1946.

Born in Bengal, where his father was a judge in the Indian Civil Service, Beveridge was educated at Balliol College, Oxford. He then taught law, wrote leaders for the *Morning Post*, and in 1908 entered the Board of Trade. Greatly concerned with the problem of unemployment, he became an authority on unemployment insurance and wrote *Unemployment: A Problem of Industry* in 1909. From 1909 to 1916 he was director of Labour Exchanges, organizing a national system of labour exchanges and a compulsory unemployment insurance scheme, and head of the small Employment Department of the Board of Trade, which eventually became the Ministry of Labour. As director of the London School of Economics (1919–37) Beveridge transformed it into an institution of international repute. He was master of University College, Oxford, from 1937 to 1944 and president of the Royal Statistical Society (1941–43).

The Beveridge Report resulted from his chairmanship of the Committee on Social Insurance and Allied Services appointed in 1941, with eleven officials from the government departments involved. It was a comprehensive scheme of social insurance, without income limits, and formed the basis of much subsequent social legislation. It provided for benefits to those experiencing unemployment, sickness, or old age, to be financed by National Insurance contributions, as well as certain income-related noncontributory benefits (later called supplementary benefits). The Report was greeted cautiously by both the government and the public but almost all its recommendations were adopted: the Family Allowances Act was passed in 1945 and the National Health Service and National Insurance Acts in 1946. In 1944 his *Full Employment in a Free Society* was published, criticizing the government's recent White Paper on full employment and containing his own definition of full employment as 3 per cent unemployed. His *Voluntary Action* (1948) was a report on the voluntary social services undertaken at the request of the National Deposit Friendly Society, and his *Power and Influence* (1953) was an autobiographical work.

Beveridge chaired the Development Corporations of Aycliffe and Peterlee, County Durham, and was for a short period (1944–45) MP for Berwick upon Tweed, though his political acumen seemed to fall a long way short of his administrative skill.

Bevin, Ernest (1881–1951) *British trade unionist and politician.*

Originally a farm worker in his native Somerset, he formed a branch of the Docker's Union in Bristol in 1910 and was one of the founders of the Transport and General Workers' Union, serving as its first general secretary (1921–40). A leading organizer of the General Strike in 1926, he later attacked the Labour prime minister Ramsay MacDONALD for failing to do enough to alleviate social distress. In 1937, Bevin was elected chairman of the Trades Union Congress, and in 1940, after the outbreak of World War II, became an MP. As wartime minister of labour (1940–45), he was responsible for the highly successful mobilization of personnel to serve Britain's war industry. Bevin was foreign secretary (1945–51) in Clement ATTLEE's postwar government.

He worked for a Europe united against communism, negotiating the Brussels treaty between Britain, France, Belgium, the Netherlands, and Luxembourg and helping to form the Organization for European Economic Cooperation (both in 1948). He was instrumental in the implementation of the Marshall Plan for Europe's postwar recovery (see MARSHALL, GEORGE C.) and in the formation of the North Atlantic Treaty Organization (1949), but his association with the USA against the Soviet Union was bitterly resented by the left wing of the Labour Party. The 1951 Colombo Plan for economic cooperation among the Commonwealth countries of South and Southeast Asia was largely Bevin's work. In the Middle East he was unsuccessful in finding a solution acceptable to both Zionists and Arabs, and he surrendered the British mandate in Palestine to the United Nations. He managed to make himself unpopular with both sides in this dispute, as well as with intellectuals in the UK. Ill-health forced his resignation in 1951. He was appointed lord privy seal but died five weeks later.

Bhutto, Benazir (1953–) *Prime minister (1988–90) of Pakistan.*

Daughter of Zulfikar Ali BHUTTO, Benazir Bhutto was born in Karachi and educated at Harvard University and Oxford, where she received a degree in philosophy, politics, and economics (1977). In the same year her father was deposed: as a result she spent much of her time subsequently under house arrest (1977–84) or in exile (1984–86), under the orders of President ZIA UL-HAQ. However, following Zia's death in an air crash in 1988, she was elected head of her father's party, the Pakistan People's Party, and became the first woman prime minister of a Muslim country. Promising radical reform of Pakistani society, including the protection of women's rights and the liberalization of the press and trade unions, she failed nonetheless to win support from other parties and her government gradually lost popular support. In the wake of increasing ethnic violence, she was dismissed as prime minister by President Ghulam Ishaq Khan in 1990; defeated in the ensuing election, she was subsequently tried on charges of corruption.

Bhutto, Zulfikar Ali (1928–79) *President (1971–73) and subsequently prime minister (1973–77) of Pakistan. The first civilian president of Pakistan, he was an outspoken defender of Pakistani interests, who became internationally known for his anti-Indian views and the rapprochement he instigated with China.*

Born in Larkana of Rajput descent, Bhutto was educated at a school in Bombay before studying at the University of California (1950), Christ Church, Oxford (1952), and Lincoln's Inn (1953), where he qualified as a barrister. He taught international law at the University of Southampton (1952–53) but returned to Pakistan in 1953 to practise and teach law at the University of Sind (1953–58).

Bhutto entered politics in 1957, when he was appointed leader of Pakistan's delegation to the United Nations General Assembly. The following year he became minister of commerce under President AYUB KHAN, the first of several portfolios. In 1962 he became deputy head of the Muslim League and by 1963, when he was made foreign minister, he had established himself as a diplomat and international speaker for Pakistan. He resigned as foreign minister in 1966, over opposition to the Indo-Pakistan settlement reached at Tashkent, and in 1967 formed the Pakistan People's Party. Imprisoned (1968–69) for opposing Ayub Khan's regime, he became deputy prime minister and foreign minister (1971) under General Agha Yahya Khan (1917–80), who overthrew Ayub Khan in 1969. When Pakistan was defeated by India in 1971 and East Pakistan (now Bangladesh) seceded, he took over from Agha Yahya Khan as president. He served in this position and then as prime minister (after constitutional changes in 1973) until 1977, when he was deposed in a military coup led by ZIA UL-HAQ. Arrested for conspiracy to murder shortly afterwards, he was sen-

tenced to death (1978) and hanged (1979) despite pleas for clemency from the international community. He wrote several books, including *The Myth of Independence* (1969) and *The Great Tragedy* (1971).

Bidault, Georges (1899–1983) *French statesman; foreign minister (1947–48, 1953–54) and prime minister (1949–50).*

Born in Moulins, the son of an insurance company director, Bidault was educated at a French Jesuit school in northern Italy and at the Sorbonne in Paris. He served with the French army of occupation in the Ruhr in 1919. Beginning a career as a history teacher in Valenciennes (1925–26), he taught in Reims (1926–31) and at the prestigious Louis-le-Grand lycée in Paris (1931–39).

Bidault first became involved in politics in 1932 as founder of *L'Aube*, a Catholic daily newspaper. Outspoken in foreign affairs, particularly against German appeasement policies, he became prominent among French intellectuals in the 1930s. At the outbreak of World War II he enlisted in the French army but was taken prisoner by the Germans. Following his release in 1941, he resumed his teaching post in Lyons, at the same time working for the resistance. In 1943 he was arrested, but escaped to become the head of the National Council of Resistance, the ruling body of the resistance movement. When DE GAULLE formed the provisional government in 1944, Bidault was named minister of foreign affairs. Helping to form the Mouvement Républicain Populaire (MRP) the same year, he was elected to the constituent assembly in 1945, succeeding de Gaulle as head of government for a brief period in 1946. He was a member of parliament throughout the Fourth Republic, becoming internationally known during his years as foreign minister and prime minister. In a move to the right he founded the Mouvement Démocratique Chrétien 1958 and was elected to the first national assembly of the Fifth Republic. As an advocate of French retention of Algeria, he was strongly opposed to de Gaulle's Algerian policy and actively sympathized with the OAS (Organisation de l'Armée Secrète), a terrorist group in France. Charged with treason in 1963, he took refuge in Brazil, where he remained until 1967. He returned to France in 1968 when the warrant for his arrest was suspended.

Bidault's published works include *D'Une Résistance à l'autre* (1965) and *Le Point* (1968).

Biko, Stephan Bantu (1946–77) *Black radical leader in South Africa, who became the martyred figurehead of a generation of black students after his brutal death while in the custody of the South African security police.*

Born in Kingwilliamstown, Biko was educated at the Catholic mission school at Mariannhill, before entering the University of Natal to study medicine in 1966. Active in the student representative council, he was a delegate to the University Christian Movement Conference in Grahamstown in 1967 and to several congresses of the National Union of South African Students.

Biko first achieved prominence when he founded and became president of the South African Students Organization (SASO) in 1968. Dropping his medical studies in 1972, he helped to form the Black People's Convention, an outgrowth of SASO, which was based on the slogan of 'black consciousness' and aimed to make the black community more aware of its oppression and to develop a sense of pride. Banned in 1973, he was arrested in 1977 and died while in the custody of security police, before he could be brought to trial. His story was told in 1987 in Sir Richard ATTEN-BOROUGH's film *Cry Freedom*.

Binet, Alfred (1857–1911) *French psychologist who made important contributions to the measurement of intelligence and educational achievement, particularly by devising the Binet scale for assessing the mental age of a subject.*

Binet began his career in law but his fascination with the work of Jean Charcot on hypnosis led him to take up medical studies at the Salpetrière Hospital, Paris. He was strongly influenced by the theories of associationism expounded by John Stuart Mill and Herbert Spencer and attempted to provide sound experimental evidence in their support. He developed techniques for measuring reasoning ability and other higher mental processes using simple tests involving pencil and paper and pictures.

In 1895 Binet founded the journal *L'Année psychologique*, which was primarily a medium for publishing his own work and that of his followers. He also established a laboratory in Paris for the study and experimental teaching of children. In 1903 he published one of his most important studies, *L'Étude expérimentale de l'intelligence* ('The Experimental Study of Intelligence'), in which he investigated the mental characteristics of his two young daughters as a systematic study of contrasting personalities. Binet pioneered the use of projective testing in which a subject's response to pictures, inkblots, and other visual material provided psychological information.

Birkhoff, George David (1884–1944) *US mathematician, who made great contributions to dynamics and other aspects of mathematics.*

The son of a physician, Birkhoff was born in Overisl, Michigan, and educated at the University of Chicago, where he obtained his PhD in 1907. After an initial teaching appointment at Princeton, Birkhoff moved to Harvard in 1912, becoming professor of mathematics there in 1919, a post he continued to occupy for the rest of his life.

A powerful mathematician with wide interests, Birkhoff contributed at some time or other to most major areas of his subject. He established his reputation with some drama. The dying Henri Poincaré (1854–1912), the leading mathematician of his generation, had published without proof an important conjecture on the three-body problem. Known variously as Poincaré's unfinished symphony and Poincaré's last theorem, it was proved by Birkhoff shortly after its publication in 1912 and is known as the Poincaré–Birkhoff fixed-point theorem. Birkhoff continued to work in dynamics. He also made significant contributions to the study of differential equations and probability theory.

To nonmathematicians Birkhoff is probably best known as the author of a curious work, *Aesthetic Measure* (1933). He identified two elements in a work of art, complexity (C) and order or symmetry (O), and went on to define the aesthetic measure (M) of such works. If M, O, and C are treated as measurable variables, he concluded that $M = O/C$ – 'the conjecture that the aesthetic measure is determined by the density of order relations in the aesthetic object'. This proposition did not find many supporters, perhaps because it seems that the square turns out to have the highest M value.

Birtwhistle, Sir Harrison (1934–) *British avant-garde composer and clarinettist. He was knighted in 1988.*

Birtwhistle was born at Accrington, Lancashire. Winning a scholarship to the Royal Manchester College of Music (1952), he studied composition with Richard Hall and, with his fellow students Peter Maxwell DAVIES, Alexander Goehr (1932–), and John Ogdon (1937–89), formed the New Music Manchester Group for the performance of contemporary music and the then rarely heard works of the Second Viennese School. After National

Service in an army band, he studied clarinet with Reginald Kell at the Royal Academy of Music. From 1962 to 1965 Birtwhistle was on the staff of Cranbourne Chase School, Dorset; in 1975 he was made musical director of the newly established South Bank National Theatre. He was connected with the Pierrot Players, which he founded jointly with Peter Maxwell Davies (1967), and with Matrix, Alan Hacker's group formed when the Pierrot Players became the Fires of London (1970).

As a composer, Birtwhistle's seriousness of purpose led him from the Stravinskian *Refrains and Choruses* (1957) to two wholly idiomatic works (1965): the instrumental *Tragoedia* and the vocal and instrumental *Ring a Dumb Carillon*. *Tragoedia* is in effect an abstract of Greek drama and it led to the theatrical works of the late 1960s, including *Punch and Judy* (1968) and *Down by the Greenwood Side* (1969), both ritualistic and often dramatically and instrumentally violent. His themes stress the cycle of nature in contrast with the passing of man. In *Medusa* (1970; revised 1978), for instrumental ensemble and tape, the evolving development of the music follows the pattern of jellyfish (medusae), who breed by subdivision. Subsequent scores, for instance *Nenia on the Death of Orpheus* (1970) for soprano and instrumental ensemble, have a more homogeneous texture. *The Triumph of Time* (1972) demonstrated his interest in textures while in *...agm...* (1979) he experimented with fragmentary material. More recent works include the opera *The Mask of Orpheus* (1973–84), *Quintet* (1981), *Still Movement* (1984), *Earth Dances* (1986), *Endless Parade* (1987), and his latest opera *Gawain* (1991). Birtwhistle's music has a sculptural, sometimes static, quality, an intense imaginative expressiveness, and an explicit purposefulness.

Black, Sir James Whyte (1924–   ) *British biochemist, born in Scotland, who shared the Nobel Prize for Physiology or Medicine in 1988 for his development of two important drugs, propranolol and cimetidine. He was knighted in 1981.*

Born in Uddingston, Scotland, he graduated in medicine at the University of St Andrews in 1946 and subsequently occupied a series of university teaching posts before joining ICI as a senior pharmacologist in 1958. In 1964 he was appointed head of biological research at Smith Kline and French Laboratories, moving in 1978 to the Wellcome Research Laboratories as director of therapeutic research; since

1984 he has been professor of analytical pharmacology at King's College Hospital Medical School at the University of London.

In the course of his research, Black sought a drug to relieve angina pectoris. His first success was to isolate a substance that would block the beta-receptors of the heart muscle, to prevent their stimulation by adrenalin or noradrenalin. The result of this work was the drug propranalol, which was the first of a now quite large group of so-called beta-blockers. These are widely used to control hypertension, angina, and other serious heart conditions. Based on the success of this work, Black then sought a drug to assist in the control of stomach and duodenal ulcers. He succeeded in producing cimetidine, which has been highly successful in blocking the histamine receptors, which stimulate the secretion of the stomach acids, oversecretion of which are the basic cause of gastric ulcers. Both these drugs have been instrumental in saving many lives and in improving the quality of life for countless sufferers of angina and gastric ulcers.

Blackett, Patrick Maynard Stuart, Baron (1897–1974) *British physicist who shared in the discovery of the positron and made valuable studies of cosmic rays and rock magnetism, for which he was awarded the 1948 Nobel Prize for Physics. A scientific adviser to the services during World War II and the government in later years, he was made a CH in 1963, awarded the OM in 1967, and created a life peer in 1969.*

The son of a stockbroker, Blackett was educated at the naval colleges of Osborne and Dartmouth. He served with the Royal Navy during World War I and saw action at the Battle of Jutland (1916). After the war Blackett resigned from the navy and studied physics at Cambridge under Lord RUTHERFORD. He remained at the Cavendish until 1933, when he was appointed professor of physics at Birkbeck College, London. While at the Cavendish Blackett carried out the work for which he was awarded his Nobel Prize. Using a cloud chamber to photograph cosmic rays, and working with G. P. S. Occhialini (1907–   ), Blackett was able in 1933 to detect the positron independently of C. D. Anderson (1905–   ) in America.

In 1937 Blackett moved to Manchester University as Langworthy Professor of Physics. By this time Blackett, with his military and scientific background, had begun to advise the government on the defence of the nation and found himself caught up in the power struggle between Lord CHERWELL and Sir Henry TIZARD, with a tendency to support Tizard. He spent much of the war in operations research at the Admiralty. After the war Blackett continued to play a role in public affairs. He argued against a too heavy commitment to nuclear weapons and warned successive governments of the technological backwardness of Britain. At Manchester Blackett's scientific interests had turned to exploring the magnetic history of the earth. With the aid of newly designed sensitive magnetometers the surveys inspired by Blackett were one of the strands leading to the wide acceptance given to the current theory of continental drift. He returned to London in 1953 to the chair of physics at Imperial College, remaining there until his retirement in 1963. As president of the Royal Society (1965–70) and as a special adviser to the 1964 Labour government, Blackett was well placed to make his views known and to influence policy.

Blackwell, Sir Basil Henry (1889–1984) *British publisher and president of Basil H. Blackwell Limited. He received a knighthood in 1956.*

It was Basil Blackwell's father, Benjamin Henry Blackwell, who founded the bookshop in Broad Street, Oxford, in 1879. After studying humanities at Merton College, Oxford, Basil Blackwell spent some time (1911–12) with Oxford University Press in London before joining his father in the business. He became chairman in 1924 following his father's death. Blackwell's keen interest in poetry led him to publish early work by such authors as Aldous HUXLEY, Edith SITWELL, and Dorothy L. SAYERS (who briefly worked for Blackwell as an editorial assistant). In 1921, Blackwell relaunched an ailing press as Shakespeare Head Press, which became known for its high-quality literary classics. In partnership with Adrian Mott, Basil Blackwell and Mott Limited was formed in 1922 and the following year the first *Joy Street* children's annual was published by Blackwell to introduce children to such authors as Hilaire BELLOC and A. A. MILNE. During the 1920s and 1930s, Blackwell steadily expanded the publishing side of the business, concentrating especially on educational books; in 1939 Blackwell Scientific Publications was formed. Blackwell served as president of the Antiquarian Booksellers' Association (1925–26) and as president of the Council of the Association of Booksellers (1934–36).

Blériot, Louis (1872–1936) *French aviator who, in 1909, made the first journey by aeroplane across the English Channel. He was made a Commandeur de la Légion d'honneur in recognition of this achievement.*

The son of a manufacturer, Blériot took a degree in engineering prior to his military service. He designed and patented an automobile lamp, which earned him a fortune. This he spent in the early 1900s attempting to build aircraft, founding the first French aircraft factory in 1906. On 17 September 1907 he flew 186 metres in one of his own monoplanes. In subsequent flights he covered up to 42 km overland. On 25 July 1909 he took off from Les Barques, Calais, at 4.41 am and landed near Dover Castle 36 minutes later after a flight of some 38 km. His number XI monoplane was powered by a 25 hp engine. Blériot won a prize of £1000 offered by the *Daily Mail* and received many orders for his planes as a result. He became one of the leading aircraft designers and manufacturers. In 1911, a Blériot plane was the first to be used for military purposes (for reconnaissance by the Italians against the Turks) and his planes were used by the French army during World War I. In 1929 he made an anniversary flight over the Channel in one of his own craft to commemorate his historic achievement.

Bliss, Sir Arthur Edward Drummond (1891–1975) *British composer of US descent. He was knighted in 1950 and served as Master of the Queen's Music from 1953 until his death.*

Born in London, Bliss took a music degree at Cambridge and studied briefly at the Royal College of Music with Charles Stanford (1852–1924), VAUGHAN WILLIAMS, and HOLST. After army service in World War I, he was attracted by the lively innovations of Parisian music of the 1920s, but his natural style of post-Elgarian romanticism (without ELGAR's introspection) prevailed. His most deeply felt work was *Morning Heroes* (1930), a choral symphony with orator, sublimating the experiences of the war years. The *Colour Symphony* is a work of his early maturity, written at the suggestion of Elgar for the Three Choirs Festival of 1922. He wrote two excellent ballet scores for Sadler's Wells: *Checkmate* (1937) and *Miracle of the Gorbals* (1944); also a powerful score for Alexander KORDA's film *Things to Come* (1934–35). Perhaps his best-known work was, however, his concerto for piano (1938). His one opera, *The Olympians* (1949), with libretto by J. B. PRIESTLEY, failed to stay in the repertoire. Bliss was Director of Music at the BBC (1942–44).

Bloch, Ernest (1880–1959) *Jewish composer, a naturalized American born in Switzerland.*

Bloch studied at the Brussels Conservatory under Ysaÿe and also in Germany. His opera *Macbeth* was produced at the Opéra-Comique in Paris in 1910 to great acclaim. In 1916 Bloch went to the USA as conductor for the Maud Allen Dance Company. Settling in New York, where he taught at the Mannes School of Music, he conducted the Boston Symphoy Orchestra in his *Three Jewish Poems* (1913), won the Elizabeth Sprague Coolidge award for his *Suite for Viola and Orchestra* (1919), and was appointed director of the Cleveland Institute of Music (1920) and director of the San Francisco Conservatory (1925), where Roger SESSIONS was among his students. A ten-year grant from a wealthy patron enabled him to give up teaching; after visits to Switzerland and Italy, he returned to the USA and spent his last years at Agate Beach, Oregon, overlooking the Pacific Ocean.

Bloch's musical language is in direct line of descent from the late nineteenth-century romanticism of Liszt and Richard STRAUSS and imbued with the potency of his Jewish heritage, apparent in the long sinuous melodies suggesting Jewish cantillation in their oriental improvisatory style. 'I aspire to write Jewish music because it is the only way in which I can produce music of vitality – if I can do such a thing at all', he wrote. Among his specifically Jewish works, the rhapsody for cello and orchestra, *Schelomo* (1916; *Solomon*), is outstanding; more classically conceived works include the concerto grosso (1925) for strings and piano obbligato, which Bloch wrote as a model for his students.

Bloch, Felix (1905–83) *Swiss-born US physicist, who won the Nobel Prize for Physics in 1952 for his work on the magnetic properties of atomic nuclei. After the war he developed the technique of nuclear magnetic resonance.*

Bloch, the son of a wholesale grain dealer, was educated at the Federal Institute of Technology in Zürich and the University of Leipzig, where he obtained his PhD under Werner HEISENBERG in 1928. Bloch taught briefly in Germany, but as a Jew left Germany for the USA in 1934 to join the physics faculty at the University of Stanford, California. In 1936 he was appointed professor of physics there, a post he held until his retirement in 1971, except for a period of a year (from 1954) when he became director of CERN in Geneva.

Bloch is best known for his work on the magnetic properties of solids and his measurements of the magnetic moments of atomic nuclei. In 1939 he measured the magnetic moment of the neutron. Shortly after World War II, Bloch and his co-workers developed the technique of nuclear magnetic resonance (NMR), which has become a widely used technique in chemistry, biochemistry, and medicine.

Bloch, Marc (1886–1944) *French historian, who died fighting with the Resistance in World War II.*

Of French-Alsatian Jewish descent, Block was born in Lyons, where his father taught ancient history. He had moved to Paris, passed the necessary examinations, travelled widely, and taught in various lycées before 'four years of fighting idleness' interrupted his studies in 1914; for his activities during World War I he was admitted to the Légion d'honneur and awarded a Croix de Guerre. After acquiring his doctorate in 1920, he taught first in Strasbourg and from 1936 as professor of economic history in Paris.

Bloch's prodigious output, including probing reviews in the magazine *Annales*, which he co-founded with Lucien FEBVRE in 1929, was as remarkable as the width of his interests, both in history, particularly comparative history, and in neighbouring disciplines. His first book, *Les Rois thaumaturges* (1924; translated as *The Royal Touch*, 1973), dealt with the healing attributes of kings. Later works included his two-volume work on feudal societies *La Société féodale* (1935; translated as *Feudal Society*, 1961); *L'Étrange Défaite* (1940; translated as *Strange Defeat*, 1949), an unforgettable essay on the fall of France, seen as a failure of character and of intelligence; and unfinished notes on history (1942–43), published and providing perhaps the best introduction to the 'craft of the historian'. As brave in World War II as in World War I, Bloch joined the army, then the Resistance. He was tortured and killed by the Germans.

Blok, Aleksandr Aleksandrovich (1880–1921) *Russian poet.*

Blok was born in St Petersburg into a prominent academic and intellectual family. His parents separated while he was still a child and Blok was brought up by his mother in the intensely literary atmosphere of the Beketov family, which encouraged his youthful talent for verse. He was educated in St Petersburg, taking a degree in philology in 1906. In the meantime he had married (in 1903) his childhood sweetheart, Lyubov, daughter of the chemist Mendeleyev. He began publishing verse at this time. In *Stikhi o prekrasnoy dame* (1904; 'Lines on the Beautiful Lady') the poet is presented as a servant of a female divinity vaguely identifiable as *sophia*, or divine wisdom, a theme that recurs in a number of his early poems. The poem reflects the influence of symbolism, then being expounded in the journal *Vessy* (1904–1910). Although Blok rejected the quasi-religious aspect of symbolist theory – his play *Balaganchik* (1906; translated as *Puppet Show*, 1963) satirizes the mysticism fashionable in St Petersburg – both his early work and the lyrics of his maturity (1907–1916), including the cycles *Plyasmi smerti* ('Dances of Death') and *Chornaya krov* ('Black Blood'), established him as the leading Russian symbolist poet.

The inspiration of some of Blok's work was personal; for example, *Zemlya v snegu* (1907; 'The Earth in Snow') and *Carmen* (1913). But he also treated the subject of Russia and its national destiny with equal intensity. *Stikhi o Rossii* (1915; 'Lines on Russia', also called *Rodina*, 'Native Land') was immediately recognized by all factions as a major work. The essay 'Narod i intelligensiya' (1909; The People and the Intelligentsia) is an important statement of his views on this subject. Among other long poems written just before the revolution is *Soloviny sad* (1915; translated as *Nightingale Garden*). In 1916 Blok served briefly in the army but returned to Petrograd in 1917. He enthusiastically welcomed the revolution and in 1918 produced his last important work: 'Dvenadtsat' ('The Twelve'), his best-known poem, in which Christ leads a new apostolate of twelve Red Guards into the revolution, and 'Skify' ('The Scythians'), on Russia and the West. In his final years Blok served on government commissions and headed the Petrograd Union of Poets, but his creative life was over. He suffered a mental decline and died of heart disease.

Of his several dramatic works the best known is *Roza i krest* (1913; translated as *The Rose and the Cross*, 1936), a symbolist play on knightly quest set in medieval France.

Blondel, Maurice Édouard (1861–1959) *French Catholic philosopher, best known for his theology of action.*

Born at Dijon and educated at the École Normale Supérieure, Blondel taught at the University of Aix-en-Provence from 1896

until growing blindness forced him to retire in 1927.

In his major work, *L'Action* (1893), Blondel emphasized the importance of action as opposed to thought. Action was defined in a wide sense and thus included thinking, feeling, and willing as well as its causes and effects. For Blondel, action therefore had meaning. But, he argued, in any action, whether it be self-regarding action, social action, or moral action, the human will can never completely satisfy itself. In this inevitable gap between action and realization Blondel saw an approach to God. Such an approach was felt by some of Blondel's contemporaries to be unorthodox enough to warrant papal rejection. In fact Blondel found his work, after some initial coolness, accepted by Leo XIII and Pius X; it is also thought to have deeply influenced the work of JOHN PAUL II.

Bloomfield, Leonard (1887–1949) *US linguist, regarded as the most important structural linguist of his generation.*

Born in Chicago, Bloomfield was educated at Harvard and subsequently taught first Germanic philology and later linguistics at the universities of Wisconsin, Illinois, Chicago, and Yale.

Bloomfield's main aim was to show that linguistics was an autonomous and, above all, a scientific discipline. This meant that only measurable and observable data could be admitted into linguistics, an assumption that led Bloomfield to favour a behaviourist account of meaning. This, however, was the weakest part of Bloomfield's work; it was his account of syntax and phonology, expressed in his extremely influential textbook *Language* (1933), that was to have the greater impact. He argued that much linguistic analysis could be pursued with only a minimal dependence upon semantic consideration. It was only necessary to know whether two forms were the same or different forms. In this way the distinctive units of a language could be identified and thereafter analysed in terms of their phonemes (phonological units) and morphemes (syntactical units).

Blum, Léon (1872–1950) *French socialist leader (1925–50) and prime minister (1936–37; 1938; 1946–47).*

Born in Paris, the son of a textile merchant, Blum studied law, becoming government attorney on the Council of the State. As a young man he attracted attention as a literary and drama critic, making regular contributions to the intellectual review *La Revue Blanche* between 1893 and 1903.

Blum joined the Socialist Party in 1902. As a Jew he was influenced by the Dreyfus affair as well as the socialist theories of Jean Juarès, (1859–1914), to whom he became private secretary. Following Jaurès's assassination in 1914, he gradually took over the party leadership, becoming editor of *Le Populaire*. As leader of the Socialist Party in opposition (1925), he worked to build the party after the communist split of 1920. During the 1930s he led the Popular Front, an anti-fascist coalition of socialists, communists, liberals, and labour organizations. In 1936 he was elected prime minister and hastily introduced several important labour reforms (collective bargaining, a forty-hour week, and holiday pay) but was forced to resign in 1937 because of a financial crisis. He returned briefly to office in 1938. After the fall of France in 1940 he was arrested by the Vichy government and tried on charges of war guilt in 1942. He was deported to Germany in 1943 but released in 1945 by the USA. On his return to France, he presided over the socialist caretaker cabinet (1946–47), headed an economic mission to the USA (1946), and served as vice-premier in 1948. He retained the leadership of the Socialist Party until his death in 1950.

A persuasive and lucid speaker and writer, Blum was the first socialist and Jewish prime minister of France. His publications include *À l'échelle humaine* (1945), a summary of his socialist-humanist philosophy, which he wrote in captivity during World War II.

Blyton, Enid Mary (1897–1968) *British writer of books for children, who created the enormously successful character Noddy.*

Brought up in Beckenham, Enid Blyton trained initially as a pianist in fulfilment of her father's ambition for her, but in 1916 she began training as a teacher, studying the Froebel and MONTESSORI techniques. After a spell as a governess and then as the proprietress of a small school, she abandoned teaching as her success as a writer grew. Her first book was a collection of poems, *Child Whispers* (1922).

In 1924 she married Hugh Pollock, who promoted her writing career. With his encouragement she wrote and edited the new children's weekly *Sunny Stories* and increased her output of longer stories. The Pollocks were divorced in 1942, but Enid Blyton remarried the following year, and the 1940s and 1950s saw the peak of her phenomenal success. A shrewd businesswoman, she made excellent use of her facility as a writer and her abundant energy. By 1965 she had published over four hundred books for children aged between five and fifteen, many of which were translated into a score of languages. Noddy, who first appeared in 1949, was her most successful creation for young children, selling millions of copies of Noddy stories and spawning pantomimes, puppet films, and commercial products. The *Famous Five* and *Secret Seven* adventure stories appealed to older children. She also ran her own fan club, conducting a huge correspondence with her readers. The facile language, characterization, and storyline of her books drew hostile criticism from teachers and librarians but children, uninfluenced by her detractors, undoubtedly enjoyed the books and bought them in very large numbers.

Boas, Franz (1858–1942) *German-born US anthropologist, who was the principal founder of the culture-history school of cultural anthropology that arose in the USA in the early twentieth century.*

Boas attended the universities of Heidelberg, Bonn, and Kiel; after receiving a doctorate in physics from Kiel in 1881 he switched to geography. He met leading German anthropologists, including Rudolf Virchow, and in 1886 went to British Columbia to study the native Indian tribes of the region, studies he continued throughout his life. After this he decided to stay in New York and became an assistant editor of *Science*. He taught at Clarke University (1888–92) and then acted as chief anthropology assistant to the World's Columbian Exposition held in Chicago (1892–93). In 1895 he joined the American Museum of Natural History, New York, where he became assistant curator (1896) and curator (1901–05). Meanwhile he joined Columbia University in 1896 as a lecturer in physical anthropology and was subsequently made professor (1899–1936).

Boas was among the first to distinguish the basic elements of modern anthropology, particularly the linguistic and cultural components of ethnology. He held that many previous studies of societies had been based on the criteria of western observers and were therefore largely subjective and invalid. Boas viewed each society and culture as the result of a unique historical development and saw his task as primarily one of description and recording rather than constructing generalizations applicable to all. His studies of Indians in northwestern America concentrated especially on their folklore and art and pioneered the use of trained native speakers to record the largely nonwritten languages. His work in physical anthropology included a study of growth in US citizens that was the first to separate the effects of heredity and physiology. His books include *The Mind of Primitive Man* (1911), *Primitive Art* (1927), and *Race, Language and Culture* (1940). Boas helped found the American Anthropological Association in 1902 and served as its president (1907–09). He was also president of the American Association for the Advancement of Science in 1931. But above all, he was the inspiration for an entire generation of US anthropologists.

Boccioni, Umberto (1882–1916) *Italian futurist painter and sculptor.*
When Boccioni settled in Milan in 1907, he was impressed and excited by its rapid industrialization. He also met the writer F. T. Marinetti (1876–1944), who believed that contemporary Italian culture was weighed down by a past that prevented progress and originality. Stimulated by Marinetti's glorification of modern technical civilization, Boccioni agreed that artists should express its dynamism, speed, vitality, and violence in their paintings. The resulting futurist manifestos appeared in 1910 and Boccioni became the most forceful member of the futurist group, devoting himself to putting the manifestos into practice. His works of 1910–11, such as *The Rising City* and *Riot in the Gallery*, painted in a seminaturalistic style and made up of dots and whirling strokes of vibrant colour, attempt to express not merely people moving but movement itself and the collective emotion of the crowd.

After 1911, when he was introduced to the cubist style, Boccioni's paintings became more rigorous in their formal construction. In 1912 he wrote the manifesto of futurist sculpture and in 1914 his book *Pittura-scultura futurista*. Only four pieces of his sculpture are known to have survived but Boccioni's innovations in this field are an important legacy to modern art. His rejection of exclusively traditional materials, his combination of different materials (e g glass, cement) in a single work, and the importance he gave to the space around an object are exemplified in his *Development of a Bottle in Space* (1912). Boccioni was killed in World War I, having volunteered in 1915.

Bogarde, Dirk (Derek van den Bogaerde; 1921–) *British actor and author.*
The son of a Dutch-born art editor on the London *Times*, Bogarde was educated at University College, London, and won a scholarship to the Royal Academy of Art. He joined the Amersham Repertory Company in 1940 and London appearances soon followed. After war service with the King's Royal Regiment in Europe and the Far East and with Air Photographic Intelligence, he returned to the stage. His film debut came in *Esther Waters* (1947). This was followed by *The Blue Lamp* (1950), *The Sea Shall Not Have Them* (1954), *Ill Met By Moonlight* (1957), and, in sharp contrast, the part of Simon Sparrow in *Doctor in the House* (1953), *Doctor at Sea* (1955), *Doctor at Large* (1956), and *Doctor in Distress* (1964). The 'Doctor' films were popular, but more demanding roles were offered to Bogarde, beginning with Basil Dearden's socially relevant *Victim* (1961), in which he played a homosexual. Joseph LOSEY's *The Servant* (1963) and *Darling* (1965) earned Bogarde British Film Awards. He went on to make two films for VISCONTI: *The Damned* (1969) and *Death in Venice* (1970), in which he gave one of his most distinguished performances.

He subsequently appeared in *The Night Porter* (1974), *A Bridge Too Far* (1977), *Providence* (1978), *Despair* (1978), and *These Foolish Things* (1991), as well as in various television and stage roles, and has collected a number of prestigious acting awards. Since the seventies he has lived in the South of France, shunning publicity, and has concentrated on writing, producing five volumes of autobiography, *A Postillion Struck by Lightning* (1977), *Snakes and Ladders* (1978), *An Orderly Man* (1983), *Backcloth* (1986), and *A Particular Friendship* (1989). His novels *A Gentle Occupation* (1980), *Voices in the Garden* (1981), and *West of Sunset* (1984) have also been bestsellers.

Bogart, Humphrey (1899–1957) *US actor. An Oscar-winning film star, usually portraying American tough guys who often turn out to have hearts of gold, his films have acquired cult status since his early death.*
The son of a doctor, Bogart was born in New York. He began his career as manager of a touring company, having served in the navy during World War I, but in 1922 made the first of many stage appearances in *Drifting*. His stage success as Duke Mantee in *The Petrified Forest* was to be repeated in the screen version (1936) of the play and thus his future in films was secure.

They Drive by Night (1940) and *High Sierra* (1941) were among the many memorable gangster films that followed; his portrayal of the cynical yet moral gangster-hero, with the barely audible lisp, won him vast audiences on both sides of the Atlantic. Perhaps even more enduring, however, were *The Maltese Falcon* (1941), in which he played the private detective Sam Spade, and the immortal *Casablanca* (1942), in which he played opposite Ingrid BERGMAN as the tough expatriate American with a soft spot for the tune 'As Time Goes By'. Other classics include *To Have and to Have Not* (1945) and the thriller *The Big Sleep* (1946), in which he played opposite his fourth wife, Lauren Bacall (1924–). Bogart's versatility was endorsed by his performance as an alcoholic trader in *The African Queen* (1952), for which he won an Oscar.

Bohrs, Niels Hendrik David (1885–1962) *Danish physicist, who was one of the most highly respected physicists of the century. His work on quantum theory and atomic structure won him the 1922 Nobel Prize for Physics.*

The son of a professor of physiology, Bohr was educated at Copenhagen University, where he gained his PhD in 1911. Supported by Carlsberg, the brewers, Bohr went to study with J. J. THOMSON in Cambridge and to work (1912–16) with Ernest RUTHERFORD at Manchester University. In 1916 he returned to Denmark, spending the rest of his life as professor of physics at Copenhagen University and, from 1920 onwards, as director of the Institute of Theoretical Physics, apart from a period during the German occupation of Denmark when, having a Jewish mother, he escaped to America, where he worked on the atom bomb.

Bohr established his reputation as a theoretical physicist with his work in Manchester in 1913 on the structure of the hydrogen atom. The Bohr atom (as it came to be known) was regarded as a miniature solar system with a central nucleus surrounded by orbiting electrons. The atomic electrons were permitted to occupy only a limited number of fixed orbits and their angular momentum was a function of Planck's constant. This model of the atom would soon prove to be too simple, but Bohr's introduction of quantum effects into atomic structure proved immensely useful in the development of subsequent theories. It also provided a good explanation of the line spectrum of hydrogen.

On his return to Denmark in 1916 Bohr created a thriving centre for the study of physics. In the 1920s and 1930s leading figures from all over the world went to Copenhagen to work with him. Bohr also developed the concept of complementarity (also known as the Copenhagen interpretation), in which he argued that no single model could adequately explain atomic phenomena. The wave–particle duality thus became not an embarrassing contradiction, but an instance of complementary phenomena.

In the late 1930s Bohr developed a model of the nucleus, the so-called liquid-drop model, which was useful in developing the concept of nuclear fission and enabled Bohr to suggest that the rarer isotope, uranium-235, would prove more fissionable than the common uranium-238. Bohr further contributed to the development of nuclear weapons when, after his escape from Denmark, he spent some time at Los Alamos. He returned to Copenhagen after the war, devoting much of his time to campaigning for adequate international control over the use of nuclear weapons. In 1955 he organized the first Atoms for Peace conference. His son Aage Bohr (1922–) also shared in a Nobel Prize for Physics (1975).

Böll, Heinrich Theodor (1917–85) *German writer. He was awarded the Nobel Prize for Literature in 1972.*

Böll was born in Cologne and educated at the Gymnasium (grammar school) there. On leaving school he worked for a bookseller for a short time, but was then called up for military service (1938) and subsequently spent six years as a private in the German army. Military life provided him with the material for his earliest publications: the short stories in *Der Zug war Pünktlich* (1949; translated as *The Train was on Time*, 1956) and the novel *Wo warst du, Adam?* (1951; translated as *Adam, Where Art Thou?*, 1955).

After World War II Böll settled in Cologne, where he continued to write novels about the changing German scene after the turmoil of the war. His style, too, moved away from the realism of his earliest writings, as is apparent in the symbolism of *Billard um halb zehn* (1959; translated as *Billiards at Half Past Nine*, 1962) and the structural innovations of *Ansichten eines Clowns* (1963; translated as *The Clown*, 1965). A Roman Catholic himself, Böll attacked organized Catholicism in *Brief an einen jungen Katholiken* (1958; 'Letter to a Young Catholic') and found other targets in the political and business ethos of contemporary Germany. His *Gruppenbild mit Dame* (1971; translated as *Group Portrait with Lady*, 1973) is a panorama of twentieth-century Germany society, and *Die verlorene Ehre der Katharina Blum* (1974; translated as *The Lost Honour of Katharina Blum*, 1975) is an attack on the way the press distorts the truth and invades the privacy of individuals. *Was soll aus dem Jungen bloss werden?, oder, Irgend was mit Büchern* (1981; translated as *What's to Become of the Boy?, or Something to do with Books*, 1981) recalled the years 1933–37. From 1971 to 1974 Böll was president of PEN (Poets, Playwrights, Editors, Essayists, Novelists) International and his works are widely translated and respected. A collected edition of his works was published in German in 1977 (vols 1–5) and 1978 (vols 6–10).

Bolt, Robert Oxton (1924–) *British playwright. He was made a CBE in 1972.*

Born and educated in Manchester, Bolt went to the university there (interrupted by a spell (1943–46) in the RAF) before teaching for eight years (1950–58) in Somerset. The success of his first play, *Flowering Cherry* (1958), encouraged him to devote himself to writing. He enjoyed a major triumph with his play about Sir Thomas More, *A Man for All Seasons* (1960), which was filmed in 1967 and won Bolt an Oscar. Other plays include *The Tiger and the Horse* (1960), *Gentle Jack* (1963), *Vivat! Vivat! Regina* (1970), *State of Revolution* (1977), and the children's play *The Thwarting of Baron Bolligrew* (1966). He also wrote scripts for films, including *Lawrence of Arabia* (1962), *Dr Zhivago* (1965), and *Ryan's Daughter* (1970), all directed by David LEAN, and *The Mission* (1986). He is married to the actress Sarah Miles (1941–); he suffered a severe stroke but continues to work.

Bond, Edward (1934–) *British dramatist whose experimental plays, often depicting scenes of violence, have aroused much controversy.*

Born in London, Bond first attracted attention with his plays in the sixties. At the Royal Court his first play, *The Pope's Wedding* (1962), was followed by *Saved* (1965), which caused a sensation because of a scene depicting the stoning to death of a baby, and was the subject of a court action because of its violence and blasphemy. Bond continued to confront his audiences with scenes of violence and cruelty deriving from such themes as imperialism, economic exploitation, war, apartheid, and social responsibility in such plays as *Narrow Road to the Deep North* (1968), *Black Mass* (1970), *Passion* (1971), a rewrite of Shakespeare's *Lear* (1971), *Bingo* (1974), and the two-part *A-A-merica!* (*Grandma Faust* and *The Swing*; 1976). In 1978 the Royal Shakespeare Company performed *The Bundle* at the Warehouse and in the same year Bond directed and produced *The Woman* at the National

Theatre. Subsequent works include *Restoration* (1981), *Summer: A play for Europe* (1982), *Derek* (1983), the trilogy *The War Plays* (1985), and *September* (1990). Several of his plays have been produced in New York, including *Saved* and *Early Morning* (1968).

As well as plays Bond has written libretti for Hans Werner HENZE's operas *We Come to the River* (1976) and *The English Cat* (1982) and the ballet *Orpheus* (1979). In addition, he wrote the screenplay for *Blow-Up* (1967) and *Laughter in the Dark* (1969) and the radio play *Badger by Owl-Light* (1975). His *Theatre Poems and Songs* was published in 1978 and *Collected Poems* in 1987.

Bondi, Sir Hermann (1919–) *Austrian-born British mathematician and cosmologist, who is joint author of the steady-state theory of the universe. He was knighted in 1973.*

Educated at the Realgymnasium in Vienna and Cambridge University, Bondi joined the research staff of the Admiralty shortly after graduating. After World War II he returned to academic life, teaching first at Cambridge and, after 1954, at King's College, London, where he became professor of mathematics. In 1948 Bondi published his best-known work in cosmology, arguing the case for the steady-state theory in similar terms to those of Fred HOYLE.

In 1967 Bondi was appointed director of the European Space Research Organization. He returned to Whitehall in 1971 and spent the next decade advising various government ministries. He was master of Churchill College, Cambridge (1983–90).

Bonhoeffer, Dietrich (1906–45) *German Protestant pastor and theologian. His arguments for a 'worldly' Christianity, pertinent to man's condition in the twentieth century, fuelled the debate on the secularization of Christianity that figured prominently in the 1960s and thereafter.*

Bonhoeffer studied theology at the universities of Tübingen and Berlin but was increasingly influenced by the neo-orthodox movement led by Karl BARTH and others. Between working as a pastor in Barcelona (1928–29) and London (1933–35), he studied at Union Theological Seminary, New York (1929–30), and was lecturer in systematic theology at Berlin University. He was a signatory of the Barmen Declaration in 1934 that founded the Confessing Church as a reaction to the increasingly pro-Nazi German Protestant Church, and in 1935 Bonhoeffer became head of its theological seminary at Finkenwalde, which he continued to operate even after its proscription by the

Nazis in 1937. During the early years of World War II, several leading members of the resistance, including Hans von Dohnanyi, Count von Moltke, and Admiral Canaris, worked with Bonhoeffer in the Military Intelligence Department. In May 1942, Bonhoeffer went to Stockholm on an abortive mission to convey the conspirators' proposals to the Allies via the Bishop of Chichester. Returning to Berlin, Bonhoeffer was arrested on 5 April 1943 and imprisoned, initially in Berlin but later in Flossenburg concentration camp, where he was executed by the Gestapo just before the liberation of the camp by US troops.

Bonhoeffer's *Nachfolge* (1937; translated as *The Cost of Discipleship*, 1948) emphasized his concern with genuine discipleship of Christ and the consequent sacrifices this involved, and attacked the notion of 'cheap grace'. In his later works, Bonhoeffer moved away from the biblical dogmatism of Barth to the realization of a 'worldly' Christianity that did not appeal to man's weaknesses but embraced his maturity and related to social and political problems of the twentieth century. Many of these ideas were contained in letters from prison to his friend and later biographer, Eberhard Bethge; they were published as *Widerstand und Ergebund* (1951; translated as *Letters and Papers from Prison*, 1955).

Bonnard, Pierre (1867–1947) *French painter and graphic artist, regarded as one of the greatest colourists of modern art.*
Having failed as a law student in Paris, Bonnard studied art against family opposition and in 1889 sold his first poster design. Until the early years of this century he was a leading member of the Nabis (a group that included Édouard Vuillard (1868–1940)), producing posters, screens, theatre designs, and illustrations as well as lithographs of Parisian figures and street life, which used simplified flat forms and decorative line and colour.

Around 1905 Bonnard began to concentrate on painting. His intimate paintings of figures in sunlit domestic interiors, such as *The Breakfast* (1907), are vivid in colour with echoes of impressionism. Together with Vuillard's works they were largely responsible for popularizing postimpressionism throughout Europe. From 1910 Bonnard made regular visits to the south of France, where he painted landscapes in less sumptuous colours that still owed much to the impressionist tradition. He also gave greater attention to the structural quality of his work.

Bonnard refused the Légion d'honneur in 1912. From the 1920s until the end of his life he exhibited widely in Europe and the USA, working mainly as a painter with occasional commissions for illustrations. His *Getting Out of the Bath* (1930) and *Nude in the Bath* (1946) are characteristic of his later work, in which colours are heavier and brighter, producing even richer and more sumptuous colour relationships.

Borden, Sir Robert Laird (1854–1937) *Canadian statesman; leader of the Conservative Party (1901–20) and prime minister (1911–20). He was knighted in 1914.*
Born in Grand Pré, Nova Scotia, Borden was educated at a private school, where he became an assistant master at the age of fourteen. He went on to study law in Halifax and was called to the bar in 1878; by the early 1890s he had established his position as a prominent figure in Halifax legal circles. He entered politics in 1896 as the Conservative member for Halifax in the Canadian House of Commons and was re-elected in 1900. The following year he became leader of the Conservative Party in opposition to the Liberal prime minister, Sir Wilfrid Laurier (1841–1919).

In 1911 Borden spoke out against the government's proposals for reciprocal trade agreements with the USA; the Liberals were defeated on this issue and Borden came to power as prime minister later that year. A staunch advocate of Canadian involvement in the development of British policy and of Anglo-Canadian cooperation in general, he pledged his country's support for Britain in World War I and represented Canada at sessions of the Imperial War Cabinet in 1917 and 1918. At the same time, however, he sought to change the status of his country from that of a British colony to that of an independent nation and was later to insist on separate membership for Canada in the League of Nations. To maintain the strength of the Canadian fighting forces he introduced compulsory military service, having formed a coalition government for this purpose in 1917. Only the Liberals of Quebec refused to enter the new Union government, largely because Borden had never managed to win the support of the French Canadians, who felt alienated by his pro-British policies.

In 1920 Borden was obliged to resign from the premiership through ill health. He represented Canada at the Washington Disarmament Conference in 1921, served as chancellor of Queen's University, Kingston

(1924–30), and devoted much of the remainder of his life to academic work, publishing two volumes of lectures, *Canadian Constitutional Studies* (1922) and *Canada in the Commonwealth* (1929).

Bordet, Jules Jean Baptiste Vincent (1870–1961) *Belgian bacteriologist, a pioneer of modern immunology who discovered complement, a complex of proteins in the blood that causes the destruction of foreign cells in an immune response. He received the Nobel Prize for Physiology or Medicine in 1919.*

Bordet graduated in medicine from the University of Brussels in 1892 and after two years clinical work moved to the Pasteur Institute, Paris. Studying the breakdown (lysis) of bacteria in immunized animals, Bordet found that two factors were involved: a specific antibody that occurred only in immunized individuals and a nonspecific component present in all individuals. (This latter he termed 'alexin', later renamed complement.) The interaction between the antibody and the foreign cell caused the complement to bind, or 'fix' to, the cell and disrupt its membrane. In 1901 Bordet was appointed to found and direct the Institut Pasteur du Brabant in Brussels. In collaboration with his brother-in-law, Octave Gengou, he developed the phenomenon of complement fixation as a means of detecting the presence of specific antibodies in blood, a valuable tool in the monitoring of disease and the basis of the Wassermann test for syphilis. In 1906 Bordet, with Gengou, discovered the whooping cough bacterium (subsequently named after him as *Bordetella pertussis*) and prepared the first vaccine. Bordet became professor of bacteriology at the Free University of Brussels (1907–35) and continued his researches on pathogenic bacteria, bacteriophages, and blood coagulation.

Borel, Émile Felix-Edouard-Justin (1871–1956) *French mathematician and politician, well known for his work in probability theory.*

The son of a clergyman, Borel was born at St Affrique and educated at the École Normale Supérieure. After teaching at the University of Lille (1893–96) he returned to Paris, where for the following forty years he held a number of appointments at the École Normale and the Sorbonne.

In his *Éléments de la théorie des probabilités* (1909) Borel sought to bring probability theory into the mainstream of mathematics. His key insight was to note that rules for calculating probabilities were similar in form to those used in the calculation of areas in geometry. It was only necessary to change the term 'set' for 'event' and 'area' for 'probability'. The introduction of measure theory to probability allowed the contemplation and even the solution of problems long considered too difficult. Borel also worked in analysis. In the 1920s he anticipated several results in the field of game theory, which were later published by J. von Neumann. In particular, he proved a restricted version of the mini-max theorem.

From 1924 to 1936 Borel served as a radical socialist member of the Chamber of Deputies. In 1925 he was appointed navy minister. In addition to his political activities Borel did much to change the structure of French science. He helped found the Institut Henri Poincaré, serving as its first director from 1928 until his death; he was also responsible for the founding of the Centre National de la Recherche Scientifique. With the outbreak of war in 1939 Borel was initially arrested by the Vichy government. He was released after a few months and, despite his age, served with the Resistance.

Borg, Björn (1956–) *Swedish tennis player and the outstanding champion in men's tennis during the latter half of the 1970s.*

Borg's emerging talent was first recognized when he won the junior Wimbledon championship of 1972. Coached by the former world-class Swedish player Lennart Bergelin, the left-handed Borg displayed a superbly crafted game marked by great fitness and athleticism, accurate and reliable groundstrokes, and an equable temperament. His first major titles were the Italian and French championships in 1974 and he went on to win five consecutive Wimbledon singles titles (1976–80). However, he was thwarted four times in the finals of the US Open, losing twice to Jimmy Connors and twice to John McEnroe.

In 1980, Borg married the Romanian tennis player Mariana Simionescu (the marriage was dissolved in 1984). The following year, although reducing his playing commitments, he captured his sixth title in the French Open championships. Obviously missing his top form in 1982, Borg became involved in a dispute with the tennis authorities over obligatory playing commitments and a poor season prompted him to quit tennis. His final appearance was in the 1983 tournament in his home town of Monte Carlo. He lost in the second round, surrounded by the world's press, who had gathered to witness the parting shots of the brilliant Swede. He subsequently devoted himself to his business interests; when these failed he returned to the professional tennis circuit in 1991 amid much speculation about whether he would recapture his earlier form. Among his many other achievements, Borg won the WCT finals in 1976 and the Masters tournament in 1979 and 1980.

Borges, Jorge Luis (1899–1986) *Argentinian short-story writer, poet, and essayist.*

Borges was born and spent most of his life in Buenos Aires. From his English grandmother Borges learned English before Spanish; his early reading (Stevenson and CHESTERTON, among others) had an acknowledged influence on his work. From 1914 to 1921 Borges travelled and studied in Europe, becoming associated in Madrid with the poets calling themselves 'Grupo Ultra'. Their doctrine, ultraísmo, was a Mallarmé-inspired belief in the 'pure' poem, i.e. one without structural, narrative, or other 'unpoetic' interest. For a time after returning to Buenos Aires Borges was spokesman for ultraísmo in various, usually short-lived, reviews and journals. His first collection of poems, *Fervor de Buenos Aires* (1923), was followed by two others (1925, 1929), which he later denounced as too dogmatically ultraist; they display, however, an interest in regional subject matter and in the paradoxes and metaphysical puzzles regarding time and the ambiguity of identity that became the main themes of his mature fiction. Borges's verse is collected in *Obra poética 1923–1967* (1967), though much further work has been published since; for example, *Elogio de la sombra* (1969, translated as *In Praise of Darkness*, 1975). A bilingual edition with English versions by thirteen English-language poets appeared as *Selected Poems 1923–1967* (1972).

From 1925 Borges began contributing weekly prose pieces to the Sunday supplement of *La Prensa* and publishing stories in the intellectual review *Sur*. His life was without incident, apart from his increasing blindness and in 1946 the vindictive dismissal by PERÓN from his job as librarian. (Perón attempted to humiliate him by offering him a post as poultry inspector.) Borges, who referred to Perón only as 'the dictator', quietly devoted himself to his work; on the fall of Perón in 1955 he was appointed director of the National Library, a post he held until 1973. Although he had published short stories from 1935 onwards, he became internationally known only after sharing the Prix Formentor with Samuel BECKETT

in 1961. His major fictional work is contained in two volumes of short stories of a completely original quality that places him among the leading storytellers of the twentieth century: *Ficciones* (1944; translated as *Fictions*, 1962) and *El Aleph* (1949; translated as *The Aleph and Other Stories, 1933–1969*, 1971), which contains an autobiography written directly in English). Other important collections are *Historia universal de la infamia* (1935, revised 1954) and *El hacedor* (1960; translated as *Dreamtigers*, 1964). His last book was *Atlas* (1986), written with Maria Kodama. Borges's influence has been extensive, most obviously in the so-called 'magic realism' of such South American writers as Gabriel García Márquez (1928–).

Borglum, (John) Gutzon (1867–1941) *US sculptor, whose most famous work is the group of colossal heads of four US presidents on Mount Rushmore, South Dakota.*

Born in Idaho, Utah, Borglum studied sculpture in Paris, where he became an admirer of Auguste Rodin. Borglum's most characteristic works depicted animals or themes of the American West, such as Red Indians on horseback, to which he gave a striking sense of movement. The group of heads of Presidents Washington, Jefferson, Lincoln, and Theodore Roosevelt, which were carved into the rock of Mount Rushmore between 1927 and 1941, is one of the best known sculptures in the USA. It cost the US government over a million dollars and was one of many public commissions undertaken by Borglum.

Borlaug, Norman Ernest (1914–) *US agronomist, whose development of high-yielding cereal varieties for the third world earned him the 1970 Nobel Peace Prize. He was admitted as a foreign member to the Royal Society in 1987.*

Born of a Norwegian immigrant farming family in Cresco, Iowa, Borlaug obtained a BSc in forestry from the University of Minnesota (1937) and a PhD for work on plant pathology (1941). After three years in commercial agrochemical research, he joined a small team of agronomists who founded what is now the International Maize and Wheat Improvement Center, near Mexico City. It was their aim to apply the new scientific developments in western agriculture to improve cereal yields in underdeveloped countries. Borlaug bred new varieties of wheat that could tolerate the application of artificial fertilizers to give much higher grain yields, often under semiarid conditions. These new varieties enabled a dramatic boost in yields during the 'Green Revolution' of the 1960s and 1970s. However, the dependence on expensive inputs of fertilizers, pesticides, herbicides, and machinery has meant that such varieties are not universally appropriate.

Bormann, Martin (1900–?45) *German Nazi politician and close advisor to* HITLER.

Born in Halberstadt, central Germany, the son of a soldier who later became a civil servant, Bormann left school to work on a farm in Mecklenburg, serving in World War I in a field artillery regiment. He returned to agriculture as an inspector of farms after the war, when he became involved in several paramilitary antisemitic groups.

Bormann joined the Nazi Party in 1924. He rose rapidly in the party ranks and was appointed chief of staff to Hitler's deputy, HESS, in 1933. Following Hess's flight to Scotland in May 1941, Hitler abolished the position of deputy-führer, appointing Bormann as head of the party chancery, which gave him control of the entire party machine. In 1943 Bormann was appointed Hitler's private secretary, a position he retained until Hitler's death. His fate was uncertain after the fall of Berlin in 1945 but he was sentenced to death *in absentia* at Nuremberg in 1946. He was formally pronounced dead in 1973 after his body was found half a mile from the site of Hitler's bunker.

Known as the 'brown eminence', Bormann was considered to be Hitler's closest collaborator. A master of political intrigue, he maintained his position of power through administrative efficiency, acumen, and a total absence of any kind of moral judgment.

Born, Max (1882–1970) *German physicist, who was awarded the 1954 Nobel Prize for Physics for his statistical interpretation of wave mechanics.*

Born in Breslau, the son of a professor of anatomy at Breslau University, Born was educated at a variety of German universities before gaining his PhD from Göttingen in 1907. He remained at Göttingen, becoming professor of physics in 1921, until he was expelled by the Nazis in 1933. He then left Germany for Britain, where he first spent some time teaching at Cambridge and visited India for a year to lecture at Raman's Institute of Physics in Bangalore. On his return to Britain in 1936 Born was appointed Tait Professor of Natural Philosophy at Edinburgh University, where he remained until his retirement to Bad Pyrmont, a spa near Göttingen.

Born was originally concerned with the elastic properties of solids but his most original contribution to physics was his proposal, made in the 1920s, to replace determinism in quantum mechanics with the concept of probability waves, arguing that the position of, say, an electron cannot be given deterministically; theory can only state its most probable position. This argument was never accepted by his friend, Albert EINSTEIN, and their conflict over the issue was published posthumously in the *Born–Einstein Letters* (1971). Born was the author of several physics textbooks in English.

Bosch, Carl (1874–1940) *German industrial chemist. His development of high-pressure chemical plant enabled the laboratory Haber process to be translated into the immensely important industrial Haber–Bosch process. He was awarded the 1931 Nobel Prize for Chemistry.*

The son of an engineer, Bosch began his career in a foundry, before being allowed by his father to pursue a formal education at the University of Leipzig. After gaining his PhD in 1898, Bosch joined the research staff of Badische Anilin und Soda Fabrik (BASF) in Ludwigshafen. There he became involved in the major task facing German industry: the synthesis of ammonia, for use in both agriculture and the armaments industry. In 1907 Fritz HABER had demonstrated that, with high temperatures and pressures and appropriate catalysts, ammonia could be synthesized from atmospheric nitrogen and hydrogen. The Haber process, however, was then restricted to the laboratory. Bosch was assigned the task of transforming the process into an industrial plant; he did this at Oppau, where BASF's first high-pressure ammonia plant opened in 1909. By 1930 well over two million tons of ammonia were being produced annually. Remaining at BASF, Bosch rose to become chairman of its successor, IG Farben, and continued to hold the position until his death.

Bose, Sir Jagadis Chandra (1858–1937) *Indian plant physiologist and physicist, noted for his studies of plant sensitivity. He was knighted in 1917.*

Born at Mymensingh in what is now Bangladesh, Bose attended St Xavier's College in Calcutta before moving to London to study medicine. Obtaining a scholarship to Christ's College, Cambridge, Bose graduated in natural sciences in 1884. On his return to India he was appointed professor of physics at Presidency College, Calcutta. His earliest work was concerned with very short radio waves and yielded an improved form of co-herer – a device for their detection – and a general theory regarding the response of inorganic materials to external stimuli. Bose extended his researches to the responses of living organisms, especially plants, and designed the crescograph, an instrument for automatically detecting and recording plant movements. This demonstrated how movements change during growth and how they can be affected by injury and other stimulation. Bose published his findings in *Response in the Living and Nonliving* (1902), which was coolly received by the scientific establishment. He wrote many other papers and books on plant sensitivity, the physiology of sap flow, and photosynthesis, including *The Physiology of Photosynthesis* (1924) and *Tropic Movements of Plants* (1929). Following his official retirement in 1915, he founded the Bose Research Institute, Calcutta, two years later. He was elected a fellow of the Royal Society in 1920.

Botha, Louis (1862–1919) *First prime minister of the Union of South Africa (1910–19), who conquered German South Africa in World War I on behalf of the Allies, but had earlier opposed the British as an Afrikaner general in the Boer War.*

Born near Greytown, the son of a Voortrekker family, Botha was educated at German mission schools before travelling with his parents to the Orange Free State, where he took up farming. Botha's political involvement began as a member of the Afrikaner force that supported the Zulus during the inter-tribal wars in 1884. One of the founders of the short-lived Vryheid Republic, which became part of the larger Transvaal Republic in 1886, Botha joined the Afrikaner forces in Natal at the outbreak of the Boer War in 1899. He rose rapidly to command the southern force besieging Ladysmith and ambushed a train on which the young Winston CHURCHILL was travelling. Botha took Churchill prisoner, but he later escaped. Botha was appointed commander of the entire Transvaal army in 1900 and participated in the peace conference at Vereeniging in 1902 at which he advocated peaceful reconciliation.

In 1905 Botha helped to found the party Het Volk. Elected the first prime minister of the Transvaal in 1907, he played a leading role in the National Convention (1908–09) that drafted the constitution for the Union, of which he became prime minister in 1910. Botha supported the Allies in World War I, gaining recognition for his annexation of German South

West Africa in 1915. He attended the Versailles Peace Conference in 1919 as the representative of the Union of South Africa, but died shortly after his return home.

Botha, Pieter Willem (1916–) *South African prime minister (1978–84) and state president (1984–89). Botha was the first South African prime minister to visit the UK (in 1984) since his country left the Commonwealth.*

Born in Paul Roux, Orange Free State, the son of a farmer, Botha was educated at Bethlehem's Voortrekker secondary school, before studying law at the University of the Orange Free State. Botha first entered politics in 1936, when he abandoned his studies to become the organizer of the Nationalist Party in Cape Province. Appointed campaign manager for all four provinces in 1946, he was promoted to chief secretary of the party in the Cape in 1948, winning the parliamentary seat of George in the same year. In 1966 he became Cape Province leader of the party and minister of defence; he was elected prime minister and minister of national intelligence in 1978. In 1984 he abolished the office of prime minister, replacing it with that of state president. An authoritarian leader in the mould of his predecessors MALAN, VERWOERD, and John Vorster (1915–83), with whom he had close association, Botha failed to end apartheid, which made him extremely unpopular in liberal democracies. In response to pressure he did introduce limited reforms, including a new constitution (1984), which allowed certain classes of non-Whites a degree of political representation. His resistance to more radical reform led ultimately to his resignation in 1989, when he was replaced by F. W. DE KLERK, who promised to speed up the reform process.

Botham, Ian (Terence) (1955–) *British cricketer.*

Born in Heswall, Cheshire, he went to Milford School and made his debut for Somerset in 1974. His Test debut followed in 1977, by which time he had established his reputation as a batsman and as a medium-fast bowler. His attacking batting style made him a great favourite with cricket watchers, although his outspoken views on cricket and a drug scandal (1986) caused some controversy. He captained England in 1980–81 and led Somerset in 1983. By then he had made cricketing history by becoming the first player to score 100 runs and take 8 wickets in a Test match (1978), while in 1979 he established a new record, having taken

100 Test wickets in just 2 years 9 days. Subsequently he became the first player to score 3000 runs and take 250 wickets in Test matches (1982). In 1987 he joined the Worcestershire side; he also spent a season with Queensland (1987–88) and during the winter played football for Scunthorpe United. Defying injury and the press, he regained his place in the England team in 1991 after a lengthy layoff. He joined the Durham side after the close of the 1991 season. Botham has also undertaken charitable work, especially a series of sponsored long-distance walks.

Botvinnik, Mikhail Moiseyevich (1911–) *Soviet chess player, who won the Soviet championship seven times and was world champion for a total of thirteen years. He was among the best players in the world for nearly forty years (1931–70).*

He learned chess at the age of twelve and quickly established a reputation in his native St Petersburg. Four years later he filled a last-minute vacancy in the 1927 Soviet championship and finished equal fifth. In 1931, the year he graduated in electrical engineering, he won the first of his seven Soviet championships. During the 1930s Soviet chess became less isolated and Botvinnik managed to travel to meet many of the leading players: he beat LASKER and CAPABLANCA, who had been world champions for over thirty years between them, and Alexander Alekhine (1892–1946), who was soon to be world champion.

The war interrupted Botvinnik's progress but in 1948 he won the world championship and successfully defended it against Bronstein in 1951 and Smyslov in 1954. In 1957 Botvinnik lost the world title to Smyslov but regained it a year later; the same thing happened in 1960 with Tal. Petrosian finally robbed him of the title in 1963, but until his retirement in 1970 Botvinnik continued to win major competitions around the world.

After his retirement he worked with a team of Soviet scientists on the development of a chess computer.

Boulanger, Nadia Juliette (1887–1979) *French music teacher, conductor, and composer. She received many honours, among them Commandeur de la Légion d'honneur.*

Boulanger's father was violin professor at the Paris Conservatoire, where she herself studied composition with FAURÉ and Charles Widor (1844–1937) and organ with Louis Vierne (1870–1937) and Alexandre Guilmant (1837–1911). She came second in the 1908 Prix de Rome with her cantata *La Sirène*. Her most successful compositions were in collaboration with Raoul Pugno (1852–1914): incidental music for d'Annunzio's *Città morte* (1911) and the Verhaeren songs, *Les Heures claires* (1909–12). As a conductor, she played an important role in the Monteverdi revival and in the performance of French renaissance and baroque music. She was the first woman to conduct a complete symphony concert in London (the Royal Philharmonic Society, 1937); she also conducted the first performance of STRAVINSKY's *Dumbarton Oaks* in Washington, DC (1938).

It is as a teacher, however, that Boulanger was most influential. Her method was based on contrapuntal exercises and the close analysis of scores of all periods; she particularly stressed the examples of Fauré and Stravinsky and attracted students from all over the world (among them BERKELEY, COPLAND, Walter Piston (1894–1976), and Jean Françaix (1912–).

Boulez, Pierre (1925–) *French composer, conductor, and theorist, who is a leading figure in contemporary music.*

Boulez was born at Montbrison. His early aptitude for mathematics destined him for a career in engineering, but in 1942, against family opposition, he enrolled at the Paris Conservatoire, where he studied composition with Olivier MESSIAEN, master and pupil quickly establishing a mutual respect. As a composer, Boulez has written his music according to mathematical formulae, which he himself derived. Starting from the serialism of SCHOENBERG, he developed a system of total serialism, in which not only the pitch, but also the timbre, intensity, and duration of each note is ordered. As this precision is difficult for human performers, in the 1950s Boulez increasingly used electronic means in the search for perfection. Boulez's pieces are subject to constant revision and therefore potential development, which ensures that they have a certain vitality.

After a period working with the Renaud-Barrault theatre company, Boulez turned to conducting. In 1954 he founded the Domaine Musical Concerts of contemporary music and also wrote *Le Marteau sans maître* (revised 1957), his first publicly acclaimed work. Since then Boulez has been in demand both as a teacher (Darmstadt, Harvard University, and Basle) and as a conductor, originally of his own works, as at Cologne with *Le Visage nuptial* (1957) and *Pli selon pli* (1960). He conducted the Cologne Orchestra at Baden-Baden

in the late 1950s, BERG's *Wozzeck* at the Paris
Opéra (1963), Edinburgh Festival concerts
(1965), and Wagner's *Parsifal* at Bayreuth
(1966). He has been guest conductor of the
Cleveland Orchestra (1967) and principal conductor of both the BBC Symphony Orchestra
(1971–74) and the New York Philharmonic
Orchestra (1971–78). In 1966 he conducted the
controversial production of Wagner's *Ring*
cycle at Bayreuth, and in 1979 the first performance of the three-act version of Berg's *Lulu*
at the Paris Opéra. His own most recent compositions include *Notations* (1980), *Répons*
(1981–86), and *Dialogue de l'Ombre Double*
(1986). In 1976 Boulez became director of the
Institut de Recherche et de Coordination
Acoustique/Musique (IRCAM) in Paris, devoted to developing the use of electronic
music. He writes and lectures on his music
copiously and entertainingly.

Boult, Sir Adrian Cedric (1889–1983) *British conductor, knighted in 1937 and made CH
in 1969.*
Born in Chester and educated at Westminster
School, Oxford, and Leipzig, Boult obtained
his DMus degree in 1914 and joined the Covent Garden music staff. He taught at the Royal
College of Music (1919–30; 1962–66), at the
same time conducting opera, ballet, and choral
and orchestral concerts in London and the
provinces, including an experimental series of
the now famous Robert MAYER concerts for
children (1922). He was director of BBC
music (1930–42) and trained and conducted
the newly formed BBC Symphony Orchestra
(1931–50). Other appointments included directorships of the London Philharmonic Orchestra (1950–57) and of the City of
Birmingham Symphony Orchestra (1924–30;
1959–60).
A staunch advocate of contemporary music,
British and European, Boult was revered as a
considerate and unsensational conductor. He
wrote an autobiography, *My Own Trumpet*
(1973), and two textbooks on conducting.

Bow, Clara (1905–65) *US actress who, as the
'It' girl of the twenties, was the epitome of
Hollywood sex-appeal, gaiety, and sophistication in the era of silent films.*
Born in Brooklyn, New York, Bow worked in
an office until winning a beauty contest and,
consequently, a part in *Beyond The Rainbow*
(1922). Very much a product of the twenties,
her rise to stardom was as meteoric as it was
short-lived. She made numerous films, including *Mantrap* (1926), in which she gave what
was possibly her best performance, *Dancing*

Mothers (1926), and *The Fleet's In* (1928).
Her most memorable roles were, however, as
the flapper with cupid-bow lips of Elinor
Glyn's *It* (1927) and *The Wild Party* (1929).
Her first talkie was *Dangerous Curves* (1929),
which was followed by such films as *Her Wedding Night* (1930) and *No Limit* (1931).
At the height of her career Bow had an
enormous following but a series of highly publicized personal scandals led to a decline in her
popularity. In 1930 she retired but returned to
the screen twice more for *Call Her Savage*
(1932) and *Hoopla* (1933). In 1931 she married cowboy actor Rex Bell (1905–62), who
became lieutenant governor of Nevada
(1954–62).

Bowen, Elizabeth Dorothea Cole
(1899–1973) *British novelist and short-story
writer.*
Elizabeth Bowen was born in Dublin to Anglo-
Irish parents. Her mother separated from her
father after his mental breakdown and took the
young Elizabeth to England. There Mrs
Bowen died, leaving her daughter to be
brought up by elderly relatives. The sense of
isolation and the unsatisfactory relationships
between the mainly upper-middle-class characters in Elizabeth Bowen's novels reflect
these early experiences.
Settling in London, she published two volumes of short stories before her first novel, *The
Hotel* (1927), which portrays a typical Bowen
heroine trying to cope in an unsympathetic
environment. This was followed by a steady
output of novels and stories in the 1930s, culminating in *The Death of the Heart* (1938).
During World War II Elizabeth Bowen remained in London, working at the Ministry of
Information during the day and as an air-raid
warden at night. Her experiences of the capital
in wartime were embodied in perhaps her most
successful novel, *The Heat of the Day* (1949).
Her later novels – which include *A World of
Love* (1955), *The Little Girls* (1964), and *Eva
Trout* (1969) – show her continuing gift for
delicate characterization and acute observation. She also published two books of essays,
Collected Impressions (1950) and *After-
thought* (1962).

Bowie, David (David Jones; 1947–) *British rock singer, songwriter, and actor, noted
for his chameleon-like changes of style and
image.*
Born in London, Bowie began his musical career while still in his teens, playing in bands
and as a solo performer. He also studied mime
and dabbled in experimental theatre. In 1969

he enjoyed his first hit with 'Space Odyssey', a song that announced his characteristic themes of space fantasy and alienation. The early 1970s saw Bowie's emergence as a major star, with such albums as *The Rise and Fall of Ziggy Stardust* (1972) and *Aladdin Sane* (1973) achieving both critical and commercial success. His popular fame owed much to his disturbing androgynous image, exploited to the full in his theatrical stage performances. After several changes of style and image, Bowie moved to Berlin in the late 1970s, where he made a series of experimental albums including *Low* (1976) and *"Heroes"* (1977). In the early 1980s he moved back towards the pop mainstream, enjoying number one hits with 'Ashes to Ashes' (1980) and 'Let's Dance' (1983). His most recent albums include two featuring his band Tin Machine.

Bowie has also pursued an occasional career as an actor, starring in such films as *The Man Who Fell to Earth* (1975) and *Merry Christmas, Mr Lawrence* (1982). In 1980 he appeared on Broadway in the title role of *The Elephant Man*.

Boycott, Geoffrey (1940–) *Yorkshire and England cricketer, one of the world's greatest run-makers and among the most technically correct of defensive batsmen. There was considerable controversy in his career, during which he lost the captaincy of his county and was banned as a test player in 1982 for taking part in matches in South Africa.*

Born in the mining village of Fitzwilliam in Yorkshire, he worked for the Ministry of Pensions before becoming a county cricketer. His first game for Yorkshire was in 1962 when he was twenty-two, and two years later he was selected for England. As an opening batsman, his self-discipline and powers of concentration were unequalled in first-class cricket. Such an approach was at times seen as an excessively selfish attitude and he was never especially popular among other players or outside his native county. In Yorkshire itself, despite public reprimands and an uneasy relationship with the county (in 1983 he was reported and warned about the time he took to score his runs against Gloucestershire at Cheltenham), he continues to command remarkable loyalty from a strong faction of the club's supporters; he has been a member of the General Committee of the club since 1984.

Boycott captained Yorkshire from 1971 to 1978 but failed to obtain the captaincy of England, his great ambition. In 1977 he scored his hundredth first-class hundred at Headingley against the Australians; by the time he retired as a player in 1986, he had scored 150 centuries in first-class cricket. His many publications on cricket include *Master Class* (1982) and *Boycott, The Autobiography* (1987).

Boyer, Charles (1898–1978) *French-born US actor who gained a reputation as the screen's 'greatest lover'.*

Before going to Hollywood in the thirties Boyer had established himself on the French stage and had made his film debut in *L'Homme du large* (1920). Early popular success in the USA came with *Private Worlds* (1935) and *Mayerling* (1936). His good looks, charm, and captivating French accent made him well suited to the leading romantic roles in which he was cast and he was soon a firm favourite with audiences on both sides of the Atlantic. DIETRICH in *The Garden of Allah* (1936) and GARBO in *Maria Walewska* (1937) were two of the many international stars opposite whom he played. *Algiers* (1938) was followed by such memorable films as *All This and Heaven Too* (1940) and *Gaslight* (1944). Notable, too, were *Madame de...* (1953), *Maxime* (1962), *The Four Horsemen of the Apocalypse* (1962), *Barefoot in the Park* (1968), and *Stavisky* (1974).

Stage appearances included 'Don Juan in Hell' from SHAW's *Man and Superman* (1951) and starring roles in many of the productions of Four Star Television, which he helped establish. For his work in founding the French Research Foundation, Los Angeles, he received a Special Academy Award in 1942.

Brabham, Jack (Sir John Arthur Brabham; 1926–) *Australian motor racing driver and three times world champion (1959, 1960, and 1966). He received a knighthood in 1979.*

Born in Sydney, Brabham served with the RAAF (1944–46) and started his racing career driving Cooper cars in midget speedway events. He came to Europe in 1955 and joined the Cooper works team in 1956, driving sports cars and in Formula Two. His Formula-One debut was at Monaco in 1957 and he was Formula-Two champion in 1958, the year in which he and Stirling MOSS drove an Aston Martin to victory in the Nürburgring 100 km race. Wins in the Monaco and British Grands Prix helped Brabham to the 1959 Formula One title and in 1960 he repeated his success, winning six Grands Prix. At the end of 1961 he left Cooper to run his own team and the Brabham-Coventry Climax car made its first appearance at the 1962 German Grand Prix. Brabham cars were fitted with the Australian Repco engine for the 1966 season and Brabham won the Drivers' Championship, the first constructor-driver to do so. Brabham-Repco cars also won the Constructors' Championship in 1966 and, driven by Denny Hulme, in 1967.

Brabham retired from driving in 1970 but Brabham cars continued to race competitively throughout the 1970s and early 1980s, although the team was no longer controlled by its founder. Brabham's autobiography, *When the Flag Drops*, was published in 1971.

Bradley, Omar Nelson (1893–1981) *US army general who commanded the massive 12th army group that spearheaded the liberation of France and Germany during the closing stages of World War II.*

A graduate of West Point Military Academy, Bradley rose through staff college and war college to become commandant of the Infantry School at Fort Benning, Georgia. His World War II service began as deputy commander, later commander, of the US 2nd Corps attached to the North African expeditionary force led by General PATTON. On 7 May 1943, Bradley's men captured the Tunisian town of Bizerta, together with some 40 000 Axis troops, thus gaining a major victory in the campaign. Following the German defeat in North Africa, Bradley took part in the invasion of Sicily and in 1943 he was posted to Britain to prepare for the Normandy invasion. Commanding the US 1st Army, Bradley's forces led the assault on Utah and Omaha beaches, established a beach-head, and took part in the decisive US army advance through Normandy. On 1 August, Bradley was given command of the huge 12th Army, comprising some 1 300 000 men. They relentlessly drove the Germans back across France, entering Paris on 25 August in a memorable liberation parade. During the winter of 1944–45, the 12th Army successfully checked a German counter-offensive in the Ardennes, breached the defensive Siegfried line, and in March 1945 established a bridgehead across the Rhine at Remagen. They quickly advanced into Germany, trapping 335 000 enemy troops in the Ruhr, and on 25 April, Bradley's forces joined up with their Soviet allies on the Elbe river, thirteen days before the German surrender.

A calm resolute commander and a skilled tactician, Bradley was also well liked by his men. He returned to the USA after the war to become administrator of veterans' affairs (1945–47) and served as army chief of staff (1948–49). He was appointed first chairman of the joint chiefs of staff (1949–53), being ele-

vated to five-star general in 1950. His memoirs, *A Soldier's Story*, were published in 1951.

Bradman, Sir Donald George (1908–) *Australian cricket captain and leading batsman of the 1930s and 1940s. He never lost a test series for his country and was knighted (1949) after a brilliant career that brought him 117 centuries, 29 of them in test matches (average 99.94).*

Born at Cootamundra, New South Wales, Bradman grew up at Bowral and attended the Intermediate High School there. From 1927 to 1934 he played for New South Wales, scoring a century on his Sheffield Shield debut for that state, and after that for South Australia (1935–49). He played for Australia from 1928 until his retirement in 1948, captaining the team from 1936.

Bradman's prodigious scoring made him the most famous cricketer in the world. On his first tour of England (1930) he averaged 98.66 runs, his 334 at Leeds being the highest score for Australia in test matches against England. His innings topped 300 half a dozen times during his career and on one occasion he made 452 not out. In all he scored 28 067 runs at an extraordinary average of 95.14. His powers of concentration were unrivalled. Toiling bowlers despaired and argued, unfairly, that his style was mechanical. As a test captain he was astute and unyielding. When he retired he became a selector and respected cricket administrator – and a successful stockbroker.

Bragg, Sir (William) Lawrence (1880–1971) *British physicist, who discovered Bragg's law and its application to the structure of crystals. He shared the 1915 Nobel Prize for Physics with his father, W. J. BRAGG, was knighted in 1941, and made a CH in 1967.*

Born in Adelaide, the younger Bragg was educated at Adelaide University and, on his father's return to Britain in 1909, at Cambridge University. Shortly after graduating he published a series of papers on crystal structure and in 1912 formulated the basic law of X-ray crystallography, now known as Bragg's law. Some of his early results were included in the book he wrote jointly with his father, *X-Rays and Crystal Structure* (1915).

After service in World War I in France, in which he was awarded the OBE and the MC, Bragg was appointed Langworthy Professor of Physics at Manchester University in 1919. He remained there until 1937, when he moved to the National Physical Laboratory as its director. However, in 1938, after the sudden death of Lord RUTHERFORD, he returned to Cambridge as director of the Cavendish Laboratory and Cavendish Professor of Physics. In 1953 Bragg left the Cavendish for the Royal Institution in London, where he became Fullerian Professor of Chemistry and, following his father, the Institution's director.

Bragg's own research during the 1920s and 1930s was devoted mainly to the structure of the silicates. Equally important was his foresight in backing such colleagues as Max PER-UTZ in their investigation of the structure of proteins. Bragg continued his support for protein analysis at the Royal Institution, where in 1954 he set up a research programme for this purpose.

Bragg, Sir William Henry (1862–1942) *British physicist who pioneered the technique of X-ray crystallography, for which he was awarded (with his son Lawrence BRAGG) the 1915 Nobel Prize for Physics. He was knighted in 1920, was president of the Royal Society from 1935 to 1940, and was made an OM in 1931.*

The son of an ex-seaman turned farmer, Bragg was educated at Cambridge University, graduating in 1884. In 1886 he moved to Australia, where he became professor of mathematics and physics at Adelaide University. He returned to Britain in 1909 and held chairs of physics at Leeds University (1909–15) and University College, London (1915–23).

Bragg's research began in his forties. He first worked on the nature of alpha particles, but on moving to Leeds in 1909 his attention was directed to the related problem of the nature of X-rays. Working with his son, Lawrence BRAGG, he devised a means of exploring the structure of matter with X-rays. In 1915 he completed the first X-ray spectrometer and with his son published in the same year the results of their early investigations in *X-Rays and Crystal Structure*. After retiring from University College, London, in 1923 Bragg was appointed director of the Royal Institution, where he formed an important research centre that attracted several young X-ray crystallographers. It was in Bragg's laboratory, for example, that plans were first made to explore the structure of biological molecules with X-rays.

Brain, Dennis (1921–57) *British horn player.*
Born in London, son of the horn player Aubrey Brain (1893–1955), he studied at the Royal Academy of Music – the horn with his father and the organ with G. D. Cunningham (1878–1948). Brain made his debut in 1938 and during World War II was principal horn

with the RAF Central Band. Afterwards, he was principal horn of the Royal Philharmonic Orchestra and, later, of the Philharmonia Orchestra. His mastery of the instrument's entire compass, together with his subtleties of phrasing and variety of tone, inspired many composers to write for him: concertos by Elisabeth Lutyens (1906–83), Gordon Jacob (1895–1984), HINDEMITH, and ARNOLD and the *Serenade for Tenor, Horn and Strings* by BRITTEN, dedicated jointly to Peter PEARS and Brain. He made some outstanding recordings, particularly of the Mozart and Strauss horn concertos. He was killed in a car accident.

Branagh, Kenneth Charles (1960–) *British actor and director.*
Born in Ireland, he trained at the Royal Academy of Dramatic Art, where he won the coveted Bancroft Gold Medal. Subsequently he made his London debut in 1982 in the role of Judd in Julian Mitchell's *Another Country*, for which he won a Society of West End Theatre award for Most Promising Newcomer as well as a Plays and Players award. After further London successes he joined the Royal Shakespeare Company in 1984, with whom he appeared in such roles as Henry V. Hailed as a new OLIVIER, he later performed a wide range of Shakespearean and contemporary parts, including several film roles; his recreation of his stage success in *Henry V* (1989), which he also directed, won the Evening Standard's Best Film of the Year award. Also admired was his performance in John Osborne's *Look Back in Anger*, subsequently televised, in which he played opposite Emma Thompson (1959–), whom he married in 1989. They both reached a wide TV audience in the *Fortunes of War* series (1988), based on Olivia Manning's novels. In 1987 he revived the old tradition of the actor-manager when he founded the Renaissance Theatre Company, some of whose productions he directed himself. His film *Dead Again* (1991), which he directed and starred in, was a major box office success on both sides of the Atlantic. He published a volume of autobiography, *Beginning*, in 1989.

Brancusi, Constantin (1876–1957) *Romanian-born French sculptor and a pioneer of abstract sculpture.*
Brancusi was born into a large and prosperous peasant family in a district with a tradition of woodworking. After studying art in Bucharest he went to Munich in 1902, from where he set out on foot for Paris, arriving in 1904. The first sculptures that he exhibited were naturalistic and sufficiently impressive for Rodin to offer

him a place in his studio as a student carver. However, Brancusi had already moved away from depicting the external apparance of objects: his ambition was to capture the essence of forms by dispensing with surface details. His *Sleeping Muse* of 1909 is an early stage of this evolution towards pure form. This and many of his subjects in marble and polished bronze were repeated and refined several times in the quest for simplicity and perfection. The abstract *Bird in Space*, for example, was produced in fifteen versions between 1923 and 1940.

The second vital characteristic of Brancusi's art is the feeling for his materials that he inherited from his native culture. In contrast to the contemporary practice of employing craftsmen to do the carving, from 1910 Brancusi worked directly with the materials himself. In the stone carving *The Kiss* (first version 1910) the forms appropriate to the material are as important as the forms required by the subject. Brancusi's work, in contrast to the cerebral approach of most avant-garde sculptors at the time, has profoundly influenced much modern sculpture. *The Kiss*, like many of his carvings in stone and wood, also reflects his respect for archaic and primitive sculpture.

Brancusi became a naturalized French citizen in 1957, shortly before his death.

Brando, Marlon (1924–) *US actor who was closely identified with the rebellious fifties and was a prominent exponent of 'method' acting in both theatre and films in the USA.*

Brando was born in Omaha, Nebraska, and attended the Actors' Studio. He made his Broadway debut as Nels in *I Remember Mama* (1944) but acclaim came with his use of 'the method' in his portrayal of Stanley Kowalski in Tennessee WILLIAMS's *A Streetcar Named Desire* (1947), the film version of which, four years later, gained him an Academy Award nomination. His portrayal of a paraplegic in *The Men* (1950) first brought 'the method' to the screen.

He received Oscar nominations and British Film Awards for *Viva Zapata!* (1952) and *Julius Caesar* (1953), in which he played Mark Antony, while the film with which his name is probably most closely linked, *On the Waterfront* (1954), won a Cannes Film Festival Prize and earned him an Oscar, a British Film Award, and New York Critics Award. *The Teahouse of the August Moon* (1956) and *The Young Lions* (1958) followed.

In 1959 he founded Pennebaker Productions and produced, directed, and starred in *One-*

Eyed Jacks (1961). During the sixties he made several less successful films, including *Mutiny on the Bounty* (1962) and *Candy* (1968). In 1971, however, came his highly successful *The Godfather*. Brando refused to accept the Oscar awarded him for this film in protest at the treatment of American Indians in the USA. Next came the critically acclaimed *Last Tango in Paris* (1972). More recent films include *Apocalypse Now* (1979), *Formula* (1980), and *A Dry White Season* (1990). He retired from the screen in 1989 and subsequently became embroiled in the scandal surrounding the murder of his daughter's lover, of which his son was accused.

Brandt, Bill (William Brandt; 1904–83) *British photographer best known for his stark images of British life in the 1930s and his later surrealistic studies of nudes.*

Brandt spent his early youth in Germany and Switzerland and in 1929 studied briefly in Paris with the surrealist artist and photographer Man RAY. In 1931 he moved to England and started to work as a photojournalist, selling his pictures to magazines and producing *The English at Home* (1936) and *A Night in London* (1938). His pictorial record of London life strongly evoked the contrasting lifestyles of rich and poor. On a trip to northern England in the mid-1930s, Brandt captured the brutality of unemployment in the depression-hit regions of Durham and Tyneside, exemplified, for instance, by his pictures of coal-searchers scouring the slag heaps.

During World War II Brandt worked for the Home Office, photographing London during the Blitz. In the late 1940s, he concentrated on landscapes and portraits. Works such as *Stonehenge under Snow* (1947) and *Maiden Castle, Dorset* (1945) characteristically employ contrasting black and white tones to give a sometimes ethereal, sometimes melancholy, effect. His portraits, including Dylan THOMAS, Robert GRAVES, Alain ROBBE-GRILLET, and Francis BACON, later gave way to a series of close-ups of single eyes. *Perspective of Nudes* (1961) made use of the wide-angle lens to depict the nude as a landscape of folds and mounds, some seemingly part of the cliffs and boulders of a seashore. Although working mainly in black and white, Brandt has used colour in some later work to add a further dimension of strangeness to mundane subjects. His other books include *Shadow of Light* (1966) and *Nudes 1945–80* (1980).

Brandt, Willy (Herbert Ernst Karl Frahm; 1913–) *West German statesman and chan-*

cellor of the Federal Republic of Germany (1969–74). He was awarded the Nobel Peace Prize in 1971.

Born in Lübeck, an illegitimate child, Brandt became an apprentice shipbuilder in 1932, after leaving school. Active as an anti-Nazi socialist worker, in 1933 he emigrated to Norway, where he changed his name to Brandt and adopted Norwegian citizenship. Graduating from the University of Oslo, he became a journalist and worked as the secretary of a Norwegian charity (1933–40). During World War II he was active in the Norwegian and German resistance movements (1940–45), returning to Berlin in 1945 and resuming his German citizenship in 1947.

Brandt was first elected to the West German assembly (Bundestag) in 1949 as the Social Democratic Party (SPD) member for Berlin. Elected president of the Berlin chamber of deputies in 1955, he became mayor of West Berlin in 1957, a position he held until 1966. In 1964 he was appointed chairman of the SPD, becoming minister for foreign affairs and vice chancellor under Kiesinger in 1966. Elected chancellor in 1969, he resigned from office in 1974 following the exposure of a personal aide as an East German spy. Since 1987 he has been honorary chairman of the SPD.

Brandt achieved international recognition for his attempts at reconciliation with the Eastern bloc, known as 'Ostpolitik'. A pragmatist, he encouraged the negotiation of joint economic projects and a policy of non-aggression towards Warsaw-Pact countries. President of the Socialist International in 1976, he was chairman of the Independent Commission on Development Issues in 1977 and a member of the European Parliament (1979–83). He chaired the Brandt Commission on the state of the world economy, whose report was published in 1980. He has been awarded numerous honorary degrees and foreign decorations including the Reinhold Niebuhr Award (1972), the Aspen Institution for Humanistic Studies Prize (1973), and a Jewish award, the B'nai Brith gold medal (1981).

Braque, Georges (1882–1963) *French painter who, with PICASSO, developed cubism. He was made a Commandeur de la Légion d'honneur in 1951.*

After training as a decorator in Le Havre, Braque studied art in Paris and began painting in the new fauvist style. However, he was temperamentally unsuited to the subjective and often impulsive style of the fauvists and, impressed by the Cézanne memorial exhibi-

tion of 1907, began to paint in a style influenced by Cézanne's geometrical simplification of forms. In the same year Braque's meeting with Picasso led to a close working relationship that lasted until World War I.

In 1909 Braque's pictures were criticized as 'bizarreries cubiques' ('cubic oddities'); the name stuck and cubism was born. Developed by Braque and Picasso together over the next five years, cubism was probably the most revolutionary force in painting since the Renaissance: it freed artists from the restriction of representing objects at a fixed moment in time and from a fixed viewpoint, which had dominated painting since the fifteenth century. No longer pretending to be windows onto the visual world, these analytical cubist paintings instead depicted different aspects and views of a motif.

In 1911 Braque originated the use of collage, which opened the way for the second phase of cubism – synthetic cubism, in which the artist's materials replaced analysis of form as the starting point for the creation of flat nonillusionistic compositions.

The outbreak of war in 1914 interrupted the fertile partnership between Braque and Picasso. The following year Braque was severely wounded and in 1916 was discharged from the army. Back in Paris he continued painting, gradually introducing brighter colours and a more naturalistic approach to his work, which still remained basically cubist. He painted domestic subjects and nudes and became a master of still life. He produced ceramics, graphics, sculpture, and illustrations and decorated a ceiling in the Louvre. In 1948 Braque received the Grand Prix for painting at the Venice Biennale.

Brassaï (Gyula Halész; 1899–1984) *Hungarian photographer, artist, and writer, noted for his depictions of artists and Parisian low-life in the late 1920s and 1930s. He became a French citizen in 1948.*

Born in Brassó, Hungary (now Brasov, Romania), he was educated in Budapest and at the Berlin Academy of Arts. Embarking on a career in journalism, he settled in Paris in 1923 and took the pseudonym 'Brassaï' after his home town. He sketched aspects of the city and in the late 1920s began taking photographs, especially of life in the Montparnasse district – the artists, pimps, and prostitutes who frequented the bars and theatres. He also displayed an artist's vision in capturing on film such unlikely subjects as graffiti and masonry. His pictures of Paris were published in *Paris de nuit* (1933) and *Voluptés de Paris* (1935). In 1932, the editors of *Minotaure* commissioned Brassaï to photograph the sculpture of PICASSO. This marked the start of a long friendship between the two men and introduced Brassaï to a wide circle of artists and intellectuals living in Paris, many of whom he photographed. *Conversations avec Picasso* appeared in 1964, with text and photographs by Brassaï.

During World War II, Brassaï lived in occupied Paris but turned to drawing and sculpture as less hazardous pursuits than photography. *Trente dessins* was published in 1946. Brassaï resumed photography after the war. His range of subjects broadened until by the late 1960s he was experimenting with a mix of photography and graphics – his 'transmutations'. Of his writings, *Histoire de Marie* (1949) is based on the life of his charwoman.

Brattain, Walter Houser (1902–87) *US physicist, who in collaboration with John BAR-DEEN invented the point-contact transistor. He shared the 1956 Nobel Prize for Physics with Bardeen and W. B. SHOCKLEY.*

Brattain was educated at Whitman College, the University of Oregon, and the University of Minnesota, where he gained his PhD in 1928. In the following year he joined the staff of the Bell Telephone Laboratories as a research physicist and remained with them until his retirement in 1967. In the 1940s, Brattain's interests at Bell centred on the properties of such semiconductors as germanium and silicon. Working with John Bardeen he developed the first workable point-contact transistor in 1947 and they published their results in 1948.

Braudel, Fernand Paul (1902–85) *French historian. One of the best-known and most widely admired of European historians, Braudel was made a Commandeur de la Légion d' honneur in 1982. He was elected to the Académie Française in 1984.*

Born at Lunéville, Braudel was educated at the Lycée Voltaire and the Sorbonne. He then spent nine years teaching history in Algiers (1923–32) and served as professor of history at São Paulo, Brazil (1935–38). Interested in the relationship between history and sociology, 'that massive, rather confused science', and in large themes covering long periods of time, Braudel made his academic reputation with his magisterial two-volume study *La Méditerranée et le monde méditerranéen à l' époque de Philippe II* (1949; translated as *The Mediterranean and the Mediterranean World in the Age of Philip II*, 1972–73). In 1949 he was appointed a professor at the Collège de France,

where he served until 1972, and in 1954 he published his *Economic and Social History of France*. He wrote large numbers of shorter historical pieces in the *Annales* tradition and three substantial challenging volumes on 'material civilization' and capitalism, which drew on a wide variety of detailed evidence and ranged dramatically over space and time. He was also a founder of the Maison des Sciences de l'Homme in Paris.

Bream, Julian Alexander (1933–) *British guitarist and lutenist, whose mastery of these instruments did much to popularize them. He was made a CBE in 1985.*

Born in London and first taught the guitar by his father, Bream was later a student at the Royal College of Music, although he continued to study the guitar privately, as it was not then taught there. He made his London debut in 1950 and quickly became popular in Britain and on the continent; he made his US debut in 1958 and has toured the Far East and Southeast Asia. He began to study the renaissance lute in 1950, which led to recitals of Elizabethan lute-songs with Peter PEARS; the formation of the Julian Bream Consort has stimulated the playing of early consort music. Bream's brilliant style, taste, and intensity have inspired several composers to write works for him, including BRITTEN (*Nocturnal after John Dowland*), WALTON (*Five Bagatelles*), and RAWSTHORNE (*Elegy*).

Brecht, Bertolt (Eugen Berthold Friedrich Brecht; 1898–1956) *German playwright, poet, and theatrical reformer.*

Born at Augsburg, the son of an industrialist, Brecht went to school there and later studied medicine and natural science at the University of Munich (1917–21); his studies were interrupted by the war, during which he worked as an orderly in a military hospital. This experience confirmed his early antinationalistic and radical views, though his earliest dramatic work, which dates from 1918, was expressionistic and anarchist rather than Marxist. Brecht became *Dramaturg* (resident dramatic adviser) first at the Munich Kammerspiele (in 1921) and then (in 1924) at the Deutsches Theater, Berlin, where he worked under the influential producer Max Reinhardt (1873–1943). He soon rejected Reinhardt's poetic theatrical style as he became influenced by the proletarian, anti-illusionist, and documentary innovations being introduced in the Berlin theatre by Erwin Piscator (1893–1966), the producer to whom Brecht is indebted for his own theory of 'epic theatre'. The turning point of his early career occurred in 1928: he married the actress Helene Weigel, who later (1949) co-founded with him the Berliner Ensemble in East Berlin, and he announced his commitment to Marxism. Since 1923 Brecht's name had been fifth on the Nazi's list of enemies of the state; with HITLER's triumph in 1933 Brecht was forced to flee to Denmark, eventually settling in California (1941–47). In Hollywood he worked on films, with CHAPLIN and Charles LAUGHTON among others. Though called before the House Un-American Activities Committee in 1947 during the first wave of anticommunist hysteria, he acquitted himself satisfactorily. He left America soon afterwards, however, settling in East Berlin in 1949 (but taking Austrian citizenship in 1950). The workers' revolt in East Germany in 1953 apparently disturbed him, but he was attacked in the West for not taking a decisive stand. He was awarded the Stalin Peace Prize in 1954.

The best-known of Brecht's many dramatic works are the collaborations with the composer Kurt WEILL (with memorable performances by Weill's wife, Lotte Lenya (1900–81)): *Die Dreigroschenoper* (published 1929; translated as *The Threepenny Opera*, 1955) and *Aufstieg und Fall der Stadt Mahagonny* (published 1930; translated as *The Rise and Fall of the City of Mahagonny*, 1956). His other major plays were all written during and just after World War II: *Mutter Courage und ihre Kinder* (1941; translated as *Mother Courage and her Children*, 1948), *Der gute Mensch von Sezuan* (1943; translated as *The Good Woman of Szechwan*, 1961, and *The Good Person of Szechwan*, 1962), *Leben des Galilei* (1943; translated as *The Life of Galileo*, 1960), and *Der kaukasische Kreidekreis* (1948; translated as *The Caucasian Chalk Circle*, 1948). The theory of epic theatre, the emphasis on clear narrative structure, 'Verfremdungseffekt' ('alienation effect'), etc., are most fully explained in *Kleines Organon für das Theater* (1949; translated as *A Little Organum for the Theatre* in *Brecht on Theatre*, 1964), which sums up Brecht's lifelong concern to control the powerful illusion created by theatre in order to confront the audience finally with the real (i e political) issues at stake.

Brendel, Alfred (1931–   ) *Czech-born pianist, of Austrian, German, and Italian lineage, resident in Britain since 1972. He received an honorary knighthood in 1989.*

As a child Brendel studied in Zagreb and Graz; later he was taught principally by Edwin Fi-

scher (1886–1960) but also by Eduard Steuermann (1892–1964). In 1948 he made his debut and the following year started his international career by winning the Busoni Competition in Bolzano; since then he has been in constant demand for solo recitals and concerto concerts, playing with the world's greatest orchestras and conductors. Brendel's repertoire is large and varied, ranging from Bach to the Second Viennese School but with emphasis on Mozart, Beethoven, and Schubert, whose sonatas he has brought out of limbo. His playing is characterized by a deeply penetrating intellectuality, intense enthusiasm, impeccable technique, and the ability to keep both the architectural scope of a work and its minutest details in balance; an example of this is his reading of Liszt's B minor sonata.

Bresson, Robert (1907–   ) *French film director who has received many awards, including the appointment as Officier de la Légion d'honneur, for his work.*

Born in Bromont-Lamothe, Bresson studied philosophy and painting before becoming a scriptwriter in 1934. In World War II he spent a year as a prisoner of war but also launched himself as a director with the film *Les Anges du péché* (1943; *Angels of Sin*), in which he demonstrated his austere intellectual style.

Bresson has not made many films, but he has had a profound influence on the cinema. Perhaps most notable of his early films was *Les Dames du Bois de Boulogne* (1945), adapted from Diderot's *Jacques le Fataliste*, with dialogue by Jean COCTEAU. *Le Journal d'un curé de campagne* (1951; *Diary of a Country Priest*), *Un Condamné à mort s'est échappé* (1956), *Pickpocket* (1959), *Le Procès de Jeanne d'Arc* (1962; *The Trial of Joan of Arc*), *Au Hasard Balthazar* (1966), *Mouchette* (1967), *Une Femme douce* (1969), *Le Diable probablement* (1977), and *L'Argent* (1983) are just some of his other remarkable films, most of which have featured relatively unknown actors. His book *Notes sur le cinématographe* (1975) has been published in both French and English.

Breton, André (1896–1966) *French poet.*
Breton was born in Tinchebray, Orne. His early involvement with the dadaists revealed itself in such works as *Mont de piété* (1919); in the same year he co-founded the movement's review *Littérature*. He became leader of the newly created surrealist movement in the early 1920s and remained faithful to its philosophy throughout his life; in his *Manifeste du surréalisme* (1924) he began to define this phi-

losophy, developing it in two later manifestos (1930, 1942) as the movement evolved. His other theoretical writings include *Les Pas perdus* (1924), *Le Surréalisme et la peinture* (1928), in which he put forward his conception of art, and the essay collection *La Clé des champs* (1953).

Among Breton's best-known creative works are the poetic novel *Nadja* (1928), a partly autobiographical account of his surreal encounters with a mysterious young woman; *L'Immaculée Conception* (1930), written in collaboration with Paul ÉLUARD; *L'Amour fou* (1937), an exploration of the connection between dreams and reality; and *Arcane 17* (1945), which deals with the subject of occultism. His poetry of the years 1919 to 1948 was collected in *Poèmes* (1948). In his creative writings Breton made much use of surrealist techniques, such as automatic writing dictated by the subconscious mind, and the startling juxtaposition of images.

With other members of the surrealist movement, Breton became briefly involved with the Communist Party in the early 1930s. He spent the major part of World War II in the USA; on his return to France in 1946 he strove to revive the surrealist philosophy in an intellectual world now dominated by new ideas, notably SARTRE's existentialism.

Breuer, Marcel Lajos (1902–81) *Hungarian-born US architect and designer. He produced the first tubular-steel chair while at the Bauhaus; his most influential concrete buildings were the UNESCO secretariat in Paris and the Whitney Museum in New York.*

Born in Pécs in Hungary, Breuer studied at the Bauhaus from 1920 to 1925, when GROPIUS put him in charge of the joinery workshop. Here he produced the first design for a tubular-steel chair in 1925. In 1928 he moved to Berlin and then, with the advent of the Nazis, to London (1935). After two years he emigrated to the USA, where he joined Gropius again at Harvard. He became a US citizen in 1944 and after the war set up his own independent practice. His first commissions were for private houses in New England but in the next twenty years he worked extensively in Europe as well as America, mostly in concrete. His buildings include the UNESCO secretariat in Paris (with NERVI and Zehrfuss; 1953), a Y-shaped eight-storey essay in concrete; the Abbey Church of St John (1953–63) in Collegeville, Minnesota, with its concrete pylon bell tower; the lecture hall for New York University (1961), a rough-textured concrete mass; the IBM Research

Centre (1962) in France; and the 'brutal' Whitney Museum of American Art (1966) in New York, consisting of three concrete tiers that overhang the street.

Breuil, Henri Édouard Prosper (1877–1961) *French archaeologist noted for his work on prehistoric cave paintings. He was made a Commandeur de la Légion d'honneur and was a member of L'Institut de France.*

Breuil studied theology and was ordained as a priest in 1900. He was appointed lecturer in prehistory and ethnography at the University of Fribourg (1905–10) and in 1910 became honorary professor at the Institut de Paléontologie Humaine. He later served as professor of prehistoric art at the Collège de France (1929–47). Breuil had a lifelong interest in cave paintings of the Palaeolithic period, predominantly dating from 15 000 to 10 000 BC, and made detailed studies of examples in the Dordogne region of France and, in particular, the paintings at Altamira in Spain, which he was able to authenticate. Breuil more or less originated the system of classifying and dating such works. In the late 1940s he studied cave art in southern Africa. His surveys of the major sites are described in *Quatre cents siècles d'art pariétal* (1952; 'Four Hundred Centuries of Cave Art').

Brezhnev, Leonid Ilich (1906–82) *Soviet statesman, general secretary of the Soviet Communist Party (1966–82) and president of the Soviet Union (1977–82). He was awarded five Orders of Lenin, two Orders of the Red Banner, and the Lenin Peace Prize. In 1976 he was promoted to marshal, the Soviet Union's highest military rank.*

Born in Kamenskoye (now Dneprodzershinsk), the son of a steelworker, Brezhnev became a manual labourer after leaving the local school. In 1923, as a student at a technical college in Kursk, he joined the communist youth organization, Komsomol. After graduating in 1927 he became a surveyor, but continued his studies in agriculture and later in metallurgy (1931–35).

Brezhnev joined the Soviet Communist Party (CPSU) in 1931. After serving in the Red Army in 1935, he returned to his hometown to become director of the Metallurgical Technical College (1936–37), moving on in 1938 to the deputy headship of the regional CPSU committee. During World War II he served as a high-level political officer in the armed forces. After the war Brezhnev rose rapidly through the party hierarchy to become a full member of the central committee and a candidate member of the Presidium in 1957, being elevated to chairman of the Presidium in 1960. In 1964 he and KOSYGIN forced KHRUSHCHEV to resign and Brezhnev became first secretary of the CPSU, rising to general secretary in 1966, a position he held until his death. By the late 1960s Brezhnev was the most powerful man in the Soviet Union and in 1977 he was appointed president.

A shrewd politician, Brezhnev instigated the so-called 'Brezhnev doctrine', which emphasized the unity of the communist world under the leadership of the Soviet Union. The SALT agreements of 1969 and 1972 were the products of his policy of detente with the West.

Briand, Aristide (1862–1932) *French socialist statesman, prime minister of France eleven times between 1909 and 1929. He received the Nobel Peace Prize in 1926.*

Born in Nantes, the son of an innkeeper, Briand was educated at St Nazaire, Nantes, and Paris, where he studied law and became involved in left-wing political circles. Returning to St Nazaire to practise, he joined a syndicalist movement, which advocated the general strike as an instrument of political change. In 1893 he moved to Paris, where he worked as a journalist and unsuccessfully stood for the Chamber of Deputies (1893 and 1895).

Briand's political career began in 1902, when he was elected to the Chamber of Deputies as the socialist member for St Étienne. He first came to the public eye in 1905 over his support of legislation to separate church and state. Expelled from the Socialist Party in 1906 for accepting office in a radical–republican coalition, he held several ministerial portfolios before becoming prime minister for the first time in 1909. During World War I Briand was in charge of foreign affairs (1915–17) and signed an accord with Britain; in the 1920s he emerged as a respected international leader. As prime minister in 1921–22 he oversaw the application of the Treaty of Versailles, participating in international conferences in London (1921) and Cannes (1922). As minister of foreign affairs he signed the Locarno Pact (1925), which permitted the entry of Germany into the Concert of Powers and agreed to its western boundaries. He became very influential in the League of Nations and shared the Nobel Peace Prize in 1926 with Gustav Stresemann (1878–1929). In 1927 he signed the Kellogg–Briand Pact, a multilateral treaty renouncing war, and in 1929 proposed a coalition of governments – the United States of Europe – as a means for ensuring peace. He stood unsuccessfully for president in 1931.

Bridge, Frank (1879–1941) *British composer, conductor, and viola player, who won particular acclaim in the field of chamber music.*

Bridge was born in Brighton. At the Royal College of Music he studied violin, later changing to viola, and composition with Stanford. He became known as a chamber music player and as a conductor capable of taking over in any emergency. In 1923 he toured the USA, conducting his own compositions with the New York, Detroit, Boston, and Cleveland Symphony Orchestras. Bridge's works include piano music, songs, orchestral tone poems, an opera, and much chamber music: he won the Cobbett Prize in 1905 for his *Fantasy String Quartet* and again, in 1908, for his *Fantasy-Trio*. His style is based on a secure orthodox technique and shows constant care for performing difficulties; later, a purposeful study of BARTÓK, SCHOENBERG, and BERG expanded his own thought harmonically and chromatically and made him a creative and valuable teacher for the young Benjamin BRITTEN.

Bridges, Robert Seymour (1844–1930) *British poet. He was appointed poet laureate in 1913.*

Born at Walmer into a prosperous landowning family, Bridges was educated at Eton and at Balliol College, Oxford. While at Oxford he met Gerard Manley Hopkins, whose poems he was later to edit (1918). He then spent some time travelling in the Near East and Europe before settling seriously to his medical studies, which he completed in 1874. Dissatisfied with his early poetry, he waited until 1873 before publishing a volume of verse; his first notable work, the sonnet sequence *The Growth of Love*, appeared anonymously in 1876, while he was working as a physician in London.

A serious illness in 1881 ended Bridges's medical career. He retired to Yattendon Manor, Berkshire, where he pursued his interest in poetry and music. In 1884 he married Monica, the daughter of the architect Alfred Waterhouse. He became increasingly interested in prosody, writing an important study of Milton's technique (1893). He also wrote a long narrative poem called *Eros and Psyche* (1885) and a number of plays, of which only the masque *Demeter* (1904) was performed. He collaborated on the famous *Yattendon Hymnal* (1895–99), which played a major role in the revival of English hymnody.

After a visit to Switzerland for his wife's health, Bridges moved to Oxford (1907). In 1913 he and other eminent men of letters joined forces to found the Society for Pure English. His wartime anthology, *The Spirit of Man* (1916), was particularly well received and illustrates the authors who had a major influence on Bridges's own thinking and poetical development. *New Verse* (1925) gave hints of the direction that Bridges's poetic experiments were taking, but it was not until near the end of his life that he produced his greatest work, *The Testament of Beauty* (1929). This long philosophical poem, written in the version of alexandrine metre that Bridges perfected, distils the essence of the poet's accumulated wisdom and understanding of the creative spirit.

Bridie, James (Osborne Henry Mavor; 1888–1951) *Scottish playwright, awarded the CBE (1946) for services to the theatre.*
Born and educated in Glasgow, Mavor qualified in medicine in 1913 and served in the Royal Army Medical Corps during World War I. He afterwards practised and taught medicine in Glasgow, where his first play, *The Sunlight Sonata* (1928), was produced.

Under the name of James Bridie, Mavor wrote around forty plays. *The Anatomist* (1931), centred on the Burke and Hare bodysnatching scandal, and *The Sleeping Clergyman* (1933) were early successes based upon the author's intimate knowledge of the medical world. He also wrote witty and charming comedies on biblical themes: *Tobias and the Angel* (1930), *Jonah and the Whale* (1932), and *Susannah and the Elders* (1937). Later successes included *Mr. Bolfry* (1943), *The Forrigan Reel* (1944), *John Knox* (1947), and *Daphne Laureola* (1949). The posthumously produced *Baikie Charivari* (1952) shows Bridie's more serious vein. His depiction of Scottish characters and speech in many of his plays was particularly well received.

In 1943 he founded the Glasgow Citizens' Theatre, to which he devoted much energy. Besides his personal achievements in the Scottish theatre, he also encouraged younger playwrights. His autobiography, *One Way of Living*, was published in 1939.

Britten, (Edward) Benjamin, Baron (1913–76) *British composer and pianist. He was created a CH (1953) and awarded the OM (1965) and a life peerage (1976).*
Born in Lowestoft, Suffolk, on St Cecilia's Day (22 November), he quickly responded to his mother's teaching, composing before he

had any formal tuition. In 1927 he became a pupil of Frank BRIDGE and in 1930 entered the Royal College of Music, studying piano with Arthur Benjamin (1893–1960) and composition with John IRELAND. His *Sinfonietta* was performed at a Macnaughten-Lemare concert (1932); in Europe the ISCM festivals offered him a platform for the oboe quartet (Florence, 1934), and the violin suite (Barcelona, 1936). The *Variations on a Theme of Frank Bridge* received tremendous applause at Salzburg (1937). At this period he composed music for GPO documentary films in collaboration with W. H. AUDEN, whose social and political disenchantment led him to emigrate to the USA; Britten, with his lifelong friend the tenor Peter PEARS, followed him in 1939. Compositions during this period include the *Sinfonia da Requiem* (1940), the first string quartet (1941), *Les Illuminations* (1939), and the *Seven Sonnets of Michelangelo* (1940).

Made homesick by reading an article by E. M. FORSTER about the Suffolk poet, George Crabbe, Britten and Pears resolved to return to England; the *Hymn to St Cecilia* and *A Ceremony of Carols* were composed during their Atlantic crossing (1942). The dramatic vocal style of *Serenade for Tenor, Horn and Strings* (1943) prepared the ground for Britten's first opera, *Peter Grimes* (libretto by Montagu Slater, based on Crabbe's poem), for the reopening of the Sadler's Wells Theatre in 1945. Its masterful achievement and success acted as a catalyst to the fertile years that followed: the *Holy Sonnets of John Donne* and the second string quartet (1945), the operas *The Rape of Lucretia* (1946) and *Albert Herring* (1947), and *The Young Person's Guide to the Orchestra* (1947). Britten's formation of the English Opera Group for the performance of chamber opera led him and Pears in 1948 to start the Aldeburgh Festival, in the Suffolk seaside town in which they lived. From that time Britten divided his energies between conducting, worldwide recital tours with Pears (Britten was one of the finest pianists of his generation), organizing the annual festival, and composing. Of his other operas, *Billy Budd* (1951; revised 1960) was commissioned for the Festival of Britain; *Gloriana* (1953) for the coronation of ELIZABETH II; *The Turn of the Screw* (1954), a chamber opera constructed on a set of orchestral variations on a twelve-note theme, for the Venice Biennale; *A Midsummer Night's Dream* (1960) for the Aldeburgh Festival; and *Death in Venice* (1973) for the Maltings at Snape, near Aldeburgh. The *War Requiem* was composed for the consecration of

Coventry Cathedral (1962). He died fifteen days before the first performance of his third string quartet.

Brook, Peter Stephen Paul (1925–) *British theatre director, best known for his work as co-director of the Royal Shakespeare Theatre. He was made a CBE in 1965 and an Officier de la Légion d' honneur in 1987.*

Brook directed his first London productions while he was still an undergraduate at Oxford – Marlowe's *Dr Faustus* (1942) at the Torch and COCTEAU's *The Infernal Machine* (1945) at the Chanticleer. After the war he continued his work in London, Stratford, and elsewhere, including a notable Stratford production of *Titus Andronicus* (1955), with Laurence OLIVIER in the title role, and *The Power and the Glory* (1956), for which he composed the music. In 1962 Brook became co-director of the Royal Shakespeare Company. His first production for them was *King Lear* with Paul Scofield (1922–), whom he has directed in a number of plays, incuding *Hamlet* (1955). For his *Marat/Sade* (1964) Brook was awarded the New York Drama Critics Award for best director in 1965, and *Timon of Athens* (Paris, 1974) brought him the Grand Prix Dominique and the Brigadier Prize. Notable subsequent productions have included *Ubu Roi* (1977) at the Aldwych, *The Cherry Orchard* (Paris, 1981; New York, 1988), and *The Mahabharata* (Avignon and Paris, 1985; Glasgow, 1988). As well as composing music he has also designed many of the sets for his productions.

Not confined to plays, he has directed opera – including an award-winning *Carmen* (1981) – and films, including *The Beggar's Opera* (1952), *Moderato Cantabile* (1960), *Lord of the Flies* (1962), *King Lear* (1969), and *The Tragedy of Carmen* (1983). He also directed *King Lear* for television in New York (1953) and has written two television plays, *The Birthday Present* and *Box For One* (both 1955). In 1971 he founded the International Centre for Theatre Research in Paris and has developed new acting techniques, arising out of mime productions presented on tour in remote parts of Africa and Asia. His publications include *The Empty Space* (1968) and an autobiography, *The Shifting Point* (1988).

Brooke, Rupert Chawner (1887–1915) *British poet, probably the best known of the 'war poets'.*

Born and educated at Rugby School, where his father was a housemaster, Brooke went up to Cambridge in 1906. There he helped found the university's famous drama club, the Marlowe

Society, and acted in numerous plays. He took part in politics through the Cambridge Fabian Society and through it formed many lasting friendships. The legend of the young poet with romantic good looks and charismatic charm was already growing.

Having completed his degree (1910), Brooke stayed on at nearby Grantchester, which in 1912 became the subject one of his best loved poems. His *Poems* (1911), containing some deeply erotic verses evoked by his Cambridge attachments, met with a mixed reception. He also travelled in Europe, living in Munich for a time with Ka Cox, whom he had loved since 1909. Later he spent more time in London before travelling across America and in the Pacific (1913).

When World War I broke out Brooke joined the navy. En route for the Dardanelles in spring 1915, he died of dysentery, heatstroke, and blood poisoning near the Greek island of Skyros, where he is buried. In his last months he wrote the poems by which he is best remembered – 'The Soldier', 'The Great Lover', and 'The Dead'. These, published in his *Collected Poems* (1918), ensured his status as the spokesman for a generation of romantically patriotic young men doomed to die in a war the full horrors of which Brooke himself never lived to experience.

Broome, David (1940–) *British showjumper and world champion (1970).*

Broome started riding as a child at his home in Newport, Gwent; he was a leading pony rider for many years before graduating to senior events. He represented Britain at four Olympics, winning individual bronze medals on Sunsalve in 1960 at the age of twenty and Mister Softee in 1968. The same two horses were involved in his European championship wins in 1961 (Sunsalve) and 1967 and 1969 (Mister Softee). In 1970 Broome rode Beethoven to become the first Briton to win the world championship.

He turned professional in 1973 and in 1975 won seven events at the Horse of the Year Show, the Prince of Wales Cup, and the Aga Khan Trophy. He was the star of the Royal International Horse Show for each of the next three years. He has represented Great Britain in team events, winning the world championship, European championship, and Nations Cup from 1978 to 1984. He has also won many individual events around the world, including six major titles on Last Resort in 1983 and 1984.

Brough, Louise (1923–   ) *US tennis player, who was among the foremost of the post-World War II generation of players who effectively established the serve and volley tactic in women's tennis.*

Louise Brough won the Wimbledon singles title in 1948, 1949, 1950, and 1955, was US national champion in 1947, and won the Australian singles crown in 1950. However, her doubles record is even more impressive. Brough and her partner, Margaret Osborne du Pont, won the US national doubles title twelve times, Wimbledon five times, and the French title three times. Brough was also four times winner of the Wimbledon mixed doubles title (1946, 1947, 1948, and 1950).

Brouwer, Luitzen Egbertus Jan (1881–1966) *Dutch mathematician, best known for his development of the philosophy of mathematics known as intuitionism.*

Born in Overschie in Holland, he was educated at the University of Amsterdam, where he spent his entire professional career and served as professor of mathematics (1912–51). From about 1908 onwards Brouwer developed an approach to the basic concepts of mathematics known variously as intuitionism, constructivism, or finitism. His central claim was that mathematical objects could only be said to exist if they could actually be constructed in a finite number of steps. It was not enough merely to show that the assumption of the entity was also required. Such a radical reformulation of the conditions of mathematical existence meant that many of the proofs of classical mathematics became invalid and much of the theory of transfinite sets developed by Georg Cantor (1845–1918) had to be rejected.

In the field of pure mathematics Brouwer's name is linked with two well-known theorems. In 1912 he proved a famous fixed-point theorem, sometimes known as Brouwer's theorem. Earlier, in 1910, he had proved the remarkable theorem, hard to believe and imagine, that three countries can be represented on a map in such a way that they will touch each other at every boundary point.

Brown, Sir Arthur (1886–1948) See Alcock, Sir John William.

Brubeck, Dave (David Warren Brubeck; 1920–   ) *US jazz pianist and composer. The Dave Brubeck quartet was enormously successful from 1951 to 1967, and made a significant contribution to popularizing jazz when it was out of favour.*

Brubeck studied with Arnold Schoenberg and Darius Milhaud. In the 1940s he formed an octet, which was soon reduced to a quartet featuring alto saxophonist Paul Desmond (1924–77), bass, and drums. The group's great popularity on college campuses was largely a result of Brubeck's cerebral approach, based on such classical devices as fugue and counterpoint, together with Desmond's wistful tone. The quartet represented a new 'cool' school in contrast to the frenetic bop that preceded it. Drummer Joe Morello (1928–   ) joined the group in 1956. The group's biggest commercial success was the recording *Time Out* (1960), one track of which, 'Take Five' (composed by Desmond), was an international hit.

Since 1967 Brubeck has led other groups, sometimes including his sons David Darius (1947–   ) and Daniel (1955–   ). He has played at several Newport Jazz Festivals as well as the Monterey Jazz Festival (1980); in 1981 he was invited to play for President Reagan at the White House. He has also composed several piano concertos, which have been played by symphony orchestras. His recent compositions include the oratorio *The Voice of the Holy Spirit* (1985) and the *Lenten Triptych* (1988).

Bruce of Melbourne, Stanley Melbourne, 1st Viscount (1883–1967) *Australian statesman and prime minister (1923–29). He was appointed a CH in 1929 and created a viscount in 1947.*

Born in Melbourne, the son of a prosperous importer, Bruce was educated in Melbourne and Cambridge. He practised law in England until 1914 when he enlisted in the British army, winning the MC (1915) and the French Croix de Guerre (1916) for his service during World War I.

In 1918, after his return to Australia, Bruce was elected to the Australian parliament as a member of the Nationalist Party. After his success as Australian representative at the League of Nations (1921), he became prime minister and minister for external affairs in 1923 at the head of a Nationalist–Country Party coalition. His joint ministry with the Country Party leader Earle Page (1880–1961), with such achievements as the creation of a central banking system and the Council of Scientific and Industrial Research (which became the Commonwealth Scientific and Industrial Research Organization), did much to improve Australia's prosperity in the postwar years. Bruce lost both the election and his seat in 1929, due to his policy of abandoning the federal arbitration system. Although he was re-elected in 1931, he subsequently spent most of his time representing Australia abroad, and in 1933 accepted the position of Australian high commissioner in London, which he held until 1945. He also represented Australia in the British war cabinet (1942–45). Subsequent positions included chairman of the World Food Council (1947–51), and first chancellor of the National University at Canberra (1951–61). He was elected a fellow of the Royal Society in 1944. Respected for his keen mind and shrewd but diplomatic manner, Bruce is remembered as an outstanding figure in both Australia and Imperial politics.

Bruch, Max (1838–1920) *German composer and conductor.*

Born into a musical Cologne family, Bruch won a scholarship to the Mozart foundation of Frankfurt (1852) and extended his musical education with lengthy visits to other centres of music, such as Munich and Leipzig. Teaching and composing, he gradually made a reputation for himself and from 1867 to 1870 was kapellmeister to the Prince of Schwarzburg-Sondershaus. From 1880 to 1883 he directed the Liverpool Philharmonic Society and introduced his choral compositions, for which he was once renowned, to English audiences. Today he is best remembered for his virtuoso instrumental works: the three violin concertos (G minor, 1868; D minor, 1878; D minor, 1891), the *Scottish Fantasy* (1880) for violin and orchestra, and *Kol Nidrei* (1880) for cello and orchestra.

Buber, Martin (1878–1965) *Jewish theologian and philosopher whose approach to the man–God relationship has had considerable impact on both Jewish and Christian theology in the twentieth century.*

Born in Vienna, Buber moved to Lvov (now in the Soviet Union) after his parents separated and was raised by his grandparents. He studied philosophy at the universities of Vienna, Berlin, Leipzig, and Zürich and became a supporter of the Zionist cause. After briefly editing the Zionist weekly *Die Welt*, he split with the Zionist leader, Theodore Herzl, and formed the Zionist Democratic Faction. *Der Jude*, founded by Buber in 1916, became the leading periodical for German Jewish intellectuals. Dissatisfied with modern Judaism and what he regarded as the alienation between man and God, Buber turned to Hasidism, the Jewish sect founded in eastern Europe in the eighteenth century. In his great philosophical work, *Ich und Du* (1923; translated as *I and Thou*, 1937), Buber distinguished two basic forms of relationship between man and the world: 'I–It', in which we only partially engage with an 'object' or some abstract notion of it, and 'I–Thou', in which our entire being conducts a 'dialogue' with another 'subject'. These latter profound relationships, usually with people but also with things, point the way to the supreme I–Thou relationship, that between man and God.

Buber was co-founder of the Free Jewish House of Learning in Frankfurt and in 1930 was appointed professor of religion at Frankfurt University. Dismissed by the Nazis in 1933, he devoted his efforts to Jewish teacher-training throughout Germany. But the Nazis increasingly restricted his activities and in 1938 he emigrated to Palestine, where he became professor of social philosophy at the Hebrew University, Jerusalem. He consistently urged cooperation with the Arabs, served as first president of the Israeli Academy of Sciences and Arts, and was co-author, with Franz Rosenzweig, of a new German translation of the Bible, completed in 1961.

Buchan, John, 1st Baron Tweedsmuir (1875–1940) *Scottish author and public servant. In 1935 he was raised to the peerage and appointed governor-general of Canada.*

Born in Perth, Buchan was educated in Glasgow before winning a scholarship to Oxford, where he enjoyed a glittering career. On leaving Oxford (1899) he studied law in London but was soon taken onto Lord Milner's staff in South Africa, where he spent two years (1901–03) in the aftermath of the Boer War.

Buchan had already published several novels, and when he returned to England he left legal practice for a job in publishing. *Prester John* (1910) was the first in his long sequence of adventure stories. *The Marquis of Montrose* (1913; revised 1928) was his first attempt at serious historiography, but its unbalanced enthusiasm evoked severe criticism. During an illness shortly after the outbreak of World War I Buchan started his four-volume *History of the Great War* (1921–22) and as light relief wrote his famous thriller *The Thirty-Nine Steps* (1915), which immediately became a best-seller. On recovery he saw active service before recall to London to help run the new Department of Information (1917–18). He kept up his writing, with *Greenmantle* (1916) and *Mr. Standfast* (1919) continuing in the vein of *The Thirty-Nine Steps*. In 1919 he settled near Oxford, producing a steady stream of books on the recent war, thrillers ('yarns', as he called them), and biographies.

In 1927 Buchan was elected MP for the Scottish universities, and his excellent record in this role, coupled with his South African experience, led to his appointment as governor-general of Canada. During his arduous public life he still continued to write: his output in these years included biographies of *Julius Caesar* (1932), *Sir Walter Scott* (1932), *Oliver Cromwell* (1934), and *Augustus* (1937) as well as some of his best yarns. In Canada he travelled throughout the country and won the respect of people from British, French, and Red Indian stock. Exhausted by his exertions, particularly over the 1939 royal tour of Canada, he died at Montreal. His autobiography *Memory Hold-the-Door* appeared in 1940.

Budge, Donald (1915–   ) *US tennis player who, in 1938, achieved the first ever grand slam, winning the four major singles titles in the same year.*

Budge sprang to fame at the age of eighteen as surprise runner-up to Fred Perry in the 1934 Pacific Coast championships. A left-hander with an aggressive and determined all-court game, Budge was recognized as one of the greatest exponents of the backhand in the history of tennis. He won the singles titles in the 1937 Wimbledon and US national championships, and in 1938 his French, Wimbledon, US, and Australian titles gave him the historic grand slam.

Budge's doubles successes included the 1936 and 1938 US titles and the Wimbledon titles of 1937 and 1938, all partnered by C. G. Mako. With the brilliant US player Alice Marble, Budge also won the 1937 and 1938 Wimbledon mixed doubles titles. He turned professional in 1938, thereby disqualifying himself from the major tournaments.

Bukharin, Nikolai Ivanovich (1888–1938) *Soviet statesman, newspaper editor, and communist theoretician.*

Born in Moscow, the son of a schoolteacher, Bukharin joined an anti-tsarist political group as a university student, becoming a member of the Bolshevik faction of the Russian Social Democratic Workers' Party in 1906. In 1908 he took over the leadership of the Moscow Bolshevik organization, but in 1911, having been arrested, he escaped to Oregon, USA, and later moved to New York, where he edited the socialist paper *Novy Mir*. Returning to Europe, he worked with Lenin in Germany on the party newspaper *Pravda* and began to establish his credentials as a major political and economic theorist.

Bukharin returned to Russia in 1917. In a dispute with Lenin, who proposed to end World War I by making a separate peace with Germany (1918), he advocated that the world war should be transformed into a European revolutionary war. Despite this disagreement, Bukharin maintained his influence in the party (and government) to become editor of *Pravda* (1918–29), of the party journal, the *Bolshevik* (1924–29), and of the *Great Soviet Encyclopaedia*. Following Lenin's death in 1924, Bukharin became a full member of the Politburo and in 1926 he was appointed chairman of Comintern (the Communist International). Although he at first supported Stalin against Trotsky, he was denounced by Stalin as the leader of the 'right deviation' and was expelled from the Politburo in 1929. He was appointed editor of the official government newspaper, *Izvestia*, in 1934 but in 1937 was arrested and removed from the party as a trotskyite. He was tried, found guilty of treason, and executed in 1938.

Bukharin was one of Lenin's closest friends and colleagues. A major economic theorist, he published many outstanding works, including *World Economy and Imperialism* (1915), *The Economy of the Transitional Period* (1920), *The ABC of Communism* (1921), and *The Theory of Historical Materialism* (1921).

Bulganin, Nikolai Aleksandrovich (1895–1975) *Soviet statesman and prime minister (1955–58). He was decorated with the Order of the Red Star for his services during World War II and in 1949 was promoted to the rank of marshal of the armed forces.*

Born in Nizhnii Novgorod (now Gorkii), the son of a factory worker, Bulganin was educated at a local school before joining the Bolshevik Party in 1917, shortly after the February Revolution. Appointed to the CHEKA (secret police) in 1918, he was transferred to the Supreme Economic Council in 1922, becoming manager of an electrical equipment factory in Moscow (1927) and chairman of the Moscow Soviet (1931).

In 1937 Bulganin was appointed premier of the Russian Socialist Federal Soviet Republic (RSFSR). Rising through the party ranks, he became chairman of the Soviet Union state bank and deputy premier of the Soviet Union (1938–41). During World War II he was a member of military councils for the defence of Moscow (1941) and the Western front (1941–43), joining Stalin's state defence committee in 1944. In 1947 he succeeded Stalin as minister of the armed forces, becoming a full member of the central committee of the Politburo (1948) and deputy premier and minister of defence under MALENKOV, following Stalin's death. In the ensuing power struggle between KHRUSHCHEV and Malenkov, Bulganin supported the successful Khrushchev, who appointed him prime minister (chairman of the council of ministers) in 1955. He retained this position and his membership of the Praesidium (Politburo) until 1958, when he was replaced by Khrushchev. His association with the group that attempted to oust Khrushchev led to his dismissal and appointment to the obscure position of chairman of the Stavropol economic council (1958–60). He lost his membership of the central committee in 1961.

Bulganin is remembered as the first Soviet premier to travel abroad (to Britain, China, India, and Burma, where he advocated the Soviet Union's coexistence policy). An outstanding administrator, he oversaw the construction of the Moscow underground railway system.

Bunche, Ralph Johnson (1904–71) *US political scientist and founding member of the UN. In 1950 he became the first black to win the Nobel Peace Prize and in 1963 he received the US Medal of Freedom, the highest civilian award in the USA.*

Bunche was born in Detroit, Michigan. With the encouragement of his maternal grandmother, who brought up Ralph and his sister after their parents' premature death from tuberculosis, he studied at the University of California in Los Angeles and at Harvard. From there he went on to Howard University, Washington, where he set up and ultimately headed the department of political science. He spent the major part of World War II in the Office of Strategic Services, specializing in African affairs; in 1944 he joined the State Department and played an important role in the drafting of the United Nations Charter.

After two years as director of the Trusteeship Department at the UN Secretariat, Bunche succeeded unexpectedly to the office of chief mediator on the assassination of Count BERNADOTTE in 1948. His tireless efforts to bring about a settlement in the Arab–Israeli conflict were rewarded with the signing of armistice agreements between the warring parties in 1949 – the achievement for which Bunche was awarded his Nobel Prize. In the mid-1950s he was appointed undersecretary for political affairs by the UN secretary-general, Dag HAMMARSKJÖLD, and as the principal organizer

of Middle East operations he headed the UN peacekeeping force in the Congo in 1960 and played a similar role in Cyprus four years later.

During the late 1960s Bunche became increasingly involved in the civil rights movement and took a more prominent stand against racial discrimination. His health was already failing, however, and in 1970 he was obliged to relinquish his post as political adviser to U THANT, then secretary-general of the UN. He died the following year.

Buñuel, Luis (1900–83) *Spanish-born film director who settled in Mexico, where he made many highly acclaimed films.*

Born in Calanda, Buñuel was educated by Jesuits and at the University of Madrid. He also attended the Académie du Cinéma, Paris. His first two films, *Un Chien andalou* (1928) and *L'Âge d'or* (1930), were both produced in collaboration with Salvador DALI and established Buñuel as a surrealist film-maker *par excellence*. Remarkable for their shocking and terrifying images, his films often attacked the Establishment, the middle classes and, in particular, the church. *Las hurdes* (1932), dealing with the plight of deprived Spanish peasants, was followed by fifteen years in the 'wilderness', dubbing American films. In 1947 Buñuel settled in Mexico and re-established his reputation with *Los olvidados* (1950), which won the Grand Prix at Cannes. Also made in Mexico were *El* (1952) and *Wuthering Heights* (1953), among others. Considered more important, however, were *Nazarin* (1958), made in France, and *Viridiana* (1961). The latter was made in Spain, where it was ultimately banned, even though the script had been passed by the censors. Among the notable films that followed were *El angel exterminador* (1962; *The Exterminating Angel*), *Belle de jour* (1966), winner of the Venice Golden Lion Award, *La Voie lactée* (1969; *The Milky Way*), *Le charme discret de la bourgeoisie* (1972; *The Discreet Charm of the Bourgeoisie*), which won a best foreign language Academy Award, and *That Obscure Object of Desire* (1977).

As a gesture of reconciliation the Spanish government awarded Buñuel, who had taken Mexican nationality, the Grand Cross of the Order of Isabel la Catolica just before his death. His autobiography, *My Last Sigh*, was translated in 1983.

Burbidge, Eleanor Margaret (1922–) *British astronomer noted for her work on optical astronomy and astrophysics.*

Margaret Burbidge was born Margaret Peachey in Davenport. She graduated from the University of London in 1948 and subsequently became director of the university observatory. She has held a number of academic posts and is currently a professor of astronomy at the University of California, San Diego. In 1948 she married Geoffrey BURBIDGE and they collaborated with Fred HOYLE and William Fowler (1911–) in an important theoretical study of the synthesis of elements in stars (1957). The Burbidges also produced an early survey of quasars (*Quasi-Stellar Objects*; 1967). Margaret Burbidge was also briefly involved in 1972 in the administration of the 98-inch Isaac Newton telescope, which was then situated at the Royal Greenwich Observatory at Herstmonceaux in Sussex.

Burbidge, Geoffrey (1925–) *British astronomer and astrophysicist noted for his work on quasars.*

Burbidge was born in Chipping Norton, Oxfordshire. He graduated from the University of Bristol in 1946 and obtained a PhD in 1951 from the University of London. Subsequently he held a number of academic posts and was director of the Kitt Peak National Observatory, Arizona (1978–84). Currently he works at the University of California, San Diego.

Burbidge began his research career in particle physics but after his marriage in 1948 to the astronomer Margaret Peachey (Margaret BURBIDGE) turned to optical astronomy and astrophysics. Together they worked on the mysterious quasars (quasi-stellar objects), which had been discovered by Allan Sandage (1926–) in 1960. The Burbidges produced an early survey of the subject in 1967.

Geoffrey Burbidge has also produced papers on cosmology and, like Fred HOYLE, is noted for his sceptical approach to the orthodox big-bang theory of the origin of the universe.

Burgess, Anthony (John Anthony Burgess Wilson; 1917–) *British author.*

Born and educated in Manchester, Burgess served in the British army during World War II. He then spent a period as a teacher before becoming an education officer in Malaya and Borneo. This became the setting for his first novels, *The Malayan Trilogy* (1956–59). From 1959 he worked freelance as a writer, becoming an influential reviewer and publishing a succession of novels, notable among which was *A Clockwork Orange* (1962; filmed by Stanley KUBRICK in 1971), a zestful depiction of a London of the future riddled with violence and juvenile delinquency. Novels of the 1960s

included *Nothing Like the Sun* (1964), a colourful account of Shakespeare's life, which annoyed sober scholars almost as much as Burgess's subsequent biography of Shakespeare (1971). He also wrote an introduction to James JOYCE and published an abridgment of Joyce's *Finnegans Wake* (1966). In the 1970s Burgess accepted a series of university appointments in the USA, continuing to publish criticism and novels that are a challenging and often comic exploration of late twentieth-century dilemmas. His later fiction includes the best-selling *Earthly Powers* (1980), the series dealing with the character Enderby, running from *Inside Mr. Enderby* (1963) to *Enderby's Dark Lady* (1984), *The Kingdom of the Wicked*, (1985) an epic of the Christian struggle against Imperial Rome, *Any Old Iron* (1989), about the rediscovery of Excalibur, and *The Devil's Mode* (1989), a collection of short stories. *Little Wilson and Big God* (1988) and *You've Had Your Time* (1990) are autobiographical. He has also written several important works in the field of linguistics.

Burnet, Sir (Frank) Macfarlane (1899– 1985) *Australian physician and virologist who, with Sir Peter MEDAWAR, won the 1960 Nobel Prize for Physiology or Medicine for the discovery that immunological tolerance to tissue transplants can be acquired. He was knighted in 1951 and received the OM in 1958.*

Burnet graduated in medicine from the University of Melbourne in 1923 and after a short period as resident pathologist at the Melbourne Hospital (1923–24) he went to London to carry out research at the Lister Institute of Preventive Medicine. Back in Australia, he became assistant director of the Walter and Eliza Hall Institute of Medical Research at the Royal Melbourne Hospital (1928–31) and was later director (1944–65).

Burnet's most significant contribution to medical science, confirmed by Medawar, was his discovery of acquired immunological tolerance; i e animals and man can acquire the ability to tolerate foreign tissues. His other notable discoveries were a means of identifying bacteria by the viruses (bacteriophages) that attack them and a technique for culturing viruses in chick embryos that is now routine laboratory practice. He contributed to the current knowledge of influenza infection and isolated the organism (*Rickettsia burneti*) that causes Q fever. His publications include *Biological Aspects of Infectious Disease* (1940), *Viruses and Man* (1953), *Enzyme Antigen and Virus* (1956), *Clonal Selection Theory of Acquired*

Immunity (1959), and his autobiography, *Changing Patterns* (1968).

Burra, Edward John (1905–76) *British painter.*

Burra left school early due to chronic ill health but later studied at the Royal College of Art in London. His early work falls into the category of social realism and shows a fascination with the squalid and seedy. *Harlem* (1934), now in the Tate Gallery, London, is an example. Burra rarely left his home in Rye, Sussex, using postcards and photographs on which to base many of his paintings. He painted mainly in watercolour, using hard outlines and simply modelled forms. In the mid-1930s social context became less important in his work than grotesque and bizarre subject matter, as in *Dancing Skeletons* (1934). Skeletons and birdmen became favourite images. Some of his paintings leant towards surrealism while others, such as *Christ Mocked*, were of a religious nature. In the 1950s and 1960s he also produced landscapes of a mysterious and menacing nature.

Burroughs, William S(eward) (1914–   ) *US novelist associated with the Beat group of writers of the 1950s.*

Born in St Louis, Missouri, Burroughs was the grandson of (and bears the same name as) the inventor of the adding machine. He studied at Harvard, served briefly in World War II, then worked at various jobs in New York. In the 1940s his apartment near Columbia University was the meeting place of Allen GINSBERG, Jack KEROUAC, and others who became the leading spirits of the Beat Generation. In 1944 Burroughs became addicted to heroin. He moved eventually to Mexico, but left after accidentally shooting and killing his wife. He travelled in South America and lived for a time in Tangiers. After many failed attempts to cure his addiction, he submitted to an apomorphine treatment in London in 1957, which proved successful.

Burroughs's writings are highly controversial, with obsessive features that, for many readers, prevent his books from qualifying as strictly literary works. From the first, in *Junkie* (1964; written in Mexico and first published under a pseudonym, 1953), a relatively straightforward record, he was concerned with the experience of drug addiction. In *The Naked Lunch* (1959), which established his reputation, plot, character, and the usual requirements of fiction play no part. Instead, the condition of the addict is presented in nightmarish fragments of farcical black humour and

sadomasochism (mass hangings, sodomy, etc.). Addiction in his later novels is portrayed as an external enemy imposing itself by force or as a virus seeking total control. The conflict plunges into the past or occurs as science-fiction wars in outer space in *The Soft Machine* (1961) and *Nova Express* (1964) and is further developed, with variations, in his other novels, including *The Wild Boys* (1971), *The Place of Dead Roads* (1984), and *The Western Lands* (1988). *The Yage Letters* (1965) is a correspondence with Allen Ginsberg.

Burton, Richard (Richard Jenkins; 1925–84) *British actor, who died an alcoholic in Switzerland, after a highly publicized and successful career on stage and screen.*

Born in Pontrhydfen, South Wales, Burton, a miner's son, adopted the name of one of his schoolmasters, who guided his acting career, helping him to lose his Welsh accent and to develop his full rich voice. He attended Exeter College, Oxford, and began his career in *Druid's Rest* (1943). After serving with the RAF (1944–47) he returned to the stage, receiving critical praise for *The Lady's Not for Burning* (1949). His performance as Prince Hal in *Henry IV* (Stratford, 1951) brought further acclaim and he became the leading member of a new generation of outstanding actors. He had, however, by then made his first film, *The Last Days of Dolwyn* (1948); despite successful seasons with the Old Vic during the 1950s, films and Hollywood proved the greater attraction.

My Cousin Rachel (1952), his first American film, brought him an Oscar nomination as Best Supporting Actor. He was nominated as Best Actor for *The Robe* (1953), *Becket* (1964), *The Spy Who Came in from the Cold* (1965), *Who's Afraid of Virginia Woolf?* (1966), and *Anne of a Thousand Days* (1970) although he never actually received an Oscar. Perhaps the best-known of his British films were John OSBORNE's *Look Back in Anger* (1959), Dylan THOMAS's *Under Milk Wood* (1971), and Peter Shaffer's *Equus* (1977).

Burton's private life received enormous publicity. He made several films with the most famous of his four wives, Elizabeth TAYLOR (to whom he was married twice; 1964–70, 1975–76), including the extravagant *Cleopatra* (1962), and starred with her in his last stage appearance, *Private Lives* (1983).

Bush, Alan Dudley (1900–   ) *British composer and teacher, whose strong political beliefs influenced his music.*

Born in London, he studied at the Royal Academy of Music with Corder (composition) and Matthay (piano), winning numerous prizes. He also studied composition privately with John IRELAND, piano with Benno Moiseivich (1890–1963) and Artur Schnabel (1882–1951). and philosophy and musicology at Berlin University (1929–31). In 1925 he was appointed professor of composition at the Royal Academy of Music; in 1936 he founded the Workers' Music Association. After army service in World War II (1941–45), he toured eastern Europe as conductor, formed the London String Orchestra, and published a textbook, *Strict Counterpoint in Palestrina Style* (1948).

Bush's political affiliation with Marxist communism is reflected in his music, both in the texts of his vocal works, such as the opera *Wat Tyler* (1950, the tyranny of oppression) and the choral work *The Winter Journey* (1946, infiltration of politics into the Nativity story), and also in the self-imposed simplification of his own style after 1945 in line with Soviet criticism of SHOSTAKOVITCH and PROKOFIEV. As a composer, his method is based on a contrapuntal use of melodic material, somewhat akin to SCHOENBERG's use of the tone row, although Bush's music remains firmly diatonic.

Bush, George Herbert Walker (1924–)
US Republican statesman; forty-first president of the USA (1989–).
Born into a wealthy political family, Bush attended private schools in New England before World War II. From 1942 he served as a naval carrier pilot in the Pacific, winning the Distinguished Flying Cross. After the war he studied economics at Yale University and worked in the Texas oil industry, becoming president of his own company in 1956. Bush became active in Republican politics in the 1960s and was a member of the House of Representatives from 1967. In the 1970s he served as US ambassador to the UN (1971–72), chairman of the Republican National Committee (1972–73), chief of the US liaison office in Peking (1974–76), and director of the CIA (1976–77).

In 1980 Bush campaigned for the Republican nomination to the presidency, later withdrawing to support the eventual winner, Ronald REAGAN, who chose Bush as his running mate. Following Reagan's victory at the polls, Bush became vice-president in 1981, a position he retained during Reagan's second term of office (1984–88). Despite doubts about his populist appeal, Bush secured the Republi-

can nomination in 1988 and enjoyed a comfortable victory over his Democrat opponent Michael Dukakis (1933–).

As president, Bush has sought to consolidate the policies of the Reagan era, while presenting a more conciliatory image. In 1990 the ever-rising federal budget deficit obliged him to raise taxes, despite electoral promises that he would never do so. A 'crusade' to stem the growth of drug-related crime in US cities made little impact. As expected, Bush's main interest has been in foreign affairs. Following the collapse of communism in eastern Europe he negotiated further arms reductions with the Soviet Union (including unprecedented cuts in nuclear arms) and saw the emergence of the USA as the world's only superpower. In 1989 he ordered US troops into Panama to oust the dictator General Manuel Noriega (1938–), who was brought to the USA to face charges of drugs trafficking. Bush responded to Saddam HUSSEIN's invasion of Kuwait in 1990 by forming an international alliance with the backing of the United Nations against Iraq and ordering a massive military build-up in the region. When Iraq remained defiant, he ordered the US-led multinational force into action in 1991. The swift liberation of Kuwait, with minimal allied losses, was seen as a personal triumph for Bush and a graphic demonstration of the USA's self-imposed role as the world's policeman. The failure of the allied forces to secure the overthrow of Saddam, however, led to continued tension in the region, notably over the issues of Iraqi persecution of Kurdish people living in Iraq and Iraq's reluctance to assist in the inspection of its arsenals by UN officials. At the same time, the USA's economic problems began to escalate and Bush was accused of neglecting the USA's domestic affairs.

Busoni, Ferruccio Benvenuto (1866–1924)
Italian pianist and composer, whose ideas greatly influenced subsequent composers.
Born in Empoli, Busoni was taught by his mother and was soon able to give concerts (Hanslick praised his performance in Vienna at the age of nine). Without the benefit of well-known teachers, he developed into a virtuoso, touring Europe and the USA. However, his interest in composition took him to Leipzig Conservatory (1886), where he was a fellow student of DELIUS. Recognition as a composer was slower, and it was not until he had settled in Berlin after World War I that his works were publicly acclaimed.

As a composer, Busoni developed a monumental style; two examples from his large output are the *Fantasia contrappuntistica* (1910–12, three versions) for one and two pianos, an essay in contrapuntal style based on an unfinished fugue by Bach; and *Doktor Faust* (1916–24), his fourth and unfinished opera, an expansive retreatment of the Faust legend, eventually completed by his pupil Philipp Jarnach (1892–1982). However, he achieved less success with his compositions than with his ideas. His experiments with scales (other than major and minor) and with microtones, together with the fascinating contrapuntal textures he conceived, stimulated a whole new generation of composers. They are set out in his book *Towards a New Aesthetic in Music* (1911).

Butler, Reg(inald) (1913–81) *British metal sculptor.*

Butler was born in Hertfordshire, which remained his home between periods elsewhere. He first trained as an architect and from 1937–39 taught at the Architectural Association School, after which he worked briefly as an engineer. As a conscientious objector during World War II he was sent to work in a smithy in Sussex. When Butler took up sculpture in 1944, his technical experiments led him to work with metal. While working as an editor for architectural journals, he produced constructivist tower shapes and female figures, both of iron. His second phase consisted of open works made up of metal wire representing hybrid insect-like creatures. From 1950 to 1953 he studied sculpture full-time at Leeds University and in 1953 won the international competition for a monument to *The Unknown Political Prisoner*. He then taught in Hertfordshire and later at the Slade School, London. Meanwhile in 1954 he had abandoned iron for bronze. Although he often placed his metal female figures in a constructivist framework of bars these works retained a rich sensuality.

Butler, R(ichard) A(usten), Baron (1902–82) *British Conservative politician, who, contrary to expectations, failed to become prime minister. He was created a life peer in 1965.*

Butler was born in India, where his father worked in the Indian Civil Service. Entering parliament in 1929 as the member for Saffron Walden, he served as undersecretary of state for India (1932–37) and was largely responsible for the Government of India Act 1935, which gave a measure of self-government to India. In the 1930s, Butler crossed swords with CHURCHILL, who appointed him minister of ed-

ucation in 1941, with the intention of allowing him to stagnate in a backwater. However, Butler made a considerable impression with his Education Act 1944, which provided for the school-leaving age to be raised to fifteen (and then sixteen) and made free secondary education available for all.

After the war, he headed a new research department at the Conservative Central Office, shaping a more progressive image for the party. As chancellor of the exchequer (1951–55), he was able, because of the country's relative prosperity, to introduce a series of lenient budgets. He was then simultaneously lord privy seal (until 1959) and leader of the House of Commons (until 1961). His opposition, stated within the party and not publicly, to Anthony EDEN's Suez adventure lost him the leadership in 1957 to Harold MACMILLAN. He failed again to win the top post in 1963, when he was defeated by Sir Alec Douglas-Home (1903–). He served under both men, first as home secretary (1957–62), introducing some notable prison reforms, and then as foreign secretary (1963–64). He was master of Trinity College, Cambridge, from 1965 to 1978. His memoirs, *The Art of the Possible*, were published in 1971.

Butt, Dame Clara (1873–1936) *British contralto. She was created a DBE in 1920.*

Born at Southwick in Sussex, she studied with Daniel Rootham in Bristol and with J. H. Blower at the Royal College of Music, having gained a scholarship there in 1889. Butt made her debut at the Albert Hall singing Ursula in Sullivan's *Golden Legend* (1892): her beautiful voice and stately presence assured her a successful career. She developed her own type of ballad concert, which she performed in all parts of the British Empire. She was popular at festivals, her first engagements being at Bristol and Hanley (1893); ELGAR wrote his *Sea Pictures* for her for the Norwich Festival of 1899. In 1900 Butt married the baritone Kennerley Rumford (1870–1957), from which time they pursued their careers jointly. She sang Gluck's *Orfeo* at Covent Garden in a notable revival in 1920.

Byrd, Richard Evelyn (1888–1957) *US aviator and explorer who made the first flight over the South Pole and who later headed US expeditions to survey Antarctica.*

Byrd was commissioned as an ensign in the US navy in 1912 and in World War I served in a naval air squadron based in Canada. On 9 May 1925, Byrd, acting as navigator to Floyd Bennett, claimed to have made the first flight over the North Pole on a return trip from King's Bay, Spitsbergen. He was promoted to commander and awarded a Congressional medal of honour in recognition of this achievement, although doubts over Byrd's claim have since emerged. Following this, in 1927, he and three companions flew the direct transatlantic route between New York and France, for which Byrd received the Distinguished Flying Cross and was made a Commandeur de la Légion d'honneur. The ensuing fame enabled him, in 1928, to obtain financial backing for an expedition to Antarctica. This set out in October and established a base camp, called Little America, near the Bay of Whales. Aerial reconnaissance revealed hitherto unknown fea-tures, including Rockefeller Plateau, named after one of Byrd's sponsors, and Marie Byrd Land, named after his wife. On 29 November 1929, Byrd and three crew members made the first flight over the South Pole. Shortly afterwards, Byrd was promoted to rear-admiral.

On a second Antarctic expedition (1933–35), Byrd spent five months alone in a weather station hut on the Ross Ice Shelf enduring temperatures as low as –60°C. His book, *Alone* (1938), details this experience. Byrd conducted further surveys in Antarctica and during World War II served on US naval staff. In 1955 he was appointed head of Operation Deepfreeze, one of the US contributions to International Geophysical Year (1957–58).

C

Cadbury, George (1839–1922) *British industrialist, enlightened social reformer, and philanthropist.*

Cadbury was the son of John Cadbury (1801–89), who had founded a cocoa and chocolate manufacturing business in Birmingham. George joined the family concern in 1856 and five years later assumed control in partnership with his elder brother, Richard. A committed Quaker throughout his life, George Cadbury began teaching at adult education classes in Birmingham in 1859 and continued to do so into his seventies. This experience enabled him to understand the grim conditions then prevailing for factory workers. In 1879 Cadbury Brothers moved to a new factory on a rural site outside Birmingham, where George introduced sweeping measures to improve the welfare and education of the workforce. In the 1890s he purchased land surrounding the factory and commissioned architect W. Alexander Harvey to build a company housing estate with large gardens and generous amenities. The Bournville Village Trust was founded in 1900 and now administers various housing developments, not exclusively company housing.

Following his brother's death in 1899, George Cadbury became head of the company. As a Liberal in politics, he acquired in 1901 the *Daily News* (later the *News Chronicle*) to provide a platform for his ideas. In 1903 he donated his former home, Woodbrooke, Birmingham, to the Society of Friends as a residential centre.

Cage, John (1912–) *US composer and writer. His experimental and aleatoric compositions have aroused considerable interest and some amusement.*

The son of an inventor, Cage was born in Los Angeles but travelled extensively in Europe (1930–31), studying music, art, architecture, and poetry. On his return to America he took courses in composition, counterpoint (with Arnold SCHOENBERG), and folk, contemporary, and non-Western music (with Henry Cowell (1897–1965)). In 1938 he began composing for percussion groups; a concert of percussion music sponsored by the League of Composers at the Museum of Modern Art in New York

(1943) included three of his compositions and focused public attention on him. He was then engaged by the Merce Cunningham Dance Company as a composer and later as an accompanist. His next experiments were with the 'prepared piano', in which a number of objects are inserted into the instrument, making a wholly percussive source of various noises. His *Metamorphosis for Prepared Piano* (1938) celebrated this phase of his life. Work with audio-frequency oscillations, variable-speed turntables, and sound-effects of various kinds led to such pieces as *Imaginary Landscape no 3* (1942).

In the 1950s Cage was studying Zen Buddhism and experimenting with aleatoric composition, using the diagrams of the *I Ching* (the Chinese *Book of Changes*) and tossing three coins to determine the pitch, duration, and timbre of each note, as in *Music of Changes* (1951). Increasingly the element of chance came into Cage's music: *Music for Piano 1* (1952) is notated entirely in semibreves and performance is determined by imperfections in the paper on which it is written. Such random concepts led to *4'33"* (1952), in which performer(s) come on to the platform, sit at their instruments for the required length of time, bow, and retire, the 'music' being whatever sounds emerge during the four minutes and thirty-three seconds. *Water Music* (1952) requires the pianist to pour water, blow whistles under water, and perform various other actions. Cage's *Musicircus* (1967) is simply a gathering of artists in different media and from different backgrounds, combining in a totally unstructured event. More recent works are *Thirty Pieces for Five Orchestras* (1981) and *Europeras 3 and 4* (1990).

With an enormous number of compositions and writings behind him, Cage explains his attitude somewhat gnomically as a determination to break distinctions between art and life: 'Everything we do is music'.

Cagney, James (1899–1986) *US film actor, who made his name in gangster roles in the 1930s.*

With his wife Frances, Cagney began a successful career as a vaudeville song and dance act on Broadway. After appearing in shows, such as *Penny Arcade* (1929) with Joan Blondell (1909–79), he made his film debut in *Sinner's Holiday* (1930). Stardom came the following year with *The Public Enemy* (1931), the first of his gangster films. Small, stocky, and gentle, Cagney seemed an unlikely candidate for fame as a tough hoodlum, but his staccato delivery and cocky cheerfulness in such films as *Angels With Dirty Faces* (1938) and *White Heat* (1944) brought him stardom the world over.

A skilled dancer and comic, Cagney was by no means limited to gangster roles. The extent of his range was exemplified by such films as *A Midsummer Night's Dream* (1935), playing Bottom, the musical *Yankee Doodle Dandy* (1942), for which he received an Oscar, *Shake Hands with the Devil* (1959), and his last film, *One Two Three* (1961). He directed one film, *Short Cut to Hell* (1957), a remake of *This Gun For Hire* (1942). *Cagney by Cagney* (1975) is the title of his autobiography.

Caillaux, Joseph (1863–1944) *French statesman; prime minister (1911–12) and five times minister of finance.*
Born in Le Mans, the son of a former government minister, Caillaux was educated at the Lycée Condorcet in Paris before entering the civil service as an inspector of finance in 1882.

Caillaux was first elected as a radical-socialist deputy to the French chamber in 1898 and was appointed finance minister (1899–1902; 1906–09). In 1911, as prime minister, he settled the Agadir (Morocco) crisis, but criticism over his ceding the French Congo to the German protectorate of Kamerun led to his resignation in 1912. He was reappointed finance minister in 1913. Following a scandal in 1914, in which his wife shot the editor of *Le Figaro* for publishing a number of their personal letters (she was later acquitted), Caillaux was forced to resign. During World War I he was sent on economic missions to South Africa and Italy. His advocacy of a negotiated peace led to his arrest in 1918 for 'communication with the enemy'. Imprisoned until 1924, he had his political rights restored in 1925, enabling him to be elected to the senate. He again served as finance minister in 1925–26 and 1935. He was instrumental in the overthrow of BLUM's Popular Front government in 1937, later supporting PÉTAIN as leader of the Vichy regime in 1940. He remained a member of the senate until his death.

Caillaux's memoirs, in three volumes, were published between 1942 and 1948.

Calder, Alexander (1898–1976) *US sculptor and painter, who was one of the first artists to introduce movement into sculpture through his invention of mobiles.*
Although his grandfather and his father were both sculptors and his mother was a painter, Calder studied engineering (1915–19) and tried a variety of jobs before turning to art studies in 1923. In the mid-1920s he began to exhibit paintings and to produce wire sculptures. The miniature circus he created during his stay in Paris (1926–27) and the performances that he gave with it in his studio are typical of the humour and fantasy that went into many of his works.

By 1931, when Calder joined the Abstraction-Création association, the abstract paintings of MONDRIAN had begun to have a strong effect on his work. He wrote of his desire to make 'moving Mondrians' and created abstract constructions of brightly coloured flat metal shapes on wires, which were set in motion by hand or by small motors. After 1934 these 'mobiles', as DUCHAMP named them, tended to be free-moving, responding unpredictably to surrounding air currents. Calder described these and his nonmoving 'stabiles' as 'four-dimensional drawings'.

In later years many of his constructions were on a grander scale, such as the *Red Sun* (1967) for the Olympic stadium in Mexico, which was 24 metres high.

Callaghan, (Leonard) James, Baron (1912–) *British politician and Labour prime minister (1976–79). He was created Baron Callaghan of Cardiff in 1987.*
The son of a petty officer, Callaghan was a tax officer before becoming a full-time employee of the Trades Union Congress. He served in the navy in World War II and was elected to parliament as member for South Cardiff in 1945. He became chancellor of the exchequer in 1964, resigning in 1967, when the prime minister, Harold WILSON, insisted on the devaluation of the pound. He was then home secretary (1967–70) and, in Wilson's last government, foreign secretary (1974–76). As prime minister, Callaghan had a far more affable and open manner than his predecessor, which impressed the country and the party, and he successfully steered his minority government through a first year of extreme economic difficulty. In 1977 he was forced to negotiate a pact with the Liberal Party to ensure his government's survival, but ultimately he failed to gain either party or union support for his anti-inflation policies. In 1979 he received a vote of no

confidence in the House of Commons, the first prime minister to do so since Ramsay MAC-DONALD in 1924. The party was defeated in the subsequent election, and in the following year Callaghan was replaced as leader by Michael FOOT. He published an autobiography, *Time and Chance*, in 1987.

Callas, Maria (Maria Cecilia Sophia Anna Kalogeropoulou; 1923–77) *Greek soprano born in the USA. During a dazzling career her voice and her unusual acting ability made her the* prima donna assoluta *of the 1950s and 1960s.*
She left the USA in 1937 to study at the Athens Conservatory with Elvira de Hidalgo. In 1945 she returned to America and subsequently (1947) made a successful appearance at the Verona Arena as La Gioconda in Ponchielli's opera. This led to performances in the widely varying roles of Aida, Turandot, Isolde, Kundry, Brünnhilde, and Elvira (in *I Puritani*). Gradually, under the guidance of the conductor Serafin, she dropped the heavier roles and concentrated on earlier Italian opera and Verdi; she repeated the success of her debut at Covent Garden, singing Norma in Bellini's opera of that name (1952), with performances in Chicago and New York.
Musically and artistically Callas was a perfectionist. Her voice was strong, brilliant, and exciting but by the middle 1960s an increasing inequality of register and tremolo led to her retirement from the stage, her last Covent Garden performance being in *Tosca* in 1965. She gave masterclasses at the Juilliard School, New York (1971–72), and produced Verdi's *Sicilian Vespers* jointly with Giuseppe di Stefano (1973). She emerged from her retirement for an extensive world concert tour with di Stefano, which confirmed her undiminished artistic powers but revealed a voice past its prime.

Calles, Plutarco Elías (1877–1945) *Mexican general and president (1924–28). A dictatorial leader, he was overthrown by Lázaro CÁRDENAS.*
Born in Guaymas, Calles became a schoolteacher in 1894. In 1910 he joined the revolutionary movement of Francisco Madero (1873–1913), which overthrew Porfirio Díaz (1830–1915) in 1911. He was elevated to the rank of general after Madero was deposed in 1913 and served as governor of Sonora (1915–19).
In 1919 Calles was appointed secretary of commerce, labour, and industry under President Carranza (1859–1920). He resigned in 1920 to support Alvaro Obregón (1880–1928), who came to power later that year and under whom he served as secretary of the interior (1920–24). Calles was elected president in 1924 but after his resignation in 1928 and the assassination of the president-elect (Obregón), he remained the power behind the scenes, supporting a succession of puppet leaders. His power base lay in the Partido Revolucionario Institutional (PRI), which he founded in 1929. In 1935 he broke with President Cárdenas (whom he had supported) and was forced into exile. He remained in the USA until 1942, when he returned to Mexico.
Calles has been described as a skilled organizer, who transformed the revolutionary movement into a professional army. As president he ran into opposition from the Catholic Church over his prohibition of church schools and his limiting of the number of clergy; he was also unpopular with foreign oil companies as a result of his restrictions on land ownership and his regulation of the industry.

Campbell, Donald Malcolm (1921–67) *British holder of world speed records on land and water. He was made a CBE in 1957.*
The son of Sir Malcolm CAMPBELL, Donald started as an apprentice engineer and became the director of an engineering firm. Inspired by his father's record-breaking ventures, he made his first attempt on the water speed record in his father's *Bluebird* in 1949 on Coniston Water. Two years later it sank at the same site and Campbell was rescued. 1955 saw the launching on Ullswater of a new *Bluebird*, a jet-powered hydroplane, which in July captured the record from the USA with two runs averaging 202.32 mph. Campbell increased this to 216.2 mph later that year on Lake Mead, Nevada, and by May 1959, Campbell's record stood at 260.35 mph.
Meanwhile, Campbell's plans to break the land speed record came to fruition and by 1961 his *Bluebird* car, designed by the Norris brothers and powered by a Bristol-Siddeley Proteus turbine, was ready for testing. In 1961 on the Bonneville salt flats, Utah, Campbell crashed at 365 mph but sustained only a hairline fracture of the skull. A replacement *Bluebird* was ready in 1963 but bad weather delayed operations until 1964, when Campbell finally succeeded in breaking the longstanding record held by John Cobb (1899–1952), reaching a speed of 403.1 mph. But this soon fell to the new generation of jet-powered cars from the USA. Campbell died on Lake Coniston while attempting to break his own water speed record

of 276.33 mph. In 1984 his daughter Gina Campbell broke the women's water speed record in her boat *Agfa Bluebird II*.

Campbell, Sir Malcolm (1885–1948) *British motor racing driver who held both land and water speed records. He received a knighthood in 1931 in recognition of his achievements.*

The son of a watchmaker and jeweller, Campbell joined Lloyds of London and became a successful underwriter, enabling him to finance his hobby – racing cars and motor cycles. He first raced at Brooklands in 1908 and the first in a long series of cars named *Bluebird* appeared at Brooklands in 1910. After service in the Royal Flying Corps during World War I, Campbell continued racing with great success. In 1925 he made the first of his record-breaking speed runs, reaching over 150 mph in his Sunbeam. In 1931 he reached 246.09 mph and in 1932, in his *Bluebird* powered by a Napier Lion aeroengine, became the first driver to reach 250 mph. Three years later, in yet another *Bluebird*, he broke the 300 mph barrier at the Bonneville salt flats, Utah. Campbell turned his attentions to the water speed record and his speed of 141.74 mph established in 1939 earned him the Seagrave trophy.

Campbell stood unsuccessfully as a Conservative candidate for Deptford in 1935 and during World War II he served on the staff of combined operations. He wrote *My Greatest Adventure: Speed* (1931) and *The Romance of Motor Racing* (1936). The family tradition of record-breaking was continued by his son, Donald CAMPBELL, and his grand-daughter Gina Campbell.

Campbell, Mrs Patrick (Beatrice Stella Tanner; 1865–1940) *British actress.*

Beatrice Tanner was born in London, daughter of John Tanner and his Italian wife. In 1894 she eloped with Patrick Campbell, who died fighting in the Boer War in 1900 (their son also died in action, during World War I). She married her second husband, George Cornwallis West, in 1914.

A fine actress, Mrs Campbell was equally famous for her wit, evidenced in correspondence with her friend George Bernard SHAW, which has been made the subject of many plays and books. She made her stage debut in *Bachelors* (1888) at the Alexandra Theatre, Liverpool, and her first London appearance at the Adelphi as Helen in *The Hunchback* (1890). Her real break, however, came at St James's Theatre as Paula Tanqueray in Pinero's *The Second Mrs Tanqueray* (1893). During the re-

mainder of the decade she appeared in numerous plays, ranging from Shakespeare to Ibsen, most notably as Mélisande in *Pelléas and Mélisande* at the Prince of Wales's Theatre. This success was repeated with even greater acclaim when the play was performed in French at the Vaudeville (1904) with Sarah Bernhardt (1845–1923) as Pelléas. Mrs Campbell also created the role of Eliza Doolittle in Shaw's *Pygmalion* (1914). Later she toured the USA and Britain and appeared in minor film roles. Her daughter, Stella Patrick Campbell, also became an actress.

Campbell, (Ignatius) Roy(ston Dunnachie) (1901–57) *South African poet.*

Campbell was brought up in Durban, a boyhood lovingly described in *The Mamba's Precipice* (1953). He was sent to Oxford (1919) but considered that study interfered with poetry, reading, and writing. *The Flaming Terrapin* (1924), his poetic and spiritual manifesto, created a stir, but when Campbell returned to Natal and co-founded the magazine *Voorslag*, with William PLOMER and Laurens VAN DER POST, he found his aspirations thwarted by his fellow countrymen. Returning to England in disgust, Campbell published *The Wayzgoose* (1928), a long poem satirizing the South African cultural scene.

In 1928 he moved to Provence, where the strenuous physical life suited his taste better than the London literary milieu, which he satirized in *The Georgiad* (1931). *Adamastor* (1930) and *Poems* (1930), which contain some of his finest lyrics, were followed by *Flowering Reeds* (1933). In 1933 Campbell moved to Spain, where his right-wing politics drew him to fight on FRANCO's side in the civil war. His poems *Mithraic Emblems* (1936) were coldly received by English liberals, thus provoking the immoderate outburst of *Flowering Rifle* (1939), which was virulently pro-Franco. During World War II Campbell served in the British army in Africa but was invalided out in 1944. From 1946 to 1949 he worked for the BBC.

In 1952 Campbell moved to Portugal. Some of his best works in this period were translations of French, Spanish, and Portuguese poetry. In 1951 his second volume of autobiography, *Light on a Dark Horse*, was published; like the first (*Broken Record*, 1934), it was characterized by Campbell's love for violent and dangerous physical activities and his enthusiasm for flamboyant anecdotes. In 1957 he was killed in a car accident in Portugal.

Campbell-Bannerman, Sir Henry (1836–1908) *British statesman and Liberal prime minister (1905–08), responsible for reuniting the party after it was split by the Boer War.*

Born in Glasgow, Campbell-Bannerman became an MP in 1868 and served under Gladstone as financial secretary to the War Office (1871–74, 1880–82), parliamentary and financial secretary to the Admiralty (1882–84), and chief secretary for Ireland (1884–85). As secretary for war in 1886 (a post he held again in 1892–95), he removed the main obstacle to reform of the armed forces by persuading the commander-in-chief, the Duke of Cambridge (the queen's cousin), to resign. He became Liberal leader in the Commons in 1899, and came into conflict with the Liberal imperialists by his condemnation of the British concentration camps in South Africa. He later healed the wounds in the party and, succeeding BALFOUR as prime minister in 1905, led the Liberals to a landslide victory in the 1906 election.

Campbell-Bannerman put together a brilliant cabinet, which included ASQUITH, LLOYD GEORGE, HALDANE, and CHURCHILL. Though the House of Lords blocked much of the Liberals' proposed legislation, the government passed some important reforms – the Trade Disputes Act 1906, the Merchant Shipping Act 1907, and the Patents Act 1907. It also gave self-government to the Transvaal (1907) and the Orange River Colony (1907). Ill-health forced Campbell-Bannerman to resign in 1908, and he died seventeen days later.

Camus, Albert (1913–60) *French novelist, essayist, and dramatist. He was awarded the Nobel Prize for Literature in 1957.*

Camus was born in Mondovi, Algeria. His father, a farm labourer, was killed in World War I, leaving his wife to bring up Albert and his elder brother on her meagre income as a charwoman. Despite the deprivations of his childhood, Camus won a scholarship to the lycée in Algiers and went on to study philosophy at the university there. Prevented by tuberculosis from pursuing an academic career, he followed a variety of occupations, including those of journalist with the *Alger-Républicain* and amateur theatrical director at the left-wing Théâtre du Travail. He was to retain a patriotic bond with his native North Africa throughout his life; the strength of this attachment is revealed in his early essays *L'Envers et l'endroit* (1937) and *Noces* (1938).

In France during World War II, Camus rose rapidly to fame on the publication of his essay *Le Mythe de Sisyphe* (1942; translated as *The Myth of Sisyphus*, 1955) and his first novel *L'Étranger* (1942; translated as *The Outsider*, 1946), both of which convey his conception of the absurdity of human existence. In *L'Étranger*, a young man who has killed an Arab is condemned to death, apparently for his refusal to conform with bourgeois society rather than for the murder itself. Camus joined the French Resistance in 1942 and edited the movement's journal, *Combat* (1944–47). His experiences at this time provided the basis for his next novel, *La Peste* (1947; translated as *The Plague*, 1948). Set in Oran, North Africa, during a plague epidemic, it is an allegory of the German occupation of France and the Resistance movement. In *L'Homme révolté* (1951; translated as *The Rebel*, 1953), Camus discussed in philosophical terms the ideology of revolution, denouncing communism for more humanistic ideals; this led to a fierce public dispute with Jean-Paul SARTRE in 1952. Camus's humanism led him also to write a condemnation of the death penalty, in collaboration with Arthur KOESTLER.

Camus never lost his early interest in the theatre. His plays include *Le Malentendu* (1944; translated as *Cross Purpose*, 1948), *Caligula* (1945; translated in 1948), and *Les Justes* (1950). He also adapted for the theatre William FAULKNER's *Requiem for a Nun* (1956) and Dostoevski's *The Possessed* (1959). His other works include the semi-autobiographical *La Chute* (1956; translated as *The Fall*, 1957), the collection of short stories *L'Exil et le royaume* (1957; translated as *Exile and the Kingdom*, 1958), and three volumes of his journalistic articles, *Actuelles* (1950–58). Since his premature death in a car accident at the age of forty-six, three volumes of his *Carnets* have been published (1962, 1964, and 1966), covering the years from 1935 to 1959.

Candela (Outerino), Felix (1910–) *Spanish-born engineer and architect. He specializes in reinforced concrete shell structures, including the spectacular sports palace for the Mexico Olympics in 1968.*

Born in Madrid, Candela was educated at the University of Madrid and the architects' school there. He was a captain in the Engineers in the Spanish republican army during the Spanish civil war (1936–39), and emigrated to Mexico in 1939. In 1941 he became a Mexican citizen and in 1978 a citizen of the USA. He has been professor at the Escuela Nacional de Arquitectura, University of Mexico, since 1953 (now on leave). His two most interesting buildings, both in Mexico, are the Church of the Miraculous Virgin (1953–55) in Mexico City and the market hall at Coyacán (1956). He used similar umbrella-shaped roofs for the John Lewis warehouse at Stevenage (1963) in England.

Capa, Robert (André Friedmann; 1913–54) *Hungarian photographer, who set new precedents in the coverage of war and its aftermath.*
Capa started work as a photojournalist based in Berlin. His first published photograph was of Leon TROTSKY, taken at a meeting in Copenhagen in 1931. Moving to Paris, he invented a US photographer, 'Robert Capa', as a ploy to boost the price of his photographs and subsequently adopted the name as his pseudonym. In 1936, while covering the Spanish civil war, Capa took one of his most memorable pictures – a loyalist soldier reeling backwards at the instant of death. Capa was the first to record so vividly and at such close quarters the full horror of war and its consequences. He witnessed the Japanese invasion of China in 1938, and during World War II covered London during the Blitz, the fighting in North Africa, and the invasion of Italy. Above all, he accompanied the first wave of US assault troops onto the Normandy beaches and moved with the Allied front to record the liberation of Paris and the closing stages of the war.

In 1947, Capa, together with Henri CARTIER-BRESSON, David Seymour (1911–56), and George Rodger, founded Magnum Photos, a cooperative freelance photographic agency, through which Capa actively promoted the work of young photographers. Capa covered the Arab-Israeli conflict in Palestine during the late 1940s, but most of his attention was now devoted to the agency. In 1954 *Life* magazine commissioned him to photograph the war in Vietnam. Capa was killed by a land mine while accompanying the French forces. He was posthumously awarded the Croix de Guerre with Palm, the highest French military honour. His books include *Slightly Out of Focus* (1947) and *Images of War* (1964).

Capablanca y Graupera, José Raul (1888–1942) *Cuban chess player who dominated world chess in the 1920s. He made a considerable impact on the game, particularly opening theory.*
Taught chess at the age of four by his father, Capablanca played very successfully in Havana before going to Columbia University, New York. In 1911 he caused a surprise by defeating Bernstein and Nimzowitch, two of the world's top players. After graduating, Capablanca joined the diplomatic corps as a

commercial attaché, which enabled him to travel extensively. His career reached a peak in 1914, when he came second to the reigning world champion, Emanuel LASKER.

Over the next six years Capablanca played and beat many of the top masters. In 1921 he won the world championship from Lasker, who had held it for twenty-seven years, and was himself champion for six years before losing the title to Alexander Alekhine (1892–1946) who, contrary to practice, did not allow Capablanca a re-match.

Capablanca died in the Manhattan Chess Club in March 1942. He was so highly thought of by the world's chess community that a memorial tournament was inaugurated in 1962 in Havana, Cuba.

Čapek, Karel (1890–1938) *Czech playwright, novelist, and essayist.*
Born at Malé Svatoňovice, Bohemia, the son of a doctor, Čapek was educated at the University of Prague and later studied at Berlin and Paris. He first pursued an interest in science, which is reflected in the scientific fantasies of his best-known work, but later shifted to philosophy, which led to a close friendship with Tomáš MASARYK, a professor of philosophy who became first president of the Czech republic. He was deeply influenced by British and American writing, especially by William James's pragmatism and the science fiction of H. G. WELLS. He married the actress Olga Scheinpflugová and was responsible for bringing modern theatre to Czechoslovakia, first as a director of the National Art Theatre and subsequently at his own Vinohradsky Art Theatre, where he produced the works of the major European playwrights and introduced the plays of young native dramatists. Čapek's career coincided with the life of the Czechoslovak republic, from its inception after the Treaty of Versailles to its demise after the Munich Conference, an event that left Čapek in despair. (He died shortly afterwards of pneumonia). Totally committed to the liberal ideals of the new state, Čapek divided his day between his own writing (in the mornings) and his journalism for *Lidove Noviny*, the newspaper speaking for the government of Masaryk and BENEŠ (in the afternoons). In the evenings his house served as the meeting place for Prague intellectuals.

The play *R.U.R.* (1921; the abbreviation is for 'Rossum's Universal Robots'), a cautionary drama about the dangers of mechanization, introduced the coinage 'robot' to the world and was an international success. His other plays include *Ze života hmyzu* (1921; translated as *The Insect Play*, 1923), *Věc Makropulos* (1922; translated as *The Makropoulos Secret*, 1925), the inspiration for JANAČEK's opera (1926) of the same name, and *Matka* (1938; translated as *The Mother*, 1939). His numerous prose works include the scientific fantasies *Továrna na absolutno* (1922; translated as *The Absolute at Large*, 1927) and *Krakatit* (1924; translated 1925); a trilogy of philosophical novels, probably his best work, entitled *Hordubal* (1933), *Provětroň* (1934; translated as *Meteor*, 1935), and *Obyčejný život* (1935; translated as *An Ordinary Life*, 1936); and several travel books. His conversations with Masaryk (3 vols, 1928–35; translated 1934, 1938) form an important biographical and political record.

Capone, Al(phonse) (1899–1947) *US gangster who built up a notoriously successful criminal empire in Chicago during the 1920s.*
Capone was born in Naples but grew up in Brooklyn, New York. A member of the infamous Five Points gang, his special talents came to the attention of Chicago vice king John Torrio, whom he joined in 1919. Capone, employed as bodyguard to the leading Mafioso James Colosimo, was known as 'Scarface' because of a scar on his left cheek – the legacy of a gangland brawl.

Prohibition had opened up vast new opportunities for Chicago criminals and Capone took full advantage of them, controlling illicit stills, organizing liquor distribution, and bribing police and politicians. In 1924, Torrio was badly wounded in a gangland attack and Capone took over as head of Torrio's organization. He ruthlessly extended his 'business', buying off or eliminating competition from rival Chicago gangs. His earnings from liquor, prostitution, gambling, extortion, and other rackets rocketed to an estimated 30 million dollars per year. But gang warfare in the city intensified, and on St Valentine's Day, 1929, members of Capone's gang lined up five rival mobsters and gunned them down. A doctor and a reporter also died. Capone himself, staying at his Miami estate, was never implicated. In 1931, however, he was found guilty of nonpayment of taxes and sentenced to jail. He was released in 1939, suffering from general paralysis as a result of advanced syphilis. He spent his last years in Florida, slowly deteriorating, and was buried in Chicago.

Capra, Frank (1897–1991) *Italian-born US film director, known for his many film comedies of the 1930s.*

Born in Palermo, Sicily, Capra was taken to the USA by his parents. He studied chemical engineering at the California Institute of Technology and served in the army during World War I. A variety of jobs followed, before he managed to talk himself into directing the short film adaptation of KIPLING's *Fultah Fisher's Boarding House* (1922). Thereafter he entered the film business full-time, working in a film lab, as a propman, film editor, and gag writer until he joined comedian Harry Langdon (1884–1944) as director.

In 1928 he was taken on by Columbia Pictures, for whom he created a whole series of film comedies. Among these were *It Happened One Night* (1934), *Mr Deeds Goes to Town* (1936), and *You Can't Take It With You* (1938), all Oscar winners. *Lost Horizon* (1937) and *Mr Smith Goes to Washington* (1939) were other examples of Capra's underlying theme of human goodness triumphing against all odds. His films found large audiences during the depressed thirties.

He then formed his own company, making *Meet John Doe* (1941) and *Arsenic and Old Lace* (1944). After World War II, during which the first in his documentary series *Why We Fight* won an Oscar, he co-founded Liberty Films with William Wyler (1902–81) and others. Although this period produced *It's a Wonderful Life* (1946) and *State of the Union* (1948), Capra's career ended with less successful sentimental remakes of his own movies. Nonetheless, Capra won six Oscars in all and was the first president of the Directors' Guild; his autobiography, aptly entitled *The Name Above The Title*, was published in 1971.

Cárdenas, Lázaro (1895–1970) *Mexican revolutionary leader and president (1934–40). He was awarded the Stalin Peace Prize in 1955.*
Born in Jiquilpar into a peasant family, Cárdenas enlisted in the revolutionary army when he was eighteen (1913) and advanced rapidly through the ranks; by 1923 he had become a general and had earned a reputation as an outstanding military leader.

Cárdenas first entered politics in 1928, when he was elected governor of the state of Michoacán. Appointed president of the National Revolutionary Party in 1930, he was given the posts of interior minister (1931) and then minister of war (1933) by Plutarco CALLES. In 1934 he was elected president and the following year forced Calles into exile. He remained in this post until 1940, when he retired. In 1943 he came out of retirement to serve as minister of defence, and in 1945 briefly served as head of the army. In his later years he supported the emerging left-wing movements in Latin America, enhancing his image as a revolutionary leader and champion of the poor, particularly the peasants, in Mexico. As president he redistributed land, expropriated foreign oil companies, and nationalized the railways. He granted asylum to TROTSKY in 1937 and in 1939–40 provided a haven for many Spanish republican refugees.

Cardin, Pierre (1922–   ) *French couturier and one of Europe's leading fashion designers. He was made a Chevalier de la Légion d'honneur in 1974.*

The son of a wine merchant, Cardin started working at the age of seventeen for a gentleman's tailor in Vichy, where he learnt the craft of cutting and fitting. He joined the French Red Cross during the latter half of World War II and then moved to Paris to join the fashion designer Paquin, where he was commissioned to design costumes for the Jean COCTEAU film, *Beauty and the Beast*. Cardin subsequently joined Christian DIOR and contributed to Dior's famous inaugural collection of 1947 as the creator of the 'Bar' suit. In 1950 he established his own workshop, designing costumes and fashions for both men and women. In the mid-1950s he moved to an elegant eighteenth-century mansion in the Faubourg Saint-Honoré, where he has carried on his immensely successful business selling elegant clothes for men and women, as well as a wide range of colourful and often whimsical accessories.

Cardin was also the founder of L'Espace Pierre Cardin theatre group and since 1970 has been a director of the Ambassadeurs-Pierre Cardin Theatre. He acquired control of the legendary Parisian restaurant, Maxims, in 1982 and his book, *Fernand Léger*, was published in 1971.

Carlson, Chester (1906–68) *US physicist and inventor of xerography.*

Born in Seattle, the son of a barber, Carlson was educated at the California Institute of Technology. After working for Bell Telephones he took a law degree and ran the patent department of P. R. Mallory and Co., a manufacturer of electrical components. The constant demand for multiple copies of complex patent specifications and drawings alerted Carlson to the need for a machine to produce cheaply and quickly smudge-free copies of a variety of documents. Working in his spare time, and with the help of Otto Kornei, a refugee German

physicist, Carlson had a workable machine ready in October 1938. The process, described initially by Carlson as electrophotography, involved sensitizing a photoconductive surface to light by giving it an electrostatic charge. It was immediately protected by Carlson with an impenetrable web of patents.

Apparently easier to invent than to sell, the process was turned down by twenty leading manufacturers of office equipment. In 1947, however, a small firm of photographic paper manufacturers, the Haloid Company of Rochester, New York, became interested. Carlson sold the process to Haloid but retained a royalty interest. A professor of classics at Ohio State University offered the name 'xerography', meaning literally dry writing, to describe Carlson's process. It took Haloid (which later became the Xerox Corporation) until 1960 and the investment of a further seventy-five million dollars to produce the xerox machine that swept the world. When Carlson died, eight years later, his royalties were estimated to have brought him fifty million dollars.

Carnap, Rudolf (1891–1970) *German-born US philosopher and a leading logical positivist.*

Carnap was educated at the universities of Freiburg and Jena, where he studied under FREGE. In 1926 he moved to Vienna and joined the influential group of philosophers led by SCHLICK and known as the Vienna Circle. Carnap was one of its most influential members, and in 1930 he and Hans Reichenbach (1913–67) founded and edited the important philosophical journal *Erkenntnis*. In 1931 he was appointed to the chair of philosophy at Prague, but in 1935 he left Europe for the USA, where he served as professor of philosophy, first at the University of Chicago and from 1954 until his retirement in 1961 at the University of California, Los Angeles.

As a radical positivist Carnap's first major work was *Der logische Aufbau der Welt* (1928; translated as *The Logical Structure of the World*, 1967). In it he sought to construct the world of experience, using the primitive relation of 'recognition of similarity'. In a later work, *The Logical Syntax of Language* (1934), he proposed his famous principle of tolerance and also introduced into philosophy the important distinction between the formal and material modes of speech. Failure to distinguish between them clearly, he declared, was a common cause of philosophical confusion.

Carnap was quick to see that such purely syntactical approaches were inadequate, and in his *Introduction to Semantics* (1942) and other works of the 1940s he began to insist that such formal concepts as 'implication' and 'analyticity' could only be analysed semantically. Later he turned his attention to problems in the philosophy of science. In his most significant work, *The Logical Foundations of Probability* (1950), he argued that there were in fact two concepts of probability. The simplest could be defined in terms of relative frequency. The other, more ambitiously, he identified as the degree of confirmation assigned to a particular hypothesis on the basis of supporting evidence. This relation, he insisted, was a logical relation, weaker than but similar to the more familiar notion of logical implication. Carnap was in reality attempting to develop a new formal inductive logic. Despite his own labours of twenty years, and a further decade's work by numerous logicians, it is still far from clear whether or not Carnap's project is a viable one.

Carné, Marcel (1909–) *French film director. He received many awards, including the appointment of Officier de la Légion d'honneur.*

Born in Paris, Carné began working in films in 1928 as an assistant cameraman. He became René CLAIR's assistant director in *Sous les toits de Paris* (1930) and assisted Jacques Feyder (1888–1948) in such films as *Le Grand Jeu* (1934) and *La Kermesse héroique* (1935). His first feature as director was *Jenny* (1936), which was followed by the pessimistic *Drôle de drame* (1937), *Quai des brumes* (1938), and *Le Jour se lève* (1939), on all of which he worked with poet-screenwriter Jacques PRÉVERT. During the German occupation they made *Les Visiteurs du soir* (1942) and the most highly regarded achievement of their successful partnership, *Les Enfants du paradis* (1945). Their last film together was *Les Portes de la nuit* (1946). *La Marie du port* (1950), adapted from SIMENON, *Juliette ou la clef des songes* (1951), *Les Tricheurs* (1958), *Les Assassins de l'ordre* (1971), and the TV feature documentary *La Bible* (1976) are among the films that followed.

Carnegie, Andrew (1835–1919) *British-born US industrialist and philanthropist who used his personal fortune, derived largely from the steel industry, to finance a variety of charitable institutions.*

Carnegie was born in Dunfermline, Scotland, but his family moved to the USA in 1848 to settle in Allegheny, a suburb of Pittsburgh, Pennsylvania. At the age of thirteen, Carnegie started work in a cotton factory but soon moved to become a messenger with a local telegraph company. Joining the Pennsylvania Railroad in Pittsburgh, he rose to become superintendent (1853–65), during which time he successfully introduced the first sleeping cars. Meanwhile he was investing his money in Storey Farm at Oil Creek, Pa., which duly yielded oil and a good return for Carnegie. In 1865 he co-founded the Keystone Bridge Company to manufacture iron railroad bridges and he also became a partner in an iron foundry. From these beginnings, the Carnegie Steel Company developed into a massive industrial complex near Pittsburgh. In 1892, the Homestead mill was the scene of a bitter industrial dispute that delivered a severe blow to the labour unions. In 1901, Carnegie sold out to John P. Morgan for an estimated $500 million.

Carnegie's *Gospel of Wealth* (1889) stated his belief that personal wealth should be used to benefit society as a whole; from the 1880s to the end of his life he put this into practice. Over 2500 public libraries in the USA, Britain, and Canada were funded by Carnegie, the first beneficiary being his home town in 1882. He also endowed a host of charitable foundations, the largest of which, the Carnegie Corporation of New York, disburses about $15 million annually for the advancement and diffusion of knowledge. Other bodies include the Carnegie Institute, Washington, and the Carnegie UK Trust. The Carnegie Institute of Technology, formed in 1912 in Pittsburgh, now forms part of the Carnegie–Mellon University. His concern with world peace prompted him to finance the Hague Peace Palace as well as the Carnegie Endowment for International Peace.

Carpentier, Georges (1894–1974) *French boxer who won the world light-heavyweight title (1920–22) and had the unusual distinction of fighting in all of the major professional divisions, from flyweight to heavyweight, and being a European champion in four of them.*

Georges Carpentier was born at Lievin, Pas de Calais, and earned his first title, as a light-weight, when only fifteen. Good-looking and charming outside the ring, he always brought an element of glamour to his contests: for the first time women went to professional boxing in large numbers especially to watch him. He won the world light-heavyweight title from 'Battling' Levinsky in 1920 but lost it two years later. In 1921 he made an unsuccessful bid for the heavyweight crown against Jack

DEMPSEY at Jersey City before a record crowd of 80 000. He retired from boxing at the age of thirty-three.

Carreras, José (1946–   ) *Spanish tenor.*
Born in Barcelona, he made his opera debut in 1970 as Gennaro in *Lucrezia Borgia* after studying under J. F. Puig at the Barcelona Conservatory. Subsequently he was coached by Montserrat Caballé (1933–   ). He appeared in London in 1971 in Donizetti's *Maria Stuarda* and made his US debut the following year as Pinkerton in *Madame Butterfly*; he has since been rapturously received at Covent Garden, the Metropolitan Opera House, the Lyric Opera in Chicago, and elsewhere. He performed in front of a worldwide television audience in 1990 when he sang at a concert in Rome with DOMINGO and PAVAROTTI as part of the celebrations for football's World Cup.

Carter, Elliott Cook (1908–   ) *US composer and writer, who has evolved a style of great rhythmical complexity.*
Born in New York, Carter was educated at Harvard University, where he obtained degrees in English and in music (studying with Walter Piston). He also studied in Paris, at the École Normale de Musique and with Nadia BOULANGER. On his return to the USA he taught and composed, winning several awards, including two Guggenheim Fellowships (1945; 1950). He has been professor of composition at the Peabody Conservatory (1946–48), Columbia University (1948–50), Queen's College, New York (1955–56), and Yale University (1960–62). He has also taught at the Tanglewood and Dartington summer schools and the Juilliard School and has been composer-in-residence at the American Academy in Rome (1963; 1967) and Berlin (1964). He dislikes playing or conducting in public.

In his compositions Carter has consistently attempted to create a music (to quote him) 'beautiful, ordered, and expressive of the more important aspects of life'. In forging his own language he has made use of a variety of compositional techniques, notably those of the Renaissance madrigal, serialism, neoclassicism, and jazz. In doing so he has moved from the diatonic style of his earlier days to an increasing dissonance. His irregular scansion of phrases and constantly shifting accent achieve a complex and supple polyrhythm, which has led to his concept of 'metrical modulation', in which he moves from one metronomic speed to another by the shortening or lengthening of the basic unit (crotchet, quaver, etc.). His cello sonata (1948) is a work in which these techniques are found. In the *Double Concerto for Harpsichord and Piano with two Chamber Orchestras* (1961) he superimposes layers of dissimilar sound, which act as both structure and expressive content, and answer the problem of 'reconciling instruments with different responses to the fingers' touch'. In an output that includes music for stage, orchestra, chorus, and chamber groups, works of particular interest are *Variations for Orchestra* (1955), a virtuoso piece for the instruments of the orchestra; a piano concerto (1965); *Symphony for three Orchestras* (1976); and three string quartets (1951; 1960; 1971). Among his most recent works are *Syringa* (1979), *Penthode* (1985), and *Three Occasions for Orchestra* (1989).

Carter, Jimmy (James Earl Carter; 1924–   ) *US Democratic statesman and thirty-ninth president of the USA (1977–81).*
Carter was born in Plains, Georgia, the son of a peanut farmer and warehouser. Having graduated from the US Naval Academy in 1946, he spent seven years with the US navy. During the latter part of his service he worked on the nuclear submarine programme and was a crew member of one of the first nuclear-powered submarines. His naval career was cut short by his father's death in 1953: as the eldest child Carter felt it his duty to return to Plains and take over the family farm. Over the next twenty years he built up the business into a sizeable concern, expanding the existing interests and embarking on a number of new enterprises.

Carter's political career began in the early 1960s, when he served two terms in the Georgia State Senate. In 1970, after an unsuccessful first attempt four years earlier, he was elected governor of Georgia. At his inaugural address he won the hearts of the black community by declaring that 'the time for racial discrimination is over' and during his four years as governor he undertook a radical reorganization of state government and increased the number of blacks in state agencies. In 1974 he announced his intention to stand as a candidate for the Democratic nomination for president. After a strenuous and well-organized campaign he managed to defeat his rivals at the first ballot. With Gerald R. FORD as his Republican opponent and Senator Walter F. Mondale (1928–   ) as his running-mate, Carter embarked on his presidential campaign, promising social and economic reforms and styling himself as 'a man of the people'. In November 1976 he was elected president.

Two major achievements of Carter's administration were the signing of the Panama Canal Treaty, which undertook to transfer control of the canal from the USA to Panama by the end of the century, and the president's involvement and assistance in the peace negotiations between Egypt and Israel in 1979. His attempts to restrict oil imports and his support of the nuclear power programme met with a certain amount of resistance, as did some of his policies for domestic reform. The admittance of the deposed shah of Iran into the USA in 1979 posed serious problems for the Carter administration: a number of Americans were taken hostage in Tehran and the US embassy was seized by Iranian students. These setbacks were not sufficient to prevent Carter's renomination as Democratic candidate for president in 1980 over such rivals as Senator Edward Kennedy (1932–   ). In the presidential election itself, however, he lost to Ronald REAGAN, former governor of California.

Cartier-Bresson, Henri (1908–   ) *French photographer and film director who was one of the key figures in the development of the photograph as a documentary record.*

The son of a wealthy textile manufacturer, Cartier-Bresson studied in Paris (1927–28) with the cubist painter and critic André Lhote (1885–1962). This instilled in Cartier-Bresson a love of painting and an eye for the surreal, both of which were to influence his later work. After studying literature in Cambridge (1929) and completing his military service, he went to Africa but contracted blackwater fever and returned to Marseilles to convalesce. He now began to exploit his interest in photography, selling his pictures to magazines and agencies, and in 1933 the first exhibitions of his work were staged in Madrid and New York. Cartier-Bresson travelled widely, recording the lives of ordinary people with an instinct for the 'decisive moment' of a scene or event; a famous example is his picture of a French family picnicking on the banks of the Marne, taken in 1935. Characteristically, when covering the coronation of George VI in London, Cartier-Bresson chose to record the reactions of spectators rather than the events themselves, introducing an entirely fresh perspective.

In 1936 Cartier-Bresson began his association with the French film director Jean RENOIR, assisting in the production of *Une Partie de campagne* (1936) and *La Règle du jeu* (1939). This prompted Cartier-Bresson to direct his own documentary film about the Spanish civil war, *Return to Life* (1937).

During World War II, Cartier-Bresson served in the French army, was captured, and spent nearly three years in a POW camp before escaping. He joined the Resistance in Paris to make a photographic record of the German occupation and the retreat following the Allied invasion. His film *Le Retour* (1945) dealt with the fate of returning French POWs. In 1947, together with the photographer Robert CAPA and others, Cartier-Bresson founded the cooperative photographic agency, Magnum Photos, and later served as its president (1956–66). His collections of photographs include *The Decisive Moment* (1952), *Europeans* (1955), *People of Moscow* (1955), *Cartier-Bresson's France* (1971), *Portraits 1932–1983* (1983), and *Henri Cartier-Bresson in India* (1988). In 1989 he published *Traits pour Traits* (*Line by Line*), a book of drawings.

Caruso, Enrico (1873–1921) *Italian tenor. One of the most highly acclaimed singers of the twentieth century, he was the first major tenor to be recorded on gramophone records, although unfortunately most of them are acoustic (pre-electric) recordings.*

Born in Naples to impoverished parents, he first sang in churches. Subsequently he studied with Guglielmo Vergine from 1891 and with the conductor Vincenzo Lombardi, making his debut in Naples in 1894. However, it was not until about 1902 that his voice was fully secure technically – the previous year a poor reception in Naples caused him to vow never to sing there again. Thereafter, his career took him to all the major opera houses in Europe and the USA, but he never sang again in Italy. His Covent Garden debut (1902) was in Verdi's *Rigoletto*, which he subsequently sang there on many occasions. He was also engaged by the Metropolitan Opera, New York, from 1902 until his death from pleurisy in 1921.

Caruso was particularly impressive in Verdi, Puccini, and Massenet. His voice combined a brilliant upper register with a baritone-like warmth. He also had perfect intonation and breath control, giving him a mastery of legato phrasing and portamento.

Cary, (Arthur) Joyce (Lunel) (1888–1957) *British novelist of Anglo-Irish descent.*

Cary was born in Londonderry but brought up in England. Despite his mother's death (1898) his childhood was mainly happy, with memorable holidays spent in Ireland with his grandmother, as Cary later described in *A House of Children* (1941). He studied art at Edinburgh (1907–09) before going to Oxford to read law. In 1912 he served in the Balkan War, and then joined the Nigerian civil service until forced to retire (1920) by ill health.

During the 1920s Cary read and wrote in Oxford, honing his skills as a novelist but publishing nothing. Africa was the setting for his first novels, which included *Aissa Saved* (1932) and *Mister Johnson* (1939). His sensitivity in writing about women is shown in *The Moonlight* (1946) and *A Fearful Joy* (1949). His major works were two trilogies, each with subtly interlocking characters. The first, concerned with art, comprises *Herself Surprised* (1941), *To Be a Pilgrim* (1942), and *The Horse's Mouth* (1944); the second deals with the nature of political life and comprises *A Prisoner of Grace* (1952), *Except the Lord* (1953), and *Not Honour More* (1955). The quality of family life, the relationship between mothers and children, and an intense philosophical quest for 'the state of grace' mark all Cary's major works.

In his last years he suffered a generalized paralysis, which forced him to accept that he would never complete his projected third trilogy on religion. *The Captive and the Free* (1959), which is all that survives of this trilogy, was left unfinished at his death.

Casals, Pablo (1876–1973) *Spanish cellist, conductor, pianist, and composer. A musician of great ability and integrity, he refused to play in HITLER's Germany, FRANCO's Spain, or any country that recognized Franco.*

The son of musical parents, he was first taught the piano, organ, and violin. He later studied the cello with Garcia at the Barcelona Municipal School of Music, soon acquiring his individual style and complete mastery of the instrument. He made his debut in Barcelona in 1891. The queen regent became his patron and he was awarded a scholarship to the Madrid Conservatory. In 1895 he was engaged as a cellist with the Paris Opéra and in 1898 made his virtuoso debut, playing Lalo's cello concerto in Paris and in London. In 1905, in association with the pianist Alfred CORTOT and the violinist Jacques Thibaud (1880–1953), he formed a trio that set new standards of performance in the piano-trio repertory. He was always admired for his intellectual authority and scrupulous adherence to the composer's text.

After the Spanish civil war he exiled himself to the village of Prades on the French side of the Spanish border. From 1950 to 1968 he organized a chamber music festival in Prades. In 1956 Casals emigrated to Puerto Rico. Many honours have been awarded to him, including the United Nations Peace Prize. Of the many recordings made by Casals, two of the most outstanding are his interpretation of the unaccompanied suites for violoncello by J. S. Bach, and the 1936 version of Dvořák's cello concerto.

Casement, Sir Roger David (1864–1916) *British public official and Irish nationalist, executed as a traitor.*

Born in Kingstown, County Dublin, Casement joined the consular service in 1892 and was posted to Africa, where in 1903 he reported on the plight of native workers on the rubber plantations of the Belgian Congo. In 1910 he turned his attention to the conditions of rubber workers in Peru, his next posting. Retiring in 1912, he joined the Irish nationalist cause and, after the outbreak of World War I, went to Germany to raise an Irish legion among prisoners of war. Failing in his mission, he returned to Ireland on Good Friday 1916 in a German submarine, with orders to halt the imminent and doomed Easter Rising in Dublin. He was arrested, convicted of high treason at the Old Bailey, and hanged at Pentonville in London. Later, he received a state funeral in Ireland. His diaries, whose authenticity is doubted, describe his homosexuality and were circulated by British agents to discredit a campaign for his reprieve.

Castro, (Ruz) Fidel (1926–) *Cuban statesman; prime minister (1959–76) and president (1976–), who has evolved his own brand of Marxism and maintained a socialist government on the doorstep of the USA.*

Born in Biran, the son of a sugar-planter, Castro was educated at a Jesuit school before studying law at the University of Havana. After graduating in 1950, he worked among the poor in Havana. In 1953 he led an unsuccessful revolt against the dictator, BATISTA (later known as the '26 July movement'), for which he was imprisoned. Released under a general amnesty in 1955, he went into exile to organize a further guerrilla campaign. A second unsuccessful revolt in 1956 drove him (along with twelve others) back into the mountains of Sierra Maestra.

Castro waged a guerrilla campaign against Batista for the next three years; finally, at the beginning of 1959, he forced Batista from power and became prime minister of a socialist government. Despite opposition from the USA, which imposed economic sanctions against Cuba (1960) and supported an invasion by Cuban exiles at the Bay of Pigs (1961), Castro remained in power. However, the antagonism of the USA forced Castro to become dependent on the Soviet Union, which led KHRUSHCHEV to instal missile bases on Cuban soil. President KENNEDY called Khrushchev's bluff and for several days in 1962 the world trembled on the verge of nuclear war. After the resolution of the Cuban missile crisis, Fidel Castro became an international figure. In 1976, Castro passed a new constitution under which he became president, secretary-general of the Communist Party, and commander-in-chief of the army. His refusal to countenance the liberalizaton of Cuban society on the lines of communist nations in eastern Europe, together with his criticisms of the reforms introduced by Mikhail GORBACHOV in the Soviet Union in the late 1980s, have resulted in Cuba's increased isolation from the international community.

Cavafy, Constantine Peter (Konstantinos Petrou Kavafis; 1863–1933) *Greek poet.*

The son of a wealthy importer-exporter, Cavafy spent almost all of his uneventful life in his birthplace – Alexandria, Egypt. His father's death when Cavafy was thirteen caused an abrupt change in the family's fortune. He spent several years of his youth in England and lived briefly (1882–85) in his mother's native Constantinople (now Istanbul). In later life he made short visits to London, Paris, and Athens, but otherwise he lived alone in Alexandria, employed in the Egyptian Ministry of Public Works. His first volume of fourteen poems appeared in 1904 and an expanded version (twenty-one poems) in 1910, but both were privately printed and were scarcely known beyond the circle of his friends. From 1912 onwards he circulated his poems on broadsheets only to his friends. His collected works did not appear until 1935. Although E. M. FORSTER had published translations of his work in his collection of Alexandrian essays, *Pharos and Pharillon* (1923), Cavafy was generally unknown to English readers until 1952. A recent translation is *Collected Poems*, translated by E. Keeley and P. Sherrard (1974).

Cavafy's work, a complex mixture of purist and demotic Greek, of the archaic and colloquial, was slow to gain acceptance among Greek readers. His posthumous reputation, however, grew rapidly and his poetry has finally exerted an enormous influence (as, for example, in the work of George SEFERIS) on modern Greek verse. The poems, virtually lacking in metaphor and rhetorical frills, mainly refer to the Alexandria of the Hellenistic and Graeco-Roman period (*c.* 325 BC to AD 400) and imply a similarity between

the city and the modern world. There is a pervading sense of the rise and decay of cultures and of all human aspirations. Mythological and historical themes (as in 'The God Abandons Antony'), erotic (homosexual) elements, and allusions to decadence and change ('Waiting for the Barbarians') combine to give Cavafy's characteristic tone of irony and regret at the transience of things.

Cavell, Edith (1865–1915) *British nurse who became an international heroine after she was executed by the Germans for harbouring and assisting fugitive soldiers in Belgium during World War I.*

A vicar's daughter from Norfolk, Cavell became interested in nursing after a European trip in 1883, during which she endowed a Bavarian hospital with funds for the purchase of instruments. In 1895 she entered the London Hospital as a probationer. Qualifying as a staff nurse, she served in Highgate and Shoreditch infirmaries and in Manchester before going to Brussels in 1906 to help establish a nursing school. In 1907 she was appointed first matron of the Berkendael Medical Institute, run by Dr Depage.

With the outbreak of war in 1914, the director was called away to organize field hospitals and Cavell was left in charge. She continued in her post after the German occupation of Belgium, treating both German and Allied wounded. However, her humanitarian instincts obliged her to use the hospital as a refuge for stranded Allied soldiers. Arrested on 5 August 1915 and brought before a military tribunal, she openly admitted her actions, dictated by conscience. In spite of interventions by the US minister in Brussels, she was sentenced to death and executed by firing squad at 02.00 on 12 October. Her death provoked international condemnation of a German military code that gave no consideration to motives and showed no mercy. A statue of Edith Cavell now stands in St Martin's Place in London.

Ceauçescu, Nicolae (1918–89) *Romanian statesman, secretary-general of the Romanian Communist Party (1965–89) and the first president of the Socialist Republic of Romania (1974–89). His repressive and totalitarian regime ended with his execution.*

Born in Scornicesti-Olt, the son of a peasant, Ceauçescu was a factory worker in Bucharest at the age of eleven. In 1932 he became a member of the Workers' Movement, joining the illegal Union of Communist Youth (UCY) and the Romanian Communist Party (RCP) in 1933. Imprisoned for antifascist activities

(1936–38; 1940–44), he spent this time studying for a degree from the Academy of Economic Studies in Bucharest.

After the communists assumed power, Ceauçescu was appointed secretary of the UCY central committee (1944–45) and elected as a deputy to the Grand National Assembly (1946). He became a member of the RCP central committee (1945–48; 1952–89), deputy minister of agriculture (1948–50), and deputy minister of the armed forces (1950–54). He continued to rise through the party ranks from his appointment as secretary of the RCP central committee in 1954 to his election as RCP secretary-general in 1965. In 1967 he became chairman of the state council and supreme commander of the armed forces, converting this office into president of the Republic of Romania in 1974.

Ceauçescu maintained and developed Romanian independence from the Soviet Union by increasing trade with the West and by not participating in Warsaw Pact military manoeuvres; Romania did not take part in the Soviet-led invasion of Czechoslovakia in 1968. At the same time he sought a role in the international community: Romania was the only eastern bloc country to establish diplomatic relations with West Germany (1966), maintain them with Israel (1967), and to attend the 1984 Olympic Games in Los Angeles.

At home, Ceauçescu pursued an ineptly planned economic policy of forced growth and repayment of western debts; by the 1980s this had resulted in Romania's having the lowest standard of living of any eastern European country. His attempt at creating a uniform Romanian state caused considerable hardship to the ethnic minorities, particularly the large Hungarian population, whose villages were systematically destroyed and their inhabitants forcibly rehoused in modern urban complexes. All signs of political opposition were ruthlessly suppressed through the activities of his hated secret police, the Securitate. Ceauçescu deliberately fostered his own personality cult, maintaining his complete hold on power by the appointment of his wife Elena and other family members to politically influential positions. Widespread discontent finally erupted in December 1989 with a bloody revolution in which thousands died; Ceauçescu and his wife were captured, tried, and executed by firing squad on Christmas Day.

Céline, Louis-Ferdinand (Louis-Ferdinand Destouches; 1894–1961) *French novelist.*

Destouches was born in Courbevoie, Seine, where his mother kept a lace shop. He joined a cuirassier regiment in 1913 and was awarded the military medal for an exploit in World War I that left him badly wounded. Working for the Rockefeller Foundation in Africa at the end of the war, he decided to make medicine his career and qualified as a doctor in 1924. After a spell in Geneva with the League of Nations and visits to the USA, Canada, and Cuba, he finally settled as a dispensing physician in Paris. Here he began work on his major novel, *Voyage au bout de la nuit* (1932; translated as *Journey to the End of the Night*, 1960), adopting as a pseudonym his grandmother's Christian name, Céline. Cynical and pessimistic, the novel was inspired by the author's own experiences in World War I and in Africa and is written in a controversial antiliterary style, heavily laden with slang. *Mort à crédit* (1936; translated as *Death on the Instalment Plan*, 1938), was based on Céline's experience as a doctor and shows his bitter contempt for humanity.

In the years immediately preceding World War II Céline further compromised his reputation by launching a violent antisemitic attack in the pamphlets *Bagatelles pour un massacre* (1937) and *L'École des cadavres* (1938). Increasingly disenchanted by politics, he was nevertheless suspected of collaboration during World War II, after the publication of *Guignol's Band* (1943), and forced to flee to Denmark. After seven years in a Danish prison Céline was allowed to return to France, where he devoted himself to writing. The first part of a trilogy, *D'Un Château à l'autre* (1957; translated as *Castle to Castle*, 1963), brought him back into the public eye. *Nord* (translated as *North*, 1972) followed in 1960; the final part of the trilogy, *Rigodon* (translated as *Rigadoon*, 1974), was published posthumously in 1969.

Chabrol, Claude (1930–   ) *French film director, who gained recognition in the 1950s as one of the first directors in the vanguard of the French 'New Wave'.*

Born in Paris into a family of pharmacists, Chabrol originally studied pharmacy, but after completing his military service he began working in Fox's Paris publicity department. After this he became a writer and critic for *Arts* and *Cahiers du Cinéma*; in this period he was also coauthor with Eric Rohmer of a critical volume on the work of HITCHCOCK.

He made his debut as a director with *Le Beau Serge* (1958), which he financed himself with an inheritance from his first wife, Agnès Goute. When this won the Grand Prix at the

Locarno Festival, he was able to set up his own production company, AJYM, through which several other 'New Wave' directors, including Jacques Rivette (1928–) and Philippe de Broca (1933–), were able to channel their work. Among his own many notable films, chiefly mystery-thrillers, were *Les Cousins* (1959), which earned him the Golden Bear Award at the Berlin Festival, *Les Bonnes Femmes* (1960), *Les Biches* (1968), *La Femme infidèle* (1969), *Le Boucher* (1970), *Les Menteurs* (1979), *Coq au Vin* (1984), and *Une Affaire des Femmes* (1988). Stéphane Audran (1938–), whom he married in 1964, has appeared in several of his films, winning a César Award for her performance in his *Violette Nozière* (1978).

He has also directed plays, including *Macbeth* (1964), *L'Adieu aux dieux* (1980), and *Vladimir et les Jacques* (1980), as well as several television productions.

Chadwick, Sir James (1891–1974) *British physicist who discovered the neutron, for which he was awarded the 1935 Nobel Prize for Physics. He took an active part in the manufacture of the first atom bomb and was knighted in 1945. He was made a CH in 1970.*
Born in Manchester, Chadwick was educated at Manchester Grammar School and University where, under Ernest RUTHERFORD, he was first introduced to the newly emerging discipline of nuclear physics. In 1913 he went to Berlin to study under Rutherford's old colleague, Hans GEIGER. Chadwick misread the political situation and found himself still in Berlin after the outbreak of World War I. Consequently he spent much of the war interned in the stables of a racecourse near Spandau. Despite fairly rigorous conditions, Chadwick, using books and equipment provided by German colleagues, was able to continue with his scientific education and even to carry out some research.

Released from internment after the war, Chadwick joined Rutherford at the Cavendish Laboratory in Cambridge, serving as assistant director from 1923 until 1935. During his time at the Cavendish, Chadwick tackled one of the overriding problems then confusing nuclear physics – the difficulty of accounting for the known mass and charge of atoms on the basis of the proton and electron. It was suspected that the nucleus also contained an uncharged particle and it was Chadwick, against much international competition, who first identified the particle (later named the neutron) in 1932.

By this time a certain amount of friction was noticeable in the Rutherford–Chadwick partnership. Chadwick had begun to see that the future of nuclear physics lay in the construction of complex machines capable of accelerating elementary particles; Rutherford was reluctant to follow this course. It seemed a good time to leave the Cavendish and consequently Chadwick accepted the offer of the chair of physics at Liverpool University in 1935. While he was there, he built Britain's first cyclotron. With the outbreak of World War II and the discovery of nuclear fission, Chadwick, as Britain's foremost nuclear physicist, was called on to advise the government. In 1943 he led the delegation of British experts who worked in the USA on the development of the first atomic bomb.

After the war Chadwick opposed government policy. He objected to the setting up of the Harwell research centre and argued that fundamental research in nuclear physics should be pursued in universities. Chadwick left Liverpool in 1948 to return to Cambridge as master of Gonville and Caius College, a post he retained until his retirement in 1958.

Chadwick, Lynn Russell (1914–) *British sculptor in metal. He was awarded the CBE in 1964.*

After studying architecture in London, Chadwick worked as an architect and furniture designer until 1946, apart from war service as a pilot (1941–44). In 1947 he moved to Gloucestershire, where he took up sculpture seriously. His first works were mobiles and constructions in metal, influenced by CALDER and GONZÁLEZ, but in the early 1950s, in a number of one-man shows, he exhibited works in a new style of his own. These sculptures were made of plaster reinforced with iron filings, which could be modelled and chiselled, placed on a steel skeleton. Some of his works suggested animal or bird shapes but usually they depicted ominous or aggressive human figures with features of crustacean species, reptiles, birds, or insects; in *Winged Figures* (1955), for example, the angular but basically human forms have batlike wings. The simple partly visible steel skeleton gave a thin-legged scarecrow-like appearance to this and many of his upright figures.

Already semiabstract, Chadwick's sculpture moved away from obvious figurative associations in the 1960s and he employed smoother blocklike forms. He won the International Sculpture Prize at the Venice Biennale in 1956.

Chagall, Marc (1889–1985) *Russian-born French painter and graphic artist. Among many honours and awards, he received the Grand Cross of the Légion d' honneur.*

Born into a poor Jewish family in Vitebsk, western Russia, he managed, despite restrictions imposed on movement for Jews, to study art in St Petersburg. Russian folk painting, however, played as great a part in forming his style as formal training. In 1910 a patron enabled him to move to Paris, where the influence of contemporary avant-garde art forms liberated him from naturalism. He joined no movements but took from each whatever was useful to him. In his early masterpiece *I and the Village* (1911), inspired by his memories of Vitebsk, the intense arbitrary colours recall fauvism while the composition, which is based on circles and triangles, seems influenced by cubism. When asked why he had painted two of the houses in this picture upside down and a milkmaid apparently inside a cow's head, he replied, 'I needed that kind of shape in that place for my composition.' Later he described his work as 'pictorial arrangements of images that obsess me'. Thus his work is frequently autobiographical and, although representational, conforms to no natural laws.

In 1915 Chagall returned to Russia and married Bella Rosenfeld, who subsequently figured in many of his paintings. After the October Revolution of 1917, he was made a commissar for fine arts but was ousted in 1919. He then moved to Moscow before going to Berlin (1922) and Paris (1923), where he stayed until World War II. Book illustrations, including over a hundred illustrations for the Old Testament, occupied much of his time in the 1920s and 1930s. After his autobiography *Ma vie* was published in 1930, he travelled widely in Europe and the Middle East. He became a French citizen in 1937 but lived in the USA between 1941 and 1947. During World War II, particularly after the death of his wife in 1944, Chagall's normally bright paintings became dark and ominous.

After returning to France in 1948, he accepted numerous public commissions, including the ceiling for the Paris Opéra (1964) and two murals for the Lincoln Centre in New York (1966). Chagall was one of only three painters to have had an exhibition of his works in the Louvre (the others were BRAQUE and PICASSO).

Chain, Sir Ernst Boris (1906–79) *German-born British biochemist, who, with Howard FLOREY, first prepared penicillin in its pure form for therapeutic use as an antibiotic. For this he received the 1945 Nobel Prize for Physiology or Medicine and he knighted in 1969.*

Born in Berlin, Chain received his doctorate in chemistry from Berlin University in 1930 and joined the Institute of Pathology at the Charité Hospital, Berlin, to study biochemistry. However, after HITLER's rise to power, he left Germany in 1933 for England and a post at Cambridge University under Sir Frederick Gowland HOPKINS to work on enzymes. Two years later he joined the team led by Howard Florey at Oxford University. Here, their first project was the purification and crystallization of the antibacterial enzyme, lysozyme. This they achieved but, disappointed with its low therapeutic value, they looked for other potential antibacterial substances and decided to concentrate on penicillin – first discovered by Alexander FLEMING. Chain's role was to extract the penicillin from the culture of the mould, *Penicillium notatum*, by freeze-drying and solvent extraction. In 1941, after scaling up the operation, the first clinical trials on humans using the purified extract were successfully performed. While Florey went to the USA to promote the large-scale manufacture of penicillin, Chain, by 1943, had described the molecular structure of penicillin G. In 1949 Chain accepted the post of scientific director of the International Research Centre for Chemical Microbiology at the Instituto Superiore di Sanità, Rome, which had its own pilot plant for penicillin production. Chain pursued the development of other types of penicillin, notably penicillin V, which resisted stomach acidity and could therefore be administered orally. His researches enabled synthetic chemical modifications of natural penicillins to combat a wider range of pathogenic bacteria and to counter resistance to the original penicillin G form – due to the production by some bacteria of the penicillin-destroying enzyme, penicillinase, discovered by Chain.

In 1961 he was appointed director of the Wolfson Laboratories at Imperial College, London. He wrote *Antibiotics* (1949) with Florey and colleagues and received a host of awards and honours worldwide in recognition of his contribution to the health of humanity.

Chaliapin, Feodor Ivanovich (1873–1938) *Russian operatic bass, widely renowned as the greatest singing actor of his age.*

Born in Kazan to a peasant family, Chaliapin was initially self-taught apart from what he learnt in the church choir. Only after he had already had some success in a provincial opera company did he have singing lessons (1892–93, with Usatov). He appeared with the Imperial Opera in St Petersburg in 1894, but did not become well known until 1896, when he went to Moscow to sing with Mamontov's private opera company. Here Chaliapin excelled, particularly in Russian opera, in such parts as Boris in *Boris Godounov* (Mussorgsky), Ivan the Terrible in *The Maid of Pskov* (Rimsky-Korsakov), and Melnik in *Russalka* (Dargomizhsky). His debut (1901) at La Scala initiated an international career that took him to concert halls and opera houses all over the world, including the Metropolitan in New York (1907–08), the Paris Opéra under the management of DIAGHILEV (1908; 1910; 1913), and Covent Garden (1913–14; 1926).

In addition to bass roles, Chaliapin also sang such baritone parts as Onegin in *Eugene Onegin* (Tchaikovsky). He was a perfectionist in his costume as well as in the musical and dramatic preparation for a part and totally intolerant of musical or artistic mediocrity.

Chamberlain, Neville (1869–1940) *British statesman and Conservative prime minister (1937–40). He advocated appeasement towards the fascist powers in the 1930s.*

Born in Birmingham, of which his father, Joseph Chamberlain (1836–1914), was to become a famous mayor, Neville Chamberlain was a Conservative MP from 1918 to 1940. He is least known for his important social reforms while minister of health (1923; 1924–29), which introduced a large-scale housing programme, new engineering industries, and pensions and insurance legislation. He served as chancellor of the exchequer (1931–37) before replacing Stanley BALDWIN as prime minister in the coalition national government (1931–40), formed to combat Britain's economic difficulties.

Chamberlain's policy of appeasement towards Germany, Italy, and Japan was designed to postpone war until Britain had rearmed. He abandoned the sanctions imposed on Italy after its conquest of Ethiopia (1936), watched while Germany annexed Austria (1938), and, in the Munich Agreement (1938), recognized HITLER's claims to the Sudetenland, then in Czechoslovakia. Chamberlain returned from his meeting with Hitler in Munich waving a piece of paper, which he proclaimed would guarantee 'peace in our time'. However, he immediately ordered the acceleration of rearmament. When, in March 1939, Hitler invaded the rest of Czechoslovakia, Chamberlain rejected appeasement and introduced the first

military conscription in Britain in peacetime. After the German invasion of Poland later that year, Chamberlain issued an ultimatum to Germany, which resulted in a state of war being declared on 3 September 1939. However, the Allied reverses in Norway at the beginning of World War II revealed the inadequacy of British preparations and forced Chamberlain's resignation (1940). He served briefly in his successor CHURCHILL's cabinet, resigning because of illness in October 1940.

Chamberlain, Owen (1920–) *US physicist who discovered the antiproton. For this work he shared the 1959 Nobel Prize for Physics with Emilio Segré.*

Born in San Francisco, the son of a radiologist, Chamberlain was educated at Dartmouth College and the University of Chicago. During World War II he worked under Emilio Segré (1905–89) at Los Alamos on the development of the atomic bomb. As professor of physics, Chamberlain taught at the University of California at Berkeley from 1958 to 1989.

Soon after his arrival at Berkeley, Chamberlain, in collaboration with Segré and others, began the search for the antiproton, the existence of which had been predicted in the 1920s by Paul DIRAC. The discovery by C. D. Anderson (1905–) in the 1930s of the positron convinced him that such a particle existed and could be found. The recent opening of the Bevatron, an accelerator capable of accelerating protons to very high energies, provided Chamberlain with the opportunity and finally, in 1955, he and his co-workers discovered among 40 000 mesons produced by the accelerator a particle – the antiproton – with the same mass as the proton and a negative charge.

Following this success Chamberlain continued to work in the field of high-energy physics, contributing in particular to studies of the interaction of antiprotons with hydrogen and deuterium and the scattering of pi-mesons.

Chandler, Raymond (1888–1959) *US writer of detective novels, creator of the private detective Philip Marlowe.*

Born in Chicago, Chandler was educated in England, where he lived with his British mother from 1896 to 1912. He published some early writings in British newspapers and weeklies and served with British and Canadian forces in World War I. After the war he worked for several US oil companies and in the 1930s began to publish crime fiction in pulp magazines. With his first novel, *The Big Sleep* (1939; filmed 1946 and 1979), he was recognized as more than an ordinary crime writer. He worked in the tough urban 'hard-boiled' vein developed by his acknowledged model, Dashiell HAMMETT. Philip Marlowe, Chandler's hero, is a moral man without illusions about the utterly corrupt society in which he operates. Chandler's other novels include *Farewell My Lovely* (1940; filmed 1944 and 1975), *The High Window* (1942), *The Lady in the Lake* (1943; filmed 1946), and *The Long Goodbye* (1953; filmed 1973). *The Simple Art of Murder* (1950) contains a dozen shorter works with a commentary. From 1943 Chandler was also a screenwriter, his films including *Double Indemnity* (1944) and *Strangers on a Train* (1951).

Chandrasekhar, Subrahmanyan (1910–) *Indian-born US astrophysicist, known for his discovery of the Chandrasekhar limit.*

Chandrasekhar was educated at the Presidency College, Madras, and moved shortly after graduating in 1930 to Cambridge, where he obtained his PhD in 1933. From 1936 to 1985 Chandrasekhar worked in the USA at the University of Chicago and the Yerkes Observatory, becoming a naturalized US citizen in 1953.

Chandrasekhar's work on stellar evolution centred round the distinction between stars that evolve into white dwarfs and those that become supernovas. In 1935 Chandrasekhar suggested that there was a limiting mass, now known as the Chandrasekhar limit, above which a star's temperature begins to rise, causing it to end its life in a supernova explosion. Chandrasekhar continued to work on stellar evolution, summarizing his work in *Introduction to the Study of Stellar Structure* (1939).

Chanel, Coco (Gabrielle Bonheur Chanel; 1883–1971) *French couturière who originated the slim low-waisted style that dominated women's fashion in the 1920s and who launched the famous range of perfumes that bears her name.*

Chanel, whose nickname derives from 'cocotte', was born in Deauville and, orphaned at an early age, worked with her sister in a milliner's shop. By 1912 Chanel had opened her own shop in Deauville and, with its success, in 1914 established her fashion house in the Rue Cambon in Paris. Chanel's clothes marked the departure from the stiff corsetted style then prevalent to looser more comfortable garments. By the 1920s the Chanel style was famous all over the world, and while still at the height of her career she took on the manufacture of her own textiles, jewellery, and perfume, besides running the couture house.

Her most famous fragrance, Chanel No. 5, was introduced in 1922. Chanel became a prominent figure in fashionable society and a close friend to the Duke of Westminster. In 1933 she became engaged to one of her directors, the designer Paul Iribe, but she never married.

In the 1930s Chanel's clothes were eclipsed by the more extrovert designs of Elsa SCHIAPARELLI and others, and in 1938 her couture house had to close, leaving only the perfumery business to continue. However, her distaste for the designs of the 1940s and early 1950s prompted her to reopen: she presented her first 'comeback' collection in 1954, which again correctly judged the needs of the public and influenced the style of an entirely new generation in the late 1950s and 1960s.

Chaplin, Charlie (Sir Charles Spencer Chaplin; 1889–1977) *British actor and director. A legendary figure in his own lifetime, he was knighted in 1975.*

Born in London to music-hall parents, Charlie Chaplin and his brother Sydney were placed in an orphanage at a very early age. Becoming a vaudeville performer, he joined Fred Karno's company in 1906. On tour in the USA Chaplin was spotted by Mack SENNETT, who signed him up for the Keystone Studio in 1913. He made his film debut in *Making a Living* (1914) and introduced the famous seedy and soft-hearted gentleman-tramp routine, which became his hallmark. Numerous films for various studios brought him world fame, all based on his mastery of pathos and slapstick acrobatics.

As well as acting, Chaplin also wrote and directed and in 1919 co-founded United Artists with D. W. GRIFFITH, Douglas FAIRBANKS, and Mary PICKFORD. In the twenties came some of his best feature films, including *The Kid* (1920) and *The Gold Rush* (1925). Reluctant to come to terms with sound, he merely added music and effects to *City Lights* (1931) and *Modern Times* (1936), with a minimum concession to dialogue in *The Great Dictator* (1940). By the 1940s a silent film had a certain novelty value, but it failed to bring in the audiences, even though Chaplin was a household name throughout the world. The bowler-hatted tramp had had his day: *Monsieur Verdoux* (1947) never quite established Chaplin as a talkie star, although he was more successful with *Limelight* (1952).

In the late forties Chaplin came to the attention of the Un-American Activities Committee; despite his denials, in 1952 he was banned from the USA as a communist sympathizer. He settled in Switzerland with his fourth wife,

Oona O'Neill (1926–91; daughter of Eugene O'NEILL) but in 1973 returned to the USA to receive his second special Academy Award (his first had been awarded in 1928). His third wife was the film actress Paulette Goddard (1911–90).

Chappell, Greg(ory Stephen) (1948–) *Australian cricketer and test captain who in the final match for his country before retiring in 1984 became the first Australian to score over 7000 runs in test matches, topping the previous highest aggregate (6996) by Sir Donald BRADMAN.*

The brother of test players Ian and Trevor Chappell, Greg made his first-class debut for South Australia in 1966–67. Two seasons of county cricket in England, playing for Somerset (1968; 1969), made him a better batsman and bowler. In the 1973–74 season in Australia he changed states and joined the Queensland side, which he captained. The first time he was chosen to play for Australia he made a century (at Perth in 1970). By the time he toured in England with his country in 1972 Chappell was an elegant and mature player: his upright stance was a model for schoolboys and his stroke play was always correct. He automatically succeeded his brother Ian in 1975 as Australia's captain. Many place him second only to Bradman among the great Australian batsmen, and he created a new world record of 122 test catches.

Charles, Prince (Charles Philip Arthur George, Prince of Wales; 1948–) *Heir apparent to ELIZABETH II of the United Kingdom.*

Born at Buckingham Palace, London, Prince Charles attended Hill House School before following his father to Cheam School and Gordonstoun. He was then sent to Geelong Grammar School in Melbourne, Australia. At Trinity College, Cambridge, he read archaeology and anthropology, taking a term off to study Welsh at the University College of Wales, Aberystwyth. In 1969 he was invested by the Queen with the insignia of the Prince of Wales at Caernarvon Castle and in the following year took his seat in the House of Lords. After a period at the Royal Air Force College at Cranwell, he entered the Royal Navy (1971) and continued in the service until 1976, when he took up full-time royal duties. In 1981 he married Lady Diana Spencer (1961–), daughter of the 8th Earl Spencer; they have two children, Prince William Arthur Philip Louis (1982–) and Prince Henry Charles Albert David (1984–). In recent years Prince Charles has aroused some controversy over his

outspoken and traditionalist views on architecture and education. He is also known for his concern for environmental issues.

Charlton, Bobby (Robert Charlton; 1937–) *British Association footballer who played 106 times for his country.*

Like his brother Jack Charlton (1935–), also an England international, Bobby came from the Northumberland mining community of Ashington. He was a schoolboy international before joining Manchester United, for whom he was an outstanding player from 1954 to 1973. He was one of the survivors of the Munich air crash in 1958, which killed several of his club colleagues. His first full international appearance was against Scotland in April 1958 and his last against West Germany in June 1970. He scored 49 goals for England and in 1966 he was a member of the national side that won the World Cup. Charlton was a brilliant inside forward and his impeccable behaviour on the field was held up as an example to young players. His playing career came to an end in 1972–73 and he spent a brief period as manager of Preston North End. Since 1984 he has been a director of Manchester United.

Charnley, Sir John (1911–82) *British orthopaedic surgeon who perfected an artificial hip joint that has brought mobility to thousands of arthritic patients. He was knighted in 1977.*

Charnley spent most of his life in his native Lancashire; he graduated from Manchester University in 1935 and trained at the Royal Manchester Infirmary and Salford Royal Hospital. He began his research on hip replacement in 1954, financing his work with patent royalties from his other discoveries, including a 'walking caliper' developed for wounded soldiers in World War II. After years of experimenting, Charnley found that the best combination of materials for a replacement hip was a thick plastic socket and a small-diameter highly polished metal ball to replace the head of the thigh bone. In 1962 he opened a centre for hip surgery at Wrightington Hospital, Wigan, where his methods are taught to surgeons of all nationalities; today more than 50 000 Charnley hip replacement operations are performed annually. Charnley served as professor of orthopaedic surgery at the University of Manchester from 1972 to 1976 and was the first orthopaedic surgeon to become a fellow of the Royal Society (in 1975).

Charpentier, Gustave (1860–1956) *French composer, best known for his opera* Louise.

Born in Dieuze, he started work in a mill at fifteen and his obvious musical talents were so impressive that his employer not only took lessons from him but sponsored his entry to the Lille Conservatory of Music. Later, Charpentier studied at the Paris Conservatoire, finding the Bohemian life of Montmartre much to his liking. Although his lack of respect for authority led to clashes with his tutors, he unexpectedly won the Prix de Rome in 1887, with his cantata *Didon*. While in Rome he wrote the orchestral suite *Impressions d'Italie*, the symphonic drama *La Vie du poète*, and the first act of *Louise*, an opera using his own libretto and based on his own experiences.

The triumph of the opera's production by Carré at the Opéra-Comique (1900) changed Charpentier's life. Moving from rags to riches in one step, he was able to develop a scheme to give free music tuition to working girls in Paris and in 1902 founded the Conservatoire Populaire Mimi Pinson (after Musset's heroine). Honours and academic status were offered to him but there was little more to come in the way of musical inspiration – except the opera *Julien* (1913). He became increasingly interested in film, radio, and recorded music. Charpentier's talent lay in the attractive use of simple thematic material and in a flair for vivid orchestration.

Cherenkov, Pavel Alekseyevich (1904–90) *Soviet physicist who discovered the form of radiation now known as Cherenkov radiation. He was awarded the 1958 Nobel Prize for Physics.*

Born in the Voronezh region, of peasant parents, Cherenkov was educated at the Voronezh State University and the Physical Institute in Moscow, where he gained his doctorate in 1930. He remained on the staff of the Institute, being appointed professor of experimental physics in 1953.

In the 1930s Cherenkov was asked to study the effects produced when radiation from such a source as radium passes through various fluids. The faint blue light that regularly appears in liquids irradiated in this way was generally assumed to be caused by fluorescing impurities in the liquid. Cherenkov, using distilled water and noting that the light continued, eliminated this possibility. Further investigation showed the light to be caused by fast secondary electrons produced by the radiation. Indeed, Cherenkov created the effect by irradiating the liquid with the electrons alone. In a series of papers published between 1934 and 1937, he established the characteristics of the radiation

soon to be named after him. It was left to his colleagues Ilya Frank (1908–90) and Igor Tamm (1895–1971) to provide a theoretical explanation of the phenomena. They showed that the electrons were travelling in the liquid faster than the speed of light in the medium, though not, of course, faster than the speed of light in a vacuum. The effect produced is in some respects analogous to the familiar sonic boom produced when a body travels faster than sound in a medium. For their work on this problem Tamm and Frank shared the Nobel Prize with Cherenkov. In later years Cherenkov worked on cosmic rays and on the design of large particle accelerators.

Chernenko, Konstantin Ustinovich (1911–85) *Soviet statesman, general secretary of the Soviet Communist Party and president of the Soviet Union (1984–85).*

Born in the Krasnoyarsk region into a peasant family, Chernenko had a limited early education (later supplemented, however, with studies at the Higher Party School in Moscow and the Kishinev Pedagogical Institute in Moldavia). Joining the Soviet Communist Party (CPSU) in 1931, he spent three years in the border guards (1930–33) and the next fifteen working for the CPSU in the provinces. The turning point in his career came in 1948, when he was sent to work in the propaganda and agitation department of the central committee of the Communist Party in Moldavia. Here he met Leonid BREZHNEV, then first secretary of the party in Moldavia, with whom he immediately established good relations and who thereafter became his patron.

Chernenko became sector head of the propaganda and agitation department of the central committee in Moscow in 1956 and head of the secretariat of the Presidium in 1960, when Brezhnev became chairman of the Presidium. After Brezhnev assumed the leadership of the CPSU, Chernenko became head of the general department of the central committee (1965–83), then a full member (1971) and then secretary (1976) of the central committee; in 1978 he became a full member of the Politburo. By now Chernenko was in a powerful position and closely allied with Brezhnev, whom he accompanied abroad for meetings with foreign politicians. On Brezhnev's death in 1983 it was Yuri ANDROPOV, however, who succeeded him in the leadership, although Chernenko remained in favour and chaired Politburo and secretariat meetings that Andropov was too ill to attend. When Andropov died on 9 February 1984, Chernenko was declared general secretary of the CPSU and two months later president of the Soviet Union. Dogged by ill health, he died in office after only thirteen months in power.

Cherwell, Frederick Alexander Lindemann, Viscount (1886–1957) *German-born British physicist, who became scientific adviser to Winston CHURCHILL and was created a viscount in 1956.*

The son of a wealthy Alsatian businessman, he was educated at schools in Scotland and Germany and the University of Berlin, where he gained his PhD in 1910. As a man of private means, Lindemann was under no pressure to find work and it was not until World War I, when he joined the Royal Aircraft Establishment at Farnborough, that he became fully employed. After the war Lindemann was appointed director of the Clarendon Laboratory and professor of experimental philosophy at Oxford, posts he held until his retirement in 1956. When Lindemann moved to Oxford, the Clarendon had an academic staff of two, no mains electricity, and a single technician. Under his direction it became one of the leading research centres in Britain and, in the field of low temperature physics, it led the world. Known universally as 'the Prof,' Lindemann made a limited contribution to physics; he invented an electrometer that bears his name and derived a formula relating the melting point of a crystal to the amplitude of its atoms' vibrations. He did, however, have many interests outside science. In 1921 he met Churchill, with whom he formed a lifelong friendship. He became Churchill's scientific adviser and exerted a considerable influence over him. Some, C. P. SNOW in particular, have argued that this influence was far from benign. He has been charged with obstructing the development of radar and supporting the discredited policy of saturation bombing during World War II.

Lindemann's influence declined during the years of the ATTLEE government (1945–51), but with the return of Churchill in 1951, Lindemann was summoned to serve as paymaster-general with a place in the cabinet. Although, initially, Lindemann had been sceptical about the feasibility of the atomic bomb, by the 1950s he appreciated the need to establish a nuclear industry and it was largely due to him that the Atomic Energy Authority was set up in 1954.

Chesterton, G(ilbert) K(eith) (1874–1936) *British journalist, poet, novelist, and broadcaster.*

A Londoner, Chesterton was educated at St Paul's School, where he illustrated E. C. BENTLEY's first book of clerihews. On leaving school he first intended to study art, but instead became a journalist. His poems, *The Wild Knight* (1900), were well received, but it was as a newspaper columnist that Chesterton's reputation was made. His vigorous paradoxical essays, attacking Victorian pretensions, decadence, and a whole range of fashionable attitudes and institutions, were so popular that they were republished in such collections as *Heretics* (1905). He also wrote major critical studies of Browning (1903) and Dickens (1906).

Chesterton also enjoyed success as a fiction writer. *The Napoleon of Notting Hill* (1904), a fantasy about civil strife between different London localities, was typical of his serio-comic narratives. Although he continued to write serious verse he also branched into a brilliant vein of comedy. *Magic*, his first attempt at drama, was a theatrical triumph in 1913. His greatest and most lasting success, however, was in the field of detective fiction, with the Father Brown stories (1911–27), the central character of which is a Roman Catholic priest.

Chesterton had met Hilaire BELLOC in 1900, and they had many opinions and attitudes in common. Like Belloc, Chesterton opposed the socialism of G. B. SHAW and H. G. WELLS. In 1922 Chesterton became a Roman Catholic, and from then on many of his copious writings were on religious topics, notably an important biography of St Francis of Assisi (1923). Towards the end of his life he became a popular and extremely expert broadcaster.

Chevalier, Maurice (1888–1972) *French actor, singer, and entertainer. He was awarded the Croix de Guerre and was an Officier de la Légion d'honneur.*

After a poor childhood in his native Paris, with periods spent in childrens' homes, Chevalier drifted into a variety of jobs before turning to the theatre. He made his stage debut in 1906, and after appearing in such shows as *Le Beau Gosse* at the Eldorado gained wide popularity as MISTINGUETT's dancing partner at the Folies-Bergère (1910). His London debut came after World War I, during which he was wounded and taken prisoner, in *Hullo, America* (1919) at the Palace Theatre. As well as revue he also appeared in straight plays, beginning with *Dédé* (1921).

The advent of sound films provided ideal scope for his gaiety and warmth, distinctive voice, and debonair manner, and his Hollywood film career began with *Innocents of Paris* (1929), in which he sang one of his most popular songs, 'Louise'. The same year he made his New York debut in ZIEGFELD's *Midnight Frolic*. Notable early films included LUBITSCH's first musical, *The Love Parade* (1930), and Mamoulian's *Love Me Tonight* (1932), singing his famous song 'Mimi'. In these and other musicals he was successfully cast opposite Jeanette MacDonald (1902–65). He also made two British films, *The Beloved Vagabond* (1936) and *Break the News* (1938). During World War II he remained in France and performed in Paris as well as in prisoner-of-war camps in Germany. Exploited by the Germans, he had to face charges of collaboration after the war. He was, however, acquitted and resumed his career in 1946 in René CLAIR's *Le Silence est d'or*. Most memorable of his later films was *Gigi* (1958). A truly international star, Chevalier appeared on stage throughout the world and made many television appearances.

Chiang Kai-shek (Jiang Jie Shi; 1887–1975) *Chinese nationalist statesman; president of China (1928–38; 1943–49) and later president of the nationalist Republic of China in Taiwan (Formosa) (1950–75). Although he never succeeded in regaining power over the communist mainland, he did oversee a period of considerable economic growth and prosperity in Taiwan, due largely to its close links with the USA.*

Born in Fenghwa, Chekian province, the son of a village merchant, Chiang was educated at the Chinese Imperial Military College and at a staff college in Japan. In 1911, he returned to China to take part in revolutionary activities against the Manchu dynasty. He served as chief of SUN YAT-SEN's staff in Canton (1921–22), becoming commander-in-chief of the army of the Chinese Revolutionary National Party (Kuomintang) in south China in 1925. Breaking with the communists in 1927, he established himself as head of a national government at Nanking in 1928.

During the 1930s he sought to maintain control while fighting the Japanese, who occupied Manchuria in 1931. He also had to contend with three rebellions (1930, 1933, and 1936), all of which he crushed. When the Japanese launched a full-scale attack on Chinese strongholds in 1937 he was forced to re-establish his headquarters in Chungking, Szechwan province.

He resigned the presidency in 1938 to devote his energy to military leadership, especially to organizing resistance to the Japanese, but in 1943 was reappointed president. He met both ROOSEVELT and CHURCHILL during World War II and was declared leader of China; he presided over the Japanese surrender in 1945. Shortly after, civil war broke out again between the Kuomintang and the communists. When Peking fell to the communists in 1949, Chiang fled from the mainland and moved his government to Taiwan, from where he planned an invasion of communist China with the military support of the USA. The invasion never took place, however, and during the 1970s the growing friendship between the USA and the People's Republic of China placed some doubt upon US guarantees of Taiwan's defence. Chiang remained president in Taiwan for twenty-five years, until his death in 1975.

Chichester, Sir Francis Charles (1901–72) *British yachtsman and aviator who made the first solo round-the-world sailing voyage in 1966–67. He was made a CBE in 1964 and received a knighthood in 1967.*

The son of a clergyman, Chichester dropped out of school and worked on a farm for a year before going to New Zealand in 1919. He took a variety of jobs, including lumberjack and door-to-door salesman, until forming a land agency in partnership with Geoffrey Goodwin, which proved successful. In 1927 the partners formed a civil airline, Goodwin and Chichester Aviation. Returning to Britain in 1929, Chichester obtained his pilot's licence, and, only three months later, made a solo flight from Croydon, England, to Sydney, Australia, in a Gipsy Moth biplane. The following year he fitted the plane with floats to make the first solo flight across the Tasman Sea between Auckland and Sydney. The flights are described in his books, *Solo to Sydney* (1930) and *Seaplane Solo* (1933). A planned round-the-world flight in 1931 ended in disaster when his plane hit telegraph wires in Katsuura harbour, Japan. He took five years to recover from his injuries.

Chichester made significant contributions to developing navigational techniques and during World War II he served with the Air Ministry, teaching navigation and writing books on the subject. After the war he founded a company, Francis Chichester Ltd., to publish maps and guides. But his pursuit of fresh challenges led him in 1953 to buy his first ocean-going yacht, *Gipsy Moth II*. This was followed in 1959 by *Gipsy Moth III*, in which he won the first solo transatlantic race and the *Observer* trophy. He again entered the race in 1964 but, in spite of a personal record time of just under thirty days, he came second. However, his most famous voyage started from Plymouth on 27 August 1966. He reached Sydney in 107 days, and after seven weeks rest in Australia, set out across the Pacific on 29 January 1967. Reaching Plymouth on 28 May, Chichester had sailed 29 600 miles in 226 days – a record time for such a voyage. This exploit made him a national hero, and the Queen used Sir Francis Drake's sword to knight him. His account of the voyage was published as *Gipsy Moth Circles the World* (1967). He made a further solo transatlantic crossing in 1971 but in 1972 ill health forced his retirement from a transatlantic race and he died some weeks later. His autobiography, *The Lonely Sea and the Sky*, was published in 1964.

Chifley, Joseph Benedict (1885–1951) *Australian statesman and Labor prime minister (1945–49).*

Born in Bathurst, New South Wales, the son of a blacksmith, Chifley left school at seventeen to join the New South Wales government railways. At twenty-four he became the youngest first-class locomotive driver in the service.

Chifley became increasingly involved in trade-union and Labor politics and entered parliament in 1928, when he won the federal seat of Macquarie for the Labor Party. Losing it in 1931, he was re-elected in 1940 and became treasurer and minister for postwar reconstruction in the CURTIN government. Following Curtin's death in 1945, Chifley became prime minister. Continuing to fulfil Labor's welfare and nationalization programme in accordance with his famous 'light on the hill' of socialism, he also initiated the postwar immigration policy and the Snowy Mountains hydroelectric scheme. He was defeated in the 1949 election but remained leader of the opposition until his death in 1951.

A tough single-minded administrator, Chifley had the image of an unpretentious idealist and is remembered for his characteristic gravelly voice.

Chirico, Giorgio de (1888–1978) *Italian painter, who was the originator of metaphysical painting and a precursor of surrealism.*

Having studied in Athens and Munich, Chirico went to Paris, where he painted most of the early pictures for which he is best known. These pictures, with their ominous dreamlike atmosphere, consist mainly of open spaces in which arcades and buildings are placed spar-

ingly in deep perspective and apparently out of context. Statues and small solitary figures cast long shadows in these eerie townscapes, as in *Melancholy of a Beautiful Day* (1913). After 1914 he also peopled his paintings with tailors' dummies and included a denser arrangement of objects, such as huge rubber gloves and abstract constructions.

In 1915 Chirico returned to Italy, where he was conscripted into the army. While in hospital, suffering from a nervous breakdown, he met the painter Carlo Carrà (1881–1966), also recovering from a mental disorder; together they formed the Scuola Metafisica in 1917. However, less than two years later Chirico changed his style: admiration for the great works of the Italian classical tradition led him to explore the techniques of the Renaissance in paintings of classical landscapes, Roman villas, horses, and gladiators. Returning to Paris in 1924 he found himself hailed as a master by the recently formed surrealist group. He temporarily reverted to his earlier style before breaking with the surrealists from 1930 and returning to an increasingly academic style.

Chomsky, Noam Avram (1928–   ) *US linguist, philosopher, and political activist, famous for his creation of the Chomskyan revolution in linguistics.*

Born in Philadelphia, the son of a Hebrew scholar, Chomsky was educated at the universities of Pennsylvania and Harvard. He joined the Massachusetts Institute of Technology in 1955 and has continued to teach there ever since.

In his early work Chomsky repeatedly made two very general claims about language. Most utterances, he declared, are unique in the sense that they consist of grammatical sentences never used before and unlikely ever to be used again. Grammar must therefore be seen as a set of rules capable of generating an infinite number of acceptable sentences. Secondly, he emphasized how incredibly quickly and accurately children of all degrees of intelligence learn to speak a language. They soon show themselves to be competent not only to use sentences they had been taught but also to generate new ones and to transform existing sentences ('John hit me') into equivalent forms ('I was hit by John'). This ability suggested to Chomsky that we are not contemplating an activity learned completely from scratch, but something for which we possess innate cognitive capacities.

This led Chomsky to a further insight. It is most improbable that English babies are born with a capacity to learn English as opposed to French. Consequently, it seemed necessary to distinguish between deep and surface grammar. The innate cognitive capacity of the mind applied only to the deep grammatical structures that are common to all languages. Surface features applicable to a particular language had to be learnt like any other acquired behaviour. Chomsky's views were first expressed at length in his *Syntactic Structures* (1957) and were continued in his *Aspects of the Theory of Syntax* (1965). Some of the more general features of his linguistic theory were presented in his *Reflections on Language* (1976) and *Language and Mind* (1972).

In the 1960s, however, Chomsky was concerned with more pressing problems than those emerging from linguistics. As a radical left-wing thinker he was naturally suspicious of the claims made by the US government about the progress of the Vietnam War. He was equally unhappy about the reasons given by the Administration for their presence in Indochina. In numerous articles, lectures, and the book *American Power and the New Mandarins* (1969) Chomsky accused the Administration of lying about the war's progress. He also claimed that there was no justification, moral, legal, or even strategic, for his country's presence in Southeast Asia. Despite the polemical nature of his argument, Chomsky's case remained a closely argued one that the Administration was compelled to take seriously. In 1991 he again voiced doubts about US foreign policy, this time in relation to the Gulf War. His most recent publications include *The Chomsky Reader* (1987) and *Necessary Illusions* (1989).

Chou En-lai (Zhou En Lai; 1898–1976) *Chinese communist statesman; prime minister (1949–76) and foreign minister (1949–58) of the People's Republic of China. Widely respected for his skill as a negotiator and for his knowledge of world affairs, he was the chief agent in maintaining amicable relations between China and the West.*

Born in Huaian, Kiangsu province, the son of a government official, Chou En-lai was educated at Nankai Middle School in Tientsin before travelling to Japan and France. While studying in Paris (1920–24), he was influenced by fellow student Ho Chi Minh and became a committed communist. On his return to China he joined Sun Yat-sen's National Party (Kuomintang), becoming head of the Whampoa Military Academy under Chiang Kai-shek in 1924.

When the communists and nationalists split and Chiang Kai-shek rose to power in 1927, Chou went underground, joining Mao Tse-tung's peasant and guerrilla movement in Kiangsi province (1931). He succeeded Mao as political commissar of the Red Army in 1932 and in the ensuing civil war between the Kuomintang and the communists, played an important role as chief negotiator (he negotiated terms for the release of Chiang Kai-shek when he was kidnapped in 1936). At the end of World War II he represented the communists (1945–47) in negotiations with the USA, which was seeking to mediate in the civil war. When, in 1949, the People's Republic of China was established, he was appointed prime minister, a position he held until his death in 1976. He also served as foreign minister (1949–58).

During the 1960s and early 1970s he continued to keep open communication channels with the USA, despite the USA's refusal to recognize the communist government, and presided over the moves towards detente made by President Nixon in 1972–73. On the domestic front, he was a moderating influence during the Cultural Revolution (1966–68).

Christian X (1870–1947) *King of Denmark (1912–47), best remembered for his courage in the face of the Nazi occupation of Denmark in World War II.*

In 1898 Christian, eldest son of the future King Frederick VIII (1843–1912), was appointed chief of the royal guard and in the same year married Alexandrine of Mecklenburg-Schwerin. He was created crown prince in 1906, and was a leading opponent of the sale of the Danish West Indies to the USA in 1911. During World War I he regularly met the two other Scandinavian neutrals, Norway and Sweden, and after the war Denmark obtained Schleswig from the defeated Germans. At home, he signed in 1915 a new constitution giving equal suffrage to men and women, and in 1918 Iceland (a Danish possession since 1381) became a separate kingdom – to achieve full independence as a republic in 1944. During World War II, when Germany occupied Denmark (1940–45), Christian was often seen riding on horseback through the capital Copenhagen to underline his continuing claims for Danish sovereignty. In 1942 he rejected Nazi demands for antisemitic legislation, but was forced a year later to condemn partisan sabotage of the railways. In August 1943 he made a speech against the occupying forces and was imprisoned for the rest of the war.

Christie, Dame Agatha Mary Clarissa
(1891–1976) *British writer of detective fiction.*
She was created a DBE in 1971.

Born Agatha Miller, she married Archibald
Christie in 1914. They were divorced in 1928
but in the meanwhile she had begun her triumphant career as a detective-story writer with
The Mysterious Affair at Styles (1920), in
which she introduced Hercule Poirot, the Belgian detective who was her most successful
fictional creation. *The Murder of Roger
Ackroyd* appeared in 1926 and received critical
acclaim. The same year she was discovered at
a health resort suffering from amnesia, after a
mysterious disappearance believed by some to
be no more than a publicity stunt.

In 1930 she married the archaeologist Max
Mallowan, and some of her stories thereafter
used their travels in the Middle East as background. Her forte was her brilliantly constructed plots with ingenious psychological
twists, ensuring a loyal readership for the fifty
or more novels she wrote in a writing career of
over half a century. A number of her books
were filmed, including *Lord Edgware Dies*
(1933) in 1936, *4.50 from Paddington* (1957)
in 1962, as *Murder She Said*, *Murder on the
Orient Express* (1934) in 1974, and *Death on
the Nile* (1937) in 1978. She also turned her
hand to stage plays. *The Mousetrap* (1952)
enjoyed the longest continuous run in the history of London theatre, and *Witness for the
Prosecution* (1953) was filmed in 1958. Apart
from Poirot, she created several other fictional
detectives, notably the resourceful Miss Jane
Marple.

Christoff, Boris (1918–) *Bulgarian bass
singer renowned for his performance of Boris
in Mussorgsky's opera* Boris Godounov *and
other bass roles in Russian opera.*

He initially studied law but as a member of the
well-known Gusla Choir he was heard by King
Boris of Bulgaria, who sponsored his tuition in
Rome (1941) with Riccardo Stracciari.
Christoff later also studied with Muratti in
Salzburg. He made his debut in Italy in 1946,
first on the concert platform, then in opera (in
the part of Colline in Puccini's *La Bohème*).
His first performance as Boris Godounov was
at Covent Garden in 1949; he then sang this
role in many of the opera houses of the world,
returning with it to Covent Garden in 1974 to
mark the twenty-fifth anniversary of its first
performance there.

His large repertoire includes most of Verdi's
leading bass roles, of which Philip II in *Don
Carlos* is particularly associated with him (he

has been much praised for the soft sustained
singing of Philip's monologue). He made his
debut in America as Boris in San Francisco
(1956) and sang in Chicago from 1957 to 1963.
A great actor-singer, he is also a polished performer on the recital platform, especially in the
Russian repertory.

Churchill, Sir Winston Leonard Spencer
(1874–1965) *British statesman and prime minister during World War II. His courage and
independence of mind, which often made difficulties in his early career, came fully into their
own in war, when he demonstrated rare qualities of leadership and outstanding gifts as an
orator. Also a writer and a Sunday painter, he
was knighted in 1952.*

Winston Churchill was born at Blenheim Palace, built by a grateful nation for his ancestor,
the first Duke of Marlborough. Churchill's
mother was the daughter of a rich American,
Leonard Jerome; his father was Lord Randolph
Churchill (1849–95), whose rapid rise to political power was cut short by ill health and
premature death. The young Churchill's talents were not apparent during his unremarkable schooldays at Harrow. After Harrow,
Churchill was, at his third try, accepted for
army training at Sandhurst. He served in India
and in the Sudan, where he simultaneously
wrote for the *Morning Post*. He was present at
the battle of Omdurman (1898). His grasp of
military matters at this early stage was shown
in his book on the reconquest of the Sudan,
published in 1899. In the same year, having
resigned his commission, he went to South
Africa, where he reported again on the second
Boer War for the *Morning Post*, and was
briefly imprisoned by Louis BOTHA. His escape
from prison camp and subsequent jumping of
a train was very much to his taste.

Returning to England, he was elected Conservative MP for Oldham in 1900 and became
associated with those who opposed the party
leadership, being especially critical of BAL-
FOUR. In doing so, he felt himself to be following the tradition of Tory democracy, of which
his father had been a leading exponent; at this
time he was undoubtedly influenced by the
work he was doing on Lord Randolph's biography (published in 1905). Churchill finally
broke with the Conservatives over the party's
adoption of tariffs, and in 1904 he joined the
Liberals, whom he regarded as the true supporters of 'the cause of the left-out millions'.
In ASQUITH'S government, he established his
reputation as a Liberal reforming minister,
working in close collaboration with LLOYD

GEORGE as president of the Board of Trade (1908–10) and home secretary (1910–11). In 1911 he became first lord of the Admiralty, but his views on military and naval issues – and his actions – provoked hostility. The failure of the Dardanelles expedition (1915) in World War I (though Churchill was vindicated in the 1917 report on the venture) brought his demotion to the duchy of Lancaster, and he resigned. He then joined up and served as a colonel in France. In 1917 he was appointed minister of munitions in Lloyd George's coalition government and, though his unpopularity with the Conservatives kept him out of the cabinet, he became a close adviser of the prime minister. He followed Lloyd George in advocating development of the tank, which proved crucial in breaking the deadlock on the western front.

Churchill was transferred to the Colonial Office after the war, and in 1922 he lost his parliamentary seat. Two years later he was re-elected as a Constitutionalist but effectively represented the Conservative vote in Woodford, a constituency he held until 1964. As chancellor of the exchequer (1924–29) under Baldwin, he was responsible for the return to the gold bullion standard (1925), which set off the disastrous train of events that led to the general strike in the following year. In 1931 Churchill resigned from the Conservative shadow cabinet in protest against his party's support of self-government for India. For the rest of the 1930s he was in the political wilderness, though he made his voice heard – against the growing Nazi menace and in support of Edward VIII in the abdication crisis. In these years he published a biography of the first Duke of Marlborough (1933–38). At the outbreak of World War II he was appointed again to the Admiralty (1939) by Neville CHAMBERLAIN. But when, in 1940, Chamberlain resigned, he advised the king to call on Churchill to form a coalition government.

Churchill's genius was to communicate his conviction that the war could and must be won. His other major contribution to victory was to forge and maintain the Alliance, especially with the United States, which defeated the Axis powers. Working closely with ROOSEVELT and keeping on equal terms with STALIN, he travelled widely throughout the war and was present at the three great Allied conferences at Teheran (1943), Yalta, and Potsdam (both 1945). During the last of these, Churchill's government was defeated at the polls and he was replaced by ATTLEE.

Already at Yalta, Churchill feared Soviet intentions in Europe and in 1946, at Fulton in the United States, he spoke of an 'iron curtain' that had descended across the Continent. In his second and last government (1951–55), Churchill ordered the manufacture of the hydrogen bomb, which he saw as a deterrent to war.

Churchill's retirement, long expected, came in 1955 when he was eighty, though he remained an MP until 1964. In 1956 he was awarded the Charlemagne Prize for his service to Europe. Churchill's other books include *The Second World War* (6 vols; 1948–54) and *A History of the English-Speaking Peoples* (4 vols; 1956–58).

Ciano, Galeazzo (1903–44) *Italian fascist politician; minister of foreign affairs (1936–43) and son-in-law of* MUSSOLINI.

Born in Livorno, the son of a World War I admiral, Ciano was educated at the University of Rome, where he graduated in law (1925). He worked briefly as a journalist before joining the diplomatic corps, which took him to Latin America and China.

Ciano's marriage in 1930 to Edda Mussolini, the daughter of the fascist leader, led to the rapid advancement of his career. Appointed minister to China in 1932, he became head of the press office in 1933 and undersecretary of state for press and propaganda in 1934. He served in the air force during the Ethiopian War (1935–36). On his return to Italy, he was appointed minister of foreign affairs and a member of the Fascist Supreme Council (1936). As foreign minister he signed the 'Pact of Steel' with Germany (1939) and persuaded Mussolini not to intervene in World War II until the fall of France in 1940. After several military defeats, Ciano advocated a separate peace with the Allies in 1942. Appointed ambassador to the Vatican in early 1943, he was instrumental in the coup of July 1943 that overthrew Mussolini and brought an end to fascism in Italy. He fled to Germany shortly afterwards, but was captured by pro-Mussolini partisans and Germans in northern Italy. He was tried and executed for high treason in 1944.

Although Ciano lived in the shadow of Mussolini, some of his personal views about foreign policy differed. He had far deeper reservations about the alliance with Germany and disliked the German foreign minister RIBBENTROP. He disapproved of Mussolini's ostentatious style, yet he spent considerable time increasing the large inheritance left by his father. Ciano's diaries (1939–43), which reveal the inside machinations of the Axis alliance, were published in two volumes in 1947.

Cicero (Elyesa Bazna; 1904–70) *Turkish spy who passed Allied secrets to the Germans while working as a valet to the British ambassador in Ankara during World War II.*

Bazna was born in Pristina (now in Yugoslavia) but the family moved to Istanbul during his childhood. He took a variety of jobs and served time in a French penal camp for petty offences against the Allied forces occupying Turkey after World War I. On his return to Istanbul he resumed his series of jobs, finally working as a driver and servant to various embassies in Ankara. In 1943 he was made personal valet to the British ambassador, Sir Hughe Knatchbull-Hugesson (1886–1971).

Bazna saw this as an opportunity both to make money and to impede Turkey's entry into the war. Using duplicate keys to gain access, he photographed documents in the embassy safe. His contact in the German embassy, L. C. Moyzisch, realized the immense importance of Bazna's material, which included top-secret minutes of Allied conferences and diplomatic correspondence, most significantly relating to the Allies' second front and moves to bring Turkey into the war. However, the German foreign minister, RIB-BENTROP, refused to accept that the information was genuine, believing that 'Cicero', as he was code-named by the Germans, was a double agent. The German embassy in Ankara sent Moyzisch to Germany to convince the head of Nazi security service, Ernst Kaltenbrunner (1902–46), that Cicero was trustworthy. Kaltenbrunner provided some £200,000 in English banknotes as payment for Cicero, who hid them under the carpet of his room in the British embassy.

By August 1943, the Allies became aware that an unidentified German agent was operating in the British embassy in Ankara. Realizing this, Cicero became more circumspect and eventually ceased his activities. He left his job at the embassy in April 1944 and after a decent interval attempted to finance various building contracts with his hoarded banknotes, only to find they were counterfeit. After the war, he tried and failed to obtain remuneration from the German government for his wartime services.

Clair, René (René Lucien Chomette; 1898–1981) *French film director, producer, and scriptwriter. In 1960 he became the first film director to be elected to the Académie Française. He was also appointed Grand Officier de la Légion d' honneur.*

Born in Paris, Clair began his career as a journalist before turning to films in 1919. When he was twenty-five he directed and wrote his first film, *Paris qui dort* (1924; *The Crazy Ray*), having acted in several films and been assistant director to Jacques de Barnocelli in Brussels. Films that followed included *Entr' acte* (1924), *Le Voyage imaginaire* (1926), and *Un Chapeau de paille d' Italie* (1927; *The Italian Straw Hat*). His international reputation firmly established, he made his first sound film, *Sous les toits de Paris* (1930). *À nous la liberté* (1931), said to have inspired CHAPLIN's *Modern Times*, was among the notable films that followed. In Britain he made *The Ghost Goes West* (1935), for Alexander KORDA, and in Hollywood he made such films as *Flame of New Orleans* (1941) with Marlene DIETRICH and *It Happened Tomorrow* (1944) with Dick Powell (1904–63). Among the memorable films he made on his return to France were *Les Grandes Manoeuvres* (1955), *Porte des Lilas* (1956), and *Tout l' or du monde* (1960). As an author he published novels and memoirs as well as books on the cinema, including *Réflexion faite* (1951) and *Cinéma d' hier, cinéma d' aujourd' hui* (1970).

Clark, Jim (James Clark; 1936–68) *British motor racing driver and World Drivers' Champion in 1963 and 1965.*

A farmer's son from Kilmany, Fife, Clark started racing in local motor club events in the mid-1950s and was a member of the Border Reivers racing team. He joined Colin Chapman's Lotus team in 1960 to drive Formula Junior, being given his first Formula One drive in the Dutch Grand Prix at Zandvoort. He stayed with Lotus and in 1962 achieved his first Grand Prix victory in Belgium, although he just lost the championship to Graham HILL in the final race of the season. The following year, Clark had a string of victories and captured the Drivers' Championship. He repeated this success in 1965, winning seven Grands Prix as well as the Indianapolis 500.

Clark survived several crashes but was killed in a Formula Two race at the Hockenheim circuit while leading the world championship. His total of twenty-five Grand Prix victories beat FANGIO's record and he ranks among the greatest drivers of all time.

Clark, Joe (Charles Joseph Clark; 1939–   ) *Canadian statesman; leader of the Progressive Conservative Party (1976–83) and prime minister (1979–80).*

Clark was born in High River, Alberta, the son of a local newspaper publisher. He graduated in history from the University of Alberta and went on to study law at Halifax, Nova Scotia, and at the University of British Columbia. Having developed an interest in politics as a student, he worked for Conservative leaders in Alberta and British Columbia in the mid-1960s and became executive assistant to Robert Stanfield (1914–), the national leader of the Conservative Party, in 1967. He served as MP for the Alberta seat of Rocky Mountain in 1972 and 1974, and on Stanfield's resignation in 1976 Clark embarked on a vigorous and successful campaign for the party leadership. Following the downfall of Pierre TRUDEAU's Liberal government in 1979, Clark became the youngest prime minister in Canadian history. His reign was brief: in early 1980 his party was defeated on the issue of higher energy prices, and in the subsequent election Trudeau was returned to power. Clark's support within the party fell off, and in February 1983 he resigned as party leader. In an attempt to regain the leadership in June of that year he was defeated by Brian Mulroney (1939–) of Quebec, who subsequently became prime minister in September 1984. Since 1984 Clark has served as Secretary of State for External Affairs.

Claudel, Paul(-Louis-Charles-Marie) (1868–1955) *French poet, dramatist, and diplomat. He was elected to the Académie Française in 1946.*

Claudel was born in Villeneuve-sur-Fère-en-Tardenois, Aisne. At the age of eighteen he underwent a sudden conversion to Roman Catholicism after a religious experience in the Cathedral of Notre Dame, Paris, on Christmas Day, 1886. In the same year he discovered the poetry of Rimbaud; these two experiences were to have a profound effect on his future literary works, beginning with the play *Tête d'or* (1889).

Claudel successfully combined a distinguished career as a diplomat with his prolific output as a writer; having entered the diplomatic service in 1890 he held posts in the consulates and embassies of Europe, China, the USA, and South America. His best poetry is to be found in the *Cinq Grandes Odes* (1910); later notable works included *Corona benignatis anni Dei* (1914), *Poèmes de guerre 1914–1916* (1922), and *Feuilles de saints* (1925). He developed a free-verse style, characterized by long unrhymed lines, that was subsequently known as the *verset claudélien*.

Having composed his first play, *L'Endormie*, at the age of fourteen, Claudel began writing seriously for the theatre in the

1890s. His early plays, such as *L'Annonce faite à Marie* (1910), first performed in 1912, and the autobiographical *Partage de midi* (1906), produced by Antonin ARTAUD in 1928, are lyrical free-verse exposés of his vision of the universe, ordered and governed by God. He achieved his greatest acclaim as a dramatist with *Le Soulier de satin* (1924; translated as *The Satin Slipper*, 1931), produced by Jean-Louis BARRAULT at the Comédie-Française in 1943. Set in Renaissance Spain and South America, it is a story of unconsummated love between the two central characters, who are forced to tread their separate paths towards God; these themes of personal destiny and forbidden love recur in many of Claudel's works. Claudel's other plays include *La Jeune Fille Violaine* (1892), an early version of *L'Annonce faite à Marie*; the trilogy *L'Otage* (1911), *Le Pain dur* (1918), and *Le Père humilié* (1920); and *Jeanne au bûcher* (1934), performed in 1938 with music by Arthur HONEGGER. He also wrote literary criticism, notably *Art Poétique* (1907); his lengthy *Correspondance, 1899–1926* with André GIDE was published in 1949.

Clemenceau, Georges (1841–1929) *French statesman and prime minister (1906–09; 1917–19).*

Born in Mouilleron-en-Pareds in the Vendée, the son of a doctor, Clemenceau was educated at Nantes and Paris, where he studied medicine. In 1865 he went to the USA as the correspondent for a Paris newspaper, teaching riding and French in a Connecticut girls' academy. On his return to France in 1869, he was appointed mayor of Montmartre (1870) and was elected to the national assembly as the Paris representative in 1871. Following the Paris Commune uprising (1870–71) he resigned from both positions but took up a post on the municipal council of Paris where he remained until 1876, assuming the presidency in 1875.

Clemenceau was elected to the national chamber of deputies in 1876 as the Radical Party member for Paris. He held this position until 1893, when he returned to journalism, gaining prominence for his support of Alfred Dreyfus (1859–1935). Elected as senator for the Var in 1902, he was appointed minister for the interior in March 1906 and became prime minister seven months later. In a dispute over naval policy in 1909 he was forced out of office. In 1913 he founded *L'Homme Libre*, a daily paper that expressed his views on disarmament and the German threat. Between 1914 and 1917 he was an outspoken critic of the

French government's military inefficiency and defeatism. Appointed prime minister by President POINCARÉ in 1917, he led the French delegation to the Paris Peace Conference, playing a major role in the drafting of the Treaty of Versailles. He resigned as prime minister in 1919 and, following an unsuccessful attempt at the presidency in 1920, retired from politics.

He devoted his years of retirement to writing, completing a two-volume philosophical testament, *Au soir de la pensée* ('In the Evening of My Thought') in 1927. His memoirs, *Grandeurs et misères d'une victoire*, were published posthumously in 1930.

Cockcroft, Sir John Douglas (1897–1967)
British physicist, engineer, and scientific administrator, who, with Ernest WALTON, was responsible for the first experimental splitting of the atomic nucleus. For this work they were awarded the 1951 Nobel Prize for Physics. The first head of the Atomic Energy Research Centre at Harwell, he was knighted in 1948 and awarded the OM in 1957.

Born in Todmorden, Lancashire, the son of a cotton-mill manager, Cockcroft went from Todmorden Secondary School into World War I as a signaller. After the war, he entered Metro-Vickers as an apprentice and was sent by them to study electrical engineering at Manchester University. He did so well there that he was persuaded to go up to Cambridge to read mathematics, though he was now in his mid-twenties. After Cambridge he was taken on by Lord RUTHERFORD as a physicist in the Cavendish Laboratory, although he had had no formal training in physics. He remained there from 1924 to 1939. It was at the Cavendish that Cockcroft, working with Ernest Walton, made his major discovery that the atomic nucleus could be split. Taking seriously a suggestion of George GAMOW, in 1932 Cockcroft and Walton directed accelerated protons at a lithium target. The result was a shower of alpha particles scintillating brightly on a zinc sulphide screen behind the lithium. Atoms of lithium had thus been split into two helium nuclei (alpha particles).

Towards the end of the 1930s Cockcroft was drawn into research on radar and with the outbreak of World War II he joined the Ministry of Supply as assistant director of scientific research. In this capacity Cockcroft was a member of the TIZARD mission of 1940, which took to the USA the secrets of radar, plans for a jet engine, and several other top-secret projects. Later in the war Cockcroft directed the Anglo–Canadian atomic energy project, building at Chalk River in Ontario the first heavy water reactor to be completed outside the USA.

After the war Cockcroft was appointed director of the newly established Atomic Energy Research Station at Harwell. Under Cockcroft's direction the first British reactor was built. Cockcroft left Harwell in 1959 to become the first master of the newly founded Churchill College in Cambridge, an office he continued to hold until his death.

Cockerell, Sir Christopher Sydney
(1910–   ) *British engineer and inventor of the hovercraft, who was knighted in 1969.*
Born in Cambridge and educated at Cambridge University, Cockerell joined Marconi in 1935 as an electronics engineer, working on the development of airborne navigational equipment. He left in 1950 to concentrate on developing his own ideas. As an amateur yachtsman, Cockerell had long been aware of the slowing friction between the water and the hull of his boat. Attempts to reduce this drag had little effect. He therefore considered the more radical solution of raising the boat's hull completely out of the water. The crucial experiment was performed by Cockerell in 1954 with a set of kitchen scales, a vacuum cleaner, and empty food tins. He found that, properly arranged, an air jet could provide the required lift and that the experiment warranted the construction of a prototype.

Granted a patent in 1955, Cockerell succeeded in attracting the attention of the Ministry of Supply in 1957, who commissioned the firm of Saunders Roe to build a full-size craft. The SR.NI, weighing seven tons and capable of sixty knots, was launched in 1959. In the same year, with Cockerell aboard, it crossed the English Channel for the first time. Since then Cockerell's life has been devoted to the development and promotion of his original invention. He served as a consultant to Hovercraft Development Ltd (1958–79). During this period he saw the hovercraft enter regular cross-Channel service in 1968 and become an established means of transport in many parts of the world.

Cocteau, Jean (1889–1963) *French poet, novelist, dramatist, artist, and film-maker. He was elected to the Académie Française in 1955.*
Cocteau was born into a wealthy bourgeois family at Maisons-Laffitte, near Paris. He served as an ambulance driver in World War I; his wartime exploits are reflected in the poems of *Le Cap de Bonne-Espérance* (1919) and in

the novel *Thomas l'Imposteur* (1923). A writer of immense versatility, Cocteau was equally at ease with poets, novelists, painters, musicians, choreographers, and film-makers. He wrote sketches for DIAGHILEV's ballet *Parade* (1917) and in the same year became involved with Les Six, a group of French composers including Georges AURIC, later collaborating with them in the ballet *Les Mariés de la Tour Eiffel* (1921). *Le Boeuf sur le toit* (1920) was a farcical ballet written for the famous clowns the Fratellinis, with music by Darius MILHAUD. In 1918 Cocteau launched his protégé, the young poet and novelist Raymond Radiguet; five years later, on Radiguet's premature death at the age of twenty-one, Cocteau took to opium. Treated for his addiction, he later described his experiences in *Opium* (1930).

Cocteau saw himself essentially as a poet, describing his novels as *poésie de roman*, his plays as *poésie de théâtre*, and his critical essays as *poésie critique*. He wrote poetry throughout his life, beginning with the early volume *La Lampe d'Aladin* (1909); later poems included *Plain-Chant* (1923), *L'Ange Heurtebise* (1925), and *Clair-obscur* (1954). He also explored the role of the poet in the novel *Le Potomak* (1919) and in some of his best-known films – *Le Sang d'un poète* (1932; translated as *The Blood of a Poet*, 1936), *Orphée* (1950), and *Le Testament d'Orphée* (1960). His other notable films include *L'Éternel Retour* (1944) and *La Belle et la bête* (1945).

Cocteau's most famous novel, *Les Enfants terribles* (1929; filmed by Cocteau in 1950 and translated as *Children of the Game* in 1955), centres on the intimate relationship between an adolescent brother and sister; an earlier novel, *Le Grand Écart* (1923), also dealt with the subject of adolescence. His plays, many of which were later filmed, include *La Voix humaine* (1930), *La Machine infernale* (1934; translated as *The Infernal Machine*, 1936), which was inspired by the Oedipus legend and is generally considered to be Cocteau's greatest play, *Les Parents terribles* (1938; translated as *Intimate Relations*, 1951), and *L'Aigle à deux têtes* (1946; translated as *The Eagle with Two Heads*, 1962), a melodrama set in Ruritania.

A keen artist, Cocteau worked in a variety of media, including glass, pottery, and ceramics; he also illustrated books and decorated public buildings and churches. His other published writings include literary criticism, such as *Le Rappel à l'ordre* (1926); *Portraits-Souvenir* (1936), a collection of his weekly articles for *Le Figaro*; and *La Belle et la bête: Journal d'un film* (1946).

Coleridge-Taylor, Samuel (1875–1912)

British composer, best remembered for his Hiawatha *trilogy.*

His father was a West African doctor who abandoned his English wife in the London suburb of Croydon and returned to his native Sierra Leone on the grounds that racial prejudice prevented him from practising in London. The young Samuel showed his musical talent early and with the support of his local choirmaster entered the Royal College of Music in 1890 to study the violin. He also joined Stanford's composition class, winning a composition scholarship in 1893, and writing chamber pieces and a symphony (1896), both of which were performed. In this period he also composed *Hiawatha's Wedding Feast* (1898), the first part of his *Hiawatha* trilogy, a set of cantatas based on Longfellow's poem. The second part, *The Death of Minnehaha*, was performed at the North Staffordshire Festival in 1899; the third and last part, *Hiawatha's Departure*, was first performed at the Albert Hall in 1900.

Hiawatha was successful and popular to a degree never attained by his other compositions. These include three operas, a considerable amount of orchestral and choral music, some chamber music, and incidental music for a series of productions at His Majesty's Theatre. Coleridge-Taylor visited the USA three times (1904, 1906, and 1910) to conduct performances of his music. In England he was an active teacher and adjudicator at festivals.

Colette (Sidonie-Gabrielle Colette; 1873–1954)

French novelist. She was a member of the Belgian Royal Academy (1935), president of the Académie Goncourt (1945), and in 1953 was created a Grand Officier de la Légion d'honneur.

Colette was born in the Burgundy village of Saint-Sauveur-en-Puisaye, Yonne. In 1893 she married the writer Henri Gauthier-Villars and her first four novels, the 'Claudine' series (1900–03), were ghost-written for her husband and appeared under his pseudonym, Willy. The adventures of the young heroine, whose freshness and sensuality contributed to the success of the novels and also aroused some scandal, were partly based on Colette's own experience. Having published *Dialogues de bêtes*, the first of many engaging sketches of animal life, under the name Colette-Willy in 1904, Colette finally broke away from her husband's influence, divorcing him in 1906. A spell in the Parisian music halls was described in *La Vagabonde* (1910) and *L'Envers du music hall* (1913). During World War I she turned to journalism, writing dramatic criticism and short stories for such periodicals as *Le Matin*, whose editor, Henri de Jouvenel, she had married in 1912.

Colette firmly established her reputation as a novelist with *Chéri* (1920), which deals with the love of a young man for an older woman. This was followed by *La Maison de Claudine* (1922), inspired by childhood memories; *Le Blé en herbe* (1923; translated as *Ripening Seed*, 1955), a study of adolescent sexuality; *La Fin de Chéri* (1926); and *Sido* (1929). Her understanding of the natural world and her love of animals are displayed in such works as *La Paix chez les bêtes* (1916; translated as *Creatures Great and Small*, 1951) and *La Naissance du jour* (1928; translated as *A Lesson in Love*, 1932). *La Chatte* (1933; translated as *The Cat*, 1953) centres on a wife's jealousy for her husband's pet cat; the theme of jealousy recurs in *Duo* (1934).

In later life, happily married to the writer Maurice Goudeket, Colette retained the warm regard and great respect of the French literary world with such novels as *Le Képi* (1943), *Gigi* (1944), *L'Étoile vesper* (1947), and *Le Fanal bleu* (1949). Dying at the age of eighty-one, a legendary figure at the end of a writing career spanning fifty years, she was given a state funeral.

Collingwood, R(obin) G(eorge) (1889–1943) *British philosopher, historian, and archaeologist of Roman Britain.*

Born at Cartmel Fell, Lancashire, of poor but talented, artistic, and imaginative parents – his father was Ruskin's secretary and biographer – Collingwood was educated (with a family friend's help) at Rugby and University College, Oxford, graduating in classics. Apart from a period in the Intelligence Department of the Admiralty during World War I, Collingwood spent the whole of his academic life in Oxford, first as philosophy tutor at Pembroke College and finally, from 1935 until his early retirement in 1941, as Waynflete Professor of Metaphysics. Quite remarkably, for much of this time Collingwood also pursued an active career in the demanding field of Romano-British archaeology. He wrote numerous works on the latter subject, including, in collaboration with J. N. L. Myers, Volume I of the *Oxford History of England – Roman Britain* (1936).

Philosophically, Collingwood was something of an outsider. He managed to resist the Hegelianism current in the Oxford of his stu-

dent days without being in any way tempted by the alternative analytic tradition emerging in Cambridge from the work of Bertrand RUSSELL and G. E. MOORE. A popular lecturer and a lively writer, he paid special attention to the neglected philosophy of history.

In his early works, *Speculum Mentis* (1924) and *An Essay on Philosophical Method* (1933), Collingwood sought to present 'a critical review of experience' and to distinguish between the methods of philosophy and science. In his mature work, *An Essay on Metaphysics* (1940), Collingwood abandoned much of his earlier philosophy and adopted in its place a more historicist approach. Metaphysics, he declared, was basically a historical discipline. Different cultures of the past had adopted different 'absolute presuppositions'. It was the task of metaphysics not to prove or disprove such presuppositions but to distinguish between them and to sort out their historical relationships. Just how Collingwood proposed to pursue this programme was sketched in his two posthumous works, *The Idea of Nature* (1945) and *The Idea of History* (1946). He published his *Autobiography* in 1939 but explicitly asked for no biography.

Colman, Ronald (1891–1958) *British-born US film star, whose cultured voice and dignified appearance made him Hollywood's epitome of the English gentleman.*

Born in Richmond, Surrey, Colman had an early desire to become an actor that led him into amateur theatricals. However, his entry into the profession in 1914 was interrupted by World War I. After service with the London Scottish in France he was invalided out in 1916 and returned to the stage, making his first appearance in *The Maharani of Arakan* at the London Coliseum. Many more stage performances followed, both in London and on tour in the USA. He appeared in films in England as early as 1917 but it was in America that his film career really began. Most notable of his early films was *The White Sister* (1923) with Lillian Gish (1896–) and Herbert Brenan's silent classic *Beau Geste* (1926). His voice made him an ideal choice for the talkies and he was nominated for Best Actor Awards for *Bulldog Drummond* (1929), *Condemned* (1929), and *Random Harvest* (1942); he won this award for *A Double Life* (1947). Other notable performances were in *A Tale of Two Cities* (1935), *Lost Horizon* (1937), and *The Prisoner of Zenda* (1937). Towards the end of his life he had a cameo part in Mike Todd's *Around the World in 80 Days* (1956) and made his last screen appearance in *The Story of Mankind* (1957).

Coltrane, John William (1926–67) *Black US jazz saxophonist. He was one of the most influential and commercially successful jazz musicians of the 1960s.*

Born in Hamlet, North Carolina, Coltrane played in a US navy band before joining a rhythm and blues group. Once established as a jazz musician, he played in groups led by Dizzy GILLESPIE, Earl Bostic (1913–65), and Johnny Hodges (1906–70). In 1955–57 he was part of the Miles DAVIS sextet, where he developed his 'sheets of sound' technique, playing passionate arpeggios as though trying to play entire chords at once. In 1957 he recorded *Blue Train* under his own name. A drug addict, Coltrane gave up tobacco, alcohol, and heroin in the first six months of 1957, deciding that music was more important to him. During the second half of the year he worked for Thelonius MONK, from whose difficult music he learned a great deal about rhythm and harmony. In 1959 he returned to Davis, recording *Kind of Blue*. In mid-1960 he formed his own quartet and soon became the most influential jazz musician since Charlie PARKER. He took up the soprano saxophone as well as the tenor and explored Eastern modes and split-reed techniques in an attempt to express his religious mysticism. Long improvisations on 'My Favourite Things', an unusual vehicle for jazz in 3:4 time, included a musical conversation with drummer Elvin Jones (1927–   ). The recording *A Love Supreme* seemed to many of his followers to sum up the soul-searching culture of the decade.

Technically, his playing was a link between the harmonically dense jazz of the 1950s and the free jazz that was evolving in the 1960s. Despite his gentle introspective personality, his music sometimes sounded angry – a phenomenon his fans described as the love in him trying to get out. He died of cancer.

Colum, Padraic (Patrick Colm; 1881–1972) *Irish poet, playwright, and novelist.*

Born in Longford in a workhouse run by his father, Colum was deeply influenced by the Irish Celtic revival of the early 1900s while still an undergraduate at Trinity College, Dublin. Foremost among his early works were the plays he wrote for the nascent theatre movement when the Abbey Theatre, Dublin, was founded under the direction of Lady Gregory (1852–1932) and W. B. YEATS. These plays, for example *The Land* (1905), depict the humble lives of Irish smallholders and peasants.

This topic recurs in much of Colum's poetry, as in *Wild Earth* (1907), which he wrote while still employed as a railway clerk. In 1911 he became co-founder of the *Irish Review*. In 1914 Colum went to the USA to lecture and never settled again in Ireland, although Irish themes remained predominant in his work. In 1923 he visited Hawaii to study local folklore in order to write it up into children's stories. He subsequently wrote many books for children retelling legends not only from Hawaii but also from Greece, Ireland, and northern Europe.

Colum's *Collected Poems* appeared in 1932, but he published other verse after this, including the narative poem *The Story of Lowry Maen* (1937) and *Images of Departure* (1968). He also successfully translated poetry from the Irish. His comedy *Balloon* was produced in 1946.

Compton, Arthur Holly (1892–1962) *US physicist who discovered the Compton effect, for which he was awarded the 1927 Nobel Prize for Physics.*

Born in Wooster, Ohio, the son of a Presbyterian minister (who was also a professor of philosophy), Compton was educated at Wooster College and Princeton, where he obtained his PhD in 1916. After teaching for a year at the University of Minnesota, Compton spent World War I at the Westinghouse Corporation working on the development of aircraft with Ernest RUTHERFORD. On his return to the USA in 1920, he worked briefly at the University of Washington, St Louis, before being appointed to the chair of physics at Chicago University in 1923.

In this year Compton discovered the effect named after him. While observing the scattering of X-rays by light elements he noticed that some of the scattered radiation increased in wavelength. Compton assumed that X-rays can behave like particles and lose energy on colliding with electrons in the target; the energy lost by the X-radiation causes an equivalent increase in their wavelength. The effect thus provides evidence for the dual wave–particle nature of radiation. Shortly afterwards Compton turned to a study of cosmic rays and argued, against Robert MILLIKAN, that they were charged particles and not electromagnetic radiation.

When the USA entered World War II in 1941, Compton became an important figure in the development of the atomic bomb. In November 1941 he presented the report of the National Academy of Sciences to the US government, arguing that a fission bomb of great power could be built. Compton's own role was to set up the laboratory in Chicago that developed the reactors needed for the production of plutonium. After the war Compton returned to St Louis as chancellor of Washington University. He resigned in 1953 to become professor of natural philosophy in order to investigate problems arising from the impact of science and technology on society.

Compton, Denis Charles Scott (1918–   ) *Middlesex and England cricketer, who played 78 times for his country. An excellent all-rounder, he made 123 centuries in first-class cricket.*

Educated in London at Bell Lane School, Hendon, he played for Middlesex from 1936 to 1958 (excluding the war years). Very briefly he was joint captain of his county but was thought to lack the tougher qualities needed for the job. His batting was largely self-taught – he took risks and defied the coaching manual with relish, but his career aggregate of 38 942 runs is evidence of his success. In 1947 he scored 3816 runs, including 18 centuries; a year later he made a score of 300 during the South African tour. Compton also had some success as a left-arm bowler. His easy-going style and convivial nature fired the imagination of schoolboys everywhere; he was equally popular among the players.

Compton was also a fine soccer player, playing outside left for both Arsenal and England. It was a football injury that eventually ended his cricket career. After his retirement he turned to journalism (writing on both sports) and became a cricket commentator for BBC Television.

Compton-Burnett, Dame Ivy (1884–1969) *British novelist. She was created a DBE in 1967.*

Ivy Compton-Burnett was born in Pinner, Middlesex, the daughter of a homoeopathic doctor and his second wife. The family was deeply split, with the five children of the first wife strongly antagonistic towards their stepmother and her seven children. Similar situations involving large families are a staple of Compton-Burnett novels.

While Ivy Compton-Burnett was reading classics at Royal Holloway College (1902–06), her favourite brother, Guy, died of pneumonia (1905). Her tyrannical mother then compelled her to act as governess to her younger sisters at Hove, a period of deep frustration during which Compton-Burnett wrote her first novel, *Dolores* (1911). Even after her mother's death (1911), dissension and tragedy dogged

the family: the younger sisters, rebelling against Ivy's authoritarianism, moved to London and barred her from their new home; Noel, Ivy's other brother, was killed on the western front (1916); in 1917 the two youngest Compton-Burnett sisters committed suicide. It was not until 1919 that Ivy found a stable and happy domestic life with the writer Margaret Jourdain, with whom she lived until the latter's death (1951).

From the late 1920s until her death Ivy Compton-Burnett produced a new novel almost every two years. Their titles reveal her preoccupation with domestic scenes and family strife: *Brothers and Sisters* (1929), *More Women than Men* (1933), *A House and Its Head* (1935), *Parents and Children* (1941), *A Father and his Fate* (1957), to name just a few. Most of the action is carried on through the characters' conversation, at which Compton-Burnett excelled. This skill in writing dialogue makes her books eminently suitable for radio and many have been adapted for broadcasting.

Connors, Jimmy (1952–   ) *US tennis player.*

Connors's first major title was the 1973 Wimbledon doubles championship, which he won with the Romanian player, Ilie Nastase (1946–   ). The following year, Connors defeated the Australian Ken Rosewall (1934–   ) in the finals of both the Wimbledon and US Open championships, establishing himself as one of the world's top players. His fast deep serves and aggressive groundstrokes, especially a powerful backhand, helped him win the US title again in 1976, 1978, 1982, and 1983. Moreover, after eight years of trying, he again won the coveted Wimbledon crown in 1982 by defeating John McEnroe in the final. Among over one hundred singles titles, he can include the 1974 Australian championship, the 1978 Masters tournament, and the WCT finals of 1977 and 1980.

Conrad, Joseph (Jozef Teodor Konrad Korzeniowski; 1857–1924) *Polish-born British novelist.*

Conrad's parents were ardent Polish nationalists, and his childhood was darkened by the exile into which they were sent by the tsarist authorities. He was born in the Ukraine. His mother died in 1865 and his father four years later, leaving the orphaned boy to be cared for by an uncle. Partly inspired by love of the sea, partly to avoid conscription in the Russian army, Conrad joined the French merchant navy (1874). After an unhappy period based in Marseilles, where he attempted suicide, he joined a

British ship (1878) and took his profession seriously, achieving his master's certificate in 1886. In 1890 he went on a brief and unsuccessful expedition to the Congo and in 1894 he left the sea. Having become a British subject in 1886, he married an English girl (1896), and settled down to write.

Despite his initial ignorance of English (French was his first foreign language), Conrad wrote *Almayer's Folly* (1895) while still at sea. It was followed by *An Outcast of the Islands* (1896), *The Nigger of the 'Narcissus'* (1898), and several shorter pieces, including 'Youth', in which Conrad's favourite narrator Marlow makes his debut. 'The Heart of Darkness', published with 'Youth' (1902), draws on Conrad's Congo experience and is a powerful evocation of isolation and moral decline in a strange and hostile environment. *Lord Jim* (1900) deals with the loss of honour by an act of cowardice for which the hero ultimately atones by offering himself up to certain death. *Nostromo*, possibly the greatest of Conrad's novels, was published in 1904; a subtle and complex tale, it has as its subject the intrigue and corruption surrounding a hoard of silver in the imaginary Latin-American state of Costaguana.

In *The Secret Agent* (1907), Conrad temporarily turned away from the sea to deal with anarchists in London. This subject doubtless owes something to the atmosphere of conspiracy in which Conrad was brought up; it reappears in *Under Western Eyes* (1911). Conrad continued to intersperse his novels with volumes of short stories, but his last major novels were *Chance* (1913) and *Victory* (1915), both on the theme of honour in the relationships between men and women. Conrad also wrote memoirs – *The Mirror of the Sea* (1906) and *A Personal Record* (1912) – and dramatized *The Secret Agent* (produced 1922) and some of his short stories.

Constantine, Learie Nicholas, Baron

(1902–71) *West Indian cricket all-rounder, generally considered to have been the greatest fielder in the history of the game. After retiring from cricket he became a distinguished politician and campaigner for race relations. He was knighted in 1962 and created a life peer in 1969.*

Born in Diego Martin, Trinidad, the son of a sugar-plantation foreman who also played cricket for his country, Constantine soon showed a natural aptitude for the game. He joined the Trinidad team in 1921 and was selected for the West Indian team to tour England

in 1923. He toured England again in 1928 and the following year became a professional in English league cricket; he continued playing professional cricket in England, apart from joining the West Indian touring team of 1939, until 1940. Because so much of his cricketing career was taken up with league cricket he totalled just five centuries in a modest first-class career of only 194 innings and 18 test matches, during which he toured England three times and Australia once. He was a fine batsman, always inclined to be unorthodox, and as a fast bowler was unsparing to his opponents. His fielding, both close to the wicket and in the deep field, was consistently outstanding.

Constantine remained in England during World War II. Always concerned about colour prejudice, against which he campaigned with great energy and dignity for much of his life, he worked in the Ministry of Labour as a welfare officer (1942–47) with special responsibility for West Indian workers. In 1944 he won a case against the Imperial Hotel, London, for their failure to 'receive and lodge' him. In 1954 he was called to the bar by the Middle Temple, and the same year published his book *Colour Bar*. Back in Trinidad, he was elected an MP in his country's inaugural democratic parliament and became minister of works and transport. In 1961 he was appointed high commissioner for Trinidad and Tobago in London, a post he resigned in 1964. He subsequently practised in the English law courts, as well as writing and broadcasting on cricket. In 1966 he became a member of the Race Relations Board.

Cook, Sir Joseph

(1860–1947) *Australian statesman and Liberal prime minister (1913–14). He was knighted in 1918.*

Born in Staffordshire, the son of a coalminer, Cook began working in a coalmine at the age of nine. Emigrating to Australia in 1885, he worked in mines in New South Wales until 1891, when he was elected to the New South Wales legislative assembly as the Labor member for Hartley. Cook left the Labor Party in 1894 when he refused to pledge himself to vote as the Labor caucus directed. Elected to the first federal parliament in 1901, Cook became deputy leader of the Free Trade Party in 1905 and leader in 1909, when he joined forces with Alfred DEAKIN to form the Liberal Party. He succeeded Deakin as leader in 1913 and became prime minister later that year, with a majority of only one in the House of Representatives and a minority in the Senate. Defeated the following year, Cook joined the Nationalist

Party in 1917 and served in the ministry under William HUGHES. He retired from politics in 1921 and became Australian high commissioner in London (1921–27).

Although prime minister for only fourteen months, Cook was regarded as an efficient administrator and a great debator.

Coolidge, (John) Calvin (1872–1933) *US Republican statesman and thirtieth president of the USA (1923–29).*

Coolidge was born in Plymouth, Vermont, and studied at Amherst College, Massachusetts, before embarking on a legal career in 1895. Admitted to the bar in 1897, he set up his practice in Northampton and around the same time began to be involved in politics, soon becoming an active member of the state Republican Party. He served as mayor of Northampton in 1909 and 1910 and was elected to the state senate in 1911, rapidly rising through the ranks to become governor in 1918. His intervention in the Boston police strike of 1919 brought him nationwide acclaim. Denying the strikers' right to be reinstated in their former jobs, Coolidge declared: 'There is no right to strike against the public safety by anybody, anywhere, any time.'

A candidate for the 1920 Republican nomination for president, Coolidge lost to Warren G. Harding (1865–1923) but was selected to stand as vice-president. At the time of President Harding's sudden death in 1923 Coolidge was on holiday in his native Plymouth and the oath of office was administered to the new president by his father, a notary public. Having spent his first year of office restoring public confidence in the Republican Party after the scandals of the Harding administration, Coolidge was re-elected to serve a full term as president in 1924. A calm, taciturn, and somewhat aloof figure, he led his country through the relatively untroubled years preceding the Depression with policies of noninterference in business and foreign affairs, concentrating his efforts on achieving government efficiency and reducing income taxes and the national debt. A notable accomplishment of his administration was the Kellogg–Briand Pact of 1928, which renounced war as a means of settling national disputes.

Coolidge made known his decision not to stand for re-election in 1928 with a characteristically terse announcement that has since become a catch phrase: 'I do not choose to run.' He spent the latter years of his life writing his autobiography and contributing articles to newspapers and magazines.

Cooper, Gary (Frank James Cooper; 1901–61) *US film star.*

Born in Helena, Montana, Cooper was the son of a British-born Montana Supreme Court judge. He was educated in England at his father's old school in Bedfordshire and in the USA at Wesleyan College, Montana, and Grinnell College, Iowa. After an unsuccessful attempt to become a cartoonist, he turned to acting, working as an extra in westerns and in various small parts. In 1926 he was cast as second lead in GOLDWYN's *The Winning of Barbara Worth*, starring Ronald COLMAN. The success of the film assured Cooper's career. Such films as *It* (1927), with Clara Bow, and *Lilac Time* (1928) followed.

His slow, somewhat hesitant, delivery did not prevent his successful transfer to the talkies in *The Virginian* (1929). HEMINGWAY's *A Farewell to Arms* (1932), *Mr Deeds Goes to Town* (1936), and *Sergeant York* (1940), for which he received an Academy Award for best actor, are among his many other films. He won a second Academy Award for *High Noon* (1952). Outstanding among his later films was *Ten North Frederick* (1958). His last film, *The Naked Edge* (1961), was made in England. Shortly before his death his friend James STEWART accepted a Special Academy Award on his behalf in recognition of Cooper's 'memorable screen performance'.

Copland, Aaron (1900–90) *US composer, pianist, conductor, and writer. He was tireless in the cause of contemporary music and in creating an environment in which that music could flourish.*

Born in New York to a family of Russian-Jewish immigrants, Copland was first taught the piano by his sister. Later he studied harmony and counterpoint with Rubin Goldmark (1872–1936), a conservative teacher, who provoked in him a reaction that led to a consuming interest in avant-garde music. There followed a visit to Paris (1921) and three years' study with Nadia BOULANGER, first at the newly founded School for Americans at Fontainebleau, then privately. Copland returned to New York in 1924 with a commission from Boulanger for her forthcoming American tour. This *Symphony for Organ and Orchestra* he later reworked without the organ as his first symphony (1928); *Music for Theater* (1925) stems from the same period and is an amalgam of the Stravinskian neoclassicism learnt from Boulanger and jazz elements; the culmination of this phase of large-scale works was the *Symphonic Ode* (1929). The *Piano Variations*

(1930), the *Short Symphony* (1933), and *Statements* (1935) for orchestra are on a smaller scale with considerably sparser texture.

At this period Copland became aware of the necessity of educating an audience to listen to the new music and of writing music specifically for this audience. Incorporating folk songs and cowboy songs, Quaker hymns, and Latin-American rhythms, he developed the style of some of his best known works: these include *El Salón México* (1936), *Danzón Cubano* (1942) for two pianos, and the ballet scores *Billy the Kid* (1938), *Rodeo* (1942), and *Appalachian Spring* (1944). His film music includes scores for STEINBECK's *Of Mice and Men* (1939) and Henry James's *The Heiress* (1948). At the same time he did not neglect the more abstract forms. His other works in this period include the piano sonata (1941), sonata for violin and piano (1943), and concerto for clarinet and string orchestra (1948, commissioned by Benny GOODMAN). An opera *The Tender Land* (1954) was less successful.

Coppola, Francis Ford (1939–   ) *US film director, writer, and producer.*
Born of Italian parents in Detroit, Coppola showed an early interest in the performing arts, creating his own puppet shows while recovering from childhood polio. After studying film at the University of California he worked on low-budget sex and horror movies until the 1960s, when he began to write and direct his own features. His first success as a director was the musical *Finian's Rainbow* (1968); notable screenplays include *Patton* (1971), a biopic of the World War II general, and *The Great Gatsby* (1974).

Coppola's reputation as a significant force in the cinema rests chiefly on the trilogy *The Godfather* (1972), *The Godfather Part II* (1974), and *The Godfather Part III* (1991). The films, conceived on an epic scale and featuring powerful performances from Marlon BRANDO, Robert DE NIRO, and Al Pacino, chart the fortunes of a New York Mafia family over several generations. Between them they collected eight Academy Awards, including three for Coppola as writer and director. A still more ambitious film was *Apocalypse Now* (1979), a retelling of CONRAD's *Heart of Darkness* in a Vietnam War setting; hailed by some critics as a masterpiece, it was dismissed by others as overblown and pretentious. His other films include *The Cotton Club* (1984) and *Peggy Sue Got Married* (1986). He has also directed in the theatre.

Correns, Carl Franz Joseph Erich (1864–1933) *German plant geneticist, whose breeding experiments contributed to the vindication of Mendel's genetic principles and helped establish the science of genetics.*
Born in Munich, Correns suffered from tuberculosis in his youth and was orphaned when only seventeen. In spite of this, he graduated from the University of Munich in 1889 and three years later became botany instructor at Tubingen University. His researches were wide-ranging, covering cell wall growth, floral morphology, and vegetatve reproduction in mosses and liverworts. But it was his series of experimental crosses using different maize and pea varieties that was to prove most significant. Correns established that certain characters, e g seed colour, were inherited by successive generations according to simple ratios. By 1899 he had arrived at a hypothesis to explain his results. Surveying the literature, he found Mendel's results, which agreed with his own, and in 1900 published his own paper (independently of DE VRIES and Erich von Tschermak (1871–1962), who published similar findings in the same year) affirming the genetic principles of Mendel.

Thereafter, all Correns's energies were focused on plant genetics. In 1902 he became assistant professor of botany at Pfeffer's Institute, Leipzig, then (1909) professor of botany at Münster University, and finally, in 1913, the first director of the Kaiser Wilhelm Institute for Biology, Berlin. During this period, Correns made a notable study of variegation in plants. He discovered a strain of *Mirabilis jalapa* that showed non-Mendelian inheritance of leaf colour. He proposed that cytoplasmic hereditary determinants were responsible, a suggestion later borne out by the discovery of DNA in chloroplasts and mitochondria.

Most of Correns's work was published only in German and has been slow to gain its deserved international recognition.

Cortot, Alfred Denis (1877–1962) *French concert pianist and conductor who was regarded as an outstanding interpreter of Chopin's music.*
Cortot studied the piano at the Paris Conservatoire with Louis Diémer, winning a first prize in 1896. As a pianist he first made his name for his interpretations of Beethoven's piano concertos and for his two-piano partnership with Édouard Risler (1873–1929). From 1905 he was a part of the legendary Cortot–Thibaud–CASALS piano trio. As a conductor he was appointed choral coach at Bayreuth in

1898, then assistant conductor working under Mottl and Richter until 1901, when he returned to Paris to prepare and conduct performances of Wagner's *Twilight of the Gods* and *Tristan*. These performances marked the beginning of a successful conducting career. From 1907 to 1917 he was professor of the piano at the Paris Conservatoire but increasing demands on him as a pianist, both in Europe and the USA, made teaching schedules impossible. In 1919 he founded the École Normale de Musique, where he supervised the tuition; here his master classes in interpretation were famous.

Cortot's empathy with German music made relationships bitter in the German-occupied France of World War II and his fellow countrymen regarded him with suspicion in the years that followed. He was, however, greatly respected for his understanding of romantic music, especially that of Schumann and Chopin. He also championed new French piano music. He had a fine library and, as well as writing textbooks on interpretation, made study editions of Chopin, Liszt, and Mendelssohn.

Cosgrave, W(illiam) T(homas) (1880–1965) *Irish nationalist and first president of the Irish Free State (1922–32).*

Born in Dublin, he became an assistant publican before fighting as an Irish Volunteer in the Easter Rising in Dublin (1916). Arrested, he was sentenced to death but the sentence was later commuted to penal servitude for life. After serving a year he was released and elected to the first Dáil Éireann (parliament), becoming minister for local government (1919–22). Accepting the Anglo-Irish treaty, in 1922 he was successively president of the second Dáil, chairman of the provisional government, and, after the establishment of the Irish Free State, president of its Executive Council (and minister of finance until 1923). Cosgrave founded the pro-treaty party Cumann na nGnaedheal, the precursor of Fine Gael, which he led in opposition from 1935 to 1944. His son Liam Cosgrave (1920–) was Fine Gael prime minister from 1973 to 1977.

Courrèges, André (1923–) *French couturier and one of the world's leading fashion designers of the 1950s and 1960s.*

The son of a butler, Courrèges studied engineering and had made a promising start in the profession when, in 1948, he moved to Paris to work in a fashion house. He joined the famous designer Cristóbal BALENCIAGA, working initially as a presser and learning all aspects of the trade. In 1961 he set up his own fashion house

and his 1963–64 winter collection marked the emergence of a new talent. The Courrèges look featured kneelength hemlines worn with white midcalf boots and also the innovation of trousers for women. Courrèges inspired numerous imitators and, disillusioned, he stopped showing for two years, while he organized control over the distribution of his clothes. He subsequently designed both for haute couture and ready-to-wear, selling the latter through his own Couture Future boutiques, which opened in the USA, Japan, Canada, and Britain during the late 1960s and early 1970s.

Court, Margaret (Margaret Smith; 1942–) *Australian tennis player who dominated the women's game during the 1960s.*

Court was originally a left-hander who trained herself to play tennis with her right hand. Her first success was in mixed doubles and in 1963 she won the grand slam of all four major titles, partnered by Ken Fletcher. Court's game was characterized by a particularly fine backhand and volley but occasionally marred by attacks of nerves. She won three singles titles at Wimbledon (1963, 1965, and 1970), was French champion five times (1962, 1965, 1969, 1970, and 1973), and more or less monopolized the Australian title for over a decade (1960–66, 1969–71, and 1973). In 1970 she achieved the grand slam in singles tournaments. She retired in 1977.

Cousteau, Jacques Yves (1910–) *French oceanographer renowned for his innovative work in undersea exploration, notably the development of manned undersea stations and his invention of the aqualung. He was made a Commandeur de la Légion d'honneur in recognition of his contributions to diving.*

The son of a lawyer, Cousteau entered the naval academy at Brest in 1930, and three years later joined the French navy as a second lieutenant. Stationed at Toulon on the Mediterranean coast, he started to design underwater breathing apparatus that would allow much greater freedom of movement than existing diving suits. Collaborating with Émile Gagnon, he first tested his prototype aqualung in 1942. After the war, this went into commercial production. During World War II, Cousteau served with the French Resistance and received the Croix de Guerre with palm for his exploits. In 1946 he founded the Undersea Research Group attached to the navy to clear mines and make scientific films underwater. Four years later he set up Campagnes Océanographiques Françaises with the aim of improving all aspects of underwater work, especially diving apparatus and photographic equipment. His specially commissioned oceanographic research ship, *Calypso*, made its first voyage, to the Red Sea, in 1951–52. This was followed by a four-year survey of the world's oceans (1952–56). In 1957, Cousteau was appointed director of the Oceanographic Institute and Museum at Monaco, and he later directed the Conshelf Saturation Dive programme (1962–65), a major project to examine the feasibility of underwater colonies situated on the continental shelves of the world. Cousteau's concern with protecting as well as exploiting the marine environment are reflected in his many award-winning films, including *The Silent World* (1956), *The Golden Fish* (1959), and *World Without Sun* (1964). He also made a long-running television series, *The Undersea World of Jacques Cousteau* (1968–76), besides writing many books, such as *The Silent World* (1953), *The Living Sea* (1963), and *The Ocean World of Jacques Cousteau* (20 vols, 1973). In 1985 he embarked on a 2½ year voyage in a wind-powered boat to publicize his support for alternative energy and his opposition to nuclear weapons. In 1989 he lent his support to a campaign to protect the environment of Antarctica.

Coward, Sir Noël (1899–1973) *British playwright, composer, producer, and actor. He was knighted in 1970.*

Coward began his career as a child actor in 1910, appearing in *The Goldfish* at the Little Theatre. He then worked for some years under Charles Hawtrey (1858–1953), playing juvenile parts and learning all the skills of his profession. After World War I Coward continued to act and began to write his own plays – he co-wrote and appeared in the revues *London Calling* (1923) and *On With the Dance* (1925). His first major success came with *The Vortex* (1924), a domestic drama accurately reflecting the brittleness and desperation of London society in the postwar period. *Hay Fever* (1925) was quite definite in tone and set the pattern for the witty social comedies for which Coward is best known, notably *Private Lives* (1930; filmed 1931) starring Gertrude LAWRENCE; *Design for Living* (1933; filmed 1933); *Blithe Spirit* (1941; filmed 1945), which achieved a record run of nearly two thousand performances and was made into a musical, *High Spirits*, in 1964; and *Present Laughter* (1943). Meanwhile Coward continued to write musicals, including the popular *Bitter Sweet* (1929) and the highly patriotic *Cavalcade* (1931). During the 1930s Coward was active both in

England and in New York, where he worked with Alfred LUNT and Lynn FONTANNE and appeared with Gertrude Lawrence in several of his plays, including *Tonight at 8.30* (1936). The most highly acclaimed of his later plays were *Relative Values* (1951) and *Nude with Violin* (1956).

Coward was equally at home in the film world. He made his own screen debut in *Hearts of the World* (1918) and continued to appear in films until the late 1960s, often in wittily observed cameo parts (notably in *Our Man in Havana*, 1959). He played the leading role in *In Which We Serve* (1942), a tribute to the Royal Navy that Coward wrote, produced, and co-directed with David LEAN. He also wrote and produced Lean's *This Happy Breed* (1944) as well as the popular and moving *Brief Encounter* (1945). A truly versatile entertainer with, in his own words, 'a talent to amuse', Coward also wrote numerous songs (including the witty 'Mad Dogs and Englishmen'), novels, some verse, and two volumes of autobiography – *Present Indicative* (1937) and *Future Indefinite* (1954).

Crane, (Harold) Hart (1899–1932) *US poet who wrote some of the most highly acclaimed American poetry of the twentieth century.*

Born in Garrettsville, Ohio, Crane had an unhappy childhood, which culminated in his parents' divorce when he was seventeen. He then had a wide variety of jobs in New York and Cleveland, finally settling in New York in 1923, as his poetry began to be accepted for publication. In this period his active homosexuality, and the difficulty he had in coming to terms with it, led to the bouts of alcoholism and depression that ended with his suicide at the age of thirty-three.

His first published collection of poems, *White Buildings* (1926), reflected his enthusiasm for city life and attracted considerable attention. In some respects it also reflected the influence of T. S. ELIOT, although his poetry is less pessimistic than Eliot's. *The Bridge* (1930), his best-known work, is an epic poem in which he sought to link a dreamlike perception of the past with modern industrial reality, using Brooklyn Bridge as the central symbol. The poem was well received and on the strength of this reception Crane was awarded a Guggenheim Fellowship, which he decided to spend in Mexico City, planning to write another epic on a Mexican theme. This never materialized, although he did write another poem, 'The Broken Tower' (1932), in Mexico. On the way back to New York by sea he jumped from the ship into the Caribbean and was drowned. *The Complete Poems and Selected Letters and Prose* was published in 1966.

Cranko, John (1927–73) *British choreographer and director of the Stuttgart Ballet (1961–73).*
Born in Rustenburg, South Africa, Cranko trained at the Cape Town University Ballet School and the Cape Town Ballet Club; in 1942, at the age of sixteen, he choreographed his first ballet, STRAVINSKY's *The Soldier's Tale*. Three years later, he joined the Sadler's Wells Ballet, working with Dame Ninette DE VALOIS, and during the 1950–51 season was named resident choreographer following the success of his ballet *Pineapple Poll*. Such works as *Bonne Bouche* (1952), *The Shadow* (1953), and the popular *The Lady and the Fool* (1954), set to music by Verdi, followed. Cranko also worked during this period for the Paris Opéra and La Scala, Milan, as well as collaborating with the painter John PIPER for a season at Henley-on-Thames. The theatrical approach he adopted towards dancing also led Cranko to write a review, *Cranks* (first performed 1956), which was well received in both London and New York.

Inspired by Frederick ASHTON's three-act ballets, Cranko choreographed *The Prince of the Pagodas* (1957) – the first full-length British ballet to have a score by Benjamin BRITTEN. After taking this ballet to Germany, Cranko became director of the Stuttgart Ballet in 1961 and soon this company, which Cranko liked to refer to as 'my Stuttgart family', achieved international fame. Cranko created several full-length ballets for the company, which include some of his best-known works – *Romeo and Juliet* (1962), *Onegin* (1965), *The Taming of the Shrew* (1969), and *Carmen* (in which he introduced the ballerina Marcia Haydée). He also choreographed shorter pieces, such as *Brouillards* and the light-hearted *Jeu de Cartes*. Cranko died suddenly at the height of his career, aged forty-five, on a flight back to Germany with the Stuttgart Ballet at the end of a successful US tour.

Crawford, Joan (Lucille le Sueur; 1908–77) *US film actress who for over forty years ranked among Hollywood's leading stars.*
Born in San Antonio, Texas, she adopted her stepfather's surname and for a time was known as Billie Cassin. Dancing engagements in cafés and nightclubs subsequently led to the chorus line on Broadway and her first film, *Pretty Ladies* (1925). Soon after, in a competition organized by her studio, she adopted the name Joan Crawford. Among her early successes was *Sally, Irene and Mary* (1925) and *Our Dancing Daughters* (1928). More dramatic roles came in the thirties with *Grand Hotel* and *Rain* (both 1932), followed by a series of successful comedies. After *Mildred Pierce* (1945), for which Crawford received an Oscar, came tougher roles in such films as *Torch Song* (1953) and *Johnny Guitar* (1954). Her first British film was *The Story of Esther Costello* (1957). Among other major films in which Crawford played during the later years of her career were *Whatever Happened to Baby Jane?* (1962), which also featured Bette DAVIS, and *The Caretakers* (1963).

In addition to her film career Crawford joined the board of Pepsi-Cola in 1959, on the death of her husband of three years, Alfred Steele, who had been chairman of the board. Previously she had been married to Douglas FAIRBANKS, JR (1929–33), Franchot Tone (1935–39), and Phil Terry (1942–46).

Crick, Francis Harry Compton (1916–   )
British molecular biologist who, with James WATSON, first proposed the structure of deoxyribonucleic acid (DNA), the molecule that carries genetic information in most living organisms. This and subsequent work earned him the 1962 Nobel Prize for Physiology or Medicine.

Born in Northampton, Crick attended University College, London, and received his BSc in physics in 1937. After graduate studies, he worked for the Admiralty during World War II, developing mines. In 1947 he joined Strangeways Research Laboratory, Cambridge, and after two years moved to the Medical Research Council unit at the Cavendish Laboratory, also in Cambridge. Here he worked under Max PERUTZ, applying the technique of X-ray diffraction crystallography to study the structure of proteins. Following Watson's arrival at the Cavendish in 1951, he and Crick collaborated informally to speculate about the structure of DNA. Using the vital evidence of Ernst Chargaff (on the ratios of constituent bases) and Linus PAULING (on helix formation in polypeptide chains), and the X-ray diffraction studies of crystalline DNA performed by Maurice WILKINS and Rosalind FRANKLIN, Crick and Watson built a model of the DNA molecule that satisfied the data and fulfilled its biological requirements – the capacity for carrying the genetic information and for self-replication. This was their famous 'double helix' of two intertwined helically coiled chains of nucleotides with complementary sequences of bases linked in pairs by interchain hydrogen bonds.

Receiving his PhD in 1953, Crick spent the next two decades working out how DNA carries its genetic information and how this directs the manufacture of specific proteins. He and his colleagues established that the basic coding unit, or codon, was a sequence of three consecutive bases. Crick was also the first to suggest, in 1957, the adaptor role of transfer RNA (tRNA) in the assembly of amino acids during protein synthesis. In 1977 Crick moved from Cambridge to the Salk Institute, San Diego, and turned his attentions to brain research. In *Life Itself: Its Origin and Nature* (1981), he speculated that life on earth may have originated from an extraterrestrial source – possibly by contamination from alien spacecraft. He has also written *Of Molecules and Men* (1966), an affirmation of his belief that life has an ultimately material explanation, and *What Mad Pursuit* (1988), in which he discussed the nature of scientific discovery.

Crippen, Hawley Harvey (1862–1910) *US murderer who poisoned his wife at their London home and whose arrest in Canada after fleeing aboard a transatlantic vessel was achieved through the intervention of radio telegraphy.*

Dr Crippen worked in London for the Munyon Patent Medicine Company and was a partner in Yale Tooth Specialists. He lived with his American wife, an aspiring actress, who used the stage name Belle Elmore (her real name was Kunigunde Mackamotzi). Elmore was last seen alive on 31 January 1910, when friends visited the house at 39 Hilldrop Crescent, in north London. Two days later, the Music Hall Ladies Guild received a note from their treasurer, Belle Elmore, saying that she had resigned and returned to the USA. Suspicions were aroused when Crippen was seen escorting from his office a young typist, Ethel Le Neve, who in March moved into the Crippens' house. Chief Inspector Dew of Scotland Yard searched the house but found nothing. Crippen claimed that his wife had died of pneumonia in the USA. However, when Crippen and Le Neve disappeared, Dew again searched the house and discovered human remains buried in the cellar. They were identified as those of the missing wife and found to contain traces of the poison hyoscine.

Crippen, with Le Neve disguised as his son, had fled aboard the transatlantic steamer SS *Montrose*, bound for Montreal. The captain became suspicious and telegraphed their descriptions to Scotland Yard in London on 22 July. Inspector Dew embarked on a faster ship and arrested the fugitives as their ship docked at Father Point, Quebec, on 31 July. Crippen was tried at the Old Bailey and found guilty of murder. He was hanged on 23 November, with the request that he be buried with a photograph of his mistress. She was acquitted.

Cripps, Sir (Richard) Stafford (1889–1952) *British politician on the left of the Labour Party, who as chancellor of the exchequer (1947–50) introduced austere measures to deal with Britain's postwar foreign exchange crisis. He was knighted in 1930.*

Born in London, Cripps was a brilliant student at Winchester and University College, London, where he read chemistry; he then became a barrister (1913) and was appointed King's Counsel in 1927 and solicitor-general in 1930. He was elected to parliament in 1931 and in the following year helped found the Socialist League. In 1936, in support of the antifascists in the Spanish civil war, he advocated formation of a British united front, to include communists. When this was revived in 1938, in the form of a Liberal–Labour popular front against Neville CHAMBERLAIN's appeasement of fascism, Cripps was expelled from the Labour Party. During World War II, he served as ambassador to Moscow (1940–42) and then as lord privy seal (1942). He headed a mission to India in 1942 that attempted, unsuccessfully, to rally the leaders of the Indian National Congress against Japan.

Cripps was readmitted to the Labour Party in 1945, when he was appointed president of the Board of Trade and then chancellor of the exchequer. His programme, which included high taxation on internal consumption and overseas purchases together with wage restraint, was largely successful in combating Britain's postwar economic difficulties.

Croce, Benedetto (1866–1952) *Italian idealist philosopher and the leading Italian intellectual of his day.*

He was born in Percasseroli, the son of a rich landowner, and lost his parents in an earthquake in 1883. Educated at the University of Rome, Croce spent the rest of his life, mainly in Naples, working as an independent scholar. He published about seventy books and in a review, *Critica*, which he founded and edited for forty years, he attempted to revitalize Italian thought and to restore it to the mainstream of European philosophy. Croce was also active politically, serving as minister of education

(1920–21) and senator in the Italian government. Initially he was sympathetic to MUSSO-LINI but after 1925, with the collapse of parliamentary rule in Italy, Croce withdrew his support and began to organize the intellectual opposition against him. See GENTILE, GIOVANNI.

Croce presented his system in his major work, *Filosofia dello spirito* (4 vols, 1902–17). The various volumes have been translated into English as *Aesthetic* (1909), *Logic* (1917), *Philosophy of the Practical* (1915), and *Theory and History of Historiography* (1921). Reality, Croce argued, was spirit, and whatever could not be found in the mind or its activity must therefore be fictitious. Mental activity could be either theoretical or practical. Theoretical activity was either conceptual (studied by logic) or intuitive (the study of aesthetics). Practical activity could be concerned with the particular (the domain of economics) or the universal (the ethical domain). Philosophy itself, for Croce, could only be a description of the principles observed by the mind as it operated within the four already identified levels. How the mind actually works is the province of history; the principles that govern its operation must therefore belong to the philosophy of history. Consequently, for Croce, philosophy could only the methodological study of history.

Cronin, A(rchibald) J(oseph) (1896–1981)
Scottish writer and physician.

Cronin was born in Cardross, Scotland. During World War I he served in the naval medical service and graduated as a physician from Glasgow University in 1919. He then practised medicine in south Wales (1921–24), making a special study of industrial medicine. Between 1924 and 1930 he practised in London, but gave up medicine for writing after the success of his first novel, *Hatter's Castle* (1931), which was later (1941), like several of his other novels, made into a film. *The Stars Look Down* (1935; filmed 1939) drew on his experience of life in the Welsh coal-mining valleys. His play *Jupiter Laughs* was produced in 1940. Other publications of the 1940s were *The Keys of the Kingdom* (1942; filmed 1944), *The Green Years* (1944; filmed 1946), and *Shannon's Way* (1948), the last two about an Irish boy brought up in Scotland. His autobiography, *Adventures in Two Worlds*, was published in 1952.

Among his later publications were *The Judas Tree* (1961) and *A Pocketful of Rye* (1969). His knowledge of medical practice and

Scottish life combined to create the basis for the very popular television series *Dr. Finlay's Casebook*.

Crosby, Bing (Harry Lillis Crosby; 1904–77)
US singer and actor. Affectionately known as 'Der Bingle' and 'The Old Groaner', he was the most popular singer of his generation and the first and best-known crooner.

Born in Tacona, Washington, Crosby began his career as a singer in a vocal group with Paul Whiteman (1890–1967) and his orchestra (1926–30). However, with the advent of radio he became known to much wider audiences, which led ultimately to his spectacular success in records, film, and television. He was the first popular singer to understand the use of a microphone, into which he crooned intimately in a conversational style, giving more weight to the lyrics than to the notes. He also used slurs, sang the consonants, and borrowed other devices from jazz. His style influenced every other singer that came after him. He had about eighty hit records after 1940; his recording of the Irving BERLIN song 'White Christmas' (from the 1942 film *Holiday Inn*) became one of the best-selling records in history.

Apart from his voice, Crosby cultivated a pleasing public persona and had a perfect comedian's timing. He appeared in more than sixty films between 1930 and 1966, including the '*Road to...*' series of comedies (1940–62) with Bob HOPE, his friend and golfing companion, and *Going My Way* (1944), for which he won an Academy Award. His television series (1964) was extremely popular.

Crossman, Richard Howard Stafford
(1907–74) *British Labour politician, whose diaries made a considerable stir when they were posthumously published.*

Son of a judge of strong Conservative convictions, Crossman was educated at Winchester and Oxford. He entered parliament in 1945 and joined Aneurin BEVAN's Keep Left group. Unsympathetic to both ATTLEE and his successor as Labour leader, Hugh GAITSKELL, Crossman was an energetic supporter of Harold WILSON's bid for the leadership in 1963. In Wilson's first government, he served as minister of housing and local government (1964–66), leader of the House of Commons (1966), and minister of social security (1969–70). Throughout this period he kept a record of events in government, including details of cabinet meetings, published, in spite of Labour efforts to repress it, as the *Crossman Papers* (1975).

Crossman was also a political journalist. He joined the weekly *New Statesman* for the first time in 1937 and was its editor from 1970 to 1972. He was also a columnist for *The Times*.

Culbertson, Ely (1891–1955) *US bridge player who revolutionized the game by formalizing and writing down a system of bidding.*

Culbertson was born in Romania, the son of an American and his Russian wife. A cosmopolitan character, he spoke numerous languages and travelled widely, taking part in revolutionary activity in Mexico, Russia, and elsewhere. After 1917 he went to Paris and then settled in the USA, where he lived by playing cards and met his future wife Josephine, a bridge teacher, whom he married in 1923. Between 1926 and 1929 contract bridge became increasingly popular and Culbertson saw a great opportunity to publicize the game. He started a bridge magazine (*The Bridge World*), organized a teachers' association, and wrote his definitive 'Blue Book'.

Although he admitted that his wife was probably a better player, he was good enough to win several major tournaments. In the 1930s he played in well-publicized 'challenges', such as the individual match with Sidney Lentz or team events with his US team tour round Europe. Interest was stimulated in these 'challenges' by high stakes. Culbertson and his team won almost every event and he made and spent a fortune living in great style. In 1938 he lost interest in bridge and devoted his considerable energies to the League of Nations and world peace. He took an active part in the formation of the United Nations in the 1940s.

Cummings, E(dward) E(stlin) (1894–1962) *US poet, writer, and painter.*

Born in Cambridge, Massachusetts, he graduated from Harvard and received an MA in 1916. While serving with a volunteer ambulance corps in France before the USA entered World War I, Cummings was erroneously charged with carrying on treasonable correspondence and imprisoned by the French. This experience inspired *The Enormous Room* (1922), a work in poetic prose that introduces the theme of much of his subsequent work – commitment to all that is free, idiosyncratic, and individual and hatred of the bureaucratic, platitudinous, prudish, and overintellectualized. After the war Cummings lived in Paris for a time but eventually returned to America to spend the rest of his life, apart from one journey to the Soviet Union, in or near New York. The diary of his Russian trip, *Eimi* (1933; the title is Greek for 'I am'), is critical of Soviet suppression of the individual in favour of mass regimentation.

By 1926 Cummings had written four volumes of mostly short lyric verse and had found what was to be his characteristic style in subsequent works. This had a fiercely 'modernistic' look: inventive typographical layout, odd end-of-line breaks and punctuation, grammatical innovations, and such quirks as the well-known aversion to capital letters. Such devices, however, were employed, albeit skilfully, to support a much more conventional poetry consisting mainly of rather sentimental love lyrics and short satirical pieces. A number of cleverly bawdy poems, intended to shock the puritans of his day, have perhaps retained more of their charm than many of his love poems and satires.

Collected Poems (1938) was followed by *50 Poems* (1940), *1 × 1* (1944), *Poems: 1923–1954* (1954), *95 Poems* (1956), and the posthumous *73 Poems* (1963). *i*, 'six nonlectures' (1953), are lectures on criticism given at Harvard. Some of Cummings's art appeared in *CIOPW* (1931; the title refers to charcoal, ink, oil, pencil, and watercolour).

Curie, Marie (1867–1934) *Polish-born French chemist. She was the first woman to win a Nobel Prize and the first person to win two Nobel Prizes. She shared the 1903 prize with her husband, the physicist Pierre Curie (1859–1906), and Henri Becquerel (1852–1908) for their studies of radioactivity and won the second prize for chemistry in 1911 for her discovery of radium and polonium.*

Born Marya Sklodowska, the daughter of a Warsaw schoolteacher, she worked as a governess for six years to enable her sister Bronya to qualify in Paris as a doctor. She followed her sister there in 1891 and despite four years of poverty graduated from the Sorbonne in physics in 1893 and mathematics in 1894. The following year she married Pierre Curie, who was in charge of the laboratory of the School of Industrial Chemistry and who already had a distinguished career behind him.

Permitted to work in her husband's laboratory without a salary, in the late 1890s she chose the subject of uranium for her doctoral thesis. Becquerel had discovered that pitchblende (uranium ore) was radioactive in 1896. It was also known that pitchblende was far more radioactive than the pure uranium. Mme Curie was convinced that the enhanced radioactivity of pitchblende resulted from the presence of an unknown substance. After a considerable amount of physically demanding work she succeeded in identifying both polonium and radium from the many tonnes of pitchblende she processed. She went on to study the nature of the radiation produced by the new elements.

Pierre died in 1906 in a road accident. Marie, by then with two daughters (see JOLIOT-CURIE, IRÈNE), succeeded to her late husband's chair of physics at the Sorbonne. Much of Marie's later life was spent in raising funds to pursue her research and to establish an appropriate institution in which to pursue the work. Fortunately Marie Curie's US admirers presented her with one gram of radium in 1921, when it was worth $100,000. The Sorbonne created for her the Curie Laboratory which, though opened in 1914, had to wait for the end of World War I in 1918 before it could begin serious research. Mme Curie herself spent the war training radiologists. She continued to work throughout the 1920s in her laboratory. Inevitably for one who had been so frequently exposed to radiation, her health began to fail, and she finally died from leukaemia.

In her honour, the unit of activity of radioactive substances was named the curie in 1910. In the same year she had allowed her name to be submitted for election to the Académie des Sciences as the first plausible woman candidate. In spite of her Nobel Prize she was rejected and thereafter she refused to allow her name to be resubmitted.

Curtin, John Joseph Ambrose (1885–1945) *Australian statesman and Labor prime minister (1941–45).*

Born in Creswick, Victoria, the son of an Irish police sergeant, Curtin left school at thirteen to work in a Melbourne printing office. Active in trade union organization and the campaign against conscription, Curtin stood unsuccessfully for a Western Australian seat in 1919 but entered federal parliament in 1928 as the member for Fremantle. Defeated in 1931, he was re-elected in 1934 and became leader of the Labor Party in 1935. He was elected prime minister in 1941.

As war-time prime minister, Curtin mobilized Australian resources to meet the danger of Japanese invasion and invited General Douglas MACARTHUR to use the continent as a base to drive back the Japanese. He laid down the groundwork for the postwar economy and introduced such welfare measures as widows' pensions and unemployment and sickness benefits. A sensitive man, with extreme loyalty to the party, Curtin died in office in 1945.

Curzon, Sir Clifford Michael (1907–82) *British pianist, who was knighted in 1977.*

Curzon entered the Royal Academy of Music in 1919 and studied with Charles Reddie, winning many prizes. Later he studied with Katherine Goodson. His first public appearance, at the age of sixteen, was in J. S. Bach's *Triple Concerto* at the Queen's Hall under Sir Henry WOOD, who was influential in Curzon's early career. From 1928 to 1930 he studied with Artur SCHNABEL, a teacher of the utmost importance for him. Later, in Paris, he also studied with Wanda LANDOWSKA and with Nadia BOULANGER. After his marriage in 1931 to the harpsichordist Lucille Wallace, he toured Europe and in 1939 visited America. Thereafter his fame was world-wide, as soloist, recitalist, and in chamber music. He gradually curtailed his large repertoire and concentrated on the music of Mozart, Beethoven, and Schubert.

Cushing, Harvey Williams (1869–1939) *US neurological surgeon, best known for his work on the pituitary gland and description of a syndrome (Cushing's syndrome) caused by overproduction of cortisol by the adrenal glands.*

Born in Cleveland, Ohio, Cushing came from a family of distinguished physicians. He graduated in arts from Yale in 1891 and in medicine from Harvard Medical School in 1895. After working at the Johns Hopkins Hospital, Baltimore, in 1912 he became professor of surgery at Harvard and surgeon-in-chief at the Peter Bent Brigham Hospital, Boston, where he built an internationally renowned school of neurosurgery.

Besides his work in endocrinology, Cushing made important contributions to the classification of brain tumours and the control of blood pressure and prevention of haemorrhage during brain surgery. Many techniques devised by him remain standard surgical procedures. His work in Boston was interrupted during World War I, when he became an army surgeon before the USA entered the war and later a consultant neurosurgeon with the American Expeditionary Force (1918). He was made Military Companion of the Bath in 1919 for his service to the British army. Cushing retired from Harvard in 1932 and became Sterling Professor of Neurosurgery at Yale; in 1937 he became director of studies in the history of medicine, playing an important part in the foundation of the Historical Medical Library. He received the Pulitzer Prize in 1926 for his biography *Life of Sir William Osler* and published his autobiography, *From a Surgeon's Journal, 1915–1918*, in 1936.

D

Daladier, Édouard (1884–1970) *French statesman and prime minister (1933; 1934; 1938–40). He became a Chevalier de la Légion d'honneur for his service during World War I.*

Born in Carpentras, near Avignon, the son of a baker, Daladier was educated at the École Normale and at the Sorbonne before becoming a lecturer in history at the Lycée Condorcet in Paris in 1914. He fought at Verdun during World War I.

Daladier was elected mayor of Carpentras in 1912 (a position he held until 1958). Entering the French chamber in 1919 as a radical-socialist deputy for the Vaucluse district of Provence, he was appointed minister of the colonies in 1924, becoming prime minister in 1933 (for ten months) and again in 1934 (for two months). At the outbreak of World War II he once more became prime minister, assuming in addition the ministries of war and foreign affairs. A signatory to the Munich pact, which supported the cession of Czechoslovakia's Sudetenland to Germany in 1938, he declared war on Germany in 1939 following the German invasion of Poland. He was replaced as prime minister by Paul Reynaud (1878–1966) in 1940 but retained his portfolios until the fall of France in June of that year. A month later he was arrested and charged with war guilt by the Vichy regime. Put on trial at Riom in 1942, he bravely denounced the Vichy government but in 1943 was deported as a political prisoner to Germany, where he spent the last years of the war. He served as a member of the national assembly from 1946 until his resignation in 1958.

Daladier was known as the 'bull of Vaucluse' because of his forceful personality. He was one of the few leaders of the Third Republic able to continue in politics under the Fourth Republic.

Dalai Lama (Tenzin Gyatso; 1935–) *Tibetan Buddhist leader; spiritual and temporal ruler of Tibet (1951–59). He was awarded the Nobel Peace Prize in 1989, in recognition of his appeals for the nonviolent liberation of his homeland from Chinese rule.*

Born into a peasant family in Amdo province, Tenzin Gyatso was five when oracles pro-claimed him the reincarnation of the thirteenth Dalai Lama, who had died in 1933. He was enthroned at Lhasa in 1940 and ruled in his own right from 1950. A year later the Chinese invaded Tibet; when the sixteen-year-old Dalai Lama's pleas were ignored by the UN, the UK, and India he had no choice but to sign an agreement in which Tibet became an 'autonomous region' of China and his own powers became largely notional. In 1959 China's attempts to destroy the national and religious identity of Tibet provoked an uprising, quickly suppressed by the occupying forces. The Dalai Lama and most of his ministers fled to India, followed by thousands of refugees. Since his exile he has acted as a spokesman for the plight of his country and has written widely on Buddhism and his search for world peace.

Dale, Sir Henry Hallett (1875–1968) *British physiologist who, in 1914, isolated the chemical acetylcholine from the fungus ergot. It was later found to be the same substance as that produced by the nerve endings during the passage of a nerve impulse, discovered by Otto LOEWI in 1921. Dale and Loewi shared the 1936 Nobel Prize for Physiology or Medicine. Dale was knighted in 1932 and awarded the OM in 1944.*

Dale worked for a short time with Paul EHRLICH and E. H. STARLING. He became a highly respected experimentalist and was appointed director of the Wellcome Institute at the age of twenty-nine. His work on the chemical composition and effects of the fungus ergot in rye showed that the physiological reactions to ergot were produced by acetylcholine. Dale recognized that such substances as acetylcholine, which can control some functions of the body, are widely distributed in nature and that they could be of therapeutic value, a branch of science he termed 'autopharmacology'. Much of modern pharmacology is based on Dale's work. In 1914 Dale joined the staff of what later became the Medical Research Council and, when insulin was discovered in 1921, he travelled to Toronto to ensure that it would be adequately standardized. He became chairman of an international committee responsible for the standardization of immunological products, hormones, vitamins, and antibiotics. He became president of the Royal Society in 1940 and chairman of the Scientific Advisory Committee to the cabinet during World War II.

Dali, Salvador (1904–89) *Spanish painter, the most widely known of the surrealists.*

Dali's childhood, according to his autobiographical *My Secret Life* (1942), was punctuated by violence and hysterical fits. He was expelled from the Madrid School of Fine Arts in 1926 for extravagant behaviour and continued throughout his life to cultivate eccentricity and exhibitionism and to suffer from paranoia, which he claimed to be a source of inspiration to his work.

In the 1920s Dali experimented with a variety of styles and was particularly influenced by the Italian metaphysical painters CHIRICO and Carlo Carrà (1881–1966). In 1929, after reading FREUD's writings on dream images and their erotic significance, Dali arrived at his mature surrealist style and quickly became a spectacular member of the surrealists in Paris. His pictures were 'hand-painted dream photographs' of subconscious images, painted in realistic detail and with technical excellence against backgrounds of arid Catalan landscapes. Characteristic images are the limp flowing watch faces in *The Persistence of Memory* (1931), half-open drawers protruding from human figures, as in *Burning Giraffe* (1935), and optical illusions, such as *Mae West* (1936), giving the effect of two pictures in one. He also worked with the Spanish director Luis BUÑUEL on two surrealistic films – *Un Chien andalou* (1928) and *L'Age d'or* (1931).

In 1940 Dali moved to New York, where he designed theatre sets and up-market shop interiors and also devoted much of his time to publicity. Returning to Spain in 1955, he became a supporter of the Franco regime. Many of his works after 1950 were religious in theme; the *Crucifixion of St John of the Cross* (1951) is well known for its unusual view of the cross from above. But he continued to explore erotic subjects: *Young Virgin Auto-Sodomized by Her Own Chastity* is from the same year. In the 1960s Dali began to produce sculpture. He wrote two more volumes of autobiography – *Diary of a Genius* (1966) and *The Unspeakable Confessions of Salvador Dali* (1976). In his last years he became a bedridden recluse after being badly burned in a fire at his home.

Dallapiccola, Luigi (1904–75) *Italian composer and pianist, the leading exponent of serialism in Italy.*

Dallapiccola was born in Istria (then part of the Austrian empire) and his education was disrupted by political upheaval. After hearing Wagner's *Flying Dutchman* in 1917 he decided to become a composer and studied the piano and composition at the Cherubini Conservatory in Florence. Influences at this time were the music of DEBUSSY, the Italian madrigalists, and the encouragement and support of his teacher Alfredo Casella (1883–1947). His friendship with Alban BERG introduced him to the serial music of SCHOENBERG and the Second Viennese School. Dallapiccola's own use of twelve-tone techniques was a gradual process of introducing serial sections into a basically diatonic context, as in *Coro degli Zitti*, one of six choral pieces (1933–36). His concern for mankind in the twentieth century is reflected in his opera *Volo di notte* (1937–39; *Night Flight*, from the novel by SAINT-EXUPÉRY), in which a doomed pilot flying above the Andes glimpses the stars, for the composer the symbol of the mystery and compassion of God. His lifelong sympathy for prisoners and concern for spiritual freedom are reflected in some of his best-known and most popular works – *Canti di prigionia* (1938–41; *Songs of Prison*), prison songs of Mary Stuart, Boethius, and Savonarola, the opera *Il prigioniero* (1944–48; *The Prisoner*), and *Canti di liberazione* (1951–55; *Songs of Liberation*) for chorus and orchestra.

Dallapiccola's achievement in shaping serialism into a lyrical Italian style stabilized after the mid-1950s and is evident in many of his shorter pieces, such as *Parole di San Paolo* (1964; *Words of Saint Paul*) for mezzo-soprano, boys' voices, and eleven instruments. The most important of Dallapiccola's later works is the opera *Ulisse* (1968; *Ulysses*) to his own arrangement of Homer's text.

Dam, (Carl Peter) Henrik (1895–1976) *Danish biochemist who discovered vitamin K, for which he was awarded the 1943 Nobel Prize for Physiology or Medicine.*

The son of a chemist, Dam received his MSc in chemistry from the Polytechnic Institute, Copenhagen, in 1920. After a period at the Royal School of Agriculture and Veterinary Medicine (1920–23), he was appointed to the physiology department at Copenhagen University, becoming assistant professor and then associate professor of biochemistry. In 1934, Dam published the results of his dietary experiments involving chicks. He found that a certain nutrient was essential for proper functioning of the blood clotting mechanism:

he termed this nutrient 'Koagulation factor' (hence vitamin K). Dam went on to extract vitamin K, one of several related fat-soluble menaquinone derivatives found especially in green vegetables, and to apply it to the prevention of haemorrhaging in clinical use.

In 1940 Dam went on a lecture tour of the USA and stayed for the duration of World War II as a research associate at Strong Memorial Hospital, Rochester. In 1946 he returned home to become professor of biochemistry and head of biology at the Polytechnic Institute. Apart from vitamin K, he worked mainly on the metabolism of sterols and other lipids.

D'Annunzio, Gabriele (1863–1938) *Italian poet, novelist, and playwright: the leading Italian literary figure of the early twentieth century.*

D'Annunzio was born at Pescara in Abruzzi and educated in Prato. He published a well-received book of verse at the age of sixteen and therefore entered the University of Rome in 1881 with some reputation as a poet. During the next decade his reputation grew steadily, if controversially, with volumes of poems then considered outspoken in their sensuality and naturalistic stories about his native Abruzzi (collected as *Novelle della Pescara*, 1902; translated as *Tales of My Native Town*, 1920). His novel *Il piacere* (1891; translated as *The Child of Pleasure*, 1898) reflects his flamboyant life at this time.

In the 1890s D'Annunzio separated from his wife (he had married in 1883), left Rome, and gradually came under the influence of Nietzsche's work, especially the glorification of the instinctive and pagan, the inspiration of his best early novel, *Il trionfo della morte* (1894; translated as *The Triumph of Death*, 1896). The following year he began a long and highly publicized affair with the actress Eleonora Duse (1859–1924) during which he wrote a number of plays, including his best dramatic work, *La figlia di Iorio* (1904; translated as *The Daughter of Iorio*, 1907). His novel *Il fuoco* (1900; translated as *The Flame of Life*) created a scandal because of clear references to his affair with Duse; he also published his best verse, *Alcyone* (1903), during this period.

D'Annunzio returned a national hero from World War I, in which he had fought with distinction. In 1919 he led an attempt to seize Fiume, then still an Austrian city, for Italy, but was forced to withdraw in 1920. He spent the rest of his life at his villa, La Vittoriale, on Lake Garda, where he devoted himself to pre-

paring an edition of his complete works, for which a national institute was eventually set up (1926). La Vittoriale and its extravagant collection of art and furnishings was left to the nation on his death. D'Annunzio's literary reputation has suffered severely since then. The charge of posturing and dilettantism, first levelled against his work by Benedetto CROCE, remains to be answered, though a careful selection of his writings, *Poesie, Teatro, Prose* (1966), has been compiled by Mario Praz and F. Gerra.

Darlan, (Jean Louis Xavier) François (1881–1942) *French admiral and politician who served as vice-premier in PÉTAIN's Vichy government during the German occupation of France in World War II.*

Appointed naval chief of staff in 1936, Darlan was made commander-in-chief in 1939. While the Germans invaded France in 1940, Darlan at one stage apparently agreed with CHURCHILL that the French fleet should seek sanctuary in British ports. But following the German occupation, he accepted the office of minister of marine in Marshal Pétain's cabinet. The French ships remained in their North African harbours, where they were subsequently badly damaged by Allied action. Becoming vice-premier in 1941, Darlan attempted to negotiate some measure of independence for France but his efforts were largely spurned by HITLER. In April 1942, Darlan was replaced by the increasingly powerful Piérre LAVAL and was put in charge of remaining French armed forces, located chiefly in French African colonies. When the Allies invaded Algeria in November 1942, Darlan decided to cooperate and signed an armistice. Darlan was seen as pro-Nazi in Britain and the USA and the agreement provoked some criticism. On Christmas Eve 1942, he was assassinated by a French monarchist.

Darlington, Cyril Dean (1903–81) *British geneticist, whose work on the structure and behaviour of chromosomes in the cell nucleus helped establish the fundamental importance of cytology in genetics.*

Born in Chorley, Lancashire, Darlington received his BSc degree from Wye College, London University. In 1923 he joined the John Innes Horticultural Institute as a volunteer, rising to become head of the cytology department (1936) and ultimately director (1939). Darlington's main interest was the mechanism by which homologous chromosomes associate and exchange genetic material during cell division (meiosis). He demonstrated that the points of contact (chiasmata) between the chromosomes resulted from the prior crossing over of their duplicate strands and that in certain species these chiasmata migrated to the ends of the chromosomes, resulting in a loop. This process, which he called 'terminalization', accounted for many hitherto unresolved cytological phenomena. He published the influential *Recent Advances in Cytology* in 1932 and in *The Evolution of Genetic Systems* (1939) he examined the significance of nuclear cytology to population genetics and evolution.

In later life, Darlington increasingly sought to explain the currents of human history in terms of genetic and cultural interactions. *The Evolution of Man and Society* (1969) was greeted with fierce controversy, with what some critics saw as racist implications. Darlington was elected fellow of the Royal Society in 1941 and received its royal medal in 1946. He was a fellow of Magdalen College, Oxford (1953–71), and made emeritus professor on his retirement.

Darrow, Clarence Seward (1857–1938) *Leading US defence counsel, who came to prominence through his work in many dramatic criminal trials.*

Darrow was admitted to the Ohio bar in 1878 and nine years later came to Chicago, where he immediately tried to secure the release of rioters who had been imprisoned for murder the previous year. In 1890 he became counsel for the Chicago city corporation and, influenced by the governor of Illinois John Peter Altgeld, became interested in labour law cases. He became nationally known as a lawyer in 1895, when he undertook the unsuccessful defence of labour organizer Eugene V. Debs, president of the American Railway Union, charged with defying an injunction in the Pullman Strike of the previous year. When Pennsylvanian anthracite coal miners went on strike in 1902, Darrow was appointed as an arbitrator by President Roosevelt, notwithstanding the lawyer's stance as an Independent Democrat. In 1907 Darrow secured the acquittal of radical labour leader William ('Big Bill') Haywood on a charge of conspiring to murder former Idaho governor Frank R. Steunenberg. However, an unexpected change of plea to guilty in the trial of the McNamara brothers (1911), charged with dynamiting the *Los Angeles Times* building, prompted Darrow to abandon labour litigation and concentrate on criminal law.

After World War I he defended war protesters on sedition charges and then embarked on a series of murder trials, such as that of Nathan Leopold and Richard Loeb. Passionately opposed to capital punishment, Darrow saved in all more than fifty accused murderers from the death penalty. Darrow's most famous trial came in 1925, when he defended John T. Scopes, a Tennessee high-school teacher accused of violating a state law by teaching Darwinian theories of evolution instead of those favoured by biblical fundamentalists. An agnostic, Darrow gave a brilliant defence; although Scopes was found guilty (the conviction was later reversed by the state supreme court), Darrow's conduct in this trial, together with his writings and his fame as a public speaker and debater, assured him of a lasting place in US legal history.

Davies, Sir Peter Maxwell (1934–   ) *British composer. He was knighted in 1987.*

Born in Manchester, he studied at the Royal Manchester College of Music, where Harrison BIRTWHISTLE, Alexander Goehr (1932–   ), and John Ogdon (1937–89) were fellow students. With them he formed the New Music Manchester Group for the performance of European avant-garde music and their own compositions. After studying with Goffredo Petrassi (1904–   ) in Rome on an Italian government scholarship (1957–58), when he was awarded the Olivetti Prize for his orchestral work *Prolation* (1958), Davies returned to England to teach music at Cirencester Grammar School (1959–62). This was a period of great importance in the development of his compositional style. His formation with Birtwhistle of the Pierrot Players (founded to perform SCHOENBERG's *Pierrot lunaire*), later to reform as the Fires of London ensemble, provided him with a group of sympathetic players and singers for the performance of his works. He held a Harkness Fellowship at Princeton University (1962–64), working with Roger SESSIONS; in 1966 he was composer-in-residence at the University of Adelaide. Since 1970 Davies has lived and composed in Orkney, producing a series of scores based on Orcadian or Scottish myth and music, including (in collaboration with the Orkney poet George Mackay Brown) *Fiddlers at the Wedding* (1973–74), for voice, alto flute, mandolin, guitar, and percussion, and *An Orkney Wedding, With Sunrise* (1985) for orchestra.

Dominant in Davies's music is his use of distortion and parody of generally medieval or Renaissance music: the string quartet (1961), the *Leopardi Fragments* (1961), and the *Sinfonia* (1962) for chamber orchestra are all based on material from Monteverdi's *Vespers*. Dramatic action has been an integral part of Davies's music since the *Eight Songs for a Mad King* (1969). The opera *Taverner* (1962–68; partly revised 1970) is one of many works that explore the conflict between blasphemy and established conformity. From 1971 Davies's religious preoccupation has been expressed in more lyrical and mystical pieces, such as *Hymn to St Magnus* (1972), the chamber opera *The Martyrdom of St Magnus* (1976), *Agnus Dei* (1984), and the opera *Resurrection* (1987). Four symphonies date from 1975–76, 1980, 1985, and 1989.

Davis, Bette (Ruth Elizabeth Davis; 1908–89) *US actress who was the first woman to receive the American Film Institute's Life Achievement Award (1977).*

Born in Lowell, Massachusetts, Bette Davis began her somewhat stormy career in the theatre. She made her Broadway debut in *Broken Dishes* (1928). Although she failed an initial screen test, she made her first film, *The Bad Sister*, in 1931 and delivered her first outstanding performance in *Of Human Bondage* (1934). *Dangerous* (1935) brought her first Oscar but she had to struggle against her studio for the parts she wanted in such films as *The Petrified Forest* (1936) and *Jezebel* (1938), for which she received a second Oscar. Her commanding style was well suited to such films as *The Little Foxes* (1941) and *Mr Skeffington* (1944). Successful, too, was *All About Eve* (1950), for which the New York Film Critics named her best actress. *The Virgin Queen* (1955) saw her as Elizabeth I, a role she had played earlier in *The Private Lives of Elizabeth and Essex* (1939), but her next great triumph came in a film that utilized her abundant talent to the full, *Whatever Happened to Baby Jane?* (1962) with Joan CRAWFORD. She continued in films to the end of her life, winning acclaim in *The Whales of August* (1986).

Television work included *Strangers*, which won her an Emmy Award in 1979. Davis also wrote two autobiographies, *The Lonely Life* (1962) and *Mother Goddam* (1975).

Davis, Sir Colin Rex (1927–   ) *British conductor, who was knighted in 1980.*

He studied the clarinet with Frederick Thurston at the Royal College of Music and became a bandsman in the Household Cavalry. His lack of training as a keyboard player precluded an apprenticeship as a repetiteur in an opera house when he wanted to take up conducting. However, he took jobs wherever they occurred and quickly made a reputation for his Mozart

performances in the early 1950s with the Chelsea Opera Group. He was appointed assistant conductor of the BBC Scottish Orchestra in 1957 and was much acclaimed for his Sadler's Wells debut (1958) conducting Mozart's *The Abduction from the Seraglio*. In 1959 he took over Mozart's *Don Giovanni* from KLEMPERER and was appointed musical director of Sadler's Wells Opera in 1961. In 1971 he became musical director of Covent Garden Opera. He made his debut at the Metropolitan Opera in New York in 1966, conducting BRITTEN's *Peter Grimes*. From 1967 to 1971 he was conductor of the BBC Symphony Orchestra. Since 1983 he has been Chief Conductor of the Bavarian Radio Symphony Orchestra.

Notable opera performances have been *The Trojans* by Berlioz (Covent Garden, 1969) and the première of TIPPETT's *The Knot Garden* (Covent Garden, 1971). Davis has made many recordings, including all Berlioz's major works.

Davis, Miles Dewey (1926–91) *Black US jazz trumpeter, bandleader, and composer. He pioneered the style of music that later became known as cool jazz.*

Born into a middle-class black family in Alton, Illinois, he studied music at the Juilliard School in New York. At the same time he was playing in various bands and began recording in 1945 with Charlie PARKER, Coleman HAWKINS, and others. In the 1950s he played and recorded arrangements by Gil Evans (1912–88) featuring the tuba and French horn; these became known as the 'Birth of the Cool' sessions. His quintet of 1955–57 featured John COLTRANE and several orchestral recordings made in the late 1950s were arranged by Evans; *Kind of Blue* (1959), recorded by a sextet, is one of his most highly regarded albums. The quintet of the 1960s began the use of an electronic keyboard. In fact, *In a Silent Way* (1969) used three of them, with an electric guitar; it was the first successful 'fusion' of jazz and rock, although Davis refused to use either of these words to describe his music. After the best-selling *Bitches Brew* (1969) his music continued in an electronic and often rhythmically complicated form; he himself used electronic amplification with his trumpet and a wah-wah pedal.

A drug addict in the early 1950s, Davis broke the habit himself. He suffered prolonged bouts of ill health and considerable pain after breaking both legs in a car crash in 1972. He toured the world several times, always winning universal acclaim.

Davis, Steve (1957–) *British snooker player.*
Born in London, he became a professional snooker player in 1978 and subsequently came to dominate the game with a series of major tournament victories, winning the UK professional championship (1980–81 and 1984–87), the Masters championship (1981–82 and 1988), and the World Championship (1981, 1983–84, and 1987–89). His records include the first maximum score of 147 in a major tournament (during the Lada Classic at Oldham in 1982) and the first run of three consecutive century breaks in a major tournament (at Stoke-on-Trent in 1988). He responded to gibes that he had a boring mechanical personality by publishing the lighthearted *How to be Really Interesting* in 1988.

Dawes, Charles G(ates) (1865–1951) *US banker and statesman remembered for the Dawes Plan (1924), which effectively saved Europe from economic collapse after World War I. For this he was awarded the 1925 Nobel Peace Prize.*
Born in Marietta, Ohio, Dawes studied law at Cincinnati Law School. He practised law in Lincoln, Nebraska, for seven years and, after a further three years at Evanston, became US comptroller of the currency (1897–1902). After a period in private business he was appointed the chief purchasing agent for the USA during World War I; at the end of the war he chaired the Allied Reparations Commission and the resulting Dawes Plan led to the reorganization of Germany's finances through a large foreign loan to enable the country to make reparations to the Allies.

Dawes served as vice-president under Calvin COOLIDGE from 1925 to 1929, when he was appointed US ambassador to Great Britain. In 1932 he returned to the USA in order to steer the economy through the Great Depression and later resumed his banking business. Dawes wrote several books based upon his experiences, including *A Journal of the Great War* (1921), *Notes as Vice President* (1935), and *A Journal of Reparations* (1939).

Dayan, Moshe (1915–81) *Israeli general and politician. A controversial swashbuckling figure, he combined a flair for military leadership with a deep interest in archaeology.*
Born in Degania on Israel's first kibbutz, Dayan attended an agricultural secondary school before studying science at the Hebrew University in Jerusalem. As a youth he joined the Jaganah, the Jewish militia force, but was jailed (1939–41) by British mandate authorities. After his release during World War II he fought with the British and Free French Forces in Syria, losing his left eye in action, after which he wore the familiar black patch.

When Israel became independent in 1948, Dayan commanded Israeli forces in the Jerusalem area, defending it against several Arab incursions. He took part in the armistice talks with Jordan (1949) and as chief of staff (1953–58) commanded the 1956 Sinai invasion. Elected to the Knesset (parliament) in 1959 as a representative of the Mapai (Labour) Party, he was appointed minister for agriculture under BEN-GURION. He resigned in 1964 but was re-elected the following year as a member of the Rafi Party, a breakaway group led by Ben-Gurion. In 1967 he became minister of defence and directed Israeli forces in the Six Day War, in which he defeated the combined forces of the Egyptians, Syrians, and Jordanians, taking considerable amounts of strategically important Arab territory (Golan Heights, West Bank, and Gaza Strip). He remained in this position, overseeing the 1973 Yom Kippur War, until he resigned in 1974 as a result of criticisms over Israel's state of unreadiness at the start of the Yom Kippur War. He returned to office in 1977 as minister of foreign affairs under BEGIN and played a prominent role in the talks that led to the Israeli-Egyptian peace treaty (1979). He resigned in 1979 over disagreement with Begin on policy towards the Arabs. He wrote about his experiences as a soldier in *Diary of the Sinai Campaign* (1966) and *The Story of My Life* (1976).

Day Lewis, C(ecil) (1904–72) *British poet, novelist, and critic. He was created a CBE in 1950 and was poet laureate from 1968 until his death.*
Day Lewis was born in County Sligo, Ireland, the son of a Church of England clergyman and his Irish wife, and was educated at Sherborne School before going to Wadham College, Oxford. From 1927 to 1935 he taught at various schools and began to find his voice as a poet; *Transitional Poem* (1929) first attracted critical attention, and such publications as *A Time to Dance* (1935) and *Overtures to Death* (1938) aligned him with the modern poetic movement under the leadership of W. H. AUDEN.

During World War II Day Lewis edited material for the Ministry of Information (1941–46). In the postwar period his work was increasingly recognized, which culminated in his appointment to the laureateship. In 1946 he was invited to give the Clark Lectures at Cambridge; he was subsequently professor of poetry at Oxford (1951–56), Charles Eliot Norton Professor of Poetry at Harvard (1964–65), and vice-president of the Royal Society of Literature from 1958. His second marriage (1951) to the actress Jill Balcon increased his interest in the spoken word, and he frequently gave readings of his own and other poetry.

Day Lewis continued to publish collections of verse at regular intervals, the last volume being *The Whispering Roots* in 1970. He published translations of Virgil in verse (*Georgics*, 1941; *Aeneid*, 1952; *Eclogues*, 1963) and several volumes of criticism, including *A Hope for Poetry* (1934), *The Poetic Image* (1947), and *The Lyric Impulse* (1965). He also wrote novels, such as *The Friendly Tree* (1936), and in a more light-hearted vein produced a number of detective stories under the pseudonym Nicholas Blake. His autobiography, *The Buried Day*, appeared in 1960.

Deakin, Alfred (1856–1919) *Australian statesman and three times prime minister (1903–04; 1905–08; 1909–10), who played a major role in the early development of the Commonwealth.*

Born and educated in Melbourne, the son of an accountant, Deakin began his career as a barrister in 1878. After entering the Victoria legislative assembly (1879) he gained a reputation as a powerful advocate and travelled widely: he made a strong impression at the London Colonial Conference (1887) with his defence of Australian interests. Elected as the member for Essendon in 1889, Deakin was one of the six Australians who went to London to steer the Federation Bill through the British parliament. Appointed attorney-general by prime minister BARTON, he introduced the legislation that produced the high court in Australia.

Following Barton's retirement in 1903 he became prime minister but was defeated a year later. Re-elected in 1905 and supported by the Labor Party, he introduced some far-reaching legislation, including a protectionist tariff, old age pensions, and commercial laws affecting copyright, trade marks, and trusts. Losing the support of the Labor Party in 1908, he formed what came to be known as the 'Fusion' of 1909 with Joseph COOK and his former opponents in the Liberal Party and was prime minister for a third time. He was defeated in the election the following year. A widely read man known as 'the silver-tongued orator of Australia', Deakin retired in 1919, shortly before his death.

Dean, Dixie (William Ralph Dean; 1906–80) *British Association footballer who played for Tranmere Rovers, Everton, and England.*

In the 1927–28 season Dixie Dean scored a record 60 league goals in 39 matches, a goal-scoring record that is unlikely ever to be beaten. They included five goals in one match and four in another. During his prolific career Dean scored no fewer than 37 hat-tricks. Tall and well-built, he was a brilliant header of the ball (many of his goals were headed). He also made 16 international appearances between 1927 and 1932. After the end of his playing career he ran a public house for some years at Chester.

Dean, James (James Byron; 1931–55) *US film star, whose early death in a car accident has ensured a permanent place in the annals of Hollywood for this 1950s symbol of disaffected American youth.*

He was born in Marion, Indiana, and after attending the University of California at Los Angeles, began acting and attended the Actors' Studio for a brief period. He made a few TV commercials and played a few small parts in such films as *Sailor Beware* (1951) and *Fixed Bayonets* (1951). On Broadway he appeared in *See the Jaguar* (1952) and *The Immortalist* (1954); it was his performance in the latter that secured him a film test at Warner Brothers. His three starring roles followed: *East of Eden* (1955), *Rebel Without a Cause* (1955), and *Giant* (1956). In the eyes of his fans he became identified with the title of his film *Rebel Without a Cause* and it was this role particularly that made him into a cult figure. A film biography, *The James Dean Story*, was made in 1957.

de Beauvoir, Simone (1908–86) *French writer and feminist.*

Born in Paris, de Beauvoir attended a number of private schools before embarking on philosophy studies at the Sorbonne. It was here that her lifelong association with Jean-Paul SARTRE began in 1929; together they became leading exponents of existentialism and founded the review *Les Temps Modernes* (1945). De Beauvoir taught in Marseilles, Rouen, and Paris before turning to literature during World War II. Her first novel, *L'Invitée*, was published in 1943; subsequent novels, reflecting various aspects of existentialist thought, include *Le Sang des autres* (1944; translated as *The Blood of Others*, 1948) and *Tous les hommes sont mortels* (1946; translated as *All Men are Mortal*, 1955). Perhaps the best known is *Les Mandarins* (1954; translated in 1956), set in postwar Paris, for which de Beauvoir was awarded the Prix Goncourt.

The first part of her extended autobiography, *Mémoires d'une jeune fille rangée*, appeared in 1958. This was followed by *La Force de l'âge* (1960; translated as *The Prime of Life*, 1965); *Une Mort très douce* (1964; translated as *A Very Easy Death*, 1966), a detached account of the death of her mother from cancer; *Tout compte fait* (1972; translated as *All Said and Done*, 1974); and a moving account of Sartre's last years, *La Cérémonie des adieux* (1981). These memoirs provide a valuable insight into Parisian intellectual life in the mid-twentieth century and into the intimate relationship between de Beauvoir and Sartre.

De Beauvoir's fame rests largely on the treatise *Le Deuxième sexe* (1949; translated as *The Second Sex*, 1953), an important milestone in the feminist cause. In it she argues against marriage and motherhood, which she regards as a submission to male domination, and attempts to rationalize the predicament of the female. Her other nonfiction works include the essays *Pour une morale de l'ambiguïté* (1947) and *Faut-il brûler Sade?* (1951). She also wrote travel books, such as *La Longue Marche* (1957; translated as *The Long March*, 1958), and the play *Les Bouches inutiles* (1945).

Debré, Michel (1912–   ) *French statesman and prime minister (1959–62). He was elected to the Academie Française in 1988.*

Born in Paris, the son of a leading paediatrician, Debré began his education at a lycée in Paris, before attending the Cavalry School at Saumar. On completion of his military service in 1932, he entered the University of Paris, where he graduated in law, becoming a law clerk to a supreme court justice in 1934. In 1938 he was appointed to the staff of finance minister Reynaud. Following the outbreak of World War II he joined the cavalry, later taking an active role in the Resistance. In 1944 he was appointed special commissioner of the republic for the Anvers region. He played a major role in the reform of public administration (1945), becoming secretary-general for German and Austrian affairs in the foreign ministry in 1947.

Debré was elected to the senate in 1948 as a member of DE GAULLE's Rassemblement du Peuple Français (RPF) (later Social Republicans). Re-elected in 1955, he became minister of justice under de Gaulle in 1958 and played a major role in the drafting of the new constitution. In 1959 he became the first prime minister of the Fifth Republic, a position he held until 1962. The following year he was elected to the national assembly as the deputy for Réunion. Over the next fifteen years he held several ministerial portfolios including economics and finance (1966–68), foreign affairs (1968–69), and the army (1969–73). Elected as a deputy to the European parliament (1979–80), he stood unsuccessfully as a presidential candidate in 1981.

A fervent patriot, Debré was a loyal supporter of de Gaulle. As prime minister his most difficult task was the resolution of the Algerian civil war (1961–62); in his various ministerial roles he advocated national independence. He was awarded the Croix de Guerre and the Rosette de la Résistance, and has written several books and articles.

de Broglie, Prince Louis-Victor Pierre Raymond, Duc (1892–1987) *French physicist whose discovery of the wave nature of electrons laid the foundation for wave mechanics and won him the 1929 Nobel Prize for Physics.*

Born into a Piedmontese family, who were ennobled by Louis XIV in 1740 and awarded the title Prince for their service to the Austrians in the Seven Years War, de Broglie inherited the dukedom from his brother, also a physicist. Educated at the Sorbonne, where he studied history, he was later attracted to science by reading the popular works of the mathematician Henri Poincaré (1854–1912). He obtained his degree in science in 1913 and, after serving throughout World War I with the engineers, returned to the Sorbonne, being awarded his doctorate in 1924. Shortly afterwards de Broglie was appointed professor of physics at the newly founded Henri Poincaré Institute, a post he held until his death.

In 1923 de Broglie published two short articles suggesting that such elementary particles as electrons could behave like waves as well as particles. His supposition was later confirmed by the electron-diffraction experiments of Sir George Paget THOMSON and C. J. Davisson (1881–1958). The waves associated with elementary particles are now known as de Broglie waves.

Debussy, (Achille) Claude (1862–1918) *French composer and pianist, regarded as the originator of impressionism in music.*

The son of a shopkeeper, Debussy was first encouraged in his love of music in 1871 by Mme Mauté de Fleurville, a former pupil of Chopin and mother-in-law of Verlaine, whose poetry he was later to set. From the age of ten

Debussy was a student at the Paris Conservatoire, studying chiefly with Marmontel (piano) and Guiraud (theory). In 1884 he won the Prix de Rome with his cantata *L'Enfant prodigue*, but in 1887 he cut short his scholarship and returned to Paris to compose. There he associated with artists and writers rather than musicians but was influenced by Wagner, Mussorgsky, Grieg, and Javanese gamelan music heard at the Paris Exhibition in 1889. His marriage in 1899 to a dressmaker broke down because of lack of intellectual compatibility; they were divorced in 1905 and he married Emma Bardac, by whom he had one daughter: Claud-Emma (Chouchou), to whom he dedicated the *Children's Corner* piano suite (1906–08).

As a composer, Debussy had a great influence on the development of twentieth-century European music: he carried the ideas of impressionist art into music to create a specifically French style, evocative of moods and atmospheres. *Prélude à l'après-midi d'un faune* (1899), the first of Debussy's truly impressionist orchestral pieces, was followed by the *Nocturnes* (1899), the symphonic sketches *La Mer* (1905), and the ballet *Jeux* (1913). His opera *Pelléas et Mélisande* (1902), based on Maeterlinck's symbolist play, brought him professional recognition. His distinctive piano style is at its most impressionistic in the two books of *Préludes* (1910–13); later his style became more classically austere, as in the two books of piano *Études* (1915) and the two-piano suite *En blanc et noir* (1915). A reluctant performer in public, he conducted his own works in the Queen's Hall in London on two occasions (1907 and 1908).

De Forest, Lee (1873–1961) *US electrical engineer who invented the triode valve.*

The son of a Congregational minister, De Forest, against his father's wishes, studied science at Yale, obtaining his PhD in 1899. He began his career with the Western Electric Co., Chicago, but was soon dismissed for devoting too much time to his own research. In 1902 he set up the first of his several companies, the De Forest Wireless Telegraph Co., which failed (as the others did) after a few years. On this occasion the failure was due to the invention of the diode valve by J. A. Fleming (1849–1945) in 1904. In 1906, however, De Forest improved on Fleming's device by adding a third electrode, making a triode. Although De Forest intended to use his triode as a demodulator, he realized that its most valuable and innovatory use would be as an amplifier. This enabled him

to explore the possibilities of broadcasting, and in 1910 he broadcast the voice of Caruso for the first time. In 1916 he established a radio station, which was able to transmit the results of the 1916 presidential elections. Never happy as a businessman, De Forest eventually sold the rights to his invention for over $300,000.

After the triode valve, De Forest is best known for his work in the 1920s on the development of talking pictures. He founded a company, the De Forest Phonofilm Corporation, to develop this invention, but it collapsed in 1927 when the film industry adopted another system.

de Gaulle, Charles André Joseph Marie (1890–1970) *French general, statesman, and president (1958–69), who became a legendary embodiment of the spirit of France.*

Born in Lille, the son of a teacher, de Gaulle was educated at the École Militaire of St Cyr, where he graduated in 1912. Joining the infantry regiment, he was wounded and taken prisoner during World War I in the battle of Verdun (1916). In 1923 de Gaulle became a lecturer at the Staff College, and during the 1930s wrote several books and articles on military subjects. Appointed a commander of an armoured division at the outbreak of World War II, he briefly held the office of undersecretary of state for war in 1940 before escaping to England to organize French resistance as leader of the Free French movement. (He refused to recognize the Vichy regime led by PÉTAIN.) He became head of the French Committee of National Liberation in Algiers in 1943 and entered Paris triumphantly in 1944. He was elected president of the provisional government but resigned in 1946 following disagreement over the constitution adopted by the Fourth Republic. After limited electoral success as leader of the Rassemblement du Peuple Français, which he founded in 1947, he withdrew from politics in 1953 to concentrate on writing.

However, in 1958 de Gaulle was asked to form a government when the Fourth Republic faced a severe economic crisis at home and civil war in Algeria. Granted wide powers by the national assembly, he promulgated a new constitution, which provided for a presidential system and enabled him to take office as president in December 1958. In the early 1960s de Gaulle presided over France's economic recovery, a solution to the Algerian crisis, and the granting of independence to African colonies. Re-elected in 1965, he pursued an assertive foreign policy that opposed British entry to the EEC, withdrew French contingents from NATO (1966), and maintained the development of a French nuclear deterrent. Overcoming student-worker protests in 1968, he resigned in 1969 after proposed constitutional changes were rejected by the electorate.

De Havilland, Sir Geoffrey (1882–1965) *British aircraft designer, the creator of the Moth, Mosquito, and Comet. He was knighted in 1944 and awarded the OM in 1962.*

The son of a clergyman, De Havilland was educated at St Edward's School, Oxford, and the Crystal Palace Engineering School. He worked in industry for a few years before setting up in 1908, with an inheritance of £1000, as an aircraft designer. Although he smashed his first aircraft, his second plane was sold to the Royal Aircraft Factory at Farnborough for £400. During World War I De Havilland, working for the Aircraft Manufacturing Company in Hendon, designed a number of bombers and fighters. In 1919, however, this company collapsed and De Havilland, who was by then one of the world's most experienced aviation engineers, set up the De Havilland Aircraft Company in Hendon with a working capital of £1875.

De Havilland's greatest achievement in the inter-war years was his introduction in 1925 of the Moth range of light aircraft. In its first twelve years 2500 models (mostly Gipsy Moths) were produced, while during World War II a further 8300 were built. It was also in the inter-war years that De Havilland designed the Mosquito, a revolutionary all-wood fighter-bomber. Although the Air Ministry showed no initial interest in the plane, it became one of the fastest and most versatile planes of its day and over 7000 Mosquitoes were built during the war. After the war De Havilland's Comet of 1949, the world's first commercial jet liner, started as a triumph of British enterprise. Unfortunately the disastrous crashes of 1954 from metal fatigue led to the need to redesign the plane. In 1959, as the Comet-4, it became the first jet in transatlantic service. There were also problems arising from De Havilland's pioneering work on supersonic planes. His son, Geoffrey, died testing a new plane in 1946, and the DH 110 crashed at Farnborough in 1952, killing many spectators while breaking the sound barrier. Although De Havilland himself continued to work on the design and development of aircraft until his death, the company and his name disappeared in 1959 when, in the process of rationalization

demanded by the government, De Havilland's was taken over by Hawker Siddeley.

de Klerk, F(rederik) W(illem) (1936–) *South African statesman; as state president (1989–) he began to dismantle the apartheid system and prepare the country for multiracial democracy.*

De Klerk was born in Johannesburg and educated at Potchefstroom University. After practising law he was elected to the House of Assembly as a National Party member in 1972, holding a number of ministerial posts before becoming minister of internal justice in 1982. On the resignation (1989) of P. W. Botha, de Klerk became leader of the National Party and was elected state president later the same year.

Although de Klerk had a reputation as a cautious reformer, it was not generally supposed that his presidency would see a decisive break with past policies. In the event, his readiness to embrace change surprised both his critics and supporters. Within months of taking office he had freed leaders of the outlawed ANC, met Nelson Mandela in prison, and announced the end of petty apartheid restrictions. In February 1990 a major speech proclaiming a 'new era' in South African politics was followed by the legalizing of the ANC and Communist Party and the freeing of Mandela. Subsequently de Klerk held talks about the country's future with Mandela and other opposition leaders and lifted the state of emergency in force since 1985. Classification by race was ended in June 1991 and a new constitution giving voting powers to Black South Africans was proposed, although interfactional violence and clashes between the ANC and the government threatened the progress of further liberal reform.

de Kooning, Willem (1904–) *Dutch-born US painter. In the late 1940s he became, with Jackson Pollock, the leading figure in the abstract expressionist school, which helped to transfer ascendancy in the arts from Paris to New York.*

In 1916 de Kooning left his native Rotterdam to serve an apprenticeship with a firm of commercial decorators in Amsterdam. At the same time he began to study art and produced drawings and paintings in a traditional style before emigrating in 1926. In the USA he experimented with abstract styles derived from Kandinsky and late cubism and formed a close friendship and working relationship with the painter Arshile Gorky. His employment on the Federal Arts Project in the 1930s enabled him to become a full-time painter. He worked in several styles, ranging from greyish abstraction to detailed figurative work, and his drawings showed him to be one of the great draughtsmen of this century.

Although he was influential among the artists in New York it was not until 1948 that he had his first one-man exhibition and his leading position was confirmed. A characteristic work of this time is *Painting* (1948), consisting of black forms that, though not representational, retain echoes of the human body. De Kooning's *Excavation* (1950) signalled the return of colour to his work. In this and in most of his later work the female form, whether represented or merely hinted at, remained a central theme, particularly in the series that began with *Woman 1* in 1952. His paintings alternated between the figurative and the abstract and during the 1970s de Kooning also turned to figurative expressionistic sculpture.

de la Mare, Walter John (1873–1956) *British poet, novelist, anthologist, and critic. He became a CH in 1948 and was awarded the OM in 1953.*

The son of a Bank of England official, de la Mare was born in Charlton, Kent, and attended St Paul's Cathedral Choristers' School, London. After school he joined the Anglo-American Oil Company (1890), for which he worked until a grant from the Privy Purse enabled him to pursue a full-time literary career (1908). Meanwhile he had already published *Songs of Childhood* (1902), among other poems and stories that appeared under the pseudonym 'Walter Ramal'. Apart from lecture tours in the USA (1916–17; 1924–25) and a spell as a civil servant during World War I, de la Mare's life was one of unbroken and externally uneventful literary activity. As a poet his talent was predominantly lyrical. His first major success was 'The Listeners' (1912) and he continued to write poetry throughout his life, including the volumes *Motley and Other Poems* (1918), *Collected Rhymes and Verses* (1944), *The Traveller* (1946), and *O Lovely England* (1954).

A particular preoccupation of de la Mare's was the imaginative world of childhood, a preoccupation especially revealed in his poems for children, which have the freshness and immediacy of nursery rhymes. *Peacock Pie* (1913), his book of poems for children, was one of his most enduring successes, and his poetic anthologies for children, *Come Hither* (1923) and *Tom Tiddler's Ground* (1931), further show his sympathetic understanding of a child's world. He also wrote numerous tales for children, including *The Lord Fish* (1933)

and *The Scarecrow* (1945). His anthology *Early One Morning* (1935) takes childhood as its theme.

Like his poetry, de la Mare's prose fiction often reveals his highly individual strain of fantasy and interest in the supernatural. This is noticeable in his novels *Henry Brocken* (1904) and *The Return* (1910). He also wrote a shrewd appraisal of the work of Lewis Carroll (1932), essays published under the title *Private View* (1953), and a quasi-philosophical discourse on the Robinson Crusoe motif – *Desert Islands* (1930).

Delaunay, Robert (1885–1941) *French painter who founded orphism and produced some of the first abstract pictures.*

After training as a decorative painter in Paris, Delaunay became a full-time artist in 1904. His early work was mainly in the style of early cubism but his researches into the application of the colour theories of Eugène Chevreul led to a preoccupation with colour that was central to all his work after 1910. This 'colour cubism', which appeared in his Eiffel Tower series and the 'window pictures', was given the name orphism by Delaunay's friend, the poet and art critic Guillaume Apollinaire. Though sometimes abstract in appearance, his pictures at this time were based on observed motifs, but from about 1912 Delaunay began to find objective motifs largely unnecessary to his exploration of the effects of colour and he painted some of the earliest abstract compositions. They consisted of arrangements of squares and circular areas of colour, as in the series *Jeu des disques multicolores* and the *Rhythmes* series. Occasionally he returned to the partial use of objective forms, as in *Homage to Blériot* (1914) and his series of football players (*c.* 1926).

During his lifetime his work had a considerable impact on contemporary artists, particularly on Paul KLEE. Delaunay's creation of a sense of movement through the interplay of colours has, together with his purely abstract use of optical phenomena, continued to influence the development of modern art.

Delius, Frederick (1862–1934) *British composer of German parentage. He was created CH in 1929.*

Born in Bradford, the son of a successful naturalized wool merchant, Delius was persuaded to enter his father's business and spent some time selling wool in Europe before emigrating to Florida (1884), where he became an orange planter. During this period he decided to devote himself to music, returning to Europe in

1886 to study at the Leipzig Conservatory. A meeting with Grieg at that time undoubtedly influenced his decision. He later settled in Grez-sur-Loing, south of Paris, with his artist wife, Jelka Rosen. In 1928, after Delius had become blind and paralysed, a young English musician, Eric Fenby (1906–87), became his amanuensis and enabled him to continue composing.

Delius forged his own unique style, his empathy with nature being a source of inspiration in many of his works. The opera *Koanga* (1895–97) reflects his experiences in Florida, while the variations for chorus and orchestra *Appalachia* (1902) owes its spirit to the American forests and mountains. The tone poem *Over the Hills and Far Away* (1895) was inspired by his native Yorkshire, and *On Hearing the First Cuckoo in Spring* (1912) and other tone poems by his garden at Grez. Other works of importance are the tone poem *Nocturne, Paris: The Song of a Great City* (1899), the opera *A Village Romeo and Juliet* (1900–01) to his own libretto based on a novel by Keller, *Sea Drift* (1903) for baritone, chorus, and orchestra to a poem by Walt Whitman, and *A Mass of Life* (1904–05) for soloists, chorus, and orchestra to words of Nietzsche. His concertos (one for violin, one for cello, and one double concerto for violin and cello) owe little to traditional forms and are orchestrally less impressive. Recognition came first in Germany, but Sir Thomas BEECHAM was his tireless advocate in England. Thanks to Beecham a six-day Delius Festival was held at the Queen's Hall in 1929 in the presence of the blind composer.

Delvaux, Paul (1897–) *Belgian painter, known for his paintings of nude or semiclothed women in surrealist architectural settings.*

The son of a lawyer, Delvaux first studied architecture and then painting at the Académie des Beaux Arts in Brussels. His early works were realistic landscapes, which by 1924 had begun to show the influence of neoimpressionism. Moving towards expressionism, he later introduced nudes and other figures into his paintings. In the 1930s, after seeing pictures by MAGRITTE and CHIRICO, he adopted the style of the surrealists and by the late 1930s was contributing to major surrealist exhibitions.

During the 1940s Delvaux's work finally achieved recognition and acclaim. His paintings characteristically depicted pale ideally beautiful women, clothed or nude, dreamily standing or reclining in classical or contemporary settings with complex architecture. They conveyed a sensation of isolation and waiting, sensuality, and sadness, as in *Phases of the Moon* (1939). Elements of the grotesque became common in his work in the 1940s with his introduction of the skeleton theme. In 1950 Delvaux became a lecturer at the Brussels École Nationale Supérieure d'Art et d'Architecture and was later appointed to a chair in fine arts at the Belgian Royal Academy.

de Mille, Cecil B(lount) (1881–1959) *Film producer and director and founder of the Hollywood film industry.*

Son of the popular dramatist Henry C. de Mille and brother of the film producer, director, and writer William de Mille (1878–1955), Cecil was born in Ashfield, Massachusetts. After training at the New York Academy of Dramatic Art he made his acting debut on Broadway in 1900. He also managed his mother's theatrical company for several years and collaborated with his brother on various plays.

His film-making career began when he joined Jesse L. Lasky and Samuel Goldfish (later GOLDWYN) to form the Jesse L. Lasky Feature Play Company in 1913. De Mille chose the then little-known suburb of Hollywood for their first film, *The Squaw Man* (1914), converting a barn into a studio. With the film's success Hollywood's life as the film capital of the world began.

Among his early films were *The Girl of the Golden West* (1915), *The Trail of the Lonesome Pine* (1916), *Male and Female* (1919), and *Forbidden Fruit* (1921). De Mille is, however, best remembered for his biblical extravaganzas beginning with *The Ten Commandments* (1923), remade in 1956, and followed by *King of Kings* (1927). Outstanding, perhaps, were *The Sign of the Cross* (1932), with Claudette Colbert (1905–) and Charles LAUGHTON, and – for sheer spectacle – the destruction of the temple by Victor Mature (1915–) in *Samson and Delilah* (1949).

Dempsey, Jack (William Harrison Dempsey; 1895–1983) *US boxer and world heavyweight champion (1919–26).*

The son of an Irishman, Jack Dempsey came from Manassa, Colorado, and struggled for recognition for some years before being successfully managed by Jack 'Doc' Kearns. Dempsey's aggressive style brought him rapid progress and popularity and he became known universally as the 'Manassa Mauler'. He won the world title in 1919 by beating Jess Willard in three rounds. Dempsey attracted huge crowds to his fights and his third defence of the title, against Georges CARPENTIER in 1921, produced boxing's first million-dollar gate. Dempsey finally lost the title in 1926 to Gene Tunney on points.

After retiring from the ring he became a boxing referee and for some years he ran a famous restaurant opposite Madison Square Gardens, New York.

Deng Xiao Ping (Teng Hsiao-p'ing; 1904–) *Chinese communist statesman; vice-premier (1973–76; 1977–80) and vice-chairman of the central committee of the Communist Party (1977–80). Denounced as a capitalist during the Cultural Revolution, Deng has been seen as an independent and nonconformist figure who has become one of the most influential men in China.*

Born in Szechwan province, the son of wealthy parents, Deng was educated at a local high school before travelling to France in 1920 to study at the University of Lyon. In 1922 he joined the Chinese Socialist Youth League and two years later took up membership of the Chinese Communist Party. Returning to China in 1926, he was appointed an instructor at the Xi'an Political and Military Academy but, following the split between the nationalists and the communists in 1927, went to Shanghai to work for the Communist Party central organization.

Deng took part in the epic 'long march' led by MAO TSE-TUNG (1934–35). He served in the Red Army during the war against the Japanese and in the civil war against the nationalists. After the establishment of the People's Republic of China in 1949, he was elected to the central committee of the Communist Party. Advancing through the party ranks he became general secretary in 1956. He fell from power during the Cultural Revolution (1966–69) as a result of his individual interpretation of Maoist doctrine but was reinstated as vice-premier in 1973. He was removed from office again in 1976 by the 'Gang of Four', shortly before Mao's death, but in 1977 became vice-chairman of the central committee (a position he held until 1980) and chief of the general staff of the armed forces. As vice-chairman he shared power with the party chairman and premier, HUA GUO FENG, and supported a rapprochement with the USA, which he advanced on a visit there in 1979. He also took a firm stance against the Soviet Union, the policies of which he viewed as a serious threat to world peace. In recent years he has been regarded as an elder statesman possessing *de facto* power,

even though he does not hold any of the three top posts. He was chairman of the State and Party Military Commissions (1982–89) and in 1984 he represented China in negotiations with Britain over the future of Hong Kong. He announced his retirement in 1989 but in practice continued to dominate China's government.

De Niro, Robert (1943–) *US actor who became a leading star in the 1970s.*

Born in New York, he studied Method acting and appeared in off-Broadway theatre before winning his first screen role, in *Greetings*, in 1968. Subsequently he established his reputation playing tough but sensitive characters in a series of admired and serious-minded, often violent, films. His breakthrough into major commercial success came in 1972 with his memorable portrayal of Johnny Boy in Martin SCORSESE's *Mean Streets*. A year later he won an Oscar for Best Supporting Actor for his role in *The Godfather Part II*. Scorsese again directed him in *Taxi Driver* (1976) and *Raging Bull* (1980), in which De Niro demonstrated his unwillingness to compromise realism by gaining fifty pounds in weight to play the seedy boxer Jake La Motta: he was rewarded with an Oscar for Best Actor. He also worked under Scorsese on *The King of Comedy* (1982) and *GoodFellas* (1990); among other directors with whom De Niro has worked have been Sergio Leone (1921–89), who directed him in *Once Upon a Time in America* (1984). His other films include *The Deer Hunter* (1978), one of the most acclaimed of the Vietnam War films, *Brazil* (1985), *The Mission* (1986), *The Untouchables* (1987), in which he played Al CAPONE, *Guilty by Suspicion* (1991), about the McCARTHY witchhunts of the 1950s, and *Awakenings* (1991). In 1989 he also founded his own production company and the TriBeCa business centre in Manhattan.

Denning, Alfred Thompson, Baron (1899–) *British jurist who served as Master of the Rolls for twenty years (1962–82). Renowned as a champion of the rights of the individual, he was created Baron Denning of Whitchurch in 1957.*

After studying mathematics and law at Oxford, Denning was called to the bar in 1923. In 1938 he became a KC and in 1944 he was appointed a High Court judge. He then progressed steadily through the judiciary until, in 1957, he joined the House of Lords as a Lord of Appeal in Ordinary. However, finding that he was not in the best position to exercise influence over the development of the law, Denning chose to return to the Court of Appeal in 1962 as Master of the Rolls.

In many cases between 1962 and 1982 Denning found opportunities to apply his beliefs in the rights of the individual in beleaguered circumstances, most notably those of deserted wives, victims of unfair contracts, and those caught up in administrative bureaucracy. His stand upon these issues endeared Denning to the public, although his rejection of precedent in favour of the pursuit of justice made him a controversial figure within the profession. Denning also served on many committees and review bodies, including the enquiry into the circumstances of the resignation of the secretary of state for war, John Profumo (1915–), in 1963 and the Committee on the Legal Education for Students from Africa.

Besides fulfilling the duties of Master of the Rolls, Denning is also the author of several popular books on legal topics, such as *The Changing Law* (1953), *The Road To Justice* (1955), *The Family Story* (1981), *The Closing Chapter* (1983), and *Leaves from my Library* (1986). In his retirement Denning startled his admirers by saying that the Birmingham Six and Guildford Four were probably guilty and should have been hanged, despite their release by the court of appeal after years of wrongful imprisonment.

Derain, André (1880–1954) *French painter and prominent member of the fauve group of artists.*

The Paris exhibition of Van Gogh's paintings in 1901 had a powerful effect on Derain's development as a painter. With VLAMINCK and MATISSE, he began to treat colour as an independent decorative and expressive element, thus freeing it from its traditional descriptive use. This led to the famous exhibition at the Salon d'Automne in 1905, when the term 'fauves' (wild beasts) was coined by a critic reacting to the violent execution and experimental colours of the paintings on view. During the next two years Derain twice visited London, where he painted scenes on the Thames and in Hyde Park. These and his other landscapes, figure compositions, and portraits are full of clashing pure yellows, purples, blues, greens, and reds. He regarded his tubes of colour as 'sticks of dynamite' discharging light.

Fauvism was intense but short-lived. In 1908 Derain became increasingly influenced by Cézanne's works and the theories of cubism being developed by PICASSO and BRAQUE. His interest turned to form and the restricted use of colour. Three years later he began to develop a style based on early Renaissance masters. This 'gothic period' continued until 1920, when he settled into a sombre realistic style, which, though classical, was very much of its time and influenced artistic life in the 1920s. Derain also designed theatre sets and costumes (notably for DIAGHILEV's Ballets Russes), produced pottery and book illustrations, and at various times in his life experimented with a variety of primitive sculptural styles.

Desai, (Shri) Morarji (Ranchhodji) (1896–) *Nationalist leader and prime minister of India (1977–79).*

Born in Bhadeli, Gujarat, the son of a teacher, Desai was educated at the University of Bombay before entering the provincial civil service in Bombay (1918). In 1930 he joined the civil disobedience movement led by Mahatma GANDHI, for which he was imprisoned several times (1930–34). Desai became a member of the Indian National Congress in 1931 and was imprisoned (1942–45) for his involvement in the Quit India Movement. He was appointed to several ministerial posts in Bombay after 1946, becoming minister for commerce and industry (1956–58) and finance minister (1958–63) in the government of Jawaharlal NEHRU.

Working for the Congress organization from 1963, he was appointed both deputy prime minister and minister of finance (1967–69) in the government of Indira GANDHI. However, he broke with the Congress Party in 1969 to become leader of the opposition and was imprisoned by Mrs Gandhi during the state of emergency (1975–77). In 1977 he defeated Mrs Gandhi at the polls and at the age of eighty-one became prime minister at the head of the new Janata Party (a coalition of four noncommunist opposition parties). During his term of office he restored many of the democratic institutions suspended during the state of emergency and adopted a friendly attitude towards the West, although he retained a general policy of nonalignment. However, the political weakness of his coalition government led to his defeat in the general election of 1979, when Mrs Gandhi swept back into power.

De Sica, Vittorio (1901–74) *Italian film director and actor.*

Born in Sora and raised in Naples, De Sica turned to acting after working in an office and having completed his military service. His film career moved through three distinct phases, from light comedy as an actor, to his most creative phase as a director in the school of neorealism, and finally to big box-office successes.

His first film as an actor was *L'Affaire Clémenceau* (1922); during the twenties he made a name for himself on both stage and screen. His real break as a film actor, however, came with *Gli uomini che mascalzoni* (1932), after which he acted in numerous films right through to the 1970s.

As a director his first ventures were the comedies *Maddalena zero in condotta* and *Teresa Venerdì* (both 1941). But it was his films made in collaboration with scriptwriter Cesare Zavattini (1902–   ) that produced the best in Italian neorealism. These included *Sciuscia* (1946; *Shoeshine*) and *Ladri di biciclette* (1948; *Bicycle Thieves*), the last of which won an Academy Award and numerous other prizes. Other films in this phase were *Miracolo a Milano* (1950; *Miracle in Milan*) and *Umberto D* (1952).

During the sixties he made many box-office successes with Sophia LOREN, including the Oscar-winning *La ciociara* (1960; *Two Women*), *Yesterday, Today and Tomorrow* (1963), and *Marriage Italian Style* (1964). De Sica had his last major success with *Il giardino dei Finzi-Contini* (1971; *The Garden of the Finzi-Continis*), which won an Academy Award as best foreign language film.

De Valera, Eamon (1882–1975) *Irish statesman, prime minister (1932–48; 1951–54; 1957–59), and president (1959–73).*

Born in Manhattan, New York, the son of an Irish mother and a Spanish father, De Valera was brought up and educated in Ireland, becoming a lecturer in mathematics at several colleges in Dublin. He joined the Irish National Volunteers in 1913 and was sentenced to death in 1916 for his part in the Easter Rising. However, he was reprieved and released the following year and elected president of Sinn Féin, a position he held until 1926. In 1918 De Valera was imprisoned in Lincoln jail. He escaped in 1919 to become president of the Dáil, travelling to the USA (1919–20) to enlist support for the Irish Republic. He was opposed to the Anglo-Irish Treaty of 1921, which gave dominion status to Ireland, and lost the presidency to Arthur GRIFFITH; in the ensuing civil war (1922–23) he led the militant republicans. In 1926 De Valera founded the party Fianna Fáil ('soldiers of destiny'), which he led in the Dáil. He won the 1932 election and became president of the executive council (prime minister).

A fervent nationalist, De Valera sought to cut the links that bound Ireland to British rule by enacting the new constitution (1937) creating the sovereign state of Eire. Elected president in 1959, he was well respected abroad and at eighty-one addressed a joint session of the US Congress in 1964. A life-long interest in mathematics led to his election as a fellow of the Royal Society in 1968. He retired in 1973 at the age of ninety.

de Valois, Dame Ninette (Edris Stannus; 1898–   ) *British choreographer born in Ireland, creator of the Sadler's Wells Ballet. She was created a DBE in 1951 and made a CH in 1982.*

After studying with Enrico Cecchetti (1850–1928) and acquiring some experience as a ballerina, de Valois joined DIAGHILEV's Ballets Russes as a very young girl, becoming a soloist in 1923. In 1926 she founded her own Academy of Choreographic Art in London and went on to choreograph ballets at the Old Vic, the Festival Theatre in Cambridge, and the Abbey Theatre in Dublin. The success of her own ballet *Job* led to the formation of the Vic-Wells Ballet Company, and the Sadler's Wells School in 1931. With Frederick ASHTON as resident choreographer, the company grew into the Sadler's Wells Ballet and finally, in October 1956, the Royal Ballet. De Valois achieved outstanding success as a choreographer with the company's *The Rake's Progress* (1935), which was based upon Hogarth's engravings, *Checkmate* (1937), *The Prospect Before Us* (1940), and *Don Quixote* (1950). De Valois chronicled the history of her company in the books *Invitation to the Ballet* (1937) and *Come Dance with Me* (1957). She has also published *Step by Step* (1977) and has maintained close connections with the Royal Ballet since her retirement as a director in 1963.

Devine, George Alexander Cassady (1910–66) *British actor, manager, and director. He was made a CBE in 1957.*

Devine began his acting career while at Wadham College, Oxford; as president of the Oxford University Dramatic Society (OUDS), in 1931 he persuaded John GIELGUD to direct a production of *Romeo and Juliet* with Peggy ASHCROFT as Juliet and Devine as Mercutio. It was also at this time that he became associated with the stage design company Motley, who worked on *Romeo and Juliet* and some of his later productions. He married one of the firm's partners, Audrey Sophia Harris, in 1939. On leaving Oxford he joined the Old Vic company, after which he became manager and producer (1936–39) at the London Theatre Studio with Michel Saint-Denis (1897–1971). During World War II he served with the Royal Regiment of Artillery and was twice mentioned in dispatches. Returning to the Old Vic (1946–51), he became founding director of the Young Vic Company until 1954.

As artistic director of the English Stage Company at the Royal Court (1955–65), Devine became a driving force in contemporary British theatre. The works of new writers, such as John OSBORNE, John ARDEN, and Arnold WESKER, were often first produced at the Royal Court. Devine also directed such plays as Arthur MILLER's *The Crucible* and Jean-Paul SARTRE's *Nekrassov* and appeared himself in several others, including Samuel BECKETT's *Endgame*. The year before his death Devine appeared in Osborne's *A Patriot for Me* at the Royal Court. In his memory, the George Devine Award for the encouragement of young theatrical talent was set up in 1966.

de Vries, Hugo Marie (1848–1935) *Dutch botanist and geneticist, whose work contributed to the validation of Mendel's genetic principles and who originated the concept of genes as hereditary determinants.*

Born in Haarlem, de Vries was a talented botanist even in his teens; after studying the subject at the universities of Leiden and Heidelberg, he became a teacher of natural history in Amsterdam. Soon he joined the newly founded University of Amsterdam, becoming professor of plant physiology in 1881. Here, his research on the osmotic properties of plant cells and the effects of the constituent solutes on osmotic pressure contributed to the later formulation of a general theory of osmosis by Van't Hoff. In 1889 de Vries's *Intracellular Pangenesis* was published; in it he proposed that the determinants of heritable characteristics in living organisms are carried by 'pangenes' (hence the term 'gene', later used by the geneticist W. L. JOHANNSEN).

De Vries's theory that each daughter cell received one complete set of pangenes following division of the mother cell anticipated later discoveries. In a series of breeding experiments, he found a range of characters in different plant species that exhibited segregation in a 3:1 ratio in the second generation of a cross – a result achieved by Mendel in 1866. De Vries published his confirmation of Mendel's findings in 1900, independently of Carl CORRENS and Erich von Tschermak, who published similar results in the same year. De Vries subsequently developed his theories, in-

troducing the term 'mutation' (*Die Mutationstheorie*, 1901–03) for a sudden change in a pangene that would provide the variation in organisms required by evolutionary theory. He discovered distinct forms of the evening primrose (*Oenothera lamarkiana*), which he mistakenly cited as examples of mutant forms. De Vries also proposed that material is exchanged between pairs of chromosomes prior to gamete production – a theory later supported by the work of T. H. MORGAN.

De Wet, Christiann Rudolf (1854–1922) *Afrikaner general and leader, who supported the Afrikaner cause for independence throughout his life and advocated Dutch as the official language in South Africa.*

Born near Dewetsdorp, the son of Voortrekker parents, De Wet grew up on a farm in the Orange Free State. After the Basuto wars in the 1860s, he joined the republican forces in the Transvaal and fought in the first Boer War (1880–81). De Wet began his political career in 1891, when he was elected to the Volksraad (parliament) of the Transvaal. In 1899 he was elected to the Volksraad of the Orange Free State but interrupted his political duties to join the Free State forces as a minor commando leader at the outbreak of the second Boer War the same year. Rising to commander, he gained a high reputation for his successful use of guerrilla tactics against the British forces.

One of the participants at the Vereeniging peace conference in 1902, he re-entered politics, becoming the minister for agriculture for the Orange River Colony under the Crown Colony regime in 1907. When the Union of South Africa was formed in 1910 he supported General HERTZOG against Louis BOTHA and founded the National Party. Joining the Afrikaner rebellion in 1914, which opposed Botha's invasion of German South West Africa in support of the Allies in World War I, he fled to Bechuanaland (now Botswana), where he was captured, tried, and found guilty of high treason. He was sentenced to six years imprisonment and fined £2000. Released a year later, he retired to his farm, where he died.

Dewey, John (1859–1952) *US pragmatist philosopher and radical educational theorist.*

Born in Burlington, Vermont, the son of a grocer, Dewey was educated at the University of Vermont. He worked at the universities of Michigan and Chicago before moving in 1904 to Columbia in New York, where he served as professor of philosophy until his retirement in 1930.

Working in the pragmatic tradition of William James and C. S. Peirce, Dewey sought to develop theories of knowledge and truth. Knowledge he defined as successful practice, while in place of the concept of truth he proposed that we substitute the notion of 'warranted assertibility'. Dewey presented his mature philosophical views in two late works, *The Quest for Certainty* (1929) and *Experience and Nature* (1935).

From such philosophical principles Dewey derived the educational theory that children would learn best by doing. He published his ideas in his *The School and Society* (1899) and established in Chicago a small experimental school where his ideas could be tested. The success of Dewey's approach convinced many American educationalists that it was necessary to develop less structured, less teacher-centred, and more practical schools.

Diaghilev, Sergei (1872–1929) *Russian artistic impresario, who founded the Ballets Russes and brought the contemporary arts of Russia to Europe. He inspired musicians, artists, and dancers and revolutionized the ballet.*

Born into an aristocratic family from Novgorod, he studied law at St Petersburg and travelled around Europe, meeting such artists as Verdi and Émile Zola. His ambition was to become a patron of the arts; however, as a homosexual with no private income this was an unlikely goal in turn-of-the-century Russia. Nevertheless he did manage to launch a review in 1899, called *The World of Art*, of which he was the editor; in 1905 he organized a well-publicized exhibition of Russian art treasures in St Petersburg.

The following year he took this exhibition to Paris, where he also organized a series of concerts devoted to the music of Russian composers. The climax of these performances was the western première of Mussorgsky's *Boris Godunov*, given in Russian at the Paris Opéra in 1908. With this success behind him he was able, in 1909, to form his own ballet company, the Ballets Russes, with Anna Pavlova, Vaslav Nijinsky, and Michel Fokine as dancers and Massine and Balanchine as choreographers in addition to Nijinsky and Fokine. Following the inspiring guidance of Diaghilev, the new company drew on the innovations of Isadora Duncan as well as drama and the decorative arts to produce an entirely new tradition that broke away from the stereotyped work of the Russian Imperial Ballet. Diaghilev commissioned such composers as Stravinsky, Ravel, Debussy, and Prokofiev to provide the music for his

ballets and was thus responsible for introducing the works of these musicians to a wider audience. In their revolutionary new ballets, which include Stravinsky's *The Firebird* (1910), *Petrushka* (1911), and *The Rite of Spring* (1913), Ravel's *Daphnis and Chloë* (1912), and Falla's *The Three-Cornered Hat* (1919), the sets were designed by such avantgarde artists as Picasso, Braque, and Matisse. The Ballets Russes toured Europe and North and South America from 1909 until 1929 and produced altogether sixty-eight ballets, many of which are still included in the modern repertoires. Diaghilev continued to introduce new ideas but the diabetes from which he had suffered for several years led to his decline and death in Venice at the age of fifty-seven.

Diefenbaker, John G(eorge) (1895–1979) *Canadian statesman; leader of the Progressive Conservative Party (1956–67) and prime minister of Canada (1957–63).*

Diefenbaker was born in Grey County, Ontario, and studied at the University of Saskatchewan. After a period of overseas service during World War I he practised law in Saskatchewan for some years and in 1936 became leader of the Saskatchewan Conservative Party. He left this post in 1940 on his election to the Canadian House of Commons. In 1956, after an unsuccessful attempt eight years earlier, he won the party leadership. The long reign of the Liberal Party came to an end in 1957 and Diefenbaker became the first Conservative prime minister for twenty-two years. In a general election the following year his party gained a significant majority – 208 of the 265 seats. As prime minister, however, Diefenbaker's indecisiveness on major national issues and his excessive concern for individuals and minority groups led to a loss of confidence in his party, and in the 1962 election his government was reduced to a minority. After a major crisis erupted over the building of missiles armed with nuclear warheads, a number of Conservative ministers resigned in 1963 and Diefenbaker was forced to call a general election, in which his party was defeated. He managed to retain the party leadership until 1967 and was re-elected to the House of Commons in 1968.

Dietrich, Marlene (Maria Magdalene von Losch; ?1901–) *German-born US actress. She was made a Commandeur de la Légion d'honneur in 1990.*

Dietrich began her career in the theatre in her native Germany during the early twenties. Beginning in the chorus line of a musical review, she subsequently studied drama under Max Reinhardt (1873–1943). Stage and film roles followed, including the film that brought her international fame and with which she has become most closely identified, *The Blue Angel* (1930), directed by Josef von Sternberg (1894–1969), who became a motivating force in her career. This was Dietrich's last film before going to the USA, where she and von Sternberg made such films together as *Morocco* (1930), *Shanghai Express* (1932), and *The Devil is a Woman* (1935). Memorable films that followed included *The Scarlet Empress* (1934), in which she played Catherine the Great, *Destry Rides Again* (1939), and *A Foreign Affair* (1948).

In 1937 she became a US citizen and during World War II entertained US troops and made fund-raising appearances and propaganda broadcasts in German, for which she was awarded the US Medal of Freedom. In the fifties and sixties she continued to make films, including *Judgment at Nuremberg* (1960), but became equally well-known as an international cabaret star, who also made many recordings and television appearances. She published her autobiography, *My Life*, in 1989.

d'Indy, (Paul Marie Théodore) Vincent (1851–1931) *French composer and influential teacher.*

Descendant of an aristocratic family, d'Indy was brought up under the strict rule of his paternal grandmother. He soon showed his talent for music and at sixteen discovered Berlioz's *Handbook of Orchestration*, which strengthened his decision to become a composer. After a period of distinguished service in the National Guard during the Franco-Prussian War (1870–71), d'Indy threw himself into the musical life of Paris. He studied composition with César Franck, and was deeply influenced by Wagner's *Ring* cycle, which he saw at Bayreuth in 1876. The two influences, Germanic myth and French nationalism, seemed incompatible and some of his compositions of this period show the dichotomy: *Poème des montagnes* (1881), a symphonic poem for piano depicting a day in his native Cévennes mountains, is Germanic in structure but reflects his recent research into folksong, and *Symphonie sur un chant montagnard français* (1886) for piano and orchestra is based on folksong collected in the Ardèche, where he had a summer residence.

After 1918 and his second marriage, d'Indy's style became much less grandiose; a series of compositions in a light vein, includ-

ing the *Diptyque méditerranéan* (1925–26), show just how much he had changed. D'Indy was an important educationalist in France, being co-founder of the Schola Cantorum (1894, with Alexandre Guilmant (1837–1911) and Charles Bordes (1863–1909)), a musical institute in Paris originally for the study of church music. Later he became a professor at the Paris Conservatoire.

Dior, Christian (1905–57) *French couturier whose 'New Look' did much to rejuvenate the French fashion industry immediately after World War II. He was made a Chevalier de la Légion d'honneur in recognition of his services to French fashion.*

The son of a wealthy industrialist, Dior began by designing clothes for his sisters; in his early teens he won a prize for a fancy dress costume he designed for a ball. He studied political science at the National School of Political Science in Paris and was initially destined for a career in the diplomatic corps. However, in the early 1930s he opened an art gallery in the Rue la Boètie to promote the works of such young artists as Salvador DALI and Christian Béraud. A bout of ill health forced Dior to leave Paris to convalesce and he travelled extensively. During this period he also worked as an illustrator for *Le Figaro* and designed for Robert Piguet. At the outbreak of World War II he served in the French army (1939–40) and, following the occupation, joined his sister in the country. In 1942 he started working as a fashion designer for Lucien Lelong and four years later, in collaboration with Marcel Boussac, a millionaire textile manufacturer, Dior set up his own fashion house. His first memorable collection was presented in the spring of 1947 and introduced a highly feminine 'New Look' with narrow-waisted tightly fitted bodices and full pleated skirts. The extravagance of the style in the austere postwar period provoked some criticism but it proved very popular and, with its generous use of fabric, was welcomed by the textile industry.

Dirac, Paul Adrien Maurice (1902–84) *British physicist, who was awarded the 1933 Nobel Prize for Physics for his contributions to quantum theory. He was made an OM in 1973.*

The son of a Swiss father and a British mother, Dirac studied electrical engineering at Bristol University before obtaining a Cambridge PhD in mathematics in 1926. After a short period in the USA, Dirac returned to Cambridge, where he was Lucasian Professor of Mathematics from 1932 until his retirement in 1969. He married Margit Wigner, the sister of E. P. WIG-NER.

In 1928 Dirac established his reputation as a theoretical physicist with an original mathematical formulation of quantum mechanics. This formulation incorporated relativistic effects, enabling Dirac to explain a new range of phenomena. He also predicted the existence of antimatter and the annihilation of antiparticles on collision. The anti-electron, or positron, was discovered in 1932 by C. D. Anderson (1905–) and the discovery of other particles of antimatter followed. Dirac provided a full account of his work in his *Principles of Quantum Mechanics* (1930). He was professor emeritus of St John's College, Cambridge, and professor of physics at Florida State University.

Disney, Walt(er Elias) (1901–66) *US film cartoonist and film producer, who created Mickey Mouse and Donald Duck.*

Born in Chicago, Disney began his career as a commercial artist and animator after attending the Kansas City Art Institute and serving in France as a Red Cross ambulance driver during World War I. With his colleague Ub Iwerks (1901–71) he made cartoons called Laugh-O-Grams, which they sold to theatres, but the venture proved unsuccessful.

Moving to Hollywood in 1923, Disney went into partnership with his brother Roy and together with Iwerks began making *Alice in Cartoonland*, an animated live-action series. The cartoons met with no success and the Disney family, which now included Walt's wife, the actress Lillian Bounds, lived in extreme poverty. Success, however, began with *Oswald the Lucky Rabbit* (1927) and was established by Mortimer the Mouse, who became a household character with his new name, Mickey, in his first talkie (using Disney's own voice), *Steamboat Willie* (1928). Other characters followed, including Mickey's mate, Minnie, as well as Goofy, Pluto, and the lovable Donald Duck. Equally memorable were the *Silly Symphonies* of the 1930s, especially *The Three Little Pigs* (1933) and its hit song 'Who's Afraid of the Big Bad Wolf'. Full-length cartoons came with *Snow White and the Seven Dwarfs* (1938), followed by *Fantasia* (1940), which attempted, with mixed success, to bring classical music to the vast cinema audiences by means of cartoons. Other full-length cartoons include *Pinocchio* (1940), *Bambi* (1942), *Lady and the Tramp* (1955), and *The Jungle Book* (1967). Nature films, such as *The Living Desert* (1953), live action classics, including *Treasure*

Island (1950) and *Kidnapped* (1960), westerns, and animal stories, such as *The Incredible Journey* (1963), were successfully added to the range. A wide variety of films have continued to be released under the Disney name since Disney's death. Disney received numerous Academy Awards for his shorts and many Special Awards for full-length features. Disneyland, an amusement park in California, was opened in 1955 and similar parks were established in Florida (1971) and Tokyo (1983). A further park is planned to open in Paris in 1992.

Djilas, Milovan (1911–) *Yugoslav politician and writer, who became a vice-president of Yugoslavia under TITO until disagreement led to his imprisonment. He was given the Freedom Award (US) in 1968.*

Born in Podbisce, Montenegro, the son of an army officer, Djilas was educated at Belgrade University, where he graduated in law. Joining the Yugoslav Communist Party (YCP) in 1932, he was arrested and imprisoned (1932–35) for political activities. In 1938 he became a member of the central committee of the YCP and in 1940 he was elevated to the Politburo.

During World War II Djilas was active in the resistance movement, organizing guerrilla warfare against the Germans with Tito. After the war he became one of the leading members of Tito's cabinet, rising to become one of four vice-presidents of the Yugoslav Republic in 1953. The following year his increasing disagreement with Tito's style of leadership led to his dismissal from all government posts and his resignation from the party. Imprisoned in 1956 for publishing an article that supported the Hungarian uprising, he remained in prison (except for the period 1961–62) for the next ten years. Throughout these years he was extremely critical of the practices of communist regimes and managed to smuggle abroad several manuscripts of books and articles saying so, including *The New Class* (1957) and *Conversations with Stalin* (1962). Following his release he was permitted to travel overseas and to return to Belgrade, although relations with the Yugoslav government remained distant.

Djilas is one of Europe's foremost intellectuals. He has continued to publish works that criticize communist societies; his *Memoirs of a Revolutionary* (London, 1973), in particular, is an indictment of the regime he helped to build. Other recent publications include *Tito* (1980).

Dobzhansky, Theodosius Grigorievich (1900–75) *Russian-born US geneticist whose* *studies of genetic variation in wild populations have provided crucial insights into the mechanism of natural selection and evolution in living organisms.*

Born in Nemirov, Dobzhansky graduated from the University of Kiev in 1921 and taught zoology at the Institute of Agriculture, Kiev, until 1924, when he became a lecturer at Leningrad University. In 1927 he moved to the USA with a Rockefeller Fellowship at Columbia University under the famous geneticist T. H. MORGAN. After working with Morgan at the California Institute of Technology, Dobzhansky returned to Columbia as professor of zoology. After a period (1962–71) at the Rockefeller Institute he officially retired, but continued working at the University of California.

Dobzhansky carried out studies of genetic variation in wild populations of fruit fly (*Drosophila*) species, mainly in California, Central America, and South America. He showed that these populations had a much greater degree of genetic heterogeneity than had hitherto been expected, with a high frequency of potentially harmful recessive genes. Dobzhansky found that this heterogeneity conferred greater overall fitness, enabling the population to adapt quickly to changing environmental conditions. Indeed, he demonstrated how structural modifications of the chromosomes (polymorphic inversions) maintained this heterogeneity. His book *Genetics and the Origin of Species* (1937) dealt with the genetic mechanisms that underly the emergence of races and new species. He also wrote extensively on human genetics and anthropology. In *Mankind Evolving* (1962) he discussed the implications for human evolution of the human ability to radically change the environment. His other books include *The Biological Basis of Human Freedom* (1956) and *Genetics of the Evolutionary Process* (1970). He became a US citizen in 1937.

Doenitz, Karl (1881–1980) *German admiral and commander-in-chief of the German navy (1943–45).*

Born in Grunau near Berlin, the son of a Prussian civil servant, Doenitz was educated at a school in Weimar. He entered the German navy in 1910 and was assigned to the light cruiser *Breslau*. During World War I (1916–18) he was in charge of U-boats. After the war he was appointed inspector of torpedo boats and later commander of the cruiser *Emden*. In 1935 he became a rear admiral and head of the submarine service responsible for the building and deployment of the U-boat fleet.

During World War II Doenitz was in charge of developing tactics used by the German submarines. At first he was successful in eliminating Allied shipping, but by the end of the war, more than two-thirds of the U-boats had been destroyed. In 1943 he replaced Admiral RAEDER as commander-in-chief of the navy. He was appointed by HITLER as his successor in the final days of the war and, following Hitler's death on 1 May 1945, established a government in Flensburg. He surrendered unconditionally to the Allies on 23 May 1945. Brought to trial at Nuremberg, he was sentenced to ten years' imprisonment in Spandau for 'conspiracy against the peace'. He was released in 1956 and lived in relative obscurity in Hamburg until his death.

Doenitz received the lightest sentence of any of the major war criminals at Nuremberg. In his memoirs, published in 1959, he pleaded his innocence, claiming, like many other Germans, that he had been unaware of the atrocities committed by Hitler and the Nazis.

Dolci, Danilo (1924–) *Italian writer and social reformer. He was awarded the Lenin Peace Prize in 1958.*

Born in Sesana, Trieste, the son of a stationmaster, Dolci was educated at high schools in Milan before being imprisoned during World War II for refusing to serve in the fascist army. After the war he began a degree course in architecture at the University of Rome (later continued at the Milan Politecnico) but abandoned his studies in 1950 to work with the Roman Catholic priest, Saltini, at a refuge for orphans in Nomadelphia. In 1952 Dolci went to work among the poor in western Sicily. Beginning with a public fast to draw attention to the hunger around him, he waged a nonviolent struggle of protest and political agitation (including the celebrated strike-in-reverse in 1956, in which local unemployed people began the unauthorized mending of a road) to focus attention on the poverty and feudal living conditions of the Sicilian people. In 1958 he donated the money from his Lenin Peace Prize to establish employment centres in Sicily.

Dolci, whose life has been dedicated to nonviolence as a means of social reform, has been called the 'Sicilian Gandhi'. He has written several books about the living conditions of Sicilian peasants and their oppression by local police and Mafia rule. They include *Banditi a Partinico* (1955; 'The Outlaws of Partinico'), *Inchiesta a Palermo* (1956; 'Report from

Palermo'), *Spreco* (1960; 'Waste'), and *Verso un mondo nuovo* (1964; translated as *A New World in the Making*, 1965). More recent publications include *The World is One Creature* (1986) and *Bozza di Manifesto* (1989; 'Draft of a Manifesto').

Dollfuss, Engelbert (1892–1934) *Austrian statesman and chancellor (1932–34).*

Born in Lower Austria, the son of a farmer's unmarried daughter, Dollfuss was educated at the Hollabrunn Gymnasium before studying at the universities of Vienna and Berlin, where he received degrees in law and theology. He served as an officer in a machine-gun unit during World War I, winning eight decorations for bravery. In 1922 he gained a doctorate in economics, enabling him to further his interest in the agrarian problems of Lower Austria.

Dollfuss began his political career after the war as secretary of the Lower Austria Farmers' League. A founder of the Lower Austrian Agricultural Chamber in 1927, he was appointed president of the Austrian Federal Railways system in 1930, becoming minister of agriculture and forests in 1931. The following year he was elected leader of the Christian Socialist Party and chancellor. He suspended parliament in 1933 following strong opposition from Austrian Nazis (who were in favour of union with Germany). In an attempt to prevent the *Anschluss* he sought support from MUSSOLINI, who was willing to guarantee Austria independence provided that the Austrian constitution was altered to include a fascist style of corporate government. Promulgating a new constitution in 1934, Dollfuss ordered the Austrian army to quell a protest by socialist workers, which resulted in fierce fighting and a thousand deaths. Five months later he was assassinated in the federal chancellory by a group of Austrian Nazis attempting unsuccessfully to overthrow the government.

A devout Catholic, Dollfuss was a politically isolated figure. On the one hand he was opposed by the Nationalists (Nazis) because of his determination to maintain Austrian integrity; on the other, he was bitterly attacked by the socialists for his authoritarian mode of government.

Dolmetsch, (Eugène) Arnold (1858–1940) *French-born British musician, instrument-maker, and pioneer in reviving the performance of early music on copies of original instruments. He was made a Chevalier de la Légion d'honneur (1938).*

Born in Le Mans into a family of musicians and instrument makers, Dolmetsch soon became a sensitive craftsman. He studied music at the Brussels Conservatory (1881–83) and at the Royal College of Music, London, after which he taught the violin at Dulwich College (1885–89). From 1905 to 1911 he worked in the USA for the firm of Chickering and Sons, Boston, making harpsichords, clavichords, lutes, and viols; he was similarly employed by Étienne Gaveau (1872–1943) in Paris from 1911 to 1914. In 1917 Dolmetsch moved to Haslemere, Surrey, where friends helped him to create a centre for the performance of early music: the Haslemere Festival was first held in 1925. He was married three times, and several of his offspring continued his work, specializing in their own particular fields. His son Carl Dolmetsch (1911–), a recorder player, now supervises the festivals. Dolmetsch recorders are used in schools in all parts of the world. An intractable man, Dolmetsch met with much prejudice and lack of understanding. However, he succeeded in bringing music of the past back into the concert room.

Domagk, Gerhard (1895–1964) *German bacteriologist and pathologist awarded the Nobel Prize for Physiology or Medicine in 1939 for his discovery of the antibacterial effects of Prontosil, the first sulphonamide drug.*

Domagk was born in Brandenburg (now in Poland) and trained in medicine at the University of Kiel. After postgraduate work at the universities of Greifswald (1924) and Münster (1925) he became director of the Bayer Laboratory for Experimental Pathology and Bacteriology at Wuppertal-Elberfeld and in 1928 was made professor of medicine at the University of Münster. Following the lead of Paul EHRLICH, he spent his career searching for chemotherapeutic agents against infections and cancer. At the Bayer works he searched systematically for new dyes and drugs that might destroy infecting organisms without harming the patient. His first major success was the discovery of germanin, which was then the most effective drug against sleeping sickness. His prize-winning work was the discovery that the dye Prontosil was effective against streptococcal bacteria in mice. The active part of the dye was the sulphonamide group and modifications led to the development of drugs that drastically reduced the mortality of pneumonia, puerperal sepsis, and cerebrospinal fever. Domagk was unable to accept his prize because of the policy of the Nazi government in Germany; he was arrested and forced to renounce the award. In 1947 he was given the Gold Medal and Diploma but his achievements had since been eclipsed by the discovery of penicillin and other more potent antibiotics.

Domingo, Placido (1941–   ) *Spanish tenor who is renowned for the conviction of his acting and his sensitive musicianship.*
Born in Spain, he moved to Mexico with his family in 1950. In Mexico he studied the piano as well as conducting and singing. In 1957 he made his debut as a baritone but his first major tenor role was Alfredo in Verdi's *La traviata* in 1960. He sang with the Israeli National Opera from 1962 to 1965 and made his New York debut at the City Opera in the US première of Ginastera's *Don Rodrigo* in 1966. He first sang at the Metropolitan Opera in New York in 1968 (as Maurizio in Ciléa's *Adriana Lecouvreur*). His first appearance at La Scala was in Verdi's *Ernani* in 1969. Two years later he appeared at Covent Garden as Cavaradossi in *Tosca*. Since then he has been in great demand in all the leading opera houses. He has also taken starring roles in several films, including ZEFFIRELLI's *La Traviata* (1983) and *Otello* (1986) and Francesco Rosi's *Carmen* (1985). He reached a worldwide audience in 1990, when he appeared in an internationally televised concert in Rome with José CARRERAS and Luciano PAVAROTTI, as part of the celebrations marking the 1990 World Cup.

Donleavy, J(ames) P(atrick) (1926–   ) *US-born Irish writer, whose bawdy comic style has brought notoriety as well as recognition.*
Born in Brooklyn, New York, of Irish immigrant parents, Donleavy was educated at preparatory school in New York and Trinity College, Dublin, where he showed few signs of his later talent. His first novel, *The Ginger Man* (1955), later hailed as a work of tragicomic genius, was not published in an unexpurgated form in English-speaking countries until the mid-1960s because of its pornographic content. His subsequent novels include *A Singular Man* (1963), *The Beastly Beatitudes of Balthazar B* (1968), *The Onion Eaters* (1971), *The Destinies of Darcy Dancer, Gentleman* (1977), *Schultz* (1977), *Leila* (1983), *Are You Listening Rabbi Löw?* (1987), and *That Darcy, That Dancer, That Gentleman* (1990), as well as several plays and a collection of short stories, *Meet My Maker the Mad Molecule* (1964).

Although Donleavy's later works have been well received by critics, they have tended to repeat the basic literary formula of *The Ginger Man* and have not been regarded as breaking any significant new ground. In 1967 Donleavy became a citizen of Ireland, where he now lives and manages a cattle farm as well as continuing to write.

Dos Passos, John Roderigo (1896–1970) *US novelist, playwright, and journalist.*
Born in Chicago of Portuguese descent, Dos Passos was educated in America and abroad, graduating from Harvard in 1916. Intending to study architecture in Spain, he soon signed up in World War I, first serving with the French and subsequently in the US medical corps. *Three Soldiers* (1921), his first novel, related the war's devastating effect on the lives of three US army privates. In *Manhattan Transfer* (1925) Dos Passos first attempted a panoramic rendering of social life and problems by depicting a large number of unrelated New York characters in numerous episodes. This approach was fully developed in his major work, the trilogy *U.S.A.* (collected 1938), which consists of *The 42nd Parallel* (1930), *1919* (1932), and *The Big Money* (1936). The nation itself, from 1900 to 1930, becomes the main character of the narrative and is portrayed critically as a materialistic society in decline. Episodes in which the lives of historical figures form a part are supported by such devices as the 'Newsreel' sections, establishing an authentic period atmosphere by quoting newspaper headlines, popular songs, etc., and 'The Camera Eye' stream-of-consciousness sections, reflecting the author's interpretation of events.

Dos Passos's work after *U.S.A.* became increasingly conservative. A second trilogy, *District of Columbia* (collected 1952), is concerned with political disillusionment, and the conservatism of his essays, *The Theme is Freedom* (1956), and his historical writings, *Men Who Made the Nation* (1957), is even more pronounced. In *Mr Wilson's War* (1963), he returned, but with a thoroughly conservative point of view, to the era covered in *U.S.A.* Dos Passos's plays, *The Garbage Man* (1926), *Airways, Inc.* (1929), and *Fortune Heights* (1933), were collected in *Three Plays* (1934).

Douglas, Donald Willis (1892–1981) *US aircraft engineer, who founded the Douglas Aircraft Company.*
The son of a bank cashier, Douglas was educated at the US Naval Academy at Annapolis and the Massachusetts Institute of Technology, from which he graduated in 1914. He began his career in aviation working as chief engineer to the Glenn Martin Aircraft Company in Los Angeles. His career was interrupted in 1916 by World War I, when he was appointed chief civilian aero engineer to the US Signal Corps.

In 1920, Douglas set up his own company to build the Cloudster, a plane commissioned by a Los Angeles sportsman who planned a non-stop flight across the USA. Although it failed in this purpose, the Cloudster became the basis of the US navy's first torpedo bomber. The greatest success of the company came, however, with the DC-3, first flown in 1935 and widely known as the Dakota. Over 11 000 models were built, making it one of the most profitable planes of all time. Douglas continued building planes throughout World War II and into the jet age. In 1967 his company was taken over by the McDonnell Aircraft Company with Douglas serving as honorary chairman of the newly formed McDonnell Douglas Company until his death.

Doyle, Sir Arthur Conan (1859–1930) *British author, best known as the creator of the detective Sherlock Holmes. He was knighted in 1902.*

Born in Edinburgh, the son of a civil servant, Conan Doyle attended Stonyhurst before taking his MB degree at Edinburgh University (1881). While practising medicine at Southsea (1882–90) he published his first novel, *A Study in Scarlet* (1887), in which Sherlock Holmes made his debut. Other novels quickly followed, including the historical novels *Micah Clarke* (1889) and *The White Company* (1891) and another Holmes story, *The Sign of Four* (1890). From July 1891 to December 1893 Conan Doyle published the series of stories called *The Adventures of Sherlock Holmes* in the *Strand Magazine*. These stories won immense popularity for the great detective and his obtuse but loyal friend and foil Dr Watson. The brilliant deductive powers of Holmes were said to be modelled on those of the Edinburgh surgeon Sir Joseph Bell, under whom Conan Doyle had studied; Watson is partly a self-parody by the author. When Conan Doyle tried to kill Holmes off in his notorious plunge with his archenemy Professor Moriarty over the Reichenbach Falls, there was a public outcry that eventually resulted in Holmes's resurrection for a new series of adventures that ran intermittently from 1903 almost until the author's death. The longer Holmes story, *The Hound of the Baskervilles*, was serialized from August 1901 to April 1902.

During the Boer War (1899–1902) Conan Doyle served as a field physician and also wrote a pamphlet called *The War in South Africa* (1902), vindicating the British action there. For these services he received his knighthood. Although he failed to obtain a seat in parliament, Conan Doyle played an active role in several public causes as a propagandist, notably in obtaining a public inquiry (1914) into the case of a man wrongly convicted of murder in 1909 and in supporting the reprieve of Sir Roger CASEMENT.

He continued to write prolifically. His later novels included the series on the Napoleonic Wars featuring Brigadier Gerard and the Professor Challenger stories, including *The Lost World* (1912). On matters of current interest he wrote, among others, *The Crime of the Congo* (1910) and accounts of World War I. The subject of spiritualism particularly engaged his interest in later life, especially after the death of his son in World War I, and he published a two-volume *History of Spiritualism* (1926), besides lecturing on the subject.

Dreiser, Theodore Herman Albert (1871–1945) *US writer who applied the naturalist genre to American society of the early twentieth century.*

The ninth child of impoverished German immigrant parents, Dreiser was born in Indiana and succeeded against the considerable odds of his background in spending a year at university before taking up a career in journalism and moving to New York. He was greatly influenced by the works of Darwin and Thomas Huxley, which probably accounted for the naturalist bias of his writing. His first novel, *Sister Carrie* (1900), was largely ignored; only marginally more successful was *Jennie Gerhardt* (1911). Among his other books, *The Financier* (1912) and *The Titan* (1914) were the first two novels of a trilogy that was completed by the posthumous publication of *The Stoic* in 1947; he also published several autobiographical books, including *A Traveller at Forty* (1913) and *A Book About Myself* (1922). His best-known work, however, was *An American Tragedy* (1925), in which he used a real-life murder case to expose the darker side of lower-class US society and which reflected his growing awareness of the importance of social reform. After visiting the Soviet Union in 1927, Dreiser's life was dominated by his socialist principles, which replaced much of the pessimistic determinism of his earlier outlook.

Some think that Dreiser's writing is often rambling and clumsy, and by the time he died, his naturalist approach was regarded as anachronistic. Nevertheless the power and honesty of his best works, with their objective portrayal of a particular society at a particular time, ensure that his books are still read.

Dreyer, Carl Theodor (1889–1968) *Danish film director.*

Born in Copenhagen, Dreyer worked as a clerk before becoming a journalist on *Extrabladet*, a Copenhagen newspaper, and, in 1912, a scriptwriter. His debut as a director came with *The President* (1919), followed by the more notable *Leaves from Satan's Book* (1921), which emulated many of the techniques used by D. W. GRIFFITH in *Intolerance* (1916). His most highly acclaimed work, however, came some five films later with *La Passion de Jeanne d'Arc* (1928; *The Passion of Joan of Arc*) starring Renée Falconetti (1901–46) in her only film. Next came his first sound film, *Vampire* (1932). This did not have the critical success of *Jeanne d'Arc*, and as it also proved a commercial failure, Dreyer returned to journalism. Ten years passed before he made his next film, a short documentary, which was followed by his most famous film, *Day of Wrath* (1943). Made during the German occupation of Denmark, the film was about seventeenth-century witch hunts and was seen as an allegory of the occupation. As a result Dreyer fled to Sweden. Returning to Denmark after the war, he made numerous documentaries for the government and continued to develop his personal style of directing. Two notable feature films were made towards the end of his life, *Ordet* (1955; *The Word*), which won the Venice Golden Lion Award, and *Gertrud* (1964).

Drinkwater, John (1882–1937) *British poet, dramatist, and actor.*

Drinkwater was born in Leytonstone, Essex. His father, an actor, was determined to discourage his son's theatrical ambitions, so when the boy left Oxford High School he was sent to work for an insurance company. In its Birmingham offices he met Barry Jackson, with whom he founded the Pilgrim Players amateur dramatic society (1907). Jackson then used his considerable personal wealth to build the Birmingham Repertory Theatre (1909) and invited Drinkwater to become its manager. In the four years before the theatre became fully professional, Drinkwater gained invaluable experience there as an actor and dramatist.

Drinkwater's first plays, including *Cophetua* (1911) and *Rebellion* (1914), were in verse, but his most successful dramas were a series of prose plays on historical themes, beginning with the triumphantly successful *Abraham Lincoln* (1918), which transferred to London and instantly established Drinkwater's reputation. Other 'chronicle' plays were *Mary Stuart* (1921), *Oliver Cromwell* (1921), and *Robert E. Lee* (1923). Among his other successful plays was the comedy *Bird in Hand* (1927).

Drinkwater also wrote in other genres. His first volume of poems (1906) attested his genuine, if intellectually undemanding, talent as a lyricist, and he remained a respected figure in the Georgian Poetry movement, with his *Collected Poems* being published in 1923. He also wrote critical studies of *William Morris* (1912) and *Swinburne* (1913) and two volumes of autobiography — *Inheritance* (1931) and *Discovery* (1932).

Dubček, Alexander (1921–) *Czechoslovak statesman. He became first secretary of the Czechoslovak Communist Party (1968–69), but his attempt to increase civil liberties led to the invasion of Czechoslovakia by Warsaw Pact troops and his own political downfall. After the communists lost power in 1989 he was appointed speaker of the Federal Assembly.*

Born in Uhrovec near Topolcany, the son of a Slovakian cabinetmaker, Dubček was educated in Kughizian, Soviet Central Asia, before becoming a machine locksmith and engine fitter in Gorkii. Returning to Slovakia with his family in 1938, because of Stalinist purges, he joined the illegal Czechoslovak Communist Party (CCP) in 1939. He took part in the Slovak resistance movement during World War II while working in an armaments factory in Dubnica.

After the war Dubček progressed through the ranks of the CCP while working in a yeast factory in Trencin. Beginning as secretary of the Trencin district committee of the party in 1949, by 1958 he had been elected to both the Czechoslovak and Slovak central committees. During this period he took a law degree at the Comenius University, Bratislava, and a doctorate in political science at the Communist Party College in Moscow. In 1960 Dubček was elected secretary of the national party's central committee and a member of the Presidium two years later, thus becoming the highest ranking Slovak in the national party. He was appointed first secretary of the Slovak Communist Party in 1963 and in January 1968 succeeded Antonin NOVOTNÝ as first secretary of the CCP. As leader of the CCP Dubček relaxed censorship and made plans for a new constitution as well as legislation to increase civil liberties. However, the invasion of Czechoslovakia by Soviet and Warsaw Pact troops (August 1968) in response to this relaxation of communist orthodoxy led to his downfall. Resigning as first secretary of the party in April 1969, he was dismissed from the Presidium in September 1969, but was appointed ambassador to Turkey in December of that year. He was expelled from the CCP in June 1970 and was forced to resign from his ambassadorial post shortly after. In 1970 he became an inspector for the forestry administration in Bratislava. He took part in the pro-democracy movement that culminated in the resignation of the politburo (1989); in December he was appointed speaker of the Federal Assembly (parliament) with Vaclav HAVEL as president of the republic.

Dubuffet, Jean Philippe Arthur (1901–85) *French painter.*
The son of a wealthy wine merchant in Le Havre, he became a brilliant, original, but dilettante member of the world of art and literature in Paris, where he started his own wine business in 1930. It was not until he was over forty that he turned to full-time painting. Dubuffet's first exhibition took place in 1944, just after the liberation of Paris from four years of German occupation. Rather than follow in the somewhat bankrupt tradition of French high culture, Dubuffet led a reaction against what he regarded as museum art. His pictures, which looked to many as if they had been scored in mud with a stick, provided a revolutionary start to postwar painting. Although Dubuffet's anti-art can be traced back to the earlier dada movement, his stance was less negative in that he used types of material that had previously been ignored or despised. He collected the artistic works of psychotics, children, and the illiterate, which he valued as being free from cultural tradition. He organized exhibitions of these works and incorporated their features into his own paintings. The later tendency of artists to experiment with all kinds of materials was heralded by Dubuffet's use of plaster, glue, and putty to build up pictures that he then embedded with pebbles and broken glass. In his later work he built up pictures from discarded items, such as pieces of newspaper, dried flowers, and foil. Between 1946 and 1967 he also wrote prolifically on polemical issues.

Duchamp, Marcel (1887–1968) *French painter and sculptor and a leading influence on anti-art movements in the USA.*
Born into a family of artists, Duchamp made his living in Paris between 1905 and 1910 by drawing cartoons and illustrations for the *Courrier français* and *Le Rire*. At the same time he studied painting and developed a special interest in depicting movement. In 1912 he exhibited his now famous *Nude Descending a Staircase*, which contained elements of cubism and had much in common with the new futurist movement in Italy. The following year he began to produce three-dimensional works and 'ready-mades', such as *Bicycle Wheel* (1913) and *Bottlerack* (1914), choosing mundane objects as subject matter in an attempt to destroy the mystique of good taste and aesthetic beauty.

In 1915 Duchamp left for the USA, where he worked until 1923 on his masterpiece *Large Glass: the Bride Stripped Bare by her Bachelors Even*, a construction of metal and glass nine feet high. Like much of his art, this unfinished work invites thought rather than a conventionally aesthetic reaction. Already notorious on his arrival in America, he increased his impact by, among other things, submitting for exhibition a urinal to which he gave the title *Fountain* and, in 1920, a reproduction of the *Mona Lisa* with added beard and moustache. He virtually gave up creative work from 1923 in favour of organizing exhibitions, editing a magazine, experimenting with film, and playing chess.

Although Duchamp had a continuing influence on anti-art movements, he acknowledged a certain failure, saying in 1962, 'When I discovered ready-mades, I thought to discourage aesthetics .... I threw the bottlerack and the urinal in their faces and now they admire them for their aesthetic beauty.' However, he appears to have made a contribution to the disturbed and elusive concept of art in the twentieth century.

Dufy, Raoul (1877–1953) *French painter.*
After studying in Le Havre and Paris, he painted in a late impressionist manner until his meeting with MATISSE in 1905. Dufy was strongly influenced by Matisse's painting *Luxe, calme et volupté*, which taught him the free use of pure flat colour to heighten effect, and in the same year he adopted the style of the fauves (see DERAIN, ANDRÉ). For a short period from 1908 Dufy came under the influence of BRAQUE and the paintings of Cézanne. As a result his palette was temporarily subdued and he turned his attention to composition and structure.

His friendship with fashion designer Paul Poiret from 1910 stimulated an interest in textile design, and he became an influential figure in the fashion world. He did numerous woodcuts, lithographs, and tapestries during this period and in the 1920s developed the personal style of painting for which he is best known – such gay and pleasant scenes as racetracks and seascapes with boats. They are roughly divided into flat areas of bright colour, over which the outlines are sketched. He also carried out numerous public commissions at this time. The final development in Dufy's style occurred around 1947, when his paintings became almost monochromatic, as in *The Red Violin* (1948) and *Homage to Mozart* (1952).

Dukas, Paul Abraham (1865–1935) *French composer, critic, editor, and teacher.*

From a musical background, Dukas entered the Paris Conservatoire at sixteen. Through playing timpani in the Conservatoire he became increasingly interested in composition and orchestration and joined the composition class of Ernest Guiraud (1837–92) forming a lasting friendship with DEBUSSY; his *La Plainte, au loin, du faune …* (1920) was written in Debussy's memory. Dukas's enthusiasm for Wagner is apparent in the piece by which he is best known, *The Sorcerer's Apprentice* (1897), a scherzo for orchestra after a ballad of Goethe. Known to millions as a result of Walt DISNEY's interpretation of it in the film *Fantasia* (1940), it also inspired works by Debussy and STRAVINSKY. Other works by Dukas that are still in the repertory are *Polyeucte* (1891), an orchestral overture based on Corneille's tragedy (1891), *La Péri*, a balletic tone poem preceded by a fanfare, written for the Russian dancer Trukhanova (1911–12), and the opera *Ariane et Barbe-Bleue* (1899–1907) based on a story by MAETERLINCK.

Dukas also edited music by Beethoven, Couperin, Rameau, and Domenico Scarlatti and was a respected critic and writer on music. He was a professor at the Paris Conservatoire, first of orchestration (1909–35) and later also of composition (1913–35). Conservative and self-critical, he destroyed much of his work, leaving only a dozen published works.

Dulles, John Foster (1888–1959) *US Republican statesman and diplomat who became secretary of state (1953–59) under President EISENHOWER. He was awarded the Medal of Freedom in 1959.*

Dulles was born in Washington, DC, the son of a Presbyterian minister and the nephew and grandson of two former secretaries of state. He studied at Princeton and George Washington Universities and at the Sorbonne; on being admitted to the bar in 1911 he joined a New York law firm as a specialist in international law. His diplomatic career began as early as 1907, when he attended the Hague Conference

as secretary to his grandfather, a member of the Chinese delegation. At the Versailles Peace Conference in 1919 he served as legal adviser to the US delegation; on his return to the USA he rejoined his law firm and soon established a reputation as one of the country's most respected international lawyers. In 1945, having assisted in the preparation of the UN Charter, Dulles was appointed as consultant to the US delegation at the San Francisco United Nations conference. His value as a spokesman and adviser on foreign affairs was widely recognized and he was selected by President TRUMAN for the delicate task of negotiating the 1951 peace treaty with Japan.

Dulles finally achieved his early ambition to become secretary of state in 1953, when he was appointed to that office by President Eisenhower. A staunch anticommunist, inflexibly opposed to any attempt at negotiation with the Soviet Union, he strove to improve the position of the USA in the Cold War and strengthened the North Atlantic Treaty Organization (NATO) with a number of additional pacts between Asian countries. In his confrontation with the Soviet Union and China he developed the policy of 'brinkmanship' – the diplomatic technique of advancing to the brink of war without being forced to engage in it. Describing the potential response of the USA to Soviet aggression he coined the phrase 'massive retaliation', referring to the use of nuclear weapons. To Dulles, communism was a moral issue: always ready to support any country threatened by the Soviet bloc, he was instinctively hostile to would-be allies that had received assistance from communist sources. This attitude has been blamed for his curt refusal in 1956 to supply financial aid to General NASSER for the construction of Egypt's Aswan Dam.

Dulles remained in the office of secretary of state until he was forced to resign through illness in April 1959: he died of cancer just six weeks later. His contribution to the strength of the western alliance was appreciated by critics and supporters alike; President Eisenhower regarded him as 'one of the truly great men of our time'.

Duncan, Isadora (1878–1927) *US dancer whose controversial interpretative dancing was extremely influential in the development of modern ballet although her flamboyant lifestyle and ardent feminism made her widely unpopular. Her short life was dogged by misfortune.*

The daughter of a music teacher in San Francisco, Isadora Duncan rejected the rigidity of a formal ballet training from her earliest years, replacing it with a completely new technique based on what she regarded as natural instinctive movements. Her first public appearances were unsuccessful, however, and at the age of twenty-one she departed for England, where the patronage of Mrs Patrick CAMPBELL enabled her to perform at private receptions and parties. Dancing barefoot to the music of Beethoven, Brahms, and Wagner in flowing gowns derived from the Greek art she had observed in the British Museum, Duncan was a controversial figure in a conventional period. In 1905 she set out on a tour of Europe, where she was seen by DIAGHILEV, who was deeply influenced by her and based much of his new theory of ballet on her revolutionary dancing.

Her success in artistic circles was not, however, backed by public support. Her feminist rejection of marriage led her to live openly with her lovers, first Gordon Craig, a stage designer, and then Paris Singer, a wealthy patron of the arts. She had a child by each; both children died in 1913 in a car that ran out of control into the Seine – a tragedy from which she never fully recovered. After World War I she moved to Moscow to found a dancing school and there she met and married Sergei Yesenin, an unstable peasant lad, seventeen years her junior, who wrote poetry. The hostile reception the couple received in America in 1922 – they were accused of being Bolshevik agents – forced her to abandon America for the rest of her life.

Back in Europe, life was no better: the depressed Yesenin turned against her and finally committed suicide in 1925. She settled in Nice after this second terrible blow, meeting her own macabre end in 1927 when her scarf caught in the rear wheel of the car in which she was travelling.

du Pré, Jacqueline (1945–87) *British cellist, who was awarded the OBE in 1976.*

Jacqueline du Pré started lessons at the age of five and became a student at the London Violoncello School at six; from the age of ten she was a pupil of William Pleeth, with whom she continued to study after entering the Guildhall School of Music, where she won every major award. She also studied with TORTELIER and ROSTROPOVITCH. Her professional career began with a Wigmore Hall concert in 1961, playing a 1672 Stradivarius cello, which she had been given by an anonymous donor. In 1967 she married Daniel BARENBOIM, with whom she gave duo recitals, as well as trio recitals with the violinists Itzak Perlman (1945–) or Pinchas Zukerman (1948–). She toured America in 1965 with the BBC Symphony Orchestra.

Du Pré's deep and instinctive feeling for music and breadth of style made her an exceptional soloist, especially in the concertos of Haydn, Schumann, Dvořák, Boccherini, ELGAR, and DELIUS. Alexander Goehr (1932–) dedicated his *Romance* (1968) for cello and orchestra to her. In 1971 she was ordered a year's rest and in 1973 multiple sclerosis sadly ended her playing career. Confined to a wheelchair, she continued to teach and gave an excellent series of master classes before her tragically early death.

Duras, Marguerite (1914–) *French novelist and dramatist.*

Born in Giadinh, Indochina, Duras went to France in 1931 and studied maths, law, and political science in Paris. Her first novel, *Les Impudents*, was published in 1942, but it was not until 1950, with the publication of *Un Barrage contre le Pacifique* (translated as *A Sea of Troubles*, 1953), that she made her name as a writer. The novel deals with an old woman's futile attempts to protect her home against the ravages of the ocean; set in Indochina, it is partly autobiographical and, like Duras's other early novels, neorealist in style.

Subsequent novels, such as *Le Marin de Gibraltar* (1952), *Le Square* (1955), *Moderato cantabile* (1958), and *L'Après-midi de M. Andesmas* (1962), carry recurrent themes of love and passion, alienation, and the passage of time. Duras's writing technique gradually became more experimental; in *Le Ravissement de Lol V. Stein* (1964), *L'Amante anglaise* (1967), and *L'Amour* (1971), the line of narrative, characters, and settings are increasingly ill-defined. *L'Amant* (1984) won both the Prix Goncourt and Ritz Paris Hemingway awards; subsequent works include the short-story collection *La Douleur* (1985) and *Practicalities* (1990). Duras has also written plays, such as *Les Viaducs de Seine-et-Oise* (1960), *Les Eaux et forêts* (1965), and *Baxter Vera Baxter* (1977), and collaborated in the cinematic adaptations of some of her novels. The best known of her film scenarios is *Hiroshima mon amour*, produced by Alain RESNAIS in 1960; others are *L'Homme atlantique* (1981) and *Les Enfants* (1985).

Durkheim, Émile (1858–1917) *French sociologist and one of the founding fathers of modern sociology.*

After he graduated from the École Normale Supérieure in Paris, Durkheim taught sociology first at the University of Bordeaux and then, from 1902 until his death, at the Sorbonne. His aim was to discover the nature of society and thus the values and principles upon which education should be based; many of his views, and those of his followers, were published in the journal *L'Année Sociologique*, which he founded in 1896.

In his first major work, *De la division du travail social* (1893; translated as *The Division of Labour in Society*, 1933), Durkheim rejected the ideas of the English utilitarians and proposed the theory that social structures influenced individual behaviour, not *vice versa*. He conceived an image of social solidarity and at the same time was able to define the proper task of sociology as the study of social rather than individual facts. In his *Les Règles de la méthode sociologique* (1895; translated as *The Rules of Sociological Method*, 1938), Durkheim went on to enlarge upon this concept and successfully laid down a methodology for the science of sociology. Durkheim then focused on the 'collective currents' that he had suggested influenced the conduct of individuals in society, and embarked upon a major empirical investigation published as *Le Suicide* (1897; translated as *Suicide*, 1952). In this pioneering study of social statistics, Durkheim distinguished between three kinds of suicide and was able to associate each with different aspects of the social order. Durkheim gave the name 'anomie' to individual behaviour or the state of a society lacking moral or social standards. In his final work, *Les Formes élémentaires de la vie religieuse* (1912; translated as *The Elementary Forms of Religious Life*, 1915), he interpreted the religious bond as a type of social bond expressed through ritual and defined God simply as Society. Throughout all his writings he never diverted from his collectivist point of view and his work was thus in striking contrast to that of many of his contemporaries, including Max WEBER.

Durrell, Lawrence George (1912–90) *British poet, novelist, and travel writer.*

Durrell was born in India and educated at Darjeeling before attending St Edmund's School, Canterbury. He joined the foreign service and was posted as press officer first to Athens and then to Cairo. Subsequently press attaché in Alexandria and Belgrade, he held similar posts in Greece and Argentina before becoming director of public relations for the government of Cyprus.

Durrell's early poetry – *Private Country* (1943), *Cities, Plains and People* (1946), *On Seeming to Presume* (1948) – displayed his sensitivity to places and his ability to capture the essence of a scene in a telling phrase. His *Collected Poems* appeared in 1960, with a revised and enlarged edition in 1968. He also wrote verse plays – *Sappho* (1950) and *An Irish Faustus* (1963) – and published various translations.

In prose his most significant achievement, and the work for which he is best known, was the *Alexandria Quartet*, comprising the novels *Justine* (1957), *Balthazar* (1958), *Mountolive* (1958), and *Clea* (1960). His liking for the cryptic and symbolic became more prominent in later novels, such as *Tunc* (1968) and *Nunquam* (1970). Subsequent works included the *Avignon Quintet*, comprising the novels *Monsieur* (1974), *Livia* (1978), *Constance* (1982), *Sebastian* (1983), and *Quinx* (1985). Durrell's brilliance at descriptive prose is revealed in his highly acclaimed travel books: *Prospero's Cell* (1945) about Corfu, *Reflections on a Marine Venus* (1953) about Rhodes, and *Bitter Lemons* (1957) about Cyprus. Embassy life in Belgrade supplied the inspiration for the humorous sketches in *Esprit de Corps* (1957) and *Stiff Upper Lip* (1958).

Dürrenmatt, Friedrich (1921–90) *Swiss playwright, novelist, and critic.*

Born at Konolfingen, in the canton of Bern, Dürrenmatt studied literature, philosophy, and science at the universities of Zürich and Bern. After first working as a graphic artist, he became theatre critic for a Zürich weekly. His first three plays were inspired by historical subjects: *Es steht geschrieben* (1947), set during the Anabaptist government of Münster (1534–36); *Der Blinde* (1948), set in the period of the Thirty Years War; and *Romulus der Grosse* (1949; translated as *Romulus the Great*, 1964), a political satire set at the end of the Roman empire. These plays, the first of which is clearly indebted to BRECHT, were not performed outside Switzerland. By the time he had gained an international reputation in the mid-1950s, Dürrenmatt had developed an original view of 'tragic comedy' that inspires his mature work and is set forth in the essays in *Theater-probleme* (1955; translated as *Problems of the Theatre*, 1958). Given the circumstances of modern experience, Dürrenmatt argued that tragedy was no longer possible. A grotesque comic element enters into the ordinary modern man's isolated confrontation with impersonal powers (money, technology, etc.) over which he has no control.

Die Ehe des Herrn Mississippi (1952; translated as *The Marriage of Mr Mississippi*, 1964) concerns the impossibility of reforming human nature and was the first of Dürrenmatt's plays to be produced outside Switzerland. After the modern morality play *Ein Engel kommt nach Babylon* (1953; translated as *An Angel Comes to Babylon*, 1964), Dürrenmatt produced his most famous and successful play, *Der Besuch der alten Dame* (1956; translated as *The Visit*, 1958), on the irresistible power of money to corrupt. (The immensely rich old woman of the title returns to her native village and bribes the villagers to murder her former lover.) Destructive technology in *Die Physiker* (1962; translated as *The Physicists*, 1963) ends in the control of an insane doctor after three famous and presumably sane physicists hide themselves in an asylum. Dürrenmatt also wrote a number of radio plays, short stories, and detective novels, for example *Das Versprechen* (1958; translated as *The Pledge*, 1959). He continued writing until his death, his last works including the plays *Achterloo* (1983) and *Achterloo IV* (1988) and the novels *Minotaurus* (1985; translated as *Minotaur*) and *Der Auftrag* (1988; translated as *The Assignment*).

Duvalier, François (1907–71) *President of Haiti (1957–71). Known as 'Papa Doc', Duvalier was fascinated by voodoo rites, which he combined with Negro nationalism to win popular support for his ruthless authoritarian regime.*

Born in Port au Prince, of Negro descent, Duvalier was raised in a Roman Catholic household. He graduated in medicine from the University of Haiti in 1934 and worked as a hospital physician until 1943, when he became involved in an anti-yaws campaign sponsored by the US government. This led to his appointment as director-general of the National Health Service in 1946. He became minister of public health and labour in the 1949 administration of Dumarsais Estimé (1900–53), but returned to his medical work in 1950 when Estimé was overthrown in a military coup. Organizing opposition to the new leader, Magliore, he gave up medicine and went underground (1954). In 1956, when an amnesty was proclaimed, Duvalier emerged from hiding to stand for the presidency. Elected with an overwhelming majority in 1957, he immediately set about making arrangements to consolidate his power. He created a gangster militia called

'Tonton macoute' which forced thousands of Haitians into exile and eliminated many others. He also manipulated elections to extend his term of office, enabling him in 1964 to be proclaimed president for life. He was excommunicated by the Vatican (1960–66) for his antipathy towards the clergy and diplomatically isolated because of his dictatorial rule. He nevertheless remained in power until his death in 1971, when his son Jean-Claude Duvalier (1951–) succeeded him. The Duvalier regime ended in 1986 when a mass uprising forced Jean-Claude Duvalier to flee the country.

Du Vigneaud, Vincent (1901–78) *US biochemist noted for his discoveries concerning the chemical structure of the pituitary hormones oxytocin and vasopressin, for which he won the 1955 Nobel Prize for Chemistry.*

Du Vigneaud was born in Chicago, Illinois. After graduating in chemistry from the University of Illinois in 1923, his interest focused on biochemistry; he worked at the Philadelphia General Hospital before joining the University of Rochester in 1925. He received his PhD in 1927 for studies on the structure of the protein hormone insulin. After brief spells at Johns Hopkins University and in Europe, he joined the biochemistry department of the University of Illinois, becoming professor (1932) and then head of department. In 1938 he was appointed professor of biochemistry at Cornell University Medical College.

Du Vigneaud's insulin studies revealed that the sulphur-containing amino acid cysteine forms disulphide bonds with adjacent cysteine molecules; these bonds are a major element in determining the conformation of the protein molecule. He went on to establish the chemical structure of the B-group vitamin biotin and to investigate the structure and synthesis of penicillin. However, it was his announcement in 1953 of the first artificial synthesis of the naturally occurring hormone oxytocin that was most significant. Oxytocin is secreted by the posterior lobe of the pituitary gland and causes contraction of involuntary muscle (hence its use for inducing labour) and triggers milk flow from the mammary glands. Du Vigneaud discovered that it was a peptide comprising eight amino acids. He followed this by successful structural analysis of another peptide hormone, vasopressin, concerned with regulating water balance in the kidneys. Both are now routinely manufactured for clinical use. Du Vigneaud worked on other aspects of protein chemistry, including transmethylation. He was latterly appointed professor of chemistry at Cornell University (1967–75).

Dylan, Bob (Robert Allen Zimmerman; 1941–   ) *US singer and songwriter whose lyrics, with their fiery imagery, made him a superstar of the 1960s.*

He began as a folksinger, accompanying himself on the guitar and harmonica in coffeehouses as a student in Minnesota. While on a visit east to see his idol, folk balladeer Woody Guthrie (1912–67), in hospital, he sang in folk clubs in New York; his first records were issued in 1962. His anti-war 'protest' songs were very popular ('Blowin' in the Wind', 'The Times They are A-Changin'', 'A Hard Rain's A-Gonna Fall') but his writing soon became more personal. His fifth album, *Bringing It All Back Home* (1965), contained some of his most highly acclaimed songs (including 'Mr Tambourine Man') but on half of the record he was accompanied by a rock band, in an attempt to interest a wider audience. When he sang with a band in public later that year he was booed by the folk fraternity. In 1965 he broke his neck in a motorcycle accident; after a long recovery he produced *John Wesley Harding* (1968), perhaps his most enigmatic collection. He acted in the film *Pat Garrett and Billy the Kid* (1973) as well as writing the score: 'Knockin' on Heaven's Door' became a hit. He toured a great deal in the 1970s and in 1979 became a born-again Christian, resulting in his music gaining a peaceful simplicity. The album *Infidels* (1983) was a return to topical material. Subsequent albums include *Oh Mercy* (1989).

His lyrics have been published and studied as though they were literature, and many inflated conclusions have been drawn from these studies.

E

Earhart, Amelia (1898–1937) *US aviator who, in 1932, became the first woman to make a solo transatlantic flight.*

Born in Atchison, Kansas, Earhart served with the Canadian Red Cross during World War I. Later she studied at Columbia University and worked as a social worker in Boston. In June 1928 she became the first woman passenger to make the transatlantic flight on a trip from Newfoundland to south Wales. She married the publisher George P. Putnam in 1931. On 20 May the following year she set out from Harbour Grace, Newfoundland, on a solo flight across the Atlantic. Her single-engined Lockheed Vega landed in Londonderry 13¼ hours later and she became an instant celebrity in Europe and the USA. In September 1932 she followed this success by becoming the first woman to fly solo nonstop across the USA, taking 19 hrs 4 mins between Los Angeles and Newark, New Jersey. Earhart made several other long-distance flights and on 1 June 1937 she left Miami, Florida, accompanied by navigator Fred Noonan, in a Lockheed Electra on the first stage of a round-the-world flight. On 2 July they took off from Lac, New Guinea, heading for Howland Island in the Pacific. They were never seen again.

Eastman, George (1854–1932) *US inventor and founder of the modern photographic industry.*

With little education, Eastman began work at the age of fourteen as an insurance agent. An ambitious young man with an interest in photography, Eastman was anxious to break into the photographic business in the 1880s. At this time, the wet plate was still in general use, so that taking a photograph was so complex and messy that it was only attractive to expert enthusiasts. The coming of the dry plate in the 1870s eliminated some of the early drawbacks; however, it remained for Eastman to develop the camera and the roll film so that, in the words of his famous slogan, 'You press the button, we do the rest.'

Eastman's first innovation was to replace the cumbersome glass plate with a transparent flexible film that could be rolled to a new position when needed. By 1884 he was ready to sell his Kodak camera. The name was thought up by Eastman: it had no other significance for him apart from the 'K', which was the initial of his mother's maiden name and he thought it was easily pronounced and remembered. In 1889 the original paper film was replaced by celluloid. The early cameras took a hundred snapshots after which the camera itself was sent to the Eastman Company, where the pictures were developed and a reloaded camera returned with the developed photographs. Eastman's camera conquered the world, taking photography to the man in the street. By 1896 he was selling a hundred thousand cameras a year and producing thousands of miles of film. Eastman himself became a man of immense wealth and one of the great philanthropists of his day. To the University of Rochester he gave 54 million dollars and a further 19 million dollars to the Massachusetts Institute of Technology. In 1932, bored with philanthropy and unable to find any new challenges in his life, Eastman committed suicide.

Eccles, Sir John Carew (1903–) *Australian physiologist who was awarded the 1963 Nobel Prize for Physiology or Medicine for his discovery of the chemical changes that bring about the transmission or inhibition of impulses by nerve cells. He was knighted in 1959 and was created a Companion of the Order of Australia in 1990.*

Eccles graduated from the University of Melbourne and subsequently spent many years researching at Oxford University (1925–37) before returning to Australia as director of the Kanematsu Memorial Institute of Pathology in Sydney (1937–43). By 1952, when he had become professor of physiology at the Australian National University at Canberra (1951–66), he was able to demonstrate that a nerve impulse is transmitted when a substance (acetylcholine) is released from the ending of a nerve cell. The substance causes the pores of the cell membrane to expand and allows sodium ions to cross into the neighbouring cell, causing the polarity of the electric charge on this cell to change. This proceeds from cell to cell causing a wave of electrical charge along the nerve, which constitutes the impulse. He also found that there are nerve endings of a second, inhibitory, type. If the cell induces this ending to release a substance into the neighbouring cells, potassium ions move across the membrane and reinforce the existing electrical charge, preventing transmission of the impulse.

Eckert, John Presper, Jr (1919–) *US electrical engineer, a pioneer in the development of the modern computer.*

Eckert was educated at the University of Pennsylvania, where in 1941 he joined the faculty. With the outbreak of World War II he was working on the calculation of ballistic tables for the US Army Ordnance Department. So complex and time-consuming were these calculations that Eckert and his colleague at Pennsylvania, J. W. Mauchly (1907–80), sought mechanical help. The result, completed in 1946, was ENIAC (Electronic Numerical Integrator and Calculator), the world's first electronic computer. It was, with its 18 000 electronic valves and a weight of 30 tons, a truly monstrous device. It also had the disadvantage that a change of program involved rebuilding the machine.

By the end of the war in 1945 Eckert and Mauchly, in collaboration with J. VON NEU-MANN, were considering the alternative – the possibility of designing a computer with a stored program. In 1947 Eckert and Mauchly formed their own company, Eckert–Mauchly Computer Corporation, to tackle the problem commercially. They first designed BINAC (Binary Automatic Computer), a smaller and faster machine that used magnetic tape; only one machine was ever built. It was, however, a step on the way to UNIVAC I (Universal Automatic Computer), the first commercial computer to appear on the market. By this time (1950) their company was struggling. In the following year it was taken over by Remington Rand. Eckert was appointed vice-president of the company, now known as Sperry Rand Corporation, with responsibility for the UNIVAC division.

Eddington, Sir Arthur Stanley (1882–1944) *British astronomer known for his work on the internal constitution of stars. He was knighted in 1930.*

Eddington was educated at Owens College (which later became Manchester University) and Cambridge. He began his career in astronomy in 1906 as an observer at the Greenwich Royal Observatory, returning to Cambridge in 1913 as Plumian Professor of Astronomy, where he remained until his death in 1944. He was a Quaker throughout his life.

Eddington's most influential work was published in *The Internal Constitution of the Stars* (1926). He showed that to understand the equilibrium of a star it is necessary to take into account the pressure of radiation, as gas pressure alone could not overcome the gravitational forces that would lead to stellar collapse. Eddington, an early supporter of EINSTEIN'S general theory of relativity, was present at Principe on the 29 May 1919 to observe the historic eclipse that provided the first observational support for Einstein's theory. Thereafter Eddington did a considerable amount to introduce Einstein's work to the British public.

For much of his later life Eddington worked on his own general theory, published posthumously in his *Fundamental Theory* (1946). He argued that such fundamental constants as the speed of light, the mass of the proton, and the gravitational constant could, if properly understood, be derived a priori from certain basic assumptions about our manner of measurement. Using his fundamental theory Eddington went on to deduce that the number of particles in the universe, the so-called cosmical number, is of the order 10^{79}. Eddington also wrote a number of books popularizing science for nontechnical readers.

Eden, (Robert) Anthony, 1st Earl of Avon (1897–1977) *British statesman and Conservative prime minister, whose brief premiership (1955–57) ended with his resignation following the Suez crisis. He received an earldom in 1961.*

Decorated with the Military Cross in World War I, he subsequently read oriental languages at Oxford. He entered parliament as a Conservative member in 1923 and held office in the coalition national government, becoming foreign secretary in 1935. He resigned in 1938 in opposition to Neville CHAMBERLAIN'S policy of appeasement, acquiring considerable support and admiration in the country by doing so. In World War II, already regarded as Winston CHURCHILL'S heir, he was successively dominions secretary, war secretary (both 1940), and foreign secretary (1940–45), in which capacity he worked especially to foster Anglo-Soviet relations. Again foreign secretary from 1951 to 1955, he made a significant contribution to the Geneva Conference (1954), which resolved the Indochina conflict, and to the London agreement (also 1954) settling French objections to the European Defence Community.

In 1955 Eden became prime minister, as long expected, after Churchill's retirement. He encountered difficulties from the first as economic conditions worsened. In 1956, joined by the French and Israelis in a secret pact, contrary to almost everyone's advice, he launched an offensive on Egypt after President NASSER nationalized the Suez Canal. Politically and militarily mishandled, Eden's offensive received worldwide condemnation. Shortly afterwards, pleading ill-health, he resigned. He published three volumes of memoirs (1960–66).

Edgeworth, Francis Ysidro (1845–1926) *British economist, known mainly for his work on indifference curves and statistical methods.*
Nephew of the Irish novelist Maria Edgeworth (1767–1849), Edgeworth came from a distinguished family. In 1911 he succeeded to the family estate at Edgeworthstown, Co. Longford, Ireland, where he was born. He was educated at Trinity College, Dublin, and then Oxford University. Called to the bar in 1877, he never practised law but became professor of political economy at King's College, London, in 1888 and three years later was appointed to the same post at Oxford. He was elected a fellow of All Souls College and spent most of his life there, resigning his post in 1922 and becoming an emeritus professor. The first editor of the *Economic Journal*, he continued – for a time as co-editor with J. M. KEYNES, and as chairman of the editorial board – until 1926.

The most outstanding mathematical economist of his time, he treated economic theory in a formal and mathematical way, concerning himself mainly with the measurement of utility, probability, index numbers, statistics, and the determination of economic equilibrium. In *Mathematical Psychics* (1881) he discussed marginal utility and introduced the concept of the indifference curve and contract curve, which he used in the 'Edgeworth Box' diagram, a widely used rectangular diagram that illustrates the most efficient allocation of resources between two producers or of two goods between two consumers. He described a model of duopolistic competition, known as the Edgeworth duopoly, in his *Theory of Monopoly* (1897). Other publications were *Theory of Distribution* (1904) and *Papers Relating to Political Economy* (1925).

Edward VII (1841–1910) *King of the United Kingdom (1901–10) – a short and relatively uneventful reign.*
Eldest son of Queen Victoria, he was created Prince of Wales shortly after his birth. In his youth he became leader of the so-called 'Marlborough House set', a fashionable and pleasure-loving group of friends who met at his

London home. The prince's early indiscretions undermined the Queen's confidence in him, and he was excluded from all political responsibility – for instance, he never saw a cabinet paper until 1892. As a result he was ill-prepared for rule and had an inflated view of his own influence, especially in foreign affairs, though his visit to Paris in 1903 may have encouraged the French to accept the Anglo-French entente cordiale (1904). His charm brought him great popularity. His consort Alexandra (1844–1925) was a Danish princess whom he married in 1863. She was founder of Queen Alexandra's Imperial (now Royal) Army Nursing Corps in 1902 and also instituted the Alexandra Rose Day, on which roses are sold every year in aid of hospitals.

Edward VIII See WINDSOR, DUKE OF.

Ehrenburg, Iliya Grigorievich (1891–1967) *Soviet writer.*
Ehrenburg was born in Kiev to middle-class Jewish parents. His father, a brewery manager, moved the family to Moscow when Ehrenburg was still a child. Imprisoned in his teens for revolutionary activities, in 1908 he went to Paris as a political emigré. His first poetry was published in 1911. During World War I he acted as a war correspondent, returning to Russia in 1917. After much wavering over the next four years he eventually threw in his lot with the Bolsheviks.

During most of the 1920s and 1930s Ehrenburg was based in Paris as foreign editor for Soviet journals. He wrote the satirical novel *Neobychainy pokhozhdeniya Khuilis Khurenito* (1921; translated as *Julio Jurenito*, 1958), attacking Western values, and a volume of satirical short stories (1923), but it was as a journalist that he made his reputation. In 1941 he returned to the Soviet Union and the next year published his well-known anti-Western novel *Padenie Parizha* (1942; translated as *The Fall of Paris*, 1945), for which he won the Stalin Prize. During the war his anti-German propaganda in *Pravda* and the *Red Star* made him famous in the Soviet Union and abroad.

In the immediate postwar period Ehrenburg, who enjoyed the personal protection of STALIN in recognition of his value as a propagandist, survived successive purges of both intellectuals and Jews. He continued writing novels, and *Ottepel* (1954; translated as *The Thaw*, 1955), published soon after Stalin's death, heralded and lent its name to a temporary relaxation in the constraints upon artists in the Soviet Union, besides containing openly adverse comments on Stalinism, especially the

Zhdanov purges. Cautiously critical as he was, Ehrenburg nonetheless retained his influence as one of the Soviet literary establishment. His voluminous memoirs *Goda, Lyudi, Zhizn* (1960–64; translated as *Years, People, Life*, 1962–66) were a major publishing event in the Soviet Union. Despite hostile reaction from KHRUSHCHEV, leading to Ehrenburg's temporary eclipse in 1963, he remained an important liberalizing influence in Soviet culture, especially among younger writers.

Ehrlich, Paul (1854–1915) *German medical scientist who played a vital role in the development of haematology and immunology and was a pioneer in the field of chemotherapy. He encouraged collaboration between research and industry and in 1908 was awarded the Nobel Prize for Physiology or Medicine for his work on immunity.*

Ehrlich was born in Strehlen, Silesia (now Strzelin, Poland), and studied medicine at the University of Leipzig, where – despite having no background in experimental chemistry or applied bacteriology – he carried out independent research in these fields. As a result of his work he was invited to continue his research at the Charité Hospital, Berlin. In 1882 Ehrlich discovered a technique for staining the tuberculosis bacillus that was vital for diagnosis of TB. He also discovered the uses of methylene blue in treating nervous disorders, diagnostic tests for typhoid, and medications for fever. During this period he contracted TB and spent three years recuperating in Egypt. He returned to Berlin in 1889 to take up a post at the Robert Koch Institute for Infectious Diseases where, in collaboration with Emil von Behring, he developed a serum against diphtheria.

In 1896 Ehrlich was made director of a new institute for serum studies and in 1899 founded the Royal Institute for Experimental Therapy in Frankfurt. At this time his relationship with Behring deteriorated when he realized the limitations of serum therapy and began to concentrate his researches on chemotherapy. In 1907 he founded the Georg Speyer Haus for chemotherapy and began his important work on syphilis. With his Japanese co-worker, Sahachiro Hata, he eventually (in 1910) found a cure for the disease in Compound 606 (later called Salvarsan), which was the 606th arsenical compound he had tried.

He was given Prussia's highest honour when he was made privy councillor and was awarded honorary degrees from Oxford, Chicago, and Athens.

Eichmann, (Karl) Adolf (1906–62) *German Nazi leader responsible for carrying out Hitler's final solution to the Jewish problem.*

Born in Solingen in the Rhineland, Eichmann grew up in Linz. He attended school in Thuringia and began studies in engineering before becoming a travelling salesman. Eichmann joined the Austrian Nazi Party in 1932. When the party was banned in 1933, he moved to Berlin to work for the *Anschluss* (union of Germany and Austria). In 1934 he joined the SD (Sicherheitsdienst), the security service of the SS (Schutzstaffel, or Black Shirts). He was then appointed head of the Scientific Museum for Jewish Affairs, with responsibility for gathering information about Jews.

From 1937 Eichmann rose rapidly through the SS hierarchy to the rank of lieutenant-colonel and was given the task of expelling Jews from Austria and Bohemia, following German annexation of those countries (1938–39). He was then made chief of subsection IV-B-4 of the Reich Central Security (RHSA) and, as an authority on Jewish affairs at the Wannsee Conference in 1942, was told to effect the 'final solution' – a euphemism for the mass extermination of European Jewry. With demonic thoroughness he introduced the death camps to which six million Jews were shipped from all over Europe; nearly all of them perished. Eichmann was captured by the US army at the end of World War II, but managed to escape to South America, where he lived incognito until 1960. In that year he was traced by Israeli agents in Argentina and kidnapped. In 1962 he was tried before an Israeli court, found guilty of crimes against humanity, and hanged.

Eijkman, Christiaan (1858–1930) *Dutch physician and pathologist who discovered that beriberi is caused by a deficient diet, which later led to the discovery of vitamins. He shared the 1929 Nobel Prize for Physiology or Medicine with Sir Frederick HOPKINS.*

After graduating from the University of Amsterdam (1883), Eijkman served as medical officer in the Dutch East Indies and, following a two-year interval studying under Robert Koch (1843–1910) in Berlin, he returned to Java to carry out investigations on beriberi. While searching for a bacterial cause of the disease, in 1890 he noticed that his laboratory chickens were suffering a nervous condition similar to that associated with beriberi; in 1897 he found that this was caused by feeding them polished rice. He believed that some toxic chemical was produced in the digestion of the rice but it was later shown by his successor, Gerrit Grijns, that the symptoms were due to deficiency of some factor (later found to be vitamin B_1) which was present in unrefined rice but was removed during polishing. Eijkman returned to the Netherlands in 1896, becoming professor of public health and forensic medicine at the University of Utrecht (1898–1928).

Einstein, Albert (1879–1955) *German-born physicist, a thinker of astounding insight, author of the special and general theories of relativity, and winner of the 1921 Nobel Prize for Physics for his work on the photoelectric effect.*

The son of a manufacturer of electrical equipment, Einstein was educated at various schools in Germany and Italy, where he failed to shine as either an apt or particularly compliant pupil. At the age of fifteen, his prophetic dislike of all things German led him to persuade his father to allow him to renounce his German nationality. Consequently, after a period of statelessness the young Einstein, who was then studying at the Federal Institute of Technology in Zürich (from 1896), became a Swiss citizen (1901). As his academic prowess did not earn him the teaching job he sought, he began his career in the Swiss Patent Office in Bern (1902–08). His duties, neither onerous nor thought-provoking, allowed Einstein time to think about the physical problems that had puzzled him for many years and in 1905 he published four papers that revolutionized twentieth-century physics.

The third and most significant of these papers, 'Zur Electrodynamik bewegter Körper' ('On the Electrodynamics of Moving Bodies'), contained Einstein's first statement of the special theory of relativity. The theory was special because it dealt only with bodies at rest or moving with uniform motion. Einstein proposed two simple principles: that the speed of light was constant irrespective of the velocity of the measurer or the source, and that the laws of nature are invariant in all frames of reference moving uniformly relative to each other. These principles enabled Einstein to predict that a body's mass increases with its velocity and the phenomenon of time dilation. The first 1905 paper developed a formula for the average displacement of particles subjected to the Brownian motion and had far-reaching effects in providing evidence for the atomic theory. The second paper, for which he won his Nobel prize, was a development of quantum theory to account for the phenomenon of the photoelec-

tric effect. The last paper, a short two-page work titled with an innocent-sounding question, 'Ist die Trägheit eines Körpers von seinem Energieinhalt abhängig?' ('Does the Inertia of a Body Depend on its Energy Content?', was perhaps the most inspired of all. Concluding that if a body emits energy E as radiation, it will lose mass E/c^2 (c is the speed of light), he came to the equation $E = mc^2$, which was later to explain the devastating energy source behind the atom bomb, about which he later warned President ROOSEVELT.

Einstein's special theory of relativity was accepted with remarkable speed and in 1908 he was invited to join the faculty at the University of Bern. He did not, however, remain long in Bern and after brief periods in Zürich and Prague, returned (1914) to Germany to become director of the Kaiser Wilhelm Institute of Physics in Berlin. While in Berlin, Einstein began work on his general theory of relativity, undoubtedly his greatest achievement as a physicist. Published in 1916 as 'Die Grundlagen der allgemeinen Relativitätstheorie' ('The Foundations of the General Theory of Relativity'), it extended the special theory to include accelerated motion, tackled the problem of gravity and inertia, and led to a number of very precise predictions. One prediction was that rays of light would be deflected by a strong gravitational field. In 1919 a solar eclipse provided an opportunity to test the theory. Sir Arthur EDDINGTON headed an expedition to Principe in Africa and found that light rays did bend in the sun's gravitational field by just the amount predicted by Einstein. At this point Einstein's uniqueness as a thinker began to be recognized by the general public. He became a household name as well as a figure the world's press pursued and whose views they sought. He also became involved in political battles. As a Jew and a pacifist, Einstein inevitably came into conflict with the Nazis. They rejected his scientific work as worthless and he was dismissed from several scientific societies. When HITLER came to power in 1933 Einstein left Germany never to return and made his home in the USA at the Institute of Advanced Studies in Princeton. He became a US citizen in 1940.

The last thirty years of Einstein's scientific career were devoted to two basic problems. The first, the validity of quantum theory, he found deeply disturbing: 'God does not play dice', he insisted, referring to the indeterminacy principle. Believing that apparent indeterminacy masks a deeper causality, Einstein found himself in a minority among physicists.

He also devoted considerable time to the search for a unified field theory, to incorporate both the gravitational and electromagnetic fields. Although Einstein published several proposed solutions, none found general acceptance.

Einstein played a decisive role in the development of the atom bomb. After the discovery of nuclear fission in 1939 it soon became clear to a number of scientists, such as Leo SZILARD, that the process could be used to develop weapons of great power. Only the voice of Einstein, it was felt, carried sufficient authority to be taken seriously in Washington. Consequently, Szilard and others persuaded Einstein to write to Roosevelt in October 1939 warning him of the danger. It marked the beginning of the process that culminated in the explosion of the first atomic bomb in 1945, as a result of which Einstein became a campaigning pacifist.

Eisenhower, Dwight D(avid) (1890–1969) *US general and Republican statesman; thirty-fourth president of the USA (1953–60).*

Born in Denison, Texas, Eisenhower graduated from West Point military academy in 1915 and held senior military positions until the outbreak of World War II. During the war he served as chief of staff of the Third Army (1941), US commander in Europe (1942), and Allied commander in North Africa and Italy (1942–43). Following the success of the North African campaign, Eisenhower became supreme commander of Allied forces in Europe (1943), and in June 1944 commanded the D-Day invasion of Normandy. At the end of the war he was appointed army chief of staff in Washington and (in 1950) military commander of NATO.

In 1952 Eisenhower decided to accept the Republican invitation to run as candidate in the presidential election, and had an overwhelming victory over the Democratic candidate, Adlai STEVENSON. Although naturally conservative in his own approach, Eisenhower was restricted by a Democratic Congress; in the domestic field he presided over major developments in the field of civil rights (using troops to enforce desegregation at schools in Little Rock, Arkansas), instigated public works projects, and made stringent efforts to revitalize the depressed economy. In foreign affairs, Eisenhower's administration adopted a stern attitude against communism during the period of the Cold War, especially through the policies of his secretary of state, John Foster DULLES, and at home it endorsed the hysterical anticommunist 'witchhunt' of Senator Joseph MCCARTHY. However, Eisenhower's personal inclinations were towards a more conciliatory approach, and in 1953 he concluded a truce to end the Korean War. He also held back from intervention in the Hungarian uprising and the Suez Crisis (both in 1956). During his later years in office relations with the Soviet Union deteriorated, however, and the 'Eisenhower Doctrine' (1957) promised military aid to Middle Eastern countries to repel communist advances. Matters came to a head when the Soviets threatened to compel US withdrawal from Berlin and (in 1960) shot down a U-2 American spy plane in the Soviet Union, on the eve of a crucial summit conference in Paris.

With failing health, and unable in any case to stand for a third term of office, 'Ike' retired in 1960, his personal popularity almost as great as when he was first elected. After his retirement he retained his interest in political affairs and wrote several books on his career: *Mandate for Change* (1963), *Waging Peace* (1965), and *At Ease: Stories I Tell to Friends* (1967).

Eisenstaedt, Alfred (1898–) *German-born US photographer, who did much to develop the art of photojournalism, especially during his long association with* Life *magazine.*

Eisenstaedt was born in Dirschau, West Prussia (now Poland), into a prosperous family: his father owned a department store. They moved to Berlin in 1906, and in 1916 Eisenstaedt was drafted into the German army. He was badly wounded in both legs and invalided for the duration. Afterwards, with the family caught in the economic upheaval, Eisenstaedt took a job selling belts and buttons. Meanwhile, he became increasingly absorbed in his hobby: photography. He learned how to develop and enlarge his own pictures and began to realize the enormous potential of the camera. He sold his first picture in 1927, to a magazine, and by 1929 he was a full-time freelance. Working for Associated Press and other agencies, Eisenstaedt undertook a wide range of assignments. Influenced by the innovative German photojournalists of the 1920s, and using the newly introduced compact Leica camera, he took pictures unobtrusively and quickly 'to find and catch the storytelling moments', as he said in *The Eye of Eisenstaedt* (1969).

The early 1930s produced some of Eisenstaedt's most memorable pictures, including GOEBBELS at the 1933 League of Nations Assembly, glaring malevolently at the camera. He also made an extensive documentary record of HAILE SELASSIE's Ethiopia just before the Italian invasion in 1935. In the same year, Eisenstaedt left Germany for the USA and in 1936 joined the staff of the recently launched *Life* magazine. Over the years, his work filled its pages and often featured on the cover. His portraits of such people as Albert EINSTEIN, Ernest HEMINGWAY, Bertrand RUSSELL, and Marilyn MONROE became known to worldwide audiences.

Eisenstein, Sergei Mikhailovich (1898–1948) *Russian film director.*
Born in Riga, Eisenstein studied engineering and architecture at the Petrograd Institute. After the civil war, during which he served with the Red Army (1918–20), he became scene designer at the Proletkult Theatre and also studied with Usevolod Mayerhold, the Russian theatrical director. After directing a few plays he turned to film, in which his own theories on the use of montage to stimulate intellectual responses could be better exploited. Eisenstein's theory of montage stemmed from his interest in Japanese calligraphy, which uses combined symbols to produce meaning.

His first film as director, *Strike* (1925), was followed by the internationally acclaimed *The Battleship Potemkin* (1925), with its famous massacre sequence. After *October* (1928) Eisenstein toured Europe and the USA. In France he co-directed *Romance sentimentale* (1930) but plans to make a film in the USA fell through, as did the ill-fated Mexican project *Que viva Mexico!* (1932). Unable to fit into the new aestheticism of socialist realism on his return to the Soviet Union in 1932, he fell into disfavour. *Alexander Nevsky* (1938), made in collaboration with PROKOFIEV and a complete departure from his earlier work and theory, restored him to favour and earned him the Order of Lenin (1939). He was then commissioned to make the three-part epic *Ivan the Terrible* (1942–46). For Part I Eisenstein and his team received the Stalin Prize, first class, but Part II brought Stalin's disapproval and was not released until after both their deaths. Part III was never completed. Eisenstein explained his theories about film in his books, *The Film Sense* (1942), *Film Form* (1949), *Notes of a Film Director* (1959), and *Film Essays* (1968).

Elgar, Sir Edward William (1857–1934) *British composer, who was knighted in 1904, awarded the OM in 1911, and made Master of the King's Music in 1924. The first British composer of stature since Purcell, he was further honoured by a grateful establishment by being created KCVO (1928), baronet (1931), and GCVO (1933).*

A countryman, born and buried in Worcestershire, Elgar found inspiration in the Severn River and the Malvern Hills, where he set his early cantata *Caractacus* (1898). His father kept a music shop in Worcester and there the young Elgar educated himself, studying scores and teaching himself a variety of instruments. A proposal that he should study in Leipzig never materialized. Further experience came from playing in orchestras (including the violin in the Three Choirs Festivals), playing the bassoon in chamber ensembles, and conducting choral and orchestral groups (including that of the County Lunatic Asylum). He was also employed as organist and choirmaster at St George's Church, Worcester.

In 1889 Elgar married Caroline Alice Roberts, whose belief in his genius helped him to realize himself as a composer, although he never quite ceased to doubt that his background and Roman Catholic faith were unacceptable to the established society in which he increasingly moved. A two-year period in London proved a failure and in 1891 the Elgars returned to Malvern, where he taught the violin and composed indefatigably. Success came with his fourteen orchestral *Enigma Variations* (1899), thirteen of which are titled by the initials of his friends, and the oratorio *Dream of Gerontius* (1900), both conducted by Hans Richter, a staunch supporter. The *Sea Pictures* for contralto and orchestra were performed at the Norwich Festival in 1899. During this period Elgar had much encouragement and useful advice from his friend A. J. Jaeger (the subject of the ninth (Nimrod) variation in the *Enigma Variations*) of the publishing firm of Novello & Co.

Elgar's catalogue of works is wide-ranging, from early salon music, such as *Salut d'amore* (1889), to marches, including the five *Pomp and Circumstance* marches (1903–30; the title taken from Shakespeare's *Othello*) of which the central tune of number one was used in his *Coronation Ode* (1902) for the words 'Land of hope and glory'. There were also the overtures *Froissart* (1890), *Cockaigne* (1901), and *In the South (Alassio)* (1903), as well as the *Serenade* (1892) and *Introduction and Allegro* (1905) for strings. Other familiar works include the symphonic study *Falstaff* (1913), the two *Wand of Youth* suites (1907, 1908), and two of a projected cycle of three oratorios *The Apostles* (1903) and *The Kingdom* (1906). Concertgoers throughout the world are familiar with the violin concerto (1910) and cello concerto (1919). Finally, in 1918 he composed three chamber works, sonata for violin and piano, string quartet, and quintet for piano and strings. After Lady Elgar's death in 1920 he wrote little music. The two sides of Elgar's personality are reflected in his music: on one hand the jingoistic pomp and circumstance of the high-Edwardian façade; on the other the tender and vulnerable man, who composed the lyrical passages with which so much of his music abounds. Perhaps this was the enigma that he was unable to resolve.

Eliot, T(homas) S(tearns) (1888–1965) *US-born British poet, critic, and playwright. He was awarded the Nobel Prize for Literature in 1948 and the OM the same year.*

Eliot was born in St Louis, Missouri, where he attended Smith Academy (1898–1903) before going on to Milton Academy, Massachusetts, and then to Harvard. After a year at the Sorbonne he returned to Harvard to write a doctoral thesis on F. H. Bradley's philosophy. However, he was already writing poetry ('The Love Song of J. Alfred Prufrock' dates from 1910) and absorbed in other reading that was to influence his work, notably the poems of Jules Laforgue and the philosophies of India. In 1914 he travelled to Europe, but the outbreak of World War I caused him to divert to Oxford. Here he continued working on Bradley before realizing that his true interests lay in poetry. He was encouraged in this by Ezra POUND, whom he met in autumn 1914, and it was Pound who at last found a journal willing to publish 'Prufrock' (1915). In 1915 Eliot married Vivian Haigh-Wood and decided to settle in England.

At first Eliot supported himself and his wife by school-teaching while he completed his thesis, although he never returned to Harvard to complete his doctorate. Instead he began working as a clerk in a London branch of Lloyds Bank (1917), writing poetry and criticism in his spare time. *Prufrock and Other Observations* was published in 1917, followed by *Poems* (1919) and a collection of essays and reviews entitled *The Sacred Wood* (1920). In 1921 overwork and worry about his wife's mental health caused him to take three months' leave from work, time he used to complete *The Waste Land* (1922). Although abstruse and technically innovative, it was at once acknowledged as a major work. In 1922 Eliot founded *The Criterion*, the most influential English literary journal of its time. It survived until 1939. In 1925 he joined the publishing house of Faber and Faber, where he was responsible for the poetry list, a job in which he also exerted a crucial influence. His increasing involvement with Christianity is evident in such poems as 'The Journey of the Magi', published in the same year (1927) that he became a British subject and an Anglo-Catholic, and 'Ash Wednesday' (1930), a long religious meditation.

In 1932–33 Eliot delivered the Charles Eliot Norton Lectures at Harvard. In 1933 he also separated from his now mentally ill wife – a decision that caused him a great deal of distress, which only eased after her death in 1947 and after he had unburdened himself in his greatest work, *Four Quartets* (1944). *Burnt Norton* was the first of the quartet, published in 1935, followed by the three other poems – *East Coker* (1940), *The Dry Salvages* (1941), and *Little Gidding* (1942).

Apart from his poetry, Eliot also wrote a series of verse dramas in a form of blank verse sometimes called 'heightened prose'. The first, *Sweeney Agonistes* (1924), was not performed until 1934. *Murder in the Cathedral* (1935), on the theme of the murder of St Thomas Becket in Canterbury Cathedral, was perhaps his best known. Later plays – *The Family Reunion* (1939), *The Cocktail Party* (1950), *The Confidential Clerk* (1954), and *The Elder Statesman* (1959) – although on secular subjects, often using plots from Greek drama, also explore fundamentally religious concepts. In a playful vein he wrote *Old Possum's Book of Practical Cats* (1939; dramatized as the musical *Cats* (1981) by Andrew LLOYD WEBBER).

Eliot's criticism, which was closely linked to his own poetic practice, had a profound influence on his contemporaries. His early essays on the metaphysical poets and the Elizabethan and Jacobean dramatists almost single-handedly initiated the modern revival of interest in these writers; his 1951 selection of Kipling's verse did the same for KIPLING. His other volumes of criticism included *Homage to John Dryden* (1924), *For Lancelot Andrewes* (1928), *The Use of Poetry and the Use of Criticism* (1933), based on his Harvard lectures, *On Poetry and Poets* (1957), and *To Criticize the Critic* (1965).

After the separation from his first wife, Eliot mainly shared accommodation in London with various friends. In 1957 he married again, to Valerie Fletcher, this time happily. He died in London and was buried at East Coker, the Somerset village from which his ancestors emigrated in the seventeenth century.

Elizabeth II (Elizabeth Alexandra Mary; 1926–   ) *Queen of the United Kingdom and head of the Commonwealth (1952–   ).*

Eldest daughter of the future GEORGE VI, Elizabeth was privately educated and in World War II served briefly in the ATS (Auxiliary Territorial Service). In 1947, when in South Africa with her parents, she broadcast a pledge to serve the Commonwealth, which has remained one of her greatest concerns. Later in 1947 she married Philip Mountbatten (subsequently Prince PHILIP, Duke of Edinburgh), and the couple have four children: the heir apparent Prince CHARLES, ANNE, the Princess Royal (1950–   ), Prince Andrew, Duke of York (1960–   ), and Prince Edward (1964–   ). The Queen's duties involve holding meetings of the Privy Council and investitures to award medals and decorations, attending state occasions, such as the state opening of parliament, and making state visits to countries abroad as well as receiving visits from foreign monarchs and political leaders. She also receives copies of all important government papers. She has to some extent modernized the image of the monarchy, while retaining the formality it requires for its survival. The Queen's main personal interest is horse racing.

Ellington, Duke (Edward Kennedy Ellington; 1899–1974) *Black US jazz pianist, bandleader, and composer. One of the outstanding jazz musicians of the century, he is remembered for his many compositions, which were uniquely interpreted by his own orchestra.*

Ellington was born in Washington, DC; his sartorial elegance even at school earned him his nickname. Already an accomplished musician when he left school, he earned his living as a signwriter by day and a nightclub pianist by night. By the late 1920s, Ellington had moved to New York, where he wrote music for revues and night-club floor shows, developing a talent as a tone-painter that had no equal. He was one of the first popular musicians to write extended compositions, although he remained essentially a miniaturist.

Royalties from his hundreds of compositions ('Mood Indigo', 'Solitude', 'Caravan', 'Sophisticated Lady', to name only a few) enabled him to keep a big band together long after it was economically feasible, so that he could hear his own music as fast as he wrote it. He always employed first-class musicians, many of whom were with him for decades and for whom the music was specifically arranged.

He restlessly toured the USA and the world all his life, composing wherever he happened to be. As a pianist he developed with jazz itself. By the 1940s his early stride style had given way to a distinctive percussive technique that characterized his playing for the rest of his life.

Known for his urbane public personality, Ellington was an intensely private man; perhaps the most revealing part of his autobiography is its title: *Music is My Mistress.*

Ellis, (Henry) Havelock (1859–1939) *British writer and physician whose studies of human sexual behaviour had considerable influence on contemporary views of sex and helped to bring about more open discussion of sexual problems and sex education.*

At sixteen Ellis sailed to Australia on his father's ship but after four years of teaching returned to study medicine at St Thomas's Hospital, London. After qualifying in 1889 he took up a literary career. He was associated with George Bernard SHAW and Arthur Symons (1865–1945) in the Fellowship of the New Life and became editor of the Mermaid Series of Old Dramatists and the Contemporary Science Series, which included his own first book *The Criminal* (1890). His major work was a study of human sexual impulses, published in seven volumes (*Studies in the Psychology of Sex*, 1897–1928). After the first volume was published, Ellis was taken to court on the grounds of obscenity and subjected to much abuse. He continued his study, publishing the remaining volumes in the USA, but until 1953 they were available only to doctors.

Elton, Charles Sutherland (1900–91) *British biologist whose studies of organisms in relation to their environment contributed greatly to the establishment of ecology as a recognized discipline.*

Born in London, Elton attended Liverpool College and New College, Oxford, from which he graduated in zoology in 1922. He was a member of several university expeditions to the Arctic Circle, including one with Julian HUXLEY to Spitsbergen in 1921. This fired his enthusiasm for studying animals in their natural habitats rather than in the laboratory. He was appointed consultant to the Hudson's Bay Company, enabling him to study annual records of animal catches back to the eighteenth century. From these he saw how populations of certain species underwent cyclical fluctuations. In his pioneering book, *Animal Ecology* (1927), Elton established some of the basic concepts of ecology, including food chains, nutrient cycles, ecological niches, and the pyramid of numbers. *Animal Ecology and Evolution* (1930) described how, by migration, animals could select more favourable environments and discussed the implications of this in the process of natural selection. In 1932, Elton founded the Bureau of Animal Population at the Department of Zoological Field Studies, Oxford. This soon became a centre for the collation and interpretation of animal population data collected worldwide and has provided much valuable information on the effects of environmental changes on animal populations and the relationships between populations of predators and their prey. Elton's own study of rodent populations, *Voles, Mice, and Lemmings: Problems in Population Dynamics* (1942), helped formulate an effective control programme for rodent pests, which was particularly important during World War II. Elton also helped found the *Journal of Animal Ecology* in 1932. He was appointed senior research fellow of Corpus Christi College, Oxford (1936–67), and made an honorary fellow in 1967. He was elected fellow of the Royal Society in 1953. Elton's other works include *The Ecology of Invasions of Animals and Plants* (1958) and *The Pattern of Animal Communities* (1966).

Éluard, Paul (Eugène Grindel; 1895–1952) *French poet.*

Éluard was born in Saint-Denis, Paris. Influenced by the works of Rimbaud, Lautréamont, and Apollinaire, which he had read during a spell of illness in his youth, he produced his first poems in 1917. *Le Devoir et l'inquiétude* (1920) expresses the pacifist convictions he developed during his active service in World War I. At the end of the war he became involved with dadaism; with André BRETON, Philippe Soupault (1897–1990), and Louis ARAGON, he was a founding member of the surrealist movement and one of its finest poets. Surrealist theories, such as the relationship between dreams and reality, are developed in his poetry of this period, notably *Capitale de la douleur* (1926), one of his best-known collections; *L'Amour, la poésie* (1929); and *La Rose publique* (1934). *L'Immaculée Conception* (1930), written in collaboration with André Breton, is a verbal picture of mental disorder.

His political awareness heightened by events in the Spanish civil war, Éluard broke with the surrealists in 1938. He joined the Communist Party in 1942 and became an active member of the Resistance movement. His collection *Poésie et vérité* (1942), denouncing the German occupation, established him as one of the greatest Resistance poets; copies of these poems, together with Éluard's later works, *Au Rendez-vous allemand* (1944) and *Dignes de vivre* (1944), were secretly distributed among the movement to raise morale.

Éluard remained with the Communist Party at the end of the war, expressing his political convictions in *Poèmes politiques* (1948). Towards the end of his life, however, his poetry became more lyrical; notable works of this period include *Poésie ininterrompue* (1946), *Tout dire* (1951), and *Le Phénix* (1951).

Empson, Sir William (1906–84) *British critic and poet. He was knighted in 1979.*

Empson was educated at Winchester College and then took his BA degree in mathematics at Cambridge (1929). The same year he published his first volume of poetry, *Letter IV*, which was privately printed, as was his *Poems* (1934). With *Poems* (1935), however, he made a tremendous impression upon the English literary scene as a poet of elaborate sophistication and subtlety. His only other major collection of poetry, apart from *Collected Poems* (1955), was *The Gathering Storm* (1940), but this small corpus gave him a prestige out of all proportion to its size.

Having discovered that his vocation lay with literature, Empson taught at Tokyo National University (1931–34) and at Peking National University (1937–39; 1947–52). During World War II he worked as Chinese editor for the BBC. He became professor of English literature at Sheffield University in 1953 and remained there until his retirement (1971).

Like his poetic output, Empson's critical writings were admired and influential, although few in number. *Seven Types of Ambiguity* (1930; revised edition 1947), his first book, was instrumental in giving a new direction to literary criticism. This, together with *Some Versions of Pastoral* (1935) and *The Structure of Complex Words* (1951), were characterized by Empson's abstruse but stimulating turn of mind. He also wrote a controversial study of Milton, *Milton's God* (1961).

Enesco, Georges (Georg Enescu; 1881–1955) *Romanian violinist, composer, conductor, and teacher, who guided the young Yehudi MENUHIN through his early career.*

At seven he entered the Vienna Conservatory of Music, his great musical gifts winning him the highest awards. He later studied in Paris with Massenet (1842–1912), FAURÉ, and Martin Marsick (1848–1924); it was in Paris that he settled, launching his virtuoso career there in 1899. However, he never lost his Romanian heritage, which is evident from his use of folksong and dance rhythms in many of his compositions. Typical of his Romanian works are the *Poème roumain* (1898) and the opera *Oedipus*, based on Sophocles with libretto by Edward Fleg, which was started in 1921 and eventually performed at the Paris Opéra in 1936. His *Souvenirs* was published in 1955 and a collection of letters in 1974.

Engler, (Gustav Heinrich) Adolf (1844–1930) *German botanist noted for his major contributions to plant taxonomy.*

Born in Sagan, now in Poland, Engler received his PhD in botany from the University of Breslau in 1866 and, after a period as a natural history teacher, was appointed (1872) custodian of botanical collections at the Munich Botanical Institute. Six years later he moved to the University of Kiel as professor of botany and in 1884 took a similar post at the University of Breslau to supervise the reconstruction of the botanical garden. During this period he founded (in 1880) the *Botanische Jahrbücher* ('Botany Yearbooks'), which he edited until his death, and wrote the first part of his encyclopedia of the plant kingdom, *Die Natürlichen Pflanzenfamilien* (1887–1911; 'The Natural Plant Families'). In 1887 he was appointed professor of botany at Berlin University. Engler wrote a host of papers and monographs on plant taxonomy and the biogeography of plants, travelling widely but with a particular emphasis on the flora of German colonies in Africa. His *Syllabus der Pflanzennamen* (1892; 'Syllabus of Plant Names') and *Die Vegetation der Erde* (1896–1923; 'The Vegetation of the Earth') both remain important texts to this day.

Epstein, Sir Jacob (1880–1959) *US-born British sculptor. He was knighted in 1954.*

Epstein was born and brought up in New York, the son of Russian Jewish immigrants. In 1901 he began evening classes in drawing while working in a bronze foundry and from 1902 to 1905 he studied in Paris, where, in the Louvre, he encountered the ancient and primitive sculptures that were to influence his work so deeply.

From Paris Epstein moved to London and in 1907 became a British citizen. His first important commission, which was finished in 1908, was a group of eighteen figures for the British Medical Association in the Strand, London; it was the first of many works to arouse violent criticism, mainly because of its expressive distortions. In France in 1911 he sculpted the memorial for Oscar Wilde at the Père Lachaise

Cemetery, Paris, which, with its dissipated-looking angels, also brought bitter protests. While in Paris he met PICASSO, BRANCUSI, MODIGLIANI, and GAUDIER-BRZESKA, and this contact may have stimulated the brief period back in England in which he helped to found the Vortex group (see LEWIS, WYNDHAM) and moved towards abstract sculpture, employing hard mechanical forms as in *Rock Drill* (1913), which originally incorporated a real drill. His primitive works of the interwar years, often on religious themes, including *Risen Christ* (1919), *Ecce Homo* (1935), and *Adam* (1939), continued to attract such enormous public derision that several were acquired by a showman as freaks for seaside sideshows.

In the 1920 he also produced a number of expressive naturalistic portraits in bronze. These highly acclaimed and masterly works (including those of EINSTEIN in 1933 and SHAW in 1934) contrasted with the controversial large carved sculptures that he continued to produce, such as *Lucifer* (1945), *Christ in Majesty* (1957) for Llandaff Cathedral, and *St Michael and the Devil* (1958) for Coventry Cathedral.

Erlanger, Joseph (1874–1965) *US physiologist who, in collaboration with Herbert Gasser (1888–1963), developed techniques for recording nerve impulses using a cathode-ray oscilloscope. In 1944 they shared the Nobel Prize for Physiology or Medicine for demonstrating that different fibres in the same nerve cord can have different functions.*

Erlanger qualified at the University of California and the Johns Hopkins Medical School (1899), where he worked for a further seven years. He was appointed professor of physiology at the University of Wisconsin (1906–10) and there began a successful collaboration with his student Gasser. Erlanger moved to the Washington University, St Louis (1910–46), and Gasser joined him soon after. There they studied various means of applying electronics to physiological research. They devised a method of amplifying electric responses occurring in an individual nerve fibre and were able to record them using the oscilloscope. An amplified impulse produced a characteristic wave form on the screen, which could then be studied.

In 1932 Erlanger and Gasser found that the fibres within a nerve conduct impulses at different rates, depending on fibre thickness, and that each fibre has a different threshold of excitability. Different fibres produced different wave forms on the screen, indicating that different types of impulse were being passed.

Ernst, Max (1891–1976) *German-born painter, who became a French citizen in 1958. He was a leading member of the dada and surrealist movements and developed the technique of frottage.*

Under parental pressure, Ernst studied philosophy and psychology at Bonn University until World War I but showed more interest in painting and in the art of the insane. After service in the war Ernst wrote, 'On 1 August 1914 Max Ernst died. He was resurrected on 11 November 1918 as a young man who aspired to find the myths of his time' and, 'How to overcome the disgust and fatal boredom that military life and the horrors of war create? Howl? Blaspheme? Vomit?'. As leader of the Cologne branch of the dada movement in 1919, Ernst did indeed blaspheme against conventional culture. On one occasion he held an exhibition whose entrance was through the lavatory of a beer hall. At the same time he became one of the first artists to use collage and photomontage for expressive purposes.

Ernst is best known, however, for the paintings he did after moving to Paris in 1922. Works such as *L'Eléphant de Célébes* and *Les Hommes n'en sauront rien*, with their dreamlike arrangements of strange images in empty landscapes, are among the first masterpieces of surrealist painting. In 1924 he joined the surrealist movement at its inception and continued to paint in an essentially surrealist style for the rest of his life. The following year saw his development of frottage, which involved adapting the technique of brass rubbing to surfaces such as leaves, wood grain, and pieces of torn paper; it liberated him from traditional figurative technique in, for example, his series of forest paintings. Vegetation was a recurring theme in his work, as were birds, monstrous creatures, and petrified cities.

After the German occupation of France, Ernst was imprisoned briefly and in 1941 he left for the USA, where he married Peggy Guggenheim (1898–1979). Following his return to France in 1949, his paintings became more richly coloured and abstract. He won the Grand Prix at the Venice Biennale in 1954.

Erté (Romain de Tirtoff; 1892–1990) *Russian-born French designer and one of the leading exponents of the Art Deco style of the 1920s and 1930s. He was made a Chevalier du Mérite Artistique et Cultural in 1970 and an Officier des Arts et Lettres in 1976.*

Romain de Tirtoff was born in St Petersburg into an aristocratic family. He graduated from Kronstadt College, St Petersburg, in 1911 and enrolled as an art student at the Académie Julian in Paris in 1912. Taking his professional name from the French pronunciation of his initials, Erté began designing clothes for a leading Paris couturier, Paul Poiret, and was soon creating theatrical costumes and stage sets, including designs for the Folies-Bergère. During World War I Erté illustrations began appearing regularly in US magazines, notably *Harper's Bazaar*, and their creator became internationally famous. Broadway shows of the 1920s, including the *Ziegfeld Follies* and *George White's Scandals*, employed elaborate tableaux vivants designed by Erté, who also worked for Metro-Goldwyn-Mayer on such films as *Ben Hur* (1925) and *La Bohème* (1926). During the 1920s and early 1930s his work extended into book illustration, interior design, fabrics, and household items. 1944 saw his first operatic set for a production of Donizetti's *Don Pasquale* and later work included the design for Glyndebourne's *Der Rosenkavalier* (1980). Erté's metal and wood sculptures were given their first exhibition in Paris in 1964 and in 1968 he showed the first of a series of lithographs. A major retrospective of his work was staged at the Grosvenor Gallery, New York, in 1967. His book, *Things I Remember: An Autobiography*, was published in 1975.

Escoffier, Georges Auguste (1846–1935) *French chef who set new standards in international cuisine. He was made a Commandeur de la Légion d'honneur in 1928, the first chef to be so honoured.*

Escoffier was born in the village of Villeneuve-Loubet on the Côte d'Azur, the son of a blacksmith. In 1859 he was apprenticed in his uncle's restaurant in Nice, where he learnt all the basic kitchen tasks and how to select and buy provisions. While on the staff of the Hotel Bellevue, Nice, he met the owner of the Paris restaurant Le Petit Moulin Rouge and was engaged as commis rôtisseur. His duties were interrupted by the Franco-Prussian War of 1870 in which, as an army chef, his culinary expertise was turned to the preparation of horsemeat. Afterwards he returned to Le Petit Moulin Rouge, as head chef. By 1883, Escoffier's reputation was such that he was hired as chef by César Ritz, manager of the Grand Hotel, Monte Carlo. Thus started a lifelong partnership between the two men. The combination of Ritz and Escoffier attracted the rich and famous, both to the Grand and to the National, Lucerne, where Escoffier spent the summer season. Regular clients included Sarah Bernhardt and Richard D'Oyly Carte: the latter persuaded Ritz to manage his newly completed Savoy Hotel in London, and Escoffier was duly installed as head chef.

Escoffier surprised and delighted London society with a constant stream of culinary inventions, such as pêche melba – peaches on a bed of vanilla ice cream coated with raspberry purée – which he first served in honour of the opera singer Dame Nellie MELBA. Ritz opened his own hotel in Paris in the 1890s and the Carlton in London in 1899. Escoffier remained there until his 'retirement' in 1919. Returning to Monte Carlo, he helped manage L'Hermitage hotel in his final years.

Evans, Sir Arthur John (1851–1941) *British archaeologist who discovered remains of the ancient Minoan civilization on the island of Crete. He was made a fellow of the Royal Society in 1901 and received a knighthood in 1911.*

Evans attended the universities of Oxford and Göttingen, attaining a degree in modern history, and became a fellow of Brasenose College, Oxford. In 1884 he was appointed keeper of the Ashmolean Museum, Oxford, and in 1908 was made extraordinary professor of prehistoric archaeology at Oxford. He first visited Crete in 1894 and five years later purchased the Kephala site near Knossos. The excavations continued until 1935 and revealed the remains of a civilization that had existed between 2500 and 1200 BC and which Evans named after the legendary Cretan king, Minos. Evans uncovered a royal palace and evidence of a cultured society with flourishing arts and crafts, as he recorded in *The Palace of Minos at Knossos* (4 vols, 1921–36), and his partial reconstruction of the Minoan palace can still be seen. Also found at the site were clay tablets inscribed with two different forms of linear script, linear A and B, described by Evans in *Scripta Minoa* (1909). Linear B was deciphered in 1952 by Michael Ventris (1922–56) and found to be an early form of Greek. Linear A remains unsolved.

Evans, Dame Edith Mary Booth (1888–1976) *British actress. She was made a DBE in 1946.*

Born in London, Edith Evans started her working life as a milliner. Her acting career began under the direction of William Poel (1852–1934), and she made her London debut as Cressida in *Troilus and Cressida* (1912).

Her range as an actress was considerable, encompassing Shakespearean roles, Restoration comedies, and modern plays. She toured with Ellen TERRY (1918), appeared as Lady Utterword in the first production of SHAW's *Heartbreak House* (1921), and portrayed the Serpent and She-Ancient in the first showing of his five-part cycle, *Back to Methuselah*, at the Birmingham Repertory Theatre (1923). She joined the Old Vic (1925–26 and 1958) and had numerous successes on Broadway. During World War II she entertained the troops in India and elsewhere with ENSA.

Evans's many notable roles included the Nurse in *Romeo and Juliet*, Millamant in *The Way of the World*, Lady Pitts in *Daphne Laureola*, and the part she made particularly her own – Lady Bracknell in *The Importance of Being Earnest*, which she first portrayed on stage in 1939 and on film in 1952. She made over twenty films in all, including *The Whisperers* (1967), for which she received a best actress award at the Berlin Film Festival. Evans gave her last stage performance in *Edith Evans ... and Friends* at the Haymarket (1974).

Evans, Sir Geraint Llewellyn (1922–   ) *Welsh baritone, who was knighted in 1969.*

After studying singing in his native Cardiff and singing in many amateur performances, he did his war service with the RAF. After the war he worked with the British Forces Radio Network in Hamburg, where he was heard by the bass Theo Hermann, who gave him lessons and encouraged his career. Evans also studied with Fernando Carpi in Geneva and Walter Hyde at the Guildhall School of Music in London. In 1948 he made his debut at Covent Garden as the Nightwatchman in Wagner's *Mastersingers*, after which he established himself as one of the leading baritones of the day. He made his debut at La Scala in 1960 singing Mozart's Figaro and at the Vienna State Opera in 1961 in the same role. His first appearance at the Metropolitan Opera in New York was in 1964, singing another part he particularly made his own – Verdi's Falstaff. At Covent Garden he created the roles of Mr Flint in BRITTEN's *Billy Budd* (1951), Mountjoy in Britten's *Gloriana*, and Antenor in WALTON's *Troilus and Cressida*. Other roles for which he is renowned are Beckmesser (*Mastersingers*), Guglielmo (*Così fan tutte*), and Leporello (*Don Giovanni*). He celebrated his twenty-fifth anniversary at Covent Garden as Don Pasquale in a new production of Donizetti's opera. He has also produced opera for the Welsh National

Opera, as well as in Chicago and San Francisco. He retired from Covent Garden in 1984.

Eysenck, Hans Jürgen (1916–) *German-born British psychologist noted for his strong criticism of conventional psychotherapy, particularly Freudian psychoanalysis, and for developing alternative treatment for mental disorders in the form of behaviour therapy.*

Born in Berlin and educated at the universities of Dijon, Exeter, and London, Eysenck practised as a psychologist in London from 1942 to 1950. In 1950 he became director of the Psychological Department of the Institute of Psychiatry at the University of London and in 1955 became professor of psychology at the university. Eysenck's behaviour therapy, developed following the theories of J. B. WATSON and PAVLOV, advocates the treatment of symptoms rather than the supposed underlying causes. Believing that psychological problems are caused by errors of learning, he uses conditioning to teach the patient new behaviour, such as how to relate more successfully with other people, or to eliminate undesirable behaviour, such as alcoholism. Behaviour therapy includes treatment of phobias by exposing the patient to the feared object or situation until the fear gradually diminishes. Eysenck also devised methods for assessing intelligence and personality and published his highly controversial ideas in *Race, Intelligence and Education* (1971). Other notable publications include *The Structure of the Human Personality* (1953), *Handbook of Abnormal Psychology* (1960), *Experiments with Drugs* (1963), *Crime and Personality* (1964), *The Inequality of Man* (1973), *Personality and Individual Differences* (1985), and an autobiography, *Rebel with a Cause* (1990).

F

Fairbanks, Douglas (Julius Ullman; 1883–1939) *US actor and producer whose good looks and charm made him an idol of the silent film era.*

Before turning to films Fairbanks, who was born in Denver, Colorado, worked in the theatre and eventually managed to secure leading roles on Broadway. He made his film debut in *The Lamb* (1915) and by 1917 had founded his own production company. *In Again, Out Again* (1917), *Arizona* (1918), and *The Knickerbocker Buckaroo* (1919) were among the films he both produced and played in.

Fairbanks married Mary PICKFORD in 1920 and, together with Charlie CHAPLIN and D. W. GRIFFITH, they formed the United Artists Corporation. During the twenties Fairbanks made many of the swashbuckling films with which he became associated, such as *The Mark of Zorro* (1920), *The Three Musketeers* (1921), *The Thief of Bagdad* (1924), and *The Iron Mask* (1929). His first sound film, *The Taming of the Shrew* (1929), in which he played Petruchio to Pickford's Katharina, was not a success. Only four more films followed, including his last, *The Private Life of Don Juan* (1934), which was made in Britain. Fairbanks and Pickford were divorced in 1936 and he subsequently married Lady Sylvia Ashley.

Douglas FAIRBANKS, JR succeeded his father as an international star, playing similar romantic roles.

Fairbanks, Jr, Douglas (1909–) *US actor and television producer. Apart from his work in the theatre and in films, Fairbanks has led an active business life and has taken part in several public service missions for the USA, such as his mission to Latin America (1940–41) as presidential envoy. He was appointed honorary KBE (1949) and is also an Officier de la Légion d'honneur.*

Son of silent film star Douglas FAIRBANKS and his first wife Anna Beth Sully, Fairbanks Jr was born in New York City and made his screen debut in *Party Girl* (1920). He combined stage appearances and film making in both the USA and England; during the thirties he made such films as *Dawn Patrol* (1930), *Outward Bound* (1930), *Catherine the Great* (1934), and the highly acclaimed *The Prisoner of Zenda* (1937), playing Rupert of Hentzau. Noteworthy, too, were *The Young in Heart* (1938), *Gunga Din* (1939), *Sinbad the Sailor* (1947), and *State Secret* (1950).

During World War II he saw service with the US navy and was awarded the DSC as well as the Croix de Guerre. In the fifties and sixties Fairbanks lived in Britain, where he produced *Chase a Crooked Shadow* (1958) and appeared in his longrunning television series *Douglas Fairbanks Presents*. Later he and his second wife returned to the USA. His first wife was Joan CRAWFORD. *Present Laughter* and *Sleuth* are among his more recent performances in the theatre. He has also produced over 160 plays for television. He published an autobiography, *Salad Days*, in 1988.

Faisal Ibn Abdul Aziz (1905–75) *King of Saudi Arabia (1964–75), fourth son of IBN SAUD.*

At the early age of fourteen, Faisal was sent by his father to head a deputation to Britain to congratulate the Allies on their victory in World War I. Faisal then took part in the reconquest of his father's territories, becoming viceroy of the province of Hejaz (in which Mecca is located) in 1926. He represented Saudi Arabia at the United Nations when, in 1945, it became a member of the organization. Though his brother, King Saud, remained nominal ruler of the country until his abdication in 1964, Faisal became its effective ruler in 1958, when ill health began to afflict Saud. Faisal worked to bring about a Muslim alliance to counter the secular Arab vision of his antagonist, President NASSER of Egypt, but he joined the Arabs in the 1967 war with Israel. At home, oil production flourished, and Faisal's reign witnessed economic development. He was assassinated by his nephew.

Faldo, Nick (Nicholas Alexander Faldo; 1957–) *British golfer.*

Born in Welwyn Garden City, Faldo scored his first notable success in 1975 when he won the English amateur championship. He turned professional the following year and was first selected to play in Britain's Ryder Cup team in 1977. He won five major tournaments in 1983 but became dissatisfied with his swing and spent two years (1985–87) totally reconstructing it, during which time he won no significant titles. Over the next three years, however, he established his reputation as the country's leading golfer, with three victories in the French Open as well as victories in the US Masters event (1989 and 1990) and the UK Masters (1987 and 1990). His successes in 1990 made him the first golfer since 1982 to win two of the four grand-slam events in a single year: he only narrowly missed also winning the US Open.

Falla, Manuel de (1876–1946) *Spanish composer, a leading exponent of the new nationalism in twentieth-century Spanish music.*

Falla studied the piano in Madrid from the age of eight; later (1902–04) he studied composition with Felìpe Pedrell (1841–1922), the prime influence in the renaissance of Spanish music in the twentieth century. In 1905 he won both a prize for piano playing and an open competition for a national opera with *La vida breve* (*The Short Life*). However, the opera was not performed and Falla left for Paris, where he was befriended by DEBUSSY, RAVEL, and DUKAS. Dukas was instrumental in producing *La vida breve*, first in Nice (1913) and later that year at the Opéra-Comique in Paris. The outbreak of World War I forced Falla to return to Madrid, where his talents were now duly recognized. A man of retiring nature, over the next twenty years he was increasingly distressed by the worsening political situation in Spain and by the brutalities of the civil war; in 1939 he emigrated to Argentina, where he spent the last seven years of his life. Falla refrained from using actual folk material in his music; instead he evoked the spirit of Spain, particularly of his native Andalusia.

His most creative years were 1915 to 1925, when he composed such colourful works as the two ballets *Love, the Magician* (1915) and *The Three-Cornered Hat* (1919), the puppet opera *Master Peter's Puppet Show* (1923), *Nights in the Gardens of Spain* for piano and orchestra (completed 1915), the *Homage on the Death of Claude Debussy* for solo guitar (1921), and the *Six Popular Spanish Songs* (1922). All these works are thoroughly Spanish in idiom; his concerto for harpsichord (or piano) and flute, oboe, clarinet, violin, and cello (1926), on the other hand, has more in common with the keyboard style of Domenico Scarlatti in its restrained classical idiom.

Fangio, Juan Manuel (1911–   ) *Argentine motor racing driver and World Drivers' Champion a record five times (1951, 1954, 1955, 1956, and 1957).*

Fangio was born into an Italian immigrant family living in the town of Balcarce. He worked as a garage mechanic and chauffeur before opening a garage in his home town in 1934. He competed in road races throughout South America, driving Fords and Chevrolets. In 1948 he acquired a Maserati in Europe and returned the following year to drive this to four victories; the same year he had two wins in Simca and Ferrari cars. In 1950 he joined the Alfa Romeo works team and in 1951 gained his first championship title.

Driving for Maserati, he crashed in the 1952 Autodrome Grand Prix at Monza through exhaustion and was out of racing until the following year. Midway through the 1954 season, Fangio switched from Maserati to Mercedes-Benz but nevertheless won six Grands Prix and his second championship title. He again won the title in 1955 and also in 1956, this time with Ferrari, after Mercedes-Benz withdrew from racing at the end of 1955. In 1957, at the age of forty-seven, he captured his fifth title driving a Maserati and retired the following year, probably the greatest driver of all time, with a total of twenty-four Grand Prix victories to his credit.

Fassbinder, Rainer Werner (1946–82) *German actor, writer, and one of Germany's most influential film directors.*

Born in Bad-Worrishofen, Fassbinder joined the Munich Action Theatre in 1967, after drama training, and in 1968 founded his own Anti-Theatre. He performed some of his own plays there and continued to act throughout his career, appearing in several of his own films. Influenced by BRECHT, Marx, and FREUD, he made more than thirty films in just over sixteen years, including the memorable *Katzelmacher* (1969), one of his first feature-length films, *Why Does Herr R Run Amok?* (1970), *The Merchant of Four Seasons* (1972), and *The Bitter Tears of Petra Von Kant* (1972). Notable, too, were *Fear Eats the Soul* (1974), *Despair* (1978) starring Dirk BOGARDE, and the allegorical *The Marriage of Maria Braun* (1979), which won first prize at the Berlin Festival. Politically committed and critical of the postwar Germany in which he lived, Fassbinder saw his films as a means of confronting audiences with social concerns and ideas.

Faulkner, William Harrison (1897–1962) *US novelist whose innovatory style established him as a major writer of the twentieth century. He was awarded the Nobel Prize for Literature*

(1949), the National Book Award (1951), and the Pulitzer Prize (1954).

Faulkner was born and grew up in Mississippi, where a desultory education produced few indications of his later talent. He fought with the Canadian Flying Corps during World War I, after which he took various jobs and travelled around. It was during this period, while living in New Orleans, that he became friendly with Sherwood ANDERSON, who encouraged him to write. His first published book, a collection of verse, *The Marble Faun* (1924), attracted little attention, but his first novel, *Soldiers' Pay* (1926), received favourable reviews. It was when Faulkner turned to his homeland, however, that he found the source of his major fiction. *Sartoris* (1929) was the first of a series of novels that drew directly on Faulkner's own family history and was set in the fictional Mississippi county, Yoknapatawpha. In the same year *The Sound and the Fury*, the first example of Faulkner's 'stream of consciousness' style, was critically acclaimed although it failed to be financially rewarding because of the intellectual demands it made on the reader. Other major works followed, including *As I Lay Dying* (1930), *Light in August* (1932), and *Absalom, Absalom!* (1936); however, it was *Sanctuary* (1931), written with the express purpose of making money, that brought both public recognition and financial security. This led to an opportunity to write scripts for Hollywood, an occupation Faulkner pursued sporadically over the next twenty years. His later novels included *The Unvanquished* (1938), *Go Down Moses* (1942), *Intruder in the Dust* (1948), *A Fable* (1954), and *The Reivers* (1962).

Fauré, Gabriel Urbain (1845–1924) *French composer, best known for his lyrical songs.*

Son of a schoolmaster, he was awarded a free place at the École Niedermeyer in Paris (1855–65). In 1860 Saint-Saëns (then aged twenty-five) had joined the staff and master and pupil began a lasting and mutually rewarding friendship. From 1866 to 1870 Fauré held a position as organist at Rennes (Brittany), after which he held similar positions in Parisian churches, including assistant to Charles Widor (1844–1937) at Saint-Sulpice and deputy (later assistant) to Saint-Saëns at the Madeleine. In 1872 he returned to the École Niedermeyer as a member of staff; in 1896 he became chief organist at the Madeleine and professor of composition at the Conservatoire. He was principal of the Conservatoire from

1905 to 1920, when deafness forced him to resign.

Unlike most of his professional contemporaries, Fauré did not fall under the spell of Wagner. His talent was to create intimate forms of song as well as piano and chamber music in a style that was reticent, lyrical, and masterly in its use of harmony and modulation. His finest compositions include some hundred songs, some in cycles such as *La Bonne Chanson* (1891–92) and *Le Jardin clos* (1915–18), nocturnes, barcarolles, impromptus, and other piano pieces, including the *Dolly Suite* for piano duet (1893–96). The *Requiem* (1887) is probably his best-known work. He also wrote incidental music to Maeterlinck's *Pelléas et Mélisande* (1898), two piano quartets (1879, 1886), two piano quintets (1906, 1921), and a string quartet completed shortly before his death (1924).

Fawcett, Dame Millicent Garrett (1847–1929) *British feminist, who led the constitutional movement for women's suffrage. She was created a DBE in 1925.*

Born in Aldeburgh, Suffolk, the daughter of an East Anglian shipowner, Millicent Garrett was drawn into the women's movement by her sister Elizabeth Garrett Anderson (1836–1917), the first woman to qualify as a doctor in Britain. In 1867 Millicent married Henry Fawcett, who was professor of economics at Cambridge University and radical MP for Brighton. He was blind and depended greatly on his wife in his political affairs, an involvement that gave her the competence to write the successful *Political Economy for Beginners* (1870). In the later 1860s she worked for the first women's suffrage committee and for the Married Women's Property Commission; she was also one of the organizers of the lecture scheme that grew into Newnham College for women at Cambridge. In 1884 Henry Fawcett died; two years later Millicent helped found a new women's suffrage society, later uniting the various separate organizations working for votes for women in the National Union of Women Suffrage Societies (NUWSS).

As the NUWSS's first president (1897–1919), she worked relentlessly for the constitutional campaign for women's suffrage, in opposition to the militant tactics advocated by Emmeline PANKHURST. Fawcett also became known for her speeches against Irish home rule and for an investigation of the British concentration camps in South Africa, which has been criticized for its vindication of the British administration of the camps. With the outbreak of World War I in 1914, she concentrated on the recruitment of women to serve in the war industry, but in 1916 returned to the suffrage campaign. Two years later parliament, largely persuaded by the contribution women had made to the war effort, passed a bill enfranchising all women over the age of thirty. Fawcett continued to work for equal votes with men, which was achieved in 1928, shortly before her death.

Millicent Fawcett was also the author of a biography of Queen Victoria and of *Women's Suffrage* (1912) and *Women's Victory and After* (1918).

Febvre, Lucien (1878–1956) *French historian.*

Born at Nancy, the son of a teacher, Febvre was educated at the École Normale Supérieure in Paris. His first field of historical study was regional – Franche-Comté. Starting with the physical environment, he moved to economics and society, finally turning to the history of events. The relationship between history and geography always fascinated him. He served as an officer in World War I and from 1920 worked at Strasbourg with his friend Marc BLOCH, with whom in 1929 he founded *Annales*, the influential problem-orientated journal. During the 1930s he was president of a committee directing a new *Encyclopédie française.* During World War II he continued working in his country house, publishing three books, and after the war, when his 'new history' was becoming increasingly popular, helped to reorganize the École des Hautes Études. The first volume of his collected essays, *Combats pour l' histoire*, was published in 1953, when he was seventy-five. They dealt with widely different subjects, while insisting forcefully that history was a 'human science' and that 'structures' had to be analysed, not described.

Fellini, Federico (1920–   ) *Italian film director, who rose to international fame in the 1950s with his neorealistic films.*

Born in Rimini, Fellini worked as a cartoonist before starting on a career in films in 1940. He cooperated with Roberto ROSSELLINI on such films as *Paisà* (1946) and acted in the 'Il Miracolo' episode of *L'amore* (1948). He drew on his experience as a cartoonist for his first films as director – *Luci della varietà* (1950; *Variety Lights*) and *Lo sceicco bianco* (1952; *The White Sheikh*). His childhood fascination with the circus and his adolescent street life were reflected in *I vitelloni* (1953; *The Young and the Passionate*). The success of *I vitelloni* was followed by the Oscar-winning *La strada* (1954; *The Road*), in which his wife, Giulietta Masina (1921–   ), starred. Among the films that followed was the outstandingly successful *La dolce vita* (1960), which won the Grand Prix at Cannes. This, together with *8½* (1963), which won first prize at the Moscow Film Festival, are the films with which he has become most closely identified. Among his later films are *Giulietta degli spiriti* (1965; *Juliet of the Spirits*), *Satyricon* (1969), *Amarcord* (1974), *Casanova* (1976), *La Città delle donne* (1980; *City of Women*), *Intervista* (1987), and *The Voice of the Moon* (1990). Fellini has written about his neorealism in *Fellini on Fellini* (1976).

Fermi, Enrico (1901–54) *Italian-born US physicist, who made the first 'atomic pile', discovered the weak interaction, and was awarded the 1938 Nobel Prize for Physics for his work on nuclear reactions.*

The son of a senior government official, Fermi was educated at the University of Pisa, where he gained his PhD in 1924. After spending some time in Germany he returned to Italy to take up an appointment at the University of Florence. While at Florence he made his first major contribution to physics by working out the statistics of those particles (later called fermions) that obey PAULI's exclusion principle.

In 1927 Fermi was appointed professor of physics at the University of Rome. Fermi, probably the greatest Italian physicist since Galileo, contributed to many areas of physics. He is best known, however, for his work on nuclear physics. Following CHADWICK's discovery of the neutron in 1932, Fermi used it to probe the structure of the nucleus, discovering over forty new isotopes. He also discovered the increased activity of slow neutrons and was the first to use water, paraffin wax, etc., as moderators. By 1938 Fermi, an antifascist, became apprehensive of the safety of his Jewish wife. As this was the year in which he was awarded the Nobel Prize, he and his wife sailed directly from the Nobel ceremony in Stockholm to New York.

Fermi began his US career at Columbia University. With the outbreak of World War II, however, he moved to the University of Chicago where, working on the development of nuclear weapons, he built the world's first atomic reactor, or pile, as it was then called, in a transformed squash court at the University of Chicago. Fermi continued to work on the project and was present at Los Alamos when the first atom bomb was tested.

After the war Fermi remained at Chicago, working on new problems in nuclear physics. He was also advisor to the US government on a number of issues and served on the general advisory committee of the Atomic Energy Commission. In this capacity he spoke in defence of J. R. OPPENHEIMER against charges of disloyalty. He died of cancer.

Ferrier, Kathleen (1912–53) *British contralto, who was awarded the CBE and the Gold Medal of the Philharmonic Society just before her untimely death from cancer.*

Born in Lancashire, she won a piano competition at fifteen but worked as a telephonist until 1940, when she won a singing competition. She then began to study singing with Roy Henderson in London. During World War II she sang to troops and in factories, making her own arrangements for her accompanists. Recognition of her distinctive mellow voice and excellent musicianship soon came; at Glyndebourne she sang Lucretia in the first performance of BRITTEN's *Rape of Lucretia* (1946) and Orfeo in Gluck's *Orfeo ed Euridice* (1947). At the first Edinburgh Festival (1947) she sang in Mahler's *Song of the Earth* conducted by Bruno WALTER, and in 1949 gave a recital in Edinburgh accompanied by Walter, with whom she was associated in some of her greatest performances. In the twelve years of her short career she was increasingly loved and admired by audiences and colleagues. Her last appearances, as Orfeo at Covent Garden in 1953, were curtailed after two performances by the onset of her final illness.

Feydeau, Georges (1862–1921) *French writer of farces.*

Feydeau was born in Paris, son of the novelist Ernest Feydeau. He began writing for the theatre in 1881, and over the next thirty-five years produced some forty plays. His works are masterpieces of construction; the intricately contrived plots and fast-moving dialogue are enhanced by satirical effects derived from his keen observation of human behaviour and everyday life. At the same time, he made use of all the classic props and devices of farce, such as elaborate stage settings and improbable situations, and did not stray far from the traditional themes of adultery, misunderstanding, and mistaken identity.

Feydeau's many successes, still performed at the Comédie Française in Paris, include *Tailleur pour dames* (1888), *L'Hôtel du libre échange* (1894; translated as *Hotel Paradiso*,

1956), *La Dame de chez Maxim* (1899), and *La Puce à l'oreille* (1907). *Occupe-toi d'Amélie* (1908) was adapted by Noël COWARD for the English stage as *Look after Lulu* (1959).

Feynman, Richard Phillips (1918–88) *US physicist, who was awarded the 1965 Nobel Prize for Physics for his work as one of the founders of quantum electrodynamics (QED) and the inventor of Feynman diagrams.*

Educated at the Massachusetts Institute of Technology and Princeton, where he gained his PhD in 1942, Feynman moved immediately to Los Alamos to work on the development of the atomic bomb. After the war Feynman taught at Cornell until 1950, when he was appointed to the chair of physics at California Institute of Technology.

In the late 1940s Feynman claimed that he could not understand the version of quantum mechanics presented in the textbooks. Consequently, he developed his own much simpler version, which enabled calculations that had previously taken weeks to be completed in hours. The results were also more accurate and enabled physical processes that lay beyond the scope of conventional quantum theory to be tackled. Feynman's approach uses a number of simple diagrams and a series of rules; manipulating them correctly enables the theory to describe with great precision the behaviour of electrons and other particles in all their complex interactions.

Fields, Dame Gracie (Grace Stansfield; 1898–1979) *British singer and entertainer. She was made a DBE in 1979.*

Very much the 'Lassie from Lancashire', Gracie Fields was born in Rochdale. During her early career she worked part-time in a cotton mill, but her success in the music halls led to a part in *Mr Tower of London* (1923) at the London Alhambra. Her singing and natural vitality ensured her success. During the thirties she became the darling of English audiences and was affectionately known as 'Our Gracie'. Her highly successful first film, *Sally in Our Alley* (1931), was followed by a series of popular films, including *Sing As We Go* (1934) and *Shipyard Sally* (1939). During World War II she entertained the British Expeditionary Force, scoring an immediate hit with 'Wish Me Luck as You Wave Me Goodbye'. Her popularity, however, slumped when she left for the USA with her second husband, Italian-born Montie Banks, who became an undesirable alien when Italy entered the war. In the USA Gracie raised money for the Allied cause but her wartime departure was viewed as an act of betrayal and on her return in 1941 to give concerts she received a mixed reception. Banks died in 1950 and in 1952 Gracie Fields, then living on the isle of Capri, remarried. Although she never again reached the heights of her popularity of the thirties, she gradually won back her place in the affections of the British public. Her performance in *The Old Lady Shows Her Medals* won her a Silvania TV Award (1956) and by 1958 she had made eight Royal Variety Command Performances. Possibly the most moving of her visits to Britain was her return to her home town in 1978 to open a theatre named in her honour.

Firbank, (Arthur Annesley) Ronald (1886–1926) *British novelist.*
Firbank was the son of a railway magnate, Sir Thomas Firbank, who had an estate near Newport, Gwent. As a small child Firbank suffered sunstroke while travelling in Egypt, which left him too delicate to go to school. After a private education he attended Trinity Hall, Cambridge, but stayed only five terms and did not take a degree. His first book, *Odette d'Antrevernes*, appeared in 1905, but it was not until 1915 that he made his reputation with *Vainglory*. In the meantime he spent much of his life travelling in Spain, Italy, North Africa, and the Near East.

The ill health that constantly dogged him rendered him unfit for military service during World War I and he lived in semi-seclusion in Oxford for part of the time. There he met, among others, the Sitwell brothers and Siegfried SASSOON, who appreciated his talent and his aestheticism and tolerated his shyness and odd social mannerisms. Other characteristic Firbank comedies of manners followed *Vainglory – Inclinations* (1916), *Caprice* (1917), *Valmouth* (1919) – and ensured him a devoted readership among those who enjoyed his bizarre wit and scintillating dialogue.

After the war he resumed his peripatetic life, returning each summer to London. He continued to write steadily: *The Princess Zoubaroff* (1920), *Santal* (1921), *The Flower Beneath the Foot* (1922), *Sorrow in Sunlight* (1924; published as *Prancing Nigger* in the USA), and *Concerning the Eccentricities of Cardinal Pirelli* (1926). He died in Rome of pneumonia shortly after wintering in Egypt, leaving incomplete the novel *Lady Appledore's Mesalliance.*

Fischer, Bobby (Robert James Fischer; 1943–) *US chess player, who was the youngest grandmaster in chess history and the first American to win the world championship.*

Fischer was born in Chicago but soon after moved with his mother and sister to Brooklyn, New York. He was six when his sister taught him to play chess and in due course joined the famous Brooklyn and Manhattan Chess Club. He won his first title, the US Open, in 1957 and the following year came equal fifth in the Portorož interzonal tournament. At fifteen he became a grandmaster. Leaving school a year later to become a professional chess player, by 1964 he was seen as a world championship contender. Over the next five years Fischer twice opted out of competition chess for long periods.

On his second comeback he won the Palma interzonal in 1970 and the following year won the right to challenge SPASSKY for the world title. Fischer objected to the venue, Reykjavik, and the financial terms, repeatedly threatening to withdraw. Eventually he played, however, and after a very long match he won. While world champion Fischer played no competition chess at all. He was due to defend his world title against KARPOV's challenge in 1975 but once again he objected to the conditions, refused to play, and so forfeited the title. Since 1977 Fischer has again retired from public chess competitions.

Fischer-Dieskau, Dietrich (1925–) *German baritone, renowned for the beauty and range of his voice and for his unique powers of interpretation. He was appointed a Chevalier de la Légion d'honneur in 1990.*

Born in Berlin, Fischer-Dieskau had an early ambition to become a conductor, but he changed direction as his remarkable voice developed. At eighteen, his studies were interrupted by war service and from 1945 to 1947 he was a prisoner of war in Italy. After the war he continued to study in Berlin and quickly made a name for himself in recitals and broadcasts. In 1948 he was invited to join the Berlin State Opera. Since then he has developed as one of the greatest singing actors of the age in such varied roles as Don Giovanni, Papageno, Falstaff, Figaro, and Wozzeck. His recital repertoire is also prodigious. Of his many performances in Britain, the première of BRITTEN's *War Requiem* at the opening of the new Coventry Cathedral (1962) was perhaps the most memorable. On this occasion he took the platform with Galina Vishnevskaya from the Soviet Union and Peter PEARS from Britain. He has also conducted on many occasions, including several recordings of opera. He received the Royal Philharmonic Society's Gold Medal in 1988.

Fisher, Andrew (1862–1928) *Australian statesman and Labor prime minister (1908–09; 1910–13; 1914–15).*

Born in Ayrshire, Scotland, the son of a miner, Fisher left school early to follow in his father's footsteps. Emigrating to Queensland in 1885, he worked for several years in the mines and served as a trade union leader there. Fisher first entered politics in 1893, when he was elected to the Queensland legislative assembly. Elected to the first federal parliament in 1901 as the Labor member for Wide Bay, Fisher became leader of the party in 1907 and prime minister in 1908. Defeated in the 1909 election by the 'Fusion' government of protectionists and freetraders, he won office again in 1910. He narrowly lost the election in 1913 but was returned to power in 1914. Fisher retired in favour of William HUGHES in 1915 and became Australian high commissioner in London (1916–21).

Remembered for his declaration during the 1914 election that Australia would assist the Empire to her 'last man and last shilling', Fisher was a modest temperate man and a devout Presbyterian.

Fisher, Sir Ronald Aylmer (1890–1962) *British statistician and geneticist, whose development of experimental design and statistical techniques facilitated the investigation of biological materials. He received a knighthood in 1952.*

Born in London and educated at Gonville and Caius College, Cambridge, Fisher graduated in mathematics in 1913 and spent two years as a statistician with an investment company. Too short-sighted for military service, he taught at Rugby School for the duration of World War I. In 1919 he joined Rothamstead Experimental Station as head of the statistics department. In 1918 Fisher published the results of his statistical analysis of characters that show continuous variation, such as human stature. By his treatment (partition) of the statistical variance, he was able to distinguish between variation due to environmental factors and that due to genetic factors; the latter were confirmed as being largely determined by the cumulative effects of many separate genes, each inherited according to Mendelian principles. Fisher subsequently did much to improve experimental methodology by introducing the concept of random sampling and the technique of analysis of variance, which quantifies sources of variation in an experiment. His books, notably *Statistical Methods for Research Workers* (1925), *The Design of Experiments* (1935), and *Statis-*

tical Tables (with F. Yates; 1938), form the foundation of statistical analysis in modern biological experimentation. Fisher's mathematical study of genes and their mutations in populations demonstrated how Mendelian genetics is consistent with the Darwinian view of evolution by natural selection, making *The Genetical Theory of Natural Selection* (1930) one of the seminal works of neo-Darwinism.

Fisher became a fellow of the Royal Society in 1929. In 1933 he was appointed Galton Professor of Eugenics at University College, London, and thereafter professor of genetics at Cambridge University (1943–57) until his retirement. From 1960 he spent the remainder of his life working for the Commonwealth Scientific and Industrial Research Organisation (CSIRO) in Adelaide, Australia.

Fitzgerald, Ella (1918–) *Black US singer and composer.*

Born in Newport News, Virginia, Ella Fitzgerald was raised in New York and started on her career by winning a talent contest in Harlem. This led to engagements with the band of Tiny Bradshaw (1905–58), then with Chick Webb (1909–38). Her first big hit was her own tune 'A-tisket, A-tasket' (1938). After Webb's death she led the band for two years before branching out on her own. She had five top ten hit records between 1940 and 1948, three of them with the vocal group The Ink Spots. Then began a successful association with the impresario Norman Granz, for whom she toured the world, appearing with Count BASIE, Duke ELLINGTON, Oscar PETERSON, and other stars. She also appeared on television and in such films as *Pete Kelly's Blues* (1955) and *Let No Man Write My Epitaph* (1960).

In her songbook series of recordings of Cole PORTER, Irving BERLIN, RODGERS and HART, and several others, she proved to be unmatched as a singer of ballads; while in her scat singing (improvisation using nonsense syllables) she is the equal of the best jazz musicians.

Fitzgerald, F(rancis) Scott (Key) (1896–1940) *US writer whose fiction is one of the most eloquent expressions of the American Jazz Age of the 1920s.*

Fitzgerald was born in St Paul, Minnesota, and educated at Princeton. He joined the army during World War I and after being demobilized moved to New York where he attempted, unsuccessfully, to become a journalist. In 1920 he published his first book, *This Side of Paradise*. This novel was immediately successful, largely because of Fitzgerald's skill in portraying the world of the postwar American adolescent, and enabled him to marry Zelda Sayre, whom he had met during the war. Now launched into affluent and fashionable society, Fitzgerald earned the money necessary to stay there by writing magazine stories. His books of this period, depicting the society of which he and Zelda were now a part, included *The Beautiful and Damned* (1922), portraying a hedonistic couple not unlike the Fitzgeralds, and a collection of short stories, *Tales of the Jazz Age* (1922). During their second trip to Europe (1924–26) Fitzgerald completed the novel that is regarded as his finest achievement: *The Great Gatsby* (1925). The eponymous hero of this book epitomizes the Jazz Age – a near-criminal but idealistic financier whose love for the materialistic and self-centred Daisy leads eventually to his death. This book has been filmed three times (1926, 1949, and 1974).

Throughout his life Fitzgerald retained a complex and ambivalent attitude towards the charisma and vices of the social class to which he aspired. Although he was relatively well off, he and Zelda spent the years after the success of *The Great Gatsby* living beyond their means in both Europe and the USA. From the late 1920s Zelda Fitzgerald's mental instability became increasingly apparent and was eventually diagnosed as schizophrenia. The pressures of caring for her, together with accumulating financial problems, contributed to Fitzgerald's alcoholism; these problems provided the material for his next novel, *Tender is the Night* (1934). By 1935 Zelda's schizophrenia required hospitalization and Fitzgerald's health had broken. He returned to Hollywood in 1937 in an attempt to make money by writing for the film industry. In the three years before his death from a heart attack, he produced an unfinished novel, *The Last Tycoon*, published posthumously in 1941, and seventeen short stories collected as *The Pat Hobby Stories*, which finally appeared in print in 1962.

Flagstad, Kirsten Malfrid (1895–1962) *Norwegian dramatic soprano, who was the outstanding singer of Wagnerian roles in this century.*

Born in Hamar into a musical family, she studied singing with Ellen Schytte-Jacobsen in Oslo and with Gillis-Bratt in Stockholm. In 1913 she made her debut in Oslo as Nuri in d'Albert's *Tiefland*. During the following seventeen years she sang a large repertory of opera and operetta roles in Scandinavia. In 1933, on the verge of retiring, she was engaged at

Bayreuth to sing small parts and in 1934 sang Sieglinde in Wagner's *The Valkyrie*.

From then on she established herself as a singer of increasing power and distinction in the Wagnerian roles of Senta, Elizabeth, Kundry, Elsa, Isolde, Sieglinde, and Brünnhilde. She made her debut at the New York Metropolitan Opera in 1935, and sang in London in 1936 and 1937. In 1938 she toured Australia. After World War II her voice had acquired added control and majesty and at the age of fifty-five she sang Isolde and Brünnhilde to the greatest acclaim. She also sang Purcell's Dido (in English) at the Mermaid Theatre in London. She retired from opera in 1954 but continued to record. She was director of the Norwegian National Opera in 1959 and 1960.

Flaherty, Robert J. (1884–1951) *US film director, who is often referred to as the father of documentary film-making.*

Flaherty was born in Iron Mountain, Michigan, the son of a miner, with whom he explored much of the wilderness of Canada. Flaherty's desire to share what he had seen of the remote peoples of the north began when as a young man he led expeditions into sub-Arctic Canada and carried out surveys of the Hudson Bay area. He made his first and best-known film, *Nanook of the North* (1922), after living with the Eskimos for sixteen months. Although undeniably romantic in its approach to the Eskimo way of life, the film set a new standard in documentary film-making.

After the success of *Nanook* came the Polynesian *Moana* (1926). Also made in the South Seas was *Tabu* (1931), a joint venture with F. W. Murnau (1889–1931). Flaherty's next successes came in Britain with *Industrial Britain* (1931) and, particularly, *Man of Aran* (1934). Breaking from documentaries, he worked on the exterior shots of *Elephant Boy* (1937) for Alexander KORDA; in the USA, he made *The Land* (1942) for the US Film Unit and worked with Frank CAPRA on propaganda films, His last film was *Louisiana Story* (1948), on which, as on many of his others, he was assisted by his wife, Frances (née Hubbard), whom he married in 1914.

Flecker, (Herman) James Elroy (1884– 1915) *British poet and diplomat.*

After attending Dean Close School, Cheltenham, where his father was headmaster, Flecker went to Uppingham and then, on a classical scholarship, to Oxford, where he took his BA in 1906. His first book of poetry, *The Bridge of Fire*, appeared in 1907, and the following year

Flecker began training to enter the consular service. He was posted to Constantinople (now Istanbul) in 1910, but had to return home because of the onset of tuberculosis. On his recovery he was posted again (1911) to Constantinople and then Beirut, but his TB returned and he died in a Swiss sanatorium.

Flecker's best work is comprised in *The Golden Journey to Samarkand* (1913) and *The Old Ships* (1915). Believing that the task of the poet was to create beauty, Flecker wrote poems that rely for their effect mainly upon strikingly sensuous imagery. He also wrote two plays that were produced posthumously: *Hassan* (1922) and *Don Juan* (1925), of which the former, with its exotic oriental setting and magnificent verse, achieved considerable fame. Flecker also left some comparatively minor prose works, published as *Collected Prose* (1920), and some interesting letters.

Fleming, Sir Alexander (1881–1955) *British bacteriologist and discoverer of penicillin – the first antibiotic. He was knighted in 1944 and received the 1945 Nobel Prize for Physiology or Medicine.*

Fleming was born in Lochfield, Ayrshire, the son of a farmer. Moving to London at the age of fourteen, he worked as a clerk in a shipping office before enrolling, in 1902, to study medicine at St Mary's Hospital. After qualifying in 1906 he joined the Inoculation Department under Almroth Wright, becoming assistant director in 1921. Fleming rapidly gained expertise in the treatment of bacterial diseases by vaccines and chemotherapy, notably the use of Salvarsan to treat syphilis. During World War I Fleming and Wright served in a military hospital at Boulogne, where they made advances in antiseptics for dressing wounds. In 1921 Fleming discovered lysozyme, a bactericidal enzyme found in mucus, tears, blood serum, etc.

However, Fleming's major discovery came in 1928. He noticed that a contaminant mould growing in a culture dish containing staphylococci bacteria had evidently killed the bacteria growing in its vicinity. He identified the mould as *Penicillium notatum* and the antibacterial substance it contained he called penicillin. Fleming further demonstrated penicillin's effectiveness against many other pathogenic bacteria and its low toxicity to other living organisms but failed to produce a pure extract or demonstrate its true therapeutic value. Thereafter he used penicillin only as an isolating agent in preparing cultures and it was not until 1940 that Ernst CHAIN and Howard FLO-

REY proved its enormous value as an antibiotic. Their work paved the way for numerous other antibiotics and a revolution in the treatment of bacterial and fungal diseases.

Fleming, Ian Lancaster (1908–64) *British novelist and author, the creator of the spy James Bond ('007').*

Fleming and his brothers were brought up by their mother after the death of their father, an MP and Yeomanry officer, in World War I. With his elder brother Peter (1907–71), who became a well-known traveller and author, he was educated at Eton, where he excelled as an athlete. He then briefly attended the Royal Military College at Sandhurst, but left to travel and study in Europe. In 1931 he joined Reuters news agency. Among his assignments was the notorious Moscow trial of British engineers on espionage charges (1933). He then worked as a banker and stockbroker in London until 1939, when he joined naval intelligence for the duration of World War II, attaining the rank of commander. After the war he returned to journalism but resigned in 1959 to devote himself to writing, much of which he did at his winter home in Jamaica.

The first James Bond adventure, *Casino Royale*, appeared in 1953. Fleming's personal experience of secret intelligence work and exotic places provided an authentic background to the violent action he excelled at depicting. The simple formula of tough sexy hero, tough glamorous heroine, and diabolical villain was repeated in subsequent books and transferred successfully to the screen. The film of *Dr. No* (1958) made James Bond's name known around the world. Fleming's ingenious but unrealistic plots are counterbalanced by his attention to the minutiae of sophisticated living: fast cars, card-playing for high stakes, gourmet food, and absurd gadgets. From 1953 Fleming published a Bond volume a year until his death: *Live and Let Die* (1954), *Moonraker* (1955), *Diamonds are Forever* (1956), *From Russia with Love* (1957), *Dr. No* (1958), *Goldfinger* (1959), *For Your Eyes Only* (short stories; 1960), *Thunderball* (1961), *The Spy Who Loved Me* (1962), *On Her Majesty's Secret Service* (1963), *You Only Live Twice* (1964), and *The Man with the Golden Gun* (1965, posthumous).

Florey, Howard Walter, Baron (1898–1968) *Australian pathologist whose contribution to the development of penicillin as an antibiotic for clinical use earned him the 1945 Nobel Prize for Physiology or Medicine. He* *was knighted in 1944 and received the OM and a life peerage in 1965.*

Born in Adelaide, Florey graduated in medicine from Adelaide University in 1922 and obtained a Rhodes Scholarship to study physiology and pathology at Oxford University. After brief spells in Cambridge, touring the USA, and working at the London Hospital, Florey returned to Cambridge, where he obtained his PhD studying the physiology of mucus secretion and encountered the distinguished biochemist Sir Frederick Gowland HOPKINS. In 1931 Florey was appointed Joseph Hunter Professor of Pathology at Sheffield University and in 1935 moved to Oxford to head the William Dunn School of Pathology. Here, Florey appointed the biochemist Ernst CHAIN to lead a biochemical unit in the department. Initially they pursued Florey's interest in the antibacterial enzyme lysozyme, discovered by Alexander FLEMING in 1921. But, disappointed with its low therapeutic value, they looked for other more effective substances and selected penicillin, first discovered by Fleming in 1928. Florey and his team cultured the mould *Penicillium notatum* and eventually obtained a pure extract of the penicillin it manufactures. By scaling up production they obtained sufficient penicillin for the first clinical trials on human patients, which took place at the Radcliffe Infirmary, Oxford, in 1941 and were a great success. Florey and his colleague Norman Heatly went to the USA to promote the enormous potential of penicillin, especially to treat war casualties, and to establish large-scale production facilities. By 1943, penicillin was being used to treat troops in Europe and ultimately saved countless lives. Florey followed up with work on another antibiotic, cephalosporin C. He also made important studies of blood circulation, especially cell movement through the capillary network, and of sperm movement in the female reproductive tract. In 1962 he resigned his professorship to become provost of Queen's College, Oxford. He was president of the Royal Society (1960–65).

Florey maintained his links with Australia, advising on medical research and forming close ties with the Australian National University at Canberra, of which he became chancellor in 1965.

Flynn, Errol (Leslie Thomas Flynn; 1909–59) *Australian-born US film actor, who played many swashbuckling romantic parts in the 1930s.*

Errol Flynn was born in Tasmania. His formal education was punctuated by expulsions from schools; at sixteen he abandoned a career as a clerk in favour of more adventurous pursuits, such as pearl diving in Tahiti and looking for gold in New Guinea.

He began acting as Fletcher Christian in a short Australian film, *In the Wake of the Bounty* (1933), after which he joined the Northampton Repertory Company in England and wrote his first autobiography, *Beam Ends* (1934). In 1935 he went to Hollywood, where his major break came with the lead in *Captain Blood* (1935). *The Charge of the Light Brigade* (1936), *The Adventures of Robin Hood* (1938), *The Private Lives of Elizabeth and Essex* (1939), and *The Sea Hawk* (1940) were among the successful costume adventures that followed. *Objective Burma* (1946), which initially caused consternation, left the impression that Flynn had won the war for the Allies singlehanded. The 1950s, however, were less successful for him although they provided the most demanding roles of his career, in such films as *The Sun Also Rises* (1957) and *Too Much Too Soon* (1958).

He died at the age of fifty, his life shortened by hard living and alcohol. His second autobiography was aptly entitled *My Wicked, Wicked Ways* (1959), his racy private life having provided gossip columnists with a wealth of material for most of his career. Flynn was married three times. His only son, Sean (1941–70), who also acted in films, disappeared while working as a correspondent in Cambodia.

Foch, Ferdinand (1851–1929) *French general who supervised the final Allied offensive that led to Germany's defeat in World War I.*

Foch studied at L'École Polytechnique, Nancy, and in 1873 was commissioned in the 24th Artillery regiment. He acquired a reputation as a military strategist, becoming a lecturer and (in 1908) director at L'École Supérieure de la Guerre. In *Principes de la guerre* (1903) Foch argued that troops must be imbued with fighting spirit – élan – a notion later changed to attack at all costs ('l'offensive à l'outrance').

In August 1914, Foch demonstrated the stupidity of such tactics in modern warfare when his 20th Corps was decimated by German machine guns. However, he was given command of the newly created 9th Army and on 8 September 1914, Foch's forces managed to hold a massive German assault on Allied lines in the Battle of the Marne, suffering 35 000 casualties. Marshal Joffre (1852–1931) appointed

Foch to coordinate French, British, and Belgian armies on the northern front but his failure to deploy adequate French reserves, especially in the second Battle of Ypres fought during April and May 1915, cost the confidence of other Allied commanders. Both Joffre and Foch were replaced, Foch being given an administrative post. In 1917, Foch followed PÉTAIN as chief of general staff. The German offensive launched in March 1918 caused the collapse of the British front, commanded by HAIG, and in a desperate effort to reorganize Allied command, Foch was appointed generalissimo of all Allied forces. The Allies managed to halt and then reverse the German advance between April and July, when, bolstered by the arrival of US troops under General John Pershing (1860–1948), they started the final offensive that ended the war.

In August 1918, Foch was made marshal of the French army. He was appointed president of the Allied Military Committee at Versailles in 1920 to administer the terms of the armistice between Germany and the Allies.

Fokine, Michel (Mikhail Mikhaylovich Fokine; 1880–1942) *Russian ballet dancer and choreographer who, under the patronage of Sergei DIAGHILEV, helped to revolutionize the ballet.*

Fokine studied at the Imperial Ballet School in St Petersburg and made his debut as a dancer with the Russian Imperial Ballet on his eighteenth birthday. Six years later (1904) he wrote the scenario for his first ballet, *Daphnis et Chloë* to RAVEL's music; in 1905 he produced the famous solo *The Dying Swan* for Anna PAVLOVA. By this time he was already protesting against the rigid conventions of the Imperial Ballet repertoire, in which most ballets consisted of waltzes and galops, interspersed with gymnastic solos to show off the virtuosity of a star ballerina. Fokine maintained that all the separate facets of a ballet should come together to form a cohesive whole, with movement, music, costume, and decor all in a consistent style and of equal importance.

These revolutionary ideas coincided with those of Diaghilev, who employed him in 1909 as the first choreographer to the Ballets Russes. For this company, in collaboration with STRAVINSKY and later Ravel and also drawing upon the music of earlier composers, such as Borodin and Rimsky-Korsakov, he created innovative ballets, incuding *Prince Igor* (1909), *The Firebird* (1910), *Schéhérazade* (1910), and *Petrushka* (1911). Léon Bakst (1866–1924) and Alexandre Beno-

ise (1870–1960), friends of Diaghilev from student days, designed spectacular sets. Fokine was the first choreographer to compose a one-act ballet and in 1909 created the first mood-ballet, *Les Sylphides*, based on Chopin's piano works. In 1923 he settled in New York, working with companies in the United States and Europe, and became a naturalized US citizen in 1932.

Fonda, Henry (1905–82) *US film and stage actor.*

Born in Grand Rapids, Nebraska, Fonda had intended to become a journalist, but subsequently abandoned this ambition to join the Omaha Community Playhouse as an amateur actor and manager's assistant. After turning professional, he worked in summer reps until joining the University Players in 1928. Broadway productions followed, including the lead in *The Farmer Takes A Wife* (1934), which also provided him with his screen debut. Stardom came quickly with such notable films as *The Trail of the Lonesome Pine* (1936), *Young Mr Lincoln* (1939), and *The Grapes of Wrath* (1940), in which he gave one of his most memorable performances (as Tom Joad).

During World War II Fonda served in the US navy, receiving a Bronze Star and Presidential Citation. After the war his career reached new heights with Broadway successes and such films as *Clementine* (1946), *Twelve Angry Men* (1957), and *Fail Safe* (1964). Most successful of his postwar performances was the Broadway production of *Mister Roberts*, which he repeated on screen in 1955. Throughout the remainder of his career he appeared in numerous films, plays, and on television; he won his only Academy Award for his role in his final film, *On Golden Pond* (1981), in which Katherine HEPBURN and his daughter Jane FONDA also appeared. His son Peter Fonda (1939–) also became an actor.

Fonda, Jane (1937–) *US actress.*

Daughter of Henry FONDA, with whom she first appeared on stage in *The Country Girl* (1954) at Omaha Community Theatre, Jane Fonda was born in New York City. Early in her career she became a model, appearing on the cover of *Vogue*. She also attended the Actors' Studio and studied with Lee Strasberg (1901–82). In 1960 she made her film and Broadway debut. *Tall Story*, her first film, was followed by numerous others, including some made for her first husband, French director Roger Vadim (1927–). Her stature as an actress was recognized in *They Shoot Horses Don't They?* (1969), for which she received a New York

Critics Best Actress Award and an Oscar nomination, and *Klute* (1971), which brought her an Oscar and another New York Critics Award.

Her anti-Vietnam War activities led her to form a troupe with Donald Sutherland (1935–) and others, which toured Southeast Asia. The tour was recorded on film and released as *F.T.A.* ('Free The Army'; 1972). Some years after the end of the war she won her second Oscar for *Coming Home* (1978), which dealt with the plight of servicemen wounded in Vietnam. Other films include the award-winning *Julia* (1977) with Vanessa REDGRAVE and *The China Syndrome* (1979), *Agnes of God* (1985), and *Old Gringo* (1989).

As well as continuing to make successful films and involving herself in political issues with her second husband, the radical activist Tom Hayden, she has also become internationally known for her fitness routine, *Jane Fonda's Workout*.

Fontanne, Lynn (1887–1983) *British actress, best known for her work in partnership with her husband, the US actor Alfred LUNT.*

Born in Woodford, Essex, Fontanne began her acting career in England, where she trained under Ellen TERRY and made her debut in pantomime at Drury Lane (1905). She toured the USA with the Grossmiths' company (1910) and subsequently returned there (1916) at the request of Laurette Taylor (1884–1946). She had her first major success on Broadway in *Anna Christie* (1921). Her marriage to Alfred Lunt in 1922 marked the start of a famous theatrical partnership. Lunt and Fontanne appeared together in many successful productions in both Britain and the USA, beginning as members of the Theatre Guild (1924–29). They successfully transferred their first major success, *The Guardsman* (1924), to the screen in 1931, but although they were offered further film contracts they remained faithful to the stage. They first appeared together in Britain in 1929 and during the thirties gained wide popularity in such plays as Noël COWARD's *Design for Living* (they had spent a successful season working with Coward in New York during the early thirties) and GIRAUDOUX's *Amphitryon*. Fontanne and Lunt spent World War II working in Britain and towards the end of the war entertained Allied troops in Europe. They played themselves in the film *The Stage Door Canteen* (1943). Their first TV production was *The Great Sebastians* (1957) and they made their final appearance in DÜRRENMATT's *The Visit* (1958–60).

Fonteyn, Dame Margot (Margaret Hookham; 1919–91) *British prima ballerina. She was also president of the Royal Academy of Dancing (1954–91) and was created DBE in 1956.*

Fonteyn first studied dance with George Goncharov (1904–54), while living in Shanghai with her parents. Returning to England to study at the Sadler's Wells Ballet School, she made her debut as a snowflake in *The Nutcracker* in 1934 with the Vic-Wells Ballet and the following year, after a leading role in a revival of Frederick ASHTON's *Rio Grande*, was chosen by Ninette DE VALOIS to take over Alicia MARKOVA's roles when she left the company. Within five years Fonteyn had danced leading roles in *Giselle, Swan Lake*, and *The Sleeping Beauty*, in which her interpretation of Aurora was described as definitive. Apart from these classical works, Fonteyn created roles in Ashton's *Le Baiser de la fée* (1935), which was choreographed for her, and subsequently in many other of his productions, such as *Daphnis and Chloë* and *Ondine* (considered by many to be her finest creation). She was also widely acclaimed for her performances in revivals of FOKINE's *Firebird* and *Petrushka* and soon achieved international recognition for her musicality, exquisite style, and characterization.

Fonteyn's marriage to the Panamanian diplomat Roberto Arias in 1955 did not halt her career, and in 1962, at the age of forty-three, she began her celebrated partnership with Rudolf NUREYEV in *Giselle*, Ashton's *Marguerite and Armand*, *Romeo and Juliet*, and many other ballets. In 1979 Fonteyn was named *prima ballerina assoluta*, a title officially given only three times in the history of the Imperial Russian Ballet and its Soviet successors, in recognition of her outstanding talent and her importance in the history of British ballet. She finally retired to Panama to care for her husband when he became ill; he died in 1989. News that Fonteyn was living in somewhat reduced circumstances led to a major fundraising effort in Britain on her behalf.

Foot, Michael Mackintosh (1913–) *British politician, leader of the Labour Party (1980–83). Widely regarded as Aneurin BEVAN's successor as leader of the Labour left, Foot proved an indecisive leader of the party and failed to create a strong and united opposition.*

Son of the Liberal politician Isaac Foot (1880–1960), he became a journalist before entering parliament in 1945. He was editor of the *Tribune*, the Labour weekly, from 1948 to 1952 and again, while out of parliament, from 1955 to 1960. He then followed Bevan as MP for Ebbw Vale, in South Wales. Foot was successively opposition spokesman on the power and steel industries (1970–71) and on European affairs (1971–74). When Labour returned to power in 1974, he was appointed employment secretary (1974–76) and then, under CALLAGHAN, lord president of the Council and leader of the House of Commons (1976–79). He was deputy leader of the party from 1976 to 1980, when he succeeded Callaghan as leader of the opposition. Three days after Labour lost the election of 1983, Foot announced his resignation and was succeeded by Neil KINNOCK. He has published, among other works, a two-volume biography of Bevan (1962–73).

Ford, Ford Madox (Ford Hermann Hueffer; 1873–1939) *British novelist and critic.*
Ford's mother was the daughter of the Pre-Raphaelite painter Ford Madox Brown; his father was a German music critic, Francis Hueffer, who moved to England in 1869. Ford was educated at University College School and soon showed evidence of his talent as a writer. He wrote studies of his grandfather (1896) and of Rossetti (1902), published poems and essays, and collaborated with Joseph CONRAD on three books, including *Romance* (1903). He also founded (1908), but could not afford to retain control of, the *English Review*.

Ford is best remembered as a novelist. Among his earlier works was the fine historical trilogy about Catherine Howard – *The Fifth Queen* (1906), *Privy Seal* (1907), and *The Fifth Queen Crowned* (1908). One of his most highly regarded novels is *The Good Soldier* (1915). In World War I he was an officer in the Welch Regiment and was badly gassed in France; these experiences are drawn upon in his tetralogy *Parade's End*, comprising the so-called 'Tietjens' novels: *Some Do Not* (1924), *No More Parades* (1925), *A Man Could Stand Up* (1926), and *Last Post* (1928).

Scandal afflicted Ford in 1910 when his wife sued him for restitution of conjugal rights. Later (1931) she sued a newspaper that had called Ford's associate, Miss Violet Hunt, 'Mrs Hueffer'. After World War I Ford did not make a permanent home anywhere for long. He lived for a time in Paris where he offered much encouragement to younger writers, including HEMINGWAY, whose work he published in the *Transatlantic Review*, which he founded in 1924. Near the end of his life he lectured in the USA, returning to France to die at Deauville.

Ford, Gerald R(udolph) (Leslie Kynch King; 1913–) *US Republican statesman and thirty-eighth president of the USA (1974–77).*
Ford was born in Omaha, Nebraska. Two years later his parents were divorced, his mother remarried, and the boy was given the name of his adoptive stepfather, Gerald R. Ford. After graduating in 1935 from the University of Michigan, where he read economics and political science, he went on to study law at Yale and ultimately set up his practice in Grand Rapids, Michigan. During the latter part of World War II he served in the US navy.

Ford's political career began in 1948, when he was elected to the House of Representatives. In 1965 he became Republican minority leader in the House, and in 1973, on the resignation of Vice-President Spiro Agnew (1918–), President NIXON nominated Ford as a possible replacement. Ford was thoroughly scrutinized by Congress and sworn in later that year. The reputation for integrity and candour he had built up over his twenty-five years in the House of Representatives stood him in good stead during and after the discovery of Nixon's involvement in the Watergate scandal and the subsequent impeachment and resignation of the president. Ford succeeded to the presidency in 1974, becoming the first man to be sworn into that office without ever having won a presidential or vice-presidential election. He chose Nelson ROCKEFELLER as vice-president and radically reorganized Nixon's cabinet over the next eighteen months, retaining only Henry KISSINGER as secretary of state and two others in their former offices.

Having been welcomed for his openness and honesty in the early days of his administration, Ford aroused widespread hostility and suspicion by granting Nixon a free pardon just two months after his resignation. In October 1974 Ford appeared before Congress to justify this magnanimous but possibly premature decision. Over the next year, however, with his prompt reaction to the Cambodian seizure of the US ship *Mayaguez* and the airlift of thousands of Vietnamese refugees to the USA, President Ford regained the approval of Congress and the support of the American people. This renewed popularity was not sufficient, however, to carry him through the presidential election of 1976: he won the Republican nomination only narrowly from Ronald REAGAN and lost to the Democratic nominee, Jimmy CARTER, in the election itself.

Ford, Henry (1863–1947) *US industrialist and car designer, the founder of the Ford Motor Company.*

Born on a farm near Detroit, Ford showed an early interest in the farm machinery around him. In 1879 he became a machinist's apprentice and began to acquire a practical understanding of the design and workings of the engines of his day. He worked variously as a watch repairer and a mechanic before building his first car in 1896. Ford was confident enough to found his own company in 1899 and a few years later, in 1903, he introduced the Model A. It was, however, with the introduction of the Model T in 1908 that Ford's great success began. Some fifteen million models were produced between 1908 and the car's withdrawal in 1927. This astounding success was based partly on the production line, introduced in its fully modern form at the Highland Park factory in 1913, and partly on the existence of an enormous market for cars. Standardization, too, was important: the Model T, Ford declared, was available in any colour of the rainbow, 'so long as it is black'.

Politically, Ford was a Republican although he bitterly opposed US entry into World War I under Republican President WILSON. To publicize his opposition Ford chartered the liner *Oscar II* in 1916 to sail to Europe to campaign against militarism. It was also about this time that Ford's antisemitism appeared, crediting the cause of the war he opposed so strongly to Jewish capitalists. In 1919 Ford resigned the presidency of his company in favour of his son, Edsel, whom he outlived briefly. On his death in 1947 the company passed to his grandson, Henry Ford II.

Ford, John (Sean Aloysius O'Feeney; 1895–1973) *US film director.*

Ford was born in Cape Elizabeth, Maine, the youngest of thirteen children of Irish immigrant parents. Following his brother Francis, who had adopted the name Ford when he became a director, writer, and author, John Ford began his career in Hollywood as a general assistant. He worked as an extra in several films, one of his earliest appearances being in D. W. GRIFFITH's *The Birth of a Nation* (1915) as a Ku Klux Klansman. Directing shorts subsequently led to the full-length feature *Cameo Kirby* (1923), followed by his first success, *The Iron Horse* (1924). Notable films of the 1930s and 1940s included *The Informer* (1935), *The Grapes of Wrath* (1940), which starred Henry FONDA, and *How Green Was My Valley* (1941), all of which brought him Academy Awards.

During World War II Ford was in charge of the Field Photographic Branch of the OSS and reached the rank of rear admiral. He made several documentaries, receiving Oscars for two: *The Battle of Midway* (1942) and *December 7th* (1943). After the war he won an Academy Award for *The Quiet Man* (1952). Often unashamedly nostalgic, Ford made numerous westerns invoking the values of the old West and the frontier spirit. His many films with John WAYNE include the classic *Stagecoach* (1939), *Fort Apache* (1948), *She Wore a Yellow Ribbon* (1949), *Rio Grande* (1950), and *The Horse Soldiers* (1959).

Forester, C(ecil) S(cott) (Cecil Lewis Troughton Smith; 1899–1966) *British novelist, creator of the naval hero Horatio Hornblower.*

Forester was born in Cairo and educated at Alleyn's School and Dulwich College, London. Although he was a keen sportsman at school, a weak heart prevented his enlistment in 1917. He became a medical student, but was unhappy in the work and was drawn increasingly to writing. He first achieved success with *Payment Deferred* (1926), a novel about a murderer that was made into a play (1931) and later filmed. He also wrote a biography of Nelson (1929).

In the 1930s Forester wrote several of his best-known novels – *Death to the French* (1932), *The African Queen* (1935; filmed, very successfully, in 1952), and *The General* (1936) – all of which have a characteristic Forester theme of people thrown upon their own resources in wartime. Forester also worked as a journalist, reporting (1936–37) the Spanish civil war and HITLER's occupation of Prague (1939). In 1937 he created, in *The Happy Return*, his most famous character, hero of the British navy at the time of the Napoleonic Wars, Horatio Hornblower. In the dozen Hornblower novels published between 1937 and 1962 Forester traced his hero's rise to admiral and incidentally demonstrated his brilliant understanding of naval warfare of the time.

During World War II Forester devoted himself to writing propaganda for the Allied war effort, much of it for American publication. In 1943, at the Admiralty's invitation, he sailed on a mission in *HMS Penelope*; the resulting book, *The Ship* (1943), was distributed by the hundred thousand throughout the navy. Towards the end of the war Forester settled permanently in the USA. Ill-health dogged him from the early 1940s; in 1964 he suffered a severe stroke and died two years later in California.

Forman, Miloš (1932–) *Czechoslovakian film director, who moved to the USA in 1968.*
Born in Caslav, Forman was brought up by relatives, his parents having died in German concentration camps; after school he trained as a scriptwriter at the film college in Prague. Among other things, he worked with Alfred Radok (1914–) on *Old Man Motor Car* (1956) and 'Laterna Magica', a mixture of live action and film presentations. *Audition* (1963) and *If There Was No Music* (1963) were his first films as a director. International success came with his first feature-length film, *Peter and Pavla* (1964), which won several awards, including first prize at the Locarno Film Festival and the Czechoslovak Film Critics Prize. His next two films, *A Blonde in Love* (1965) and *The Fireman's Ball* (1967), confirmed his status as a leading international director. They were also his last Czech films – he moved to the USA in 1968. There he made the Oscar-winning *One Flew Over the Cuckoo's Nest* (1975), and also achieved success with *Taking Off* (1971), *Hair* (1979), and *Ragtime* (1981). He also contributed the episode on the decathlon event to the film of the 1972 Munich Olympics, *Visions of Eight* (1974). His film *Amadeus* (1984), from the stage play of Peter SHAFFER, was largely shot in Prague and won eight Oscars, including that for best director. Subsequent films include *Valmont* (1989).

Forster, E(dward) M(organ) (1879–1970) *British novelist and literary critic. He became a CH in 1953 and was awarded an OM in 1969.*
After a childhood pampered by his widowed mother and several adoring aunts, Forster went to Tonbridge School (1893), where he was predictably miserable. He then won a classical exhibition to King's College, Cambridge (1897), where his tutor encouraged him to write. After graduating (1901) he travelled in Greece and Italy, where he found the Mediterranean passion, beauty, and spontaneity a liberating contrast to the cold and narrow English middle classes. He began writing short stories and in 1905 published his first novel, *Where Angels Fear to Tread* (filmed 1991), which, like *A Room With a View* (1908; filmed 1985), is partly set in Italy. *The Longest Journey* (1907) is a sad tale of hopes ruined by a disastrous marriage. *Howards End* (1910; filmed 1991) is partly inspired by the beneficent aura

of his childhood home, recreated as a haven for those who follow their hearts rather than the 'life of telegrams and anger'.

In 1912 he accompanied his Cambridge friend G. Lowes Dickinson to India. On his return he lived with his mother and published very little. At this time he was working on *Maurice*, a novel depicting a homosexual relationship, such as Forster craved and eventually found for himself with the policeman Bob Buckingham, whom he met in 1930. *Maurice* was not published until 1971 (it was filmed in 1987). During World War I Forster worked for the Red Cross in Alexandria (1915–18), where he met the Greek poet CAVAFY, whose work Forster later publicized to the English-speaking world. Forster's second visit to India (1921), this time as private secretary to a maharajah, resulted in his greatest novel, *A Passage to India* (1924; filmed 1984). He was then invited back to Cambridge to deliver the prestigious Clark lectures (published in 1927 as *Aspects of the Novel*) and for a time (1927–33) was a fellow of King's. He continued to write essays, criticism, talks, and biographies; some of the shorter pieces were published in *Abinger Harvest* (1936). In 1934 he became first president of the National Council for Civil Liberties.

After his mother's death (1945) Forster was elected an honorary fellow of King's and spent the rest of his life there. Although he knew many leading writers well, he particularly valued his many friendships with undergraduates. Among his later works, *Two Cheers for Democracy* (1951) reflects the deep concern for individual liberty that dominated his thinking before and after World War II. In 1953 he published an account of his 1921 visit to India, *The Hill of Devi*, which shows his sympathy for Muslim and Hindu culture already apparent in *A Passage to India*. He also collaborated with Benjamin BRITTEN on the libretto for *Billy Budd* (1951).

Foucault, Michel (1929–84) *French philosopher and historian of ideas, best known for his work on the history of western attitudes to the insane.*
Foucault was educated at the École Normale Supérieure, where he was a student of Louis Althusser. He began his career outside France teaching at the University of Uppsala in Sweden in the 1950s and for a year he held the post of director of the Institut Français in Hamburg. Foucault returned to France in 1960 to the chair of philosophy at Clermont-Ferrand, moving in 1970 to the Collège de France,

where he held the post of professor of the history of systems of thought until his death.

Foucault first revealed his power as a historian and philosopher in his *Histoire de la folie* (1961; translated as *Madness and Civilization*, 1971). He followed this study with equally penetrating accounts of western attitudes to punishment and to sexuality (*L'Histoire de la sexualité* – 3 vols, 1976–84; translated as *The History of Sexuality*, 1978). Much of the theoretical background to this work was contained in his *Les Mots et les choses* (1966; translated as *The Order of Things*, 1970) and *L'Archéologie du savoir* (1969; translated as *The Archaeology of Knowledge*, 1972). In all his work Foucault sought to identify and describe certain 'discourses' present in such disciplines as biology, medicine, and politics. How do the 'discourses' emerge? What rules do they obey? How do they change? Foucault's answers to his central questions, always obscure, have been endowed with great insight by some; others, with equal conviction, have dismissed them as pretentious and shallow.

Fowler, H(enry) W(atson) (1858–1933) *British lexicographer and commentator on the usage of the English language.*

Fowler was the eldest child of a military tutor at Tunbridge Wells and attended Rugby School before winning a scholarship to Balliol College, Oxford (1877). Soon after obtaining his degree he took a job teaching at Sedburgh School, remaining there for seventeen years (1882–99), until he resigned over his conscientious objection to the compulsory preparation of boys for confirmation. Living quietly in London, he wrote essays for periodicals (later collected into several volumes). After moving to Guernsey in 1903, he collaborated with his brother F. G. Fowler (1870–1918) on a translation of Lucian (1905) and wrote the first of his popular works on English usage, *The King's English* (1906). In 1908 he married his Guernsey landlady.

In 1911 the first of his lexicographic works for Oxford University Press appeared: *The Concise Oxford Dictionary*, also in collaboration with F. G. Fowler. By lying about his age he managed to see active service (1915–16) in World War I, but was invalided out. After F. G. Fowler died of tuberculosis in 1918, Fowler moved to Somerset and continued to work for the Oxford University Press; *The Pocket Oxford Dictionary* appeared in 1924 and *A Dictionary of Modern English Usage*, for which he is best known, in 1926. Fowler's wife died

in 1930 and he commemorated their relationship in the *Rhymes of Darby and Joan* (1931). In 1933 Fowler himself died of pneumonia and overwork, leaving another brother, A. J. Fowler (1868–1939) to carry on his work for the University Press.

Fowles, John (1926–   ) *British novelist.*

John Fowles was born in Leigh-on-Sea, Essex, and attended Bedford School. After graduating from Oxford (1950) he taught in Britain and Europe. His first novel, *The Collector* (1963), was an immediate success and was subsequently filmed. He then wrote a philosophical study, *The Aristos* (1964), and followed this with his second novel, *The Magus* (1965), set on a Greek island. Like *The Collector*, *The Magus* deals with the theme of sexuality and power, a theme that Fowles explored further in *The French Lieutenant's Woman* (1969). This won the W. H. Smith Award (1969) and was also made into a successful film. Later publications include *Poems* (1973), a volume of short stories entitled *The Ebony Tower* (1974), and the novels *Daniel Martin* (1977), *Mantissa* (1982), and *A Maggot* (1985). He also edited the previously unpublished work of the seventeenth-century antiquarian John Aubrey, *Monumenta Britannica* (1980–82).

Foyt, A(nthony) J(oseph) (1935–   ) *US motor racing driver, who has won more races and set more records in many different types of cars and events than any other driver.*

Foyt was born in Houston Heights, Texas, and grew up in a racing environment. His father was a mechanic and Foyt soon became enthusiastic about racetracks and high-powered engines – he worked as a mechanic before becoming a driver. He raced an extraordinary variety of cars, including stocks, sports, sprints, and championship, and his reputation grew rapidly. 1960 was his first great year and first national championship; he won four of the next seven championships. He was winner of the US Auto Club Championship a record seven times and took championships on dirt tracks and in stock cars. Perhaps his most significant achievement was taking part in twenty-five Indianapolis 500 races and winning four of them. He also won the Daytona 500 and the Le Mans Twenty-Four Hour Race.

Foyt was involved in several accidents and suffered some appalling injuries, among which were a broken back, innumerable burns, and fractures. He was not popular in the racing fraternity because he was outspoken about good and bad drivers alike and about the cut-throat nature of the commercial side of the sport.

Franck, James (1882–1964) *German-born US physicist, author of the Franck Report and the first physicist to provide experimental support for the quantum theory. For this he shared with Gustav Hertz (1887–1975) the 1925 Nobel Prize for Physics.*

The son of a banker, he was educated at the universities of Heidelberg and Berlin, where he obtained his PhD in 1906. After winning the Iron Cross twice in World War I, Franck was appointed professor of experimental physics at Göttingen. He remained there until 1933, when HITLER's racial laws banned Jews from holding university appointments. Although he was Jewish, Franck was exempt from this restriction on account of his military record. He nevertheless chose to resign in protest and publicly expressed his opposition to a law that treated him as an alien in his own country. One Göttingen professor supported him, the remaining forty German Aryans did not. Franck therefore left Germany and, after spending a year in Copenhagen, emigrated to the USA. He settled eventually in Chicago, where he became professor of physical chemistry (1938–49).

Franck's most significant work in Germany, in collaboration with Gustav Hertz, provided experimental evidence in support of the quantum theory. In 1914 they found that when mercury atoms were bombarded with electrons the mercury atoms absorbed precisely 4.9 eV of the electrons' energy, neither less nor more, thus supporting the theory that energy can only exist at the submicroscopic level in discrete quanta.

During World War II Franck worked on the development of the atomic bomb. In June 1945 he was asked to prepare a report giving the scientist's views on what role the bomb should play in the war and after it. The Franck Report argued against an unannounced use of the bomb, warned against the futility of trying to keep its workings secret, and recommended a system of international control. Denied the support of J. R. OPPENHEIMER and Arthur COMPTON, the report was ignored by the politicians. In the postwar years Franck worked mainly on the physical chemistry of photosynthesis.

Franco, Francisco (1892–1975) *Spanish general, statesman, and dictator (1939–75).*

Born in Galicia, the son of a naval paymaster, Franco was educated at the Toledo Infantry Academy before entering the army in 1910. Serving mainly in Morocco (1910–27), he achieved the rank of general and was appointed director of the Saragossa Military Academy in 1927. He became chief of staff in 1935 and governor of the Canary Islands in 1936. Joining General José Sanjurjo (1872–1936) in a military revolt against the Republican government, which began the civil war, he became the undisputed leader of the insurgents following the death of Sanjurjo.

Franco became leader of the Falange (Fascist) Party in 1937. He proclaimed himself 'caudillo' (leader) of Spain and took control of the government in 1939, after the surrender of Madrid and the defeat of the republic. During World War II he sympathized with the Axis powers but declined to enter the war, preferring to concentrate on the establishment of a corporate state under the Falange (the single political party). Reorganizing the government in 1947, he enacted a law of succession making Spain a monarchy, himself head of state for life, and his successor anybody he chose to name king. In 1969 Franco nominated Prince JUAN CARLOS to succeed him, which the Prince did 'provisionally', three weeks before Franco's death in 1975.

Frank, Anne (1929–45) *German Jewish girl whose graphic account of two years spent hiding from the Nazis in occupied Amsterdam is a testament to the suffering of the Jews during World War II.*

Anne was born in Frankfurt-am-Main, the daughter of Otto Frank, a German businessman. With the rise of the Nazis during the early 1930s, the Frank family moved to Amsterdam. German forces occupied the Netherlands in 1941 and soon all Jews faced the threat of deportation. On 9 July 1942, the Franks and their two daughters went into hiding in a secret apartment at Otto Frank's warehouse, where they were joined by the van Daans and their son, Peter. Friends brought them food and news from the world outside.

Showing remarkable insight, Anne Frank recorded the emotions and conflicts that marked their lives, how they responded to the constant fear of discovery, and her own emotional evolution on the brink of womanhood. The property was sold and an architect came to do a survey; a burglary resulted in police searching the premises. Still they remained undetected and, as news of the Allied invasion reached them in June 1944, their spirits rose. The last entry in Anne's diary is for 1 August. On 4 August, they were all seized by the Gestapo and sent to concentration camps. Anne's mother died in Auschwitz; Anne and her sister in Bergen-Belsen. Only Otto Frank survived. He published his daughter's diary in 1947 as *Het Achterhuis* (translated as *The Diary of A Young Girl*, 1953). Their hiding place in Amsterdam is now a museum and Anne's diary has been translated into over thirty languages, dramatized for the stage, and filmed (1959).

Franklin, Rosalind Elsie (1920–58) *British chemist and biophysicist whose X-ray crystallography studies of deoxyribonucleic acid (DNA) contributed to the discovery of its molecular structure by James W*ATSON *and Francis* C*RICK*.

Franklin graduated from Newnham College, Cambridge, in 1941 and the following year joined the British Coal Utilization Research Association to study the physical chemistry of various types of coal. In 1947 she moved to the Laboratoire Central des Services Chimiques de l'État in Paris and used the technique of X-ray crystallography to study carbon forms. When, in 1951, she joined the Medical Research Council Unit at King's College, London, her expertise in this technique was directed towards DNA. Collaborating with Raymond Gosling, Franklin made the important discovery that increasing humidity caused the crystalline A form of the DNA sample to transform into a paracrystalline B form, which exhibited an X-ray diffraction pattern strongly indicative of a helical molecular structure. However, Franklin's cautious inductive approach and subsequent misleading X-ray results led her to argue against a helical structure. Visiting King's College in January 1953, Watson was shown, by Maurice W*ILKINS*, an X-ray photograph of B-DNA taken by Franklin. To Watson this was striking evidence of a helical structure comprising two molecular chains. Franklin correctly asserted that the phosphate–sugar backbone of each chain was external and the bases internal and she produced vital data that suggested to Crick that the two chains were antiparallel, i.e. running in opposite directions. All these ideas were embodied in Watson and Crick's successful model of DNA.

Meanwhile, Franklin had moved to Birkbeck College, London, to work on the structure of tobacco mosaic virus (TMV). Shown a copy of the Watson and Crick paper in March 1953, she and Gosling set down their own findings in support of the double helix DNA model. Both papers appeared in the April 25 issue of *Nature*, along with one by Wilkins and others. Franklin continued to work on TMV and to publish her previous work on

coals. She died of cancer at the age of thirty-seven.

Fraser, (John) Malcolm (1930–) *Australian statesman and Liberal prime minister (1975–83). He was awarded the CH in 1977 and made a Companion of the Order of Australia in 1988.*

Born near Deniliquin, Victoria, the son of a grazier, Fraser was educated in Melbourne and Oxford, where he obtained an MA degree in 1952. Fraser entered parliament in 1955, when he won the seat of Wannon for the Liberal Party. Becoming leader of the party in 1975 (in opposition), he was installed as a caretaker prime minister when the Labor prime minister Gough W*HITLAM* was dismissed by the governor-general. Winning the subsequent election by a large majority, Fraser remained in power until 1983, when his government was defeated and he retired.

Fraser, Peter (1884–1950) *New Zealand statesman and Labour prime minister (1940–49).*

Born in Fearn, Scotland, the son of a bootmaker, Fraser was educated at a boarding school in Scotland before emigrating to New Zealand in 1910. A waterside worker in Auckland, Fraser became president of the General Labourers Union in 1911 and on the formation of the Labour Party (1916) became a member of the national executive.

Fraser was first elected to parliament in 1918 as the Labour member for Wellington. As minister for health in the S*AVAGE* Labour government (1935–40), which instigated considerable social reform, Fraser was responsible for the introduction of the national health scheme. Becoming deputy prime minister in 1939 and prime minister in 1940 he guided New Zealand through World War II, giving full support to the British Commonwealth. Defeated in the 1949 election, Fraser remained leader of the opposition until his death.

An idealist, known for his courage of conviction and concern for the Maori people, Fraser firmly supported New Zealand's attachment to Britain, viewing the Commonwealth as an example of peace created through unity. Keenly interested in the formation of international organizations after the war, he was elected chairman of the UN Social, Cultural, and Humanitarian Committee in 1946; he attended the UN General Assembly in Paris in 1948 and was involved in discussions on the Atlantic Pact in 1949.

Frazer, Sir James George (1854–1941)
*British anthropologist and classical scholar.
He was knighted in 1914 and awarded the OM
in 1925.*

Frazer was born in Glasgow and took his MA
at Glasgow University (1874) before winning
a scholarship to Trinity College, Cambridge.
In 1879 he was elected to a fellowship at Trinity, which he held until his death.

The whole vast structure of *The Golden
Bough* (2 vols, 1890; third and complete edition in 12 vols, 1911–15) grew from Frazer's
attempt to explain the ancient office of priest-
king of the Arician grove near Lake Nemi in
Italy. To this end he gathered material from
primitive cultures throughout the world, and
although his conclusions may be suspect or
superseded by more recent research his great
compendium of comparative data has never
been surpassed. In 1907 Liverpool University
created a chair of social anthropology for Frazer; his inaugural lecture was published as *The
Scope of Social Anthropology* (1908). Although he retained the Liverpool chair until
1922, Frazer always preferred Cambridge and
continued to work there. He delivered the
Gifford lectures at St Andrews on two occasions (1911–12, 1924–25), and these lectures
were published as *The Belief in Immortality
and the Worship of the Dead* (1913) and *The
Worship of Nature* (1926). Other major contributions to his subject were *Psyche's Task*
(1909), *Totemism and Exogamy* (1910), and
Folklore in the Old Testament (1918). In 1936
he added an *Aftermath* to *The Golden Bough*
and in 1938 and 1939 the contents of his anthropological notebooks were published as *An-
thologia Anthropologica*.

Always a prodigious and disciplined
worker, Frazer also published in the field of
classical studies. His editions of Pausanias
(1898) and of Ovid's *Fasti* (1929) are permeated with his anthropological insights. In English literature he admired the prose stylists of
the eighteenth century and in 1912 he published an edition of the *Letters of William
Cowper*.

Frazier, Joe (1944–   ) *US boxer and world
heavyweight champion (1968–73).*
Born in Philadelphia, Frazier became an
Olympics gold medallist in 1964. He then
turned professional and built his career on his
fitness and agility, a powerful left hook, and
prowess as an in-fighter. He won the world
heavyweight title in 1968, after the championship had been taken away from Muhammad
ALI and passed to Jimmy Ellis. In 1971 he

became the first man to beat Ali in a professional fight. When the two met, as undefeated
champions, the fight was watched by 20 455
spectators at Madison Square Gardens.

Frazier held the title until 1973, when he lost
to George Foreman (1948–   ) in the second
round. Subsequently, he lost to Foreman once
more and to Ali twice before retiring in 1976.

Frege, Gottlob (1848–1925) *German mathematician, logician, and philosopher who laid
the foundations for modern investigations into
the philosophy of logic and language.*

Born in Wismar (now East Germany), the son
of a clergyman, he spent his entire career at the
University of Jena, being appointed professor
of mathematics in 1896. His first important
book, *Die Grundlagen der Arithmetik* (1884;
translated as *The Foundations of Arithmetic*,
1950), was little noticed. In it he argued that
number could be defined in terms of the more
fundamental notion of a class and went on to
argue that mathematics could be derived from
logic. For the next twenty years Frege sought
to develop such a derivation and to present it in
a completely formal manner. The task seemed
to have been triumphantly completed in 1903,
when he prepared for publication of the second
volume of his *Grundgesetze der Arithmetik*
(translated as *The Basic Laws of Arithmetic*,
1964). But before the work could appear Frege
received a letter from Bertrand RUSSELL in-
forming him that a contradiction had been
found in his system. All he could do was to add
a postscript lamenting: 'Hardly anything more
unfortunate can befall a scientific writer than
to have one of the foundations of his edifice
shaken after the work is finished.' Yet what
was at stake, he could justifiably point out, was
'not just my particular way of establishing
arithmetic, but whether arithmetic can possibly
be given a logical foundation at all'.

If Frege did not succeed in reducing mathematics to logic he had at least managed to
identify the problem. He had earlier, in his
Begriffsschrift (1879; translated as *Conceptual
Notation*, 1972) developed a workable logical
notation and shown in some detail the kind of
logic that mathematics would have to be derived from. It was one in which, for the first
time, predicates and quantifiers could be handled as readily as propositions. Frege also, in
such classic papers as *Über Sinn und
Bedeutung* (1892; translated as *Sense and Reference*, 1940) and *Begriff und Gegenstand*
(1892; translated as *Concept and Object*,
1952), began to explore the interrelations between such important concepts as meaning,

reference, truth, negation, thought, and function. The exploration has continued uninterrupted ever since.

It would be difficult to overestimate the influence of Frege on modern logic and philosophy. Since his introduction by Russell (in 1903) to a wider public his work has come to be more and more discussed by each succeeding generation.

French, John Denton Pinkstone, 1st Earl (1852–1925) *British army officer who commanded the British Expeditionary Force in France and Belgium during the early stages of World War I. He was awarded the OM in 1914, created Viscount French of Ypres in 1916, and raised to the earldom in 1922.*

The son of a naval commander, French served briefly as a midshipman before embarking on his army career in 1870. He commanded a detachment of Hussars in the Gordon relief expedition to Egypt in 1884 and, with experience in India and at the War Office, was posted to the Boer War as brigadier-general in 1899. Here French displayed his skill as a cavalry commander, clearing the Cape Province of rebels and playing an important role in the relief of Kimberley in February 1900. He was promoted to major-general and awarded the KCB.

Back home, French's seniority increased, culminating in 1912 in his appointment as chief of the imperial general staff. In 1913, however, French signed an undertaking to a group of Irish officers based at Curragh that they would not be called on in any moves to coerce Ulster into Home Rule. Disavowal of this by the British cabinet forced French's resignation. Nonetheless, with the declaration of war in August 1914, French was appointed commander-in-chief of the British Expeditionary Force to France. From the outset, the British C-in-C's relationship with his French counterpart was strained. In the German advance of 1914, elements of French's force became separated, some continuing to retreat even after the enemy had been halted, thereby jeopardizing other Allied forces. The secretary of state for war, Lord KITCHENER, met the pessimistic General French in Paris to stiffen his resolve. Following the German reverse, French's pessimism turned to extreme optimism and he ordered a series of fruitless attacks on strong enemy positions along the River Aisne. By 1915, French's leadership became increasingly suspect. A number of assaults had failed, and in September French squandered an initial advantage in the battle of

Loos, which finally proved inconclusive. In December he was replaced as C-in-C by HAIG.

French was appointed C-in-C of Home Forces, in which capacity he organized measures to counter raids over England by Zeppelin airships. He also served as lord-lieutenant of Ireland (1918–21), a job for which he was patently unsuited and from which he resigned having survived an attempt on his life.

Freud, Lucian (1922–) *British painter, born in Germany. He was made a CH in 1983.* A grandson of Sigmund FREUD, Lucian Freud left Berlin in 1932 and became a British subject in 1939. After studying at the Central School of Art in London and at Goldsmith's College, he joined the merchant navy during World War II but was invalided out in 1942. His first one-man show was in 1944. Freud painted figure subjects in a meticulously detailed style based on firm draughtsmanship. *Girl with a White Dog* (1951) is a characteristic example in the Tate Gallery, London. His sharply focused almost obsessive hyperrealism produced effects that could be merciless or whimsical but nearly always intense. During the 1960s he began to use dramatic lighting and brushwork that was more overt and expressive of form.

Freud, Sigmund (1856–1939) *Austrian psychiatrist and founder of psychoanalysis. Freud developed important theories about the structure and functioning of the mind and the desires, conflicts, and motives in human behaviour. He devised psychoanalytical techniques for analysing normal and abnormal behaviour and showed that many illnesses with no apparent organic cause could be treated by psychoanalysis. Although many of his ideas have been revised and modified since his death, they have been a major influence in psychiatry and have had wide application.*

Freud was born in Freiberg, Moravia (now Czechoslovakia), and began to study medicine in 1873 at the University of Vienna. In 1882 he gave up his research and began to work at the General Hospital of Vienna to qualify for private practice, hoping to be able to afford to marry. In the hospital psychiatric clinic he studied under Theodor Meynert; his work on hallucinatory psychosis (Meynert's amentia) led to his hypothesis on the wish fulfilment mechanism. In the same year he began collaborating with the Viennese physician Josef Breuer (1842–1925), who was treating a girl with complex psychosomatic symptoms that interested Freud. While she was recounting her symptoms to Breuer, they gradually eased and disappeared and Breuer supplemented this 'talking cure' with hypnosis. Freud discussed the case repeatedly and eventually went to Jean Charcot (1825–93), the French expert on hypnosis, to discuss it. Charcot was able to induce the same types of symptoms under hypnosis, proving that if they could be induced and removed by thought, they must have a psychogenic origin. As a result Freud formulated the principle of 'conversion', suggesting that hysterical symptoms are caused by suppression of thoughts from conscious influence so that the 'mental energy' suppressed is diverted to cause bodily disorders. In 1893, he and Breuer published *The Psychical Mechanism of Hysterical Phenomena*, later expanded into *Studies in Hysteria* (1895).

Between 1892 and 1895 Freud developed his techniques of analysis using methods of free association, i e by encouraging his patients to pursue aloud a particular train of thought. In 1897 he began to analyse himself and suggested that many of his own psychoneuroses had their origin in events in his childhood. In support he found that many of his patients stated that a parent had attempted seduction when they were children; he postulated that this was really a fantasy of the child who had sexual desires towards the parent of the opposite sex. This led him to formulate the concept of the Oedipus complex (*Three Essays on the Theory of Sexuality*, 1903), which so shocked the medical world. In 1899 he published one of his most important books, *The Interpretation of Dreams*, in which he analysed dreams in terms of unconscious experiences and desires that originated in childhood. He developed further his theory that neuroses have their origins in suppressed sexual desires and real or imagined sexual experiences in childhood. His insistence that mental disorders had a sexual aetiology rooted in infancy caused considerable controversy and was a major cause of his estrangement from many of his colleagues, including Breuer. In 1902 Freud formed the Psychological Wednesday Circle, in which a number of colleagues met at his home to discuss psychological matters. In 1910 the enlarged group became the International Psycho-Analytical Society. Many members of the group, including Alfred ADLER and Carl JUNG, became increasingly opposed to Freud's theories and left the circle to form their own highly influential schools of psychology.

Freud had become internationally renowned and was invited to take part in a lecture tour of the USA in 1909. On his return he maintained his flourishing private practice while continuing to hold the chair of neurology at the University of Vienna (1902–38). By 1923, Freud had begun to notice symptoms of the cancer that eventually killed him, but despite his illness and numerous operations he continued to work prodigiously. In 1923 he published *The Ego and the Id*, discussing his division of the mind into different levels of consciousness, and in *The Future of an Illusion* (1927) he speculated on the nature of man, religion, and God. In 1938 Austria was annexed by Nazi Germany and Freud reluctantly moved to London (where his son was already living) with his daughter Anna Freud (1895–1982), who became a noted children's psychoanalyst. He died the following year.

Friedman, Milton (1912–   ) *US economist and leading proponent of monetarism in the Chicago School. He won the Nobel Prize for Economics in 1976.*

Born in New York and educated at Chicago and Columbia universities, Friedman worked for the Natural Resources Commission in Washington, followed by research at the National Bureau of Economic Research. During World War II he served in the Tax Research Division of the US Treasury and then in the Statistical Research Group of the Division of War Research, Columbia University. He became professor of economics at Chicago in 1945 and remained there until 1976, when he joined Stanford University as senior research fellow at the Hoover Institution. Since 1981 he has been a member of the President's Economic Policy Advisory Board.

Friedman is known worldwide for his studies of the influence of the quantity of money (bank deposits and currency) in an economy on the level of production. He is a strong believer in the efficiency of the market and minimal government interference. In Friedman's view, changes in the money supply cause changes in the level of production, not the other way round, and controlling the money supply is the most effective way to tackle inflation. He also proposed a theory of permanent income, in which an individual's spending decisions depend not on his or her wealth at the time but on expected lifetime wealth.

Friedman's publications include *Taxing to Prevent Inflation* (1943), *A Theory of the Consumption Function* (1957), *Price Theory* (1962), *A Monetary History of the United States 1867–1960* (1963), *The Optimum Quantity of Money* (1969), and *A Theoretical Framework for Monetary Analysis* (1971). With Anna J. Schwartz he wrote *Monetary*

Trends in the United States and the United Kingdom (1982) and with his wife Rose (whom he married in 1938) he wrote *Free to Choose* (1980) and *Tyranny of the Status Quo* (1984).

Frisch, Karl von (1886–1982) *Austrian zoologist whose studies of animal communication contributed greatly to the founding of ethology as a distinct discipline. In recognition of this he was awarded the 1973 Nobel Prize for Physiology or Medicine (with Konrad LORENZ and Niko TINBERGEN).*

Born in Vienna, the son of a surgeon, Frisch was a keen amateur naturalist in his youth. In 1905 he enrolled at Vienna University to study medicine but transferred to the Zoological Institute, Munich. Here he undertook many field trips and worked periodically at the Trieste Marine Research Institute, studying light perception in minnows. He obtained his PhD from Vienna University in 1910 and returned to the Munich Institute to teach and continue research on colour perception in fish. In 1912 he joined the University of Munich as a lecturer. Here he showed that bees could distinguish colour but further studies were curtailed by the outbreak of war. In 1919, after working in a Vienna hospital during World War I, he returned to Munich as an assistant professor in the Zoological Institute.

Von Frisch discovered how bees returning to the hive from foraging trips can communicate to other members of the hive the direction and distance of a food source by a pattern of movements and tail wagging – the famous 'dance of the bees'. This formed the basis of his classic work *Aus dem Leben der Bienen* (1927; translated as *The Dancing Bees*, 1955). Von Frisch held posts at the universities of Rostock (1921–23) and Breslau (1923–25) before returning to Munich as director of the Zoological Institute. Apart from four years at Graz University (1946–50), he remained at Munich until his retirement in 1958. Von Frisch's other works include *Bees, Their Vision, Chemical Sense, and Language* (1950), which explains how bees navigate using the sun together with an internal biological clock as compass, and *Tanzsprache und Orientienung der Bienen* (1965; translated as *The Dance Language and Orientation of Bees*, 1967), which elaborates details of the dance movements and relates how they are inherited rather than learnt.

Frisch, Max (1911–91) *Swiss playwright and novelist. He was appointed Commandeur de l'Ordre des Arts et des Lettres in 1985.*

Born in Zürich, Frisch studied architecture and practised as an architect for a number of years before deciding in 1954 to devote himself full-time to his literary work. He wrote two novels and one successful play and published the first volume of his continuing literary diary before producing his first characteristic play, *Die chinesische Mauer* (1947; translated as *The Chinese Wall*, 1961). Showing the influence of BRECHT, the play attacks totalitarian politics in the intellectual hero's attempt to demonstrate, to an uncomprehending audience, man's irrational pursuit of power. Only the emperor of China understands his argument – and makes him his court jester. *Graf Öderland* (1951; translated as *Count Oederland*, 1962) examines the ironic result of the protagonist's wish to break out of the deadening round of his life to some form of individual freedom.

Frisch's international fame rests on his next two plays: *Biedermann und die Brandstifter* (1958; translated as *The Fire Raisers* or *Biedermann and the Firebugs*, 1962), in which the threat posed by the fire-raisers goes unchecked because of a complacent failure to confront it; and *Andorra* (1961; translated 1962), which again deals with the failure to confront a reality, in this case antisemitism. In both plays Frisch's concern is not unlike Brecht's: the powerful illusions or images of language (or art) distance or distort reality or create a complacency that corrupts and prevents action. Autobiographical elements and a concern with the idea of personal identity form a major part of Frisch's work, as in his best-known novel *Stiller* (1954; translated as *I'm Not Stiller*, 1958) – in which the protagonist, against all the evidence from his past, insists that he is not the person he is said to be – and also in *Biographie* (1967) and *Montauk* (1975). Among Frisch's other works are *Homo Faber* (1957), *Mein Name sei Gantenbein* (1964; translated as *A Wilderness of Mirrors*, 1965), *Der Mensch erscheint im Holozän* (1979; translated as *Man in the Holocene*), and *Blaubert, eine Erzählung* (1982). Some of his important diaries have been translated as *Sketchbook* (1974).

Frisch, Otto Robert (1904–79) *Austrian-born British physicist, co-discoverer with his aunt, Lise MEITNER, of nuclear fission.*

The son of a publisher, Frisch was educated at the University of Vienna, where he gained his doctorate in 1926. He worked first at the universities of Berlin and Hamburg but was dismissed under German racial laws in 1933 and moved to the Institute for Theoretical Physics in Copenhagen, before settling in Britain in 1939.

While on his way to England, Frisch paid a Christmas visit to Lise Meitner, another refugee from the Nazis, in Sweden. Discussing the recent observation of her former colleague, Otto HAHN, that uranium when bombarded by neutrons produced lighter barium nuclei, they concluded that uranium nuclei must have split into lighter barium nuclei, a process later described as nuclear fission. In 1939 Frisch moved to Birmingham where, in collaboration with Rudolf Peierls (1907–), he demonstrated that under certain circumstances the process of uranium fission could lead to an explosive chain reaction. This insight led them to conclude that, using uranium-235 as suggested by Niels BOHR, it might be possible to produce a powerful new kind of bomb. This idea was reported by Frisch to Sir Henry TIZARD, and provided the first step on the road to the atom bomb. As an alien, Frisch was excluded from any work on weapon development and he accepted a post in 1940 as lecturer in physics at the University of Liverpool. In 1943, however, he was permitted to become naturalized and to join many of his former exiled German colleagues at Los Alamos in the USA, where the first atom bomb was being developed.

After the war Frisch returned to Britain, where he worked at the Atomic Energy Research Establishment at Harwell before being appointed in 1947 to the Jackson Chair of Physics at Cambridge, a post he held until his retirement in 1972. Frisch left a personal autobiography called *What Little I Remember* (1979).

Fromm, Erich (1900–80) *German-born US psychoanalyst and social philosopher who emphasized the cultural determinants of personality. His works investigate emotional problems in free societies and advocate psychoanalysis as a cure for cultural ills and an aid to the development of a sane society.*

Born in Frankfurt, Fromm was educated at the universities of Heidelberg and Munich before establishing a private practice in psychotherapy in 1925. Fromm was initially a disciple of FREUD, combining his psychological theories with the social principles of Karl Marx in both his clinical practice and his philosophy. His methods of therapy were unique in that the analyst confronted the patient both as his therapist and as another, empathic, person, stressing that there is no factor present in the patient that is not present in everyone else.

In 1933 Fromm went to Chicago to lecture at the Psychoanalytic Institute; rather than return to HITLER's Germany, he remained to become a US citizen (in 1934). He taught at several American universities and later expanded his views to incorporate some of the principles of Zen Buddhism. In 1957 he cofounded the National Committee for Sane Nuclear Policy and in 1962 gave up his religion although acknowledging that it had profoundly influenced his philosophy.

Fromm wrote many books and articles for both academics and the general reader, including *Escape from Freedom* (1941), *Psychoanalysis and Religion* (1950), *The Art of Loving* (1956), and *The Wellbeing of Man in Society* (1978). He also published his analysis of Freud in *Sigmund Freud's Mission: An Analysis of his Personality and Influence* (1959) and *The Greatness and Limitations of Freud's Thought* (1980).

Frost, Robert Lee (1874–1963) *US poet who did much to popularize contemporary poetry. He received forty-four honorary degrees and many awards, including the Pulitzer Prize (1924, 1931, and 1937).*

Frost was born in San Francisco and on his father's death ten years later was taken to New England by his mother. His affinity for New England, where he spent most of his life, permeated his poetry, and it is as a regional poet that he is best remembered. An inauspicious education at Dartmouth College and Harvard (he never completed a college course to obtain a degree) was followed by various menial jobs and marriage at the age of twenty-one. It was not until he took his wife and four children to England in 1912, where he met several of the Georgian poets, that Frost's talent began to emerge. His first two books of verse, *A Boy's Will* (1913) and *North of Boston* (1914), were first published in England. The latter volume, in particular, brought recognition and success in his own country when he returned to the USA in 1915. The collections that followed, *Mountain Interval* (1916) and *New Hampshire* (1923), contained some of Frost's best and most popular poetry. His later books included *Collected Poems* (1930), *A Further Range* (1936), and *A Witness Tree* (1942).

The ironic tone of much of Frost's poetry, its simple language, and conversational manner make him one of the most accessible of modern poets. Throughout his life Frost successfully combined a career as poet, farmer, and university professor and is regarded as one of the masters of twentieth-century US poetry.

Fry, C(harles) B(urgess) (1872–1956) *British all-round sportsman of great distinction. He held the world long-jump record for more than twenty years, played for Southampton in an FA Cup Final, and as a county cricketer scored 30 886 runs at an average of more than 50.*

Fry was educated at Repton and at Oxford University, where he captained both the cricket and football teams. On coming down from Oxford he played county cricket first (briefly) for Surrey, then for Sussex, and finally for Hampshire. He made 26 test appearances and in 1912 captained England against Australia and South Africa. He frequently topped the national batting averages and in 1901 scored a record six hundreds in a row. Two years later he hit 232 not out for the Gentlemen against the Players. In all he scored 94 centuries during his first-class career before retiring in 1921, his fiftieth year.

Outside sport, Fry had many varied interests. A brilliant classicist and a fine writer, he contributed to numerous periodicals and published several books on cricket and other matters, including his autobiography *Life Worth Living* (1939). He also stood several times (unsuccessfully) for Parliament. For over forty years (1908–50) he commanded a training ship on the Hamble for the Royal Navy and the Merchant Navy, maintaining his interest in this until the end of his life.

Fry, Christopher (Christopher Harris; 1907–) *British dramatist.*

Born in Bristol, Fry was educated at Bedford Modern School and in 1927 gained his first professional theatrical experience as an actor in Bath. After a spell as a preparatory schoolteacher (1928–31) he returned to the theatre as director of the Tunbridge Wells Repertory Players (1932–35). His play, *The Boy with a Cart*, was published in 1939 but not produced until 1950. In the meanwhile he also directed the Oxford Repertory Players (1940; 1944–46) and worked at the Arts Theatre, London (1945 and 1947). The highly acclaimed productions of *A Phoenix Too Frequent* (1946) and *The Lady's Not for Burning* (1948) established him as a major writer of verse drama. Later plays included *Venus Observed* (1950), *A Sleep of Prisoners* (1951), *The Dark is Light Enough* (1954), *Curtmantle* (1961), and *One Thing More, or Caedmon Construed* (1986), and he also translated and adapted plays by ANOUILH, GIRAUDOUX, and Ibsen. He won the Queen's Gold Medal for Poetry in 1962.

Fry provided the commentary for the 1953 film of Queen Elizabeth II's coronation and was also involved in writing the scripts for several films, including *Ben Hur* (1959). He also wrote *The Brontës of Haworth* (1973) for television.

Fry, Roger Eliot (1866–1934) *British painter and art critic.*

After reading science at Cambridge University, Fry turned to the study of art. Although he considered himself primarily a painter, it is as a critic that he came to exert most influence. The art historian Kenneth Clark (1903–83) said of him, 'In so far as taste can be changed by one man, it was changed by Roger Fry.'

Fry worked on *The Athenaeum* and *The Burlington Magazine* and from 1905 to 1910 directed the Metropolitan Museum of Art, New York. Although orthodox to begin with, Fry's writings began to shock academic opinion when, after 'discovering' Cézanne in 1906, he began to champion the modern French schools of painting, to which he gave the name postimpressionism. He bears most of the credit for introducing postimpressionism to the British public through the exhibitions he organized in 1910 and 1912. In 1913 he founded the Omega workshops, whose purpose was to produce well-designed household items. Fry's painting style is naturalistic and reflects the aesthetic tastes of the Bloomsbury group.

Fuchs, (Emil Julius) Klaus (1911–88) *German-born British physicist who, in the 1940s, conveyed important details of the Anglo-US atom bomb programme to the Soviet Union.*

Fuchs arrived in Britain from Germany in 1933 and continued his studies in theoretical physics at Bristol and Edinburgh, proving himself to be a brilliant mathematician. He was interned in 1940 and spent some time in a Canadian detention camp but, at the urging of prominent scientists in Britain, he returned in 1941 to work on the atom bomb project at Birmingham University. Fuchs had made no secret of his communist sympathies while in Germany, but he was given security clearance by the British authorities and in August 1942 was granted British citizenship.

Taking full advantage of his access to top-secret information, Fuchs began to pass vital technical information to Soviet agents almost immediately and continued to do so in the USA after he joined the research team working on the Manhattan Project at Los Alamos in November 1943. He returned to Britain in June 1946 and was appointed head of theoretical physics at the Harwell Atomic Energy Establishment. The following year he again began passing secrets to the Soviets. Not until 1949 did Fuchs fall under suspicion, when US cipher experts managed to break Soviet intelligence codes. Fuchs eventually confessed to a senior MI5 officer, James Skardon, and was sentenced to fourteen years in prison. His evidence was used to incriminate his contact in the USA, Harry Gold, and the spies Ethel and Julius ROSENBERG. Fuchs was released in 1959 and went to East Germany, where he became deputy director of the Central Institute of Nuclear Research at Rossendorf, near Dresden.

Fuchs, Sir Vivian Ernest (1908–   ) *British geologist and explorer who led the Commonwealth Trans-Antarctic Expedition, which made the first overland crossing of Antarctica (1957–58). He was knighted in 1955 and made a fellow of the Royal Society in 1974.*

Fuchs read geology at St John's College, Cambridge. In 1929, immediately after graduating, he accompanied the Cambridge East Greenland Expedition. During the 1930s he was a member of four expeditions to east Africa and he received his PhD in 1935 for a thesis about the tectonic geology of the rift valley in that region. After serving in the Cambridgeshire Regiment during World War II, in west Africa and northwest Europe, he was demobilized as a major and in 1947 was appointed leader of the Falkland Islands Dependencies Survey, based at Stonington Island. His duties involved surveying the British sector of Antarctica whose sovereignty was disputed by Argentina. During his spell as director of the Falkland Islands Dependencies scientific bureau (1950–55), Fuchs obtained support for his proposed transantarctic expedition. Financed by HM Government, the Royal Geographical Society, and private donation, it became part of the British contribution to the International Geophysical Year (1957–58). In November 1955, Fuchs set out aboard *Theron* for Vahsel Bay on the Weddell Sea coast. His team were collaborating with a New Zealand contingent led by Sir Edmund HILLARY, whose task was to set up supply bases along the route. Using snow tractors, Fuchs's team left Shackleton base on 24 November 1957, reached the South Pole on 19 January and finally arrived at Scott base on the McMurdo Sound on 2 March, having completed the 2158 miles in 98 days. Fuchs was awarded a special gold medal by the Royal Geographical Society to mark his achievement. In 1958 he was appointed director of the British Antarctic Survey, retiring in 1973. His books include *The Crossing of Ant-*

arctica (with Hillary; 1958), *Of Ice and Men* (1982), and *A Time to Speak* (1990).

Fuller, Richard Buckminster (1895–1983) *US architect, engineer, and inventor of the geodesic dome.*

Born in Milton, Massachusetts, Fuller had a limited formal education, although it did include two years at Harvard. His early interest in boats led him into the US navy in World War I. After the war he spent several years in industry before announcing the prototype Dymaxion house in 1927. This mast-and-wire construction, reflecting Fuller's nautical experience, was intended as a technological solution to the American housing shortage of the Depression. Designed on the basis of vehicle and aircraft production lines as a 'machine for living', the Dymaxion house and its contents absorbed Fuller's energy for the next twenty years. Indeed, he only abandoned it in 1947, after he had failed to secure support for his idea of converting World War II factories into Dymaxion production lines. In some ways the failure of Dymaxion led to the success of his postwar invention, the geodesic dome. Envisaged as a solution to a variety of unlikely problems, the dome in practice turned out to be both expensive and of limited application. Nevertheless, the 76-metre geodesic dome at the US pavilion at the 1967 Montreal International Exhibition attracted considerable international attention and some 10 000 of these domes have been constructed throughout the world, from the arctic to the tropics.

However, it is perhaps as a highly stimulating talker, extemporary lecturer, and educator that Buckminster Fuller will best be remembered by generations of students. He was professor of architecture at Southern Illinois University (1959–75) and, as the 'first poet of technology' (as he was called), professor of poetry at Harvard. He was awarded the Gold Medal of the RIBA in 1983.

Fuller, Roy Broadbent (1912–91) *British poet and novelist. He was awarded the CBE in 1970.*

Born in Failsworth, Lancashire, Fuller was educated at Blackpool High School before training to become a solicitor. He qualified in 1934 and practised until World War II, when he served in the Royal Navy (1941–46), attaining the rank of lieutenant (1944). After the war he specialized in legal work relating to building societies, becoming chairman of the building societies' legal advice panel (1958–69). From 1968 to 1973 he was professor of poetry at Oxford.

Besides professional legal publications, Fuller kept up a steady output of poetry from the publication of *Poems* in 1939. *Collected Poems* was published in 1962 and subsequent volumes included *New Poems* (1968), which won the Duff Cooper Memorial Prize, *Off Course* (1969), *The Reign of Sparrows* (1980), *Subsequent to Summer* (1985), and *Available for Dreams* (1989). Fuller also wrote a number of novels, including *The Ruined Boys* (1959), *The Father's Comedy* (1961), *My Child, My Sister* (1965), *The Carnal Island* (1970), and *Stares* (1990). His *Owls and Artificers* (1972) and *Professors and Gods* (1973) contain some thought-provoking criticism. He also wrote a three-volume autobiography: *Souvenirs* (1980), *Vamp Till Ready* (1982), and *Home and Dry* (1984), collected as *The Strange and the Good* (1989).

Furtwängler, Wilhelm (1886–1954) *German conductor. Although he clashed with the Nazis on several occasions, he remained in Germany during World War II.*

Son of a Berlin professor of archaeology, he studied in Munich with Joseph Rheinberger (1839–1901) and Max von Schillings (1868–1933). He started his conducting career with appointments in Zürich, Strasbourg, Lübeck, and (from 1915 to 1920) Mannheim. Increasingly prestigious engagements followed: Vienna (1919), the Berlin State Opera (1920–22), the Leipzig Gewandhaus, and the Berlin Philharmonic (1922). He visited England in 1924, conducting the Royal Philharmonic Society's concerts and the London Symphony Orchestra. From 1924 his Wagner performances in Berlin and Paris won him great acclaim; he conducted at Bayreuth in 1931, 1936, 1937, 1943, 1944, and 1951. He made his Covent Garden debut in 1935 with Wagner's *Tristan and Isolde*, and conducted the *Ring* cycle there in 1937 and 1938. Furtwängler's position in Germany under the Nazi regime has been the subject of much controversy; although British audiences eventually accepted him again he was refused permission to conduct in the USA.

G

Gabin, Jean (Jean-Alexis Moncorgé; 1904–76) *French film and theatre actor.*

Gabin was born in Mériel. His first job was as a building worker, but encouraged by his father, who was a music-hall comedian, he joined the Folies-Bergère as a singer and dancer. He eventually became leading man at the Folies and in 1930 made his first film, *Chacun sa chance*. Soon Gabin had established himself as the most popular actor in France with forceful dramatic performances in such films as *Coeur de Lilas* (1931), *Maria Chapdelaine* (1934), which won the Grand Prix du Cinéma Français, *Pépé le Moko* (1936), and Jean RENOIR's *La Grande Illusion* (1937).

During the occupation of France, Gabin went to Hollywood, where he made his least successful films, *Moontide* and *The Imposter* (both 1942). In 1943 he joined the Free French Forces, with which he was still serving when Paris was liberated. Back in his native France after the war, he soon re-established his career and went on to make such remarkable films as CARNÉ's *La Marie du port* (1949) and to win a Cannes Award for *L'Air de Paris* (1954). His outstanding postwar performance came in *Un Singe en hiver* (1962; *A Monkey In Winter*).

With Fernandel (1903–71) he founded Gafer Films, a production company, in 1963. He continued to make films throughout the 1960s and early 1970s, his last being *L'Année sainte* (1976).

Gable, (William) Clark (1901–60) *US film actor known for thirty years as the 'King of Hollywood'.*

Born in Cadiz, Ohio, Gable worked as a lumberjack and salesman, while trying to break into acting. At first he managed only small parts in films and the theatre, for which he was coached by actress Josephine Dillon, who became his first wife (1924–30). His first Broadway lead was in *Machinal* (1928); after the film *The Painted Desert* (1931) came the lead in *Susan Lennox* (1931), followed by such memorable films as *Red Dust* (1932), remade as *Mogambo* (1953), and the Oscar-winning *It Happened One Night* (1934). Also notable among the seventy films he made were *Mutiny on the Bounty* (1935), *San Francisco* (1936), and his most famous, *Gone With the Wind* (1939), in which he played opposite Vivien LEIGH.

Married five times, he was struck by tragedy when his third wife, Carole Lombard (1908–42), was killed in an aeroplane crash in 1942. During World War II Gable served in the USAF and was decorated twice. His postwar films included *Soldier of Fortune* (1955) and his last, *The Misfits* (1961), in which he played opposite Marilyn MONROE.

Gabo, Naum (Naum Neemia Pevsner; 1890–1977) *Russian-born US sculptor, who was one of the founders of European constructivism and a pioneer of kinetic art.*

Born in Briansk, Russia, he was the brother of the painter and sculptor Antoine Pevsner (1886–1962) but he always signed his work Gabo. Gabo's training was in medicine, natural sciences, and then engineering, which he studied in Munich. At the same time he attended lectures on art history and in 1913 and 1914 he visited Paris, where his brother Antoine introduced him to avant-garde art movements, particularly cubism.

On the outbreak of World War I, the brothers went to Norway, where Gabo produced his first constructions. These heads constructed of metal and plastic sheets were superficially similar to some cubist sculpture but represented different intentions, which were later explained in the *Realistic Manifesto* of 1920. This manifesto of the principles of constructivism was published in Russia after Gabo's return there in 1917. Revolutionary Russia, however, was only interested in art with more obvious social usefulness than Gabo's abstract constructions, and so in 1922 he left for Berlin.

During his ten years in Berlin, Gabo exhibited in Europe and the USA and lectured on constructivist ideas. He was one of the first sculptors to use transparent materials (such as Perspex or Plexiglass) and he introduced the elements of time and movement into sculpture through his kinetic sculptures powered by electric motors. In 1932 he went to Paris and became a member of the Abstraction-Création association. With other constructivists Gabo moved to London in 1935 and to the USA after the beginning of World War II. He continued to lecture in the USA and to work on important commissions. He became a US citizen in 1952 and in 1971 he received an honorary KBE.

Gabor, Dennis (1900–79) *Hungarian-born British electrical engineer and inventor of the hologram. He was awarded the 1971 Nobel Prize for Physics.*

Born in Budapest, the son of a businessman, Gabor was educated in Budapest and at the Technische Hochschule, Charlottenburg. After working in industry for several years, Gabor decided to leave Germany for England in 1933 on account of the Nazis. In the following year he accepted a position with British Thomson-Houston in Rugby, for whom he worked on the development of the electron microscope. With the outbreak of World War II in 1939 Gabor, as an alien, was excluded from the secret work being undertaken by British Thomson-Houston. Consequently he turned to problems that were sufficiently theoretical to remain unclassified. One of the problems he tackled arose out of attempts to improve the resolving power of the electronic microscope. Rather than attempt to obtain a better picture, Gabor chose to consider how more information could be extracted from the existing picture. The result, first obtained in 1948, was the three-dimensional picture, now known as a hologram, reconstituted from the diffraction patterns made by the illuminated object.

In 1948 Gabor left British Thomson-Houston for Imperial College, London, where he became professor of electron physics (1958–67). In later years he began to speculate about the future and the ways in which it could be influenced by scientific innovation. He expressed his views in the popular work *Inventing the Future* (1963).

Gaddafi, Mu'ammer Muhammad al (1942–) *Libyan statesman and colonel, chairman of the Revolutionary Council (1969–77) and president of Libya (1977–).*

Born in Serte, the son of a nomadic family, Gaddafi received a traditionally Islamic education at preparatory school in Fezzan and attended secondary school in Misurata, from which he was expelled for political agitation. He started to read history at the University of Libya in 1962, but gave up the following year to join the Benghazi Military Academy. Inspired by General NASSER, who symbolized Arab renaissance to him, Gaddafi formed the Free Officers Movement, a group modelled on the organization set up by Nasser for revolution in Egypt. Commissioned as a second lieutenant in the Signal Corps in 1965, he was sent on a training course to England in 1966 to learn English and advanced signals procedures. In 1969, following his failure to receive a promotion to captain, Gaddafi used the Free Officers Movement to execute a carefully planned bloodless coup, overthrowing the regime of King Idris (1890–1983) and proclaiming the Libyan Arab Republic. Assuming the role of chairman of the ten-member revolutionary council, he promoted himself to colonel and became commander-in-chief of the armed forces. He was promoted to the rank of major-general in 1976, but preferred to keep the title of colonel. He was appointed president in 1977.

Preoccupied with a dream of Arab unity, Gaddafi has been critical of governments he claimed were opposed to this goal while actively supporting dissident groups within these countries. This policy has led to military conflict against Chad (1982–88) and Libyan involvement in various terrorist campaigns, especially against Israel and various western powers. In retribution for Gaddafi's support for terrorism, US planes bombed Tripoli and Benghazi from British bases in 1986. Gaddafi subsequently toned down his rhetoric against the West, signalling a new era of pragmatism, although he also continued to repress dissidents within Libya.

Gagarin, Yuri Alekseevich (1934–68) *Soviet cosmonaut who, in 1961, made the first manned space flight. He received the Order of Lenin and was made Hero of the Soviet Union in recognition of this achievement.*

Gagarin was born in Gzhatsk, a village in the Smolensk region, the son of a carpenter. His childhood was disrupted by the Nazi invasion during World War II. After the war he trained as a metalworker at the Lybertsky plant, Moscow, before entering trade school at Saratov-on-Volga, where he received a foundryman's certificate. In 1955 he graduated in technical science from Saratov technical college and in the same year took a course at Saratov Aeroclub. This shaped the course of Gagarin's career and, after training at the Soviet Air Force School, he joined the Soviet Air Force in 1957.

Gagarin was one of several selected for the cosmonaut training programme in the late 1950s and on 12 April 1961 he was launched into space aboard *Vostok 1* from Tyuratam in Kazakstan. He completed a single orbit of the earth at a maximum altitude of 327 km and a maximum speed of 28 096 km per hour. His craft parachuted safely down 108 minutes later near the village of Smelovka in the Saratov region of the USSR, and Gagarin became an international hero. He made no further space flights but was involved with the cosmonaut training programme. He was killed in a plane crash while on a training flight.

Gaitskell, Hugh Todd Naylor (1906–63) *British politician and leader of the Labour Party in opposition (1955–63).*

A middle-class London intellectual, Gaitskell taught political economy at London University; after the war he was elected (1945) MP for South Leeds, a seat he held until his death. Under ATTLEE he was successively minister of fuel and power (1947–50), minister of economic affairs (1950), and chancellor of the exchequer (1950–51). His budget in 1951 proposed charges on dentures and spectacles, formerly provided free by the National Health Service, and provoked the resignation of Aneurin BEVAN and Harold WILSON. Gaitskell, who had considerable union support, subsequently defeated Bevan for the leadership in succession to Attlee. The loss of three elections in a row (1951, 1955, 1959) gave rise to severe self-questioning within the party, but Gaitskell's proposal to modify its image by altering clause 4 of the constitution (concerning 'common ownership of the means of production') was rejected.

At the 1960 party conference, his opponents to the left of the party carried a motion in favour of unilateral nuclear disarmament, which he and the executive opposed. For the next year, beginning with his famous speech 'Fight and fight and fight again', Gaitskell campaigned for the decision to be reversed and was successful at the 1961 conference. The party was thereby largely reunited and its image in the country considerably restored. Gaitskell also campaigned against British membership of the European Economic Community. He died suddenly, at the height of his powers and prestige.

Galbraith, John Kenneth (1908–) *US economist, popularly known for his criticism of western society's preoccupation with growth for its own sake.*

Born in Ontario, Galbraith graduated in agriculture at Toronto and obtained a doctorate at the University of California; in 1937 he went to Cambridge as a social science research fellow and was later a visiting fellow of Trinity College. In 1939 he was appointed assistant professor of economics at Princeton and from 1949–75 he was professor of economics at Harvard. He held various government posts, including that of US ambassador to India

(1961–63) under President KENNEDY, and became president of the American Economic Association in 1972.

In *American Capitalism: The Concept of Countervailing Power* (1952) Galbraith discussed monopolistic power systems among buyers and sellers (including labour) bred by capitalism, so that bargaining strength is replacing competition in the determination of prices and wages. His most widely read book is perhaps *The Affluent Society* (1958), in which he criticized the creation of false consumer needs by the advertising practices of large corporations and the consequent waste of resources, which could have been spent on social needs. In *The New Industrial State* (1967) Galbraith examined the growth of giant companies, necessitated partly by modern technology, and the separation of ownership and management that they entail. In his view they have undermined the assumptions of consumer sovereignty and profit maximization, on which conventional microeconomic theory is based. Other works include *A Theory of Price Control* (1952), *The Great Crash 1929* (1955), *Economics, Peace and Laughter* (1971), *The Nature of Mass Poverty* (1979), *The Anatomy of Power* (1983), and *A History of Economics* (1987). Galbraith's books have a very wide readership and popular appeal because of their challenging real-world subject matter and their readability.

Galli-Curci, Amelita (1882–1963) *Italian soprano who went to the USA in 1916 and remained there for the rest of her life.*

She studied the piano at the Conservatory in her native Milan, graduating with the first prize in 1903. As a singer she was largely self-taught, making her debut in opera as Gilda in Verdi's *Rigoletto* in 1906. In 1916 she joined the Chicago Opera Company and from 1921 to 1930 she sang with the Metropolitan Opera in New York. Galli-Curci made an extensive concert tour of Britain in 1924 but never sang in opera in London. A throat illness terminated her career, although after an operation in 1935 she unsuccessfully attempted a come-back singing Mimi in Puccini's *La Bohème* (Chicago, 1936).

The lyrical freshness of Galli-Curci's voice and her vocal agility won her much applause in such roles as Gilda, Elvira (*I puritani*), and Violetta (*La traviata*). That she frequently sang out of tune and made little attempt at characterization does not seem to have worried audiences of her day. She retired to California in 1940.

Gallup, George Horace (1901–84) *US stat-istician who pioneered public opinion polls.*

The son of a land speculator, Gallup attended Iowa State University, where he edited the college newspaper and graduated in 1923. He stayed on to teach journalism and in 1928 was awarded his PhD for a thesis concerning techniques for measuring readers' opinions. Gallup served as head of the department of journalism at Drake University (1929–31) and as professor at Northwestern University (1931–32), during which time he was developing his opinion-sampling techniques. In 1932 he joined the Young and Rubicam advertising agency in New York, where he used his techniques to survey consumer response to different advertising methods. With the founding of the American Institute of Public Opinion in 1935, Gallup introduced his opinion surveys into many areas of marketing and commercial and social planning. He also polled the electorate before the 1936 presidential election and correctly predicted ROOSEVELT's victory.

Gallup's basic technique was to interview a random sample of people representing a cross-section of society. Although useful in many areas, the use of opinion polls in politics has been criticized for influencing the decisions of both voters and politicians, especially before elections. However, Gallup's organization flourished and now conducts opinion surveys worldwide. His books include *Public Opinion in a Democracy* (1939), *The Pulse of Democracy* (with Saul Forbes Rae; 1940), and *The Sophisticated Poll Watcher's Guide* (1972; 1976).

Galsworthy, John (1867–1933) *British novelist and playwright. He was awarded the 1932 Nobel Prize for Literature, having been appointed to the OM in 1929.*

The son of a prosperous lawyer, Galsworthy was born in Coombe, Surrey, educated at Harrow, and took a degree in law at New College, Oxford (1889). Called to the bar in 1890, he specialized in marine jurisprudence. To gain experience he travelled (1893) to the Far East on a merchant ship, encountering Joseph CONRAD, who later became his friend.

Galsworthy's first novels and stories were published under the pseudonym John Sinjohn: *From the Four Winds* (1897), *Jocelyn* (1898), *Villa Rubein* (1900), and *The Island Pharisees* (1904). His first major success, however, was with *The Man of Property* (1906), which later became the first volume in *The Forsyte Saga*, first published in its complete form in 1922 (the other volumes are *In Chancery*, 1920; and

To Let, 1921). In the Forsyte novels Galsworthy cast a critical eye over the new monied class in Victorian and Edwardian society. The trilogy *A Modern Comedy* (1929), comprising *The White Monkey* (1924), *The Silver Spoon* (1926), and *Swan Song* (1928), embodied a more sympathetic view.

In 1906 Galsworthy enjoyed a triumph with his first play, *The Silver Box*. Like the plays that followed it – *Strife* (1909), *Justice* (1910), and *The Skin Game* (1920) among them – *The Silver Box* succeeded by reason of its naturalistic dialogue and straightforward construction, with the dramatist content to let his characters act out their own destinies without his obviously taking sides or manipulating them to expound a message. Galsworthy's *Collected Plays* appeared in 1929. Among his posthumously published works was a volume of poems (1934). The theme of *Justice* had demonstrated Galsworthy's concern with the plight of prisoners. He also took an active interest in various literary causes, becoming the first president of PEN (International Association of Poets, Playwrights, Editors, Essayists, and Novelists) when it was founded in 1921 and president of the English Association (1924).

Galway, James (1939–   ) *British flautist. His many concert and television appearances have brought renewed popularity to the flute and its music. He was appointed Officier des Arts et des Lettres in 1987.*

Born in Belfast, Galway studied in London on a scholarship with John Francis at the Royal College of Music (1956–59) and with Geoffrey Gilbert at the Guildhall School of Music (1959–60). A further scholarship enabled him to study at the Paris Conservatoire with Jean-Pierre Rampal; he later worked with Marcel Moyse in the USA. For fifteen years Galway worked as an orchestral player with Sadler's Wells Opera (1961–66), Covent Garden Opera (1965), the London Symphony Orchestra (1966–67), the Royal Philharmonic Orchestra (1967–69), and the Berlin Philharmonic Orchestra under KARAJAN (1969–75). His outstanding brilliance led to increasing solo and chamber music engagements; he also became professor of flute at Eastman School of Music in the USA.

Galway plays on an A. K. Cooper 14-carat gold flute with a virtuosity that makes the most difficult music appear easy. He has a large repertoire of classical and preclassical works, but he also readily accepts the challenge of interpreting contemporary works. Among those written for him are MUSGRAVE's *Orpheus* and Schroeder's *Variations for Flute and Orchestra.* Galway published *James Galway: an Autobiography* in 1978, *Flute* in 1982, and *James Galway's Music in Time* in 1983.

Gamow, George Antony (1904–68) *Russian-born US physicist, best known for his work on the big-bang theory and the genetic code.*

The son of a teacher, Gamow was educated at Leningrad University, where he obtained his PhD in 1928. After postdoctoral study in Copenhagen, Cambridge, and Göttingen, Gamow emigrated to the USA in 1934 and subsequently held chairs of physics at George Washington University (1934–58) and the University of Colorado (1958–68).

In 1948 Gamow published, in collaboration with Ralph ALPHER and Hans BETHE, one of the best-known papers in modern cosmology (often referred to as the Alpher–Bethe–Gamow paper). It gave a theoretical background to the big-bang theory of Abbé Georges Lemaître (1894–1966), proposing realistic mechanisms by which the elements could have been formed in the original cosmic explosion.

Gamow also made contributions to the theory of the origin of life and the significance of the genetic code, which was identified in 1953. Gamow quickly saw the essential nature of the code, arguing in 1954 that it was a triplet code with a sequence of three consecutive bases required to encode each protein. Gamow was known to a wider public through his thirty or so books popularizing modern science.

Gandhi, Indira (1917–84) *Indian prime minister (1966–77; 1980–84). Regarded as the strong-willed matriarch of India, Mrs Gandhi dominated her country's affairs both nationally and internationally for nearly eighteen years, until her assassination by Sikh extremists.*

Born in Allahabad, the daughter of Jawaharlal NEHRU, she was educated at the University of Bengal and at Somerville College, Oxford. Following in her father's footsteps, she entered politics early, joining the All-India Congress Party in 1938 at the age of twenty-one. In 1942 she married Feroze Gandhi (unrelated to Mahatma GANDHI; he died in 1960). When India achieved independence in 1947 Mrs Gandhi began to serve as political hostess for her father, often travelling with him abroad and gaining an insight into Indian politics. In 1959 she was elected president of the Congress Party, and on Nehru's death in 1964 became

minister of information and broadcasting in the government of Lal Bahadur Shastri (1904–66); she was elected leader of the Congress Party and prime minister after Shastri's death in 1966.

Mrs Gandhi first demonstrated the extent of her power when she intervened decisively in the Pakistani civil war (1971). During the struggle of the Bengalis in East Pakistan to achieve independence from central government in the west, thousands of Bengali refugees poured into India; with Indian support, East Pakistan finally achieved independence as Bangladesh (1972) and Mrs Gandhi was praised for her handling of the situation both in India and abroad. In international affairs she pursued a policy of nonalignment and 1974 saw the emergence of India as a nuclear power with the explosion of a nuclear device in the Rajasthan desert. However, droughts in 1972 and 1974 led to a worsening in India's economy and widespread political unrest. In response to what she saw as a threat to national security, Mrs Gandhi instituted a state of emergency in June 1975, in which political opponents were imprisoned, the press was censored, and civil liberties were restricted. At the same time she introduced measures to control population growth (by sterilization) and clear slum dwellings in Delhi, which caused widespread unrest. Her younger son, Sanjay Gandhi (1948–80), assumed a prominent role in these activities. Aware of the unpopularity of her emergency measures, Mrs Gandhi called a general election in 1977 and was defeated by Morarji DESAI at the head of the Janata Party. However, the new government proved indecisive and in the 1979 general election she was returned to power. Mrs Gandhi's second term of office was plagued by religious disturbance. Between 1982 and 1984 some three thousand people died in ethnic clashes in Assam, and there were terrorist activities among extremist Sikhs of the Punjab, who were demanding autonomy. In November 1984 Mrs Gandhi was assassinated by her own Sikh bodyguard, months after ordering the army to storm the Sikh base at the Golden Temple at Amritsar, during which over three hundred people were killed.

She was succeeded as prime minister by her elder son, Rajiv Gandhi (1944–91), who – after his brother's death in a plane crash in 1980 – became her close adviser. His position as Mrs Gandhi's successor was confirmed with an overwhelming victory at the general election in December 1984. Continuing ethnic violence, however, eventually led to his resignation in 1989. During the election campaign of 1991, Rajiv Gandhi was killed in a terrorist bomb attack; his widow Sonia Gandhi (1949–) declined to succeed him as leader of the Congress Party. This murder therefore brought to an end the Nehru dynasty, which had dominated the Congress Party and Indian politics since partition.

Gandhi, Mahatma (Mohandas Karamchand Gandhi; 1869–1948) *Indian nationalist leader, known as the Mahatma ('Great Soul'). Although he never held government office, Gandhi was regarded as the supreme political and spiritual leader of India and the main force in achieving his country's independence.*

Born in Porbandar, the son of an office worker who belonged to the merchant caste, Gandhi went to England at the age of nineteen to study law (1888). He was called to the bar at the Inner Temple in 1891 and practised as a barrister in Bombay before travelling to South Africa in search of clerical work in 1893. During the Boer War (1899–1902) and the Zulu uprising (1906) he worked with a volunteer medical corps. He came to prominence in the period 1907–14, during which he challenged through passive resistance the Transvaal government's discrimination against Indian settlers. He was jailed several times for his opposition, but managed to achieve some concessions for the Indian minority.

Gandhi returned to India in 1915. After protesting at Britain's sedition laws and the massacre at Amritsar (1919), when hundreds of Indian nationalists were killed by British-controlled troops, he emerged as the leader of the Indian National Congress and instituted his policy of noncooperation and nonviolent civil disobedience against the British in order to achieve Indian independence. His measures included the boycott of British products in order to encourage India's village industries and often took the form of hunger strikes. He also travelled round India campaigning against the degradation of the untouchables and urging friendship between the Hindu and Muslim communities. In 1930 he made his famous walk from Ahmedabad to the sea, publicly distilling salt from the seawater in protest against the government's salt monopoly. He was imprisoned for this and for various other activities between 1922 and 1942.

During World War II, in the face of a possible Japanese invasion of India, the British government offered complete independence to India after the war in return for their cooperation in repelling the Japanese. Gandhi, how-ever, refused to go against his principles of nonviolence and noncooperation and was imprisoned until 1944, when negotiations for Indian independence were resumed. Gandhi (with Jawaharlal NEHRU) played a crucial role in these negotiations and India was declared an independent republic in 1947. Gandhi reluctantly agreed to the partition creating the separate state of Pakistan for the Muslim minority. Fighting subsequently broke out between the Hindus and Muslims of Bengal, and Gandhi was assassinated by a young Hindu fanatic in 1948.

Garbo, Greta (Greta Louisa Gustafsson; 1905–90) *Swedish-born US film star, whose perverse 'I want to be alone' combined with her quite exceptional beauty to create one of Hollywood's most lucrative and enigmatic legends.*

Born in Stockholm, Garbo won a scholarship to the Royal Dramatic Theatre training school in Stockholm, where she was discovered by the Swedish film director Mauritz Stiller (1883–1928). He chose her for a part in the Swedish film *Gösta Berlings Saga* (1924; *The Story of Gosta Berling*) and in the following year took her to Hollywood, where MGM put her under contract. Her natural diffidence offscreen immediately gave her an air of mystery which, with her success in *Torrent* (1927), created the beginnings of the legend.

Many silent films followed, including *Flesh and the Devil* (1927) and *Love* (1927), an adaptation of Tolstoy's *Anna Karenina*, playing opposite John Gilbert (1895–1936) in both. With the advent of talkies, her husky voice and charming accent added a new dimension to the myth. 'Garbo talks' proclaimed the advertising for *Anna Christie* (1930). Other beads in the Garbo rosary include *Mata Hari* (1931), *Grand Hotel* (1932), and *Queen Christina* (1933). For *Anna Karenina* (1935) and *Camille* (1937) she was named best actress by the New York film critics. Her last films were the comedies *Ninotchka* (1939), directed by Ernst LUBITSCH, and *Two-Faced Woman* (1941). The second was not as successful as the first and Garbo thereafter refused all inducements to return to the screen. For over forty years she lived a totally private life that never ceased to excite the curiosity of the world's press. In 1956 she was awarded a special Oscar for her 'unforgettable screen performances'.

García Lorca, Federico (1898–1936) *Spanish poet and dramatist, regarded as one of the*

García Márquez, Gabriel

most prominent figures of Spanish literary life in the twentieth century.

Born at Fuente Vaqueros, Granada, the son of a wealthy landowner, García Lorca studied literature and law and planned a career as a musician before he turned to writing. Whilst at Madrid in 1919 he met such influential figures of the avant-garde movement as Salvador DALI, Rafael ALBERTI, Luis BUÑUEL, and Pablo NERUDA. García Lorca's first play, *El maleficio de la mariposa* (1920; 'The Evildoing of the Butterfly'), was not a success, but his first volume of verse, *Libro de poemas* (1921; 'Book of Poems'), was an early indication of his talent. García Lorca became the centre of a group of poets known as 'The Generation of 1927' and in 1928 he published his most celebrated book of poems, *Romancero gitano* (translated as *Gypsy Ballads*, 1953). His reputation was consolidated with *Poema del Cante Jondo* and *Llanto por Ignacio Sánchez Mejías* (1934; translated as *Lament for the Death of a Bullfighter*, 1937); all three drew on life in his native Andalusia and were characterized by a highly original imagery. His work during a six-month visit to New York (1929) showed a marked change in tone; a sense of mortality and rebellious disgust is evident in his book of surrealistic poems *Poeta en Nueva York* (1940; translated as *Poet in New York*, 1955).

A spirit of revolt also showed itself in García Lorca's plays, for example *La zapatera prodigiosa* (1930; translated as *The Prodigious Wife*, 1941), which attacked the fashionable realism of the Spanish theatre and showed the influence of COCTEAU and Valle-Inclán. In 1932 he became co-director of the La Barraca theatrical company and toured Spain. García Lorca wrote a number of farces for the troupe but it is his trilogy of plays, *Bodas de sangre* (1933; translated as *Blood Wedding*, 1939), *Yerma* (1934; translated 1941), and *La casa de Bernarda Alba* (1936; translated as *The House of Bernarda Alba*, 1947) on which his reputation as a playwright rests. These 'folk tragedies', dealing with the subject of frustrated womanhood, combine savage passion with a strong dramatic construction. By this time García Lorca was recognized as supreme master of the Spanish language and unequalled in his expression of the dark forces underlying life and death and love and hate. However, when he was at the peak of his creativity, the Spanish civil war broke out and García Lorca, arrested for his republican sympathies, was shot by a nationalist firing squad and buried in an unmarked grave.

García Márquez, Gabriel (1928–) *Colombian novelist and short-story writer, one of the founding figures of modern South American literature. He was awarded the Nobel Prize in 1982.*

Brought up in poverty in the remote village of Aracataca, García Márquez nevertheless studied law and journalism at the universities of Colombia and Bogotá. He followed a journalistic career from 1948 and began to write the short fictions collected in his first book, *Leaf Storm* (1955). The title story introduced the imaginary village of Macondo, loosely based on his birthplace, which appears throughout his work as a symbolic focus of Colombian history and identity. From 1955 García Márquez worked in Paris and Rome as foreign correspondent of El Espectador, a liberal daily; enthusiasm for the 1959 Cuban revolution then led him to spend two years with the Cuban press agency in Havana and New York. His first major novel, *In Evil Hour*, appeared in 1962.

Owing to his left-wing sympathies, García Márquez came into disfavour with the Colombian government and spent the 1960s and 1970s in voluntary exile in Mexico and Spain. In 1967 he produced his masterpiece, *One Hundred Years of Solitude*, a dense work tracing the history and ethos of Macondo through the lives of several generations of its leading family. The book is celebrated for its 'magic realism', an intermingling of realism and fantasy that has been much imitated in recent writing. *The Autumn of the Patriarch* (1975) is a complex satire on South American dictatorship.

Following the award of the Nobel Prize, García Márquez was formally invited back to Colombia where he has since lived. Recent novels such as *A Chronicle of a Death Foretold* (1982), *Love in a Time of Cholera* (1985), and *The General in His Labyrinth* (1990), a novel about Bolivar, have enjoyed considerable success in Europe and the USA.

Garland, Judy (Frances Gumm; 1922–69) *US film actress, singer, and superstar, whose life ended tragically and prematurely as a result of drink and drug abuse.*

Born in Grand Rapids, Minnesota, the daughter of vaudeville entertainers, Garland had appeared on stage before she was five. She and her two older sisters formed the 'Gumm Sisters Kiddie Act' but eventually she went solo, changing her name to Judy Garland. Thanks to the ambition of her mother, at thirteen she was under contract to MGM.

Her first film was *Every Sunday* (1936), with Deanna Durbin (1921–). *Broadway Melody of 1938* (1937), in which she sang her famous 'Dear Mr Gable', brought her to the public's attention, after which she made nine films with the 1930s Hollywood juvenile lead Mickey Rooney (1920–). It was, however, as Dorothy in *The Wizard of Oz* (1939), a role originally intended for Shirley TEMPLE, that she achieved international stardom. For this she received a special Oscar and became identified with the film's hit song 'Over the Rainbow'.

Other successful films followed, including *For Me and My Gal* (1942), *Meet Me In St Louis* (1944), *The Clock* (1945), *Easter Parade* (1948), and *A Star is Born* (1954), which gained her an Oscar nomination. But an unhappy private life, psychiatric illness, and drink and drug dependence finally destroyed her. By the sixties ill health began to affect her concert and cabaret appearances, although she continued to make films, including *Judgment at Nuremberg* (1960) and her final film, *I Could Go On Singing* (1963). She was married five times; her daughter Liza Minnelli (1946–), by her second husband Vincente Minnelli (1910–86), emerged as a star in the 1970s.

Gaudier-Brzeska, Henri (Henri Gaudier; 1891–1915) *French sculptor.*
Born at St Jean de Braye, near Orléans, Gaudier studied both commerce and painting in London before turning to sculpture in Paris (1910). Here he met Sophie Brzeska from Poland, with whom he lived from that time and whose name he added to his own. The following year they settled in London, where they lived in poverty.

From 1912 Henri Gaudier-Brzeska immersed himself in London's artistic and literary life, meeting Katherine MANSFIELD, Wyndham LEWIS, Jacob EPSTEIN, Ezra POUND, Roger FRY, and others. He signed the vorticist manifesto and experimented with many other avant-garde styles. On the outbreak of World War I he enlisted in the French army and was killed in 1915 at the age of twenty-four. It is widely recognized that had his artistic development not been cut short at this eclectic experimental stage he might have been one of the twentieth century's greatest sculptors.

Gavaskar, Sunil Manohar (1949–) *Indian cricketer who in 1983, at the age of thirty-four, scored his thirtieth test hundred to beat Sir Donald BRADMAN's world record.*
Educated at St Xavier's High School and Bombay University, he made his mark as an out-

standing schoolboy and university player – he once scored 327 in his last year at university. As a twenty-year-old he made his test debut in the West Indies; although he missed the first test through injury he still ended with an aggregate of 774 runs for the series. Gavaskar returned to India a national hero and went on to make several centuries against England, Australia, New Zealand, and Pakistan. He captained India in 1978–82 and 1984–85. In 1980 he had one season of county cricket with Somerset and was happiest in the three-day matches. Although of small stature, Gavaskar did not wear a protective helmet – rare among modern players. In 1987 he became the first batsman to score 10 000 runs in Test cricket. Outside his playing career, he has worked as a public relations officer and is the author of widely read cricketing books.

Geiger, Hans Wilhelm (1882–1945) *German physicist, who invented the Geiger counter.*
The son of a philologist, Geiger was educated at the universities of Munich and Erlangen, where he obtained his PhD in 1906. His first academic appointment took him to Manchester University as assistant to Professor Arthur Schuster (1851–1934). In the following year Schuster was succeeded by Ernest RUTHERFORD. In 1908, in cooperation with Rutherford, Geiger investigated the nature of the alpha particle, showing that it had a double positive charge. Geiger also designed instruments capable of detecting and counting alpha particles. These were the prototypes of the counter Geiger developed in the 1920s with W. Müller, which has since become widely known as the Geiger counter (or Geiger–Müller counter).

Geiger returned to Germany in 1912 to direct the Physikalisch-Technische Reichanstalt in Berlin. He later held chairs of physics at the universities of Kiel (1925–29) and Tübingen (1929–36). In 1936 he was appointed head of physics at the Technical University, Charlottenburg.

Geldof, Bob (1954–) *Irish singer, guitarist, and songwriter who organized massive fundraising events for famine relief in the 1980s; in 1986 he was awarded an honorary KBE and nominated for the Nobel Peace Prize.*
Born to middle-class parents in Dublin, Geldof was educated by Jesuits at Blackrock College. After a rebellious adolescence he led an unsettled life in Britain and Canada, working through a succession of jobs, including spells as a journalist. In 1975 he helped to found the Boomtown Rats, a Dublin-based rock band,

becoming their lead singer. Owing partly to Geldof's flair for publicity, the band became one of the most successful to emerge from the punk scene, enjoying number one hits with 'Rat Trap' (1978) and 'I don't like Mondays' (1979). By the 1980s, however, their popularity was declining.

In 1984, deeply affected by television images of the famine in Ethiopia, Geldof organized the recording of a charity single by an all-star rock group calling themselves Band Aid. 'Do They Know It's Christmas', written by Geldof and the singer Midge Ure, became the biggest-selling UK single ever, raising some £8 million for famine relief. This success led to the ambitious Live Aid project, two marathon rock concerts held simultaneously in London and Philadelphia on 13 July 1985. Broadcast worldwide to an estimated one and a half billion viewers, the concerts raised upwards of £50 million, a triumph owing much to Geldof's determination, organizational ability, and skill in handling the media. Sports Aid, organized in 1986, raised a further £50 million.

As chairman of the Band Aid trust, Geldof remained closely involved with the administration of the funds, visiting many relief projects in the field. His sincerity and abrasive outspoken style made him a popular hero and his autobiography *Is That It?*, became a bestseller in 1986. Attempts to resume his musical career as a solo artist have met with only moderate success.

Gell-Mann, Murray (1929–) *US physicist, who proposed the concepts of strangeness and quarks and predicted the existence of the omega-minus particle. For his contributions to particle physics he was awarded the 1969 Nobel Prize for Physics.*

The son of Austrian immigrants, Gell-Mann was born in New York and educated at Yale and the Massachusetts Institute of Technology, where he gained his PhD in 1951. He taught at Chicago for four years before moving in 1955 to the California Institute of Technology, where he had been appointed professor of physics at the age of twenty-six. His first major contribution to high-energy physics was made in 1953, when he demonstrated how some puzzling features of hadrons (particles responsive to the strong force) could be explained by a new quantum number, which he called 'strangeness'.

He also made extensive use of group theory in a most original manner to create order among the increasing number of known particles, forming them into a number of natural families. Gell-Mann's insight was supported by the discovery of the omega-minus particle and others, which his groupings had predicted in the previous year. Having introduced some order into an increasingly complex field, Gell-Mann attempted a measure of simplification by assuming that all hadrons consisted of groups of fractionally charged particles, which he called quarks (a name taken from a quotation in James Joyce's *Finnegans Wake*: 'Three quarks for Muster Mark'). Gell-Mann proposed that all particles could be reduced to combinations of at most three quarks, although the number of quarks and their antiparticles has since been increased. The quark hypothesis is still widely accepted.

Genet, Jean (1910–86) *French novelist and dramatist.*

Genet was born in Paris, the unwanted illegitimate son of a prostitute. Convicted of petty theft at an early age, he spent part of his youth in the reform school at Mettray; his experiences there inspired part of the novel *Miracle de la rose* (1946; translated as *Miracle of the Rose*, 1965). In the 1930s Genet made his living as a pickpocket and male prostitute in various European cities and was frequently imprisoned; the autobiographical *Journal du voleur* (1949; translated as *The Thief's Journal*, 1965) describes this period of his life. While in Fresnes prison in 1942 Genet began to write poetry, and the novel *Notre-Dame des Fleurs* (translated as *Our Lady of the Flowers*, 1964) followed in 1946. This vivid account of crime and prostitution in the Montmartre underworld brought Genet to the attention of such literary figures as Cocteau and Sartre; when the writer was convicted of burglary again in 1947 an appeal from his supporters in the literary world secured a reprieve. Other novels of this era include *Pompes funèbres* (1947) and *Querelle de Brest* (1947).

The eroticism and unsavoury subject-matter of Genet's novels prevented their official publication in Britain and the USA until the more tolerant 1960s. Meanwhile, Genet had begun to write for the theatre; the influence of Sartre is evident in his early avant-garde plays. *Les Bonnes* (1947; translated as *The Maids*, 1954), a complicated drama of false identities and impostures, ranked him alongside contemporary dramatists of the Theatre of the Absurd. Later plays, dealing with contemporary prejudices, were intended to shock his audience and encountered censorship problems in France. These include *Le Balcon* (1956; translated as *The Balcony*, 1958), set in a brothel; *Les*

Nègres (1958; translated as *The Blacks*, 1960), a play within a play about racial killings; and *Les Paravents* (1961; translated as *The Screens*, 1962). Vividly symbolic, they combine a skilful use of imagery with a total rejection of social, moral, and political values.

Gentile, Giovanni (1875–1944) *Italian idealist philosopher, supporter of MUSSOLINI and proponent of the philosophy of 'actualism'.*
Gentile was born in Sicily and held academic appointments at the universities of Palermo, Pisa, and Rome. A supporter of fascism, he served as minister of education (1922–24) under Mussolini and as president of the Fascist Institute of Culture. Between 1929 and 1937 Gentile held the demanding post of director of the 35-volume *Enciclopedia italiana*. He was assassinated in Florence at the hands of the partisans after Mussolini's overthrow.

Benedetto CROCE, the leading Italian philosopher of Gentile's day, had distinguished between theoretical and practical activities of the mind. In his most important work, *Teoria generale dello spirito come atto puro* (1916; translated as *The Theory of Mind as Pure Act*, 1922), Gentile rejected Croce's distinction and argued instead that only the pure act of thought was real. Nature itself thus became what was thought. Any other approach, Gentile argued, would lead to the conclusion that nature must remain something forever unknown.

George V (1865–1936) *King of the United Kingdom and Emperor of India (1910–36).*
He was born at Marlborough House in London, the second son of EDWARD VII, with whom he had excellent relations. He served in the navy from 1877 to 1892, when his elder brother, the Duke of Clarence, died and George became heir apparent. In the following year he married Clarence's fiancée, Mary of Teck (1867–1953), and in 1901 was created Prince of Wales. A year after his accession to the throne he became the only Emperor of India to hold a coronation durbar (court) in the subcontinent – at Delhi in December 1911. In 1915, during World War I, while visiting his troops in France, he was thrown from his horse and severely injured, never fully recovering his strength. The inappropriateness of the royal surname – Saxe-Coburg – at a time of war with Germany caused him to adopt the name Windsor in 1917. George became enormously popular after the war and was to prove an excellent radio speaker, recording frequent messages, which were widely appreciated. The warmth of his 1935 Jubilee celebrations testified to the affection and respect he commanded. George

was succeeded briefly by his eldest son Edward VIII (see WINDSOR, DUKE OF) and then by his second son, GEORGE VI.

George VI (1894–1952) *King of the United Kingdom (1936–52) and the last Emperor of India (until 1947).*
Second son of GEORGE V, he was born at Sandringham House in Norfolk and was known as Prince Albert until his accession to the throne. He served in the navy from 1909 to 1917 and was present, in *HMS Collingwood*, at the Battle of Jutland in World War I. He then spent a year in the air force. Created Duke of York in 1920, he became involved especially in youth work and established an annual camp attended by young men from different social backgrounds. In 1923 he married Lady Elizabeth Bowes-Lyon (1900–   ; now the Queen Mother), and their two daughters, the future ELIZABETH II and Princess Margaret, were born in 1926 and 1930 respectively. George became king after the abdication of his brother (see WINDSOR, DUKE OF) in 1936 and coped remarkably with a role for which he was not prepared and had little natural inclination – he overcame a severe stammer in order to perform his duties. During World War II he made a great contribution to public morale, especially in London during the Blitz. He died at Sandringham.

Gershwin, George (Jacob Gershovitz; 1898–1937) *US composer and pianist, known for his popular songs, musicals, and his folk opera* Porgy and Bess.
Born in Brooklyn of impoverished Russian-Jewish immigrant parents, he grew up on the tough East Side of New York City. By the age of ten he had managed to arrange for piano lessons and at sixteen began his professional career as a song plugger for a music publisher. In 1916 his first song was published and in 1919 he made his name with 'Swanee', the song popularized by Al JOLSON. In the same year his first musical, *La, La Lucille*, appeared. In 1924 Gershwin had his first really successful musical *Lady Be Good* (including 'Fascinating Rhythm', 'Oh, Lady Be Good', and 'The Man I Love'), which was also the first show written in collaboration with his brother, lyricist Ira Gershwin (1896–1983). This partnership continued until George's death and included such shows as *Funny Face* (1927), *Girl Crazy* (1930), and *Of Thee I Sing* (1931).

In the 1920s, as well as keeping up a flow of songs and shows, Gershwin spent a considerable amount of time studying composition with avant-garde composers, such as Henry

Cowell (1897–1965). In 1924 he expanded into the world of orchestral music with *Rhapsody in Blue* (1924), a blend of jazz and Lisztian romantic pianism. This work was introduced to the public with fantastic success by Paul Whiteman at a Town Hall concert with the composer at the piano. A subsequent piano concerto (1925) was not so well received and the orchestral *An American in Paris* (1931), although popular, shows the limitations of Gershwin's symphonic technique. The negro folk opera *Porgy and Bess* (1935), with Ira Gershwin as librettist, has had more success posthumously than in the composer's lifetime. At the peak of his career a brain tumour was discovered and Gershwin did not survive the operation.

Getty, J(ean) Paul (1892–1976) *US industrialist whose oil companies and other interests, reputed to be worth over $500 million, made him one of the world's wealthiest individuals.*
Getty attended the universities of Southern California and Oxford before joining his father's successful oil-prospecting company. A dollar millionaire at the age of twenty-four, Getty became president of George F. Getty Incorporated in 1930, on the death of his father, and inherited the estate estimated at $15 million. By judicious share acquisition during the Depression, he took control of Tidewater Associated Oil Company in 1937 and ten years later he became president of Mission Corporation and Getty Oil Company. This formed the base of his business empire, which eventually included over a hundred companies worldwide.

Getty, a reclusive man, married five times and was also a noted art collector. In 1953 he founded the J. Paul Getty Museum at Malibu, California, which now houses a remarkable collection of paintings, drawings, sculptures, and books in a facsimile Roman villa. His British home, the Tudor manor house Sutton Place in Surrey, is now an arts centre.

Giacometti, Alberto (1901–66) *Swiss sculptor, painter, and draughtsman, who is best known for the thin elongated figures that he produced after World War II.*
Son of the postimpressionist painter Giovanni Giacometti (d. 1883), Alberto was born in an Alpine village in the Italian part of Switzerland. He studied painting and sculpture in Geneva and Italy before settling in Paris in 1922. His early sculpture was impressionistic but he became interested in primitive sculpture and after a two-year cubist period began in 1927 to produce highly simplified figures reminiscent

of primitive Mediterranean idols. Giacometti produced a large number of paintings and brilliant drawings throughout his life but during the period 1929–35, when he was associated with the surrealist movement, these began to take second place. The most famous of his imaginary surrealist constructions is his open skeletal construction of glass, wood, wire, and string, *The Palace at 4 am.*

In 1935 he broke with the surrealists, having determined to concentrate on the expression of human reality as he perceived it. The heads and figures that he produced became increasingly minute after 1937. He spent World War II in Geneva, returning in 1945 to Paris, where two years later he found the mature style upon which his fame now largely depends. The tall volumeless figures of this period were modelled with a thin roughly textured coating of plaster of Paris on wire. A great deal has been said and written about the meaning and intentions of Giacometti's art – however, much of it was apparently as enigmatic to him as it was to others. His view of man was influenced by the philosophy of Jean-Paul SARTRE; this existential view of life as a losing game without the possibility of real communication is reflected in his emaciated, apparently distant, figures and in his own dissatisfaction with much of his work. His portrait paintings, usually executed in dull colours or grey, and his drawings have many of the qualities of his sculpture.

His style continued to develop: the 1950s saw the appearance of his painted narrow heads on massive shoulders and in the 1960s he produced more monumental heads and figures. He won the Carnegie Sculpture Prize in 1961, the Grand Prix for sculpture at the Venice Biennale in 1962, and the Guggenheim International Award for Painting in 1964.

Gibbon, Lewis Grassic (James Leslie Mitchell; 1901–35) *Scottish novelist.*
Gibbon was born near Auchterless, Aberdeenshire, and brought up at Arbuthnott, Kincardineshire, where he attended the local school. He tried his hand at journalism (1917–18), joined the army (1918–22), and then pursued his interests in archaeology and exploration, which brought him his first commission for a book, *Hanno* (1928). Encouraged by H. G. WELLS, he began writing short stories, which were regularly published from 1927 in the *Cornhill Magazine*; his *Calends of Cairo*, a collection of twelve stories, appeared in 1931.

In the meantime Gibbon had published his first novel, *Stained Radiance* (1930), which he followed with *The Thirteenth Disciple* (1931), *Image and Superscription* (1933), and *Spartacus* (1933), all under his true name. His greatest achievement, however, was the trilogy *A Scots Quair*, written under the name of Lewis Grassic Gibbon: *Sunset Song* (1932), *Cloud Howe* (1933), and *Grey Granite* (1934), the first set in the 'Mearns' country of his childhood, the last in a great industrial city. He also published in 1934 a biography of Mungo Park, two books on exploration (*The Conquest of the Maya*, *Nine Against the Unknown*), and a celebration of Scotland in prose and verse entitled *Scottish Scene*, in which he collaborated with Hugh MACDIARMID.

Gide, André-Paul-Guillaume (1869–1951) *French novelist and moralist. He was awarded the Nobel Prize for Literature in 1947.*

Gide was born in Paris, the son of a university professor and a Norman heiress. His uncle was the political economist Charles Gide. He was a delicate child whose education was interrupted by frequent bouts of ill health. His wealthy family background enabled him to devote his life to writing; he published *Les Cahiers d'André Walter*, his first prose work, anonymously in 1891. This was followed by poems and other narrative works inspired by the symbolist movement. In 1895 he married his cousin Madeleine Rondeaux, the culmination of a devotion that had begun some thirteen years before, in his adolescence, and was to inspire a number of his literary works.

Les Nourritures terrestres (1897; translated as *Fruits of the Earth*, 1949), a blend of prose and verse, was written during two visits to North Africa in 1893 and 1894. It was here that Gide met Oscar Wilde and Lord Alfred Douglas and became aware of his own homosexuality; *Les Nourritures terrestres* is an encouragement to cast aside moral constraint and yield to impulse and desire. The book passed relatively unnoticed when it first appeared in 1897, but was highly acclaimed on its republication after World War I. In 1908, with other contemporary writers, Gide founded the literary journal *La Nouvelle Revue Française*.

During the early 1900s Gide produced three narrative works, which he classified as *récits*: *L'Immmoraliste* (1902; translated in 1930), a development of the themes introduced in *Les Nourritures terrestres*; *La Porte étroite* (1909; translated as *Strait is the Gate*, 1924); and *Isabelle* (1911; translated in 1931). *La Porte étroite* is a reflection of Gide's austere Protestant upbringing, in which the young hero and heroine are prevented by religious scruples from consummating their love. *Les Caves du Vatican* (1914; translated as *The Vatican Swindle*, 1925), a satirical tale (or *sotie*), was the last of Gide's prewar works.

In the years following World War I, Gide became increasingly introspective in his writings. He questioned the sincerity and validity of religious faith, ultimately pronouncing himself an agnostic, and struggled to come to terms with the problems arising from his ill-suppressed homosexuality. Notable works of this period include *La Symphonie pastorale* (1919; translated in 1949); *Corydon* (1924), a frank defence of homosexuality; and *Les Faux-Monnayeurs* (1926; translated as *The Counterfeiters*, 1927), which Gide considered to be his only real novel (on a much grander scale than his earlier *récits*). *Si le Grain ne meurt* (1926; translated as *If it die...*, 1935) is an autobiographical work dealing with the author's early life, in which he seeks to resolve some of his moral and religious conflicts. During the latter part of 1926 Gide visited French Equatorial Africa, a trip that helped to broaden his outlook and turn his attention towards social problems in the world outside. On his return he published *Voyage au Congo* (1927) and *Retour du Chad* (1928), both of which attacked French colonialism and the exploitation of African natives. A brief involvement with the Communist Party followed, ending in disillusion after a visit to the Soviet Union in 1936. Gide's last narrative work, *Thésée* (1946; translated in 1950), was written in North Africa, where he spent the second half of World War II.

Gide also wrote literary criticism, essays, and translated some of Shakespeare's tragedies. Perhaps his most outstanding work is his *Journal* (1939–50), which catalogues his thoughts, attitudes, and experiences up to the age of eighty and provides a valuable insight into one of the greatest literary minds of the twentieth century. The *Journal* is supplemented by a collection of random notes, *Ainsi soit-il, les jeux sont faits*, published posthumously in 1952.

Gielgud, Sir (Arthur) John (1904–) *British actor and director. He was knighted in 1953 and made a CH in 1977. He is also a Companion of the Légion d'honneur.*

Born in London, the great-nephew of actress Ellen TERRY, Gielgud was educated at Westminster School and studied drama at Lady Benson's School and the Royal Academy of Dramatic Art, of which he was president

(1977–89). His long and distinguished career began at the Old Vic, where he made his debut as the Herald in *Henry V* (1921). As well as acting he also became manager of the Queen's Theatre for a season in 1937. He toured military installations as a member of ENSA during World War II and, immediately after, played to SEAC forces in Burma.

One of the truly outstanding Shakespearean actors of the twentieth century, noted for the diction and tone of his fine speaking voice, he has performed Hamlet on many occasions since 1929 and his Shakespearean one-man show, *Ages of Man*, in many parts of the world. His highly successful non-Shakespearean roles include John Worthing in *The Importance of Being Earnest* and Harry in *Home* (1970), which earned him the Evening Standard best actor award and the New York Tony.

He has appeared on television and in many films, being nominated for an Oscar for his King Louis VII of France in *Becket* (1964) and winning an Oscar for *Arthur* (1980). His most recent films include *The Shooting Party* (1985), *The Whistle Blower* (1987), and *Prospero's Books* (1991). He also produced Benjamin BRITTEN's opera, *A Midsummer Night's Dream* (1961), at Covent Garden. His publications include his autobiography, *An Actor and His Times* (1979). In 1985 he received a special Olivier award for services to theatre.

Gierek, Edward (1913–   ) *Polish statesman. First secretary of the Polish United Workers Party (1970–80), he was forced to resign after a heart attack and interned and discredited by General Jaruzelski's military government. As a compliant Warsaw-Pact leader he was awarded the Order of Lenin in 1973.*

Born in Porabka, the son of a coalminer, Gierek left school at thirteen to work in the mines. After moving to France in 1923 he became active in the French trade-union movement and joined the Communist Party there in 1931. In 1937 he emigrated to Belgium, where he again worked as a coalminer. As a member of the Belgian Communist Party, he organized and led Belgian resistance among Polish miners and factory workers during World War II. After the war he founded the Union of Polish Patriots and the Polish Workers Party (1945), becoming chairman of the National Council of Poles in Belgium (1946).

Gierek returned to Poland in 1948 and joined the Polish United Workers Party in the same year; he was elected as a deputy to the national Sejm (parliament) in 1952 and gained a degree in mining engineering in 1954 from Kraków School of Mining. He gradually worked his way through the party ranks to become director of the department of heavy industry, a member of the central committee of the Politburo (1956), and first secretary of Katowice province, Poland's most industrialized region (1957). As head of this wealthy and relatively autonomous region, Gierek developed a strong power base. When, in 1970, there were riots by industrial workers over a substantial rise in food prices, Gierek took over the position of first secretary of the party from GOMUŁKA. Ten years later he was forced out of office following a heart attack, when a similiar economic crisis led to conflict between Solidarity, a coalition of independent trade unions, and the government. He was interned by the military government under General Jaruzelski (1981–82) and discredited as a party member.

Gigli, Beniamino (1890–1957) *Italian tenor, often regarded as the natural successor to Enrico CARUSO.*

He was born in Recanati near Ancona. Poverty and lack of schooling made his early life hard and frustrating, but he eventually won a scholarship to the Santa Cecilia Conservatory in Rome, where he studied singing with Enrico Rosati. In 1914 he won first prize in an international singing competition in Parma and later in the year made his debut at Rovigo as Enzo in Ponchielli's *La Gioconda*. The following year he sang Faust in Boito's *Mefistofele* to great acclaim in both Bologna and Naples. He also sang this role again under TOSCANINI at La Scala in 1918 and later that year at the Metropolitan Opera in New York, where he was engaged from 1920 to 1932 and again from 1938 to 1939. He sang frequently at Covent Garden in these periods and appeared there again in 1946 opposite his daughter Rina in Puccini's *La Bohème*.

Although his acting was rudimentary, Gigli's sumptuous voice earned him a fortune and in 1928 he built himself a palatial home outside his native town. His technique was as secure at sixty as it had been during his prime; he fortunately made many records of his most memorable roles.

Gill, (Arthur) Eric Rowton (1882–1940) *British sculptor, illustrator, typographer, and writer.*

The son of a Church of England clergyman, Gill became a Roman Catholic at the age of thirty-one and throughout his career attempted to revive a religious attitude to art. He was trained at the Chichester Art School and then at the Central School of Art in London, where he studied art and typography after being apprenticed to an architect in 1900. From 1903 he worked as a letter cutter and in 1910 began to carve figures. In 1914, the year after his conversion, he began work on fourteen relief carvings, *The Stations of the Cross*, for Westminster Cathedral. This commission was finished in 1918. From 1924 Gill illustrated such books as *The Canterbury Tales* (1927) and *The Four Gospels* (1931) for the Golden Cockerel Press and he became a major figure in the revival of typography, designing new typefaces including Perpetua and Gill Sanserif. His second most famous sculpture, *Prospero and Ariel*, was finished in 1931 for Broadcasting House, London. His best-known books are *Art* (1934) and his *Autobiography* (1940).

Gillespie, Dizzy (John Birks Gillespie; 1917–) *Black US jazz trumpeter, bandleader, and one of the founders of modern jazz in the mid-1940s.*

Born in Cheraw, South Carolina, he learnt to play the trumpet from his father. By 1955 he had become a professional musician, replacing his idol Roy Eldridge (1911–89) in Teddy Hill's band in 1937. In 1939 he was engaged by Cab Calloway as composer and arranger as well as trumpet player. From 1941 to 1943 he worked in several bands, also playing in jam sessions at Minton's Playhouse, where Teddy Hill led the house band and where the new jazz style of bop was emerging. In 1944 he was appointed musical director of the first big bop band, led by singer Billy Eckstine. This influential band did not record because of a musicians' strike but the next year Gillespie made some of the first bop records, with Charlie PARKER. His own first band failed but the second was more successful (1946–50). Since then he has toured the world almost annually, often with a quintet. However, whenever it is possible he forms a big band to play at the major jazz festivals.

Ginsberg, Allen (1926–) *US poet and leading figure of the Beat group of writers, who first came to prominence in the 1950s.*

The son of poor Jewish immigrants from Russia, Ginsberg was born in Newark, New Jersey, and educated at Columbia University, where he studied under Lionel Trilling and Mark Van Doren and won prizes for his poetry. At William BURROUGHS's apartment near Columbia he met Jack KEROUAC and others with whom he was later associated, and in 1948 had

several mystical experiences that had a lasting effect. He moved to San Francisco and in 1956 published *Howl and Other Poems*. This volume, successfully defended against a charge of obscenity, epitomized the Beats' attack on the smugness of the 1950s. Owing much to Blake and Whitman, it signalled a break with the academically respectable verse of the day and the dominant influence of the New Criticism, inspiring an alternative culture of visionary experience, drugs, and oriental forms of religious meditation. *Kaddish and Other Poems* (1961), considered his best work, and *Reality Sandwiches* (1963) followed. A talented performer, Ginsberg won an immense international following in the 1960s. He was instrumental in reviving poetry as a popular spoken art and played an effective role in various movements of the time – for civil rights, peace, gay liberation, and so on. His later work, including *The Fall of America* (1972), *Mind Breaks* (1977), and *White Shroud, Poems 1980–85* (1986), is relatively tranquil and hardly compares in intensity to his earlier books. Together with a number of former Beat poets, he is now associated with the Naropa Institute of Buddhist Studies and the Arts at Boulder, Colorado.

Giraudoux, (Hyppolyte-)Jean (1882–1944) *French dramatist, novelist, and diplomat. He was admitted to the Légion d'honneur for his service in World War I.*

Born at Bellac, Haute-Vienne, Giraudoux embarked at the age of eighteen on a long and successful career in the diplomatic service. He published his first novel, *Provinciales*, in 1909; this was followed by a series of novels written in a precious and impressionistic style, such as *Suzanne et le Pacifique* (1921), *Siegfried et le Limousin* (1922), and the gently satirical *Bella* (1926).

Giraudoux began writing for the theatre in 1928, encouraged by the actor and director Louis Jouvet; his first play, *Siegfried*, was an adaptation of his own novel. Many of his subsequent plays were derived from mythological or biblical tradition and, like his novels, contain a delicate blend of reality and fantasy. *Amphitryon 38* (1929) was so named, according to Giraudoux, because it was the thirty-eighth dramatization of that particular myth. In *La Guerre de Troie n'aura pas lieu* (1935; translated as *Tiger at the Gates*, 1955), a war-weary Hector returns to Troy determined on peace; the ensuing conflict, in this case between the peacemakers and the warmongers, is a recurring theme in Giraudoux's works. His women are at once sophisticated and disarmingly innocent; they dominate many of his plays and their love transcends all. Such is the case with Alcmène, in *Amphitryon 38*, and the gentle Isabelle in *Intermezzo* (1933; translated as *The Enchanted*, 1950).

Giraudoux's other plays include *Électra* (1937), *L'Impromptu de Paris* (1937), *Sodome et Gomorrhe* (1943), and *La Folle de Chaillot* (1945; translated as *The Madwoman of Chaillot*, 1949). The remainder of his literary output consists of critical essays, notably *Les Cinq Tentations de La Fontaine* (1938) and *Littérature* (1941), short stories, memoirs of his diplomatic career, and two film scripts.

Giscard D'Estaing, Valéry (1926–) *French statesman and president (1974–81), regarded as the principal architect of France's economic growth in the 1960s and early 1970s.*

Born in Koblenz, the son of a prominent economist, Giscard began his education at the École Polytechnique in Paris. Interrupting his studies in 1944, he joined the French army, serving in North Africa and Germany during World War II. After the war he completed his degree then attended the École Nationale d'Administration (1949–51), which provided training for senior positions in the civil service.

Giscard entered the civil service in 1952 as an assistant in the ministry of finance and economic affairs, rising to the position of inspector within two years. An active member of the conservative National Centre of Independents and Peasants, he was elected as a Gaullist deputy (for Puy-de-Dôme) to the national assembly in 1956. Following his re-election in 1958 as the member for Clermont, he became secretary of state for finance under DE GAULLE (1959) and minister in 1962. In 1965 he founded and became president of the Républicains Indépendants (changed to Parti Républicain in 1977), a pro-Gaullist faction that split from the National Centre of Independents and Peasants. Sacked from the ministry by de Gaulle in 1967 because of mounting opposition from business and labour to his stringent economic policies, he regained the finance portfolio under President POMPIDOU. Pompidou's death in 1974 paved the way for Giscard's election the same year as president, the youngest head of state in France since Louis Napoleon in 1848. He lost office in 1981 when he was defeated at the polls by MITTERRAND. Since 1989 he has been a member of the European Parliament.

Glashow, Sheldon Lee (1932–) *US particle physicist, who shared the 1979 Nobel Prize for Physics with Steven WEINBERG and Abdus SALAM for his discovery of the charmed quark and the unification of the electromagnetic and weak forces.*

Glashow was educated at the famous Bronx School for Science, where he was a classmate of Steven Weinberg. He continued his education at Cornell and Harvard, gaining his PhD there in 1959. After short periods at the Bohr Institute in Copenhagen, CERN in Geneva, and the California Institute of Technology, Glashow spent the years 1961–66 teaching at the University of Stanford and the University of California (Berkeley), before returning to Harvard in 1967 as professor of physics.

In 1964 Glashow postulated the existence of a fourth quark. Murray GELL-MANN, who had introduced the quark concept into physics, had initially recognized three quarks. Glashow, however, was struck by the lack of symmetry between quarks and leptons, as the latter group consisted of two pairs, electron and electron neutrino together with the muon and muon neutrino, interacting with each other through the weak force. If a fourth quark were to be added, a lepton–quark symmetry would result. The fourth quark, given by Glashow a quantum number to represent the property he called charm, was ignored by many physicists as speculative. Confirmation of Glashow's proposal came in 1974 with the discovery of the J particle, whose unexpectedly long life is believed to be due to its composition from a charmed quark and a charmed antiquark.

Glashow's charmed quark also had implications for the work of Weinberg and Abdus Salam, in their attempts to unite the electromagnetic and the weak force. In their original formulation the theory only applied to leptons. Glashow's contribution was to show that it could be applied to all elementary particles.

Gobbi, Tito (1915–84) *Italian baritone, who was one of the finest singing actors of his day.*

After reading law at Padua University, Gobbi studied singing in Rome with Giulio Crimi, having first to overcome the two most dreadful afflictions of a singer – a stammer and asthma. He made his opera debut at Gubbio as Rodolfo in Bellini's *La sonnambula* in 1935 and the following year won an international singing competition in Vienna and sang Germont in Verdi's *La traviata* at the Teatro Adriano in Rome. His performance of the title role in BERG's *Wozzeck* in its first Italian performance (1942) was an outstanding success. The same year he sang Belcore in Donizetti's *L'Elisir d'amore* at La Scala and made his Covent Garden debut in the same role in 1951. Gobbi particularly excelled in Verdi's baritone roles as well as in Scarpia in Puccini's *Tosca*, which he sang with Maria CALLAS in memorable performances in 1964. In 1965 he produced Verdi's *Simon Boccanegra* in Chicago and at Covent Garden. He made twenty-six films and his autobiography was published in 1979.

Godard, Jean-Luc (1930–) *French film director of the New Wave, noted for his experimental style.*

Born in Paris, Godard spent his childhood in Switzerland, where his father had a wealthy medical practice. After studying ethnology at the University of Paris he began to write criticism for *Cahiers du Cinéma*, the main organ of the French New Wave, and to make short films. Godard's feature films of the 1960s, of which *Breathless* (1960) was the first, are remarkable for their improvised dialogue, dislocated narratives, and unconventional techniques of cutting and shooting. Thematically, they deal with existentialist problems, such as the meaning of personal identity, and are much concerned with the idea of betrayal. Made cheaply, for the most part using television equipment, they were shown outside the main distribution channels on the art-house circuit.

In the late 1960s Godard's work became didactically Marxist, such films as *Le Gai savoir* (1968) and *Wind from the East* (1969) making even fewer concessions to mainstream taste than his earlier work. This political commitment continued in the 1970s, which saw Godard experimenting with television and video, mainly in Switzerland. At the turn of the decade he began to make more conventional narrative films, achieving some commercial success with *Every Man for Himself* (1979) and *Slow Motion* (1980). In 1985 *Je vous salue Marie*, a controversial updating of the Annunciation story, provoked violent protests by some Christian groups.

Over the years Godard's influence has been profound, many of his once-revolutionary techniques having been absorbed, in less extreme forms, by commercial film makers.

Goddard, Robert Hutchings (1882–1945) *US rocket engineer, who produced the first successful liquid-fuel rocket.*

Goddard was educated in his home town of Worcester, Massachusetts, at Clark University, where he gained his PhD in 1911 and became professor of physics in 1919. It soon

became apparent to Goddard that solid-fuel propellants were too heavy, bulky, and inefficient to power a space flight that he dreamed of. In his classic paper of 1919, *A Method of Reaching Extremely High Altitudes*, he therefore proposed the use of liquid fuels. He went on to design practical rockets and in 1926 actually succeeded in launching a rocket powered with petrol and liquid oxygen to a height of twelve metres. He also envisaged a multistage rocket.

Although he gained financial support from the Smithsonian, the military authorities showed no interest. Goddard therefore continued on his own and in 1937 one of his rockets attained a height of 37 kilometres. Again, with the outbreak of World War II, Goddard tried to interest the War Department in his work but rockets that could fly only a few kilometres were of no interest to a country fighting enemies thousands of miles away. Goddard did, however, live long enough to see a captured German V-2 rocket in 1945. Although it was larger and vastly more powerful than his own models, he was gratified to note the similarity in the design to his own plans. Some fifteen years after his death Goddard's widow received one million dollars from the US government for the use they had made in the space programme of the hundred or so patents Goddard had prudently taken out.

Gödel, Kurt (1906–78) *Austrian-born US logician and mathematician, who discovered one of the most important mathematical results – the undecidability of mathematics.*

Born in Brno (now in Czechoslovakia), the son of a businessman, Gödel was educated at the University of Vienna, where he gained his PhD in 1930 and joined the faculty. He left Austria in 1938 and emigrated to the USA, where he joined the staff of the Institute for Advanced Studies at Princeton, becoming professor of mathematics from 1953 to 1976. Gödel became an American citizen in 1948. An extremely cautious and distrustful man, he became intensely hostile to Austria and consistently rejected the many honours they wished to bestow on him in later life.

A previous generation of logicians, especially Gottlob FREGE and Bertrand RUSSELL, had sought to derive the whole of mathematics from exclusively logical principles and concepts. While they and their followers had achieved considerable success, it remained a limited success. It is possible to derive parts of mathematics from logic but no one had yet proved that all mathematical truths could be so derived. In 1931 the twenty-five-year-old Gödel published his *Über formal unentscheidbare Sätze der Principia Mathematica und verwandter Systeme* (translated as *On Formally Undecidable Propositions of Principia Mathematica and Related Systems*, 1962). Gödel demonstrated that in any system, such as Russell's *Principia Mathematica*, that was consistent and rich enough to contain the laws of simple arithmetic, there would always be true propositions of the system that could never be proved or disproved within the system. They were undecidable. No tinkering with the system, no additions, and no new axioms could overcome this startling defect. There was therefore no way of completing the Frege–Russell programme.

In 1938 Gödel published the solution to a second major mathematical problem. Georg Cantor, towards the end of the nineteenth century, had claimed that there is no set greater than the natural numbers but smaller than the set of real numbers. Known as the continuum hypothesis, it resisted all attempts at proof or disproof until 1938. Gödel then succeeded in showing that the hypothesis was consistent with the axioms, including the axiom of choice, of set theory. Gödel's other important results include the first proof of the completeness of first-order logic in 1930 (his doctoral thesis) and, in 1933, a new axiomatization of modal logic together with a suggestive analysis of the intuitionist logic of BROUWER. In the field of cosmology Gödel proposed a number of models of the universe including, in 1949, one that satisfied the equations of general relativity theory without incorporating Mach's postulate.

On his death Gödel left some eighty scientific notebooks running to 5000 pages written in an old-fashioned German shorthand. Although they have yet to be fully analysed it has been reported that they deal at some length with topics as diverse as theology and demonology.

Goebbels, Paul Joseph (1897–1945) *German minister of enlightenment and propaganda (1933–45) during the Nazi period.*
Born in Rheydt in the Rhineland, the son of a factory foreman, Goebbels was educated at eight universities before receiving a doctorate of philosophy from the University of Heidelberg in 1921. He intended to become a writer but became involved in Nazi politics, joining the party in 1924 and editing the Nazi newspaper *Voelkisch Freiheit*. Appointed leader of the party (Gauleiter) in Berlin in 1926, he was put

in charge of the party propaganda machine in 1929.

Goebbels was elected to the Reichstag in 1933. He became minister of enlightenment and propaganda under HITLER, controlling the whole range of communication from the press to educational and cultural activities. During World War II he developed his techniques of mass communication, using the machinery of his office to control public opinion and manipulate information about the progress of the war. By 1943 he had achieved such influence that he was virtually running the country. He supported Hitler until the end, remaining in the Führer's bunker during Hitler's suicide. Afterwards he committed suicide himself, having shot his wife and six children.

Goebbels was considered one of the best educated of Hitler's followers. An adroit administrator and powerful orator, he was completely cynical and a fervent antisemite. His diaries, published in 1948, disclose his reverence for Hitler and his contempt for the general populace.

Goering, Hermann Wilhelm (1893–1946) *German Nazi leader, creator of the Luftwaffe and the Gestapo.*
Born in Rosenheim, Bavaria, Goering fought in World War I as a pilot commanding the famous Richthofen squadron, which earned him the highest military decoration in Germany. After the war he began a degree course in economics at the University of Munich but abandoned his studies to join the Nazi Party (1922). A year later he was put in charge of the SA (stormtroopers) participating in the Munich putsch, an unsuccessful attempt to overthrow the government in which he was wounded (1923).

In 1928 Goering was elected to the Reichstag as one of twelve Nazi deputies. He became president of the Reichstag in 1932 and was appointed Prussian prime minister, minister of the interior, and air minister in 1933 after HITLER came to power. As air minister he ordered the secret rebuilding of the Luftwaffe (German air force), becoming commander-in-chief of the Luftwaffe in 1935. From the Prussian secret police, which he commanded, he established the Gestapo, setting up the first concentration camps. In 1936 he became director of the Four Year Plan for armament. He reached the height of his career in 1940 when the position of Reichsmarshal, Germany's highest-ranking officer, was created for him. Thereafter he lost favour with Hitler as a result of military defeats (particularly the Battle of

Britain) that discredited the previously prestigious Luftwaffe. Expelled from the Nazi Party and arrested in 1945 for attempting to take over the leadership, he escaped and was later captured by US troops. He was condemned to death at the Nuremberg war trials, but committed suicide by poison a few hours before his execution should have taken place.

Golding, Sir William Gerald (1911–)
British novelist. He was awarded the CBE in 1966, won the Nobel Prize for Literature in 1983, and was knighted in 1988.

William Golding was born near Newquay, Cornwall, and educated at Marlborough Grammar School and Brasenose College, Oxford. After graduating (1935) he became a schoolmaster in Salisbury until joining the Royal Navy (1940) in World War II. His war service included taking part in the hunt for the *Bismarck* and in the Normandy landings. Afterwards he returned to schoolteaching and wrote three unpublished novels before achieving instant fame with *Lord of the Flies* (1954; filmed 1963), a disturbing moral fable about marooned schoolboys slipping back into savagery.

The human capacity for evil and guilt is a predominant preoccupation in Golding's work, exemplified in his next two novels – *The Inheritors* (1955), about the extermination of Neanderthal man by *Homo sapiens*, and *Pincher Martin* (1956), about the torturing guilt suffered by a naval officer facing imminent death. Later works include *Free Fall* (1959), *The Spire* (1964), *The Scorpion God* (1971), *Darkness Visible* (1979), *Rites of Passage* (1980; which won the Booker McConnell prize), and *The Paper Men* (1984); *Close Quarters* (1987) and *Fire Down Below* (1989) were both sequels to *Rites of Passage*. His play *The Brass Butterfly* (1958) is based on his own short story 'Envoy Extraordinary'.

Goldschmidt, Victor Moritz (1888–1947)
Swiss-born Norwegian geochemist who was known as the father of geochemistry.

The son of a noted physical chemist, he was educated at Christiania (now Oslo) University, where he obtained his PhD in 1911. He became director of the Mineralogical Institute but in 1929 moved to a similar post in Göttingen. As a Jew, Goldschmidt was expelled from Germany in 1933; he consequently returned to Norway, but the Nazi invasion of Norway in 1939 made him a refugee until he managed to escape to England in 1942. He worked first at the Institute for Soil Research in Aberdeen and later at the Rothamstead Experimental Station,

Harpenden. Shortly after the war Goldschmidt returned to Oslo.

Goldschmidt sought to make use of advances in chemistry, geology, and mineralogy to develop a rigorous and predictive new science of geochemistry. He aimed to do more than describe the distribution and frequency of the various elements and compounds in the earth's crust, wishing to establish the laws that control the observed pattern. Goldschmidt published his results in his eight-volume *Geochemische Verteilungsgesetze der Elemente* (1923–38; 'The Geochemical Laws of the Distribution of the Elements').

Goldwyn, Samuel (Samuel Goldfish; 1882–1974) *Polish-born US film producer who became one of the legendary names of Hollywood.*

Goldwyn was born in Warsaw and emigrated to the USA via England at the age of thirteen. Beginning as a glovemaker's apprentice, he became a successful salesman and then entered the emerging film business, with his brother-in-law, Jesse Lasky (1880–1958), and Cecil B. DE MILLE. Their first film, *Squaw Man* (1914), was highly successful and numerous films followed. Subsequently he joined Edgar Selwyn in forming the Goldwyn Company, a name made by combining the 'Gold' of Goldfish with the 'wyn' of Selwyn. In 1918 he adopted the name Goldwyn as his own. After the creation of Metro-Goldwyn-Mayer in 1924 he became an independent producer and released his films through the United Artists Corporation. His second wife, Frances Howard, whom he married in 1925, worked with him. Their son, Samuel Goldwyn Jr (1926–), also became a producer.

An astute businessman, Goldwyn brought the best talents together: photographer Gregg Toland (1904–48), such writers as Lillian HELLMAN, Robert Sherwood (1896–1955), and Maxwell Anderson (1888–1959), and directors of the calibre of King Vidor (1894–1982), Howard HAWKS, and William Wyler (1902–81). His association with Wyler was particularly successful, resulting in the *The Best Years of Our Lives* (1946), which won a special Academy Award. Goldwyn was also very strict about the moral standpoint of his films and effectively set the standard that all Hollywood film-makers had to follow for thirty years. Ronald COLMAN, Gary COOPER, and David NIVEN were among the numerous stars he signed up. Memorable films included *Bulldog Drummond* (1929), *Wuthering*

Heights (1939), *The Westerner* (1940), and *The Little Foxes* (1941) with Bette DAVIS.

Goldwyn was famous for his unique use of English – 'A verbal contract isn't worth the paper it's written on' is one of the many Goldwynisms that have become as legendary as the man himself.

Gollancz, Sir Victor (1893–1967) *British publisher and founder of Victor Gollancz Ltd, noted also for his humanitarian work. He received a knighthood in 1965.*

Gollancz came from a Jewish family and attended New College, Oxford, until the outbreak of World War I interrupted his studies. Commissioned in the Northumberland Fusiliers as a second lieutenant, Gollancz was assigned to assist officer training at Repton School on account of his poor eyesight. He concentrated instead on teaching political philosophy to the pupils and, following the publication of a faintly antiwar article in a school magazine, Gollancz was posted elsewhere. After his discharge from the army in 1919, he joined the publisher Ernest Benn the following year; in 1928 he formed his own publishing company. Gollancz concentrated on fiction (especially crime fiction), philosophy, politics, and music, publishing such authors as Daphne du Maurier (1907–89) and Dorothy L. SAYERS. He established the distinctive yellow jackets as a familiar feature of bookshelves.

In 1937 Gollancz helped to found the Left Book Club and campaigned to expose the growing menace of fascism in Europe. After World War II he organized the Save Europe Now campaign to alleviate starvation, especially in Germany, and in 1951 he established the Association for World Peace – the forerunner of War On Want. Gollancz was a staunch supporter and sometime chairman of the National Campaign for the Abolition of Capital Punishment and also an advocate of nuclear disarmament. He served as governor of the Hebrew University of Jerusalem (1944–52). His autobiographical memoirs include *My Dear Timothy* (1952) and *More for Timothy* (1953), while his interest in religious matters resulted in *A Year of Grace* (1950) and *From Darkness to Light* (1959).

Gómez, Juan Vicente (1857–1935) *Venezuelan dictator, who was elected president in 1910 and remained in power until his death. A ruthless tyrant, he became immensely rich, earning the reputation of being the wealthiest man in South America.*

Born in Tachira state, of almost pure Indian blood, Gómez had little education and spent his early years working on ranches and farms. In 1899 he joined the successful revolutionary movement of Cipriano Castro (1858–1924) and became vice-president in the new regime established by Castro in 1902.

In 1908, when Castro was in Europe for medical treatment, Gómez seized power with the support of the army. For the next twenty-seven years he ruled Venezuela either as president or through puppets. During this period Venezuela emerged as one of the world's major oil producers, following the discovery of oil in 1918. Negotiating favourable contracts with foreign companies, Gómez was able to finance the payment of public debts and the construction of extensive public works. Throughout his years in power he managed to maintain internal stability, increase agricultural production, and build up the wealth of the country. He achieved this by making use of the army to suppress any opposition and by imprisoning thousands of opponents. Numerous unsuccessful plots were mounted against Gómez, but he remained in power until his death in 1935.

Gomułka, Władysław (1905–82) *Polish statesman. Secretary-general of the Polish Workers Party (1943–48; 1956–70), he was imprisoned for his criticism of STALIN but later reinstated. As a leader he attempted to find a Polish form of socialism.*

Born in Krosno, Gomułka worked as a locksmith and in an oil refinery before joining a socialist youth movement at the age of sixteen. Becoming a member of the Communist Party in 1926, he was active in trade-union affairs throughout the late 1920s and 1930s, being imprisoned several times. During World War II he worked with the resistance movement as a member of the illegal Polish Workers Party, becoming its secretary-general in 1943.

After the war Gomułka was appointed deputy premier and minister for the territories taken from Germany in the new government of National Unity. However, his opposition to Stalin and criticism of the Soviet Union led to his arrest and imprisonment (1951–56) as well as his expulsion from the Polish Workers Party. He was readmitted to the party in 1956, however, after the denunciation of Stalin at the Twentieth Soviet Party Congress; as a result of the Poznań riots, which gave Poland more independence from the Soviet Union, Gomułka was re-elected secretary-general. He held this office until 1970, when he was forced to resign over riots against the increases in food prices.

In defying the Soviet Union, Gomułka sought a specifically Polish form of socialism. To this end he allowed a considerable amount of land to be owned privately, halted the collectivization of farms, relaxed opposition to the Roman Catholic Church, and limited the powers of the secret police. However, the defence commitment to the Soviet Union, which he retained, limited his ability to enact popular reforms and thus contributed to his downfall.

González, Julio (1876–1942) *Spanish sculptor and painter, who pioneered sculpture in metal.*
González came from a family of metalworkers in Barcelona. He originally intended to be a painter, and after studying drawing and painting in Barcelona moved in 1900 with his brother Joan to Paris; here he became a close friend of his compatriot Picasso and produced paintings and masks in hammered bronze.

The death of his brother in 1908 shattered González and he lived a more or less solitary life as an unsuccessful painter until 1927, when he began instead to work in wrought iron. Applying the lessons he had learned while working as an apprentice welder for the Renault car company in 1918, he first produced metal masks reminiscent of negro sculpture. He worked with Picasso and taught him his technique; in turn Picasso's influence on him inspired González to move towards abstraction. He produced more open skeleton-like structures, which he thought of as 'like drawings in space', and then surrealist constructions of welded pipes and sheets. His mature style is represented by such works as *Angel* (1933), *Woman Combing her Hair* (1936), and *Cactus People* (1930–40). During World War II he began to work in plaster owing to the shortage of iron. He died in Arceuil.

Goodman, Benny (Benjamin David Goodman; 1909–86) *US jazz clarinettist and bandleader, known as the 'King of Swing'.*
Born and raised in Chicago, he joined the musician's union there at the age of thirteen. After many years of freelance work, he formed his first regular band in 1934, which was featured on a network radio show from New York. On tour the next year, the band was suddenly a huge success in Los Angeles and elsewhere with ballroom dancers. Big-band jazz, which had been played by black bands for years, became a nationwide craze.

His band remained faithful to the jazz concept. An excellent musician and jazz soloist, he insisted on adequate rehearsal time and precise playing by each of the sections. He also employed the best arrangers, such as Edgar Sampson (1907–73) and Fletcher HENDERSON. As well as the drummer Gene Krupa (1909–73) and the trumpeter Harry James (1916–83), Goodman also featured the black pianist Teddy Wilson (1912–   ) and Lionel HAMPTON – a brave innovation at that time. The band appeared in several films, such as *The Big Broadcast of 1937*, and played in the first jazz concert at Carnegie Hall in 1938. In 1939 Goodman published *The Kingdom of Swing* (with music critic Irving Kolodin). His sextet of 1939–41 was a particularly popular jazz group, with Cootie Williams (1908–85) on the trumpet and Charlie Christian (1916–42) on the electric guitar.

As an elder statesman of American music, he led many bands, large and small, all over the world, including the Soviet Union in 1962. A biographical film, *The Benny Goodman Story*, was released in 1955, taking the usual fictional liberties. Goodman also played classical music, commissioning works by COPLAND and HINDEMITH, but it will undoubtedly be for his jazz that he will be remembered.

Goossens,   Sir   (Aynsley)   Eugene (1893–1962) *British conductor and composer of Belgian descent. He became a Chevalier de la Légion d'honneur in 1934 and was knighted for services in Australia in 1955.*
Goossens was born in London, into a musical family, but was sent to school in Bruges and became a student at the Bruges Conservatory of Music in 1903. In 1904 he entered the Liverpool College of Music and from there won a scholarship to the Royal College of Music in London, where he studied the violin with Achille Rivarde (1865–1940) and composition with Sir Charles Stanford (1852–1924). Stanford arranged for a performance of Goossens's *Chinese Variations* to be conducted by the composer at the college in 1912 and later at a Promenade Concert. Goossens played the violin in a number of orchestras and chamber ensembles, including the Queen's Hall Orchestra (1912–15).

In 1915 he decided to concentrate on the career of conductor. His much acclaimed performance of STRAVINSKY's *Rite of Spring* in 1921 led to other concerts of contemporary music and he earned the reputation of being able to deputize for conductors in concerts of modern or complex music at short notice. In 1923 Goossens went to America as conductor of the newly formed Rochester Philharmonic Orchestra, the start of a long association with

US orchestras. In 1947 he was appointed director of the New South Wales Conservatory. His technical excellence on the rostrum is also apparent in his many compositions, but perhaps because compositions do not survive on technical expertise alone they are now rarely played.

His brother Leon Goossens (1897–1988) was a well-known virtuoso oboist.

Gorbachov, Mikhail Sergeevich (1931–) *Soviet statesman; general secretary of the Soviet Communist Party (1985–91) and president of the Soviet Union (1988–). He was awarded the Nobel Peace Prize in 1990 in acknowledgment of his influence in defusing East–West tensions.*

Born into a peasant family in the agrarian district of Stavropol, Gorbachov was a farm worker in his youth. Being clever and ambitious, however, he went to Moscow University in 1953 to study law, returning to Stavropol after graduating to begin his political career. He became head of the Komsomol (Young Communist League), took charge of collective farms in the area, and in 1970 became first secretary of the Communist Party in Stavropol. In 1978 he moved to Moscow to become agriculture secretary of the central committee of the Soviet Communist Party (CPSV), in which position he introduced the contract system for team payment by results; two years later he was elected a full member of the Politburo. In the four years before he succeeded Konstantin CHERNENKO as general secretary Gorbachov acquired responsibility for economic, ideological, and party matters as well as for some aspects of foreign affairs. In 1984 he made an official visit to London, where he was well received.

A dynamic and forceful character, as general secretary he embarked on a radical programme of economic and social reform, summarized as *glasnost* (openness) and *perestroika* (restructuring). Political changes introduced under the programme included the toleration of dissident views, with the establishment in 1989 of a Congress of People's Deputies to enable such views to be heard, and culminated in 1990 in the Communist Party losing its monopoly of power. Social reforms included religious toleration, increased artistic freedom, and easing of travel restrictions. In foreign policy, Gorbachov agreed major new arms-limitation treaties (1987 and 1990) with the USA, effectively bringing about the end of the Cold War. Hailed as a hero in the West, at home he was regarded with some hostility by both conserva-

tives within the Communist Party and by reformers who considered the pace of liberalization too slow. In 1989 he acquiesced in the dismantling of communist regimes in eastern Europe, culminating in the reunification of Germany in 1990. At the same time, however, he steadfastly opposed the break-up of the Soviet Union into its constituent republics, using the army to suppress nationalists in such dissident regions as Lithuania. His attempts to transform the Soviet Union's outdated industrial base by encouraging a more free-market approach seem only to have exacerbated the country's economic problems, undermining any remaining support for him among ordinary people.

In the face of gathering opposition from critics, especially Boris YELTSIN, he pushed through legislation in 1990 greatly strengthening his personal powers as president. Signs that Gorbachov may be backing away from the principles of *glasnost* caused some disquiet in the West and prompted speculation about his continuing political survival. These fears proved well-founded in August 1991 when, on the eve of the signing of a new Union treaty, he was overthrown in a coup led by hardliners fearing the break-up of the Soviet Union. The coup collapsed after several days, largely through the open resistance of Boris Yeltsin, enabling Gorbachov to return from a brief exile in the Crimea, where he had been held under house arrest. To remain in power, he was obliged to agree to a new political understanding with Yeltsin and his supporters and to dissociate himself from the discredited Communist Party, resigning as its general secretary. He went on to promise de facto independence to the Baltic republics, new parliamentary elections, more rapid progress to a market economy, and other liberal reforms, hoping to preserve the Soviet Union as a union of sovereign states with a single army and a single economy. His writings include *Perestroika* (1987) and *The August Coup* (1991).

Gordimer, Nadine (1923–) *South African novelist and short-story writer, known as an outspoken critic of apartheid. She was awarded the Nobel Prize for Literature in 1991.*

Gordimer was born in the mining town of Springs, Transvaal, the daughter of Jewish immigrants from London and Latvia. The contrast between her own privileged background – her father was a wealthy jeweller – and the conditions of the Black mine workers stirred her political conscience at an early age.

Largely self-educated, she began writing at the age 11, when her parents removed her from school after she was diagnosed as having a weak heart; her first short story was published when she was 15. She later briefly attended the University of Witwatersrand.

Gordimer published her first mature works, the stories in *The Soft Voice of the Serpent* and the novel *The Lying Days*, in 1953. Like later work, these writings deal with the corrupting effect of the apartheid system on the privileged and the oppressed alike. Novels of the 1960s and 70s included *A Guest of Honour* (1970), the Booker-prize-winning *The Conservationist* (1974), and *Burger's Daughter* (1979), a story of Communist activists that was banned in South Africa. Recent works include the novella *July's People* (1981), set in a war-torn South Africa of the near future, and *My Son's Story* (1990). Gordimer is also a prolific and much admired writer of short stories.

Although a committed advocate of Black majority rule (and member of the ANC), Gordimer eschews overt polemic, treating characters from all part of the political spectrum with the same mixture of sympathy and severity. A recurring theme is the predicament of the White liberal, caught between the claims of political commitment on the one hand and a sense of his or her impotence on the other. Gordimer's integrity has made her one of the most universally respected figures in South Africa: the award of the Nobel Prize was welcomed by both the government of DE KLERK and the ANC leadership.

Gorki, Maksim (Aleksei Maksimovich Peshkov; 1868–1936) *Russian writer and playwright.*
Born at Nizhnii-Novgorod (now Gorkii) on the Volga, Gorki was orphaned as a child and went to work at the age of eight. He read and wrote in his spare time and by 1898 had established a reputation with realistic stories dealing sympathetically with outcasts and the oppressed. His first novel, *Foma Gordeyev* (1899), dealt with the rising merchant class. Encouraged by Chekhov, Gorki wrote the play for which he is best known, *Na dne* (1901; translated as *The Lower Depths*, 1912). Set in a doss-house, it was produced by STANISLAVSKY at the Moscow Art Theatre to great acclaim, a success soon repeated in the West. In 1902 Gorki was elected to the Imperial Academy of Science and Letters, an honour instantly cancelled by the tsarist government. At first a supporter of the Social Democrats, Gorki soon came to side with the Bolsheviks. He was briefly arrested in

the revolution of 1905 but freed after intervention by western writers. In exile in the USA, he hoped to raise funds for the Bolsheviks; the tsarist embassy, however, publicized the fact that his travelling companion was not his wife and Gorki was snubbed everywhere. The novel *Mat'* (1907; translated as *Mother*), an account of how the simple uneducated mother of a young striker is converted to Marxism, was written in America and became the model of the 'socialist realism' later instituted by STALIN. From 1907 to 1913 Gorki remained abroad, living mainly in Capri. He began a lengthy series of memoirs that are of unique importance in spanning pre- and post-revolutionary periods in Russian history. They were eventually published as *Detstvo* (1913–14; translated as *My Childhood*, 1915), *V lyudyakh* (1915–16; translated as *In the World*, 1917), and *Moi universitety* (1923; translated as *My Universities*, 1952).

He returned to Russia in 1913 and supported the Bolshevik revolution. In 1917–18, however, he attacked the brutality of the Bolsheviks in his journal *Novaya Zhizn* ('New Life') and these criticisms (translated as *Untimely Thoughts*, 1968) continued until LENIN, an old friend of Gorki's, forced the journal to close. Lenin reached a compromise with him by supporting Gorki's efforts to preserve cultural life during the period of revolutionary devastation. From 1918 to 1921 he devoted himself wholeheartedly to saving individuals from arrest or starvation and initiated a massive translation project of the world's classics in order to provide employment. Exhausted and ill with tuberculosis, Gorki was persuaded by Lenin to seek treatment abroad in 1921. He maintained his Russian contacts while writing novels and publishing *Zametki iz dnevnika* (1924; translated as *Fragments from My Diary*). He made two visits to Russia (1928, 1929) but was not persuaded to return until 1931, when he received every possible honour and (in 1932) was made chairman of the Union of Soviet Writers. Although Gorki appeared to acquiesce in the Stalinist regime, his reasons for returning are uncertain, as are the circumstances of his death. Yagoda, former head of the NKVD, was accused of causing his death and executed in 1938, but it is possible that Stalin ordered his murder or that he died of natural causes.

Gorky, Arshile (Vosdanig Manoog Adoian; 1905–58) *Armenian-born US painter, who has been described as the last of the great surrealists and the first of the abstract expressionists.*

Gorky arrived in the USA in 1920 and in 1925 moved to New York in order to study and later teach at the Grand Central School of Art. He deliberately set out to emulate and master the styles of old and modern masters: the influence of PICASSO, in particular, can be seen in his early work. His first one-man show was in 1934 after he had given up teaching, and the following year he began work on an abstract mural at Newark airport for the Federal Arts Project.

The personal style for which he is best known appeared in the early 1940s in such paintings as *Garden in Sochi* (1941) and *The Liver is the Cock's Comb* (1944). It was an abstract style using biomorphic forms not unlike those of MIRÓ, characterized by thin painting, bright colours, and sinuous black lines. The surrealistic element in his work is evident in the ambiguity of its hybrid forms, which replaced the literal idiom of earlier surrealism. The accidental destruction by fire of much of Gorky's late work in 1946 began a series of misfortunes that included illness and injury and ended in the artist's suicide.

Gorton, Sir John Grey (1911–) *Australian statesman and Liberal prime minister (1968–71). He was knighted in 1977 and created Companion of the Order of Australia in 1988.*

Born in Melbourne, the son of a fruit grower, Gorton was educated in Geelong and Oxford, where he graduated in 1935 with an MA degree. He was severely wounded during World War II, during which he served in the RAAF (1940–44). Gorton joined the Liberal Party in 1949 and was elected as a senator for Victoria. He was government leader in the Senate from 1967 to 1968, when he transferred to the House of Representatives to stand for the leadership of the party, which he won. He was the first senator to be appointed prime minister.

Branded a centralist, Gorton practised a style of leadership that was presidential rather than prime ministerial. In 1971 he lost the support of Liberal Party members and resigned as prime minister. He stood unsuccessfully as an independent in 1975.

Graf, Steffi (Stefanie Maria Graf; 1969–) *German tennis player, who dominated women's tennis in the late 1980s.*

Born in the former West Germany, she was encouraged to make tennis her career by her father, who coached her for many years. Her introduction to the professional circuit was rewarded with immediate success; by 1987 she was ranked top women's player in the world

(deposing Martina NAVRATILOVA), winning eleven of the thirteen tournaments she contested that year, as well as the French Open title. Her devastating forehand was perhaps the secret of her subsequent victories in the Australian, French, Wimbledon, and US Open singles championships in 1988, making her only the fourth woman to achieve the coveted grand slam of the four major titles. She also added to her triumphs an Olympic gold medal in the women's singles. In 1989 she repeated her victories in the Australian, Wimbledon, and US Open tournaments. Her career subsequently declined as a result of the strong competition from a new generation of rising young stars and the disruption of her own private life but in 1991 she captured her third Wimbledon singles title.

Graham, Billy (William Franklin Graham; 1918–) *US evangelist who has conducted large religious meetings throughout the world.*

Born into a farming family in Charlotte, North Carolina, Billy Graham showed an early interest in Christianity. After studying for a short time at Bob Jones University, he moved to Florida Bible Institute and in 1939 he was ordained as a Southern Baptist minister. The following year he returned to college and three years later became pastor of a church in Western Springs. In the 1940s he worked with the Youth for Christ movement as an evangelist. The turning point in his career came in 1950 when, as the result of a series of highly successful evangelistic meetings in Los Angeles, the Billy Graham Evangelistic Association was formed.

Graham has conducted numerous evangelistic 'crusades' throughout the world, including several in Britain (most recently in 1989), and also in eastern Europe, South Korea, South Africa, and the Soviet Union. He is said to have preached about Jesus Christ to more people than any other person in history. Graham's style of preaching is simple but powerful; it combines a highly developed sense of theatre with the techniques of the pop festival to teach the Bible as the word of God. As a climax to his meetings, he is renowned for asking people 'to make a decision for Christ'. To make a public demonstration of their decision they are invited to get up out of their seats and give their names to members of his supporting team. There is then an efficient follow-up through their local churches.

Graham, a persuasive and convincing preacher, has undoubtedly made many converts. He also knows personally a great number of world leaders, notably US presidents, has broadcast widely, and has written many books, including *Peace with God* (1954), *World Aflame* (1965), *Angels: God's Secret Agents* (1975), and *Answers to Life's Problems* (1988).

Graham, Martha (1893–1991) *US dancer and choreographer and one of the most influential teachers of modern dance. She was appointed a Chevalier de la Légion d'honneur in 1984.*

Graham became interested in dance after seeing a performance by Ruth St Denis (1878–1968) and in 1916 she went to the Denishawn School in order to study with her husband Ted Shawn (1891–1972). Her first leading role was in Shawn's Aztec ballet *Xochitl*, and she travelled the country with the troupe until 1923, when she began to experiment in dance and teach at the Eastman School of Music. In 1927 Graham set up her own studio, the Martha Graham School of Contemporary Dance, and – with the help of musician-composer Louis Horst (1884–1964) – evolved a new dance language using flexible movements intended to express psychological complexities and emotional power.

During the 1930s Graham produced works on the theme of American roots and values, such as *Primitive Mysteries* (1931), *Frontier* (1935), and *Appalachian Spring* (1944), drawing on sources as varying as Emily Dickinson and the Brontë sisters. In subsequent years her work concentrated on making dance reveal 'the inner man' and her ballets became increasingly psychological in approach. *Care of the Heart* (1946) and *Errand into the Maze* (1947) were two of the most notable pieces from this period and Graham's reputation for controversy was gradually eclipsed by the recognition of the quality and intelligence of her troupe's performances. Until she retired as a dancer in 1970, Graham danced in most of her own works with her company and was reluctant to pass on her roles. However, renewed interest in her early works and her own concentration on lecturing and coaching her company resulted in a revival of popular interest in the Graham technique. She choreographed almost one hundred and seventy pieces, both solo and ensemble, including three films and various television performances.

Grahame, Kenneth (1859–1932) *British author of books for children, most notably* The Wind in the Willows *(1908).*

Born in Edinburgh, Grahame was orphaned at an early age and went to live with his grand-

mother in England. He was educated at St Edward's School, Oxford, before going to work in his uncle's offices in London. He joined the Bank of England as a clerk (1879) and rose to the office of secretary to the Bank (1898–1908). He narrowly escaped death in an assassination attempt by a lunatic in 1903.

Grahame began writing essays and reviews in the 1880s and in the 1890s contributed to *The Yellow Book*. His first book, a volume of essays entitled *Pagan Papers* (1893), was followed by *The Golden Age* (1895). Both collections of essays were much admired, particularly for their insight into the world of children, and in 1898 *Dream Days* appeared as a sequel to *The Golden Age*. But Grahame's claim to immortality was established with *The Wind in the Willows*, with its magical world of the riverbank and its characters Rat, Mole, Badger, and Toad. The story was drafted in letter form for Grahame's only child Alastair, who later died tragically while an undergraduate at Oxford (1920). The story was successfully dramatized by A. A. MILNE as *Toad of Toad Hall* (1929).

Grainger, Percy Aldridge (1882–1961) *Australian-born US-naturalized composer, pianist, editor, folksong collector, writer, and teacher.*

Born near Melbourne, he was brought by his mother to study in Frankfurt (1895–99) with Kwast (piano) and Knorr (composition). He later studied the piano with BUSONI in Berlin. By 1901 he had settled in London as a concert pianist, taking audiences by storm with his panache, handsome features, and easy charm. He toured in Scandinavia (to which he felt a great affinity), central Europe, Australia, New Zealand, and South Africa. His overpossessive mother travelled everywhere with him and accompanied him to America in 1914, where he spent two years as an army bandsman (1917–19), taking US nationality in 1918. His mother's suicide in 1922 so severely disorientated him that he returned to Australia in an effort to rediscover his roots. In 1928, having returned to the USA, he and the Swedish poet and artist Ella Viola Ström celebrated a spectacular wedding in the Hollywood Bowl with a concert of music composed and conducted by Grainger before an audience of some twenty thousand.

As early as 1905 Grainger had joined the English Folk Song Society and in 1906 had started to use the wax-cylinder phonograph in the collection and transcription of folksongs. This enabled him to introduce his friend

DELIUS to the folksong 'Brigg Fair', used by the latter as the theme for a set of orchestral variations (1908). His friendship with Grieg, dating from 1906, led to close professional partnership on the concert platform and in the editing of certain of Grieg's works, including the piano concerto. Most of Grainger's original compositions are neglected in favour of some of his slighter but charming pieces: *Country Gardens, Handel in the Strand, Molly on the Shore*, are pieces he originally conducted at the Balfour Gardiner Concerts at the Queen's Hall in 1912 and 1913. In later life he became deeply resentful at this public repudiation of his prodigious output for what he called 'my fripperies'.

Gramsci, Antonio (1891–1937) *Italian intellectual and founder of the Italian Communist Party.*

Born in Alès, Sardinia, Gramsci was educated at the University of Turin, where he studied history and philosophy. As a student he participated in the Socialist Youth Federation and in 1913 joined the Socialist Party. During World War I (1914–18) he studied Marxist thought and became a leading speaker and theoretician for the left. In 1919 he founded *L'Ordine Nuovo*, a paper that was influential among left-wing intellectuals and asserted that the benefits of the Russian revolution were not necessarily limited to Soviet Russia.

In 1921 Gramsci, with TOGLIATTI and several others, broke away from the Socialist Party during the Livorno conference, to form the Communist Party. He became leader of the party and was elected to the chamber of deputies three years later. However, he was arrested and imprisoned by the fascists when the party was banned in 1926. He spent the next eleven years in prison and died in Rome shortly after his release in 1937.

Gramsci is best remembered for his political and philosophical writings. Written in prison and published posthumously, they include *Lettere dal Carcere* (1947) and *Opere* (nine vols, 1947–54), which deal with the nature of the political process and the role of intellectuals in that process. His theoretical ideas and practical criticism are still influential among left-wing thinkers.

Grant, Cary (Archibald Alexander Leach; 1904–86) *British-born US film star, renowned for his performances as the handsome, suave, and slightly bemused man-about-town in a host of films.*

In his teens Grant ran away from his native Bristol to join a troupe of touring tumblers,

with whom he travelled to the USA in 1920. Three years later he returned to England, playing fairly unrewarding parts in musical comedies. When Arthur Hammerstein offered him a part in the musical *Golden Dawn*, he returned to New York. Other musicals followed, including Broadway appearances and a season with the Municipal Opera Company in St Louis, Missouri.

He made his screen debut in *This is the Night* (1932), which was followed by *Blonde Venus* (1932) with Marlene DIETRICH, and *She Done Him Wrong* (1933), in which he was the object of Mae WEST's immortal misquoted invitation 'Come up and see me sometime'. Straight romantic roles gave way to comedy, with his distinctive, slightly clipped, English accent enhancing his excellent sense of timing. *Topper* (1937), *Bringing Up Baby* (1938), *My Favourite Wife* (1940), and *Arsenic and Old Lace* (1944) were typical of his many comedy films. Memorable, too, were his performances in Alfred HITCHCOCK's *Suspicion* (1941), *To Catch a Thief* (1955), and *North by Northwest* (1959). His last film was *Walk Don't Run* (1966) and in 1970 he received a special Academy Award. On retiring from the screen Grant became an executive of a large cosmetics firm.

Granville-Barker, Harley (1877–1946) *British actor, producer, playwright, and critic.*

Born in London, Granville-Barker began his career in Harrogate and in 1891 joined Sarah Thorne's repertory company in Margate. He made his London debut at the Comedy Theatre in 1892 and subsequently joined the Stage Society, where his friendship with G. B. SHAW began. He pioneered the establishment of permanent repertory companies and a national theatre, co-writing with William Archer (1856–1924) *A Scheme and Estimates for a National Theatre* (1907). As a co-manager of the Royal Court Theatre with J. E. Vedrenne (1904–07), he presented plays incorporating social comment and greater realism, including works of Shaw, GALSWORTHY, MASEFIELD, and Ibsen, as well as his own plays. Lillah McCarthy, a member of the company, became his wife in 1906 but the marriage ended in divorce.

Granville-Barker's influence on the staging and interpretation of Shakespeare in particular was profound, beginning with his productions at the Savoy of *The Winter's Tale* and *Twelfth Night* (both 1912) and *A Midsummer Night's Dream* (1914); his *Prefaces to Shakespeare* (1927–46) remains a permanent and illuminating guide for theatrical practitioners and students alike. After World War I, during which

Granville-Barker served first with the Red Cross and then in military intelligence, he married poet and novelist Helen (Huntington) Gates (d. 1950). In the thirties he spent most of his time lecturing at Oxford and Cambridge, and in 1937 he became director of the British Institute in Paris. In 1939 he went to the USA as visiting professor at Yale and Harvard, returning to Paris after the war, where he died.

Grappelli, Stephane (1908–) *French jazz violinist, one of the few successful jazz musicians to play the violin.*

After early classical training, he turned to jazz in about 1927. In 1934 he formed the Quintette du Hot Club de Paris with the Belgian guitarist Django REINHARDT, subsequently making many recordings with the group, as well as writing arrangements and occasionally playing the piano. In 1940 he moved to England, returning to Paris in 1948 and maintaining his international career ever since. His best-known performances are the recordings with Reinhardt, in which his lyricism and sweet tone are perfectly balanced with the guitar. He also made some recordings with Yehudi MENUHIN in 1973. A charming conversationalist, Grappelli has been a popular guest on television talk shows.

Grass, Günter Wilhelm (1927–) *German novelist, poet, and playwright, recognized as one of the leading figures in contemporary German literature.*

Grass was born and educated in Danzig (now Gdańsk, in Poland). From the Hitler Youth Movement he was drafted into the army at the age of sixteen, was wounded, became a prisoner of war, and was released by the Americans in 1946. After trying various manual jobs he studied sculpture in Frankfurt and at the Düsseldorf Kunstakademie. He then spent four years in Paris, where he wrote his immensely successful first novel *Die Blechtrommel* (1959; translated as *The Tin Drum*, 1962), a picaresque tale of a dwarf whose experiences reflect Grass's own as a youth in Nazi Germany. (It was filmed in 1979.) In 1960 Grass settled in Berlin, although he subsequently undertook many lecture tours abroad.

In addition to *Die Blechtrommel*, Grass also wrote two volumes of poems and the symbolical play *Hochwasser* (1957; translated as *Flood*, 1968) before his return to Germany. *Katz und Maus* (1961; translated as *Cat and Mouse*, 1963), a short novel that was filmed in 1967, was followed by the epic *Hundejahre* (1963; translated as *Dog Years*, 1965), set in the same period as *Die Blechtrommel*. Subse-

quent novels include *Der Butt* (1977; translated as *The Flounder*, 1978) and *Kopfgeburten* (1980; translated as *Headbirths*, 1982). Other prose works include *Die Rättin* (1986; translated as *The Rat*, 1987). Of his plays, perhaps the best known is *Die Plebejer proben den Aufstand* (1966; translated as *The Plebeians Rehearse the Uprising*, 1967), a searching critique of Brechtian artistic detachment set against the background of the Berlin rising of July 1953. He also published further collections of poetry, much of which has been translated. Grass's moral concern was not confined to his writings, as his involvement in politics proved, and he gained a considerable reputation as a crusader for causes that interested him.

Graves, Robert Ranke (1895–1985) *British poet, author, and critic.*

Graves was the son of the Irish writer A. P. Graves (1846–1931) and his second wife, a niece of the German historian Leopold von Ranke. On leaving Charterhouse in 1914 he immediately enlisted in the army; his experiences in the Royal Welch Fusiliers during World War I form the basis for his brilliant memoir *Goodbye to All That* (1929; revised edition 1957). In 1919 he went up to St John's College, Oxford, where contact with other poets encouraged his determination to be a writer. Despite a young family and financial difficulties, Graves published a volume of poetry every year from 1920 to 1925. After a year (1926) teaching English at the Egyptian University, Cairo, Graves devoted himself to writing. In 1929, having parted from his first wife, he made his home in Majorca, where he has lived ever since, apart from an enforced absence during the Spanish civil war and World War II. He was Clark Lecturer at Trinity College, Cambridge, in 1954 and professor of poetry at Oxford from 1961 to 1966.

Graves's poetic publications are very numerous and astonishingly consistent in the quality of the erudite and often witty lyrics at which he excelled. In 1968 he won the Gold Medal in the Cultural Olympics and the Queen's Gold Medal for Poetry. Among his prose writings, his fictional recreations of the classical world in *I, Claudius* (1934), *Claudius the God* (1934), and *Count Belisarius* (1938) won Graves an international reputation. His interest in mythology resulted in the novel *The Golden Fleece* (1944) and in some controversial studies, notably *The White Goddess* (1948); he also published a major survey of *The Greek Myths* (1955). In *The Nazarene*

Gospel Restored (1954; with Joshua Podro) he turned his attention to primitive Christianity. Graves also published collections of essays, including *The Common Asphodel* (1949).

Graves had a distinguished record as a translator of classical authors, with versions of Apuleius' *Golden Ass* (1949), Lucan's *Civil Wars* (1956), Suetonius' *Twelve Caesars* (1956), and Terence's *Comedies* (1962). His translation and adaptation of the *Iliad, The Anger of Achilles* (1959), and a new version of the *Rubaiyat of Omar Khayyam* (1968; with Omar Ali-Shah) were more controversial.

Greene, (Henry) Graham (1904–91) *British novelist. He was made a CH in 1966, a Chevalier de la Légion d'honneur in 1969, and was awarded the OM in 1986.*

Greene was educated at Berkhamsted School, where his father was headmaster. After Balliol College, Oxford, he joined the staff of *The Times* (1926–30). A key event of this period was his conversion to Roman Catholicism (1926); the profound moral paradoxes in his adopted faith underlie most of his work, even the apparently light-hearted 'entertainments'. After the publication of his first novel, *The Man Within* (1929), Greene concentrated on his writing, apart from a visit to Liberia (1935), described in *Journey Without Maps* (1936), and a fact-finding tour of Mexico (1938) reporting on religious persecution. During World War II Greene worked for the Foreign Office, part of the time in Sierra Leone (1941–43).

Greene's first success was the 'entertainment' *Stamboul Train* (1932), and *Brighton Rock* (1938) established him as a major novelist. Other 'entertainments' included *The Confidential Agent* (1939) and *The Ministry of Fear* (1943). *The Lawless Roads* (1939) and *The Power and the Glory* (1940) drew on his Latin American experiences, while *The Heart of the Matter* (1948) is the tragedy of a colonial police officer in West Africa. A characteristic Greene preoccupation is the motives that make people commit themselves to a cause or ideology, whether religion, patriotism, or sex; this is worked out with brilliant comic effect in *Our Man in Havana* (1958) and more sombrely in *The End of the Affair* (1951), *The Quiet American* (1955), *A Burnt-Out Case* (1961), *The Comedians* (1966), *The Honorary Consul* (1973), and *The Human Factor* (1978). His last novel was *The Captain and the Enemy* (1988). Many of Greene's novels have been filmed, often using his own scripts; *The Third Man* (1949) was written as a film, directed by Carol REED.

Greene also wrote short stories and books for children. His plays, beginning with *The Living Room* (1953), are deeply concerned with religious and ethical problems; of them, his comedy *The Complaisant Lover* (1959) was perhaps the most successful. His fascination with South America (he was a member of the Panamanian delegation to Washington at the signing of the Canal Treaty in 1977) is reflected in the semibiographical *Getting to Know the General* (1984).

Greer, Germaine (1939–) *Australian-born feminist writer.*

Germaine Greer was born in Melbourne and went to Melbourne and Sydney universities before moving to England to study at Cambridge. She was appointed as English lecturer at Warwick University (1968–73), and soon afterwards achieved celebrity with the publication of *The Female Eunuch* (1970), a devastating exposé of women's frustration and subservience in a male-dominated society. Hailed as a leading figure in the women's movement of the 1970s, she wrote numerous feminist articles and gave talks and TV appearances to propound her case, including a well-publicized public debate with Norman MAILER. *Sex and Destiny* (1984) to some extent retracts her earlier optimism about the sexual revolution, presenting sex as a destructive rather than a liberating force in society. Subsequent publications include *Shakespeare* (1986), the autobiographical *Daddy, We Hardly Knew You* (1989), and *The Change* (1991).

Gregory, Jack (John Morrison Gregory; 1895–1973) *Australian cricketer widely regarded as one of the finest all-rounders in the world in the years after World War I. A fast bowler, noted slip fielder, and aggressive batsman, he played in 24 tests.*

He came from a cricketing family – his father had played for New South Wales, and his cousin, Syd Gregory (1870–1929), played 58 times for his country. Jack was a fine athlete, who also played rugby and tennis with some success. He first played for New South Wales in 1920 and made a great impact in the 1920–21 series against England. His bowling partnership with Ted McDonald became one of the most famous – and feared – in the history of the game. Gregory's bowling action was distinctive for the leap he gave before delivering the ball. He took 504 first-class wickets and scored more than 5000 runs. He toured England twice and South Africa once.

Gregory's cricketing career ended prematurely when he was only thirty-two, due to a serious knee injury that caused him to break down during a test match in 1928–29.

Grierson, John (1898–1972) *British film producer who was the founding force in the British documentary film movement.*

Born in Kilmadock, Scotland, Grierson served with the RNVR (1915–19) and gained an MA at Glasgow University (1923). He became interested in film while studying mass communications in the USA, where he met the father of documentary film-making, Robert FLAHERTY. It was while writing about Flaherty's work that Grierson coined the word 'documentary'. He also saw, and was influenced by, EISENSTEIN'S techniques in *The Battleship Potemkin* (1925). Returning to England in 1927, he became Films Officer with the Empire Marketing Board, where he directed the highly successful *Drifters* (1929), about the herring fleet in the North Sea. After this he devoted himself to producing films for the Board, including *Industrial Britain* (1933), directed by Flaherty. In 1933 Grierson moved to the GPO Film Unit, where he brought together such young talents as W. H. AUDEN and Benjamin BRITTEN. With Alberto Cavalcanti (1897–1982) as director, he produced a number of highly praised documentaries, including *Song of Ceylon* (1934), *Coalface* (1935), and *Night Mail* (1936). Moving to Canada, he was instrumental in setting up the National Film Board (1937), and he remained in Canada throughout the war. Of his many postwar ventures, which included working at UNESCO in Paris (1946) and the Central Office of Information in London (1948) and co-directing the Group Three Production Company (1950–55), perhaps the most rewarding was the Scottish TV series *This Wonderful World* (1957–68). He then returned to Canada as professor at McGill University, Montreal.

Griffith, Arthur (1871–1922) *Irish nationalist leader; founder of Sinn Féin and president of the Irish Free State (1922).*

Born in Dublin, Griffith was educated at the Christian Brothers Schools before working as a printer and in the South African gold mines (1896–98). Following his return to Ireland in 1898, he founded *The United Irishman*, a nationalist paper advocating a new status for Ireland.

In 1905 Griffith founded and became president of the Irish nationalist party Sinn Féin. Starting a newspaper of the same name in 1906 to replace *The United Irishman*, he supported the Irish National Volunteers but did not participate in the Easter Rising of 1916. Between 1916 and 1921 he was imprisoned several times but was twice elected as the Sinn Féin member for East Cavan. Becoming vice-president of the newly declared republic in 1919, he was acting president while DE VALERA was in the USA (1919–20). In 1921 Griffith led the Irish delegation that negotiated the Anglo-Irish Treaty to establish the Irish Free State. Elected president of the Dáil in January 1922, following the resignation of De Valera, who opposed the treaty, he died in office several months later.

Griffith, D(avid) W(ark) (1875–1948) *US film director, who was one of the most influential figures in the early history of the film industry.*

After a somewhat inauspicious beginning as an actor and writer, Kentucky-born Griffith turned to films in 1907. His first film as director, *The Adventures of Dollie* (1908), was followed by over 150 one-reel films. Then came the more ambitious *Enoch Arden* (1911; his first two-reel film), *Man's Genesis* (1912), and his most famous epic, *The Birth of a Nation* (1915). Though highly successful, the last of these was attacked for its apparent colour prejudice. A good deal of controversy raged around its release, to which Griffith responded with the four-part *Intolerance* (1916). Griffith not only developed and exploited cinematic techniques to the full but also worked with actors such as Mary PICKFORD, Donald Crisp (1880–1974), Dorothy Gish (1898–1968), and her sister Lillian Gish (1899–   ), to produce a style of acting less theatrical and more suitable for the camera.

During World War I Griffith made *Hearts of the World* (1918) in France and England. Subsequently he joined Douglas FAIRBANKS, Mary Pickford, and Charlie CHAPLIN in the formation of United Artists. Notable postwar films included *Broken Blossoms* (1919), *Way Down East* (1922), *Orphans of the Storm* (1922), and *Isn't Life Wonderful* (1924).

Griffith's sentimental Victorian outlook eventually put him out of business in the 1930s but, in recognition of his achievements in motion pictures, he was presented with an Honorary Oscar in 1935.

Grimond, Jo(seph), Baron (1913–   ) *British politician and leader of the Liberal Party (1956–67). He was created a life peer in 1983.*

An officer in the British army in World War II, and then secretary of the National Trust, Grimond entered parliament as MP for Orkney and Shetland (islands off the north coast of Scotland) in 1950. As Liberal leader, he advocated British membership of the European Economic Community (the Liberals were the first political party to do so) and his idea of 'copartnership in industry' between management and labour. He was opposed to Britain's maintaining an independent nuclear deterrent. The Liberals performed well in the elections of 1959 (when they more than doubled their vote and yet won only six seats) and 1964, but reverses in 1966 persuaded Grimond to resign the leadership. He retired from politics in 1983. His books include *The Liberal Future* (1959), *The Liberal Challenge* (1963), *The Common Welfare* (1978), *Memoirs* (1979), and *Britain – a view from Westminster* (1986).

Gris, Juan (José Victoriano Gonzales; 1887–1927) *Spanish painter, who was a pioneer of synthetic cubism.*

The son of a businessman, Gris studied at the School of Industrial and Fine Arts in Madrid before moving at the age of nineteen to Paris, where he arrived with a mere sixteen francs, his father having gone bankrupt. He lived in the same house in Montmartre as his compatriot PICASSO, supporting himself by producing satirical drawings for periodicals. Around 1910 he began painting seriously, adopting the cubist style of Picasso and BRAQUE in still lifes and protraits of friends. Within two years he was producing important work and exhibiting in Paris and Barcelona. He was, however, less interested in the analytical treatment of form, which was the essential quality of cubism at that time. Instead of reducing objects to geometric shapes he tended to begin with simple fragmented shapes and build up a picture from them, using brighter more decorative colour and collage. Gris thus bears much of the credit for the development of cubism's second stage – synthetic cubism.

Unlike Picasso, Gris continued to develop the cubist style throughout his artistic life. The two painters also differed in their approach to painting: Gris left nothing to improvisation but imposed severe almost scientific disciplines upon his works. In 1920 Gris had a serious attack of pleurisy and suffered from increasingly bad health until his death seven years later at the age of forty. His paintings in the 1920s are characterized by softer more descriptive lines and occasionally lack the vitality of his earlier work. Gris also created a large number of book illustrations, designed stage sets and costumes for DIAGHILEV's Ballets

Russes in the early 1920s, and produced writings on the theory of cubism.

Grivas, George Theodorus (1898–1974) *Greek-Cypriot patriot. He was proclaimed a national hero by the Greeks after the Cyprus settlement and promoted to lieutenant-general, the highest rank in the Greek army.*

Born in Trikomo, Cyprus (then part of the Ottoman Turkish empire), the son of a grain merchant, Grivas was educated at schools in Nicosia and Athens. In 1919 he entered the Royal Hellenic military academy as a cadet, joining a Greek nationalist group that favoured union of Cyprus with Greece ('enosis'). He adopted Greek nationality in 1919 and fought with the Greek army in the Asia Minor campaign, during which he was decorated for bravery. Between 1925 and 1933 he studied military techniques in France, returning to Greece to lecture at the Salonika training school.

At the outbreak of World War II Grivas joined the general staff of the Greek army. He commanded a division in the Albanian campaign (1940–41) and established an underground resistance group called X, which developed into an extreme anticommunist nationalist movement in 1945. During the 1950s Grivas led an underground campaign against British rule in Cyprus, forming the guerrilla army EOKA (National Organization for Cyprus Struggle) in 1954 and calling himself 'Dighenis Akritas' after the legendary Greek hero. He left Cyprus in 1959 after the agreement for independence was reached but returned in 1964 to command (until 1967) the Cypriot National Guard. In 1971 he returned secretly to Cyprus to organize supporters against President MAKARIOS, whom he regarded as a traitor for accepting Commonwealth membership rather than enosis. He died in hiding in 1974.

Gromyko, Andrei Andreievich (1909–89) *Soviet minister of foreign affairs (1957–85), member of the Politburo (1973–88), and president of the Soviet Union (1985–88). He was the longest serving foreign minister in Soviet history and was awarded the Order of Lenin six times.*

Born in Starye Gromyky, Byelorussia, the son of a peasant, Gromyko was educated at the Minsk Agricultural Institute and the Moscow Institute of Economics, from which he graduated in 1936. He was then a senior research scientist at the Academy of Sciences (1936–39) and at the same time edited *Voprosy Ekonomiki*.

In 1939 Gromyko joined the diplomatic service, as head of the American Division of the National Council of Foreign Affairs. He was appointed minister at the Soviet embassy in Washington (1939–43), before becoming ambassador to the USA (1943–46). Leading the Soviet delegation to the Dumbarton Oaks Conference on postwar security (1944), he participated in the Tehran, Yalta, and Potsdam conferences and was a delegate to the United Nations conference in San Francisco (1945); he was also the principal Soviet delegate to the United Nations (1946–49). In 1949 he was made deputy foreign minister, and appointed ambassador to Britain (1952–53). In 1957 he became foreign minister. He was elected a member of the central committee of the Soviet Communist Party (CPSU) in 1956 and became a member of the ruling Politburo in 1973. His appointment as president, then largely a formal position, in 1985 was widely interpreted as a manoeuvre by the new general secretary of the Communist Party, Mikhail GORBACHOV, to reduce Gromyko's influence on Soviet affairs. The process was completed in 1988, when he was stripped of all his posts and obliged to retire, one of the last of the 'old guard' to be swept away by the impact of *glasnost* and *perestroika*.

Gropius, Walter (1883–1969) *German-born architect who lived in the USA for the last thirty-two years of his life. As director of the Bauhaus and, later, professor of architecture at Harvard, he exerted an enormous influence on the development of modern architecture. The Bauhaus building at Dessau was his main contribution to the corpus of influential twentieth-century buildings.*

Born in Berlin, Gropius soon decided to follow his father into architecture. After attending technical high schools in Charlottenburg and Munich, he became an assistant to the prestigious Peter BEHRENS. In 1910 he set up his own practice, his first major commission being the Fagus shoe factory (1911) at Alfeld. This revolutionary industrial building, with its cubic outline and vast expanses of glass, aroused considerable interest. Gropius's last prewar building was the Machinery Hall at the 1914 Cologne Exhibition. War service as an airborne observer, rewarded with an Iron Cross, interrupted his architectural career. However, this was resumed after the war; in 1919 he was appointed to set up the Staatliches Bauhaus in Weimar. In 1925 this influential school moved into a remarkable building, designed by Gropius, in the town of Dessau. Here, with

KANDINSKY, MOHOLY-NAGY, and Paul KLEE, he built up a distinguished team that deeply influenced all aspects of design in Europe and America.

This bud of German culture was nipped with characteristic stupidity by the Nazis. The Bauhaus was closed down in 1933 and the following year Gropius left for England. His three years of private practice in London did not prove particularly fruitful and in 1937 he accepted the chair of architecture at Harvard University. During this period he teamed up with a group of young architects, directing them in the design of a number of buildings, including the Harvard University Graduate Centre (1950) and the US Embassy in Athens (1961). Gropius was certainly one of the most influential architects of the century, chiefly because of his understanding of the role of the architect in modern society and the use he made of modern materials of construction.

Grosz, George (1893–1959) *German-born US draughtsman and painter, who is best known as a caricaturist and satirist.*

After studying art in Dresden and Berlin, Grosz drew for satirical reviews, depicting social corruption in Germany, particularly of the Prussian military caste. His first books of drawings appeared in 1915 and 1916. His experience of military and civilian life in World War I strengthened his feeling of revulsion at contemporary bourgeois society and he became a prominent member of the Berlin dada movement after the war. In the mid-1920s he became a leading figure in the Neue Sachlichkeit (New Objectivity) movement, which combined the German postwar feeling of resignation and cynicism with an enthusiasm for social realism in art. Grosz published several books of satirical drawings in the 1920s, including *The Face of the Ruling Class* and *Ecce Homo*, for which he suffered numerous prosecutions.

In 1932 his international reputation as a left-wing satirical artist led to an invitation to teach at the US Art Students' League in New York and he settled in the USA the following year, becoming a US citizen in 1938. Although he continued to satirize bourgeois materialism, his later work, much of it in oils, was more romantic and lyrical, consisting largely of landscapes and still lifes and occasionally of apocalyptic visions. Returning to Germany in 1959, Grosz died soon after his arrival in Berlin.

Guevara (de la Serna), Che (Ernesto Guevara; 1928–67) *Argentine revolutionary*

who became a legend in his own lifetime and a model for radical students all over the world.

Born in Rosario, of Spanish-Irish descent, Guevara completed his medical studies in 1953. After travelling throughout Latin America, he became convinced of the need for violent revolution to overcome the poverty endured by the mass of the people. In 1956 he joined Fidel CASTRO in the campaign to overthrow the regime of BATISTA in Cuba. Active as a guerrilla leader, he became one of Castro's principal lieutenants, eventually serving as a diplomat and administrator in the revolutionary procommunist Cuban government, which won power in 1959. Attracted by the cause of liberation from poverty in other third world countries, he went to Africa and fought with LUMUMBA's supporters in the Congo war. In 1966 he emerged in Bolivia, where he trained and led a team of guerrillas in the Santa Cruz region. Here he was captured and executed by troops from the Bolivian army in 1967, before he could bring about the insurrection he had planned.

Guevara wrote several books about his experiences as a guerrilla, the tactics of guerrilla warfare, and his theories of revolutionary change. They include *Guerrilla Warfare* (1961), *Reminiscences of the Cuban Revolutionary War* (1963), and *Bolivian Diary* (1969).

Guinness, Sir Alec (1914–   ) *British actor, one of the most versatile actors on stage and screen. He was knighted in 1959.*

Guinness was born in London and attended the Fay Compton Studio of Dramatic Arts, making his stage debut at the Playhouse in *Libel* (1934). He went on to play a wide variety of roles, including Sir Andrew Aguecheek in *Twelfth Night*, Hamlet (in modern dress), and Jonathan Swift in *Yahoo*. For the part of T. E. LAWRENCE in *Ross* (1960) he received an Evening Standard Award and for the title role in *Dylan* (New York, 1964) a Tony Award. Subsequent stage successes included *The Old Country* (1977) and *A Walk in the Woods* (1989). He made his New York debut in 1942 in Arthur MILLER's *Flare Path*. After the war, during which he served in the Royal Navy, Guinness made his screen debut in David LEAN's *Great Expectations* (1946) playing Herbert Pocket, a role he had portrayed in his own stage adaptation of the book in 1939. Other major successes on screen include *Oliver Twist* (1948) as Fagin and the unforgettable *Kind Hearts and Coronets* (1949), in which he played eight parts. David Lean's outstanding

success, *The Bridge on the River Kwai* (1957), brought him both a British Film Academy Award and an Oscar. In 1979 he was awarded a Special Oscar for his contribution to films. His most notable television success was his portrayal of the enigmatic George Smiley in John LE CARRÉ's *Tinker, Tailor, Soldier, Spy* (1979) and *Smiley's People* (1981–82), for both of which he won BAFTA Awards. He published an autobiography, *Blessings in Disguise*, in 1985.

Guitry, Sacha (1885–1957) *French actor, director, and dramatist, who was appointed an Officier de la Légion d' honneur in 1936.*

Son of the French actor and dramatist Lucien Germain Guitry (1860–1925), Sacha was born in St Petersburg, where his parents were working at the time and where he made his stage debut at the age of five. Returning to France, he attended a succession of schools before writing his first play, *Le Page*, at sixteen; the following year he became a member of the Renaissance Theatre, Paris. His first success came five years later with the play *Nono* and this was the start of a long and successful career, during which he wrote well over a hundred plays and numerous filmscripts, often casting himself in the lead opposite his current wife (he was married five times). Early theatre successes included *Le Veilleur de nuit* (1911) and *La Jalousie* (1915). In 1920 he and his company appeared in London with such plays as *L'Illusioniste* and *Le Grand Duc*, having already established his own theatre in Paris. Other plays included *Histoire de France* (1929), *Mémoires d'un tricheur* (1935), and *N'écoutez pas, mesdames* (1942).

Guitry's first film role was in *Roman d'amour et d'aventure* (1917), after which he made many more, either as actor, director, or screenwriter, or as all three, often taking several parts, as in *Remontons les Champs-Elysées* (1938). Other notable films included *Quadrille* (1938), *Napoléon* (1955), and *La Vie à deux* (1957).

Gulbenkian, Calouste Sarkis (1869–1955) *Armenian oil magnate and philanthropist who played a major role in the initial exploitation of Middle Eastern oilfields.*

Born in Scutari, near Istanbul, Gulbenkian fled from Turkey in 1896 to escape persecution of Armenians; after living in various European countries, he settled in London in 1900. He was appointed financial counsellor to the Ottoman Embassy and also worked for the Royal Dutch Shell Oil Company, negotiating contracts with the French and Italian governments and handling various share deals. In the early 1900s Gulbenkian was instrumental in the formation of the Turkish Petroleum Company, in which he held a 40 per cent stake. He helped to negotiate the deal in 1912 and 1914, in which control was split between the Anglo-Persian Oil Company (now British Petroleum), Royal Dutch Shell, and German interests. Gulbenkian retained a 5 per cent lifetime holding in the company, which in the 1920s became the Iraq Petroleum Company. This was the origin of his nickname 'Mr Five Percent'. He also negotiated the entry of US oil companies into the Iraq and Iranian oil fields and arranged a concession for Royal Dutch Shell in the Venezuelan oil field.

Gulbenkian refused decorations offered by both the British and French governments and bequeathed his immense fortune and valuable art collection to the Calouste Gulbenkian Foundation. This has its headquarters in Lisbon and disburses funds for social and cultural projects in Portugal and elsewhere.

Gunn, Thom(son William) (1929–) British poet.

After leaving Trinity College, Cambridge, Gunn went to California (1954) and taught English at the University of California at Berkeley (1958–66), becoming senior lecturer there in 1990. Gunn was quickly recognized as one of the leading poets of his generation with the publication of *Fighting Terms* (1954); this was followed by *The Sense of Movement* (1957), *My Sad Captains* (1961), *Positives* (1966; with photographs by his brother Ander Gunn), *Touch* (1967), *Moly* (1971), *Jack Straw's Castle and other poems* (1976), and *The Passages of Joy* (1982). Much of his poetry is concerned with themes of energy, power, and movement. *Poems 1950–75; A Selection* was published in 1979, and he has also published a selection of Ben Jonson's poems (1974).

Gurdjieff, George Ivanovich (1877–1949) Russian sage and occultist, who was also accused of being a charlatan.

Gurdjieff was born in Alexandropol (the present Leninakan), the son of a Greek carpenter and his Armenian wife. His early life is somewhat obscure. There are tales of an organization known as the Seekers of Truth and travels in Tibet, where he is supposed to have stayed from 1900 to 1902, and of studies in Sufi monasteries in central Asia. In 1905 he was practising tentatively and humbly as a hypnotist and faith healer, but by 1912 he seemed to have worked out his ideas sufficiently precisely to prepare to open in Moscow his grandly titled Institute for the Harmonious Development of Man. During this period his principal advocate was P. D. OUSPENSKY. But with the outbreak of World War I and the Russian revolution, Gurdjieff's plans were delayed. He left Russia after 1917 and after some time in Constantinople and Berlin finally settled in France, where in 1922 he opened his Institute at Le Prieuré, Fontainebleau.

It proved to be highly successful, and a trip to the USA in 1924 aroused considerable American interest in his work. On examination, however, Gurdjieff's Institute offered nothing that could not be obtained from numerous other comparable regimes. Basically he provided hard physical labour for his disciples, together with lectures and specially designed exercises and dances. If these were faithfully pursued, the student would rise from the lowest level on the evolutionary ladder, the 'instinctive man', to become, at the highest seventh level, the 'complete man'. It is reported that the writer Katherine MANSFIELD did try the course in 1923 but died within a few weeks from tuberculosis.

Gurdjieff's ideas were partially revealed in two obscure works published after his death, *All and Everything* (1950) and *Meetings with Remarkable Men* (1963). He does, however, still have a considerable following.

Guthrie, Sir (William) Tyrone (1900–71) British theatre actor and director. He was knighted in 1961.

Born in Tunbridge Wells, Guthrie, whose maternal great-grandfather was the Irish actor Tyrone Power (1797–1841), made his debut at the Oxford Playhouse in 1924. After a period in radio he became director at the Festival Theatre, Cambridge (1929–30), and in 1931 directed his first play in London, *The Anatomist* at the Westminster Theatre. His long connection with the Old Vic and Sadler's Wells began in 1933 as play director; Flora ROBSON, Laurence OLIVIER, and John GIELGUD were among the actors he directed during this period. On the death of Lilian BAYLIS (1937) he became administrator of both theatres. He also directed the Shakespeare Repertory Company (1933–34; 1936–45) and scored major successes with *Measure for Measure* (1933), with Charles LAUGHTON, and a modern-dress *Hamlet* (1939), with Alec GUINNESS.

After World War II Guthrie began to develop his ideas for a closer relationship between audience and actors, which he felt was essential for productions of Shakespeare in

H

Haber, Fritz (1868–1934) *German chemist who invented the process for the synthesis of ammonia that bears his name. For this work he was awarded the 1918 Nobel Prize for Chemistry.*

The son of a merchant, Haber was educated at the universities of Berlin, Heidelberg, and Jena. He taught in Karlsruhe from 1894 until 1906, when he moved to Berlin to become director of the Kaiser Wilhelm Institute of Physical Chemistry.

Aware that no modern state could survive long without a supply of ammonia for agriculture and the explosives industry, Haber also realized that in a war Germany would find itself cut off from its supplies of natural ammonia. He therefore devised a process in which ammonia is synthesized from the elements nitrogen and hydrogen under pressure. The Haber process was essentially a laboratory process, but with the help of the industrial chemist Carl BOSCH commercial production of ammonia by the Haber–Bosch process began in 1913. Haber also worked, during World War I, on the development of poison gas for military use.

In the 1920s Haber attempted to repay the massive reparations imposed on Germany at Versailles by extracting sufficient gold from the oceans. This venture, however, failed. Haber was a patriotic German, but his service to his country over many years meant little to the Nazis as he was also a Jew. After most of his staff had either been dismissed or forced out of office, Haber himself resigned in 1933, protesting that he should be treated no differently from his fellow Jews. Later the same year he dissociated himself from the Institute and left for England. Before he could reach his destination, he died in Switzerland.

Hahn, Kurt (1886–1974) *German progressive educationalist who founded Gordonstoun public school in Scotland.*

Hahn was educated at the Wilhelms Gymnasium in Berlin and then at the universities of Oxford, Heidelberg, Freiburg, and Göttingen, where he studied classics. At the outbreak of World War I Hahn joined the German Foreign Office and in 1917 was appointed to the post of private secretary to the chancellor, Prince Max von Baden. At the end of hostilities, the prince sponsored Hahn's ambition of establishing a German boarding school at Salem, modelled on the English public school but upholding the tenets of Plato's educational philosophy, with the aim of instilling into its pupils the qualities of self-confidence and self-reliance combined with service to the community. Hahn himself served as headmaster of this coeducational school from 1920 to 1933, when he was arrested by the Nazis and imprisoned in a concentration camp for his opposition to HITLER's policies.

When he was released, on the personal intervention of Ramsay MACDONALD, Hahn fled to the UK and set up a new school, which eventually made its home at Gordonstoun in Scotland. As the school's headmaster for the next twenty years, Hahn was able to develop the educational principles he had applied at Salem and acquired a reputation as an eccentric and idealist. He was elected a member of the Headmasters' Conference in 1944, and after World War II the school expanded rapidly and became a pattern for similar schools in other parts of the world. Among the school's more illustrious pupils was Prince PHILIP, Duke of Edinburgh, who – impressed by the system – saw that his sons went there also. Hahn was instrumental in the establishment of the Duke of Edinburgh's Award Scheme, the Outward Bound schools, the Atlantic College, and the Trevelyan Scholarships.

Hahn, Otto (1879–1968) *German chemist, one of the discoverers of nuclear fission, for which he was awarded the 1944 Nobel Prize for Chemistry.*

The son of a successful merchant, Hahn had to overcome considerable family opposition before he was allowed to pursue a scientific career. He was educated at the University of Marburg, where he received his doctorate in 1901. After some years abroad working with William Ramsay (1916–) in London and Ernest RUTHERFORD in Canada, Hahn returned to Germany in 1907 to join the chemistry department of Berlin University. Although he became professor of chemistry in 1910, Hahn left in 1912 for the Kaiser Wilhelm Institute of Chemistry where he spent the rest of his career.

Although he had intended to become an industrial chemist, Hahn was so engrossed in radioactivity that he turned instead to academic chemistry. For some thirty years, in collaboration with the physicist Lise MEITNER, he explored the chemistry of the newly discovered radioactive elements. Together they discovered (1917) the new element protactinium. Their most important work, in collaboration with Fritz Strassmann (1902–80), was carried out in the 1930s on uranium. When uranium was bombarded with slow neutrons they repeatedly found barium in the decay products. It was left to Meitner, and her nephew Otto FRISCH, to interpret the reaction as nuclear fission.

Hahn himself contributed little to the German war effort and would have nothing to do with nuclear weapons. In 1957, with seventeen other leading German atomic scientists, he signed a declaration stating that he would never work on the production or testing of atomic weapons.

Haig, Douglas, 1st Earl (1861–1928) *British army general who commanded British forces on the western front in the latter half of World War I. He also founded the British Legion. He was awarded the OM and created Earl Haig of Bemersyde in 1919.*

The son of a landed family from the Scottish Borders, Haig attended Brasenose College, Oxford, and the Royal Military College, Sandhurst, and was commissioned into the 7th Hussars in 1885. He served under KITCHENER in the Sudan (1898) and commanded a cavalry column in the later stages of the Boer War, being promoted to colonel. Thereafter, Haig was posted to India as inspector-general of cavalry and was chief of staff of the Indian army, with an intervening spell at the War Office (1906–09). In 1912 he was appointed GOC at Aldershot. At the outbreak of World War I, Haig's 1st Corps spearheaded the British Expeditionary Force in France and in 1915 he took command of the British 1st Army. He replaced General FRENCH as commander-in-chief of British forces in December. Between July and November 1916, Haig carried out a war of attrition against German positions along the Somme. His largely civilian volunteer force sustained terrible losses (400 000 casualties, including 90 000 dead), which provoked criticism at home, particularly from LLOYD GEORGE. Haig used similar tactics the following year at Arras and at Ypres, where he captured the town of Passchendaele on 6 November. But in March 1918, the British

front collapsed following a German offensive and Allied command was restructured under General FOCH. Haig remained British C-in-C largely through his personal connections with the Royal Family.

After the war, Haig was appointed C-in-C of Home Forces. Two years later, in 1921, he helped found the British Legion, to improve the welfare of ex-servicemen and their families. He served as its first president and was also chairman of the United Services Fund.

Haile Selassie I (1892–1975) *Emperor of Ethiopia (1930–36; 1941–74), who made his country a prominent force in Africa and in world affairs.*

Son of the chief adviser to Emperor Menelik II, Ras (Prince) Tafari Makonnen (as Haile Selassie was born) became a provincial governor and was associated with the movement for reform. When Menelik's daughter Zauditu became empress in 1917, Ras Tafari was appointed regent and heir apparent. In 1923 he negotiated his country's membership of the League of Nations and was the first Ethiopian ruler to travel abroad. On Zauditu's death in 1930, he was crowned Emperor Haile Selassie (Might of the Trinity). He established a personal autocracy but simultaneously took some steps towards modernizing his country. In 1935 Italy invaded Ethiopia and Haile Selassie, having himself, as the warrior-emperor, taken up arms on the battlefield, fled to England. The Allies restored him in 1941, and he set about rebuilding his former autocracy and the country's prominence. After the establishment of the Organization of African Unity in 1963, Addis Ababa became its centre. An attempted coup was aborted in 1960, but a second military rising in 1974 deposed Haile Selassie, and he was imprisoned in his palace. The monarchy was then abolished, and in the following year the emperor died in unexplained circumstances. From the mid-1950s he became the focus of the Rastafarian religious cult, which regarded him as a divine being who would lead the world's black population back to an earthly paradise in Africa. The name Rastafarian is derived from Ras Tafari.

Haitink, Bernard Johann Herman (1929–) *Dutch conductor who has established an international reputation.*

After studying the violin at the Amsterdam Conservatory and conducting with Felix Hupka, he became a violinist with the Netherlands Radio Philharmonic. During this time he attended conductors' courses led by Ferdinand Leitner, organized by the Netherlands Radio

Union (1954; 1955). In 1955 Haitink was appointed second conductor with the Radio Union and in 1957 principal conductor of the Netherlands Radio Philharmonic. In 1956 he made his name overnight as a last-minute substitute conductor with the Concertgebouw Orchestra in a performance of Cherubini's *Requiem*. This was followed by his US debut with the Los Angeles Symphony Orchestra (1958) and his British debut with the Concertgebouw (1959). In 1961 Haitink became the Concertgebouw's youngest-ever principal conductor, in joint appointment with Eugen Jochum (1902–87); in 1964 he took over sole responsibility. His connection with Britain continued as guest conductor with the London Philharmonic Orchestra (1964) and its principal conductor and artistic adviser (1967–79). He subsequently became music director at Glyndebourne (1978–88) and at Covent Garden (1987–   ).

Haldane, J(ohn) B(urdon) S(anderson) (1892–1964) *British geneticist, biometrician, and philosopher who made valuable contributions to the physiology of respiration and chromosome mapping but above all to popularizing science and emphasizing its political and social context.*

The son of the eminent physiologist John Scott Haldane (1860–1936), whom he assisted as a child, and nephew of Richard Burdon HALDANE, J. B. S. Haldane was born in Oxford and educated at Eton and New College, Oxford, until World War I interrupted his studies. During his war service he observed first-hand the effects of gas warfare and was involved in designing a more effective gas mask. In 1919 he became a fellow of New College and taught physiology. He caused a stir by using himself as guinea pig while investigating the changes in blood composition associated with respiration. Two years later he was appointed reader in biochemistry at Cambridge University, working under Frederick Gowland HOPKINS on enzyme reaction kinetics. Haldane's innovative mathematical treatment of the subject is described in *Enzymes* (1930). Concurrent with this, he was studying genetic linkage and formulating a mathematical theory of natural selection – *The Causes of Evolution* (1932). Haldane was appointed professor of genetics at University College, London, in 1933 and later became professor of biometry (1937–57). He worked on linkage maps of chromosomes in various species, including the human X-chromosome, in which he pinpointed the relative positions of deleterious mutants such as those

responsible for haemophilia and colour blindness.

The outbreak of the Spanish civil war prompted Haldane to join the Communist Party in 1938 and he was for several years chairman of the editorial board of the *Daily Worker*, for which he also wrote numerous articles about science. *The Marxist Philosophy and the Sciences* (1938) describes Haldane's position at this time. During World War II, he worked for the British Admiralty, investigating the physiological effects of deep-sea submersion, often at great personal risk. In protest at the Anglo-French invasion of Suez, he emigrated to India in 1957, working for a time with the Indian Statistical Office, Calcutta, before establishing a genetics and biometry laboratory in Bhubaneswar, Orissa province.

Haldane, Richard Burdon, 1st Viscount (1856–1928) *British lawyer and politician, distinguished for his military reforms before World War I. He was created a viscount in 1911.*

Born and educated in Edinburgh, Haldane was called to the English bar in 1879 and became a queen's counsel in 1890. A Liberal MP from 1885 to 1911, Haldane achieved his most important work while he was secretary of war (1905–12). In 1907 he created the Territorial Force (later the Territorial Army), which facilitated the immediate dispatch of the British Expeditionary Force to France at the start of World War I. In 1909 he set up the Imperial General Staff, having visited Germany three years earlier to study the workings of the German general staff. As Anglo-German relations deteriorated, Haldane again visited Germany (1912) in a vain attempt to ease tensions. Appointed lord chancellor in 1912, he created additional lords of appeal to speed up the legal system, but was dismissed in 1915 for allegedly pro-German views. By the end of the war he had joined the Labour Party, and in 1924 Ramsay MACDONALD appointed him again to the lord chancellorship. Haldane was also a philosopher, the author of *The Reign of Relativity* (1921). An associate of Sidney and Beatrice WEBB, he helped found the London School of Economics in 1895.

Hall, Sir Peter Reginald Frederick (1930–   ) *British theatre manager and director. He was knighted in 1977.*

Born in Bury St Edmunds, Suffolk, Hall was educated at St Catherine's College, Cambridge, where he directed a number of university productions. He began his career as director at the Theatre Royal, Windsor, in

1953. The following year he was appointed assistant director of the Arts Theatre, London, and became director there in 1955, directing the English premiere of BECKETT's *Waiting for Godot* (1955). Hall then founded his own company, the International Playwrights' Theatre at the Phoenix (1957). In the same year he directed his first opera at Sadler's Wells, *Cymbeline* at the Shakespeare Memorial Theatre, and *The Rope Dancers*, his first Broadway production. He became managing director of the Royal Shakespeare Company (1960–68), after which he joined its board of directors. He directed numerous Shakespearean and non-Shakespearean plays for the company at Stratford, the Aldwych in London, and on tour abroad. From 1973 to 1988 he worked with the National Theatre, first as assistant director and then director.

Hall has directed a number of films, most notably *A Midsummer Night's Dream* (1968), *Akenfield* (1974), and *She's Been Away* (1989) as well as the television productions of *The Wars of the Roses* (1965) and several operas, including *Carmen* (1985) and *The Marriage of Figaro* (1989). He was artistic director of the Glyndebourne Festival (1984–90) and has received many awards for his work in both theatre and opera. In 1988 he left the National Theatre to found the Peter Hall Company, whose productions have included *Orpheus Descending* (1988) and *The Wild Duck* (1990). *Peter Hall's Diaries* (edited by John Goodwin) were published in 1983.

Hammarskjöld, Dag Hjalmar Agne Carl (1905–61) *Swedish secretary-general of the United Nations (1953–61). He was awarded the Nobel Peace Prize posthumously in 1961.*

Born in Jönköping, Sweden, the son of a judge and former prime minister, Hammarskjöld was educated at Uppsala University, where he graduated with arts and law degrees in 1928 and 1930. In 1933 he became an assistant professor at Stockholm University, gaining a doctorate in 1934. He established a reputation for expertise in economic matters after working as secretary of the Bank of Sweden (1935–36). He later became chairman of the bank (1941–48) and undersecretary to the ministry of finance (1936–45).

In 1946 Hammarskjöld became economic advisor to the ministry of foreign affairs. He participated at the Paris Conference in 1947, which established the MARSHALL Plan, and was the chief Swedish delegate to the Organization for European Economic Cooperation (1948–49), becoming foreign affairs minister

in 1951. After serving as Swedish delegate to the Council of Europe and the United Nations General Assembly (1951–53), he was elected secretary-general of the United Nations in 1953. As secretary-general he was influential in the establishment of a UN emergency force in Sinai and Gaza (1956); he also initiated peace moves in the Middle East (1957–58) and sent observers to Lebanon (1958). He died in an air crash in Zambia in 1961 while negotiating a resolution of the Congo crisis. His journal *Markings*, which contains personal reflections and meditations about his role as secretary-general, was published posthumously.

Hammerstein II, Oscar Greeley Clendinning (1895–1960) *US lyricist, playwright, and one of the most influential writers of musical plays.*

Son of a prominent American opera impresario, he became a stage manager in the theatre and began writing plays in 1917. He wrote the book or lyrics, or both, collaborating with composers Vincent Youmans (1898–1946), Rudolf Friml (1879–1972), Sigmund Romberg (1887–1951), Jerome KERN, and, most successfully, Richard RODGERS. The Rodgers and Hammerstein musicals include *Oklahoma* (1943), *South Pacific* (1949), and *The Sound of Music* (1959), many of which were subsequently made into films.

Hammett, (Samuel) Dashiell (1894–1961) *US writer of detective novels that are the original and outstanding examples of the tough realistic type of crime fiction.*

Hammett was born in Maryland and spent his youth in Baltimore and Philadelphia. After leaving school at fourteen, he eventually found work in the famous Pinkerton's Detective Agency, where he acquired the experience that distinguishes his novels: an unerring sense for authentic speech and character and a habit of acute observation. In World War I he served as a sergeant. Afterwards he continued to work for a time at Pinkerton's, married, and had two children. He eventually left his family, however, and started writing: in 1930 he met Lillian HELLMAN, who remained his friend for the rest of his life. Hammett's major novels were all published between 1929 and 1932. His fictional heroes are dedicated to seeing justice done in a society whose corruption is rendered with detachment and in convincing detail. *Red Harvest* (1929), *The Dain Curse* (1929), *The Maltese Falcon* (1930; filmed 1941 with Humphrey BOGART), and *The Glass Key* (1931; filmed 1935 and 1942) were followed by the comic *The Thin Man* (1932; filmed 1934).

Hammett became a celebrity in New York and Hollywood, where he wrote film scripts for a time, but his writing career was virtually finished by the time he was forty. He suffered from heavy drinking but managed to control it after 1948. A victim of the McCARTHY witchhunt of the 1950s, he spent six months in prison because of his political views; although he resolutely refused to answer questions regarding communist affiliations, it is likely that he was a party member. Certainly one cause of the writing block of the final decades of his life was his refusal to suppress deeply held political beliefs and adjust his writing to the bland conformity of the EISENHOWER years. He tried to write an autobiographical novel but failed to complete it before his death. This unfinished work was printed in *The Big Knockover and Other Stories* (1966), edited by Hellman.

Hammond, Dame Joan (1912–) *New Zealand soprano. She was created DBE in 1974.*

After studying singing and the violin at Sydney Conservatory, she played the violin for three years with the Sydney Philharmonic Orchestra. She made her operatic debut in 1929 with the Williamson Imperial Grand Opera Company in Sydney and thereafter studied singing in London (with Dino Borgioli) and also in Vienna. Hammond made her London debut in a recital in 1938. From 1942 to 1945 she sang with the Carl Rosa Opera, her roles including Violetta, Mimi, Tosca, and Cio-cio-san; she made her Covent Garden debut in 1948 singing Leonora in Verdi's *Il trovatore*. At Sadler's Wells in 1951 she sang Elizabeth in Verdi's *Don Carlos* and in 1959 took on the title role in Dvořák's *Rusalka* there (the first professional British stage performance of that opera). She also sang in New York, with the City Centre Opera (1949), and in Moscow and Leningrad (now St Petersburg), singing (in Russian) Tatyana in Tchaikovsky's *Eugene Onegin*. She sang Dido in Purcell's *Dido and Aeneas* at the 1959 Bath Festival. As well as opera, Hammond's repertory included a wide range of choral works and oratorio; her strong vibrant voice, excellent projection, and expressive interpretation can be heard in many recordings, of which the aria *O mio babbino caro* from Puccini's *Gianni Schicchi* sold over a million copies and won her EMI's Golden Disc of 1969. She retired in 1965 and published her autobiography, *A Voice, A Life*, in 1970.

Hammond, Wally (Walter Reginald Hammond; 1903–65) *Gloucestershire and England cricketer who scored more than 50 000 first-class runs, including 167 centuries. He played in 85 test matches, 20 of them as captain, and many would nominate him as England's greatest batsman of all time.*

Born in Dover, Kent, Hammond was the son of a regular soldier and spent his early childhood in Malta, where his father was stationed before World War I. When the family returned to England in 1914, young Hammond attended the grammar school at Cirencester and in 1926 started to play for Gloucestershire; he remained with that county until 1946, apart from the war years and an isolated appearance in 1951. He first played for England in 1927 and his spectacular progress as a test cricketer earned him the captaincy of England in 1938 and again after the war until his retirement in 1947. In 1932 against the New Zealanders at Auckland he scored an undefeated 336 and he was also a fine bowler as well as one of the most brilliant slip fielders in the history of the game. In all, he scored 7249 runs and 22 centuries for his country.

Hampton, Lionel (1913–) *US black jazz vibraphonist, drummer, pianist, singer, and bandleader. A highly extrovert performer, he has been one of the best-loved American musicians for nearly fifty years.*

Born in Louisville, Kentucky, he was brought up in Chicago and moved to California in 1927. He became a professional musician at sixteen, playing the drums on records and in films with the bands of Les Hite (1903–62) and Louis ARMSTRONG. In 1930 he made his first record playing the vibraphone, but his real success came as the vibraphonist in Benny GOODMAN's quartet (1936–40). Besides recording with Goodman, Hampton made about a hundred small-group jazz records for Victor (1937–41), hiring the best people who happened to be available for each recording session and often using men from the bands of Goodman, Count BASIE, or Duke ELLINGTON. These classics were subsequently reissued.

Since 1940 Hampton has led his own big bands on tours all over the world. He has appeared in many films, played himself in *The Benny Goodman Story* (1955), and was featured at President CARTER's White House Jazz Party in 1978.

Hancock, Tony (Anthony John Hancock; 1924–68) *British comedian.*

Born in Birmingham, Hancock began his acting career in the Royal Air Force (1942–46), when he appeared with ENSA and Ralph Reader's Gang Show. After the war he worked in pantomime, summer shows, and at the Windmill Theatre before appearing in such BBC radio shows as *Workers' Playtime* and *Educating Archie* (1951–53). Success came in 1954 with his highly popular radio series *Hancock's Half Hour*, scripted by Alan Simpson and Ray Galton, in which he played opposite Sid James (1913–76). The sardonic wit of the apparently materialistic, yet essentially lonely, misfit he portrayed, 'Anthony Aloysius St John Hancock' of 23 Railway Cuttings, East Cheam, found a ready audience in the fifties. The series readily adapted to television – the image of the pretentiously grand black homburg and fur-collared overcoat became his hallmark and the Hancock catch phrase, 'Stone me', became part of the language. In the early sixties he made a couple of films, *The Rebel* (1961) and *The Punch and Judy Man* (1963), and had cameo parts in *The Wrong Box* (1966) and *Those Magnificent Men in Their Flying Machines* (1965).

A deeply unhappy man, both of whose marriages ended in divorce, he gradually succumbed to alcohol and depression. In 1968 he took his own life in Sydney, New South Wales, where he had gone to make an Australian TV series.

Handley Page, Sir Frederick (1885–1962) *British aeronautical engineer, founder of Handley Page Ltd. and builder of the Halifax bomber. He was knighted in 1942.*

Born in Cheltenham, Handley Page was educated at the City and Guilds Engineering College in London, where he studied electrical engineering. Fascinated by aviation, he turned down an offer from Westinghouse in the USA and in 1909 founded his own aircraft company, Handley Page Ltd. His first major triumph came in World War I with his 0/400 in 1918, a two-engined bomber capable of delivering a bomb load of 1800 lb. With the end of the war Handley Page formed a transport company, converting his military planes for use as airliners to be flown around Europe and the Empire. In 1924 he joined with several other airlines to form Imperial Airways. It was for Imperial Airways that Handley Page produced the Heracles in 1930, the first forty-seater airliner. With the start of World War II Handley Page Ltd. reverted to the production of military aircraft. Their most successful plane was the Halifax bomber, of which nearly seven thousand were built. In the postwar years Handley Page extended their range of military planes with the introduction in 1952 of the Victor, a jet bomber.

The days of Handley Page Ltd. were numbered, however. During the 1950s the policy of the various governments of the day was to urge aircraft manufacturers to amalgamate in order to meet the huge developmental costs of new models. Only Handley Page, of the large independent producers, refused to comply. Consequently, with government funds directed to the newly formed Hawker Siddeley group and the British Aircraft Corporation (BAC), Handley Page were unable to survive. The founder died in 1962 and in 1970 the firm went into liquidation.

Hardie, (James) Keir (1856–1915) *British politician and chief founder of the Labour Party.*

A coalminer in his native Lanarkshire, Scotland, from the age of ten, Hardie was active in miners' organizations before becoming an independent Labour MP in 1892. When the Independent Labour Party was formed in 1893, he was its first chairman, and in 1900 he helped create the Labour Representation Committee, precursor of the Labour Party. Hardie lost his parliamentary seat in 1895 but was re-elected in 1900 as MP for Merthyr Tydfil in Wales. From 1906 to 1907 he was the first leader of the parliamentary Labour Party. Before the outbreak of World War I, he tried, unsuccessfully, to persuade the Second International to declare a general strike in all countries, should war occur. His pacifism received another blow when the bulk of Labour MPs gave their support to British participation in World War I, and he withdrew from Labour politics. He remained an MP until his death the following year.

Hardy, Godfrey Harold (1877–1947) *British mathematician, best known for his work in analysis and number theory.*

The son of a schoolteacher, Hardy was educated at Winchester and Trinity College, Cambridge. He remained at Cambridge as a fellow of Trinity until 1919, when he moved to Oxford as Savilian Professor of Geometry. He returned to Cambridge in 1931 and occupied the Sadleirian Chair of Pure Mathematics until his retirement in 1942.

Hardy set himself the major task of introducing much-needed rigour into British mathematics. To this end such works as his *Course of Pure Mathematics* (1908) did much to revitalize and reform the archaic system of Cambridge mathematics. An ally in his campaign was J. E. Littlewood (1885–1977) with whom, beginning in 1911, he collaborated on nearly a hundred papers. Two years later Hardy discovered the great natural talent of the largely self-taught mathematician, Srinivasa Ramanujan (1887–1920). 'All my best work ... has been bound up with theirs', he later wrote. They tackled, among other work, problems in Diophantine analysis, inequalities, convergent series, definite integrals, and number theory. They also mounted a major campaign against the still unproved Riemann hypothesis. In 1908 he formulated the Hardy–Weinberg law, an important principle in the field of population genetics, but later dismissed it as mathematically trivial.

A radical in politics and an atheist in religion, Hardy was also a cricket fanatic. In his later life he published two remarkable and stylish works. The first, *A Mathematician's Apology* (1940), presents Hardy's views on the nature and value of mathematics; the second, *Bertrand Russell and Trinity* (1942), gives a fascinating account of the dismissal in 1916 of Russell from his Trinity College lectureship. Something of Hardy's style and aspirations can be seen in a mock-serious list of new-year wishes recorded in the 1920s. At the top of the list was his hope of proving the Riemann hypothesis followed by wishes to score 211 in the final innings of the last test at the Oval, to discover a conclusive argument for the nonexistence of God, and to be the first president of the USSR in Britain.

Harlow, Jean (Harlean Carpenter; 1911–37)
US film star, the platinum blonde wise-cracking Hollywood sex symbol of the 1930s.

Born in Kansas City, Missouri, she eloped at sixteen into a short marriage. Her film career began inauspiciously as an extra, followed by appearances in a few Laurel and Hardy pictures. She featured in *The Saturday Night Kid* (1929) but the real breakthrough came in Howard HUGHES's *Hell's Angels* (1930). *The Public Enemy* and *Platinum Blonde* (both 1931) were among those that followed. In 1932 she joined MGM, where her image was further sharpened. With Clark GABLE she made *Red Dust* (1932), *Hold Your Man* (1933), *China Seas* (1935), *Wife vs Secretary* (1936), and her last film, *Saratoga* (1937). Other films included *Riffraff* and *Libelled Lady* (both 1936) with Spencer TRACY.

Harlow's personal life was as tempestuous as it was short. She and her career survived her marriage (1932) to director Paul Bern (1889–1932). He did not; his suicide caused a considerable stir in Hollywood. Her marriage (1933) to lighting director Harold Rosson (1895–1960) lasted twelve months. Finally,

she formed a close relationship (1934) with William Powell (1892–1984), with whom she played in such films as *Reckless* (1935). Her death from kidney failure, while making *Saratoga*, brought to an end a glamorous but very unhappy life. She was the subject of *Harlow*, starring Carroll Baker (1931–), and a TV film starring Carol Lynley (1942–), both made in 1965.

Harriman, W(illiam) Averell (1891–1986)
US diplomat.

Harriman was born in New York City, the son and heir of the financier and railway magnate Edward Henry Harriman. He became vice-president of the Union Pacific Railroad Company in 1915 and served as chairman of the board from 1932 to 1946. In 1920 he set up a private banking firm, which merged with Brown Brothers in 1931. An active Democrat during Franklin D. ROOSEVELT's administration, Harriman made his first diplomatic mission in 1941, visiting Britain and the Soviet Union to coordinate lend-lease aid. As ambassador to the Soviet Union (1943–46), Harriman was involved in a number of major wartime conferences and adopted a realistic and unsentimental approach to postwar relations with eastern Europe. He went on to serve as ambassador to Britain in 1946 and secretary of commerce to President TRUMAN (1947–48). In 1955 he was elected governor of New York, a post he held until 1958. During President KENNEDY's administration Harriman was appointed ambassador-at-large (1961) and served as assistant secretary of state for Far Eastern affairs (1961–63). He was one of the chief negotiators of the Nuclear Test-Ban Treaty, signed in 1963 by Britain, the USA, and the Soviet Union. In 1965 he entered the service of a fourth US president, Lyndon B. JOHNSON, and was appointed to lead the US delegation at the Paris peace talks on Vietnam (1968–69). Having retired from active politics in 1969, Harriman set down his personal observations on international affairs in such works as *America and Russia in a Changing World* (1971).

Harris, Sir Arthur Travers (1892–1984)
British marshal of the RAF and head of Bomber Command during World War II. Known as 'Bomber' Harris, he instituted the controversial policy of area bombing over German cities. He was created a baronet in 1953.

The son of an Indian civil servant, Harris was in southern Africa at the outbreak of World War I and joined the Rhodesia regiment as a bugler. In 1915 he joined the Royal Flying Corps and ended the war in the rank of major. Now with the newly formed RAF, he was posted to India, Iraq, and Egypt, was promoted to group-captain in 1933, and served as deputy director of plans at the Air Ministry (1934–37). In 1939, Air Vice-Marshal Harris was appointed air officer commanding for 5 Bomber Group and the following year he became deputy chief of air staff. At this stage in the war, the targets for Britain's bomber aircraft were principally well-defined strategic installations, such as rail termini, oil installations, and steel works. After heading an RAF mission to the USA and Canada in 1941 to coordinate air operations, Harris was appointed commander-in-chief of Bomber Command. The limited effects of precision attacks led to a change of policy from late 1941. Now German towns and cities became the targets, thereby hitting the morale of the civilian population as well as inflicting major economic damage. Enthusiastically supported by CHURCHILL, Harris presided over a massive increase in the strength of Bomber Command, mobilizing huge numbers of aircraft in night-time area-bombing raids. On 30 May 1942, the first '1000 raid' took place, when 1046 aircraft took off for Cologne. Reconnaissance was greatly improved and bomb size and destructive power increased. However, as the scale of German civilian casualties became known in Britain, the morality of Harris's policy was repeatedly questioned. Both he and Churchill remained firmly committed.

After the war, Harris was one of the few top military leaders not to have been offered a peerage, which was perhaps a reflection of the national guilt still attached to his wartime policies.

Harrison, George (1943–   ) *British guitarist, singer, and songwriter. He played the lead guitar and sang in the Beatles (1962–70).*
Born in Liverpol, Harrison played in local skiffle groups from about 1956. One of these, the Quarrymen, with John LENNON and Paul McCARTNEY (and later Ringo STARR), became the enormously successful Beatles. That the group was reluctant to record Harrison's songs was one of the reasons it broke up in 1970, although a Harrison song ('Something') was a Beatles hit in 1969. In his solo career, his top-of-the-charts song 'My Sweet Lord' (1970, from the album *All Things Must Pass*) led to a lawsuit; the court found that he had unconsciously borrowed the tune from a 1963 song called 'He's So Fine'. In 1971 he pro-

duced benefit concerts for the people of Bangladesh; the resulting three-record set, with many guest stars, won an EMI award. More than $10 million was raised for Bangladesh but most of it was held up until 1981, while the Internal Revenue investigated the Beatle company, Apple. His 1974 recording, *Dark Horse*, was a best-seller, but subsequent recordings have been regarded as disappointing, with the possible exception of *Cloud Nine* (1989).

Harrison has produced records for other artists, and played on some of them. He also became an executive producer of Monty Python films and eventually founded his own film company, Handmade Films; he has also published an autobiography, *I Me Mine* (1982).

Harrod, Sir (Henry) Roy Forbes (1900–78) *British economist, whose major contribution to economics was his model of economic growth. He was knighted in 1959.*
Educated at New College, Oxford, Harrod began lecturing in economics at Christ Church, Oxford, and in 1952 was appointed Nuffield Reader of International Economics. During World War II he served as statistical adviser to the Admiralty, and from 1945 to 1961 he was joint editor of the *Economic Journal*.

The Harrod–Domar model, so called because it was formulated independently by the American Evsey D. Domar (1914–   ), was an analysis of economic growth bringing the time dimension to the Keynesian income-expenditure model. He studied the effect on growth of the interaction of the multiplier and accelerator and examined the conditions under which an economy grows continuously at a constant rate as well as those under which cyclical fluctuations occur as a result of sector imbalances. He analysed the level of growth necessary to keep pace with growth in the labour force and established that an equilibrium growth rate is possible only if savings and investment remain consistently equal. Harrod also developed the concept of 'neutral technical progress', which has the equivalent effect of increasing the supply or productivity of labour.

Harrod's major works include *The Trade Cycle* (1936), *Towards a Dynamic Economics* (1948), *The Life of John Maynard Keynes* (1951), *Money* (1969), and *Economic Dynamics* (1973).

Hart, Herbert Lionel Adolphus (1907–   ) *British jurisprudent and philosopher.*
Hart was born in Harrogate, Yorkshire, and educated at New College, Oxford, where he

read Greats (classics and philosophy). After graduation (1929), he read for the bar and practised subsequently at the Chancery Bar from 1932 to 1940. During World War II Hart worked in intelligence. He returned to Oxford in 1945 and served initially as a philosophy tutor until his election in 1952 to the chair of jurisprudence, a post he held until 1968.

As both philosopher and lawyer Hart was favourably placed to illuminate both disciplines from his unique position. A major influence on his early thought was J. L. AUSTIN, with whom he conducted for several years a memorable seminar on the topic of excuses. From Austin he learned that we may use 'a sharpened awareness of words to sharpen our perception of phenomena'. Such a 'sharpened awareness' was seen in his first major study, *Causation in the Law* (1959), written in collaboration with Antony Honoré (1921–   ). Hart's main work, however, was directed towards the central concept of jurisprudence, the nature of law itself. In *The Concept of Law* (1961) he attempted to show how laws could be distinguished from orders backed by threats and precisely how legal and moral obligation differed. He argued in terms of primary rules, which impose duties, and secondary rules, which confer powers. 'Most of the features of law', Hart argued, 'can best be rendered clear, if these two types of rule and the interplay between them are understood.'

Hart's *Law, Liberty, and Morality* (1963) and *Punishment and Responsibility* (1968), though less fundamental in their approach, were nevertheless important contributions to the debates of the 1960s and 1970s. On such issues as capital punishment, the enforcement of morals, and human responsibility, Hart presented a humane, tolerant, and closely argued position. His other publications include writings on Jeremy Bentham and *Essays in Jurisprudence and Philosophy* (1983). He was principal of Brasenose College, Oxford (1973–78) and became a QC in 1984.

Hart, Lorenz Milton (1895–1943) *US lyricist. He collaborated with only one composer, Richard RODGERS, with whom he wrote many successful shows and songs.*

Hart and Rodgers met while they were both students at Columbia University in their native New York City. Influenced by the new sophisticated musical shows being written by Jerome KERN and others, their first success was a Columbia University show in 1920. In all, they wrote songs for about thirty shows and films, including *The Garrick Gaieties* (1925), *The*

Boys from Syracuse (1938), and *Pal Joey* (1940). Hart's lyrics had a poetic quality and a wit unusual in a writer of popular songs. However, his increasing dependence on alcohol made him difficult to work with. In fact Rodgers is said to have described him as 'a partner, a best friend, and a source of permanent irritation'.

Hartnell, Sir Norman Bishop (1901–79) *British fashion designer and dressmaker to the British Royal Family. He received a knighthood in 1977.*

Born in Streatham, London, the son of a grocer, Hartnell attended Magdalene College, Cambridge, but left after two years to become a clerk in a London dress shop. In 1923 he opened his own business in Bruton Street, Mayfair, and started designing and making dresses for his friends and acquaintances. His first Paris show in 1927 was shunned by the French but provided him with an opening into the US market. By the early 1930s his reputation as a leading young designer was well established and his imaginative use of embroidery and sequins featured in designs for leading actresses, such as Gertrude LAWRENCE. A commission to design the Duchess of Gloucester's wedding dress in 1935 was followed by a request from Queen Mary for some Hartnell dresses. In 1938 he became dressmaker by appointment to Queen Elizabeth, now Elizabeth the Queen Mother. World War II saw Hartnell's talent turned to the design of utility garments, making economical use of scarce materials. He designed the wedding gown for the marriage of Princess Elizabeth in 1947 and in 1953 was responsible for her coronation dress and the new coronation robes for many peers and peeresses. He was retained as royal dressmaker by Queen ELIZABETH II. His autobiography, *Silver and Gold*, was published in 1955.

Hašek, Jaroslav (1883–1923) *Czech novelist and short-story writer.*

The son of a journalist and a teacher, Hašek was born and grew up in Prague, where he became locally famous as an anarchistic and satirical personality in bohemian circles. In World War I he served in the Austrian army but soon deserted, joining Czech patriots in Russia. He joined the Russian Communist Party in 1918 and was a political commissar with the Red Army in Siberia for two years before returning to Prague, where he devoted the rest of his life – shortened by alcoholism – to the long rambling uncompleted work for which he is best known.

Osudy dobrého vojáka Švejka za světové války appeared in four volumes published between 1921 and 1923. Purged of its many vulgarities and coarseness, an English version, *The Good Soldier Schweik*, was published in 1930; a full unbowdlerized translation, *The Good Soldier Švejk*, was not published until 1974. The original was unimpressively completed by the Czech writer Karel Vaněk. The Czech government of MASARYK, embarrassed by the book's vulgar humour, found it difficult to admit that Hašek had produced a comic masterpiece, but the character Schweik clearly had universal appeal and gained an international following. The hero appears to be an amiable fool, though he overcomes everything authoritarian and pompous in the military life in which he has to survive. Given an order, he carries it out with a lunatic thoroughness that amounts to sabotage. Much of the interest of the episodes lies in a carefully maintained ambiguity: one does not know whether Schweik is supremely stupid or devilishly cunning in pretending to be so. As the archetypal story of the little man against the system, *The Good Soldier Schweik* has had a wide literary influence; its spirit can be seen in Joseph Heller's *Catch-22* (1961), to name just one example.

Havel, Václav (1936–) *Czechoslovak playwright who was twice imprisoned as a dissident; following the collapse of communist rule, he was elected as the country's president (1989–).*

Havel came from a bourgeois family known for its hostility to the ruling communists – a background that prevented him from studying drama at university. However, he found work as a stagehand and in the 1960s became resident playwright at an avant-garde theatre in Prague. His early plays, such as the *The Garden Party* (1963) and *The Memorandum* (1965), used absurdist techniques to scrutinize totalitarianism, especially its corrupting effects on language and personal integrity. In 1968 he was active in the movement towards liberalization but following the Soviet-led clampdown of 1968–69 his work was banned from the stage for over twenty years.

In the 1970s Havel became the leading spokesman for Charter 77 and other human-rights groups and was subjected to repeated harassment and house arrest by the authorities. In 1979 he was sentenced to four and a half years in prison for sedition. After his release he wrote such plays as *Largo Desolato* (1983) and *Temptation* (1985), which were successfully staged in the West.

In 1989 Havel was jailed for a second time following anti-government demonstrations, a sentence that sparked a further wave of protests. On his release he founded the opposition group Civic Forum and led a renewed campaign for political change. In November mass protests and the threat of a general strike led to the resignation of the entire ruling politburo and the formation of a majority non-communist government. Havel, the most popular man in the country, was elected to the post of president the following month.

Hawke, Robert James Lee (1929–) *Australian statesman and Labor prime minister (1983–). He became a Companion of the Order of Australia in 1979.*

Born in Bordertown, South Australia, the son of a Congregational minister, Hawke was educated in Perth, where he graduated from the University of Western Australia with arts and law degrees. Awarded a Rhodes scholarship in 1952, Hawke became an advocate for the ACTU (Australian Council of Trade Unions) in 1958 and was elected president in 1970, a position he held for ten years. Having stood unsuccessfully for the federal seat of Corio in 1963, Hawke became a member of the International Labour Organization governing body (1972–80) and then of the Reserve Bank Board (1973–80). An articulate advocate, he established a reputation as a skilful negotiator in industrial disputes. He eventually entered federal parliament in 1980 as the member for Wills and was elected leader of the Labor Party and prime minister within a month in 1983, when Malcolm FRASER's Liberal government was defeated in the election.

As prime minister, he pushed forward a radical programme of privatization; he won a record fourth election victory in 1990. In 1991 he defeated a challenge to his leadership of the Labor Party by his finance minister, Paul Keating.

Hawking, Stephen William (1943–) *British theoretical physicist whose main work has been on black holes and quantum gravity. He was created a Companion of Honour in 1989.*

Hawking was educated at the universities of Oxford and Cambridge; after working in various institutions he was appointed Lucasian Professor of Mathematics at Cambridge in 1979. Since 1962 he has suffered from amyotrophic lateral sclerosis, a rare crippling disease that has left him confined to a wheelchair, unable to write, and capable of such distorted speech that only a few intimates can understand him. Hawking's mind, however, has remained quite unaffected by his devastating disease. Able to work out complex equations without external aid and capable of retaining much technical data in his powerful memory, Hawking has been able, with the help of family and colleagues, to lead an active intellectual life and is indeed one of the world's most productive and creative theoretical physicists.

Through his work on black holes Hawking is widely known. In 1971 he argued that black holes could be formed other than by a star's gravitational collapse. They could have been produced, in the form of mini-black-holes, in the original big bang. These objects, if they exist, still await discovery. Hawking went on in 1974 to describe a process by which black holes could quite unexpectedly emit radiation at a steady rate; this is known as 'Hawking radiation'. Hawking has also considered the problem of the quantization of gravity, although he has not yet reached any generally accepted conclusions. He explained his scientific theories in a popular book, *A Brief History of Time* (1987), which has been in the bestseller list for nearly four years.

Hawkins, Coleman (1904–69) *Black US jazz tenor saxophonist, also called 'Hawk' or 'Bean'. He was one of the most influential saxophone players in jazz.*

Born in St Joseph, Missouri, he studied music in Topeka, Kansas, making his recording debut in 1923. Until then, the saxophone had a moaning sound with comical connotations; as a member of the Fletcher HENDERSON band (1923–34), Hawkins played with a stiff reed in order to increase the volume so that he could be heard as a soloist over the band. He also developed a large deep rich tone, which he used as a means of personal expression. From 1934 to 1939 he resided in Europe. After returning to the USA he recorded 'Body and Soul' (1939), which is often regarded as the definitive summary of his style. For the rest of his life he freelanced with small groups, also touring the USA and Europe several times with Jazz at the Philharmonic.

Hawkins was one of the few jazz masters who was always willing to listen to new ideas, playing and recording with younger men, such as Thelonius MONK and Sonny Rollins, twenty-five years his junior. For the last two decades of his life, Hawkins was one of the leading figures in mainstream jazz.

Hawks, Howard (1896–1977) *US film director, screenwriter, and producer who special-* ized in making popular westerns, gangster films, comedies, and musicals.

Born in Goshen, Indiana, Hawks was educated at Philips-Exeter Academy and Cornell University. After World War I, during which he served as a pilot, he became a racing-car driver before entering the film business in 1922 and progressing through the various departments until he finally became a director. His first feature was *The Road to Glory* (1926), which he also wrote. Throughout his career he continued to write and to contribute to his film scripts. Although he made several silent films, he only became known with his talkies *Scarface* (1931), *Barbary Coast* (1935), and *Only Angels Have Wings* (1939). Rita Hayworth (1918–87), Katherine HEPBURN, and Cary GRANT were among the great stars he directed. So, too, were Humphrey BOGART and his wife Lauren Bacall (1924–), who appeared together in *To Have and Have Not* (1944) and again in one of the screen's best-known films, *The Big Sleep* (1946).

Hawks then turned to westerns, producing, among others, *Red River* (1948) and *Rio Bravo* (1959). In sharp contrast were *Monkey Business* (1952) and *Gentlemen Prefer Blondes* (1953) with Jane Russell (1921–) and Marilyn MONROE. His last film was *Rio Lobo* (1970); in 1974 he was awarded an Honorary Oscar for his work.

Hawthorne, Mike (John Michael Hawthorne; 1929–59) *British motor racing driver and 1958 World Drivers' Champion.*

Hawthorne's father was a motor engineer and raced motor cycles at Brooklands. With parental encouragement, Hawthorne studied engineering and started racing motor cycles and then motor cars, competing with a Riley in 1951. The following year he started his first international season with a brilliant display at the Goodwood Easter meeting, driving a Cooper-Bristol. Enzo Ferrari signed him for the 1953 season and Hawthorne narrowly beat the great FANGIO in the French Grand Prix. His 1954 season was marred by a bad crash and public controversy surrounding his ineligibility for national service. Above all, his father was killed in a road accident. In spite of this, Hawthorne won the Spanish Grand Prix and achieved two second places. In 1955 he drove for Vanwall and then returned to Ferrari, but Formula One success eluded him. However, co-driving a Jaguar D-type with Ivor Bueb, he won the Le Mans Twenty-Four Hour race, an event overshadowed by the disastrous crash in which a car hurtled into the crowd killing eighty-two spectators. Hawthorne drove for BRM in the 1956 Formula One season but was more successful in sports cars, winning the Monza 1000 km race with his friend, Peter Collins. An indifferent 1957 season with Ferrari was followed by his best-ever performance. Driving the new V-6 Dino, Hawthorne won the French Grand Prix, gained five second places, and took the title. However, the death of Collins during the German Grand Prix prompted Hawthorne to announce his retirement at the end of 1958. He was killed in a road accident shortly afterwards.

Hay, Will (1888–1949) *British music-hall comedian, film actor, and director.*

Born in Stockton-on-Tees, Hay began his stage career in 1909, having earlier trained as an engineer. Between 1923 and 1924 he toured the world, after which he performed in the USA (1927) and South Africa (1928–29). He made numerous appearances on radio and in 1933 made his screen debut in the film short *Know Your Apples.* Among the many memorable films that followed were *Boys Will Be Boys* (1935), *Oh, Mr Porter!* (1937), *Convict 99* (1938), *Ask a Policeman* (1939), and *The Goose Steps Out* (1942). He also collaborated on the scripts of many of his films, often working with Basil Dearden (1911–71) at Ealing, where his later films were made, including his last, *My Learned Friend* (1944). In many of his films he appeared as a somewhat down-at-heel schoolmaster.

During World War II he served as an instructor to the Sea Cadet Corps, reaching the rank of lieutenant. A keen amateur astronomer, he was a fellow of the Royal Astronomical Society and the author of *Through My Telescope: astronomy for all* (1935).

Headley, George Alphonso (1909–83) *West Indian cricketer and his country's first black captain. One of the finest West Indian batsmen, he had a career record of 33 centuries and a batting average of just under 70.*

Born in Panama, he lived for a short time in Cuba before moving to Jamaica (his mother was Jamaican) at the age of nine. The intention was for him to attend an English school. Here he took immediately to cricket and was playing for Jamaica when he was eighteen. In his second match, against an English touring side, he scored 211. At this point he gave up his plan of going to the USA to study dentistry.

Known as 'the black Bradman', Headley was the idol of West Indian cricket between the two world wars. In a test career of 22 matches his highest score was 270 not out against England at Kingston in 1935. He invariably did well against the England bowlers and his test record was impressive. After World War II, he made his first appearance as a test captain, in England in 1947. He last played in a test, against England at Kingston in 1954, at the age of forty-four. From 1950 Headley played league cricket with success in England, and in 1955 he was appointed Jamaica's national cricket coach, a post he held until 1963. One of his nine children, Ron Headley (1939–), played county cricket for Worcestershire and made two test appearances.

Healey, Denis Winston (1917–) *British politician, deputy leader of the Labour Party (1980–83).*

Healey served in the army in World War II and was then secretary of the International Department of the Labour Party (1945–52), gaining early experience in foreign affairs. He was elected MP for Leeds, South-East in 1952 (Leeds, East from 1955), and was on the executive of the Fabian Society (1956–61). Under Harold WILSON he was defence secretary (1964–70) and then opposition spokesman on foreign affairs (1970–72) and Treasury matters (1971–74). After Wilson's retirement, Healey, a moderate, failed three times to win the party leadership (in 1976, 1980, and 1983). As Chancellor of the Exchequer (1974–79), he presented a record number of budgets. Healey was subsequently, with Labour again in opposition, spokesman on Treasury matters (1979–81) and foreign affairs (1981–87). He left the shadow cabinet in 1987 and published an autobiography, *The Time of My Life*, in 1989.

Heaney, Seamus (Justin) (1939–) *Irish poet and critic.*

Born into an Ulster farming family, Heaney was educated at St Columb's College, Londonderry, and Queen's University, Belfast. In the 1960s he taught and lectured in Belfast, emerging as the leading figure among a group of young poets based in the city. His early poetry, published in *Death of a Naturalist* (1966) and *Door into the Dark* (1969), is best known for its evocation of the author's childhood in rural Ulster. In 1972 Heaney moved south of the border and took Irish citizenship, a gesture to which he has ascribed considerable symbolic importance. The same year saw a marked change in his poetry, which began to deal with wider social and cultural themes, notably the role of language and its relation to politics and history. In *North* (1975), his most acclaimed volume, Heaney dealt more directly than hitherto with the present-day conflict in Northern Ireland. These poems, together with those in *Field Work* (1979), led to Heaney being recognized as the most important Irish poet since Yeats. Later collections include *Station Island* (1984), *The Haw Lantern* (1987), and *Seeing Things* (1991); his critical essays and lectures have been published in *Preoccupations* (1980) and *The Government of the Tongue* (1988). In 1989 he was appointed Professor of Poetry at Oxford University.

Hearst, William Randolph (1863–1951) *US newspaper tycoon and politician.*

The son of a mine-owner and rancher, Hearst was expelled from Harvard University for a prank. In 1887 he took charge of the San Francisco *Examiner*, owned by his father, and he substantially increased its circulation with journalistic techniques modelled on those of Joseph Pulitzer's New York *World*. In 1895 Hearst bought the New York *Journal* and embarked on a fierce struggle for circulation with the rival *World*, using dramatic headlines and exaggerated stories to boost sales. From the early 1900s, Hearst began his acquisition of newspapers in Chicago, Los Angeles, and elsewhere; by 1927 he owned twenty-five dailies with a combined circulation of over five million. He also owned magazines, such as *Cosmopolitan* and *Harper's Bazaar*, news agencies, a film newsreel, and the Cosmopolitan Movie Company, which made several features starring Marion Davies (1897–1961), a former Ziegfeld Follies dancer and lifelong companion to Hearst.

Hearst's political career was less successful. Elected congressman for New York (1903–07), he narrowly missed the Democratic presidential nomination in 1904, ran unsuccessfully for mayor of New York in 1905, and the following year lost his fight to become governor of New York. He was again defeated in the 1909 elections for mayor.

Immensely wealthy, Hearst spent lavishly, notably on the construction of a castle at San Simeon, California, and a vast collection of art treasures. The Depression hit Hearst's empire and forced the sale of his collections but Hearst survived to become one of the archetypal figures of US society, embodied in Orson WELLES's film *Citizen Kane* (1941).

Heath, Edward Richard George (1916–) *British politician and Conservative prime minister (1970–74), who took Britain into the European Community.*

Born in Broadstairs and educated at Oxford, Heath became MP for Bexley in 1950. He was successively minister of labour (1959–60), lord privy seal (1960–63), and secretary of state for trade and industry (1963–64) in the MACMILLAN and Douglas-Home administrations. In 1965 he became the first Conservative leader to be elected by ballot. As prime minister, Heath came into conflict with the trade unions, who opposed his Industrial Relations Act (1971); twice his government was challenged by miners' strikes. In 1972, after the country was reduced by energy shortages to working a three-day week, Heath submitted to the miners' demands. Heath's long-standing commitment to European unity came to fruition in his negotiation of Britain's entry into the Community in 1973. In 1974 he went to the country and lost both elections that year (in February and October). In the subsequent battle for the leadership the rattled Conservative party voted him out in favour of Margaret THATCHER. He later became a focus of disaffection within the party as a result of some of Mrs Thatcher's economic policies. He is known also for his interests in music and yachting. In 1990 he embarked on a widely publicized and controversial trip to Iraq to secure the release of British hostages held by Saddam HUSSEIN.

Heaviside, Oliver (1850–1925) *British electrical engineer, who proposed the existence of the ionosphere in 1902.*

The son of an engraver, Heaviside was born partially deaf and consequently received no systematic formal education. He did, however, gain a fairly comprehensive understanding of the science of his day through private study. His first and only regular job, to which he was appointed in 1870, was as a telegraph operator with the Great Northern Telegraph Company in Newcastle. Heaviside's poor hearing made it difficult to carry out his duties and he consequently resigned in 1874, spending the next fifteen years in private study and research at his parents' home in Camden Town in north London. In 1889 Heaviside moved with his parents to Devonshire to live with his elder brother Charles. On the death of his parents a few years later, Heaviside left his brother and spent the rest of his life living in seclusion in various Devon towns. He was supported financially by magazine writing, gifts from friends, and, in his later years, a civil list pension.

Heaviside is remembered for his proposal in 1902 that a layer of electrically charged particles in the upper atmosphere is capable of reflecting back to earth radio waves broadcast from the surface of the earth. This would account for the transmission across the Atlantic by MARCONI in the previous year of messages carried by radio waves. In 1911 this layer was given the name Heaviside–Kennelly layer after Heaviside and Arthur KENNELLY, who independently proposed its existence. After Sir Edward APPLETON proved its existence in 1924, it became known as the ionosphere, although the E region of the ionosphere is still sometimes known as the Heaviside–Kennelly layer. This proposal was only one aspect of Heaviside's work contained in the three volumes of his *Electromagnetic Theory* (1893–1912). Expressed in a strikingly original mathematical form, Heaviside's work soon became accepted by practical engineers. Told that his work was difficult to read, Heaviside replied that it had been harder to write.

Heidegger, Martin (1889–1976) *German existential philosopher, author of* Sein und Zeit *(1927).*

Born a Catholic, the son of a sexton in Baden-Wurttemberg, Heidegger originally intended to become a Jesuit priest. He was educated at the University of Freiburg, where he studied under HUSSERL. After several years of teaching at the University of Marburg, Heidegger returned to Freiburg in 1928 as professor of philosophy. In 1945 he was sacked from his post by the Allied rulers of Germany for his support of the Nazis. He was also refused permission to teach elsewhere. When the ban was lifted in 1951 Heidegger chose to continue living on his own and working in isolation for the rest of his life.

In his major work, *Sein und Zeit* (1927; translated as *Being and Time*, 1962), Heidegger sought to offer a comprehensive account of being. He distinguished between *Dasein*, the being of humans; *Vorhanden*, the being of objects; and *Zuhanden*, the being of tools. Objects in the world are merely in the world. They can, of course, be given a use by human beings. The being (*Dasein*) of human beings is more complex. It involves, among other features, an awareness of the future, the necessity of choice, and the ultimate reality of death. In another work, *Was ist Metaphysik?* (1929; translated in *Existence and Being*, 1949), Heidegger attacked logic for its inability to cope with nothingness and the totality of existence.

Much of this later work was often held up to ridicule by positivistically minded philosophers, such as A. J. AYER. How, they asked, can 'Nothing' annihilate itself? Heidegger, however, did not respond to such criticism and consequently his work has been largely ignored by analytical philosophers.

Heifetz, Jascha (1901–87) *Russian-born US-naturalized violinist. He was made a Commandeur de la Légion d'honneur in 1957.*

Born in Vilnius, Heifetz was first taught by his father, an accomplished violinist. Later he studied at the Vilnius School of Music under Elias Malkin. At the early age of eight he graduated and started performing in public. In 1910 he entered the Imperial Conservatory in St Petersburg as the youngest member of Auer's prestigious class; at twelve he started on his long virtuoso career, touring Russia, Scandinavia, and Germany. The turmoil of the revolution in Russia (1917) disrupted his career and eventually he emigrated to the USA, becoming a naturalized citizen in 1925.

Heifetz first performed in London in 1920 at the Queen's Hall. He toured Australia and the East, revisiting England in 1922 and 1925. His 1926 engagements included a visit to Palestine, where he gave a series of free concerts to large audiences of the Jewish colony. His reputation as a sensational violinist preceded him wherever he went, and he became known the world over through the large number of recordings he made. He commissioned several important works in the repertory, including WALTON's violin concerto (1939), of which he gave the first performance that year with the Cleveland Symphony Orchestra under Artur Rodzinski (1892–1958). After a long performing career he became a teacher in California.

Heisenberg, Werner Karl (1901–76) *German theoretical physicist best known for his work on the quantum theory, including matrix mechanics and the uncertainty principle, for which he was awarded the 1932 Nobel Prize for Physics.*

The son of a professor of history, Heisenberg was educated at the universities of Munich and Göttingen. He subsequently spent two years in Copenhagen working with Niels BOHR before returning to Germany in 1926 to take up an appointment as professor of physics at Leipzig. There, in the mid-1920s, he developed a formalism, known as matrix mechanics, that provided a plausible basis for quantum mechanics. This insight was followed by the uncertainty principle, according to which it is not possible to determine exactly and simultaneously both the position and momentum of elementary particles. No improvement in instruments or techniques can overcome what is a theoretical limitation on the knowledge we can gain of the material world. This remains a fundamental tenet of modern physics.

With the outbreak of World War II Heisenberg was called to Berlin to direct the German nuclear programme. Although he had defended EINSTEIN against attack from the Nazis, and had himself been investigated by the secret police, as an ardent nationalist he was convinced that it was his duty to fight for his country. In 1941 he visited occupied Copenhagen and contacted his old teacher and colleague, Niels Bohr, to warn him, so Heisenberg later reported, that Germany could make an atomic bomb. However, Heisenberg's team made little progress in overcoming either the theoretical or the industrial problems implicit in such a programme. Whether the lack of progress was due to the banishment of the best physicists to England and America because they were nearly all Jewish, or to a reluctance of the remaining ones to provide the Nazis with such a weapon, remains a matter of dispute. Heisenberg's role in the affair is still controversial; when he first heard of the Hiroshima bomb he is said to have dismissed it as 20 000 tons of high explosive.

After the war Heisenberg helped to establish the Max Planck Institute for Physics, serving as its director, first at Göttingen and after 1955 in Munich. Although he campaigned strongly against the ADENAUER government's failure to back the construction of nuclear reactors, Heisenberg also made his own position clear on nuclear weapons, declaring in 1957 that he would take no part in their testing or their production.

Hellman, Lillian (1905–84) *US playwright, most of whose work had a political theme.*

Born in New Orleans of Southern Jewish parents, she was taken to New York at the age of five and thereafter attended schools in both cities. She studied at New York University and later at Columbia but did not take a degree. After working in publishing and as a reader for a theatrical agent, she wrote her first play, *The Children's Hour* (1934; filmed 1936 (as *These Three*), 1962, and 1979 (as *The Loudest Whisper*)), about a girl who maliciously accuses her teachers of lesbianism. The play caused an uproar and was highly successful, running for 691 performances. *The Little Foxes* (1939; filmed 1941) concerns a Southern family and its attempt to hang on to its position of dominance. *Watch on the Rhine* (1941), an effective anti-Nazi play filmed in 1943, suggested that the country would soon join the war against HITLER. Among her other plays are *Another Part of the Forest* (1946; filmed 1948), *The Autumn Garden* (1951), and *Toys in the Attic*

(1960; filmed 1963). Hellman dramatized novels for the stage and with the poet Richard Wilbur worked on the musical adaptation of Voltaire's *Candide* (1957), with a score by Leonard BERNSTEIN. She was called before the House Committee on Un-American Activities during the MCCARTHY witch-hunt of the 1950s. Although she herself was by this time anti-communist, she refused to give names or betray friends. This period is covered in *Scoundrel Time* (1976). Other autobiographical volumes, which include an account of her long friendship with Dashiell HAMMETT, are *An Unfinished Woman* (1969), *Pentimento* (1973), and *Maybe* (1980). Parts of *An Unfinished Woman* and *Pentimento* formed the basis of the film *Julia* (1977).

Helpmann, Sir Robert (1909–86) *Australian ballet dancer, choreographer, actor, and director, who played a major role in the early years of the Sadler's Wells Ballet and is credited with bringing classical ballet to his native land. He was knighted in 1968.*

When Anna PAVLOVA toured Australia and New Zealand in 1923, Helpmann joined her company and then spent five years touring Australia with J. C. Williamson's troupe. In 1933 he came to England and joined the Vic-Wells (later the Sadler's Wells) Ballet; the same year he was given his first major role, as Satan in DE VALOIS's *Job*. He became the company's principal dancer in 1934 and partnered Alicia MARKOVA in *The Haunted Ballroom*. The following year marked the start of a long partnership between Helpmann and the new prima ballerina, Margot FONTEYN, with whom he danced a large classical repertoire. Helpmann's dramatic ability was used to great effect in such ballets as *Checkmate* (1937), *The Prospect Before Us* (1940), and the immensely popular *Cinderella* (1948), which was choreographed by Frederick ASHTON and featured Helpmann as one of the ugly sisters. Helpmann also began to choreograph ballets himself and achieved great success in *Comus* (1942), *Hamlet* (1942), and *Miracle in the Gorbals* (1944). He also appeared in the ballet films *The Red Shoes* (1948) and *Tales of Hoffmann* (1951).

In 1950 Helpmann resigned from the Sadler's Wells Ballet, although he still appeared as a guest artist. He continued to work as a choreographer, dancer, director, and actor, appearing as Oberon in *A Midsummer Night's Dream*, in which he played opposite Vivien LEIGH, and in the title role in *Hamlet*. In 1963 Helpmann became a co-director of the Australian Ballet with Dame Peggy Van Praagh

(1910–90) and produced such works as *Display*, based on Australian legend, and *Yugen*, based on Japanese legend. In 1976, after the success of his *The Merry Widow* (1975), he retired.

Hemingway, Ernest Miller (1898–1961) *US writer, whose novels and short stories include some of the best-known works of this century. He was awarded the Nobel Prize for Literature in 1954.*

Hemingway was born and grew up in the Great Lakes region of Illinois. He began his working life in journalism but during World War I served as an ambulance driver in Italy, where he was badly wounded and invalided back to the USA in 1919. After recuperating, he travelled extensively as a foreign correspondent before settling in Paris as an expatriate. Here he had his first major literary success with *The Sun Also Rises* (1926), which was followed three years later by the highly acclaimed *A Farewell to Arms*. Subsequent books include *For Whom the Bell Tolls* (1940), *Across the River and Into the Trees* (1950), and *The Old Man and the Sea* (1952), as well as the nonfiction works *Death in the Afternoon* (1932), in which he revealed his passion for bullfighting, and *The Green Hills of Africa* (1955).

The popular image of both writer and books is of rugged toughness, an exterior that masked the sensitive side of Hemingway's character and the complexity of his work. Nevertheless it is by the economy of his style and the starkness of his writing that he is known. Although his base remained in the USA, Hemingway continued to travel in Europe, serving as a war correspondent in France during World War II and in Spain at the beginning of the Spanish civil war. As he grew older, he became increasingly tormented by his failing powers, both physically to pursue the sporting outdoor life he had always led and intellectually to write. He committed suicide at his home in Idaho.

Henderson, (James) Fletcher (1897–1952) *Black US jazz pianist, bandleader, and arranger. He was one of the leaders of the big-band style of the Swing Era.*

Born in Cuthbert, Georgia, he obtained a degree in chemistry but drifted into bandleading in 1921. Always a great judge of talent, in the period 1923–24 he hired Louis ARMSTRONG, Coleman HAWKINS, and Don Redman (1900–64), with whom he developed the new style of jazz for larger groups with brass and reeds in separate sections playing from written charts. Forced to do this arranging work himself after Redman left in 1929, Henderson and his overshadowed brother Horace (1904–   ) refined the style to such an extent that it suddenly became a nationwide success in 1935 for the white band of Benny GOODMAN. Thereafter Goodman played many Henderson arrangements, always giving him full credit. Henderson also wrote for other bandleaders, as well as playing the piano in Goodman's sextet for most of 1939.

Henderson was handicapped not only by being black but also because he was an indifferent businessman, especially after being badly injured in a 1928 car crash. Nevertheless, at one time or another he employed virtually all the best black musicians in the country, and his band was considered one of the best of its kind.

Hendrix, Jimi (James Marshall Hendrix; 1945–70) *Black US rock guitarist and singer. He was one of the pop 'superstars' of the 1960s.*

The son of a gardener, he was born in Seattle, Washington, his antecedents being part negro, part Cherokee Indian, and part Mexican. When he left the US army in 1963 he started touring the southern states as a backing musician for rhythm and blues stars. By 1966 he had formed his own band in New York called Jimmy James and his Blue Flames. Here he was heard by the British rock musician Chas Chandler, who urged him to come to London. Once in London he formed a trio called the Jimi Hendrix Experience. The group's first recording, 'Hey Joe', was number 6 in the British pop charts in 1967; it was followed by 'Purple Haze' and several other successes. The wild appearance of the group, their deafening amplification, and their ability to improvise made them a great success at the Monterey Pop Festival in 1967 and the Woodstock Festival in 1969. Back in the USA after his great success in Britain, a planned concert tour was banned by the Daughters of the American Revolution, who regarded his performing antics as obscene. After a year's isolation, Hendrix formed another group, the Band of Gypsies, which performed at the Isle of Wight Festival and made a recording in 1970.

Hendrix was the first to explore the potential of the electric guitar as an electronic sound source. Admired by Miles DAVIS and Bob DYLAN he had a profound effect on the pop music that came after him. He was building his own studio in New York when he died of an accidental overdose of barbiturates.

Henze, Hans Werner (1926–) *German composer. One of the most prolific of contemporary composers, he has a world-wide reputation as a conductor of his own works.*

Parental opposition to music as a career and to the arts in general made Henze's early days difficult and frustrating, but eventually he was allowed to study at the local music school in Brunswick (mainly piano and percussion). From 1944 to 1945 he was in the army, for a time a prisoner-of-war in England. After repatriation he went to study with the composer Wolfgang Fortner (1907–) at the Heidelberg Institute of Church Music, where he received a concentrated grounding in contrapuntal techniques; later he studied with René Leibowitz (1913–72) in Paris. For several years he worked in various provincial opera houses, including two years as musical adviser for ballet at the Wiesbaden Opera. In 1951 his piano concerto won the Robert Schumann Prize at Düsseldorf. In 1953 Henze moved to Italy, first living on the island of Ischia, then in Naples, and later near Rome. He is currently artistic director of the Philharmonic Academy, Rome (since 1981) and holds the chair of composition at both the Hochschule für Musik, Cologne (since 1980) and the Royal Academy of Music, London (since 1987). He also makes frequent international conducting tours.

Henze's music derives from the serial techniques of the second Viennese School (see SCHOENBERG, ARNOLD FRANZ WALTER), particularly those of BERG. His deep love and respect for Mozart and other classical composers is manifest in his use of classical structures, such as sonata and passacaglia, in many of his works. The influence of STRAVINSKY is apparent in his use of rhythm, and Henze's wide knowledge and love of Italian opera is apparent from the beauty and lyricism of much of his writing. Increasingly, in the late 1960s, Henze's left-wing Marxist ideals influenced his music: *Das Floss der 'Medusa'*, a requiem for CHE GUEVARA based on the same story as Gericault's painting, caused an uproar at its first abortive Hamburg performance (1968). Henze's operas include the early *Boulevard Solitude* (1951), an updating of the Manon Lescaut story, *Elegy for Young Lovers* (1959–61) and *The Bassarids* (1965), both with libretti by W. H. AUDEN, and *We Come to the River* (1974–76) and *The English Cat* (1982), both with libretti by Edward BOND, the former an indictment of social injustice in which several episodes are portrayed simultaneously, the music of one counterpointing the music of another. Other works include seven symphonies, concertos, chamber music, and ballet music, including *Undine* (1959), a work that helped to establish Henze's reputation in both Britain and the USA.

Hepburn, Katherine (1909–) *US film and stage actress. Her characteristic voice and dazzling personality have been a feature of show business for over fifty years.*

Hepburn was born in Hartford, Connecticut, into a well-to-do family; after graduating from Bryn Mawr in 1928 she secured a small part in a local theatrical production of *Czarina*. This was followed by a successful, but somewhat stormy, Broadway career and her screen debut in *A Bill of Divorcement* (1932).

The recipient of four Oscars for *Morning Glory* (1933), *Guess Who's Coming to Dinner* (1967), *The Lion in Winter* (1968), and *On Golden Pond* (1981), Hepburn has been nominated for no fewer than eight other Academy Awards during her long and distinguished career. These were for *Alice Adams* (1935), *The Philadelphia Story* (1940), *Woman of the Year* (1942), *The African Queen* (1951), *Summertime* (1955), *The Rainmaker* (1956), *Suddenly Last Summer* (1959), and *Long Day's Journey Into Night* (1962). *Woman of the Year* (1942) was the first of many films with Spencer TRACY, with whom she was to have a long relationship that ended with his death shortly after their last film, *Guess Who's Coming to Dinner* (1967). Their romance, which has become a Hollywood legend, was the subject of the best-selling book *Tracy and Hepburn: An Intimate Affair* (1971) by Garson Kanin.

As well as films, Hepburn has continued to appear on Broadway and in television films, such as *The Corn is Green* (1979). In 1987 she published the autobiographical *The Making of The African Queen*.

Hepworth, Dame (Jocelyn) Barbara (1903–75) *British sculptor who, with Ben NICHOLSON and Henry MOORE, led the abstract movement in Britain in the 1930s. She was made a DBE in 1965.*

Born in Wakefield, Yorkshire, she trained at the Leeds School of Art and at the Royal College of Art, London (1921–24). Here she was a contemporary of the sculptor Henry Moore, who came from the same mining district in Yorkshire and whose artistic development ran parallel to hers until the 1930s. She studied in Florence and Rome before returning to London in 1926.

In 1931 she met Ben Nicholson, who later became her second husband, and she joined the Seven and Five society of artists. In the same year she produced her first *Pierced Form*, introducing holes into British sculpture. At this time she was working increasingly towards abstraction in carved stone and wood. Using forms suggested by the natural qualities of these materials and other natural forms, she assimilated these to the human figure or to free organic forms. In touch with BRAQUE, BRANCUSI, MONDRIAN, and PICASSO, in 1933 she became a member of the Abstraction-Création association and of the British Unit One group. Her work from 1934 included nonorganic geometric abstracts. Colour was sometimes incorporated to heighten the effect of interior spaces, which became larger and more complex and expressive in the 1940s after her move from Hampstead to St Ives in Cornwall at the beginning of World War II.

Her marriage to Ben Nicholson ended in 1951. In the late 1950s she began to work in bronze, producing large works of monumental simplicity for landscape and architectural setting. She won the Grand Prix at the São Paulo Biennale in 1959. In the 1960s she worked on a broad range of sculptures in bronze and in stone, both geometric constructions and organic forms.

Herbert, Sir A(lan) P(atrick) (1890–1971)
British writer, reformer, and politician. He was knighted in 1945 and made CH in 1970.

Herbert was born at Ashstead, Surrey, the son of a civil servant. He was educated at Winchester College and New College, Oxford. Despite taking a first-class degree in law (1914), he never practised; after serving in World War I with the Royal Naval Division in Gallipoli and France, during which he was wounded, he became private secretary to an MP. His war experiences resulted in two volumes of war poetry (1916, 1918) and the novel *The Secret Battle* (1919). Before the war he had already been writing for *Punch*, and in 1924 he joined the magazine's staff. The same year he published the first of his famous mock law reports, later collected as *Misleading Cases in the Common Law* (1927) and its successor volumes. In 1924 he also wrote his first stage play, a Christmas piece called *King of the Castle*.

From 1935 until the abolition of university seats in 1950, Herbert was Independent MP for Oxford University, using his position to achieve numerous reforms. Of these perhaps the most famous was his Marriage Bill (1937), which brought about a much-needed overhaul of divorce law, the anomalies of which he had exposed in the novel *Holy Deadlock* (1934). He related the stormy battle over this Bill in

The Ayes Have It (1937) and described his time as an MP in *Independent Member* (1950). During World War II he served in the Naval Auxiliary Patrol on the Thames in his boat *Water Gipsy*.

Herbert wrote the libretti for a number of successful musicals, including the revue-type *Riverside Nights* (1926), the adaptations from Offenbach *La Vie Parisienne* (1929) and *Helen* (1932), *Tantivy Towers* (1931), and *Bless the Bride* (1947). In *The Thames* (1966) he celebrated his long association with the river. His many interests and campaigns, the most prominent of which was his crusade to give authors some benefit when their books are borrowed from public libraries, are described in his autobiographical *A.P.H.: His Life and Times* (1970).

Hertzog, James Barry Munnik (1886–1942) *South African statesman and prime minister (1924–39). A firm supporter of the Dutch language and culture, Hertzog consistently opposed British influence in South Africa.*

Born in Wellington, Cape Colony, the son of a farmer, Hertzog was educated at Victoria College, Stellenbosch, and at the University of Amsterdam, where he obtained a law degree in 1892. Opening a legal practice in Pretoria in 1893, he became a judge of the Supreme Court of the Orange Free State in 1895, remaining in office for four years. Hertzog first rose to prominence in 1900, as a Free State general during the second Boer War. An active participant in the Vereeniging peace negotiations in 1902, Hertzog became the unchallenged Free State political leader, co-founding the Orangia Unie Party in 1906. Elected to the Crown Colony cabinet in 1907, he participated in the national convention that drafted the constitution for the Union of South Africa, becoming a minister in the first Union cabinet under Louis BOTHA in 1910.

Breaking with Botha in 1914, he formed the exclusively Afrikaner National Party and became prime minister in 1924, following a pact with the Labour Party. In 1933 he entered into a coalition with the South African Party led by General SMUTS, retaining the position of prime minister when the two parties fused in 1934. However, he lost office in 1939 when disagreement over entry into World War II led the two parties to split up. Briefly joining MALAN's National Party in 1940, Hertzog resigned in 1941 to found the Afrikaner Party and retired the following year to his farm in Pretoria, where he died.

Hertzsprung, Ejnar (1873–1967) *Danish astronomer, best known for his independent discovery of the Hertzsprung–Russell diagram.*

The son of a senior civil servant, Hertzsprung trained as a chemical engineer at the Copenhagen polytechnic, as his father believed that he would be unable to earn his living as an astronomer. Hertzsprung therefore worked as a chemist for some years before his first astronomical appointment at the Potsdam Observatory (1909). At the end of World War I Hertzsprung moved to Leiden University, where he was appointed director of the Leiden Observatory (1935–44).

Quite independently of H. N. RUSSELL, Hertzsprung published in 1911 the 'Giant and Dwarf' theory of stellar evolution, which he represented in the type of diagram now known as a Hertzsprung–Russell diagram. Hertzsprung also made important calculations of the distance from earth of certain Cepheids (a group of variable stars) in the small magellanic cloud. Although now regarded as an underestimate, it was one of the first measurements of an extragalactic distance.

Hess, Dame Myra (1890–1965) *British pianist who was created CBE in 1936 and DBE in 1941 in recognition of the lunchtime concerts she organized at the National Gallery during World War II.*

She began her musical education at the Guildhall School of Music under Julian Pascal and Orlando Morgan, winning a scholarship to the Royal Academy of Music (1902), where she was one of the earliest and most distinguished pupils of Tobias Matthay (1858–1945; author of a method of pianism based on relaxation). Her debut under BEECHAM at the Queen's Hall in London (1907), playing Beethoven's fourth piano concerto, was enthusiastically acclaimed. She quickly became known for her interpretations of the major concertos and for her playing in recitals and in chamber ensembles. She also formed a piano duo with her cousin Irene Scharrer (1888–1971), another Matthay pupil. In the 1920s she toured Europe and the USA, where she became a welcome and regular visitor. Her repertoire was wide and, in her early days, adventurous in the field of contemporary music.

From 1939 to 1946 she inaugurated, organized, and frequently played at the National Gallery lunchtime concerts as a regular musical feature of wartime London. Dame Myra Hess made numerous piano transcriptions of baroque music, of which 'Jesu, Joy of Man's Desiring' from Bach's *Cantata 147* achieved great popularity. Her playing combined a brilliant technique with great warmth of feeling. Among her pupils were Stephen Bishop-Kovacevich and Yonty Solomon.

Hess, (Walther Richard) Rudolf (1894–87) *Deputy leader of the German Nazi Party (1934–41).*

Born in Alexandria, Egypt, the son of a German importer, Hess was educated at Alexandria, Bad Godesberg, Neuchâtel, and Hamburg. After serving in the same regiment as HITLER during World War I, in 1920 he became Hitler's political secretary. He was a participant in the abortive Munich putsch (an attempt to overthrow the government) in 1923 and was imprisoned for seven months in Landsberg prison, where he took down *Mein Kampf* from Hitler's dictation.

Hess remained closely associated with Hitler during the 1930s. Appointed chairman of the central political commission of the National Socialist (Nazi) Party in 1932, he became deputy leader of the party in 1934 and was named successor to Hitler (after GOERING) as head of state in 1939. In 1941 Hess flew alone from Augsburg to Scotland in an attempt to negotiate peace with the British government. He was interned until 1945 and then tried as a war criminal at Nuremberg in 1946, where he was sentenced to life imprisonment at Spandau prison. From 1966 he was the sole prisoner at Spandau, the Soviets refusing to agree to his release, despite proposals to do so from the French, British, and Americans. After his suicide in 1987 the prison was demolished.

Hess, Victor Francis (1883–1964) *Austrian-born US physicist who was awarded the Nobel Prize for Physics in 1936 for his discovery of cosmic rays.*

The son of a forester, Hess was educated at the University of Graz, where he obtained his PhD in 1906. Hess worked at the newly founded Institute for Radium Research in Vienna from 1910 until 1919, when he was appointed to the chair of physics at the University of Vienna. In 1938 Hess, as a Roman Catholic, was summarily dismissed from his post by the invading Nazis and he emigrated to the USA. He was naturalized in 1944 and became professor of physics at Fordham University, New York, from 1938 until his retirement in 1956.

When Hess joined the radium institute in 1910 he became interested in the source of penetrating ionizing radiation that appeared to occur on the surface of the earth. Initially it was thought to come from the earth and consequently it was assumed that its effect would fall off with increasing altitude. Earlier attempts to confirm this conjecture proved inconclusive. In 1911, therefore, Hess made a number of balloon ascents with his measuring instruments, including one to a height of over 5000 metres. He found that there was an initial decrease in ionization up to a height of about 150 metres, but thereafter activity increased markedly with altitude. As the same effect was observed at night, the source of the radiation could not be attributed to the sun. Hess concluded that the radiation came from outside the solar system and it later became known as cosmic radiation. This subject continued to absorb him for the rest of his scientific life.

Hesse, Hermann (1877–1962) *German-born Swiss novelist. He was awarded the Nobel Prize for Literature in 1946 and shortly afterwards the Goethe Prize in Germany.*

Born at Calw near Württemberg, Hesse was the son of Christian missionaries who had worked in India and grandson of a distinguished expert on Indian languages. An emphasis upon spiritual values, especially as formulated in Indian mysticism, was to mark his most important later writings, particularly after his own visit to India in 1911. As a child he lived in Basel (1881–86) and had Swiss citizenship. He spent a short period studying at a seminary in Germany but soon left to work as a bookseller in Switzerland. From 1904 he devoted himself entirely to writing. After a first volume of verse (1899), Hesse established his reputation with a series of lyrical romantic novels – *Peter Camenzind* (1904; translated 1961), *Unterm Rad* (1906; translated as *Beneath the Wheel*, 1968), and *Gertrud* (1910; translated 1915) – and the short story *Knulp* (1915; translated 1971). After his journey to India in 1911, he lived for a time in Bern. Having proclaimed his pacifism in 1914, Hesse was denounced in Germany (he took up permanent Swiss citizenship in 1923) and his reputation suffered there until after World War II. In 1916–17 he underwent Jungian analysis and the experience had a profound effect on his subsequent work. In 1919 he left his family and published *Demian*, a novel that broke with the rather provincial romanticism of his earlier work and shows the impact of JUNG's psychology in a pattern of conflicting oppositions. The novel ends with a disaster to civilization (i e World War I) but looks forward to a future rebirth.

Hesse's major work was published between the wars and enjoyed not only an initial success but a very influential revival in the 1960s and 1970s with the popular interest in oriental religion, meditation, and mysticism. *Siddhartha* (1922; translated 1954) is a novelistic rendering of a Brahmin's search for and attainment of the spiritual goal. *Der Steppenwolf* (1927; translated 1929) refers to the human and bestial elements in the character of the protagonist, Harry Haller, and is a surrealistic tale of how the hero resolves this conflict. *Narziss und Goldmund* (1930; translated as *Death and the Love*, 1932), set in the Middle Ages, again explores conflicting aspects of human nature in the two contrasting central characters. Hesse's last notable work and the one that attempts to summarize his views is *Das Glasperlenspiel* (1943; translated as *The Glass Bead Game* or *Magister Ludi*, 1970). Set in the future (in twenty-third century Castalia), the novel is Hesse's final statement on the need to discover an as yet untried spiritual solution to the problems and contradictions of human nature and culture.

Hevesy, George Charles von (1885–1966) *Hungarian-born Swedish chemist who was awarded the 1943 Nobel Prize for Chemistry for his work on radioactive tracers.*

The son of a wealthy industrialist, Hevesy was educated at the University of Freiburg, where he obtained his PhD in 1908. Thereafter, in a career much interrupted by war and politics, Hevesy worked in seven different countries. After brief periods in Zürich and Karlsruhe, Hevesy joined RUTHERFORD in Manchester. There he was given the task of separating radioactive radium D from lead. As radium D is an isotope of lead, the chemical methods used by Hevesy proved unsuccessful. However, Hevesy realized that his apparent failure could be utilized to make an entirely new type of tracer. If radioactive lead and ordinary lead could not be distinguished chemically, the radioisotope could be used to monitor the path of lead through a complex system. By 1923 he had shown how radioactive lead could be used to label salts taken up by plants in solution; in 1934, using radioactive phosphorus, Hevesy applied his tracer technique for the first time to animals. This use of artificial radioisotopes enabled the technique to find very wide applications.

Hevesy left Manchester in 1913 for the University of Vienna but, with the outbreak of World War I in 1914, he returned to his native Budapest. After the war he worked in Copenhagen from 1920 to 1926, when he accepted the chair in physical chemistry at the University of Freiburg. Hevesy, however, abandoned Germany with the rise of HITLER and returned in 1934 to Denmark. Several years later the Germans caught up with him once again and in 1942 Hevesy left Denmark for the safety of Sweden, where he completed his academic career. Hevesy is also known for his discovery, in 1923, in collaboration with Dirk Coster (1889–1950), of the new element hafnium.

Heyerdahl, Thor (1914–) *Norwegian anthropologist noted for his ocean voyages in primitive craft that demonstrated the possibility of cultural contact between widely separated early civilizations.*

After reading zoology and geography at Oslo University (1933–36), Heyerdahl became interested in Polynesian culture as a result of visiting Fatu Hiva in the Pacific Marquesas Islands. He recognized a parallel between the legendary Polynesian figure, Tiki, and the pre-Incan hero, Kon-Tiki, who, according to legend, had fled from South America around 500 AD to escape massacre. After serving with the free Norwegian armed forces' parachute unit during World War II, Heyerdahl set out to show that Polynesia could have received migrants from South America. He supervised the construction of a primitive balsa-log raft and, with five companions, set sail on 28 April 1947 from the Peruvian coast. After drifting with Pacific currents, they made land on Raroia Reef, south of the Marquesas, after a voyage of 101 days covering 6880 km. Heyerdahl wrote a book, *The Kon-Tiki Expedition* (1948), and also made an Oscar-winning documentary film, *Kon-Tiki*, based on the voyage. In 1949 he founded the Kon-Tiki Museum, Oslo.

Heyerdahl led further expeditions – to the Galapagos Islands (1953) and to Easter Island (1955–56). In the mid-1960s he constructed a 50-foot papyrus reed raft, *Ra* (named after the Egyptian sun god), according to ancient Egyptian design to prove that such a vessel could have made the transatlantic crossing. Heyerdahl and his six-member crew set out from Morocco in May 1969 but were forced to abandon the attempt only 600 miles short of their target. Nevertheless, the following year, Heyerdahl successfully completed the crossing in *Ra II*, reaching Barbados on 12 July after a 56-day voyage. Both trips are described in *The Ra Expeditions* (1970). In 1978, he set sail from Qurna, Iraq, in a reed boat of ancient Sumerian design bound for India. However, in protest against the war in the Horn of Africa, he burnt his craft, *Tigris*, in Djibouti. *The Tigris Expedition* (1979) is Heyerdahl's ac-

count of the event. Subsequently he led expeditions to the Maldives (1982–84), Easter Island (1986–88), and Peru (1988).

Hilbert, David (1863–1943) *German mathematician, a major contributor to most branches of modern mathematics.*

Born in Königsberg (East Prussia; now Kaliningrad, USSR), the son of a judge, Hilbert was educated at the universities of Königsberg and Heidelberg. He taught briefly at Königsberg before moving to the University of Göttingen, where he was professor of mathematics from 1895 until his retirement in 1930.

Hilbert established his reputation as a mathematician by his work in the 1880s on the theory of invariants. This was followed by his *Grundlagen der Geometrie* (1899; 'The Foundations of Geometry') in which, some two millennia after its initial appearance, Hilbert offered the first rigorous treatment of Euclidean geometry. In the following year Hilbert became known to a much larger audience by his presentation at the International Congress of Mathematicians in Paris of twenty-three outstanding unsolved problems. In maths, he declared, 'there is no Ignorabimus', all problems are solvable. The first problem (on the continuum hypothesis) was solved only in 1963, while the second (on the consistency of arithmetic) and the eighth (on the Riemann hypothesis) remain, with several others, essentially unsolved. On the principle that 'physics is too difficult to be left to physicists,' Hilbert spent much of his time on his sixth problem, the axiomatization of physics. Although this may now look somewhat presumptuous, it yielded concepts and results that have profoundly influenced the development of modern physics. The notion of Hilbert space, for example, was an important constituent in the evolution of quantum mechanics.

Hilbert also tackled the problem of the foundations of mathematics. In opposition to the intuition of BROUWER and the logic of Bertrand RUSSELL he developed a formalist approach in which the axiomatization and consistency of all mathematics was sought. The programme was most fully developed, in collaboration with Paul Bernays (1888–1977), in their *Grundlagen der Mathematik* (2 vols, 1934–39). The work of GÖDEL, however, has thrown doubt on the viability of the formalist approach.

Hill, (Norman) Graham (1929–75) *British motor racing driver and World Drivers' Champion in 1962 and 1968.*

The son of a stockbroker's clerk, Hill studied engineering at Hendon Technical College (1942–45) and served a five-year apprenticeship with an engineering company before entering the Royal Navy in 1950 for two years' national service. In 1953, in response to an advertisement, Hill paid to drive a racing car for a few laps of Brands Hatch. This made him determined to become a racing driver and he resigned his job in engineering to offer his services as a mechanic in return for occasional drives. He joined Colin Chapman's Lotus team as a mechanic and by 1958 was a regular driver. In 1960 he joined BRM and two years later, driving the new V-8 BRM, won four Grands Prix and the Drivers' Championship. Runner-up in 1963, 1964, and 1965, in 1966 he won the Indianapolis 500 driving a Lola. 1967 saw Hill's return to Lotus and the following year he drove the highly competitive Lotus 49 to three wins and his second championship title. During the 1969 US Grand Prix, Hill crashed and broke both legs but, with extreme tenacity and courage, he returned to racing in 1970. However, apart from victory in the 1972 Le Mans 24-hour race driving a Matra with Henri Pescarola, he never matched his former success.

Hill became a well-known television personality and a noted campaigner for disabled drivers. In 1975, he and five members of his recently formed Embassy-Hill racing team were killed in a light aircraft crash after returning from a test session in France. The Embassy-Hill cars never raced after Hill's death.

Hillary, Sir Edmund Percival (1919–) *New Zealand explorer and mountaineer who, in 1953 with the Sherpa mountaineer Tenzing Norgay (c. 1914–86), first climbed to the summit of Mount Everest. He received a knighthood in the same year in recognition of his achievement.*

Born in Auckland, Hillary spent two years at Auckland University College and worked on his father's farm before joining the Royal New Zealand Air Force in 1943. He served as a navigator aboard flying boats in the Pacific during the war. Afterwards he worked as an apiarist in partnership with his brother, an interest he has maintained between expeditions. The first of these was in 1951 – the New Zealand Garwhal Expedition to the Himalayas – after which he joined the British Everest Reconnaissance, the British Cho Oyu Expedition (1952), and the British Expedition (1953) led by Sir John HUNT. With Tenzing, Hillary reached the 8848 m peak on 29 May 1953. The

following year, Hillary led the New Zealand Himalayan expedition to the Barun Valley, east of Everest, and in 1955 he headed the New Zealand contingent of the Trans-Antarctic Expedition organized by Vivian FUCHS. During this, Hillary made an overland journey to the South Pole, reaching it in January 1958.

Hillary made further expeditions to the Everest region and in 1967 led the first ascent of Mount Herschel in Antarctica. He has also navigated the Sun Kosi river in the Himalayas (1968) and in 1977 led a powerboat expedition up the Ganges river from its mouth, continuing to its source on foot. From 1984 to 1989 he was New Zealand high commissioner in New Delhi. His books include *High Adventure* (1956), *From the Ocean to the Sky* (1979), his autobiography *Nothing Venture, Nothing Win* (1975), and (with Peter Hillary) *Two Generations* (1983).

Hilton, James (1900–54) *British novelist, best known as the author of* Lost Horizon *and* Goodbye Mr. Chips.

The son of a Walthamstow schoolmaster, Hilton attended schools in his home town before being sent to Cambridge, first to the Leys School and then to Christ's College. He graduated in 1921. His first novel, *Catherine Herself* (1920), had already been published; this encouraged Hilton to spend the next decade writing novels, none of which achieved any real success. However, *Lost Horizon* (1933) had a world-wide triumph and gave the word 'shangri-la' to the language to denote a distant or unattainable paradise, named after the fabulous valley in the Himalayas described in the book. Hilton immediately followed *Lost Horizon* with *Goodbye Mr. Chips* (1934), another instant success.

He was invited to Hollywood to create the film scripts of these two stories and lived there from 1935 until his death. As a scriptwriter he gained considerable esteem, winning a Hollywood Motion Picture Academy award for his script of *Mrs. Miniver* (1942). He also continued to produce competent and highly popular novels, among them *Random Harvest* (1941; filmed in 1942), *So Well Remembered* (1947), and *Time and Time Again* (1953).

Himmler, Heinrich (1900–45) *German Nazi leader and head of the SS.*

Born in Munich, the son of a schoolmaster, Himmler was educated at a high school in Landshut, Bavaria, before joining the Bavarian infantry as a cadet-clerk in 1917. After World War I he studied agriculture at Munich Technical College, becoming a poultry farmer in

Bavaria and associating with a right-wing paramilitary nationalist organization. A participant in the Beer Hall putsch in 1923, which attempted to overthrow the government, he became progressively more involved with the Nazis. In 1929 he was chosen by HITLER as head of the SS, the Führer's personal bodyguard. In 1933 he became chief of police in Munich and a year later was given command of the Prussian secret police (Gestapo). By 1936 he was in charge of the entire German police system.

During World War II Himmler's powers expanded. Appointed Reich Commissar for the consolidation of German Nationhood in 1939, he was responsible for organizing the mass genocide of non-Nordic people, especially Jews. In 1943 he became minister of the interior, gaining control of the civil service and the courts. Throughout these years he built up the influence of the SS, enlarged concentration and extermination camps, and authorized pseudomedical experiments on prisoners.

Towards the end of the war Himmler began to lose influence. His approach to the Allies through Count BERNADOTTE (April 1945), calling for capitulation, angered Hitler, who ordered his arrest. Himmler fled, only to be caught by British troops near Bremen in May 1945. He committed suicide shortly afterwards.

Hindemith, Paul (1895–1963) *German composer and viola player.*

Hindemith decided on a musical career very young and despite parental opposition, lack of means, and minimal schooling left home at eleven to earn a living playing in café and theatre orchestras. Eventually he studied at the Hoch Conservatory in Frankfurt and later under Arnold Mendelssohn (1855–1933) in Darmstadt. At twenty he was appointed leader of the Frankfurt Opera orchestra, a post he held for eight years. In 1921, with the Turkish violinist Licco Amar, he formed the Amar Quartet, which toured extensively in central Europe and was known for its performance of contemporary works. Hindemith also appeared increasingly often as a soloist (he gave the first performance of WALTON's viola concerto at a Promenade Concert in London in 1929). In 1927 he was appointed professor at the Berlin Hochschule für Musik. During this time he was becoming well known as a composer; he was among the founders of the Donaueschingen festivals of the 1920s and later of those at Baden-Baden. In the 1930s Hindemith incurred the disfavour of the Nazi regime, both

for his music and for his philosophical concepts. He joined his former colleague Amar in Turkey, undertaking a plan for the entire redevelopment of Turkish musical life. In 1939 he settled in the USA, where he was frequently engaged as soloist, becoming head of the music department of Yale University and also teaching at the Tanglewood Music Center. His Charles Eliot Norton lectures given at Harvard University (1949–50) were later expanded into his book *A Composer's World* (1952). In 1953 he returned to Europe and settled in Zürich, while continuing to tour as conductor and soloist.

Hindemith's music is basically contrapuntal in texture, grounded in the tradition of Lutheran German baroque and dictated by his philosophy of music as 'an agent of moral elevation'. His early experiments were iconoclastic (he was one of the first composers to incorporate jazz elements) but soon gave way to a neoclassical style in which he used the established forms of concerto grosso, passacaglia, toccata and fugue, and occasionally sonata. His idea of himself of a master craftsman, rather than a virtuoso genius, is manifest in his *Gebrauchsmusik* ('utility music'): compositions especially written for amateurs to play at home. Hindemith was a prolific composer in all genres, his works including the operas *Cardillac* (1926) and *Mathis der Maler* (1938; *Mathis the Painter*), the ballet *Nobilissima visione* (1936), orchestral and chamber music, and choral works. He was also the author of textbooks and other writings, including *The Craft of Musical Composition* (1934–36).

Hindenburg, Paul Ludvig von Beneckendorf und (1847–1934) *German field-marshal who rose to prominence in the later stages of World War I to become second president of the Weimar Republic.*

The son of a Prussian aristocrat, Hindenburg was commissioned into the 3rd Foot Guards in 1866. He served with distinction in the Austro-Prussian War of 1866 and the Franco-Prussian War (1870–71). His subsequent career was unremarkable, promotion duly occurred, and he retired a general in 1911. Recalled at the outbreak of World War I, Hindenburg was sent to the East German front, where, with the brilliant tactician Major-General Erich von LUDENDORFF, he headed off the Russian invasion with celebrated victories at Tannenburg and Masurian Lakes in August and September 1914. Hindenburg took the credit and was elevated to field-marshal, commanding all forces on the eastern front. In 1915 the Russians were pushed back some three hundred miles. The following year, Hindenburg, with Ludendorff as his chief of staff, was given command of the western front. Together they constructed the Siegfried–Stelling defensive line – known as the Hindenburg line – in France, and introduced unrestricted submarine warfare at sea – a move that eventually led to US involvement on the Allied side. The duo were by now a powerful political as well as military force, contriving the replacement of Chancellor Theobald Bethmann Hollweg (1856–1921) by the more compliant Georg Michaelis (1857–1936). However, following Germany's defeat in 1918, Hindenburg largely retired from public life until 1925, when he succeeded Friedrich Ebert (1871–1925) as president of the Weimar Republic. He used his presidential power to implement expenditure on the armed forces instead of much-needed social measures but was re-elected president in 1932. With the growing prominence of HITLER, Hindenburg was persuaded to sack the moderate Heinrich Brüning (1885–1970) in favour of the right-wing Franz von Papen (1879–1969) as chancellor. In January 1933, Hitler finally became chancellor and subsequently banned all opposition parties. Hindenburg's authority was now totally eclipsed and when he died, Hitler assumed the title of Führer (Leader).

Hines, Earl (Kenneth) (1903–83) *Black US jazz pianist often known as 'Fatha' Hines. One of the most influential of jazz musicians, he retained his talent and vitality throughout his long life.*

Born in Duquesne, Pennsylvania, he grew up in a musical family and studied piano playing in Pittsburgh. While still a student he began playing in clubs before moving to Chicago in 1922. By 1927 he had recorded with Louis ARMSTRONG and in 1928 joined his Hot Five. In that year he also recorded eight of his piano solos, as well as several immortal numbers with groups led by Louis Armstrong and the legendary Chicago clarinettist Jimmy Noone. His piano style called for octaves and tremolo effects in the right hand, with a startling single-note virtuosity, while using the left for far more than just bass lines. He was the first jazz pianist to make the keyboard sound like an orchestra.

From 1930 to 1947 he led big bands, usually called the Grand Terrace Band (after his home base in a Chicago night club). In the mid-1940s the band contained several young modernists, such as Charlie PARKER and Dizzy GILLESPIE, who were making innovations of their own. In

1948 the ensemble was disbanded and Hines rejoined Armstrong and his All Stars for three years. In the 1950s Hines slipped into obscurity, but was rediscovered by public and critics alike at concerts in New York in 1964. From then on he toured the world and recorded many solos, both at concerts and in the studio. His technical command and the sheer joy in his music-making lasted until his death in hospital following a heart attack.

Hinshelwood, Sir Cyril Norman (1897–1967) *British physical chemist, who was awarded a half-share in the Nobel Prize for Chemistry in 1956 for his work on chemical kinetics. He was knighted in 1948 and awarded the OM in 1960.*

The son of an accountant, Hinshelwood was educated at Balliol College, Oxford, having first spent the years of World War I working as a chemist in the explosives supply factory at Queensferry. He spent most of his active career at Oxford, becoming Dr Lee's Professor of Chemistry (1937–64). Thereafter he held a senior research fellowship at Imperial College.

Hinshelwood established his reputation with *The Kinetics of Chemical Change in Gaseous Systems* (1926), which contained an investigation of chemical chain reactions. He continued to work in the field of chemical kinetics, seeking in *The Chemical Kinetics of the Bacterial Cell* (1946) to open up new areas of research. In 1950 he came close to solving the problem of DNA, just three years before the genetic code was broken by WATSON and CRICK. He was also a man of considerable learning, well known as a connoisseur and collector of Chinese porcelain and as a formidable linguist. Indeed, for a period he had the unique distinction of serving as president of the Royal Society (1955–60) and of the Classical Association at the same time.

Hirohito (1901–89) *Emperor of Japan (1926–89), who ruled as a divinity until Japan's defeat in World War II.*

Born in Tokyo, 124th direct descendant of Jimmu, Japan's first emperor, he was the first Japanese crown prince to travel abroad. On his return from a visit to Europe in 1921, Hirohito became regent, when his father Yoshihito was declared insane. In 1924 he married Princess Nagako Kuai, and in the same year survived an assassination attempt. In 1926 he became emperor, after his father had died. He remained aloof from politics but was unsympathetic to his government's prewar expansionism. In 1945 he ordered the Japanese forces to accept unconditional surrender from the Allies. In so doing he broke with the precedent of imperial silence and spoke to his nation personally over the radio. Hirohito became a constitutional monarch by the terms of the new constitution, imposed by the Americans in 1946. His main interest was in marine biology, on which he wrote several books. He was succeeded by his son, Akihito (1933–), in 1989.

Hiss, Alger (1904–) *US public servant who was allegedly part of a communist intelligence network operating in the USA during the 1930s.*

Hiss graduated in political science from Johns Hopkins University in 1926 and then entered Harvard Law School (1926–29). After a spell in private legal practice, he joined the Department of Agriculture in 1933 and three years later moved to the State Department as assistant to Francis Sayre, assistant secretary of state. In 1944, Hiss joined the Office of Special Political Affairs, preparing policy and legal briefs for United Nations affairs. He attended the Yalta conference in February 1945, as one of President ROOSEVELT's advisers, and served as temporary acting secretary-general for the 1945 UN conference held in San Francisco. The following year he was elected president of the Carnegie Endowment for International Peace.

It was against this impeccable record that, on 3 August 1948, Whittaker Chambers – a senior editor at *Time* magazine and self-confessed former member of an underground communist network – denounced Hiss as an accomplice to the House Committee on Un-American Activities. Chambers's assertions, upheld by congressman Richard M. NIXON, were that Hiss had copied and filmed State Department documents. Hiss's typewriter and also microfilm that Chambers claimed to have hidden inside a pumpkin – the so-called 'pumpkin papers' – were produced as evidence. Hiss's first trial in 1949 resulted in a hung jury, but the following January he was convicted of perjury and sentenced to jail. The case led to an upsurge of anticommunist feeling and the rise to prominence of Senator Joseph MCCARTHY. Hiss was released in 1954, still maintaining his innocence.

Hitchcock, Sir Alfred (1899–1980) *British film director who lived in Hollywood after 1940. Known as the 'Master of Suspense', he made his reputation with a unique brand of thriller. He was knighted in 1980.*

Born in London and educated by Jesuits, Hitchcock entered films after having worked as a commercial artist. He made his debut as a

director with *The Pleasure Garden* (1925), but more typical of his early work was *The Lodger* (1926). Alma Reville, whom he married in 1926, collaborated on many of his filmscripts. His first venture in sound was *Blackmail* (1929), originally silent and hastily converted to meet the new trend. Notable films of the 1930s included *The Man Who Knew Too Much* (1934), *The Thirty-Nine Steps* (1935), and *The Lady Vanishes* (1938).

Rebecca (1940), his first American film, won a best picture Oscar and earned him a best director nomination; other notable films of the 1940s were *Spellbound* (1945) and *Notorious* (1946). Some of Hitchcock's finest work, however, came in the 1950s and 1960s, including *Strangers on a Train* (1951), *Rear Window* (1954), *Vertigo* (1958), *North by Northwest* (1959), *Psycho* (1960), and *The Birds* (1963). His last films included *Frenzy* (1972) and *Family Plot* (1976). His films, in all of which he appeared in crowd scenes or tiny parts, had in common the ability to make audiences sit on the edges of their seats. In 1979 he was presented with the American Film Institute's Life Achievement Award.

Hitler, Adolf (1889–1945) *German chancellor and Nazi leader (1933–45). One of the most powerful and evil leaders in the history of the world, he achieved a following in his native Germany and Austria that enabled him to make a serious bid for world domination.*

Born in Braunau, the son of a customs official, Hitler left school in 1905. He had an ambition to become an artist but, after failing to be admitted to the Academy of Fine Arts in Vienna in 1907, he became an architect's draughtsman. During World War I he served as a lance corporal in the Bavarian army, winning the Iron Cross.

In 1919 Hitler co-founded the National Socialist Workers' Party, which later became known as the Nazi Party. As one of the organizers of the unsuccessful 'Beer Hall' putsch in Munich in 1923, which attempted to overthrow the government, he served thirteen months in Landsberg Prison, where he dictated his political testament *Mein Kampf* to his secretary, Rudolf HESS. Following his release in 1924 he worked to build up the Nazi Party, which increased its numbers in the Reichstag from 12 in 1928 to 232 in 1932. In 1933 he was elected chancellor and by means of devious political intrigue acquired dictatorial powers for an initial period of four years. On the death of HINDENBURG in 1934 he combined the offices of president and chancellor to become

Hobbs, Jack

'Der Führer' (leader). By then he had acquired considerable support from right-wing industrialists and bankers throughout Germany.

Over the next few years Hitler embarked on a massive programme of rearmament. At home, he destroyed German democracy by murdering his political rivals (in 1934 one hundred people were executed in what was known as the 'night of the long knives'), introduced a fanatical antisemitic policy, and whipped up emotional support for his regime through elaborate propaganda and mass meetings at which he and his audiences attained a degree of hysteria that now make old film clips of these events appear comic. In a series of military moves on the way to world domination, he reoccupied the Rhineland (1936), established the Rome–Berlin Axis (1936), and annexed Austria (1938), Sudetenland (1938), and Czechoslovakia (1939). A nonaggression pact with the Soviet Union (1939), negotiated by his foreign minister RIBBENTROP, enabled him to invade Poland in 1939. This brought Germany into a state of war with Britain and France. France soon fell (1940) and the rest of continental Europe succumbed to his jack-booted troops. However, when he attempted to invade the Soviet Union (1941) he was forced to retreat after disastrous losses at Stalingrad in 1943. As the tide of the war turned against him, so did factions in Germany. He survived an assassination attempt in 1944 but in the final days of the war, with British, US, and Soviet armies advancing on Berlin, he committed suicide (1945) in his bunker, shortly after marrying his mistress, Eva Braun.

Ironically, Hitler's failure to achieve world domination can be attributed to his fanatical antisemitism, which caused his Jewish physicists to flee to Britain and the USA, where they created the atomic weapons that would have enabled him to succeed. In the end, he and his followers achieved the most devastating war in the history of man, the genocide of six million people, and the ineradicable disgrace and humiliation of Germany.

Hobbs, Jack (Sir John Berry Hobbs; 1882–1963) *Surrey and England opening batsman who scored 61 237 runs and completed 197 centuries in an illustrious career from 1905 to 1934 (excluding the war years). He was knighted in 1953.*

Hobbs was born in Cambridge but qualified to play for Surrey. In his first match for the county, against a side led by W. G. Grace, he scored 88; two weeks later, in his championship debut against Essex at the Oval, he made

a century before lunch. From then on his great reputation was quickly established. He was selected to go to Australia in 1907–08 and went on to make 61 test appearances, with 15 centuries and a batting average of 56.94. His most notable partner was the Yorkshire batsman Herbert SUTCLIFFE.

Seventeen times he exceeded 2000 runs in a season and in 1926, at Lord's, he made 316 not out against Middlesex. He was a man of unaffected charm, dignity, and good nature; England probably never had a more popular cricketer. After giving up the game he ran a sports shop in London.

Hochhuth, Rolf (1931–) *German playwright.*

Born at Eschwege near Kassel, Hochhuth first worked as a bookseller and in publishing after leaving school. Since 1963 he has lived in Switzerland. His reputation was firmly established by his first two highly controversial plays, both written in free verse – *Der Stellvertreter* (1963; translated as *The Deputy* or *The Representative*, 1964) and *Soldaten* (1967; translated as *Soldiers*, 1968). Both plays deal with issues arising from World War II, have historical figures as protagonists, and were based on meticulous research, which led to Hochhuth's being classified as a 'documentary dramatist'. The label, though accurately reflecting a careful gathering and presentation of evidence, does not do justice to the dramatic and poetic power of much of Hochhuth's writing, in which the factual or documentary element is secondary to his artistic purpose. Against those playwrights who believe that tragedy in the classical sense is no longer possible in the theatre, Hochhuth has argued that a freedom of choice does exist in times of historical crisis and that tragic errors of judgment occur.

Der Stellvertreter portrayed PIUS XII as too concerned with diplomatic manoeuvres to take a decisive stand against the Nazi policy of exterminating the Jews. The pope thus forfeits his role as the deputy, or vicar, of Christ, which is fulfilled instead by the Jesuit priest Riccardo. The violent controversy aroused by this accusation had hardly settled before another arose over *Soldaten*. In it CHURCHILL is portrayed as arranging the death of General Sikorski (1881–1943) for political reasons and backing a brutal policy of bombing German cities despite a forceful argument against it (presented in the play by the Bishop of Chichester). A misinterpretation of evidence may have been involved in the first charge; in any case, Sikorski's Czech pilot brought a libel action and was awarded damages against Hochhuth in 1972.

It is perhaps inevitable that Hochhuth's more recent work has not met with such passionate or anxious responses. It includes *Guerrillas* (1970), the comedies *Die Hebamme* (1971; 'The Midwife') and *Lysistrate oder die NATO* (1974), *Tod eines Jägers* (1976; 'Death of a Hunter'), on HEMINGWAY's suicide, and *Judith* (1984). He has also published a book of essays, *Krieg und Klassenkrieg* (1971; 'War and Class War'), two novellas, and a novel, *Eine Liebe in Deutschland* (1978; translated as *German Love Story*, 1980).

Ho Chi Minh (Nguyem That Thanh; 1890–1969) *First president of the Democratic Republic of Vietnam (1945–54) and of North Vietnam (1954–69). One of the foremost leaders of the procommunist and anticolonial movements in South-East Asia, Ho Chi Minh successfully led his country to independence from the French.*

Born in Kim-lien, North Vietnam, the son of a minor official, Ho Chi Minh was educated at a Franco-Annamite school. For a number of years he worked as a teacher (1907–11) and a seaman (1912–16) before he moved to England (1915) and then to France (1917). Agitating for the end of colonialism in Indochina, he attended the Versailles Peace Conference (1919) and in 1920 became a founder member of the French Communist Party. Three years later he moved to Moscow, where he studied at the East University (1923–24). As a translator with the Soviet consulate in Canton, he travelled to China in 1924 and began to organize the Vietnamese nationalist movement. In 1927 he returned to Moscow after the split between the Chinese nationalists and the communists.

Ho Chi Minh founded the Indochinese Communist Party in 1930. Arrested by the French as a revolutionary, he escaped to Moscow in 1932 and in 1938 returned to China. Three years later he formed the communist-dominated Vietnam League for Independence, known as the Viet Minh. During World War II he organized the Viet Minh in guerrilla raids against the Japanese, who had overthrown the French in Vietnam, and when the Japanese were defeated by the Allies in 1945 Ho Chi Minh entered Hanoi and declared Vietnam an independent republic with himself its first president. The return of the French to Vietnam after the war led to eight years of bitter conflict (1946–54) culminating in the final defeat of the French at Dien Bien Phu. At the Geneva conference (1954) Ho agreed to the partition of Vietnam along the 17th parallel and assumed the presidency of North Vietnam. When the second Indochina war began in 1959 he gave support to the Viet Cong guerrilla movement in South Vietnam, although he became less actively involved in running the government. When Saigon fell to the communists in 1975, it was renamed Ho Chi Minh City in his honour.

Hockney, David (1937–) *British painter, etcher, and draughtsman. Hockney won international success while still a student at the Royal College of Art, London. His charismatic personality and life style have also contributed to his wide public recognition.*

Hockney's initial training was at his local art school in Bradford. At the Royal College (1959–62) he developed an individual style that was representational but deliberately naive, showing the influence of abstract art, children's drawings, and graffiti. A light ironic humour characterizes these pictures, whose themes are usually either autobiographical, as in *Flight into Italy – Swiss Landscape*, or, in his own words, 'propaganda for homosexuality', as in *We Two Boys Together Clinging*.

Between 1964 and 1967 he taught in the USA and developed his Californian style. Rejecting abstraction, these paintings depict figures amidst flat, almost shadowless, architecture, lawns, and swimming pools. A characteristic painting, *A Bigger Splash*, is one of many that reveal Hockney's delight in rendering moving water. After 1968 Hockney began to create deeper space in his pictures without abandoning his striving towards clarity, as in the many closely observed double portraits. His etched illustrations for *Grimm's Fairy Tales* in 1969 were a perfect vehicle for his wit and sense of fantasy. His many drawings of friends in this period began as a form of note taking on his frequent travels but in the early seventies, during his Paris period, drawing became an end in itself. A similar development took place with photography in the eighties, as seen in *David Hockney: Cameraworks* (1984) and *Hockney on Photography* (1988). From the mid-seventies Hockney moved constantly between homes in Paris, London, New York, and Los Angeles. He became an Associate of the Royal Academy in 1985.

Hodgkin, Dorothy Mary Crowfoot (1910–) *British chemist noted for her work on the structures of penicillin, vitamin B₁₂, and* *zinc insulin. She was the first British woman to be awarded the Nobel Prize for Chemistry (1964) and in 1965 she was awarded the OM.*

Dorothy Crowfoot was born in Egypt, as her father worked in the Egyptian Ministry of Education, and was educated at Somerville College, Oxford. After graduating in 1932, she spent two years as a research student at Cambridge under J. D. BERNAL before returning to Oxford in 1934; she has since spent most of her career there, first as a fellow of Somerville, from 1960 until 1977 as Wolfson Research Professor of the Royal Society, and as a fellow of Wolfson College (1977–82). She was also Chancellor of Bristol University (1970–88). In 1937 she married Thomas Hodgkin (1910–82), the distinguished Africanist; the couple had three children and campaigned together in many political battles. Despite the demands of her family and her politics, Dorothy Hodgkin has managed to lead a scientific career noted both for its productivity and its creativity.

The most decisive intellectual influence on Hodgkin was undoubtedly J. D. Bernal. When she joined him he was just beginning to explore the possibility of determining the three-dimensional structure of such complex biological molecules as proteins. The most talented of the group was building up was Dorothy Hodgkin. The work involved much patience and enormous effort; for example, the elucidation of the structure of insulin was only completed in the 1970s. Her first major breakthrough came in 1949, when she published the structure of penicillin; it was followed in 1956 with the structure of vitamin B$_{12}$. It was for these two latter achievements that Dorothy Hodgkin was awarded her Nobel Prize.

Hoffman, Dustin Lee (1937–) *US actor. Very much a product of the theatre and the changing face of Hollywood in the 1960s, Hoffman has always been more of an actor than a film star.*

Born in Los Angeles, Hoffman studied music at Santa Monica City College before turning to acting. From the Pasadena Playhouse he went on to take a variety of jobs, including working in a psychiatric hospital. When possible he appeared in local theatres and on television. By 1966 he had won an Obie Award as best off-Broadway actor for *The Journey of the Fifth House* and in the same year he also appeared in the successful farce *Eh?*.

He made his screen debut in *Madigan's Millions*, which was not released until 1969, after his major success in *The Graduate* (1967).

This was followed by fine character performances in John Schlesinger's *Midnight Cowboy* (1969), *Little Big Man* (1970), in which he aged from teenager to centenarian, *Papillon* (1973), *Lenny* (1974), as comedian Lenny Bruce, and *All The President's Men* (1976). He was nominated for Academy Awards for *The Graduate*, *Midnight Cowboy*, and *Lenny* and won an Oscar for his performance in *Kramer vs Kramer* (1979), the outstanding film in which he fights his estranged wife for custody of their seven-year-old son. He won another Oscar and a Golden Globe award for his performance in *Rain Man* (1989). Other memorable films include *Straw Dogs* (1971), *Marathon Man* (1976), and *Tootsie* (1982), in which he played the part of a man masquerading as a woman. He returned to the stage in 1984 in a Broadway production of *Death of a Salesman* and made an acclaimed London debut in 1989 in *The Merchant of Venice*.

Hogan, Ben (William Benjamin Hogan; 1912–) *US professional golfer whose total of sixty-two US tournament victories has only twice been surpassed.*

Hogan's success is attributable to innumerable hours of practice and a tough, relentless, and competitive spirit. In 1948 he had eleven tournament wins, including the US Open. The following year Hogan suffered multiple fractures in a motor accident. The doctors said he would never play golf again but he proved them wrong by winning the US Open in 1950 and both the US Open and the Masters tournaments in 1951. In 1953 he achieved the unique feat of winning the US Open, the US Masters, and the British Open.

Holiday, Billie (Eleanora Fagan Holiday; 1915–59) *Black US jazz singer, often described as the greatest true jazz singer of all time.*

Born in Baltimore, she was called 'Billie' after a film star. Her father, Clarence Holiday (1900–37), played the guitar in jazz bands, but she was not introduced to jazz until hearing recordings of Bessie SMITH and Louis ARMSTRONG in the brothel in which she worked. Escaping from the brothel, she became a singer in obscure Harlem clubs, where she was discovered by record producer John Hammond; she made her first records with Benny GOODMAN's band in 1933.

During the period 1935–42 she made well over a hundred records with small jazz groups, usually led by pianist Teddy Wilson. Made quickly and cheaply for sale in the 'race' market, they are all now regarded as jazz classics.

Her style was inimitable: she subtly altered a melody and phrased a lyric as if she was a jazz musician playing a solo; singing slightly behind the beat, she created an illusion of langour, resignation, or wistfulness. Having had no training, and possessing no technical knowledge, she sang the way she did instinctively. Asked about her skill, she protested that she didn't know any other way to do it. She also sang briefly with the bands of Count BASIE, Artie SHAW, and others, appeared in films (notably *New Orleans*, 1947), and gave concerts at Carnegie Hall. She appeared on television and toured Europe in the 1950s; her acerbic ghosted autobiography, *Lady Sings the Blues*, was published in 1956. (A biographical film of the same name, made in 1973, is not highly regarded.)

Deeply scarred by her early poverty and humiliation, she battled in later life against the heroin addiction from which she died – sadly and prematurely. As a *coup de grâce* she was arrested for possession of narcotics on her hospital deathbed; the charge was widely believed to be false.

Holland, Sir Sidney George (1893–1961) *New Zealand statesman and National Party prime minister (1949–57). He was knighted in 1957.*

Born and educated in Canterbury, Holland worked for a hardware firm on leaving school and joined his father's haulage business in 1912, when the latter became mayor of Christchurch. Enlisting in the Field Artillery in 1914, he was invalided home in 1918 and after the war began several commercial ventures, including his own engineering company.

Holland first entered parliament in 1935, when he was elected as the member for Christchurch North, replacing his father – who was incapacitated by an accident two weeks before the election – as candidate. Elected leader of the National Party in opposition in 1940, Holland developed and consolidated the National Party, fighting three elections before finally winning office in 1949. Holland's ministry was characterized by the greatest possible encouragement to private enterprise; he was a great advocate of Britain and supported the British government's Suez policy in 1956. Holland retired from politics in 1957 to breed sheep and cattle on his farm in North Canterbury.

Holst, Gustav Theodore (1874–1934) *British composer and teacher.*

Of Swedish lineage, Holst was born in Cheltenham and entered the Royal College of Music in 1893, winning a composition scholarship in 1895 and studying with Charles Stanford (1852–1924). He also studied the organ and piano but because of neuritis in his hand switched to the trombone, which gave him the opportunity for orchestral experience. While still a student he met Ralph VAUGHAN WILLIAMS, who became a valued colleague and lifelong friend. In 1898 Holst joined the Carl Rosa Company as first trombone and répétiteur, the first of several orchestral appointments. In 1903 he gave up orchestral playing to concentrate on composition; after many of his scripts had been rejected, a chance request to deputize as a music teacher in a Dulwich school opened up a new and fruitful career that led to his appointment as music master at St Paul's Girls' School (1905) and musical director of Morley College (1907), positions he held concurrently to the end of his life. At St Paul's he had the opportunity to write music undisturbed in his spare time and the stimulus of composing for enthusiastic pupils; at Morley he was able to rehearse choirs and ensembles twice a week. In 1919 he was also appointed professor of composition at the Royal College of Music; he held a similar post at Reading College from 1919 to 1923. Overwork led to a breakdown in health and a period of enforced rest in 1924, the year he was elected fellow of the Royal College of Music.

Holst had little ambition to compose large works but wrote a number of choral works for his choirs and the *St Paul's Suite* (1913) for the string orchestra at St Paul's. His study of folksong led to such pieces as the *Somerset Rhapsody* (1907), and his interest in Sanskrit resulted in a series of works including the four sets of *Rig-Veda Hymns* (1908–12) and the chamber opera *Savitri* (1908). With the orchestral suite *The Planets* (1914–16; performed at a Philharmonic concert under Sir Adrian BOULT in 1919) Holst became a celebrity; his *Hymn of Jesus* (1917) confirmed his success when it was performed by the Royal Philharmonic Society and Choir in 1920. Festivals were now eager to commission new works and his operas were performed: *The Perfect Fool* (1921) at Covent Garden in 1923; *Savatri* and *At the Boar's Head* (1924) in Manchester in 1925. Holst's music is full of melody, often in the folksong idiom. His own preferred composition was the orchestral *Egdon Heath* (1927). Biographies of Holst were written by his daughter Imogen Holst (1907–84), who was also a composer and teacher.

Holt, Harold (1908–1967) *Australian statesman and Liberal prime minister (1966–67).*

Born in Sydney, the son of a theatrical businessman, Holt was educated in Melbourne and graduated in law from Melbourne University in 1930. He first entered politics in 1935, when he won the safe federal seat of Fawkner for the United Australia Party. In 1956 he became deputy leader of the Liberal Party and was elected unopposed as leader and prime minister in 1966 following the retirement of MENZIES.

As prime minister, Holt was preoccupied with the Vietnam War. Remembered for his statement 'all the way with LBJ', he gave an open-ended commitment of military support to the USA in Vietnam. He also encouraged friendship with Asia and relaxed Australia's restricted immigration laws. Known for political loyalty, Holt was a less dominant leader than his predecessor. He died in a swimming accident at Portsea, Victoria, while in office.

Holyoake, Sir Keith Jacka (1904–83) *New Zealand statesman and National Party prime minister (1957; 1960–72). He became a CH in 1963 and was knighted in 1978.*

Born in Scarborough, near Pahiatua, the son of a storekeeper, Holyoake was educated in Tauranga, Hastings, and Motueka before working on the family farm, which he eventually bought and diversified. He first entered politics in 1932, winning a by-election as the Reform Party candidate for Motueka and becoming the youngest member in the House of Representatives. Elected as the member for Pahiatua in 1943, he became deputy leader of the opposition (National Party) in 1946, deputy prime minister and minister for agriculture in 1949, and leader and prime minister for a brief period in 1957, following the retirement of Sir Sydney HOLLAND. Elected prime minister a second time in 1960, he remained in office until his retirement (1972). Resigning from parliament in 1977, he was appointed governor-general, a position he held until 1980.

One of New Zealand's longest serving statesmen, Holyoake favoured a property-owning democracy and was known for his attachment to farming and the outdoors. Appointed chairman of the General Council of the FAO in 1955, he was awarded honorary degrees from the universities of South Korea and Victoria, Wellington, and received the freedom of the City of London.

Honegger, Arthur (1892–1955) *French-born composer of Swiss parentage.*

Born in Le Havre into a comfortable bourgeois environment, Honegger was not only musically talented but also a first-class athlete with a passion for engines and boats. He studied music for two years at the Zürich Conservatory, absorbing the German classics; when he entered the Paris Conservatoire in 1912 a new world opened to him in the music of DEBUSSY on the one hand and that of the young composers who were to constitute the group Les Six (including AURIC and MILHAUD) on the other. During World War I he served in the Swiss army for a time, returning to Paris in 1916. Although he was a member of Les Six, it was an allegiance of friendship rather than ideology: 'I do not follow the cult of the fair and the music-hall but, on the contrary, that of chamber music and symphony in their most serious and most austere aspect'. Honegger visited the USA in 1929 on a concert tour of his works, his wife, the pianist Andrée Vaurabourg (1894–), playing his *Concertino* for piano and orchestra (1925). During the last decade of his life he was professor of composition at the École Normale de Musique in Paris. His book of reminiscences *I Am a Composer* (1951) suggests personal disenchantment with his career.

Honegger's double nationality was a prime factor in his development, combining a Protestant seriousness with the flair, opportunism, and outrageous showmanship of Parisian life in the twenties. His style is basically polyphonic, often polytonal, and of great complexity. His avowed model was J. S. Bach. Honegger considered melody to be 'the touchstone of all successful works' and underlaid it with massive harmonic structures. Central to his work are the five symphonies dating from 1930 to 1951. His catalogue of works includes operas, oratorios, such as *Le Roi David* (1921; *King David*) and *Jeanne d'Arc au bûcher* (1935; *Joan of Arc at the Stake), ballet scores, including Skating Rink* (1921), in line with the twenties vogue for sports ballets, incidental music for plays, radio and film music, choral and vocal works, orchestral works, of which *Pacific 231* (1924) reflects both his passion for trains and his virtuosic use of motor rhythms, and much chamber, piano, and instrumental music.

Hoover, Herbert C(lark) (1874–1964) *US Republican statesman and thirty-first president of the USA (1929–33).*

Hoover was born in West Branch, Iowa, to Quaker parents, who were both dead by the time he was nine years old. Herbert went to live with a maternal uncle in Newburg, Oregon, where he was educated at the Quaker academy. He graduated from Stanford University as a mining engineer in 1895 and his subsequent engineering career, which took him to Australia, China, India, Africa, Canada, and Britain, won him a worldwide reputation and considerable wealth. In China at the time of the Boxer Rebellion (1900), Hoover became involved in war relief work, an experience that stood him in good stead as head of Allied relief operations in London during World War I. On the entry of the USA into the war in 1917 President Woodrow WILSON appointed Hoover national food administrator. In the immediate postwar period he was responsible for the distribution of food to the starving millions in European countries hit by famine.

Hoover entered politics in 1921 as secretary of commerce to President Warren G. Harding (1865–1923) and continued in this position under President COOLIDGE. During his extended period in office (1921–29) he radically reorganized the department and served as chairman on commissions that led to the construction of the Hoover Dam and the St Lawrence Seaway. On Coolidge's decision not to run for re-election in 1928, Hoover campaigned successfully for the Republican nomination and won a massive victory over his Democratic opponent, Alfred E. Smith, in the presidential election. Just seven months after his inauguration in March 1929, Hoover found himself faced with the immediate and long-term problems resulting from the stock market crash and the Great Depression. His decision to supply federal aid only to public works and financial institutions, leaving the support of the unemployed masses to private charities and local government, earned him widespread criticism and severely marred the humanitarian reputation he had built up through his wartime relief work. In the 1932 election he was soundly defeated by the Democratic candidate, Franklin D. ROOSEVELT.

During World War II, Hoover became involved in relief work once again; in 1946 President TRUMAN appointed him coordinator of food supplies to avert the threat of postwar famine. He subsequently led the two Hoover commissions (1947–49; 1953) aimed at streamlining federal government procedures, eliminating waste, and improving efficiency. Having retired from active politics at the age of eighty, he devoted the rest of his life to writing and public speaking on international and national affairs.

Hoover, J(ohn) Edgar (1895–1972) *US lawyer and director of the Federal Bureau of Investigation (FBI) from 1924 until his death in 1972.*

Born in Washington, DC, Hoover helped support his family by working as a messenger at the Library of Congress after the death of his father. At the same time he took evening classes in law at George Washington University, receiving his bachelor's degree in 1916 and his master's degree the following year. In 1917 he entered the Department of Justice as an attorney, rising to the office of special assistant to the attorney-general in 1919 and assistant director of the FBI in 1921. On his appointment as director of the FBI in 1924, Hoover set about reorganizing the bureau along more efficient lines. He established better methods for recruiting and training personnel, expanded and consolidated the fingerprint files, instituted a crime-detection laboratory, and founded the FBI National Academy for the training of police officers from all parts of the USA. During Hoover's term of office the FBI dealt successfully with the gangsters of the twenties and thirties, the spy rings of World War II, the communist sympathizers and subversives hunted out in the MCCARTHY era, and the civil rights disturbances of the sixties. Throughout this period Hoover strove to maintain an apolitical stance and to shield the FBI from political control, concentrating his efforts on creating one of the most efficient law-enforcement machines of all time.

Hope, Bob (Leslie Townes Hope; 1903–   ) *British-born US comedian and film actor, whose dry wit and mastery of the wisecrack have made him an international superstar.*

Hope, who was born in Eltham, London, and taken to the USA at the age of four, reached stardom and films through vaudeville, Broadway, and radio. After several shorts he appeared in *The Big Broadcast of 1938* (1938), in which he sang the number that was to become his theme song, 'Thanks for the Memory'. The success of *The Cat and the Canary* (1939) was followed by *Road to Singapore* (1940), the first of the seven popular 'Road' films that brought Bob Hope together with Bing CROSBY and Dorothy Lamour (1914–   ). The rivalry between Crosby and Hope on the screen and on the golfcourse became a standard part of both their repertoires.

Among Hope's many films were *My Favourite Blonde* (1942), *The Paleface* (1948), *Fancy Pants* (1950), and *The Facts of Life* (1960). His numerous television and club appearances, as well as his shows to entertain the troops, have ensured his continuing popularity. In addition to several volumes of autobiography, including *The Road to Hollywood* (1977), he has published several other humorous books. In recognition of his charitable works and contributions to the film industry he has received several Special Academy Awards.

Hopkins, Sir Frederick Gowland (1861–1947) *British biochemist and one of the principal founders of biochemistry, whose discovery of vitamins earned him the 1929 Nobel Prize for Physiology or Medicine. He was knighted in 1925 and received the OM in 1935.*

Born in Eastbourne, Sussex, Hopkins worked briefly after leaving school in an insurance office before training as an analytical chemist. He studied for an external degree at London University and obtained an assistant's post with a Home Office forensic specialist. In 1888 he enrolled at Guy's Hospital, London, to study medicine. Qualifying in 1894, he continued at Guy's until 1898, when he joined the University of Cambridge to teach physiology and anatomy. In 1901, Hopkins isolated the amino acid tryptophan. He subsequently demonstrated that this and various other amino acids – the essential amino acids – were essential dietary components for many animals, including humans. In 1912, Hopkins published the results of a series of feeding trials in which young rats had been fed the necessary protein, fat, carbohydrate, and salts in purified form. They failed to grow normally, causing Hopkins to conclude that certain essential 'accessory food factors' were missing. These were later termed vitamins. Also at this time, he was proving, with Walter Fletcher, that muscle contraction produces lactic acid. He also isolated the tripeptide glutathione, which functions as a hydrogen acceptor and coenzyme in metabolism. Hopkins was a consultant on nutrition during World War I and advised on the manufacture of vitaminized margarine, first retailed in 1926. But he was foremost the founder of a world-famous school of biochemistry and the inspiration for an entire generation of biochemists.

Hopkins, Harry Lloyd (1890–1946) *US Democratic administrator, who had a prominent role in advising on and administering Franklin D. ROOSEVELT's New Deal relief programme.*

Hopkins was born in Sioux City, Iowa, and graduated from Grinnell College in 1912. During the next fifteen years he worked on a number of social welfare projects in New York, spending some years with the Association for Improving the Condition of the Poor and serving with the Red Cross in the latter part of World War I. In 1931, at the height of the Depression, he was appointed director of the Temporary Emergency Relief Administration by Roosevelt, then governor of New York, and on Roosevelt's inauguration as president in 1933 Hopkins became federal emergency relief administrator. He advised and encouraged the president on the New Deal relief programmes of the mid-1930s, aimed at providing work for the unemployed on public projects, food and clothing for the needy, and a variety of social and economic reforms. Ignoring the accusations of waste and extravagance hurled at him by his opponents, as director of the Works Progress Administration (WPA) Hopkins spent some 8500 million dollars on unemployment aid and brought relief to more than fifteen million Americans.

By the beginning of Roosevelt's second term as president Hopkins had become one of his closest advisers, serving as secretary of commerce (1938–40). Prevented by ill health from pursuing his own political ambitions, he travelled to London and Moscow during World War II as the president's personal representative, negotiating with CHURCHILL and STALIN on such questions as arms supplies and military strategy. In 1941 he was appointed head of the lend-lease programme to provide aid and support to the Allies. After Roosevelt's death in April 1945, Hopkins continued for some months to serve as presidential assistant, visiting Moscow on behalf of the new president, Harry S. TRUMAN.

Hopper, Edward (1882–1967) *US realist painter of contemporary American life.*

Hopper was born in Nyack, New York, and studied at the New York School of Art (1900–06). During the period 1906–10 he made several visits to Europe, after which he was obliged for financial reasons to work as a commercial illustrator for several periodicals (1913–23).

Recognition of Hopper's paintings came in the 1920s, when he exhibited at the Whitney Studio Club in a one-man show and in group exhibitions. In 1924 he married the painter Josephine Nivison. Hopper was eventually able to concentrate his efforts on painting; his mature work is known for its presentation of a certain mood in pictures of American city life, such as *Early Sunday Morning* (1930) and *Nighthawks* (1942). Everyday scenes are portrayed in stark reality, conveying an atmosphere of desolation or loneliness. Still figures are shown in exterior and interior settings in introspective isolation. From the 1930s Hopper's work underwent no major changes. He became a recognized exponent of American realism but always denied alliance to any school or having any social message. His work portrayed a personal and emotive view of the American scene.

Horowitz, Vladimir (1904–89) *Russian-born pianist, resident in the USA.*

After studying with Felix Blumenfeld (1863–1931) at the Kiev Conservatory, Horowitz began giving concerts at the age of twenty: his prodigious talent amazed even the most sophisticated audiences. His first visit to London was in 1928; a 1930 performance of the Rachmaninoff D minor concerto with Willem Mengelberg (1871–1951) conducting the London Symphony Orchestra has become legendary. In 1928 he first toured the USA, giving recitals and playing concertos with the leading orchestras; he decided to settle there, and in 1933 he married TOSCANINI's daughter Wanda. Horowitz's technical perfection and stylish interpretation made him the king of players. In 1936 an illness that started with appendicitis and extended into a series of nervous complications threatened to shorten his career. However, he made a comeback in Paris and London in 1939 with renewed mastery. He officially retired in 1950 from the concert platform, but continued to give occasional recitals and to record. In 1986 he returned to the Soviet Union to give a concert. His many awards included the Royal Philharmonic Society Gold Medal (1972).

Horthy de Nagybánya, Miklós (1868–1957) *Hungarian naval officer who became head of state after World War I.*

Descended from a long line of Protestant aristocrats, Horthy entered the Imperial Naval Academy at Fiume (now Rijeka, Yugoslavia) at the age of fourteen. He joined the Austro-Hungarian navy in 1886, rising to lieutenant by 1900. While he was commanding the *Taurus*, stationed off Constantinople (now Istanbul), his diplomatic activities found favour with Emperor Franz-Joseph, who made Horthy his aide-de-camp in 1909. During World War I, Horthy commanded the battleship *Hapsburg* and the cruiser *Norvara*. In 1917 he became a national hero after breaking through the Allied blockade across the Straits of Otranto and returning safely to his Adriatic port. The following year he was promoted to rear-admiral and later supervised the transfer of the Austro-Hungarian fleet to Yugoslavia.

1919 saw a communist revolution in Hungary led by Béla Kun (1886–?1939). Horthy organized an army to crush the communists, entering Budapest in November. He then presided over the White Terror, in which any suspected communist sympathizers, including many Jews, were either interned or shot. The Hungarian parliament appointed Horthy regent to the monarch, but when the exiled Charles I returned in 1921, Horthy prevented him from reclaiming the throne. In the 1930s, Horthy played an active role in his country's attempts to accommodate HITLER, who ceded parts of Czechoslovakia and Romania to Hungary, which, in November 1940, joined the Axis powers. But Horthy soon became an obstacle to the Germans and they abducted him to a concentration camp in 1944. The Allies held him in Nuremberg prison but he was released without trial in 1946 to spend his remaining years in Portugal.

Hounsfield, Sir Godfrey Newbold (1919–) *British electronics engineer who invented the computerized axial tomography (CAT) X-ray scanner for use in clinical diagnosis. In recognition of this he was awarded the 1979 Nobel Prize for Physiology or Medicine. He was knighted in 1981.*

Hounsfield was born in Newark, Nottinghamshire, and attended the City and Guilds College, London, where he qualified in radio communications. During World War II he served in the Royal Air Force, spending some time at Cranwell Radar School as a lecturer. In 1947 he enrolled at Faraday House Electrical Engineering College, receiving his diploma in 1951 and in the same year joining EMI. He worked initially on radar systems and then computer design, becoming a project engineer for the first large solid-state computer to be manufactured in Britain, the EMIDEC 1100. Expertise in computer systems led him in 1967 to start work on the CAT scanner. This entailed using a computer to construct a cross-sectional planar image of an organ or body using the information from a series of axial transverse X-ray scans – a form of tomography. In this way a much more detailed picture of soft tissues could be obtained, compared with conventional X-radiography, with improved detection and resolution of tumours, blood clots, and other features. The first CAT scanner was installed in 1971, for taking brain scans, and four years later the first whole-body scanners were introduced. Hounsfield was senior staff scientist (1977–85) and subsequently consultant at Thorn EMI and continued development of the CAT scanner as well as investigating the potential of nuclear magnetic resonance as a diagnostic tool. He became a fellow of the Royal Society in 1975.

Housman, A(lfred) E(dward) (1859–1936) *British poet and classical scholar.*

The eldest son of a Worcestershire solicitor, Housman was educated at Bromsgrove School. Despite the death of their mother (1870), the family was in general a close and happy one. Housman showed an early talent for nonsense verse for family entertainments. He won a scholarship to St John's College, Oxford (1877), but failed his final examination there, partly perhaps through intellectual arrogance but also on account of emotional turmoil over his love for a fellow-student, Moses Jackson.

In 1882 Housman took a post in the Patent Office, studying Greek and Latin in his spare time to redeem his Oxford failure. His classical publications in the 1880s so impressed the academic world that in 1892 he was appointed professor of Latin at University College, London. In 1896 his collection of poems, *A Shropshire Lad*, was published; it gradually became one of the best-loved volumes of English poetry. In 1903 the first volume of Housman's classical magnum opus, his edition of Manilius' *Astronomica*, appeared (the last came out in 1930), and in 1905 his edition of *Juvenal* was published.

In 1911 Housman became professor of Latin at Cambridge, where he remained until his death. In addition to his work on Manilius, he published an edition of *Lucan* (1926), and his lectures on *The Application of Thought to Textual Criticism* (1921) and *The Name and Nature of Poetry* (1933) attracted considerable attention. *Last Poems* (1922) was the only other volume of poetry to appear in his lifetime. A few weeks before his death Housman refused the office of poet laureate, as he had earlier refused the Order of Merit (1929). *More Poems* (1936) and *Collected Poems* (1939) contain new material published at the discretion of Housman's brother, the writer Laurence Housman (1865–1959). Some of these poems deal more explicitly with the poet's homosexuality than Housman cared to publish during his lifetime.

Howard, Sir Ebenezer (1850–1928) *British town planner and founder of the garden-city movement. He was knighted in 1927.*

Born in London, Howard started work in the City as a clerk but later became a shorthand writer at the law courts. In the period 1872–77 he visited the USA, where he was influenced by the ideas of Walt Whitman, Ralph Waldo Emerson, and Edward Bellamy. From them he conceived the idea of a garden city, an independent unit built in the countryside, where each house had its own grounds and all the other urban land was in public ownership. Howard envisaged a unit limited to a population of 30 000 surrounded by a green belt of agricultural land. These ideas, an attempt to curb the unplanned growth of the suburbs of large cities, were published in his *Tomorrow: a Peaceful Path to Real Reform* (1898) and republished four years later as *Garden Cities of Tomorrow*. The book was immediately effective and in 1902 the Garden City Association was founded. By 1903 the first garden city, Letchworth in Hertfordshire, was begun. Welwyn Garden City followed in 1920. The concept has been used throughout the world and was of considerable influence in the postwar towns created in the UK.

Howard, Leslie (Leslie Howard Stainer; 1893–1943) *British actor, director, and producer, who portrayed the archetype of the English gentleman in a series of films.*

The son of Hungarian Jewish immigrants, Howard was born in London and educated at Dulwich College. Shell-shocked during World War I, he joined a touring company in 1917 and made his first film, *The Happy Warrior*, the same year. It was in the USA, however, that his career began to flourish. Plays included *Her Cardboard Lover*, with Tallulah Bankhead, and *Outward Bound*, made into a film in 1930. With his impeccable English accent, natural charm, and youthful good looks, Howard is remembered best for his portrayal of *The Scarlet Pimpernel* (1935). Other memorable films include *Berkeley Square* (1933), *Of Human Bondage* (1934), *The Petrified Forest* (1936), *Pygmalion* (1938), which he co-directed with Anthony Asquith (1902–69), the epic *Gone With the Wind* (1939), and *Intermezzo* (1939).

With the outbreak of World War II he returned to England and made several suitably patriotic films, including *Pimpernel Smith* (1941), in which his son Ronald Howard (1918–) also appeared, *49th Parallel* (1941), and the story of the Spitfire, *The First of the Few* (1942), which he also wrote and directed. He died when his plane was shot down by German aircraft on the return flight from a lecture tour in Spain and Portugal during World War II.

Howard, Trevor Wallace (1916–88) *British actor, whose many roles included a relaxed English gentleman, a Nottinghamshire miner, and a distressed priest.*

The son of an English father and Canadian mother, Howard was educated at Clifton College and the Royal Academy of Dramatic Art, making his stage debut in 1936 at the Shakespeare Festival, Stratford-on-Avon. During World War II he served in the First Airborne Division, returning to the stage in 1944, in which year he made his film debut in Carol REED's *The Way Ahead*.

Notable among his early films was David LEAN's *Brief Encounter* (1945) with Celia Johnson (1908–82), whom he was to play opposite again thirty-five years later in the television play *Staying On*. Other exceptional films included *The Third Man* (1949), *Outcast of the Islands* (1951), and *The Heart of the Matter* (1953). He received a British Film Academy Award for *The Key* (1958) and was nominated for an Oscar for *Sons and Lovers* (1960) and *The Count of Monte Cristo* (1976). Memorable, too, were his portrayals of Captain Bligh in the remake of *Mutiny on the Bounty* (1962) and Lord Cardigan in *The Charge of the Light Brigade* (1968). His later films included *Sir Henry at Rawlinson End* (1980) and *The Missionary* (1983). Stage and television appearances continued until his death. He received the Emmy Award for *The Invincible Mr Disraeli* (1963) and was nominated for an award for *Napoleon at St Helena* (1966).

Hoyle, Sir Fred (1915–) *British astronomer, writer, and broadcaster, one of the authors of the steady-state theory. He was knighted in 1972.*

Hoyle was educated at Bingley Grammar School and the University of Cambridge, becoming lecturer in mathematics in 1945. He soon established his reputation as an original thinker with his proposal in 1948 that the universe has remained basically unchanged, the observed expansion of the universe being compensated by the continuous creation of matter. A similar theory was suggested also by Hermann BONDI and Thomas Gold (1920–). Hoyle's second major contribution to cosmology came in 1957 when, with Margaret BURBIDGE, Geoffrey BURBIDGE, and William Fowler (1911–), he proposed how elements could be produced in the interior of stars.

Boyle became Plumian Professor of Astronomy (1958"72) and in 1962 he was appointed director of a newly created Institute of Theoretical Astronomy. Hoyle found it impossible to work harmoniously with the Cambridge radio astronomers, however, and when an application for a £50,000 grant for a new computer was turned down by the Science Research Council he decided to leave Cambridge. Since 1972 he has held no permanent academic appointment, working as Visiting Professor at a number of institutions in both Britain and the USA.

In later years, mainly in collaboration with Chandra Wickramansinghe (1939–), Hoyle has tackled the problem of the origin of life on Earth. In such works as *Lifecloud* (1978), *Origin of Life* (1980), and *Cosmic Life Force* (1988), they have argued for an extraterrestrial origin. Hoyle has continued to defend his modified steady-state theory against the evidence of radio astronomers that the microwave background is more compatible with its main rival, the big-bang theory. He has also written science fiction and children's books. He published an autobiography, *The Small World of Fred Hoyle*, in 1986.

Hua Guo Feng (Hua Kuo-feng; 1920–) *Chinese prime minister (1976–80) and chairman of the Communist Party (1976–81). Known for his ability to combine pragmatism with ideology, he attempted to follow in MAO TSE-TUNG's footsteps but failed to find support with other Chinese leaders.*

Born in Shansi province, the son of a peasant, Hua joined the Red Army when he was fifteen and took part in the epic 'long march' led by Mao (1934–35). He fought against the Japanese during World War II and against the Nationalist Party under CHIANG KAI-SHEK, emerging as party secretary of the Shansi province at the end of the civil war (1949).

During the 1950s Hua advanced through the ranks of the Communist Party organization. In 1958 he was appointed vice-governor of Hunan province, a position he held until 1967. Two years later he was elected to the central committee of the party; in 1971 he was appointed to lead the enquiry into the abortive coup of Lin Biao (1905–71), abortive coup of Lin Biao, becoming a member of the Politburo in 1973. In 1975 he became deputy prime minister under CHOU EN-LAI and consolidated his position as designated heir to the premiership in his role as chief editor of Mao's writings. After the death of Chou En-lai in 1976 Hua was named prime minister; when Mao died later in the same year he was also appointed chairman of the Communist Party and of the

Military Affairs Commission, becoming the first person to hold all three posts at the same time.

Hua was quick to denounce the so-called 'Gang of Four' when he came to power, having Mao's widow and other radicals arrested. However, Hua was increasingly criticized by his powerful deputy premier DENG XIAO PING; by 1982 Hua had been stripped of all three crucial posts.

Hubble, Edwin Powell (1889–1953) *US astronomer, who discovered Hubble's law, and established the reality of the expanding universe on experimental evidence.*

The son of a lawyer, Hubble studied law at the University of Chicago and Oxford University. An excellent athlete, he fought against the French boxer Georges CARPENTIER and was offered a match with the world champion, Jack JOHNSON. Hubble began to practise law in 1914 but the attraction of astronomy became too strong and he returned to the University of Chicago, where he gained his PhD in 1917. After a year's military service in France, Hubble joined the Mount Wilson Observatory in California, where he spent the rest of his life.

Using measurements of galactic red shifts obtained by his colleagues at Mount Wilson, Hubble found that there was a linear relationship between the velocity of the receding galaxies and their distance from the earth. Hubble thus demonstrated the reality of the expanding universe. He went on to use this linear relationship, since known as Hubble's law, to measure the universe. He assigned to the knowable universe a radius of about 18 billion light-years and an age of about 2 billion years. Both figures have been revised by a factor of about 10 since Hubble's death but the principles he laid down remain unchanged.

Hughes, Howard Robard (1905–76) *US industrialist, aviator, and film producer whose obsessive privacy in later life led to considerable speculation and a notorious fraud case.*

When his father died in 1924, Hughes took control of Hughes Tool Company, which held the patent on a drilling bit used in oil and gas exploration. This was the basis of Hughes's fortune. After he had attended the Rice Institute of Technology, Houston, and also the California Institute of Technology, in 1926 Hughes made his debut as a Hollywood film producer. *Hell's Angels* (1930), featuring Jean HARLOW, was a box-office success. Later Hughes films included *Scarface* (1932) and *The Outlaw* (1941), with Jane Russell (1921–   ) in her first starring role.

The founding of the Hughes Aircraft Corporation enabled Hughes to indulge his passion for flying and in 1935 he set a new record speed of 352 mph in a plane of his own design. Three years later he made a record round-the-world flight in 91 hr 14 min piloting a Lockheed 14. Hughes was less successful with his massive eight-engined flying boat designed for 750 passengers. It managed a flight of just one mile, in 1947.

Hughes shunned public life for the last twenty-five years of his life, living in almost total seclusion except for his bodyguard. In 1971, author Clifford Irving sold the publication rights for what he claimed to be Hughes's personal memoirs. These were subsequently discovered to be a fake and Irving and his wife were convicted of fraud. Hughes died on a flight from Acapulco to Houston for medical treatment.

Hughes, Ted (Edward James Hughes; 1930–   ) *British poet. He was appointed poet laureate in 1984.*

Hughes was born in Yorkshire and took his degree at Pembroke College, Cambridge. In 1956 he married the American poet Sylvia PLATH. His first volume of poetry, *The Hawk in the Rain* (1957), made an immense impact and at once Hughes was acclaimed as a major new poetic talent. He won a Guinness Poetry Award (1958), was awarded a Guggenheim Fellowship (1959–60), and in 1961 won the Hawthornden Prize. Throughout the 1960s Hughes was a leading poetic influence, publishing *Lupercal* (1960) and *Wodwo* (1967), as well as several children's books. *Crow* (1970) contains perhaps his finest achievement in the creation of the sinister and violent anti-hero of the title. Hughes's unique vision of the natural world as terrible and violent is continued in later collections, such as *Cave Birds* (1978) and *River* (1983). He has also published a number of stories, written radio plays, and edited the poems of Keith Douglas (1964) and Emily Dickinson (1968). His most recent publications include *Wolfwatching* (1988) and *Moortown Diary* (1989).

Hughes, William Morris (1862–1952) *Australian statesman and prime minister (1915–23). He was awarded the CH in 1941.*

Born in London, of Welsh parents, Hughes was educated in London as a student teacher before emigrating to Australia at the age of twenty. He studied law at Sydney University and was admitted to the bar in 1903.

Hughes first entered politics when he was elected to the New South Wales parliament in

1894. Transferring to federal parliament in 1901, he became leader of the Labor Party and prime minister in 1915. Leaving the party in 1917 over the issue of conscription, which he strongly supported, Hughes continued in office as the head of the newly formed Nationalist Party. In 1919 he attended the Paris Peace Conference, where he secured an Australian mandate for part of New Guinea and other German colonies in the Pacific. Resigning as prime minister in 1923 after the break-up of his coalition government, he remained politically active and was appointed attorney general (1939–41) and a member of the War Advisory Council (1941–45). His death, at eighty-eight, ended a political life spanning fifty-eight years.

Hunt, (Henry Cecil) John, Baron (1910–) *British mountaineer and explorer who led the 1953 expedition on which Edmund* HILLARY *and Tenzing Norgay (c. 1914–86) made the first ascent of Mount Everest. Hunt received a knighthood in 1953 and was created Baron Hunt of Llanfair Waterdine in 1966.*

Hunt was born in India, the son of a British army officer. He was educated at Marlborough College and the Royal Military College, Sandhurst, after which, in 1930, he was commissioned in the King's Royal Rifle Corps as a second lieutenant. He later served with the Indian police and spent his vacations mountaineering, joining three Himalayan expeditions before the outbreak of World War II. Serving with the KRRC in Italy, he was awarded the Military Cross for gallantry and achieved the rank of brigadier. In 1944 he took command of the 11th Indian Infantry Brigade. After the war he became a staff officer and in 1951 moved to Supreme Headquarters Allied Powers Europe (SHAPE) with the rank of colonel. The following year he was appointed leader of the successful British Everest Expedition. Hillary and Tenzing reached the summit on 29 May 1953.

After retiring from the army in 1956, Hunt served as director of the Duke of Edinburgh's Award Scheme (1956–66) and was rector of Aberdeen University (1963–66). During the Biafran crisis in Nigeria, Hunt acted as personal advisor to the prime minister and led the government relief mission to the stricken area. He has served on many other committees and boards. In 1981 he gave his support to the newly formed Social Democratic Party; subsequently he allied himself with the Social and Liberal Democrats. His books include *The Ascent of Everest* (1953) and *The Red Snows*

(1959). Among his many other awards and honours, he was made a CBE in 1945 and a KG in 1979. He also won the DSO in 1944.

Hussein (ibn Talal) (1935–) *King of Jordan (1953–). A controversial Middle-Eastern figure, he has maintained an ambivalent attitude to the Arab League, to Egypt, and to Israel, which has made his reign both precarious and dangerous.*

Born in Ammar, the son of King Talal and claiming descent from the prophet Mohammed, Hussein was educated at Victoria College, Alexandria, before attending Harrow School and Sandhurst Military Academy in England. In 1953 at the age of eighteen he succeeded his father, who was deposed by parliament as a result of mental illness.

While his English education has encouraged him to maintain close ties with Britain, financial aid from the USA and his membership of the Arab League have given him an invidious position in Middle-East politics. In the 1948–49 Arab-Israeli War he gained the West Bank for Jordan, but lost it to Israel by his ill-advised participation in the Six Day War (1967). When large numbers of Palestinians from the West Bank then crossed the river into Jordan, al-Fatah guerrillas established themselves on the East Bank, from which they launched raids on Israel. Hussein chose to use his army to subdue these guerrillas in order to avoid Israeli retaliations, which brought him into disfavour with other Arab states. The uneasy peace among the Arabs was maintained by Hussein's participation in the Yom Kippur War (1973) and his recognition of the PLO as the body entitled to govern the West Bank if it is recovered from Israel. In 1990 he was heavily involved in negotiations between Saddam HUSSEIN of Iraq and the leaders of the nations allied against him following Iraq's invasion of Kuwait. His lack of support for the US-led alliance during the ensuing Gulf War (1991) made him unpopular in the UK and USA, as well as with many of the Arab members of the alliance.

Hussein has been married four times; his second wife was an Englishwoman, Toni Gardiner (married 1961–72). His present wife, who he married in 1978, is the American Elizabeth Halaby (known as Nur el Hussein). He has survived several assassination attempts and has reflected upon his career in his books *Uneasy Lies the Head* (1962) and *My War with Israel* (1963).

Hussein, Saddam (1937–   ) *Iraqi dictator, whose attempts to make Iraq a regional superpower led to the Gulf War of 1991.*

Born into a Sunni Muslim peasant family, Saddam joined the radical Ba'ath party in 1957. In 1959 he received a death sentence for the attempted murder of Abdul Kassim, prime minister of Iraq, but escaped abroad. After studying law in Cairo, he returned to Iraq when the Ba'athists seized power in 1963. Following a second coup later that year, he joined a plot against the new regime, leading to his imprisonment in 1964. Escaping once more, he worked in the Ba'athist underground and took a leading role in the coup that brought the party back to power in 1968. As deputy chair of the Revolutionary Command Council, Saddam became effective ruler of Iraq, assuming absolute power in 1979 when he appointed himself president, prime minister, and head of the armed forces.

Domestically, Saddam's rule has been characterized by the systematic use of terror against political opponents and the creation of an extravagant personality cult. There has been widespread nationalization, notably of the oil industry in 1972. Internationally, his goals have been dominance in the Gulf region and the leadership of the Arab world. In 1980 Iraqi forces seized oil fields inside Iranian borders; resistance proved stiff and the ensuing Iran–Iraq war settled into a long stalemate, vastly expensive in both lives and resources. The conflict ended in 1988 with no significant gain on either side. During this period Saddam also used poison gas against rebellious Kurdish villages in northern Iraq, killing and injuring thousands and provoking widespread international protests.

The expansion of Iraq's armed forces continued unabated after the war, as did Saddam's stockpiling of sophisticated modern weapons. In 1990 he ordered the invasion of Kuwait and announced its annexation, in defiance of UN resolutions calling for Iraq's withdrawal. A US-led multinational force, authorized by the UN, was assembled in Saudi Arabia and, when Saddam still refused to comply, massive air attacks were launched on targets in Iraq and Kuwait, followed by a ground operation in which Iraqi forces were totally overwhelmed. Despite Saddam's personal humiliation, the destruction of much of his army, and the devastation of Iraq's economy, rebellions in several parts of the country after the war failed to dislodge him from power.

Husserl, Edmund (1859–1938) *German philosopher, logician, and founder of the modern school of phenomenology.*

Born at Prossnitz (now Prostějov, Czechoslovakia), Husserl studied at the universities of Leipzig and Berlin, where he read mathematics under Karl Weierstrass, and at the University of Vienna, where he came under the influence of the psychologist Franz Brentano. Husserl held appointments at Halle (1887–1901), Göttingen (1901–16), and finally Freiberg, retiring in 1928.

Husserl's first work, *Philosophie der Arithmetik* (1891), sought to derive arithmetical concepts from psychological principles. It was savagely criticized by FREGE, and in his next work, *Logische Untersuchungen* (2 vols, 1900–01; translated as *Philosophical Investigations*, 1970), Husserl rejected all such psychological approaches to logic. Logical laws, he declared, were necessary and therefore could not possibly be seen as empirical generalizations. He consequently tried to construct a 'pure' logic dealing with the concepts common to all sciences. To complete his ambitious programme Husserl proposed a phenomenological approach, i e a scrupulous attention to the abstract entities, be they propositions, meanings, or essences, arising from conscious (especially intellectual) processes.

In later works, such as his *Ideen zu einer reinen Phänomenologischen Philosophie* (1913; translated as *Ideas*, 1931), Husserl argued, in a most Cartesian way, that consciousness was the one absolute thing that cannot be thought away. This allowed him to attempt to develop a 'transcendental phenomenology' through which the world of objects can be approached.

Huston, John (1906–87) *US film director, screenwriter, and actor.*

Son of actor Walter Huston (1884–1950), John Huston was born in Nevada, Missouri. His early life was spent touring with his parents, after which his activities included acting, writing, and becoming a cavalryman in Mexico. For a time he collaborated on screenplays but in 1941 he had his first opportunity to direct a film, the classic detective film *The Maltese Falcon*, starring Humphrey BOGART, Mary Astor (1906–87), Peter Lorre (1904–64), and Sidney Greenstreet (1879–1954). During World War II he saw service with the US army and directed outstanding documentaries, including *The Battle of San Pietro* (1944) and *Let There Be Light* (1945). Postwar success quickly followed with *The Treasure of the Si-*

erra Madre (1948), which won Oscars for best director, best screenplay, and best supporting actor (his father). Notable among the many films that followed were *The Asphalt Jungle* (1950), *The Red Badge of Courage* (1951), *The African Queen* (1952), *Moby Dick* (1956), *The Misfits* (1961), *The Night of the Iguana* (1964), *Fat City* (1972), *The Man Who Would Be King* (1975), the screen version of the musical *Annie* (1982), and *Prizzi's Honour* (1985).

Huston also acted in his own and other people's films, including *The Bible* (1966), as Noah, and *Winter Kills* (1977). His colourful life – including several marriages – was the subject of William Nolan's book *John Huston: King Rebel* (1965) and his autobiography, *John Huston: An Open Book* (1980).

Hutton, Sir Leonard (1916–90) *Yorkshire and England cricketer who scored a record 364 against the Australians at the Oval in 1938. The first professional to captain the England team, he is also one of the limited number of players to have scored a century of centuries. He was the second cricketer to be knighted (in 1956) for services to the game.*

He was born and brought up at Pudsey, Yorkshire, and played for his native county from 1934 to 1955. During the war he served with the Royal Artillery and the Army Physical Training Corps, returning to cricket in 1946 with a right arm shortened because of a wartime accident. He first played for England in 1937 – he reached a century in his test debut against the Australians (1938) – and in all represented his country 79 times. He finished with a test average of 56.67 and a career average that was only marginally less. Hutton had the distinction of being the first professional to be appointed captain of England on a regular basis (1952–54); he was also made a member of the MCC while still a professional player.

Huxley, Aldous Leonard (1894–1963) *British novelist. He spent the last twenty-five years of his life in California, among several other British emigrés with an interest in mysticism.*

Besides his father's distinguished Huxley connections (he was the son of the biologist T. H. Huxley), Aldous Huxley was related through his mother to the Arnolds of Rugby. His brother was the biologist Julian HUXLEY. He first attended a school his mother had established and then from a preparatory school won a scholarship to Eton (1908). However, a serious eye infection left him nearly blind and he was soon forced to leave. He learnt braille, continued his education by private tuition, and eventually an improvement in his eyesight enabled him to go up to Balliol College, Oxford (1913–16). There he scored both academic and social successes and at nearby Garsington met many of the significant writers of the period.

After a period of financial struggle, Huxley's witty and daring novels of the 1920s made him a fashionable figure in the febrile society that he satirized in *Crome Yellow* (1921), *Antic Hay* (1923), and *Those Barren Leaves* (1925). *Point Counter Point* (1928) established his reputation in both Britain and the USA. His vision of a nightmare Utopia, *Brave New World*, appeared in 1932, and *Eyeless in Gaza* in 1936. Huxley also wrote short stories, essays, and travel books.

With increasing financial security Huxley and his wife (whom he had married in 1919) travelled abroad. In 1938 they decided to settle in California, an overriding factor in their decision being Huxley's belief that the Californian sunshine would help his eyesight. In this he was correct; with the aid of eye exercises his sight improved dramatically by mid-1939 and in 1942 he published a book on his recovery, *The Art of Seeing*. The novels of Huxley's Californian years are more deeply understanding of the human condition than those of the 1920s. They include *After Many a Summer* (1939), which won the James Tait Black Memorial Prize, *Time Must Have a Stop* (1944), *Ape and Essence* (1948), *The Devils of Loudun* (1952), and *Island* (1962), which portrays an optimistic Utopia to contrast with *Brave New World*. Huxley developed his own brand of philosophy and was in the forefront of the band of Californian emigrés interested in oriental mysticism, especially Zen. He propounded his views in such books as *The Perennial Philosophy* (1941). In 1953 he experimented with psychedelic drugs, writing of his experiences in *The Doors of Perception* (1954) and *Heaven and Hell* (1956). In 1955 Huxley's wife died of cancer and the following year he married again, but in 1960 he himself developed the cancer from which he eventually died.

Huxley, Sir Julian Sorell (1887–1975) *British biologist and scientific administrator who contributed much to the beneficial use of science in society. He was knighted in 1958.*

The grandson of the famous biologist T. H. Huxley (1825–95), and brother of the writer Aldous HUXLEY, Julian Huxley studied at Eton and Balliol College, Oxford, receiving his zoology degree in 1909. He spent some time studying marine sponges at the Naples Zoological Station before returning to Oxford in 1910 as a zoology lecturer. Two years later he moved to the Rice Institute, Houston, Texas, as a research associate to establish the biology department. In 1916 he returned home to enlist in the Intelligence Corps. After the war he was made a fellow of New College, Oxford, during which time he organized the University expedition to Spitsbergen (1921). Huxley was appointed professor of zoology at King's College, London, in 1925 but resigned two years later to allow more time for research. His notable studies of the differential growth of different body parts, *Problems of Relative Growth* (1932), were but one facet of his wide-ranging interests. He wrote many popular articles and essays, especially on ornithology and evolution, and co-produced several history films, including the *Private Life of the Gannet* (1934). He adopted a firmly humanistic philosophical stance, as evidenced by *Religion Without Revelation* (1927).

Huxley served (1935–42) as secretary of the Zoological Society of London and instigated an ambitious programme of rebuilding, unfortunately never realized because of the war. He was elected a fellow of the Royal Society in 1938. He became widely known through his appearances on the BBC programme *Brains Trust*.

In 1946 Huxley was appointed as the first director-general of the newly founded United Nations Economic and Scientific Organization (UNESCO), during which time he travelled widely and identified the growing problems of population expansion and environmental destruction. No stranger to controversy, Huxley supported the contentious view that the human race could benefit from planned parenthood using artificial insemination by donors of 'superior characteristics'.

I

Ibert, Jacques François Antoine
(1890–1962) *French composer and administrator.*

Ibert, the quintessential Parisian, studied composition at the Paris Conservatoire with Paul Vidal (1863–1931), winning a number of prizes. After war service he was awarded the Prix de Rome (1919) for his cantata *Le Poète et la fée* and during his residence in Rome wrote several of his best works, including a ballet based on Oscar Wilde's *Ballad of Reading Gaol* (1920) and the symphonic suite *Escales* (1922; *Ports of Call*). Later works from his large output include the *Concertino da camera* (1934) for alto saxophone and small orchestra, the comedy opera *Angélique* (1927), the most successful of his six operas; the popular *Divertissement* (1930) for chamber orchestra; ballets; incidental theatre music and film scores; songs; characteristic piano pieces such as *The Little White Donkey*, the second of ten *Histoires*; and chamber works including the string quartet (1937–42), one of his most deeply felt works.

Ibert's style is the blend of neoclassic techniques and impressionism popular in the Paris of his heyday: 'all systems are valid, providing one derives music from them', he wrote. As administrator he was director of the French Academy in Rome from 1937 to 1960; he also administered the Réunion des Théâtres Lyriques Nationaux (1955–56) and was elected a member of the Institut de France in 1956.

Ibn Saud (*c.* 1880–1953) *The first king of Saudi Arabia (1932–53), who was largely responsible for his country's fabulous wealth as an oil producer.*

Born in exile because his father was excluded by Ibn Saud's uncle from rule of his country, he fought for over twenty years to regain his possessions – from the conquest of the sultanate of Najd in 1902 to that of Hejaz in 1924. The reconquest of the state that in 1932 became Saudi Arabia was facilitated by the Arabian tribesmen called al-Ikhwan (the Brethren), who later formed the backbone of the National Guard of Saudi Arabia. In 1933 Ibn Saud negotiated an agreement with the Standard Oil Company of California, selling them prospecting rights for oil. The discovery of oil in 1936 brought immense wealth to the country and enabled Ibn Saud to initiate a modernization programme. Unenthusiastic about the Arab League, he eventually overcame his distrust of its Hashimite leadership enough to give it grudging support. Ibn Saud had about 150 wives and many children.

Ikeda Hayato (1899–1965) *Japanese finance minister (1949–52; 1956–57) and prime minister (1960–64). Widely respected for his financial and administrative expertise, he initiated anti-inflationary policies that assisted the Japanese economy in its recovery after World War II.*

Born in Yoshina, the son of a sake distiller, Ikeda was educated at the Kyoto Imperial University, where he graduated in law and economics in 1925. In 1927 he entered the civil service as an administrative officer in the finance ministry. He worked his way through the ranks of the ministry over the next twenty-three years, rising to the post of chief of the national taxation office in 1945. Two years later he was appointed vice-minister of finance in the first YOSHIDA cabinet.

Ikeda was first elected to the lower house of parliament (Diet) in 1949 as the representative for Hiroshima. As finance minister under Yoshida, he went to Washington in 1950 to negotiate a peace treaty with the USA and attended the San Francisco peace conference as a member of the Japanese delegation (1951). In 1952 he became secretary-general of the Liberal Party (subsequently the Liberal-Democratic Party) and international trade and industry minister. He served in various other posts in the successive cabinets of Ishibashi (1956–57) and Kishi (1957–60) and also became chairman of the party's political affairs research committee. When Kishi resigned in 1960, Ikeda became president of the party and the new prime minister. He then launched a high-rate economic growth policy, eased internal political tensions, and strengthened Japanese ties with Europe, Asia, and the USA. He remained in office until 1964, when he retired because of ill health. He wrote several books of a financial nature, including a detailed explanation of Japan's taxation and property taxation laws.

Ilyushin, Sergei Vladimirovich (1894–1977) *Soviet aeronautical engineer and designer of the numerous Ilyushin planes.*

Born in the Vologda province into a large peasant family, Ilyushin began his working life as a labourer in St Petersburg. His interest in aviation was aroused while preparing the site in 1910 for Russia's first international aviation meeting. To pursue this interest he worked for some time as a security guard at the St Petersburg Kommandantiski aerodrome. He also managed, during World War I, to qualify as a pilot. With these qualifications he was admitted to the Air Force Academy after the Revolution; after graduating in 1926 he was employed on the design and development of military aircraft. He eventually became a Red Army general and professor at the Air Force Academy.

Of the many planes designed by Ilyushin, three in particular became widely known beyond the Soviet borders. The first, the IL-2 attack aircraft, known to the Germans as the 'Schwarze Tod' (black death), did much with the 36 163 models produced to repel the German invasion. The second, the IL-18, the four-engined Moskva, was one of the first turbo-prop liners, widely used in and sold to the Third World in the 1960s. The third is the IL-62 turbojet passenger aircraft.

Inönü, Ismet (1884–1973) *Turkish prime minister (1923–37; 1961–65) and president (1938–50). Although for many years considered a dictator, he emerged as a defender of democracy during his period as leader of opposition and introduced several social and political reforms to westernize Turkey.*

Born in Smyrna, the son of a lawyer, Inönü was commissioned into the Ottoman army in 1904. He served in Macedonia and Yemen; during World War I he fought at Gallipoli and commanded forces in Syria (1916). He was undersecretary of war in Istanbul when the Ottoman empire surrendered (1918).

Inönü joined Kemal ATATÜRK's forces in 1920 and became commander-in-chief of the western army fighting the Greeks in Anatolia (1921–22). He took his name from the two battles of Inönü fought in January and April 1921. A close associate of Atatürk, he was appointed minister of foreign affairs in 1922 and signed the Treaty of Lausanne on behalf of Turkey in 1923. When Turkey was proclaimed a republic in October of that year he became prime minister, remaining in this post until

1937. On Atatürk's death in 1938 he was elected president and chairman of the Republican People's Party (RPP). He continued as president throughout World War II but was defeated in 1950 by the Democrat Party, which he had permitted to be formed in 1946. He then led the opposition until the 1960 military coup; between 1961 and 1965 he served as prime minister of three coalition governments. Defeated in 1965, he remained leader of the RPP until 1972.

Inukai Tsuyoki (1855–1932) *Japanese prime minister (1931–32). His sympathy towards liberation movements in Asia, in particular the Chinese revolution, alienated the militaristic faction within the Japanese government and led to his assassination.*

Born in Okayama of samurai origin, Inukai was educated at a school in Tokyo. During the Seinan War he worked as a correspondent for the newspaper *Yubin Hochi* before co-founding the magazine *Takai Keizai Shimpo* (1880), in which he gave support to protectionist trade policies. Throughout the 1880s he worked in a variety of positions, from government official at the Institute of Statistics (1881) to chief editor of the paper *Akista Nippo* and special correspondent for *Yubin Hochi* in Korea at the time of the Seoul incident (1884).

Inukai was elected to the Tokyo Prefectural Assembly in 1885. In 1890 he entered the lower house of parliament (Diet) as the representative from Okayama, the first of nineteen terms. Appointed minister of education in the first Okuna cabinet (1898), he founded and became president of the Rikken Kokuminto (Constitutional National Party) in 1910. In 1913 he led a popular movement against the autocratic ruling regime and opened the way for more democratic cabinet elections. Nine years later the Kokuminto was disbanded and Inukai formed the Kakushin Karabu (Reformation Party) in its stead, becoming minister of communications in the Takaaki cabinet (1923). He left the government in 1924 to join the Rikken Seiyūkai (Friends of Constitutional Government), of which he became president in 1929. He became prime minister in December 1931, at the time of the Manchurian crisis, and attempted to curb the armed forces but was assassinated six months later by a group of military officers.

Ionesco, Eugène (1912–) *Romanian-born French dramatist, a pioneer of the Theatre of the Absurd.*

Born in Slatina, Romania, son of a Romanian father and a French mother, Ionesco spent most of his early childhood in France, returning to Romania at the age of thirteen. He took his first degree at the University of Bucharest; in his mid-twenties he won a scholarship to study for a doctorate in Paris. After the end of World War II he became a naturalized French citizen and settled in Paris.

While attempting to learn English, Ionesco was struck by the inadequacy of language as a valid medium for the communication of ideas; this led to his awareness of the alienation of man and the sterility of human relationships. These themes are developed in his first play, *La Cantatrice chauve* (1950; translated as *The Bald Prima Donna*, 1958); the meaningless platitudes of the dialogue, inspired by an English-language textbook, are blended with absurd logic and surrealist effects. Similar themes pervade Ionesco's subsequent plays, such as *La Leçon* (1951; translated as *The Lesson*, 1958), in which language becomes a lethal weapon, *Les Chaises* (1952; translated as *The Chairs*, 1958), *Amédée, ou comment s'en débarrasser* (1954; translated in 1958), which features an ever-growing corpse, and *Tueur sans gages* (1959; translated as *The Killer*, 1960).

With *Rhinocéros* (1959; translated in 1960) Ionesco entered the field of political satire: the play depicts a totalitarian society to which even the more rebellious characters gradually conform by turning into rhinoceroses. The horror of death, an underlying theme in many of Ionesco's earlier plays, is at its most evident in *Le Roi se meurt* (1962; translated as *Exit the King*, 1963). Later plays include the fantasy *Le Piéton de l'air* (1963; translated as *A Stroll in the Air*, 1965), *Le Soif et la faim* (1966), *Jeux de massacre* (1970), *L'Homme aux valises* (1975; translated as *Man with Bags*, 1977), and *Journeys Among the Dead* (1986). Ionesco also wrote theatrical criticism in *Notes et contre-notes* (1962), a collection of short stories, *La Photo du colonel* (1962), and revealed his strong right-wing political stance in *Journal en miettes* (1967).

Ireland, John Nicholson (1879–1962) *British composer.*

Born into a literary family, but not a happy one, Ireland grew up sensitive to music and poetry but with feelings of insecurity that lasted all his life; a marriage in 1927 was immediately annulled. Entering the Royal College of Music at fourteen, he studied piano with Frederick Cliffe (1893–97) and composition with Charles Stanford (1897–1901). In 1905 he obtained a music degree at Durham,

the university that honoured him with a doctorate in 1932. His life was spent mainly in teaching and composing, although from 1904 to 1926 he was also organist and choirmaster at St Luke's, Chelsea. Among his pupils at the Royal College of Music, where he taught composition from 1923 to 1939, were Alan BUSH, E. J. Moeran (1894–1950), Humphrey Searle (1915–82), and Benjamin BRITTEN, whose genius Ireland was among the first to recognize. The writings of Arthur Machen stimulated an enduring interest in pagan mysticism, what Ireland himself called 'racial memory'. This inspired a series of compositions including the orchestral tone poem *The Forgotten Rite* (1913) and *Legend* (1933) for piano and orchestra. It also moved him to visit places of antique mystery, such as Chanctonbury Ring and locations in the Channel Isles. His sojourn in Guernsey was cut short by the German occupation in 1940, and he spent the rest of his life in West Sussex.

Ireland's music, founded on Brahms and the teaching of Stanford, was later influenced by that of STRAVINSKY and the impressionism of DEBUSSY; it owes nothing to the English folksong revival. Ireland's output was not large and is uneven in inspiration, but some works are of outstanding interest. His early successes were in the field of chamber music, including the second violin sonata (1915–17), the cello sonata (1923), the piano sonata (1918–20) and the piano concerto (1930). His lyricism and richly melodic imagination are at their best in such works as the song cycle *Land of Lost Content* (1920–21) and in piano pieces, such as the impressionistic *Island Spell* and *Amberley Wild Brooks*. A boisterous vein is evident in *A London Overture* (1936) and such songs as 'I Have Twelve Oxen' (1918). One of his last compositions was a vivid film score for *The Overlanders* (1946–47).

Ironside, William Edmund, 1st Baron (1880–1959) *British general whose exploits in the Boer War inspired the character of Richard Hannay in John BUCHAN's novels. He received a knighthood in 1919, a peerage in 1941, and was awarded the grand cross of the Legion of Honour in 1946.*

Ironside attended the Royal Military Academy, Woolwich, before joining the Royal Artillery in 1899. Proficient in seven languages and of tall muscular build, 'Tiny' Ironside served as an intelligence officer in the Boer War, at one stage joining a German military expedition to southwest Africa disguised as a Boer driver. Promoted to captain in 1908, at the outbreak of World War I he was a major attached to the 6th Division in France. He fought in the battles of Vimy Ridge and Passchendaele in 1917 and the following year, as acting colonel, he commanded a machine gun corps on the Somme, subsequently assuming command of the 99th Infantry Brigade. In September 1918 he went to northern Russia as chief of general staff of the abortive Allied expeditionary force charged with the ludicrous objective of countering the Bolshevik revolution. His experiences as commander of the Russian town of Archangel are recounted in *Archangel 1918–1919* (1953).

After serving as a military aide to Admiral HORTHY in Hungary and as commandant of Camberley Staff College, Ironside spent time in India, was promoted to general, and in 1936 took over Eastern Command, Home Forces. He consistently urged the need to re-equip and at the start of World War II was appointed chief of general staff at the War Office. After temporarily commanding Home Forces following the Dunkirk evacuation, he ended his career as a field marshal.

Isherwood, Christopher William Bradshaw (1904–86) *British-born US-naturalized novelist and playwright.*

Isherwood was educated at Repton School and Corpus Christi College, Cambridge, from which he was sent down. His first novel, *All the Conspirators* (1928), received poor reviews and, after briefly studying medicine (1928–29), he went to teach English in Berlin (1930–33). His novels *Mr. Norris Changes Trains* (1935) and *Goodbye to Berlin* (1939) memorably recreate the atmosphere in Germany at the time. Back in London in 1934 he was a leading figure in the group of left-wing intellectuals that included AUDEN and SPENDER. He collaborated with Auden on the plays *The Dog Beneath the Skin* (1935), *The Ascent of F6* (1936), and *On the Frontier* (1938), and their visit to China in 1938 was recorded in the travel-diary *Journey to a War* (1939). Isherwood's first autobiography, *Lions and Shadows*, appeared in 1938.

Isherwood emigrated with Auden to the USA in 1939 and settled in California, eventually becoming a US citizen (1946). In 1947–48 he travelled in South America, describing his experiences in *The Condor and the Cows* (1949). His output of novels continued steadily, if not abundantly: *Prater Violet* (1945), *The World in the Evening* (1954), *Down There on a Visit* (1962), *A Single Man* (1964), *A Meeting by the River* (1967), and *Kathleen and*

Frank (1971). The autobiographical *Christopher and His Kind* (1977) treats the author's homosexuality frankly.

In the 1940s Isherwood became much interested in oriental religion. With Swami Prabhavananda he produced a fine translation of the Hindu religious classic *The Bhagavad-Gita* (1944) and a collection of the aphorisms of Patanjali (1953).

Issigonis, Sir Alec Arnold Constantine (1906–88) *British car designer, who was responsible for both the Morris Minor and the Mini. He was knighted in 1969.*

Of Greek descent (his father was a naturalized Briton), Issigonis came to England in 1918 when, with his parents, he left his home in Smyrna to escape the impending Turkish invasion. Educated at the Battersea Polytechnic, Issigonis began his working life in 1933 as a draughtsman at Rootes, the Coventry car manufacturer. In 1936 he moved to Morris Motors at Cowley and stayed with them throughout their later transformations into the British Motor Corporation and British Leyland, serving variously as chief engineer (1957–61), technical director (1961–72), and as advanced design consultant (since 1972).

Issigonis's first major triumph came with his Morris Minor in 1948. Starkly functional, it was greeted with little enthusiasm by the Morris sales staff. It went on, however, to sell 1 293 331 models before it was withdrawn in 1971. Issigonis took his ideas further in 1959 with his most famous design, the Mini Minor. Despite some initial design faults there was never any doubt that Issigonis's model would sweep the world.

Ives, Charles Edward (1874–1954) *US composer. His experimental techniques were eventually recognized as an important contribution to twentieth-century American music.*

Ives was born in Danbury, Connecticut. His father, George Ives, was an ex-bandmaster with an extraordinary flair for experimenting with sounds. From this musical background Charles Ives went to Yale University (1894–98), where he studied composition with Horatio Parker (1863–1919), a traditional teacher. Well aware that the career of composer would be unlikely to support a family, Ives entered the insurance business; by the age of forty he was head of the largest agency in the country. During this period he spent all his spare time composing. His experiments predated many techniques used by later composers, such as BARTÓK's percussive use of dissonant chords, SCHOENBERG's use of inversion and rhythmic augmentation and diminution, and STRAVINSKY's flexible metre, as well as polytonality and atonality. These radical techniques precluded Ives from publication or performance and, apart from a few private ventures, his music remained unknown.

In 1918 his health broke down and from then on he did very little composing. However, he eventually arranged for the printing of *114 Songs* and his *Concord Sonata* (1911–15), with an elaborate programme note *Essays before a Sonata*. He distributed these to libraries and music critics, creating some interest among the younger generation of American composers. The conductor Nicolas Slonimsky (1894–   ) saw some of Ives's scores and eventually conducted his *Three Places in New England* (1911–14) with the Boston Chamber Orchestra in the USA, Havana, and Paris in 1931. The following year he conducted *The Fourth of July* (1911–13) in Paris, Berlin, and Budapest. A performance of the *Concord Sonata* by the pianist John Kirkpatrick in 1939 brought Ives's music to a wider public; *Concord* itself was hailed as 'the greatest music composed by an American'. Ives was then sixty-five. The rest of his life was spent putting in order the work of a lifetime, including the four symphonies (1896–98; 1897–1901; 1901–04), a host of orchestral works, including *The Unanswered Question* (1908), and choral, vocal, keyboard, and chamber works.

J

Jackson, Glenda (1936–) *British actress who has made her name on stage, screen, and television.*

Glenda Jackson was born in Birkenhead; after working in a chemist's shop she attended the Royal Academy of Dramatic Art in London before working with various repertory companies (1957–63). In 1963 she made her film debut in *This Sporting Life* and joined the Royal Shakespeare Company. Throughout her career Jackson has combined film and stage work. After the Royal Shakespeare Company's Theatre of Cruelty Season at the London Academy of Music and Dramatic Art (1964), she played in *Marat/Sade* in London, New York, and Paris (1965) as well as performing in the film version (1967). She also appeared in the stage and film versions of both *Hedda Gabler* (1975; filmed 1976) and *Stevie* (1977; filmed 1978), about the poet Stevie SMITH. Other plays include *Hamlet* (1965), *Antony and Cleopatra* (1978), *Great and Small* (1982), *The House of Bernarda Alba* (1986), and *Mother Courage* (1990).

Of her films, *Women in Love* (1969) earned her an Academy Award and awards from the New York Film Critics and National Board of Review. A second Oscar came with the comedy *A Touch of Class* (1973). Other memorable films include *Sunday Bloody Sunday* (1971), for which she received a British Academy Award, and *The Romantic Englishwoman* (1974). She also starred in *The Incredible Sarah* (1976), as Sarah Bernhardt, *Health* and *Giro City* (both 1982), *Turtle Diary* (1985), and *The Rainbow* (1989). On the small screen she has proved no less successful, her distinctive style earning her an Emmy Award for the title role in the television series *Elizabeth R* (1971), after she had appeared in the same role in the film *Mary Queen of Scots* in the same year.

Long noted for her left-wing views, in 1990 she was adopted as the Labour parliamentary candidate for Hampstead.

Jackson, Michael (1958–) *US singer and songwriter, who became the topselling pop artist of the 1980s.*

Born in Gary, Indiana, he was the youngest of the original Jackson Five, a successful black pop group of the 1970s consisting of the five (later six) Jackson brothers. He sang on most of the group's recordings but released songs under his own name as early as 1971 when 'Got to Be There' became a major hit. However, his full-time solo career did not start until 1979, when *Off the Wall* became the bestselling album by a black artist to date; it included songs written by Paul McCARTNEY as well as songs by Jackson himself. The up-tempo album *Thriller* (1983) was even more successful and Jackson's skill as a dancer was exploited to the full in spectacular accompanying videos. Songs from the album and its successor, *Bad* (1987), confirmed Jackson's status as the most commercial US star of the decade. The success of the albums and videos was further enhanced by a plethora of stories about Jackson's reclusive lifestyle, especially such eccentricities as his exotic pets, extreme vegetarianism, and rumoured cosmetic surgery.

Jacobsen, Arne (1902–71) *Danish architect and furniture designer, known for many buildings in the international modern style, including the SAS tower in Copenhagen and the new St Catherine's College in Oxford.*

Born in Copenhagen, Jacobsen studied architecture at the Copenhagen Academy of Arts. In the late 1920s he was deeply influenced by the works of LE CORBUSIER and MIES VAN DER ROHE, an influence that is abundantly evident in his Bellavista housing estate (1933) and Bellevue Theatre, both in Copenhagen. His prewar town halls, for example Arhus (1937) and Søllerød (1940), are also influenced by these trends, while his postwar town hall at Rødøvre (1955) and his SAS Tower in Copenhagen (1960) make use of curtain walling in the American style. The group of buildings that he designed for the new St Catherine's College in Oxford (1959) aroused considerable controversy and are thought by some to be too severe.

Jacobsen was professor of architecture at the Copenhagen Academy of Arts from 1956 and was made an honorary DLitt of Oxford University in 1961.

Jagger, Mick (Michael Phillip Jagger; 1943–) *British rock singer and songwriter. As leader of the Rolling Stones he is an inter-*

national celebrity and the most enduring pop superstar of the 1960s.

While still a student at the London School of Economics in 1960 he became an enthusiastic devotee of black American rhythm and blues. The Rolling Stones released their first record in 1963 but they did not have an American number one hit until they began writing their own material ('Satisfaction') in 1965. Billed as a mad and bad alternative to the Beatles, the Stones had an image centred around Jagger's blatant sexuality. Arrested on a drugs charge, Jagger was the subject of an editorial in the London *Times* in 1967, which spoke of 'breaking a butterfly on a wheel', and he was given a suspended jail sentence. The album *Beggar's Banquet* (1968) contained 'Sympathy for the Devil', a sarcastic commentary on human behaviour, taken by some as an endorsement of evil, which did little harm to their image. More serious was a murder that took place as the band played at a pop festival in California in 1969; this was recorded in the film *Gimme Shelter*. Jagger also acted in two films, *Performance* and *Ned Kelly* (both 1970). Jagger's energy as a performer, even now that he is in his late forties, and the band's record over the last three decades, ensure that their recordings and tours are enormously successful.

His private life has always been a favourite subject of the tabloid press. In the 1960s his girlfriend Chrissy Shrimpton (model and sister of Jean Shrimpton) was replaced by pop star Marianne Faithfull (whose hit song 'As Tears Go By' Jagger co-wrote); in the 1970s he and his wife, Nicaraguan model Bianca Perez Morena de Macias, were jet-set celebrities. After their divorce, he married his girlfriend of many years, Texas-born model Jerry Hall.

Janáček, Leoš (1854–1928) *Czech composer. Best known for his operas, Janáček was deeply influenced by native folk music and the rhythms of Czech speech.*

One of fourteen children of a village schoolmaster, Janáček was born in Hukvaldy (Silesia). At the age of eleven he was sent to the Augustine monastery at Brno, where for eight years he was taught as a chorister by Pavel Křižkovský (1820–85), who also conducted the school orchestra. When Křižkovský left, Janáček was given his post, but, recognizing the inadequacy of his training, he went to the Prague Organ School in 1874, where he survived in the utmost penury. He returned to Brno a year later. In 1876 he was appointed conductor of the Brno Philharmonic Society, but as he still lacked confidence in his musical

education he spent a further period studying in Leipzig and Vienna (1879–80). In 1881 he was appointed director of the newly formed Brno school of music. In the same year he married one of his students, with whom he had a son and a daughter. After both children had died tragically, the marriage broke up and Janáček was left grief-striken.

During this period he became increasingly active as a composer; his first work to be published professionally was a set of four choral pieces for male voices (1886). With the establishment of a Czech theatre in Brno, Janáček became interested in opera, and in 1887 completed *Šarka* (not performed until 1925). In 1904 he resigned his teaching appointment in order to devote more time to composing. In 1916 his opera *Jenufa* (1894–1903) was performed at the National Theatre in Prague to wide acclaim; Janáček was then over sixty. In the next twelve years he produced another five operas: *Mr Brouček* (completed 1917), *Katya Kabanová* (1919–21), *The Cunning Little Vixen* (1921–23; inspired by a strip cartoon in the local paper), *The Makropoulos Affair* (1923–25), and *From the House of the Dead* (1928; based on Dostoyevsky). Janáček's other compositions include the choral *Glagolitic Mass* (1926), the Slavonic rhapsody *Taras Bulba* (1918), and the popular *Sinfonietta* (1926). Of his chamber works, the two string quartets (1923; 1927–28) are outstanding. The two sets of pieces, *By Overgrown Paths*, are examples of his piano music.

Jansky, Karl Guthe (1905–50) *US radio engineer and the founder of radio astronomy.*

Jansky was educated at the University of Wisconsin, where his father held the chair of electrical engineering. After graduating he joined the staff of Bell Telephone Laboratories and was given the task of investigating the source of the interference encountered by long-distance short-wave radio signals. In addition to such obvious candidates as thunderstorms, Jansky also noticed a persistent hissing sound. At first he suspected that this interference originated in the sun; he found, however, that the strongest source of interference was related to sidereal rather than solar time. With further observations he was able to identify the centre of our galaxy as the site of maximum radiation. Jansky published his results in 1932 but neither Bell nor any professional astronomers seemed interested enough in Jansky's work to pursue the matter further. It was not until after World War II that astronomers turned to the results first disclosed by Jansky, and the science of radio astronomy and the building of radio telescopes did not begin in earnest until the 1950s.

Jansky's own interest in radio astronomy was also short-lived. After publishing his results in 1932, he returned to more straightforward engineering pursuits before his early death from a heart attack in 1950.

Jeans, Sir James Hapgood (1877–1946) *British mathematical physicist and astronomer, known for his popular writings on scientific subjects. He was knighted in 1928.*

The son of a journalist, Jeans was a precocious child. He was educated at Cambridge University, from which he graduated in 1900. Suffering from tuberculosis he was admitted to a sanatorium from 1902 until 1904, during which period he wrote his first book, *The Dynamical Theory of Gases* (1904). After his recovery Jeans taught briefly at Cambridge before being appointed (in 1905) professor of applied mathematics at Princeton University in the USA. While in America Jeans worked mainly in the field of mathematical physics and published the two textbooks, *Theoretical Mechanics* (1906) and *The Mathematical Theory of Electricity and Magnetism* (1908). Jeans returned to Britain in 1909 and again taught at Cambridge (1910–12). Thereafter he never accepted a full-time academic appointment; his scientific career was supported by the wealth of his first wife and the earnings from his numerous books. In later life he was connected with a number of institutions, including the Mount Wilson Observatory at Pasadena, California, where he was a research associate (1923–44), and the Royal Institution in London, where he held the title of professor of astronomy (1935–46). He was also secretary (1919–24) of the Royal Society.

As an astronomer Jeans is best remembered for his account of the origin of the solar system. In his last serious scientific work, *Astronomy and Cosmogony* (1928), he rejected the view that the solar system had condensed from a cloud of interstellar gas and dust, arguing instead for the tidal theory, i e that matter was attracted from a passing star by the sun's gravitational field.

For the rest of his life Jeans devoted himself to the writing of several popular works, such as *The Mysterious Universe* (1930) and *The Stars in their Courses* (1931), which were at one time widely read.

Jellicoe, John Rushworth, 1st Earl (1859–1935) *British admiral who commanded the Imperial Fleet at the Battle of Jutland in 1916. He was awarded the OM in 1916, cre-*

ated Viscount Jellicoe of Scapa Flow in 1918, and raised to the earldom in 1925.

The son of a merchant navy captain, Jellicoe passed out of training school in 1874 as a midshipman. He entered the Royal Naval College, Greenwich, in 1878, becoming a lieutenant two years later. He gained distinction in gunnery and torpedo classes and was made gunnery-lieutenant in 1884. Jellicoe was among the survivors of the *Camperdown* tragedy in 1893, when the ship *Victoria* sank off Tripoli after colliding with the *Camperdown*. Some four hundred lives were lost. He also survived serious wounds sustained as a member of a naval task force that went ashore during the 1900 Boxer Rebellion in China. In 1904, Jellicoe was posted to the Admiralty to supervise the modernization of naval ordnance and the introduction of dreadnought-class vessels. By 1910 he was vice-admiral of the Atlantic fleet and in August 1914, on the outbreak of World War I, was appointed commander-in-chief of the British fleet.

Based in Scapa Flow, Jellicoe waited for the German fleet to show itself. However, apart from hit-and-run raids on Britain's east coast, there was little action until 31 May 1916 – the Battle of Jutland. An advance squadron of British cruisers, commanded by Admiral BEATTY, engaged German cruisers in the late afternoon. The main body of Jellicoe's fleet sighted the enemy in the early evening and what ensued was probably the last great set-piece naval battle in history. Eventually, after several hours of fighting, the German ships managed to slip through the British fleet in the confusion and darkness and return to port. Whether Jutland was a victory or a defeat for Britain is arguable: certainly Jellicoe lost more ships and men than his opponents, but it undoubtedly impressed the strength of the British fleet on the Germans.

Jellicoe was posted to the Admiralty as first sea lord in November 1916, charged with improving the defence of the merchant fleet. However, his apparent reluctance to introduce the convoy system, favoured by LLOYD GEORGE, led to his dismissal on Christmas Eve 1917. After the war (1920) he was appointed governor-general of New Zealand and later served as president of the British Legion (1928–32).

Jenkins, Roy Harris, Baron (1920–) *British politician and one of the founders of the Social Democratic Party in 1981. He was created Baron Jenkins of Hillhead in 1987.*

Son of a Labour MP, Jenkins was born in Wales and educated at Oxford. After war service in the army he was elected a Labour MP in 1948. Chairman of the Fabian Society (1957–58), he became minister of aviation (1964–65), home secretary (1965–67), chancellor of the exchequer (1967–70), and then deputy leader of the Labour Party (1970–72). He resigned in 1972 over a disagreement with Harold WILSON regarding British entry into the European Community. He was again home secretary (1974–76) in Wilson's last government, and in 1977 was appointed president of the EEC Commission. On completing his term of office in Europe, he returned to British politics as a joint founder of the new Social Democratic Party, becoming its first leader (1982–83) and SDP MP for Glasgow, Hillhead (1982–87). He was subsequently elected Chancellor of Oxford University (1987–   ). Among many books, he has written biographies of Attlee (1948), Asquith (1964), and Truman (1986).

Jinnah, Mohammad Ali (1876–1948) *Indian nationalist, Muslim leader, and founder of Pakistan.*

Born in Karachi, the son of a merchant, Jinnah was educated at schools in Karachi and Bombay before qualifying as a barrister at Lincoln's Inn, London (1895). He returned to India in 1896 and established a legal practice in Bombay (1897–1906). Jinnah joined the Indian National Congress in 1906. Elected to the Imperial legislative council in 1910, he became a member of the Muslim League in 1913 and an organizer of the Indian Home Rule League. As an advocate of Muslim rights, he was instrumental in the signing of the Lucknow Pact (1916), which in principle provided for separate Muslim and Hindu electorates on Indian independence.

In 1919 he was appointed to represent Bombay Muslims in the Imperial legislative council but resigned from the council later that year in protest over legislation to prevent 'revolutionary activity'. In 1920 he left the Indian National Congress and the Home Rule League because he disagreed with Mahatma GANDHI's Hindu politics and campaign of civil disobedience. He continued to work for the Muslim League as president and remained hopeful of Hindu–Muslim unity well into the 1930s. However, relations between the League and the Congress deteriorated and in 1940 the League resolved to form a separate Muslim state of Pakistan on independence. Jinnah participated in constitutional meetings (1942,

1945, and 1946) that culminated in partition and the establishment of an independent Pakistan (1947). He was appointed the first governor-general of Pakistan and officially given the title 'Qaid-i-Azam' ('great leader'), but died in office the following year.

Jodl, Alfred (1890–1946) *German general who, as HITLER's chief of staff of the armed forces, was the principal architect of German military campaigns throughout World War II.*
Jodl served in the Bavarian artillery during World War I, was wounded in 1914, and continued as a staff officer. Appointed by Hitler in 1935 to head the department of national defence at the War Ministry, in 1939 Jodl was given the key post of operations chief in the armed forces high command – the Oberkommando der Wehrmacht (OKW) – headed by Wilhelm KEITEL. On the outbreak of World War II, Jodl was thus Hitler's closest military adviser, and remained so for the duration, implementing the Führer's policy decisions in terms of military reality, often against the wishes of the army high command.

Jodl was an intelligent and shrewd staff officer but with limited experience of command in the field. He was occasionally critical of Hitler, particularly over the Führer's inept handling of the Soviet campaign in 1943, but generally Jodl merely reinforced his leader's increasingly idiosyncratic and ill-considered directives. Thus, his support for Hitler's totally unrealistic objectives in the Ardennes offensive of December 1944 served only to hasten the final German collapse. On 7 May 1945, Jodl surrendered the German army to the Allies at Reims. Jodl's diaries revealed their author's complicity in many of Hitler's war crimes, including the murder of hostages. Tried by the International Military Tribunal at Nuremberg after the war, he was convicted and hanged.

Johannsen, Wilhelm Ludvig (1857–1927) *Danish geneticist who introduced the terms phenotype, genotype, and gene, and was one of the founders of modern genetics.*
Born in Copenhagen, the son of an army officer, Johannsen was apprenticed to a pharmacist in 1872 and worked in Denmark and Germany, passing his pharmacist's exam in 1879. Two years later he was appointed assistant in the chemistry department at the Carlsberg laboratory under the famous chemist Johan Kjeldahl (1849–1900). Here Johannsen investigated the metabolism of dormancy and germination in seeds, tubers, and buds. In 1892 he was appointed lecturer at Copenhagen Ag-

ricultural College and eventually became professor of botany and plant physiology.

Johannsen's most notable experiments concerned his so-called 'pure lines' of the self-fertile princess bean, *Phaseolus vulgaris*. Studying the progeny of self-fertilized plants, he selected the character of bean weight and found that both the lightest and the heaviest beans produced progeny with the same distribution of bean weights, i e they were genetically identical. He concluded that the variations in bean weight were due to environmental factors and he introduced the terms genotype (for the genetic constitution of an organism) and phenotype (for the characteristics of an organism that result from the interaction of its genotype with the environment). Johannsen favoured the view of DE VRIES that inheritance was determined by discrete particulate elements and abbreviated de Vries's term 'pangenes' to 'genes'. Johannsen's *Arvelighedslaerens elementer* (1905; 'The Elements of Heredity') was later (1909) rewritten, enlarged, and translated into German to become one of the founding texts of genetics. In 1905 Johannsen was appointed professor of plant physiology at Copenhagen University, becoming rector in 1917.

John XXIII (Angelo Giuseppe Roncalli; 1881–1963) *Pope (1958–63) who, by convening the Second Vatican Council (1962–65), set in motion the process of 'aggiornamento' – bringing the Catholic Church up to date to meet the demands of the twentieth century.*

Roncalli was born in the Lombardy region of Italy, the son of a tenant farmer. After attending a seminary in Bergamo, he went to Rome in 1900 and four years later was ordained as a priest. He subsequently received a doctorate in canon law and became secretary to the Bishop of Bergamo. In 1920 he worked in the Vatican and, five years later, was appointed apostolic visitor to Bulgaria. After occupying a similar position in Greece and Turkey (1935–45) Roncalli was given the much more prestigious job of papal nuncio to newly liberated France. Here he skilfully performed the delicate task of reconciliation between the state and those bishops who were alleged to have collaborated with the Vichy government. Made a cardinal in 1953, Roncalli was then appointed Archbishop of Venice.

When, at the age of seventy-seven, Roncalli succeeded PIUS XII as Pope John XXIII, he was considered an 'interim pope'. Therefore, many were surprised when the following year (1959) he announced his plans for the Second

Vatican Council, ninety-four years after the first. He presided over the first session in October 1962, when 2300 bishops gathered to discuss the modernization of the Church, especially how power could be devolved to the bishops and their congregations. 'Vatican II' continued under Pope John's successor, PAUL VI, who implemented many of its proposals. Pope John XXIII's major encyclical, *Pacem in Terris* ('Peace on Earth'), urged the need for peaceful coexistence, especially between communist east and capitalist west. The Cuban missile crisis of 1962 prompted a personal appeal for restraint from the pope. In his short reign, during which he met representatives of many other religions and faiths, John XXIII became respected throughout the world.

John, Augustus Edwin (1878–1961) *British painter and draughtsman, who is best known for his portraits and for the appealing spontaneity of his style. He was admitted to the OM in 1942.*

Born in Tenby, the son of a Welsh solicitor, John entered the Slade School in London at the age of seventeen. Here his individual style of draughtsmanship soon became apparent. Having begun life at the Slade as a quiet and diligent student, he underwent a sudden metamorphosis in 1897, becoming a flamboyant bohemian and rebelling against nineteenth-century artistic traditions.

John had his first exhibition in 1899, married in 1901, and in 1903 met Dorothy McNeill (Dorelia), who joined his household in 1904 and after the death of his wife (1907) became his lifelong companion. She was the model for his portrait *The Smiling Woman* (1908), in which he portrayed a robust gypsy type of beauty. The gypsies of his native Wales were frequent subjects of his work and he lived an intermittent caravan life with his large family until World War I. Around 1910 he flirted with postimpressionism and visited Provence. In the following year he produced many fine landscapes, and by 1914 he had secured an international reputation. After the war John's creative vitality seemed to diminish and he became a portraitist of the wealthy and famous.

His sister Gwen John (1876–1939) was also a painter, noted for her sensitive portraits of women. After studying at the Slade she worked in Paris as a model and painter, becoming the student and mistress of RODIN. Her later years were marked by growing reclusiveness and a conversion to Roman Catholicism. Although she achieved little recognition in her own lifetime, her reputation has since come to rival that of her brother.

John, Elton Hercules (Reginald Kenneth Dwight; 1947–   ) *British singer, pianist, and songwriter. With lyricist Bernie Taupin (1950–   ), he wrote many of the hit songs of the 1970s.*

John and Taupin began their collaboration in 1966, writing songs for other singers and groups. In 1968 John began to record them himself. Taupin wrote lyrics quickly and easily, John added tunes without changing a word, and the two rarely needed to meet. As a songwriting team, like John Lennon and Paul McCartney in the 1960s, they captured the feeling of a decade. John's public performances, including handstands at the piano and a large collection of outrageous eyeglasses, were well received by his teenage audiences. Since 1977 he has collaborated with other lyricists and in 1979 he became the first pop star to tour the Soviet Union. His most recent albums include *Leather Jackets* (1986) and *Sleeping with the Past* (1989).

A wealthy man, John maintains lavish homes in three countries. He became Chairman (1976–90) and subsequently Life President (1990–   ) of Watford Football Club, of which he has been a follower since childhood.

John Paul II (Karol Jozef Wojtyla; 1920–   ) *Polish-born pope (1978–   ) whose many travels abroad included a visit to Britain (1982) and tours of his native country (1979, 1987, and 1991).*

Wojtyla was born in Wadowice, Poland, the son of a junior army officer. In 1938 he moved with his father to Kraców and enrolled in the Jagiellonian University to take Polish studies and philosophy. Here he also developed a keen interest in drama and poetry. During the Nazi occupation of Poland in World War II, he joined the underground student resistance movement to pursue his studies, working in a stone quarry by day. In 1942 he started to study theology in secret at the university and was finally ordained a priest in 1946. After studying philosophy and moral theology at the Angelicum Institute, Rome, he returned home in 1948 to a climate of oppression and growing conflict between Church and state. He served as a priest to the village of Nieogowić near Kraców and then in a city parish. Meanwhile, in 1949, he obtained the first of two doctorates from the Jagiellonian University and in 1954 was appointed professor of theology at the Catholic University of Lublin. Made a bishop in 1958, Wojtyla attended the Second Vatican Council (1962–65) and in 1964 was appointed Archbishop of Kraców. He became a cardinal three years later. Under the Polish primate, Cardinal Wyszynski, Wojtyla helped steer Poland's Catholics through the 1970 food riots, becoming increasingly outspoken against the communist authorities.

On 16 October 1978, the election of the first Polish pope was greeted with jubilation in Poland. In spite of a series of worldwide visits in which he has confronted poverty, hunger, and distress, especially in Central and South America, he has upheld staunchly and uncompromisingly the church's traditional opposition to artificial means of contraception and abortion. This has led his critics to assert that the size of the new generation of Catholics is more important to him than measures to alleviate hunger and deprivation. He has reaffirmed his conservative attitude to other social issues by his condemnation of homosexuality (1986) and his opposition to the ordination of women and the relaxation of the rule of celibacy for priests. He has also had to grapple with the Church's dilemma over the politicization of the priesthood. In May 1981, John Paul II survived an assassination attempt in St Peter's Square, Rome.

Johns, Jasper (1930–   ) *US painter, sculptor, and printmaker, who was a forerunner of pop art.*

Johns was brought up in South Carolina and attended the University of South Carolina. After brief service in the army he settled in New York (1952), where he painted while supporting himself by doing display work for large stores. Some of his best-known paintings – the series of *Flags*, *Targets*, and *Numbers* – were produced in the mid-1950s. Using mainly traditional painting techniques, Johns depicted completely flat, commonplace, and universally recognized images, such as the US flag. This search for an impersonal art can be seen as a radical reaction to abstract expressionism, which was the current avant-garde movement, as well as his questioning of the nature of art. His use of rich encaustic (wax-based) paint helped to focus attention on the act of painting, as did his later inclusion of objects from his studio, which he affixed to the canvas. At the same time visual ambiguity played an important part in his work: for example, the word 'red' would appear in a painting but the letters of the word would be blue. In the late 1950s Johns began to produce sculptures consisting of exact reproductions of such objects as beer cans and light bulbs cast in bronze. He has exhibited widely in Europe, the USA, and Japan. He has also collaborated with Merce Cunningham (1919–) and John CAGE on various dance projects.

Johnson, Amy (1903–41) *British aviator who, in 1930, made a celebrated solo flight from Britain to Australia. In recognition of this achievement she was made a CBE.*

The daughter of a herring importer from Kingston upon Hull, Amy Johnson attended Sheffield University and received her BA in 1925. While working in London as a secretary she joined the London Aeroplane Club and became the first woman recipient of a ground engineer's licence. Having obtained her pilot's certificate in December 1929, she set out on 5 May of the following year from London bound for Australia in an attempt to break the solo record. Her de Havilland Moth powered by a 90 hp Gipsy engine arrived in Port Darwin on 24 May in spite of several stops for repairs. Although not a record, her achievement was received enthusiastically in Britain and elsewhere. The *Daily Mail* gave her £10 000 and she became a national heroine. She made further long-distance flights both alone and with her husband, the pilot James Mollison (1905–59), whom she married in 1932. In June 1933 their de Havilland biplane crashed on landing at Bridgeport, Connecticut, after a nonstop flight from Britain and in May 1936 Johnson broke the London to Cape Town solo record. The marriage was dissolved in 1938 and in 1939 Johnson joined the Auxiliary Air Force. The following year her plane disappeared on a flight over the Thames estuary and she was presumed dead.

Johnson, Jack (1878–1946) *US boxer who became the first black to hold the world heavyweight title (1908–15).*

Born in Galveston, Texas, the son of a bareknuckle fighter, Jack Johnson ran away from home to work in racing stables before deciding to become a professional boxer. In 1908 he took the world title by beating the Canadian Tommy Burns. He lost it in 1915 in Havana, Cuba, to Jess Willard, when he was knocked out in the twenty-sixth round. Apart from Muhammad ALI Johnson was arguably the greatest heavyweight of all time. Outside the ring he lacked popularity, largely because of his flamboyant and arrogant lifestyle. After one comeback, he retired officially in 1928. He died in a car crash in 1946.

Johnson, Lyndon B(aines) (1908–73) *US Democratic statesman and thirty-sixth president of the USA (1963–69).*

Johnson was born into a poor but politically active farming family in Gillespie County, Texas, and became interested in politics at an early age. After graduating from high school in 1924 he worked at a variety of odd jobs before enrolling at the Texas State Teachers' College in 1927; he subsequently taught public speaking and debating in a Houston school. In 1932 he went to Washington as secretary to the newly elected Congressman Richard M. Kleberg, having assisted with his congressional campaign. He came to the notice of President ROOSEVELT, who appointed him director of the National Youth Administration in Texas (1935), and after two years in this office Johnson successfully campaigned for a seat in Congress. His relationship with Roosevelt was based on mutual admiration and support, but this proved insufficient to win him a seat in the US Senate in 1941; it was a further seven years before he was able to take this next step up the political ladder. Elected to the Senate in 1948, he became Democratic Whip in 1951 and party leader two years later. At that time the Democrats were the minority party in Senate but by 1955, after a series of vigorous campaigns, Johnson had established himself as majority leader, a position of considerable importance and power. He led the Democrats with tact, resourcefulness, and determination, insisting on cooperation with the Republican President EISENHOWER and achieving some measure of unity in a party that was frequently divided on major issues, notably the civil rights legislation of 1957 and 1960.

In 1960 Johnson campaigned unsuccessfully for the Democratic nomination for president, losing to John F. KENNEDY, who surprisingly invited Johnson to be his running mate. As vice-president, Johnson represented Kennedy on visits to southeast Asia and Berlin and served as chairman of a committee set up to provide equal employment opportunities for blacks. On his sudden accession to the presidency after Kennedy's assassination in 1963, he reassured the nation by carrying on the policies of the Kennedy administration and encouraging Congress to pass delayed legislation on civil rights and tax reduction. Re-elected by a massive majority in 1964, Johnson began to develop his ambitious Great Society programme for wide-ranging domestic reform. An assortment of legislative proposals were passed by Congress, including the provision of medical care for the elderly, federal aid for schools, housing and urban development, improved civil rights legislation, and a number of antipoverty measures. The increasing US involvement in the Vietnam War, however, undermined much of the support won by Johnson through his domestic reforms. Rather than hasten a negotiated settlement of the hostilities, he escalated the crisis by increasing US military aid to South Vietnam in their renewed struggle against the communists. By the end of 1965 the US presence in Vietnam comprised some 180 000 troops; this number was nearly doubled the following year and continued to rise steadily. Johnson found himself faced by increasing opposition at home and finally gave way to demands to halt the bombing of North Vietnam in March 1968. At the same time he announced his decision not to stand for re-election in the forthcoming presidential campaign. In 1969 he retired to his ranch in Texas where he compiled his memoirs of the presidency, published in 1971 as *The Vantage Point*.

Joliot-Curie, Irène (1897–1956) *French physicist who, in collaboration with her husband, Frédéric Joliot-Curie (1900–58), discovered artificial radioactivity. For this they were awarded the 1935 Nobel Prize for Chemistry.*

The daughter of the early Nobel laureates Pierre and Marie CURIE, Irène Curie obtained a doctorate from the Sorbonne in 1925 and married her mother's research assistant, Frédéric Joliot, a year later. Together they worked under Marie Curie at the Radium Institute in Paris – Irène eventually becoming director of the institute in 1946. She was succeeded on her death in 1956 by her husband, who held the office until his own death in 1958.

In 1931 the Joliot-Curies began a classic series of experiments. They bombarded aluminium with alpha particles and noted that the aluminium emitted protons under the bombardment. To their surprise, however, the aluminium continued to emit protons after the alpha-particle source was removed. It soon became clear that the stable aluminium atoms had absorbed alpha particles and been transmuted into radio isotopes of silicon. The Joliot-Curies' later work on the neutron bombardment of uranium was of great importance in the discovery of nuclear fission. After World War II Mme Joliot-Curie was involved in the foundation of the French Atomic Energy Commission. Like her mother, Irène Joliot-Curie died of leukaemia.

Jolson, Al (Asa Yoelson; 1886–1950) *US singer. Billed as 'The World's Greatest Entertainer', he was extremely popular on the stage for several decades.*

Born in Lithuania, Jolson was brought to the USA in 1894 and brought up in Washington, DC, where he made his first appearance on the stage. After a period in the circus and in vaudeville, he made his Broadway debut in 1899; his first recording was made in 1911. He derived his style from the black-face minstrels of Lew Dockstader's minstrel troupe, which he joined in 1909. In this now unacceptable guise he popularized the sentimental 'Mammy' song, often delivered on one knee with emotional declamation almost regardless of the melody. Between 1918 and 1921 he was famous for his performances of 'Avalon', 'Swanee', 'April Showers', 'My Mammy', 'California, Here I Come', 'Toot, Toot, Tootsie, Goo'bye', and 'Rockaby Your Baby With a Dixie Melody'.

He featured in the first full-length talking motion picture, *The Jazz Singer*, in 1927 (though not a jazz singer by any definition), and can be heard on the sound tracks of two biographical films, *The Jolson Story* (1946) and *Jolson Sings Again* (1949).

Jones, Bobby (Robert Tyre Jones; 1902–71) *US amateur golfer, whose record of thirteen major titles out of twenty-seven championships in eight years (1923–30) is unlikely to be surpassed by amateur or professional.*

Jones's impressive performance began when he won the US Open championship at the age of twenty-one. He went on to win three more US Opens, three British Open championships, five US amateur championships, and one British amateur championship. In 1926 he became the first person to hold both the British and US Open titles simultaneously. The greatest year of his career was 1930, when within five months he won the British amateur, the British Open, the US Open, and the US amateur.

During these eight years he also obtained degrees in English literature and engineering and qualified as a barrister. He retired from tournament golf in 1930 in order to make a series of golf instruction films for Warner Brothers for a reputed fee of $250,000. Although he refused to play as a professional, he continued to take an interest in golf and helped many young players, including Jack NICKLAUS.

Jones, Daniel (1881–1967) *British linguist and phonetician who developed the idea of an international phonetic alphabet and described the influential Received Pronunciation.*

After obtaining a degree in mathematics at Cambridge University, Jones read law and finally became interested in phonetics, which he studied in Paris (1905–06) under Paul Passy (1859–1940). Passy, with A. J. Ellis (1814–90) and Henry Sweet (1845–1912), had been working on the concept of an international phonetic alphabet, and when Jones returned to England in 1907 he continued to develop and refine the principles for such a system at the first British department of phonetics, at University College, London.

A key step forward in Jones's work came with his invention of a system of 'cardinal vowels', used as reference points for transcribing all vowel sounds. In his *An English Pronouncing Dictionary* (1917), Jones set out his description of Received Pronunciation – the socially prestigious dialect that is 'most usually heard in everyday speech in the families of Southern English persons whose men-folk have been educated at the great public boarding-schools'. This guide was widely used and much revised, although Jones never intended it as a practical reference book that would influence the future development of the language. Other major works produced by Jones include *An Outline of English Phonetics* (1918), contributions to *The Principles of the International Phonetics Association* (1949), and *The Phoneme: its Nature and Use* (1950), in which he tried to classify the nature of phonemes (sound units of a language).

Joplin, Scott (1868–1917) *Black US ragtime pianist and composer, the first to write down his compositions.*

Born in Texarkana, Texas, Joplin won several local piano contests before turning his attention exclusively to the syncopated piano style known as ragtime. A strong influence on the stride piano style of Fats WALLER, ragtime became a precursor of jazz. The first pieces called rags were written in 1897–98; two of Joplin's best known, 'Original Rags' and 'Maple Leaf Rag', were written in 1899. The latter was so successful that a publishing company was formed on the strength of it, and a million copies of the sheet music were soon sold. Ragtime became nationally popular and for a time Joplin achieved his ambition of wealth and fame for a black musician. However, he aspired to create a more serious school of ragtime composition although the style does not sustain extended forms. He also wrote two operas, *A Guest of Honor* (*c.* 1903; now lost) and *Treemonisha* (1911), and started an opera company and a symphony based on ragtime. None of these ventures succeeded; he was particularly distressed by the failure of *Treemonisha*. These failures, the ravages of

syphilis, and declining interest in ragtime, combined to lead to his early death in a mental home.

He wrote about fifty piano rags, of which many are subtle and stylish compositions as well as delightful period pieces. The tendency to play them at breakneck speed, in spite of the instruction 'Do not play fast', was reversed when they were revived in the late 1960s; Joplin's music was featured in a popular film, *The Sting*, in 1973.

Josephson, Brian David (1940–   ) *British physicist, who was awarded the 1973 Nobel Prize for Physics for the discovery of the Josephson effect.*

Josephson was educated at Cardiff High School and the University of Cambridge, where he obtained his PhD in 1964. After a year at the University of Illinois, Josephson returned to Cambridge in 1967, becoming professor of physics in 1974.

Josephson's name is associated with the effect he discovered in 1962 while still a graduate student. He predicted that at a superconductor junction, under certain circumstances, a current can cross the junction even when there is no potential difference across it. Shortly afterwards the zero-voltage current predicted by Josephson was detected experimentally. At the same time the oscillating currents produced when a voltage was applied were also found, as predicted by Josephson. Josephson has also lent some support to the study of parapsychology. He has admitted to having been influenced by the Maharishi Mahesh Yogi and in 1980 he called for a new approach to physics in which Krishna consciousness has a place.

Jouvet, Louis (1887–1951) *French actor and producer.*

Jouvet was born in Brittany and began his career with touring companies; he made his Paris debut in *The Brothers Karamazov* in 1910. Subsequently he became involved in management, joining Jacques Copeau (1878–1949) as actor and stage manager at the opening of the Vieux-Colombier in 1913. After war service (1914–17) Jouvet spent two years with Copeau and his company at the Garrick, New York. After an attempt to set up his own theatre he became director of the Comédie des Champs-Élysées (1924–34). He produced several plays by Jules Romains, most notably *Knock, ou le triomphe de la médicine*, and all of GIRAUDOUX's, beginning with *Siegfried* (1928). In 1934 he assumed the management of l'Athénée and two years later was appointed a professor at the Conservatoire. He also became one of a group of occasional producers at the Comédie Française. Throughout his distinguished career Jouvet continued to act and during World War II he toured South America. Arnolphe in *École des femmes* and Geronté in *Les Fourberies de Scapin* are among the many Molière roles he portrayed. Jouvet made his film debut in Pagnol's *Topaze* (1933). Subsequent films in which he appeared included *La Kermesse héroïque* (1935), Jean RENOIR's *Les Bas-fonds* (1936), Marcel CARNÉ's *Drôle de drame* (1937) and *Hôtel du Nord* (1938), *Quai des Orfèvres* (1947), and *Une Histoire d'amour* (1951). He also wrote *Réflexiones du comédien* (1939) and *Témoignages sur le théâtre* (published posthumously in 1952).

Joyce, James Augustine (1882–1941) *Irish novelist and poet. His major novel,* Ulysses, *developed the much discussed, and subsequently much used, unexpurgated 'stream-of-consciousness' style in twentieth-century novel writing.*

Joyce was born at Rathgar, Dublin, and received his early education from the Jesuits. He then studied modern languages at Dublin's Catholic University, graduating in 1902. In 1904 Joyce began working on a long autobiographical novel, *Stephen Hero*. The material of this was eventually reworked and published as *A Portrait of the Artist as a Young Man* (1916), a book that admirably captures the flavour of his early Dublin years. But chafing against the parochialism of Irish culture, then dominated by YEATS and the Celtic revival, Joyce left Ireland in 1904 to live and write abroad. Accompanied by Nora Barnacle (whom he did not marry until 1931), Joyce settled in Trieste until 1915. His first volume of verse, *Chamber Music*, appeared in 1907, its title reflecting Joyce's early interest in music. During World War I he moved to Zürich before making a permanent home in Paris (1920–40).

His first major publication was *Dubliners* (1914), a collection of short stories of his native city. He also published a play, *Exiles* (1918), heavily influenced by Ibsen. In 1922 *Ulysses* was published in Paris in defiance of the moral scruples of the publishing establishment. Despite being banned for many years in English-speaking countries, it gained acceptance as an individual masterpiece, using the 'stream-of-consciousness' technique. Structured loosely along the lines of Homer's *Odyssey*, *Ulysses* chronicles a day in the life of three Dubliners, Leopold and Molly Bloom and Stephen Dedalus. *Finnegans Wake* (1939) is even more difficult and allusive, being conceived as a dream-sequence. In 1937 Joyce published his *Collected Poems*. Since his death Joyce's work has attracted an immense amount of critical exegesis and has exerted an influence upon twentieth-century prose out of all proportion to its size.

Joyce, William (1906–46) *Anglo-US political activist who broadcast Nazi propaganda to Britain during World War II. Known as 'Lord Haw Haw', he was hanged as a traitor after the war.*

Born in Brooklyn, New York, Joyce was the son of a successful building contractor, Michael Joyce, who in 1909 returned to his Irish homeland. A Protestant and ardent Anglophile, Michael Joyce had a profound influence on his son, who acted as an informer for the Black and Tans. Following Irish Independence (1921), the family was persecuted and fled to England, where their claim for restitution was rejected by the authorities. Joyce took first class honours at Birkbeck College, London, in English and history. In pursuit of his cherished ideal of the 'English gentleman', he joined Oswald MOSLEY's British Union of Fascists and was soon displaying his zeal as an organizer and skill as an orator. However, Mosley's admiration for MUSSOLINI conflicted with Joyce's conviction that HITLER's Nazis offered the solution to Britain's malaise. Joyce quit Mosley's organization and formed his own National Socialist League. A summons for his arrest was issued in August 1939, but – obtaining a British passport – he managed to flee to Berlin.

The Nazi Ministry of Propaganda employed Joyce to make radio broadcasts in English. Soon he was writing his own scripts and employing every possible device to deride the British government and exaggerate German military success. 'Lord Haw Haw' made his final broadcast on 30 April 1945 from Hamburg; he was subsequently captured after talking to two British officers who recognized his distinctive voice. He was charged with treason, but when his US citizenship was revealed the prosecution were forced to resort to a legal judgement of 1608 to make the charge stick. Nevertheless, Joyce was found guilty and hanged.

Juan Carlos I (1938–) *King of Spain (1975–). Named by FRANCO as his successor, Juan Carlos has steered his country from dictatorship to democracy.*

Son of Don Juan de Bourbon, Count of Barcelona, and grandson of Alfonso XIII of Spain, Juan Carlos was born in Rome and educated privately in Switzerland and later in Spain. He then received training in all three of the Spanish armed forces (1957–59). In 1962 he married Sophia (1938–   ), daughter of King Paul of Greece, and they have three children: Crown Prince Felipe, Princess Elena, and Princess Cristina. The transition to democracy in Spain has not been entirely smooth, and in February 1981 a military coup was staged but successfully aborted. Juan Carlos was awarded the Charlemagne Prize for his contribution to Europe in 1982 and subsequently UNESCO's Bolivar Prize (1983) and the Nansen Medal (1987).

Jung, Carl Gustav (1875–1961) *Swiss psychiatrist who founded analytical psychology and who proposed the concepts of extrovert and introvert personalities, archetypes, and the collective unconscious. He also developed valuable methods of psychiatric therapy.*

After qualifying in medicine at the University of Basle (1900), Jung became an assistant at the Burghölzi Asylum, Zürich, and later senior staff physician and an instructor in psychiatry to the University of Zürich. He was appointed professor of psychology at the Federal Polytechnical University, Zürich (1933–41), and from 1943 was professor of medical psychology at the University of Basle. Jung made his name as a psychiatrist with a study of schizophrenia, using both physiological and psychological methods. He showed that word association tests were useful indicators of emotional activity and that such activity was accompanied by physiological changes – in heart rate and respiration. In 1906 Jung met FREUD and they collaborated closely until 1912, when their differences could no longer be reconciled. Jung opposed Freud's theory of a sexual basis for neurosis and completed their break when he published *Psychology of the Unconscious* (1913), which expressed his conflicting views. Jung analysed the causes of the

estrangement and concluded that he and Freud were constitutionally different types and therefore approached problems from opposing viewpoints. He expanded his theories of typology (*Psychological Types*, 1923), first differentiating two major types of people, according to their attitudes, as either extrovert or introvert. Later he differentiated four functions of the mind (thinking, feeling, sensations, and intuition), one or more of which predominate in any particular person.

Jung's major interest was the unconscious background underlying the conscious life and he attached great importance to dreams. He had always been subject to dreams and fantasies and allowed them free expression so that he could study them scientifically by detailing his thoughts and experiences. He considered that a dream is a fact and discovered in some patients some ancient material that could not be explained in terms of their own history. This led him to study some less advanced peoples in an attempt to learn about human thought. He eventually developed his hypothesis of the collective unconscious, in which he proposed that the mind, like the body, has a very long ancestry so that an individual's unconscious contains both his own personal experiences and the common inherited cultural experiences consisting of symbols and ideas shared by all humans. The patterns of shared ideas that he found he called archetypes, and he spent the rest of his life developing these ideas, particularly in relation to religion and psychology.

Jung's methods of therapy emphasized normal and healthy psychology, pointing out that symptoms were disturbances of normal processes and not entities themselves. He suggested that neuroses were interferences with the healthy mind and that dreams were unrealized potentialities. His approach was to concentrate on the patient's present and his failure to cope with life at that particular time. He pioneered psychotherapy for the middle-aged and the elderly, especially those who considered that their lives were no longer of value.

K

Kádár, János (1912–89) *Hungarian statesman. First secretary of the Hungarian Socialist Workers Party (1956–88), he was appointed by the Soviet Union to form a new government after the 1956 revolution.*

Born in Fiume (now Rijeka, Yugoslavia), the son of a peasant, Kádár was trained as a mechanic. His political career began at nineteen, when he joined the Young Communist Workers Federation and the illegal Communist Party. Active in the resistance movement during World War II, he was elected to the national assembly in 1945, becoming a member of the Politburo of the Communist Party in the same year, assistant general secretary in 1946, and minister of internal affairs in 1948. In 1950 he was dismissed from his government and party posts and imprisoned (1951–54) for opposition to Stalinism.

He was rehabilitated in 1954 and during the October Revolution (1956) joined the government of Imre NAGY. However, Nagy was opposed to the revolution and it was Kádár that the Soviets chose to form a new government after the Red Army had crushed the uprising. He served as premier and chairman of the council of ministers (1956–58; 1961–65) but his most influential role was as first secretary of the Hungarian Socialist Workers Party, a position he held from 1956 to 1988 and which effectively made him head of state.

Kadar was regarded as a traitor by many Hungarians for his support of the Soviet intervention in 1956; in the aftermath of the uprising many of its leaders and supporters were imprisoned or executed. In the 1960s he appeared to modify his severity and a number of political prisoners were released under an amnesty (1963). He consistently supported the Soviet Union (Hungarian troops participated in the 1968 invasion of Czechoslovakia), but also permitted a degree of decentralization of the economy, making Hungary the most prosperous state in eastern Europe. His resistance to the political reforms that swept the communist world in the 1980s, however, resulted in his eventual removal as first secretary and his appointment as president of the party, a largely honorary title; in 1989 he was dismissed from all his political offices.

Kafka, Franz (1883–1924) *Austrian novelist and short-story writer.*

Born in Prague (then in Bohemia), Kafka was the only son of a successful Jewish businessman. He was educated at German-speaking schools and, in obedience to his father's wishes, studied law at the University of Prague, graduating in 1906. He was employed by a workers' insurance company. In 1917 he was diagnosed as suffering from tuberculosis and spent a year (1919) in various sanatoriums. Using the disease as a respectable means of leaving his restricting office job, he resigned in 1922. He died two years later at a sanatorium near Vienna, having published only a small part of his writings. Kafka had rather unsatisfactory relationships with several women, but it was his domineering father who appears to have had the greatest effect on his work.

According to his will, Kafka's unpublished manuscripts were to be destroyed, but the executor, the writer Max Brod (1884–1968), decided to ignore the instruction and began publishing major works soon after his death. Brod edited the *Gesammelte Schriften* ('Collected Works') in 1935–37, but it was not until the 1950s, when an expanded edition appeared, that Kafka was recognized as one of the major German writers of the twentieth century. Kafka's characteristic tone results from presenting subjects, which have the frightening dramatic and enigmatic quality of nightmares, in a prose that is lucid and detached. A sense of guilt haunts the stories, but they are not philosophical or psychological allegories; the reason for things going wrong remains ambiguous or inexplicable.

In *Die Verwandlung* (1915; translated as *Metamorphosis*, in *Metamorphosis and Other Stories*, 1961) the protagonist Gregor Samsa awakes one morning to find himself transformed into a gigantic beetle. His attempts to adjust are described in detail; after his death his family returns, with relief, to its normal existence. *Der Prozess* (1925; translated as *The Trial*, 1939) recounts the arrest (for uncertain reasons) and interrogation of Joseph K. The legal process is extended into an absurd parody before the hero is taken away and stabbed to death by two executioners in black. In *Das Schloss* (1926; translated as *The Castle*, 1930) a surveyor, K., takes up winter quarters in a suspicious village in order to contact the nearby castle, but all attempts (including his relationships to the villagers) come to nothing and a nagging suspicion remains that the castle may be uninhabited. Of Kafka's many brilliant shorter works one of the finest is *In der Strafkolonie* (1919; translated as *In the Penal Colony*, 1948), which concerns an ingenious form of execution. The instrument, a creation of the long departed Old Commandant, slowly inscribes the names of the crimes into the flesh of the accused, who can thus read it in a final moment of enlightenment. In the story the neglected machine falls apart while horribly killing the man who has preserved it and enthusiastically volunteered himself for a final demonstration of its use.

Kahn, Louis Isadore (1901–74) *Estonian-born US architect who acquired a considerable international reputation for his dramatic geometrical buildings and his talents as a teacher.*
Born on the Estonian island of Ösel, he emigrated to America with his family at the age of four and became a US citizen at the age of fourteen. After a traditional training as an architect at the University of Pennsylvania he became a pioneer of the modern international movement with a well-established firm of US architects. In 1948 he returned to academic life, first as professor of architecture at Yale (1948–57) and then at the University of Pennsylvania (1957–74). His international reputation did not emerge until the 1950s, with his Yale University Art Gallery (1951–53) and the University of Pennsylvania Medical Research Building (1958–60). His striking Salk Institute Laboratories (1959–65) in La Jolla, California, led to several commissions outside America. The most important of these were the Institute of Management (1963–74) in Ahmedabad, India, and the Assembly Building (1962–74) in Dhaka, Bangladesh. By the 1970s his influence was reflected in the awards of Gold Medals by the American Institute of Architects in 1971 and the RIBA two years later.

Kaldor, Nicholas, Baron (1908–86) *Hungarian-born British economist. He was created a life peer in 1974.*
Born in Budapest, Kaldor moved to Britain in the 1920s to study, graduating at the London School of Economics in 1930 and lecturing there until 1947, when he was appointed director of the Research and Planning Division of the Economic Commission for Europe in Geneva. He was a member of the Royal Commis-

sion on Taxation of Profits and Incomes (1951–55) and later special adviser to the chancellor of the exchequer on the social and economic aspects of taxation policy. He advised on tax reform for the government of India and subsequently for Ceylon, Mexico, Ghana, British Guiana, Turkey, Iran, and Venezuela. In 1966 he was appointed professor of economics at Cambridge University.
An antimonetarist, Kaldor was involved in the discussions of the Radcliffe Committee in the 1950s, whose report in 1959 stated that in ordinary times monetary policy should be subordinate to other measures in economic policy. Attacking the government's monetarist policies in the early 1980s, he wrote *The Scourge of Monetarism* (1982) and *The Economic Consequences of Mrs Thatcher* (1983). In welfare economics he put forward the compensation principle: with Sir John Hicks (1904–89) he postulated that state A is preferable to state B if those who gain from moving to A can compensate those who lose and still be better-off (whether or not compensation is actually implemented). This is known as the Kaldor–Hicks test and may be compared with PARETO's system, which requires that no-one is made worse-off by such a move.
Among Kaldor's most important publications were essays on *Economic Stability and Growth* (1960) and on *Value and Distribution* (1960), *Capital Accumulation and Economic Growth* (1961), *Causes of the Slow Rate of Growth of the UK* (1966), and *Conflicts in Policy Objectives* (1971). Kaldor was a leading figure, along with Joan ROBINSON, in the postwar Cambridge school, which emphasized a macroeconomic approach following J. M. KEYNES.

Kandinsky, Wassily (1866–1944) *Russian-born painter who pioneered abstract painting.*
It was not until the age of thirty that Kandinsky decided to abandon an academic career that had begun with his appointment as lecturer in the faculty of law at the University of Moscow in 1893. He left his native city in 1897 to study painting in Munich. The current avant-garde movement there was art nouveau, and Kandinsky's early work is characterized by the broad flat areas of colour and the rhythmic lines of this style. He travelled between 1903 and 1908 but continued to be based in Munich until 1914.
Most historians credit Kandinsky with the first abstract painting, dating it as early as 1910. In his treatise *On the Spiritual in Art* (1912), he argued that art depended on inherent

aesthetic principles, not on resemblance to the outside world. He wanted to produce a 'spiritual vibration' by expressing inner and essential feelings rather than the surface appearances of the natural world. His emphasis on spiritual qualities in art reflected the intensely mystical side of his personality and the influence of such theories as those of theosophy. The two years before World War I also saw the foundation of the Blaue Reiter group of artists led by Kandinsky, all of whom were interested in nonobjective painting and the correspondence between different art forms. The latter conception was explored in Kandinsky's stage play *The Yellow Sound*.

The outbreak of war in 1914 forced Kandinsky along with other Russians to leave Germany. He returned to Russia and for six years became closely involved in cultural policy after the revolution. Increasing hostility to his ideas, however, prompted his acceptance in 1921 of the post of professor at the Bauhaus in Weimar, Germany, where he taught from 1922. During his Bauhaus period, Kandinsky's pictures, which until then had often contained echoes of external reality, became for the most part wholly abstract. They were full of energy and movement conveyed purely by colour, line, and shape. They also displayed a mastery of composition, which has provided the basis for almost all nonspontaneous expressive abstract painting in this century. In 1926 the Bauhaus published his most influential treatise, *Point, Line and Plane*. When the Bauhaus was closed by the Nazis in 1933, Kandinsky, who had become a German citizen in 1927, went to live in Paris and in 1939 became a French citizen. During this period he continued to create masterpieces, which became less harshly geometrical with softer and less angular shapes.

Kapitza, Peter Leonidovich (1894–1984) *Soviet physicist noted for his work in the field of low temperature physics. He was awarded the Nobel Prize for Physics in 1978.*

The son of an engineer, Kapitza was educated at the Petrograd Polytechnic. He graduated in 1918 and in 1921 came to England to study physics under Ernest RUTHERFORD at the Cavendish Laboratory in Cambridge. While in Cambridge, Kapitza developed his interest in low-temperature physics, his first major success being the invention of a simple method to liquefy helium. As a result liquid helium, once a rare and expensive commodity, became readily available in laboratories.

Kapitza first published details of his helium liquefier in 1934. In the same year he returned to Moscow to visit his family, but was refused permission to return to Cambridge. Instead he was appointed director of the Vavilov Institute for Physical Problems in Moscow. In the previous year a special laboratory had been opened for Kapitza in Cambridge, stocked with specialized equipment. On hearing that Kapitza would not be returning to Cambridge, Rutherford arranged for the new equipment to be purchased by the Soviet authorities for his former pupil. Kapitza continued his low-temperature research in Moscow and in 1938 announced his discovery of the phenomena of superfluidity. He found that below 2.2 K, helium flows through fine channels with no apparent friction and will defy gravity and flow up the walls of its container.

Kapitza's career in Stalinist Russia was not without its problems. He is reported to have been under arrest from 1945 to 1953 for, according to one account, refusing to work on nuclear weapons. Clearly a man of independence and courage, he defended Lev LANDAU in 1938, when he had been arrested as a German spy, and many years later, in 1970, he publicly protested at the detention of the biologist Zhores Medvedev.

Karajan, Herbert von (1908–89) *Austrian conductor. A conductor of major symphony orchestras and operas, who said he sought 'TOSCANINI's precision with FURTWÄNGLER's fantasy', he managed to survive his association with the Nazis.*

Born in Salzburg, Karajan, a child prodigy pianist, studied at the Salzburg Mozarteum while at school; later he studied conducting with Franz Schalk (1863–1931) at the Vienna Academy. His first appointment was at the Ulm Städtisches Theater; in 1934 he became general music director at Aachen, where he joined the Nazi Party. A highly acclaimed performance of Wagner's *Tristan and Isolde* at the Berlin State Opera in 1937 made his name as an opera conductor.

At the end of World War II his Nazi affiliations blocked his career until 1947, when the Vienna Symphony Orchestra appointed him conductor and two years later made him concert director for life. 1947 was also the year of Karajan's London debut with the Philharmonic Orchestra, with whom he made its first European tour in 1952. His US debut was in 1955, with the Berlin Philharmonic Orchestra; the same year he became its principal conductor in succession to Furtwängler, remaining in the post until his death. Karajan was artistic director of the Salzburg Festival (1956–60; 1964) and founded the Salzburg Easter Festival in 1967. His association with the Vienna State Opera (1957–64) ended in his resignation after conflict. However, he rejoined in 1977.

Karamanlis, Konstantinos (1907–)
Greek statesman. He was prime minister (1955–63; 1974–80) and subsequently became president (1980–85; 1990–) after the fall of the monarchy.

Born in Prote, the son of a schoolmaster, Karamanlis was educated at the University of Athens, where he graduated in law in 1932, and subsequently set up a legal practice in Serrai. In 1935 he was elected to parliament as the populist member for Serrai but as a result of the closure of parliament by METAXAS (1936) and the Axis occupation during World War II he returned to his home town. He joined the Free Greek forces in Cairo in 1944.

After the war Karamanlis was re-elected to parliament and was appointed minister of labour in the Tsaldaris government. Between 1947 and 1955 he held several ministerial portfolios including transport, social welfare, national defence, public works, and communications. Elected prime minister in 1955 as head of the right-wing Greek Rally (having left the Populist Party in 1950), he founded the National Radical Union the following year and led it to victory. He remained in power until 1963, when he resigned over conflict with the monarchy and fled to Paris and a self-imposed exile of eleven years. He returned to Greece in 1974 to lead a national emergency government after the collapse of the military junta, when war with Turkey was imminent. Establishing the New Democracy Party, he was again elected prime minister and remained in power until 1980. As prime minister he was responsible for the referendum abolishing the monarchy and establishing a republic (1974) with a new constitution (passed in 1975). This constitution enhanced the powers of the head of state and provided for the entry of Greece into the European Economic Community. Karamanlis became president in 1980 but resigned in 1985, just before the presidential election, when the prime minister, Andreas Papandreou (1919–), withdrew his support and announced constitutional proposals to reduce the powers of the president. He was re-elected president in 1990, Papandreou having lost power in 1989.

Karpov, Anatoly Yevgenyevich (1951–)
Soviet chess player and world chess champion (1975–85).

Karpov was only four when he learned to play chess. A Soviet candidate master at the age of eleven, he became full master at fifteen, international master at eighteen, and grandmaster at nineteen. In his mid-teens Karpov was brilliant at 'lightning' chess and greatly feared, even by grandmasters.

In 1969 Karpov won the world junior championship. Within two years he finished equal first with KORCHNOI in the Leningrad interzonal tournament and thus qualified for the Candidates Series – the eliminating competition to find the best challenger to the reigning world champion. Karpov beat former champions SPASSKY and Polugayevsky easily but Korchnoi made him fight hard. In all Karpov played forty-three qualifiers before establishing his right to challenge Bobby FISCHER for the world championship. Fischer argued about the conditions and in the end did not defend his title. Karpov thus became the twelfth world champion (the first by default) in April 1975, aged twenty-three.

Karpov successfully defended his title against Korchnoi in 1978 and 1981. The 1984–85 world championship tournament, between Karpov and KASPAROV, was abandoned after a record five months with Karpov needing one more win to retain the title but suffering from nervous exhaustion. He lost the title to Kasparov later that year and failed to regain the championship in challenges in 1986, 1987, and 1990. He was elected to the Soviet Congress of People's Deputies in 1989.

Kasparov, Gary (Garri Weinstein; 1963–)
Soviet chess player and world chess champion since 1985.

Of Armenian-Jewish origin, he was born in Baku, Azerbaidzhan, and played his first game of chess at the age of six. At the age of thirteen he captured the Soviet youth championship, going on to win his first international tournament in 1979, having studied under BOTVINNIK from 1973 to 1978. He attained the rank of international grandmaster in 1980.

The 1984–85 world championship tournament, between Kasparov and KARPOV, became an epic contest lasting 48 games and ended without a result when the International Chess Federation stepped in, despite protests from Kasparov that he wished to continue. Kasparov won the rematch in 1985, becoming, at twenty-two, the youngest-ever world champion. Volatile and energetic, he successfully defended his title against Karpov in 1986, 1987, and 1990.

Kaunda, Kenneth David (1924–) *Zambian politician and first president (1964–91).*
Born in Lubwe (Northern Rhodesia, now Zambia) of Malawi descent, the son of a teacher and missionary, Kaunda was educated at a mission school and at Munali Training School, where he qualified as a teacher. Returning to Lubwe to teach in 1943, he served as headmaster (1944–47), before teaching at schools in Tanganyika (1947–48) and Mufulina (1948).

Kaunda's political career began in 1950, when he became founder-secretary of the Lubwe branch of the African National Congress (ANC). Elected ANC secretary-general in 1953, he subsequently became editor of the *Congress News*, the journal of the ANC. In 1953 he was arrested but not imprisoned; however in 1955 he was imprisoned for two months for possessing banned political literature. In 1958 Kaunda broke away from the ANC to form the Zambia African National Congress. Arrested again in 1959 (when the party was also banned), Kaunda was released in 1960 and elected president of the United National Independence Party (UNIP). This party formed a coalition government with the ANC in 1962, in which Kaunda was minister for local government. Leading the UNIP to a resounding electoral victory in 1964, Kaunda became prime minister and was appointed the first president of independent Zambia, a position he held until electoral defeat in 1991. Kaunda was appointed chairman of the Organization of African Unity (1970–71; 1987–88) and of the Non-Aligned Nations Conference (1970–74). During the 1980s he played a key role in the negotiations leading to Namibian independence (achieved in 1990).

Kawabata Yasunari (1899–1972) *Japanese novelist. He won the 1968 Nobel Prize for Literature, the first Japanese to do so.*
Kawabata was born in Osaka and had a lonely childhood after the deaths of his parents and grandparents when he was still very young. His literary talent emerged while he was still at high school in Tokyo and developed during his years at Tokyo Imperial University. As co-founder of the journal *Bungei Jidai* ('The Artistic Age'), he was a leading member of the neoimpressionist group who used the journal to express their artistic manifesto opposing the preceding realist movement. In 1925 he published his first successful novel, *Izu no odoriko* (translated as *The Izu Dancer*, 1955), which is semiautobiographical and deals with the romance between a student and an itinerant girl dancer. *Yukiguni*, perhaps his most famous novel, was begun in 1935 but not completed until twelve years later; it was translated as *Snow Country* (1956). *Sembazuru* (translated as *Thousand Cranes*, 1959) was begun in 1949 but its intended second part was never completed. *Yama no oto* (1949–54; translated as *The Sound of the Mountain*, 1970) was another of Kawabata's characteristically melancholic novels. In 1972 he committed suicide.

Kazan, Elia (Elia Kazanjoglous; 1909–) *Turkish-born US film and stage director, who was a co-founder of the Actors' Studio and whose 1950s 'method' films were highly acclaimed.*
Born in Istanbul, Kazan emigrated with his family to the USA in 1913 and began his career with Lee Strasberg (1901–82) at the Group Theater in 1932. Most of his early career was spent in the radical theatre and for a time he was a member of the Communist Party. His first wife was playwright Molly Kazan and his second, the actress-director Barbara Loden.

Kazan first made a name for himself on Broadway. Successful stage productions included *All My Sons* and *A Streetcar Named Desire* (both 1947). As well as directing, he also acted in several plays and films and worked on a few documentaries before directing his first feature film, *A Tree Grows in Brooklyn* (1945). His first Oscar as director came with *Gentleman's Agreement* (1947), an indictment of antisemitism. Socially relevant, too, was *Boomerang* (1947), dealing with judicial corruption. In the same year Kazan became a co-founder of the Actors' Studio, the home of 'method' acting, and there followed a series of films featuring Marlon BRANDO, the chief exponent of the 'method'. These included *A Streetcar Named Desire* (1951), *Viva Zapata!* (1952), and the Oscar-winning *On the Waterfront* (1954). After the Brando films Kazan made *East of Eden* (1955), starring James DEAN. Subsequent films included *Baby Doll* (1956), *A Face in the Crowd* (1957), film versions of two of his novels, *America America* (1963) and *The Arrangement* (1969), and *The Last Tycoon* (1977). He published his autobiography *Elia Kazan, A Life*, in 1988.

Kazantzakis, Nikos (1885–1957) *Greek novelist, poet, and playwright.*
Born in Herakleion, Crete, Kazantzakis grew up during a period of violent rebellions in Crete against the Ottoman government, an experience reflected in a vein of stark realism that runs through his work. His family moved to

Naxos, where he went to a French school; he later studied law in Athens and philosophy at Paris, where he was influenced by the books of Nietzsche and especially by the works of Henri BERGSON, under whom he studied. His own writings, though not adhering to any formal religion, are basically religious in emphasis and deeply concerned with metaphysical issues. Human existence is seen as essentially tragic but capable of being transformed into spiritual value. This theme informs Kazantzakis's most substantial work, *Odusseia* (1938; translated as *The Odyssey: A Modern Sequel*, 1958), a philosophical epic poem extending to 24 books and 33 333 verses.

His international reputation is due largely to the success of his novels, though these account for only a part of his work, which covers a wide range of genres. In addition to his two best-known novels, *Vios kai politeia tou Alexi Zorba* (1946; translated as *Zorba the Greek*, 1952) and *O Christos xanastavronetai* (1948; translated as *Christ Recrucified*, or *The Greek Passion*, 1954), he wrote *O teleftaios peirasmos* (1955; translated as *The Last Temptation of Christ*, 1959) and *O ftochoulis tou Theou* (1956; translated as *God's Pauper – St. Francis of Assisi*, 1962). He wrote a number of plays in which the philosophical interest predominates. *Melissa* (1936), *Kouros* (1955), and *Christophoros Kolomvos* (1956) have been collected in English as *Three Plays* (1969). He made translations of Dante's *Divine Comedy* (1934) and Goethe's *Faust* (1937) as well as modern renderings of the *Iliad* (1955) and *Odyssey* (1965). An enthusiastic traveller, Kazantzakis published books on Britain, Spain, China, and Japan.

Keaton, Buster (Joseph Francis Keaton; 1895–1966) *US actor and director. His straight face and daring acrobatic skills made him one of the most popular mime artists of silent films.*

Given the name 'Buster' by Houdini, after he had fallen unscathed down a flight of stairs as a baby, Keaton made his first stage appearance at the age of three in his parents' acrobatic and comedy act 'The Three Keatons'. He made his screen debut in *The Butcher Boy* (1917), after Roscoe 'Fatty' Arbuckle (1887–1933) had introduced him to films. In 1920 the first films to come from the Buster Keaton Studio were released, including *The Saphead*, which established him as a star. Among the many memorable films that followed were *One Week* (1920), *The Playhouse* (1921), *The Boat* (1921), and *Cops* (1922). One of his finest

feature-length films was *Our Hospitality* (1923). Notable, too, were *Sherlock, Jr* (1924), *The Navigator* (1924), a huge commercial success, and *The General* (1927). From 1921 to 1932 he was married to Natalie Talmadge (1898–1969), who appeared in a few of his films.

With the coming of sound Keaton's heyday was over. He did, however, make later appearances in a number of films, including Charlie CHAPLIN's *Limelight* (1953) and Mike Todd's *Around the World in 80 Days* (1956). He was given a special Academy Award for his screen comedies in 1959. An autobiography, *My Wonderful World of Slapstick*, was published in 1960. Shortly before his death *Film* (1965), a silent short, brought a rapturous reception and ovation at the Venice Film Festival.

Keitel, Wilhelm Bodewin Johann Gustav (1882–1946) *German field-marshal who served as supreme commander of HITLER's armed forces throughout World War II.*

From a middle-class landowning family, Keitel served in the artillery and as a staff officer during World War I. After the war he found his true métier as a punctilious and conscientious administrator and in 1935 was appointed chief of staff to the war minister, Werner von Blomberg (1878–1946). When in 1938, Hitler replaced the war ministry with a high command of the armed forces – the Oberkommando der Wehrmacht (OKW) – he appointed Keitel its head.

Keitel was one of the few Hitler appointees to retain his post throughout the war – an indication of his unquestioning loyalty to the Führer. But the armed service chiefs and even Hitler himself regarded Keitel with contempt and he was dubbed 'Lakeitel' – 'lackey Keitel'. Indeed, military matters were largely delegated to his chief of staff, Alfred JODL. Together, Keitel and Jodl contrived to supplant the existing army high command with Hitler's OKW. Promoted to field-marshal in July 1940, Keitel offered his resignation in August after Hitler had rejected a memorandum opposing the opening of an eastern front. His offer was sternly refused and thereafter Keitel stifled all criticism of the Führer's proposals. In the wake of the attempted assassination of Hitler on 20 July 1944, Keitel was one of those chosen to investigate fellow officers arrested on suspicion of complicity. After the war, he was convicted of crimes against humanity by the International Military Tribunal at Nuremberg and hanged.

Keller, Helen Adams (1880–1968) *US writer and academic who, deaf and blind herself, championed the cause of blind, deaf, and dumb people throughout the world.*

The daughter of a newspaper editor, Helen Keller contracted scarlet fever at the age of nineteen months, which left her blind and deaf. When nearly seven, she came under the care of Anne Sullivan, herself partially sighted, who undertook the onerous and often frustrating task of teaching Keller the manual alphabet, tapped onto the palm of her hand, and how to lip-read by placing her thumb and forefingers on the speaker's face. These skills enabled Keller to demonstrate her remarkable intelligence. She learnt braille, how to type, and even how to speak. With her mentor and friend constantly at her side, Keller attended courses at Radcliffe College and in 1904 received her AB degree *cum laude*. She developed a flair for languages and specialized in philosophy, receiving doctorates from universities around the world. But above all, Helen Keller is remembered for her campaigning in aid of the American Foundation for the Blind, through which she raised some two million dollars and increased public awareness of the cause.

Anne Sullivan's death in 1936 caused a crisis in Keller's life; how she coped with this is related in *Helen Keller's Journal* (1938). The role of mentor was gradually filled by Polly Thompson, who had been her secretary since 1914. But it was Keller's perseverance, determination, and humour that surmounted her immense handicaps and inspired countless others. She wrote several other books, including *The Story of My Life* (1902), *The World I Live In* (1909), *The Song of The Stone Wall* (1910), and *Teacher* (1956).

Kelly, Grace (1928–82) *US film actress, whose short career came to an end when she married Prince Rainier III (1923–   ) of Monaco in 1956.*

Born into a wealthy family in Philadelphia, daughter of self-made businessman and Olympic sculling champion John Kelly and niece of playwright George Kelly, Grace attended the Academy of Dramatic Arts and for a short time worked as a model and appeared in television commercials. She made her Broadway debut in *The Father* (1949) and her first screen appearance in a bit part in *Fourteen Hours* (1951).

Her first starring role came in the classic *High Noon* (1952), opposite Gary COOPER, followed by *Mogambo* (1953), for which she was nominated for an Oscar as best supporting ac-

tress. *The Country Girl* (1954) brought her an Academy Award for best actress and the New York Critics Award, and in *High Society* (1956) she appeared with Bing CROSBY and Louis ARMSTRONG. Memorable too were three HITCHCOCK films: *Dial M for Murder* (1954), *Rear Window* (1954), and *To Catch a Thief* (1955), during the shooting of which she met her future husband.

After her marriage in 1956 Kelly retired from show business, devoting herself to charitable causes, life in Monaco, and her three children until her untimely death in a road accident.

Kempe, Rudolf (1910–76) *German conductor, who was much admired for his Wagner* Ring *cycles.*

Kempe studied at the Dresden Music High School and in 1928 joined the Dortmund Opera Orchestra as principal oboe, later that year moving to the Leipzig Gewandhaus Orchestra, where he remained for seven years. In 1935 he made his conducting debut at the Leipzig Opera, directing Lortzing's *Der Wildschütz*.

After war service near Chemnitz, Kempe was appointed to the opera house there; he was subsequently director of music at the Dresden State Orchestra (1949–52) and general music director at the Munich State Opera (1952–54). In 1953 he made his Covent Garden debut, conducting operas by Richard STRAUSS, a composer with whose works he was closely associated. Thereafter, Kempe was a regular guest conductor at Covent Garden. In 1954 he conducted at the Metropolitan Opera, New York, and in 1960 at the Bayreuth Wagner Festival. From 1960 he was associate conductor (and from 1961 principal conductor) of the Royal Philharmonic Orchestra, with memorable success at the Bradford Delius Centenary concerts in 1962. After the Royal Philharmonic Orchestra was reorganized in 1970 he was made conductor for life, but resigned in 1975. He was principal conductor of the BBC Symphony Orchestra from 1975 – his unexpected death a year later cut short an ambitious series of programmes.

Kendall, Edward Calvin (1886–1972) *US biochemist noted for his discovery of the adrenal hormone cortisone (later used to treat rheumatoid arthritis), for which he received the 1950 Nobel Prize for Physiology or Medicine.*

Kendall gained his PhD from Columbia University in 1910 and worked briefly as a research chemist before joining St Luke's

Hospital, New York City, in 1911. Three years later he was appointed professor of physiological chemistry at the Mayo Foundation, Rochester, Minnesota. Here he was the first to isolate the principal hormone secreted by the thyroid gland – thyroxine. Following this, Kendall's work on the coenzyme glutathione led to the determination of its tripeptide chemical structure. In the 1930s, Kendall concentrated on isolating and identifying the hormones produced by the cortex of adrenal glands. He discovered six steroid hormones (known as corticosteroids), one of which, cortisone, was used by Philip Hench (1896–1965) to relieve pain and swelling in patients suffering from rheumatoid arthritis. By 1951 the artificial synthesis of cortisone was achieved, making it a widely used drug. Kendall retired in 1951 to become visiting professor of chemistry at Princeton University (1952–72).

Kendrew, Sir John Cowdery (1917–)
British molecular biologist who determined the structure of the muscle protein myoglobin. For this achievement Kendrew shared the 1962 Nobel Prize for Chemistry with Max Perutz. He was knighted in 1974.
The son of a well-known climatologist, Kendrew graduated from Cambridge in 1939 shortly before the start of World War II. He spent the early part of the war working on the development of radar and, after 1940, in operations research. While in operations research he met J. D. Bernal, from whom he first heard of the challenge of working out the three-dimensional structures of biological molecules by X-ray diffraction methods. At the end of the war in 1945 Kendrew returned to Cambridge and began working with Max Perutz on the structure of haemoglobin. In 1948 he set out on his own to tackle the structure of myoglobin, a protein found in muscle. Working on sperm-whale myoglobin, in ten years Kendrew succeeded in locating the exact position of almost all the molecule's 2600 atoms. It was the first protein to reveal its full three-dimensional structure.

From 1959 to 1987 Kendrew served as editor-in-chief of *The Journal of Molecular Biology*. In 1975 he became the first director-general of the European Molecular Biology Laboratory in Heidelberg and continued to hold the post until 1982; he was president of St John's College, Oxford, from 1981 to 1987.

Kennedy, John F(itzgerald) (1917–63) *US Democratic statesman and thirty-fifth president of the USA (1961–63). His brief term of office, cut short by assassination, presented the world with a vision of freedom and social justice whose impact has far outlived the legislative achievements of his administration.*
The second son of Joseph P. Kennedy (1888–1969), millionaire banker, business tycoon, and US ambassador to Britain in the late 1930s, John F. Kennedy was instilled with a sense of public duty from an early age. Graduating from Harvard University in 1940, he entered the navy the following year. In 1943, the motor torpedo boat he was commanding was hit and sunk in the Solomon Islands. In spite of his injuries, Kennedy ensured the escape of his crew and subsequently received an award for gallantry. However, the incident left him with a chronic legacy of back pain. Influential Kennedy family connections in his home state of Massachusetts helped him to a Congressional seat in 1946 and in 1952 he was elected to the Senate. Kennedy campaigned vigorously for labour law reform and civil rights improvements and, as a member of the Senate foreign relations committee, advocated increased US aid for underdeveloped countries. While recovering from surgery to his back, he wrote *Profiles of Courage* (1956), which won a Pulitzer Prize. He became increasingly prominent in the national Democratic Party, his overwhelming victory in the 1958 Senate elections marking the start of his campaign for the White House. After a successful campaign for the 1960 Democratic nomination for president, Kennedy declared: 'We stand today on the edge of a new frontier.' Finally beating Republican Richard Nixon by a narrow margin in the presidential election, Kennedy was sworn in as president in January 1961, the youngest man ever elected to the office and the first Roman Catholic.

The new president's administration was characterized by youth, brilliance, and glamour, undoubtedly enhanced by the charm and good looks of his wife Jacqueline (1929–). Nonetheless Kennedy was soon embarrassed when CIA-backed anti-Castro Cuban insurgents were crushed after landing in Cuba at the Bay of Pigs in the spring of 1961. He was criticized on the one hand for supporting an attack on a neighbouring country and on the other for failing to give sufficient military support to the rebels. Following his meeting with Kennedy in June 1961, Khrushchev threatened the West's access to Berlin but later backed down when Kennedy mobilized US reserves. A far more serious confrontation occurred in October 1962, when Kennedy ordered the removal of Soviet missiles from

Cuba. He blockaded the island and the world teetered on the brink of a US–Soviet conflict until, thirteen days later, Khrushchev complied with US demands. The crisis marked a thawing in East-West relations so that the following June, Kennedy, Khrushchev, and Harold MAC-MILLAN signed the nuclear test-ban treaty that halted atmospheric weapons testing.

In domestic policy, Kennedy was closely assisted by his brother Robert Kennedy (1925–68) who, as attorney general, championed the civil rights movement and supervised a crackdown on organized crimes. (However, civil rights reforms were delayed by Congress until 1964, following the president's death.) On 22 November 1963, Kennedy was shot and mortally wounded while riding in a motorcade through Dallas, Texas. Lee Harvey OSWALD was arrested for the attack but was himself murdered two days later. The brief 'Camelot years' were over. Robert Kennedy was also assassinated, during his campaign for the Democratic presidential nomination in 1968. The youngest brother, Edward Kennedy (1932–   ), was elected senator in 1962 and remains a prominent spokesman for the liberal wing of the Democrats.

Kennelly, Arthur Edwin (1861–1939) *British-born US electrical engineer, who proposed the existence of the ionosphere in 1902.*

Born in Bombay, the son of an Irish naval officer in the employ of the East India Company, Kennelly was educated in Britain. He left school at thirteen and worked for some years with a variety of telegraph companies. In 1887 he emigrated to the USA, where he worked until 1894 as an assistant to Thomas Edison. Kennelly left Edison to set up as a consulting engineer on his own. However, in 1902 he began an academic career as professor of electrical engineering at Harvard, a post he occupied until his retirement in 1930.

Although Kennelly made many contributions to the theory of alternating currents and worked extensively on the accurate measurement and standardization of electrical units, he is best known for his suggestion in 1902 that a layer of electrically charged particles in the upper atmosphere is capable of reflecting back to earth radio waves broadcast from the surface of the earth. This would explain how radio waves were first transmitted across the Atlantic by MARCONI in 1901. The layer was initially known as the Heaviside–Kennelly layer after Kennelly and Oliver HEAVISIDE, who independently proposed its existence at the same time. Its existence was confirmed in 1924 by Sir Edward APPLETON, after which it became known as the ionosphere, although the E region of the ionosphere is still sometimes known as the Heaviside–Kennelly layer.

Kenyatta, Jomo (Kamau Ngengi Kenyatta; *c.* 1891–1978) *Nationalist leader and first president of Kenya (1964–78). Recognizable all over the world as the African leader with the fly-swat, Kenyatta used the slogan 'harambee' (pull together) to build racial and tribal harmony in the divided Kenyan nation.*

Born near Gatundu in Central Kenya, of mixed Kikuyu and Masai descent, Kenyatta was educated by the Church of Scotland Mission, before working as a translator and interpreter in the early 1920s. His political involvement began when he joined the Kenya Central Association in 1922, of which he became general secretary in 1928. Travelling to London in 1929, Kenyatta represented the Kikuyu people before the Hilton Young Commission on land use. Two years later he visited England again to study anthropology under Bronislaw MALINOWSKI at London University, writing *Facing Mount Kenya* (1938), an analysis of the colonial impact on Kikuyu life. Having formed a pan-African organization in 1945 with NKRUMAH and other future African leaders, Kenyatta was hailed as the leader of the nationalist movement on his return to Kenya in 1946. Having transformed the Kikuyu-dominated Kenya African Union (KAU) into a national party, he was accused of complicity in the 'Mau Mau' uprising. In 1952 he was arrested and imprisoned for nine years by the colonial government (although he persistently denied having managed the rebellion).

In 1960, the newly formed Kenya African National Union (KANU) named Kenyatta as president *in absentia*. He was released from prison in 1961 and became leader of the parliamentary opposition in 1962. Following a further victory by KANU in 1963, he was allowed to form a government and became the first prime minister on independence later that year. When Kenya declared itself a republic in 1964, Kenyatta became the first president, an office that he held until his death in 1978.

Kerenski, Alexander Feodorovich (1881–1970) *Russian statesman. He was prime minister of the Russian provisional government (1917) but lost power to the Bolsheviks after the October Revolution (1917).*

Born in Simbirsk (now Ulyanovsk), the son of a headmaster, Kerenski was educated at his father's school before attending the University of St Petersburg, where he graduated in law

(1904). Joining the Socialist Revolutionary Party in 1905, he was known at the St Petersburg bar for defending socialists accused of political crimes.

In 1912 Kerenski was elected to the Fourth Duma as the Trudoviki (Labour Group) delegate from Volsk. An advocate of Russian involvement in World War I, he became disillusioned with the tsarist war effort and following the 1917 February Revolution came out in favour of the abolition of the monarchy. On the formation of the provisional government in March 1917, he was appointed minister of justice, taking over the war and admiralty portfolios in May 1917. He was elected prime minister in July 1917 but his mismanagement of economic policy made it impossible for him to muster support for his government against the Bolsheviks. Moreover, his support of Russian involvement in World War I, when troops were weary and demoralized, and his failure to unite the various political factions in Russia enabled the Bolsheviks to seize power in October 1917 (the October Revolution). He remained in hiding for several months before escaping to western Europe, where he lived in exile until 1939. In 1940 he emigrated to the USA, where he spent the rest of his life. In exile, he devoted his time to writing about the revolution and editing emigré newspapers and journals. His major works in English include *The Prelude to Bolshevism* (1919), *The Catastrophe* (1927), and *The Crucifixion of Liberty* (1933).

Kern, Jerome David (1885–1945) *US composer and songwriter. He wrote many of the century's best-loved songs and played a leading part in replacing the European operetta by the American musical.*

He studied music in his native New York and briefly in Heidelberg. By 1904 he was working as a song plugger and rehearsal pianist in New York and also had a brief spell in London (1906). From 1915 he began to transform the American musical theatre by integrating songs with the dramatic action, rather than simply dropping them in. *Very Good, Eddie* was his first success that year, and deeply influenced Richard RODGERS.

Showboat (1927) was probably the most influential show in the history of American musical plays; it included the hit songs 'Ol' Man River', 'Can't Help Lovin' Dat Man', and 'Why do I Love You?'. It was filmed several times and entered the repertory of the New York City Opera Company in 1954. Other Kern successes included *Roberta* (1933;

'Smoke Gets in Your Eyes'), *Swing Time* (1936; 'The Way You Look Tonight'), and *Very Warm for May* (1939; 'All the Things You Are'). In 1939 he moved to Hollywood and wrote exclusively for films, including *Lady Be Good* (1941; 'The Last Time I Saw Paris'), and *Cover Girl* (1944; 'Long Ago and Far Away').

Altogether he wrote more than a thousand songs for over a hundred plays and films, working with lyricists P. G. WODEHOUSE, Johnny Mercer (1909–76), Oscar HAMMERSTEIN II, and many others.

Kerouac, Jack (Jean-Louis Kerouac; 1922–69) *US novelist and poet, a leading figure of the Beat group of writers.*

Born in Massachusetts into a French-Canadian family, Kerouac studied at Columbia University (1940–42). A decisive event in his short career was his meeting with Allen GINSBERG at William BURROUGHS's apartment near Columbia: their names became indissolubly linked in the public mind as leaders of the Beats, a term supposedly coined by Kerouac from 'beatific'. After leaving university and travelling about for a time, Kerouac wrote his first novel, *The Town and the City* (1950), drawing on the experiences of his family up to World War II. With *On the Road* (1957), written in a loose 'spontaneous' style, Kerouac was established as a hero of the Beat movement. An autobiographical novel, it relates the constant and aimless wanderings of characters whose only motive is self-gratification (chiefly through sex and drugs), though much is made of the spiritual quest that is somehow involved. In *Dharma Bums* (1958) the subject is similar but with an explicitly Buddhist theme of a search for truth (*dharma*). This was followed by *The Subterraneans* (1958), *Doctor Sax* (1959), *Big Sur* (1962), *Desolation Angels* (1965), and other novels. *Mexico City Blues* (1959) is a collection of poems. Long before his early death, caused by drink, Kerouac withdrew from the Beat scene; none of his later work equalled or had the influence of his first novels.

Kertész, André (1894–1985) *Hungarian-born US photographer whose pictures of Paris in the 1920s were a major influence on the development of photojournalism.*

After attending the Hungarian Academy of Commerce, Kertész worked as a clerk in the Budapest Stock Exchange. During World War I he served with the Austro-Hungarian army and was wounded in 1915. He had begun taking pictures as a hobby, and after the war magazines started to accept his work for publication. His early pictures display a remarkable instinct for humanity and warmth in a scene, usually of everyday life in Budapest.

In 1925 Kertész moved to Paris. His love for the city is evident in numerous pictures, often showing features of the city from new and unexpected vantage points, with unusual juxtaposition of form and light. He also took portraits of friends, who included artists and writers, such as Piet MONDRIAN, Marc CHAGALL, and COLETTE. Kertész owned one of the first compact Leica cameras and took full advantage of the greater freedom and flexibility it allowed. His first one-man show was held in 1927 and the following year he exhibited in the prestigious First Independent Salon of Photography. Besides contributing to various European magazines, in the early 1930s he published collections of photographs with the themes of children, Paris, and animals and made over 150 surrealistically distorted studies of nudes.

In 1936, Kertész took a one-year contract with the Keystone picture agency in the USA. With war looming in Europe, he decided to stay on in New York, becoming a US citizen in 1944. He freelanced for several years until in 1949 he was contracted full-time to photograph fashions and design for a US magazine house. His early work underwent reappraisal in the 1960s and 1970s, with the publication of such retrospective works as *André Kertész: Sixty Years of Photography, 1912–1972* (1972) and *J'aime Paris: Photographs since the Twenties* (1974). His eye for the telling detail, such as the small signs of nature in the environs of his New York apartment, is still obvious in *Of New York* (1976).

Kesselring, Albrecht (1885–1960) *German air force chief and field-marshal who demonstrated great skill as an army commander in World War II, particularly during the defensive campaign in Italy.*

Commissioned in the Bavarian artillery in 1906, Kesselring served as a captain during World War I and remained in the army until, in 1933, he joined HITLER's Air Ministry to help create the Luftwaffe – the German air force. He was later appointed chief of general staff. At the outbreak of war, Kesselring was given command of Luftflotte 1, operating in Poland, and in 1940 was moved to head of Luftflotte 2, comprising about half the German air force.

Directing German air attacks on Britain in 1940, Kesselring supported the tactical error made by GOERING in switching from raids on

RAF bases to civilian targets and cities. This enabled the RAF to recover and eventually regain control of British air space. On 1 December 1941, after some six months directing Luftflotte 2 in the Soviet offensive, Kesselring was promoted to commander-in-chief (south) with control over both air and land forces. Kesselring saw the forces of his colleague, Erwin ROMMEL, take Tobruk from the British in 1942, then falter, starved of adequate supplies. By May 1943, the Axis forces had been defeated in North Africa. The retreat of Kesselring's forces from Sicily in July marked the start of a great tactical defensive campaign on the Italian mainland. Using the mountainous terrain to great advantage, Kesselring ensured that the Allied advance was a slow and costly business. In March 1945, Hitler appointed Kesselring commander-in-chief (west), replacing von Runstedt in an attempt to salvage the already hopeless position. Kesselring surrendered to US forces on 6 May 1945. He was tried by a British military court in Venice in 1947 and sentenced to death for his culpability regarding the shooting of 335 Italian civilian hostages. The sentence was commuted to life imprisonment and he was released in 1952 because of ill health. He later served as president of the ex-serviceman's association, Stahlhelm.

Keynes, John Maynard, 1st Baron
(1883–1946) *British economist who laid the foundations of modern macroeconomics with his* General Theory of Employment, Interest and Money *(1936). He was a member of the Bloomsbury group and was created Baron Keynes of Tilton in 1942.*

Born in Cambridge, the son of John Neville Keynes (1852–1949), a logician and political economist, he was educated at King's College, Cambridge, where he studied under Alfred MARSHALL and Arthur PIGOU, and subsequently lectured in economics. Entering the civil service in 1906 he worked at the India Office and served as a member of the Royal Commission on Indian Finance and Currency. In World War I he was an adviser to the Treasury and attended the Versailles Peace Conference but resigned his post in strong opposition to the terms of the treaty; in 1919 he wrote *The Economic Consequences of the Peace*, in which he expressed the view that the high war reparation payments being imposed on Germany would have disastrous consequences. Further academic work followed, notably his *Treatise on Probability* (1921), *Tract on Monetary Reform* (1923), in which he opposed the return to the gold standard, and *Treatise on Money* (1930). Meanwhile Keynes was amassing a personal fortune on the money markets. During World War II he again worked for the Treasury and was prominent at the Bretton Woods Conference (1944) in New Hampshire, where his plan (the 'Keynes Plan') for world monetary reform by the creation of a world clearing bank for the settlement of international debts and a new international currency (which he termed 'bancor') was rejected in favour of the US plan for the establishment of the International Monetary Fund and the World Bank. Keynes also handled negotiations with the USA over lend-lease payments.

The principal economic problem of his time was that of unemployment, and Keynes's *General Theory of Employment, Interest and Money* presented a revolutionary approach to the problem. Traditional economic theory held that market forces would eventually solve the problems if wages and the rate of interest were allowed to fall. Keynes, however, believed that an equilibrium level of national income would not automatically equate with full employment. He stressed effective demand as determining the level of employment and advocated stimulating this by government spending on public works. His main innovative analytical tools were the consumption function and the speculative demand for money. His analysis of economic national aggregates, particularly savings and investment, has won wide acceptance.

In 1925 Keynes married Lydia Lopokova, a dancer with DIAGHILEV's Ballets Russes.

Khachaturian, Aram Ilich (1903–78) *Soviet composer of Armenian origin, whose work was deeply influenced by Armenian folk music.* Born in Tiflis, Georgia, as a child Khachaturian played in a band, taught himself the piano, and became interested in his native folk music, which later was to characterize his compositions. His talent for music was discovered when he joined the Gnesin Music Academy in Moscow at the age of eighteen; after this his progress was rapid. *Tants* for violin and piano and *Poema* for piano (1926–27) were both successfully performed. In 1929 he entered the Moscow Conservatory and by the time he had finished postgraduate study (1937) he was already known internationally as a composer for such works as the piano concerto (1936), the first of three symphonies (1935), and the *Song of Stalin* for chorus and orchestra (1937, expanded 1938).

As an administrator he was active in the newly formed Union of Soviet Composers, although this did not prevent censure for so-called 'formalism'. He also admitted to over-preoccupation with technique and the desire to make his music more 'cosmopolitan'. Once the accusations and confessions were over he was appointed professor at the Moscow Conservatory and became a successful conductor of his own works; in 1954 he was named People's Artist of the USSR. Other works for which he is known include the ballets *Gayaneh* (1942), with its popular 'Sabre Dance', and *Spartacus* (1954; revised 1968) and the expansive violin concerto (1940). He married the Soviet composer Nina Makarova (1908–   ).

Khomeini, Ayatollah Ruhollah (Ruhollah Hendi; 1900–89) *Iranian religious leader, who, after sixteen years in exile, returned to Iran to lead the Islamic Revolution following the deposition of the shah (see PAHLAVI, MOHAMMED REZA). He was seen by his supporters in Iran as a saviour and liberator; in the West he was more often regarded as a savage and fanatical theologian who had dragged his country back into the Middle Ages.*

Born in Khomein, the son of a religious leader (ayatollah), Khomeini was educated at Islamic schools in Khomein and Qom. He taught in the theological school in Qom and wrote many works on Islamic philosophy, law, and ethics. Like his father and grandfather, he too became a religious leader. In 1963 he was arrested for speaking out against the shah's land reforms; the following year he was forced out of the country.

In exile Khomeini lived in Najaf, Iraq (1964–78), and Neauphlé-le-Château, France (1978–79). During this period he worked to overthrow the shah and to create a Shiite Islamic republic. Particularly active in France, he exercised considerable influence over the revolution (1978–79) that forced the shah to leave Iran in January 1979. Khomeini returned to Iran in February 1979 and was hailed as the leader of the Islamic Revolution. After appointing a new government, he returned to the theological seminary in Qom, although he continued to wield power behind the scenes as leader of the Islamic movement. Under the new constitution that promulgated the Islamic Republic of Iran (1979) he was officially declared a religious leader (Velayay Faghih).

Khomeini presided over the introduction of revolutionary tribunals to conduct summary trials, which ordered many executions; he also caused considerable outrage when he supported the seizure of the US embassy (1979) by militant Iranian students, showing scant regard for international law and creating further divisions between Iran and the West. He pursued the war with Iraq (1980–88) with relentless religious fervour, regarding the Sunni Muslim Iraqi president, Saddam HUSSEIN, as the infidel leader of the country that expelled him in 1978, executed the Iraqi Shiite leader Ayatollah Baqir al-Sadr in 1980, and finally declared war on the Islamic Revolution. International criticism of Khomeini reached a new height in 1989, when he announced that Iran would pay a large reward to anyone who killed the British author Salman RUSHDIE, whose novel *The Satanic Verses* (1988) had been judged blasphemous. The affair culminated in a rift in diplomatic relations between the UK and Iran, which continued after Khomeini's death.

Khomeini published numerous religious and political books, including *The Government of Theologians*, a series of lectures he gave in exile.

Khrushchev, Nikita Sergeevich (1894– 1971) *Soviet statesman, first secretary of the Soviet Communist Party (1953–64) and prime minister (1958–64). Khrushchev was a close associate of STALIN and the leader to emerge after his death. He was ousted by BREZHNEV and KOSYGIN, largely as a result of his antagonism to China.*

Born in Kalmkova, the son of a miner, Khrushchev was a shepherd before becoming a metalworker. Joining the Bolsheviks in 1918, he fought against the Germans in World War I and against the White Army during the civil war. After the civil war he worked for the Communist Party in Kiev and then in Moscow, where he graduated from the Industrial Academy in 1931. By 1935 he had become first secretary of the Moscow region, earning recognition for the building of the Moscow metro.

In 1938 Khrushchev became first secretary of the Ukraine region and the following year became a member of the Politburo. He served on the southern front at Stalingrad during World War II before returning to the Ukraine to continue as first secretary and chairman of the council of ministers (1944–49). After the power struggle following Stalin's death in 1953, he emerged as first secretary of the party and gradually built up enough support to launch an attack on Stalinism and the 'cult of personality' at the Twentieth Party Congress (1956). Two years later he succeeded BULGANIN as chairman of the council of minis-

ters (prime minister) but resigned in 1964, officially for reasons of ill health.

During his years in power he instigated the notion of 'peaceful coexistence', travelling to China, India, western Europe, and the USA. But he came close to war with the USA over the Cuban missile crisis in 1962 and as a result of his climbdown before Kennedy's resoluteness lost some of his ebullience and popularity both at home and abroad. His quarrel with China over economic aid, philosophy, and borders finally led to his downfall. He lived outside Moscow, in virtual isolation, for the last years of his life.

Kim Il Sung (Kim Song Ju; 1912–) *Korean statesman; first premier (1948–72) and president (1972–) of North Korea.*

Born near Pyongyang, Kim Song Ju was a devoted communist from adolescence, serving a prison sentence for his activities in 1929–30. In the 1930s he led the armed resistance to Japanese domination of Korea, taking the name of a legendary national hero. During World War II he continued this struggle as commander of a Korean contingent in the Soviet army; on the defeat of the Japanese he was installed as the leader of Soviet-occupied North Korea. Talks aimed at reuniting the north with the US-occupied south failed in 1947 and the Democratic People's Republic of Korea became an independent state with Kim at its head.

In 1950 Kim, who has never accepted partition, ordered the military invasion of South Korea, precipitating the Korean War. A US-led United Nations force came to the defence of the South, while China intervened on the Northern side. The fighting proved inconclusive and the eventual armistice (1953) left the border little altered. Since the war, Kim has repeatedly pressed for the peaceful reunion of the Koreas, proposing a new federation plan in 1980.

Domestically, Kim has maintained a one-party communist state and created an extraordinary personality cult around himself and his family. All industry is nationalized and since the 1950s most agricultural land has been collectivized. There have been notable successes in the provision of medical care and education. In 1980 he named his son, Kim Chong Il, as his successor.

King, Billie Jean (Billie Jean Moffitt; 1943–) *US tennis player and one of the greatest players of all time, who won a record twenty Wimbledon titles.*

In a long tennis career, King competed in over one hundred singles matches at Wimbledon alone. Her powerful serve and volley game gained her a total of six singles titles (1966–68, 1972, 1973, and 1975), ten doubles titles, and four mixed doubles titles. In 1983, at the age of forty, she retired after losing to Andrea Jaeger in the semi-finals. King's other notable singles titles included the 1968 Australian championship, the US Open in 1971, 1972, and 1974, and the 1972 French championship. She also won a string of doubles titles, usually partnered by Rosemary Casals; in 1980, she won her fourth US Open doubles title, playing with Martina NAVRATILOVA.

In 1973 King played before the largest ever tennis crowd of over 30 000 people in a challenge match against the 1939 Wimbledon men's champion, Bobby Riggs, at the Houston Astrodome, Texas. King won 6-4, 6-3, 6-3 and collected record prize money of $100,000. She played a prominent part in improving the status and prize money for women's professional tennis following the introduction of open tournaments in the late 1960s. Her writings on tennis include *We Have Come a Long Way* (with Cynthia Starr; 1989).

King, (William Lyon) Mackenzie (1874–1950) *Canadian statesman; leader of the Liberal Party (1919–48) and prime minister (1921–26; 1926–30; 1935–48).*

King was born in Berlin (now Kitchener), Ontario, the son of John King, a noted lawyer, and the grandson of William Lyon Mackenzie (1795–1861), leader of the 1837 rebellion in Upper Canada. He studied at the University of Toronto before embarking on postgraduate work at Chicago and Harvard universities. Having declined an academic position at Harvard, he entered the civil service in 1900 as deputy minister of labour in Ottawa. Encouraged by the Liberal prime minister, Sir Wilfrid Laurier (1841–1919), he campaigned successfully for a seat in Parliament and became minister of labour in 1909. On the defeat of the government in 1911, King remained an active member of the Liberal Party and retained his interest in labour relations, conducting an investigation on this subject for the Rockefeller Foundation in the USA (1914–18) and publishing his results in *Industry and Humanity* (1918).

On Laurier's death in 1919, King was elected leader of the Liberal Party and in 1921 he became prime minister for the first time. Having tried and failed to gain a majority in Parliament he resigned early in 1926, returning to power with a significant majority later that year. He represented Canada at the imperial conferences in London (1923, 1926, 1927), where he played an important role in establishing the status of the self-governing nations of the Commonwealth. Defeated by R. B. Bennett (1870–1947) in the 1930 election, King remained at the head of the opposition throughout the Great Depression. In 1935 he was re-elected prime minister and served for a further thirteen years until his retirement in 1948. Through his intuitive and shrewd approach to the French-Canadian problem, he succeeded in uniting a country frequently split on major issues; his government survived two potentially divisive crises over the conscription issue to make a valuable contribution to World War II and led the country smoothly into the postwar period. King went on to strengthen Canada's ties with the USA and Britain and made significant advances in the economic and social development of the country. He died just eighteen months after his resignation.

King, Jr, Martin Luther (1929–68) *US black civil-rights leader and Baptist minister. He was awarded the Nobel Peace Prize in 1964.*

Born in Atlanta, Georgia, the son and grandson of Baptist preachers, King studied at Morehouse College, Atlanta, graduating in sociology at the age of nineteen. Encouraged by his father to enter the Baptist ministry, he abandoned his earlier ambitions to become a doctor or lawyer and went on to Crozer Theological Seminary in Chester, Pennsylvania. He completed his education at Boston University, where he was awarded a PhD in 1955. While studying for his doctorate, King was appointed pastor of a Baptist church in Montgomery, Alabama. In December 1955 he became involved with the Montgomery Improvement Association, a black activist group protesting against racial segregation on public transport, and led a year-long black boycott of the local bus company. Inspired by the principles of Mahatma GANDHI, notably that of nonviolent resistance, King led his followers with courage and determination to a successful conclusion of the dispute and went on to found the Southern Christian Leadership Conference (SCLC) in 1957.

During the next three years King lectured throughout the country on civil rights and discussed with religious and state leaders the problems faced by blacks in their struggle for freedom. In 1960 he returned to his native

Atlanta to share the pastorship of Ebenezer Baptist Church with his father and to devote more time to the civil rights movement: in the early sixties he led numerous nonviolent demonstrations against segregation and was arrested several times. His attempt in May 1963 to desegregate the hotels and restaurants of Birmingham, Alabama, met with violent resistance from the police, who used dogs and fire hoses to disperse the crowds of blacks. King was among those imprisoned for the incident and from Birmingham jail he wrote a celebrated letter expressing his moral philosophy in eloquent and forceful terms. Later that year, with other leaders of the civil rights movement, King marched to the Lincoln Memorial in Washington at the head of some 200 000 Americans of all races and creeds in the peaceful demonstration known as the March on Washington. A brilliant orator, King made a profound impression on all those present on that occasion with the now-famous speech beginning 'I have a dream…'. In 1964 King's tireless efforts were rewarded with the Nobel Peace Prize and, earlier in the year, by the passing of the Civil Rights Act. In 1965 a demonstration in Selma, Alabama, led to the passage of the Voting Rights Act, but King's nonviolent approach was increasingly challenged by younger and more militant black activists. He responded by attacking a broader range of issues, including the Vietnam War and the conditions of the poor, and in 1968 he announced plans for a Poor People's March on Washington. While these plans were being finalized, King travelled to Memphis, Tennessee, to lend his support to a largely black group of striking sanitation workers. As he talked with his associates on the balcony of their motel he was shot and killed by a hired assassin, James Earl Ray, a white man who was convicted of the murder in 1969 and jailed for ninety-nine years.

Kinnock, Neil Gordon (1942–) *British politician, leader of the Labour Party since 1983.*

A trades union tutor with the Workers' Educational Association, Kinnock was elected to parliament in 1970, representing Bedwellty in Wales until 1983 and then the new constituency of Islwyn. He became a member of the Tribune Group, on the left of the party. He was parliamentary private secretary to the employment secretary, Michael Foot (1974–75), and chief opposition spokesman on education (1979–83). Sponsored by the Transport and General Workers' Union, Kinnock succeeded

Foot as leader of the Labour Party. As leader of the opposition, he promoted party unity by ousting members of the extreme left-wing organization Militant Tendency. Since Labour's defeat in the general election of 1987 Kinnock has realigned Labour policies on a more moderate basis.

Kinsey, Alfred Charles (1894–1956) *US zoologist and sexologist, noted for his surveys of human sexual behaviour in the USA.*

Born in Hoboken, New Jersey, Kinsey graduated in 1916 with a BS degree from Bowdoin College and received his ScD in life sciences from Harvard University in 1920. He joined Indiana University as an assistant professor of zoology, becoming associate professor (1922) and full professor (1929). Between 1922 and 1936, Kinsey published numerous papers on gall wasps as well as several high-school biology texts. In 1938 he started his survey of human sexual behaviour, interviewing people and compiling data. Based on over five thousand case histories, *Sexual Behaviour in the Human Male* (1948) revealed distinct correlations between sexual habits and social class and disclosed a hitherto unsuspected diversity of sexual practices and mores. It became known as the 'Kinsey Report' and aroused much controversy. This was followed by *Sexual Behaviour of the Human Female* (1953). Both books became best-sellers and profits went to finance Kinsey's Institute of Sex Research, which he established at Indiana University in 1947.

Kipling, (Joseph) Rudyard (1865–1936) *British poet, short-story writer, and novelist. He was awarded the Nobel Prize for Literature in 1907 but declined both the poet laureateship and the OM.*

Kipling was born in Bombay, where his father was teaching at the School of Art. Sent back to England as a child, Kipling endured callous treatment from the family with which he boarded; this period of unhappiness is recalled in the story 'Baa, Baa, Black Sheep'. In 1878 he was sent to the United Services College, Westward Ho!, where the camaraderie and fun Kipling enjoyed were to be the basis of the school stories published as *Stalky and Co* (1899).

On returning to India (1882), Kipling became a reporter on the Lahore *Civil and Military Gazette*. In this job he saw the diversity of life in India in the heyday of the British raj, which is reflected in the poems and short stories by which he soon gained a considerable reputation. *Departmental Ditties* appeared in

1886, and several short-story collections, among them *Plain Tales from the Hills*, were published in 1888. He developed a special sympathy and affection for the unsung heroes of the empire: the ordinary soldiers whom he celebrates in one of his earliest volumes of stories, *Soldiers Three* (1888), and in *Barrack-Room Ballads* (1892). In 1889 Kipling left India to return to England via Japan and the USA. In London he met Wolcott Balestier, an American with whom he collaborated on the novel *The Naulakha* (1892) and whose sister Caroline he married in 1892. After further travels, Kipling settled temporarily on his wife's property in Vermont; although he disliked the USA he continued to write, completing *Many Inventions* (1893) and the two *Jungle Books* (1894, 1895) before returning to England.

Anti-imperialist feeling engendered by the Boer War (1899–1902) harmed Kipling's popularity and in the early 1900s he concentrated on books for children: *Kim* (1901), *Just So Stories* (1902), *Puck of Pook's Hill* (1906), and *Rewards and Fairies* (1910). In 1902 he bought a house in Sussex, which became the focus of the traditional English life (expressed in his poem of the same year, 'Sussex'). During World War I Kipling threw himself into patriotic writing but in a generally more sombre tone than that of his earlier work. His only son was killed in action (1915) and in later life Kipling devoted much energy to the War Graves Commission. He also suffered chronic ill health on account of an undiagnosed duodenal ulcer. He died after an operation, leaving his autobiography, *Something of Myself* (1937), unfinished.

Kirchner, Ernst Ludwig (1880–1938) *German expressionist painter who was the dominant personality in the Brücke group.*

A compulsive draughtsman from childhood, Kirchner became a full-time artist in 1905 after studying painting and architecture. In the same year he became a founder member of the first organized group of German expressionists, Die Brücke. In reaction to the academic and pompous nature of official German art, the Brücke artists determined to paint real life and to express real emotions. They rented an empty butcher's shop in a working-class area of Dresden and sought inspiration from Gothic and medieval German art, primitive art, and Van Gogh and Gauguin. The revolutionary style that resulted is exemplified in Kirchner's nudes and portraits with their bright and boldly contrasting colours and simple forms and in his graphic works, which utilize the crudest methods of woodblock and linocut with powerful effect.

In 1911 Kirchner settled in Berlin, where he established an international reputation. During his Berlin period he produced psychological portraits and claustrophobic street scenes of dandies and prostitutes. His increasingly frenzied style reflected both his vision of modern life and his impending mental crisis. This materialized in the form of a serious nervous and physical breakdown soon after his mobilization at the beginning of World War I. After a period in a sanatorium Kirchner spent the rest of his life in the quiet and solitude of the Swiss mountains near Davos, where his paintings of mountain landscapes and peasants expressed a new serenity. In 1928 his style turned towards the abstract as he experimented with his 'hieroglyphic' forms. In 1938 the condemnation of his work by the Nazis worsened his depression and he took his own life.

Kissinger, Henry Alfred (1923–   ) *German-born US statesman and academic, who masterminded his country's foreign policy during the NIXON and FORD administrations.* He shared the 1973 Nobel Peace Prize with the North Vietnamese negotiator, Le Duc Tho, for their treaty enabling the withdrawal of US troops from South Vietnam.

Originally from Fürth, near Nuremberg, Kissinger's family emigrated to the USA in 1938 to escape Nazi antisemitism. Kissinger attended City College, New York, and also worked in a shaving-brush factory. Gaining US citizenship in 1943, he was drafted into the army and served in Europe. After the war he entered Harvard University, received his BA in 1950, and stayed on as a fellow, specializing in international relations. His research programme on US–Soviet relations for the Council on Foreign Relations resulted in *Nuclear Weapons and Foreign Policy* (1957). This introduced the concept of 'flexible response' in a nuclear confrontation and was followed by *The Necessity for Choice; Prospects of American Foreign Policy* (1961). In 1957 he was appointed lecturer in government at Harvard's Centre for International Affairs, becoming associate professor (1959) and subsequently professor (1962). For ten years (1959–69) he directed the Harvard defence studies programme while serving as defence consultant to the KENNEDY administration and to the Arms Control and Disarmament Agency (1961–67). When Nixon was elected president in 1968, he appointed Kissinger presidential assistant for national security affairs. Kissinger's pragmatism bore fruit in improved relations with both China and the Soviet Union, culminating in the presidential summits of 1972 and the SALT treaty. But his handling of the Vietnam War was more contentious, especially the covert US bombing of Cambodia in 1969 and 1970. Secret negotiations with the North Vietnamese began in Paris in 1969. Meanwhile, US and South Vietnamese forces entered Cambodia and Laos in 1970. A North Vietnamese offensive in spring 1972 prompted US bombing of the North and an attempt to blockade North Vietnamese ports. Further Kissinger diplomacy in August and September brought a ceasefire closer but, following a further breakdown in negotiations and Nixon's re-election, a massive airstrike against the North was ordered over Christmas. On 23 January 1973, Kissinger and Tho signed the cease-fire agreement and US troops subsequently pulled out. A communist takeover followed in 1975 and Cambodia suffered a catastrophic civil war. US policy had failed utterly.

Appointed secretary of state in September 1973, Kissinger embarked on a frantic round of 'shuttle diplomacy' following the Yom Kippur War between Egypt and Israel in October 1973 and achieved disengagement between the warring parties. He remained in office after Nixon's resignation until 1977, when he accepted the chair of diplomacy at Georgetown University. His publications include two volumes of memoirs, *The White House Years* (1979) and *Years of Upheaval* (1982).

Kitchener of Khartoum, (Horatio) Herbert, 1st Earl (1850–1916) *British field-marshal who led British and Egyptian forces to victory in the Battle of Omdurman in 1898, served as commander-in-chief during the Boer War, and was war minister during World War I. He was created a baron in 1898, elevated to the viscountcy in 1902, and granted an earldom in 1914.*

The son of an army officer, Kitchener was educated in Switzerland and attended the Royal Military Academy, Woolwich. Commissioned in the Royal Engineers in 1871, he undertook survey work in Palestine and Cyprus before being appointed second in command of the Egyptian cavalry in 1883. The following year, during the revolt by followers of the Mahdi in the Sudan, he performed intelligence work and accompanied the unsuccessful expedition to rescue General Gordon from Khartoum. After serving as governor-general for east Sudan, Kitchener was appointed sirdar

of the Egyptian army (1892). Following his reorganization of forces, in 1896 Kitchener embarked on a campaign to quell the Dervishes in the Sudan. The town of Dongola was captured and Kitchener was promoted to major-general. On 2 September 1898, Kitchener's combined British and Egyptian force routed the Dervishes at Omdurman. Khartoum was recaptured two days later. This victory made Kitchener a national hero in Britain and, after briefly returning to the Sudan as governor-general, he was posted to South Africa as army chief of staff to Lord Roberts, whom he succeeded as commander-in-chief in November 1900.

Kitchener ordered the destruction of Boer homesteads and the internment of Boer families in camps, where many suffered and died. This typified his cold-hearted, often ruthless, approach. Kitchener's war of attrition against the Boers finally ended in May 1902 and he was transferred to India as army C-in-C. Here he quarrelled with the viceroy, Lord Curzon (1859–1925), who gave priority to the civil service over the military, and Curzon resigned. However, ASQUITH failed to appoint Kitchener as his successor; instead, in 1911, he became consul-general of Egypt.

At the outset of World War I in August 1914, Kitchener was appointed secretary of state for war. He warned that the war would last years rather than months and pressed for massive expansion of the army. Throughout the country, his commanding image appeared on recruiting posters urging 'Your country needs you!' But relations with his cabinet colleagues became increasingly strained and in 1915 he was relieved of responsibility for military strategy, although his resignation was rejected by the prime minister. On 5 June 1916, he was aboard the HMS *Hampshire* bound for Russia on a diplomatic mission when the vessel hit a German mine off Orkney and sank. Kitchener was not among the few survivors.

Klee, Paul (1879–1940) *Swiss painter and graphic artist, one of the most imaginative and prolific of twentieth-century masters. Working in many styles, both representational and abstract, by his teaching and his work he profoundly influenced innovative art in this century.*

The son of a Swiss mother and a German father, Klee left Bern to study art in Munich at the age of nineteen. His early drawings and etchings were in an expressionist style containing elements of fantasy and satire. In the early 1900s he came into contact with many of the influential artists of the period, including KANDINSKY, who involved him in the Blaue Reiter group in 1912. In 1914 a visit to Tunisia stimulated a new awareness of colour and he began to concentrate on painting. His paintings, mainly on paper or small canvases, have a childlike quality that masks the sophistication of his work. Though the rhythms and forms of his pictures were often based on remembered impressions, his was an art of free fantasy. He once described his drawing method as 'taking a line for a walk.'

After serving in the German army in World War I, Klee achieved an international reputation by 1920 and in that year was invited to join the faculty of the Bauhaus. In 1924 he had his first US exhibition in New York and in 1925 the Bauhaus published his influential *Pedagogical Sketchbook*. He left the Bauhaus in 1931 to teach at the Düsseldorf Academy but in 1933 the Nazis forced him out of this post and in 1937 they included his works in their notorious Exhibition of Degenerate Art. Klee by this time had moved to Bern, where he suffered from depression and became increasingly ill. Although not all his paintings lost their earlier lightness of tone, the themes of corruption and malevolence together with darker colours and heavier lines predominated in the last seven years of his life.

Klein, Melanie (1882–1960) *Austrian psychoanalyst who was a major influence in child psychology and psychiatry, devising techniques of analysis that gave insight into the depths of a child's mind.*

Klein began her training in child psychoanalysis in Budapest with Sándor Ferenczi (1873–1933), a close associate of FREUD, and later studied in Berlin with Karl Abraham (1877–1925). She developed a method of analysis using children's free play with small toys and their spontaneous communication, which she believed provided an insight into their desires and anxieties, their relationships with parents, and their fantasies and psychological defences during the early years of life. Her methods initially provoked much criticism: although Freud's work on adults showed that many fantasies and anxieties were rooted in childhood, Klein suggested that they were already established early in infancy and provided evidence of superego and an Oedipus complex in a two-and-a-half year old child. This, and the idea that sadism and aggression are present in infants, were found shocking but her theories and methods of therapy later gained wide acceptance.

In 1926 Klein moved to England to work at the British Psychoanalytical Society and spent most of her working life in London. In *The Psychoanalysis of Children* (1932), she presented her techniques and observations of anxiety situations in infancy and their effect on normal growth and the development of emotional disorders. Klein believed that a child could be protected from later mental disorders by psychoanalysis. After 1934 she worked with adults to extend her studies of child anxiety situations and made valuable contributions to the psychopathology of depressive and schizoid-paranoid illnesses that have their origins in childhood.

Klemperer, Otto (1885–1973) *German-born conductor, who became a US citizen in 1937 and an Israeli citizen in 1970. A legendary figure in his own lifetime, he was renowned for his strict observance of the composer's intentions.*

Klemperer grew up in Hamburg and at the age of sixteen entered the Hoch Conservatory in Frankfurt; he later studied in Berlin, specializing in composition, under Hans Erich Pfitzner (1869–1949) and piano. Deeply influenced by Mahler's music, he met the composer in Berlin in 1905 and they became close friends; on Mahler's recommendation Klemperer was appointed conductor of the German Opera in Prague, making his debut there in 1907 conducting Weber's *Der Freischütz*. Klemperer was then successively conductor at the Municipal Theatre in Hamburg (1910–14), the Strasbourg Opera (1914–17), the Cologne Opera (1917–24), and the State Opera, Wiesbaden (1924–27). By the time he took over the musical directorship of the Kroll Theatre, Berlin, in 1927, he had an extremely high reputation for the classical opera and concert repertoire. He was also known for his eagerness to present contemporary works; at the Kroll Opera he introduced such works as JANÁČEK's *From the House of the Dead*, SCHOENBERG's *Erwartung* and *Die glückliche Hand*, STRAVINSKY's *The Soldier's Tale* and *Oedipus Rex*, and HINDEMITH's *Cardillac* and *Neues vom Tage*.

Increasingly conservative opposition to his methods, however, led to the closing of the theatre in 1931. Moreover, as a Jew, Klemperer was persona non grata in Germany in the 1930s. When his contract with the Berlin State Opera was summarily cancelled in 1933, he left Gemany to become the director of the Los Angeles Symphony Orchestra. After the war he returned to Europe to conduct the Budapest Opera (1947–50). Despite a series of accidents and operations that left him partially paralysed, he continued to be very much in demand as a conductor. In the 1950s, for example, under the auspices of Walter Legge, he made a series of recordings in London with the Philharmonia Orchestra, including the highly acclaimed Beethoven cycles.

Kline, Franz (1910–62) *US painter.*

Born in Wilkes-Barre, Pennsylvania, Kline studied at Boston University (1931–35) and in London at the Heatherley School of Art. After returning to the USA in 1939 he painted mainly urban landscapes in traditional style. By 1950, however, he had developed a personal style of expressive abstraction and in that year he had his first one-man show. In such works as *Chief* (1950), which like a number of his paintings was named after a famous train, and *Two Horizontals* (1954), thick bold black lines stand out like oriental calligraphy against a white background. But unlike calligraphy the white areas also play a part in the picture. Throughout the 1950s Kline combined painting with teaching. His style remained constant, although in the last years of his life he began to use vivid colour as well as black and white.

Knox, Ronald Arbuthnot (1888–1957) *British Catholic theologian noted for his modern English translation of the Bible.*

The son of an Anglican clergyman, Knox read classics at Balliol College, Oxford, and was ordained as an Anglican priest. In 1912 he was appointed Anglican chaplain of Trinity College, Oxford. A member of the ritualistic movement, Knox opposed modernist trends within the Church of England. His disenchantment grew during World War I, when he taught at Shrewsbury School and also worked at the War Office (1916–18), and in 1917 he joined the Catholic Church. Two years later, following his studies at St Edmund's College, he was ordained a Catholic priest, becoming Catholic chaplain at Oxford in 1926.

During this time, Knox was the author of a succession of works, ranging from satirical fantasy to translations of Virgil, and in 1939 he resigned the chaplaincy to devote himself entirely to a new translation of the Bible. Completed in 1950, his version of the fourth-century Vulgate Latin text in modern English superseded the Douay translation of 1610, thus representing a major landmark in contemporary biblical scholarship. Knox's other works include *Enthusiasm* (1950), an examination of the phenomenon of divine revelation.

Kodály, Zoltán (1882–1967) *Hungarian composer. He made a deep study of Hungarian folk music and his vocal music makes remarkable use of the Hungarian language.*

The son of a railway official, Kodály grew up in a musical atmosphere, his father being a keen violinist and quartet player. In 1900 a scholarship took him to Budapest to study at the university and in 1902 he entered the Budapest Academy of Music. Four years later he was awarded a PhD for a thesis on 'Strophic Construction in Hungarian Folksong'. In the same year he embarked on his remarkable collaboration with BARTÓK, collecting Hungarian folk tunes.

As a composer, Kodály was influenced by the classical repertoire and by the music he sang as a choirboy. However, a visit to Paris in 1907, where he attended the classes of Charles Widor (1844–1937), introduced him to the music of DEBUSSY and the impressionists. This broadening of his horizons was reflected in his music, which after 1910 began to be played outside Hungary, particularly at the festivals of the International Society of Contemporary Music. The powerful *Psalmus Hungaricus* (1923), which was composed in celebration of the union of Buda and Pest, has become part of the international repertoire. Two operas followed – *Háry János* (1925–27), from which a suite was made, and *The Spinning Room* (1924–32). Perhaps Kodály's most vivid music is to be found in two sets of dances, the *Marossek Dances* (1930) and the *Galánta Dances* (1933). In 1938 the Amsterdam Concertgebouw Orchestra commissioned the *Peacock Variations* for its fiftieth anniversary; a year later the Chicago Philharmonic Orchestra commissioned the *Concerto for Orchestra*.

Kodály taught at the Budapest Academy of Music from 1907, an association that lasted for the greater part of his life, especially as he was recalled from retirement in 1945 to become director. Kodály was particularly interested in originating methods for teaching young children to participiate in music-making; much of the music in the four volumes of his *Bicinia Hungarica*, a collection of folksongs and original compositions for two voices (published between 1937 and 1942), was written for this purpose.

Koestler, Arthur (1905–83) *Hungarian-born British writer whose works, on a variety of topics, reflect his humane, versatile, and inquiring mind.*

Koestler was born in Budapest and studied engineering at Vienna. He worked as a journalist in the late 1920s and 1930s, visiting the Soviet Union, the Middle East, Paris, London, Zürich, and Spain. While in Berlin (1930–32) he joined the Communist Party, but was subsequently disillusioned and resigned (1938). He reported the Spanish civil war and was taken prisoner by FRANCO, narrowly escaping death; this experience is related in *Ein spanisches Testament* (1937; translated as *Spanish Testament*, 1938) and also drawn upon in his novel about the Stalinist purge trials, *Darkness at Noon* (1940). In 1940 he made Britain his permanent home and thereafter wrote in English. From 1941 to 1942 he served in the British Pioneer Corps.

Koestler's interests covered a wide range of subjects. The novel *Thieves in the Night* (1946) about the Jews in Palestine and the political study *The Yogi and the Commissar* (1945) both had strong topical relevance, as did *Reflections on Hanging* (1956) during the anti-capital punishment debate. The workings of science and the scientific mind enthralled him and he published a number of books on the subject, including *The Act of Creation* (1964), *The Ghost in the Machine* (1967), and *The Case of the Midwife Toad* (1971). Parapsychology attracted his attention in his later years and he left money in his will to found a chair in the subject, which he wrote about in *The Roots of Coincidence* (1972). He also wrote autobiography: *Scum of the Earth* (1941), *Arrow in the Blue* (1952), and *The Invisible Writing* (1954). In 1983 he and his third wife Cynthia committed suicide together when Koestler could no longer cope with his terminal illness – Parkinson's disease.

Kohl, Helmut (1930–   ) *German statesman, chancellor of West Germany (1982–90) and of Germany since 1990.*

The product of a conservative Roman Catholic background, Kohl showed an early interest in politics, joining the Christian Democratic Union (CDU) in 1947. He studied at the universities of Frankfurt and Heidelberg, obtaining a doctorate in political science in 1958. From 1959 to 1976 he served on the state legislature of Rhineland-Palatinate, becoming minister-president in 1969 and gaining a reputation as an administrator. He became chairman of the CDU in 1973 and the conservative candidate for the chancellorship in 1976, when he lost narrowly to the incumbent, Helmut SCHMIDT.

Kohl led the opposition in the Bundestag until 1982, when the ruling coalition collapsed and he became chancellor, retaining the post

with a much increased majority after elections the following year. As chancellor he has pursued centre-right policies, reducing government spending and showing a firm commitment to NATO: he has also been prominent in moves towards greater European integration. Although he was re-elected in 1987, the results showed a serious decline in his standing. In 1990, however, he presided skilfully over the reunification of the two Germanies, greatly increasing his stature at home and abroad. He was elected chancellor of the united Germany later the same year.

Köhler, Wolfgang (1887–1967) *US psychologist, born in Estonia, who was a founder of the Gestalt school of psychology.*

After graduating at the Friedrich-Wilhelm University in Berlin, Köhler carried out research on hearing and in 1911 became assistant lecturer at the University of Frankfurt. In 1912, he and fellow scientist Kurt Koffka (1886–1941) were used as subjects for experiments on perception by Max Wertheimer (1880–1943), and the three of them proposed a new explanation of mental processes, together founding the Gestalt school. This regards mental processes, such as learning and perception, as wholes (gestalts) that cannot be analysed into smaller components. According to Gestalt theory each time something is learnt the individual's total perception of the environment changes to incorporate the new information. From 1913 to 1920 Köhler was director of the anthropoid research station of the Prussian Academy of Sciences at Tenerife, where he worked on chimpanzee behaviour, including problem solving and the use of tools. His resulting publication *Intelligenzprüfungen an Menschenaffen* (1917; 'The Mentality of Apes') radically altered the current views of learning. As head of the Psychological Institute and professor of philosophy at the University of Berlin (1921–35), Köhler continued to explore many aspects of the Gestalt theory, publishing *Gestalt Psychology* in 1929.

Köhler was outspoken in his opposition to the HITLER government and in 1935 left Germany to live in the USA, where he became professor of psychology at Swarthmore College, Pennsylvania, and then (in 1958) research professor at Dartmouth College, New Hampshire. He continued to work on the physiological basis of perception and suggested that learning does not result from simple associations but involves more complex mental organization.

Kokoschka, Oskar (1886–1980) *Austrian-born expressionist painter, best known for his searching portraits.*

Kokoschka studied at the Vienna School of Arts (1904–09) at the time when Sigmund FREUD was developing his theories of psychoanalysis in Vienna. Other early influences were French symbolism and the prevailing art nouveau style. But in Kokoschka's paintings the decorative lines of this style soon gave way to more expressive drawing in portraits that reveal the artist's quest for the subject's innermost soul. In *Portrait of Auguste Forrel* (1910), for example, the subject's head and hands, upon which the picture concentrates intensely, are a complex of exaggerated linear details. In 1910 Kokoschka moved to Berlin, where he played an important part in the development of German expressionism. During this period he did graphic illustrations and his paintings became richer in colour. In 1914 he produced the first of his allegorical pictures, *The Bride of the Wind*.

After World War I, in which he was wounded, Kokoschka moved to Dresden, where he became a professor at the Dresden Academy in 1920 but left abruptly in 1924 to spend seven years travelling. Contact with impressionist paintings in Paris led to a more exuberant use of colour in the landscapes and townscapes of this period. In 1937 he painted *Self Portrait of a Degenerate Artist* after the Nazis had condemned his work. He fled to London in 1938, becoming a British citizen in 1947, but left in 1953 to settle in Switzerland.

Kolff, Willem Johan (1911–) *US physician, born in the Netherlands, who pioneered the science of biomedical engineering. During World War II he invented the kidney dialysis machine and in 1982 he and his colleagues performed the first heart transplant using an artificial heart.*

Born in Leiden, the son of a doctor, Kolff was educated at the University of Leiden Medical School. He did postgraduate research at Groningen University and worked there until the German occupation of the Netherlands, when he moved to Kampen rather than work under a Nazi director. During this time he developed crude versions of his kidney dialysis machine and later supplied researchers in Britain, Canada, and the USA with his successful design.

Kolff emigrated to the USA in 1950 to join the staff of the Cleveland Clinic Foundation (1950–67), where he studied cardiovascular problems. In collaboration with a student he designed a successful artificial heart–lung machine that made open heart surgery possible; he also invented the intra-aortic balloon pump to help circulation during heart attacks (1961). His most ambitious idea was to design an artificial heart and in 1957 he implanted one in a dog, which survived for ninety minutes. In 1967 Kolff moved to the University of Utah as director of the Institute for Biomedical Engineering and head of the Division of Artificial Organs, heading a team developing new prostheses and artificial organs. He has championed the trend towards home dialysis and in 1975 produced the wearable artificial kidney. In 1982, twenty-five years after his operation on the dog, Kolff and his colleagues used an aluminium and plastic heart to replace the diseased heart of Dr Barney Clark, who survived for 112 days.

Konoe Fumimaro, Prince (1891–1946) *Japanese prime minister (1937–39; 1940–41; 1941). He tried unsuccessfully to prevent Japan's war with China from becoming a world conflict.*

Born in Kyoto, the son of a prominent politician, Konoe was educated at the Tokyo Imperial University, where he studied philosophy, and at the Kyoto Imperial University, where he studied law. While still a student he wrote magazine articles advocating a 'new order' in southeast Asia and in 1916 was given a seat in the upper house of the parliament (Diet). Upon his graduation a year later he was appointed to the ministry of home affairs and worked towards the expansion of parliamentary politics.

Konoe's political career began in earnest in 1933, when he was elected president of the upper house. Regarded as a promising political figure by prominent military leaders, he was recommended for the premiership in 1936 but did not accept the post until 1937, when the Hayashi government fell. Shortly afterwards Japan's conflict with China escalated; finding himself unable to contain the military he resigned (1939). He was re-elected in 1940 and attempted to prevent a widening of the Sino-Japanese conflict. During this period he also advocated a reform of the political system, in which political parties were replaced by the newly formed Imperial Rule Assistance Association, of which he was president. In 1941 he established a third cabinet, but his attempts to negotiate with the Americans were opposed by the war minister TOJO HIDEKI and he was again forced to relinquish office. Three years later he helped to bring down the Tojo cabinet and, after the Japanese surrender to the Allied forces in 1945, became deputy minister of national affairs. He committed suicide in 1946, after an order for his arrest as a war criminal had been issued.

Korda, Sir Alexander (Sándor Kellner; 1893–1956) *Hungarian-born British film director, producer, and executive. He was knighted in 1942 and was made an Officier de la Légion d' honneur in 1950.*

Korda was born in Turkeve and began his career as a journalist before becoming a film director in 1916. After working in Austria, Germany, the USA, and France he settled in Britain in 1930, where he made the successful *Service for Ladies* (1932) at Elstree before setting up his own studios at Isleworth and Denham and founding London Films. As a producer, *The Private Life of Henry VIII* (1933) with Charles LAUGHTON, *Things To Come* (1936) with Raymond MASSEY, and *The Four Feathers* (1939) were among his many British prewar successes. In 1939 he made *The Lion Has Wings* for the British government. *The Fallen Idol* (1948), *The Third Man* (1949), *The Sound Barrier* (1951), and *Hobson's Choice* (1954) were a few of his best-known postwar films, which established him as a major force in the British film industry.

Korda married three times, his second wife, Merle Oberon (1911–79), appearing in several of his films. His brothers, Zoltan and Vincent, also worked on many of his productions.

Kornberg, Arthur (1918–) *US biochemist who discovered how deoxyribonucleic acid (DNA) is replicated in the cell and reproduced the reaction in the test tube. He was awarded the 1959 Nobel Prize for Physiology or Medicine.*

Kornberg was born in New York and educated at the City College, New York, and the University of Rochester. He became a commissioned officer in the US Public Health Service (1941–42) and worked at the National Institutes of Health from 1942 to 1952, directing a programme of research on enzymes and intermediary metabolism (the chemical reactions maintaining the life of the cell). He helped to discover the mechanism of formation of the coenzymes flavine adenine dinucleotide (FAD) and nicotinamide adenine dinucleotide (NAD), which have an important role as hydrogen carriers in cell respiration.

From 1953 to 1959 Kornberg was professor and chairman of the Department of Microbiology of the Washington University School of Medicine, St Louis, where he studied the mechanisms of manufacture of nucleotides

and how they combine to build molecules of DNA and RNA. By adding labelled nucleotides to extracts from cultures of the bacterium *Escherichia coli* he found evidence of a polymerization reaction catalysed by an enzyme. He isolated and purified the enzyme (DNA polymerase) and was able to use it to produce short replicas of DNA molecules by adding it to nucleotides. The results were published in *The Enzymatic Synthesis of DNA* in 1961. Kornberg was head of the Department of Biochemistry at Stanford University (1959–69).

Kosygin, Aleksei Nikolaevich (1904–80) *Soviet statesman. He was prime minister (1964–80), sharing power with BREZHNEV, but his power declined in the later 1960s. He was awarded the Order of Lenin six times.*

Born in St Petersburg, the son of a lathe operator, Kosygin was educated at the St Petersburg Cooperative Technicum, where he graduated in 1924. He became an instructor of cooperative groups in Siberia (1924–30) before returning to St Petersburg (renamed Leningrad in 1924) to study engineering at the Leningrad Textile Institute. From 1935 to 1938 he worked in textile mills in Leningrad, of which he was elected mayor in 1938.

Kosygin joined the Communist Party in 1927. Elected to the central committee twelve years later, he became deputy prime minister in 1940. In 1946 he was elected a candidate member of the Politburo, attaining full membership (1948–52) and being appointed minister of finance and of light industry (1948–53). He supported KHRUSHCHEV after STALIN's death and was elected deputy prime minister in 1953, becoming minister of economic planning (1956–57) and chairman of the state economic planning commission in 1960. The same year he was re-elected a full member of the Politburo. He succeeded Khrushchev in 1964 as chairman of the council of ministers (prime minister), a position he held until 1980 when he retired due to ill health.

Koussevitsky, Sergei Alexandrovich (1874–1951) *Russian conductor and double-bass player. He became one of the leading champions of contemporary music, commissioning and playing works by many young American and European composers.*

Koussevitsky studied at the Philharmonic School in Moscow and by 1896 had made his reputation as a virtuoso double-bass player. In 1907 he gave recitals at the Wigmore (then the Bechstein) Hall in London and in 1908 conducted the London Symphony Orchestra in a Beethoven programme. He had also studied conducting in Berlin with Arthur Nikisch (1855–1922), building a reputation for his interpretation of Beethoven and of the Russian and Finnish repertoire. In Russia he toured with his own orchestra and founded a music-publishing house in Berlin, commissioning works from SCRIABIN and STRAVINSKY, among others. After the revolution Koussevitsky left his homeland (1920), conducting in London, Rome, Berlin, Paris, and Barcelona. In 1924 he became conductor of the Boston Symphony Orchestra and in the next twenty-five years made it one of the great orchestras of the world. In 1940 he founded the Berkshire Music Center at Tanglewood, Massachusetts, which he continued to direct after his retirement from the Boston Symphony Orchestra in 1949. He also set up the S & N Koussevitsky Music Foundation for commissioning works by little-known musicians.

Krebs, Sir Hans Adolf (1900–81) *German-born British biochemist who discovered the tricarboxylic acid, or Krebs, cycle – the series of chemical reactions that are fundamental to the metabolism of living organisms. For this he was awarded the 1953 Nobel Prize for Physiology or Medicine. He was knighted in 1958.*

Born in Hildesheim, Lower Saxony, Krebs studied medicine at the universities of Göttingen, Freiburg, and Munich before receiving his MD from Hamburg University in 1925. Thereafter he worked as an assistant in Otto WARBURG's department at the Kaiser Wilhelm Institute (1926–30), as an assistant at Hamburg Municipal Hospital, and at Freiburg University. Growing antisemitism forced him to flee Germany in 1933 and he obtained a post at Cambridge University under Sir Frederick Gowland HOPKINS, receiving his MA in 1934. The following year he was appointed lecturer in pharmacology at Sheffield University, becoming head of the biochemistry department (1938) and professor of biochemistry (1945) as well as director of the Medical Research Council's cell metabolism unit at Sheffield.

In 1932, at Freiburg, Krebs established the series of chemical reactions by which ammonia, produced by the breakdown of amino acids, is converted to the much less toxic compound, urea, prior to excretion – the urea cycle. In 1937 he went on to unravel the chemical pathway by which food is converted to usable energy. Using pigeon liver and breast muscle, Krebs found that the pyruvic acid resulting from the initial breakdown of carbohydrate (glycolysis) is oxidized to carbon dioxide and water via nine intermediate reactions involv-

ing a series of carboxylic acids together with various enzymes and cofactors. The Krebs cycle not only drives the production of the metabolic fuel needed for muscle contraction and other energy-demanding processes but also forms a metabolic crossroads for the synthesis and breakdown of lipids and proteins as well as carbohydrates.

Krebs became a fellow of the Royal Society in 1947 and was appointed Whitley Professor of Biochemistry and fellow of Trinity College, Oxford (1954–67). With Hans Kornberg (1929–   ), he wrote *Energy Transformations in Living Matter* (1957).

Kreisler, Fritz (1875–1962) *Austrian violinist. One of the most popular performers of his day, he frequently included his own short solo compositions in his recitals.*

Born in Vienna, the son of an eminent physician, Kreisler had prodigious talent, which was encouraged by his music-loving father, and he entered the Vienna Conservatory at the age of seven. One of the youngest students ever to be admitted, he studied with Leopold Auer (1845–1930) and won the gold medal for violin playing at ten. At twelve, having moved to the Paris Conservatoire, where he studied with Joseph Massart (1811–92), he again won the gold medal against fierce competition. Kreisler's first public appearances were in America, where he toured with the pianist Moriz Rosenthal (1862–1946) in 1889. Feeling the need to develop other interests, Kreisler studied first medicine and then art in Rome and Paris. Finally he spent a year as an officer in the Austrian army. After these diversions he returned to music and made his mature debut in Berlin in 1899. In 1901 he played in London under Hans Richter (1843–1916) and won the affection of English audiences, particularly with his interpretations of the Beethoven and Brahms concertos. In 1904 he was awarded the gold medal of the Royal Philharmonic Society and in 1910 he gave the first performance of ELGAR's violin concerto (dedicated to him), afterwards playing it in St Petersburg, Moscow, Vienna, Berlin, Dresden, Munich, and Amsterdam. During World War I he served with the Austrian army, was wounded, and eventually went to America, where he settled.

Kreisler's playing combined a distinctive sweetness of tone with great vigour. Many of his own compositions were believed at the time to be arrangements of eighteenth-century trifles; only in later years were they disclosed as his own work.

Krupp (von Bohlen und Halbach), Alfried (1907–67) *German industrialist who was the last of the family to have sole control over the vast manufacturing and armaments empire founded by Friedrich Krupp (1787–1826).*

Alfried was the eldest son of Gustav Krupp von Bohlen and Halbach (1870–1950), husband of the founder's great-granddaughter. He studied engineering at Aachen Technical Institute before joining the family business. He was made a member of the board in 1936 and two years later joined the Nazi Party. The Krupp empire comprised an international conglomerate of companies involved in iron and steel manufacture, armaments, munitions, banking, property, and hotels. In 1943 Alfried took control from his ailing father (who had been imprisoned after World War I) by special dispensation from HITLER and immediately embarked upon a ruthless programme of plant seizure in German-occupied countries and the use of concentration-camp inmates as forced labour. An Allied bombing raid devastated the main Krupp factories at Essen in 1945, and after the war Krupp was tried at Nuremberg and sentenced to twelve years' imprisonment and the confiscation of his assets. However, in 1951 his release was secured by the intervention of the US high commissioner in Germany on the political grounds that a change in circumstances necessitated a strengthening of German industry. Krupp's estate was restored to him and he once more prospered with diversification into such areas as industrial equipment, food, and nuclear power. However, the economic recession of the mid-1960s led to a financial crisis and Krupp was forced to relinquish personal control in return for financial help from the Federal German government. Krupp died shortly before an anticipated announcement accepting the deal. His son, Arndt, renounced rights of inheritance and the firm is now a corporation.

Kubelík, (Jeronym) Rafael (1914–) *Czech-born conductor and composer. He left his native country after the war and in 1973 became a Swiss national. He has been a great champion of Czech music, especially of the operas of JANÁČEK.*

The son of Jan Kubelík (1880–1940), the Czech violinist, he studied conducting and composition at the Prague Conservatory. In 1934 he made his conducting debut with the Czech Philharmonic Orchestra and two years later he was appointed conductor; in 1938 he toured Britain with the orchestra, giving a series of widely acclaimed concerts. After two years as director of the Brno Opera, Kubelík returned to the Philharmonic. In 1945 he emigrated to England; later he moved his residence to Switzerland, eventually becoming a Swiss national.

Kubelík's international career has included appointments with the Chicago Symphony Orchestra (1950–53), Covent Garden Opera in London (1955–58), and the Bavarian Radio Symphony Orchestra in Munich (1961–79), with a two-year break (1972–74) at the Metropolitan Opera, New York. As a composer, his works include symphonies, concertos, three requiems, chamber music, and five operas. He is married to the soprano Elsie Morison.

Kubrick, Stanley (1928–) *US film director, producer, and writer.*

Born in New York City, where he attended the City College, Kubrick began his career as a photographer on *Look* magazine. He entered the film business in 1950 with his independently produced short documentary *Day of the Fight*. *Flying Padre* (1951), another documentary, followed, after which he made his first feature, *Fear and Desire* (1953), and his first major critical success, *The Killing* (1956).

Often presenting a somewhat pessimistic view of man's place in the scheme of things, Kubrick has since made such successful films as *Paths of Glory* (1957), *Spartacus* (1960), *Lolita* (1962), and *Dr Strangelove* (1964), which won a New York Critics Award. His remarkable *2001: A Space Odyssey* (1968), which he produced, directed, and co-scripted with Arthur C. Clarke (1917–), was awarded an Oscar for best special visual effects, while the very different *A Clockwork Orange* (1971) brought a second New York Critics Award. More recent films include *Barry Lyndon* (1975), *The Shining* (1980), and *Full Metal Jacket* (1987).

Kundera, Milan (1929–) *Czech novelist and writer living in France, who is noted for his poignant sexual comedies.*

Kundera was born in Brno, the son of the pianist and musicologist Ludvik Kundera. He studied film at the Prague Academy of Music and Dramatic Arts, where he later (1958–69) followed an academic career. In his twenties he published several volumes of poetry, which incurred official displeasure for their sexual content and irreverent attitude. His first novel, *The Joke* (1967), was still more daring, being an ironic exploration of the private lives of Czechs during the years of Stalinism; its publication was a sign of the increasing liberalization of Czech life in 1967–68, a movement in which Kundera participated keenly. Following the Soviet-led invasion of 1968, however, he was removed from his teaching posts and saw the proscription of all his works. His next novel, *Life is Elsewhere*, was a satirical treatment of events following the communist takeover of 1948; it was published in France in 1973.

In 1975 Kundera was permitted to emigrate to France, where he took up teaching posts in Rennes and Paris. Stripped of his Czech citizenship in 1979, he became a French national two years later. Recent novels, such as the semi-autobiographical *The Book of Laughter and Forgetting* (1979) and *The Unbearable Lightness of Being* (1984; filmed 1987), have enjoyed great success in the West but remained banned in Czechoslovakia until the fall of communism. He published *Immortality* in 1991.

Kurosawa Akira (1910–) *Japanese film director, whose ability to blend the traditional with the modern has made him one of the most widely known Japanese film-makers to both western and eastern audiences.*

Kurosawa was born in Tokyo, the youngest of eight children of an army officer. After training at a western-style art school he abandoned a career as a professional artist to enter films as an assistant director to Kajiro Yamamoto (1902–74) in 1936. He made his debut as a director with *Sugata Sanshiro* (1943) and at the Venice Film Festival achieved international acclaim with the prize-winning *Rashomon* (1951), on which the American-made *The Outrage* (1964) was based. *Ikiru* (1952; *Living*), *Seven Samurai* (1954), remade in Hollywood as *The Magnificent Seven*, *Throne of Blood* (1957), based on Shakespeare's *Macbeth*, *Yojimbo* (1961), and *Sanjuro* (1962), are among his many memorable films. *Dersu Uzala* (1975) won a best for-

eign film Oscar and *Kagemusha* (1980) won the Golden Palm award at the Cannes Film Festival. *Ran* (1985) is a Japanese version of *King Lear*. His most recent films include *Dreams* (1990) and *Rhapsody in August* (1991).

Kuznets, Simon (1901–85) *Russian-born US economist and statistician, who won the Nobel Prize for Economics in 1971 for his research on social change and economic growth.*

Born in Kharkov, Kuznets emigrated to the USA as a young man. Educated at Columbia University, he joined the staff of the National Bureau of Economic Research in New York in 1927 and continued this work until 1961. Meanwhile he held teaching posts at the University of Pennsylvania (1936–54), Johns Hopkins (1954–60), and Harvard (1960–71), where he eventually become professor emeritus.

Kuznets's major field of study has been the relationships between population, per capita incomes, and economic growth, and in particular why stable populations are related to high rates of economic growth. He noted that the developed countries were more advanced than the rest of the world when their industrial development began. In *National Product Since 1869* (1946) Kuznets describes a pronounced cyclical movement in the economy of 16–22 years' duration (the 'Kuznets cycle'), one factor in which he believed to be changes in the rate of population growth. He found a long-term stability in the proportion of income saved in industrial countries. In *National Income and its Composition 1919–38* (1941) Kuznets discussed national income statistics, particularly the measurement of gross national product. Other important publications are *Income and Wealth in the US: Trends and Structures* (1952), *Economic Growth of Nations* (1971), and *Growth, Population and Income Distribution: Selected Essays* (1979).

L

La Guardia, Fiorello Henry (1882–1947)
*US politician and mayor of New York
(1933–45), who was responsible for wide-
ranging civic improvements in the city.*
The son of an Italian bandmaster in the US
army, La Guardia was born in New York City
and spent his early years in South Dakota and
Arizona. He subsequently moved with his father to Florida, where he worked on a St Louis
newspaper. After his father's death in 1898 he
accompanied his mother to Budapest, Hungary, to join her relatives. In the early 1900s he
served as US consul in Budapest and Fiume,
returning to New York in 1906 to study law.
Admitted to the bar in 1910, he became deputy
attorney-general for New York in 1915 and the
following year was elected to Congress as a
Republican. His first term in Congress was
interrupted by a period of service with the air
force in World War I; re-elected in 1922, he
served for a further ten years. During this period he was involved in the passage of the
Norris–La Guardia Act (1932), which outlawed anti-union contracts and protested the
unions' right to strike and picket peacefully.

In 1933, after an unsuccessful attempt four
years earlier, La Guardia became mayor of
New York, an office he was to hold for three
consecutive four-year terms. He rapidly won
the respect and affection of the people of New
York, who dubbed him 'the Little Flower',
from his first name. Under La Guardia's administration the city received a facelift, with
massive slum-clearance projects and ambitious building programmes for schools, bridges, playgrounds, and parks. He also improved
the efficiency of the municipal government,
fighting corruption at all levels, and obtained a
new city charter (1938). During World War II
he served as director of the US Office of Civilian Defense (1941–42) and in 1946, having
declined to stand for re-election in the mayoral
campaign of the previous year, he was appointed director general of UNRRA (the
United Nations Relief and Rehabilitation Administration). One of New York's main air-
ports is named after him.

Laing, R(onald) D(avid) (1927–89) *British
psychiatrist whose controversial views on*
*madness and family life influenced some radi-
cal psychiatric movements of the late 1960s.*
Laing was educated at the University of Glasgow and trained at the West of Scotland Neu-
rosurgical Unit. He began his career at the
Army Psychiatric Unit, Netley (1951–52), and
the Psychiatric Unit of the Military Hospital,
Catterick (1952–53). This was followed by
three years in the Department of Psychological
Medicine at the University of Glasgow, after
which Laing moved to Tavistock (1956),
working initially at the Tavistock Clinic and
then at the Institute of Human Relations. In
1964 he became the principal investigator at
the Schizophrenia and Family Research Unit.
His major interest was in exploring the minds
of severely disturbed patients and he proposed
that schizophrenia is a defensive façade and
that what society calls madness is in fact a
journey towards self-realization (*The Divided
Self*, 1960). In *Sanity, Madness and the Family*
(1965) and *The Politics of the Family* (1971)
his questioning of society's definitions of sanity and insanity and his view of madness as an
escape from the tensions of the close-knit nuclear family aroused considerable controversy
and is not now taken too seriously. Outside his
professional career, Laing published several
volumes of poetry, including *Knots* (1970).

Lalique, René Jules (1860–1945) *French
designer and producer of art nouveau jewellery and glassware, whose name is particularly
associated with a style of glassware decorated
with figures in relief.*
Born in Ay on the Marne, Lalique was apprenticed to a jeweller at the age of sixteen and then
studied at art schools in Paris and London. He
founded his own company in Paris in 1885 and
by 1900, the year of the Paris International
Exhibition, his patrons included royalty. Many
of his best pieces were created for the actress
Sarah Bernhardt. The most characteristic of
them combined the use of gold with enamel-
ling and relatively few gemstones. They often
depicted the female figure nude or with flowing hair and drapery, sometimes with butterfly
or dragonfly wings. He frequently employed
insect motifs and other forms from nature.

By 1914, however, his experiments with
glass had completely taken over from the cre-

ation of jewellery. The style of glass that he developed, with its moulded surfaces and patterns in relief, became the height of fashion in the 1920s. As well as luxury items, such as scent bottles, Lalique produced glass for architectural use and lighting equipment. His son, Marc, continued his work after his death.

Lamb, Jr, Willis Eugene (1913–) *US physicist, who discovered the Lamb shift, for which he was awarded the 1955 Nobel Prize for Physics.*

The son of a telephone engineer, Lamb was educated at the University of California, where he gained his PhD in 1938 under the supervision of J. R. OPPENHEIMER. He has since held chairs of physics at Columbia (1948–51), Stanford (1951–56), Oxford (1956–62), Yale (1962–74), and the University of Arizona (since 1974).

In 1947 Lamb used radiofrequency resonance techniques to examine the fine structure of the hydrogen spectrum. In particular he was checking Paul DIRAC's prediction that the hydrogen atom could exist in either of two equal energy states. Lamb's work revealed that in fact the states were not of exactly equal energy and the small Lamb shift, as it came to be called, led to a fundamental revision of quantum theory at the hands of Richard FEYNMAN, Julian Schwinger (1918–) and Sin-Itiro Tomonaga (1906–79).

Lambert, Constant (1905–51) *British composer and conductor. He had a lifelong association with the music of the ballet.*

Son of the Australian painter George Washington Lambert, he was educated in England at Christ's Hospital and the Royal College of Music, where he studied under Ralph VAUGHAN WILLIAMS. While still a student, Lambert was commissioned by DIAGHILEV to compose for his Ballets Russes. Lambert's *Romeo and Juliet*, a work in two tableaux, choreographed by Bronislava Nijinska (1891–1972), depicting the lovers as members of a ballet school who elope in an aeroplane, was first produced in Monte Carlo in 1926. It was the first of his neoclassical scores, amalgamating the movements of a classical suite with twentieth-century jazz and polytonality.

Following his performance as reciter of Edith SITWELL's poems in WALTON's *Façade*, Lambert became friendly with the Sitwell family, which led to the ballet *Pomona* (1926). Other ballet scores include *Horoscope* (1937) and *Tiresias* (1950) as well as numerous arrangements for ballet made during his time as conductor of the Carmargo Society (later the

Sadler's Wells Ballet), of which he was musical director (1930–47). Of Lambert's other works, the *Elegiac Blues* (1927) and *Rio Grande* (1927), a setting of a poem by Sacheverell Sitwell (1897–1988) for piano, chorus, and orchestra (excluding woodwind), reflect his increasing interest in jazz; other choral works are the setting of Nashe's *Summer's Last Will and Testament* (1932–35) and *Dirge* (1940) to words of Shakespeare. The sonata for piano (1928–29) and the concerto for piano and nine instruments (1930–31) are works of a more serious nature; the *Aubade héroique* (1942) for chamber orchestra was inspired from the Sadler's Wells Ballet's hasty escape from Holland after the German invasion in 1940. His numerous transcriptions include works by Boyce, Handel, and Roseingrave. Lambert's witty survey of contemporary music, *Music Ho!*, was published in 1934.

Land, Edwin Herbert (1909–91) *US inventor best known for his development of the Polaroid camera.*

Land was educated at Harvard but left before graduating to develop a new kind of polarizing filter. In 1937 he founded the Polaroid Company to exploit his discovery. By reducing sun glare, this material found wide use in the manufacture of glasses, binoculars, prisms and other optical instruments. During World War II, Land was in charge of research into the development of weapons. He first demonstrated his instant camera in 1947 and in the following year the Polaroid Land Camera was on sale to the public. Land was chairman of the Polaroid Company from 1937 until his retirement in 1980.

He also pursued over the years a continued interest in the mechanism of colour vision, arguing in his retinex (retina-and-cortex) theory that the nature of the perceived image is not determined solely by the flux of radiant energy reaching the eye.

Landau, Lev Davidovich (1908–68) *Soviet physicist, who was awarded the 1962 Nobel Prize for Physics for his work on low-temperature physics.*

Born into a Jewish family in Baku (his father was an engineer, his mother a physician), Landau was educated at Baku University and the University of Leningrad, where he graduated in 1927. A mathematical prodigy, he travelled throughout Europe visiting most of the leading research centres before being appointed head of the theoretical physics department at the Physical-Technical Institute in 1932. In 1937, at the suggestion of Peter KAPITZA, Landau was invited to join his Institute for Physical Problems in Moscow; in 1943 he became professor of physics at Moscow University.

At the institute Landau was asked by Kapitza to investigate the recently discovered phenomenon of superfluidity. In the early 1940s Landau worked out the mathematical theory of superfluidity, introducing a number of new concepts into the work. Involved in a serious car accident in 1962, Landau was unable to attend the Nobel ceremony, and although he lived another six years was unable to do any creative work.

In addition to his theoretical work, Landau is well known in the West for his much-translated multi-volume encyclopedic *Course in Theoretical Physics*, written in collaboration with his pupil E. M. Lifshitz and first published in 1938.

Landowska, Wanda (1877–1959) *Polish-born pianist and harpsichordist. She was a leading figure in the twentieth-century revival of baroque harpsichord music.*

Landowska was born in Warsaw and studied at the Warsaw Conservatory; in 1896 she went to Berlin to study composition with Urban, but was (as she put it) 'refractory to rules'. Beginning to make a reputation as a pianist, Landowska moved to Paris in 1900, where she married Henry Lew (an expert in Hebrew folklore) who, until his death in a car crash in 1919, collaborated with her in research into seventeenth- and eighteenth-century music and its interpretation. This research culminated in 1909 in the publication of their book *Musique ancienne*.

During her association with the Schola Cantorum in Paris in the first decade of the twentieth century, Landowska gave up playing Bach on the piano in favour of the harpsichord, first playing this instrument in public in 1903; in 1912 she had a harpsichord built for her by Pleyel to her own specifications. During World War I she was detained in Berlin, where she was teaching the harpsichord at the Royal High School of Music. In 1919 she played the continuo of Bach's *St Matthew Passion* on the harpsichord in a concert in Switzerland, the first twentieth-century performance of its kind. In 1925 Landowska settled in Saint-Leu-la-Forêt, near Paris, where she founded the École de Musique Ancienne. Here she divided her time between teaching and extensive recital tours of Europe and the USA. At the beginning of World War II, she had to abandon her school with its valuable library and collection of in-

struments in face of the German invasion and she eventually settled in the USA in Lakeville, Connecticut. There she made a complete recording on the harpsichord of Bach's *Well-tempered Clavier* (1951).

Many works were written for her, including POULENC's *Concert champêtre* for harpsichord and orchestra (1927–28) and FALLA's concerto (1923–26).

Landsteiner, Karl (1868–1943) *Austrian immunologist who discovered the major human blood groups and developed the ABO system of blood typing, which enabled blood transfusions to become routine practice. He was awarded the Nobel Prize for Physiology or Medicine in 1930.*

Landsteiner was a research assistant at the Vienna Pathological Institute (1898–1908) when he first discovered some basic differences in human blood that made indiscriminate transfusions of blood potentially lethal. In 1901 he demonstrated that there were at least three types, which he labelled A, B, and O, characterized by the types of antigens attached to the red blood cells. The AB group was found a year later. By matching the blood groups of the donor and the patient, the rejection of the transfused blood by the recipient could be avoided. The classification of the blood groups was useful not only in transfusions but also as evidence in trials for paternity and murder; the finding that blood groups are inherited through particular genes was useful in genetics and anthropology.

Landsteiner became professor of pathology at the University of Vienna (1909–15) and later moved to the Rockefeller Institute for Medical Research, New York (1922–43). In 1927 he discovered two more blood groups, M and N, and in 1940 discovered the Rhesus (Rh) factor. His book *The Specificity of Serological Reaction* (1936) is an important text that laid the foundation for the science of immunochemistry.

Lane, Sir Allen (1903–70) *British publisher and founder of Penguin Books Limited. He was knighted in 1952 and made a CH in 1969.*

After leaving Bristol Grammar School, Lane was apprenticed to his uncle, a publisher at The Bodley Head, whom he succeeded as managing director in 1925. Ten years later the impending collapse of The Bodley Head led Lane into a personal gamble that he had been considering for some time – the publication of good-quality sixpenny paperbacks of well-established hardback books. The success of this venture led to the formation of Penguin Books

Limited, which with the introduction of Penguin Specials (original books on topical subjects) established the new company as market leaders in the expanding field of paperback publishing. Lane's publication of the unabridged version of D. H. LAWRENCE's *Lady Chatterley's Lover* in 1960, and his subsequent triumph in the law courts, where the publication was clumsily challenged for obscenity, won him considerable favourable publicity, enormous sales of the book, and a handsome oversubscription when he floated the company a few months later. A man of limited intellectual stature himself, he is remembered for his commercial sense and his wise judgment in surrounding himself with intellectuals of great ability.

Lang, Fritz (1890–1976) *Austrian-born film director, who left Germany for France when the Nazis came to power and settled in the USA in 1935.*

Son of an architect, Lang was born in Vienna and educated at the College of Technical Sciences and Academy of Graphic Arts there. He also studied painting and travelled extensively abroad until World War I, when he joined the Austrian army. Wounded and decorated several times, he was discharged in 1916. Screenplays written during convalescence led to his film career, which proved to be one of the most influential in the early years of the German cinema. During the pioneering days of silent films he directed such notable films as *Die Spinnen* (1919–20; *The Spiders*), *Der müde Tod* (1921; *Destiny*), and *Dr Mabuse der Spieler* (1922). After *Die Nibelungen* (1924) came his silent classic *Metropolis* (1927).

Lang's first talkie was the remarkable thriller *M* (1931), his personal favourite. This was followed by *Das Testament des Dr Mabuse* (1933), which was banned in Germany. Nazi propaganda minister GOEBBELS suggested that he should direct films for the government but Lang decided to leave the country instead. His wife, Thea von Harbou, who collaborated with him on many of his early scripts and who remained in Germany working on official films, divorced him. In France Lang made *Liliom* (1934), after which he went to the USA, where he made such films as *Fury* (1936), *The Woman in the Window* (1944), *Rancho Notorious* (1952), and *The Big Heat* (1953).

Langmuir, Irving (1881–1957) *US chemist, who made an early contribution to the design of light bulbs and was awarded the 1932 Nobel Prize for Chemistry for his work on surface phenomena.*

Langmuir was born in New York and educated at the Columbia School of Mines and in Germany at Göttingen University, where he gained his PhD in 1906. Langmuir taught briefly on his return home, before joining the General Electric Company at their research laboratory in Schenectady, New York, in 1909. He remained with them until his retirement in 1950.

In 1913 Langmuir found that the life of the tungsten vacuum bulbs then in use could be extended considerably if they were filled with a mixture of nitrogen and argon. So successful and profitable did his innovation prove to be that General Electric granted Langmuir the freedom to pursue his own research. This enabled him to make numerous contributions to chemical theory and practice. His most significant work was concerned with atomic structure and the surface properties of liquids and solids.

Lansbury, George (1859–1940) *British Labour politician, widely respected for his commitment to socialist ideals.*

Born in Suffolk and a railway worker at the age of fourteen, Lansbury became a poor law guardian in 1892 and a borough councillor in 1903. He was elected to parliament in 1910 but resigned two years later to fight a by-election, which he lost, on a women's suffrage ticket. Mayor of Poplar (1919–20), in 1921 he and other councillors were imprisoned for six weeks for refusing to raise the county rate because Poplar was too poor to pay it. He edited the socialist *Daily Herald* (1919–22) and then his own *Lansbury's Labour Weekly* (1925–27). Entering parliament again in 1922, he served under Ramsay MACDONALD as first commissioner of works (1929–31), in which capacity he created the Lido on the Serpentine in Hyde Park, London. When MacDonald abandoned Labour to head the national government (1931), Lansbury became leader of the Labour Party and of the opposition. As a committed pacifist, he resigned when the Labour Party conference voted, despite the risk of war, to support sanctions imposed on Italy for its attack on Ethiopia (1935). In a vain attempt to avert conflict with the fascist powers, he visited both HITLER and MUSSOLINI in 1937.

Larkin, Philip Arthur (1922–85) *British poet and novelist.*

Larkin was educated at King Henry VIII School, Coventry, and St John's College, Oxford. From 1943 he worked as a librarian, becoming librarian of the University of Hull in 1955. *The North Ship* (1945) was Larkin's first volume of poems, but his individuality and poetic craftsmanship were not recognized until *The Less Deceived* was published in 1955 and only fully acknowledged with *The Whitsun Weddings* (1964). The tone is pessimistic but there is a lurking recognition of humorous possibilities in the banalities of everyday life. *High Windows* (1974), even more sombre in tone, secured Larkin's position as a leading modern poet, despite the slenderness of his *oeuvre*. He also wrote novels – *Jill* (1946; revised 1964) and *A Girl in Winter* (1947) – and a book of essays on jazz (*All What Jazz?*, 1970) and edited *The Oxford Book of Twentieth Century Verse* (1973). His *Collected Poems* were published in 1988.

Larwood, Harold (1904–) *Nottinghamshire and England cricketer who was the central figure in the controversy over bodyline bowling during the 1932–33 tour of Australia. During his cricketing career he took 1427 first-class wickets and played in 21 test matches.*
A miner at the age of fourteen, Larwood first played for Nottinghamshire in 1924 and had his last match for them in 1938. Not big for a fast bowler, he had a rhythmic action that gave him a pace none of his contemporaries could match. He was a quiet and modest man, unhappy to find himself enmeshed in the bad feeling engendered by the bodyline controversy. In this ploy the English bowlers delivered very fast short-pitched balls and relied on a packed leg-side field to catch their victims. Larwood took 33 wickets in that series and his short-pitched deliveries made him a feared and unpopular figure in Australia. He never represented his country again after that contentious tour; his career was cut short by an injury to his left foot. 'Lol', as he was known to everyone in cricket, ironically emigrated to Australia after World War II and settled very happily in Sydney.

Lasdun, Sir Denys Louis (1914–) *British architect best known for his designs for the National Theatre in London. He was knighted in 1976 and awarded the Gold Medal of the RIBA in 1977.*
Educated at Rugby School and the Architectural Association, he went into practice after serving in the army in World War II. Early buildings include Fitzwilliam College in Cambridge (1959), some of the buildings for the University of East Anglia (1963), and the Royal College of Physicians in Regent's Park,

London (1961–64). The innovative but much criticized National Theatre (1965–76) was followed by the School of Oriental and African Studies (1970–73) for London University. He has also designed buildings in Israel and Luxembourg.

Lasker, Emanuel (1868–1941) *German chess player and world champion from 1894 to 1921, longer than anyone else has ever held the title.*
Lasker was twenty years old when he began to interest chess circles in Germany. Within five years his reputation had spread throughout Europe and the USA; in 1894 he beat Steinitz to become world champion. In the next six years Lasker won major tournaments in St Petersburg, Nuremberg, Paris, and London; he also returned to his university studies and completed his doctorate in 1902. In 1914 Lasker had to face his most serious challenge when he fought to retain his title against the young Cuban, CAPABLANCA. In 1921 Capablanca finally wrested the world title from Lasker after twenty-seven years. His playing tactics were frequently under discussion and during his long career he was accused of hypnotizing opponents, using 'psychological devices', and of having extraordinary luck. In fact he was a determined fighter who could be both forceful and subtle. After losing the title Lasker devoted his time to philosophy, writing, and teaching. He continued to play chess for financial reasons and went on doing well in international competitions until 1936.

Laski, Harold Joseph (1893–1950) *British political scientist, author, and educator who influenced socialist thinking in Britain in the 1930s and 1940s.*
Born in Manchester, Laski was educated at Oxford and went on to teach at McGill University (1914–16) and Harvard (1916–20), before holding the chair of political science at the London School of Economics (1926–50). In the early stages of his career Laski argued the case for political pluralism in such writings as *Authority in the Modern State* (1919) and *The Foundation of Sovereignty and Other Essays* (1921), but his views became gradually more collectivist – in his *Grammar of Politics* (1925) he saw the state as 'the fundamental instrument of society' – and, by the 1930s, Marxist.
The growth of fascism in Europe encouraged Laski to promote socialism as the only viable political alternative in such books as *The Rise of European Liberalism* (1936) and he became active as a member of the Labour

331 Laughton, Charles

Party's national executive. During World War II he was assistant to the deputy prime minister, Clement ATTLEE, and in 1945, when Labour came to power, he became chairman of the Labour Party. However, Laski's Marxist views were not favoured by the parliamentary leaders of the party and there was a marked clash between the elected parliamentary members and the national executive. Although Laski continued his calls for economic reform in such publications as *Reflections on the Constitution* (1951), he was unable to exercise much influence upon the new government.

Laski's other writings included a long and controversial study of US politics, *The American Democracy* (1948), and a selection of letters between him and US Justice Oliver Wendell Holmes, Jr (1841–1935) were published as *Holmes–Laski Letters* (1953).

Lauda, Niki (Nikolaus Andreas Lauda; 1949–) *Austrian motor racing driver who won the World Drivers' Championship in 1975, 1977, and 1984.*

Lauda started racing Formula V cars in 1969, progressed through Formula Three and sports cars, and by 1971 was driving in Formula Two. In 1974 he joined the Ferrari works team and gained his first Grand Prix victory in Spain. In 1975, five 'firsts' helped him to his first World Championship title but the following year he crashed during the German Grand Prix at the Nürburgring and sustained severe burns. However, he staged a remarkable recovery and returned to racing the same year only to lose the championship to James Hunt after refusing to drive in treacherous conditions at the Japanese Grand Prix. After winning his second championship in 1977, Lauda quit Ferrari and joined the Brabham team. He retired in 1979 to concentrate on developing his air charter company but in 1982 he was back, driving for the McLaren team. He won the US Grand Prix at Long Beach and achieved fifth place in the championship. Lauda stayed with McLaren for 1983 and 1984 and won the 1984 World Drivers' Championship by the narrowest of margins from his team-mate, Alain PROST. He finally retired in 1985.

Lauder, Sir Harry MacLennan (1870–1950) *Scottish comedian, one of the most popular of Britain's music-hall artists.*

Lauder was born in Portobello, near Edinburgh, and worked in a flax mill and as a miner before becoming a professional entertainer. He toured with concert parties and made appearances in provincial music halls before his London debut at Gatti's in 1900.

The stocky Scotsman, dressed in a kilt and glengarry and brandishing a crooked walking stick, became highly popular with London audiences. 'I Love a Lassie', 'Roamin' in the Gloamin'', 'A Wee Deoch-an'-Doris', 'Stop Your Tickling Jock', and the more serious yet rousing 'The End of the Road', were among the songs that endeared him to audiences both in Britain and on his many tours of the British Empire and the USA. During both world wars (he lost his son in the first) he entertained troops at home and abroad. As well as writing many of his own songs he also published *Harry Lauder at Home and on Tour* and *Harry Lauder's Logic* (1917).

Laughton, Charles (1899–1962) *British-born actor who became a US naturalized citizen in 1950. He is remembered for a series of outstanding character performances in the cinema.*

The son of a hotelier, Laughton was born in Scarborough and was originally destined to follow in his father's footsteps. After World War I, during which he served on the western front, he became interested in the stage and in 1925 enrolled at the Royal Academy of Dramatic Art, where he won a gold medal. In 1926 he made his professional debut in *The Government Inspector* at Barnes and by 1931 was appearing in New York in *Payment Deferred*. His work in the theatre included a season with the Old Vic in 1933, an appearance with the Comédie Française in 1936, and performances in *A Midsummer Night's Dream* and *King Lear* at Stratford-on-Avon in 1959. His film career began with three silent shorts with Elsa Lanchester (1902–86), whom he married in 1929. In the same year he made *Piccadilly* and three years later made his first Hollywood film, *The Old Dark House* (1932).

A large man with an expressively mobile face, Laughton played his many character parts with an exuberance that some of his peers regarded as overacting. *The Private Life of Henry VIII* (1933), which won him an Oscar, *The Barretts of Wimpole Street* (1934), in which he played Elizabeth Barrett Browning's father, *The Mutiny on the Bounty* (1935), playing the irascible Captain Bligh, *Ruggles of Red Gap* (1935), and *Rembrandt* (1937) are among his memorable films. His make-up and performance as Quasimodo in *The Hunchback of Notre Dame* (1939) were widely acclaimed, as was his portrayal of the inebriated bootseller in *Hobson's Choice* (1954). His last film was *Advise and Consent* (1961).

Laval, Pierre (1883–1945) *French states-man and prime minister (1931–32; 1935–36; 1942–44), whose collaboration with the Germans in World War II led to his execution after the war as a traitor.*

Born in Châteldon, the son of a café owner, Laval attended school at the Lycée Saint-Louis in Paris. He studied law and became a well-known advocate for the wellbeing of the working classes, joining the Socialist Party in 1903.

Laval entered politics in 1914 when he was elected as the socialist deputy for Aubervilliers, a working-class district in north-east Paris. He was opposed to World War I and narrowly escaped imprisonment in 1918 for sedition. Defeated at the polls in 1919, he returned to his legal practice and journalism but re-entered the chamber as an independent leftist in 1924, following his resignation from the Socialist Party. He became mayor of Aubervilliers in 1923 (a post he held until 1944) and again represented Aubervilliers in the chamber (1924–27) before winning a seat in the senate (1927). Continuing a move to the right, he became prime minister briefly in 1931–32 and 1935–36 but was best known as foreign minister (1934–36) when, with the British foreign secretary Sir Samuel Hoare (1880–1959), he sought the support of Musso-LINI against the growing influence of Germany by granting concessions to Italy in Abyssinia (the Hoare–Laval Pact).

In June 1940 after the fall of France, Laval became chief minister in Pétain's government, advocating collaboration with the Third Reich. Ousted from power in December 1940, he was reinstated in 1942 following German pressure. In 1944 he fled France after liberation but was arrested in Austria and returned to France. Convicted and sentenced to death for treason, he was executed in 1945, after trying to poison himself.

Law, (Andrew) Bonar (1858–1923) *British statesman and Conservative prime minister (1922–23).*

Born in Canada and brought up in Scotland, Bonar Law entered parliament as a Conservative member in 1900 and became party leader in 1911. He supported Ulster opposition to home rule (his father had been of Ulster descent), and in Asquith's wartime coalition government served as colonial secretary (1915–16). Under the influence of Sir Max Aitken (later Lord Beaverbrook), he supported the efforts of Lloyd George to remove Asquith and in Lloyd George's coalition became leader of the House of Commons

(1916–21). As chancellor of the exchequer (1916–19) he introduced National Savings, his only innovation. In 1919 he became lord privy seal, retaining the post until ill health forced his retirement in 1921. In the following year he returned to the fray to help manoeuvre the resignation of Lloyd George. Many leading Conservatives opposed the action, and Bonar Law became prime minister in what Chur-chill described as a 'second eleven' Conservative government. In 1923 he resigned, again because of illness, and died shortly afterwards. When his ashes were interred in Westminster Abbey, Asquith commented: 'It is fitting that we should have buried the Unknown Prime Minister by the side of the Unknown Soldier.'

Lawrence, D(avid) H(erbert) (1885–1930) *British novelist and poet.*

Lawrence was born in the coalmining village of Eastwood, Nottinghamshire, the fourth child of a miner and a schoolteacher. The conflict between his coarse violent father and refined socially ambitious mother shaped all Lawrence's subsequent attitudes to relations between men and women. He was a sickly boy and his mother made considerable sacrifices to enable him to gain an education at Nottingham High School and University to equip him for life outside the colliery. He became a schoolteacher and, encouraged by Ford Madox Ford and Edward Garnett (1869–1937), began to write. After his first novel, *The White Peacock* (1911), he abandoned teaching for writing.

In 1912 his second novel, *The Trespasser*, was published and in 1913 *Love Poems and Others* was followed by one of Lawrence's most successful novels, *Sons and Lovers*, which reflects his own tortured relationship with his mother. During this period he travelled in Germany and Italy, having eloped with Frieda Weekley, cousin of the German air ace von Richthofen and wife of his former teacher at Nottingham University; they married in 1914 after Frieda's divorce. Lawrence's loathing for war caused his profound distress during World War I; this distress was compounded by a prosecution for indecency brought against *The Rainbow* (1915) and by official harassment that forced him and Frieda to leave (1917) their cottage in Cornwall and lead a nomadic existence. Nonetheless he published two more volumes of poetry – *Amores* (1916) and *Look, We Have Come Through* (1917) – and worked on *Women in Love* (privately printed in New York, 1920), which is perhaps Lawrence's most complete fictional exploration of his views on sex.

In 1919 the Lawrences left England for good, moving first to Italy, where Lawrence wrote the travelogue *Sea and Sardinia* (1921) and at least part of the novels *The Lost Girl* (1920) and *Aaron's Rod* (1922). Other works of this period include *Fantasia of the Unconscious* (1922), *Studies in Classical American Literature* (1923), and the poems *Birds, Beasts and Flowers* (1923). In 1922 the Lawrences travelled to the USA via Australia, the setting for his novel *Kangaroo* (1923). They lived first at Taos, New Mexico, then at Oaxaca, Mexico. *The Plumed Serpent* (1926) celebrates the victory of what Lawrence perceived as the dark primeval forces of ancient Mexican religion over the pallid bourgeois values of Christianity. He also wrote the essays published as *Mornings in Mexico* (1927). A nearly fatal bout of malaria (1925) forced Lawrence back to Italy. While living near Florence he wrote *Lady Chatterley's Lover*, which he published there privately in 1928. The notoriety of this novel brought him financial gain but also attracted considerable scandal and other vexations, including the confiscation of the manuscript of his poems *Pansies* (1929). It was not until 1960 that Sir Allen LANE's Penguin Books received a judgment in the High Court that allowed *Lady Chatterley's Lover* to be published in an unexpurgated edition in Britain.

Lawrence's health, which was never good, declined sharply in 1930, and in March of that year he died of tuberculosis in a clinic at Vence, France. His *Letters*, edited by Aldous HUXLEY, and *Last Poems* were published in 1932.

Lawrence, Ernest Orlando (1901–58) *US physicist, who was awarded the 1939 Nobel Prize for Physics for his invention of the cyclotron. Element 103 was named lawrencium in his honour.*

The son of a school superintendent in rural South Dakota, he was educated at the universities of South Dakota, Minnesota, Chicago, and Yale, where he gained his PhD in 1925. In 1928 Lawrence joined the University of California at Berkeley, remaining there for the rest of his career as professor of physics. After 1936, he also became director of the Radiation Laboratory, later to be known as the Lawrence Radiation Laboratory.

In 1929 Lawrence began the development of one of the key devices of high-energy physics, the cyclotron. Although Lawrence's first model produced only a few thousand electronvolts it was the ancestor of modern accel-erators that are capable of producing energies of many billions of electronvolts. Under Lawrence's direction the Radiation Laboratory became one of the leading centres in the world for research into high-energy physics and a number of new radioisotopes were discovered there, including tritium, carbon-14, uranium-233, and plutonium. Many synthetic elements were also created and, for the first time, antiparticles were produced in a laboratory.

With the entry of the USA into World War II, Lawrence diverted his laboratory's resources to the preparation of uranium-235 for use in the atom bomb. However, despite a massive investment, the amount of uranium produced was negligible and had little impact on the construction of the bomb. In 1949, when news came that the Soviets had exploded an atomic bomb, Lawrence and Edward TELLER actively campaigned for the development of the hydrogen bomb. Lawrence was also concerned with his Materials Testing Accelerator (MTA), a device generally agreed to be ill-conceived. The project with its intractable problems did much to wear Lawrence out and probably hastened his death from ulcerative colitis in 1958.

Lawrence, Gertrude (Gertrud Alexandra Dagmar Lawrence Klasen; 1898–1952) *British actress, best known for her performances in Noël COWARD's plays.*

Gertrude Lawrence was born in London into a theatrical family: her mother was a small-part actress and her Danish father was a music-hall singer. After making her stage debut in *Dick Whittington* at Brixton in 1910 she toured in musicals for several years. She first met Noël Coward in 1913 and her first success came in Coward's revue *London Calling* (1923). She subsequently appeared in other revues and in plays before her outstanding performance opposite Coward in *Private Lives* (1930) as Amanda, a part Coward wrote especially for her. This was followed by appearances in other Coward plays, notably *Tonight at 8.30* (1936). Many of Gertrude Lawrence's London productions were repeated in the USA, where she toured in *Susan and God* (1937) and appeared in New York in *Lady in the Dark* (1941) and SHAW's *Pygmalion* (1945). During World War II she entertained troops in Europe and the Pacific and in 1948 starred in the London production of Daphne du Maurier's *September Tide* (1948). Lawrence was married twice, first (1917) to playwright and producer Francis Howley, by whom she had a daughter, and then

(1940) to American producer and manager Richard Aldrich.

She made a handful of films (1929–50), the best being *The Glass Menagerie* (1950), and gave her last performance in New York in the musical *The King and I* (1951).

Lawrence, T(homas) E(dward) (1888– 1935) *British soldier and author, known also as 'Lawrence of Arabia'.*

Lawrence was born at Tremadoc, North Wales, and educated at Oxford High School and Jesus College, Oxford. An excellent scholar, he wrote a thesis on medieval military architecture (published as *Crusader Castles*, 1936), which enabled him to travel in the Middle East, where he learnt to live like an Arab. At the outbreak of World War I he was assisting the archaeologist Leonard WOOLLEY in excavating the Hittite city of Carchemish.

Rejected as being too short for active service, Lawrence was assigned to military intelligence in Egypt. In 1916 he went to Jidda where he became adviser to Prince Faisal (1885–1933), who was then uniting the Arabs in revolt against their Turkish rulers. Winning the Arabs' confidence, Lawrence helped Faisal seize the important port of Wajh and he himself, with the Howaitat tribe, routed the Turks near Ma'an before capturing Aqaba (1917). The campaign of guerrilla raids and pitched battles then turned north, culminating in the Arabs' occupation of Damascus (1918). At the peace conference and subsequently in the Colonial Office (1921–22) Lawrence strove to obtain a just settlement for the Arabs, but was disillusioned by the great powers' dismissal of their claims now their help was no longer needed. He was also occupied with the writing of his most important book, *Seven Pillars of Wisdom* (1935; private edition 1926), of which he would allow only an abridged version, *The Revolt in the Desert* (1927), to be made publicly available during his lifetime.

To avoid the fame his exploits had attracted, Lawrence joined the Royal Air Force in 1922 under an assumed name (first J. H. Ross, later T. E. Shaw) and for most of the rest of his life remained obscurely in the ranks of that service. Lawrence's experience of life in the ranks was published as *The Mint* in 1955. Shortly after his retirement in 1935 he was killed in a motorcycle accident in Dorset.

Leach, Bernard Howell (1887–1979) *British potter, who was the major figure in modern studio pottery in Britain. He became a CH in 1973.*

The son of a colonial judge, Leach spent the first ten years of his life in Hong Kong, Singapore, and Japan before going to school in Britain in 1897. He attended the Slade School of Art (1903–08) and in 1909 returned to Japan, where in 1911 he took up pottery, apprenticing himself to the sixth generation of Kenzan potters. Apart from two visits to China, Leach remained in Japan until 1920, when he returned to Britain with fellow potter Hamada Shōji and set up his studio pottery in St Ives, Cornwall.

Leach's style – and the public response to his work – developed slowly. His products were designed to be aesthetically pleasing as well as functional and at their best they combined the best qualities of both traditional English and eastern pottery. He visited Japan and China regularly in the 1920s and 1930s. The first edition of his *A Potter's Book*, which outlined his philosophy and methods, was published in 1940: it rapidly became a potters' bible and influenced potters all over the world. His ideas were also spread by the pupils he took. In the 1950s he paid a number of visits to the USA, where he met his third wife, and in the sixties and seventies he wrote two more books: *Kenzan and his Tradition* (1966) and *Hamada, Potter* (1975).

Leadbelly (Huddie Ledbetter; 1888–1949) *Black US blues and folk singer and guitar player.*

His early life was spent in poverty, picking cotton, learning songs that dated back to slavery, and playing the guitar. He was discovered by the folklorists Fred and Alan Lomax in the Louisiana State Prison Farm in 1933, where he was serving time for attempted murder. The Lomaxes wrote a book about him in 1936 and brought him to New York. He subsequently recorded folk songs as well as his own songs, such as 'Midnight Special' and 'Rock Island Line'. It is not known whether he collected and adapted them or wrote them himself.

Leadbelly was a fluent player of the twelve-string guitar and, influenced by his friend Blind Lemon Jefferson, he helped to keep alive interest in folk music and the blues during a period in which it was being neglected. He was in prison on four separate occasions for violent crimes. A few months after he died, his song 'Goodnight Irene' was a number one hit in the USA.

Leakey, Louis Seymour Bazett (1903–72) *British-born Kenyan palaeontologist and anthropologist whose discoveries of fossil homi-*

nids in east Africa established that region as the likely origin of Homo sapiens.

Leakey was born in Kenya, where his parents were missionaries, and attended Weymouth College and St John's College, Cambridge. Following an injury while playing rugby, he took leave from his studies and joined an archaeological expedition to east Africa, the first of many. His PhD in 1929 was for a thesis about a Stone Age site in Kenya. He later spent two years studying the customs of the Kikuyu tribe, among which he had lived as a boy. He worked for British Intelligence in Nairobi during World War II and afterwards was appointed curator of the Coryndon Memorial Museum, Nairobi (1945–61). Working on Rusinga Island on Lake Victoria in 1948, Leakey discovered a skull of *Proconsul africanus*, an apelike ancestor of modern primates, which lived 25–40 million years ago. Leakey concentrated his search for further fossil remains in east Africa, especially in the Olduvai Gorge in what is now northern Tanzania. In 1959, his wife Mary Leakey (1913–) discovered the upper jaw and palate of a hominid, named *Zinjanthropus* by Leakey and now called *Australopithecus boisei*. In 1964 his son, Richard Leakey (1944–), found a lower jaw of the same type. The Leakeys also unearthed fragments of another hominid, *Homo habilis*, which had a larger brain than the australopithecines. Leakey maintained that this was ancestral to *Homo sapiens* while *Zinjanthropus*, which lived at the same time, eventually died out, a view currently favoured by palaeontologists. Leakey gave many lectures throughout the world and wrote many books, including *Adam's Ancestors* (1934), *Olduvai Gorge* (1952), and *Unveiling Man's Origins* (1968). Although Leakey's interpretation of his finds was often questioned, their value remains undisputed.

Lean, Sir David (1908–91) *British film director, who made a number of outstanding British films, perhaps the most highly acclaimed of which is* The Bridge on the River Kwai. *He was knighted in 1984.*

Born in Croydon, Lean began his career with Gaumont Pictures in 1928 as a number board boy and messenger. He progressed to director, having worked on the editing and commentary for Gaumont and British Movietone News. He also edited such films as *Pygmalion* (1938) and *The 49th Parallel* (1941) before making his debut as director with *In Which We Serve* (1942), co-directed with Noël COWARD. Subsequently he directed such notable films as

Coward's *Blithe Spirit* (1945) with Margaret RUTHERFORD and *Brief Encounter* (1946) with Trevor HOWARD and Celia Johnson (1908–82), outstanding adaptations of Dickens's *Great Expectations* (1946) and *Oliver Twist* (1948), *The Sound Barrier* (1952) with Ralph RICHARDSON, which won a British Film Academy Award, and *Hobson's Choice* (1954) with John MILLS and Brenda de Banzie (1915–81). Also in *The Sound Barrier* was Lean's second wife, Ann Todd (1909–), who appeared in two of his other films, *The Passionate Friends* (1949) and *Madeleine* (1950).

Summer Madness (1955) with Katherine HEPBURN earned Lean a New York Critics Award, as did *The Bridge on the River Kwai* (1957) with Alec GUINNESS, which also brought his first Oscar. A second Oscar, as well as the Italian Silver Ribbon award, came with the spectacular *Lawrence of Arabia* (1962) starring Peter O'Toole (1932–). Continuing with lavish productions he went on to direct *Doctor Zhivago* (1965), *Ryan's Daughter* (1970), and *A Passage to India* (1984).

Leavis, F(rank) R(aymond) (1895–1978) *British literary critic and university teacher. He was made a CH in 1978.*

Leavis was educated at Cambridge, first at the Perse School and then at Emmanuel College, where he taught from 1925. In 1929 he married Queenie Dorothy Roth, herself a notable critic and author of *Fiction and the Reading Public* (1932). Leavis's own first major book, *New Bearings in English Poetry*, also appeared in 1932, the year he founded the periodical *Scrutiny*, which was to run until 1953 and was the means of disseminating much of the new criticism written by Leavis and his followers. At this period Leavis moved to Downing College, Cambridge, where he became a fellow (1936) and created his own distinctive school of literary studies. In addition to his profound influence upon his pupils, Leavis wrote several major critical works, among them *Revaluation* (1936), *The Great Tradition* (1948), and *The Common Pursuit* (1952), which won him an international following. Some of his criticism was notorious, notably his debunking of Milton, but, more positively, he was an ardent advocate for D. H. LAWRENCE (in *D. H. Lawrence, Novelist*, 1955) and led the way to a more serious appreciation of Dickens (in *Dickens the Novelist*, 1970).

After his retirement (1962) academic honours were heaped upon him. The same year he became embroiled in an infamous and ill-natured controversy with C. P. SNOW over the 'two cultures' issue. His Clark Lectures (1967) at Trinity College, Cambridge, were published as *English Literature in Our Time and the University* (1969). Besides his Dickens study, Leavis also wrote in retirement *Anna Karenina and Other Essays* (1967), *Lectures in America* (1969; with his wife), *Nor Shall My Sword* (1972), and *The Living Principle* (1975).

Le Carré, John (David John Moore Cornwell; 1931–   ) *British writer of espionage thrillers.*

Le Carré was educated at Sherborne before attending Bern University and Lincoln College, Oxford. He taught at Eton (1956–58) before joining the British Foreign Sevice (1960–64). The latter gave him the background and material for the series of espionage novels that made him an international best-seller. His first two books, *Call for the Dead* (1961) and *A Murder of Quality* (1962), were modest successes but he made his reputation with *The Spy Who Came in from the Cold* (1963). The story of Alec Leamas, a seedy disillusioned middle-aged spy, with its combination of psychological realism, astute plotting, and the minutiae of international espionage, is typical of Le Carré's best work. The same mood and ambience are sustained in *The Looking-Glass War* (1965) and *A Small Town in Germany* (1968). In later books, which include *Tinker, Tailor, Soldier, Spy* (1974), *The Honourable Schoolboy* (1977), and *Smiley's People* (1980), he develops the character of George Smiley, brilliant but self-doubting *éminence grise* of the British intelligence organization nicknamed 'the Circus'. In *The Little Drummer Girl* (1983) the espionage theme is maintained against a background of the Israeli intelligence service in the Middle East conflict. His most recent spy thrillers include *A Perfect Spy* (1986) and *The Secret Pilgrim* (1991).

Le Corbusier (Charles-Édouard Jeanneret-Gris; 1887–1965) *Swiss-born French architect, engineer, painter, and writer. As a pioneer of modern architecture, a designer of many outstanding buildings, a town planner on the grand scale, and an architectural theorist, Le Corbusier stands head and shoulders above his contemporaries. He was also an abstract painter of considerable talent.*

Born at La Chaux-de-Fonds (Neuchâtel), Le Corbusier (a name he borrowed from his maternal grandfather) came from a family of watchmakers. After attending a local school and technical college he spent a few months in 1905 in the Vienna studios of the architect Josef Hoffmann (1870–1956). This was followed by a stay in Paris (1908–09) to study the use of ferroconcrete under Auguste Perret (1874–1954), and during 1910 he spent a few months in the Berlin workshops of Peter BEHRENS. For the next three years (1911–14) Le Corbusier practised as an interior designer in Switzerland, then, moving to Paris, he worked as a factory manager for seven years, moonlighting as a painter and an architect. However, it was not until 1922 that he and his cousin, Pierre Jeanneret (1896–   ), could afford to set up an architectural practice on their own. It was then that Le Corbusier adopted his grandfather's name as an architect, retaining his own name as a painter.

In 1923 the partnership produced plans for the 'Citrohan' House, a residence designed as if it was a car, 'a machine for living in', which had many features in common with an earlier design of Le Corbusier's, his 1914 Dom-ino standard concrete house. In the following years Le Corbusier and his cousin produced a number of private houses built on concrete pillars (pilotis), which supported the structure and freed the walls from their traditional load-bearing function. Several of these cubist-style houses, such as the Villa Savoye (1931), aroused considerable architectural interest. In 1922 Le Corbusier and Jeanneret won a competition for the Palace of the League of Nations in Geneva, but the scheme was later rejected and the partnership suffered a number of other setbacks. In 1940 they broke up, Jeanneret settling in Grenoble and Le Corbusier returning to Paris. During this unsettled period (1936–45), Le Corbusier had been designing one of his most original buildings, the Ministry of Education in Rio de Janeiro, in cooperation with Lúcio Costa (1902–63) and NIEMEYER. After the war he redesigned the bombed town of La Pallice and embarked on one of his most famous buildings, the block of flats in Marseilles, called *unité d'habitation*, which he described as 'a town for 1600 people under one roof'. In this building he began his departure from the functional glass-and-metal façade, introducing innovative sculptural effects and unusual rooflines. On a different scale and in a different idiom, his chapel at Ronchamp (1950–55), which abandons the strict functionalism of his earlier buildings, made use of a number of novel and irrational features, such as randomly placed and irregularly shaped windows to create what he called a religious ambience. The monastery and church at Eveux-sur-Arbreste, near Lyons, is another ecclesiastical exercise in reinforced concrete.

Although many of Le Corbusier's early town-planning ideas were unexecuted, he did create a highly successful city, the new capital of the Punjab, Chandigarh. This was planned by Le Corbusier for a population of 150 000 in the 1950s, expanding to half a million by the end of the century. For this work Le Corbusier prepared the outline plan and the design for many of the public buildings. Some of these structures have a severity that foreshadows the brutalism of his last works, the Museum of Modern Art in Tokyo (1957) and the Dominican Friary of La Tourette (1957–60).

Le Corbusier was also a prolific writer; his books include *Vers une architecture* (1923; translated as *Towards a New Architecture*, 1946), *La Ville radieure* (1935), *Le Modulor I* (1948; translated 1954), and *L'Unité d'habitation de Marseilles* (1952).

Lederberg, Joshua (1925–) *US geneticist who discovered genetic recombination in bacteria, for which he received the 1958 Nobel Prize for Physiology or Medicine.*

Born in Montclair, New Jersey, Lederberg graduated in 1944 from Columbia College and in 1946 joined Edward TATUM at Yale University. Here they discovered that a mixture of two mutant strains of the bacterium *Escherichia coli* produced limited numbers of normal (wild-type) individuals. By a process called conjugation, DNA was transferred through minute hairlike structures (pili) from one individual to another, a form of sexual reproduction. Lederberg received his PhD in 1948 and was appointed assistant professor of genetics at Wisconsin University, becoming associate professor in 1950 and professor in 1954. In 1952 he published his discovery of transduction in the bacterium *Salmonella typhimurium*. A virus (bacteriophage P22) infective to the bacterium incorporated part of the bacterial chromosome into its outer coat and transferred this to another individual bacterium. These findings contributed much to understanding the mechanisms of recombination in DNA that are crucial to genetic engineering.

In 1959 Lederberg was appointed professor of genetics at Stanford University's newly founded department. He was president of Rockefeller University, New York, (1978–1990). He received the US National Medal of Science in 1989.

Lee, Tsun-Dao (1926–) *Chinese-born US physicist who, with C. N. YANG, was awarded the 1957 Nobel Prize for Physics for their discovery that parity is not conserved in the weak interaction.*

Born in Shanghai, Lee fled the Japanese invasion in 1945, abandoned his early studies in Kweichow, and entered the National Southwest Associated University in Yunnan, where he first met Yang. With the aid of scholarships, Lee continued his education in the USA, at the University of Chicago. After gaining his PhD in 1950 and spending a year at the University of California, Lee was appointed a member of the Institute for Advanced Studies, Princeton, in 1951. Already recognized by the institute's director, J. R. OPPENHEIMER, as a brilliant theoretical physicist, Lee joined the physics faculty at Columbia University in 1953, where he has since remained.

According to the conservation of parity, a principle introduced into physics by E. P. WIGNER, there can be no fundamental distinction in nature between left and right. In 1956 Lee and Yang proposed that the conservation law does not apply to the weak interaction. They suggested ways of testing the proposal and experiments performed shortly afterwards at Columbia confirmed their theory.

Lee Kwan Yew (1923–) *First prime minister of the Republic of Singapore (1965–90). The longest-serving leader in southeast Asia in modern times, he was criticized for repressing political opposition and praised for Singapore's economic progress, which was achieved through his encouragement of foreign investment.*

Born in Singapore, the son of wealthy Straits Chinese parents, Lee was educated at Raffles College, Singapore, before attending Cambridge University, England, where he studied law. He was admitted to the bar in 1950 but returned to Singapore and became a legal adviser to the trade union movement.

Lee founded the moderate left People's Action Party (PAP) in 1954. Winning a seat in the legislative assembly the following year, he became leader of the opposition and in 1959, after helping to negotiate a new constitution, was elected head of the government. Lee led Singapore into the federation of Malaysia in 1963 and out of it in 1965, when the Malay-dominated federal government opposed his policy of multiracialism. When, in 1965, Singapore was declared an independent republic, he was elected its first prime minister. Lee sought to promote Singapore's trade and industry and maintain independence, adopting a policy of nonalignment and regional cooperation.

Léger, Fernand (1881–1955) *French painter, who was one of the leading cubists.*

The son of a prosperous Normandy farmer, Léger arrived in Paris in 1900 having served an apprenticeship to an architect in Caen. In Paris he worked in an architect's office and as a photographic retoucher, while studying art in his spare time. From 1909 he was in close touch with cubist painters but, unlike PICASSO and BRAQUE, who broke down objects into rectilinear forms, Léger favoured cylindrical forms. He became the first of the cubists to experiment with abstraction, which resulted in his *Contrast of Forms* series (1913). In these brightly coloured pictures the contrast is between flat and tubular forms.

On the outbreak of World War I he enlisted in the engineers. Léger shared the enthusiasm of the futurists for mechanization and for modern industrial society and his war experiences strengthened that enthusiasm. They also opened his eyes to life outside his own social class, and the paintings that followed his discharge as a result of gas poisoning often had proletarian subjects, such as *The Mechanic* (1920). As well as figure compositions he painted still lifes and semi-abstract brightly coloured cityscapes. In 1924 he made an experimental film, *Ballet mécanique*. The 1930s saw a looser and more curvilinear style of painting with elements of surrealism.

Léger spent World War II in the USA, where he taught at Yale and in California and painted figure compositions, mainly of acrobats and cyclists. On his return to France in 1945, he joined the French Communist Party, and monumental working-class subjects became still more prominent in his painting. This characteristic of Léger's art made him especially suited to public commissions, such as the stained-glass windows for the church at Audincourt (1951).

The final development in Léger's style occurred in the 1950s, when he began to separate colour from outline. In *Two Women with Flowers* (1954), for example, the robust black outlines define the figures but no longer act as boundaries for the areas of pure flat colour. In 1955, the year of his death, Léger won the Grand Prix at the São Paolo Biennale.

Lehmann, Lotte (1888–1976) *German-born naturalized US singer, who was one of the great lyric sopranos of her era.*

Born into a family of musicians, she studied singing in Berlin, chiefly with Mathilde Mallinger (1847–1920), Wagner's original Eva in *The Mastersingers*. She made her professional debut in Hamburg in 1910 as Aennchen in Nicolai's *Merry Wives of Windsor* and in 1914 BEECHAM engaged her to sing at the Drury Lane Theatre, London, as Sophia in Richard STRAUSS's *Die Rosenkavalier*. She excelled in Strauss's music and also sang the Composer and Ariadne in *Ariadne auf Naxos*, Arabella in *Die Frau ohne Schatten*, and Christine in *Intermezzo*. Lehmann also sang Wagnerian roles in addition to Italian and French opera. The Nazis forced her to leave Vienna in 1938 and she eventually settled in the USA, where she became a naturalized citizen. She made her farewell performances singing Strauss in New York (1945) and Chicago (1946). In retirement she turned to teaching, painting, and writing.

Leigh, Vivien (Vivien Mary Hartley; 1913–67) *British actress.*

Born under the Raj in India, Vivien Leigh trained at the Royal Academy of Dramatic Art and gained early recognition on stage in *The Mask of Virtue* (1934). She made her film debut in *Things are Looking Up* (1934), which was followed by such films as *A Yank at Oxford* (1938) with Robert Taylor (1911–69) and *Fire Over England* (1937) with Laurence OLIVIER, whom she married in 1940.

In 1939 Leigh shot to international fame playing Scarlett O'Hara in *Gone With the Wind*, winning an Oscar and a New York Critics Award, after which she returned to England and, with Olivier, appeared in numerous plays. In this period she also made *Waterloo Bridge* (1940), *Lady Hamilton* (1941), and *Caesar and Cleopatra* (1945). They toured Australia and New Zealand with the Old Vic (1947) and Leigh's stage success as Blanche du Bois in *A Streetcar Named Desire*, directed by Olivier, was repeated in KAZAN's film version (1951), bringing a second Oscar and New York Critics Award. Divorced in 1961, she again toured with the Old Vic (1961) and in 1963 made her first appearance in a musical on Broadway in *Tovarich*, winning a Tony Award. Ill health dogged the remainder of her career. Her last film was *Ship of Fools* (1965).

Lenin, Vladimir Ilich (Vladimir Ilich Ulyanov; 1870–1924) *Russian Revolutionary and founder of the Bolshevik Party. A national hero, he was the first leader of the Soviet Russia (1918–24). His body is preserved in the mausoleum in Red Square, Moscow.*

Born in Simbirsk, the son of a schoolmaster, Lenin was educated at the University of Kazan, from which he graduated in law (1891), and at the University of St Petersburg (now

Leningrad), where he studied Marxism and began his involvement with revolutionary politics. Arrested in 1895 and sent to Siberia in 1897, he spent three years in exile, during which he wrote *The Development of Capitalism in Russia*, a study of the Russian economy.

In 1900 Lenin moved to western Europe. Largely as a result of the publication of pamphlets, including *What is to be Done?* (1902), and his newspapers *Iskra* ('The Spark') and *Vperyod* ('Forward'), he emerged as leader of the Bolshevik faction of the Russian Social Democratic Workers' Party (1903). He participated in the 1905 revolution but returned to Europe after its failure, settling in Switzerland in 1907. In March 1917 he re-entered Russia, passing through Germany in a 'sealed train', and in the October Revolution that year led the Bolsheviks to power, overthrowing the provisional government headed by KERENSKI. Concluding peace with Germany in 1918, he commanded the Bolsheviks in the civil war (1918–21), fighting off opposition from both the White Army and foreign intervention. As chairman of the Council of People's Commissars (head of state), he founded the Third (Communist) International (1919) and established the authority of the Bolshevik (Communist) Party in Russia. He was also responsible for instituting the New Economic Policy in 1921, which emphasized economic reconstruction and education and allowed a small amount of free enterprise. He survived an assassination attempt in 1918 but suffered a stroke in 1922 and died prematurely in 1924.

A powerful orator, fluent in many languages, he wrote several major studies of socialist theory, including *Materialism and Empirico-criticism* (1909) and *Imperialism: The Highest Stage of Capitalism* (1916). He was revered as the architect of the Soviet Union for over sixty-five years, but the mystique surrounding his name has diminished considerably in recent years, during which the basic tenets of communism have been renounced by liberal reformers in many formerly communist states; in the Soviet Union itself the Communist Party was stripped of its monopoly of power in 1990. The queues that once formed every day to file past his preserved body in Red Square had by 1991 slowed to a trickle. Also in 1991, the citizens of Leningrad, so named in his honour, voted to change the name of their city back to St Petersburg.

Lennon, John (1940–80) *British singer, guitarist, and songwriter: the most important member of the Beatles (1962–70).*

Born in Liverpool, he formed an amateur skiffle group, the Quarrymen, in 1957, with Paul McCARTNEY, George HARRISON, and later Ringo STARR. It evolved into the Beatles, whose 1964 single 'Can't Buy Me Love' was the first record to top both the US and UK charts simultaneously. The group's musical sound profoundly influenced not only popular music but a whole generation of teenagers. With McCartney, Lennon formed a highly successful songwriting team. In 1969, frustrated by the restrictions implied by being a Beatle, Lennon announced that he was quitting. In the same year he and his Japanese-born second wife, Yoko Ono (1933–), embarked on a programme for the promotion of world peace; their international 'bed-in' was a field day for the media. His post-Beatle musical career was a mixture of rock and roll, public exhibition of his 'primal scream' psychotherapy, and naive peacemongering. His recordings *Imagine* (1971) and *Walls and Bridges* (1974) were highly regarded, especially in the USA. The title song from the former was a top-ten hit in both the USA and the UK and is probably the song for which his post-Beatle career will be best remembered. 'Whatever Gets You Through the Night' (a duet with Elton JOHN from *Walls and Bridges*) was Lennon's only post-Beatle record to top the charts in the USA.

In 1975, after a long battle, Lennon was granted legal residency in the USA, and Yoko gave birth to their son (with Elton John as godfather). Thereafter Lennon became virtually a recluse for five years, concentrating on being a father. Just as he and Yoko were about to resume a public career, Lennon was murdered by a psychopath on the steps of his apartment building.

Leontief, Wassily (1906–) *Russian-born US economist, who won the Nobel Prize for Economics in 1973 for his development of input-output analysis.*

Born in St Petersburg, Leontief was educated at the universities there and in Berlin. He worked at the Kiel Economic Research Institute and in 1929 served as economic adviser to the Chinese government in Nanking. In 1931 he moved to Harvard, where he was appointed professor of economics in 1946. He was director (1975–85) of the Institute for Economic Analysis at New York University.

Leontief developed an input-output matrix for the US economy to show the interrelationships of the various sectors of the economy and the effect on other sectors resulting from a change in one of them: this became known as the 'Leontief model'. Each sector uses outputs of other sectors and itself produces commodities or services that are used as factors of production in other sectors or for final consumption. All flows of goods and services are set out in a rectangular table so that the relationship between total inputs (raw materials, labour, manufacturing and related services) and total demand for final goods and services can be seen. This technique is widely used as the basis for economic planning, making extensive use of computers to obtain the data. The 'Leontief paradox' is Leontief's theoretical refutation (1954) of the Heckscher–Ohlin model of international trade using input-output analysis: imports to the USA appear to be capital-intensive and US exports labour-intensive using this system.

Leontief's principal works include *The Structure of the American Economy 1919–29* (1941), *Studies in the Structure of the American Economy* (1953), *Input-Output Economics* (1966), *The Future of Non-Fuel Minerals in the US and the World Economy* (1983), and *The Impact of Automation on Workers* (with F. Duchin; 1986).

Lessing, Doris (1919–) *British novelist, brought up in Rhodesia (now Zimbabwe). She was an active communist and a feminist in her youth, and social and political conflicts, especially as they affect women, are a dominant feature of her work.*

Mrs Lessing was born Doris May Tayler in Kermanshah, Persia (now Iran), where her father was a captain in the British army. In 1924 the family settled in Rhodesia and Doris Lessing was educated at schools in Salisbury (now Harare). Between 1939 and 1949 she had two short-lived marriages and in the latter year moved to England. The novel *The Grass is Singing* (1950) and the short stories *This Was the Old Chief's Country* (1951) reflect her experiences of Africa. *Martha Quest* (1952) was the first in a series of five novels published under the general title *Children of Violence*. The last was *The Four Gated City* (1969). Among her novels *The Golden Notebook* (1962) was critically acclaimed and *The Good Terrorist* (1985) won the W. H. Smith literary award in 1986 as well as the Palermo Prize and Premio Internazionale Mondello in 1987. Subsequent novels include *The Fifth Child* (1988). Her philosophical preoccupations were explored in an ambitious series of novels with a science fiction setting called *Canopus in Argos: Archives – Shikasta* (1979), *The Marriages between Zones Three, Four and Five*

(1980), *The Sirian Experiments* (1981), *The Making of the Representative for Planet 8* (1982), and *The Sentimental Agents in the Volyen Empire* (1983).

Besides her full-length novels, Doris Lessing also wrote short novels published under the title *Five* (1953), which won the 1954 Somerset Maugham Award. Her several collections of short stories include *A Man and Two Women* (1965) and *The Story of a Non-Marrying Man* (1972). *In Pursuit of the English* (1960) describes her impressions on first coming to England. She has also published poetry and wrote the play *Play with a Tiger* (1962).

Lévi-Strauss, Claude Gustave (1908–) *Belgian-born French anthropologist, who was one of the leading exponents of the structuralist movement in social anthropology that arose in the 1950s and 1960s. He was awarded the Legion d'honneur in 1991.*

Lévi-Strauss studied philosophy at the Sorbonne, Paris, and was appointed professor of sociology at the University of São Paulo (1935–39). During World War II he served in the French army (1939–41) before moving to the USA, first as a teacher at the New School for Social Research, New York (1941–45), and then as a cultural attaché to the French embassy in Washington (1946–47). In 1948 he became professor of ethnology at the University of Paris and in 1959 was appointed professor of anthropology at the Collège de France. He was made honorary professor in 1983.

The approach of Lévi-Strauss and others assumes an underlying essential structure common to humans in different societies which, if adequately described, can be used to account for and predict behaviour. Lévi-Strauss saw language as such an essential common denominator underlying cultural phenomena and this formed the basis of his theories concerning the relationship of such societal elements as kinship, religion, and myth. His books include *Les Structures élémentaires de la parenté* (1949; translated as *The Elementary Structures of Kinship*, 1969), *Anthropologie structurale* (2 vols, 1958 and 1973; translated as *Structural Anthropology*, 1964 and 1977), *Mythologiques* (4 vols, 1964–72), and *La Potière jalouse* (1985; translated as *The Jealous Potter*, 1988).

Lewis, Carl (Frederick Carleton Lewis; 1961–) *US black athlete, who in the 1980s was acclaimed as the greatest track and field competitor since Jesse OWENS.*

Born in Birmingham, Alabama, Lewis became nationally known in 1981, when he was given the Sullivan Award for the top US amateur athlete. In 1983 he set new world records for the javelin with a throw of 8.79 metres (28 ft 10¼ in) and in the 200 metres event with a time of 19.75 seconds. In the 1984 Olympics at Los Angeles he won four gold medals – in the 100 metres, 200 metres, long jump, and (setting a new world record) the 4 x 100 metres relay event; these were the same events for which Owens won gold medals in 1936. In the 1988 Seoul Olympics he repeated his victory in the 100 metres and long jump events and won a silver medal for the 200 metres race; his time of 9.92 seconds in the 100 metres established another world record. In 1989 he set yet another world record as part of a 4 x 200 metres team in Koblenz. He captured his third world championship title in the 100 metres event in 1991, establishing a new record of 9.86 seconds.

Lewis, C(live) S(taples) (1898–1963) *British author and scholar, many of whose works deal with religious and moral themes reflecting his Christian faith.*

Born in Belfast, Lewis was sent after his mother's early death to various preparatory schools and then to Malvern College. In 1916 he won a classical scholarship to University College, Oxford. However, he first joined the army, fought in France, and was wounded at Arras (1918). While convalescing from this wound he began his long friendship with the mother of one of his friends who had been killed in the war, and he shared her house in Oxford until her death in 1951. At Oxford Lewis enjoyed academic success in the fields of classics and English and in 1925 Magdalen College appointed him to a tutorial post, which he held for nearly three decades. His rooms there were for many years the meeting-place of a group of friends, the 'Inklings', which included J. R. R. TOLKIEN. Even when he became professor of medieval and Renaissance English at Cambridge (1954–63) he kept his Oxford connections very much alive.

Lewis's first major critical work was *The Allegory of Love* (1936). *A Preface to Paradise Lost* (1942), *The Abolition of Man* (1943), *English Literature in the Sixteenth Century* (1954), and *The Discarded Image* (1964) were all based on his lectures. He became a convinced Christian and wrote several popular works on Christian ethics, the most famous of these being *The Problem of Pain* (1940), *The Screwtape Letters* (1942), and *Mere Christianity* (1952). He also wrote a trilogy of allegorical science-fiction novels (beginning with *Out of the Silent Planet*, 1938), the seven *Chronicles of Narnia* (1950–56) for children, and a volume of autobiography, *Surprised by Joy* (1955). *Till We Have Faces* (1956) is a sequel to *The Allegory of Love* in the form of a novel set in an antique world. *A Grief Observed* (1961) expresses his distress over the illness and death of his wife, whom he married in 1956.

Lewis, (Harry) Sinclair (1885–1951) *US novelist and the first American to win the Nobel Prize for Literature (1930).*

Lewis was born in Sauk Centre, Minnesota, and completed his education at Yale, where he first began to write. After working in publishing and journalism, he published his first book, *Our Mr Wrenn*, in 1914. His subsequent novels until *Main Street* (1920) were unremarkable, but this book brought him instant recognition and acclaim for its use of caricature and satire to attack many aspects of small-town life in the American midwest. *Babbitt* (1922), *Arrowsmith* (1925), *Elmer Gantry* (1927), and *Dodsworth* (1929) continued in a similar vein, making such institutions as the urban middle class and the church targets for his satirical prose. Lewis was awarded the Pulitzer Prize for *Arrowsmith* in 1926 but refused it, claiming there were worthier contenders than himself, a self-deprecating disclaimer he repeated in his speech accepting the Nobel Prize in Stockholm in 1930. Although he was regarded by many during the 1920s as the leading US novelist, Lewis's reputation has since steadily waned. He continued to perplex many of his admirers by producing several novels that were banal and superficial interspersed with others that were satirical and provocative. He offered no explanation for this unevenness as a writer and in his later years appeared to become increasingly conservative in both his life and his work.

Lewis, (Percy) Wyndham (1882–1957) *British writer and painter; the founder of vorticism, he is also known for his portraits. In his books he attacked and satirized the cultural establishment.*

Lewis was born on his father's yacht in the bay of Fundy, Maine, USA (his father was American, his mother British). He attended Rugby School, studied art at the Slade School in London, and then studied and lived on the Continent, mainly in France (1901–09). Back in England he had stories published by *The English Review* and exhibited his paintings with the Camden Town Group (1911) and at Roger FRY's postimpressionist exhibition (1912). Breaking with Fry (Lewis was notoriously quarrelsome), he founded his own movement, known as vorticism, which aimed to reflect the pace and technology of modern life. He publicized vorticism in a review entitled *Blast*, of which just two issues (1914–15) appeared; he also organized the sole British exhibition of vorticist art (1915).

In 1916 Lewis enlisted in the army and served in France both as soldier and as official war artist. His first novel, *Tarr*, appeared in book form in 1918. During the 1920s and 1930s Lewis maintained a prolific output of books and articles, besides holding several exhibitions of his paintings. *The Childermass* (1928) became the first part of his trilogy *The Human Age*, which he completed with *Monstre Gai* and *Malign Fiesta* (both 1955), and *The Apes of God* (1930) satirizes art and literature in contemporary London. As a painter he was particularly successful as a portraitist. Among his other writings, *Hitler* (1931) aroused much hostility by its adulation of the Führer, hostility that was not overcome by his recantation in *The Hitler Cult* (1939).

Lewis and his wife, whom he had married in 1929, spent World War II in straitened circumstances in the USA and Canada, both of which Lewis detested. He returned to England in 1945 and continued writing and painting; he was art critic on *The Listener* (1946–51). By 1954 he was totally blind. Among the books of his later years was the autobiographical *Rude Assignment* (1950), which followed his earlier autobiography *Blasting and Bombardiering* (1937).

Libby, Willard Frank (1908–80) *US chemist, who was awarded the 1960 Nobel Prize for Chemistry for his discovery of radiocarbon dating.*

The son of a farmer, Libby was educated at the University of California, where he gained his PhD in 1933. He remained at Berkeley until the outbreak of World War II, when he moved to Columbia University to work on the separation of uranium isotopes for use in the atom bomb. In 1945 Libby joined the faculty at the University of Chicago as professor of chemistry.

In 1939 Serge Korff discovered that carbon has a rare radioisotope, carbon-14, which has a half-life of about 5600 years. In 1947 Libby suggested that as carbon dioxide in the atmosphere would contain some carbon-14, the proportion of carbon-14 found in biological specimens could be used to date objects no older than 50 000 years. Although it took some years to work out the test procedures and meth-

ods of calibration, carbon-14 dating was soon shown to be both reliable and versatile. Libby's radiocarbon clock therefore provided archaeology with the precise instrument it had been seeking for centuries. Libby returned to the University of California in 1959 as director of the Institute of Geophysics at Los Angeles, where he continued to refine and develop methods of radioactive dating.

Lichtenstein, Roy (1923–   ) *US painter and sculptor and a leading exponent of pop art.*

Lichtenstein was born in New York and began to study art there in 1939. He then moved to Ohio State University, but his studies were interrupted by service in the army and he did not graduate until 1949. In the 1950s Lichtenstein worked in Cleveland, Ohio, as a freelance commercial artist, painting at the same time in the abstract expressionist style. About 1957, after his return to New York to teach, he rejected the subjectivism of this style and began, like Claes OLDENBURG, to use subjects from everyday life. His paintings were based on images from commercial art, such as cartoons from bubble-gum wrappers and pictures from advertisements and travel posters (*Girl With a Ball*, 1961). Some of his best-known works are blown-up pictures from comic strips (*Whaam!*, 1963). Though conscious of the crudity of these images, Lichtenstein intended no social comment and the paintings have high formal quality. His first one-man exhibition of work in this new pop art style was in 1962. During the 1960s he produced ceramics, including *Dinnerware Objects*, and also worked with enamel on steel; in his paintings he parodied earlier painters and familiar images, such as the likeness of George Washington. In 1977 he exhibited his first sculptures in bronze, again representing banal objects from everyday life. In the 1980s he produced still lifes influenced by such artists as DALÍ, MATISSE, and PICASSO.

Liddell Hart, Sir Basil Henry (1895–1970) *British military theorist and historian who originated many aspects of strategy for mechanized warfare that were first used during World War II. He received a knighthood in 1966.*

Born in Paris, the son of an English minister, Liddell Hart's studies at Corpus Christi College, Cambridge, were interrupted by World War I. Commissioned in the King's Own Yorkshire Light Infantry, he was wounded at Ypres in 1915 and gassed on the Somme the following year. He joined the army education corps in 1921 but the legacy of his combat injuries forced him to retire from the army in 1927 at the rank of captain.

Liddell Hart worked as defence correspondent of the *Daily Telegraph* (1925–35) and *The Times* (1935–39), at the same time establishing his reputation as an innovator of military strategy. He formulated the theory of the 'expanding torrent' – a spearhead thrust through enemy lines that is immediately reinforced in depth to outflank the enemy. His outstanding work, *Strategy – The Indirect Approach* (six editions; 1929–67) emphasizes the need to destroy enemy command centres, communications, and supply lines to secure victory at minimum cost. In addition, Liddell Hart's biographies of military leaders, including Scipio (1926), Sherman (1930), FOCH (1931), and T. E. LAWRENCE (1934), were acclaimed as the work of a fine military historian. However, his role as unofficial adviser to war minister Hore-Belisha (1937–38) was more controversial. Besides pressing for increased mechanization of the army, Liddell Hart sought to identify 'dead wood' among the military hierarchies, thus causing considerable resentment in establishment circles. During World War II, his advocacy of a defensive campaign went unheeded by CHURCHILL, although certain German commanders, especially Guderian and ROMMEL, ascribed their military successes to reading Liddell Hart.

After the war, Liddell Hart's interviews with German generals formed the basis of *The Other Side of The Hill* (1948). His other works include *The Rommel Papers* (1953), *The Tanks* (two vols, 1959), and *History of The Second World War* (1970). His immensely valuable archives are now housed at King's College, London.

Lie, Trygve Halvdan (1896–1968) *Norwegian Labour politician and first secretary-general of the United Nations (1946–52).*

Born in Oslo, the son of a carpenter, Lie was educated at the University of Oslo, where he graduated in law in 1919, becoming a member of the Norwegian Labour Party secretariat the same year. In 1922 he was appointed the legal adviser to the Norwegian Trade Unions Federation, a position he held for thirteen years.

Lie was first elected as a Labour member to parliament in 1935. Initially minister of justice, he became minister of shipping and supply at the outbreak of World War II in 1939. In 1940 he fled from Norway to England with the government, where he acted as foreign minister until 1945. The same year he led the Norwegian delegation to the United Nations

conference in San Francisco, playing a major role in the drafting of the charter of the Security Council. He was elected as the first secretary-general of the United Nations in 1946 with the support of the USA and the Soviet Union. He resigned in 1953 following Soviet refusal to recognize him as secretary-general after he had advocated UN military action against North Korea. On his return to Norway he resumed ministerial duties in the government and wrote his memoirs.

Lifar, Serge (1905–86) *Russian choreographer and dancer, who revitalized French ballet and re-established the Paris Opéra as one of the world's leading companies.*

Having studied dance in Russia, Lifar was brought to Paris by Bronislava Nijinska (1891–1972) to join DIAGHILEV'S Ballets Russes. In 1925 he became the company's principal dancer and four years later scored a great triumph in the title role of BALANCHINE'S *Prodigal Son* (1929). Lyrical, virile, and athletic as a dancer, Lifar first tried his hand at choreography in 1929 with his ballet *Le Renard*, in which acrobats doubled for the principal dancers. After Diaghilev's death in the same year and the break-up of his company, Lifar became principal dancer and ballet master at the Paris Opéra, the original home of ballet. Over the next twenty-eight years he restored the reputation of the company as a major balletic influence and not only staged and danced in such classics as *Giselle* but also presented more than fifty new works, including *Prométhée* (1929) and *Icare* (1935), in which he emphasized the importance of rhythm in ballet.

Dismissed as ballet master in 1944, for entertaining the Germans during their occupation of France, Lifar returned to the Paris Opéra in 1947 and scored major successes with *Phèdre* (1950), *Snow-White* (1951), and *Daphnis and Chloë* (1958). In 1956 he retired as a dancer and two years later left the Paris Opéra to work with companies all over the world. He appeared in a film, *Le Testament d'Orphée* (1960), and was the author of twenty-five books, in which he described his own career and his theories of experimental choreography.

Ligeti, György Sándor (1923–   ) *Hungarian composer, who was only able to develop his more radical musical ideas after his escape to western Europe.*

Born in Romania, Ligeti completed his musical education at the end of World War II at the Budapest Academy of Music under Ferenc Farkas (1905–   ) during the period in which

Lillee, Dennis Keith

Otto KLEMPERER was conducting at the Budapest Opera. In 1948 communications with the West were cut and Ligeti had to earn his living conforming musically in a teaching post at the Academy (from 1950). However, he was clandestinely listening to music broadcasts from the West and composing private compositions that he did not hear in performance until his escape in 1956 to Vienna, where he made contact with the European musical avant garde. He was invited by STOCKHAUSEN to work at the Cologne Radio electronic studios and he also absorbed the total serialism of BOULEZ. Although he rejected both as compositional methods, each left its mark upon him. After the performances of his *Apparitions* (1956–59) and *Atmosphères* (1961), Ligeti was accepted as an important composer. In his work, a distinct melody, harmony, and rhythm are replaced by complexes of polyphonic sound that are particularly effective in such choral works as the *Requiem* (1963–65) and *Lux aeterna* (1966). As a humorist, Ligeti has made musical fun of his more eccentric colleagues; the *Trois Bagatelles* (1961) satirizes CAGE and his followers, the *Poème symphonique* for a hundred metronomes (1962) ironically explores the implications of mechanical music. Ligeti's catalogue of compositions is large and includes a cello concerto (1966), string quartets, and a piano concerto (1986). He was Professor of Composition at the Hamburg Academy of Music (1973–89).

Lillee, Dennis Keith (1949–) *Australian cricketer, a formidable fast bowler who took more test wickets (355) than any other bowler. He announced his retirement from test cricket in 1984, at the age of thirty-four, to devote more time to his family.*

Born in Perth, he first played for Western Australia in 1969–70 and at the end of the summer went to New Zealand with an Australian B side. His test debut came against England at Adelaide in 1971 and he went on to obtain his 351 wickets for Australia at an average of 25.94. At times he was accused of intimidatory methods: various public incidents, one involving injury to Javed Miandad, captain of Pakistan, lost Lillee some support in cricketing circles.

Lindbergh, Charles Augustus (1902–74) *US aviator who, in 1927, made the first solo transatlantic flight. In recognition of his achievement, Congress awarded him its medal of honour.*

The son of a congressman, Lindbergh gave up his engineering studies to take flying lessons and bought his first plane for $500 in 1923. He was a flying cadet in the US Air Service Reserve before joining the US Air Mail Service as a pilot in 1926, flying the Chicago to St Louis route. Backed by a group of St Louis businessmen, Lindbergh planned his attempt on the solo crossing of the Atlantic and the prize of $25 000 that was on offer. On 20 May 1927 his craft, *Spirit of St Louis*, left Roosevelt Field, New York, and after a 33½-hour flight he landed in Paris to a hero's reception. On his return to the USA he was promoted to colonel in the Air Reserve and made several goodwill trips on behalf of his government.

In 1929, Lindbergh married Anne Morrow, an ambassador's daughter, and the following year Charles Augustus Jr was born. Tragedy struck in March 1932 when the Lindberghs' son was kidnapped and later found dead. The murder hunt and arrest in 1934 of Bruno Richard Hauptmann preceded one of the most widely publicized criminal trials in US legal history. To escape the glare of publicity, the Lindberghs and their second son, Jon Morrow, left the USA in December 1935 to live temporarily in Britain. Hauptmann was convicted and, in April 1936, executed.

Following a tour of Germany in 1936, Lindbergh gave warning of growing German air power but in 1938 he aroused controversy by accepting a German decoration from GOERING. In 1939 he made the first of several speeches urging US neutrality in World War II. However, he served the US war effort as a consultant to the United Aircraft Corporation and the Ford Motor Company. After the war he continued as a consultant to Pan American Airlines and served on various aeronautical boards and committees. He was made a brigadier-general in 1954. His book, *Spirit of St Louis* (1953), describes his epic flight and won a Pulitzer Prize.

Lindemann, Frederick Alexander See CHERWELL, FREDERICK ALEXANDER LINDEMANN, VISCOUNT.

Linklater, Eric Robert Russell (1899–1974) *British writer. He was made a CBE in 1954.*

Linklater was born in Orkney, the son of a sea captain, and educated at Aberdeen Grammar School; in the latter part of World War I he served as a private in the Black Watch (1917–19). After the war he went to Aberdeen University and studied medicine before going to Bombay as the assistant editor of *The Times of India* (1925–27). On his return he taught English at Aberdeen (1927–28) and then went to the USA as a Commonwealth Fellow (1928–30). During World War II he served as a major in the Royal Engineers, first holding a command in the Orkneys (1939–41) before moving to the War Office directorate of public relations (1941–45). He was rector of Aberdeen University (1945–48), served briefly with the army in Korea (1951), and was deputy lieutenant in Ross and Cromarty from 1968 to 1973.

From 1920 onwards Linklater published a prolific flow of work in several genres. Some of his best-known fiction reflects his travels, including *Juan in America* (1931) and *Juan in China* (1937). He also wrote plays – among them, *The Devil's in the News* (1934), *Love in Albania* (1949), and *Breakspear in Gascony* (1958) – and several volumes of autobiography. Among his works of nonfiction were his official history of the Italian campaign (1951) and numerous biographical and historical studies; many of these have a Scottish theme or interest, beginning with *Ben Jonson and King James* (1931), *Mary Queen of Scots* (1933), and *Robert the Bruce* (1934) and continuing through to *The Royal House of Scotland* (1970).

Lipchitz, Jacques (1891–1973) *Lithuanian-born French innovative sculptor.*

Against the wishes of his father, a Jewish building contractor who wanted his son to study engineering, Lipchitz went in 1909 to Paris, where he studied sculpture and developed a life-long interest in ancient, medieval, and primitive art. His work was vital, mannered, and representational. He was recalled to Russia for military service in 1912 but later discharged because of ill health. In 1915 he began to produce constructions of geometric forms, which were some of the first sculptures to apply the principles of cubism in three dimensions. He also anticipated a technique of later sculptors, such as Henry MOORE and Barbara HEPWORTH, when in 1916 he bored a hole in his *Man with a Guitar* to bring space within the sculpture.

Lipchitz became a French citizen in 1925. The same year he abandoned the discipline of cubism and returned to a more expressive and exuberant style in such works as the monumental *Joie de Vivre* (1927), which was probably the earliest revolving sculpture. He also began to produce more abstract open-work constructions of strips and bands of metal; for example, *The Couples* (1929).

Mythological, often violent, themes and more solid monumental forms, as in *Pro-metheus* (1937), are typical of his work after 1930. In 1940 he fled from Paris, arriving in New York in 1941. Increasingly rich in symbolism, his postwar work expressed the suffering and pathos of the previous decade in European history. His later years brought him numerous public commissions and saw continued experimental work, such as the semiautomatics of 1956, which were modelled by touch alone.

Littlewood, (Maud) Joan (1914–   ) *British theatre director, co-founder of the Theatre Workshop.*

Born in London, Littlewood studied at the Royal Academy of Dramatic Art. Somewhat unfashionably, but typical of her unorthodox experimental approach to drama, she began her career in Manchester, where she founded a street theatre, the Theatre of Action (1931–37), and the Theatre Union (1937–39). After World War II she returned to London as co-founder, with Gerry Raffles, of the Theatre Workshop, which moved to the Theatre Royal, Stratford, in 1953. Between 1945 and 1953 the company toured the British Isles and abroad. Among the many plays Littlewood directed at the Theatre Royal a number transferred to the West End, including *The Quare Fellow* (1956), *A Taste of Honey* (1948), and, perhaps the one with which her name became most closely associated, *Oh, What A Lovely War* (1963).

Littlewood has often worked abroad, her first Broadway production being in 1960. After the Theatre Workshop disbanded in 1964 she worked at the Centre Culturel in Hammamet, Tunisia (1965–67), and at Image India in Calcutta (1968). The Theatre Workshop was revived and reformed in the early 1970s but in 1975 she left England to work in France.

Littlewood's other activities have included creating the Children's Environments around the Theatre Royal (1968–75) and directing the film *Sparrers Can't Sing* (1962). She has also appeared on stage from time to time. She received a special Society of West End Theatre award in 1983.

Litvinov, Maksim Maksimovich (1876–1951) *Soviet diplomat and foreign minister (1930–39). His long-established ties with STALIN enabled him to survive the purges of the 1930s and become one of the few Jews to retain high office in the Soviet Union.*

Born in Bialystok in Poland into a poor Jewish family, Litvinov joined the Russian Social Democratic Labour Party in 1898. A supporter of the Bolshevik faction, he moved to western Europe in 1902 but was deported from France on conspiracy charges in 1908. He then moved to London, where he worked in a publishing house and married (1916) an English girl, Ivy Low.

After the October (Bolshevik) Revolution in 1917, Litvinov attempted to muster support for the Bolsheviks in London but was deported in 1918. On his return to Moscow he was appointed to the Commissariat of Foreign Affairs, becoming deputy commissar in 1921. During the 1920s he led several Soviet delegations to disarmament conferences. By 1930 he had become commissar of foreign affairs, gaining prominence throughout the 1930s for his policy of collective security; he established diplomatic ties with the USA (1933), joined the League of Nations (1934), and concluded a mutual defence pact with France (1935). When Stalin signed the German-Soviet nonagression pact in 1939, Litvinov (as a Jew he was strongly anti-Nazi) was replaced by MOLOTOV. However, following the German invasion of the Soviet Union in 1941, he was appointed ambassador to the USA (1941–43). He continued to serve in the foreign affairs commissariat until his retirement in 1946.

Lloyd, Clive Hubert (1944–   ) *West Indian cricketer and captain of the West Indies team (1974–78; 1979–85), renowned as a hard-hitting batsman and for his quiet diplomacy and authority as a leader.*

Clive Lloyd was born in Georgetown, British Guiana (now Guyana), and was educated at Chatham High School there. When his father, a doctor's chauffeur, died he left school early to support the family, working as a clerk in a hospital. He made his cricketing debut for Guyana in 1963 and played his first test match (against India) in 1966. In 1968 he joined Lancashire, the start of a lengthy relationship during which he captained the county side. He captained the West Indies in a total of 74 tests from 1974 to 1978 and again – despite several knee operations – from 1979 until his retirement in 1985; his side was beaten in only two of his 18 test series. He was manager of the West Indies team that toured Australia (1988–89). Since 1987 he has been a member of the Commission for Racial Equality and has also worked to help the unemployed.

Lloyd, Harold (1893–1971) *US film comedian, who made hundreds of shorts as the bemused and bespectacled American man in the street.*

Born in Burchard, Nebraska, Lloyd trained at the San Diego Dramatic School and made his screen debut in 1912. Much of his early work

with Hal Roach (1892–   ) and Mack SENNETT at Keystone was highly Chaplinesque; although he gained wide popularity as Lonesome Luke it was from 1917, as the rather ordinary-looking individual in glasses and straw hat, that he was to achieve real fame. *Grandma's Boy* (1922), *Safety Last* (1923), *Girl Shy* (1924), and *The Kid Brother* (1927) are a few of the hundreds of films he made. Not only did audiences warm to the new character and his ability to triumph in the end but they also thrilled at Lloyd's many stunts, most of which were performed by Lloyd himself.

In many of his films he played opposite Mildred Davis (1900–69), whom he married in 1923. *Welcome Danger* (1929) was his last silent film and although he made several talkies, including the notable *Movie Crazy* (1932), the triumphs of his silent days were never repeated. However, his two compilations, *Harold Lloyd's World of Comedy* (1962) and *The Funny Side of Life* (1963), were highly successful. In 1952 he was honoured with a Special Oscar.

Lloyd George, David, 1st Earl (1863–1945) *British statesman and Liberal prime minister (1916–22), renowned as an orator. He was created an earl in 1945.*

He was born in Manchester, of Welsh parents. His father, a schoolmaster surnamed George, died in 1864; the young David was brought up in North Wales by his mother and her brother Richard Lloyd. Disliking the name David George, he later incorporated his uncle's name and required that he be addressed as David Lloyd George. He qualified as a solicitor before entering parliament in 1890, as Liberal member for Caernarfon Borough, a constituency he represented until 1945.

Lloyd George became known as a gifted and outspoken orator during the second Boer War (1899–1902), when he supported the Boers' claims to independence. As president of the Board of Trade under CAMPBELL-BANNERMAN, he introduced the Merchant Shipping Act 1906 and the Patents Act 1907 and established the Port of London Authority (1908). When ASQUITH replaced Campbell-Bannerman in 1908, Lloyd George became chancellor of the exchequer. His famous 'people's budget' of 1909 proposed a redistribution of wealth on the basis of a new liberalism designed to foster the interests of employers and workers at the expense of landowners. The budget included a supertax on incomes above £5000, higher death duties, and an income tax allowance for children under sixteen. Most of the revenue raised was to finance old-age pensions (introduced in 1908); a part was also allocated to the Development Commission to experiment in farming, forestry, and land reclamation. The budget was rejected by the Lords, and the consequent constitutional crisis led to passage of the Parliament Act 1911. Lloyd George was also responsible for the National Insurance Act 1911, which with the Pensions Act formed the basis of the welfare state.

In 1912 it was revealed that he had bought shares in the Marconi Company, which the government planned to employ to set up an 'imperial wireless chain'. A parliamentary investigation found the purchase 'imprudent' rather than 'corrupt', but the scandal tarnished his reputation.

In World War I Lloyd George advocated all-out pursuit of victory and served successfully as minister of munitions (1915–16). Appointed secretary for war (1916), he found his powers limited and became increasingly disenchanted with Asquith's leadership. Supported by Bonar LAW and the Conservatives, he manoeuvred Asquith's resignation. Thereafter the Liberals were split between the two men and Lloyd George became prime minister of a coalition dominated by the Conservatives. He assumed dictatorial control of war policy, conflicting with the general staff and with the Admiralty over his introduction of convoys, which successfully evaded the German U-boat blockade. The war won, Lloyd George went to the country (December 1918) and the coalition received an overwhelming mandate. He performed with distinction at the Paris Peace Conference, advocating moderate peace terms and proposing measures to revive the postwar European economies. At home, economic difficulties were aggravated by the return of four million ex-servicemen requiring work. The establishment of the Irish Free State in 1921 lost Lloyd George the support of many Conservatives (the 'Diehards'), who were further alienated by a confrontation between Turkish and British troops at Chanak in 1922, which almost led to a war. The Conservatives withdrew their support from the coalition and Lloyd George resigned.

Lloyd George is widely regarded as one of the most brilliant British statesmen of the twentieth century, although he was also generally distrusted. In addition to the Marconi scandal, there were rumours that he had sold honours for personal and party gain. The irregularities of his private life also caused public concern – he lived openly with his secretary Frances Stevenson (1888–1972), whom he

married after the death of his first wife, Dame Margaret Lloyd George (1864–1941). Though he led the Liberals once more, between 1926 and 1931, he never again held office.

Lloyd Webber, Andrew (1948–   ) *British composer. His musical plays have been international hits for over twenty years.*
After studying at the Royal College of Music he was determined to create a successful British musical theatre. He met lyricist Tim Rice (1944–   ) in 1965 and together they wrote a few pop songs; their *Joseph and his Amazing Technicolour Dreamcoat*, an attempt to present pop music in an oratorio style, was first produced in 1968. The extremely successful musical *Jesus Christ Superstar* was unusual in that it was released first on a recording in 1970, staged in 1971, and filmed in 1973. His film scores include *Gumshoe* (1971) and *The Odessa File* (1974); the musical show *Jeeves* (1975) was based on P. G. WODEHOUSE with lyrics by Alan Ayckbourn. *Evita* (1978), with Rice again, was an international hit, as was *Cats* (1982), adapted from T. S. ELIOT's *Old Possum's Book of Practical Cats*. *Starlight Express* opened in London in 1984; with the theatre rebuilt and the entire cast on roller skates, it was said to be the most expensive musical then mounted. His *Requiem*, premiered in New York in 1985, represented a departure from the genre of musicals. *Phantom of the Opera* (1986), based on the old horror story, was another enormous hit, as was *Aspects of Love* (1989). Members of his Really Useful Theatre Company, based at London's Palace Theatre, included Prince Edward. Lloyd Webber's enormous success is reflected in his wealth, assessed at several tens of millions of pounds. His brother Julian Lloyd Webber (1951–   ) is a cellist and composer.

Lodge, Henry Cabot (1850–1924) *US statesman and Republican senator (1893–1924).*
Lodge was born into a wealthy family in Boston and graduated from Harvard in 1874 with a degree in law. Admitted to the bar in 1876, he returned to Harvard to take up an academic post in the American History department before entering politics in 1879. After two years in the Massachusetts House of Representatives, he moved to the US House in 1887 and to the Senate in 1893, where he served until his death. He soon became a prominent Republican leader, with a particular interest in international affairs, rising in 1918 to a position of considerable power as Republican floor leader in the Senate and chairman of the Foreign

Relations Committee. Although he had backed President Woodrow WILSON on the issue of the USA's entry into World War I, Lodge was bitterly opposed to certain aspects of the Treaty of Versailles and the League of Nations Covenant, which Wilson submitted to the Senate for approval in 1919. Supported by fellow isolationists in the Senate, Lodge drew up a list of objections known as the 'Lodge reservations'. Wilson refused to compromise, however, and the treaty was ultimately rejected by the Senate.

In 1921, under the new Republican administration of President Warren G. Harding (1865–1923), Lodge served as delegate to the Washington Conference on the Limitations of Armaments. He also wrote prolifically on historical and political issues, notably *Daniel Webster* (1882), *George Washington* (1888), and *The Senate and the League of Nations* (1925).

His grandson, also called Henry Cabot Lodge (1902–85), was a prominent diplomat.

Loewe, Frederick (1904–88) *US songwriter. With lyricist Alan Jay Lerner (1918–86), he wrote the music for some of America's most popular musicals.*

Born in Austria, Loewe was trained as a concert pianist and went to the USA in 1924. He began writing songs while playing the piano in restaurants in the 1930s. His earliest compositions were in an outmoded operetta style, but after meeting Lerner (1942) he changed his style and together they evolved a highly sophisticated form of musical play. The most successful of these were *Brigadoon* (1947), *Paint Your Wagon* (1951), *My Fair Lady* (1956), *Gigi* (1958), and *Camelot* (1960). *My Fair Lady* won a Pulitzer Prize, and all these musicals have been filmed.

Loewi, Otto (1873–1961) *German-born US physiologist who showed that the passage of a nerve impulse is associated with the release of a chemical at the nerve endings. This chemical, acetylcholine, was later isolated by DALE, and Loewi and Dale shared the 1936 Nobel Prize for Physiology or Medicine.*

After obtaining his medical degree at the University of Strasbourg (1896), Loewi studied chemistry, physiology, and pharmacology and was appointed professor of pharmacology at Graz University in 1909 – a position he held until he was expelled by the Nazis in 1938. He emigrated to the USA and in 1940 became a research professor at New York University's College of Medicine.

Loewi's most important findings, in 1920, were concerned with the mechanism of nerve impulse transmission. By electrically stimulating the nerves of a frog's heart, he slowed its rate of contraction. The fluid bathing this heart was then allowed to perfuse a second heart, which was not electrically stimulated. However, the contraction rate of the second heart was also decreased, demonstrating that a chemical released by the first heart into the perfusing fluid was responsible for this action. The chemical was later identified as acetylcholine – the first substance demonstrated to be a chemical neurotransmitter.

Lonsdale, Dame Kathleen (1903–71) *British physicist, known for her work in X-ray crystallography. She was the first woman to be elected a fellow of the Royal Society (in 1945) and was made a DBE in 1956.*

Born Kathleen Yardley, the daughter of a postman in Newbridge, Ireland, she was educated at Bedford College, London, where she graduated in 1922. She then joined the research staff of the Royal Institution under Sir William BRAGG and over the following twenty years, with such colleagues as Dorothy HODGKIN and J. D. BERNAL, worked as an X-ray crystallographer. Her first major success came in 1929 with her publication of the structure of benzene. Lonsdale continued to work on the structure of a number of organic molecules, producing in 1948 a survey of the discipline in her *Crystals and X-Rays*.

As a Quaker, Lonsdale refused to register for employment at the outbreak of World War II. As the mother of three small children, she would have been exempt from any government service, but was fined £2 for refusing to register. She refused to pay and spent one month in Holloway prison. In 1948 Lonsdale moved to University College, London, where she became professor of crystallography (1948–68).

Loos, Adolf (1870–1933) *Austrian architect, whose rejection of all ornament, curves, and decorative features had a profound influence on a generation of architects. He was not, however, himself a successful architect.*

Born in Brno, Moravia (now in Czechoslovakia), he studied in Dresden and then spent the years 1893–96 in the USA, where he was strongly influenced by Louis Sullivan (1856–1924). On returning to Vienna he spent the rest of his life there, except for six years (1922–28) in Paris. In 1908 he wrote his famous article *Ornament and Crime*, in which he argued that the use of ornament in architecture was now obsolete and that its presence in modern buildings was degenerate. His Steiner House in Vienna (1910) used the new material, reinforced concrete, in severe curveless cubic forms without decoration of any kind. The result deeply influenced GROPIUS and through him the Bauhaus movement, with the result that an architectural fashion was set for a whole generation. The severity that led to brutalism owes much of its roots to this single early essay in reinforced concrete. It is perhaps ironic that several of Loos's later buildings make free use of classical decoration.

Loren, Sophia (Sofia Scicolone; 1934–) *Italian-born French film actress and international superstar. She was awarded the Legion d'honneur in 1991.*

Born into poverty in Rome, Sofia Scicolone was raised in the slums of Naples. In her teens she began to enter beauty competitions and by the time she was fifteen had met the man who was to groom her for stardom and become her future husband, Italian producer Carlo Ponti (1910–).

She began her film career inauspiciously as an extra in *Quo Vadis* (made 1949, released 1951). By 1953, however, she was appearing in the title role of *Aïda* and subsequently attracted considerable attention for *La donna del fiume* (1955; *Woman of the River*). After making several other Italian films she went to the USA, where she appeared in such films as *Boy on a Dolphin* (1957), *The Black Orchid* (1959), for which she received a Venice Festival Award, and *That Kind of Woman* (1959). Her range has extended from the comedy of Anthony Asquith's *The Millionairess* (1960) with Peter SELLERS to the award-winning dramatic performance of *La ciociara* (1961; *Two Women*). For the latter she won, among other honours, Academy, Cannes Festival, and British Film Awards. She also received a Moscow Festival Award for *Marriage Italian Style* (1964). Later films include *Man of La Mancha* (1972), *The Cassandra Crossing* (1977), and *Firepower* (1979). A biography, *Sophia – Living and Loving: Her Own Story*, was published in 1979. Charges of tax evasion resulted in her spending a widely publicized month in prison in 1982.

Lorenz, Konrad Zacharias (1903–89) *Austrian ethologist whose studies of animal behaviour in its natural environment helped establish ethology as a distinct discipline and earned him the 1973 Nobel Prize for Physiology or Medicine.*

The son of a Viennese surgeon, Lorenz was an ardent collector of animals from childhood. He studied medicine at Columbia University and Vienna University, from which he received his MD in 1928 and PhD in zoology in 1933. Lorenz kept a variety of animals at the family home in Altenburg, near Vienna, and from his observations of jackdaws and geese established the phenomenon of imprinting. He found that for a short time after hatching, chicks are genetically predisposed to identify their mother's sound and appearance and thereby form a permanent bond with her. In 1937, Lorenz was appointed lecturer in comparative anatomy and animal psychology at Vienna and later (1940–42) became professor of psychology at Königsberg University. He served as a doctor in the German army until his capture by the Russians in 1944. Released in 1948, he returned to Altenberg and wrote the popular account of his work, translated as *King Solomon's Ring* (1949), which was followed by *Man Meets Dog* (1950). Funded by the Max Planck Institute, Lorenz co-founded the Institute for the Study of Comparative Behaviour at Seewiesen in 1955, later becoming its sole director (1961–73).

Lorenz gained major insights into the genetic basis of behaviour patterns and how they are triggered, how they develop through learning, their social significance, and their evolution. *On Aggression* (1963) aroused controversy because of Lorenz's extrapolation of his work to the possible implications for human society. He argued that with their sophisticated weaponry, often remotely controlled, humans lack the inhibitory mechanisms that largely prevent serious injury or death arising through social conflict in other species.

Los Angeles, Victoria de (1923–) *Spanish soprano. She sings a wide variety of operatic roles, including those of Wagner, Puccini, and Purcell, as well as Spanish folk songs; in recitals she sometimes accompanies herself on the guitar.*

Born in Barcelona into a musical family, she soon learnt to sing and play the guitar. She went on to study at the Barcelona Conservatory, where she made her recital debut in 1944; her professional opera debut came the following year at the Teatro del Liceo as the Countess in Mozart's *Marriage of Figaro*. After winning an international singing contest in Geneva in 1947 she became widely known, making tours of Europe and South America. In 1950 she made her US debut at the Metropolitan Opera, New York, as well as her first appearance in Britain, in *La Bohème* at Covent Garden, where she appeared regularly over the next decade.

Losey, Joseph (1909–84) *US-born film director, who took up residence in the UK after being blacklisted as a communist in the 1950s.*

Losey was born in La Crosse, Wisconsin, and educated at Dartmouth College and Harvard. Medical studies gave way to the theatre, beginning with bit parts and writing reviews. He was involved in setting up the Brechtian-inspired *Living Newspaper* and after war service achieved some success as a theatre director, most notably with BRECHT's *Galileo* (1947), starring Charles LAUGHTON.

Losey's work in films began in 1938, supervising documentaries for the Rockefeller Foundation. *The Boy With Green Hair* (1948) was his first feature. After making *M*, a remake of Fritz LANG's original, and *The Prowler* (both 1951), his problems with Senator McCARTHY and the Un-American Activities Committee began. Finding himself blacklisted and unable to work, Losey moved to Britain, where his first film, directed under the pseudonym Victor Hanbury, was *The Sleeping Tiger* (1954) with Dirk BOGARDE, with whom he went on to make such successful films as *The Servant* (1963) and *Accident* (1967). These last two, the scripts for which were written by Harold PINTER, made Losey a cult figure in Europe. *Accident* also starred Stanley Baker (1928–76), with whom Losey successfully worked in *The Criminal* (1960) and *Eve* (1962). Losey and Pinter came together again for the popular adaptation of L. P. Hartley's novel *The Go-Between* (1971), which won the Cannes Golden Palm Award. These, together with such films as *King and Country* (1964), *Modesty Blaise* (1966), and Nell Dunn's *Steaming*, completed just before Losey's death, are an indication of his range and the substantial contribution he made to Britain's film industry.

Louis, Joe (Joseph Louis Barrow; 1914–81) *US boxer, known as 'The Brown Bomber', and world heavyweight champion (1937–49).*

Born at Lexington, Alabama, Joe Louis was a Golden Gloves champion and turned professional in 1934. Within three years he had become the second black boxer, after Jack JOHNSON, to hold the world title when he defeated James J. Braddock. By any standards he was a great boxer, noted for his powerful left jab and knock-out punch. He successfully defended his title twenty-five times and knocked out four other heavyweight champions – Max

Schmeling, Jack Sharkey, Primo Carnera, and Max Baer.

He retired from the ring in 1949 but shortage of money forced him to make a comeback. After his third defeat, against Rocky MARCIANO in 1951, he gave up for good.

Lovell, Sir (Alfred Charles) Bernard (1913–) *British radio astronomer, who created and was first director of the observatory at Jodrell Bank in Cheshire. He was knighted in 1961.*

After receiving his PhD from Bristol University in 1936, Lovell was appointed to a lectureship in physics at Manchester University. He returned to Manchester in 1945 after having spent World War II working on the development of radar. His first postwar research made use of abandoned military radar equipment to study cosmic rays and meteors. Realizing that something more elaborate was needed to study radio sources, Lovell spent ten years in winning financial support and gaining academic backing for the 250-foot steerable parabolic reflector he planned. It was completed in 1955, with a deficit of £250,000. The success of the telescope in tracking the first Sputnik in 1957 finally persuaded the Treasury to pay all outstanding debts in 1960. Lovell was appointed professor of radio astronomy and director of the Jodrell Bank Observatory in 1951, appointments that he held until his retirement in 1981. His publications include *The Exploration of Outer Space* (1961), *Out of the Zenith* (1973), and *Astronomer by Chance* (1990).

Lowell, Jr, Robert Traill Spence (1917–77) *US poet.*

Born in Boston, Massachusetts, Lowell was the son of a naval officer and a member of a prominent New England family, a great-grandnephew of the poet and diplomat James Russell Lowell (1819–91) and cousin of the poet Amy Lowell (1874–1925). He was educated at St Mark's School, Harvard, and Kenyon College, Ohio, where he studied under the critic and poet John Crowe Ransom. After graduating in 1940, he married his first wife, the writer Jean Stafford. He later married (and remarried after one divorce) the writer Elizabeth Hardwick. Lowell taught at Louisiana State University and worked in publishing for a time. During World War II he was imprisoned for five months, not strictly as a conscientious objector – a classification he refused to accept – but for his opposition to the bombing of civilians. In 1947–48 he was appointed poetry consultant at the Library of Congress.

In 1940 Lowell had converted to Roman Catholicism (later abandoned) and his first two volumes, *The Land of Unlikeness* (1944) and *Lord Weary's Castle* (1946), reflect this in their treatment of historical themes and criticism of modern life. Influenced by the Metaphysical poets, the books are formally structured and written in an intensely dramatic and allusive iambic verse. *Lord Weary's Castle*, which won a Pulitzer Prize in 1947, was followed by *The Mills of the Kavanaughs* (1951). By this time Lowell was acknowledged as a major poet; *Poems 1939–1949* collected most of his work to this date.

With the publication of *Life Studies* (1959; revised edition 1968), which won the National Book Award, his work took a new turn. The verse was loosely structured and dealt with disturbing episodes from his personal life, including the madness that forced Lowell to spend time in institutions for periods throughout his career. Although it was labelled 'confessional' poetry at the time and extremely influential (for example, on the work of Sylvia PLATH), scenes that seem shockingly intimate were rendered with controlled art and are not merely autobiographical. *Imitations* (1961), a book of verse translations, was also highly influential. Not strict translations – the language occasionally departs radically from the originals – they are paraphrases intended to function as poems in their own right. Lowell's later work, less intense and innovative than the earlier poetry, includes *For the Union Dead* (1964), *Old Glory* (1965), three plays based on stories by Hawthorne and Melville, *Prometheus Bound* (1969), a prose rendering of Aeschylus, *Notebook* (1969), and *For Lizzie and Harriet* (1973).

Lowry, L(aurence) S(tephen) (1887–1976) *British painter of industrial and urban landscapes.*

Lowry's father was an estate agent in Manchester but in 1909 the family moved to an industrial part of Salford. Lowry had private lessons in painting and studied intermittently at art schools in Manchester and Salford between 1905 and 1925. The personal style that he developed around the end of World War I remained essentially the same throughout his life. Against backgrounds of factories, chimneys, and streets melting into the white haze characteristic of industrial cities in Lancashire, Lowry painted groups of small dark busily moving figures. He also painted a number of deserted landscapes and seascapes.

The popularity of his apparently naive pictures grew following his first London exhibition in 1939, but the importance of his work was underrated until the mid 1960s. Lowry's work has no niche in the development of twentieth-century painting and the diversity of opinion about his artistic stature was highlighted during the large retrospective exhibition of his paintings at the Royal Academy in 1976.

Lowry, (Clarence) Malcolm (1909–57) *British novelist, whose life was a long and unsuccessful battle against alcoholism. Although his most important book was published during his lifetime, much of his work was published posthumously.*

The son of a prosperous cotton broker, Lowry went to the Leys School, Cambridge. He then made a voyage to the China Seas (1927) and visited the American writer Conrad Aiken (1899–1973) in Massachusetts before returning to Cambridge to study English at St Catharine's College. His first novel, *Ultramarine* (1933), was based on his experiences during his 1927 voyage. Even as an undergraduate Lowry had a serious alcohol problem, to which his marriage (1934) to Jan Gabrial did not provide a solution. In 1934 Lowry went to the USA and thence to Mexico (1936). Drinking heavily again, he was deserted by his wife. In 1938, after an interlude in gaol, Lowry left Mexico and went to Los Angeles. Nonetheless he had gathered the material that was to form the body of his most highly acclaimed book, *Under the Volcano* (1947), a symbolic semi-autobiographical novel, set in a Mexican town, that follows the decline of the alcoholic British consul. This was mainly written while Lowry was living in a shanty at Dollarton, British Columbia, with his second wife Margerie Bonner, whom he married in Canada in 1940. A second visit to Mexico (1945–46) furnished him with material for two more novels, which he left unfinished at his death.

In 1955 Lowry and his wife settled at Ripe, Sussex, where Lowry died of barbiturate poisoning. Among his posthumously published works were a collection of short stories entitled *Hear Us O Lord From Heaven Thy Dwelling Place* (1961), *Selected Poems* (1962), *Lunar Caustic* (1963), a novella based on his treatment for alcoholism in New York in the mid-1930s, *Dark as the Grave Wherein my Friend is Laid* (1968), his second Mexican novel completed and edited by his wife and Douglas Day, and *October Ferry to Gabriola* (1970).

Lubitsch, Ernst (1892–1947) *German-born US film director, particularly known for his witty and sophisticated comedies. He was made an Officier de la Légion d'honneur in 1938.*

Lubitsch was born in Berlin and began his career as an actor, first on stage with Max Reinhardt (1873–1943) at the Deutsches Theater and then in films. Early success in his native Germany came in a series of comedy shorts in which he played a small Jewish businessman named Meyer, but it was as a director that he achieved international acclaim, with such films as *Carmen* (1918; *Gypsy Blood*) and the spectacular *Madame Dubarry* (1919; *Passion*). Shortly afterwards, he went to the USA, where he made an inauspicious start with *Rosita* (1923), starring Mary PICKFORD. The silent films that followed, however, were highly regarded, including *The Marriage Circle* (1924) and *Lady Windermere's Fan* (1925). Success continued when he moved on to talkies, beginning with the musical *The Love Parade* (1930). His particular brand of satire was both successful and influential, and many directors tried to imitate the 'Lubitsch touch'. *Trouble in Paradise* (1932), *Ninotchka* (1939), with Greta GARBO, *To Be or Not To Be* (1942), with Jack Benny (1894–1974), and *Heaven Can Wait* (1943) were among his many notable films. In recognition of his contribution to motion pictures he was awarded a Special Oscar in 1937.

Ludendorff, Erich von (1865–1937) *German general and one of the principal military strategists of World War I.*

Ludendorff was commissioned in the infantry in 1883 and joined the general staff in 1894, becoming head of the deployment section in 1908. He held the view that peace was merely the interval between wars and that the nation's prime duty was to provide the means to wage war. He contributed to the Schlieffen Plan for invading France, a crucial element of which was the capture of the Belgian fortress at Liège. His active lobbying for increased military spending irritated the establishment; nevertheless, on the outbreak of World War I, he was appointed quartermaster-in-chief of von Bulow's 2nd Army and personally led the assault and capture of Liège. Called to the eastern front as chief of staff to the elderly General von HINDENBURG, Ludendorff ordered a regrouping of forces that resulted in defeats for the Russians at Tannenburg (August 1914) and Masurian Lakes (September 1914). He remained in the east until August 1916 when

Hindenburg was appointed supreme commander on the western front. With Hindenburg's authority, Ludendorff began to wield immense political as well as military influence. He instituted unrestricted submarine warfare against the Allies and contrived the dismissal of the 'defeatist' chancellor von Bethmann-Hollweg (1856–1921). March 1918 saw the start of the final German offensive in the west, concentrated against the British 5th Army near Amiens. Their initial success was never consolidated and despite continued attacks in the spring and summer, the Allied counterattack (July–September) finally overwhelmed the Germans. Ludendorff suffered a nervous collapse and Emperor William II accepted his resignation on 26 October.

After a brief exile in Sweden, Ludendorff returned to Germany and in 1923 participated in HITLER's attempted putsch. He contested the 1925 presidential election against his former commander, Hindenburg, while sitting as one of Hitler's National Socialist MPs (1924–28). In his later years, Ludendorff espoused eccentric notions of a Germanic 'divine faith' that alienated most of his former colleagues, even Hitler.

Lukacs, Giorgi Szegedy von (1885–1971) *Hungarian philosopher and critic, probably the most original and influential western Marxist of the century.*

Born in Budapest, the son of a wealthy Jewish banker, Lukacs studied philosophy and law at the universities of Berlin and Heidelberg. In 1918 he joined the Communist Party and served briefly in the short-lived government of Béla Kun (1886–?1939). When it collapsed in 1919 Lukacs left Hungary and settled for several years in Germany. He later moved to the Soviet Union but returned to Hungary in 1944 to take up the post of professor of aesthetics at the University of Budapest. In 1951 Lukacs withdrew from public life but returned in 1956 as minister of culture in the NAGY government. After the Russian invasion of Hungary (1956) and the fall of the Nagy government Lukacs was deported to Romania and expelled from the party. He was allowed to return to Hungary shortly afterwards and was readmitted to the party in 1967.

Lukacs's most important work is undoubtedly his *Geschichte und Klassenbewusstsein* (1923; translated as *History and Class Consciousness*, 1971). It is a work that was repudiated by the Comintern and by Lukacs himself on more than one occasion. The essence of Marxism, he declared, lay in the dialectic rather than in any specific set of beliefs. Lukacs also went on to stress the central role played by the concept of alienation in Marxist thought. As a critic, two of Lukacs's most influential works were *Die Theorie des Romans* (1916; translated as *The Theory of the Novel*, 1971) and *Der Historische Roman* (1955; translated as *The Historical Novel*, 1962). His standpoint was unqualifiedly realist, although his notion of realism was far more subtle and demanding than that frequently adopted by Marxist critics. He also produced a number of works on individual writers, among which were studies of Balzac, Thomas MANN, and Goethe. Not surprisingly, Lukacs had little time for such writers as James JOYCE and Franz KAFKA.

Lumumba, Patrice (1925–61) *First prime minister of the Belgian Congo (now Zaïre) (1960). To the Congolese he was a leader who freed them from imperialism, to the Belgians he was a communist and a murderer, to the world at large he was an unstable and dangerous charmer.*

Born in Onalua, a member of the Batetela tribe, Lumumba was educated at Catholic and Protestant schools before joining the colonial civil service to work as an assistant postmaster in Stanleyville (now Kisangani). He was dismissed from this post after conviction for embezzlement in 1956. In 1955 he was elected president of a Congolese trade union in the Orientale Province and he became a director of a brewery in Leopoldville (now Kinshasa) in 1957.

Lumumba's political involvement began as a member of the Belgian Liberal Party in the Congo. Petitioning the government for independence in 1958, he was one of the founders of the Mouvement National Congolais (MNC) and led several mass demonstrations in an effort to establish the MNC as a national party. He was imprisoned on a charge of inciting a riot in 1959, but was released in 1960, gaining at the same time nationwide support for this party in the 1960 elections. A participant in the conference on independence in Brussels in the same year, he emerged then as the only important national leader and was asked to form the first government in the independent Congo. Following the army mutiny five days later and the secession of Katanga province led by TSHOMBE, Lumumba requested UN and later Soviet assistance to oust the Belgians from the region and to bring the country under control. In 1961 he was deposed by the army led by Mobutu and he was forced to seek UN protec-

tion in Leopoldville while trying to re-establish his government. However, he was recaptured and murdered several months later.

Lunt, Alfred Davis (1893–1977) *US actor and director.*

Born in Milwaukee, Lunt began his career with the Castle Square Theatre repertory company, Boston (1912). He made his Broadway debut in 1917 and two years later achieved a notable success in the title role of *Clarence*. However, it was in partnership with Lynn FONTANNE, whom he married in 1922, that he was most enthusiastically received. As members of the Theatre Guild (1924–29) they had their first major success together in MOLNÁR'S *The Guardsman* (1924), the screen version of which they made in 1931. Lunt made his London debut with Fontanne in Vara's *Caprice* at St James's Theatre (1929) and during the thirties they became popular in Britain for their highly sophisticated handling of comedy in such plays as *Amphitryon* (1938). This was something they had already perfected in New York when they worked closely with Noël COWARD, who wrote a number of roles for them. The three of them appeared together in several productions, including the highly successful *Design for Living* (1933). During World War II, they worked in Britain, visiting many hospitals and military installations.

The old Globe Theatre in New York was renamed the Lunt–Fontanne in their honour and they appeared in the opening production of DÜRRENMATT's *The Visit* (1958), which they successfully repeated at the opening of the new Royalty Theatre, London (1960). They also appeared on TV and Lunt directed several plays and opera.

Luthuli, Albert John (c. 1898–1967) *South African nationalist leader. He was awarded the Nobel Peace Prize in 1960 for his commitment to nonviolence as a means of opposing apartheid.*

A Zulu, born in Groutville, Natal, Luthuli was educated at mission schools and at Adams College, where he qualified as a teacher. Teaching for fifteen years in a mission college, he assumed an inherited Zulu chieftaincy in 1935 and served on the Native Representative Council until it was abolished in 1946. Joining the African National Congress (ANC) the same year, he became president-general in 1952 (a position he held until his death) but was dismissed from his chieftaincy because of his political activities during the Defiance Campaign (civil disobedience campaign). In 1952, he was restricted to his rural home by the government and in 1956 was among the Africans arrested and charged in the notorious treason trial. Released in 1957, he was permanently banned in 1959. Luthuli was elected rector of Glasgow University in 1962 but was not permitted to attend the installation. He was killed by a train while crossing the tracks near his home in 1967.

Lutyens, Sir Edwin Landseer (1869–1944) *British architect, often regarded as the last great designer of traditional buildings. He is known for his plans for New Delhi (especially the Viceroy's House), the Cenotaph in London, and the Roman Catholic Cathedral in Liverpool. He was knighted in 1918 and awarded the OM in 1942.*

Born in London, Lutyens was a self-taught architect who after a brief apprenticeship set up in practice on his own in 1889. His designs for Munstead Wood (1896) for the garden designer Gertrude Jekyll helped him to become established very rapidly. Many Surrey houses in the arts-and-crafts style were created by Lutyens in the years before World War I. His range also extended to the grander palladian Heathcote (1906) in Yorkshire and the neogothic Castle Drogo (1910) in Devonshire. In 1912 he was selected as adviser on the plans for New Delhi, where he introduced an open garden-city layout. His Viceroy's House succeeds in combining the dominant features of classical architecture with decoration in the Indian idiom. After World War I he was much in demand as a designer of memorials to the fallen. The Cenotaph (1919) in Whitehall is the best known in the UK and there were several in France. Other buildings designed by him include the head office of the Midland Bank in the City (1924) and the British Embassy in Washington (1927). In 1929 he designed a Roman Catholic cathedral for Liverpool, but only the crypt and sacristy were built to his plan before work stopped owing to World War II: the cathedral was subsequently completed to a different design.

Luxemburg, Rosa (1871–1919) *Polish-born German revolutionary leader.*

Born in Zamosc in Russian Poland, the daughter of a timber merchant, Luxemburg attended high school in Warsaw before emigrating to Switzerland in 1889, where she studied law and political economy at the University of Zürich. Gaining a doctorate in 1898, she became active in the international socialist movement, founding the Polish Social Democratic Party, a forerunner of the Polish Communist Party. In 1898 she obtained German citizen-

ship through marriage and settled in Berlin, where she joined the German Social Democratic Party and participated in political activities of the extreme left.

Between 1907 and 1914 Luxemburg taught at the Social Democratic Party school in Berlin. Imprisoned in 1915 for opposing World War I, she founded with Karl Liebknecht (1871–1919) the Spartacus League, a revolutionary group dedicated to ending the war and establishing a proletarian government. In 1918 she was released from prison and co-founded the German Communist Party. The following year she and Liebknecht were assassinated by troops in Berlin after she had organized a revolt by the Spartacists.

Known as 'Red Rosa', Luxemburg was an impressive speaker and incisive political writer. She stressed democracy and mass action as a means to achieve socialism in many of her publications but also wrote about imperialism in the developing world in *Die Akkumulation des Kapitals* (1913). She is regarded by many socialists as a martyr and heroine.

Lyons, Joseph Aloysius (1879–1939) *Australian statesman and United Australian Party prime minister (1931–39). He was appointed a CH in 1936.*

Born near Stanley, Tasmania, the son of a farmer, Lyons qualified as a teacher at the age of seventeen. He began his political career in 1909, when he was elected as the Labor member for Wilmot in the Tasmanian House of Assembly. He became leader of the party in 1916 and was premier of Tasmania from 1923 until 1928. In 1929 Lyons was elected for the Wilmot seat in the federal House of Representatives. Resigning from the cabinet over financial policy in 1931, he joined forces with the opposition to form the United Australia Party and was elected as leader. He became prime minister on Labor's defeat in the same year and remained so until his death in 1939.

Seen as a temporary prime minister, Lyons proved to be a sound leader with considerable administrative ability. His wife, Dame Enid Lyons, was the first woman member of the House of Representatives and of the federal cabinet. Lyons was given an honorary LLD degree by Cambridge University in 1937.

Lysenko, Trofim Denisovich (1898–1976) *Soviet biologist whose unorthodox ideas, notably his antagonism towards widely accepted genetic principles, profoundly influenced Soviet biology for over two decades.*

Born in Karlovka, of a peasant family, Lysenko graduated from the Uman School of Horticulture in 1921 and received his doctorate in agricultural science from the Kiev Agricultural Institute in 1925. Following work at Gandzha Experimental Station, he joined the All-Union Genetics Institute at Odessa in 1929 and began a series of experiments to investigate the technique of vernalization, the artificial cold treatment of partly germinated seeds to accelerate their development. Lysenko claimed that the changes in the seeds caused by vernalization, i e environmentally induced changes, were inherited by the next generation, thereby reviving Lamarck's long-discredited doctrine of the inheritance of acquired characteristics and contradicting the generally accepted theories of Mendel, T. H. MORGAN, and other geneticists.

This belief influenced much of Lysenko's later work and conflicted with the conventional views of other Soviet biologists, especially the eminent director of the Institute of Genetics of the USSR Academy of Sciences, N. I. VAVILOV. However, Lysenko's ideas supported Stalinist dogma and with STALIN's backing his influence grew until in 1940 he succeeded Vavilov, who was subsequently arrested and exiled. Under Lysenko's instruc-tions, Soviet biology was overturned, adherents to western 'bourgeois' science dismissed, and the textbooks rewritten in accordance with Lysenkoism. In the 1950s, after Stalin's death, his power waned but it was not until KHRUSHCHEV's death in 1964 that Lysenko was finally stripped of his authority and Soviet biology allowed to return to the mainstream of western science.

Lyttelton, Humphrey (1921–   ) *British jazz trumpeter, bandleader, broadcaster, and author.*

After leaving Eton, Lyttelton joined George Webb's Dixielanders and formed his own band in 1948, playing in traditional jazz style. His own composition, 'Bad Penny Blues', was a British top twenty hit in 1956. He first appeared in the USA in 1959, and has often toured Europe. In the 1950s he evolved a mainstream style influenced by Duke ELLINGTON and has remained an effective catalyst for much fine jazz playing.

He is a well-known broadcaster, on both radio and television, and has published several books, including *I Play As I Please* (1954), *Second Chorus* (1958), *Take it From the Top* (1975), and histories of jazz.

M

MacArthur, Douglas (1880–1964) *US general who commanded US forces in the Pacific during World War II and UN forces in the Korean War.*

Son of the Civil War hero and governor of the Philippines, Arthur MacArthur, Douglas graduated from West Point military academy in 1903. He continued the family link with the Philippines throughout his military career besides serving as aide de camp to president Theodore Roosevelt (1906–07), chief of staff of the US 42nd Division in France and the Rhinelands (1917–19), superintendent at West Point after the war, and US army chief of staff (1930–35). He retired from the army in 1937 with the rank of general to continue as military adviser to the Philippines authorities.

In 1941, he was recalled by the USA to command US and Filipino troops defending the Philippines against the Japanese. MacArthur's troops fought a stubborn strategic withdrawal, holding out in Bataan and Corregidor until May 1942. Meanwhile, MacArthur left for Australia in February with the famous pledge: 'I came through and I shall return'. Now supreme Allied commander in the southwest Pacific, he planned the Pacific offensive launched in autumn 1942 following the massive naval and air battle at Midway Island in June. Skilfully coordinating air cover, naval bombardment, and amphibious landings, MacArthur embarked on a strategy of island hopping, capturing the smaller islands and bypassing those that were larger and more heavily defended. By the spring of 1944, northern New Guinea and the Solomon Islands had been captured from the Japanese. In October, MacArthur commenced the reconquest of the Philippines and on 4 February 1945, US troops entered Manila. MacArthur's command was extended to all US army forces in the Pacific and soon southern Philippines and Borneo were liberated. It was MacArthur who accepted the Japanese surrender on 2 September 1945 following the explosion of US atomic bombs over Hiroshima and Nagasaki on August 6 and 9.

MacArthur headed the occupation forces in Japan after the war until June 1950, when he was appointed commander of UN troops in South Korea following the insurgence of Northern Korean troops across the border. He masterminded the amphibious landings at Inchon, behind the enemy lines, thereby retrieving the situation. However, in November, North Koreans supported by Chinese communists again pushed south over the border. MacArthur's desire to extend the war by bombing mainland China and blockading Chinese ports led to his dismissal by President TRUMAN.

Seemingly aloof and arrogant, he was nevertheless an intelligent and gifted strategist and a dedicated soldier; his personal staff testified to a warmth and courtesy kept hidden from public view.

Macartney, Charles George (1886–1958) *Australian cricketer of great ability, rated with TRUMPER and BRADMAN as one of Australia's greatest batsmen. He played 35 times for his country, and in 1921 scored 345 in less than four hours against Nottinghamshire.*

Born in Maitland, New South Wales, he played for New South Wales from 1905–06 to 1926–27 and for his country from 1909 to 1926–27. He toured England four times, the last tour in 1926. At home he was known as the 'Governor General'; he made his own rules at the wicket. His dislike for slow scoring made him bat, at times, with a cavalier disregard for the bowlers.

Macaulay, Dame (Emilie) Rose (1881–1958) *British writer, whose works display her skill as a social satirist and reflect her Christian faith. She was created a DBE in 1958.*

Rose Macaulay was the daughter of a master at Rugby School. During her childhood she lived with her family for some years in Italy, which she loved. On returning to England she went to Oxford High School and then to Somerville College, Oxford (1900–03). She soon began to publish novels, beginning with *Abbots Verney* (1906), and collections of poetry. *What Not* (1918) showed a new satirical strain in her fiction, and she followed this up with a number of highly popular novels that readily appealed to the 1920s novel-reading public: they included *Potterism* (1920), *Dangerous Ages* (1921), *Told by an Idiot* (1923), *Crewe Train* (1926), and *Keeping Up Appearances* (1928). While working at the Ministry of Information in 1918 she fell in love with a married ex-priest, Gerald O'Donovan, with whom she carried on a secret affair until his death in 1942. This undoubtedly influenced her subsequent work, which took a more serious turn in the 1930s with her only historical novel, *They Were Defeated* (1932), set in the English Civil War period. She also wrote a biography of Milton (1934) and in 1938 published *The Writings of E. M. Forster*. During World War II she worked as an ambulance driver in London, and it was some time before she returned to fiction writing with *The World My Wilderness* (1950). Her last and possibly best novel, the semi-autobiographical *The Towers of Trebizond* (1956), reflects Macaulay's love and mourning for O'Donovan, her guilt over their affair, and her spiritual development during this time. *Letters to a Friend* (1961), her posthumously published correspondence with an Anglican priest, Father Hamilton Johnson, records her feelings about the affair and her gradual return to the Anglican Church (from which she had been estranged for many years). Throughout her life she wrote prolifically for periodicals and she published several volumes of essays and also some travel books – *They Went to Portugal* (1946) and *Pleasure of Ruins* (1953).

MacDiarmid, Hugh (Christopher Murray Grieve; 1892–1978) *Scottish poet and writer.*

The son of a postman, Hugh MacDiarmid was born in Langholm and attended Langholm Academy before going on to Edinburgh University. He served in the army during World War I and then became a journalist in Montrose, Angus. There he founded and edited *Northern Numbers* (1921–23), an anthology of Scottish verse, and in 1922 he founded the monthly *Scottish Chapbook*, which promoted the Scottish poetic renaissance of which MacDiarmid himself was a leading figure. His lyrics in *Sangschaw* (1925) and *Penny Wheep* (1926) proved that the Scots language was capable of being rescued from the dialectical quaintness to which it had succumbed since Burns's heyday. His extended philosophical poem *A Drunk Man Looks at the Thistle* (1926) attracted considerable attention.

MacDiarmid was a moving spirit behind the formation of the Scottish Nationalist Party (1927). He was also an active Marxist, publishing his *First Hymn to Lenin* in 1931. In the 1930s he became increasingly dissatisfied with the limitations of the Scots tongue as a medium for the expression of twentieth-century scientific and philosophical concepts; he used archaic Scots in *Scots Unbound* (1932), but

subsequently employed his own idiosyncratic brand of English in such poems as those in *Stony Limits* (1934) and *Second Hymn to Lenin* (1935). In 1934 he collaborated with Lewis Grassic GIBBON on *Scottish Scene* and his later publications include several works on Scotland, as well as the anthology *The Golden Treasury of Scottish Poetry* (1940). His autobiography, *Lucky Poet*, was published in 1943. *Collected Poems 1920–61* (1962) was brought out in a revised edition in 1967 and *More Collected Poems* followed in 1970.

MacDonald, (James) Ramsay (1866–1937) *British statesman and the first Labour prime minister (1924; 1929–31; 1931–35).*

MacDonald was born in Scotland, the illegitimate son of a maidservant, and became a journalist. He joined the Independent Labour Party in 1894 and was the first secretary of the Labour Representation Committee (1900–05) and then (1906–12) of its successor the Labour Party. Elected to parliament in 1906, he led the parliamentary Labour Party from 1911 until resigning in 1914 in protest against World War I. He lost his seat in 1918 but was re-elected in 1922. He became the first Labour prime minister in January 1924; however, his minority government lasted only until November, when the Liberals withdrew their support. Aiming to prove that Labour was a realistic alternative to the Liberals and Conservatives, MacDonald's government did no more domestically than introduce Wheatley's Housing Act, which assisted the building of council houses, and some reforms in secondary education. MacDonald's main interest was in foreign affairs and, acting as his own foreign secretary, he was instrumental in obtaining acceptance of the DAWES Plan to enable Germany to pay war reparations. At the League of Nations he made a powerful plea for collective security and disarmament, to be achieved by means of the Geneva Protocol, a document that was later rejected.

In his second government (1929–31), MacDonald again concentrated on foreign affairs, at the expense of Britain's worsening economic difficulties. His proposal, in response to the international financial crisis in the summer of 1931, to make cuts in unemployment benefits was rejected by Labour's General Council, and MacDonald resigned. The next day he agreed to the king's request that he form a government with the Liberals and the Conservatives, for which he was denounced as a betrayer of the Labour movement and expelled from the party. He remained prime minister of the coalition national government until succeeded by Stanley BALDWIN in 1935. He died at sea, on his way to seek a health cure in South America.

Mackenzie, Sir (Edward Montague) Compton (1883–1972) *British novelist. He was knighted in 1952 and received many other civil and military awards.*

Mackenzie was born at West Hartlepool, where his actor parents were on tour. He was educated at St Paul's School, London, and then read history at Magdalen College, Oxford (1901–04). His earliest publications were *Poems* (1907) and a period novel, *The Passionate Elopement* (1911). *Carnival* (1912), drawing upon his theatrical background, and the semiautobiographical *Sinister Street* (1913–14), relating to his time at St Paul's and at Oxford, established him as a novelist. In 1914 he converted to Roman Catholicism.

During World War I Mackenzie joined the Royal Marines, served on the Dardanelles Expedition (1915), and, after being wounded, worked in intelligence in Greece. These experiences furnished material for a number of novels and memoirs, including *Greek Memories* (1932), which was prosecuted under the Official Secrets Act. In revenge he satirized the intelligence service in *Water on the Brain* (1933). His most successful fictional vein was light comedy, beginning with *Poor Relations* (1919). He also wrote a trilogy on clerical life, comprising *The Altar Steps* (1922), *The Parson's Progress* (1923), and *The Heavenly Ladder* (1924), and a partly autobiographical sextet entitled *The Four Winds of Love* (1937–45).

In the early 1930s Mackenzie strongly associated himself with the Scottish nationalist cause and made his home on Barra in the Outer Hebrides. Scottish life inspired a number of his works, including the famous comic novel *Whisky Galore* (1947; filmed 1948). Among his nonfiction were biographies of ROOSEVELT (1943) and BENEŠ (1946). At the invitation of the Indian government he visited the battlefields on which the Indian army had fought in the period 1939–47 and wrote his account of its exploits in *Eastern Epic* (1951). In later life Mackenzie divided his time between Scotland and France. He wrote about his interests (music and cats) and between 1963 and 1971 published a ten-volume autobiography, *My Life and Times*.

Mackintosh, Charles Rennie (1868–1928) *British architect, who is considered an important figure in the evolution of modern architec-*ture. *One of the exponents of art nouveau, he established a severe version of the style; nearly all his best buildings are in Glasgow.*

Born in Glasgow, Mackintosh trained as an architect as an evening student at the Glasgow School of Art, after which he was articled to a Glaswegian architect, John Hutchinson. In 1896 he won a competition for the new building of the Glasgow School of Art (1898–1909), which was widely acclaimed when the building was completed. For Catherine Cranston he designed four unusual tearooms (1897–1912), including the decor and the furniture. These were much admired by Viennese designers, who invited him to submit a design for the interior of a flat for an exhibition in Vienna in 1900. This too attracted considerable attention abroad; at home, however, he was considered difficult and unreliable. In 1923 he moved south, first to Walberswick in Suffolk, then to London. Apart from a brief spell in France, he remained in London until he died. However, he never succeeded in establishing an architectural practice in London.

Maclean, Donald (1913–83) *British Foreign Office official who, following his defection to the Soviet Union in 1951, was revealed to be a Soviet agent.*

The son of the former Liberal cabinet minister Sir Donald Maclean, Maclean went up to Cambridge in the early 1930s, where he met Kim PHILBY, Guy Burgess (1911–63), and Anthony Blunt (1907–83), all of whom held ardent procommunist views. Maclean joined the Foreign Office and was posted in 1938 to the British Embassy in Paris, where he passed intelligence to the Soviets. Following a spell in London, he was promoted to first secretary to the British ambassador in Washington in 1944. While maintaining the appearance of a hard-working conscientious career diplomat, Maclean made regular trips to his Soviet contact in New York, conveying information, especially Anglo-US strategy relating to the development of the atomic bomb. However, in the late 1940s, US cipher experts had cracked Soviet intelligence codes and were trying to identify the Soviet agent operating inside the British Embassy. Maclean returned to London in 1948 where his old acquaintance, Guy Burgess, kept him informed of progress in the security investigation. When he was posted to Cairo as head of chancery at the British Embassy, Maclean's drinking led to several embarrassing episodes and he was recalled on medical grounds. He resumed work at the Foreign Office in November 1950, where he recommenced his espionage activities. On 25 May 1951, Burgess, who had earlier returned from Washington at the insistence of Philby, got wind of impending proceedings against the long-suspected Maclean. They promptly caught a cross-Channel ferry and made their way to the Soviet Union. Adapting to his new life in Moscow, Maclean worked at the Foreign Ministry before taking on the British desk at the Institute of World Economic and International Relations.

Mac Liammóir, Micheál (1899–1978) *Irish actor and dramatist, co-founder of the Gate Theatre in Dublin.*

Mac Liammóir was born in Cork and, under the stage name of Alfred Willmore, made his London debut in 1911 in the title role of *Oliver Twist*. Having worked under Sir Herbert Beerbohm Tree (1853–1917) at His Majesty's Theatre, playing such roles as the son of Macduff in *Macbeth* and John Darling in *Peter Pan*, he travelled abroad and studied art at the Slade before returning to Ireland to tour with his brother-in-law Anew McMaster's Shakespeare Company.

In 1928 Mac Liammóir co-founded the Dublin Gate Theatre and the Galway Gaelic Theatre with Hilton Edwards (1903–82) and established himself as one of Ireland's leading actors. As well as playing a wide variety of Shakespearean and non-Shakespearean roles, Mac Liammóir staged a number of one-man shows, most notably *The Importance of Being Oscar* (1960–61). He wrote in both English and Gaelic, translating many of the world's classics into Irish; his own plays included *Ill Met By Moonlight* (1957) and *Where Stars Walk* (1961).

MacMillan, Sir Kenneth (1929–) *British choreographer. Principal choreographer to the Royal Ballet (since 1977), he was knighted in 1983.*

MacMillan began his career as a dancer, studying at the Sadler's Wells Ballet School and becoming a member of the Sadler's Wells Theatre Ballet in 1946. Two years later he joined the Sadler's Wells Ballet, which became the Royal Ballet in 1956, and came under the influence of the principal choreographer, Frederick ASHTON. In 1953 MacMillan returned to the Sadler's Wells Theatre Ballet and choreographed his first works, which included *Danses Concertantes* (1955). In succeeding years he choreographed for both Sadler's Wells companies and his successes of the 1960s include *Le Baiser de la fée* (1960), *The Invitation* (1960), *Le Sacre du printemps*

(1962), *Las Hermanas* (1963), and *Das Lied von der Erde* (1965).

From 1966 to 1969 MacMillan directed the Deutsche Oper Ballet in West Berlin. He created a new one-act ballet, *Anastasia* (1967), for the ballerina Lynn Seymour (1939–) as well as interpreting such classics as *The Sleeping Beauty* (1967) and *Swan Lake* (1969). In 1970 he succeeded Frederick Ashton as director of the Royal Ballet; he gave up this post in 1977 in order to concentrate upon his role as principal choreographer for the company. Since then MacMillan has continued to choreograph ballets, combining an exuberance reminiscent of Ashton with a more melancholy atmosphere in a number of compact dramatic pieces for the Ballet Rambert, theatre, television, and cinema as well as for his own company. Since 1984 he has also been an Artistic Associate of the American Ballet Theatre. Works in recent years have included the full-length narrative ballet *Romeo and Juliet* (1967), an expanded version of *Anastasia* (1971), *Manon* (1974), *Mayerling* (1978), *Isadora* (1981), *Dance of Death* (1983), and *The Prince of the Pagodas* (1989); he has also directed IONESCO's plays.

Macmillan, (Maurice) Harold, Earl of Stockton (1894–1986) *British statesman and Conservative prime minister (1957–63). He was awarded the OM in 1976 and accepted an earldom in 1984.*

Macmillan was educated at Eton and Balliol College, Oxford. After service in World War I, he worked in the publishing house founded by his grandfather Daniel Macmillan. An MP from 1924 to 1929 and from 1931 to 1964, he opposed Neville CHAMBERLAIN's policy of appeasement in the 1930s and did not hold office until World War II. He was then parliamentary secretary to the Ministry of Supply (1940–42), colonial undersecretary (1942), the government's representative in North Africa (1942–45), and finally secretary of state for air (1945). In CHURCHILL's postwar government, Macmillan was a very successful housing minister (1951–54), subsequently serving as minister of defence (1954–55) and foreign secretary (1955). As chancellor of the exchequer (1955–57), he introduced premium bonds (1956).

Succeeding Sir Anthony EDEN as prime minister in 1957, Macmillan restored good relations with the USA (damaged by Eden's Suez policy). He also visited the Soviet leader KHRUSHCHEV (1959), hoping to serve as a moderating influence between East and West. In 1963 he played an important part in the negotiation of the Nuclear Test Ban Treaty. Macmillan advocated the granting of independence to British colonies, notably in his famous 'wind of change' speech in Cape Town in 1958. At home, though the government won the 1959 election on the slogan 'You've never had it so good', its attempts to deal with inflation were both unpopular (especially the 1961 'wages pause') and ineffective. Macmillan's dismissal of seven cabinet ministers in 1962 earned him the sobriquet 'Mac the Knife'. His attempts to take Britain into the European Economic Community were frustrated in 1963 by DE GAULLE, who was angered by the Nassau agreement in which the USA agreed to supply nuclear missiles for British submarines. The failure in Europe, together with the scandal of the war secretary John Profumo's affair with Christine Keeler, simultaneously the mistress of a Soviet naval attaché, undermined the government and Macmillan resigned during a period of ill health.

Chancellor of Oxford University from 1960 until his death, Macmillan wrote several volumes of memoirs.

MacNeice, (Frederick) Louis (1907–63) *British poet. He was created a CBE in 1958.*

MacNeice was born in Belfast, where his father (later a bishop) was rector of Holy Trinity church. His mother's death (1914) cast a shadow on his childhood, but he was happy at Sherborne preparatory school and later at Marlborough (1921–26), where he made several lifelong friends, among them John BETJEMAN. At Merton College, Oxford, his friends included the poets AUDEN, SPENDER, and C. DAY LEWIS. He also enjoyed academic success and published his first poetic collection, *Blind Fireworks* (1929). In the same year he married and took up a lectureship in classics at Birmingham, where he wrote *Eclogue for Christmas* (1933) and *Poems* (1935). After his wife left him (1935) he took a post as lecturer in Greek at Bedford College, London. But before this he had travelled to Spain and with Auden to Iceland, an experience they recounted in *Letters from Iceland* (1937). In London MacNeice became increasingly interested in left-wing politics, but he continued to write poetry – *The Earth Compels* (1938) and *Autumn Journal* (1939) – as well as several prose works.

In 1941, after a period of restlessness involving visits to Spain and the USA, MacNeice joined the BBC, where he remained for twenty happy and productive years. He wrote a critical study of YEATS (1941), radio drama, notably the fantasy *The Dark Tower* (1947),

inspired by World War II, and several more volumes of poetry. He also married again (1942). *Collected Poems 1925–1948* appeared in 1949. During the 1950s MacNeice's work caused him to travel all over the world and his poetry appeared less regularly: *Ten Burnt Offerings* (1952), *Autumn Sequel* (1954), and *Visitations* (1957). In 1960 he and his second wife parted. The following year MacNeice resigned from the BBC to concentrate on his own work, the first fruits of this being *Solstices* (1961). In summer 1963 while supervising the sound engineers making a BBC programme called *Persons from Porlock*, he caught a chill working underground and died from viral pneumonia. His autobiography, *The Strings Are False*, was published in 1965.

Madariaga, Salvador de (1886–1978) *Spanish writer and diplomat, whose chief works are political essays that reflect his liberal and humanistic values and his commitment to internationalism.*

Born at La Coruña, Madariaga studied engineering at the École Polytechnique in Paris but made his career in politics and diplomacy. Fluent in French and English as well as Spanish, Madariaga moved to London in 1916, working for several years as a journalist before joining the secretariat of the League of Nations in 1921; he was director of the disarmament section until 1927. For three years he held the chair of Spanish studies at Oxford, then served as ambassador of the Spanish Republic to the USA (1931) and France (1932–34). After the civil war, he continued in exile in England until 1976, when he returned to Spain.

His political essays, all translated by himself, include *Disarmament* (1929), *Anarquía o jerarquía: ideario para la constitución de la tercera república* (1935; translated as *Anarchy or Hierarchy*, 1937), *Bosquejo de Europa* (1951; translated as *Portrait of Europe*, 1952), *The Anatomy of Cold War* (1955), and *Memorias 1921–36: Amanecer sin mediodía* (1974; translated as *Morning Without Noon: Memoirs*, 1974). He also published two volumes of poetry, several plays, and a number of novels, among them *La jirafa sagrada* (1924; translated as *The Sacred Giraffe*, 1925), *Ramo de errores* (1952; translated as *A Bunch of Errors*, 1954), and *Sanco Panco* (1964). His creative work, however, is overshadowed by his critical and literary essays, which include *The Genius of Spain* (1923), a guide to *Don Quixote* (1934), and lives of Columbus (1939), Cortés (1941), and Bolívar (1951).

Madonna (Madonna Louise Veronica Ciccone; 1958–   ) *US singer and film actress, who has sold more records worldwide than any other female performer.*

Born into an Italian family in Rochester, Michigan, Madonna Ciccone studied performing arts at the University of Michigan and dance at the Alvin Ailey studios in New York. From 1979 she worked as a dancer in Paris and played in New York rock bands. Her first recordings as a singer, made in 1982, enjoyed some success in US dance clubs; the following year 'Holiday' became a national hit and she released her first album.

Although her records had a certain infectious appeal, Madonna's rise to stardom in 1984–85 owed more to her skilful use of video, which allowed her to exploit her striking looks, talent as a dancer, and ability to project a strong identifiable image. In 1985 the album *Like a Virgin* became a million seller in both the USA and Europe, spawning no less than five best-selling singles. The same year saw her first major screen role in the comedy *Desperately Seeking Susan*; subsequent films, which have been less well received, include *Who's That Girl* (1987) and *Dick Tracy* (1990). Her success as a singer has continued with such albums as *True Blue* (1986), which topped the charts in twenty-eight countries, and *Like a Prayer* (1989); recent hit singles include 'Vogue' (1990) and 'Justify My Love' (1991). An astute self-publicist, Madonna has caused controversy with recent videos and stage performances, which have provoked charges of blasphemy and indecency. Her knowing, sometimes ironic, exploitation of her status as a sex symbol has provoked much discussion about her significance as a role model for women in the 1980s and 1990s. A deal in 1991 made her the highest-paid entertainer of all time.

Maeterlinck, Maurice Polydore-Marie-Bernard (1862–1949) *Belgian dramatist, poet, and essayist, many of whose works reflect his interest in mysticism. He was awarded the Nobel Prize for Literature in 1911.*

Maeterlinck was born in Ghent and studied law at university there. In 1885 he travelled to Paris, where he became involved with leading members of the symbolist movement; after 1890 he spent most of his life in France. His first collection of poems, *Serres chaudes*, was published in 1889; this was followed in 1896 by *Douze Chansons*.

Maeterlinck's tendency towards mysticism, later examined in the essay *La Sagesse et la* *destinée* (1898), first became evident in his early plays, notably *La Princesse Maleine* (1889) and the immensely popular *Pelléas et Mélisande* (1892), on which Debussy's opera (1902) was based. Subsequent plays included *Monna Vanna* (1902), a historical drama; *L'Oiseau bleu* (1908; translated as *The Blue Bird*, 1909), set in a fairy world of fantasy; *Le Bourgmestre de Stilmonde* (1918), a patriotic drama of World War I; and *La Puissance des morts* (1926), in which Maeterlinck developed the idea of death, another recurrent theme in his works.

In 1901 he produced the first of a series of philosophical prose works on natural history subjects, *La Vie des abeilles*. This was followed some years later by *La Vie des termites* (1926), *La Vie des fourmis* (1930), and *L'Araignée de verre* (1932). He also published several essay collections, notably *Le Double Jardin* (1904), *L'Intelligence des fleurs* (1907), and *La Mort* (1913). In his later years Maeterlinck produced few works of note. He went to the USA in 1940 and lived there for seven years, returning to France two years before his death.

Magritte, Réne-François-Ghislain (1898–1967) *Belgian surrealist painter.*

Magritte's career as a painter began early with private lessons at the age of ten. When he was fourteen he experienced the tragedy of his mother's suicide. At eighteen he was a student in Brussels, where he came under the influence of futurism and the metaphysical painter Giorgio de Chirico, whose work he first saw in 1922. In the same year he married Georgette Berger, who modelled for a number of his later pictures, including one of his first surrealist paintings, *The Two Sisters*, painted in 1925. Before 1925 he had made his living from commercial work but from this point he began to paint full time.

After an unsuccessful first exhibition of surrealist paintings in 1927, he left for France, where he made personal contact with the Paris surrealist group. Magritte painted prolifically and soon began to exhibit widely in Europe and in the USA. He worked in a style that has been described as 'magic realism' because of the highly realistic manner in which it depicted impossible disturbing juxtapositions of familiar objects in incongruous settings; for example, in *Threatening Weather* (1928) a chair, a table, and a torso hover like clouds over the sea. Many of his pictures explored ambiguities, such as that between picture and landscape in *Human Condition* (1935). In the

1950s and 1960s he received a number of commissions for murals. His clear simple style and his exploration of surrealistic ideas and images remained unchanged apart from brief experiments with other styles during World War II and in 1948. Magritte's influence among his contemporaries as well as on younger artists, such as Claes OLDENBURG, was considerable.

Mahfouz, Naguib (1911–) *Egyptian novelist and writer; the first Arabic writer to receive the Nobel Prize for Literature (1988).*
Educated at Cairo University, Mahfouz followed his father into the Egyptian civil service in 1934, serving in the department of arts and censorship and becoming director of the state cinema organization. His early novels, published in the 1940s, were historical works set in ancient Egypt but he soon turned to contemporary themes. In 1956–57 he produced his most celebrated work, the Cairo Trilogy of novels, comprising *Palace Walk*, *The Palace of Desire*, and *The Sugar Bowl*. The books, which cover a time span from World War I to the 1950s, describe the interlocking destiny of several Cairo families against a background of momentous historical and social change and have won their author comparisons with Galsworthy, Proust, and Tolstoy.

In 1959 Mahfouz published his most controversial work, *The Children of Gebelawi*; owing to its treatment of Mohammed and other religious figures the book was banned by the Muslim authorities in Egypt for many years. His later novels include *The Thief and the Dogs* (1961), *The Beggar* (1965), amd *Miramar* (1967). He has also produced numerous short stories and screenplays.

Mailer, Norman (1923–) *US novelist, journalist, and essayist whose life and work have provoked considerable interest in recent years because of their preoccupation with violence.*
Mailer was born in New Jersey and grew up in New York. He graduated from Harvard in 1943 and spent two years in the Pacific with the US army during World War II. During his career as a writer, he has assumed several roles, including commentator on US politics, poet, and film writer and producer.

His first novel was the highly successful *The Naked and the Dead* (1948), drawing on his own war experiences; it was followed by *Barbary Shore* (1951), his Hollywood novel *The Deer Park* (1955), and a series of existentialist essays, *Advertisements for Myself* (1959). A recurrent theme of Mailer's work is war and the effects of violence on the relationships be-

tween younger and older men. In his early writing, loyalty and comradeship provided a positive view of humanity, but these relationships are depicted with increasing pessimism in such later books as *Why Are We in Vietnam?* (1967). Equally apparent is a consistent strain of social criticism that centres on the inevitable sacrifice of personal integrity in the pursuit of material success, notably in the black comedy *An American Dream* (1965). *The Armies of the Night* (1968), which won the 1969 Pulitzer Prize, is a personal account of the 1967 peace march on the Pentagon. His later books include *Of a Fire on the Moon* (1970), concerning the Apollo II moon landing; *The Prisoner of Sex* (1971), his response to the women's liberation movement; *Marilyn* (1973), on the life and death of Marilyn MONROE; *The Executioner's Song* (1979), a fictionalized account of the execution of a real-life murderer that won him a second Pulitzer Prize (1980); *Ancient Evenings* (1983); *Tough Guy Don't Dance* (1984; filmed 1988), and *Harlot's Ghost* (1991).

Major, John (Roy) (1943–) *British Conservative politician; prime minister from 1990.*
The son of a music-hall performer turned designer of garden ornaments, Major was born and brought up in south London. After the collapse of the family business the Majors lived in some poverty; John left school at sixteen and worked in labouring jobs before starting a banking career in 1963. His rise to a senior position at the Standard Chartered Bank coincided with a growing interest in politics and he served in local government from the late 1960s.

Elected to parliament in 1979, he became minister of state for social security (1986–87) and chief secretary to the Treasury (1987–89). In 1989 he was unexpectedly promoted to foreign secretary, a position he held for only three months before being moved to the Exchequer on the resignation of Nigel Lawson (1932–). As Chancellor, Major followed a tough policy of maintaining high interest rates as a measure to curb inflation. Following the enforced resignation of Margaret THATCHER in November 1990, he emerged as the candidate best able to maintain party unity and, with Mrs Thatcher's backing, was chosen as her successor.

The youngest prime minister of the century, Major had been in the Cabinet for only three years and was little known by the public. Although his diffident and apparently modest style led some detractors to label him as 'grey', it made a favourable impression in the country; this was enhanced by his calm handling of the

Gulf War with Iraq (1991) only weeks into his premiership and his competent hosting of the G7 conference of leading industrial nations (1991). Although Major signalled a break with Thatcherite policies by abolishing the hated poll tax, taking a more conciliatory line on Europe, and announcing the Citizen's Charter to protect the rights of individuals, his party remained in trouble over the economy and planned reforms of the NHS.

Makarios III (Mikhail Khristodolou Mouskos; 1913–77) *President of the Republic of Cyprus (1959–77). A well-known figure on the international scene in the 1950s, he combined the roles of benevolent orthodox patriarch with that of cunning political leader.*

Born in Pano Panayia, the son of a peasant, Makarios was educated at a monastery before attending the University of Athens, where he studied divinity. In 1946 he was ordained as a priest, shortly before beginning studies in sociology and theology at the University of Boston in Massachusetts, USA. Elected a bishop two years later, he became archbishop of the Orthodox Church in Cyprus in 1950. In this role he assumed the mantle of national and spiritual leader of his people.

Makarios became involved in politics as a supporter of EOKA, the movement for *enosis*, the union of Cyprus with Greece. Opposed to both British proposals for Cypriot independence within the Commonwealth and Turkish pressures for partition, he secured Greek support for *enosis* in 1954. He was arrested in 1956 by the British on suspicion of terrorist activities and exiled to the Seychelles until 1957, when he was permitted to move to Athens (but banned from Cyprus). When, in 1959, he gave up the quest for *enosis*, EOKA was disbanded and Cyprus was granted independence; he was then elected president. Although he was criticized for his change of policy, he then worked to achieve the integration of the two communities. Despite several assassination attempts by *enosis* agitators and continual conflict between Greeks and Turks on the island, he remained in power until 1974. In July of that year an attempted coup by the Greek Cypriot National Guard forced him to flee to Malta and then to London. Shortly after, Turkey invaded Cyprus and proclaimed a separate state for Turkish Cypriots in the north. On his return to Cyprus five months later Makarios faced difficulties in reasserting Greek dominance, but was determined to oppose partition. He died of a heart attack in 1977.

Malamud, Bernard (1914–86) *US novelist and short-story writer, one of the leading Jewish writers to emerge in America in the 1950s.*

Malamud was born in Brooklyn, the son of Russian Jewish immigrants, and lived there until his marriage in 1945. His father was a small grocer who made a precarious living and the family frequently had to move, according to where business was available. In 1936 he graduated from City College of New York and worked in various jobs for the next four years. He started writing seriously in 1940 while studying for an MA at Columbia. From 1949 to 1961 he and his family lived in Oregon where he taught at Oregon State College. After returning east, he settled in New England.

Malamud's work, though based in solid realistic detail, has a moral intensity and often fabulous or allegorical qualities that are unmistakably his own. His first novel, *The Natural* (1952; filmed 1984), is the story of a baseball player that is heightened by allusions to the legend of the Holy Grail. *The Assistant* (1957) concerns a poor Jewish grocer and his Gentile assistant, who had earlier robbed him. The assistant finally succeeds to the business and is converted to Judaism. The relations between victim and vitimizer (or parasite and host) and the redeeming value of suffering are themes further developed in the short stories of *The Magic Barrel* (1958), which won the National Book Award. *A New Life* (1961) concerns Levin, who leaves the east to teach in the northwest. The story collection *Idiots First* (1963) was followed by *The Fixer* (1966), a novel based on an incident in tsarist Russia in which a Jew was accused of the ritual murder of a Christian child. *Pictures of Fidelman* (1967) and *Rembrandt's Hat* (1973) are collections of short stories; the novel *The Tenant* (1971) examines the relationship between a black and a white writer. His last writings included the novels *Dubin's Lives* (1979) and *God's Grace* (1982).

Malan, Daniel François (1874–1959) *South African prime minister (1948–54). A right-wing Afrikaner nationalist, Malan introduced the policy of apartheid into South Africa, incurring great displeasure among western nations.*

Born near Riebeck West, Cape Province, the son of a farmer, Malan was educated at Victoria College, Stellenbosch, where he studied for the Dutch Reformed Church, receiving a doctorate in divinity at the University of Utrecht, Holland, in 1905. After working as a minister at the Cape, he toured and wrote a short book

(*Na Congoland*) about Dutch Reformed missions in Rhodesia and the Belgian Congo. Malan's political involvement began in 1915 as editor of *Die Burger*, the Nationalist Party newspaper. Entering parliament in 1919, he became a member of General HERTZOG's cabinet in 1924. In 1933 he broke with Hertzog over the coalition with SMUTS and formed the Gesuiwerde Nasionale Party ('Purified' National Party), which gained much of its support from the Afrikaner Broederbond. Appointed leader of the opposition in the Assembly, Malan gradually gained electoral support, defeating the United Party in the 1948 election to become prime minister and minister for external affairs. His administration is notorious for the introduction of apartheid. He retired from office in 1954.

Malcolm X (Malcolm Little; 1925–65) *US black militant leader.*

The son of a Baptist minister, Malcolm X was born in Omaha, Nebraska, and grew up in Lansing, Michigan. At the age of sixteen he moved to his sister's home in Boston, where he soon became involved in crime. In 1946 he was imprisoned for burglary and while in jail he was converted to the Nation of Islam or Black Muslim faith, led by Elijah Muhammad. On his release in 1953 he became an active member of the sect and at the same time abandoned his surname, which he saw as a relic of slavery, and became known as Malcolm X. During the next ten years he travelled around the country on behalf of the movement, speaking out against the exploitation of black people and founding Muslim temples in a number of US cities.

Having referred in public to the assassination of President John F. KENNEDY as a case of 'chickens coming home to roost', Malcolm found himself suspended from the Black Muslim movement in 1963. The following year he announced the formation of a rival organization, the Muslim Mosque, Inc., which encouraged blacks to arm themselves and use violence in the fight against social injustice. After visits to the Middle East and Africa in the same year, he founded the Organization of Afro-American Unity. The violent feud between the Black Muslims and followers of Malcolm X continued to grow in intensity, and in February 1965 Malcolm's home was destroyed by a fire bomb. Later that month, while addressing a rally in a New York ballroom, Malcolm was shot and killed by three blacks, at least two of whom were Black Muslims.

Malcolm was widely condemned in his lifetime by orthodox civil-rights leaders who advocated nonviolent methods of achieving racial equality. His personal philosophy, however, set out in his autobiography (1965), exerted considerable influence on the new wave of black consciousness in the USA, culminating in the Black Power movement of the late 1960s.

Malenkov, Georgi Maksimilianovich (1903–88) *Soviet statesman. As a protégé of STALIN he became prime minister (1953–55) after Stalin's death, although the leadership of the party was taken from him by KHRUSHCHEV. He was awarded the Order of Lenin twice.*

Born in Orenburg (now Chkalov) into a middle-class family, Malenkov interrupted his studies to join the Bolshevik army. During the civil war he served as political commissar for the Eastern and Turkestan districts, joining the Soviet Communist Party (CPSU) in 1920. After the war he entered the Moscow Technical College, from which he graduated in engineering in 1925. As secretary of the Bolshevik Students Organization, he attracted the attention of the central committee of the CPSU and was appointed to Stalin's personal secretariat (1925–30).

Malenkov became organizing secretary of the Moscow section of the CPSU in 1930. Rising rapidly through the party ranks, he was promoted to the inner secretariat of the party in 1939 and became a candidate member of the Politburo in 1941. During World War II he was one of the five members of the Committee of State Defence (war cabinet) (1942–44); he was also a leading member of the Committee for the Economic Rehabilitation of Liberated Districts. Appointed deputy prime minister and a full member of the Politburo in 1946, he became head of government as prime minister and first secretary of the CPSU in 1953, following the death of Stalin. Within days he was ousted from the secretaryship by Khrushchev (who formally secured the post six months later) but retained the premiership until 1955, when he was forced to resign over his inability to solve problems of agriculture and industrialization. He served as minister for electrical energy for the next two years but was dismissed from all his government and party offices in 1957 for attempting to overthrow Khrushchev. He was employed as manager of a hydroelectric power station in Kazakhstan until his retirement in 1963, but was expelled from the Communist Party in 1961.

Malinowski, Bronislaw Kaspar (1884–1942) *Polish-born British anthropologist who developed the functionalist approach to social anthropology, which sought to explain social phenomena in terms of their functional significance.*

Malinowski attended the Jagellonian University, Kraców, receiving a degree in physics and mathematics in 1908. While at Leipzig University he was considerably influenced by the psychologist Wilhelm Wundt (1832–1920). Moving to the London School of Economics, Malinowski studied anthropology, received his DSc in 1916, and remained as lecturer, becoming reader in anthropology (1924) and then professor (1927). In 1939 he was appointed visiting professor of anthropology at Yale University.

Based in Australia during World War I, Malinowski conducted his celebrated studies of native society on the Trobriand Islands in the southwest Pacific region. He learnt their language and participated in daily life, one of the first anthropologists to use such a participant-observation method. He described his findings in *Argonauts of the Western Pacific* (1922). He also toured the USA and Mexico, studying the Pueblo Indians, and observed Bantu tribes in Africa, examining the significance of customs, ceremonies, religion, taboos, and other cultural elements not in a historical context but in relation to the functioning of the society. His other books include *The Father in Primitive Psychology* (1927) and *Scientific Theory of Culture* (1944).

Malraux, André (1901–76) *French novelist, essayist, art critic, and politician.*

Born in Paris, Malraux travelled to the Far East on an archaeological expedition in the 1920s and became involved with revolutionary groups in Indochina, Canton, and Shanghai. His taste for action and adventure, which he saw as the only means of preserving human dignity, is evident in such novels as *Les Conquérants* (1928; translated as *The Conquerors*, 1929), *La Voie royale* (1930; translated as *The Royal Way*, 1935), and *La Condition humaine* (1933; translated as *Man's Fate*, 1934). The three novels are set in Hong Kong, Indochina, and Shanghai respectively and are based on the author's own experiences. *La Condition humaine* won the Prix Goncourt and worldwide acclaim for Malraux; set at the time of CHIANG KAI-SHEK's struggle against the communists, its somewhat pessimistic view of life is countered by the more optimistic reassertion of human dignity and solidarity.

Returning to Europe, Malraux joined wholeheartedly in the fight against fascism. *Le Temps du mépris* (1935; translated as *Days of Contempt*, 1936) condemns the rise of Nazism in HITLER's Germany, and *L'Espoir* (1937; translated as *Days of Hope*, 1938) is based on Malraux's active experience on the republican side in the Spanish civil war. During World War II he served as a private soldier with a French tank unit and later in the Resistance movement. Captured by the Germans and subsequently liberated, he founded and took command of the Alsace-Lorraine brigade. His last novel, *Les Noyers d'Altenburg*, first published as *La Lutte avec l'ange*, appeared in 1943. In the postwar period he turned to politics, participating in General DE GAULLE's Rassemblement du Peuple Français (1947) and becoming minister for cultural affairs (1959–69) after de Gaulle's return to power in 1958.

In the meantime Malraux wrote extensively on art: *Les Voix du silence* (1951; translated as *The Voices of Silence*, 1953) a comprehensive and philosophical history of art, was followed by the three volumes of *La Musée imaginaire de la sculpture mondiale* (1952–54; translated as *Museum Without Walls*, 1967). The autobiographical elements of Malraux's novels seem to have obviated the need for an autobiography as such; in his *Antimémoires* (1967) the author reflects on the ethical and aesthetic value of his life rather than on specific more personal events, such as the tragic death of his two sons in a car accident. His other writings include *Les Chênes qu'on abat* (1971; translated as *Fallen Oaks*, 1972) and a collection of speeches, *Oraisons funèbres* (1971).

Mandela, Nelson (Rolihlahia) (1918–) *Black South African political leader, whose long imprisonment made him an international symbol of the struggle against apartheid; president of the African National Congress.*

Mandela was born in Transkei, the son of a tribal chief. After attending university in Johannesburg he practised as a lawyer, setting up the country's first black legal practice. An ANC activist from his twenties, he responded to the banning of the organization in 1960 by inciting a wave of strikes; when nonviolent means made little impact, he formed the Spear of the Nation movement to undertake a campaign of sabotage and guerrilla activity. Mandela evaded arrest until 1962, when he received a five-year sentence for incitement; in 1964 this became a life sentence, following a second trial at which he was found guilty of sabotage and treason.

Mandela spent the first part of his sentence on Robben Island, a notorious high-security prison. A campaign for his release was spearheaded by his second wife, Winnie Mandela (1934–), whom he had married in 1958; she became a political figure in her own right, suffering imprisonment (1969–70) and harassment at the hands of the authorities. By the late 1970s he had become an internationally famous figure, showered with honours and tributes from sympathizers worldwide. His refusal to gain his own freedom by making a political deal with his captors had by this time invested him with an almost mythical status in the eyes of many black South Africans. In 1988 his seventieth birthday was marked by renewed calls for his release and much international publicity; later that year he was moved to more comfortable quarters following an alarming decline in his health.

Mandela was finally released in 1990, to worldwide acclaim, on the intervention of the new state president, F. W. DE KLERK. He has since engaged in talks about the country's future with de Klerk and other government figures and travelled widely to argue the case for continued international pressure on South Africa. His triumph has been marred by continuing violence between ANC supporters and the Zulu Inkatha movement and by the trial of his wife on charges arising from the abduction and beating of several youths, one of whom died, by members of her bodyguard. Winnie Mandela received a six-year prison sentence in 1991, against which she has appealed.

Mandelshtam, Osip Emilyevich (1891–1938) *Russian poet. The subjects of Mandelshtam's work, which is ranked among the greatest Russian poetry of the twentieth century, are uncompromisingly literary and belong to the mainstream of European culture. This, and the poet's commitment to his art above all else, proved intolerable to the Stalinist regime.*

The son of nonreligious Jewish parents, Mandelshtam was born in Warsaw but grew up in St Petersburg, where he attended the Tenishev School. He studied at the Sorbonne, Heidelberg, and the University of St Petersburg but without taking a degree. His first poetry (1910) was published in the journal *Apollon*. He became associated with the Acmeists, a group of poets favouring concrete detail and clarity and precision of language against the mystical fuzziness to which the symbolist poets had succumbed. The group disintegrated after World War I. Mandelshtam, having pub-

lished one volume, *Kamen'* (1913; 'Stone'), managed to avoid the war and the most difficult period after the revolution, probably by moving to the south. By 1922 he had returned, however, and published his second volume, *Tristiya*, which assured his reputation as a major poet. Even so, much of his time was spent on routine journalistic and translating work by which he had to support himself. In 1928 he published another collection, *Stikhotvoreniya* ('Poems'), a book of prose, *Yegipetskaya marka* (translated as *The Egyptian Stamp*, 1965), and a critical volume, *O poezii* ('On Poetry'), none of which allayed official disapproval of his work, which clearly gave no support to the new regime.

Advised to leave Petrograd at the end of the 1920s, he found journalistic work in the provinces. An account of these years, *Puteshestviya v Armeniyu* (1933; translated as *Journey to Armenia*, 1973), was subjected to scathing criticism and no work by Mandelshtam was allowed in the Soviet press for more than three decades. He was arrested in 1934, having composed an unflattering poem on STALIN, interrogated, and exiled to Voronezh until 1937. He continued to write in exile despite worsening health. He was released briefly, during which time he and his wife, Nadezhda, were forced to move from place to place in a desperate attempt to find shelter and work; but he was rearrested in 1938 and sent to a camp near Vladivostok. His death was officially recorded as occurring on 27 December 1938. His wife wrote two books describing their lives and the suffering imposed on them by the regime: *Hope against Hope* (1971) and *Hope Abandoned* (1973).

Manley, Michael (1923–) *Jamaican prime minister (1972–80; 1989–), who became a charismatic leader representing the views of poor people of the third world.*

Born in Kingston, the son of former prime minister Norman Manley, Michael Manley was educated at a college in Jamaica before briefly attending McGill University, Montreal. He enlisted in the Royal Canadian Air Force in 1942 and after the war attended the London School of Economics (1945–49).

Manley first came to prominence as the principal sugar organizer for the National Workers Union (1953–55). Demonstrating skills in negotiation, he was appointed island supervisor and first vice-president in 1955. In 1962 he was elected to the senate and in 1967 transferred to the house of representatives. Two years later he became president of the People's

National Party. He was elected prime minister in 1972 on a moderate socialist platform. As prime minister he sought to strengthen Jamaica's economy through the expansion of public works and the encouragement of local industry. He also introduced a free education system and training programmes for the unemployed. He was defeated at the polls in 1980 by Edward Seaga, following a downturn in the economy, but was re-elected in 1989. His many publications include *The Politics of Change* (1974), *A Voice at the Workplace* (1975), *The Search for Solutions* (1976), and *A History of West Indies Cricket* (1988).

Mann, Thomas (1875–1955) *German novelist and essayist. He was awarded the Nobel Prize for Literature in 1929.*

Born in Lübeck, Mann was the son of a well-to-do merchant; his mother came from Portuguese Creole stock. His elder brother, Heinrich Mann (1871–1950), also became a novelist. After the death of his father in 1891, the family moved to Munich, where Mann lived until 1933. He began writing stories in the 1890s and established a major reputation with an epic two-volume novel about the decline of an aristocratic Lübeck family, *Buddenbrooks* (1901; translated 1924). The ironic style of this novel was further refined in two novellas of 1903, *Tonio Kröger* and *Tristan* (both translated in *Three Tales*, 1929). In these he developed an opposition between the bourgeois and the artistic, or *Bürger* and *Künstler*: the one obtuse and dull but healthy, the other sensitive and brilliant but diseased and decadent. In 1905 Mann married the daughter of a leading authority on Wagner. His second novel, *Königliche Hoheit* (1909; translated as *Royal Highness*, 1916), was followed by the highly regarded novella *Der Tod in Venedig* (1912; translated as *Death in Venice*, 1925), in which the protagonist, Gustav von Aschenbach, a dignified writer of serious works, becomes fatally infatuated with a fourteen-year-old youth. In the complex symbolism of the story, Mann used a literary equivalent of the Wagnerian leitmotiv (for example, in the recurrence of a devil figure with red hair). The novella was made into a highly acclaimed film (1970) by Visconti. In the novel *Der Zauberberg* (1924; translated as *The Magic Mountain*, 1927), the setting is a sanatorium, a microcosm of contemporary Europe with its intellectual ferment and disorders.

Though conservative and pro-German in World War I, Mann openly opposed fascism in Germany and attacked it in such stories as *Mario und der Zauberer* (1930; translated as *Mario and the Magician*). While he was abroad in 1933, his property was confiscated by the Nazis. He lived in Switzerland until 1939, having warned Europe of the impending danger (*Achtung Europa!*, 1938). Emigrating to America, he became a US citizen in 1944 but returned to Switzerland in 1952. He was awarded the Goethe Prize in 1949. Of his later works three are outstanding. The tetralogy *Joseph und seine Brüder* (1933–43; translated as *Joseph and his Brothers*, 1948) is an archaeologically detailed study of the biblical story and an exploration into the essence of civilized life. In *Doktor Faustus* (1947) the emergence of demonic forces in the work of the composer and genius who is the hero is paralleled by the apocalyptic collapse of Germany and Europe in the war. The unfinished picaresque novel *Die Bekenntnisse des Hochstaplers Felix Krull* (1954; translated as *Confessions of Felix Krull, Confidence Man*, 1955) strikes a note of comic resignation as the charming trickster, Krull, triumphs over everything. Important critical work by Mann is collected in *Essays of Three Decades* (1947).

Mansfield, Katherine (Kathleen Mansfield Beauchamp; 1888–1923) *New Zealand short-story writer.*

Katherine Mansfield was born in Wellington, New Zealand, and educated there until she went (1903) to London to study music. Failing to settle after her return to New Zealand in 1906, she came back to London where she married George Bowden (1909). This marriage was a failure and the pair separated. In 1911 she met John Middleton Murry, with whom she lived from 1912 until her divorce from Bowden (1918) enabled them to marry.

Meanwhile she became a regular contributor to various journals and her first collection of short stories, showing her mastery of the form, appeared in 1911 under the title *In a German Pension*. In 1915 she, Murry, and D. H. Lawrence collaborated in producing a magazine called *The Signature*. The death of her only brother in France (1915), added to the general strain of the war, undermined her health. In 1917 tuberculosis was diagnosed and she spent much of the rest of her life travelling in search of a cure in Italy, Switzerland, and France. Nonetheless she published two more collections of stories: *Bliss* (1920), which included the stories 'Prelude' and 'Je ne parle pas français' and *The Garden Party* (1922). She died at the Gurdjieff Institute near Fontainebleau early in 1923. Two further collections of stories, *The Dove's Nest* and *Something Childish* (both 1924), were published posthumously, as were collected poems (1923); her journal, edited by her husband, was published in 1927, and her letters in 1928.

Mao Tse-tung (Mao Ze Dong; 1893–1976) *First chairman and chief architect of the People's Republic of China (1949–59) and the Chinese Communist Party (1949–76). He has been described as the 'father of the Chinese revolution' and was one of the most important revolutionary figures of the twentieth century.*

Born in Shaoshan, Hunan province, the son of a farmer, Mao enlisted in the revolutionary army in Hunan during the first rebellion against the Manchu dynasty (1911). Already influenced by the nationalistic ideas of SUN YAT-SEN, he became converted to Marxism and was involved in political activity as a student at Peking University (1919); in 1921 he cofounded the Chinese Communist Party. He organized communist guerrilla army units as part of the revolutionary forces (1924–25), and it was at this time that he first became aware of the revolutionary potential of the peasant classes. Appointed chief of propaganda and publicity under Sun Yat-sen, he also publicly supported a united front with the Kuomintang (Nationalist Party) under CHIANG KAI-SHEK until 1927, when the two parties split and he was dismissed from his office. With Chiang Kai-shek purging the communists, Mao – now leader of the communists – retreated to southeastern China, where he established the Chinese Soviet Republic in a portion of Kiangsi province (1931–34). In 1934 Kuomintang troops forced Mao to evacuate Kiangsi and he led his army on the epic 'long march' to Shensi in northern China, a distance of 6000 miles. During World War II an alliance was negotiated between the communists and the Kuomintang against the Japanese (1937–45), during which time Mao wrote major works on Marxist Leninism. In 1945 the civil war resumed and eventually resulted in victory for the communists; in 1949 Mao became chairman of the newly established People's Republic of China.

During the 1950s Mao abandoned the Soviet economic model and began to introduce his own reforms, designed to rejuvenate the economy. In 1958 he initiated the Great Leap Forward – a campaign to increase industrial production by mobilizing China's enormous manpower into rural 'people's communes'. However, these new policies disrupted Chinese industry and opposition mounted; in 1959

Mao retired as chairman of the republic, but continued to influence policy as chairman of the Communist Party. He launched the Cultural Revolution (1966–69) against party bureaucracy, during which revolutionary attitudes were encouraged, officials dismissed, and all aspects of culture were destroyed. Schools and universities were closed, the students being organized into Red Guards who attacked nonconformists and destroyed property. Mao's 'thoughts' in the 'Little Red Book' became doctrine for the mass of Chinese people, inspiring a cult based on his own personality that lasted until his death.

In the struggle for power that followed Mao's death, a radical group known as the 'Gang of Four', which included Mao's widow, Chiang Ch'ing (Jiang Qing; 1912–91), was finally ousted by HUA GUO FENG.

Maradona, Diego (Armando) (1960–   ) *Controversial Argentinian footballer, who led Argentina to victory in the 1986 World Cup.*

The product of a Buenos Aries shanty-town, Maradona became a child celebrity for his foot-juggling tricks, performed on national television and during half-time at football matches. He became a professional footballer at fifteen, quickly establishing himself as the nation's favourite by his flamboyance and skills. In 1982 he was sold to Barcelona for a then-record fee of about £5 million, but developed a reputation for temperamental behaviour; he was sold again, for an even higher fee, in 1984. This time the investment paid off, as Maradona hit peak form, taking his new club, Napoli, to a series of national and European triumphs. His presence dominated the 1986 World Cup as player, captain of the victorious Argentinian side, and scorer of a notorious fisted goal in the quarter-final against England (ascribed by its author to 'the hand of God'). In 1991 Maradona's future looked uncertain when traces of cocaine were found in his urine and he was banned from world football for fifteen months; at the same time rumours of his links with drug dealers, organized crime, and prostitution became the subject of an official enquiry in Italy.

Marc, Franz (1880–1916) *German expressionist painter associated with the Blaue Reiter group of artists.*

The son of a Munich painter, Marc became a student at the Munich Academy in 1900. He travelled to Italy and to Paris – the centre of postimpressionism, the influence of which is seen in his early works. From 1907 Marc's paintings most frequently depicted animals, especially horses. In 1910 he met KANDINSKY and they formed the Blaue Reiter group, holding an exhibition in 1911 that also included works by August Macke (1887–1914) and Henri Rousseau (1844–1910). The influence of the group on Marc's work is apparent in his increased use of brilliant colour and strong forms, as in *Red Horses* (1911).

In 1912 his work again altered under the cubist influence of DELAUNAY. With the use of intermingled shapes and colours, the animals began to merge with the background, in *Deer in the Forest* (1913–14), for example. Marc, an intensely religious man, portrayed idealized rather than realistic animals, to which he ascribed a spiritual purity lacking in man. He saw his painting as an attempt to reach and interpret the source of reality. Increasingly his work tended towards abstraction. The paintings immediately preceding World War I convey drama and strong emotion, in particular *The Fate of the Animals* (1913), which presages catastrophe. Marc was called up in 1914 and killed in action near Verdun two years later.

Marceau, Marcel (1923–   ) *French mime artist. He is a Chevalier de la Légion d'honneur and has received many other honours.*

Born in Strasbourg, Marceau began his career as a member of Jean-Louis BARRAULT's company in 1945. In 1946, however, he began to devote himself entirely to mime, appearing for the first time as the white-faced Bip, a character he developed from the French nineteenth-century Pierrot, at the Théâtre de Poche, Paris. As Bip, the central character of many of his sketches, he gained universal recognition and acclaim. He directed his own company (1948–64) and toured the world on numerous occasions.

Marceau has also created several mime-dramas, including *Mort avant l'aube* (1947) and *Le Manteau* (1951), based on Gogol, and several pantomimes. His *Don Juan* was produced at the Théâtre de la Renaissance in Paris (1964) and the ballet *Candide* at the Hamburg Opera House (1971). Other notable performances included *Jardin public* (1949), in which he played ten characters. In 1978 he founded the École de Mimodrame de Paris at the Théâtre de la Porte-Saint-Martin, but he remains most famous for his solo mimes, of which he sometimes gives three hundred performances a year. His publications include *L'Histoire de Bip*.

Marciano, Rocky (Rocco Marchegiano; 1923–69) *US boxer and world heavyweight champion (1952–56).*

Born in Brockton, Massachusetts, Rocky Marciano earned a living as a labourer, working for a time in a shoe factory. He turned professional in 1947. He was short for a heavyweight and many considered his style to be crude. However, he ensured that few opponents could stand up to him. In 1951 he knocked out Joe LOUIS and the following year took the title by beating 'Jersey Joe' Walcott. Marciano successfully defended his title six times before retiring, undefeated, in 1956: he was the only world heavyweight champion in the history of boxing who never lost a professional fight. He was killed in an air crash in 1969, just before his forty-sixth birthday.

Marconi, Guglielmo (1874–1937) *Italian electrical engineer, the best known and most successful pioneer of radio telegraphy. He was awarded the Nobel Prize for Physics (jointly) in 1909.*

Marconi was born in Bologna, the son of a wealthy landowner and his Irish wife, Anne Jameson, a member of the well-known distilling family. He was educated first privately and then at the Leghorn Technical Institute and the University of Bologna. Marconi's interest in radio was aroused in 1894 when he first heard of Hertzian waves. He began, in collaboration with Augusto Righi (1850–1920), to investigate the possibilities of transmitting these waves over greater distances. Experimenting on his father's estate, he managed to broadcast a signal over a distance of a mile but, finding the Italian government uninterested in his work, Marconi left for Britain in 1896; he first demonstrated his wireless apparatus for the British Post Office the same year.

The crucial question to be decided was whether radio waves would follow the curvature of the earth's surface or travel no further than the horizon. Convinced of their ability to travel around the earth, Marconi took out the necessary patents and on 12 December 1901 apparently proved his point by transmitting a signal, the letter 'S', in Morse code from Poldhu in Cornwall to Newfoundland in Canada. The explanation of the phenomenon was not as Marconi supposed but, as Arthur KENNELLY later showed, due to reflection by the ionosphere. Nevertheless, the critics were silenced and Marconi became instantly famous. He also proved to be an astute businessman. The Marconi Wireless Telegraph Co. Ltd, founded in 1900, set about establishing radio stations on land and at sea throughout the world to receive and transmit radio waves; in 1901 messages were first sent across the Atlantic. Initially Marconi worked with long waves but in the 1920s he investigated the advantages to be gained from using short waves. Much of the 1920s was therefore spent by Marconi aboard his 700-ton yacht, *Elettra*, testing short-wave reception and transmission. By the end of the decade Marconi had set up a worldwide system of short-wave stations.

With his numerous triumphs and the success of his company, Marconi in the 1930s had become one of the best-known Italians in the world. Having been a member of MUSSOLINI'S Fascist Party since 1923, Marconi was given the title of Marchese (Marquis) in 1929. He was consequently, despite declining health, chosen in 1935 to make a tour of Latin America and Europe to defend his country's invasion of Abyssinia. Nevertheless he feared political pressure and hoped that radio waves would serve as an instrument of peace. He died shortly afterwards from a series of increasingly serious heart attacks. Although Marconi was not a pioneer of broadcasting, he recognized the importance of radio telephony and all BBC stations were silent for two minutes on the day of his funeral.

Marcuse, Herbert (1898–1979) *German-born US philosopher, famous in the 1960s as the theorist of the new revolutionary left.*

Born in Berlin, Marcuse was educated at the University of Freiburg, where he gained his doctorate in 1922; he then became an associate at the influential Institute of Social Research in Frankfurt. With the rise of HITLER, however, Marcuse left Europe for the USA, where he remained for the rest of his life. He taught at Columbia, Brandeis, and the University of California, La Jolla. During World War II Marcuse worked for military intelligence.

In 1964, with the publication of his *One-Dimensional Man*, the sixty-six-year-old Hegelian social philosopher became an international celebrity. Like other Frankfurt colleagues he argued that while the modern industrial society had satisfied the material needs of man it had done so only by ignoring his true needs and by restricting his liberty. The apparent freedom present in many industrial societies was discounted by Marcuse as 'repressive tolerance'. There was, therefore, no way in which a 'liberated' man could ever come to terms with capitalism. In addition to the repressions of capitalism, Marcuse also identified, in his *Eros and Civilization* (1955), the repressions im-

posed on us by the unconscious mind. In this area he recognized fairly orthodox Freudian solutions. In the political domain, however, as he argued in his *Soviet Marxism* (1958), Marcuse rejected the approach of bureaucratic communism. Revolutionary change, it seemed, could therefore only come from an alienated elite, such as the students of Paris and Berkeley. Although such views seemed highly plausible at one time, they failed to survive the student movement of the 1960s.

Marini, Marino (1901–80) *Italian sculptor, painter, and graphic artist, who belonged to no artistic movements or groups but is recognized as an outstanding creative sculptor.*

Marini was born in Pistoia. He studied painting and sculpture at the Academy of Fine Arts in Florence but before 1930 limited his output mainly to painting and graphic art. From 1929 he travelled widely and taught in Monza until 1940, when he became professor of sculpture at the Brera Academy in Milan.

When he returned to sculpture in 1930 Marini produced single human figures of great vigour and intensity, usually either female figures or entertainers (such as jugglers or acrobats) in bronze. But in 1935 he discovered the theme that was to dominate his mature work: that of the horse and rider. Frequently showing the influence of archaic sculpture, these usually motionless figures have a powerful inner vitality and seem to bear an obscure tragic symbolism. These are the works for which he is best known, but he also did a large number of graphic works and portraits in bronze, of which the best is perhaps that of Igor STRAVINSKY (1951).

Marini took up painting again in 1948 while living in Switzerland. Many of these later pictures approach abstraction as did some of his sculptures, for example *Monumental Rider* (1958), which became more dramatic and distorted, with rougher surfaces. He obtained numerous prizes during this period, including one at the Venice Biennale in 1952.

Maritain, Jacques (1882–1973) *French neo-Thomist philosopher.*

Born in Paris, a Protestant, Maritain was educated at the Sorbonne and converted to Roman Catholicism in 1906. After further study in Germany, where (among other pursuits) he studied biology under Hans Driesch (1867–1941), he was appointed in 1914 as professor of modern philosophy at the Institut Catholique, Paris. In later life Maritain taught mainly in North America, first in Toronto and from 1948 to 1956 at Princeton.

Maritain began his philosophical life as a follower of Henri BERGSON. In 1908, however, he began to study the work of St Thomas Aquinas and found, to his surprise, that he had been a Thomist for many years without being aware of it. Thomist thinking, Maritain realized, had become uncritical, inward-looking, and absurdly traditional. He therefore saw it as his task to instil into Thomism a greater intellectual rigour and to open it up to the twentieth century. Thus in such works as *Art et scolastique* (1920; translated as *Art and Scholasticism*, 1962) and *Scholasticism in Politics* (1940), Maritain attempted to deploy Thomist principles in new areas. He also, in his *Les Degrés du savoir* (1932; translated as *The Degrees of Knowledge*, 1959), made an ambitious attempt to incorporate within the traditional Thomist scheme the concepts and processes of modern physics.

Maritain was more than an academic philosopher. He served as French ambassador to the Vatican (1945–48) and later, a strong opponent of the Vatican Council, argued against the neomodernist movement.

Markova, Dame Alicia (Lilian Alicia Marks; 1910–) *British prima ballerina, director, teacher, and co-founder of the London Festival Ballet. She was made a DBE in 1963.*

Born in London, Markova studied with Astafyeva, Cecchetti, Legat, and Celli before making her debut with DIAGHILEV's Ballets Russes at the age of fourteen. Her first major success with the company came in George BALANCHINE's *Le Rossignol* (1926), and when the Ballets Russes was disbanded on the death of Diaghilev in 1929 she joined the Vic-Wells Ballet and became its first prima ballerina (1933–35). She subsequently danced leading roles in Frederick ASHTON's *La Péri, Foyer de Danse, Les Masques,* and *Mephisto Valse,* Antony Tudor's *Lysistrata,* and Ninette DE VALOIS's *Bar aux Folies-Bergère* and also appeared in the Camargo Society's performances. With the Vic-Wells Ballet she became the first English dancer to take the lead in *Giselle* and the full-length *Swan Lake,* as well as creating roles in *Les Rendez-vous* (1933), *The Haunted Ballroom* (1934), and *The Rake's Progress* (1935).

In 1935 Markova formed the Markova–Dolin Company with Anton Dolin (1904–83) and appeared as its prima ballerina. She went on to dance with the René Blum–Léonide Massine Ballet Russe de Monte Carlo (1938–41), with whom she made a successful US debut in *Giselle* (1938), and

with the American Ballet Theatre (1941–44; 1945–46), with whom she created the role of Juliet in Antony Tudor's *Romeo and Juliet* (1943). Over the next twenty years she made guest appearances with many companies and won great acclaim for her dancing in such works as *Les Sylphides, Pas de Quatre,* and *The Nutcracker.*

In 1950 Markova became a co-founder of the London Festival Ballet with Dolin and later, on retiring from the stage in 1963, became successively a director of the Metropolitan Opera Ballet (1963–69), professor of ballet at the University of Cincinnati's Conservatory of Music (1969–74), and finally a teacher at the Royal Ballet School. She became president of the London Festival Ballet in 1986. She is the author of *Giselle and I* (1960) and *Markova Remembers* (1986).

Marshall, Alfred (1842–1924) *British economist, regarded as one of the founders of the neoclassical school in economics.*

Marshall was born in London and graduated in mathematics from St John's College, Cambridge. He began lecturing in moral science at Cambridge in 1868. In 1882 he took the chair of political economy at Bristol and was the first principal of University College, Bristol. In 1885 he became professor at Cambridge University, retiring in 1908.

Marshall's influence on microeconomics was profound, both in teaching and policy-making. He developed downward-sloping demand curves from marginal utility theory and upward-sloping supply curves (higher prices leading to greater output), at the intersection of which equilibrium prices are determined. He coined the term 'elasticity' to denote responsiveness of demand or supply to small changes in price, and introduced the concepts of consumer surplus, quasi-rent, and external economies. Marshall described the time periods over which consumption and production decisions can change:

(1) very short run, when supply is fixed and prices are determined by market forces;

(2) the short period, when supply can be increased up to the maximum capacity of the existing capital stock;

(3) the long period, when supply can be varied given the existing state of technology, and firms enter or leave the industry; and

(4) the very long period, when technology also changes.

Although his main contribution was his work on the theory of value and the theory of the firm, Marshall also wrote on *The Pure*

Marshall, George C(atlett)

372

Theory of Foreign Trade (1879), *Industry and Trade* (1919), and *Money, Credit and Commerce* (1923).

Marshall, George C(atlett) (1880–1959) *US general, administrator, and diplomat, who organized the massive expansion in US armed forces during World War II and implemented the programme of economic aid to postwar Europe known as the 'Marshall Plan'. He was awarded the 1953 Nobel Peace Prize.*

Graduating from Virginia Military Institute in 1901, Marshall joined the infantry as a second-lieutenant. He served in the Philippines (1902–03) and in World War I was chief of operations of the US 1st Army, then chief of staff of the 8th Army Corps. After serving as aide to General Pershing (1919–24), and a posting to China with the 15th Infantry, in 1927 Marshall started a six-year spell as assistant commander of instruction at Fort Benning infantry school. On the outbreak of World War II in Europe, he was appointed army chief of staff, responsible for building an army of less than 200 000 men into the formidable fighting force that joined the Allies after the attack on Pearl Harbor in 1941.

With his effective political lobbying, Marshall won government approval for his ambitious plans and in 1942 entirely regrouped the US army command structures into ground, air, and supply forces. Appointed chairman of the joint chiefs of staff committee, he was thus in close contact with the president, accompanying ROOSEVELT to all the major Allied conferences and winning great respect for his immense ability and self-effacing manner. Instead of commanding the Allied invasion of Europe, as many had anticipated, Marshall stayed on in Washington as top presidential adviser. In December 1944 he was promoted to five-star general.

After the war, TRUMAN appointed him special ambassador to China in a vain attempt to mediate in the civil war. Appointed secretary of state in 1947, Marshall instituted a programme of emergency economic aid to European countries, which played a crucial role in helping them to reconstruct their shattered economies. However, his period as defence secretary (1950–51) was marred by unfounded attacks on Marshall by the bête noir of postwar US politics, Senator Joseph McCARTHY.

Martin, (Basil) Kingsley (1897–1969) *British journalist and influential editor of the* New Statesman.

Martin was the son of a dissenting minister. He was educated first at Hereford Cathedral School and later at Mill Hill School, London. A conscientious objector, he served in World War I in the Friends' Ambulance Unit. In 1919 he went to Magdalene College, Cambridge, and in 1924 obtained a lectureship in politics at the London School of Economics. In 1927 he became a leader writer on the *Manchester Guardian*.

At the instigation of J. M. KEYNES, in 1931 he became editor of the *New Statesman*, which shortly afterwards combined with the *Nation* and later (1934) the *Week-End Review*. In this post, which he retained until 1960, he was outstandingly successful, and the *New Statesman* became the country's most influential journal on the whole spectrum of left-wing issues and literary criticism. Martin himself was a brilliant writer well able to command the respect of the intelligentsia, who formed his principal readership.

Besides his journalistic work, Martin wrote two major studies in political history: *The Triumph of Lord Palmerston* (1924) and *French Liberal Thought in the Eighteenth Century* (1929). After his retirement he published two autobiographical books: *Father Figures* (1966) and *Editor* (1968). He died in Cairo on one of his frequent and extensive travels abroad.

Marx Brothers *A legendary US comedy team consisting of Chico (Leonard Marx; 1886–1961), Harpo (Adolph, later Arthur, Marx; 1888–1964), Groucho (Julius Henry Marx; 1890–1977), and Zeppo (Herbert Marx; 1901–79).*

Originally five brothers, the fifth being Gummo (Milton Marx; 1893–77), the Marx Brothers were launched on their showbusiness career by their mother, Minnie Marx (née Schoenberg), whose own family had a vaudeville background. In the early days the members of the team varied, with Minnie herself sometimes taking part. At one time they appeared as 'The Six Musical Mascots', at others 'The Three Nightingales' or 'The Four Nightingales'; finally they emerged as 'The Marx Brothers' in the twenties, Gummo having left the act by the time they made their Broadway debut in the musical *I'll Say She Is* (1924). *The Cocoanuts* (1925) and *Animal Crackers* (1928) were among the shows that followed, both being transferred to the screen in 1929 and 1930, respectively. Zeppo, the least zany of the four, left the act in 1933 but not before taking part in one of their funniest films, *Duck Soup* (1933).

All of them were accomplished musicians: Chico, with his Italian appearance and fractured English, played the piano; the silent Harpo, who communicated by means of a motor horn but was not in fact dumb, was a harpist; and the moustachioed wise-cracking Groucho played the guitar. Other Marx Brothers classics were *A Night at the Opera* (1935), *A Day at the Races* (1937), and *A Night in Casablanca* (1947); *Love Happy* (1950) was their last film as a team. Chico and Harpo subsequently retired but Groucho went on, appearing on television and making his last film, *Skidoo*, in 1968.

Masaryk, Tomáš Garrigue (1850–1937) *Czechoslovakian statesman, first president (1918–35) and co-founder of Czechoslovakia.* Born in Hodonin, Moravia, the son of a coachman, Masaryk was educated at the University of Vienna, where he gained a doctorate in 1876 and was appointed a lecturer in philosophy in 1879. In 1882 he became a professor at the University of Prague, a position he held until 1914.

Masaryk first entered politics when he was elected to the Hapsburg Parliament (1891–93; 1907–14). After the outbreak of World War I he escaped to London where he became a lecturer in Slavonic history at King's College, at the same time campaigning for an independent Czechoslovakia; he co-founded with BENEŠ the Czechoslovakian National Council. In 1918 he travelled to the USA, where he received the support of President WILSON for the establishment of a Czechoslovak Republic. In June 1918, when Czechoslovakia was recognized as an Allied power, he became president-elect; in December 1918, Masaryk returned to Czechoslovakia as president. He was re-elected three times (1920, 1927, and 1934) but resigned on account of his age in 1935 in favour of Beneš. A philosopher with a considerable reputation, he published several works, including *Marxism* (1898) and a survey of the social and intellectual history of Russia, *Russia and Europe* (1913).

Mascagni, Pietro (1863–1945) *Italian composer. His one success,* Cavalleria rusticana *(known to opera buffs as* Cav*), is an enduring part of the repertoire of many opera houses.* The son of a Livorno baker, Mascagni was forbidden to waste his time learning about music and had to study in secret. When he was found out by his father he was for a time adopted by a sympathetic uncle; however, after his early efforts at composition, a symphony for small orchestra and a *Kyrie*, were per-formed at the local music school, parental opposition ceased and Mascagni returned home (1881). A successful performance of a setting of Schiller's *Ode to Joy* encouraged a wealthy patron, Count Florestano de Larderel, to pay to send Mascagni to the Milan Conservatory. However, Mascagni found the academic discipline unacceptable and he left to join a travelling opera company.

Conducting and composing with a series of such companies, Mascagni made a living until 1889, when he won a publisher's competition with his one-act opera *Cavalleria rusticana*, based on a story of Verga. It was performed in Italy, France, England, and the USA, spawning a crop of similar works in the verismo style. Having achieved fame and fortune by the age of twenty-six, Mascagni went on to write several more operas, including *Iris* (1898), *Le maschere* (1900), and *Nerone* (1935). None of these, however, achieved any real success.

Masefield, John Edward (1878–1967) *British poet and novelist. He was appointed poet laureate in 1930 and received the OM in 1935.* Orphaned at an early age, Masefield was brought up on a farm at Ledbury, Herefordshire, where he gained his abiding love of the English countryside. He attended the King's School, Warwick, before going to sea at the age of thirteen. After some years in the USA he returned to England (1897) and settled in London, where he began to contribute to various journals. *Salt Water Ballads* (1902), *Ballads* (1903), and *Ballads and Poems* (1910) show his growing mastery of narrative verse forms. He also wrote histories and edited poetic selections, sometimes collaborating with his wife, whom he married in 1903. Of his critical works, his *William Shakespeare* (1911) was particularly acclaimed. As a playwright he was less successful, although *The Tragedy of Nan* (1908) had considerable merit. *The Everlasting Mercy* (1911) and *Dauber* (1913), both long narrative poems, were Masefield's first major poems.

During World War I Masefield served in the Red Cross in France and the Dardanelles; after the war he settled near Oxford, where he lived for the rest of his life. *Reynard the Fox* (1919) and *Right Royal* (1920) celebrate the courage of man and beast in the traditional rural pursuits of foxhunting and steeplechasing. Courage and endurance also feature as themes in Masefield's prose fiction, for instance *Lost Endeavour* (1910) and *The Bird of Dawning* (1933). His admiration for these qualities and his patriotism underlie his prose piece on

Dunkirk, *The Nine Days Wonder* (1941). Although much of the poetry he wrote as poet laureate did not enhance his reputation, he fulfilled the post's duties conscientiously for thirty-seven years. *So Long to Learn* (1952) and *Grace Before Ploughing* (1966) are autobiographical.

Massey, Raymond (1896–1984) *Canadian-born actor and producer, who became a US citizen in 1944.*

Born in Toronto, Massey was educated at Toronto University and Balliol College, Oxford. During World War I he served with the Canadian army in France and Siberia and in World War II was a major on the staff of the adjutant general.

Massey's distinguished acting career began in London in the early twenties at the Everyman Theatre in Hampstead, where he subsequently joined the management and became a producer. His film debut came in *The Speckled Band* (1931), as Sherlock Holmes. Throughout his career he combined theatre and film work, receiving acclaim for his fine character performances. His imposing bearing, rich voice, and compelling eyes contributed to a distinctive style that could be either benevolent or vicious. The part with which his name became most closely associated was that of Abraham Lincoln in *Abe Lincoln in Illinois*, a role he played first at the Plymouth Theatre, New York (1938–39), then on a tour of the USA (1939–40), and finally in the film (1940). Other notable films included *The Old Dark House* (1932), *Things to Come* (1936), *The Prisoner of Zenda* (1937), *49th Parallel* (1941), *Arsenic and Old Lace* (1944), *Mourning Becomes Electra* (1947), and *East of Eden* (1955). He also gained wide popularity as Dr Gillespie in the television series *Dr Kildare* (1961–65).

An autobiography, *When I Was Young*, was published in 1977. The children of his marriage (1929–39) to actress Adrianne Allen (1907–   ), Daniel Massey (1933–   ) and Anna Massey (1937–   ), have become notable actors in their own right.

Massey, William Ferguson (1856–1925) *New Zealand statesman and Conservative prime minister (1912–25).*

Born in Limavady, Northern Ireland, the son of a tenant farmer, Massey was educated in Ireland before emigrating to New Zealand with his family in 1870. A successful farmer, Massey was elected president of the Auckland Agricultural Association (later the New Zealand Agricultural and Pastoral Association) in 1891.

Massey entered parliament in 1894, when he was elected as the member for Waitemata. He became the member for Franklin in 1896 (a seat he held until his death), leader of the Conservative Party in opposition (1903), and prime minister in 1912, heading the first Conservative government for twenty-two years. A fervent supporter of the Empire, Massey pledged the loyalty and solidarity of New Zealand to Britain during World War I, stating 'All we have and all we are are at the service of the King.' Twice attending meetings of the war cabinet, he represented New Zealand at the Paris Peace Conference in 1919 and visited London five times during his period of office.

An avid reader with a keen memory, Massey was an astute parliamentarian who, as a farmer, enjoyed the continuous support of the farming community. He was awarded honorary degrees from the universities of Oxford, Edinburgh, and Belfast and received the freedom of several cities, including London, Edinburgh, and Glasgow.

Massine, Léonide (Leonid Fyodorovitch Miassin; 1896–1979) *Russian dancer and choreographer. He became a naturalized French citizen in 1944.*

The son of a horn player and a singer in the chorus at the Bolshoi Theatre in Moscow, Massine attended the theatre's school, where he studied ballet. As a result of his lack of technical excellence he was about to settle for a career as an actor, when he was persuaded by DIAGHILEV in 1914 to join the Ballets Russes; the same year he danced the lead role (intended for NIJINSKY, who had been dismissed) in *The Legend of Joseph*. After this success Diaghilev took him seriously in hand and trained him as both dancer and choreographer; his first work to be performed, *Le Soleil de nuit*, was staged in New York in 1915.

During the next few years Massine produced some of his most famous ballets. He collaborated with COCTEAU, PICASSO, and SATIE on *Parade* (1917), he produced *Le Tricorne* and *La Boutique fantasque* for the Alhambra Theatre in London in 1919, and in 1920 created *Le Chant du Rossignol*, *Pulcinella*, and a new version of *The Rite of Spring*, all with music by STRAVINSKY. In 1921 he married an English girl in the Ballets Russes and was dismissed from the company. He continued to create ballets for Diaghilev, however, including *Les Matelots* (1925) and *Ode* (1928), as well as touring with his own company. In 1932 he joined the Ballets Russes de Monte Carlo, for whom he produced such comedy ballets as *Jeux d'enfants* and *Gaîté Parisienne*. During this period he choreographed ballets for existing symphonies, including *Les Présages* (1933) using Tchaikovsky's fifth symphony and *Symphonie fantastique* (1936) using Berlioz's work: these ballets were controversial and are no longer performed in the modern repertory, but the idea has influenced subsequent choreographers.

After the war Massine concentrated on reviving his older, more popular, works for companies throughout the world; he also choreographed and danced in the films *The Red Shoes* (1948) and *Tales of Hoffmann* (1951). In 1960 he formed a new company, the Balletto Europeo, and continued to dance until well past the age of sixty.

Mastroianni, Marcello (1924–) *Italian actor, who has played the leading roles in a great variety of Italian films.*

Mastroianni worked as a draughtsman and clerk before becoming an actor. As an amateur he appeared with a theatrical group at the University of Rome and made his professional debut in films in an adaptation of Victor Hugo's *Les Misérables* – *I miserabili* (1947). His theatrical debut came the following year, after which he appeared in a number of plays, including *Death of a Salesman*, *A Streetcar Named Desire*, and in various Shakespearean roles. He subsequently achieved international recognition in such films as Luchino VISCONTI's *Le notti bianche* (1957; *White Nights*) and, particularly, Federico FELLINI's *La dolce vita* (1960). Other notable films include *Il bell' Antonio* (1960), *La notte* (1961), *Divorce Italian Style* (1961), for which he won a British Film Award, *8½* (1963), and *Leo the Last* (1969). Successful, too, were *Yesterday, Today and Tomorrow* (1963), for which he received his second British Film Award, and *Marriage Italian Style* (1964), in both of which he played opposite Sophia LOREN. His films since then include *Drama of Jealousy* (1970), for which he received a best actor award at Cannes, *What?* (1972), *A Special Day* (1977), *Blood Feud* (1981), *Ginger and Fred* (1985), *Intervista* (1987), and *The Beekeeper* (1988).

Matisse, Henri Émile Benoît (1869–1954) *French painter and sculptor, who was the leader of the fauve group of painters in the early 1900s and is acknowledged as the master of colour in this century. He won first prize at the Carnegie International at Pittsburg, in 1927 and at the Venice Biennale in 1950. In 1925 he was made Chevalier de la Légion d'honneur and in 1947 was elevated to Commandeur.*

The son of a grain merchant, Matisse studied law in Paris and worked in a lawyer's office before turning to painting during convalescence from appendicitis in 1890. He subsequently abandoned law and studied art in Paris. His early works were unadventurous but highly competent still lifes and interiors, executed while working part-time as a decorator. Around 1897 he became acquainted with the paintings of the impressionists and from this point colour became the most important element in his pictures. The energy and radical nature of his experiments with colour in such pictures as *Portrait of Madame Matisse* (1905) and *Open Window Collioure* (1905) ensured his leadership of the fauves. His aims, however, were different from those of the impressionists and of the short-lived fauve movement: in 1908 he wrote in his *Notes d'un peintre*, 'I dream of an art of balance, purity and calm, without troubling or depressing themes, that will offer … a soothing influence.' This aim is illustrated in the large figure compositions, increasingly simple in form, of the period 1905–10, including *Luxe, calme et volupté* (1905), *Joie de vivre* (1906), and *La Danse* (1909). During this period Matisse began to spend the summers painting in the south of France. He travelled widely between 1907 and 1914 and developed an admiration for Near East art, which inspired the exotic decorations that feature in many of his paintings.

After a less lyrical cubist-influenced phase during World War I, Matisse continued to celebrate the joys of life and, in contrast to his previous practice, incorporated modelling and perspective into his *Odalisque* paintings of the 1920s. He also produced some sculpture and from 1948 produced abstract pictures with cut-out coloured paper, such as *The Snail* (1953). Two of the greatest works of his maturity were commissions: the *Dance* murals for the Barnes Foundation, Pennsylvania (1930–32), and the decoration of the chapel at Vence (1949–51). In 1952 the Musée Matisse opened at Le Cateau-Cambrésis, where he was born.

Matthews, Sir Stanley (1915–) *British Association footballer with Stoke City, Blackpool, and England. He was knighted in 1965.*

Matthews was born at Hanley, near Stoke-on-Trent, where his father was a barber and boxer. At the age of fourteen he signed as an amateur for Stoke City, making his senior debut three years later. He went on to play in 886 first-class matches, 54 of them for England. He was dubbed 'the Wizard of Dribble', for few players have had better balance or shown more deceptive ball control. At the age of thirty-eight he achieved a long-standing ambition by winning an FA Cup medal for Blackpool against Bolton. His remarkable career lasted for thirty-three years and he made his last appearance as a player in 1965, five days after his fiftieth birthday.

After retiring from playing, he had a brief spell in management and later coached youngsters in various parts of the world. He made his home in Malta.

Maugham, W(illiam) Somerset (1874–1965) *British novelist, short-story writer, and playwright. He was made a CH in 1954.*
Maugham was born in Paris, where his father was legal adviser to the British embassy, and he spent his childhood in France, speaking French as his first language. When his parents died he was sent to England and King's School, Canterbury: his delicate health and pronounced stammer made this an unhappy time for him. After a spell at Heidelberg University and a holiday on the French Riviera, recovering from suspected tuberculosis, he became a medical student at St Thomas's Hospital, London (1892). His contact with London slum life through his hospital work inspired his first novel, *Liza of Lambeth* (1897); after graduating he abandoned medicine, more on the strength of two legacies than the success of his first novel.

He next wrote some undistinguished novels, but in 1903 his first play, *A Man of Honour*, was staged, and in 1907 the phenomenal success of *Lady Frederick* made him the country's most popular playwright. In 1908 he had four successful plays running in London. His pre-eminence as a writer of comedy continued through the 1920s. In the years before World War I he worked on his major novel, *Of Human Bondage* (1915), and during the war he worked for British Intelligence in France and Russia. His affair with Syrie Wellcome led to her divorce from the pharmaceutical manufacturer Sir Henry Wellcome and they were married in 1916; the relationship engendered considerable acrimony, even after Maugham and Syrie were divorced in 1927. A principal factor in the breakdown of the marriage was Maugham's relationship with a young American, Gerald Haxton, with whom he travelled and eventually settled (1928), after Haxton was barred from Britain, in the south of France. Meanwhile Maugham continued his output of best-selling novels: *The Moon and Sixpence* (1919), *The Painted Veil* (1925), and *Cakes and Ale* (1930) among them. He also wrote short stories, which were collected in such volumes as *The Casuarina Tree* (1926); a number of these stories were filmed.

Maugham had to flee from France in 1940 to escape the Nazis. He lived quietly in the USA until the war ended, when he returned to France with a new companion, Alan Searle, Haxton having died in 1944. Maugham's last important novel, *The Razor's Edge*, was published in 1944, but he continued to write essays, short stories, and historical fiction. He also supplemented his earlier autobiographical works, *The Summing Up* (1938) and *Strictly Personal* (1941), with a notorious memoir called *Looking Back* (1962), reviving the old bitterness between him and Syrie.

Mauriac, François (1885–1970) *French novelist. He was elected to the Académie Française in 1933 and received the Nobel Prize for Literature in 1952.*
Mauriac was born in Bordeaux, the setting for many of his novels, into a pious family of the upper middle classes. His strict Catholic upbringing had a marked effect on his literary works, the first of which was a collection of poems, *Les Mains jointes* (1909). In his early novels Mauriac began to develop the themes that were to recur throughout his works of fiction: the conflict between religion and passion, the problems of human relationships, and the narrowness and oppression of life in the provinces. He made his name with *Le Baiser au lépreux* (1922; translated as *The Kiss to the Leper*, 1923) a study of marital incompatibility and unsatisfied yearning for love. This was followed by *Génitrix* (1923), which centres on a mother's possessive love, and *Le Désert de l'amour* (1925; translated as *The Desert of Love*, 1949), awarded the Grand Prix du Roman of the Académie Française. *Thérèse Desqueyroux* (1927; translated in 1928), in which the young heroine, stifled by provincial life, finds herself driven to poison her husband, and *Le Noeud de vipères* (1932; translated as *Viper's Tangle*, 1933), a tale of avarice and family conflict, are generally considered to be Mauriac's finest novels. Later works of fiction include *Le Mystère Frontenac* (1933; translated as *The Frontenac Mystery*, 1951), which paints a somewhat rosier picture of family life in the provinces, and *La Pharisienne* (1941; translated as *A Woman of the Pharisees*, 1946).

Mauriac also wrote for the theatre, his most notable play being *Asmodée* (1938). Later plays, such as *Les Mals Aimés* (1945) and *Passage du Malin* (1948) were less successful. Events of the 1930s prompted him to polemical writings; he condemned totalitarianism and fascism, worked with the Resistance movement in World War II, writing under the clandestine pseudonym 'Forez', and openly supported General DE GAULLE in the early 1960s. Sensitive to criticism, he justified his moral attitudes and literary intentions in such studies as *Dieu et Mammon* (1929) and *Le Romancier et ses personnages* (1933). Further insight into his methods, values, and political views may be gleaned from his *Journal* (1934–51) and *Mémoires* (1959–67). His *Bloc-Notes* (1958–71) are collections of journalistic articles written for *Le Figaro* and *L'Express*.

Maurois, André (Émile Herzog; 1885–1967) *French biographer, novelist, and essayist. In 1938 he was elected to the Académie Française.*

Maurois was born in Elbeuf, Normandy. His father owned a textiles factory, where Maurois worked for a time as director before turning to literature. In this decision he was influenced by the philosopher Alain (1868–1951), who had taught him in Rouen and whose biography Maurois published in 1949. During World War I Maurois served with the British army as liaison officer and was awarded the DCM and later the KBE. *Les Silences du Colonel Bramble* (1918; translated as *The Silence of Colonel Bramble*, 1919) is a humorous recollection of life in an English officers' mess.

The first of a long series of romanticized biographies, *Ariel ou la Vie de Shelley*, appeared in 1923. This was followed by biographies of Disraeli (1927), Byron (1930), and Voltaire (1935), *À la Recherche de Marcel Proust* (1949), *Lélia*, a biography of George Sand (1952), *Olympio ou la Vie de Victor Hugo* (1954), and *Prométhée ou la Vie de Balzac* (1965). The remainder of Maurois's literary output consists of novels, such as the semiautobiographical *Bernard Quesnay* (1926), based on his experiences as director of his father's factory, and *Climats* (1928); short stories, such as *La Machine à lire les pensées* (1937); histories of England (1937) and the USA (1962); and critical and philosophical essays.

Maxwell, (Ian) Robert (Jan Ludvik Hoch; 1923–91) *British publisher, newspaper proprietor, and entrepreneur, born in Czechoslo-*

vakia; chief executive of the Maxwell Communications Corporation.

Maxwell was born into a poor Jewish family in the remote mountain village of Solotvino, then in Czechoslovakia (now in the Soviet Union). He received little formal education, dropping his studies for the rabbinate to become an itinerant salesman. The Nazi invasion of Czechoslovakia in March 1939 found him in Budapest, where he joined an underground resistance group. He later fought with Czech units in France before being evacuated to Britain in 1940. Joining the Pioneer Corps of the British Army, he was later transferred to fighting regiments; after D-Day he was commissioned in the field for bravery in Normandy, later becoming a captain and being awarded the Military Cross. In 1945 he adopted the name Maxwell, becoming a British citizen the following year. Most of his close family, including both parents, had died at the hands of the Nazis.

After the war, Maxwell served in military intelligence in Germany, working in the press department in Berlin from 1946. During this period he became involved in the export of German scientific journals to London and in 1951 he took over several of the companies concerned, forming Pergamon Press, the foundation of his publishing empire. Over the next fifteen years Pergamon expanded rapidly, acquiring many other small companies and making its founder a millionaire. A professed socialist, Maxwell became a Labour Party parliamentary candidate in 1959 and was MP for Buckingham from 1964 to 1970.

Maxwell's business methods had always aroused some controversy and in 1971 his conduct during the takeover of Pergamon by a US company led to an investigation by the Department of Trade and Industry. In a highly critical report the inspectors concluded that Maxwell could not be relied upon "to exercise proper stewardship of a publicly quoted company". By 1974 Maxwell was almost bankrupt, but managed to regain a controlling interest in Pergamon, whose share price had collapsed. With the rise in price of these shares during the ensuing years, Maxwell's fortune was restored and indeed increased sufficiently to enable him to become the proprietor of the Mirror Group of newspapers in 1984. The 1980s saw a further expansion of his business interests into computing, cable television, and many other areas, in all of which he adopted a high public profile. He was chairman of Derby County FC (1987–   ) and was formerly chairman of Oxford United FC (1982–87). His sudden death, in unclear circumstances while on his luxury yacht off Tenerife, prompted speculation about his links with Mossad, the Israeli secret service, and threatened the future of his many business enterprises. He was buried in Jerusalem.

Mayakovsky, Vladimir Vladimirovich (1893–1930) *Russian poet and playwright, a leading exponent of futurism and enthusiastic supporter of the Bolsheviks.*

Born at Bagdadi in Georgia, Mayakovsky was the son of an impoverished forester who belonged to the lower gentry. While he was a child the family moved to Moscow where they attempted to live on an inadequate pension. Mayakovsky first showed an interest in painting, but his talent for language and drama soon asserted itself. He aligned himself with the Russian futurists, rejecting traditional forms and genres as outmoded expressions of a pastoral society and irrelevant to modern life. New styles were necessary to express the specific qualities (mechanized, urban, etc.) of the twentieth century. Mayakovsky, who had been arrested several times for political (Social Democratic) activities, dropped politics and devoted himself to avant-garde creative work. His prerevolutionary pieces – for example, *Vladimir Mayakovsky Tragediya* (1913), a play written, directed, and acted by him, and the poems in *Oblako v shtanakh* (1914–15; 'A Cloud in Trousers') and *Fleyta-pozvonochnik* (1915; 'The Backbone Flute') – were calculated to shock the bourgeoisie and focus attention on the poet himself. They were not conceived as political pieces.

From 1917, however, Mayakovsky saw a natural role for futurism in the Bolshevik revolution and the new society. He designed brilliantly innovative propaganda posters and altered his aggressive avant-garde style to have a witty popular mass appeal, as in the play *Misteriya-buff* (1918) and the poem *150,000,000* (1919–20), both revolutionary classics. In 1923 he founded LEF (Left Arts Front) and edited its review (followed later by *Noviy LEF*) and throughout the 1920s played an active public role, giving poetry readings, designing posters for state shops, and writing film scripts and journalism in verse. He also produced appropriate pieces for public occasions, including a poem on the death of LENIN (1924) and *Khorosho!* (1927; 'Good!'), on the tenth anniversary of the October Revolution.

Futurism, however, did not suit the stuffy conservative tastes the regime revealed once it established itself. Mayakovsky's work was in-creasingly criticized and he found himself slighted by party officials. The plays *Klop* (1929; translated as *The Bedbug*, 1960) and *Banya* (1930; translated as *The Bath House*, 1963) reflect an anxiety at the betrayal of the revolution by officialdom. It is not known whether his suicide – he shot himself in his flat – was due to these causes or to some personal crisis. His reputation in the Soviet Union was ironically restored in 1935 when STALIN remarked that Mayakovsky remained the most talented Soviet poet and 'indifference to his memory and works is a crime'. His birthplace Bagdadi was renamed Mayakovsky in his honour but reverted to its original name in 1991.

Mayer, Louis B. (Eliezer Mayer; 1885–1957) *Russian-born US film executive who became one of Hollywood's most powerful film magnates.*

Born in Minsk, Mayer emigrated as a child with his family to Canada, where he began his working life as a scrap-metal dealer. His entry into the motion picture business came in 1907, when he acquired the first of a chain of picture houses in New England. He later became a distributor, securing the highly profitable New England distribution rights for D. W. GRIFFITH's *The Birth of a Nation* (1915). Production interests soon followed and by 1924 the amalgamation of the Metro, GOLDWYN, and Mayer companies, to form MGM, had been completed. 'Paternalistic', 'despotic', 'sentimental', are some of the conflicting adjectives that have been used to describe Mayer, who became the highest paid person in the USA and wielded almost unfettered power for more than two decades. Under his control such stars as the BARRYMORES, Greta GARBO, Clark GABLE, Judy GARLAND, Spencer TRACY, Katherine HEPBURN, and Elizabeth TAYLOR were to appear in a long string of highly successful films.

Mayer was also actively involved in setting up the Academy of Motion Picture Arts and Sciences (1927) and was president of the Association of Motion Picture Producers (1931–36). In 1950 the Motion Picture Academy presented him with a special award and the following year he left MGM, after being forced to retire by the board.

Mayer, Sir Robert (1879–1985) *German-born British patron of music. He was knighted in 1939 in recognition of his contribution to London's musical life and became a CH in 1973.*

Born in Frankfurt, the son of wealthy and musical German-Jewish parents, he was a gifted pianist and began to study at the Mannheim Conservatory at the age of six. At eleven Brahms heard him play and encouraged him to follow a musical career. However, he went into business and made a fortune in the metal industry in America and on the stock exchange. In 1896 he settled in England, becoming a naturalized British citizen in 1902 and marrying the singer Dorothy Moulton in 1919. Seeking to make a lasting contribution to musical life, Mayer instituted the Robert Mayer Children's Concerts, the first one taking place in 1923 under the baton of Adrian BOULT with Malcolm Sargent (1895–1967) taking over the following season. In 1929 Mayer retired from business in order to devote himself to extending the scope of his musical ventures. In 1932 he co-founded (with Sir Thomas BEECHAM) the London Philharmonic Orchestra and in 1954 he founded the Youth and Music concerts, catering for an older age group. His memoirs, *My First Hundred Years*, were published to mark his hundredth birthday in 1979, which was also marked by a concert at the Royal Festival Hall in his honour, attended by the Queen. In 1980, at the age of 101, he remarried.

Mboya, Tom (Thomas Joseph Mboya; 1930–69) *Kenyan politician and labour organizer. A fervent pan-Africanist, he is remembered for his anti-white slogan 'Scram out of Africa'.*

Born a member of the Luo tribe in western Kenya, the son of a sisal worker, Mboya was educated at Catholic mission schools. Qualifying as a sanitary inspector in 1951, he worked for the Nairobi City Council until 1952, when he became active in nationalist politics and trade union affairs. In 1952, when political parties were banned and many nationalist leaders were detained by the government following the Mau Mau uprising, Mboya began to build the trade union movement, becoming general secretary of the Kenyan Federation of Labour in 1953 (a post he held until 1962). After spending a year at Ruskin College, Oxford, on a scholarship (1955–56), Mboya was elected to the legislative council in 1957, where he led an unofficial opposition of eight African members. In 1960 he was elected general secretary of the newly created Kenya African National Union (KANU), which formed the independent government in 1963 under KENYATTA. Mboya was minister of labour (1962–64), minister of justice (1963–64), and minister of economic planning (1964–69). Although Mboya was considered a leading contender to succeed

Kenyatta, he never did so as he was assassinated in Nairobi in 1969 by a member of the Kikuyu tribe. His book, *Freedom and After*, was published in 1963.

McBride, Willie John (1941–) *Irish rugby player, who won sixty-three Irish and seventeen British Lions caps.*

McBride was only four years old when his father – a farmer – died. His tough childhood gave him a physical and mental courage that was to be invaluable in his rugby. He did not start playing the game until he was sixteen, but his outstanding talents were soon recognized. A lock forward, he first played for Ireland in 1962 and went on Lions tours to South Africa (1962, 1968) and New Zealand (1966). The heavy defeats of these tours were redressed in 1971, when the Lions pack under McBride's leadership won the series against the All Blacks.

Under McBride's captaincy in 1974 Ireland won the international championship for the first time for twenty-three years. His leadership of the 1974 Lions tour in South Africa inspired the team to win twenty-one out of twenty-two games (with one drawn), scoring ten tries in four tests. McBride was a superb tourist off the field as well – a fine ballad singer and all-round entertainer and one of the game's most endearing personalities.

McCarthy, Joseph R(aymond) (1908–57) *US politician whose startling series of allegations concerning 'communist subversion' in US public life in the early 1950s triggered widespread unsubstantiated accusations, a phenomenon that became known as the 'McCarthy witchhunt' or 'McCarthyism'.*

A farmer's son from Wisconsin, McCarthy qualified in law at Marquette University in 1935 and became a lawyer, later serving as a circuit judge (1940–42). Service in the marine corps earned him the rank of captain and a DFC and in 1946 he was elected a Republican senator. He first hit the headlines in February 1950 with a speech in which he claimed to know of 205 communists or communist sympathizers working for the State Department, although he later failed to substantiate his claims before a committee chaired by Senator Millard Tydings. However, McCarthy's strident campaign attracted popular support and leading Democrats, including the secretary of state Dean ACHESON and defence secretary George MARSHALL, became targets for McCarthy. He received endorsement from EISENHOWER during the latter's 1952 presidential campaign but the following year, after his appointment to the Senate's permanent subcommittee on investigations, McCarthy even turned on fellow Republicans. 'McCarthyism' became widespread; 'subversive' literature was banned and many innocent people were vilified. The climax came in May and June 1954 when, in televised hearings, McCarthy cross-examined US army personnel following further unfounded allegations. The true nature of his demagoguery was exposed to the nation and in December he was censured by a Senate committee. The McCarthy witchhunts were over.

McCarthy, Mary Therese (1912–89) *US novelist, critic, and journalist.*

Born in Seattle of Irish and Jewish ancestry, she and her three brothers, one of whom is the actor Kevin McCarthy (1914–), were brought up by relatives after the death of their parents in the influenza epidemic of 1918. This early period of her life is covered in *Memories of a Catholic Girlhood* (1957). A grandfather saw to it that she had a good education and she graduated from Vassar College (1933). She worked as a book reviewer for the weeklies *The Nation* and *New Republic* and was associated with the group of radical noncommunist and Trotskyite writers of *The Partisan Review*, of which she became drama critic and an editor. McCarthy married, but later divorced, the critic Edmund Wilson (1895–1972), by whom she had one son in 1948. In subsequent years she lived much of the time in Paris.

McCarthy's novels are witty, often devastating, satires that draw on her experience of intellectual circles and academic life. *The Company She Keeps* (1942) concerns a young woman's introduction to life in urban bohemia. *The Oasis* (1949) satirizes the attempt at an intellectual utopian community in New England, and *The Groves of Academe* (1952) is an outstanding example of the satirical academic novel. *The Group* (1963; filmed 1966) traces the various careers of eight college alumnae some thirty years after their graduation. Towards the end of her career McCarthy argued for a return to the novel of ideas, of which *Birds of America* (1971) and *Cannibals and Missionaries* (1979) are examples. Her early short stories were collected in *Cast a Cold Eye* (1950).

McCarthy was a distinguished critic, political and cultural commentator, and travel writer. Among her many nonfiction works are *Sights and Spectacles* (1956), a collection of theatre criticism, *Venice Observed* (1956), *The Stones of Florence* (1959), *On the Contrary*

(1961), *Vietnam* (1967), *Hanoi* (1968), *The Writing on the Wall and Other Literary Essays* (1970), *The Mask of State: Watergate Portraits* (1973), and the autobiographical *How I Grew* (1987).

McCartney, Paul (1942–) *British guitarist, singer, and songwriter. The bass guitarist in the Beatles (1962–70), he has had the most successful post-Beatle career of the four.*

Born in Liverpool, he bought a guitar in 1956 and a year later joined the Quarrymen, a skiffle group led by John Lennon. This group later became the phenomenally successful Beatles. The combination of Lennon's acerbity and McCartney's gentleness made them by far the most influential songwriters of the 1960s. After the Beatles broke up in 1970, McCartney formed the band Wings, which was in existence longer than the Beatles (although the personnel changed), finally disbanding in 1981. His later music has not been highly praised by critics but McCartney's talent for creating music people want to hear has ensured that more than two dozen records have been hits in the USA. Among his most successful albums since 1981 have been *Pipes of Peace* (1983), *Give My Regards to Broad Street* (1984), which was the score to a film he wrote and financed himself and in which he and his wife Linda starred, and *Flowers in the Dirt* (1989). He also wrote the score for and produced the film *Rupert and the Frog Song* (1984), which won a BAFTA award for Best Animated Film. His first classical composition, the *Liverpool Oratorio*, was co-written with Carl Davis and had its first performance in 1991. His many honours include several Ivor Novello awards.

McCartney has invested wisely in song copyrights, adding to his wealth accumulated from the Beatles era, and is now one of the world's richest men. He has been arrested several times on drugs charges and publicly favours the legalization of marijuana.

McCullers, Carson Smith (1917–67) *US novelist and playwright.*

Born in Columbus, Georgia, she had intended to study music at the Juilliard School in New York but, having lost the tuition money, she turned to writing instead. Her first book, *The Heart is a Lonely Hunter* (1940; filmed 1968), has some of the bizarre features of the 'Southern gothic' novel (the central character is a deaf-mute), but the distinction of the writing and the treatment of the implied subject of fascism led to great critical acclaim. Her second novel, *Reflections in a Golden Eye* (1941;

filmed 1967), again made use of gothic elements in relating the story of a murder in an army camp before World War II. *The Member of the Wedding* (1946), her best and most successful work, concerns the loneliness of a young girl at the time of her brother's marriage. The novel was dramatized (1950), with an extremely successful run, and filmed (1952). The title story of the collection *The Ballad of the Sad Café* (1951) was also successfully dramatized, by Edward ALBEE (1963). *Clock Without Hands* (1961), her last novel, dealt with Southern racial problems.

Troubled by recurrent ill-health, McCullers suffered a number of strokes and finally developed cancer. She married, was divorced in 1940, then after the war remarried the same man, who had returned from the war wounded, alcoholic, and addicted to drugs. He committed suicide after their second separation. These traumatic experiences she attempted to transform in the comedy *The Square Root of Wonderful* (1958), the heroine of which is twice divorced from the same husband.

McEnroe, John Patrick (1959–   ) *US tennis player who dominated the game in the early 1980s.*

McEnroe came to prominence in 1977 by reaching the semi-finals at Wimbledon in his first appearance. Subsequently his powerful serve and volley game produced an almost invincible combination. McEnroe lost his first Wimbledon singles final in 1980 to Björn BORG in a memorable five-set struggle, but the following year he beat Borg to capture the title. However, his mercurial temperament led to disputes with the tournament officials and a $10,000 fine (later revoked).

McEnroe's major singles successes include the Wimbledon title in 1981, 1983, and 1984, and the US Open championship in 1979, 1980, 1981, and 1984. He also won the WCT final in 1979, 1981, and 1983 and the Masters tournament in 1979 and 1983. Usually partnered by the US player Peter Fleming, McEnroe's doubles victories have included four Wimbledon titles (1979, 1981, 1983, and 1984), the US Open in 1979 and 1981, and the Masters six times (1978–83). In the mid-1980s, in the wake of a series of defeats, he announced his retirement from the game. After his return towards the end of the decade he struggled to repeat his earlier successes. He married the US film actress Tatum O'Neal (1962–   ) in 1986.

McIndoe, Sir Archibald Hector (1900–60) *New Zealand plastic surgeon who pioneered techniques of treating airmen badly burned during World War II, rebuilding their faces and limbs and helping them to overcome the psychological problems caused by their mutilations.*

McIndoe graduated from the University of Otago Medical School (1924) and after several years in the USA at the Mayo Clinic, Rochester, he went to London and trained as a plastic surgeon. By 1939 he was a renowned expert in this field and became consultant to the RAF, which arranged for him to organize a centre at East Grinstead for the treatment of severely burned airmen and air-raid victims. McIndoe created a world-famous centre for reconstructive plastic surgery and followed the progress of his patients, fighting on their behalf for better pay and conditions until they were rehabilitated. He personally operated on six hundred men, who formed their own club, 'McIndoe's Guinea Pigs', which met annually and through which he was able to follow the fortunes of his patients. After the war McIndoe improved the facilities at East Grinstead and trained plastic surgeons from all over the world.

McKellen, Sir Ian Murray (1939–   ) *British actor who made his name in a variety of classical and modern stage roles. He was knighted in 1991.*

Educated at Wigan and Bolton, he made his first stage appearance at the Belgrade Theatre in Coventry in 1961. After further work in the provinces, he made his London debut in 1964, winning the Clarence Derwent award for his performance in *A Scent of Flowers*. He joined the National Theatre in 1965 and the Prospect Theatre Company in 1968, establishing his reputation as a major contemporary actor with his portrayals of Shakespeare's Richard II and Marlowe's Edward II. In 1972 he became a founder member of the Actors' Company, with whom he continued to win praise in classical roles, and in 1974 he made his first appearance with the Royal Shakespeare Company. His most famous roles include Faustus (1974), Macbeth (1976), Iago (1989), for which he won Evening Standard and London Critics' awards, Coriolanus (1984), for which he won a London Standard award, and King Lear (1990) and Richard III (1990), which were two of his main successes with the Royal National Theatre. Other acclaimed performances have been in plays by such varied authors as Chekhov, Otway, and Stoppard. He won a Laurence Olivier award and other honours for *Wild Honey* in 1986 and numerous accolades, including a Tony award, as Salieri in *Amadeus*

(1980). He has toured internationally and was an associate director with the National Theatre (1985–86). His films include *Priest of Love* (1981), *Plenty* (1985), and *Scandal* (1989). His is also known for his outspoken campaigns on behalf of homosexuals.

McLuhan, (Herbert) Marshall (1911–80) *Canadian sociologist, literary critic, and commentator upon communications technology.*

McLuhan was educated at the universities of Manitoba and Cambridge and from 1946 to 1966 taught in the English department at the University of Toronto. He first examined the effects of technology and the media on society in his book *The Mechanical Bride: Folklore of Industrial Man* (1952), in which he analysed the sociopsychological impact of mass communication and interpreted it as an extension of the human nervous system. This thesis was reinforced and expanded by McLuhan in *The Gutenberg Galaxy* (1962), *Understanding Media: The Extensions of Man* (1964), and *The Medium Is the Massage* (1967), a title he adapted from his own slogan 'the medium is the message'.

In 1966, McLuhan became director of the Centre for Culture and Technology at the University of Toronto and his controversial views received wider publication. He became one of the most popular prophets of the decade and, in 1967, was appointed Albert Schweitzer Professor in the Humanities at Fordham University, New York. He predicted that, as society switched its attention away from the printed word towards the more technological media, books would disappear and it would be the characteristics of a particular medium rather than its content that would influence the individual. He also argued that electronic communication would create a world of instant awareness in which a sense of private identity would be impossible.

McLuhan's other publications included *War and Peace in the Global Village* (1968) and a selection of his literary criticism, *The Interior Landscape* (1969).

McMahon, Sir William (1908–88) *Australian statesman and Liberal prime minister (1971–72). He received the CH in 1972 and was knighted in 1977.*

Born and educated in Sydney, the son of a lawyer, McMahon practised as a solicitor until his service in the war (1939–45). McMahon joined the Liberal Party in 1949, winning the federal seat of Lowe in New South Wales. He became deputy leader in 1966 and leader in 1971, when he assumed the prime ministership

on GORTON's resignation. A determined leader with a great capacity for work, McMahon lacked the charisma needed to be prime minister. In 1972 he lost the federal election that brought the Labor Party to power for the first time in twenty-three years. McMahon remained a member of parliament until his resignation in 1982.

Mead, Margaret (1901–78) *US anthropologist noted for her studies of so-called 'primitive' societies and her work on the role of culture in character development.*

Mead studied psychology at Barnard College, where she encountered the distinguished anthropologist Franz BOAS. In 1925 she travelled to Samoa to study the transition of native girls from adolescence to adulthood and published her findings in *Coming of Age in Samoa* (1928). In 1926 she was appointed assistant curator of ethnology at the American Museum of Natural History, New York, becoming associate curator in 1942 and curator in 1964. She was made adjunct professor of anthropology at Columbia University in 1954.

In the late 1920s a trip to New Guinea enabled her to examine intellectual development in young children in relation to their cultural environment (*Growing Up in New Guinea*, 1930) and later, on the Indonesian island of Bali, she made innovative use of film to record aspects of the society and compiled the photographic study *Balinese Character* (1942). During World War II, Mead conducted a survey of eating habits in the USA and was also concerned with the social impact of US troops stationed in Britain. Her postwar work concentrated more on contemporary US society, especially the influence of cultural phenomena in psychiatry, mental health, child development, and education. She served on several government committees and lectured widely. Critics pointed to her tendency to disregard established sociological methods in favour of a more subjective approach. In the late 1960s, her concern with the disillusionment among the young and the problems of overpopulation and environmental crisis led to such works as *Culture and Commitment* (1970) and *A Way of Seeing* (1970).

Meade, Richard (1939–) *British three-day event rider who was among the best in the world for nearly twenty years. A member of the teams winning Olympic gold medals and the world and European championships, he has also won individual medals in these events.*

Meade grew up in Monmouthshire, where his parents had a stud farm. After school he joined the Hussars and then studied engineering at Cambridge University before making a career as a business consultant. He first represented Great Britain in the 1964 Tokyo Olympics. In 1966 Meade won the first of his two individual silver medals in the world championships and in 1968 won the team gold medal in the Olympics. From 1970 to 1972 with his horse The Poacher, he won Badminton, his second individual silver medal in the world championships, and team gold medals in the world and European championships. Perhaps the highlight of his career came at the 1972 Munich Olympics, where he took both the team and the individual gold medals on Laurieston. Again in 1981 and 1982 he won Badminton and world and European championship team gold medals. He became president of the British Equestrian Federation in 1989.

Meads, Colin Earl (1936–) *New Zealand rugby player. He was on the winning side in forty-one of his fifty-five international appearances.*

Meads had represented the New Zealand Under-23 team as both flanker and number eight by the age of nineteen. He won his first full cap in 1957 and was in the All Black team that beat the 1959 British Lions by three matches to one. In 1961 he was joined for the first time in the All Black pack by his brother, Stan Meads. They regularly locked the New Zealand scrum until 1966, when Stan retired.

Known as 'Pinetree', Colin Meads symbolized the New Zealand pack of the 1960s. His great skills, combined with a great will to win, made the All Blacks an unbeatable side and enabled the heavy defeat of the British Lions. The period 1963–71 saw several titanic struggles in the second row of the scrum between Meads and the Ireland and British Lion lock, Willie John McBRIDE.

After his retirement in 1972 Meads continued to influence New Zealand rugby through coaching his province, King County. He continues to run the farm that was also his training ground for many years.

Medawar, Sir Peter Brian (1915–87) *British immunologist who shared with Macfarlane BURNET the 1960 Nobel Prize for Physiology or Medicine for his work on acquired immunological tolerance and the development of the immune system in embryos and young animals. He was knighted in 1965, became a CH in 1972, and was awarded the OM in 1981.*

Born in Brazil, the son of a Lebanese businessman, Medawar was educated in England and obtained a zoology degree from Magdalen

College, Oxford. After lecturing and researching at Oxford he became professor of zoology first at Birmingham University (1947–51) and then at University College, London (1951–62). He was director of the National Institute for Medical Research (1962–71) and subsequently head of the Transplantation Biology Section of the Clinical Research Centre.

Medawar's interest in immunology arose from his research after World War II into the problems of skin grafting for burns. In 1949 Burnet suggested that during embryonic life and the early postnatal period cells gradually acquire the ability to distinguish their own tissues from foreign material, and Medawar found evidence to support this idea when he discovered that fraternal cattle twins accept skin grafts from each other. This indicated that antigens 'leak' from the yolk sac of each embryo to that of the other, giving them immunity to each other's tissues before their own system is fully developed. In a series of experiments on mice, Medawar found evidence that each cell contains genetically determined markers (antigens) important to the immunity process. An individual injected with a donor's cells while still an embryo will later accept tissues from all parts of the donor's body and from its twin. Medawar's work changed the basis of immunological research from attempting to treat the fully developed immune system to altering the mechanism itself. It has been invaluable in research to find methods of preventing the rejection of transplanted organs. In addition to his own research, Medawar was much admired for his essays on the philosophy of science, especially for his Reith lectures *The Future of Man* (1960) and his book *The Art of the Soluble* (1967).

Meir, Golda (Goldie Mabovitch; 1898–1978) *Israeli prime minister (1969–74). As a matriarchal Israeli figure, she was a fighter for peace in the Middle East.*

Born in Kiev, Russia, Meir emigrated with her parents to Milwaukee, USA, in 1907. She was educated at a teachers' seminary and in 1921 moved to Palestine, where she worked on a kibbutz. She served her political apprenticeship as secretary of the Histradut (Federation of Labour) Women's Labour Council (1928–32) and on the Histradut executive committee (1934–39).

Meir became active in national politics at the end of World War II. As head of the Political Department of the Jewish Agency (1947–48), she worked for the release of many people detained by the British authorities controlling

Palestine. Following the proclamation of Israel's independence in 1948, she became ambassador to the Soviet Union. Elected to the Knesset (parliament) in 1949, she was appointed minister for labour (1949–56) and minister for foreign affairs (1956–66). She retired from the government in 1966 and spent the next three years building up the Israeli Labour Party from disparate socialist factions (including the former Mapai Party, of which she was a co-founder). In 1969 she was elected prime minister, a position she managed to retain, through coalition rule, until 1974, when she retired after having seen her country through the Yom Kippur War. She published *My Life*, her autobiography, in 1975 and died in 1978 after a twelve-year struggle against leukaemia.

Meitner, Lise (1878–1968) *Austrian-born Swedish physicist, discoverer with her nephew Otto Frisch of nuclear fission.*

The daughter of a lawyer, she became interested in physics when she was told as a small girl that the colours produced by oil-stained puddles were due to the interference of light waves. Her father insisted that she first qualify as a teacher before he allowed her to study science at the University of Vienna. She duly obtained her doctorate in 1906 and moved to Berlin, where she began a remarkably fruitful research partnership with Otto Hahn that lasted more than thirty years.

Initially, Meitner was received somewhat churlishly and exiled to an old carpentry shop, on the grounds that her hair was liable to catch fire in a laboratory. However, apart from the period of World War I, which she spent nursing, Meitner remained in Berlin working mainly on establishing some of the basic features of the radioactive elements. In 1935, with her collaborators Hahn and Fritz Strassmann (1902–80), her work included a study of the decay products of irradiated uranium. It proved to be a complex field, but before their work could be completed Meitner, as a Jewess, fled from Berlin in 1938. After brief periods in Holland and Denmark, Meitner settled in Sweden, where she worked at the Nobel Institute of Physics in Stockholm until her retirement in 1960. Her last years were spent in Cambridge, England, where Otto Frisch and other relatives had settled.

It was with Frisch in 1938, while he was visiting her in Gothenburg, that she discussed the latest results of her collaborators, Hahn and Strassmann. They had found that isotopes of barium were produced when uranium was irradiated with neutrons. After a few calculations Meitner and Frisch concluded that the results could be explained by assuming the uranium nucleus had split into smaller parts, a process they later called nuclear fission. It was this process that led to the development of nuclear weapons, work that Meitner refused to pursue. Although she was not awarded a Nobel Prize, Meitner did share with Hahn and Strassmann the prestigious Fermi Prize in 1965.

Melba, Dame Nellie (Helen Armstrong; 1861–1931) *Australian operatic soprano, created a DBE in 1918. Her popularity was so great that Escoffier named his icecream dish pêche melba after her. Melba toast is also named after her, tradition has it because she so enjoyed a piece of toast a young waiter had burnt, while she was staying at the Savoy Hotel.*

Nellie Mitchell was born in Melbourne – her father was Scottish and her mother of Spanish descent. Although she showed evidence of her musical talent as a child, she did not study singing until after her marriage to Charles Nesbitt Armstrong in 1882. In 1886 she went to Paris, where she was taught by Mathilde Marchesi (1821–1913), and the following year made her debut at the Théâtre de la Monnaie, Brussels, as Gilda in *Rigoletto*; she sung under the name of Melba (after her native city). The success of this performance led to her debuts in London (1888, in the title role of *Lucia di Lammermoor*), Paris (1889), and New York (1893), and for the next twenty years she made regular appearances in all the major opera houses in Europe and the USA. Her voice – a light lyrical soprano of great purity – was ideally suited to such roles as Violetta in Verdi's *La traviata*, the title role of Delibes's *Lakmé*, and Mimi in Puccini's *La Bohème*. She gave her last performance at Covent Garden in 1926, after which she returned to Australia to become president of the Melbourne Conservatory. She published *Melodies and Memories* in 1925.

Melchior, Lauritz (1890–1973) *Danish-born naturalized US tenor, regarded as the greatest Wagnerian heroic tenor of his day. Among other awards, he became a Chevalier de la Légion d'honneur.*

He studied with Paul Bang at the Royal Opera School, Copenhagen, making his debut in the baritone role of Silvio in Leoncavallo's *I pagliacci* in 1913; for the next four years he continued to sing other baritone parts. Further study with Vilhelm Herold revealed the tenor quality of his voice, and after intensive study with eminent European teachers, notably Mildenburg, Melchior made his Covent Garden debut in 1924, as Siegmund in Wagner's *Ring* cycle. This marked the start of his international career. He sang at Bayreuth later the same year and made his debut at the Metropolitan Opera, New York, in 1926; from 1929 his career centred on this house until 1950. During this time he also appeared in a few Hollywood films and made frequent guest appearances in Europe and in Buenos Aires.

Melchior's repertoire was limited to predominantly Wagnerian roles – Lohengrin, Parsifal, Tristan (which he sang over two hundred times), Siegfried, and Siegmund – for which his powerful physique, vocal stamina, and ringing tenor were ideally suited.

Mendès-France, Pierre (1907–82) *French statesman and prime minister (1954–55).*

Born in Paris, the son of a dress manufacturer, Mendès-France was educated at the University of Paris, where he graduated in law and political science in 1927. He was admitted to the bar in 1928 at the age of twenty-one, the youngest lawyer in France at that time.

Mendès-France was elected to the national assembly as the Radical-Socialist deputy for Eure in 1932. At the outbreak of World War II he enlisted in the air force, but after the Vichy regime was set up in 1940, he joined the Free French forces in London. Appointed commissioner of finance in the Committee of National Liberation in 1943, he led the French delegation at the Bretton Woods Monetary Conference in 1944, later representing France on the governing boards of the IMF and IBRD (1946–58).

After France was liberated in 1944, Mendès-France was appointed minister of the national economy in DE GAULLE's provisional government. He resigned the next year over the rejection of his economic plan but later became permanent French representative to the United Nations Economic and Social Council (1947–50). Elected prime minister in 1954, he negotiated with Vietnam to end the Indochinese war, believing French withdrawal to be essential to its resolution and to the French economy. He was defeated in the 1955 election but retained a ministerial post without portfolio in 1956, eventually losing his seat in 1958. He briefly held a seat in the national assembly (1967–68).

Mendès-France was seen as a nonconformist who supported causes that were often unpopular with the general electorate. He became known as the 'Frenchman who drank milk'

because of his crusade for teetotalism, which antagonized the wine industry. His lack of wholehearted support for the European Defence Community also aggravated prominent French leaders in favour of European unity.

Menotti, Gian Carlo (1911–) *Italian composer resident in the USA, who has made his name with a series of highly acclaimed operas.*

Menotti was born near Lake Lugano into a prosperous and musical family, the sixth of ten children. When he entered the Milan Conservatory at the age of thirteen, he had already written two operas. In 1928 he crossed the Atlantic to study at the Curtis Institute of Music, Philadelphia, under Rosario Scalero. *Amelia goes to the Ball* (1936), a one-act opera buffa, was first performed there in 1937 and transferred to the Metropolitan Opera, New York, the following season with enormous success. This was the first of a long line of operatic successes for which Menotti has been his own librettist, writing generally in English.

Menotti's operas nevertheless derive from his Italian background, particularly from the works of Puccini and MASCAGNI. The operas are sectional, the arias connected by expressive declamation, and intimate, often of chamber proportions – Menotti himself calls them 'plays with music'. Among the best known are *The Medium* (1945), of which Menotti also made a successful film version, *The Telephone* (1946), a short opera buffa, *The Consul* (1949), *Amahl and the Night Visitors* (1951), *The Saint of Bleeker Street* (1954), and *The Most Important Man* (1971). More recent works include *St Teresa* (1982) and *Goya* (1986). Menotti has also written symphonic and choral works, as well as librettos for operas by Samuel BARBER and Lukas Foss (1922–). His foundation and organization of the Festival of Two Worlds, at Spoleto, is aimed at bringing young artists from the New World into contact with those of the Old.

Menuhin, Yehudi (1916–) *US violinist resident in the UK. A member of a distinguished family of musicians, he has received many honours, including the KBE (hon.) in 1965, the OM in 1987, and Grand Officier de la Légion d'honneur. He is also widely known as a writer and broadcaster on a number of humanitarian subjects.*

A child prodigy, Menuhin had his first violin lessons with Sigmund Anker in San Francisco, later studying with Persinger. His debut in San Francisco (1924), playing the Mendelssohn concerto, caused a minor sensation. Similar concerts in New York (1926) and Paris (1927) led Menuhin to become a pupil of Georges ENESCO, who was an enduring influence on his musical development. At twelve, after a performance of the Beethoven concerto under Fritz Busch (1890–1951) in New York, Menuhin's status as a world celebrity was assured, not only because of his technical brilliance and spontaneity but also because of the maturity of his interpretations of Bach, Mozart, and Beethoven. A notable event was the performance in 1932 of ELGAR'S violin concerto, for which the seventy-five-year-old composer coached Menuhin and conducted the orchestra.

In addition to Menuhin's long and successful career as a soloist, he has been active in a large number of other musical spheres: as director of the Bath, Windsor, and Gstaad festivals; as a conductor, notably of his own chamber orchestra (founded 1958) and since 1988 of the English String Orchestra; in duets with the Indian musician Ravi SHANKAR and the jazz violinist Stephane GRAPPELLI; and in the foundation of a boarding school at Stoke d'Abernon, near London (1962), for musically gifted children. Menuhin has also appeared with other members of his family, notably his sisters Hephzibah Menuhin (1920–80) and Yaltah Menuhin (1922–) and his son Jeremy Menuhin (1951–), all of whom are pianists.

Menzies, Sir Robert Gordon (1894–1978) *Australian statesman and longest serving prime minister as leader of the Liberal Party (1949–66). He was knighted in 1963.*

Born in Japarit, Victoria, the son of a farmer who later became a politician, Menzies was educated in Melbourne and graduated in 1916 with a first class honours degree in law from the University of Melbourne. Menzies's political career began in 1928, when he was elected to the Victorian Upper House. He entered federal parliament as a member of the United Australia Party in 1934 and became attorney-general and deputy leader. He earned the nickname 'Pig Iron Bob' when, in 1938, he was instrumental in the government's decision to sell pig iron to Japan, which was then at war with China. Resigning from the cabinet in 1939, he was elected leader of the party and prime minister within a month, following the death of LYONS. Two years later, in 1941, he resigned after clashes with party members and attacks from the press.

Considered finished in politics, Menzies, as leader of the newly formed Liberal Party, made a triumphant come-back in 1949 to again become prime minister. He remained so for the next sixteen years in what for many Australians was an era of material prosperity. Seen as a master of political manoeuvre, Menzies had a reputation for arrogant dominance of his fellow ministers. He was generally conservative in domestic politics, but progressive in his policy of developing Australian universities. Pro-American in many respects, he nevertheless retained an emotional attachment to the British Commonwealth throughout his career. In 1965 Menzies was the first Australian to become Lord Warden and Admiral of the Cinque Ports and Constable of Dover Castle; following his retirement in 1966 he became a Knight of the Thistle.

Mercouri, Melina (1925–) *Greek actress and politician, who has achieved considerable success as a film star and as a minister.*

Mercouri was born in Athens into a politically active family, her father having been a minister of state and deputy mayor of Athens. After training with the Greek National Theatre Company, she worked successfully in the theatre for a time before making her screen debut in *Stella* (1955). This was followed by *Celui qui doit mourir* (1957; *He Who Must Die*), directed by Jules Dassin (1911–), whom she married in 1966. He also directed *Never on Sunday* (1960), the film with which she has become most closely identified and which brought international recognition, a Best Actress Cannes Festival Award, and an Oscar nomination. Mercouri also appeared in the Broadway musical version, *Illya Darling* (1967). Other successes include *Topkapi* (1964), another Dassin film, and several Greek dramas including *The Medea*, performed on Mount Likavittos, and *The Oresteia*.

In the sixties Mercouri campaigned against the ruling junta and was forced into exile until a civilian government was restored in 1974. Returning to Greece, she was elected Member of Parliament for the Port of Piraeus in 1977 and appointed minister of culture and science (1981–85), in which role she appealed repeatedly for the return of the Elgin Marbles from Britain. She was minister of culture, youth, and sports (1985–89). An autobiography, *I Was Born Greek*, was published in 1971.

Merleau-Ponty, Maurice (1908–61) *French philosopher and phenomenologist.*

Educated at the École Normale Supérieure, Merleau-Ponty graduated in 1931. Before World War II, in which he served in the army, Merleau-Ponty taught in a number of lycées. After the liberation of France in 1945 he held chairs at the University of Lyon (1945–49), the

Sorbonne (1949–52), and, from 1952 until his death, the Collège de France.

Merleau-Ponty is best known outside France for his two major works, *Phénomenologie de la perception* (1945; translated as *The Phenomenology of Perception*, 1962) and *La Structure du comportement* (1942; translated as *The Structure of Behaviour*, 1963), in which he sought to unite modern psychological theory, the phenomenology of HUSSERL, and the existentialism of Jean-Paul SARTRE. In France he was also known for his left-wing political activities. With Sartre he had founded and edited the journal *Les Temps Modernes*, in which many of the left-wing debates of the period were pursued. With the Korean War (1950–53), however, Merleau-Ponty abandoned his commitment to communism, broke with Sartre, and adopted a more independent position.

Messerschmitt, Willy (1898–1978) *German aircraft designer and manufacturer, responsible for many German warplanes including the Me 262, the first operational jet fighter.*

The son of a wine merchant, Messerschmitt was educated at schools in Frankfurt and Bamberg and at the Technische Hochschule, Munich. An early interest in aviation, stimulated by the WRIGHT brothers' achievements, led Messerschmitt to design, produce, and eventually sell a long line of sophisticated gliders. In 1923 he set up his own business at Bamberg and turned to the production of powered aircraft. Although Messerschmitt went bankrupt in 1931 he was able to resume business in 1933. Shortly afterwards the Luftwaffe ceased to conceal its rebuilding programme and Messerschmitt was commissioned to plan a new fighter plane. The result was the Me 109 (1935), the most successful German plane of World War II, 33 675 of which were built for the Luftwaffe. Messerschmitt's most original plane, however, was the Me 262, the first jet fighter. Work began on the twin-engined jet in 1938 and for its construction a large underground plant free from the effects of allied bombing was built at Kahla. Fortunately only 1294 models of this plane, with a top speed of 600 mph, were built. Hitler insisted that priority be given to jet bombers and in the resulting confusion insufficient numbers of both were produced.

With the collapse of Germany in 1945, Messerschmitt was taken prisoner, first by the British and then by US forces. He was held in custody by the USA for two years. On his release an Augsburg court decided that the 44 186 planes he had built between 1939 and 1945 had been produced against his will. As Germany was banned from building planes, Messerschmitt turned to the construction of low-cost houses and sewing machines. Free once more to build planes in 1958, he began to produce over the next decade a range of satellites, missiles, helicopters, and aircraft. After a series of mergers in the late 1960s, Messerschmitt's company began work in the 1970s on the airbus and the Tornado, a fighter for NATO.

Messiaen, Olivier (1908–) *French composer who has been immensely influential not only as a result of the modal harmony and rhythmic complexity of his own music but also as a teacher. He is a Grand Officier de la Légion d'honneur.*

Born in Avignon, Messiaen was the son of a professor of literature and the poet Cécile Sauvage. He entered the Paris Conservatoire at the age of eleven, winning every major prize during his time there; among his teachers were Marcel Dupré (1886–1971) and Paul DUKAS. In 1930 Messiaen became the principal organist at La Trinité, Paris, a post he held for over forty years, and in 1936 joined the staff of the École Normale de Musique and the Scola Cantorum. During this period in Paris, he and several other young composers formed the group 'La Jeune France', to reaffirm the human and spiritual values of music.

When World War II began, Messiaen was drafted into the army and later captured by the Germans. In a prison camp in Silesia he composed and performed (to an audience of five thousand prisoners) his *Quatuor pour la fin du temps* (*Quartet for the End of Time*) for violin, clarinet, cello, and piano (1941). On his release, Messiaen became professor of harmony at the Paris Conservatoire, where his class in musical analysis later attracted students from all over the world. In his *Technique de mon langage* (1944), Messiaen identifies the sources of his work as medieval modes and Gregorian chant, Hindu rhythm, Debussyan impressionism, and the sounds of nature, especially birdsong.

Much of Messiaen's work is written or influenced by the organ, and his Catholic faith and vision of death are recurring themes. From his long catalogue of works, the culmination of his early compositions is the Hindu-influenced *Turangalîla-symphonie* (1946–48), expressed in chord clusters and birdsong using a large orchestra. The *Livre d'orgue* (1951) is an example from a more intellectual period, while

Couleurs de la cité céleste (1963) is a later synthesis of his musical language. *Et exspecto resurrectionem mortuorum* (1964) was commissioned by the French government in memory of the dead of two world wars. Later works include *La Transfiguration* (1969), *Des Canyons aux étoiles* (1974), the opera *Saint François d'Assise* (1975–83), and *La Ville d'En-Haut* (for piano and orchestra; 1987). Messiaen is married to the pianist Yvonne Loriod, his former student and a brilliant exponent of his works.

Meštrović, Ivan (1883–1962) *Yugoslav-born US monumental sculptor.*

Meštrović was born into a peasant family in Vrpolje, Croatia, and grew up in the mountain village of Otavica. After being apprenticed to a stonemason in Split in 1899 he studied sculpture in Vienna (1900–04) and from 1902 exhibited with the Vienna Secession group. After a visit to Paris (1907–08) he returned to Yugoslavia, where he began to establish himself as a monumental sculptor. Back in Yugoslavia after World War I, spent in Italy, France, Switzerland, and England, he was made head of the Academy in Zagreb and given numerous public commissions. His colossal sculptures in bronze or marble usually reflected his strong patriotism and employed simplified elongated classical or cubistic forms. One of his best-known works is the massive mausoleum for the unknown soldier outside Belgrade (1934).

Following imprisonment by the Gestapo (1941–42), Meštrović spent World War II first in Rome, where he accepted commissions from the Vatican, and then in Switzerland, emigrating to the USA in 1946. He was professor of sculpture first at Syracuse University (from 1947) and then at the University of Notre Dame, Indiana (from 1955).

Metaxas, Ioannis (1871–1941) *Greek general and dictator (1936–41). A staunch monarchist throughout his life, Metaxas was known for his pro-German sympathies until Greek independence was threatened, when he declared war on the Axis powers.*

Born in Ithaca, Metaxas joined the army in 1890. He fought against the Turks in 1897 before completing his military training in Germany. During the Balkan War (1912–13) he served as assistant chief of staff, becoming principal military adviser to King Constantine, who shared his opposition to Greece's entry into World War I. Promoted to general in 1916, he lived in exile in Italy (1917–20) when King Constantine was deposed.

On the restoration of the monarchy, Metaxas returned to Greece but departed again briefly in 1923, following its fall. During the late 1920s and early 1930s he led a monarchist party in opposition to the republican government. Eventually, in 1935, the monarchy was restored with his help and under King George II he became minister of war and prime minister (1936); he formally instituted a dictatorship in August of that year. His style of leadership had many of the attributes of fascism, such as the suppression of political opposition, but he introduced several social and economic reforms. When, in 1940, Italy invaded Greece, he led his country into World War II and joined the Allies. He died in 1941, only three months before the Germans occupied Greece.

Mies van der Rohe, Ludwig (1886–1969) *German-born architect and director of the Bauhaus, who emigrated to America in 1937. His pavilion for the 1929 International Exhibition in Barcelona was regarded as a masterpiece of twentieth-century architecture, and his New York Seagram Tower is one of his most elegant edifices.*

Born in Aachen, Mies was apprenticed to his father, a stonecutter, and learnt to draw as a designer of stucco decorations. He had no formal architectural training. However, in 1908 he joined the workshop of the illustrious Peter BEHRENS as a junior assistant, at the same time as GROPIUS was a senior assistant. After World War I Mies emerged as a highly original architect in his own right with a great talent for making exciting use of new materials: this was especially evident in his unexecuted designs for glass skyscrapers. His commission to design the German pavilion at the 1929 International Exhibition in Barcelona, at the age of forty-three, was an appropriate recognition of his achievements. That the building no longer exists has done little to detract from its importance to the evolution of modern architecture. Every detail of the building was designed personally by Mies (including the famous upholstered metal Barcelona chair). In 1930 Mies was appointed to succeed Gropius as director of the Bauhaus in Dessau. Under Mies, this institution continued to have an enormous influence on all aspects of design in Europe until it was closed by the Nazis in 1933.

Mies, as unable to contemplate life in Nazi Germany as the Nazis were to accommodate his original mind, emigrated to the USA in 1937. Here he became director of architecture at the Chicago Armour Institute, which later became the Illinois Institute of Technology, occupying a remarkable campus designed by Mies himself in 1940. Two glass and metal tower apartment blocks on Lake Shore Drive, Chicago (1948–51), led to commissions in many other US towns and cities. The most admired of these is undoubtedly the bronze tower of the Seagram building in New York (1954–58). Although there are no Mies buildings in the UK, his design for the Mansion House Square development in London was his last set of plans. These were completed in 1969 but were rejected by the City of London Corporation in 1985 as a result of public controversy.

Miles, Bernard James, Baron (1907–91) *British actor and director, founder of the Mermaid Theatre. He was knighted in 1969 and created a life peer in 1979.*

Miles was born in Uxbridge and educated at Pembroke College, Oxford. After working as a schoolteacher he made his first appearance on the stage in *Richard III* at the New Theatre in 1930. Several years in repertory were followed by appearances at the Old Vic and elsewhere. After the war he appeared in radio shows, specializing in rustic monologues, and in 1950 he appeared at the London Palladium; the following year he and his wife, Josephine Wilson, founded the first Mermaid Theatre in the garden of their house in St John's Wood in London. Designed as an Elizabethan playhouse, it moved in 1953 to the Royal Exchange and finally, in 1959, to Puddle Dock in Blackfriars. Miles directed and appeared in many Mermaid productions, including his own dramatization of Stevenson's *Treasure Island*, playing Long John Silver. The opening production at Puddle Dock, the musical *Lock Up Your Daughters*, was another of his adaptations, this time from Fielding's *Rape Upon Rape*.

Miles made his film debut in *Channel Crossing* (1933). Among the many memorable films that followed was *Great Expectations* (1946), in which he produced the definitive Joe Gargery. Others include *In Which We Serve* (1942), *Tawny Pipit* (1944), which he produced, co-directed, and starred in, *The Guinea Pig* (1948), and *Heavens Above* (1965). He was the author of several books, including *The British Theatre* (1947).

Milhaud, Darius (1892–1974) *French composer. A member of the group known as Les Six, Milhaud wrote a considerable amount of polytonal music and was deeply influenced by jazz.*

Born into a wealthy and cultured Jewish family from Provence, Milhaud was encouraged in his musical ambitions. He entered the Paris Conservatoire at the age of seventeen as a violinist, but soon decided to become a composer. His teachers at the Conservatorie, including Paul DUKAS and Charles Widor (1844–1937), influenced him less than his contacts with the artistic and literary personalities of contemporary Paris, notably the writers Francis Jammes (1868–1939) and Paul CLAUDEL, both of whom later collaborated with him as librettists. Jammes wrote the text for *La Brebis égarée* (1910–15), Milhaud's first opera, and Claudel was the librettist of *Christophe Colomb* (1928). Claudel also adapted three of the dramas of Aeschylus, for which Milhaud wrote incidental music: *Agamemnon* (1913), *Les Choëphores* (1915), and *Les Euménides* (1922). During World War I, Claudel was appointed French minister to Brazil and as Milhaud was unfit for military service Claudel invited him to be his secretary. This two-year period in Brazil provided the inspiration for such works as the two dance suites *Saudades do Brasil* (1920–21). On his return to Paris in 1918, Milhaud was drawn into the COCTEAU circle of writers and musicians, incuding Erik SATIE, and became known as one of Les Six. *Le Boeuf sur le toit* (1919), a ballet with scenario by Cocteau, was written at this time.

In 1920 Milhaud visited London and first heard jazz, which became another important influence on his compositions. Two years later he toured the USA and wrote the ballet *La Création du monde* (1923) based on the revelation of Harlem jazz. Milhaud both travelled and composed compulsively during the 1920s and 1930s, although he was frequently confined to a wheelchair as a result of rheumatoid arthritis. In 1940, after the fall of France, he and his wife sought refuge in the USA. There he was offered a teaching post at Mills College, Oakland, California, a connection that he maintained after his return to France in 1947. He then combined teaching at the Paris Conservatoire and in the USA with his other activities, only resigning from Mills College in 1971.

In spite of all this teaching and travelling, Milhaud had a large output, including twelve symphonies, chamber music, and piano works. His last work, the cantata *Ani maamim, un chant perdu et retrouvé*, was written for the 1973 Israel Festival.

Miller, Arthur (1915–   ) *US dramatist, regarded as one of the leading American playwrights of the twentieth century.*

The son of a Jewish manufacturer, Miller was born in New York City, where he suffered at first hand the effects of the Depression when it hit his hitherto prosperous family. This taught him at an early age to recognize the destructive effects of poverty, a theme he later explored in his plays. After leaving school, he held various odd jobs and attended a course in journalism at Michigan University. His first play was *The Man Who Had All the Luck* (1944), but it was *All My Sons* (1947) and more notably his masterpiece *Death of a Salesman* (1947), which won him the 1949 Pulitzer Prize, that brought both recognition and fame. *The Crucible* (1953) drew an unmistakable parallel between the Salem witch trials of 1692 and the McCarthy witchhunts of the 1950s, during which Miller refused to name suspected communists. Subsequent plays included *A View from the Bridge* and *A Memory of Two Mondays* (both 1955), *The Price* (1968), *The Creation of the World and Other Business* (1972), *Playing for Time* (1981), *The Archbishop's Ceiling* (1986), and *The Ride down Mt Morgan* (1991). *After the Fall* (1964) examined with typical honesty his unhappy relationship with Marilyn Monroe, whom he married in 1955; he also wrote the screenplay for Monroe's last film, *The Misfits* (1961).

A fine distinction can be drawn between Miller's early and later plays. All concern relationships, but the earlier works focus primarily on the individual in relation to the outside world and the need for self-knowledge as a means of coming to terms with reality; the later plays emphasize relationships between individuals. He published his autobiography, *Timebends: A Life*, in 1987.

Miller, (Alton) Glenn (1904–44) *US trombonist, arranger, and bandleader. He led one of the most popular dance bands in the world, whose records are still selling after nearly fifty years.*

He was born in Iowa into a middle-class family and began playing the trombone professionally while still at college. In 1926–28 he worked for Ben Pollack, who led one of the best white bands of the period. Miller played in theatre bands on Broadway and wrote arrangements for others before forming his own band in 1937. His first venture went broke but a new band, formed the following year, rapidly became one of the most successful in the business. A competent but not outstanding trombone player, he hired musicians like himself, who could do what they were told. His arrangement of 'Moonlight Serenade', which

became his signature tune, was known throughout the world – in a sense it became the signature tune of the whole swing era. 'In the Mood', 'Tuxedo Junction', and 'Moonlight Cocktail' were Miller's great recording successes. The band also starred in two films, *Sun Valley Serenade* (1941) and *Orchestra Wives* (1942). Miller joined the US army in 1942; his all-star service band was being posted to France, six months after the invasion of Normandy, when his aircraft disappeared in a fog over the English Channel.

A biographical film, *The Glenn Miller Story* (1953), played somewhat loosely with the facts.

Miller, Henry (1891–1980) *US novelist, whose works achieved notoriety for their use of sexually explicit and obscene language.*

Born and brought up in Brooklyn, Miller studied very briefly at City College of New York and then had various jobs in New York. He determined in 1924 to devote himself to writing and in 1930 moved to Paris, where he lived until 1940. On returning to America, he toured the country and published a highly critical account of what he found, *The Air-Conditioned Nightmare* (1945). He settled in Big Sur, California.

Although classed as fiction, Miller's writing is autobiographical. Incidents may be selected, arranged, and otherwise fictionalized, but the identity of the first-person narrator is never in doubt. *Tropic of Cancer* (1934) deals with Miller's life in Paris and is distinguished by a bawdy sense of humour, a frank depiction of sex, and a use of obscenities as they are actually used, especially in all-male company. These qualities prevented US publication of the book until 1961, while assuring an enormous underground readership. *Black Spring* (1936; US edition, 1963) is a collection of pieces intended as a companion to *Tropic of Cancer*, and *Tropic of Capricorn* (1939; US edition, 1962) deals with Miller's youth in New York. *The Colossus of Maroussi* (1941) records a visit to Greece in 1939, *The Time of the Assassins* (1956) is a critical work on Rimbaud, and *Big Sur and the Oranges of Hieronymus Bosch* (1958) concerns Miller's life in Big Sur. A trilogy of novels written in California and entitled *The Rosy Crucifixion* (*Sexus*, 1945; *Plexus*, 1949; *Nexus*, 1960) suggests an American equivalent to the early Paris books, but it lacks the comic sense that enlivens and redeems *Tropic of Cancer* and *Tropic of Capricorn*. Miller published several volumes of nonfiction as well as collections of correspondence with Lawrence DURRELL (1962) and Anaïs Nin (1965).

Miller, Keith Ross (1919–) *Australian cricketer renowned for his entertaining attitude to the game. He played 55 times for his country (1946–56), scoring 7 centuries and taking 170 wickets, although these figures do not reflect the impact of his personality.*

Born in Melbourne, he began his career with Victoria in 1937–38 with a dynamic innings of 181 against Tasmania. After World War II, during which he served with the RAAF, he stirred the imagination of the British public with his hard hitting in the 1945 Victory Tests at Lord's. He then returned to domestic cricket, this time for New South Wales for whom he played from 1947–48 to 1955–56, part of the time as captain. For Australia, BRADMAN valued Miller more as a bowler: the combination with Ray Lindwall (1921–) was highly successful. In his bowling, as in his batting, Miller was unpredictable, but spectators always enjoyed his performance on the field.

Millikan, Robert Andrews (1868–1953) *US physicist, who was awarded the 1923 Nobel Prize for Physics for his determination of the charge on the electron.*

Born in Illinois, the son of a Congregational minister, Millikan was educated at Oberlin College and Columbia University, where he gained his PhD in 1895. He spent a year in Europe at Göttingen and Berlin before taking up an appointment in 1896 at the University of Chicago. Millikan left Chicago in 1921 to become director of the Norman Bridge Laboratory of the California Institute of Technology, a post he held until his retirement in 1945.

Although J. J. THOMSON had identified the electron in 1897, the magnitude of its charge was still uncertain when Millikan carried out a series of classic experiments in 1909. In a paper published in 1913, based on 58 observations with charged oil drops, Millikan showed that the charge on the oil drops was always an integral multiple of 1.6×10^{-19} coulomb, a figure close to the currently accepted figure. Later research has shown that Millikan's results were carefully selected from a larger list of 140 observations. Those observations that did not agree with Millikan's conclusions were omitted.

Millikan also worked for many years on the nature of the cosmic rays first identified in 1912 by Victor HESS. In a series of ingenious observations begun in the 1920s Millikan conclusively demonstrated that they originated beyond the earth's atmosphere. He was less

satisfactory on the nature of the rays, however, insisting in a lengthy controversy with Arthur COMPTON that they were electromagnetic radiation and not charged particles. Compton turned out to be right.

Mills, Sir John (Lewis Ernest Watts; 1908–) *British actor, producer, and director. He was knighted in 1976.*

Born in North Elmham, Norfolk, Mills began his stage career as a song and dance man in 1929. Subsequently he appeared in productions of *Cavalcade* and *Of Mice and Men*, as well as with the Old Vic in its 1938 season. Although he made his film debut in *The Midshipman* (1932), it was not until after war service with the army that he began to be noticed in such patriotic films as *In Which We Serve* (1942), *This Happy Breed* (1944), and *Scott of the Antarctic* (1948). He also played the adult Pip in David LEAN's *Great Expectations* (1946) and the definitive Mr Polly in H. G. WELLS's *The History of Mr Polly* (1949), which he also produced.

Memorable films of the fifties included *Hobson's Choice* (1954) and *Ice Cold in Alex* (1958), followed by *Tunes of Glory* (1960). Starring roles, however, eventually gave way to cameo parts in such films as *Oh! What a Lovely War* (1969), *Young Winston* (1972), and *Gandhi* (1982), all directed by his friend Richard ATTENBOROUGH. For his supporting role as the village idiot in *Ryan's Daughter* (1970) he received an Academy Award. He has continued to combine film and theatre work including, most recently, *Pygmalion* (1987) and *When the Wind Blows* (1987), as well as various television series.

Mills is married to playwright Mary Hayley Bell and their daughters, Hayley Mills (1946–) and Juliet Mills (1941–), are also actors. Hayley appeared in *Sky West and Crooked* (1965), which Mills directed.

Milne, A(lan) A(lexander) (1882–1956) *British writer, noted especially for his ever-popular children's books and his plays.*

After attending Westminster School and reading mathematics at Trinity College, Cambridge, Milne found his métier as assistant editor of *Punch* (1906–19). During World War I he served as a signalling officer until invalided out of the army. In 1917 his first play, *Wurzel Flummery*, was produced and was followed by a string of successful light comedies: *Mr. Pim Passes By* (1919), *The Truth About Blayds* (1921), *The Dover Road* (1922), and *The Great Broxopp* (1923). *The Fourth Wall* (1928) was a clever murder mystery and in

1929 came the first production of *Toad of Toad Hall*, his dramatization of *The Wind in the Willows* by Kenneth GRAHAME.

Milne's most famous work, however, was the series of nursery stories written for his young son Christopher Robin, who was born in 1920: *Winnie-the-Pooh* (1926) and *The House at Pooh Corner* (1928). He also wrote two collections of verses for children; *When We Were Very Young* (1924) and *Now We Are Six* (1927). In the 1930s Milne turned to essays and short stories. *Peace With Honour* (1934) is a strong plea against war. His autobiography *It's Too Late Now* came out in 1939. He also wrote the novels *Two People* (1931) and *Chloe Marr* (1946). In 1952, the year that his last work, *Year In, Year Out*, appeared, he suffered a stroke and retired to his Sussex home, where he died.

Mingus, Charles (1922–79) *Black US jazz bass player, bandleader, and composer. He was the complete master of the string bass and of considerable importance as a jazz composer.*

Mingus was born in Arizona near the Mexican border. The family soon moved to Los Angeles, and he was one of the few great jazz musicians to grow up on the west coast. He had Chinese, Swedish, and British ancestry as well as American Negro; his father was very light-skinned and tried to teach him not to associate with darker people. Mingus, however, rejected this advice and remained, throughout his life, a passionate opponent of racism.

He began playing professionally while still a teenager and was one of the first and most successful black musicians to try to retain control of his own work. He formed Debut records (with drummer Max Roach) in 1952 and the Jazz Composers Workshop in 1953. In 1955 he combined a RACHMANINOV prelude with a song by Jerome KERN to create 'All the Things You C-sharp', in which the melody is combined with a powerful swinging melancholy. 'Pithecanthropus Erectus' (1956), an attempt to describe the ages of man, was an early extended composition. From 1959 such works as 'Better git it in your soul' and 'Ecclusiastics' introduced the passion of the black 'holiness' church into his music. Mingus discovered and employed many young musicians in his bands, always exhorting them to apply their own creativity to his music. His concert recordings, made in Europe and the USA during his last two decades, have established him as one of the best-known and best-loved of jazz musicians. His autobiography, *Beneath the Under-*

dog (1971), is not always reliable, but it is as vivid as any of his compositions.

Mintoff, Dom(inic) (1916–) *Prime minister of Malta (1955–58; 1971–84). A figure of some controversy in the international sphere, he was the dominant personality in Maltese affairs for thirty-five years.*

Born in Cospicua, the son of a naval cook, Mintoff studied engineering and architecture at the University of Malta before studying English at Oxford University on a Rhodes Scholarship (1939). He worked as a civil engineer in Britain (1941–43) and then returned to Malta to find employment as an architect (1943).

Mintoff joined the Maltese Labour Party in 1944. Elected to the Council of Government in 1945, he became a member of the legislative assembly, minister of works, deputy leader of the Labour Party, and deputy prime minister in 1947 (at the age of thirty-one). Two years later he was elected leader of the Labour Party, becoming prime minister when he led his party to victory in 1955. However, he was forced into opposition in 1958, over his failure to achieve Malta's integration with Britain. He returned to power in 1971 as prime minister of a now independent Malta (proclaimed 1964) and retained the leadership for the next thirteen years. In 1974 he abolished the monarchy and instituted a republic. During the 1970s he strained relations with Britain when he placed restrictions on British naval facilities on Malta, closing the base completely in 1979, and called for an end to Malta's association with NATO. Mintoff cultivated relations with Libya early in his term of office, turning to Italy for support in maintaining Malta's neutrality in 1980, when relations with Libya soured. He also fostered ties with the Soviet Union in the hope of increased Soviet investment in the country, making a visit to Moscow shortly before he finally retired from office in 1984. In the last few months of his premiership Malta was plunged into violence over a dispute between the government and the Roman Catholic Church on the control of Catholic schools.

Miró, Joan (1893–1983) *Spanish semi-abstract painter and graphic artist.*

After overcoming opposition from his father, who, like his father before him, was a craftsman in Barcelona, Miró attended the local school of fine arts. His first portraits and landscapes combined elements of Catalan folklore with the colourful technique of the fauves, recently current in Paris. Miró visited Paris in 1919 and became friendly with his compatriot

Mistinguett

PICASSO, whose influence can be seen in Miró's paintings of that period. Miró settled in Paris in 1920 and joined the dada group there but was later associated with the surrealists. He contributed to the first exhibition of the surrealists in 1926, by which time he had evolved the completely personal style that he continued to develop throughout his career. His paintings represented a brightly coloured fantasy world of spiky calligraphic forms against plain backgrounds. The lines and semifigurative forms were variously angular, amoeba-like, or reminiscent of cave paintings. André BRETON, the chief theorist of surrealism, described Miró's work as 'pure psychic automatism' and Miró himself said, 'I begin painting and as I paint the picture begins to assert itself, or suggest itself, under my brush. The form becomes the sign for a woman or a bird as I work.'

Collage, assemblage, and graphics began to interest him in the 1930s and ceramics in the 1940s. During the Spanish civil war he left Spain and lived in Paris, where the element of gaiety disappeared from his work; such pictures as *Head of a Woman* (1938) expressed his sense of impending horror. He returned to Spain in 1940 to escape World War II. In the 1950s he produced two large ceramic murals for the UNESCO building in Paris. In 1954 he won the Grand Prix for graphic art at the Venice Biennale.

Mishima, Yukio (Kimitake Hiraoka; 1925–70) *Japanese novelist, whose obsession with the loss of traditional Japanese military values led to his suicide.*

Born into an upper-class family in Tokyo, Mishima was educated at the Peers' School, where he developed an interest in the literature of Japan's classical (pre-Meiji) period, and graduated from the University of Tokyo in law. He resigned from a civil service position after a short time to devote himself to writing. His reputation was made with his first roughly autobiographical novel, *Kamen no kokuhaku* (1949; translated as *Confessions of a Mask*, 1960), in which there is a suggestion of the abnormal in the hero's (unfulfilled) desire to die young in some great conflagration. After a visit to Greece in 1952, Mishima published *Shiosai* (1954; translated as *The Sound of Waves*, 1956), an adaptation, transposed to Japan, of the classical Greek romance *Daphnis and Chloe*. His experience in Greece also led him to undertake a strenuous physical regime that transformed his appearance from that of an introspective intellectual into something approximating the ideal classical male. (He liked

to publish photographs of himself, a favourite pose – among many of somewhat erotic interest – depicting the martyrdom of St Sebastian.)

Kinkakuji (1956; translated as *The Temple of the Golden Pavilion*, 1959), perhaps his finest novel, was based on a real event in which a disturbed young apprentice priest burnt down a famous temple. In the novella *Yūkoku* (1960; 'Patriotism'), concerned with a conspiracy in the 1930s to seize political power and restore it to the emperor, Mishima took up a theme that was to become of obsessive interest. On the same day that he completed his final work, the tetralogy *Hōjō no umi* (1969–71; translated as *The Sea of Fertility*), Mishima and a few colleagues belonging to his private army, the Shield Society, seized the headquarters of the Japanese Self-Defence Force. After delivering a speech, Mishima committed suicide with a companion in a traditional manner (seppuku). He appeared to hope that this act would publicize his views on, and perhaps lead to a military uprising in defence of, Japan's traditional heritage; later speculation as to what other motives might be involved probably derives from assumptions about the pathological subject matter of some of his novels. Mishima also wrote a number of plays, both on modern subjects and on themes from the classical nō drama.

Mistinguett (Jeanne-Marie Bourgeois; 1875–1956) *French revue artist, singer, and actress, best remembered for her spectacular dance numbers, often partnered by Maurice* CHEVALIER.

A Parisian of Belgian and French extraction, Mistinguett made her stage debut in music hall in her late teens and subsequently performed at the Eldorado for some eight years. She appeared in light comedy and musical plays and for a time was part-owner of the Moulin Rouge, where she also gave many memorable performances. Mistinguett and Chevalier first became partners at the Folies-Bergère in 1910 and they appeared together many times during the next ten years. Fabulous costumes and large elaborate hats became her hallmark and her beautiful legs were said to carry the highest insurance in the business. She also appeared at the Casino de Paris on numerous occasions, singing such songs as 'Mon Homme'. Returning briefly to drama, she starred in Sardou's *Madame Sans-Gêne* (1921). Most of Mistinguett's career was spent in Paris. She appeared in the USA in 1911 and 1951 but her only appearance in London came late in her career (1947). She wrote two volumes of autobiography: *Mistinguett and her Confessions* (1938) and *Mistinguett* (1954).

Mitchell, Reginald Joseph (1895–1937) *British aircraft designer who created the Spitfire.*
Mitchell was born in Talke, Staffordshire, the son of a schoolteacher. In 1911 he entered into an apprenticeship with a firm of locomotive manufacturers in Stoke-on-Trent but as a consequence of his long interest in aviation he moved to Southampton to work as a designer for the Supermarine Aviation Works in 1916. He was appointed chief designer in 1919 and remained with the company for the rest of his life.

Mitchell initially worked on seaplanes. His first success was the Sea Lion, winner of the Schneider Trophy in 1922. This was followed by the various Supermarines that continued to win Schneider trophies and speed records throughout the 1920s and early 1930s. The turning point in Mitchell's life came with the publication in 1931 by the Air Ministry of their specifications for a new fighter to replace the ageing Bristol Bulldog. For the Supermarine entry Mitchell designed a low-winged monoplane of all-metal construction, powered by the 660 horsepower Rolls-Royce Goshawk engine. The first design from Mitchell, ready for testing in early 1934, impressed neither Mitchell nor the Air Ministry. He tried again with a second prototype, powered this time by the newly designed Rolls-Royce Merlin engine. It began its flight trials in March 1936. By this time it had already been given what Mitchell called its 'bloody silly name' – Spitfire. Mitchell, dying of cancer, lived long enough to see his plane accepted by the RAF. The first model was delivered in July 1938. The 19 000 that followed before the end of World War II were largely responsible for the defeat of the Luftwaffe and the consequent collapse of the Nazis. Mitchell's life and the story of the Spitfire were related in the wartime film *First of the Few* (1942), written, directed, and acted in by Leslie HOWARD.

Mitterrand, François Maurice Marie (1916–) *French statesman and the first socialist president (1981–) for thirty years.*
Born in Jarnac, southwestern France, the son of a stationmaster, Mitterrand was educated at the University of Paris, where he received degrees in law, arts, and political science. During World War II he fought in the colonial infantry but was wounded and captured by the Germans at Verdun (1940). In 1941 he escaped from his prison camp (at the third attempt) and returned to France, where he was active in prisoner-of-war and resistance movements. He worked briefly as a minor official for the Vichy regime before becoming secretary-general of the Organization for Prisoners of War, War Victims, and Refugees (1944–46) in DE GAULLE's provisional government.

Mitterrand was elected to the national assembly in 1946 as the deputy for Nièvres, representing the Democratic and Socialist Resistance Union (UDSR). During the Fourth Republic he held a number of ministerial posts including secretary of state for information (1948–49), minister for overseas territories (1950–51), minister of state (1952–53), minister of the interior (1954–55), and minister for justice (1956–57). He lost his seat in the national assembly in 1958 but was returned in 1962, having served as a senator in the intervening years. In 1965 he stood unsuccessfully as the presidential candidate for the Federation of the Democratic and Socialist Left (FGDS). As president of the FGDS (1965–68) he sought to strengthen the left alliance but the landslide victory of de Gaulle in 1968, following the student-worker protest, led to its demise. Nominated the first secretary of the Socialist Party in 1971, he again stood unsuccessfully as the socialist candidate for the presidency in 1974. In 1981 he finally won, a victory that brought to power the first socialist government of the Fifth Republic. As president Mitterrand has introduced several major reforms, including the nationalization of French banking and a decentralization programme. He has also played a leading role in the formation of EC policy. In 1986 he asked the right-wing politician Jacques Chirac to serve as prime minister in an unprecedented power-sharing arrangement; he defeated Chirac in the presidential elections of 1988 to secure a second seven-year term. His authorization of lavish expenditure to celebrate the bicentenary of the French Revolution in 1989 attracted criticism from some quarters.

Mizoguchi, Kenji (1898–1956) *Japanese film director.*
Born in Tokyo, Mizoguchi studied art and worked on a newspaper before entering films as an actor. Beginning at the Nikkatsu Studios in 1922, he soon abandoned acting for directing. With more than eighty films to his credit, he became one of Japan's most respected directors, his work, the central theme of which was a concern for women, being admired for its beauty and strong visual sense. *Osaka Elegy* (1936), *Sisters of the Gion* (1936), and *The*

Story of the Last Chrysanthemum (1939) are three of his many notable early works. As a figure in world cinema, however, he is particularly remembered for *Saikaku Ichidai Onna* (1952; *The Life of Oharu*), *Ugetsu Monogatari* (1953), *Sansho Dayu* (1954; *Sansho the Bailiff*), and *Yokihi* (1955; *The Princess Yang Kwei Fei*).

Modigliani, Amedeo (1884–1920) *Italian painter, sculptor, and draughtsman known for his melancholy elongated portraits.*

The fourth son of a wealthy Italian Jewish banker who lost his money, Modigliani suffered during his short life from both poverty and chronic illness. Pleurisy at the age of eleven was followed by tuberculosis at fourteen, after which he left school. He studied painting in his home town of Livorno and later in Florence and Venice before moving to Paris in 1906. Here, although he immersed himself in the life of the highly stimulating artists' quarter of Paris, his style remained uniquely his own. Despite the simplifications of form and the traces of cubism and expressionism, his paintings were influenced as much by fourteenth-century Italian artists, such as Botticelli, as by his contemporaries. The subjects of his portraits, with their elongated forms, frequently appear tired, sad, and vulnerable and they display the artist's sensitivity to character. The nudes, painted in the same linear style, are enlivened by an element of eroticism. He exhibited infrequently during his lifetime but was supported by two patrons who bought most of his work. Although there is little certain knowledge of the details of Modigliani's life in Paris, he has traditionally been regarded as the typical romantic genius: handsome, starving, obsessively painting and carving while destroying himself with drink and drugs and finally dying young from tuberculosis.

Moholy-Nagy, László (1895–1946) *Hungarian-born US painter, kinetic sculptor, and teacher, who pioneered the use of new materials and has been described as the prototype of the modern experimental artist.*

Moholy-Nagy studied law at Budapest and served in World War I before becoming a full-time artist. His first paintings were landscapes and portraits but in 1919 his growing interest in avant-garde experimentation was quickened by contact with the Russian artists Kazimir Malevich (1878–1935), El Lissitsky (1890–1941), and GABO in Vienna. In 1921 he was in Berlin experimenting with collage and photomontage. From 1923 he taught at the influential Bauhaus school, where he continued to paint and worked with photography, films, theatre design, typography, industrial design, and experiments with light, as well as co-editing Bauhaus publications. On leaving the Bauhaus in 1928 he moved to Berlin, Amsterdam, and London, where he was a member of the constructivist group, and finally to Chicago in 1937. Here he took charge of the New Bauhaus, founded the Chicago School of Design, and continued to produce his space modulations, which he had begun to develop in London. These three-dimensional constructions of glass, metal, and plastic produced optical effects by the play of light on their moving surfaces. Moholy-Nagy saw the manipulation of light as the art form of the future. His belief in this, and in the reintegration of art with the environment, are expounded in his two books, published in 1932 and 1947.

Molnár, Ferenc (1878–1952) *Hungarian playwright, whose ingeniously constructed boulevard dramas effectively combine romantic feeling with lightly cynical realism.*

Born at Budapest, Molnár studied law but in 1896 became a journalist and was soon drawn to the theatre. His first farces were popular locally and were followed by *Az ördög* (1907; translated as *The Devil*), a treatment of the Faust story that was produced in New York the following year. *Liliom* (1909), his best play, was a failure when first performed in Budapest, but it was revived after World War I and was an outstanding success in New York (1921) and London (1926) and again later in the form of the Broadway musical *Carousel* (1945). During the war, Molnár was a war correspondent and during the 1920s lived mainly in western Europe. He eventually settled in America, becoming a US citizen in 1940. He died in New York.

Among his many other plays are *A testör* (1910; translated as *The Guardsman*), which was adapted for radio by Arthur MILLER (1947); *The Glass Slipper* (1924), based on Cinderella; and *The Play in the Castle* (1924). The last was shown in New York as *The Play's the Thing* (1926), adapted by P. G. WODE-HOUSE, and more recently in London formed the basis of *Rough Crossing* (1984), an adaptation by Tom STOPPARD. Because he had lived most of his life abroad and perhaps because of his western success, Molnár was neglected in his native land until 1965, when his third wife, Lili Darvas, performed *Olimpia* (1927) there and revived his Hungarian reputation. His novel *A Pál utcai fiuk* (1907; translated as *The Paul Street Boys*, 1927) was widely translated.

Molotov, Vyacheslav Mikhailovich (Vyacheslav Mikhailovich Scriabin; 1890–1986) *Soviet statesman and diplomat, prime minister (1930–41), and foreign minister (1939–49; 1953–56). A staunch Stalinist, his opposition to* KHRUSHCHEV *resulted in his dismissal from office (1957) and expulsion from the Soviet Communist Party (1962).*

Born in Kukarka (now Sovetsk), he joined the Bolshevik Party and assumed the name of Molotov ('hammer') in 1906, while still a student, having participated in the antitsarist revolution of the previous year. After a two-year period in exile following his arrest in 1909, he worked for the party and was editor of the Bolshevik newspaper *Pravda* (1912–17). After the Bolsheviks' success in the October Revolution (1917), Molotov worked his way through the ranks of the party until in 1922 he became second secretary of the central committee and STALIN's right-hand man. In 1925 he became a member of the Politburo and in 1930 prime minister. At the outbreak of World War II, as foreign secretary, he negotiated the Molotov–Ribbentrop Pact (1939) – the treaty of nonaggression between the Nazis and the Soviets – and after the Germans invaded the Soviet Union he served in the war cabinet, arranging alliances with the Allies and attending the Tehran (1943), Yalta (1945), and Potsdam (1945) conferences. After the war, as Soviet representative at the United Nations Security Council, his frequent exercise of the Soviet Union's veto contributed to the prolongation of the Cold War.

After Stalin's death (1953) he resumed the position of foreign secretary but fell out with Khruschev over the latter's agricultural and industrial policies. Dismissed from his office in 1956, he joined the 'antiparty group', which attempted unsuccessfully to oust Khrushchev, and in 1957 was stripped of all his other offices, including membership of the central committee. He subsequently served as ambassador to Mongolia (1957–60) and Soviet representative to the International Atomic Energy Agency in Vienna (1960–61); in 1962 he was expelled from the Communist Party. He was rehabilitated in the mid-1980s, shortly before his death.

Mondrian, Piet (Pieter Cornelis Mondriaan; 1872–1944) *Dutch painter, who was a founder of the* De Stijl *movement and the chief exponent of neoplasticism, one of the earliest and strictest forms of geometric abstraction.*

Mondrian's Calvinist parentage and upbringing may partly account for the purity and single-mindedness of his later artistic life. A strong mystical element in his personality was also well developed by his early twenties. After studying in Amsterdam, his early subdued landscapes soon gave way to brightly coloured pictures, such as *The Red Tree* (1908); in Paris from 1911 he then went through a symbolist phase and finally a cubist phase. Feeling, however, that cubism was 'not developing abstraction towards its ultimate goal', Mondrian progressed from cubist-related works, such as *Still Life with Ginger Pot* (1912), towards more autonomous compositions consisting of lines and rectangular shapes with virtually no suggestion of depth, such as *Composition in Grey, Blue and Pink* (1913).

Before the end of World War I, which he spent in the Netherlands, he had, with Theo van Doesburg (1883–1931) and Jacobus OUD, formed the *De Stijl* group. Its members held that natural forms obscured 'pure reality'. Thus they 'denaturalized' their art by eliminating signs of brushstrokes and personal technique and by limiting themselves to the irreducible elements of form: vertical and horizontal lines, rectangular shapes, and primary colours, which they regarded as symbols of natural forces and underlying universal reality. Their style was also known as neoplasticism. In Paris in the 1920s and 1930s Mondrian's pictures became increasingly sparse. He broke with van Doesburg after the latter had introduced diagonal lines into his pictures. Mondrian left Paris for London in 1938 and moved to New York in 1940. The influence of New York life inspired a less ascetic style in which the vertical and horizontal lines were broken up into small squares of bright colour. He called these pictures the *Boogie-Woogie* series.

Monk, Thelonius Sphere (Thelious Junior Monk; 1920–82) *Black US jazz pianist, leader, and composer. His unique harmonic sense and original compositions made him a founding father of postwar jazz.*

He was born in Rocky Mount, North Carolina; by the time he was twenty he had become the house pianist at Minton's Club in New York, where most of the young innovators were playing. Later he led small groups and occasionally big bands. During the 1960s he led and recorded with a quartet, while in 1971–72 he toured the world with The Giants of Jazz. He was often unable to work through illness, but was honoured at President CARTER's White House Jazz Party in 1978.

His angular miniaturist piano style led some people to believe that he couldn't play well, but his compositions, such as 'Round Midnight', 'Straight, No Chaser', and 'Well, You Needn't', became favourite and challenging vehicles for improvisation, and his recordings have dated rather less than those of most of his contemporaries.

Monnet, Jean (1888–1979) *French economist and politician, first president of the European Coal and Steel Community and a leading figure in the formation of the European Economic Community.*

Born at Cognac, Charente, Monnet was educated in Cognac and in 1914 entered the Ministry of Commerce, becoming an expert in finance. He was appointed the first deputy secretary-general of the League of Nations in 1918. He had many public posts, including the chairmanship of the Franco-British Economic Coordination Committee in 1939, but he is mainly known for drawing up the French Modernization Plan (the Monnet Plan) in 1947 and later the Schuman Plan on European resources, which led to the formation of the ECSC. From 1956 to 1975 he chaired an action committee for a United States of Europe. *Les États Unis d'Europe ont commencé* consists of extracts from Monnet's speeches; he also published his *Mémoires* (1977).

Monnet was awarded the Charlemagne Prize in 1953, the Schuman Prize in 1966, and the title Honorary Citizen of Europe in 1976.

Monod, Jacques Lucien (1910–76) *French biochemist who, with François Jacob (1920–), first proposed the concept of a functional gene cluster (operon) to explain how gene expression is regulated in microorganisms. For this work they were awarded the 1965 Nobel Prize for Physiology or Medicine.*

Monod was born in Paris and graduated in science from the University of Paris in 1931, receiving his doctorate ten years later. He was appointed assistant professor of zoology (1934–45), during which time he spent a year at the California Institute of Technology working under the geneticist T. H. MORGAN. During World War II he fought in the French Resistance and in 1945 joined the Pasteur Institute, becoming head of the cellular biochemistry department (1954) and ultimately its director in 1971.

Working with the bacterium *Escherichia coli*, Monod and Jacob investigated the genes that code for three enzymes required by the bacterium to utilize the sugar lactose, and in 1961 they proposed a cluster (operon) comprising five neighbouring genes – a regulator, an operator, and three structural genes. The

regulator gene codes for a repressor protein. In the absence of lactose in the medium, this repressor binds to the adjacent operator gene, preventing transcription of the three structural genes. However, if present in the medium, lactose (acting as an inducer) binds to the repressor molecule so removing it from the operator and allowing transcription to proceed. Many examples of operons have been found in bacteria but not in higher organisms.

Monod's best-selling *Le Hazard et la nécessité* (1970; translated as *Chance and Necessity*, 1971) expounded his belief that life arose from a chance assembly of molecules and is directed only by the forces of natural selection. Man alone must 'choose between the kingdom and the darkness'.

Monroe, Marilyn (Norma Jean Baker/Mortenson; 1926–62) *US film star and sex symbol.*

An unhappy childhood in Los Angeles orphanages and foster homes, her mother's mental illness, an unknown father, a short broken teenage marriage, and early struggles as a photographer's model form the background to a Hollywood saga that ended, predictably, in a drug overdose. Monroe made her debut in a small part in *Scudda-Hoo! Scudda-Hay!* and subsequently appeared in a better part in the B-picture *Ladies of the Chorus* (both 1948). She first attracted attention in John HUSTON's *The Asphalt Jungle* (1950) and her first starring role came in the minor thriller *Don't Bother to Knock* (1952). The disclosure, at this crucial point in her career, that she had posed nude for a calendar photograph in her modelling days, and her protestation that she had done so because she was hungry, was skilfully used by Hollywood to convert the emerging starlet into a world-class sex symbol. The studios cashed in and during 1953 she made three highly successful films – *Niagara*, *Gentlemen Prefer Blondes*, and *How to Marry a Millionaire*. She also married the baseball player Joe DiMaggio, but this second marriage lasted less than a year.

In spite of a steadily deteriorating personal life, Monroe then became the darling of the critics, who decided that sexy and beautiful as she was, she was also an accomplished actress with an unusual talent for comedy. The ensuing series of films, *The Seven Year Itch* (1955), *Bus Stop* (1956), *The Prince and the Showgirl* (1957), in which some thought she outshone her co-star (Laurence OLIVIER), *Some Like it Hot* (1959), and *Let's Make Love* (1960), were all highly acclaimed. In this period she married

Arthur MILLER, the intellectual playwright. Before this odd misalliance ended in divorce (1960), Miller wrote for her the screenplay of her last film, the appropriately named *The Misfits* (1961).

Monsarrat, Nicholas John Turney
(1910–79) *British writer.*

Monsarrat was educated at Winchester College and Trinity College, Cambridge, where he graduated in 1931. His first publications appeared in the 1930s, but it was his war service, in which he served in the Royal Navy (1940–46) and reached the rank of lieutenant-commander, that furnished him with the experiences that enabled him to create his international best-seller *The Cruel Sea* (1951; filmed in 1953). Other novels relying upon his skill at depicting men and ships in action include *HMS Marlborough Will Enter Harbour* (1952) and *The Ship That Died of Shame* (1959; filmed in 1955).

After the war Monsarrat lived for a time in South Africa (1946–53) before moving to Canada. Later books include publications on Canada, but the sea and adventure continued to dominate his writing. Novels of this period include *The Tribe that Lost Its Head* (1956) and *Richer than All His Tribe* (1968). He also wrote a two-volume autobiography, *Life is a Four-Letter Word* (1966, 1970).

Montale, Eugenio
(1896–1981) *Italian poet, critic, and translator. He was awarded the Nobel Prize for Literature in 1975.*

Born in Genoa, Montale lived there or in a nearby town for the first thirty years of his life. Because of poor health, his education was interrupted when he was fourteen and he was mainly self-taught. His distinguished first volume of poems, *Ossi di seppia* (1925; 'Cuttlefish Bones'), was starkly unrhetorical compared to the work of other poets fashionable at the time. Montale eventually moved to Florence, where in 1929 he was appointed a director of the lending library Gabinetto Vieusseux. (The first English translation of his work 'Arsenio' had been published in *Criterion* by T. S. ELIOT the previous year.) In 1938 Montale lost his job for refusing to join the Fascist Party and his second volume of collected poems, *Le occasioni* (1939), was attacked for being difficult, a charge he contested. A third volume of collected poems, *La bufera* (1956), appeared after the war; the title ('The Storm') refers to World War II.

In 1948 Montale was appointed literary editor of one of Italy's leading newspapers, *Corriere della Sera* in Milan. Over the next two decades he published in the newspaper a series of autobiographical pieces that form a commentary on his creative work. These were collected as *Farfalla di Dinard* (1956; translated as *The Butterfly of Dinard*, 1971). He subsequently published a book of critical essays, *Auto-da-fé* (1966), and a volume of important longer critical essays on his own poetic activity, *Sulla poesia* (1976). He also published his correspondence with Italo Svevo (1861–1928), *Montale–Svevo. Lettere* (1966). In 1967 Montale was given a life appointment as senator. In addition to his poetry, Montale produced an exceptional body of translations from English that includes works by Shakespeare, Hawthorne, Melville, Mark Twain, O'NEILL, FAULKNER, STEINBECK, and Eliot.

Montessori, Maria
(1870–1952) *Italian physician and educationalist who revolutionized the teaching of infants by devising the Montessori method.*

Born in Ancona into a noble family, Montessori studied medicine at the University of Rome. In 1896 she became the first woman in Italy to receive a medical degree, after which she began to work with retarded children in the psychiatric clinic of the university. This experience, coupled with her studies in philosophy, psychology, and education, prompted her in 1907 to open the first Casa dei Bambini ('children's house'), a school in which she applied her own ideas about teaching to children of normal intelligence from the San Lorenzo slum district of Rome. After this successful experiment she devoted herself to the education of normal children.

In 1909 Montessori set out her educational system in a book, *Il metodo della pedagogia scientifica* (translated as *The Montessori Method*, 1912), to encourage others to adopt the same approach. Developing the ideas of Jean Itard and Edouard Séguin, she advocated the use of a 'prepared environment' in which the child would be provided with a variety of sensory materials and be allowed to progress at its own pace. In this way the child's self-confidence and self-discipline would increase and the teacher would have only a limited role. The emphasis would be placed on the child's natural creative potential, rather than compulsion, and the reading and writing process would commence only when the child was ready, which was usually sooner under the Montessori method than under other regimes.

The success of her system led Montessori to open schools in Italy, Spain, south Asia, and the Netherlands and she expanded her theories in several later publications. With some modifications, her ideas have become an integral part of modern nursery- and infant-school education.

Monteux, Pierre (1875–1964) *French conductor. A musician of worldwide acclaim, he insisted on thorough rehearsals and a minimum of gesture during performance. He was appointed Commandeur de la Légion d'honneur.*

While still a student of the violin at the Paris Conservatoire, Monteux played the viola in the orchestra of the Opéra-Comique and with the Geloso Quartet. In 1894 he became assistant conductor and choirmaster of the Colonne concerts. After a period as conductor of the Orchestre du Casino in Dieppe, Monteux was appointed conductor of DIAGHILEV's Ballets Russes (1911). In this capacity, between 1911 and 1914, he conducted the first performances of some of the most original works of the twentieth century, including STRAVINSKY's *Petrushka* and *The Nightingale*, RAVEL's *Daphnis et Chloë*, and DEBUSSY's *Jeux*. He also presided over the uproar at the first night of Stravinsky's *Rite of Spring* (1913), when it was his imperturbable conducting that kept the ballet moving through to the end.

Released from war service in 1916, Monteux travelled to the USA in charge of the French repertory at the Metropolitan Opera, New York (1917–19). In 1920 he decided to stay in America and joined the Boston Symphony Orchestra. He returned to Europe in 1924 as joint conductor, with Willem Mengelberg (1871–1951), of the Amsterdam Concertgebouw Orchestra. In 1932 Monteux founded the École Monteux in Paris for students of conducting. He returned to the USA in 1936 as permanent conductor of the San Francisco Symphony Orchestra, taking US citizenship in 1942. In 1961, at the age of eighty-six, the indefatigable Monteux became conductor-in-chief of the London Symphony Orchestra.

Montgomery of Alamein, Bernard Law, 1st Viscount (1887–1970) *British field-marshal who commanded Allied ground forces during the liberation of France in World War II but who is perhaps best known for his victory over ROMMEL's Afrika Korps at El Alamein in 1942. He was created a viscount in 1946.*

A bishop's son, Montgomery was educated at St Paul's School before entering the Royal Military College, Sandhurst. He joined the Royal Warwickshire regiment and while a platoon leader in World War I, was wounded; after various staff postings he became a battal-

ion commander. At the outset of World War II he commanded the 3rd Division in France until their evacuation from Dunkirk. Heading the 5th Corps and later the 12th Corps, as part of home defences, Montgomery was given command of the 8th Army in Egypt in August 1942. With General ALEXANDER as his commander-in-chief, Montgomery swiftly re-equipped and restored the morale of his battered forces. On 31 August at Alam al Halfa, they repulsed an attack by combined Italian and German forces led by Rommel. On 23 October, Montgomery launched the Battle of El Alamein against Rommel's by now numerically weaker forces. The week-long battle resulted in the first significant Allied victory of the war. Montgomery's forces pursued their opponents, joining with EISENHOWER's expeditionary force from the west, achieving victory in North Africa by May 1943. In July, Montgomery's 8th Army invaded Sicily with the US 7th Army under General PATTON. In December, after spearheading the assault on the Italian mainland, Montgomery was recalled to Britain for the D-Day preparations.

Montgomery was given command of all Allied ground forces during the June 1944 invasion of Normandy. Following the German retreat across France, opinions differed about the next move. Montgomery advocated a narrow pencil-line thrust deep into Germany, while Eisenhower favoured a steady advance on a broader front. As a result, Eisenhower was given overall control of Allied ground forces with Montgomery commanding the 21st Army group fighting through Belgium and Holland. He was normally a cautious strategist; the parachute assault on the Rhine bridge at Arnhem in September 1944 was one of Montgomery's few military disasters. In December, the Germans broke through Allied lines in the Ardennes and Montgomery was given command of all forces north of the 'bulge' to check and eventually repel the enemy. His army soon swept down into Germany from the north while US troops crossed the Rhine in the south. German forces surrendered in May 1945.

After the war he served as chief of general staff (1946–48) and was deputy commander of NATO forces (1951–58). He wrote his *Memoirs* (1958) and *The Path to Leadership* (1961). Known to his men, and almost everyone else, as 'Monty', he developed a style of leadership that was more endearing to his men than to his superiors; the battledress and tank beret, the caravan, the clipped gnomic utterances, and the austere image of the clean-living

Christian soldier provoked Winston CHUR-CHILL's epithet 'indomitable in retreat; invincible in advance; insufferable in victory'.

Montherlant, Henry(-Marie-Joseph-Millon) de (1896–1972) *French novelist and dramatist. He was elected to the Académie Française in 1960.*

Montherlant was born in Paris of aristocratic Catholic parents. He saw active service in the latter half of World War I, which inspired the semiautobiographical novel *Le Songe* (1922). His early works exalt physical prowess and manly courage: *Les Olympiques* (1924) centres on the competitive atmosphere of the athletics track and *Les Bestiaires* (1926) is about bullfighting. In 1934 Montherlant published what was to become one of his best-known novels, *Les Célibataires* (translated as *Lament for the Death of an Upper Class*, 1935), a satire based on the inability of two impoverished aristocrats to adapt to modern society. This was followed by a series of four novels, *Les Jeunes Filles* (1936–39; translated as *The Girls: a Tetralogy of Novels*, 1968), in which Montherlant's misogynistic outlook and the arrogant virility of his hero alienated the author from female contemporaries, such as Simone DE BEAUVOIR. *Le Chaos et la nuit* (1963), the finest of Montherlant's late novels, returns to the subject of bullfighting.

Montherlant began writing for the theatre in 1942. His plays are set at various points in history and have elements of classical tragedy; his heroes and heroines illustrate recurrent themes of austere isolation, personal pride, and refusal to compromise or surrender. *La Reine morte* (1942) takes place in fourteenth-century Spain, *Malatesta* (1946) in Renaissance Italy, *Le Maître de Santiago* (1947) in Spain's Golden Age, *Port-Royal* (1954) in seventeenth-century France, and *La Guerre civile* (1965) in ancient Rome. For *La Ville dont le prince est un enfant* (1951) Montherlant recreated the Catholic college of his own schooldays. His dramas of contemporary life were less successful. In 1972, afraid that he was losing his sight, Montherlant committed suicide.

Moore, Bobby (Robert Frederick Moore; 1941–) *British Association footballer with West Ham United and England and the captain of the national team that won the World Cup in 1966.*

Bobby Moore joined West Ham United in 1958 and remained with the team for the next sixteen years. He soon became a brilliant defensive player. Having led his country at under-23 level, Moore won his first full international cap in May 1962, against Peru. This was followed by three great successes: in 1964 he captained West Ham when they won the FA Cup; in the following year they won the European Cup Winners' Cup; and in 1966 he led England to victory in the World Cup. Moore played for England a record number of 108 times and finished his career with Fulham (1974–76).

Moore, G(eorge) E(dward) (1873–1958) *British philosopher, a leading figure in modern analytic philosophy.*

Born in London, the son of a doctor, Moore was educated at Dulwich College and Trinity College, Cambridge, where he served as a fellow from 1898 to 1904. As a man of independent means he worked privately at philosophy before returning to Cambridge as a lecturer (1911–25) and later as professor (1925–39) of philosophy.

It was Moore, Bertrand RUSSELL declared, who in 1903 forced him to change his mind on 'fundamental questions'. He led the revolt against the prevalent Hegelianism of their student days, objecting that it was inapplicable to the familiar world of 'tables and chairs'. With Moore, philosophical problems arose not from his contemplation of the world, from the work of scientists, or even from his own imagination but, he confessed, exclusively from the puzzling writings of other philosophers. What could they have meant, he repeatedly asked himself, when, like J. E. McTaggart (1866–1925), they proclaimed time to be unreal? Or, like George Berkeley (1685–1753), when they insisted that matter was unreal? In response Moore offered two of his best-known papers, 'A Defence of Common Sense' (1923) and 'A Proof of an External World' (1939). Moore pioneered the techniques of philosophical analysis, in which he was followed by many of his younger colleagues. He also wrote extensively on problems of perception, insisting throughout his life that sense data are directly perceived.

Moore's best-known work, *Principia Ethica* (1903), exercised, through his contact with the Bloomsbury group, more than a strictly philosophical influence. He argued that good was a simple, indefinable, unanalysable, and non-natural property. Thus, any attempt to define it would lead to some form of the naturalistic fallacy. It was still possible, he insisted, to identify certain things as pre-eminently good. These he declared to be 'personal affection and aesthetic enjoyments', values seized upon and deployed in unexpected ways by several of his more worldly Bloomsbury friends.

Moore's personality was marked by a passionate commitment to philosophy, a remarkable honesty, and a simplicity bordering on innocence. Only once, Russell declared, had he known Moore to lie, when he had asked him if he always spoke the truth. 'No', Moore replied. Russell also described the familiar sight of Moore unavailingly attempting to light his pipe while engaging in philosophical discussion. Match after match would be struck, leaving Moore eventually still talking but holding an unlit pipe, an empty matchbox, and several burnt fingers.

Moore, Henry (1898–1986) *British sculptor and draughtsman. He was made a CH in 1955 and received the OM in 1963.*

The son of a Yorkshire miner, Moore studied at Leeds School of Art after doing his military service and then at the Royal College of Art, London (1921–24). He assimilated a variety of influences, including medieval and primitive sculpture and the pre-Columbian art of Mexico, as can be seen in his *Mother and Child* (1924–25). After travelling in Italy Moore taught at the RCA and in 1926 he completed the first of his many public commissions, the figure of *North Wind* for the London Transport headquarters.

In 1930 Moore joined the abstraction-oriented Seven and Five Society and from 1932 to 1939 he taught at the Chelsea School of Art. Between 1930 and 1936 he explored surrealism and nonobjective geometrical sculpture before returning to organic and human forms, particularly the reclining figure. In each of his works Moore aimed at vitality and power of expression, rather than beauty – in his own words 'a pent-up energy, an intense life of its own, independent of the object it may represent'. To achieve this motionless vitality he took inspiration from natural and organic forms, such as rocks, shells, and bones, and also from the natural qualities of the stone and wood that he carved. These he assimilated to the human form, which remained the dominant theme throughout all his work apart from the stringed pieces and helmet heads of the late 1930s.

As a war artist (1940–42) Moore's drawings of figures sheltering in the London Underground led to wider popular recognition. Following the bombing of his London studio in 1940, he moved to Hertfordshire. Already an influential figure in British art, he achieved international fame after World War II and re-ceived a large number of commissions from various countries as well as numerous honours and international prizes for sculpture, including that of the Venice Biennale in 1948. Major commissions include sculptures of reclining figures for the Time-Life building in London (1952–53) and the UNESCO building in Paris (1956–57). Additional themes in his postwar work were upright figures, internal and external forms, family groups, and two- or three-piece semiabstract reclining figures, as well as the abstract pieces of the 1960s, notably *Atom Piece* (1964–66) for the University of Chicago. Though based in Hertfordshire, Moore also worked in Forte dei Marmi, Italy, from 1977.

Moore, Marianne Craig (1887–1972) *US poet and prominent figure in the US avant-garde literary scene of the mid-twentieth century. She received several prizes for her poetry, including the Pulitzer Prize for Poetry (1951).*

Moore was born near St Louis, Missouri, and studied biology at Bryn Mawr College, where the rigours of objective analysis undoubtedly affected her later poetic style. After teaching shorthand for four years, she moved to Greenwich Village, New York, in 1918 and worked over the next eleven years both as a schoolteacher and a library assistant. Recognition first came in 1915 with the publication by T. S. ELIOT of some of her poetry in *The Egoist*, a London imagist journal. In 1921, unknown to her, a collection of her poetry was published in England by friends under the title *Poems*; the same collection, with some additions, was reissued three years later in the USA, retitled *Observations*. It won the *Dial* Award in 1925 and Moore became editor of that publication from then until it closed in 1929. Subsequent works include *The Pangolin and Other Verse* (1936), *What Are Years?* (1941), and *Collected Poems* (1951). A major undertaking was the translation of the fables of La Fontaine (1954); her later work included *Like a Bulwark* (1956), *The Arctic Fox* (1964), and *Tell Me, Tell Me* (1966).

Morandi, Giorgio (1890–1964) *Italian painter of still lifes.*

Morandi studied at the Academy of Fine Arts in his native Bologna from 1907 to 1913. He was influenced by the masters of the early Renaissance, whom he studied on a visit to Florence, and by the impressionist and postimpressionist masters, particularly Cézanne. In addition he was loosely involved with contemporary movements such as futurism and the

Scuola Metafisica. Apart from a few deserted landscapes his pictures were all still lifes. These were sparse and meticulously arranged constructions, made up of a few everyday objects, whose effect depended on the artist's mastery of architectonic form. As well as paintings he produced graphic works and was professor of graphic art at the Bologna Academy of Fine Arts from 1920. He received the Grand Prix for painting at the Venice Biennale in 1948.

Moravia, Alberto (Alberto Pincherle; 1907–90) *Italian writer, whose novels and short stories display his narrative skill and psychological insight.*

Moravia was born of Jewish stock in Rome, the setting for most of his stories. He began to write after he became ill with tuberculosis at the age of sixteen and had to spend two years in various sanatoria. On recovering he became a journalist, first in Turin and then in London. His first novel, *Gli indifferenti* (1929; translated as *The Time of Indifference*, 1953), was a sensational success; like most of his work it deals with the moral degeneracy and spiritual aridity of the middle classes. Such themes were not well received in fascist Italy and during World War II Moravia and his wife, the novelist Elsa Morante, spent some time in hiding from the fascists, an experience mirrored in Moravia's novel *La ciociara* (1957; translated as *Two Women*, 1958).

Disillusionment and alienation, especially in sexual relations, are constant topics in Moravia's novels and short stories. Of the novels, some of the best known are *La mascherata* (1942; translated as *The Fancy Dress Party*, 1948), *La romana* (1947; translated as *The Woman of Rome*, 1949), *La disubbidienza* (1948; translated as *Disobedience*, 1950), *Il conformista* (1951; translated as *The Conformist*, 1952), *La noia* (1961; translated as *The Empty Canvas*, 1961), and *L'attenzione* (1965; translated as *The Lie*, 1966). His short stories include *Racconti romani* (1954; translated as *Roman Tales*, 1956), which won the Marzotto Prize in its year of publication, *Nuovi racconti romani* (1959), *L'automa* (1961), *Una cosa è una cosa* (1966) and *Io e lui* (1971; translated as *The Two of Us*, 1971). In 1961 he was awarded the Viareggio Prize for his work as a whole. Among his other writings are the collection of essays on life and literature entitled *L'uomo come fine e altri saggi* (1963; translated as *Man as an End*, 1966), several plays, and a book on the Cultural Revolution in China

(1967). He became a member of the European Parliament in 1984.

Morgan, Charles Langbridge (1894–1958) *British novelist, playwright, and drama critic.*

Morgan joined the Royal Navy in 1907 but resigned in 1913, intending to go up to Oxford. When World War I broke out he rejoined the navy, but was captured and interned in Holland until 1917, spending some of this time on parole at Rosendaal Castle, which he later used as the setting for his best-known novel, *The Fountain* (1932). In 1919 he went up to Brasenose College, Oxford, and published his first novel, *The Gunroom*, about the unhappy experiences he suffered as a young midshipman.

In 1921 he joined *The Times* and was for many years its drama critic (1926–39). *My Name is Legion* (1925) was his second novel, but he only achieved recognition with *Portrait in a Mirror* (1929). In the 1930s, in addition to the novels *The Fountain*, *Sparkenbroke* (1936), and *The Voyage* (1940), Morgan wrote the biographical sketch *Epitaph on George Moore* (1935) and a successful play, *The Flashing Stream* (1938).

During World War II he worked for the Admiralty, but also wrote a series of essays later published as *Reflections in a Mirror* (1944–46). *The Judge's Story* (1947) is a characteristic Morgan fable on the struggle between spiritual and material values, a theme to which he returned in the essays *Liberties of the Mind* (1951) and the play *The Burning Glass* (1953). In 1948 Morgan, an ardent Francophile, visited France, where his work was much admired, and used it for the setting of *The River Line* (1949; dramatized in 1952). His last novel was *Challenge to Venus* (1957), set in Italy.

Morgan, Thomas Hunt (1866–1945) *US geneticist, who confirmed the theory first proposed by Hugo DE VRIES that exchange of genetic material between homologous (paired) chromosomes takes place during cell division (meiosis) prior to the formation of gametes in living organisms. He received the 1933 Nobel Prize for Physiology or Medicine.*

Born in Lexington, Kentucky, Morgan graduated from Kentucky State College with a BS degree in zoology (1886) and obtained his PhD from Johns Hopkins University four years later. He became a teacher at Bryn Mawr College, Philadelphia, during which time he made several visits to the Naples Zoological Station and began his own work in experimental embryology. In 1904 he was appointed professor of experimental zoology at Columbia University and in 1908 started his series of breeding experiments using the fruit fly *Drosophila melanogaster*. Morgan and his team, A. H. Sturtevant, C. B. Bridges, and H. J. Muller, found a recessive mutant character, white eye, that was only manifest in male progeny. Morgan proposed that the gene responsible was carried by the sex chromosome, i e it was sex-linked. Previously sceptical, Morgan was convinced by this evidence that chromosomes did carry the genes. Going on to work with two different sex-linked characters, Morgan discovered that they were not always inherited together and so, in 1911, postulated exchange between regions of homologous chromosomes during meiosis – a phenomenon he termed crossing-over. In 1913, A. H. Sturtevant produced the first linkage map, in which he related the frequency of crossing over between any two genes to their notional distance apart on the chromosome. Morgan's team made many further discoveries, all of which firmly established the link between inheritance of characters and chromosomes.

In 1928, Morgan joined the California Institute of Technology to found the biology division. Here he resumed his research in embryology. He served as president of the National Academy of Sciences (1927–31) and wrote *The Mechanism of Mendelian Heredity* (1915) and *The Theory of the Gene* (1926).

Moro, Aldo (1916–78) *Italian statesman and Christian Democrat prime minister (1963–68; 1974–76), who was assassinated by the Red Brigades in 1978.*

Born in Maglie, the son of a teacher, Moro was educated at the University of Bari where, as a student, he was active in the Catholic University Federation. He graduated with a doctorate in law in 1940 and taught criminal law at the university, becoming professor at the age of twenty-four.

Towards the end of World War II Moro helped to establish a Christian Democrat (CD) organization in Apulia. Elected to the constituent assembly as a member of the CDP in 1946, he was appointed undersecretary of state for foreign affairs in 1948, becoming minister of justice in 1955 and minister of education in 1957. Made political secretary of the CDP two years later, he succeeded Amintore Fanfani (1908–) as prime minister in 1963, leading the government until 1968 and again from 1974 to 1976. In 1963 Moro assembled a centre-left coalition of Christian Democrats, So-

cialists, Social Democrats, and Republicans, and in 1976 he laid the ground work for parliamentary cooperation between the Christian Democrats and the Communists.

Morris, Desmond John (1928–) *British zoologist, whose work has brought to the fore developments in the study of animal behaviour and their possible implications for the human condition.*

Interested in animals from childhood, Morris received a BSc in zoology from Birmingham University in 1951. Moving to Oxford University, he studied the courtship and other behaviour of the ten-spined stickleback under the tutelage of the famous ethologist Niko TINBERGEN and received his DPhil in 1954. In the late 1950s, as presenter of the popular television series *Zoo Time*, Morris became a well-known personality; subsequent TV series included *Life* in the 1960s and *The Animals Roadshow* in the 1980s. He was appointed curator of mammals for the Zoological Society of London (1959–67), served as director of the Institute of Contemporary Arts (1967–68), and was research fellow at Wolfson College, Oxford (1973–81).

Morris is best known for his books. *The Naked Ape* (1967) aroused controversy because of Morris's interpretation of many aspects of human behaviour in terms of that exhibited by our ape relatives, including such phenomena as hunting instincts, pair bonding, mutual grooming, and territoriality. Contentious, provoking, and best-selling, it was seen by some critics as simplistic. It was followed by *The Human Zoo* (1968), in which Morris drew parallels between the symptoms of stress shown by animals in overcrowded conditions and the social problems associated with life in crowded cities. His other books include *Intimate Behaviour* (1971), *Manwatching* (1977), *The Book of Ages* (1983), *Bodywatching* (1985), and *Animal-Watching* (1990).

Morrison, Herbert Stanley, Baron (1888–1965) *British Labour politician, prominent on the London County Council (LCC) and in national politics.*

Born in Brixton, London, the son of a policeman, Morrison was earning a living as an errand boy at the age of fourteen. After working as a journalist he became secretary of the London Labour Party in 1915. He was elected mayor of Hackney (1919) before election to the LCC (1922–45) and to parliament (1923–24; 1929–31; 1935–59). As minister of transport (1929–31) under Ramsay MACDONALD, he devised the London Passenger Transport Board (to be set up in 1933). In 1939 he became leader of the LCC. During World War II he was minister of supply (1940) and then home secretary (1940–45), serving in the war cabinet from 1942. Morrison was largely responsible for writing the manifestoes that took Labour to victory in the elections of 1945 (*Let Us Face the Future*) and 1950 (*Labour Believes in Britain*). Under Clement ATTLEE, he served as leader of the House of Commons and lord president of the Council (1945–51) and then as foreign secretary (1951). Inexperienced in foreign policy, he was unpopular for his handling of a dispute with Persia, which led to the expulsion of British personnel in the Anglo-Iranian Oil Company, and was blamed for the unfavourable economic effects of the Korean War. After the 1951 election defeat, Morrison became deputy leader of the party until 1955, when he lost to GAITSKELL the contest to succeed Attlee as leader.

Morton, Jelly Roll (Ferdinand Joseph Lemott; 1885–1941) *Black US jazz pianist, composer, and bandleader. He was an important link between ragtime and New Orleans jazz.*

Born in New Orleans, he was disowned by his family for playing the piano in brothels. In the 1920s he moved to Chicago and in 1923–24 recorded nineteen piano solos, nearly all his own compositions; in 1926–30 he made a long series of recordings for Victor with small bands called Jelly Roll Morton and his Red Hot Peppers. His arrangements were revolutionary in that they retained the blend of individual instrumental voices and the excitement of their cross-talk within formal orchestration. However, Morton's style was unable to change with the times, and he saw his popularity wane, although his composition 'King Porter Stomp' was a hit in the Swing Era that followed. In 1938 he recorded hours of conversation and reminiscence of turn-of-the-century New Orleans, as well as playing and singing, for the Library of Congress. The following year he recorded a set of piano solos and blues.

His piano playing was technically complex and harmonically advanced for its time; his music contained Spanish and operatic elements, as well as the blues and ragtime. His recordings provide an invaluable insight into the evolution of early jazz.

Moseley, Henry Gwyn Jeffries (1887–1915) *British physicist, best known for his law relating the atomic number and the X-ray emission of an atom.*

After graduating from Oxford in 1910 Moseley joined Ernest RUTHERFORD in Manchester, with whom he worked for three years. He returned to Oxford in 1913 but a year later enlisted in the army with the start of World War I. A sniper's bullet at Suvla Bay during the Gallipoli campaign brought to an end a brilliant scientific life.

In work that would have undoubtedly earned him a Nobel Prize had he survived the war, Moseley made the crucial connection between the emerging discipline of nuclear physics and the long-established periodic table of the chemists. Measuring the X-rays emitted from a series of elements when bombarded with electrons, he found that the square root of the frequency of the X-rays was proportional to the atomic number for certain groups of elements. This result is known as Moseley's law. Moseley concluded that some property of the atom grew as the atomic weight increased. He suspected that the factor was in fact the charge on the nucleus and he went on to assign the simplest atom, hydrogen, a charge of +1, the next simplest, helium, a charge of +2, and so on throughout the periodic table. These numbers were called atomic numbers by Moseley, and this is the name that has remained.

Mosley, Sir Oswald Ernald (1896–1980) *British politician whose enviable political connections and early brilliance were entirely wasted on the short-lived British Union of Fascists.*

Born in London, son of Sir Oswald Mosley, the 5th Baronet, Mosley became an MP after World War I, first as a Conservative (1918–22), then as an Independent (1922–24), lastly as a member of the Labour Party (1924; 1929–31). He was chancellor of the Duchy of Lancaster from 1929 to 1930. In 1931 he founded the socialist New Party, which gained no seats in the election of that year. He created the British Union of Fascists (BUF) in 1932, dressing himself and his followers in black uniforms. The Blackshirts, as they were called, provoked antisemitic violence, especially in the East End of London. Numbering 20 000 members at its height, the BUF – essentially a collection of right-wing thugs – was effectively destroyed by the Public Order Act 1936. Mosley was interned (1940–43) during World War II. In 1948 he founded the right-wing Union Movement, but never made an effective return to politics. His third wife was Lady Diana Mitford (1910–   ), whose sister Unity (1914–48) was another supporter of HITLER. Mosley's papers were made available in 1983.

Moss, Stirling (1929–   ) *British motor racing driver who achieved a string of victories in the 1950s and early 1960s but failed to win a World Championship.*

The son of a dentist, Moss started racing in 1948, driving a Cooper 500, and in 1950 joined Peter Heath's HWM team, in whose cars he achieved his first major successes. From 1952 onwards Moss drove a variety of cars in Grands Prix, including Coopers and a Maserati, as well as Jaguars in sports car events. Briefly joining the Maserati works team in 1954, he switched to Mercedes-Benz the following year and came second in the Drivers' Championship. This was the year of his famous victories in the gruelling Mille Miglia and Targa Floria events, driving a Mercedes 300 SLR sports car. In 1956 he rejoined Maserati and clinched another second place in the Drivers' Championship. In 1957 and 1958, driving for Vanwall, Moss was again runner-up, by only a single point in 1958, the year in which he and Jack BRABHAM won the Nürburgring 1000 km race in an Aston Martin DBR1 sports car. Moss won the event again the following year.

Moss's successes in both Formula One and sports cars continued, driving Lotus and Cooper cars for the Rob Walker team. He broke both legs in a crash at Spa in Belgium in 1960 but returned to racing the same year. However, in 1962 he was seriously injured after crashing in a Formula Two race at Goodwood and retired from racing. Moss has since pursued various interests, especially in design and motoring journalism.

Mössbauer, Rudolf Ludwig (1929–   ) *German physicist who was awarded the 1961 Nobel Prize for Physics for his discovery of the Mössbauer effect.*

Mössbauer was educated at the Munich Technische Hochschule, where he gained his doctorate in 1959. In 1956, while working for his doctorate, Mössbauer made the observations that led to the formation of his effect. When a gamma ray is emitted by an atomic nucleus, normally the recoil of the atom affects the wavelength of the emission. Mössbauer found, however, that, at very low temperatures the nucleus is locked into the crystal lattice and the recoil is spread over the whole lattice, with a negligible effect on the wavelength of the emission. In 1960 the effect was used to provide experimental confirmation of EINSTEIN's theory of general relativity. It has also been widely used in spectrographic work (Mössbauer spectroscopy).

After completing his doctoral studies in Munich, Mössbauer moved to the California Institute of Technology, where he was professor of physics (1962–64), before being appointed to a similar post in Munich in 1964.

Mountbatten of Burma, Louis, 1st Earl (1900–79) *British member of the royal family, statesman, and admiral, who as the last viceroy of India supervised the transfer of power to India and Pakistan. He was created a viscount in 1946 and an earl in 1947.*

Son of Prince Louis of Battenberg (who took the name Mountbatten during World War I) and of Princess Victoria of Hesse-Darmstadt, a granddaughter of Queen Victoria, Mountbatten was a second cousin of GEORGE VI and an uncle of Prince PHILIP. He entered the Royal Navy as a cadet in 1913 and thereafter held various naval positions. In 1922 he married the heiress Edwina Ashley, whose grandfather, Sir Ernest Cassel, had been financial adviser to EDWARD VII; they had two daughters. During World War II, after commanding destroyers, he became adviser and later chief of Combined Operations (1941–43) and then Supreme Allied Commander in South-East Asia (1943–45), when the British retook Burma. In 1947 he was appointed viceroy of India with the task of negotiating the partition of the subcontinent into the two independent nations of India and Pakistan. He was then made governor-general of India (1947–48). Commander-in-chief of the Mediterranean Fleet (1952–54), he was successively first sea lord (1955–59) during the Suez crisis (1956) and chief of UK Defence Staff (1959–65); he became an admiral in 1956. From 1953 he was personal aide-de-camp to the Queen. Mountbatten was murdered by an IRA bomb, which exploded on his boat in Donegal Bay, Northern Ireland.

Mubarak, (Muhammad) Hosni Said (1928–   ) *Egyptian statesman and president (1981–   ), who stepped into the breach created by SADAT's assassination.*

Born at Kafr El-Moseilha in the Nile delta, he studied at the Egyptian military academy at Cairo and joined the Egyptian air force in 1950. He rose to become Director-General of the Air Academy (1967–69) and then Air Force Chief of Staff (1969–72) before being appointed head of the air force by Sadat in 1972. In this role he won praise for his direction of Egyptian air operations in the war (1973) with Israel and was subsequently promoted to air marshal. In 1975 Sadat made Mubarak his vice-president, in which post he took an important role in the complex negotiations between the various Arab powers.

Following his appointment as president after Sadat's assassination, Mubarak did much to establish closer links between Egypt and other Arab nations and distanced himself from Israel when the latter invaded Lebanon in 1982. A moderate, he also worked to caution fundamentalists within Egypt and to improve the country's technological base. He was rewarded with both the respect of foreign leaders and increased popularity at home, being endorsed for a second six year-term as president in 1987. During the Gulf War of 1991 he aligned Egypt against the regime of Saddam HUSSEIN, risking division with other Arab states.

Mugabe, Robert Gabriel (1924–) *Prime minister (1980–87) and president (1987–) of Zimbabwe, who has managed to improve the lot of the black peoples of Zimbabwe without provoking an economically disastrous exodus of the whites.*

Born in Kutama, the son of a labourer, Mugabe was educated at Catholic mission schools and at the Fort Hare University College, where he obtained degrees in education and economics. He began his teaching career in local mission schools in 1952 and later taught in Northern Rhodesia (now Zambia) (1955) and Ghana (1956–60). Mugabe's political career began in 1960 when he was appointed publicity secretary of the National Democratic Party (NDP) in Southern Rhodesia. After serving as the deputy secretary-general of the Zimbabwe African People's Union (ZAPU) under Joshua Nkomo (1917–), following the banning of the NDP, Mugabe was arrested in 1962 and 1963 but managed to escape to Tanzania. In 1963 he broke with Nkomo and co-founded (with Sithole) the Zimbabwe African National Union (ZANU). He was arrested again in 1964 and detained until 1974, when he was among the political prisoners released in Ian SMITH's general amnesty. Going into exile in Mozambique, Mugabe became leader of ZANU, forming the Patriotic Front with Nkomo in 1976. He contested the February 1980 elections as leader of the ZANU (PF) party and was declared prime minister in April 1980. He subsequently (1982) ousted Nkomo from the cabinet and in 1987 became executive president with ZANU and ZAPU merged as ZANU (PF) in a one-party state.

Muir, Edwin (1887–1959) *British poet, critic, and translator. He was awarded the CBE in 1953.*

Muir was born on Orkney and was brought up in rural Scotland until he was fourteen, when his family moved to Glasgow. There Muir worked unhappily as a clerk. In World War I he was rejected as physically unfit for active service. Although he had contributed to *New Age* from 1913, Muir's literary career really began with his marriage to Willa Anderson (1919). They left Glasgow for London, where Muir became assistant editor on *New Age*, and between 1921 and 1927 spent much time on the Continent. During this period they collaborated on translations of several distinguished European novelists, notably KAFKA, making them available for the first time to an English readership. Muir also wrote some important critical studies: *Latitudes* (1924) and *Transition* (1927), which contain high praise for the works of D. H. LAWRENCE.

From 1942 Muir worked for the British Council, first in Edinburgh (1942–45), then in Prague (1945–48), and finally in Rome (1949–50). He was then warden of Newbattle Abbey, near Edinburgh (1950–55), and became Charles Eliot Norton Professor of Poetry at Harvard (1955–56), before retiring to live near Cambridge. His autobiography (1954) contains some of the visionary qualities of his best poetry, most of which was written when he was over fifty. *The Voyage* (1946) and *The Labyrinth* (1949) were his first major collections, followed by *Collected Poems* (1952) and *One Foot in Eden* (1956).

Mujibur Rahman, Sheik (1920–75) *First prime minister (1972–75) and president (1975) of Bangladesh. Popularly known as Sheik Mujib, he was considered the father of Bangladesh at the time of independence, but his assumption of dictatorial powers in 1975 led to his assassination.*

Born in Tungipana, the son of a middle-class landowner, Sheik Mujib attended a mission school in Gopalganj before studying at Islamia College, Calcutta, and at the University of Dacca, from which he was expelled for his political activities in 1948. In 1949 he co-founded the Awami (People's) League, a political party that advocated autonomy for East Pakistan. He was appointed general secretary of the party in 1953.

Sheik Mujib was elected to the East Pakistan legislative assembly (1954) and to Pakistan's national assembly (1955). Imprisoned during the 1960s on numerous occasions for his support of a separate Bengali state, he became a national hero and led the Awami League to victory in the elections of 1970, winning a majority in the national assembly. When civil war broke out in 1971 he was again imprisoned but was released in 1972 by BHUTTO after East Pakistan (now Bangladesh) had been formally recognized, with Sheik Mujib declared president in his absence. Returning to Bangladesh he gave up his presidency for the post of prime minister. Faced with mass starvation, economic chaos, and ineffective government, he proclaimed a state of emergency (1974) and suspended fundamental civil liberties. In January 1975 he introduced legislation that enabled him to become president once more, and effectively dictator; he was overthrown and assassinated in August of that year during a military coup.

Muldoon, Sir Robert David (1921–)
New Zealand statesman and National Party prime minister (1975–84). He became a CH in 1977 and was knighted in 1984.

Born in Auckland, the son of an accountant, Muldoon was educated at Mount Albert Grammar School before enlisting in the services during World War II. Studying accountancy throughout the war, he won a scholarship to England in 1946, returning to New Zealand in 1947 to join a firm of accountants. He became president of the New Zealand Institute of Cost Accountants in 1956.

Muldoon's political career began when he won the seat of Tamaki in 1960 for the National Party. Deputy prime minister for a brief period in 1972, he was deputy leader of the opposition (1973–74), leader of the opposition (1973–74), and finally prime minister in December 1975, when the Labour administration was defeated. Known for his pugnacious style, Muldoon developed a reputation for firm and determined leadership. He was chairman of the board of governors for the IMF and World Bank (1979–80) and became chairman of the ministerial council for the OECD in 1982. His government fell to Labour in the general election of 1984, David Lange succeeding him as prime minister. In 1986 he was appointed shadow minister for foreign affairs.

Mulliken, Robert Sanderson (1896–1986)
US chemist, who was awarded the 1966 Nobel Prize for Chemistry for his work on molecules.

The son of an organic chemist, Mulliken was educated at the Massachusetts Institute of Technology and the University of Chicago, where he was awarded his PhD in 1921. Apart from a brief period in New York, Mulliken spent the whole of his academic career at Chicago, serving from 1931 until his retirement in 1961 as professor of physics.

Mulliken's work in chemistry was concerned with the development of molecular orbital theory. According to his theory, electrons move in orbitals, which encircle several atomic nuclei. The shapes of these orbitals can be calculated, enabling conclusions to be drawn about the energy of bonds, etc. Mulliken also made major contributions to the interpretation of molecular spectra.

Munch, Charles (1891–1968) *French violinist and conductor, who was well known to concert-goers on both sides of the Atlantic.*

Munch studied the violin at the Strasbourg Conservatory, where his father was a professor, and later with Carl Flesch in Berlin and with Lucien Capet in Paris. As a resident of Alsace, Munch was conscripted into the German army during World War I, after which he returned to Strasbourg to become professor of the violin at the Conservatory. In 1926 he was appointed leader of the Gewandhaus Orchestra in Leipzig, under FURTWÄNGLER, remaining there until 1933, when he made his debut as conductor in Paris. For the next fifteen years he played a leading role in the musical life of the city as conductor of the Lamoureux concerts, the Orchestre Symphonique de Paris, the Société Philharmonique de Paris (1935), and the Société des Concerts du Conservatoire (1937). Munch also toured the main European cities and made his debut in the USA in 1946. In 1948 he succeeded KOUSSEVITSKY as chief conductor of the Boston Symphony Orchestra, where he introduced many contemporary French and US works to the public. In 1962 he returned to France and was instrumental in the foundation of the Orchestre de Paris (1967). He died on a concert tour of America.

Munthe, Axel Martin Frederik (1857–1949) *Swedish physician, psychiatrist, and writer.*

Born at Oskarshamn, Kalmar, Munthe studied at the universities of Uppsala and Montpellier and at the Salpetrière in Paris, where he worked under the renowned neurologist Jean-Martin Charcot (1825–93); he later parted company with him after disagreeing with his theory of hypnotism. He practised for twelve years in Paris as a gynaecologist and then carried on an equally lucrative practice in Rome, where he lived for a time in the house formerly occupied by Keats. He was honoured by Italy for his services after an earthquake at Messina and in 1908 was appointed physician to the Swedish royal family. His one book of note, *The Story of San Michele* (1929), was written in English and became a worldwide best-seller in forty-four languages. Referring to the Villa San Michele, which he built on Capri, it is an account of his experiences as a doctor and reveals a sympathetic and humane character who had led a full and dedicated life.

Murdoch, Dame (Jean) Iris (1919–)
British novelist and university teacher. She was created a DBE in 1987.

Iris Murdoch was born in Dublin and educated in London and Bristol before reading classics at Somerville College, Oxford (1938). After a spell as a civil servant (1942–44) she worked for the United Nations Relief and Rehabilitation Administration (1944–46). She then studied philosophy at Newnham College, Cambridge, and from 1948 taught philosophy at St Anne's College, Oxford. She married the critic John Bayley in 1956.

Besides her philosophical writings, including *Sartre, Romantic Rationalist* (1953), *The Sovereignty of Good* (1970), and her study of Plato, *The Fire and the Sun* (1977), Iris Murdoch made a considerable reputation as a novelist with a regular succession of subtle and entertaining narratives beginning with *Under the Net* (1954) and including *The Sandcastle* (1957) and *The Bell* (1958). One of her early successes, *A Severed Head* (1961), was made into an equally successful stage play. The disastrous havoc caused by unruly passion in the lives of her characters is a recurrent theme, treated in a highly inventive vein of black comedy. At the same time she explores such themes as the struggle of the artist (*The Black Prince*, 1973) and the quest for the spiritual life, as in the Booker Prize-winning *The Sea, The Sea* (1978) and *Nuns and Soldiers* (1980). 'Heroes' typical of Murdoch's novels are egotistical and destructive, such as Austin Gibson Grey in *An Accidental Man* (1971), Hilary Burde in *A Word Child* (1975), and George McCaffrey in *The Philosopher's Pupil* (1983). Her most recent novels include *The Good Apprentice* (1985), *The Book and the Brotherhood* (1987), and *The Message to the Planet* (1989).

Murdoch, (Keith) Rupert (1931–) *US publisher and media entrepreneur, born in Australia; founder and head of the News International Communications empire.*

Murdoch was born in Melbourne, the son of the celebrated war correspondent and newspaper publisher Sir Keith Murdoch (1886–1952). After taking a degree at Oxford he worked as a subeditor on the *Daily Express* until 1952, when he returned to Australia to run the family newspapers. As owner of the *Adelaide News*

he quickly hit on the formula that would become his hallmark, taking the paper radically downmarket with a heavy emphasis on scandal, show-business gossip, and sport. Sales rocketed, allowing him to obtain further titles in Sidney and Perth, which he subjected to a similar transformation.

Murdoch became a major figure in the British newspaper business with the acquisition of the *News of the World* (1969) and *The Sun* (1970). The latter became Britain's best-selling daily but in doing so incurred considerable disapproval for its aggressive right-wing populism and emphasis on sex. In 1981 Murdoch acquired the Times Newspaper Group amid much controversy. Five years later he pulled off an audacious coup by switching the papers' production to a new high-tech plant in east London without obtaining a union manning agreement. Although a violent dispute ensued, the power of the print unions was effectively broken and the British newspaper industry underwent a technological revolution that many considered long overdue.

In the 1980s Murdoch began to move into other areas of the communications industry, building up holdings in radio, TV, film, and publishing companies, chiefly in the USA. As the owner of Harper Collins he has a substantial stake in book publishing on both sides of the Atlantic. He became a US citizen for business purposes in 1985, the year he acquired 20th Century Fox and a number of US television stations. In 1989 he launched the satellite television channel Sky (known as British Sky Broadcasting since its absorption of its main rival in 1990).

While some argue that Murdoch's effect on the newspaper industry has been bracing and salutary, others blame him for declining standards of taste and accuracy in popular journalism. There is also anxiety about the sheer number of communications outlets he now controls, a situation that some see as near-monopolistic.

Murry, (John) Middleton (1889–1957) *British author and critic.*

Murry was born into a poor family at Peckham, London, and received his education at Christ's Hospital, London, and Brasenose College, Oxford, by winning scholarships. He began writing for the *Westminster Gazette* in 1912, and later contributed to the *Times Literary Supplement*. During World War I he worked in Intelligence, becoming chief censor (1919), and afterwards edited the *Athenaeum* (1919–21).

He had already published his first novel *Still Life* (1917), but his forte was as a critic.

The fatal illness of his first wife, Katherine MANSFIELD, led to his resignation from the *Athenaeum*, but after her death in 1923 he founded the *Adelphi*, which he continued to edit until 1948. His domestic life was dogged by misfortune and unhappiness and his literary squabbles were often bitter (notably those with D. H. LAWRENCE). Nevertheless he wrote a number of influential books on literary and philosophical issues; Blake, Swift, Keats, and Shakespeare were among the writers on whom he produced studies. *The Necessity of Communism* (1932), *The Necessity of Pacifism* (1937), and *The Free Society* (1948) chart three stages in his thinking on important issues of his time. During World War II he ran a farming community staffed by conscientious objectors and edited the pacifist journal *Peace News*. *Love, Freedom and Society* (1957), a comparative study of D. H. Lawrence and Albert SCHWEITZER, was Murry's last important work. His early life is described in his autobiography *Between Two Worlds* (1935).

Musgrave, Thea (1928–) *British composer, whose earlier diatonic compositions were later superseded by her serial music.*

After a three-year course at Edinburgh University and lessons with Hans Gál, Musgrave studied (1950–54) with Nadia BOULANGER in Paris. In 1953 she received her first commission, *The Suite o' Bairnsangs* for the Scottish Festival at Braemar; the following year the BBC (Scotland) commissioned *Cantata for a Summer's Day*. The ballet *A Tale for Thieves* and the chamber opera *The Abbot of Drimock* also date from this period. These early works are generally diatonic in style, but by 1960 Musgrave was using serial techniques, as in the trio for flute, oboe, and piano (1960). Musgrave's opera *The Decision* (1964–65), with a libretto about a mining disaster, was staged at Sadler's Wells in 1967. In 1966 her first experiments in nonsynchronized but fully notated instrumental parts appeared in the second and third chamber concertos (1966; 1967). These techniques were used vocally in the opera *The Voice of Ariadne*, commissioned for the 1974 Aldeburgh Festival. Her viola concerto, a BBC commission, was given its first performance in 1973 by her husband Peter Mark and the Scottish National Orchestra, conducted by Musgrave. More recent works include the opera *Mary, Queen of Scots* (1976–77), *Black Tambourine* (for women's chorus and piano; 1985), and *The Seasons* (for orchestra, 1988).

Mussolini, Benito Amilcare Andrea (1883–1945) *Italian fascist dictator (1921–45), whose dream of building a new Roman Empire was thwarted by his disastrous military alliance with the Germans.*

Born in Dovia, the son of a blacksmith, Mussolini was educated at a school in Forlimpopoli, where he gained a diploma in primary education (1901). Teaching briefly in Gualtieri Emilia, he emigrated to Switzerland (1902) but was forced to return to Italy (owing to a false passport) in 1904. Over the next few years he became involved in socialist politics as a labour leader and editor of *La Lotta di Classe* ('The Class Struggle'), a provincial newspaper. He was elected director of the Socialist Party and editor of its newspaper *Avanti* in 1912, but was expelled from the party in 1914 because he advocated that Italy should enter World War I. He then founded an independent paper *Popolo d'Italia*, supported by the Autonomous Fascist Party.

In 1919 Mussolini formed a new fascist party – the fighting fascists, or Blackshirts – and was elected to parliament in 1921 as the head of a National Fascist Party. He led the march on Rome the following year, establishing himself as a dictator (known as 'il Duce'). After he had developed a powerful militia and secret police, he reorganized the state economy into a 'corporate state'. In 1929 he signed the Lateran Treaty establishing the independent Vatican state to mollify Roman Catholic opinion and by the early 1930s he had crushed virtually all opposition to his regime. Although three assassination attempts were made on him, he had at this time a large measure of popular support. However, this support began to wane when their theatrical and arrogant dictator began to involve the Italians in a series of military measures. In 1936 he invaded Ethiopia, provoking worldwide protests and a feeble attempt by the League of Nations to impose sanctions on Italy. Initially antagonistic towards HITLER, he attempted to prevent him from annexing Austria by blocking the Brenner Pass (1934). However, in 1937 he formed the Rome–Berlin Axis with Hitler to show his people that they did not stand alone in the world. To emulate the Nazis, in the hope that some of their success might rub off on him, he introduced antisemitism and a number of other measures that attracted only the hooligan element in Italy (unlike Germany) and alienated the mass of the peace-loving Italian people. At the beginning of World War II he invaded Albania (1939) and finally declared war on the Allies after the German invasion of France (1940). From that point onwards his fortunes declined rapidly. By 1943 he had suffered disastrous military losses and was forced to resign. He was executed by Italian partisans at Donego in 1945 and his body ended up hanging upside down from a lamp-post in a Milan square.

Myrdal, (Karl) Gunnar (1898–1987) *Swedish economist, winner of the 1974 Nobel Prize for Economics jointly with Friedrich August von Hayek (1899–) for his work in broadening the scope of economics.*

Myrdal graduated from Stockholm University in law in 1923 and entered private practice. He later graduated in economics and lectured at Stockholm; in 1933 he took the chair of political economy and financial science at Stockholm University. Myrdal held several public posts (member of parliament, economic adviser to the Swedish legation in the USA, minister of commerce, and secretary-general of the United Nations Economic Commission for Europe) before being appointed professor at the Institute for International Economic Studies of Stockholm University in 1957.

Myrdal's book *Monetary Equilibrium* (1931) was a forerunner in some ways to KEYNES's *General Theory*. He emphasized the dynamic nature of macroeconomic processes and in 1927 began the usage of the terms 'ex post' and 'ex ante' with particular reference to actual and planned expenditure and savings in an economy. He was well known for his work on poverty in the developing countries. *Asian Drama* (1968) deals with the poverty, lack of education, ill health, and underemployment in these countries; arguing that direct economic aid from the developed countries was counterproductive, Myrdal advocated instead their independent development. He wrote as an institutional economist, i e he stressed the importance of the noneconomic environment (political and social institutions and customs) of the country in which economic theory is being applied: without these factors being taken into account, the theory becomes irrelevant. Myrdal criticized the use of sophisticated mathematical techniques and believed that social, legal, health, and educational factors were necessary for the realistic study of underdeveloped economies.

Other important works include *An International Economy: Problems and Prospects* (1956), *Economic Theory and Underdeveloped Regions* (1957), *Challenge to Affluence* (1963), *The Challenge of World Poverty* (1970), and *Against the Stream* (1973).

N

Nabokov, Vladimir (1899–1977) *Russian-born US novelist.*

Born in St Petersburg, Nabokov left Russia with his aristocratic family in 1919, having already published two verse collections (1916 and 1918). In England he became a student of zoology at Trinity College, Cambridge, but later switched to literature. While in England he wrote two further volumes of Russian poetry, published in 1923. From 1922 to 1940 Nabokov lived in Germany and France, establishing himself as an émigré writer of Russian novels with *Mashenka* (1926; translated 1970), an autobiographical love story, and *Korol-dama-valet* (1929; translated as *King, Queen, Knave*, 1928). Subsequent works included his chess novel *Zashchita luzhina* ('The Luzhin Defence', 1930; called *The Defence* in the 1964 English translation), *Camera Obscura* (1933; translated 1936 and called *Laughter in the Dark* in the 1938 US translation), and *Priglasheniye na kazn* (1935; translated as *Invitation to a Beheading*, 1959). The last of his Russian novels, *Dar* (1937; translated as *The Gift*, 1963), introduced his mastery of literary parody of which he was to make skilful use in his later English books. All these Russian-language novels were published under the pseudonym V. Sirin, derived from the name of an admired Russian publisher.

In 1940 the penurious Russian émigré moved with his wife Véra and son Dimitri to the USA; here he established himself as a teacher of Russian subjects at a variety of colleges, became a US citizen in 1945, and in 1948 settled down for eleven years as professor of Russian literature at Cornell University. In this period he wrote under his own name in a consummately polished English that gave no hint of being a second language. The first of his English novels, *The Real Life of Sebastian Knight* (1941), was followed by *Bend Sinister* (1947) and *Pnin* (1957), a portrait of the New World as seen through the lugubrious eyes of an Old-World émigré professor of entomology. The contrast between the brash new and the intellectual old is again the theme of *Lolita* (1958; Paris publication 1955), a hilarious and highly sophisticated account of a middle-aged European's physical obsession with a monstrous twelve-year old American girl. As por-nography, *Lolita* brought Nabokov instant notoriety and considerable wealth. As allegorical literature, he had to wait a little longer for acclaim and for the interest that it awakened in his earlier (and later) work. Most notable of his novels after *Lolita* are *Pale Fire* (1962), written in the form of a long poem accompanied by a parody of a critical commentary, and *Ada: A Family Chronicle* (1969), a long and sensual novel written in his uniquely erudite style.

Apart from his eighteen papers on entomological subjects (he was an ardent amateur lepidopterist), Nabokov wrote an autobiography, *Speak, Memory* (1967), a translation of *Alice in Wonderland* into Russian (1923), and a translation of *Eugene Onegin* (1964) into English.

Nader, Ralph (1934–) *US lawyer and leading figure in the US consumer protection movement.*

Educated at Princeton and Harvard Law School, Nader was admitted to the Connecticut bar in 1958 and took up private practice in Hertford. He rose to prominence after testifying before Congress on motor-car safety in 1966. His views on defective car design, set out in his book *Unsafe at Any Speed* (1965), were an important factor in the passing of the National Traffic and Motor Vehicle Safety Act (1966), which established federal control over car design.

Nader then turned his attention to the protection of American Indians, as well as such consumer issues as safety in food and drugs. Various pieces of important legislation resulted from these efforts, including the Wholesale Meat Act (1967) and the Occupational Safety and Health Act (1970). In order to increase the effectiveness of the consumer lobby, Nader mobilized college students in groups, known as 'Nader's Raiders', to study the activities of government regulatory agencies. In 1969 these groups were organized as part of Nader's Center for the Study of Responsive Law. Other organizations established by Nader include the Public Interest Research Group and the Tax Reform Group. These organizations put pressure on the authorities to improve their regulation of a wide variety of manufacturers and processes and also pressed for the establishment of a federal office to represent consumer interests. By acquiring single shares in large industrial companies Nader was able to observe and criticize from within, without recourse to the courts. The accident at the Three-Mile Island nuclear power plant in Pennsylvania provided further ammunition for Nader and his organizations in 1979, the year he published his *Menace of Atomic Energy*. Subsequent publications in which he continued to tackle questions of universal concern have included *Who's Poisoning America?* (1981) and *The Big Boys: Power and Position in American Business* (1986).

Nagy, Imre (1896–1958) *Hungarian statesman. He became prime minister in 1953 but resigned in 1955. In the Hungarian revolution (1956) he again became prime minister and was executed by the Soviet Union for his part in it.*

Born in Kaposvar into a peasant family, Nagy worked as an apprentice locksmith before serving in the Austro-Hungarian army during World War I. Taken prisoner by the Russians, he joined the Bolsheviks and fought with the Red Army, taking part in the Bolshevik revolution (1917). He returned to Hungary in 1921 but his membership of the illegal Hungarian Communist Party forced him to escape to the Soviet Union in 1929. He remained there until 1944, studying agriculture and working for the Institute of Agrarian Sciences in Moscow.

Nagy returned to Hungary towards the end of World War II, when his country was under Soviet occupation. He became minister of agriculture (1945–46) in the new postwar government and briefly held the posts of interior minister and speaker of parliament; he was however, excluded from the government of Rakosi in 1949. When MALENKOV came to power in the Soviet Union (1953), Nagy was appointed prime minister but was forced to resign in 1955 over his liberal policies (e g more consumer goods, less collectivization). The next year he was called upon by the central committee of the Communist Party to resume the premiership in the hope of quelling the national uprising that had broken out. When the Red Army moved in to crush the revolt, Nagy sought refuge in the Yugoslav embassy. On an understanding that he had been granted safe passage he left the embassy, but was deported to Romania before returning to Hungary, where he was abducted by Soviet troops and subsequently executed.

Nagy was a popular leader who, as minister for agriculture, was responsible for significant land reforms; as prime minister he sought to liberalize economic policies by supporting the manufacture of consumer goods and relaxing agricultural collective programmes. He is principally remembered for his independent attitudes in the face of communist orthodoxy and he became a hero of the new liberal republic declared in 1989.

Naipaul, Sir V(idiadhar) S(urajprasad) (1932–   ) *Trinidadian novelist of Indian parentage. He was knighted in 1990.*

Naipaul was educated at Queen's Royal College, Trinidad, and University College, Oxford. After Oxford he settled in Britain. His first novel, *The Mystic Masseur* (1957), was awarded the 1958 John Llewelyn Rhys Memorial Prize. It maps out typical Naipaul territory: the Indian subculture of the West Indies, observed with a satirical but sympathetic eye. It was followed by *The Suffrage of Elvira* (1958), *Miguel Street* (1959), and *A House for Mr. Biswas* (1961), the last based partly upon the life of Naipaul's own father. *The Middle Passage* (1962) and *an Area of Darkness* (1964) are travel books about Naipaul's visits, respectively, to his native Caribbean and his ancestral India, and *The Loss of El Dorado* (1969) is a history of Trinidad. He has also written collections of short stories: *A Flag on the Island* (1967) and *The Overcrowded Barracoon* (1972). Later novels include *The Mimic Men* (1967), *In a Free State* (1971), *Guerrillas* (1975), *A Bend in the River* (1979), and *The Enigma of Arrival* (1987). *Among the Believers: An Islamic Journey* appeared in 1981; other recent nonfiction includes *Finding the Centre* (1984) and *A Turn in the South* (1989).

Namier, Sir Lewis Bernstein (1888–1960) *Polish-born British historian. He was knighted in 1952.*

Born in Poland of Jewish parents, Namier emigrated to England in 1906 and was awarded a first-class honours degree at Balliol College, Oxford, in 1911. He became a naturalized British citizen in 1913 and after a year in the army at the start of World War I spent five years working in the Foreign Office. Before the war he had already turned to the study of British parliamentary history during the eighteenth century, and in 1920 he returned to what became his major academic preoccupation. Although he lectured for a year (1920–21) in Oxford, it was Manchester University, where he was professor of modern history from 1931 to 1953, that was his academic base.

Collecting and analysing a mass of detailed information about individual Members of Parliament, Namier reshaped the interpretation of British eighteenth-century politics. Disciples and critics alike called it 'namierizing'. He himself deemed as 'definitive' his *The Structure of Politics at the Accession of George III* (1929) and *England in the Age of the American Revolution* (1930); he followed the same approach to research in the last nine years of his life as editorial board member of the *History of Parliament*, given special responsibility for the later eighteenth century.

Namier's other historical writings on European nineteenth- and twentieth-century history were outstanding, and he was a powerful (if controversial) personality, a conservative, and a Zionist.

Nansen, Fridtjof (1861–1930) *Norwegian explorer, oceanographer, and statesman who led the first expedition to cross the Greenland icecap. For his humanitarian work for casualties of World War I he received the 1922 Nobel Peace Prize.*

Nansen studied zoology at Christiania (now Oslo) University and in 1883 was appointed curator of zoology at the Bergen Museum. In May 1888, Nansen and five companions made the first crossing of the Greenland interior and returned triumphantly to Norway in 1889. In the same year, Nansen was appointed curator of zoology at Oslo Museum. He devised a scheme to test his theory that the Arctic icepack flowed from Siberian waters over the polar region and down the eastern Greenland coast. His ship, *Fram*, was built to withstand the pressures of pack ice and on 24 June 1893 they set sail. By September *Fram* was enclosed by the ice and started its northwesterly drift. But it failed to pass near the Pole, and in March 1895, Nansen and Frederick Johansen left the *Fram* and headed northwards by dog-sledge. Although reaching a record latitude, they were forced to turn back on 8 April and spent the winter on Franz Joseph Land in an improvised shelter. In August 1896 they fortunately encountered the explorer Frederick Jackson and returned with him to Norway. The *Fram* meanwhile emerged from the ice in 1896 and returned home a few days after Nansen! *Farthest North* (2 vols, 1897) is Nansen's account of this epic journey.

Appointed professor of zoology at Christiania University in 1896, Nansen turned his attention to oceanography and became professor of the subject in 1908. He organized four oceanographic expeditions to the North Atlantic and Arctic Oceans and in 1901 helped found the International Council for the Exploration of the Sea. He became increasingly involved in affairs of state and served as Norwegian minister in London (1906–08). From 1920 he headed the Norwegian delegation to the League of Nations and in the same year was appointed high commissioner responsible for the repatriation of 500 000 prisoners-of-war. The following year he headed an International Red Cross mission to relieve famine in the Soviet Union, and in 1922 introduced the 'Nansen passport', an identification card for stateless persons.

Narayan, R(asipuram) K(rishnaswamy) (1906–) *Indian novelist. He won the Padma Bushan award for services to literature.*

Narayan was born in Madras, the son of a headmaster, and educated at the Maharaja's College, Mysore. On leaving college (1934) he was appointed teacher in a small village school, an environment that he found so constricting that he turned to writing. The result was his first novel, *Swami and Friends* (1935), about a group of schoolboys. Writing with elegance and humour in his adopted language, English, he built up in his subsequent novels an affectionate if ironical picture of life in a small southern Indian town, a place that he named Malgudi. Among his successful novels are *The English Teacher* (1945), *The Financial Expert* (1952), *The Guide* (1958), *The Man-Eater of Malgudi* (1961), *The Sweet Vendor* (1967), *The Painter of Signs* (1977), *A Tiger for Malgudi* (1983), *The Talkative Man* (1989), and *The World of Nagaraj* (1990). He also published a large number of short stories, notably *Under the Banyan Tree and Other Stories,* (1985), collections of essays, such as *The Reluctant Guru* (1974), and the autobiography *My Days* (1975). He edited an abbreviated modern prose version of the Hindu epic the *Ramayana* (1973) and followed this with an edition of the *Mahabharata* (1978).

Nash, Paul (1889–1946) *British landscape painter.*

Nash was born in London and educated at St Paul's School; unlike his brother John Nash (1893–1977), who became an artist without formal training, he went on to study at the Slade School of Art. In his final year there (1912) he had his first one-man exhibition. His water-colour landscapes had a poetic character, which derived from a powerful sense of the personality of place. Their style seemed to be influenced by Cézanne but was completely individual. During World War I he served in the Artists' Rifles and in 1917 as an official war artist. His direct and powerful pictures of the frightful landscapes of the front created an enormous impression. Nash also began to paint in oils from this time. From 1928 the influence of surrealism reinforced his sense of the ominous mystery of landscape and of objects. As a war artist again in 1940 he depicted combat in the air. His final landscapes, such as *Vernal Equinox* (1943) and the *Sunflower* series, are remarkable for their strange visionary quality.

Nash, Sir Walter (1882–1968) *New Zealand statesman and Labour prime minister (1957–60). He became a CH in 1959 and was knighted in 1965.*

Born in Kidderminster, the son of a clerk, Nash emigrated from England in 1909, having worked in a legal office and cycle manufacturing firm in Birmingham after leaving school. Taking an interest in labour issues, Nash was elected to the national executive of the New Zealand Labour Party in 1919 and was subsequently secretary (1922–32) and president (1935).

Nash was first elected to parliament in 1929 as the member for Hutt. An idealist who was devoted to the alleviation of social and economic injustice, Nash accomplished his most important achievements as Labour minister for finance and customs (1935–49). He initiated several important economic and social reforms, including the controversial social security legislation in 1939, which laid the basis for a welfare state in New Zealand. Becoming deputy prime minister in 1940, Nash was appointed New Zealand minister in the USA and a member of the Pacific War Council (1942–44). He was leader of the New Zealand delegation to the UN Monetary and Financial Conference at Bretton Woods (1944), at which the IMF (International Monetary Fund) and the IBRD (International Bank for Reconstruction and Development) were created. He became leader of the opposition in 1950 and prime minister in 1957, with a working majority of only two seats. Losing the 1960 election, he resigned from the party leadership in 1963 but retained his parliamentary seat until his death.

Nasser, Gamal Abdel (1918–70) *Egyptian statesman; prime minister (1954–56) and president (1956–70). His domestic reforms and nationalization of the Suez Canal made him a national hero and the acknowledged leader of the Arab world.*

Born in Alexandria, the son of a postal clerk, Nasser was educated at the Cairo Military Academy. There he helped to establish a nationalist movement among cadets and junior officers, known as the Free Officers Movement. After the unsuccessful war against Israel (1948–49), in which he was wounded, the movement grew quickly; by 1952 Nasser (now a colonel) was sufficiently strong to lead other Free Officers in a successful coup against the regime of King Farouk (1920–65). In the new Egyptian republic Nasser was appointed minister of the interior and deputy to President Mohammed Neguib (1901–84).

Becoming prime minister in 1954, he replaced Neguib as president in 1956 when he promulgated a new constitution. In the same year the world became aware of Colonel Nasser, after his announcement that he was about to nationalize the Suez Canal. The ensuing Suez Crisis, in which Anthony EDEN, in collaboration with France and Israel, led Britain in an attack on Egypt, destroyed Eden's political reputation and greatly enhanced Nasser's prestige in the Arab world. His prestige also rose among third world and nonaligned nations, although later he moved closer to the Soviet Union. His international reputation brought him new strength at home, enabling him to initiate substantial political and social reforms by redistributing land, encouraging industrialization, expanding public works (including the construction of the Aswan High Dam) with Soviet financing, and providing more equitable educational opportunities. He led Egypt in two unsuccessful wars against Israel (1956 and 1967), resigning briefly after Egypt's humiliating defeat in the Six Day War (1967). He remained in power until 1970, when he died of a heart attack.

Navratilova, Martina (1956–   ) *US tennis player, born in Czechoslovakia, and the most successful woman player since Billy Jean KING.*

Navratilova defected to the USA in 1975 and became a US citizen in 1981. During this time she developed her powerful game to a pitch rarely seen in women's tennis: in 1982 and 1983, for example, she lost only 4 out of 180 matches and won 31 of the 35 tournaments she entered. In 1983 she won both the Wimbledon and US Open singles titles without conceding a set and her earnings for that year – over $1,400,000 – broke all records for a sportswoman. Going on to take the Australian championship and the 1984 French singles title, Navratilova thereby achieved the coveted grand slam of all four major titles in succession.

Her major successes include nine Wimbledon titles (1978–79, 1982–87, and 1990), and victories in the US Open, the French championship, and the Australian championship. She has been equally successful in doubles, with a record eight successive grand slam titles in partnership with the US player Pam Shriver. She won the doubles title at Wimbledon in 1976, 1979, and 1981–86. Her records include an unbroken run of 74 consecutive wins. By the end of 1989 her earnings totalled more than $15 million.

Needham, Joseph (1900–) *British zoologist and historian, noted for a mammoth scholarly work on science and civilization in China.*

Born in London, Needham studied medicine at Cambridge University and in 1924 was appointed fellow of Gonville and Caius College, marking the start of a long association. He later became demonstrator in biochemistry (1928–33), then reader (1933–66) and emeritus professor (1966–). Needham studied under the eminent biochemist Frederick Gowland HOPKINS and specialized in embryology, writing *Chemical Embryology* (3 vols, 1931). However, his interests were wide-ranging, spanning politics, history, religion, and philosophy, and in particular the role of science in society. His interest in China arose out of his first visit in 1942 as head of the British scientific mission in Chungking and he was a staunch supporter of the communist revolution that finally achieved power in 1949. He embarked on a massive survey of scientific achievement in China and its impact on Chinese civilization – *Science and Civilization in China* (7 vols, 1954–). His other books include *The Sceptical Biologist* (1929), *Time, The Refreshing River* (1943), *Celestial Lancets* (1980), about acupuncture, and *The Hall of Heavenly Records* (1986), about Korean astronomical instruments and clocks.

Nehru, Jawaharlal (1889–1964) *First prime minister of India (1947–64). He was a central figure in the creation of an independent India and established a political dynasty.*

Born in Allahabad, the son of a lawyer, Nehru was educated in England at Harrow School and Trinity College, Cambridge, before qualifying as a barrister at the Inner Temple, London. In 1912 he returned to India and became involved in the Indian nationalist movement. He joined the Indian National Congress led by Mahatma GANDHI after the massacre at Amritsar (1919). Nehru was elected leader of the Congress Party in succession to his father, (Pandit) Motilal Nehru (1861–1931), in 1929. Imprisoned nine times by the British during the 1930s and 1940s, he worked closely with Gandhi in the

campaign for Indian independence and participated in negotiations with the British authorities. When India and Pakistan became separate nations in 1947, he was elected India's first prime minister.

One of Nehru's first acts during his long premiership was to decide to keep India as an independent republic within the Commonwealth, which undoubtedly helped to determine its evolution into a multiracial association of equals. He was the architect of India's foreign policy, which was based on the principle of nonalignment, although he was obliged to ask for US and British military aid against the Chinese during the border dispute of 1962. Popular with his people, he was more of a pragmatist than Gandhi and tried to mitigate the conflict between the various religious groups on the Indian subcontinent.

He remained in office until his death at the age of seventy-four. His daughter Indira GANDHI followed in his footsteps and was elected prime minister in 1966.

Nernst, Walther Hermann (1864–1941) *German physical chemist, who was awarded the 1920 Nobel Prize for Chemistry for his work in the field of thermodynamics.*

The son of a Prussian judge, Nernst was educated at the universities of Zürich, Berlin, Würzburg, and Graz. He taught briefly at Leipzig before being appointed in 1890 to the chair of chemistry at Göttingen. In 1904 Nernst moved to Berlin, where he remained in the post of professor of physical chemistry until his retirement in 1933. In this year Nernst, despising the Nazis, retired to his estates in Prussia. These had been acquired quite early in his life with the fortune he had made selling his rights in a form of electric light he had invented. Offered a royalty, Nernst prudently took a lump sum shortly before his light was replaced by the more efficient tungsten-filament lamp.

Nernst's early work in electrochemistry was followed in 1906 by the major advance in thermodynamics he made with his statement of the Nernst heat theorem, which predicts that there would be no change in entropy in a reaction between crystalline solids at absolute zero. This can be generalized into a statement of the third law of thermodynamics, which states that absolute zero cannot be attained in a finite number of steps. Nernst is also remembered for his *Theoretische Chemie vom Standpunkte der Avogadroschen Regel und der Thermodynamik* (1893; translated as *Theoretical Chemistry from the Standpoint of Avogadro's*

Rule and Thermodynamics, 1895), a much used and translated textbook of the period.

Neruda, Pablo (Ricardo Eliezer Neftalí Reyes; 1904–73) *Chilean poet and diplomat. He was awarded the Nobel Prize for Literature in 1971.*

Neruda adopted the pseudonym in 1920 in honour of the Czech poet Jan Neruda and changed his name by deed poll in 1946. The son of a train-driver, he moved with his father to Temuco in southern Chile after the death of his mother. Here he was educated and met the poet Gabriela Mistral (1889–1957), then a teacher (she won a Nobel Prize in 1945). He went to Santiago in 1920 or 1921 as a student and published three books of poems (1921, 1923, and 1924), which earned him a reputation throughout Latin America. From 1927 to 1943 Neruda lived abroad, for six years serving as the Chilean consul in Rangoon and Java and from 1934 living in Spain, where he founded the poetry magazine *Caballo verde para la poesía* in Madrid in 1935. It printed works by GARCÍA LORCA, Guillén, and other major writers until it ceased publication at the start of the civil war. As editor, Neruda attacked the idea of 'pure' detached poetry and argued instead for an 'impure' poetry committed to the realities of life, a principle adopted in his own work, *Residencia en la tierra, 1925–31* (expanded in 1935; translated as *Residence on Earth and Other Poems*, 1946). In 1939 he fled to Paris, where he was involved in war refugee work; he later joined the Communist Party.

After his return to Chile he was elected senator (1945) and published the collection *Tercera residencia, 1935–1945* (1947), notably plain and straightforward in style, and began publishing a major work of epic length, the *Canto general* (completed 1950). Originally conceived as an epic on Chile, this poem was expanded in many parts to cover all the Americas, their natural history, ancient civilizations (as in *Alturas de Macchu Picchu*; translated as *The Heights of Macchu Picchu*, 1966), modern wars of liberation, etc. Forced to leave Chile in 1948, Neruda travelled to eastern Europe, the Soviet Union (where he was awarded the World Peace Prize, 1950), and China. He returned to Chile in 1952. *Odas elementales* (1954; translated as *The Elemental Odes*, 1961) and *Tercer libro de las odas* (1959) were followed by an edition of his complete poems, *Obras completas* (1957; revised 1968), though he produced several further volumes afterwards. He was presidential candidate in 1970 but withdrew in favour of ALLENDE, the same year he was appointed ambassador to France. From 1961 Neruda lived, when possible, at his house on Isla Negra, which he bequeathed to the copper workers of Chile. A verse autobiography, *Memorial de Isla Negra* (1964), was supplemented by one in prose, *Confieso que he vivido* (1974).

Nervi, Pier Luigi (1891–1949) *Italian engineer and architect. A pioneer in the use of reinforced concrete, Nervi combined the roles of architect, engineer, entrepreneur, and academic to an unusual extent. His many famous buildings include the Pirelli skyscraper in Milan.*

Born in Sondrio, Lombardy, he obtained his civil engineering degree at the University of Bologna and then worked in a firm of architects who specialized in reinforced concrete. After serving as an officer in the Italian Engineering Corps during World War I, he set up his own partnership as a consulting engineer. He became a professor in the faculty of architecture at the University of Rome in 1947 and remained there until 1961. His first major building was the communal stadium in Florence (1930–32), with a cantilever roof and unusual spiral staircase; he then won a competition for a series of aircraft hangars, for which he developed his lattice vault of precast concrete beams (1936–40). In 1948 he designed the Turin Exhibition Hall, using corrugated concrete to form the vault lattice. The grid vault was also a feature of the Chianciano spa (1952) and of the two circular Olympic stadiums in Rome (1960). In 1953 he cooperated with Marcel BREUER and Zehrfuss in the construction of the UNESCO secretariat in Paris. Perhaps his best-known building is the elegant Pirelli skyscraper in Milan (1958), with its thirty-two cantilevered floors. His last building was the San Francisco cathedral (1970).

Newman, Paul (1925–   ) *US film actor, director, and producer, who is also politically active.*

Born in Cleveland, Ohio, Newman turned to acting after World War II, in which he served with the navy air corps. After spending some time at Kenyon College, Ohio, studying economics, he moved on to the Yale Drama School and the New York Actors' Studio. In 1953 he made his successful Broadway debut in *Picnic* and two years later he made his first film, *The Silver Chalice* (1955). This was quickly followed by such notable films as *The Long Hot Summer* (1958), for which he received a Cannes Festival Award, and *Cat on a*

Hot Tin Roof (1958), which earned him his first Academy Award nomination. *The Hustler* (1961), *Hud* (1963), and *Cool Hand Luke* (1967) brought further Oscar nominations. In both *Butch Cassidy and the Sundance Kid* (1969) and *The Sting* (1973), he played opposite Robert REDFORD. Successes since then have included *The Verdict* (1982), *The Color of Money* (1986), for which he won an Oscar, and *Blaze* (1990). In 1986 he received an honorary Oscar for career achievement.

His first film as director was *Rachel, Rachel* (1968), starring his wife Joanne Woodward (1930–). Newman and Woodward have co-starred in many films since 1958, when they made *The Long Hot Summer* and *Rally Round the Flag Boys!* together. With Sidney Poitier (1924–), Steve McQueen (1930–80), Barbra STREISAND, and others, he founded the production company First Artists. Politically active, Newman served as US delegate to the UN conference on disarmament in 1978. After his son died of a drug overdose, he set up the Scott Newman Foundation to make films highlighting the dangers of drug abuse. He is also chairman of a company making his own brand of salad dressing, the profits of which go to charity.

Ngo Dinh Diem (1901–63) *First president of South Vietnam (1955–63). A devout Roman Catholic, he was a fervent nationalist who gradually alienated both domestic and international support for his regime.*

Born in Hué, of royal descent (his family were among the first Vietnamese to be converted to Roman Catholicism in the seventeenth century), Ngo Dinh Diem spent much of his youth associated with the imperial family. Following in his father's footsteps, he worked for the government and during the 1930s served as minister of the interior for the emperor, Bao Dai. In 1945 Ngo Dinh Diem was taken prisoner by communist forces led by HO CHI MINH. Asked by Ho to join his government in the north (in the hope that Ngo would secure Catholic support for the communists), he refused and fled to the USA, where he remained in exile for almost a decade. He returned to South Vietnam as prime minister in 1954 under the sponsorship of the USA. In 1955 he ousted the monarchy, declared South Vietnam a republic, and became president.

Pitting himself against the communists from the start, he also spurned the French and was fiercely independent of the Americans who supported him. While he promised land reforms and the elimination of corruption, he lost much support because of his nepotism, his promotion of fellow-Catholics and members of his own family, and his ruthless suppression of opposition. His persecution of the Buddhists, who made up an overwhelming majority in South Vietnam, led to the withdrawal of US support and a military coup (1963), in which he was assassinated.

Nicholas II (1868–1918) *The last tsar of Russia (1895–1917), an autocratic ruler who believed in the divine right of kings.*

Son of Alexander III, in 1894 Nicholas married ALEXANDRA of Hesse-Darmstadt, whose strong personality exerted a fatal influence over her husband. She encouraged his ambitions in Asia, where his occupation of Port Arthur (1896) and Manchuria (1900) aroused the antagonism of Japan. The subsequent Russo-Japanese War (1904–05) ended with Russia's surrender. At home, Nicholas's autocratic method of government was a prime cause of the revolution of 1905. Forced to accept the establishment of a representative assembly (the Duma), he nevertheless continued to try to rule absolutely. After the outbreak of World War I he took command of the Russian forces, leaving Alexandra to govern in his absence. Her disastrous administration, which was largely in the hands of her mentor RASPUTIN, was a significant immediate cause of the Bolshevik revolution in 1917, in which the monarchy was overthrown. In 1918 Nicholas and his family were assassinated at Ekaterinburg.

Nicholson, Ben (1894–1982) *British abstract painter and the first winner of the Guggenheim International Prize. He received the OM in 1968.*

The eldest son of the painter Sir William Nicholson (1872–1949), he received formal training in art for only one term at the Slade School in London in 1911. He was briefly connected with the vorticist movement at the beginning of World War I (see LEWIS, WYNDHAM), but his first one-man exhibition in 1922 revealed the cubist influence that had arisen from his visit to Paris the previous year and was to provide the basis for much of his future work. The paintings of the 1920s were cubist still-lifes and landscapes as well as abstract paintings.

In 1930 he began to work with the sculptress Barbara HEPWORTH, who became his second wife. From 1933 he produced painted reliefs with circular and rectangular motifs. These became his main output, together with austere geometric compositions on canvas. During the 1930s he was a member of the Seven and Five group, Unit One, and the Abstraction-Création Association and was in contact with many of the major artists of the period, particularly MONDRIAN. From 1939 he lived in St Ives in Cornwall, where his second marriage ended in 1951. A year after his third marriage, in 1958, he moved to Ticino in Switzerland. During the 1950s he won first prizes at the Carnegie International, the Guggenheim International, and the São Paolo Biennale.

Nicklaus, Jack William (1940–) *US golfer and winner of more tournaments than any golfer in history, including nineteen major championships.*

Nicklaus was born in Columbus, Ohio. Encouraged by his father, an excellent all-round sportsman, he started playing golf at the age of ten at the Scioto Country Club in Columbus, where Bobby JONES had won the 1926 Open. He adopted the less fashionable interlocking grip, which he has retained throughout his career. Nicklaus won the US amateur championship twice while a student at Ohio State University (1959–62). He turned professional in 1962 and won the first of four US Open championships that year. He has also won the US Masters six times, the PGA championship six times, and the British Open championship three times (including 1978 at St Andrews, perhaps his greatest victory). Nicklaus has over 70 other tournament victories around the world to his credit. When not actually playing the game he finds time to design and build golf courses.

Nielsen, Carl August (1865–1931) *Danish composer who has been called the co-founder, with SIBELIUS, of modern Scandinavian music.*

Nielsen was born into a poor but musical family, the seventh of twelve children; his early musical experiences were restricted to his mother's singing and his father's playing of the violin or cornet. Nielsen taught himself these instruments and although he had little formal education, his natural inquisitiveness enabled him to acquire a basic musical background. His musical career started in 1879 in a military band, from which he gained a scholarship in 1886 to study at the Copenhagen Conservatory. His first major success as a composer was the performance of his *Little Suite* for string orchestra by the Tivoli Orchestra (1888). In 1889 he became a second violinist in the orchestra of the Royal Chapel, a post he held for sixteen years, gaining a wide knowledge of music through its comprehensive repertoire. In 1891, during a journey through Europe, Nielsen met and married the sculptor Anne Marie

Brodersen and ten years later was given a state pension enabling him to compose without financial worries. From 1916 until 1931 Nielsen was connected with the Copenhagen Conservatory as teacher, examiner, and member of the board of governors, and in 1931 he was appointed director. During this period he increasingly conducted his own works at home and abroad and acquired an international reputation.

The core of Nielsen's achievement is the six symphonies written between 1890 and 1925. These illustrate the composer's development from his early classical style to a personal idiomatic language using polyphony, not only in the instrumental lines, but as a means of structural cohesion. Other works of particular importance are the two operas *Saul and David* (1899–1901) and *Maskerade* (1904–06), concertos for violin (1911), flute (1926), and clarinet (1928), the wind quintet (1922), and his last major work, *Commotio* for organ (1931).

Niemeyer, Oscar (1907–) *Brazilian architect. Deeply influenced by LE CORBUSIER, Niemeyer created his own flamboyant style, especially in the public buildings for Brasília, for which he was director of architecture.*

Born in Rio de Janeiro, Niemeyer was educated at the Escola Nacional de Belas Artes and the University of Brazil. He joined the office of Lúcio Costa (1902–63) in 1935 and in 1936 was part of the team of Brazilian architects who worked with Le Corbusier on the new Ministry of Education in Rio de Janeiro. In 1939 he collaborated with Costa on the Brazilian Pavilion for the New York World Fair (1939). Thereafter he worked largely on his own. The casino, club, and church designed for Pamphulha, Belo Horizonte (1941–43), with its liberal approach to the relationship between design and function, attracted considerable attention and he was appointed architectural director of the new capital, Brasília, which was created in the 1950s. The president's palace, the cathedral, the law courts, and the Square of the Three Powers in Brasília have all been highly praised. Indeed the whole enterprise has been greeted as a triumph of modern planning as well as a highly commendable vision on the part of the Brazilian government. The plan and the vision have only been partly marred by reality in the form of the dismal shanty town that has grown up around Niemeyer's opulently planned city.

Niemöller, Martin (1892–1984) *German Lutheran pastor, who became internationally known for his courageous resistance to Nazism*

in the 1930s and later for his support of the ecumenical and peace movements.

Born at Lippstadt (Westphalia), the son of a pastor, Niemöller enrolled in the German navy, becoming a much-decorated U-boat commander in World War I. He was ordained in 1924 and became pastor of the Berlin parish of Dahlem in 1931. His book, *From U-boat to Pulpit* (1934), became a best-seller. Although he was an ardent nationalist he quickly foresaw the dangers of Nazism and two years later formed the Pastors' Emergency League, out of which grew the Confessing Church, in opposition to HITLER's German Christian Movement (See BARTH, KARL). Summoned to a personal interview with Hitler, Niemöller stood firm and as a consequence was prohibited from preaching. His decision to ignore this prohibition led to his arrest and subsequent imprisonment (1937–45) in concentration camps at Sachsenhausen and Dachau, much of the time in solitary confinement. During World War II, Niemöller became the symbolic figure of Protestant opposition to Nazism. In his last days, Hitler ordered his execution – an order that was not carried out.

After the war, Niemöller took a leading part in the 'Declaration of Guilt' at Stuttgart, which confessed the war guilt of Germany and especially of German churchmen – a guilt he insisted upon sharing. A member of the German delegation that formed the World Council of Churches, he became the first president of the German Protestant Church's Office for Foreign Relations. Niemöller was an ardent advocate of a united neutral Germany and in 1952 made a much-publicized visit to the Soviet Union in support of reconciliation with the East. By 1954 he had become a Christian pacifist and a vociferous supporter of nuclear disarmament.

Niemöller retired in 1964 from the presidency of the Church in Hesse and Nassau, devoting the remaining years of his life to the Christian peace movement.

Nijinsky, Vaslav (1888–1950) *Russian ballet dancer and choreographer regarded as one of the greatest dancers in the history of ballet. He was greatly encouraged by Sergei DIAGHILEV, who built up his enormous reputation. His creative life was short, however, as a result of his mental illness.*

Nijinsky was dancing with his parents' troupe in Kiev by the age of three. With the amazing ability to perform ten entrechats (crossing and uncrossing the legs ten times while still in the air), he entered the Imperial Ballet School in

1907. Having gained a reputation as an outstanding dancer with a virtuoso technique, he joined the Imperial Ballet after graduation and was at once given solo roles. After meeting Sergei Diaghilev, whose protégé he became, he toured Europe with Diaghilev's Ballets Russes: his performances, including the poet in *Les Sylphides* and the golden slave in *Schéhérazade*, caused a sensation in Paris, where he created a vogue for all things Russian and associated with ballet. Nijinsky joined the Ballets Russes on a permanent basis in 1911, having been dismissed from the Imperial Ballet in a dispute over costume. Diaghilev then commissioned especially for him the ballets *Petrushka* (music by STRAVINSKY) and *Le Spectre de la Rose* (adapted from music by Weber), both choreographed by FOKINE. At this stage in his career Diaghilev encouraged Nijinsky to choreograph his own ballets: *L'Après-midi d'un faune* (1912) and *Jeux* (1913) both used DEBUSSY's music, and *Le Sacre du Printemps* (1913; *The Rite of Spring*), with music by Stravinsky, caused an uproar when it was first performed. Nijinsky's unexpected marriage in 1913 to a Hungarian countess, Romola de Pulszka, provoked the jealous and betrayed Diaghilev to dismiss him from the company. Four years later, however, Diaghilev produced Nijinsky's last ballet, *Till Eulenspiegel*, based on Richard STRAUSS's tone poem, in which he danced the title role. Thereafter he became increasingly ill with schizophrenia and was forced to retire. His wife took him to Switzerland in 1919, where for many years he was in and out of psychiatric hospitals. The last three years of his life were spent in London.

Nilsson, (Märta) Birgit (1918–   ) *Swedish singer, known for her soprano roles, especially in the operas of Wagner and Richard STRAUSS.*

Nilsson entered the Royal Academy of Music, Stockholm, in 1941, where she studied with Joseph Hislop. Her debut (1946), at the Stockholm Royal Opera in the role of Agathe in Weber's *Der Freischütz*, was as a last-minute replacement. Five years later she had her first success outside Sweden, as Electra in Mozart's *Idomeneo* at Glyndebourne. By 1955 Nilsson had established herself as a world-class singer, making her US debut in San Francisco (1956) and singing Wagner's Isolde at the Metropolitan Opera, New York, in 1959. She is also well known for her interpretation of Leonora in Beethoven's *Fidelio*.

Niven, (James) David (Graham) (1909–83) *British film actor and producer, who later became a best-selling author.*

Niven was brought up in Kirriemuir, Scotland, and educated at Stowe School. He arrived in the USA in 1935 after Sandhurst, the Highland Light Infantry (1929–32), and a variety of occupations in various parts of the world. Beginning as an extra (appropriately classed 'Anglo-Saxon Type No. 2008'), he progressed to larger parts in such films as *The Charge of the Light Brigade* (1936), with his friend Errol FLYNN. His first starring role came in *Bachelor Mother* (1939).

During World War II he served with the Rifle Brigade and the Phantom Reconnaissance Regiment, reaching the rank of colonel. On leave, he made *The First of the Few* (1942), with Leslie HOWARD, and *The Way Ahead* (1944). His memorable postwar films included *The Moon is Blue* (1953), *Around the World in 80 Days* (1956), and *Separate Tables* (1958), which brought him an Oscar and a New York Critics Award. Always urbane and debonair, Niven fitted Hollywood's image of the aristocratic English gentleman exactly. Notable of his later films were *The Guns of Navarone* (1961), *55 Days at Peking* (1963), and *The Pink Panther* (1964). Successful, too, was the Four Star Television Company he co-founded with Charles BOYER and Dick Powell (1904–63).

A natural raconteur, he achieved huge success with his autobiographies, *The Moon's a Balloon* (1971) and *Bring on the Empty Horses* (1975), and several novels. Niven's first wife, Primula, died tragically young in an accident in 1946. His second wife was Swedish model Hjordis Tersmeden, whom he married in 1948. He died bravely fighting an illness that deprived him of his voice.

Nixon, Richard Milhous (1913–) *US Republican statesman and thirty-seventh president (1969–74) of the USA. His success in improving US relations with China and the Soviet Union was overshadowed by the revelations of the Watergate scandal, after which he became the first president to resign the office.*

Nixon graduated from Whittier College, California, in 1934 and three years later, having qualified as a lawyer at Duke University, entered legal practice in Whittier. He served in the US navy during World War II and was elected to Congress in 1947. His staunch anticommunist views received wide publicity, especially in the Alger HISS case. After serving as a senator for California (1950–53), Nixon

was appointed vice-president by EISENHOWER. But his defeat by KENNEDY in the 1960 presidential election seemed to mark the end of Nixon's political career. However, during the 1960s, he staged a remarkable comeback, won the 1968 Republican nomination, and defeated Hubert Humphrey (1911–78) to enter the White House.

On the domestic front, Nixon instituted conservative social policies with emphasis on 'self-reliance'. After bouts of inflation and recession, strict wage and price controls were introduced and the dollar devalued to boost trade. From the outset however, one of Nixon's main preoccupations was the continuing presence of US troops in South Vietnam. His policy of 'peace with honour', which envisaged progressive US disengagement while reinforcing South Vietnamese forces, was principally conducted by his national security adviser, Henry KISSINGER. As the war dragged on, domestic unrest grew, especially among the young. Nixon's popularity was boosted in February 1972 by his historic visit to China, which led to the restoration of the Sino–US diplomatic relations, and in May by his agreement with Soviet leaders in Moscow on arms limitations (SALT 1) and trade.

Nixon overwhelmingly beat George McGovern (1922–) in the November 1972 presidential election. A ceasefire agreement was signed with the North Vietnamese on 23 January 1973 following intensification of US bombing raids over the region, and US troops were finally withdrawn. But Nixon was soon embroiled in a major domestic scandal. Seven men found guilty of illegally entering Democratic Party offices in Washington's Watergate building were found to have links with Nixon's re-election campaign committee. The affair was probed by both a Senate committee and the Grand Jury and in April three of Nixon's top aides – John Ehrlichman, H. R. Haldeman, and John Dean – resigned following their implication in the affair. Taped transcripts of White House conversations, regarded as crucial evidence, became the focus of a legal and constitutional battle between the judiciary and the president so that in October a Congressional judicial committee began studying the grounds for Nixon's impeachment. Many of Nixon's statements concerning Watergate were subsequently exposed as mendacious and in July 1974 the House committee voted to impeach the president. On 9 August, Nixon resigned. His gross abuse of power had greatly diminished respect for the presidency and left America bewildered and cynical. Nixon was granted an unconditional pardon by his successor, Gerald FORD. He has subsequently been involved in a number of informal political missions, some overseas; his publications include such volumes of memoirs and political commentaries as *Memoirs* (1978), *No More Vietnams* (1986), and *In the Arena* (1990).

Nkrumah, Kwame (Francis Nwia Kofi; *c*. 1909–72) *The first president of independent Ghana (1957–66) and the first popular African leader to achieve independence for his country. Unfortunately the national hero turned into a despotic tyrant, who was deposed by a military coup.*

Born in Nekroful, western Gold Coast (now Ghana), the son of a goldsmith, Nkrumah was educated at Catholic mission schools before training to be a teacher in Accra in 1926. Travelling to the USA in 1935, he took degrees at Lincoln and Pennsylvania universities (later teaching political science at Lincoln University) before studying law in London in 1945. In London Nkrumah was active in the West African Students Union and when he returned to the Gold Coast in 1947 he became leader of the newly formed United Gold Coast Convention (UGCC) Party. Forced to resign in 1949 over his 'Positive Action' campaign (a programme of civil disobedience, agitation, and widespread propaganda), Nkrumah launched his own political party, the Convention People's Party (CPP), the first mass-appeal party to emerge in black Africa. In 1950 he was imprisoned by the British as a subversive but was released in 1951, following a landslide victory by the CPP. He was elected the first prime minister of the Gold Coast by the legislative assembly in 1952. Six years later in 1957, the Gold Coast became the first black African colony to achieve independence, adopting the name of Ghana. Declaring Ghana a republic in 1960, Nkrumah decreed himself president for life in 1964, banning all opposition parties. He was deposed in 1966, while on a visit to Peking; the reaction of the Ghanaian people to this was jubilation and the destruction of the statue of himself that Nkrumah had erected. He died in exile.

Nolan, Sir Sydney Robert (1917–) *Australian painter, whose best-known pictures were inspired by Australian history and landscapes. He was knighted in 1981 and became a Companion of the Order of Australia in 1988.*

Born in Melbourne, the son of an Irish tram driver, Nolan grew up an athletic young man with a keen interest in cycle racing. In 1938,

after attending evening classes in painting, he became a full-time artist, moving in Bohemian avant-garde circles and producing pictures that shocked the public.

During his service in the Australian army in World War II Nolan produced his first backwoods landscapes. In 1946 he began his famous *Ned Kelly* series, celebrating the notorious folk hero in a deliberately naive style. Nolan travelled widely in Australia in the 1940s, painting pictures inspired by folk history and legend, such as the *Fraser* series and the paintings of explorers set in desert landscapes. From 1949, the year of his exhibition at UNESCO in Paris, he began to travel in Europe. Back in Australia three years later he painted scenes of the drought of 1952 and the following year worked on the film *Back of Beyond*, which won the Grand Prix at the Venice film festival in 1954. From that year he lived in London. He did a second series of Ned Kelly paintings in 1955 and thereafter continued to produce pictures of Australian subjects, as well as works that reflected his experiences of other countries visited.

Nolde, Emil (Emil Hansen; 1867–1956) *German expressionist painter and graphic artist.*

The son of a farmer, Emil Hansen changed his name in 1904 to that of his birthplace, Nolde, in Schleswig. After teaching at the School of Industrial Design in St Gallen (1882–88), he worked and studied in Munich, Dachau, and (from 1900) Paris and Copenhagen. The influence of Van Gogh's works can be seen in the bright colours and thick expressive brush strokes in his series of garden paintings (1907–08), but in his figure compositions his violent handling of colour and characterization reflects his inclination towards the grotesque. Although he was briefly a member of the Brücke group of artists after settling in Berlin in 1906, Nolde has been described as 'the great solitary of German expressionism'. His most outstanding works are his large religious pictures of 1909–12 (for example, *Dance Round the Golden Calf*, 1910); these were followed by figure compositions of rowdy low-life scenes painted with a barbaric sensuality reflecting his passionate temperament and belief in expression above accuracy.

Nolde's admiration of primitive art's 'concentrated often grotesque expression of force and vitality in the simplest possible form' led him to Polynesia in 1913–14, and primitive 'stylized, rhythmical, and decorative' qualities are apparent in such works as *The Dancers* (1920). Though best known for his figure compositions, Nolde also painted landscapes, and during World War II – when he was forbidden to paint by the Nazis – he produced small comparatively mild water-colours in secret.

Noriega, Manuel Antonio (1938–) *Panamanian general, dictator of Panama (1983–89) whose corrupt regime was eventually overthrown by a US invasion force.*

Educated at the University of Panama, he subsequently studied at the Military Academy in Peru before becoming a lieutenant in the Panama National Guard in 1962. By 1970 he was head of the Panama Intelligence Services and was establishing his control of the country's affairs. In 1983 he was made commander-in-chief of the Panama Defence Forces and the effective ruler of the state, installing and deposing the country's presidents. During the 1960s he was recruited by the CIA but US support for his regime became increasingly strained and was finally withdrawn in 1987. In 1988 he was charged in the US courts with drug trafficking and racketeering; as a result economic sanctions were imposed on Panama. Noriega refused to respond to this pressure and survived an attempted coup in 1989. When Guillermo Endara Gallimany won a presidential election in 1989, Noriega annulled the result and had himself formally declared head of the government. His declaration of a state of war between Panama and the USA proved the last straw and on 20 December 1989, in Operation Just Cause, 23 000 US troops with air cover seized control of Panama, installing Endara as president; 230 people died. Noriega took refuge in the Papal Nunciature but, after the building was bombarded for ten days with rock music and other psychological measures, he gave himself up and was flown to the USA to await trial on drugs charges. In 1990 progress towards his conviction was threatened when tapped telephone calls made by Noriega were publicized, leading to protestations that Noriega's rights under the US constitution had been contravened. Nonetheless, legal proceedings began in 1991, as planned.

Norman, Jessye (1945–) *US soprano, who has won wide acclaim for both her operatic and concert performances.*

Born in Augusta, Georgia, she won a scholarship to Howard University, Washington D.C., to study music and subsequently continued her studies at the University of Michigan. In 1968 she won the Bavarian Radio Corporation International Music Competition and a year later she made her opera debut in Wagner's *Tannhäuser* in Berlin. This success led to a contract with the Berlin Opera and to the growth of her reputation in Europe, where she was widely admired for her powerful stage presence. She made her debut at Covent Garden and La Scala in 1972, finally appearing in New York for the first time in 1973. Her performances in the works of Wagner, Schubert, and Mahler increased her prestige in the 1970s and 1980s. In association with Colin Davis and the BBC Symphony Orchestra she has made many notable recordings.

Novello, Ivor (David Ivor Davies; 1893–1951) *British actor-manager, composer, and dramatist, best known for his romantic musicals and popular songs.*

Novello was born in Cardiff. Encouraged by his mother, Clara Novello Davies, an accomplished musician and singing teacher, he began writing songs at an early age. While still a chorister at Magdalen College School, Oxford (1905–11), he published his first song, 'Spring of the Year'. His most popular early song, especially among World War I troops, was 'Keep the Home Fires Burning'; others included 'We'll Gather Lilacs'. During World War I Novello served in the Royal Naval Air Service (1914–18). He made his acting debut as Armand Duval in *Duburau* (1921) at the Ambassadors and subsequently became an actor-manager, the first of his many productions being *The Rat* (1924), which he wrote with Constance Collier (1878–1955). He wrote many other plays including *Symphony in Two Flats* (1929) and *The Truth Game*, performed in New York (1930), both of which were successfully transferred to the screen. Novello is best remembered, however, for his musicals, which include *Glamorous Night* (1935), *Careless Rapture* (1936), *Crest of a Wave* (1937), *The Dancing Years* (1939), *Perchance to Dream* (1945), and *King's Rhapsody* (1949). He also made a number of films, some written by himself or adapted from his plays. His first film as actor was the silent *The Call of Blood* (1919) and he spent some time in 1922 working with D. W. GRIFFITH. Other notable films include *The Rat* (1925), *Sleeping Car* (1933), and *Autumn Crocus* (1934).

Novotný, Antonín (1904–75) *Czechoslovak statesman. He was president of Czechoslovakia (1957–68) until his Stalinist sympathies and loyalty to Moscow culminated in his replacement by the reformists.*

Born in Prague into a working-class family, Novotný attended a local school before serving an apprenticeship as a blacksmith. At seventeen he became a founding member of the

Czechoslovak Communist Party (CCP), when the Social Democratic Party split (1921). Novotný began working for the CCP in 1929. By 1937 he had gained a prominent position in the party organization, moving from Prague to Moravia, where he took over the regional organization and edited a newspaper. During World War II he was active in the resistance movement until he was captured by the Germans, who imprisoned him (1941–45) at Mauthausen concentration camp. After the war he continued to rise in the party organization, becoming a member of the central committee (1946) and a member of the Politburo (1951). In 1953 he was elevated to first secretary of the party, becoming president of Czechoslovakia in 1957. He held both these offices until 1968, when the reform movement ousted him and he was replaced by DUBČEK as first secretary and General Suoboden as president. He was suspended from the party in 1968 but at the party congress in 1971, after the fall of Dubček, he was reinstated.

Nu, U (Thakin Nu; 1907–   ) *First prime minister of the Union of Burma (1948–56; 1957–58; 1960–62). He adopted the epithet 'U' (which means uncle) in 1952 at the height of his political career. Respected for his integrity and lack of pretension, he nevertheless failed to exert the authority of his office in the face of numerous difficulties.*

Born in Wakema, Lower Burma, Nu attended schools in Wakema and Rangoon, where he graduated in 1929. For several years he taught at the National High School in Pantanaw but in 1934 returned to university to study law. He was expelled in 1936 for participating in a student strike, as a result of which he became leader of the nationalist Thakin Party. At the outbreak of World War II he was imprisoned by the British for belonging to the anti-imperialist Dobhana Asiaijore ('Our Burma') Organization. In 1942 he was released when Burma was invaded and joined the Japanese-sponsored government as foreign minister under Ba Maw. Disillusioned with the Japanese, however, he cooperated with Aung San's anti-fascist organization and, when the war ended, was elected vice-president of a nationalist coalition known as the Anti-Fascist People's Freedom League.

In 1947 he announced his retirement from politics, but, upon the assassination of Aung San and most of his cabinet, was persuaded to lead the party and head a new government. Elected speaker of the constituent assembly in 1947, he signed the Anglo-Burmese Treaty in London, which paved the way for Burmese independence. The following year he became the first prime minister of the Union of Burma, a position he retained (with a brief interlude out of office in 1956–57) until 1958, when he gave in to pressure from the army in order to prevent civil war and resigned. He was returned to power in 1960 but was again frustrated by the extent of Burma's war damage; after a military coup under General Ne Win (1911–   ) in 1962, he was taken into custody (1962–66). He left Burma in 1969 to organize opposition to the military regime from Thailand until forced to leave. He returned to Burma in 1980 and in 1988 once more emerged as an opposition leader.

Nuffield, William Richard Morris, Viscount (1877–1963) *British industrialist and philanthropist who founded Morris Motors and, among numerous benefactions, financed the Nuffield Foundation. He was made a baronet in 1929, created Baron Nuffield in 1934, and elevated to the viscountcy in 1938. He was made a CH in 1958.*

William Morris was born in Worcester, the son of a draper's assistant. The family moved to Cowley, near Oxford, and Morris began work in a local cycle firm. In 1893 he started his own business with just £4, hiring and repairing cycles. Soon he was making his own models and in 1902 he switched to motor cycles and then to motor cars. A Morris Garage was set up in 1904 and his growing interest in construction led to the opening of the first Morris factory in 1912. The following year the first Morris car appeared: an 8.9 hp two-seater capable of 50 mph and selling for £165. The post-World War I Morris Cowley and Morris Oxford models established Morris's reputation as a manufacturer and sales increased rapidly to over 50 000 cars per year in 1925. Morris Motors Limited was formed in 1926, the year of the first MG (Morris Garage) models, and in 1927 Morris bought Wolseley. The company's growth continued with the acquisition of various component manufacturers and, in 1938, Riley Motors was bought. During World War II, Lord Nuffield, as he was by then, served as director-general of maintenance at the Air Ministry, while his factories turned out tanks and aircraft. 1948 saw the start of Nuffield tractors and also an association between the company and Austin Motors that led, in 1952, to their merger as the British Motor Corporation.

From 1926, Nuffield gave large sums to various organizations in Britain and throughout the world, usually in the fields of medicine or academic research. In 1937 he provided land and money for the foundation of Nuffield College, Oxford, as a graduate college specializing in social and political studies. The Nuffield Foundation was created in 1943, endowed with some £10 million to fund medical and social projects.

Nureyev, Rudolf Hametovich (1939–)
Soviet-born ballet dancer and choreographer, whose spectacular dancing has been compared to that of NIJINSKY. He defected from the Soviet Union in 1961 and became a naturalized Austrian citizen in 1982. He became a Chevalier de la Légion d'Honneur in 1988.

Born into a farming family in Ufa, Siberia, Nureyev studied at the Kiev Ballet School and made his first appearance with the Kiev Company in 1959. A rebellious student, he refused to join the Communist Youth Organization and while touring in May 1961 he sought political asylum in Paris. He claimed that the rigid organization of Soviet ballet restricted the number of roles available to him.

Immediately after his defection he joined the Grand Ballet du Marquis de Cuevas and frequently performed abroad. In 1962 he became permanent guest artist at the Royal Ballet in London, where he enjoyed a remarkable partnership with Margot FONTEYN, dancing with her in such ballets as *Giselle* and *Swan Lake*. In 1964 he reworked *Swan Lake*, giving the dominant role to the male dancer, following a precedent set by Nijinsky. Other works that he has choreographed include *La Bayadère, Raymonda, Tancredi, The Nutcracker, Don Quixote, Romeo and Juliet,* and *The Tempest.* He has also starred in several films, among them *Valentino* (1977) and *Exposed* (1982). From 1983 to 1989 he was artistic director of the Paris Opéra Ballet. Nureyev's explosive temperament and versatile acrobatic dancing have helped to make him the best-known contemporary male dancer. Unfortunately, how-ever, a series of tours undertaken in 1991 was less than highly acclaimed.

Nyerere, Julius Kambarage (1922–)
President of Tanzania (1964–85). Known as 'Mwalimu' (teacher) by Tanzanians, he is regarded as the father of the nation.

Born in Butiama near Lake Victoria, the son of Chief Nyerere Burito of the Zanaki tribe, Nyerere was educated at a government school before attending Makerere University College (1943–45), where he received a diploma in education. Beginning teaching at a mission school in Tabora (1946–49), he spent a period at Edinburgh University, before returning in 1953 to teach at St Francis College near Dar es Salaam.

Nyerere's political career began in 1954, when he founded the Tanganyika African National Union (TANU). Campaigning for the nationalist movement, he addressed the Trusteeship Council of the United Nations in 1955 and the Committee of the United Nations General Assembly in 1956. In 1957 he was nominated as the TANU member for Tanganyika in the legislative council but resigned in protest at its lack of progress. He was re-elected in 1958 as the Eastern Province member. In 1960, he was appointed chief minister, becoming prime minister in 1961 and president of the Tanganyika Republic in 1962. In 1964 he became president of Tanzania, announcing the introduction of a one-party system of government. Subsequently he was re-elected at regular intervals with a substantial majority on each occasion. He finally resigned as president in 1985, by which time Tanzania was one of the most politically stable nations in Africa (with one of the highest literacy rates); he stepped down as party chairman in 1990. The author of several publications on Ujamaa socialism, which contends that Tanzanian economic development should be based on self-help and self-reliance, he has also translated several Shakespeare plays into Swahili.

O

Oates, Lawrence Edward Grace (1880–1912) See SCOTT, ROBERT FALCON.

Obote, Milton Apollo (1924–   ) *President of Uganda (1966–71; 1980–85), who was deposed by Idi AMIN but returned as president after Amin's defeat.*

Born in the Lango district of Uganda, the son of a farmer, Obote was educated at a Protestant mission school in Lango before attending high school in Jinja and graduating with an arts degree from Makerere College, Kampala, in 1949. After working in a variety of jobs (including labourer, clerk, and salesman) he migrated to Kenya in 1950.

Obote first became involved in politics when he worked for the Nairobi District African Congress. Later he joined Tom MBOYA's rival People's Convention Party and became a founding member of the Kenya African Union. In 1955 he returned to Uganda, where he established a Lango branch of the Uganda National Congress (UNC). Entering the legislative council in 1957, he founded and became leader of the Uganda People's Congress, a party that emerged from a split in the UNC in 1960. In 1961 he was elected parliamentary leader of the opposition and he became prime minister in 1962, leading Uganda to independence. As leader of the Ugandan revolution in 1966 he suspended the constitution, becoming interim president until 1967, when he was elected first executive president under the republican constitution. Obote was deposed in a military coup led by Idi Amin in 1971 and he remained in exile in Tanzania from 1971 to 1980. Returning to Uganda in May 1980, he was re-elected president in December and established a multiparty democracy. Subsequently he set about improving the country's economic base. The army remained unwilling, however, to submit to his control, and in 1985 he was deposed in a second military coup.

O'Casey, Sean (Shaun O'Cathasaigh; 1880–1964) *Irish playwright and author. With the help of W. B. YEATS, he overcame the disadvantages of his background to become a leading Irish writer of the twentieth century.*

O'Casey was brought up in poverty in his native Dublin, the last of thirteen children, eight of whom died in infancy. Deprived of formal education by a painful disease of the eyes, he taught himself to read in his teens but could only find employment as a casual labourer. From 1916 he began writing plays with the encouragement of W. B. Yeats and Lady Gregory of the Abbey Theatre, and after several rejections his first play, *The Shadow of a Gunman* (1923), was successfully staged at the Abbey. Like all his best work, the play is based on the lives of the Irish poor during the political troubles in which O'Casey himself had played a minor but active part. His best-known play, *Juno and the Paycock*, followed in 1924, but *The Plough and the Stars* (1926) aroused such hostility in Dublin that O'Casey decided to move to England.

O'Casey's subsequent experience with the theatre was almost uniformly discouraging. The anti-war play *The Silver Tassie* (1929), which had been rejected by the Abbey Theatre and caused O'Casey to break with Yeats, was a critical success in London but a financial failure. *Within the Gates* (1934) was even less successful. Only *Red Roses for Me* (1946) achieved any critical recognition. *Purple Dust* (1940), *Cock-a-Doodle Dandy* (1949), and *Bishop's Bonfire* (1955) were poorly received and are now rarely performed. *The Drums of Father Ned* (1960), which O'Casey wrote for the 1958 Dublin International Festival, had to be withdrawn after objections were raised about its production. From 1938 O'Casey lived in Devon, where he wrote his six-volume autobiography starting with *I Knock at the Door* (1939) and ending with *Sunset and Evening Star* (1954).

O'Connor, Frank (Michael O'Donovan; 1903–66) *Irish writer, best known for his short stories.*

Frank O'Connor was born and brought up in Cork. As a teenager he became a republican activist and was interned during the civil war (1921–22). He worked as a librarian, first in Cork and later in Dublin, where he was friendly with several prominent figures in the Irish literary world, notably George Russell (A.E.; 1867–1935) and W. B. YEATS. In the 1930s he was involved in running the Abbey Theatre, Dublin, for which he wrote several plays. When his short-story collection *Guests of the Nation* (1931) was published he was instantly acclaimed as a major new writer. He followed this success with *Bones of Contention* (1936) and a biography of Michael Collins, *The Big Fellow* (1937).

During World War II O'Connor worked for the Ministry of Information in London. He tried his hand at novel writing, but his métier was really the short story: subsequent collections include *Crab Apple Jelly* (1944), *The Common Chord* (1947), *Travellers' Samples* (1950), and *Domestic Relations* (1957).

In the postwar period O'Connor taught for a time in the USA and expanded into other fields of literature. He published two significant critical studies, one on Shakespeare (*The Road to Stratford*, 1948) and one on the novel (*Mirror in the Roadway*, 1956). He also wrote the traveller's guide *Irish Miles* (1947) and two volumes of autobiography, *An Only Child* (1961) and *My Father's Son* (1969). His fine verse translations from Gaelic were a feature of his later years, beginning with *The Midnight Court* (1945) and continuing with *Kings, Lords, and Commons* (1961) and *The Little Monasteries* (1963). *The Book of Ireland* (1958) was a notable anthology of Irish prose and poetry, and he also worked on *A Golden Treasury of Irish Poetry, A.D. 600–1200* (with David Greene; 1967).

Odets, Clifford (1906–63) *US dramatist and a founder member of the influential Group Theatre of the 1930s.*

Born into a wealthy middle-class Jewish family in Philadelphia, Odets grew up in New York. An inauspicious education was followed by an equally unremarkable career as a radio actor. However, this occupation gave him time to write and he became a founding member of the avant-garde Group Theatre. In 1934 he also joined the Communist Party. *Waiting for Lefty* (1935), written for a one-act play competition, was staged by the Group Theatre and brought instant success, with Odets being hailed as the most revolutionary discovery of the decade. The same year that and his first play, *Awake and Sing!*, were both staged on Broadway, as was *Till the Day I Die*. The vehement left-wing political message of Odets's work earned him both adulation and criticism from the public. *Paradise Lost* (1935) did not fulfil expectations, however, and Odets took up script writing in Hollywood, where he also embarked upon a tumultuous but brief marriage with the actress Luise Rainer (1909–). Subsequent plays met with a mixed critical reception. They include *Golden Boy* (1937), *The Big Knife*

(1949), and *The Country Girl* (1950), probably the biggest commercial success of his career. Despite his rapid rise to fame, Odets failed to hold his position among the elite of US dramatists. As his work became more personal than political, it lost the excitement and enthusiasm that had been the key to his popularity.

O'Hara, John Henry (1905–70) *US novelist whose rapid rise to fame made him one of the most popular writers of the 1930s and 1940s.*

O'Hara was born in Pottsville, Pennsylvania, the eldest child of a large family. An unsatisfactory education (he was expelled from school twice) was followed by a multiplicity of short-lived jobs, including gas-meter reader and soda clerk, resulting in a lifestyle that was both impecunious and insecure. A venture into journalism provided him with a more stable income and a vocational opening. From 1934 he also wrote scripts for Hollywood.

O'Hara's first two novels, *Appointment in Samarra* (1934) and *Butterfield 8* (1935), are considered his best work, together with some of the short stories in the collection *The Doctor's Son* (1935) and *Files on Parade* (1939). Perhaps his most dramatic rise to fame, however, came with the successful transformation of his *Pal Joey* sketches (1940) into a Broadway musical. Of his later novels, the best-known is *Ourselves to Know* (1960).

Literary critics have been divided about O'Hara's work. He has been compared to HEMINGWAY and F. Scott FITZGERALD and praised for his ability to recreate realistically the atmosphere and mores of a particular era.

Oistrakh, David Fyodorovich (1908–74) *Soviet violinist with an international reputation. PROKOFIEV, SHOSTAKOVICH, and KHACHATURIAN all dedicated works to him.*

Born in Odessa of musical parents, Oistrakh studied the violin and viola from the age of five and graduated from the Odessa Conservatory in 1926. He then toured the Soviet Union as a soloist, playing Glazunov's concerto at the composer's invitation in 1927, and in the following years won prizes at the major Soviet festivals. In 1934 he was appointed professor at the Moscow Conservatory and achieved international recognition when he won the Violin Competition in Brussels (1937). In 1951 he helped to ease postwar political tensions when he toured western Europe, playing in the USA in 1955. His appearances with Yehudi MENUHIN playing the Bach *Double Concerto* in Moscow, directly after World War II, and again in London, in the late 1950s and 1960s, were memorable occasions. Oistrakh also played and recorded with his son, Igor Oistrakh (1931–), whom he himself had trained at the Moscow Conservatory. A chess enthusiast, Oistrakh frequently played with Prokofiev. He died suddenly in Amsterdam, while on a tour of Europe.

Oldenburg, Claes Thure (1929–) *Swedish-born US pop sculptor.*
The son of a diplomat, Oldenburg spent his childhood in the USA, Sweden, and Norway before going to live in Chicago in 1937. After studying English literature and art at Yale University, he decided to become an artist, and while working for a newspaper studied at the Chicago Art Institute. He became a US citizen in 1953. Having moved to New York, Oldenburg produced figure paintings and graphic works until 1958, when he began to organize 'happenings'. 'I am for an art that does something other than sit on its ass in a museum' he later explained. During the 1960s Oldenburg became one of the leaders of the pop art movement. In 1961 he opened his store, where he sold plaster replicas of food and other commonplace items. 1962 saw his 'giant objects', such as *Giant Hamburger*, and the soft sculptures for which he is most famous, such as *Soft Typewriter* (1963). Oldenburg was fascinated by the values attached to size and in the late 1960s produced numerous 'projects for colossal monuments', which included an enormous lipstick and a half-peeled banana for Times Square. He subsequently re-explored, in different media, motifs used in his previous period and exhibited widely in Europe and the USA, as well as undertaking commissions in Nevada (1978–80), Des Moines, Iowa (1978–79), and Rotterdam (1977–80).

Olivier, Laurence Kerr, Baron (1907–89) *British actor, director, and manager. He was knighted in 1947, created Baron Olivier of Brighton in 1970, and received the OM in 1981.*
Olivier was born in Dorking, the son of an Anglican clergyman, and first appeared on stage as Katherine in a schools' production of *The Taming of the Shrew* (1922) at the Festival Theatre, Stratford-on-Avon. After studying at the Central School of Dramatic Art he made his professional debut in London as the Suliot Officer in *Byron* (1924) at the Century Theatre and subsequently joined the Birmingham Repertory Company (1926–28). In 1929 he made his New York debut as Hugh Bromilow in *Murder on the Second Floor*. Returning to England, Olivier appeared in Noël COWARD's *Private Lives* (1930) and alternated the parts of

Romeo and Mercutio with John GIELGUD in a London production of *Romeo and Juliet* (1935).

In 1937 he joined the Old Vic and in his first season portrayed a memorable Hamlet; he again performed *Hamlet* at Elsinore with Vivien LEIGH, who became his second wife in 1940, as Ophelia. He made his name in the USA in a series of successful films, including *Wuthering Heights* (1939), *Rebecca* (1939), and *Pride and Prejudice* (1940), before returning to serve in the Fleet Air Arm during World War II. In 1944 he became co-director of the Old Vic with Ralph RICHARDSON and produced some of his most powerful performances – the title role in *Richard III* (filmed 1955), Astrov in Chekhov's *Uncle Vanya*, Hotspur in *Henry IV Part I*, and Oedipus in Sophocles's *Oedipus Rex*. During this period he also produced, directed, and starred in the film version of *Henry V* (1945) and *Hamlet* (1948), which were internationally successful. During the 1950s he acted in and managed his own company and gave an outstanding performance in the title role of Peter BROOK's Stratford production of *Titus Andronicus* (1955). At the same time Olivier did not neglect the work of modern playwrights, and gave a notable portrayal of Archie Rice in John OSBORNE's *The Entertainer* (1957; filmed 1960).

In 1961 his marriage to Vivien Leigh was dissolved – because of her mental instability the couple had lived apart for some time – and he married the actress Joan Plowright (1929–). The same year he became first director of the Chichester Festival and in 1962 founder-director of the National Theatre Company (the Royal National Theatre from 1988). During his eleven years with the National Theatre, which under his direction became a major international company, Olivier continued to give some memorable performances, notably the title role in *Othello* (filmed 1965) and James Tyrone in Eugene O'NEILL's *Long Day's Journey into Night*. Ill health forced him to leave the National Theatre in 1973 but he continued to act in films and on television. His autobiography, *Confessions of an Actor*, was published in 1982.

Olivier is usually regarded as the greatest actor of the century; for his outstanding work in the theatre, especially in establishing the Royal National Theatre, he was the first theatrical knight to be raised to the peerage. The Society of West End Theatre presents the Laurence Olivier awards annually in his honour.

Onassis, Aristotle Socrates (1906–75)
Greek shipping magnate who married the widow of President KENNEDY.

Onassis was born in Smyrna (now Izmir), Turkey, the son of a Greek tobacco merchant. In 1922 the family fled from Turkish hostility in the region to Athens. The young Onassis decided to try his luck in South America and in 1923 arrived in Buenos Aires with just $60. His first job was as a telephone operator on a nightshift. Meanwhile, during the daytime, he built up his own tobacco importing business and in due course started to manufacture cigarettes. In 1928 he negotiated a trade treaty with Argentina on behalf of the Greek government. Extending his business into other commodities, Onassis was reckoned to be a dollar millionaire by the age of twenty-five. In 1932, in the depths of the economic recession, he bought six Canadian freighters at a bargain price and subsequently, as the international freight market picked up, put them into service. His fleet grew steadily and in the 1950s he invested heavily in oil tankers and bulk carriers to make it one of the world's largest privately owned merchant fleets and Onassis one of the world's wealthiest individuals. He disposed of his whaling fleet to the Japanese in 1956, the year in which Onassis was awarded the contract to operate the Greek national airline. Olympic Airways started in 1957.

With his 1800-ton yacht, *Christina*, and numerous houses throughout the world, Onassis entertained many of the world's leaders, including Winston CHURCHILL. His first marriage was dissolved in 1960 and, after a much-reported friendship with the opera singer Maria CALLAS, he married Jacqueline Bouvier Kennedy (1929–   ), the widow of John F. Kennedy, in 1968.

O'Neill, Eugene Gladstone (1888–1953)
US dramatist regarded as one of the leading playwrights of the twentieth century. He was awarded the Pulitzer Prize three times (1920, 1921, and 1928) and the Nobel Prize for Literature (1936).

O'Neill was born in New York City; his childhood was divided between Catholic boarding schools and accompanying his actor parents on tour. He began a career in journalism but interrupted it by spending six years at sea. In 1912, following a brief and unsuccessful marriage, O'Neill spent six months in a sanatorium recovering from tuberculosis, during which he realized he wanted to become a writer. His first plays, staged by the Provincetown Players, were the one-act *Bound East for Cardiff* (1916)

followed by *Beyond the Horizon* (1920). Their success led O'Neill to continue writing for the theatre and among his plays of the 1920s – his most prolific period – were *The Emperor Jones* (1920), *The Great God Brown* (1926), and *Strange Interlude* (1928). Only three plays appeared in the 1930s: *Mourning Becomes Electra* (1931), *Ah, Wilderness* (1933), and *Days Without End* (1934). *The Iceman Cometh* (1939) and *Long Day's Journey Into Night* (1940–41) were written while O'Neill was suffering from Parkinsonism and alcoholism. Although they were not performed until 1946 and 1956, respectively, they are now regarded as two of the most powerful plays in the modern theatre. Posthumous productions included a 1973 revival of the hitherto unsuccessful *A Moon for the Misbegotten* (originally staged in 1947).

Oppenheimer, Julius Robert (1904–67)
US physicist, who directed the US atomic bomb project in Los Alamos and was subsequently deemed to be a security risk because he had had communist connections.

Oppenheimer came from a wealthy family of Jewish immigrants. He was born in New York and educated at Harvard, Cambridge, and Göttingen, where he gained his PhD in 1927. On his return to the USA, Oppenheimer joined the faculty of the University of California, teaching at both the Berkeley and the Pasadena campuses. Throughout the 1930s Oppenheimer built up a reputation as a theoretical physicist, making several important contributions to quantum theory. Outside science his interests included literature and politics.

When, in 1942, the US government decided to build an atom bomb, Oppenheimer was chosen to direct the project. He helped to select the site for the laboratory in Los Alamos and proved to be a director with sufficient intellectual authority to command the support of several hundred scientists and sufficient diplomatic skill to deal with the politicians and generals. Oppenheimer was undoubtedly successful, for on 16 July 1945 an atom bomb was exploded in the nearby New Mexico desert. Oppenheimer was involved in and supported the decision to drop the first bomb on a Japanese town rather than on an uninhabited area as a threat. However, as a known left-wing sympathizer, Oppenheimer was regarded by the Los Alamos security forces as a risk. In 1942 it was suspected that Oppenheimer had been approached by a Soviet agent and that, although he did not hand over any secrets, his failure to inform the security forces of the approach was suspicious. After several interviews and threats to force his dismissal, late in 1943 Oppenheimer identified Haakon Chevalier, an old colleague and friend, as his contact. Security issues were again raised when Oppenheimer was chairman of the Advisory Committee of the Atomic Energy Commission (1946–52). In 1949 the Committee recommended that the US government should not attempt to build the hydrogen bomb. In the hysteria that developed after the explosion of the first Russian atom bomb in 1949 this decision appeared misguided. In 1953 Oppenheimer's security clearance was withdrawn and his appeal dismissed on the grounds that he had not provided the 'enthusiastic support' for the hydrogen bomb expected by someone in his position. His clearance was never restored.

Exiled from power, Oppenheimer retired to the Institute of Advanced Studies at Princeton, where he had been appointed director in 1947. Towards the end of his life the US government made gestures taken to indicate that it may have been mistaken about his loyalty. In 1963 he was given the Fermi Award by the AEC and in 1964 he was invited to Los Alamos to lecture to a packed audience. He died of cancer of the throat.

Orff, Carl (1895–1982) *German composer who won international recognition for his modern operas and innovative approach to musical education.*

Born in Munich, Orff studied composition at the Music Academy before working as a professional musician (1915–19). Gradually he evolved a system of musical education for children, beginners, and amateurs based on Dalcroze eurhythmics, enabling anyone to participate creatively in music-making. He set out his theory and method in *Schulwerk* (1930–33; 'Schoolwork'), stressing the value of simplicity and controlled improvisation in music. He conducted early experiments in this system at the Günther School for gymnastics, dance, and music, which he co-founded in Munich (1924). After World War II German radio based a five-year series of children's broadcasts on his ideas, the material for the series being set out in *Music for Children* (1950–54) and translated into many languages. Orff's interest in the dramatic presentation and unity of words and music featured strongly in his best-known work, *Carmina Burana* (1937), a cantata based on a collection of medieval Latin poems found in a Bavarian monastery in 1803. Other works, mostly choral or dramatic, included *Der Mond* (1939), *Die Kluge* (1943), and *Antigone* (1949).

Ormandy, Eugene (Eugene Blau; 1899–1985) *Hungarian-born US conductor, whose name became almost synonymous with that of the Philadelphia Orchestra.*

At the age of five Ormandy became a violin student at the Budapest Royal Academy and toured Hungary as a child prodigy. He was appointed professor at seventeen, but left Hungary in 1921 and emigrated to the USA when he was offered a US concert tour. The tour did not materialize, however, so he took work with the Capitol Theater orchestra in New York, accompanying silent films, becoming conductor in 1924. He then conducted a number of different orchestras before assuming leadership of the Minneapolis Symphony Orchestra (1931–36). He joined the Philadelphia Orchestra in 1936 as associate conductor and eventually succeeded STOKOWSKI as music director and conductor (1938–80). During his long association with the orchestra, Ormandy maintained a high standard of excellence and undertook several successful tours of Europe, Australia, South America, and Japan. Famous for conducting entirely from memory, he appeared as guest conductor of many other orchestras.

Orozco, José Clemente (1883–1949) *Mexican mural painter.*

While still a schoolboy in Mexico City, Orozco lost his left hand and the sight of one eye in an accident. However, he studied art at the National University of Mexico and intermittently at various art schools although as a painter he was self-taught to a great extent. His early work consisted of paintings of schoolgirls and prostitutes in a postimpressionist style.

The first of his large frescos for public buildings was painted in 1922 for the Mexican National Training School; it showed his respect for Renaissance painting as well as his strong commitment to the twelve-year-old Mexican revolution. In the monumental works that followed, Orozco expressed the cause of the peasants and workers in his depictions of the life and history of the Mexican people. His style, well suited to this purpose, was dramatic, realistic, and expressionist, with strong outlines and colours. In the 1930s, after his stay in the USA (1927–34) and a visit to Europe (1934), his themes widened to include suppression and rebellion throughout the world. Probably his best murals were done in the late 1930s in Guadalajara. His work influenced realist painting in the USA, where he painted numerous murals, including one in the Museum of Modern Art, New York (1940).

Ortega y Gasset, José (1883–1955) *Spanish philosopher, opponent of FRANCO and critic of modern democracy.*
Born in Madrid of wealthy parents, Ortega was educated at the University of Madrid, where he obtained his doctorate in 1904. He studied philosophy in Germany for five years, becoming a neo-Kantian in the process, before returning to Madrid in 1910 as professor of metaphysics. Politically active, he played a role in the overthrow of Alfonso XIII in 1931, served as a deputy for Leon, and became civil governor of Madrid. With the rise of Franco, however, Ortega chose exile in Argentina and Portugal. When he finally returned to Spain in 1946 he founded in Madrid the Institute of Humanities, where he taught and wrote for the last years of his life.

Philosophically Ortega argued for what he called 'perspectivism'. The world, he insisted, can be known only from a specific point of view. He consequently took as the fundamental feature of his theory of knowledge not mind or matter but 'perspective'. All perspectives were equally valid. The only indubitably false perspective was that which claimed to be the one and only true perspective. He is best remembered, however, for a work of political analysis: *La rebelión de las masas* (1930; translated as *The Rebellion of the Masses*, 1932). Democracy, he warned, could all too easily lead to tyranny, both of the left and the right. It was the duty of the intellectual elites of Europe to fight against this threatened tyranny of the masses.

Orwell, George (Eric Arthur Blair; 1903–50) *British novelist and essayist.*
Orwell was born in India, where his father was in the Bengal Civil Service. After leaving Eton (1921), he joined the Imperial Police in Burma for five years. On his return to Europe he determined to reject his class and to develop his own brand of social philosophy by experiencing poverty at first hand; the result was his first book, *Down and Out in Paris and London* (1933), written under the new name he had chosen for himself. He followed this with his first novel, *Burmese Days* (1934). He also wrote provocative and influential essays and reviews, many of them published in the *New Statesman*.

Orwell was a passionate crusader for socialism in such books as *The Road to Wigan Pier* (1937), a trenchant study of the plight of the unemployed. *Homage to Catalonia* (1938) records his experiences in the Spanish civil war, when he fought on the republican side and was severely wounded. He also wrote the novels *Keep the Aspidistra Flying* (1936) and *Coming Up for Air* (1939), the latter a typical defence of the individual against the impersonal forces of big business and industrialized society.

His dislike of totalitarianism, whether of the left or the right, is explicit in his two best-known books, the political novels *Animal Farm* (1945) and *Nineteen Eighty Four* (1949; filmed 1984). The former is a satire on communism as it developed in the Soviet Union under STALIN; the latter describes a nightmare Utopia of the future. Orwell died from tuberculosis.

Osborne, John James (1929–) *British playwright and actor, who first came to prominence as an 'angry young man' of the 1950s.*
Osborne was educated at Belmont College, Devon, and began his acting career in 1948. He did not, however, make his mark as a dramatist until *Look Back In Anger* was produced at the Royal Court Theatre in London in 1956. Its 'hero' Jimmy Porter was immediately recognized as the archetype of contemporary disillusioned youth, and his creator was labelled an 'angry young man' (along with Kingsley AMIS and several other writers who satirized the ills of postwar British society). This play established a landmark in the British theatre; thereafter a new and more potent kitchen-sink realism became not only acceptable but indispensable. In 1957 *The Entertainer* was produced with Laurence OLIVIER in the part of the seedy down-at-heel music-hall artist, trying to pretend, like England, that he still had an important role to play. Both *Look Back in Anger* and *The Entertainer* were subsequently filmed (in 1958 and 1959 repectively).

Later plays include *Luther* (1960), *Inadmissible Evidence* (1964), *A Patriot for Me* (1965), *The Hotel in Amsterdam* (1967), *West of Suez* (1971), and *Déja Vu (1991), as well as plays for television, including You're Not Watching Me, Mummy* (1978) and *God Rot Tunbridge Wells!* (1985). In 1964 his screenplay for *Tom Jones* was awarded an Oscar. Besides writing and acting he also directed a number of plays both on stage and for television, the latter including his own *The Right Prospectus* (1969). In 1989 he published his own translation and adaptation of Strindberg's play *The Father*. In 1981 he published an autobiography up to 1955, *A Better Class of Person*, in which he revealed his less than sympathetic relationship with his mother. A second volume appeared in 1991.

Oswald, Lee Harvey (1939–63) *US gunman who was arrested and charged with the assassination of President John F. KENNEDY in 1963. He was himself murdered two days after his arrest.*

Oswald was raised by his widowed mother in various parts of the USA. He developed an interest in Soviet culture and Marxist ideology and in 1959, following his dishonourable discharge from the Marine Corps, he travelled to the Soviet Union and requested citizenship. He was granted resident alien status in 1960 and moved to Minsk, taking a job in a factory. The following year he married Marina Prusakova, a Soviet citizen, and the following June they returned to the USA to settle in Fort Worth, Texas.

Oswald achieved international notoriety when he was arrested in Dallas, Texas, following the assassination of Kennedy at 12.30 pm on 22 November 1963. The official commission of inquiry – the Warren Commission – found that Oswald had fired two bullets from an upper window of the Texas School Book Depository, which overlooked the presidential motorcade. His second bullet fatally wounded the president. Oswald also shot and killed a police patrolman who apprehended him shortly before his capture, at around 2 pm in the Texas Theatre. However, Oswald never stood trial. On 24 November, he was shot and killed by a Dallas night-club owner, Jack L. Ruby, while being moved by police out of Dallas police station.

Many have disputed the Warren Commission's conclusion that Oswald was acting alone. Other theories have been advanced, including the propositions that Oswald was a Soviet agent or that he was part of a much larger domestic conspiracy organized by forces hostile to Kennedy. That Oswald was murdered to ensure his silence was always denied by Ruby, who died in prison in 1967 while awaiting a retrial for the murder.

Oud, Jacobus Johannes Pieter (1890–1963) *Dutch architect, who was a founder member of the De Stijl group and municipal architect for Rotterdam.*

Born in Purmerend, Oud worked for a period in Germany under the town-planning expert Theodor Fischer (1862–1938). In 1917 he joined the painters Theo van Doesburg (1883–1931) and Piet MONDRIAN in publishing the periodical *De Stijl* and forming part of the associated group. In architecture, this group

promoted a severe cubist style in which ornament was not used. However, after disagreement with Doesburg, Oud left the group in 1921. In 1918 Oud was appointed architect to the City of Rotterdam and between the wars built several housing estates in the *De Stijl* idiom, which were influential among architects interested in mass-produced low-cost housing. Oud resigned as architect to Rotterdam in 1927 and thereafter his style became lighter and somewhat decorative, his only important building in this style being the Shell building in The Hague (1938–42).

Ouspensky, Peter Demianovich (1878–1947) *Russian sage and occultist.*

Born in Moscow, the son of an army officer, Ouspensky was educated at Moscow University, where he studied the natural sciences. A reading of theosophical literature convinced him that scientific knowledge could never be more than partial knowledge and started him on his search for a deeper wisdom. He first travelled in the East but it was not until 1915, when he met GURDJIEFF in Moscow, that Ouspensky found the intellectual insight he was seeking. He remained with Gurdjieff until 1921, when he moved to England at the invitation of Lord Rothermere (1868–1940), the proprietor of the *Daily Mail*. Ouspensky settled in London, where he lectured to his disciples for the rest of his life.

Ouspensky published his views in such works as *A New Model of the Universe* (1931) and *In Search of the Miraculous* (1949). Although he made considerable play with the idea of the fourth dimension taken from contemporary cosmology, Ouspensky actually did nothing more than reformulate in a more modern terminology the traditional hermetic philosophy of the Renaissance. Man, he insisted, is really asleep – only self-study would enable him to awake to a higher consciousness. Further, Ouspensky insisted, man consisted of a multiplicity of conflicting egos that could only be harmoniously integrated by practising his specially designed Gurdjieff-type psychological exercises.

Owen, David Anthony Llewellyn (1938–   ) *British politician, leader of the Social Democratic Party (1983–87; 1988–90).*

Owen qualified as a medical doctor and worked as a neurologist before entering parliament in 1966 as Labour MP for Plymouth, Sutton (Plymouth, Devonport, since 1974). In 1972 he resigned as opposition spokesman on defence because he disagreed with Labour's stand against British membership of the European Economic Community. Under CALLAGHAN, he was health minister (1974–76) and then foreign secretary (1977–79). He became increasingly discontented with Labour policies, especially the decision, which he regarded as undemocratic, that MPs should have only 10 per cent of the votes in the election of the Labour leader. He left the party in 1981 to become a founding member of the Social Democratic Party. He became its leader in 1983, but opposed the merging of the SDP with the Liberal Party and resigned in 1987. After the parties joined in 1988, Owen revived the SDP but failed to win back substantial support and finally wound the party up in 1990, to end his parliamentary career as an independent social democrat.

Owen, Wilfred (1893–1918) *British poet, whose best-known poems were inspired by his experiences during World War I.*

Owen was born at Oswestry, Shropshire, and educated at the Birkenhead Institute, Liverpool, before going on to London University (1910). From his early teens Owen had been fascinated by poetry and by this time he was writing competent but not strikingly original verse, much under the influence of Keats. In 1913 he was sent to France to convalesce from a serious illness; here his rapidly maturing talent was encouraged by the French poet Laurent Tailhade (1854–1919).

In 1915 Owen enlisted in the Artists' Rifles. His letters home show how deeply the horrors of trench warfare – the gas, the cold, the ugliness – oppressed him and altered his perceptions. In 1917 he was invalided back to Britain from the Somme, and while he was in the Craiglockhart War Hospital near Edinburgh he became friendly with Siegfried SASSOON. This friendship boosted Owen's self-confidence as a poet and during the remaining months of his life he wrote most of the poems for which he became famous. In August 1918 he went back to France and in November was killed trying to lead his men across the Sambre Canal. He left

incomplete the draft of a preface for his proposed book of poems ('My subject is War, and the pity of War') and the poem 'Strange Meeting'. His poems were collected and edited by Sassoon (1920).

Owens, Jesse (James Cleveland Owens; 1913–80) *US black athlete, numbered among the greatest track and field competitors in the first half of the twentieth century.*

Born at Danville, Alabama, the son of a poor farmer, Owens soon showed an aptitude for athletics and became renowned for his natural grace of movement. In May 1935, at an event in Michigan, he beat or equalled six world records. His long jump of 8.13 metres (26 ft 8¼ in) was not beaten for another twenty-five years. Owens will always be associated with the Berlin Olympics of 1936, when he won four gold medals in defiance of HITLER's racist views – in the 100 metres, 200 metres, long jump, and 4 × 100 metres relay. Soon after those Games he turned professional so no-one was really able to measure what he might have ultimately achieved.

Ozu, Yasujiro (1903–63) *Japanese film director.*

Born in Tokyo, Ozu spent his career, which began in 1923 as a scriptwriter and assistant editor, at the Shochiku studio. He made his debut as a director with *Zange no Yaiba* (1927; *Sword of Penitence*), which was followed by such films as *Hitori Musoko* (1936; *The Only Son*). His work was interrupted by periods of military service between 1937 and the mid-forties, although he did manage to direct two films during this period. Traditional in both theme and technique, Ozu has often been cited as the most Japanese of Japan's film directors. The theme of middle-class family life runs throughout his films. From a canon of fifty-four films, perhaps the most notable are *Tokyo Monogatari* (1953; *Tokyo Story*), *Soshun* (1956; *Early Spring*), and *Samma no Aji* (1962; *An Autumn Afternoon*).

P

Paderewski, Ignacy Jan (1860–1941) *Polish pianist, composer, and statesman. Briefly prime minister of Poland, he was acknowledged as the finest pianist of his time. His many honours included the Grand Cross of the Légion d'honneur.*

Paderewski studied and taught at the Warsaw Conservatory (1872–84) before going to Vienna for further piano study with Theodor Leschetizky (1830–1915). Between 1887 and 1891 he made his first public appearances as a pianist and received critical acclaim, most notably from George Bernard SHAW. He played in London and New York (1890) and then toured South America, Australia, New Zealand, and South Africa. In 1889 he settled in Switzerland, where he composed his opera *Manru* (1901). Appointed director of the Warsaw Conservatory in 1909, he raised money for Polish war victims throughout World War I.

In 1918 he gave up music and served as the Polish representative in Washington. The following year he was elected prime minister and minister of foreign affairs, so becoming Poland's signatory to the Treaty of Versailles. Resigning after ten months in office, he returned to concert tours, composition, and the editing of Chopin's works. In 1939 he served briefly as head of the Polish government in exile before emigrating to the USA. Upon his death, he was given a state burial by order of President ROOSEVELT. Of his compositions, only the *Minuet in G* and the piano concerto are still performed.

Pahlavi, Mohammed Reza (1919–80) *Shah of Iran (1941–79). Considered one of the most stable leaders in the Middle East until he was swept from power by a sudden wave of revolution, he introduced substantial reforms in Iran despite criticism from Islamic fundamentalists.*

The son of Reza Khan Pahlavi (shah of Iran 1925–41), Mohammed Reza Pahlavi was educated in Switzerland. In 1941 he succeeded his father when the latter was forced into exile after the occupation of Iran by British and Soviet forces. His early years as shah were dominated by conflict with the prime minister Mohammed Mosaddeq (1880–1967), a fervent nationalist. He was forced to leave the country

briefly in 1953 but returned to power and curtailed Mosaddeq's planned nationalization of British oil interests. He then assumed direct control over all aspects of Iranian life and, with the financial assistance of the USA, embarked on a national development plan, known as the 'White Revolution'. This promoted the expansion of public works, the development of industry, and the introduction of land and social reforms. At the same time he built up an enormous stock of armaments, bought with the country's oil revenue. He also attempted to establish closer relations with the eastern bloc countries during the 1960s and 1970s, even though his country's position in the Persian Gulf ensured financial and military support from the USA. Opposition to his regime, which focused upon his autocratic style of government and allegations of corruption, gradually increased despite the efforts of his brutal secret police, the SAVAK, to quell it. By 1978 his antagonists were united under the leadership of the Ayatollah KHOMEINI, who was then living in exile in France. When a number of concessions failed to prevent riots breaking out in January 1979, the shah left Iran; in April 1979 the Islamic Republic of Iran was proclaimed. Although demands were made for his extradition in return for the release of fifty American hostages seized at the US embassy in Tehran by Iranian students, he remained in exile until his death from cancer.

Paisley, Ian Richard Kyle (1926–   ) *Northern Irish Presbyterian minister and politician. Paisley is relentlessly committed to the continuing union of Northern Ireland and the United Kingdom and opposes power-sharing in the province between Roman Catholics and Protestants.*

Born in Armagh, Paisley was ordained as a minister of the Free Presbyterian Church in 1946, becoming its moderator (leader) in 1951. He founded the Ulster Protestant Volunteers in 1966 and was twice imprisoned (in 1966 and 1969) for militant anti-Catholic activities. He was elected a member of Stormont (the Northern Irish Parliament) in 1970 and campaigned for its restoration after it was suspended following the imposition of direct rule from Westminster in 1972. Founder of the Protestant Unionist Party, in 1974 Paisley joined with other Protestant leaders to form the Ulster Democratic Unionist Party. He became Westminster MP for North Antrim in 1970 and Member of the European Parliament for Northern Ireland in 1979. In 1985 he resigned his seat as MP for North Antrim in protest at the Anglo-Irish Agreement, which formally allowed the Republic of Ireland a say in Northern Irish affairs, but was re-elected to the same seat the following year. In 1991 he was a key figure in inconclusive talks, with other Northern Ireland party leaders, aimed at deciding the future of government in Northern Ireland.

Pankhurst, Emmeline (1858–1928) *British feminist and leader of the militant campaign for women's suffrage in the years before World War I.*

Emmeline Goulden was one of ten children of a well-to-do Manchester printer. Her parents held progressive views, and she was introduced to feminism by her mother. After schooling in Manchester and then Paris, she married (1874) Dr Richard Pankhurst, a barrister, who drafted the earliest British women's suffrage bill in the late 1860s and the Married Women's Property Acts 1870 and 1882. They both worked for the Manchester Women's Suffrage Committee and the Married Women's Property Commission before moving to London in 1885, where Emmeline managed a shop in Tottenham Court Road. In 1889 they founded the Women's Franchise League and, after abandoning the Liberal Party when women's suffrage was omitted from the Reform Act 1884, they joined the Independent Labour Party (1893).

After her husband's death in 1898, Emmeline returned to Manchester to become registrar of births and deaths in the Rusholme district of the city. In 1903, with her daughter Christabel Pankhurst (1880–1958), she founded the Women's Social and Political Union (WSPU). Two years later, Christabel was arrested for assaulting the policeman who removed her from a Liberal election meeting in the Free Trade Hall, Manchester, which she had interrupted to call for votes for women. It was then that the *Daily Mail* coined the term 'suffragette' to describe the women campaigners for the vote, and from then on the WSPU became increasingly militant.

Emmeline moved back to London, where she lived with her second daughter Sylvia Pankhurst (1882–1960), then an art student, and rallied, with brilliant oratory, more and more women to the cause. She was frequently

jailed, and in 1913, after the bombing by the WSPU of LLOYD GEORGE's home, was arrested for inciting violence and reimprisoned thirteen times under the notorious 'Cat and Mouse Act', whereby a prisoner made ill by hunger-striking could be released to regain her health and then rearrested. She twice, in 1909 and 1911, travelled to the USA to raise funds, but a visit in 1913 ended almost as soon as it began, with her detention on Ellis Island. At the outbreak of World War I in 1914, Emmeline called a halt to the militant campaign and directed her energies to the recruitment of women for the war industry. After the war she adopted four orphans and lived for a time in the USA. In 1926 she joined the Conservative Party and was a prospective candidate for Whitechapel.

A third daughter, Adela Constantia Pankhurst (1885–1961), was active in the women's movement in Australia.

Papadopoulos, Georgios (1919–) *Greek colonel. He headed the military junta that ruled Greece from 1967 until 1973, becoming prime minister (1967–73) and president (1973).*

Born in Eleochorion, Achaia, the son of a village schoolmaster, Papadopoulos was educated at the War Academy, from which he graduated in 1940 as a second lieutenant. He served on the Albanian front during the Greek–Italian war (1940), joining the national resistance units when Greece was occupied by the Germans.

After an early career as an artillery officer, Papadopoulos was promoted to the central intelligence service, where he worked as a liaison officer between Greek and American intelligence agencies (1959–64), being promoted to colonel in 1960. In April 1967 he led a military coup to overthrow the government, ostensibly to prevent a leftist electoral victory predicted later that year. He was sworn in as prime minister in December 1967, after an unsuccessful counter-coup by King Constantine forced the king into exile. Following the introduction of martial law he banned political parties, imposed restrictions on the press, and is alleged to have imprisoned thousands of political opponents. In 1973 Papadopoulos established a republic with himself as president and lifted martial law. However, loss of support from other senior military officers led to his arrest by Ioannidis and to his replacement as president by General Phaedon Ghizikis in November 1973. Ghizikis invited KARAMANLIS to return from exile to assume office in July

1974, following the military intervention by Turkey in Cyprus and the collapse of the dictatorship. In August 1975 Papadopoulos was tried and given a death sentence for high treason and insurrection, which was later commuted to life imprisonment.

Pareto, Vilfredo Frederico Damaso (1848–1923) *Italian economist, mathematician, and sociologist, whose work formed the foundation of modern welfare economics.*

Born in Paris, Pareto studied at Turin and became an engineer. He was director of the Italian railways and a superintendent of mines before being appointed professor of economics at Lausanne University in 1892, succeeding Léon Walras (1834–1910), whose mathematical approach he continued. In *Cours d'économie politique* (1896–97) he set out mathematically the conditions for general equilibrium in an economy. He also postulated a law of income distribution based on an attempt to prove mathematically that the distribution of income within any society follows a universal pattern, though this has been found to hold only for the highest levels of income. However, Pareto justified income inequality on this reasoning and rejected socialism. His *Trattato di sociologia generale* (1916; translated as *Mind and Society*, 1963) is said to have formed the basis of Italian fascism. He believed that the different social grouings of which all societies are composed reflect the differing abilities of their members: those of superior ability actively try to improve their social position, which results in a 'circulation of elites' as the cleverest of the lower class challenge the position of the upper class, some of whom are replaced.

Pareto's most important work, however, is contained in his *Manuale di economia politica* (1906; translated as *Course of Political Economy*, 1972), in which he analysed consumer demand using the concepts of ordinal utility (as opposed to measurable or cardinal utility, which he rejected) and indifference curve analysis (invented earlier by EDGEWORTH). Exchange takes place where the ratio of consumers' marginal utilities equals the price ratio. 'Pareto optimality' is a condition where no-one is worse-off in one state than another but someone is better-off, and there is no state 'Pareto-superior' to it (i e in which more people would be better-off without anyone being worse-off). The compensation principle of KALDOR and Hicks is a development of this.

Parker, Charlie (Charles Christopher Parker; 1920–55) *Black US jazz saxophonist,*

often nicknamed 'Bird'. In spite of his short life, he was one of the most important jazz musicians of his generation.

He grew up in Kansas City, during the time it was incubating new jazz talent. After humiliation on the bandstand, he taught himself to modulate instantly from any key to any other, an accomplishment that stood him in good stead for the rest of his life. In the late 1930s he played and recorded with the band of Jay McShann (1909–), and played the tenor saxophone with Earl HINES, but in 1944 he moved to New York and freelanced or led his own small groups. A new jazz style was being born, led by Thelonius MONK, Dizzy GILLESPIE, and others; by the time Parker took part in the now highly regarded recording sessions in 1945, he was already a legend – one of the most creative and technically fluent musicians of the period.

Unfortunately Charlie Parker was a heroin addict all his adult life; he tried, but often failed, to persuade younger hero-worshipping musicians not to imitate him in that respect. He also drank to excess. When he died he had pneumonia, perforated ulcers, cirrhosis of the liver, and a weak heart. A medical examiner estimated his age at fifty-three – he was thirty-four.

Parker, Dorothy Rothschild (1893–1967) *US humorist, short-story writer, and poet. A member of the famous Algonquin Round Table, she was noted for her witty repartee (her canary was called Onan – because he spilled his seed on the ground!).*

Born in New Jersey, she started work in 1916 as a theatre reviewer for *Vanity Fair*, marrying her first husband, Edward Parker, in 1917 (they were divorced in 1928). She had been writing satirical verse for some years when she became a book reviewer for *The New Yorker* in 1927. *Enough Rope* (1927), her first book of poems, was published shortly after this appointment and became a best-seller; she accordingly resigned her job to work freelance. Parker remained a leading contributor to *The New Yorker* and was one of its legendary wits, a member of the circle that met regularly at the Algonquin Hotel. *Enough Rope*, together with *Sunset Gun* (1928) and *Death and Taxes* (1931), was collected in *Not So Deep as a Well* (1936). Her short prose pieces printed in *Laments for the Living* (1930) and *After Such Pleasures* (1933) were collected in *Here Lies* (1939). Her style of wit in these books – cynical, sardonic, with no illusions left – had a lasting influence.

In 1933 Parker married Alan Campbell, a film actor; they were divorced in 1947 but shortly afterwards remarried. They collaborated on film scripts and other projects. A lifelong supporter of left-wing causes, Parker worked as a newspaper correspondent in Spain during the civil war. With hundreds of other writers in the 1950s, she was named before the House Committee on Un-American Activities as having left-wing political affiliations.

Parker's repartees were famous almost as soon as they were uttered and have attained the status of classic witticisms.

Parsons, Sir Charles Algernon (1854–1931) *British engineer and inventor of the multistage steam turbine. He was knighted in 1911.*

Born in London, the youngest son of a noted astronomer, the 3rd Earl of Rosse (1800–67), Parsons was educated at home as a boy. He later attended Trinity College, Dublin, and St John's College, Cambridge. After graduating in 1877, Parsons served an apprenticeship with W. G. Armstrong & Co. at Elswick for four years and subsequently spent two years gaining practical experience in the workshops of Kitson & Co. in Leeds. In the following year Parsons joined the Gateshead firm of Clarke, Chapman & Co. as a junior partner.

Presented with the problem of designing a machine to turn an electric generator, he experimented with turbines of various kinds, as an alternative to the slow-turning steam engines then available. In 1884 he took out his first patent for an axial-flow reaction turbine, which generated 7.5 kilowatts at a turbine speed of 18 000 r.p.m. Following disputes with his partners, Parsons left Gateshead in 1889 and set up his own company in Newcastle. Although his first turbines were sold to power stations, in the 1890s he founded a separate company, Parsons Marine Steam Turbine Company, at Wallsend to develop turbines capable of propelling ships. To prove the value of his engines, Parsons fitted his ship Turbinia with one of his 2000 h.p. engines and took it to the 1897 Naval Review. With a maximum speed well over 30 knots it easily outpaced the fastest of the Royal Navy's destroyers. By 1905 it had been decided that all warships would be powered by turbines. Parsons's turbines spread throughout the world and have continued to power ships and power stations long after his death. Such famous ships as the *Queen Mary, United States, France,* and both *Queen Elizabeth*s were all driven by turbines produced by one of Parsons's companies.

Parsons, Talcott (1902–79) *US sociologist whose controversial theories strongly influenced US sociology in the 1950s and 1960s.*

Having studied at Amherst College, the London School of Economics, and Heidelberg University, Parsons became an instructor in economics at Harvard until, in 1931, he switched to sociology. Six years later he produced his first book, *The Structure of Social Action* (1937), in which he began his search for a general theory of society that would encompass all dimensions of human behaviour. In this first study Parsons stressed the theory of social action being based on a voluntaristic principle and gathered together the arguments of the European writers DURKHEIM, PARETO, WEBER, and Alfred MARSHALL to support this thesis.

In 1946 Parsons founded the Department of Social Relations at Harvard and taught his theory there. However, when *The Social System* (1951) was published it was clear that his views had changed and that Parsons now favoured a more functionalist approach. He argued that no part of society could be understood without reference to the whole and that the entire social framework depended upon the interaction of its many units.

Parsons continued to expand this functional analysis of social stratification in such works as *Essays in Sociological Theory* (1954), *Social Structure and Personality* (1964), and *Politics and Social Structure* (1969). The theory was also made to incorporate evolutionism and cybernetics, partly in response to widespread criticism that Parsons had not addressed himself to the problems of conflict, power, and deviance. Nonetheless, although in many respects the theory was abstract and vulnerable to intellectual attack, it did exercise a considerable influence on contemporary anthropology, psychology, and history as well as sociology.

Pasmore, (Edwin John) Victor (1908–) *British painter. He became a CH in 1981.*

On leaving Harrow School, Pasmore worked as a clerk for the London County Council while attending evening classes at the Central School of Art. His early work was mainly naturalistic: views of the Thames were among the most characteristic of his pictures in the 1930s and early 1940s. A change of direction took place in 1947, and it is for the abstract paintings and reliefs of this later period that he is best known. In the early 1950s Pasmore exhibited with Ben NICHOLSON and Barbara HEPWORTH and worked on mural reliefs and other architectural projects. An influential teacher, he taught at the Central School of Arts and Crafts (1949–53) and at Durham University (1954–61); his teaching method, based on his studies of the principles of abstract art and known as basic design, came to be widely adopted in British art schools. In 1984 he executed stage designs for the Royal Ballet.

Pasolini, Pier Paolo (1922–75) *Italian film director and writer, who caused considerable controversy with his films.*

Pasolini was born in Bologna and educated at the university there. He began in films as a scriptwriter in the early 1950s, having already published poems, novels, and several essays. During the 1940s he had lived in the slums of Rome among petty criminals and prostitutes, and it was these experiences that he drew upon for his novels and films, including his first film as director, *Accattone!* (1961). Never predictable, his contribution to *Rogopag* (1962), which was greeted with controversy and charges of defamation, was followed by the biblical *The Gospel According to St Matthew* (1964). Here, as in many of his other films, he used family and friends, rather than professional actors, for his cast. *Oedipus Rex* (1967), *Pigsty* (1969), *Medea* (1970), *The Canterbury Tales* (1972), and *The Arabian Nights* (1974), were among the other films he made before being murdered in a Rome suburb.

Pasternak, Boris Leonidovich (1890–1960) *Russian poet and novelist. Under pressure from the Soviet authorities, he was obliged to refuse the 1958 Nobel Prize for Literature after it had been awarded to him.*

Pasternak was born and spent most of his life in Moscow. He was the son of Jewish parents prominent in the cultural life of the city: his father Leonid Pasternak (1862–1945) was a famous portrait painter and his mother, Rosa, a concert pianist. Encouraged by the composer SCRIABIN, a family friend, the young Pasternak thought of becoming a musician, but after studying philosophy at Marburg (1912) and later graduating from the University of Moscow, he published his first book of poems in 1914. On the eve of the revolution in 1917 he wrote the highly original lyric poems that established his reputation when they were published several years later. These were *Sestra moya zhizn* (1922; 'My Sister, Life') and *Temy i variatsii* (1923; 'Theme and Variations'). These lyrics referred only indirectly to the historical events of the time. Pasternak therefore took this historical subject up directly in his next two long poems, *Devyatsot pyaty god* (1926; 'The Year Nineteen Five') and *Lyuten-*

ant Schmidt (1926–27). The lyrics of his next volume, *Vtoroye rozhdeniye* (1932; 'The Second Birth') were characterized by a simple classical style that contrasted with his earlier more experimental work. Despite this, he was officially suspected of aestheticism, and in the 1930s found it impossible to conform to the simple-minded standards of 'socialist realism'. He published nothing of his own but turned instead to translating major foreign authors, including Goethe, Schiller, Keats, Shelley, and Verlaine. He made important translations of eight Shakespearean plays, including *King Lear, Hamlet, Othello, Romeo and Juliet*, and the two parts of *Henry IV*.

At some time in the 1930s he also started writing a long fictional, but essentially autobiographical, narrative that was eventually published as *Doctor Zhivago*. After a new wave of criticism against writers in 1946, Pasternak devoted his time wholly to finishing his novel. In the year following Stalin's death in 1953 there was an apparent cultural thaw that encouraged Pasternak to submit his completed manuscript to the literary journal *Novy Mir*. It was turned down, however. In 1957 a publisher in Milan who had earlier obtained the Italian rights published the novel, and French, German, and English translations immediately followed. Pasternak was awarded the Nobel Prize the following year, but under intense pressure in the Soviet Union – he was expelled from the Writers' Union – he refused it. He died at Peredelkino, the writers' colony near Moscow. *Doctor Zhivago*, which has Christian overtones and is extremely critical of Marxism, contains some of Pasternak's finest verse in an appendix. A large selection of his poetry appeared briefly in Russian in 1965, but when it was learned that the editor, Andrey Sinyavsky, was identical with the dissident underground writer Abram Tertz, the volume was quickly withdrawn. *Doctor Zhivago* was finally made available in the Soviet Union during the liberal reforms of the 1980s.

Paton, Alan Stewart (1903–88) *South African writer and politician.*

Paton was born and educated in Pietermaritzburg, Natal. From 1924 he taught at various schools in Natal, until appointed principal of Diepkloof Reformatory for young blacks, near Johannesburg. During his tenure at Diepkloof (1936–48) he struggled to make the reformatory a more humane institution; the early part of his life, including his Diepkloof years, is described in his autobiography *Towards the Mountain* (1980). In 1947, on a tour of reformatories in Britain, Europe, and North America, he began writing *Cry, the Beloved Country* (1948). The worldwide success of this novel, which was subsequently dramatized and filmed, decided Paton to continue his war on social injustice by means of writing, and he resigned from Diepkloof.

His second novel, *Too Late the Phalarope*, was published in 1953; in this year, alarmed by the breakdown of race relations in South Africa under Nationalist Party rule, he helped to found the Liberal Party. He later became its president and played an active part in promoting its policies until it was banned in 1968. In 1969 he started the magazine *Reality*. Besides his writings on the South African political scene, Paton wrote several biographies – *Hofmeyr* (1965) about the statesman Jan H. Hofmeyr (1845–1909) and *Apartheid and the Archbishop* (1973) about Archbishop Geoffrey Clayton – a memoir of his first wife, and *Kontakion for You Departed* (1969). Other works of fiction include short stories, published as *Debbie Go Home* (1961), and the heavily autobiographical novel *Ah, But Your Land is Beautiful* (1981). A second volume of autobiography, *Journey Continued*, was published shortly after his death.

Patton, George Smith (1885–1945) *US general whose armoured divisions swept across France following the Allied invasion of Normandy in World War II.*

A graduate of West Point military academy in 1909, Patton joined the 15th Cavalry as a second-lieutenant. He fought under General Pershing in Mexico (1916–17) and led the 1st US Tank Brigade during the 1918 Allied offensive in northern France. Patton became a staunch advocate of armoured mechanized warfare throughout his various staff and command postings between the wars. In late 1942, now a major-general, Patton was chosen to lead ground forces in the Allied invasion of northwest Africa – Operation Torch – under the command of EISENHOWER. Patton's 2nd Corps achieved decisive victory in Tunisia in May 1943 and he went on to lead the US 7th Army in the invasion of Sicily two months later, racing MONTGOMERY's 8th Army to capture Palermo. Here, Patton received much publicity for reportedly striking a shell-shocked soldier, whom he accused of cowardice. This was one of many controversial incidents in the flamboyant general's career and caused a few months of official disfavour. In July 1944, however, he was given command of the US 3rd Army in the Normandy invasion. His tanks

broke through German lines from the beachhead and began a rapid advance, encircling part of the enemy force and pursuing the remainder across France.

A complete contrast to his British counterpart, Montgomery, the gun-toting Patton displayed a reckless disregard for conventional military wisdom; in spite of his frequent disregard for the wishes of his superiors, he acquired a reputation for achieving results. At one point, when his tanks ran out of fuel at the River Meuse, he was not above taking supplies from other sections of the army. His forces held the German Ardennes offensive in the winter of 1944–45, relieved the garrison at Bastogne, and eventually rolled on to the Czech border. In October 1945 he was appointed commander of the 15th Army in France but he died after a car crash near Mannheim on 21 December. His memoirs, *War As I Knew It*, were published in 1947.

Paul VI (Giovanni Battista Montini; 1897–1978) *Pope (1963–78) who undertook the process of modernization of the Catholic Church initiated by the Second Vatican Council (1962–65).*

The son of a middle-class lawyer and editor from the Brescia region of Italy, Montini was ordained a priest in 1920. After graduate studies in Rome at the Gregorian University and the University of Rome, he joined the training school for Vatican diplomats. He served briefly as apostolic nuncio in Warsaw in 1923 until ill health forced his return to Rome, where he entered the Vatican civil service. As spiritual adviser to Catholic students in Rome, Montini formed friendships with many of Italy's future statesmen, including Aldo MORO. Appointed papal undersecretary of state in 1939, Montini became a close personal adviser and confidant to Pope PIUS XII. He was appointed Archbishop of Milan in 1953 and was made a cardinal in 1958.

Five years later, Montini was elected successor to JOHN XXIII and, as Pope Paul VI, presided over the remaining sessions of the Second Vatican Council. Thereafter he proceeded to implement, albeit cautiously, its decisions. Communications between the Vatican and laity were improved and power devolved to the bishops through the biennial synods. The traditional Latin mass was phased out in favour of the vernacular and, through his meetings with other religious leaders, Pope Paul gave impetus to the ecumenical movement. However, his 1968 encyclical, *Humanae Vitae* ('Of Human Life'), caused considerable controversy and widespread dismay by reasserting the Vatican's total opposition to artificial means of contraception. Throughout his reign, Pope Paul travelled widely, although his critics found his repeated appeals for universal justice and liberty and an end to hunger and misery somewhat ambivalent in view of the contents of the reactionary *Humanae Vitae*.

Pauli, Wolfgang (1900–58) *Austrian-born US physicist who discovered the exclusion principle. For this work he was awarded the 1945 Nobel Prize for Physics.*

The son of a physician, Pauli was educated at the University of Munich where, under the supervision of Arnold SOMMERFELD, he gained his PhD in 1921. After a year in Copenhagen with Niels BOHR, Pauli taught briefly at the University of Hamburg before moving in 1928 to the chair of physics at the Zürich Federal Institute of Technology. Apart from the war years, spent in the USA at the Institute of Advanced Studies, Princeton, and despite becoming a US citizen in 1946, Pauli remained in Zürich for the rest of his life.

Pauli established his reputation as a physicist while still in his early twenties. An early book on relativity published in 1922 was extravagantly praised by Albert EINSTEIN. It was followed, three years later, by Pauli's exclusion principle, stating that no two electrons can be in the same quantum state. This principle produced an enormous simplification in the foundations of physics. In 1930 Pauli solved the problem of the missing energy in beta decay by proposing that a small neutral particle is emitted with the electron or positron. This particle, later named the neutrino by FERMI, was first detected by Frederick Reines (1918–) in 1953.

Pauling, Linus Carl (1901–) *US chemist, who was awarded the 1954 Nobel Prize for Chemistry for his work on the nature of the chemical bond as well as the 1962 Nobel Peace Prize for his work on banning nuclear testing.*

The son of a pharmacist, Pauling was educated at the Oregon State Agricultural College and the California Institute of Technology, where he gained his PhD in 1925. After two years on a postdoctoral fellowship in Europe, Pauling returned to CIT, where he served as professor of chemistry (1931–64). He went on to hold chairs at San Diego and Stanford, before taking up the directorship of the Linus Pauling Institute at Menlo Park in 1974.

Pauling's chemical reputation was established with his work during the 1920s and the 1930s on chemical bonding, which was summarized in his book *The Nature of the Chemical Bond and the Structure of Molecules and Crystals* (1939). Pauling's other important work includes the study of such complex biological molecules as proteins and haemoglobin. In the former case he revealed, with R. B. Corey, in 1950 that in certain configurations protein molecules adopt a helical form; in the latter case, Pauling identified the abnormal haemoglobin present in patients with sickle-cell anaemia.

In 1962 Pauling received the Nobel Peace Prize. It was awarded for a petition organized by Pauling in which 11 021 scientists signed a document proposing a multilateral ban on nuclear testing. Pauling has also attained some renown as a result of his publication *Vitamin C and the Common Cold* (1970), in which he suggests that a daily dose of ten grams of vitamin C provides protection against colds. More recent publications include *How to Live Longer and Feel Better* (1986).

Pavarotti, Luciano (1935–) *Italian operatic tenor. He is widely known throughout the world on account of his many popular recordings.*

After studying with Pola and Campogalliani, he made his debut at the Teatro Municipale in Reggio Emilia in 1961 as Rodolfo in Puccini's *La Bohème*. In 1963 he sang in Amsterdam and repeated his success in the role of Rodolfo at Covent Garden as a last-minute substitute for Giuseppe Di Stefano (1921–). In 1964 he sang Idamante in Mozart's *Idomeneo* at Glyndebourne after which he toured Australia (1965) with Joan SUTHERLAND's company, singing opposite her in Donizetti's *Lucia di Lammermoor*, which they later recorded together. Pavarotti made his debut at La Scala, Milan, as the Duke of Mantua in Verdi's *Rigoletto* in 1965, and in 1968 he made his US debut in San Francisco and at the Metropolitan Opera. Other notable performances have included Nemorino in Donizetti's *L'Elisir d'amore*. In 1990 he became familiar to a worldwide audience when he sang (alongside CARRERAS and DOMINGO) at an internationally televised concert in Rome to mark the World Cup. His recording of 'Nessun Dorma' also topped pop charts throughout the world in 1990.

Pavese, Cesare (1908–50) *Italian novelist, poet, and critic. The elegance and clarity of his prose goes far beyond simple realism; the quietly disturbing imagery of deserted roads and towns and of emptiness in nature has had not* *only a literary influence but has been felt by artists in other media, as for example in the films of* ANTONIONI.

Born in the Piedmont village of Santo Stefano Belbo, Pavese studied English and American literature at the University of Turin. After graduating he published one volume of verse, a novel, and a shorter book of fiction and made a number of important translations from English of works by Defoe, Melville, Dickens, Sinclair LEWIS, Gertrude STEIN, James JOYCE, and William FAULKNER. After writing an antifascist article in the review *La Cultura*, which was printed by the influential publisher Einaudi, Pavese was imprisoned for a time. On being released, he became a director of Einaudi in Rome. He spent two years (1943–45) in a Piedmont village when the area was a centre of partisan activity against the Germans, and in 1945 returned to Turin as editorial director of Einaudi. He was briefly a member of the Communist Party and a contributor to its newspaper, *L'Unità*. He committed suicide on 27 August 1950, an act that, as was revealed in his posthumous diary *Il mestiere di vivere* (1952; translated as *The Burning Brand*, 1961), he had seriously considered since the 1930s. In the last few years of his life he wrote five novels, several novellas, and a number of short stories. He was awarded the Strega Prize for literature shortly before his death.

Pavese's work is much concerned with isolation and the failure of relationships and communication. *Paesi tuoi* (1941; translated as *The Harvesters*, 1961) and *Il carcere* (1949; translated as *The Political Prisoner*, 1955) both dramatize the breakdown of relationships through the experience or fear of violence, a theme also explored in *Dialoghi con Leucò* (1947; translated as *Dialogues with Leucò*, 1965). This alienation is not merely one of individual tragic lives but is finally seen as part of a general human predicament, as in *La luna e i falò* (1950; translated as *The Moon and the Bonfires*, 1953), perhaps his best novel. Other works are *Il diavolo sulle colline* (1948; translated as *The Devil in the Hills*, 1954) and a posthumous book of poems, *Verrà la morte e avrà i tuoi occhi* (1951; 'Death Will Come and its Eyes Will be Yours').

Pavlov, Ivan Petrovich (1849–1936) *Russian physiologist, best known for his classic experiments with dogs through which he developed the concept of the conditioned reflex. He was awarded the 1904 Nobel Prize for Physiology or Medicine.*

Pavlov was born at Ryazan in central Russia, a son of the village priest. After studying physiology and chemistry at the University of St Petersburg and medicine at the Imperial Medical Academy, graduating in 1879, he worked under Carl Ludwig (1816–95) in Leipzig and Rudolf Heidenhain in Breslau (now Wroclaw, Poland). Back in St Petersburg (1888), Pavlov began his first independent research – on heart physiology and the regulation of blood pressure in dogs. It was during this time that he developed his surgical skills and established the practice of taking repeated measurements from normal unanaesthetized animals that was to be so crucial to his later work. In 1890 Pavlov became professor of physiology at the Imperial Medical Academy and started his researches into the physiology of digestion. Using a skilful surgical technique that enabled part of the stomach to be isolated from pancreatic and salivary secretions while maintaining its nerve supply intact, Pavlov showed that gastric secretion in dogs could be stimulated by the sight of food and by nerve stimulation via the brain. It was this work, published in translation as *Lectures on the Work of the Digestive Glands* (1902), that earned Pavlov his Nobel Prize.

In 1898 Pavlov began his work on conditioned reflexes, using unanaesthetized dogs with externalized salivary glands, which was to occupy him for the next thirty years. He found that dogs that salivated in response to the sight of food accompanied by the beat of a metronome would eventually salivate in response to the beating metronome alone: they had learnt to associate the beat of the metronome with the appearance of food. By measuring the amount of saliva produced, he was able to quantify the strength of the conditioned reflex (salivation). Pavlov thus pioneered techniques for studying higher mental activities (in this case learning) in terms of measurable physiological processes, and he went on to explore the idea that many aspects of human and animal behaviour (normal and abnormal) may be explained in terms of conditioned responses.

Always an outspoken critic of communism and the Soviet regime, who denied his request to move his laboratory abroad after the Bolshevik revolution, Pavlov resigned his chair of physiology in 1924, on hearing that priests' sons had been expelled from the Medical Academy. Nevertheless, the government continued to fund his experiments and treat him with respect, and Pavlovian theories have always been popular in the Soviet Union.

Pavlova, Anna (1882–1931) *Russian ballerina who, with her famous tours, popularized ballet throughout the world.*

Pavlova was born in St Petersburg. Her family was poor, as her father died when she was two, and even though she was a sickly child she passed the entrance examination to the Ballet School in St Petersburg (now the Kirov Theatre) in 1891, and by 1906 was the prima ballerina at the Russian Imperial Ballet, for whom she toured Russia and northern Europe in 1907 and 1908. After brief appearances with DIAGHILEV's Ballets Russes in Paris, she performed independently in New York and London. She left the Imperial Ballet in 1913 and the following year began to tour with her own company. These tours, which often involved travel to remote parts of the world, were managed by Victor Dandré, whom she married in 1914.

Pavlova's repertoire was largely classical: her most famous performances were the solo parts in *Giselle* and in *The Dying Swan*, which was choreographed especially for her by Michel FOKINE. Although she did little to change the art of dancing, she was responsible for introducing ballet to people in all parts of the world. She also helped the renaissance of the dance in India and made a special study of oriental dance techniques.

In 1912 she settled in London and established a dance school in Hampstead. During the next decade her tours abroad of these performances were too much for her and she died of pneumonia at the age of forty-nine. An Anglo-Soviet film of her life, *Pavlova*, appeared in 1985.

Peake, Mervyn Laurence (1911–68) *British writer and artist.*

Peake was born and received his early education in China, where his father was a missionary doctor. He then attended Eltham College, Kent, and the Royal Academy Schools before joining an artists' colony on Sark. From 1935 to 1939 he taught drawing at the Westminster School of Art. Invalided out of the army with a nervous breakdown (1943), Peake spent a short period as an official war artist at the end of the war before returning to Sark (1946). Family commitments forced him to take up teaching again, until increasing ill health forced his retirement (1960).

Peake was acknowledged as an outstanding illustrator of books; besides illustrating works by Lewis Carroll and R. L. Stevenson, he also wrote and illustrated his own children's books, such as *Captain Slaughterboard Drops An-*

chor (1939). He published collections of sketches in his lifetime and more were collected after his death. Peake also made his reputation as a poet, writing both for adults (as in *The Glassblowers*, 1950) and for children. He is, however, best remembered for his three Kafkaesque novels *Titus Groan* (1946), *Gormenghast* (1950), and *Titus Alone* (1959), set in the strange surreal world of Gormenghast castle.

Pears, Sir Peter Neville Luard (1910–86) *British tenor, who co-founded (with Benjamin BRITTEN) the Aldeburgh Festival and created the tenor roles in all Britten's operas. He was knighted in 1978.*

Pears studied at the Royal College of Music on an opera scholarship and later with Elena Gerhardt (1883–1961). In 1934 he joined the BBC Singers and toured in the USA with them, becoming a member of the Glyndebourne chorus in 1938. In 1939 he returned to America to give a number of recitals with Benjamin Britten, with whom he formed a lifelong partnership. Returning to Britain in 1943, Pears joined the Sadler's Wells Opera Company, where he created the title role in Britten's *Peter Grimes* in 1945. In the following years Pears sang the leading tenor roles in Britten's operas, including *The Rape of Lucretia* (1946), *Albert Herring* (1947), *Billy Budd* (1951), *Gloriana* (1953), *The Turn of the Screw* (1954), *A Midsummer Night's Dream* (which he co-adapted with Britten) (1960), the dramatic church parable *Curlew River* (1964), *Owen Wingrave* (1970), and *Death in Venice* (1973). He was also renowned for his role as the Evangelist in Bach's *St Matthew Passion* and for his interpretation of Elizabethan lute songs in partnership with Julian BREAM. With Britten he founded and organized the Aldeburgh Festival (1948) and the Britten–Pears School for advanced studies in music.

Pearson, Lester B(owles) (1897–1972) *Canadian statesman and diplomat who became leader of the Liberal Party (1958–68) and prime minister (1963–68). He was awarded the Nobel Peace Prize in 1957.*

Pearson was born in Toronto, the son of a Methodist minister. His studies at Toronto University were interrupted by a period of service in World War I, after which he graduated in history and returned to Toronto University as a lecturer (1924–28). He joined the diplomatic service (1928) in the Department of External Affairs, serving at the office of the Canadian high commissioner in London (1935–41) and as ambassador to the USA

(1945–46). As undersecretary of state for external affairs (1946–48) he worked with Louis St Laurent; when St Laurent succeeded to the Liberal party leadership in 1948, Pearson entered parliament as secretary of state for external affairs, an office he was to hold until 1957. During this time he headed the Canadian delegation to the United Nations, served as chairman of NATO (1951), and displayed his skill as a mediator in the resolution of the Suez crisis (1956), for which he received the Nobel Peace Prize. After the retirement of St Laurent in 1957, Pearson succeeded to the leadership of the Liberal Party and became prime minister, at the head of a minority government, in 1963. Beset by scandals of bribery and corruption among high-ranking government officials, Pearson was criticized for his often blind loyalty to his cabinet colleagues and for his handling of internal problems with the Quebec separatists. He resigned in 1968.

Pei, I(eoh) M(ing) (1917–) *US architect, born in China, best known for his monumental public buildings.*

Born in Canton, Pei studied in the USA from 1935, attending the University of Philadelphia and the Massachusetts Institute of Technology. When World War II prevented his return to China he worked on architectural contracts in several cities and served on the National Defense Committee. In 1948 he became architectural director of Webb and Knapp Inc., in New York, where he worked with the developer William Zeckendorf on a number of large urban projects, including the Mile High Center, Denver (1952–56), a masterpiece of the International Style. Pei became a US citizen in 1954, founding his own architectural firm a year later. His buildings of the 1960s and 1970s include the John F. Kennedy Memorial Library at Harvard University (1964), the National Center for Atmospheric Research, Boulder, Colorado (1967), the John Hancock Tower, Boston (1973), and the East Wing of the National Gallery of Art, Washington DC (1971–78). These major works are all characterized by the dramatic juxtaposition of simple geometric forms that has become the architect's hallmark. Similar principles governed Pei's design for a glass and steel pyramid erected in the forecourt of the Louvre Museum, Paris, in 1989 – a work that has caused some controversy.

Peierls, Sir Rudolph Ernst (1907–) *German-born British physicist, best known for his work on the quantum theory of solids and nuclear physics. He and Otto FRISCH were the first physicists to calculate that an atom bomb could be made. He was knighted in 1968.*

Peierls was born in Berlin and educated at the universities of Berlin, Munich, and Leipzig. After teaching for a short period in Zürich he moved to Britain, where he first taught at Manchester; in 1937 he was appointed professor of mathematical physics at Birmingham University. While at Birmingham he was joined in 1940 by another refugee from Nazi Germany, Otto Frisch. Their calculations, based on the newly discovered phenomenon of nuclear fission led them to conclude that two or three pounds of uranium-235 would be sufficient to produce a devastating bomb. Frisch and Peierls wrote a memorandum entitled *On the Construction of a Super-bomb*, which they sent in March 1940 to Sir Henry TIZARD. Peierls and Frisch were therefore the first scientists to demonstrate the practicality of nuclear weapons. In 1943 Peierls joined the staff of Hans BETHE at Los Alamos to work on the preparation of the bomb. In 1946 he returned to Birmingham, where he remained until 1963 working on atomic energy projects. In 1963 he moved to Oxford as Wykeham Professor of Physics, a post he held until his retirement in 1974.

Pelé (Edson Arantes do Nascimento; 1940–) *Brazilian Association footballer, regarded as the greatest inside forward of all time.*

Born in humble circumstances, Pelé was only fifteen when he joined Santos (São Paulo). The following year he scored his first senior goal and in 1958, at the age of seventeen, he played for Brazil when they won the World Cup for the first time. Pelé became a national hero and a world-class star. He appeared in four World Cup finals and by the time he retired from international football in 1971 had played for Brazil 110 times, scoring a total of 97 goals. He stayed with Santos until 1974 and then played for New York Cosmos (1975–77). In all, he scored a total of over 1200 goals in 1363 matches.

Penney, William George, Baron (1909–91) *British mathematician and leading figure in the development of British nuclear weapons. He was knighted in 1952, made a life peer in 1967, and awarded the OM in 1969.*

Born in Sheerness, Penney was educated at Imperial College, London, Wisconsin University, and Cambridge, where he obtained a PhD in 1935. He taught mathematics at Imperial College until, at the outbreak of World War II, he was engaged in work that eventually led to the development of the atomic bomb. He worked for some time at Los Alamos in the USA and was one of the official British observers at Nagasaki in 1945, when the second atomic bomb was dropped.

After the war Penney remained in government service with the Ministry of Supply (1946–52). During this period he was concerned with the development of the first British atomic bomb, which was tested on 3 October 1952 in the Monte Bello Islands. With the success of this venture behind him, Penney was appointed director of the Atomic Weapons Research Establishment, Aldermaston, with the instruction to develop Britain's hydrogen bomb. The result was successfully tested on Christmas Island in 1957. Shortly afterwards Penney left Aldermaston to concentrate on his duties as a board member of the United Kingdom Atomic Energy Authority (UKAEA), serving as deputy chairman (1961–64) and then chairman (1964–67). In 1967 Penney returned to Imperial College as rector, a position he held until his retirement in 1973.

Perelman, S(idney) J(oseph) (1904–79) *US humorist. His stories in* The New Yorker *constitute his important work and have assured him a place among the leading twentieth-century writers of humour.*

Born in Brooklyn, Perelman attended Brown University and began writing for the university magazine. Among his undergraduate friends was the writer Nathanael WEST, whose sister Perelman later married. He graduated in 1925 and published humorous pieces in a number of magazines. *Dawn Ginsbergh's Revenge* (1929), his first book, brought immediate recognition and introduced the extravagant comic style that made him famous. Accepting an offer from Hollywood, Perelman worked there for a time writing scripts and jokes for the MARX BROTHERS films, but from 1934 he was associated with *The New Yorker* and almost all of his short prose pieces were first published in that magazine. He also collaborated successfully on films and plays – with his wife on the film *Ambush* (1939) and the play *The Night Before Christmas* (1941) and with Ogden Nash (1902–71) on the comedy *One Touch of Venus* (1943). For years Perelman lived in rural Bucks County, Pennsylvania. Seeking a change of scene for his work, he moved to London in the 1970s, but returned to America before his death.

Perelman's very literary style has been described as surrealist and absurdist and comparisons have been made with Poe and JOYCE. A faultless command of the vernacular combines with parody, puns, jokes, and all manner of linguistic play in wild lampoons and fantasies. The pieces are often constructed around short excerpts from newspapers, which lead into sketches or playlets in which the wise-cracking author is usually defeated by some absurdity of modern life. Among his many collections are *Strictly from Hunger* (1937), *Look Who's Talking* (1940), *Crazy Like a Fox* (1944), *Keep It Crisp* (1946), *Westward Ha!* (1948), *Listen to the Mocking Bird* (1949), *The Ill-Tempered Clavichord* (1953), *The Road to Miltown; or, Under the Spreading Atrophy* (1957), *The Rising Gorge* (1961), and *Baby, It's Cold Inside* (1970).

Peres, Shimon (Shimon Perski; 1923–) *Polish-born Israeli prime mininster (1984–86).*

Born in Wołozyn, Poland, he emigrated with his family to Palestine in 1934 and joined the Zionist Haganah movement in 1947. He soon established his political reputation and in 1948, at the age of twenty-five, he was appointed by David BEN-GURION the first head of Israel's navy. Thereafter he held a series of posts at the ministry of defence, including that of director-general (1953–59).

In 1959 Peres was elected to the Knesset as a member of Mapai, the leading left-wing party, but resigned to join Ben-Gurion's breakaway Rafi party in 1965. Following the creation of the merged Labour Party in 1968 he held a number of ministerial posts, including that of defence minister (1974–77). In 1977 the sudden resignation of prime minister Itzhak Rabin (1922–) led to Peres being appointed acting prime minister pending a general election later that year, in which Labour was defeated.

As leader of the opposition (1977–84), Peres consistently advocated a more conciliatory approach to the Palestinian problem than that followed by the Likud government. In 1982 he led protests over the massacre of Palestinians in refugee camps in Beirut, allegedly with Israeli connivance. Peres became prime minister of a coalition government in 1984 as part of a power-sharing deal with the Likud leader Yitzhak SHAMIR; after two years in office (during which Israeli forces withdrew from Lebanon (1985) and the soaring rate of inflation was reduced) he handed over to Shamir and served under him as deputy prime minister and foreign minister. The coalition was renewed following Labour's narrow defeat in the general election of 1988, with Peres once more in the role of deputy premier. In 1990, however, the coalition collapsed when Peres was sacked for lending his support to US proposals for an Israeli-Palestinian peace conference. He has since led the opposition to the right-wing government of Shamir.

Pérez de Cuéllar, Javier (1920–   ) *Peruvian secretary-general of the UN (1982–91).*
Born in Lima, Peru, he joined the foreign ministry in 1940 and entered diplomatic service in 1944, working at Peru's embassies in France, the UK, Bolivia, and Brazil and becoming a member of the Peruvian delegation to the UN. After several more years in diplomatic posts, including those of ambassador to Switzerland and the Soviet Union, he was appointed Peru's representative at the UN in 1971. In 1982 he succeeded WALDHEIM as secretary-general of the UN, soon winning respect for his negotiating skills. In particular, he played a key role in healing the rifts following the Falklands conflict (1982), and in ending the Iran-Iraq War in 1988. He was re-elected for a further five-year term as secretary-general in 1986. In 1990 his determined efforts to avert the Gulf War, following Iraq's invasion of Kuwait, increased his international standing, even though his last-minute visit to Saddam HUSSEIN was unsuccessful. He was, however, to a great extent responsible for maintaining UN support for the American-led coalition that liberated Kuwait in 1991. He also played a key role in negotiating the release of Western hostages held in the Middle East.

Perón, Eva (Maria Eva Duarte de Perón; 1919–52) *Argentine actress and politician. As wife of President Juan PERÓN, she became a legendary – if somewhat controversial – figure, idolized by the poor (who called her Evita) but feared and opposed by the army, who saw in her a dangerous socialist threat to its reactionary power base.*
Born in Los Toldos, the youngest of a large poor family, Maria Duarte was educated at a school in Junin before leaving for Buenos Aires with theatrical aspirations. During the 1930s and 1940s she worked as a radio and later film, actress, primarily for Radio Belgrano.
Her initiation into politics came in 1944, as a member of the radio workers union. Through various benefactors she met many of the leading personalities in the Argentine hierarchy, including Juan Perón, then minister of war, whom she married in 1945. When he was elected president the following year, she devoted much of her time, energy, and influence

to social welfare and the alleviation of poverty. Exercising increasing influence within her husband's regime, she became *de facto* minister of health and labour, organized female workers, secured the vote for women, and directed large amounts of government funds to social welfare. She was nominated for the vice-presidency in 1951 but was forced by the army to withdraw. She died the following year from cancer. At the height of her influence her book *La razón de mi vida* (1951; 'The Purpose of My Life') was made compulsory reading in Argentine schools.

Perón, Juan Domingo (1895–1974) *Argentine statesman and president (1946–55; 1973–74), who succeeded in winning the support of the working classes through his powers of oratory and programme of social reform.*
Born in Lobos into a modest Creole family, Peron entered a military college in 1911 and in 1914 was commissioned as a lieutenant in an infantry regiment. He spent three years at the Argentine Staff College (1926–29) before being appointed professor of military history there in 1930. Six years later he became military attaché in Chile. During the late 1930s and early 1940s he visited Italy and Germany to familiarize himself with the practice of fascism.
Perón's rise to power began in 1943, when he took part in the coup organized by pro-fascist army officers. Appointed vice-president, minister of war, and secretary of labour in 1944, he was forced to resign all his posts and briefly imprisoned in 1945. The following year he was elected president. With the support and popularity of his second wife, Eva PERÓN, he secured re-election in 1951 but after her death in 1952, the faltering economy, military uprisings, (1953, 1955), and conflict with the Roman Catholic Church caused him to lose ground. In September 1955 a military coup forced him into exile. He finally settled in Spain (1960), where he remained until 1972, when he was allowed to return to Argentina. The Peróniste movement, which he created, provided support for him in exile, as well as in power, and enabled him to be re-elected president in 1973. However, he died in office the following year to be succeeded by his third wife, Isabel Perón (1930–   ), who was deposed by the army in 1976.
As a pro-fascist president, Perón improved the lot of the working class by speeding up industrialization and introducing a programme of social reform, known as 'justicialismo'. He also allowed his wife considerable influence in

the departments of health and labour. Internationally, he antagonized Great Britain by nationalizing the British-owned railways and also made the unpopular decision of granting an American oil company a concession in Patagonia.

Perry, Fred (1909–) *British tennis player, who dominated the game during the 1930s.*
Perry was a world-class table-tennis player before he embarked on a career in lawn tennis. His agility and style combined with a supremely confident approach and a characteristic running forehand, which he used to gain position at the net for an attacking forecourt game, earned him three consecutive Wimbledon singles titles in 1934–36. His other singles titles included the US national championship (1933, 1934, and 1936), the Australian championship in 1934, and the French title in 1935. He also won the French (1933) and Australian (1934) doubles titles, partnered in both by G. P. Hughes (1902–). Perry and Dorothy Round (1909–82) won the mixed doubles championship at Wimbledon in 1935 and 1936.

Perry turned professional in 1936 and has remained a familiar figure on the tennis circuits, especially as a commentator for radio and television. His name also adorns a successful range of sports and leisure wear.

Perutz, Max Ferdinand (1914–) *Austrian-born British biochemist, who discovered the molecular structure of the blood pigment, haemoglobin. In recognition of this he received the 1962 Nobel Prize for Chemistry, which he shared with John C. KENDREW. He was awarded the CBE in 1963, the CH in 1975, and the OM in 1988.*
Perutz was born in Vienna and studied chemistry at the University of Vienna before moving to Cambridge University in 1936, where he received his PhD in 1940. Working at the Cavendish Laboratory under J. D. BERNAL, Perutz gained expertise in the technique of X-ray crystallography and in 1937 used it to elucidate the structure of haemoglobin. Progress was slow and so Pertuz and Kendrew, who was working on myoglobin, developed the technique of isomorphous replacement in which heavy metals, such as mercury or gold, are used to label specific sites in the molecule. This enabled more information to be obtained from the complex diffraction patterns produced by the X-rays scattered through such a large molecule as haemoglobin (molecular weight about 64 000). Eventually, in 1960, Perutz published his results: haemoglobin

comprised four polypeptide subunits, each binding a haem group that reversibly binds with oxygen. He went on to show how its three-dimensional structure alters as oxygen is bound and how its function is impaired by slight structural variations, as in sickle cell anaemia.

Perutz served as director of the Medical Research Council's unit for molecular biology at the Cavendish (1947–62) and was chairman of the European Molecular Biology Organization (1963–69). He became a fellow of the Royal Society in 1954. He wrote *Proteins and Nucleic Acids, Structure and Function* (1962) and was co-author of *Atlas of Haemoglobin and Myoglobin* (1981); subsequent publications include *Is Science Necessary?* (1988) and *Mechanisms of Co-operativity and Allosteric Control in Proteins* (1990).

Pétain, (Henri) Philippe (Omer) (1856–1951) *French general and statesman, head of the Vichy regime (1940–42). A national hero in World War I, he was sentenced to death as a traitor for his submission to the Germans in World War II.*

Born in Cauchy-à-la-Tour, the son of a peasant family, Pétain attended St Cyr, where he graduated in 1878. In 1888 he became an instructor at the École de Guerre, a position he held until he was almost sixty.

Promoted to general in 1915 during World War I, he became a national hero (1916) for his defence of Verdun, succeeding General Robert Nivelle (1857–1924) as commander-in-chief in 1917. In 1918 he was named marshal of France, becoming vice-president of the Supreme War Council in 1920 and inspector-general of the army in 1922. An advocate of defensive military policy, he retired from the army in 1931 but entered politics as minister for war in 1934 under Doumergue. Many consider that the weakness of the French army at the start of World War II was directly attributable to Pétain's outdated notions of how a modern army should be organized. Nevertheless, shortly before the fall of France in 1940 he succeeded Raynaud as prime minister and immediately sought to negotiate an armistice with the Germans. The government that he subsequently established in Vichy was a puppet regime for the Third Reich, until German occupation in 1942. Pétain was arrested in 1944 and brought to trial by DE GAULLE's provisional government in 1945. His subsequent death sentence was commuted to life imprisonment on the Île de Yeux, where he died at the age of ninety-six.

Peterson, Oscar Emmanuel (1925–)
Black Canadian jazz pianist, one of the most popular and accomplished jazz musicians of the second half of the twentieth century.

A successful instrumentalist in Canadian bands in the mid-1940s, Peterson went to the USA in 1949 for a Norman Granz concert. As a result, he toured with Granz's Jazz-at-the-Philharmonic concerts throughout the 1950s, moving onto the international circuit in the 1960s, when he often appeared with Ella FITZ-GERALD. During this period he usually led a trio with a bass and guitar, replacing the guitar with drums in 1959. Peterson is also a composer and a singer: he won praise for his *Canadian Suite* in 1965 and recorded the vocal *With Respect to Nat* in the same year. In the early 1970s he often played the piano solo, subsequently dividing a concert time between a solo performance and trio; a popular solo recording is *My Favourite Instrument* (1973). He has hosted his own television series, *Oscar Peterson Presents* and *Oscar Peterson's Piano Party* in the 1970s and *Oscar Peterson and Friends* in 1980; he has also published books of jazz exercises and piano solos.

Although he is not an innovator, Peterson is a serious and articulate man with complete technical command of the instrument, who acknowledges his debt to Nat 'King' Cole (1919–65), Earl HINES, and especially Art TATUM.

Petit, Roland (1924–) *French dancer and choreographer, whose innovative and influential ballets combine fantasy with contemporary realism. He is a Chevalier de la Légion d'honneur.*

Petit trained at the Paris Opéra Ballet School under Gustave Ricaux and joined the Paris Opéra Ballet in 1939. During the next five years he danced in a variety of works and achieved his first success as a choreographer with *Orphée, Rêve d'Amour*. In 1944 Petit concentrated on performing his own works at the Théâtre Sarah Bernhardt in Paris with Irène Lidova and support from such artists as Christian Bérard (1902–49) and Jean COCTEAU. With financial help from his father, Petit founded the Ballets des Champs-Elysées the following year, acting as principal dancer, ballet-master, and choreographer. Secure in his own company, his dancing began to develop an angular acrobatic style; with *Les Forains* (1945) he had his greatest success to date. The company also performed *Le Jeune Homme et la mort* (1946) to considerable critical acclaim, proving that

ballet could still be used as a contemporary medium.

In 1948 Petit formed the Ballets de Paris de Roland Petit, which presented his works *Les Demoiselles de la nuit* (1948), *L'Oeuf à la coque* (1949), *Carmen* (1949), and *La Croqueuse de diamants* (1950), among others. Several important dancers emerged from Petit's company, including Renée (Zizi) Jeanmaire (1925–), who attracted attention with her passionate dancing in the popular *Carmen* and whom Petit married in 1954. They subsequently worked together in Hollywood on several films, including *Hans Christian Andersen* (1952), *Daddy Long Legs* (1955), and *Anything Goes* (1956). In 1960 *Black Tights*, a film incorporating several Petit dances, was released and since then many new choreographic opportunities have been offered to Petit with leading companies around the world. Recent ballets by Petit include *Die Fledermaus* (1980) and *Soirée Debussy* (1985).

Petty, Richard Lee (1937–) *US stock-car driver and winner of a record seven grand national circuit championships. His numerous victories include a record seven wins in the prestigious Daytona 500 race.*

Petty was born into a North Carolina car-racing family. His father, Lee Petty (1914–), rebuilt and raced secondhand cars in his spare time and later became one of the first National Champions as stock-car racing became popular after World War II. Petty worked as an apprentice to his father and began driving when he was twenty-one. Just two years later in 1960 he won his first grand national circuit victory. His father retired from driving in 1962 to devote his energy to working in the pits, determined to ensure Petty's success. In 1972 Petty won his fourth national championship and in doing so not only broke his father's record of three championships but also passed an unprecedented 100 000 miles of racing.

Petty Enterprises, the family business, has always been very safety-conscious, pioneering such devices as roll bars, nylon windscreens, cooled helmets, and two-way radios. Petty is a keen supporter of his son Kyle, who is a successful racing driver in his own right.

Philby, Kim (Harold Adrian Russell Philby; 1912–88) *British secret-service officer who, in 1963, was revealed as a longstanding Soviet agent.*

The son of the explorer and civil servant Harry St John Philby, Kim Philby studied history and economics at Trinity College, Cambridge, where he first met Donald MACLEAN, Guy Burgess (1911–63), and Anthony Blunt (1907–83), all communist sympathizers. Philby was recruited by Soviet Intelligence in London in 1933 and the following year married a militant communist, Litzi Kohlmann, in Vienna. Following the outbreak of the Spanish civil war, Philby went to Spain in 1937 as a freelance journalist and also as a Soviet agent. He was later employed by *The Times* as their correspondent, both in Spain and with the Allied western front during the initial stages of World War II. He returned to London in 1940 and was recruited into the British Secret Intelligence Service. In this role he was able to maintain a regular flow of information to his Soviet contacts. He became head of Soviet counter-intelligence in the latter stages of the war, which gave him even greater opportunities to further Soviet interests. He intervened in 1945 to obstruct the defection to the Allies of a KGB official in Istanbul, Konstantin Volkov. Philby remained undetected and was even awarded the OBE for his wartime services.

After serving as intelligence officer at the British embassy in Istanbul, Philby was given the delicate job of chief liaison officer between the British and US intelligence services at Britain's embassy in Washington DC. In 1950, his friend and collaborator, Burgess, was posted to Washington by the Foreign Office and even lodged with Philby. Aware of the tightening security net around Maclean, Philby sent Burgess home to warn him. Following the defection of both men in 1951, Philby was recalled to London for interrogation but, in the absence of firm evidence against him, was merely asked to resign. He resumed working as a journalist and was exonerated in a Commons statement made by Harold MACMILLAN, then foreign secretary, in 1955.

On 23 January 1963, Philby disappeared from his Beirut home, bound for Moscow. Five months later the British government officially admitted that he was a spy – one of the most successful and damaging of the twentieth century.

Philip, Prince, Duke of Edinburgh
(1921–) *Consort of ELIZABETH II of the United Kingdom. Prince Philip is known and respected for his outspoken views and the asperity with which he often expresses them.*

Son of Prince Andrew of Greece and Denmark and of Princess Alice (1885–1969), daughter of the 1st Marquess of Milford Haven, he was born in Corfu. He attended Cheam School in England, Salem at Baden in Germany,

Gordonstoun in Scotland, and then the Royal Naval College at Dartmouth. In 1947, the year of his marriage to Princess Elizabeth, he became a naturalized British subject and took the surname of his maternal grandfather, Mountbatten. He also renounced his right to the Greek and Danish thrones and, on the eve of his marriage, was created Duke of Edinburgh. He became a prince of the realm in 1957. The Duke continued to serve in the navy, commanding the frigate *Magpie*, until Elizabeth's succession in 1952, when he became fully involved in royal duties. He was made an admiral of the fleet in 1953. In 1956 he founded the Duke of Edinburgh's Award Scheme for the encouragement of creative activities among young people, which now operates in over forty countries. A keen environmentalist, he is president of many bodies, including the WWF (Worldwide Fund for Nature) International (1981–). His publications include *Wildlife Crisis* (with James Fisher; 1970), *The Environmental Revolution: Speeches on Conservation 1962–77* (1978), and *Men, Machines and Sacred Cows* (1984).

Piaf, Edith (Edith Giovanna Gassion; 1915–63) *French singer, cabaret artiste, and songwriter. Her nostalgic style and irrepressible star quality made her France's best-loved singer and an international entertainer.*

She was born in Paris; her mother was a café singer and her father a well-known acrobat. She decided early to be an entertainer and by the time she was fifteen had become a street singer. A cabaret impresario named her 'la môme piaf' (the little sparrow) in 1935. She made her radio debut in 1936 and was successful in films (including Jean RENOIR's *French Cancan*), on stage (*Le Bel Indifférent*, specially written for her by COCTEAU), as well as on television. She first toured the USA in 1947. Piaf wrote about thirty songs herself, the best-known of which is the much-recorded 'La vie en rose'. Many of the songs with which she is associated ('Mon Légionnaire', 'Je m'en fou pas mal', and 'Je ne regrette rien') evoke an image of defiance and despair reflected in her own tragic abuse of alcohol and drugs. The sad and private death of the heroin-ravaged little old lady (of only forty-eight) was in marked contrast to her public funeral a few days later, at which all of Paris seemed to throng the streets. Two volumes of her memoirs were published in Paris in 1958 and 1964.

Piaget, Jean (1896–1980) *Swiss psychologist noted for his studies of thought processes in children and widely regarded as one of the*

most important psychologists of this century. His descriptions of the development of perception, reason, and logic changed the current views of children's intelligence and greatly influenced methods of child education, particularly in the USA.

Piaget began to observe behaviour patterns at an early age, publishing his observations of an albino sparrow when he was ten years old. At fifteen his writings on molluscs were known internationally and at twenty-two he obtained his doctorate from the University of Neuchâtel. After two years at the Sorbonne he was appointed director of the Institut J.-J. Rousseau in Geneva and in 1929 became the professor of psychology at the University of Geneva; in 1955 he was made director of the International Centre of Genetic Epistemology, Geneva.

Piaget had intended to study the development of thought processes in children in order to elucidate the inherent mental structures of humans. His earliest research, which focused on why children fail reasoning tests, led to the long-term study of child intelligence. He suggested that mental growth was determined by interplay of both developing innate structures and environmental influences, an interaction he termed 'equilibration'. Equilibration supposes that when a new experience is assimilated into a child's concept of the world, the concept becomes inadequate and a new, more complex, concept must be invented to accommodate the new information. Equilibrium is then maintained until further experiences require another change of concept. Such a precept requires the existence of logic from early infancy, with intelligence being developed by progressive refinement of cognitive ability by a flexible process of trial and error. Piaget defined the development of children's thinking as a four-stage process, beginning with the sensorimotor stage in infants, who learn from experience by connecting new with older experiences. In the preoperational stage (two to seven years), a child can use words and manipulate them mentally. From seven to twelve years a child begins to think logically and can compare and differentiate, and from twelve to adulthood begins to experiment with formal logic and can think flexibly.

In his later work, Piaget attempted to describe the interactions of cognitive and emotional factors within his four-stage framework. He was a prolific writer, publishing many articles and over thirty books, including *The Origin of Intelligence in Children* (1954) and *The Early Growth of Logic in the Child* (1964).

Picasso, Pablo (Ruiz y) (1881–1973) *Spanish artist. Because of his prolific inventiveness and his technical and expressive brilliance he dominated avant-garde art in the first half of the twentieth century; his name is now more widely known than that of any other modern artist.*

Picasso was born in Malaga, the son of José Ruiz Blanco, an art teacher, and Maria Picasso. He signed his earliest paintings with both surnames but dropped his father's name after about 1900. Showing a prodigious talent from an early age, he ran through a wide variety of styles in Barcelona until he visited Paris in 1901. The paintings of his 'blue period' (1901–04) used melancholy blue tones to depict the poor, the suffering, and the outcast. In 1904 he settled in Montmartre, and began to paint the less austere pictures of his 'rose period' (1905–06) in which acrobats, dancers, and harlequins were represented in pinks and greys.

The Cézanne memorial exhibition of 1907 and a growing interest in African masks and carvings inspired him to concentrate on the analysis and simplification of form during the years 1907–09, sometimes called his 'negro period'. This tendency can be seen as early as 1906 but it was in 1907 that he shocked even the most avant-garde artists of his circle by the violent simplification and distortion of the human form in his *Les Demoiselles d'Avignon*, which can now be seen as a turning point in European art. It heralded the cubist movement, which from 1909 Picasso developed in close collaboration with Georges BRAQUE, whom he had met in 1907, and with Juan GRIS after about 1912. This was the most revolutionary aesthetic movement of the century. Its first stage – analytical cubism – provided artists with new ways of representing visual reality; its second stage – synthetic cubism – included the use of collage and established the concept that a picture no longer had to mirror the world but could exist in its own right.

The partnership with Braque ended in 1914 when Braque enlisted in the army, and in 1917 Picasso left for Rome to design costumes and scenery for DIAGHILEV's Ballets Russes. The following year he married the ballerina Olga Koklova, who bore his son Paul in 1921. As well as pictures in the cubist manner, Picasso produced many neoclassical figure paintings. After 1925 the calmness of his earlier pictures gave way to a sense of convulsive movement, as in *The Three Dancers* (1925), and he began to exhibit with surrealists. Images of violence and anguish became increasingly dominant from 1928 and culminated in Picasso's huge masterpiece, *Guernica* (1937). This painting expressed protest and horror at the bombing of a Basque town by fascist forces in 1937. Picasso never returned to Spain after that year. Meanwhile he had bought a chateau at Boisgeloup in 1932, where he produced large iron sculptures and plaster heads. He had left his wife in 1935 and his new mistress, Marie-Thérèse Walter, bore him a daughter, Maïa.

Throughout World War II Picasso remained in Paris, although the occupying Nazis forbade him to exhibit. After the liberation of Paris he joined the Communist Party. He then lived with Françoise Gilot, first in Antibes and then in Vallauris, until she left him in 1953. They had a son, Claude, and a daughter, Paloma. These were stormy but happy years, and at Vallauris Picasso developed a passion for ceramics and also for bullfighting. His last companion, from 1953, was his model Jacqueline Roque, whom he married in 1961. He continued to produce illustrations, metal sculptures, ceramics, pictures (including a large series of variations on classical paintings), and graphics, which included the famous series of 347 etchings, many of them erotic. He continued to live in the south of France and died at Mougins, near Cannes.

Pickford, Mary (Gladys Mary Smith; 1893–1979) *Canadian-born US film star of the silent screen.*

Pickford was born in Toronto and at the age of five was touring on stage as 'Baby Gladys'. At fourteen she was starring on Broadway in *The Warrens of Virginia*, with the new name of Mary Pickford, and two years later began her film career with D. W. GRIFFITH. Moving from company to company, she made many films, including *The Little American* (1917), directed by Cecil B. DE MILLE, *Rebecca of Sunnybrook Farm* (1917), and *Daddy-Long-Legs* (1917). She married Douglas FAIRBANKS in 1919 (having divorced Owen Moore) and together with Charlie CHAPLIN and Griffith established United Artists. This resulted in such Pickford successes as *Pollyanna* (1920) and *Little Lord Fauntleroy* (1921).

The image of curly-headed innocence and unsophisticated charm that made her famous in silent films did not transfer easily to sound. Although her first talkie, *Coquette* (1929), won her an Academy Award, this was followed by the less successful trio, *The Taming of the Shrew* (1929), *Kiki* (1931), and her last film, *Secrets* (1933). After retiring from films she did some broadcasting and wrote several books, including her autobiography, *Sunshine and Shadow* (1955). Her marriage to Fairbanks ended in divorce; in 1937 she married Charles 'Buddy' Rogers (1904–).

Piggott, Lester Keith (1935–) *British flat-racing jockey. He won more Classics (including a record nine Derbys) than any other jockey and his successes outside Great Britain included three Prix de l'Arc de Triomphe.*

Piggott was born into a family with two hundred years of racing history. His father was a successful jockey and trainer, his mother the daughter of a trainer. Apprenticed to his father, he rode his first winner in August 1948 at the age of twelve. In 1954 he won his first Derby on Never Say Die and in 1960 was champion jockey for the first time (he was subsequently champion jockey from 1964 to 1971 and in 1981 and 1982). For twelve years he was first jockey to Noel Murless's stable but in 1967 became freelance. By 1985 he had achieved a total of twenty-nine Classic victories. The Irish trainer Vincent O'Brien provided Piggott with many winners, such as Sir Ivor, Roberto, and Nijinsky (possibly the best horse Piggott ever rode). His determination to win sometimes brought him into conflict with the stewards and the Jockey Club and he was suspended on several occasions. However, he was a superb judge of a horse's strengths and weaknesses. He announced his retirement as a jockey in 1985. In 1987 he was found guilty of a £3.25 million tax fraud and imprisoned. After his release in 1988 he worked for a time as a trainer before staging a comeback as a jockey in 1990.

Pigou, Arthur Cecil (1877–1959) *British economist regarded as one of the last of the classical school of economists and a major contributor to the field of welfare economics.*

Descended from a Huguenot family, Pigou was born on the Isle of Wight and educated at King's College, Cambridge, where he began by studying history. In 1901 he started lecturing at King's and at the age of thirty succeeded Alfred MARSHALL to the chair of political economy. He retired in 1944 but continued to live at King's. He served on various government committees but was best suited to academic life.

Pigou lectured brilliantly on Marshallian economics and provided the essential ideas and techniques of philosophical reasoning for the Cambridge school of economic thought. His principal work was *Wealth and Welfare* (1912; later renamed *The Economics of Welfare*), which created welfare economics, examining the conditions for maximizing satisfaction and analysing the distinction between private and social utilities. He argued that a redistribution of income would increase total welfare on the basis of marginal utility being a decreasing function of income, but was challenged by many for treating welfare as though it could be measured, compared, and aggregated for the formulation of practical policies.

In his work on monetary theory and national income, Pigou wrote a severely critical review of KEYNES's *General Theory*, which attacked the classical viewpoint, but he later came to agree publicly with much of it (Keynes and Pigou were friends). He introduced a concept known as the 'Pigou effect' (or 'real balance effect') to describe a means by which falling prices and wages might restore a full employment equilibrium: when employment is less than full, falling prices would increase the value of money balances held, causing increased demand for commodities and thereby stimulating employment.

Pinter, Harold (1930–) *British playwright, director, and actor. He was awarded the CBE in 1966.*
Pinter was educated at Hackney Downs Grammar School and in 1949 became an actor, working mainly in the provinces. In 1956 he married the actress Vivien Merchant (1929–82). *The Birthday Party* (1958) was his first full-length play, but his first success was with *The Caretaker* (1960); both plays were subsequently filmed. These plays, like *The Homecoming* (1964), the one-act *Landscape* and *Silence* (both 1969), *No Man's Land* (1975), the one-act *Mountain Language* (1988), and *Party Time* (1991), depend for their dramatic tension upon the nervous suspicious relationships between the principal characters, a theatrical style that has attracted the label 'comedy of menace', or upon breakdowns in personal communication.

In addition to his stage plays, Pinter has written drama for television and radio. He has also written the screenplays for several distinguished films, beginning with *The Servant* (1962) and including *The Pumpkin Eater* (1963), *The Go-Between* (1969), *The French Lieutenant's Woman* (1981), *Turtle Diary* (1985), *The Comfort of Strangers* (1989), and the Proust films. Among the stage productions that he has directed are a number of plays written by Simon Gray (1936–), including *Butley* (1971) and *The Common Pursuit* (1984). In 1980, after his first marriage was

dissolved, Pinter married the historian Lady Antonia Fraser (1932–).

Piper, John Egerton Christmas (1903–) *British painter, designer, illustrator, and author. He was made a CH in 1972.*
The son of a solicitor in Epsom, Surrey, Piper was trained as a solicitor (1922–26) in his father's law firm. At the same time he studied art at the Richmond School of Art and the Royal College of Art. His work in the late 1920s and the 1930s included discerning critical reviews and abstract paintings in subdued colours. He also produced topographical paintings of coastal and rural scenes, which took over from his abstract works towards the end of the 1930s.

His pictures of buildings in picturesque decay highlighted the romantic nature of Piper's art and in 1942 he wrote *British Romantic Artists*. The previous year, as official war artist, he had completed drawings of bomb-damaged London buildings, which were widely acclaimed. After the war his paintings continued to provide a comprehensive record of the buildings of Britain; at the same time he broadened his scope by producing stained glass windows, such as the one in Coventry Cathedral (1962), tapestries, and theatre designs. A recipient of numerous honorary degrees, Piper published books on art, architecture, and topography, some in collaboration with John BETJEMAN.

Pirandello, Luigi (1867–1936) *Italian playwright. He was awarded the Nobel Prize for Literature in 1934.*
Born at Agrigento, the son of a well-to-do mine owner, Pirandello studied Romance philology and took a doctorate at the University of Bonn. In 1893 he moved to Rome and embarked on an active literary life, publishing collections of poetry, translations, and fiction, including his best-known early work, the novel *Il fu Mattia Pascal* (1904; translated as *The Late Mattia Pascal*, 1923). In 1894 he married the daughter of one of his father's business associates. He and his wife, who had been raised traditionally in strict seclusion from the world, were dependent on their families for support. Soon after his wife had borne their third child, the family fortunes were destroyed by a flood. Pirandello's wife suffered a breakdown in 1904 and gradually became violently insane. To protect the children, he was eventually forced to commit her to an asylum in 1918. During this critical period of his life, between 1910 and 1918, Pirandello wrote the stories, many inspired by dreams, that form the bases

of more than half of his forty-three plays. His first play was produced in 1910 and his first three-act play (unsuccessfully) in 1915; he won international acclaim with *Sei personaggi in cerca d'autore* (1921; translated as *Six Characters in Search of an Author*, 1922). He formed his own theatrical company in Rome and directed and acted in his own plays in Europe and America.

The effect of Pirandello's work was extremely important for the modern theatre both within Italy and abroad. He restored the prestige of serious drama in Italy, where there had long been nothing to challenge the dominance of opera. Although his work is not so systematic as it may at first appear, his plays undermined the safe conventions of realistic drama and paved the way for the later absurdist theatre of IONESCO, BECKETT, and others. Individual identity in Pirandello's plays is at risk; distinctions between actor and role, theatre and reality, and form and content are blurred, and normal expectations are sabotaged. *Six Characters*, in which characters from another play take over the rehearsal in progress on the stage, forms a trilogy with two other works concerned with theatre-within-theatre: *Ciascuno a suo modo* (1924; translated as *Each in His Own Way*, 1925), in which the usual art-and-reality relationship is reversed; and *Questa sera si recita a soggetto* (1930; translated as *Tonight We Improvise*, 1932), in which the role takes over the actor (rather than vice versa). Among Pirandello's many other plays are *Così è (se vi pare)* (1917; translated as *Right You Are, If You Think You Are*, 1922), *Enrico IV* (1922; translated as *Henry IV*, 1922), and *Lazzaro* (1929; translated as *Lazarus*, 1952).

Pius XII (Eugenio Pacelli; 1876–1958) *Pope (1939–58) who maintained the neutrality of the Catholic Church during World War II but encouraged efforts to counter the rise of communism in postwar Italy.*

Pacelli came from an aristocratic family that held longstanding connections with the Vatican. He studied at the Gregorian University in Rome and the Collegio Romano, receiving his doctorate in theology in 1898, and was ordained a priest the following year. Pacelli proved himself an outstanding scholar and, after obtaining a doctorate in civil and canon law in 1902, he joined the Vatican civil service. He was appointed undersecretary of state in 1914 and three years later, now an archbishop, was sent to Munich as papal nuncio. He later served in Berlin and in 1929 became a cardinal. As Vatican secretary of state, Pacelli

toured South America and the USA in the 1930s and in 1939, on the death of Pius XI, was elected Pope Pius XII.

During the 1930s, MUSSOLINI became increasingly hostile to the Catholic Church and the Vatican in Rome. Thus, during World War II, Pius's alleged distaste for fascism was rarely expressed and then only guardedly; he contrived a studied neutrality, maintaining diplomatic relations with both Allied and Axis governments. After the war there was mounting criticism for the lack of condemnation from the pope of Nazi atrocities and the ambivalent attitude of the Vatican towards Italian and German antisemitism. However, the postwar years brought active Vatican support for pro-Catholic parties in meeting the challenge of the strong communist movement. Pius XII threatened to excommunicate priests who joined the Communist Party and he encouraged the formation of anticommunist Catholic Action groups in every parish. Poor health in the later years of his reign resulted in power residing more and more with the Vatican bureaucracy.

Planck, Max Karl Ernst Ludwig (1858–1947) *German physicist, who was awarded the 1918 Nobel Prize for Physics for his discovery of the quantum theory.*

The son of a professor of law, Planck was educated at the universities of Munich and Berlin, where he obtained his PhD in 1879. Planck taught briefly in Munich and Kiel before moving to Berlin in 1889 to succeed Gustav Kirchhoff (1824–87) as professor of physics, a post Planck held until his retirement in 1926.

Planck initially established his reputation with his work on thermodynamics. In 1900, however, he published a paper, 'Zur Theorie des Gesetzes der Energieverteilung im Normal-Spektrum' ('On the Theory of the Law of Energy Distribution in the Continuous Spectrum') that contained, Planck is reported to have told his son, discoveries as great as any made by Newton. It certainly transformed the physics of the twentieth century. As classical physics was unable to explain the manner in which he found that a black body radiated energy, he suggested in this paper that energy was radiated in discrete quanta of magnitude hf, where f is the frequency of the radiation emitted and h is a constant, later called Planck's constant. After 1900 a considerable amount of research was devoted to using Planck's quanta to explain such varied phenomena as the photoelectric effect, the scattering of X-rays, the orbit of electrons around the atomic nucleus, and the Zeeman effect. By the time of Planck's death quantum theory had become the best-tested theory in the history of science. Planck's tombstone at Göttingen records his achievement by displaying the value of h, Planck's constant, as 6.6×10^{-27} erg second.

By the time of his retirement in 1927 Planck was the most respected German scientist and consequently many scientists looked to him to defend German science against the racial laws of 1933 banning Jews from holding government office. Although Planck did not speak out publicly against the Nazis and their laws, he did raise his fears with HITLER in May 1933. It is reported that Hitler's response was that 'the annihilation of contemporary German science was acceptable for a few years'. Neither of them foresaw that so many of the expelled German Jewish scientists would produce the atom bomb in the USA instead of in Germany.

Plath, Sylvia (1932–63) *US poet, whose career was cut short by her suicide at the age of thirty.*

Plath was born in Boston, Massachusetts, the daughter of a professor of biology at Boston University. Her father died in 1940, an event that could have contributed to the periods of suicidal depression that recurred throughout her life. She began writing verse as a child and in 1952, as an undergraduate at Smith College, won a fiction contest that involved acting as guest editor for a national magazine. On returning home after this interval of literary celebrity, she suffered from a severe depression, attempted suicide, and had to undergo therapy. She recovered fully after a time and graduated from Smith *summa cum laude* in 1955. The following year she married the British poet Ted HUGHES. Plath taught at Smith for a year in 1957 and then spent a year in Boston devoting herself to her work. She also attended lectures given by Robert LOWELL at Boston University. In 1959 she and Hughes left America to settle permanently in Britain. Plath published her first book of poems, *The Colossus* (1960), while they were living in London, where she also gave birth to their first child, a daughter. (A son was born in 1962.) Deciding to live in the country, they bought a house and moved to Devon in 1961. At the end of the following year Hughes went to live with another woman and Plath returned to London with the two children. The remaining months of her life were ones of intense creativity and declining mental stability. While working occasionally

for the BBC, she published a novel, *The Bell Jar* (1963), under the pseudonym Victoria Lucas. A month after its publication she committed suicide by gassing herself.

Plath's major work, the posthumous collection *Ariel* (1965), consists of poems written mainly during the final months of her life. While often compared to the 'confessional' poetry of Lowell's *Life Studies* (1959), Plath's work has an extreme and painful quality that is entirely original. Other poems written during the final period of her life have been posthumously collected in *Crossing the Water* (1971) and *Winter Trees* (1971). *Letters Home*, a volume of letters written to her mother, appeared in 1975. Her *Collected Poems*, which won a Pulitzer Prize, was published in 1981.

Plomer, William Charles Franklyn (1903–73) *South African writer. He was awarded the CBE in 1968.*

William Plomer was born at Pietersburg, northern Transvaal, and educated at Rugby School in England. He returned to South Africa to farm and then became a trader in Natal. His first publication, the novel *Turbott Wolfe* (1925), raised a storm of hostility and his editorship (with Roy CAMPBELL and Laurens VAN DER POST) of the satirical monthly review *Voorslaf* (1926) caused further controversy. In 1927 he travelled to Japan and then to Greece, before returning to England as a publisher's reader. His short stories, *I Speak of Africa*, came out in 1927, followed by his first collections of poetry, *Notes for Poems* (1928), *The Family Tree* (1929), and *The Fivefold Screen* (1932).

In the 1930s Plomer branched out from fiction and poetry into other literary genres. Among other projects, he wrote biographies – *Cecil Rhodes* (1933) and *Ali the Lion* (1936) – and undertook the editing of the diaries of the nineteenth-century parson Francis Kilvert (1938–40). The verse collections *Visiting the Caves* and *Selected Poems* came out respectively in 1936 and 1940. During World War II Plomer worked for the Admiralty but also found time to write an autobiography – *Double Lives* (1943). *The Dorking Thigh* (1945) contains some of the satirical ballads that Plomer rated among his best work.

The libretto *Gloriana* (1953) was the first of Plomer's collaborations with Benjamin BRITTEN. His other libretti for Britten are *Curlew River* (1964), *The Burning Fiery Furnace* (1966), and *The Prodigal Son* (1968). A second satirical collection, *A Shot in the Park* (1955), contributed with Plomer's other verse publications to his *Collected Poems* (1960). Other postwar publications include the memoir *At Home* (1958).

Poincaré, Raymond Nicolas Landry (1860–1934) *French statesman; prime minister (1912–13; 1922–24; 1926–29) and president (1913–20).*

Born in Bar-le-Duc, the son of a meteorologist and civil servant, Poincaré was educated at lycées in Bar-le-Duc and Paris before studying law at the Sorbonne. He entered politics in 1887 when he was elected to the chamber of deputies as the Moderate-Republican member for Meuse. After serving as a cabinet minister (1893–94), he was appointed vice-president of the chamber in 1895, a position he held until 1897. Between 1899 and 1912 he devoted his energies to his legal practice, briefly holding the post of finance minister in 1906 and accepting a seat in the senate. From January 1912 to January 1913 he was prime minister of a Republican–Radical coalition. He accepted the post of president in February 1913. After his term expired in 1920 he resumed the premiership (1922) and was elected as a senator the same year. Retaining his senate seat, he resigned from the premiership in 1924 but was recalled in 1926 to lead the National Union government. He retired from office in 1929 due to ill health.

Poincaré is remembered for sending French troops to occupy the Ruhr (1923) in response to Germany's failure to make reparations payments. As president he sought to assert the authority of his office but conflicted with CLEMENCEAU during World War I and the ensuing Paris Peace Conference.

Polanski, Roman (1933–) *Polish-born French film director.*

Born in Paris to Polish parents, Polanski grew up in Poland (his parents having returned there when he was three) but later became a French citizen. His films undoubtedly reflect the fear and insecurity of his traumatic childhood. His parents were incarcerated in German concentration camps, where his mother subsequently died, and the eight-year-old Polanski was left to fend for himself, drifting from family to family. After the war, reunited with his father, he returned to Kraców and at fourteen began his acting career in the theatre. He attended the film school of Łódź and appeared in such films as *Pokolenie* (1954; *A Generation*) and *Lotna* (1959). He also directed several shorts, one of which, *Two Men and a Wardrobe* (1958), won the competition for experimental films at the World's Fair, Brussels.

His first feature film as director was *Knife in the Water* (1962), which won a prize at the Venice Film Festival. Since then he has worked abroad, directing such notable films as *Repulsion* (1965) and *Cul de Sac* (1966), both of which won prizes at the Berlin Film Festival, *Rosemary's Baby* (1968), and *Macbeth* (1971). For *Chinatown* (1974) he received a Society of Film and TV Arts best director award and for *Tess* (1979) the Golden Globe. He left the USA for France in 1977 after being charged with a sexual offence against an underaged girl. His most recent films include *Pirates* (1985) and *Frantic* (1988). In 1981 he directed and starred in his own Warsaw production of Peter SHAFFER's play *Amadeus*.

Polanski's second wife, actress Sharon Tate (1943–69), whom he married in 1968, died tragically, the victim of a bizarre murder.

Pollock, Jackson (1912–56) *US abstract expressionist painter, who developed the style known as action painting.*

Pollock was born in Cody, Wyoming, and moved to southern California when he was thirteen. He studied painting at the Art Students' League in New York and in the 1930s produced realistic work, mainly in the manner of the regionalists. From 1938 to 1942 Pollock was employed on the Federal Arts Project and in 1940 he had his first exhibition, together with Willem DE KOONING.

Pollock's style moved away from realism in the 1940s but it was not until 1947 that he developed the method of painting that, over the next four years, brought him recognition as the leading figure in the abstract expressionist movement. This method, known as 'action painting', involved tacking a large sheet of canvas to the floor or wall, then pouring or spattering paint over it in rhythmic movements, covering the entire canvas and avoiding any point of emphasis in the picture (known as the 'all over' style). Occasionally the paint was thickened with sand or broken glass. Brushes were replaced by sticks, trowels, and knives. Thus the canvas became an arena in which the artist's unconscious emotions could be acted out. Pollock continued to paint in this style until his death in a motor accident.

Pol Pot (1925–) *Cambodian communist leader, whose rule (1975–79) led to the deaths of over two million people from execution, forced labour, starvation, and disease.*

Born into a peasant family in Kompong Thom province, Pol Pot spent much of his youth in a Buddhist monastery, where he seems to have become a monk. In the 1940s he attended technical college in Phnom Penh and became active in the movement against French rule, joining the Communist Party in 1946. From 1949 he studied electronics in Paris until his scholarship was withdrawn, owing to poor academic performance – an experience sometimes offered as an explanation for his hatred of intellectuals. Returning to Cambodia in 1953, he became a leading communist organizer, rising to the position of party secretary in 1963. In the same year he left his job as a schoolteacher to join the underground guerrilla movement.

By 1968 Pol Pot was the recognized leader of the Khmer Rouge insurgents, who waged an effective campaign against the governments of Prince Sihanouk (1923–   ) and General Lon Nol (1923–85). Having established themselves in the countryside, the Khmer Rouge took over the government in Phnom Penh in 1975. Renaming the country Democratic Kampuchea, the new regime set about translating Pol Pot's fanatical vision of a wholly peasant and agrarian society into reality. The urban population was stripped of its property and driven into forced labour camps in the countryside, where many thousands died of ill-treatment (in the so-called 'killing fields'). The educated middle class was systematically eradicated.

In 1978–79 Vietnamese forces invaded Cambodia, swiftly overcoming the Khmer Rouge, who retreated to the Thai border spreading further devastation as they went. After the Vietnamese victory, the scale of the Khmer Rouge brutality was first revealed to a horrified world. Pol Pot himself was sentenced to death for genocide but had by this time escaped to Thailand, whence he continued to direct guerrilla warfare against the occupiers.

In 1982 the Khmer Rouge joined with other exiled Cambodian factions, including supporters of Sihanouk, to form the Coalition Government of Democratic Kampuchea. Despite the dominant role of the Khmer Rouge, this coalition gained UN recognition as the legitimate Cambodian government, winning considerable western support. Although Pol Pot officially withdrew from the Khmer Rouge leadership in 1985, he appears to remain firmly in control. Vietnamese withdrawal from Cambodia in 1988–89 led to renewed civil war and widespread fears that he would return to power. In 1990 the warring factions accepted a UN peace formula partly devised to prevent such an outcome.

Pompidou, Georges Jean Raymond (1911–74) *French statesman, prime minister (1962–68), and president (1969–74).*

Born in Montboudif, Auvergne, the son of a university professor, Pompidou was educated at the École Normale Supérieure in Paris and at the École Libre des Sciences Politiques, where he gained a diploma in administration. In 1935 he began a teaching career, which was briefly interrupted in 1939 when he was conscripted into the army. He returned to the classroom in 1940 after the fall of France.

In 1944 Pompidou became an adviser to DE GAULLE, with whom he maintained a close association throughout his political exile. He was appointed to the Council of State in 1946 but resigned to become director-general of the Rothschild Bank in 1954. When de Gaulle returned to power in 1958, Pompidou was appointed chief of cabinet to assist in the drafting of the new constitution. In 1961 he established a dialogue with Algerian nationalists and was instrumental in the signing of the Evian Agreements to end the war. He succeeded DEBRÉ as prime minister in 1962, but was forced to resign in 1968 over his handling of the student-worker protest. Elected president in 1969, he died while in office in 1974.

Popper, Sir Karl Raimund (1902–   ) *Austrian-born British philosopher who developed a philosophy of science based on a theory of falsifiability. He was knighted in 1965 and made a CH in 1982.*

The son of a lawyer, Popper was born and educated in Vienna. After graduation he worked as a schoolteacher for some years while continuing his philosophical researches. Although associated with the Vienna Circle centred around SCHLICK, he was never a member and was in fact highly critical of many of their most distinctive doctrines, as his first book, *Logik der Forschung* (1934; translated as *The Logic of Scientific Discovery*, 1959), clearly revealed. Popper objected to the positivists' obsession with meaning and the role they assigned to the verification principle. Instead, he proposed that scientific statements, unlike the propositions of metaphysics, were falsifiable, and that consequently scientific theories were acceptable only in so far as they managed to survive the challenge of frequent attempts to falsify them. Scientific hypotheses should not be judged as true, but rather as having survived all stringent tests so far applied to them. It was a view that many scientists found congenial and it has influenced the work of such figures as the biologist Sir Peter

MEDAWAR, the art historian Sir Ernst Gombrich (1909–), and the economist Friedrich von Hayek (1899–1938).

Popper left Austria before the *Anschluss* to accept, in 1937, a post at the University of New Zealand, Canterbury. He became a naturalized British citizen in 1945. While there he published *The Open Society and its Enemies* (2 vols, 1945), a work he described as his war effort. In this, and his later *The Poverty of Historicism* (1957), Popper launched a strong attack against the view that there are laws of historical development. Plato, Hegel, and Marx were all subjected to a lengthy and rigorous analysis and finally dismissed with other historicists as enemies of the open society.

In 1947 Popper moved from New Zealand to the London School of Economics, where he held the chair of logic and scientific method from 1949 until his retirement in 1969. He was appointed Senior Research Fellow at the Hoover Institution, Stanford University, in 1986. In his later works, *Conjectures and Refutations* (1963) and *Objective Knowledge* (1972), Popper explored some new themes. He sought to develop an 'evolutionary approach' to the problem of knowledge, while also seeking to identify a third world distinct from the familiar worlds of things and of experiences. This is the world of statements, theories, problems, and arguments – entities for which Popper has offered a realist analysis.

Popper is one of the few modern philosophers to have produced anything like a comprehensive body of work. In addition to his work on the philosophy of science, social theory, and epistemology, Popper has also published important work on probability, truth, indeterminism, quantum physics, and the foundations of logic. Much else remains unpublished. His most recent publications include *Popper Selections* (1985) and *A World of Propensities* (1990).

Porter, Cole (1891–1964) *US songwriter. His technical ability as a composer and his wit as a lyricist combined to make him one of the best songwriters of the century.*

Born in Peru, Indiana, to an affluent family, he started studying music at the age of six and began writing songs as a law student. After the failure of his first comic opera, *See America First* (1916), he went to France in 1917 and served briefly in the French Foreign Legion. Although he was a homosexual, in 1919 he married another socialite, Linda Lee Thomas, with whom he lived a highly sophisticated life in Europe but continued to study music and

write songs, some of which appeared in musical reviews. From 1929 onwards, some of the best songs of the century appeared in song-and-dance musicals: 'Let's Do It', 'Night and Day', 'Begin the Beguine', 'Just One of Those Things', 'What is This Thing Called Love?', and many more. He wrote songs for several films, including *Born to Dance* (1936) and *Rosalie* (1937). His songs appeared in twenty-seven shows and revues, of which the best-known are *Kiss Me, Kate* (1948) and *Can-Can* (1953). A film biography, *Night and Day*, was made in 1946.

In 1937 he suffered a riding accident in which both legs were crushed by a horse; he was in pain for the rest of his life and one leg eventually was amputated. After his wife died, he became virtually a recluse for the last ten years of his life, but a late song, 'True Love' (from the film *High Society*, 1956), was an international hit sung by Bing CROSBY and Grace KELLY. *The Complete Lyrics of Cole Porter* was published in 1983.

Porter, George, Baron Porter of Luddenham (1920–) *British physical chemist who with Ronald Norrish (1897–1978) invented the technique of flash photolysis. They shared the 1967 Nobel Prize for Chemistry with Manfred Eigen (1927–). Porter was knighted in 1972, awarded the OM in 1989, and created a baron in 1990.*

He was educated at Leeds University, graduating in 1941. During World War II he was a radar officer in the Royal Navy Volunteer Reserve. After the war, Porter moved to Cambridge where, from 1949, he and his supervisor Norrish developed flash photolysis – a method of studying extremely short-lived chemical species, thought to be intermediates in certain chemical reactions.

The method used by Norrish and Porter was to enclose a sample of gas in a long glass quartz tube. The sample was subjected to an intense flash of light from surrounding flash tubes, causing photochemical reactions. Excited molecules and free radicals produced in the tube could be investigated spectroscopically by directing a beam of light down the tube and measuring the absorption spectrum of the gas. The technique gives information on the nature of excited species and radicals and is also a way of studying very fast reactions. From 1955 until 1963, Porter was Professor of Chemistry at the University of Sheffield. From 1966 to 1985 he was the Fullerian Professor and director of the Royal Institution, where he worked particularly on solar energy. He was president of the Royal Society from 1985 until 1990 and is currently chairman of the Centre for Photochemistry and Photosynthesis at Imperial College, London.

Porter is noted for his interest in popularizing science for non-scientists, especially young people. Indeed he has suggested that people who have no taste for science should be dragged "kicking and screaming into the twenty-first century".

Potter, (Helen) Beatrix (1866–1943) *British writer of children's books.*

Born in London, Beatrix Potter had a dull and solitary childhood, regulated by governesses. On holidays in Scotland and the Lake District she was enthralled by the wild animals and flowers, and one of her governesses, recognizing her artistic talent, encouraged her to draw and study them. She kept small animals as pets to compensate for the domination of her authoritarian parents.

Her books grew out of letters written in her late twenties to entertain a sick child. Their enduring popularity owes as much to the delicate watercolour illustrations, drawn by Beatrix Potter herself, as to the economical unsentimental prose of the animal stories. The first two, *Peter Rabbit* and *The Tailor of Gloucester*, were privately printed in 1900 and 1902, but she then found a publisher in Frederick Warne, who published most of her twenty-eight or so books. In 1905 she became engaged to his son Norman, but he died a few months later and Miss Potter, now financially independent through the sale of her books, bought a farm at Sawrey in the Lake District. She spent increasing periods of time there learning about farming, especially Herdwick sheep, on which she became an authority. In 1913 she married the Ambleside solicitor William Heelis and virtually ceased writing; she spent the rest of her life caring for her expanding estates in the Lake District, all of which were left to the National Trust at her death.

Poulenc, Francis Jean Marcel (1899–1963) *French composer and pianist, who became a member of Les Six.*

Poulenc studied the piano with Ricardo Viñes (1875–1943) and composition with Charles Koechlin (1867–1950). After a period in the army (1918–21), his talent, recognized in the *Rhapsodie nègre* (1917, revised 1933), made him a part of postwar Parisian artistic life and he became one of the group known as Les Six. As a Roman Catholic, he experienced a strong renewal of faith during a visit to Notre Dame de Rocamadour, the outcome of which was *Litanies à la vierge noire* (1936), the first of a series of lyrical and contrapuntal sacred choral pieces. Poulenc's lyrical gift is also apparent in his large collection of solo songs, which the tenor Pierre Bernac (1899–1979) sang, accompanied by Poulenc at the piano, in a number of recitals in Europe and the USA. For texts, Poulenc generally restricted himself to a limited number of poets, especially Apollinaire (*Quartre poèms*, 1931), Éluard (*Tel jour, telle nuit*, 1937, and *Le Travail du peintre*, 1956), Max Jacob (*Cinq poèmes*, 1931), and de Vilmorin (*Trois poèmes*, 1937).

Poulenc's operas include the burlesque *Les Mamelles de Tirésias*, with a libretto by Apollinaire, *Les Dialogues des Carmélites* (1953–55), a tragedy of a group of Carmelite nuns during the French Revolution, and *La Voix humaine* (1948), a one-act monologue. Orchestral works include the highly successful *Concerto champêtre* for harpsichord and orchestra (1928) and the concerto for organ, strings, and percussion (1936). His best-known ballet score is *Les Biches* (1923), which was commissioned by Diaghilev for the Ballets Russes. Of his chamber works, Poulenc's sonatas for various wind instruments and piano are now part of the contemporary repertoire.

Pound, Ezra Weston Loomis (1885–1972) *US poet, critic, and translator. A key figure in the development of contemporary poetry, who received the Dial award (1928) for contributions to American letters and the Bollingen Prize (1949).*

Born in Idaho, Pound spent his childhood near Philadelphia and was educated at Hamilton College, New York, and the University of Pennsylvania, where he gained a wide knowledge of the classical and Romance languages. A small scholarship enabled him to travel in Europe and he left the USA in 1908. Pound's first collection of poems, *A Lume Spento* (1908), was published at his own expense in Italy. While here he met the violinist Olga Rudge, who became his mistress and bore him a daughter, Mary. Soon after, in London, he founded (with Richard Aldington and H.D.) the imagist movement, which rejected the lingering influence of Victorianism on poetry and emphasized the importance of the language of common speech, the creation of new poetic rhythms, and complete freedom of choice of subject matter. Various collections of his poetry were published, including *Exultations* (1909) and *Ripostes* (1912), while *Cathay* (1915) illustrated his increasing preoccupation with translation as a poetic medium.

The impact of World War I probably stimulated Pound's greatest poetry, of which *Homage to Sextus Propertius* (1917) and *Hugh Selwyn Mauberley* (1920) are often seen as the culminating achievements. His most ambitious poetry, however, was encompassed in *The Cantos*, conceived as early as 1913 and never completed, primarily because the poet's ideas and aims were in a constant state of flux. The work was ultimately regarded as a failure by Pound himself.

Pound left London for Paris in 1921 and three years later moved to Italy. There he remained throughout World War II, expressing admiration for MUSSOLINI and broadcasting fascist propaganda in English. This involvement damaged his reputation and alienated him from many of his friends. He was arrested for treason in 1945, but found insane and never brought to trial. Instead he was committed to St Elizabeth's Hospital, Washington, DC, where he remained for ten years. During this period, he wrote the *Pisan Cantos* (1948), often regarded as the most successful section of the larger work. Pound was released in 1958 and returned to Italy, where he died at the age of eighty-seven.

Powell, Anthony Dymoke (1905–) *British novelist. He was awarded the CBE in 1956 and made a CH in 1988.*

Powell was educated at Eton and Balliol College, Oxford. His talent as a novelist was apparent in his first novel, *Afternoon Men* (1931; adapted as a play, 1963). In 1934 he married Lady Violet Pakenham, daughter of the 5th Earl of Longford. Other books of the 1930s include *Agents and Patients* (1936) and *What's Become of Waring?* (1939). During World War II he served in the Welch Regiment and the Intelligence Corps.

After two volumes on the seventeenth-century scholar John Aubrey in the late 1940s, Powell embarked on his major novel sequence, *A Dance to the Music of Time*. The first volume in this twelve-volume series was *A Question of Upbringing* (1951) and the last *Hearing Secret Harmonies* (1975). They constitute a panoramic and satirical view of the changes in the lives and fortunes of the English upper middle classes during Powell's lifetime. Subsequent publications included the novel *The Fisher King* (1986) and *Miscellaneous Verdicts* (1990), a collection of his criticism. His four volumes of autobiography *To Keep the Ball Rolling* were published between 1976 and 1982.

Powell, Cecil Frank (1903–69) *British physicist, who was awarded the 1951 Nobel Prize for Physics for his discovery in 1947 of the pi-meson (pion).*

The son of a gunsmith, Powell was educated at Cambridge University where he gained his PhD in 1927. In the same year he joined the physics department of Bristol University and spent the rest of his career there, becoming Wills Professor of Physics in 1948 and director of the Wills Physics Laboratory in 1964.

In the 1930s Powell began his study of the tracks of nuclear particles on photographic plates. The standard instrument of his day, Wilson's cloud chamber, was unwieldy and limited in its use. Powell therefore developed photographic emulsions some hundred times thicker than those normally used and eventually used them to study cosmic rays by sending them up in balloons to heights of 30 000 metres. The lengthy analysis of these plates was rewarded in 1947 by Powell's discovery of the tracks of the pi-meson, predicted by H. YUKAWA in 1935. Powell also, in collaboration with Giuseppe Occhialini (1907–), published one of the earliest textbooks on these new methods, *Nuclear Physics in Photography* (1947). Powell was also involved in setting up CERN in 1953 and was present at the Pugwash meeting in Nova Scotia in 1957 that initiated the Pugwash movement, enabling scientists to influence world affairs.

Preminger, Otto Ludwig (1906–86) *Austrian-born US film and theatre director and producer.*

The holder of a doctorate in law from the university in his native Vienna, Preminger did not pursue a legal career; instead, he entered the theatre as an actor with Max Reinhardt's company in 1923. He subsequently became a director and took over responsibility for the Josefstadt Theatre. His New York debut came with *Libel* (1935), after which he produced several other Broadway successes. He became an associate professor at Yale University (1938–41) and then joined Twentieth Century Fox as producer and director (1941–51). Notable among his early films were *Laura* (1944), *Fallen Angel* (1945), *Forever Amber* (1947), and *Daisy Kenyon* (1948).

Some of his films as an independent producer were to prove milestones in the history of the cinema: *The Moon is Blue* (1953) broke the censorship barrier of the day with its mature and frank dialogue, while *The Man with the Golden Arm* (1955) uncovered new ground with its handling of drug addiction. He also

Presley, Elvis Aaron   458

produced adaptations of Bizet's *Carmen, Carmen Jones* (1954), SHAW's *Saint Joan* (1957), and Françoise SAGAN's *Bonjour Tristesse* (1958), the last two starring Jean Seberg (1938–79). *Anatomy of a Murder* (1959), *Exodus* (1960), *Advise and Consent* (1962), and *The Cardinal* (1963) were among his other successes. *Preminger: An Autobiography* was published in 1977.

Presley, Elvis Aaron (1935–77) *US pop singer. The first rock and roll star, he profoundly changed popular music with his fusion of country music and rhythm and blues.*

Born in East Tupelo, Missouri, of a poor family, he attended an Assembly of God church, where the hymn singing was an early influence. He was still a truck driver when he made his first recordings for Sun records in Memphis, Tennessee, in July 1954. He also appeared on country-music radio shows, his early records becoming hits in the country charts. In 1955 his contract and all his Sun recordings were purchased by RCA for $35,000, considered a large sum at the time. His first recording for RCA, 'Heartbreak Hotel', went to the top of all three charts – country, pop, and rhythm and blues – in 1956. All his recordings of that year became rock classics, including 'Money Honey', 'I Got a Woman', and 'Blue Suede Shoes'. All were originally other people's songs, but his controversial eccentric improvisational style caught the imagination of the world's teenagers.

After two years (1958–60) stationed in Germany in the US army, he appeared in more than thirty musical films between 1960 and 1969; none achieved critical acclaim. In 1969 he resumed his personal appearances, mostly in Las Vegas. Presley spent much of his life secluded in his Memphis mansion, called Graceland after his mother, surrounded by friends. He became an accomplished singer of ballads and his musical backing was usually first-rate. He remained the biggest pop star in the world until he died of an accidental overdose of drugs.

Prévert, Jacques (1900–77) *French poet.*
Prévert was born in Neuilly-sur-Seine. His involvement with the surrealists lasted only a few years (1925–29) but its effects, particularly the use of surprising and unexpected images, may be seen throughout his work. Prévert's poetry, in which he used a variety of linguistic and literary devices, ranges from lyrical and often melancholy pictures of life on the streets of Paris to humorous anarchic attacks on all aspects of the establishment. His

first major success was the satirical poem *Tentative de description d'un dîner de têtes à Paris-France* (1931). A number of poems from the collection *Paroles* (1946), set to music by Josef Korma, became popular songs and helped to bring Prévert's poetry to a wider audience; the more satirical antiestablishment poems appealed particularly to young people. Among his later poetical works are *Spectacle* (1951), *La Pluie et le beau temps* (1955), and *Fatras* (1966).

Prévert also wrote for the theatre and cinema. His film scenarios include *Le Crime de Monsieur Lange* (1935), directed by Jean RENOIR, and *Les Visiteurs du soir* (1942) and *Les Enfants du paradis* (1945), both directed by Marcel CARNÉ.

Priestley, J(ohn) B(oynton) (1894–1984) *British novelist, critic, playwright, and broadcaster. He was awarded the OM in 1977.*

Priestley was born and educated in Bradford, where his father was a schoolmaster. Throughout World War I he served in the army and then went up to Trinity Hall, Cambridge, where he began his career as a writer. In the 1920s he published several volumes of criticism, including studies of Meredith (1926) and Peacock (1927), as well as several novels. However, his first major success came with the picaresque novel *The Good Companions* (1929). He consolidated this success with *Angel Pavement* in the following year. The 1930s saw Priestley's debut as a playwright, first with the dramatization of *The Good Companions* (1931), and then with *Dangerous Corner* (1932) and *Laburnum Grove* (1934). In addition to fiction and drama, he also wrote essays, such as *English Journey* (1934) and *Rain upon Godshill* (1939). In 1937 he became president of the London PEN Club (Poets, Playwrights, Editors, Essayists, Novelists).

During World War II Priestley was a popular broadcaster and in his plays and other works accurately reflected the wartime mood of the British public. After the war he was a delegate to the UNESCO conferences (1946–47) and subsequently chairman of the international theatre conferences in Paris (1947) and Prague (1948). The 1940s, a particularly prolific period in his theatrical output, saw the production of such plays as *An Inspector Calls* (1946) and *The Linden Tree* (1947). He also wrote the libretto for BLISS's opera *The Olympians* (1949).

In 1953 Priestley married his third wife, the archaeologist Jacquetta Hawkes (1910–), with whom he collaborated on the play

Dragon's Mouth (1952) and wrote *Journey Down a Rainbow* (1955). His output continued almost as varied and prolific as ever, and he added writing for television to his repertoire. Turning again to criticism, he wrote *The Art of the Dramatist* (1957) and *Literature and Western Man* (1960); his *Essays of Five Decades* appeared in 1969. Among his postwar novels were *Festival at Farbridge* (1951), *Saturn Over the Water* (1961), and *It's an Old Country* (1967), but he published an increasing proportion of nonfiction, including the autobiographical volumes *Margin Released* (1962) and *Instead of the Trees* (1977). His patriotic delight in national traditions and history is the motivating force in much of this work, notably *The English* (1973).

Pritchett, Sir V(ictor) S(awdon) (1900–   ) *British writer and critic. He was knighted in 1975.*

V. S. Pritchett was born in Ipswich and educated at Alleyn's School, London, and other schools. He then worked in the leather trade and various other businesses before becoming a newspaper correspondent in France in the 1920s. From there he moved to Spain, about which he wrote articles and travel books. He also began writing fiction, including *Claire Drummer* (1929), *Elopement in Exile* (1932), *Nothing Like Leather* (1935), *Dead Man Leading* (1937), and *Mr Beluncle* (1951). He was literary editor of the *New Statesman and Nation* for a time and afterwards continued his association with the journal, contributing a regular column and eventually becoming a director. His critical essays have been collected in several volumes – among them *In My Good Books* (1942), *The Living Novel* (1946), *Books in General* (1953), *The Myth Makers* (1979), and *A Man of Letters* (1985).

From the 1950s onwards, Pritchett fulfilled lecturing engagements at several US universities. He also delivered the Clark Lectures at Cambridge (1969). He wrote a number of memorable short stories, of which collected editions were published in 1956, 1982, and 1983; *The Camberwell Beauty* (1974) and *A Careless Widow and Other Stories* (1989) contain some of his later stories, and in 1981 he edited the *Oxford Book of Short Stories*. Among his studies of other writers are *George Meredith and English Comedy* (1970) and biographies of Balzac (1973), Turgenev (1977), and Chekhov (1988). His two volumes of autobiography are *A Cab at the Door* (1968) and *Midnight Oil* (1971).

Prokofiev, Sergei Sergeievich (1891–1953) *Soviet composer and pianist. His earlier works express his mocking sense of humour, often underlined by dissonance, while his later works, composed after his return to the Soviet Union, are more relaxed and melodic.*

From earliest childhood Prokofiev composed music; by the time he entered the St Petersburg Conservatory at the age of thirteen he had written four operas, two sonatas, a symphony, and numerous piano pieces. Prokofiev's student days were somewhat disturbed both by the current political unrest and by his own impatience with the teaching routine; however, he graduated in 1914 winning the first prize for the piano with his own first piano concerto (1911–12). In 1918, shortly after the successful first performance of his *Classical Symphony* (1916–17), Prokofiev emigrated to the USA, where he was well received as a pianist but met considerable criticism as a composer. However, his opera *The Love for Three Oranges* (1919) was eventually produced in Chicago in 1921. In 1922 Prokofiev moved to Paris, where DIAGHILEV commissioned him to write the ballet *Le Pas d'acier* (1925).

By 1936 Prokofiev was drawn back to his native Soviet Union, where he established his home with his wife, the Spanish-born singer Lina Llubera, and their two sons. Having no interest in politics, Prokofiev was ill-prepared for state criticism that, from time to time, accused him of 'formalistic tendencies'. Nevertheless he managed to settle into the communist state, continuing his prolific flow of compositions until 1948, when ill health brought his creative life to an end.

Many of Prokofiev's works have found a regular niche in the repertoire. Among his principal works are the suite *Lieutenant Kije* (1934), *Peter and the Wolf* (1936), a 'symphonic fairy-tale for young and old', the ballets *Romeo and Juliet* (1935) and *Cinderella* (1944), the third piano concerto (1921), the two violin concertos (1914–35) and the opera *War and Peace* (1941–43; revised 1946–52).

Prost, Alain (1955–   ) *French motor racing driver and World Drivers' Champion (1985, 1986, and 1989), who ranks with F*ANGIO *as the most successful Formula One driver.*

Born in St Chamond in southern France, he drove his first Formula One race in 1980 in Argentina and scored his first Grand Prix success (in France) in 1981, driving a Renault. In 1984 he switched to the Malboro-MacLaren team and a year later won his first world championship, becoming the first Frenchman to do so. Remaining with Malboro-MacLaren, he repeated his success in 1986 and, driving for Malboro-Honda, again in 1989. He was runner-up in 1983, 1984, 1988, and 1990. In 1987 he passed Jackie STEWART's record of twenty-seven Grand Prix victories; by 1990 he had scored forty-three wins. In 1990 he also established a new record with his career total of Grand Prix points.

Proust, Marcel (1871–1922) *French novelist.*

Proust was born in Auteuil, the son of an eminent Catholic physician and a wealthy Jewess. At the age of nine he suffered his first attack of the asthma that was to torment him for the rest of his life. After a period of military service (1889–90) he studied law and literature in Paris and became an habitué of the fashionable salons of the Parisian aristocracy. During this period he published a volume of short stories, *Les Plaisirs et les jours* (1896), but the auto-biographical novel *Jean Santeuil* (written 1895–99), a preparatory sketch for his masterpiece *À la recherche du temps perdu*, did not appear until 1952, many years after his death. In the early 1900s he produced translations of Ruskin – *La Bible d'Amiens* (1904) and *Sésame et les lys* (1906) – and wrote a lengthy critical essay, *Contre Sainte-Beuve* (published posthumously in 1954), in which he expounded his theory of literary criticism, attacking Sainte-Beuve's biographical approach.

Proust's parents died within two years of each other (1903, 1905), leaving him alone in the world but with ample financial resources to devote himself to writing. Retiring to a cork-lined room fumigated with asthma inhalants in the Boulevard Haussmann, where he lived as a virtual recluse for the remainder of his life, he embarked on his major novel, *À la recherche du temps perdu* (1913–27; translated as *Remembrance of Things Past*, 1922–30). The first volume, *Du côté de chez Swann* (translated as *Swann's Way*), was refused by a number of publishers; Proust financed its publication himself in 1913. During World War I he revised and enriched his early draft for the remainder of the novel, increasing its proposed length from three to seven volumes; *À l'ombre des jeunes filles en fleurs* (translated as *Within a Budding Grove*) was published by the *Nouvelle Revue Française* in 1919 and won the Prix Goncourt and worldwide fame for its author. His health rapidly deteriorating, Proust managed to publish two more sections of the novel before his death from pneumonia: *Le Côté de Guermantes* (1920–21; translated as

The Guermantes Way) and *Sodome et Gomorrhe* (1921–22; translated as *Cities of the Plain*). The remaining three volumes, completed but not fully revised, were published posthumously: *La Prisonnière* (1923; translated as *The Captive*), *Albertine disparue* (1925; translated as *The Sweet Cheat Gone*), and *Le Temps retrouvé* (1927; translated as *Time Regained*).

The novel is a complex reconstruction of the life of the narrator, Marcel, from childhood to middle age. Marcel is in some respects Proust himself – a delicate child from a wealthy middle-class family, he moves through the fashionable salons frequented by Proust in his twenties – but the novel is an allegory of the author's life rather than pure autobiography. Characters, places, and events are altered but not invented or falsified. The central theme of the novel is the recovery of the lost past through the stimulation of unconscious memory; in the early part of the novel the narrator, tasting a madeleine dipped in tea, is suddenly reminded of childhood breakfasts and finds he can gain access to memories he thought he had lost forever. The novel also analyses the decadence of the aristocracy, the emptiness of love, whether homosexual or heterosexual, and the vices and depravity to be found at all levels of society, represented by a cast of some two hundred characters. It ends, however, on a note of optimism: in *Le Temps retrouvé* the disillusioned Marcel, having renounced his ambition to be a writer, suddenly discovers that the beauty of his past experiences is not irretrievable; the novel comes full circle as he settles down to write the story of his life.

Puccini, Giacomo (1858–1924) *Italian composer, the most important figure in Italian opera since Verdi.*

Puccini was born in Lucca, where his family had supplied several generations of musical directors to the cathedral of S. Martino. Giacomo was schooled to follow in their footsteps and worked as a church organist from the age of fourteen. By the time he had begun his studies at the Milan Conservatory in 1880, however, a performance of Verdi's *Aida* had convinced him that he should devote his talents to opera rather than sacred music. His first opera, *Le Villi*, was produced shortly after his graduation in 1884, achieving moderate success. A second opera, *Edgar* (1889), was poorly received, partly because of its unsuitable libretto.

By this time Puccini had caused a scandal in his home town by eloping with Elvira

Gemignani, a married woman. Although Elvira bore him a son, the couple were not able to marry until the death of her husband in 1904. After some wandering they settled in Torre del Lago, a lakeside village in Tuscany that remained the composer's home until the last years of his life. It was here that Puccini composed *Manon Lescaut*, the opera that established his reputation; it was produced at Turin in 1893, to scenes of wild enthusiasm, and went on to enjoy international success. The composer used the same librettists, Luigi Illica and Giuseppe Giacosa, for his next three works; *La Bohème* (1896), *Tosca* (1900), and *Madama Butterfly* (1904). Like *Manon Lescaut*, these enduringly popular works are love stories, centring on the plight of a female protagonist, who meets a tragic end in the last act. Puccini's dramatic power, skilful use of the orchestra, and gift for original lyrical melodies are displayed to the full. Puccini's tempestuous relationship with his wife erupted into national scandal in 1908, following the suicide of a female servant whom Elvira had accused of having an affair with her husband. When the girl's innocence was established, her relatives brought a successful prosecution. Puccini was horrified by the publicity, which seems to have affected his ability to work. His next opera, *La fanciulla del West* (The Girl of the Golden West), was produced in New York in 1910 with CARUSO in the tenor part and TOSCANINI conducting. Its shows the influence of such contemporaries as DEBUSSY and STRAUSS and the composer's attempts to rebut criticisms that his style and subject matter had become formulaic. This is even more true of *Turandot*, an opera on a fantastic fairy-tale theme that Puccini was working on at the time of his death. The score was completed by a pupil and produced in Milan in 1926. Puccini had by then become a national figure in Italy, his death being marked both by public honours and a genuine outpouring of popular grief.

Q

Quant, Mary (1934–) *British fashion designer and one of the principal creators of the 'sixties look'. She was awarded an OBE in 1966.*

While studying for a teaching diploma at Goldsmith's College, London, Mary Quant met her collaborator and future husband, Alexander Plunkett-Greene. After working as a milliner, she opened her first boutique, called Bazaar, in the mid-1950s in Chelsea's King's Road in partnership with Plunkett-Greene and Archie McNair. Quant's choice of clothes and accessories caught the imagination of young Londoners and this success inspired her to design and make garments herself. Beginning on a single sewing machine in her bedsitter, she expanded rapidly and in the early 1960s branched out into furs, rainwear, underwear, and cosmetics. Her most famous innovation was the miniskirt, launched in 1966, which took hemlines to unprecedented heights. By 1967 two further Quant boutiques had opened in London and their founder held a host of design contracts and retail concessions with other stores. In the early 1970s another Quant creation – hotpants – made a brief appearance, and at the same time she produced designs for household furnishings and linen.

Quine, Willard van Orman (1908–) *US philosopher, logician, and radical critic of modern empiricism.*

Born at Akron, Ohio, Quine was educated at Oberlin and Harvard, where he completed his PhD under the supervision of A. N. WHITE-HEAD. After spending some time in Vienna, Warsaw, and Prague he returned to Harvard as Edgar Pierce Professor of Philosophy (1948–78).

Quine has worked extensively in the field of set theory, publishing his mature views in *Set Theory and its Logic* (1963). He is best known, however, for his work on the philosophy of language. In his classic paper 'Two dogmas of empiricism' (1951) he challenged the basic empiricist assumption that there is a clear distinction between analytic and synthetic propositions. Attempts to define the notion of analyticity invariably presupposed an understanding of some intentional concept like syn-

onymy, which was in fact no clearer than analyticity itself. Further, Quine argued, 'no statement is immune from revision': even the principles of logic themselves could be questioned and replaced.

In his later book, *Word and Object* (1961), Quine argued for the indeterminacy of translation. This amounted to the claim that there was in principle no satisfactory way to ensure that a sentence P was a correct translation into language A of sentence Q in language B. He further conceded his commitment to the existence of abstract entities on the grounds that mathematics demands the use of sets with which to define such essential concepts as number.

His most recent publications include *The Time of My Life* (1985), *Quiddities* (1987), and *The Logic of Sequences* (1990).

Quisling, Vidkun Abraham Lauritz Jonsson (1887–1945) *Norwegian head of state under the German occupation (1940–45). His name has come to mean any national traitor who collaborates with the enemy.*

Born in Fyresdal in Telemark, the son of a Lutheran pastor, Quisling was educated at the Norwegian Military Academy, where he became a general staff officer at the age of twenty-four before joining the diplomatic and intelligence service in 1918. He observed the Russo-Finnish conflicts (1918–21) as a military attaché and represented British interests at the Norwegian legation in the Soviet Union (1927–29). In 1931 he entered the Norwegian parliament (Storting) as a member of the Agrarian Party and became minister of defence (1931–33).

Quisling founded the Nasjonal Samling (National Party) in 1933, a Norwegian version of the German National Socialist (Nazi) Party. Following a visit to Berlin in 1939, he collaborated with the Germans in the invasion of Norway (1940), becoming head of state under the German occupation, when he attempted to employ Nazi practices in Norway. He remained puppet leader until the liberation of Norway in 1945, when he was tried as a traitor and executed.

R

Rabi, Isidor Isaac (1898–1988) *Austrian-born US physicist who invented magnetic resonance spectroscopy. He was awarded the Nobel Prize for Physics in 1944.*

Rabi was born in Rymanov in Austria, now in the Soviet Union. His parents emigrated to America when he was young and he grew up in a Yiddish-speaking community in New York, where his father ran a grocery store. He was educated at Cornell University, graduating in chemistry in 1919, and at the University of Columbia, where he gained his PhD in physics in 1927.

Rabi then moved to Germany, where he worked for Otto STERN. He had been impressed by the famous experiment by Stern and Walter Gerlach (1920–21), which showed that some atoms have a magnetic moment. When he returned to Columbia two years later he began a research program that led to the development of molecular-beam magnetic resonance spectroscopy.

This is an extremely precise method of measuring the magnetic properties of atomic nuclei. It opened up a number of subsequent applications, including the atomic clock for precise measurement of time, the maser, the laser, and the use of magnetic resonance for studies of molecules and nuclei and, more recently, for medical tomography.

Rabi became Professor of Physics at Columbia in 1937. During World War II he led a group working on the development of radar in the US. From 1946 to 1956 he was a member of the General Advisory Committee of the Atomic Energy Commission, succeeding J. Robert OPPENHEIMER as its chairman in 1952. Rabi also helped set up the CERN centre for high-energy physics in Geneva and the Brookhaven National Laboratory, New York. He remained at Columbia until his retirement in 1967.

Rachmaninov, Sergei (1873–1943) *Russian composer, pianist, and conductor. His sumptuous and lyrical music, strictly in the Russian romantic tradition, is familiar all over the world. It is clear from his records that he was a virtuoso pianist.*

Coming from an affluent but somewhat improvident family, Rachmaninov had to learn self-reliance early. He studied first at the St Petersburg Conservatory and later in Moscow, where his teachers were the rigid disciplinarian Zverev, Ziloti (piano), Taneyev (counterpoint), and Arensky (composition). Graduating in 1892 with the Gold Medal, Rachmaninov soon established himself as a brilliant pianist. However, he was by then increasingly interested in composition: his first piano concerto (1890–91, revised 1917), the one-act opera *Aleko* (1892; his graduation exercise), and the famous *Prelude* in C sharp minor for piano (1892) were among his early works. However, the poor reception given to a performance of his first symphony (1897) distressed him to such an extent that he stopped composing for three years, until the success of the second piano concerto (1900–01) launched him internationally as a composer. In 1909 he had his debut in the USA as a pianist and in the same year composed the tone poem, *The Isle of the Dead*. The choral symphony *The Bells* (1913) followed. After the 1917 revolution, Rachmaninov and his wife (his cousin Natalya, whom he had married in 1902) left Russia for Switzerland. In 1936 he emigrated to the USA, where he died of cancer at his home in Beverly Hills.

Rachmaninov's works include three symphonies, four piano concertos, three operas, and the *Rhapsody on a Theme of Paganini* (1934), as well as much chamber and piano music and many songs.

Rackham, Arthur (1867–1939) *British book illustrator.*

The son of an admiralty marshal, Rackham left school at sixteen because of ill health; after a voyage to Australia, he attended an evening art course in London while working as a clerk. He took a job with *The Westminster Budget* in 1892 and began to illustrate books the following year.

By the end of the 1890s Rackham had developed his personal style. A sensitive linear method using black ink and a muted range of watercolours, it was well suited to the newly invented four-colour printing process that was revolutionizing the art of book illustration. Some of the best-known examples of his work, with its rich fantasy combined with naturalistic detail and a slightly gothic flavour, are his illustrations for *Peter Pan in Kensington Gardens* (1906), *Alice's Adventures in Wonderland* (1907), and *Hans Andersen's Fairy Tales* (1932).

Radhakrishnan, Sir Sarvepalli (1888–1975) *Indian philosopher and statesman. His many honours included a knighthood (1931) and an honorary OM (1963).*

Radhakrishnan was born in Tirutani, near Madras. Although his family was poor, he was a Brahmin, which – together with his natural gifts – ensured him an education. After graduating from Madras Christian College, Radhakrishnan taught philosophy at universities in Madras, Mysore, and Calcutta, during which time he began to write. His growing international reputation led to his appointment in 1936 as the first Spanding Professor of Eastern Religion and Ethics at Oxford University. In such works as *Indian Philosophy* (2 vols, 1923–27), *The Hindu View of Life* (1926), and *Eastern Religion and Thought* (1939), Radhakrishnan did much to present in an authoritative form the main ideas of classical Indian philosophy and religion to the West. Human history, he declared, was not just a series of secular happenings but a 'meaningful process'. He also argued that despite the obvious differences between western and eastern thought, such differences were negligible and there were in addition many striking agreements.

Radhakrishnan's later life was devoted to public affairs. In 1949 NEHRU appointed him ambassador to the Soviet Union. He returned to India in 1952 to become vice-president and, from 1962 to 1967, president of the republic.

Radzinowicz, Sir Leon (1906–   ) *Polish-born British criminologist and educator. He was knighted in 1970.*

Radzinowicz was born in Łódź and educated in Paris (at the Sorbonne), Geneva, and Rome, where he studied under the controversial criminologist Enrico Ferri. In 1932 he returned to Poland to teach the positivist criminology that Ferri favoured and became a professor at the Free University of Warsaw. His views made him unpopular, however, and after coming to England in 1938 to report on the working of the English penal system he settled in Cambridge on the outbreak of World War II.

In 1946 Radzinowicz was made director of the Department of Criminal Science at Cambridge and in 1948 became a fellow of Trinity College. In 1959 he took the major step of setting up the Institute of Criminology, which

played a significant role in the development of the discipline and lent the subject a new legitimacy in the eyes of the establishment. Radzinowicz became the first Wolfson Professor of Criminology in the same year and was recognized as a leading authority on penal affairs. In this capacity he became an influential and charismatic member of many top-level committees and commissions and took a leading part in the debates on capital punishment and the treatment of dangerous prisoners. He also contributed to the scholarship of the field through numerous lectures and a number of books, most notably in his fifty-two-volume series *English Studies in Criminal Science* (later retitled *Cambridge Studies in Criminology*), which he began in 1948 and continued to enlarge throughout his subsequent career.

Raeder, Erich (1876–1960) *German admiral who commanded his country's navy in the first half of World War II.*

Raeder served as chief of staff to Admiral Franz von Hipper (1863–1932) during World War I, taking part in the naval battles of Dogger Bank and Jutland. After the war he was promoted to rear-admiral (1922) and in 1928 became admiral and commander-in-chief of the navy. Raeder started to build up the small fleet, ordering the construction of pocket battleships and, after HITLER's accession to power in 1933, a new generation of submarines, both contrary to the terms of the Treaty of Versailles. By the outbreak of World War II, Raeder's efforts had restored the German navy as a significant force. Hitler promoted him to grand admiral in 1939.

In April 1940, Germany invaded Denmark and Norway, partly at Raeder's urging, to provide the German fleet with deep water fjords from which to operate. By the summer, Germany controlled the entire northern coastline of mainland Europe. Raeder argued for air supremacy before any invasion of Britain. Meanwhile his U-boats mounted a wide-ranging campaign against vessels supplying Britain in an attempt to starve the island into submission. When plans for the invasion were shelved in 1941–42, Raeder pressed for the major campaign to be focused on gaining control of the Mediterranean. However, Hitler, who consistently underrated the navy, insisted on invading the Soviet Union. Differences between them culminated in Raeder's resignation on 30 January 1943 and his replacement by the submarine commander DOENITZ.

After the war, Raeder was tried and sentenced to life imprisonment by the International Military Tribunal at Nuremberg but was released in 1955 because of ill health.

Rafsanjani, Ali Akbar Hashemi

(1934–) *Iranian statesman and religious leader, who became president of Iran (1989–) on the death of Ayatollah* KHOMEINI. Rafsanjani was born near Kerman, the son of a pistachio-nut-farming family of moderate wealth. From the age of fourteen he attended theological college at Qom, where he studied Islamic jurisprudence under Ruhollah Khomeini, becoming a mullah. Adopting Khomeini's teachings in both politics and religion, Rafsanjani became a committed enemy of the Shah's westernizing regime and an advocate of theocratic government. When Khomeini was exiled in 1963, Rafsanjani worked with the revolutionary underground; the following year he was gaoled for alleged complicity in the murder of the prime minister, Hassan Ali Mansour. Despite repeated imprisonment he managed to build up the family business and amassed a considerable personal fortune from land speculation.

Released from prison for the last time in 1978, Rafsanjani helped to orchestrate the mass demonstrations that led to the downfall of the Shah in January 1979. Following the triumphant return of Khomeini, he became prominent in the new Islamic government, acquiring a reputation for his ability and unswerving loyalty to the Ayatollah. As acting minister of the interior he was involved in the purging of dissidents in the early 1980s. Rafsanjani's own doubts about aspects of the regime seem to date from the mid-1980s, when he became privately convinced of the futility of the war with Iraq. Appointed acting commander-in-chief of the armed forces in 1988, he successfully persuaded Khomeini to accept UN peace terms. By this time he had also decided that Iran's blanket hostility to the West was unsustainable and in 1986 he became involved in the secret diplomacy with US negotiators that erupted as the Irangate scandal.

In the power struggle that followed Khomeini's death Rafsanjani outwitted more extreme figures and emerged at the head of his country's affairs. As president he has performed a careful balancing act, on the one hand working patiently to improve Iran's relations with the West and on the other appeasing hardline Muslim opinion on such issues as the RUSHDIE affair. He kept his country studiedly neutral during the Gulf War of 1991.

Raman, Sir Chandrasekhara Venkata

(1888–1970) *Indian physicist, who was* awarded the 1930 Nobel Prize for Physics for his discovery of the Raman effect. He was knighted in 1929.

Raman was educated at the Hindu College, Vishakapatam, where his father taught mathematics and physics, and at the Presidency College, Madras. Unable to find employment as a scientist, Raman worked for ten years as an auditor in the Indian Finance Department in Calcutta. However, he continued his private research, even managing to publish a substantial amount of work. In 1917 Raman was offered the newly endowed Palit Professorship of Physics at Calcutta University. He remained there until 1933, trying to attract other talented Indian physicists and to create, despite appalling financial difficulties, a centre for scientific research in India. Raman continued his work after 1933 at the Indian Institute of Science, Bangalore, where he directed the physics section. In 1947 he set up his own Raman Institute, also in Bangalore.

Raman's research interests were very wide; his best-known work, however, is concerned with the scattering of light. He found that when monochromatic light is passed through a transparent medium some of it is scattered. Examining the spectra of the scattered light revealed not only light of the initial wavelength but some weaker lines. Now known as Raman lines, they were explained by Raman in 1928 as the result of interactions between the passing photons and the vibrating molecules of the medium. Raman's insight has proved to be of considerable value in the study of molecular energy levels and also as an experimental proof of quantum theory.

Rambert, Dame Marie

(Cyvia Rambam; 1888–1982) *Polish-born British ballet producer, director, and teacher who played a leading part in the establishment of English ballet. She was made a DBE in 1962.*

Born in Warsaw, Rambert studied eurhythmic dance with Émile Jacques-Dalcroze (1865–1950) and came to the attention of Sergei DIAGHILEV in 1913. Joining Diaghilev's Ballets Russes as a teacher of rhythmic technique shortly afterwards, she became an assistant to Vaslav NIJINSKY and helped him choreograph *L'Après-midi d'un faune* and *Le Sacre du printemps*. After further study with Enrico Cecchetti (1850–1928), Rambert joined the corps de ballet of the Ballets Russes. She came to London to continue her ballet training, staged her first ballet in 1917, and became a British citizen in 1918 following her marriage to playwright and producer Ashley Dukes.

In 1920 Rambert opened a ballet school in London, based upon Cecchetti's teaching methods, and in 1926 she produced her pupil Frederick ASHTON's first ballet. Four years later she formed the first permanent English company, the Ballet Club, which later became the Ballet Rambert, and helped to found the influential Camargo Society with Ninette DE VALOIS. Over the next fifty years Rambert built up her reputation as the mother of English ballet and introduced many leading performers and designers, such as Sophie Fedorovitch, Hugh Stevenson, and William Chappell. Among the choreographers she encouraged besides Ashton were Antony Tudor, Andrée Howard, Frank Staff, Walter Gore, and Norman Morrice. Notable dancers of the Ballet Rambert were Pearl Argyle, Diana Gould, Elisabeth Schooling, Prudence Hyman, Maude Lloyd, Celia France, Sally Gilmour, Lucette Aldous, Harold Turner, Peggy van Praagh, Hugh Laing, and Leo Kersley.

The emphasis Rambert placed on British performers did much to revitalize the national ballet and in 1966 the company abandoned its classical repertory in order to concentrate on more innovative works, a trend that continued well after Rambert's retirement as a director. She was a co-author of *Dancers of Mercury: The Story of Ballet Rambert* (1960) and translator of *Ulanova: Her Childhood and Schooldays* (1962).

Ramsey, Sir Alf(red) (1922–   ) *British Association footballer who became manager of the national team that won the World Cup in 1966. He was knighted in 1967.*

Alf Ramsey began his playing career with Southampton and joined Tottenham Hotspur in 1949, for whom he won championship winners' medals in the second and first divisions (1950 and 1951). He was a stylish full back who played thirty-one times for England. In 1955 he was appointed manager of Ipswich Town and during the next eight years took them from the third to the first division championship. He was team manager of England from 1963 to 1974.

Ranjitsinhji Vibhaji, Kumar Shri, Maharajah Jam Sahib of Nawanagar *(1872–1933) Indian cricketer and statesman who played for England and captained Sussex (1899–1903).*

He was educated at Rajkumar College, India, and Trinity College, Cambridge, where he did not win his blue until his final year. In his first match for Sussex, only two summers later (1895), he scored a century at Lord's. In all, he scored 72 centuries and 24 692 runs. In 1900 he scored more than 3000 runs in English cricket including five double centuries, one of them reached in only three hours against Middlesex. His top score was an undefeated 285 at Taunton against Somerset in 1901.

'Ranji' was never free of ill health. Asthma restricted his sleep and in 1907 he almost died after an attack of typhoid. During World War I he fought as a colonel with the Indian troops in France. During a period of convalescence from ill health in England, he was accidentally hit in the face by some shot while shooting on the Yorkshire moors and lost an eye. A popular figure and a progressive ruler (he succeeded his cousin as maharajah in 1907), he served on the Indian Council of Princes and the League of Nations.

Rank, J(oseph) Arthur, 1st Baron (1888–1972) *British industrialist who, as founder of the Rank Organization, played a major part in establishing the British film industry. He was created Baron Rank of Sutton Scotney in 1957.*

J. Arthur Rank was the youngest son of Joseph Rank, who had built up a thriving flour-milling company and was also a leading Methodist. Arthur was apprenticed to his father's business after leaving school. He served in France during World War I and in the early 1930s became interested in producing instructional films for Methodist Sunday schools. The Religious Film Society was founded in 1933 to promote the use of film for Christian purposes. This led in 1934 to the formation of the British National Film Company. Its first film, released in 1935, was *Turn of The Tide*, about life in a Yorkshire fishing village. In the same year, Rank, in partnership with Charles Woolf, established General Film Distributors to control distribution of his films. As chairman of the Rank Organization (1941–62), Rank invested heavily in the British film industry, acquiring several film studios, including Denham and Pinewood, controlling the Gaumont and Odeon cinema chains, and also making many films, such as the memorable screen adaptations of Shakespeare starring Laurence OLIVIER. The Rank Organization later diversified its business into hotels, leisure, motorway services, and Xerox copying machines. Its founder also became chairman of Rank Hovis McDougall (1952–69), formed by amalgamation of the family's milling and animal feeds concern with the other companies.

Rasmussen, Knud Johan Victor (1879–1933) *Danish explorer and ethnologist noted for his studies of Eskimo culture.*

Rasmussen, born in Greenland of part-Eskimo ancestry, was brought up to be fluent in the Eskimo language and familiar with their culture. After education in Denmark he returned to Greenland and was a member of a Danish expedition to the Thule district of Greenland (1902–04). In 1910 he established a trading post in Thule to provide Eskimos with goods and equipment in exchange for skins, carvings, and other items. Thule became the base for a series of expeditions. In 1912, Rasmussen led a team over the ice cap to Independence Fjord; in a later expedition in northern Greenland (1916–18), two of his companions perished. A visit to Angmagssalik on the eastern coast in 1919 gave Rasmussen the opportunity to study Eskimo folklore and formed the basis of his book, *Myter og sagn fra Grønland* ('Myths and Legends from Greenland'; 3 vols, 1921–25). In 1921 he started a three-year study of Eskimo culture throughout the American Arctic, during which time he and two Eskimo companions travelled the breadth of the continent to Alaska, an expedition he described in *Fra Grønland til Stillehavet* (2 vols, 1925–26; translated as *Across Arctic America*, 1927).

Rasputin, Grigori Yefimovich (1871–1916) *Russian monk and charlatan, who acquired such a pernicious influence over the household of Tsar NICHOLAS II that he was assassinated by loyal royalists.*

Born in Pokrovskoye, Tobolsk, into a peasant family, Rasputin gained a reputation as a libertine in his youth, earning the name Rasputin ('the dissolute') from the local villagers. After becoming a farmer at the age of twenty, he joined a religious sect, the Khlysty (flagellants), with whom he went on a pilgrimage to Mount Athos in Greece. He reappeared two years later as a Siberian mystic and 'holy monk', alleging that he possessed unusual abilities to heal the sick.

In 1903 Rasputin visited St Petersburg, where he charmed his way into the houses of the aristocracy. When he was presented at court in 1907, he exercised his hypnotic powers over Alexis, the haemophiliac crown prince, gaining the confidence of the imperial family, particularly Empress ALEXANDRA, who believed he could cure her son. Through his manipulation of the imperial family Rasputin gained great national influence, arranging the appointment and dismissal of both political and clerical leaders. He was particularly influential at the beginning of World War I, when the tsar was away at the front and the empress was left in control of domestic affairs. His corrupt and incompetent sycophants held such important posts that he was widely believed to be receiving payment from the Germans. At the same time his notoriously profligate behaviour outside the imperial circle caused great scandal and further discredited the imperial family. He was assassinated in 1916 by a group of aristocratic monarchists, which included Prince Felix Yusupov, a nephew by marriage, of the tsar. Their attempt to poison the 'mad monk', as he was often called, failed and they finally shot him and disposed of his body in the River Neva.

Rattigan, Sir Terence Mervyn (1911–77) *British dramatist. He was knighted in 1971.*

Rattigan was educated at Harrow School and Trinity College, Oxford. His earliest play, *First Episode*, was produced in 1934, but he made his mark as a playwright with the farce *French Without Tears* (1936), which, like several of his other plays, was produced in both London and New York and also filmed. After writing a number of successful comedies he turned to more serious themes, as in *The Winslow Boy* (1946), in which the characters' belief in the suspect's guilt or innocence and the operations of justice create acute dramatic tension. *The Browning Version* (1951), *The Deep Blue Sea* (1952), and *Separate Tables* (1954) are even more sombre examinations of the fate of passion in a cold convention-ridden society. The plays *Ross* (1960) and *A Bequest to the Nation* (1970) are constructed around T. E. LAWRENCE and Lord Nelson, respectively. In addition to his plays, Rattigan also wrote a number of screenplays, including *The Yellow Rolls Royce* (1965) and *Goodbye Mr Chips* (a 1969 remake of the James HILTON novel originally filmed in 1939).

Rattle, Simon (Denis) (1955–) *British conductor; principal conductor of the City of Birmingham Symphony Orchestra (1980–).*
Born in Liverpool, Rattle was eighteen when he won the Bournemouth John Player Conducting Competition in 1973. He subsequently studied at the Royal Academy of Music, conducting the Bournemouth Symphony Orchestra and Sinfonietta from 1974. Other appointments have included associate conductor of the Royal Liverpool Philharmonic Orchestra (1977–80) and principal conductor of the London Choral Society (1979–84). By the time Rattle began his association with the City of Birmingham Symphony Orchestra in 1980

he had already gained considerable respect for his lucid readings of difficult early twentieth-century works, such as those of Mahler, and was well known to a larger public for his youth and flamboyant style. Despite lucrative offers from abroad he has remained in Birmingham, where he is regarded as one of the city's greatest cultural assets.

Rauschenberg, Robert (1925–) *US painter and experimental artist, who was a pioneer of pop art.*
Rauschenberg was born in Port Arthur, Texas. He studied pharmacy at Texas University and served in World War II before studying painting in Kansas City, Paris, and (1948–49) at Black Mountain College under Joseph ALBERS. After a further period of study at the Art Students' League in New York, in 1951 he exhibited a series of paintings depicting a few black numbers or symbols on an all-white background. These impersonal pictures represented a reaction against abstract expressionism.

During the 1950s and 1960s Rauschenberg produced the 'combine paintings' for which he is best known. These pictures, such as *Charlene* (1954), *Bed* (1955), and *Rebus* (1955), combined painting with collage and assemblage of objects, including nails, rags, bottles, etc. They displayed what Rauschenberg described as his desire to 'act in the gap between art and life'. During the 1960s he experimented with lithographs and silk-screen printing as well as three-dimensional constructions. He won first prize at the Venice Biennale in 1964 and in 1966 founded EAT (Experiments in Art and Technology) – a collaboration of artists, scientists, and technicians to investigate the utilization of new technology in works of art. It resulted in works of art such as *Soundings* (1968) – an environmental piece whose workings were stimulated by the sound of spectators' voices.

In the 1970s Rauschenberg explored the use of fragile materials, such as fabric and cardboard, and continued his involvement with dance companies, which had begun in the 1950s. He has subsequently worked extensively with photographic techniques.

Ravel, (Joseph) Maurice (1875–1937) *French composer, who was influenced by DEBUSSY and whose richly orchestrated music and chamber music form part of the modern repertoire.*
Ravel studied at the Paris Conservatoire (1889–1905), where he joined FAURÉ's composition class. His lack of success in the Prix de

Rome, when he was already a composer of some renown, caused something of a scandal and the resignation of the director of the Conservatoire. This battle with the establishment led Ravel to refuse the Légion d'honneur in later life. During World War I Ravel joined the motor transport corps but was invalided out just before the death of his mother in 1916. This was a great shock to him, and it seems that after his mother's death he was never able to establish any other emotional relationship.

He was, however, a prolific composer. Among his most important works were *Ma mère d'oye* (1908; *Mother Goose*), originally a suite for piano duet, later orchestrated (1912); *Daphnis et Chloë* (1909–12) and *Bolero* (1928), both ballets; and *Gaspard de la nuit* (1908), an impressionistic piano suite. Ravel also had a strong feeling for the exotic and the antique, which was expressed in such works as the orchestral *Rapsodie espagnole* (1907–08; *Spanish Rhapsody*) as well as the piano music *Sonatine* (1903–05) and *Le Tombeau de Couperin* (1914–17). He wrote two fantasy operas, *L'Heure espagnole* (1907–09; *The Spanish Hour*) and *L'Enfant et les sortilèges* (1920–25; *The Child and the Spells*), as well as a considerable amount of other music for the human voice, including the *Histoires naturelles* (1906) for voice and piano, *Shéhérazade* (1903) for voice and orchestra, and *Chansons madécasses* (1925–26) for voice and chamber ensemble.

Rawsthorne, Alan (1905–71) *British composer, who is known both for his film music and his 'serious' compositions.*
Rawsthorne studied at the Royal Manchester College of Music (1926–30) and later with Egon Petri (1930–31) in Berlin. After a period as director (1932–35) of the School of Dance and Mime associated with Dartington Hall, Rawsthorne worked as a composer, producing three symphonies, four concertos, and much chamber and instrumental music, as well as scores for twenty-two films and four plays.

Basically a tonal composer, Rawsthorne interpolated his own brand of serialism into his later works, as in the quintet for piano and wind (1962–63). Recognition came in 1938 with the performance in London of the *Theme and Variations* (1937) for two violins. The *Symphonic Studies* (1938), the first piano concerto (1939), and the more popular second piano concerto (1951) confirmed Rawsthorne's status as a composer of great originality.

Ray, Man (1890–1976) *US photographer, painter, draughtsman, and sculptor. His reputation is principally based on his photographic interpretations of the surreal.*

After studying painting in New York, Ray and his wife moved to New Jersey in 1913 with the intention of creating an artists' community. Particular influences in European art of the time were synthetic cubism and surrealism. 1915 saw the beginning both of his friendship with DUCHAMP and of his experimentation in photographic images. He helped to found the New York dada movement and two magazines followed: *The Blind Man* and *Rongwrong* (both 1917). A year later came the spraygun paintings intended to imitate photographic effects, a technique also used in the fifties and sixties by pop artists.

In 1921 Man Ray went to Paris, where he became involved with European dadaists and the surrealist movement. He pioneered the photogram (a photographic impression made without a camera on a sensitized plate), which he named the 'Rayograph'. His work was shown in the first international dada exhibition in 1922 and a collection of his photographs was published. He lived in Paris until 1940 and during this period he also made surrealist films, including *Le Retour à la raison* (1923) and *L'Étoile de mer* (1928). After leaving Paris he eventually arrived in Hollywood. Throughout the 1940s his work was widely exhibited and he continued to work right up to his death.

Ray, Satyajit (1921–) *Indian film director, composer, and writer. He was appointed a Chevalier de la Légion d' honneur in 1989.*

Ray was born in Calcutta and attended the university there, before studying art and poetry at Rabindranath TAGORE's university. He then became a commercial artist (1943–56), making his first film, *Pather Panchali* (1955), on a part-time basis and founding Calcutta's first film society in 1947. The success of *Pather Panchali* brought Indian films to the attention of western audiences and also established Ray's own reputation. Based on a novel and filmed in neorealist style, the film was self-financed. Although Ray had no real experience of film-making, he had been able to learn something from Jean RENOIR, who was in India for the filming of *The River* (1951). Nevertheless *Panchali* won a prize at the Cannes Festival and its success ensured the completion of the two remaining parts of the trilogy: *Aparajito* (1956), which won the Venice Golden Lion, and *Apur Sansar* (1959; *The World of Apu*). He also completed *Paras Pathar* (1957) and *Jalsaghar* (1958; *The Music Room*). Notable films that followed included *Kanchenjunga* (1962), for which he wrote the music, *Charulata* (1964; *The Lonely Wife*), *Ashanti Sanket* (1973; *Distant Thunder*), which won the Berlin Golden Bear Award, *The Chess Players* (1977), *The Home and The World* (1984), *Ganashatru* (1989; *Public Enemy*), and *Branches of the Tree* (1990).

Ray is also the editor of *Sandesh* children's magazine and the author of *Our Films, Their Films* (1976) and *Stories* (1987).

Reagan, Ronald W(ilson) (1911–) *US actor and Republican statesman who became the fortieth president of the USA (1981–89).*

Reagan was born in Tampico, Illinois, the son of a shoe salesman. After his graduation from Eureka College, Illinois, in 1932 he became a radio sports announcer in Des Moines, Iowa, until 1937, when he signed a contract with Warner Brothers and went to Hollywood. During the next twenty-five years he appeared in some fifty films, including *King's Row* (1941) and *The Voice of the Turtle* (1948), served two terms as president of the Screen Actors Guild (1947–52; 1959–60), and hosted two television series, *General Electric Theatre* (1954–62) and *Death Valley Days* (1962–65).

Having joined the Republican Party in 1962, Reagan made his mark on the political world with his contribution to the campaign of Republican presidential nominee Barry Goldwater (1909–) in 1964. Three years later he won the governorship of California from Edward G. Brown and was re-elected in 1970 to serve a second term. Although popular with the Californian electorate, he failed to gain the Republican presidential nomination in 1968 and 1976. Finally successful in 1980, he went on to defeat Jimmy CARTER in the presidential election and became, at the age of sixty-nine, the oldest-ever president of the USA. On 30 March 1981, less than two months after his inauguration, he survived an assassination attempt.

Reagan had fought his campaign on promises of economic reform: on taking office he immediately put into action deflationary policies aimed at reducing taxes and government spending. In his foreign policy he expressed the need to take a firm stand against the Soviet Union, increasing the defence budget to maintain the position of the USA in the nuclear arms race and to fund the Strategic Defence Initiative, which envisaged a new network of military satellites in space; as a result, the federal deficit continued to grow. Relations with the Soviet Union deteriorated sharply, culminating in the Korean airliner disaster of 1983, in which a Korean passenger aircraft strayed off course and was shot down by Soviet missiles. Later that year, he authorised US intervention in Grenada and approved CIA operations against Nicaragua, provoking opposition from Congress. Despite these problems, Reagan retained his popularity with the American people and entered the 1984 presidential campaign with the wholehearted support of his party. He was re-elected president with the biggest political landslide in US history, winning every state of the USA except Minnesota, the home state of his defeated Democratic opponent, Walter F. Mondale (1928–).

His second term of office witnessed a dramatic improvement in East–West relations with the signing (with Mikhail GORBACHOV) of the Intermediate Nuclear Forces (INF) treaty in 1987 and significant progress in other arms control talks. He also undertook reform of the US tax system and late in his presidency weathered controversy over the sale of US arms to Iran, some of the profits of which were, it was reported, secretly diverted to fund the Contras, opponents of the Sandinista government of Nicaragua. He failed, however, to make much impact on the still-worsening federal deficit, which became a problem he bequeathed to his successor, George BUSH, who had been his vice-president.

Redford, Robert (1936–) *US film actor and director.*

Redford was born in Santa Monica, California; he turned to acting after attending the University of Colorado on a baseball scholarship. He also attempted to become a painter after a trip to Europe and studied art at the Pratt Institute in Brooklyn. Having settled for the stage, he took up drama studies at the American Academy of Dramatic Arts. *Tall Story* (1959) provided him with his Broadway debut; by 1963 he was starring in the Broadway production of *Barefoot in the Park*. He also starred in the film version of this play in 1967.

He appeared in several TV series and dramas in the early 1960s and made his film debut in *War Hunt* (1962). Early films included *The Chase* and *Inside Daisy Clover* (both 1966) but it was *Butch Cassidy and the Sundance Kid* (1969), playing opposite Paul NEWMAN, that first brought him international recognition. He and Newman were successfully to co-star again in *The Sting* (1973), a widely acclaimed

film that revived the music of Scott JOPLIN and brought Redford an Oscar nomination. *Downhill Racer* (1969), *The Candidate* (1972), *The Great Gatsby* (1974), *The Natural* (1984), and *Out of Africa* (1986) are among his other successful films. Redford's own production company, Wildwood, was responsible for *All The President's Men* (1976), about the Watergate investigation, in which he played the reporter Bob Woodward. As the director, he won an Academy Award for *Ordinary People* (1980).

Redgrave, Sir Michael Scudamore (1908–85) *British actor and director. He was knighted in 1959.*

Although he eventually followed his parents onto the stage – his father was G. E. ('Roy') Redgrave and his mother Margaret Scudamore – Redgrave began his career as a modern language master at Cranleigh School, after graduating from Magdalene College, Cambridge. Subsequently he joined the Liverpool Repertory Theatre (1934–36), the Old Vic (1936–37; 1949–50), and the National Theatre, appearing in numerous Shakespearean roles and other plays. Two of his most memorable performances were as Antony in *Antony and Cleopatra* (1953) and the title role in *Uncle Vanya* (1963).

Equally at home on the screen, Redgrave made his film debut in *The Lady Vanishes* (1938), followed by *The Stars Look Down* (1939) and a memorable performance in *Kipps* (1941). He has played a variety of roles, including the disturbed ventriloquist in *Dead of Night* (1945), Ernest Worthing in the film version of *The Importance of Being Earnest* (1952) with Edith EVANS, Barnes WALLIS in *The Dambusters* (1954), and the alcoholic father in *Time Without Pity* (1957). Other films include *Mourning Becomes Electra* (1947), for which he was nominated for an Oscar, *The Browning Version* (1951), and *The Quiet American* (1958). After the peak of his career in the 1940s and 1950s he appeared in such films as *The Loneliness of the Long Distance Runner* (1962), the remake of *Goodbye Mr Chips* (1969), and *Nicholas and Alexandra* (1971).

His publications include *The Aspern Papers* (play, 1959), *The Mountebank's Tale* (novel, 1959), *Actor's Ways and Means* (1953), *Mask or Face* (1958), and *In My Mind's Eye* (autobiography, 1983). He married the actress Rachel Kempson (1910–) in 1935; their children, Lynn (1943–), Corin (1939–), and Vanessa REDGRAVE, are also actors.

Redgrave, Vanessa (1937–) *British actress, who is known for both her films and her political activity.*

The daughter of Sir Michael REDGRAVE, whose own parents were actors, Vanessa Redgrave trained at the Central School of Speech and Drama and joined the Frinton Summer Repertory in 1957. Since then she has appeared in a variety of roles and has been a member of the Royal Shakespeare Company.

Her film debut came in *Behind the Mask* (1958), playing opposite her father. Among the many memorable films that followed were *Morgan – A Suitable Case for Treatment* (1966), which earned her the Cannes Festival Award as Best Actress and an Oscar nomination, *Isadora* (1968) and *Mary Queen of Scots* (1972), both of which brought further Academy Award nominations, and *Julia* (1976), for which she received both an Academy Award and a Golden Globe Award. She is also remembered for her roles as Queen Guinevere in *Camelot* and Anne Boleyn in *A Man for All Seasons* (both 1967). Other films include *Agatha* (1978), the television film *Playing for Time* (1980), in which she gave a commanding performance as the Auschwitz-incarcerated French entertainer, Fania Fenelon, *The Bostonians* (1984), *Comrades* (1987), and *Young Catherine* (1990). In the film *Wetherby* (1985) she appeared with Joely Richardson, her daughter by her marriage (1962–67) to director Tony Richardson (1928–91).

Politically active, Redgrave has stood unsuccessfully as the parliamentary candidate for the Workers' Revolutionary Party. She has also produced a children's anthology, *Pussies and Tigers* (1963).

Reed, Sir Carol (1906–76) *British film director who made some exceptional films in the postwar years. He was knighted in 1952.*

Reed, who was born in London and educated at King's School, Canterbury, began his career as an actor in 1924. In 1927 he joined Edgar WALLACE as stage director but in the early 1930s joined Ealing Studios under Basil Dean (1888–1978). His first film as director was *Midshipman Easy* (1934). *Talk of the Devil* (1937), *Bank Holiday* (1938), the highly successful comedy *A Girl Must Live* (1939), *The Stars Look Down* (1939), *Night Train to Munich* (1940), *Kipps* (1941), and *The Young Mr Pitt* (1942) were some of his early films. During World War II he worked with the Army Film Unit, where his success led to *The Way Ahead* (1944). Next came the Oscar-winning *The True Glory* (1945), co-directed with Garson Kanin (1912–).

Some of Reed's most highly acclaimed films were made during the 1940s. *Odd Man Out* (1947), with James Mason (1909–84) playing a dying gunman, was followed by *The Fallen Idol* (1948) with Ralph RICHARDSON. *The Third Man* (1949) has been called one of the best films ever made, largely due to Reed's skill in bringing together an outstanding cast (Joseph Cotten (1905–), Trevor HOWARD, Alida Valli (1921–), and Orson WELLES), a crisp screenplay (by Graham GREENE), and the original sound of the zither (played by Anton Karas), all photographed and directed with exceptional sensitivity against the haunting background of postwar Vienna. Notable, too, at about this time were *Outcast of the Islands* (1952) and *The Man Between* (1953). Later films included *A Kid for Two Farthings* (1955), *The Running Man* (1963), and the musical *Oliver!* (1968), for which he won an Oscar as best director.

Reid, Sir George Houston (1845–1918) *Australian statesman and prime minister (1904–05). He was awarded the KCMG in 1909 and the GCMG in 1916.*

Born near Paisley, Scotland, the son of a Presbyterian minister, Reid was educated in Melbourne, where he studied law. He was admitted to practice in 1879 and became a QC in 1898. Reid entered politics when he was elected to the New South Wales legislative assembly in 1880. He became premier of New South Wales in 1894 and remained in office until 1899. Entering the first federal parliament in 1901 he became leader of the opposition and in 1904 combined with the Labor Party to defeat the DEAKIN protectionist government. Later the same year, he supported a group of protectionists to defeat Labor, forming a coalition government with himself as prime minister. Defeated in 1905 by Deakin, who was supported by the Labor Party, Reid retired from Australian politics in 1908.

Considered an incisive public speaker and an astute politician, Reid was the first Australian high commissioner in London (1910–16) and on completion of his term, won a seat in the British House of Commons (1916–18). His wife, Dame Flora Reid, was awarded the Grand Cross of the British Empire for her work with Australian soldiers during World War I.

Reinhardt, Django (Jean Baptiste Reinhardt; 1910–53) *Belgian gypsy jazz guitarist.*

Born in Liberchies in Belgium, a French-speaking gypsy, Reinhardt spent his early life in a caravan near Paris. As a teenager he played the banjo, guitar, and violin. A caravan fire in 1928 badly damaged his left hand, but he managed to develop his own unorthodox fingering and a superbly fluent guitar style, without learning to read either words or music. In 1934 he formed the Quintette du Hot Club de Paris with the jazz violinist Stephane GRAPPELLI. The band was unique among jazz groups for being a string quintet, with two rhythm guitars and a string bass. It became world-famous, making many recordings in both France and Britain. Reinhardt was also a composer; his 'Nuages' was an international hit and was recorded many times. He also played and recorded with visiting American jazz musicians, such as Coleman HAWKINS. After World War II, during which he worked in occupied France, he was reunited with Grappelli for one recording session. However later that year he switched permanently to the electric guitar and went to the USA to tour with Duke ELLINGTON. In 1953 he co-starred with Dizzy GILLESPIE, and was scheduled to tour with Jazz at the Philharmonic, when he died.

Reinhardt was an extravagant, romantic, and illiterate man – a gambler and notoriously unreliable. All he cared about was music, and his ideas were so original that his recordings still bring pleasure to many people.

Reith, John Charles Walsham, 1st Baron (1889–1971) *British administrator, who was first general manager of the British Broadcasting Company (1922–26) and first director-general of the new public corporation, the BBC (1927–38). He was knighted in 1927 and created a baron in 1940.*

Born in Stonehaven, Scotland, a son of the manse, Reith had already served with distinction as a soldier in World War I and managed a large engineering company in Glasgow before he joined the British Broadcasting Company. His influence on its operations, its growth, and its philosophy was immense and was internationally recognized. Refusing to treat broadcasting simply as a means of entertainment, Reith saw it in the right hands as 'contributing consistently and cumulatively to the intellectual and moral happiness of the community'.

He always had his critics outside the BBC yet no one doubted his power, his dedication, and his sense of mission. He was never so successful again after 1937, when he became chairman of Imperial Airways, yet he held ministerial posts during World War II, being appointed minister of information in 1940 and

subsequently minister of transport and minister of works. He lost his post in February 1942 and never forgave CHURCHILL thereafter. Continuing hopes of high office were not realized after the war, although from 1950 to 1959 he was chairman of the Colonial Development Corporation.

Reith is now remembered as much for his writings, including his diaries (published in 1975), as for his activities in the BBC. His first volume of autobiography (1949) he called with apt irony *Into the Wind*; the second volume, *Wearing Spurs*, was published in 1966. In 1947 the BBC established the highly prestigious Reith Lectures, broadcast annually, in his honour.

Remarque, Erich Maria (Erich Paul Remark; 1898–1970) *German novelist.*

Remarque, the son of a bookbinder, was born and educated in Osnabrück. After World War I, in which he was twice wounded, he worked at various jobs before becoming a journalist for a sports magazine and writing in his spare time. His first novel, *Im Westen nichts Neues* (1929; translated as *All Quiet on the Western Front*, 1929), a brilliantly realistic and unpartisan tale of a common soldier's experiences in the war, was a huge and immediate international success, selling millions of copies in twenty-five languages. After publishing a sequel entitled *Der Weg zurück* (1931; translated as *The Road Back*, 1931), Remarque was increasingly attacked by the emerging Fascist Party. His first novel was burned in the bonfire of books in 1933 and his works were banned.

He emigrated to America in 1939, becoming a US citizen in 1947 (he refused an offer to restore his German citizenship after World War II). His later books are well-constructed and realistically observed stories of action in times of crisis, often set in wartime, but none achieved the wide audiences of his first novel. *Drei Kameraden* (1938) was first published in its English translation, *Three Comrades* (1937). *Flotsam* (1941) was followed by his next most successful novel, *Arc de Triomphe* (1946), set in wartime Paris. His other works include *Der Funke Leben* (1952; translated as *Spark of Life*), *Zeit zu leben, Zeit zu sterben* (1954; translated as *A Time to Live and a Time to Die*), *Der schwarze Obelisk* (1956), *Schatten in Paradies* (1971; translated as *Shadows in Paradise*, 1972), and a play, *Die letzte Station* (1974; translated as *Full Circle*).

Renoir, Jean (1894–1979) *French-born film director, who went to the USA during World War II but subsequently returned to Europe.*

Born in Paris, the son of impressionist painter Pierre Auguste Renoir (1841–1919), he was brought up in France. After World War I he worked for a time as a potter but by 1924 he had begun his career in films. His early silent films starred his wife, Catherine Hessling, who had been his father's model. Notable among them was Émile Zola's *Nana* (1926). Renoir also appeared with his wife in a few Cavalcanti films, including *Le P'tite Lille* (1929) and *Le Petit Chaperon rouge* (1930; *Little Red Riding Hood*), shortly after which they parted.

With the coming of sound, Renoir enhanced his reputation as a director with such films as *Madame Bovary* (1934), *Toni* (1935), and *Les Bas-Fonds* (1936). He achieved international acclaim with *La Grande Illusion* (1937), and *La Bête humaine* (1938). *La Règle du jeu* (1939), which failed when it first appeared, was subsequently recognized as a masterpiece. In the USA Renoir directed such films as *The Southerner* (1945) and *The Woman on the Beach* (1947). *The River* (1951), his first film in colour, was shot in India with his nephew, Claude Renoir (1914–), as cameraman. Returning to Europe he had a profound effect on the New Wave film-makers of the 1960s. In this period he made several films, including *Le Déjeuner sur l'herbe* (1959), *Le Caporal épinglé* (1962), and his last, a TV film, *Le Petit Théâtre de Jean Renoir* (1971).

Renoir also wrote and directed plays and was the author of a novel, a volume of autobiography, *My Life and My Films* (1974), and *Renoir, My Father* (1962).

Resnais, Alain (1922–) *French film director, who became one of the New Wave of filmmakers.*

Born in Vannes, Resnais studied acting and attended the Institut des Hautes Études Cinématographiques. He first made several short documentaries, including *Van Gogh* (1948), *Guernica* (1950), in collaboration with Robert Hessens, and *Nuit et brouillard* (1955), a study of concentration camps. He then embarked on his first feature, the moving *Hiroshima mon amour* (1959), about a romance between a French woman and a Japanese man. This was followed by *L'Année dernière à Marienbad* (1961), which won the Venice Grand Prix. Throughout, Resnais has collaborated closely with such writers as Marguerite DURAS (*Hiroshima mon amour* and *Providence*, 1977), Alain ROBBE-GRILLET (*Marienbad*), Jorge Semprun (*La Guerre est finie*, 1966), and Jacques Sternberg (*Je t'aime, je t'aime*, 1968, and *Stavisky*, 1974). He is much

admired for the distinctive techniques he has developed to explore the thematic preoccupations of his films – memory and time. Resnais has won several awards, including a special prize awarded for *Mon oncle d'Amérique* (1980) at the Cannes Film Festival. Recent films include *La Vie est un roman* (1983), the highly acclaimed *L'Amour à mort* (1985), and *Melo* (1987).

Respighi, Ottorino (1879–1936) *Italian composer, who is remembered for his brilliantly orchestrated tone poems.*

Respighi began his musical career as a string player, studying the violin and viola at the Liceo Musicale in Bologna (1891–99). In 1900 he went to Russia as first viola in the St Petersburg Opera Orchestra; there he met Rimsky-Korsakov, from whom he took composition lessons. Although he continued to earn his living as a string player, from then on he also worked as a composer, especially after further study with Max BRUCH in Berlin (1902).

Respighi made his name in Italy as a composer of operas (of which he composed nine between 1904 and 1935), but his reputation rests on the series of impressionistic tone poems that owe much to Rimsky-Korsakov in the brilliance of their orchestration. These include *The Fountains of Rome* (1917), *The Pines of Rome* (1924), *Church Windows* (1927), *Three Botticelli Pictures* (1927), and *Roman Festivals* (1929). In addition to these colourful pieces, the three sets of *Ancient Airs and Dances* (1917; 1924; 1932), which are orchestral transcriptions of lute pieces, reflect Respighi's strong affinity with Italian music of the seventeenth and eighteenth centuries. His wife, a former pupil, later arranged them as a ballet (1937). Respighi was also commissioned by DIAGHILEV to arrange and orchestrate the ballet *La Boutique fantasque* (1919) from Rossini's original music; it ends with the famous 'Can-Can'.

Rhee, Syngman (1875–1965) *First president of the Republic of Korea (South Korea) (1948–60). Devoted to his country's independence, Rhee was initially a popular leader who introduced land reform and a literacy programme, but was forced from office after uprisings against his repressive style of government.*

Born in Whanghai province, a descendent of the Ri dynasty, which governed Korea (1456–1910), Rhee received a classical Confucian education before attending a Methodist missionary school in Seoul. In 1896 he joined the Independence Club, which advocated political reform and Korean nationalism. Imprisoned for his political activities in 1898, he was released in 1904 when a general amnesty was announced. A converted Christian, Rhee translated several English books into Korean while in prison and wrote *Spirit of Independence* (1904). He spent the next six years in the USA, where he gained degrees from George Washington, Harvard, and Princeton universities before returning to Korea in 1910 as a teacher for the Methodist Mission Board. Korea was annexed by the Japanese later that same year and Rhee eventually went back to the USA (1912). He spent the next thirty years working for the Korean Methodist Church and directing the activities of the Korean independence movement. He also served as the elected president of the Korean provisional government in exile for twenty years (1919–39).

Rhee returned to Korea in 1945 following the defeat of Japan at the end of World War II, and the country was divided into communist North Korea and noncommunist South Korea. When the Republic of Korea was proclaimed in the south in 1948, Rhee was elected president with the sanction of the USA. Having outlawed the opposition, he attempted to establish control over the entire country, but at the end of the indecisive Korean War (1950–53) the country remained divided. In his later years, an authoritarian streak, revealed in electoral malpractice, corruption, and police repression, lost him considerable support and provoked the demonstration that led to his downfall in 1960.

Rhine, Joseph Banks (1895–1980) *US parapsychologist noted for his investigations into extrasensory perception (ESP) and psychokinesis, which he claimed proved the existence of such phenomena.*

Rhine studied at the University of Chicago, receiving a BS degree in 1922, a MS degree the following year, and a DPhil in 1925. He worked as a botany teacher at the University of West Virginia (1924–26), during which time he attended a lecture on spiritualism given by Sir Arthur Conan DOYLE. Much impressed, Rhine became research assistant to the psychologist William McDougall (1871–1938) at Duke University. In 1930 they established a parapsychology laboratory and in 1934 Rhine published *Extra-Sensory Perception*, an account of his experiments in telepathy. Using special ESP cards each bearing one of a set of symbols, Rhine claimed statistical proof that subjects could identify unseen the card in the dealer's hand by thought transference. Critics

held that Rhine was predisposed to accept the existence of such phenomena, thereby increasing the likelihood of success. This was followed by *New Frontiers of the Mind* (1937), a best-selling popular work on the subject.

Rhine became professor of psychology in 1937 and director of the parapsychology laboratory (1940–65) where, in a further series of experiments, he claimed evidence to support the existence of psychokinesis – the influencing of physical objects by thought processes. He also wrote *New World of the Mind* (1953) and edited *Progress in Parapsychology* (1970).

Rhodes, Wilfred (1877–1973) *Yorkshire and England cricketer who took 4187 first-class wickets, more than any other player. He was an all-rounder who played fifty-eight times for his country, making his final appearance at the age of forty-eight.*

Born at Kirkheaton, near Huddersfield, he first played for Yorkshire in 1898; his last match was in 1930. In that time he scored almost 40 000 runs, including 58 centuries and one score of 267 not out. But it was as a slow left-arm bowler that he is best remembered. In only his second match for Yorkshire he took 13 wickets against Somerset. His test debut was in 1899 and his last match for England came in 1926, when he was recalled to play against the Australians. He took 7 for 17 against the Australians at Birmingham in 1902 and two years later, at Melbourne, his match analysis was 15 wickets for 124 runs.

A quiet but popular figure, he became a coach at Harrow School after he give up playing first-class cricket.

Rhys, Jean (Ellen Gwendolen Rees Williams; 1894–1979) *British novelist and short-story writer.*

Jean Rhys was born in Dominica to a Welsh doctor and his Creole wife. She was educated at a convent school until she was sixteen. She then went to England to study acting and lived the kind of bohemian life that features in her short stories. The love affair that forms the starting point for *After Leaving Mr Mackenzie* (1931) draws on her own experience at this time.

In 1919 she married her first husband, Jean Lenglet, and moved to Paris with him. She showed some of her short stories to Ford Madox FORD, who encouraged her to publish, and *The Left Bank and Other Stories* appeared in 1927. Her novels *Voyage in the Dark* (1934) and *Good Morning, Midnight* (1939) made a considerable impact, as did some of her journalism in Paris, on account of the freedom with which she handled controversial feminist topics. In 1932 she and Lenglet were divorced, and she subsequently married again, first Leslie Tidden Smith, who died in 1945, and then Max Hamar.

At the beginning of World War II Jean Rhys stopped writing and it was not until the 1950s that she began working on another novel. This was her *Wide Sargasso Sea* (1966), which won immense critical acclaim. Drawing on her childhood memories of the West Indies, it tells the story of Antoinette Cosway, Mr Rochester's mad wife in Charlotte Brontë's *Jane Eyre*. She later published two further collections of stories and left an unfinished autobiography, *Smile Please*, published posthumously in 1979.

Ribbentrop, Joachim von (1893–1946) *German Nazi diplomat and foreign minister (1938–45), who ended up in the dock at Nuremberg and was hanged for his war crimes.*

Born in Wesel, the son of an army officer, Ribbentrop was educated at schools in Kassel and Metz before going to Canada in 1910 as a sales representative. He returned to Germany at the outbreak of World War I and served in the cavalry (1914–18). After the war he was, for several years, a wine merchant.

Ribbentrop joined the National Socialist (Nazi) Party in 1932. Becoming HITLER's adviser on foreign affairs when the Nazis came to power (1933), he was appointed successively Reich commissar for disarmament at Geneva (1934), ambassador at large (1935), and ambassador to Great Britain (1936–38). In these positions he negotiated the Anglo-German naval agreement, which provided for the rearmament of the German navy (1935) and the German-Japanese anti-Comintern pact (1936). In 1938 he was appointed minister of foreign affairs, a position he held until the fall of the Third Reich in 1945. His most significant diplomatic move as foreign minister was the signing in 1939 of the German-Soviet Treaty of Nonaggression (MOLOTOV–Ribbentrop Pact), which cleared the way for the German invasion of Poland and resulted in World War II. In 1940 he negotiated the Tripartite Pact with Japan and Italy, which provided for mutual support against the USA. Thereafter his influence waned and in 1945 he was dismissed from office by Admiral DOENITZ. He was tried and convicted as a war criminal at Nuremberg and hanged in 1946.

Richard, Cliff (Harry Roger Webb; 1940–) *British pop singer and actor.*

Born in India, he returned with his family to Britain in 1947. While still working as a clerk in a television factory he played and sang in skiffle groups; in 1958 he had his first hit record, 'Move It'. His backing group, originally called the Drifters and later renamed the Shadows, stayed with him until 1968. He made his film debut in *Serious Charge* (1959) and appeared in several subsequent films, mainly musicals. In *Expresso Bongo* (1960) he was cast as a pop singer; in *Two a Penny* (1968) he played a straight dramatic role. He has toured widely, appearing in eastern Europe and on German television in 1970; he also made his stage debut that year in *Five Finger Exercise* and in 1986 starred in the musical *Time* in London.

Richard's music has evolved from an early identification with teenagers' rock and roll to the category admired by pop fans when they reach their middle-age. Now over fifty himself, he maintains a clean-living boyish appearance, no doubt not totally unconnected with his conversion as a born-again Christian in the 1970s; he has continued to enjoy success in the pop single charts. A ghost-written autobiography, *Which One's Cliff?*, was published in 1977.

Richards, Sir Gordon (1904–86) *British flat-racing jockey. Knighted in 1953, he was champion jockey 26 times (1925–53), winning 4870 races, including 14 Classics.*

Born in Shropshire, Richards was a miner's son, one of twelve children. He was apprenticed to Martin Hartigan's stable when he was sixteen and rode his first winner in March 1921. In 1925 he rode 118 winners and was champion jockey for the first time. In 1933 he rode a total of 253 winners to break Fred Archer's record of 246 set in 1885. Fourteen years later he had 269 winners in the season.

Richards was knighted for his services to racing in the Coronation Honours List in June 1953. A few days later he won the only great prize that had eluded him – the Derby. A year later, after a fall when he sustained serious injuries, he retired. In 1955 he started training with thirty two-year-olds. His stable jockey from 1956 to 1969 was the Australian Scobie Breasley, with whom he shared many notable successes on the horses of Stavros NIARCHOS and others. Soon after his retirement from training in 1970, Richards was made an honorary member of the Jockey Club.

Richards, Viv (Isaac Vivian Alexander Richards; 1952–) *West Indian cricketer and test captain regarded as one of the best batsmen in recent cricket history.*

Born at St John's, Antigua, he made his debut for Leeward Islands in 1971–72 and for the West Indies in 1974–75. In England he played for Somerset (1974–86). As captain of the West Indies (1985–91) he confirmed the team's dominance of international cricket with a series of victorious tours. His own tally of test runs has passed 6000. It was his influence that brought test cricket to Antigua. Apart from his natural grace and power as a batsman, Richards was also a brilliant fielder in almost any position, latterly more in the slips than the covers, and he also bowled off-breaks with modest success. He retired from test cricket in 1991.

Richardson, Sir Ralph David (1902–83) *British actor. He was knighted in 1947.*

Born in Cheltenham, Richardson first worked in an office before beginning his distinguished acting career as Lorenzo in *The Merchant of Venice* (1921) at Lowestoft. He subsequently joined the Birmingham Repertory Company and made his London debut in *Oedipus at Colonus* (1926). His long association with the Old Vic began in 1930 and he soon established his reputation as a fine actor of Shakespearean and other classical roles. It was at the Old Vic that he and John GIELGUD acted together for the first time; their later plays together included N. C. Hunter's *A Day by the Sea* (1953), David Storey's *Home* (1970), which they also performed in New York, and Harold PINTER's *No Man's Land* (1976). Among Richardson's most notable roles were Shakespeare's Falstaff, the title role in *The Amazing Dr Clitterhouse*, Dr Sloper in *The Heiress*, the title role in *Peer Gynt*, Cherry in Robert BOLT's *Flowering Cherry*, Wyatt Gilman in John OSBORNE's *West of Suez*, and Cecil, a part he inspired, in William Douglas Home's *The Kingfisher*. After World War II, during which he served in the Fleet Air Arm and became a lieutenant-commander, Richardson was co-director of the Old Vic with Laurence OLIVIER.

Although primarily a stage actor, Richardson appeared in many films portraying a variety of characters. These included *The Fallen Idol* (1948), *The Heiress* (1949), *The Sound Barrier* (1952), *Richard III* (1955), as Buckingham, *Long Day's Journey into Night* (1962), *Alice's Adventures in Wonderland* (1972), as the Caterpillar, and his last film,

Greystoke (1984), in which his performance was highly acclaimed.

Richthofen, Manfred Freiherr von (1892–1918) *German aviator, nicknamed the 'Red Baron', who shot down eighty Allied aircraft during World War I.*

From an aristocratic Prussian family, Richthofen joined the Prussian cadet corps and, in 1912, became a lieutenant in the Uhlan 1st Regiment. Initially he fought as a cavalry officer in World War I, heading a troop of lancers, but dissatisfaction led him to transfer to the flying corps. At first he was trained as an observer and in August 1915 was assigned to a squadron of twin-engined bombers – the Ostend Carrier Pigeons. Transferred to Metz in late 1915, he learnt to fly and the following March was posted to a squadron in France. In August he joined the newly formed crack fighter squadron, or Jagdstaffel, headed by air ace Oswald Boelcke.

Richthofen relished the aerial dogfights with Allied planes in much the same way as he enjoyed hunting game on the family estates. He scored his first 'kill' on 17 September 1916 in the skies above the Somme. Thereafter he dedicated himself to developing his combat skills, largely abstaining from the revelries of the officers' mess. His craft, an Albatross D II, was painted a distinctive bright red, hence his nickname, and on 16 January 1917 the Red Baron received the coveted 'Blue Max', or 'Pour le Mérite', for destroying sixteen enemy planes. By now acknowledged as the top German pilot, he was given command of his own squadron, Jagdstaffel 11, based at Douai, and then appointed head of the combined squadrons comprising Jagdgeschwader 1. On 6 July 1917 he narrowly escaped death from a serious bullet wound in the head. The following year, on 21 April, luck finally deserted him when he was killed, probably by a shot from Australian infantry, near Vaux sur Somme. His brother, Lothar, survived the war credited with forty 'kills'.

Rilke, Rainer Maria (René Karl Wilhelm Josef Maria Rilke; 1875–1926) *Austrian poet.*
Born in Prague, Rilke was the son of an unsuccessful army officer and a devoted mother whose ambitions centred on her son. As a boy he was sent to a military academy, then to a business school, but both proved uncongenial. In 1895, financed by an uncle, he studied philosophy at the University of Prague but left to go to Munich the following year, allegedly to continue his studies but actually to write. He had written several volumes of verse by 1899,

the last few giving hints of a distinctive style. The turning point of his early career occurred on two trips to Russia (1899–1900), accompanied by Lou Andreas-Salomé (1861–1937) – former companion of Nietzsche, later a member of FREUD's circle, and the first of many intellectual or aristocratic women who supported Rilke as friends or patrons. During these visits Rilke met Tolstoy and the painter Leonid Pasternak (father of Boris PASTERNAK) and was stirred by the vastness of Russia and its devoutly religious people. *Das Stundenbuch* (1899, published 1905, translated as *The Book of Hours*), written as if by a Russian monk, reflects this deep impression and shows the emergence of Rilke's conception of art as a (quasi-)religious vocation.

After leaving Russia he lived in Worpswede, an artists' community near Bremen. In 1901 he married Clara Westhoff, a sculptor; they had one daughter, Ruth, but separated the following year, when he published *Das Buch der Bilder* (1902; translated as *The Book of Images*). From 1903 to 1909 Rilke lived mainly in Paris. He acted briefly as secretary to the sculptor Auguste Rodin, whose committed craftsman-like approach influenced Rilke's attempt to rid his own work of subjectivity. This is seen in the sharply visual prose of *Die Aufzeichnungen des Malte Laurids Brigge* (1910; translated as *The Notebooks of Malte Laurids Brigge*, 1958) and in the completely individual mature style of *Neue Gedichte* (two vols, 1907–08; translated as *New Poems*), which focus intensely on tangible things, as in 'Das Karussell', 'Der Panther', and 'Archaïscher Torso Apollos', to cite well-known examples.

In 1911–12 Rilke twice visited Duino castle on the Dalmatian coast as the guest of Princess Marie von Thurn und Taxis-Hohenlohe. Here he began a series of elegiac free-verse hymnic poems, but this work was interrupted by World War I. In 1922 a Swiss patron offered him the use of the castle of Muzot (Valais) and in an astonishing three-week burst of creativity Rilke completed the *Duineser Elegien* (1923; translated as *Duino Elegies*, 1963) and *Die Sonette an Orpheus* (1923; translated as *Sonnets to Orpheus*, 1936). Concerned with the struggle against transience and the journey into death, the *Elegies* are the culmination of Rilke's career. He died of leukaemia at Val-Mont near Montreux, Switzerland.

Rivera, Diego Maria (1886–1957) *Mexican painter, whose murals pioneered monumental social realism in Mexico.*

Rivera showed early talent and in 1907, after studying art in Mexico, received a grant to study in Madrid. After travelling widely in Europe, from 1911 he lived mainly in Paris, where he was influenced by cubism. On his return to his home country in 1921 he abandoned European avant-garde styles and techniques in order to create an art of the people appropriate to Mexico after the revolution. Adopting traditions from Mexican folklore and Aztec and Mayan art, he produced huge murals for public buildings depicting the social and political history of Mexico from a Marxist viewpoint. These works are characterized by strong decorative areas of colour and flat simplified forms. His best works were done in the 1920s. In 1929 Rivera became director of the Central School of Fine Arts in Mexico and in the early 1930s, his fame having spread to the USA, he painted murals in San Francisco, Detroit, and New York. He was also invited to the Soviet Union in 1927 and 1955. Rivera continued to undertake public commissions well into the 1950s and, as a member of the Communist Party, remained active in politics.

Robbe-Grillet, Alain (1922–) *French novelist, a leading exponent of the* nouveau roman.

Born in Brest, Robbe-Grillet worked as statistician in the National Institute of Statistics and as an agronomist at an establishment researching tropical fruits before turning to writing. The *nouveau roman* was conceived as a form of anti-novel that rejects accepted devices such as characterization, narrative, and plot, avoiding psychological analysis and moral judgment and presenting reality as a series of objective impressions. *Les Gommes* (1953; translated as *The Erasers*, 1964) was an early example of the new form; it centres on a detective who eventually murders the victim whose death he is investigating. It was followed by such novels as *Le Voyeur* (1955; translated as *The Voyeur*, 1958), which won the Prix des Critiques, *La Jalousie* (1957; translated as *Jealousy*, 1959), a tale of jealousy in which the subjective observations and impressions of the suspicious husband are interwoven with an objective view of the situation, *Dans le labyrinthe* (1959), *La Maison de rendez-vous* (1966; translated as *The House of Assignation*, 1970), *Un Régicide* (1978), *Djinn* (1981), and *Angélique ou l'enchantement* (1988).

Robbe-Grillet set out his revolutionary theories in *Pour un nouveau roman* (1963; translated as *Towards a New Novel*, 1965), a collection of essays. He also explored the visual potential of his techniques in film scenarios, notably in RESNAIS's *L'Année dernière à Marienbad* (1961), later published as a ciné-roman (translated as *Last Year at Marienbad*, 1962), and has directed such films as *L'Immortelle* (1963), *L'Eden et après* (1970), and *La belle captive* (1983).

Robeson, Paul (1898–1976) *Black US singer and actor.*
Robeson was born in Princeton, New Jersey, the son of a Presbyterian minister and school-teacher, and was a leading athlete at Rutgers University. Although he attended Columbia University Law School, he abandoned his career as a lawyer and became an actor. Early stage appearances included *Voodoo* (1922) with Mrs Patrick CAMPBELL. Critical acclaim, however, came with Eugene O'NEILL's *All God's Chillun Got Wings* (1924), followed by *The Emperor Jones* (1925), in which his reputation as a singer as well as an actor was established. The screen version of this play provided his first film role in 1933. This was followed by several films in Britain, including the memorable *Sanders of the River* (1935), *Song of Freedom* (1936), and *Proud Valley* (1940).
Returning to the USA he toured successfully in *Othello* throughout the 1940s. However his outspoken political comments brought him to the attention of the Un-American Activities Committee and his passport was revoked in 1950. It was returned in 1958, and the following year he came to London to sing at the Albert Hall and to appear at Stratford in *Othello*. After touring Europe he returned to the USA, where he lived in relative seclusion until his death. As well as his film and stage successes, Robeson is remembered for his remarkable singing of 'Ole Man River' from Jerome KERN's *Showboat*, filmed in 1936.

Robey, Sir George (George Edward Wade; 1869–1954) *British music-hall comedian and actor. He was knighted in 1954.*
Making his music-hall debut in 1891, Robey went on to be billed as 'the Prime Minister of Mirth', appearing at all the top halls throughout the country. During World War I he served with the Motor Transport Service and raised funds for war-time charities. He also appeared in reviews, including *The Bing Boys Are Here* (1916), *Zig-Zag* (1917) and, after the war, *Bits and Pieces* (1927), which he wrote. Among his pantomime dames was Dame Trott in *Jack and the Beanstalk* (1921).
He appeared as Falstaff on stage in *Henry IV, Part I* (1935) and on film in OLIVIER's

Henry V (1944). Other films included the second version of Feodor CHALIAPIN's *Don Quixote*. During World War II he entertained the troops and worked on war-savings drives. A man of many talents, Robey featured on radio and television and was an exhibiting member of the Royal Academy Institute of Painters in Water Colours. Nonetheless the bushy eyebrows, bowler hat, and long frock coat of the music-hall artist remained his most enduring image. His publications included *My Life Up To Now* (1908) and *Looking Back on Life* (1933).

Robinson, Edward G. (Emanuel Goldenberg; 1893–1972) *Romanian-born US film star, renowned throughout the world for his gangster roles and his collection of paintings.*
Born in Bucharest, Robinson emigrated with his family in 1903 to the USA, where they settled in New York. After attending City College he won a scholarship to the American Academy of Dramatic Arts; he made his professional debut in 1913. Having carved out a notable stage career on Broadway, he made his screen debut in *The Bright Shawl* (1923). Short and stocky, with a distinctive face and twinkling eyes, Robinson first became known as Rico Bandello in the gangster classic *Little Caesar* (1931). Although he went on to make many other gangster films, the extent of his range was soon recognized with such film biographies as *Dr Ehrlich's Magic Bullet* and *A Dispatch from Reuters* (both 1940), as well as Billy Wilder's *Double Indemnity* (1944), Fritz LANG's *The Woman in the Window* (1944), and Arthur MILLER's *All My Sons* (1948). In the 1950s Robinson was summoned to appear before the Un-American Activities Committee and his film career went into temporary decline. For a time he returned to the stage and although the peak of his film career had passed he made several films in the 1960s, including *Seven Thieves* (1960), *Sammy Going South* (1963), and the highly successful *The Cincinnati Kid* (1965).
In recognition of his work for the cinema he was posthumously awarded a Special Academy Award. A noted art collector, Robinson at one time owned one of the finest collections of paintings in private hands. An autobiography, *All My Yesterdays*, was published in 1973.

Robinson, Joan Violet (1903–83) *British economist and leading proponent of the Cambridge school of economic thought.*
Born in Camberley, Surrey, Joan Robinson was educated at Girton College, Cambridge. She began lecturing at Cambridge in 1931,

becoming a reader in 1949 and professor of economics in 1965, on the retirement of her husband, Professor Sir E. A. G. Robinson (1897–), who was himself a distinguished economist.

In 1933 she published *Economics of Imperfect Competition*, which introduced the concept of a market situation in between pure monopoly and perfect competition, in which each firm has a monopoly in its own products despite close substitutes produced by other firms, because of consumer preference for particular brand names, etc. She was critical of the free market's ability to allocate resources efficiently and of neoclassical theory. During the 1950s and 1960s a major debate took place between the Cambridge school and US neoclassical economists about the meaning of capital, and particularly the possibility of 'capital re-switching', or switching back from more to less capital-intensive production methods, in response to required rates of return. The Cambridge school developed macroeconomic theories of growth and distribution based partly on KEYNES but using analytical techniques resembling those of the earlier classical economists.

Joan Robinson's other major works include *Essay on Marxian Economics* (1942), *Accumulation of Capital* (1956), *Essays in the Theory of Economic Growth* (1963), *Economic Heresies* (1971), and *Introduction to Modern Economics* (1973). She was a prolific writer and one of the leading economic theorists in the post-Keynesian era.

Robinson, John Arthur Thomas (1919–83) *Anglican clergyman and theologian whose book* Honest to God *(1963) sparked a heated debate about modern, more secular, interpretations of Christianity.*

Robinson grew up in the ecclesiastical milieu of Canterbury, where his father was a clergyman. He read classics at Jesus College, Cambridge, but in 1941 entered Westcott House theological college, Cambridge, graduating in 1942. His postgraduate studies in the philosophy of religion earned him an MA (1945) and a PhD in 1946, when he was also ordained a priest. He served in a Bristol parish as curate to Mervyn STOCKWOOD and then moved to Wells Theological College as chaplain and lecturer (1948–51) before returning to Cambridge as dean and fellow of Clare College. He also lectured in divinity and wrote the first of a series of books that established him as a notable New Testament scholar. In 1959 he was appointed suffragan Bishop of Woolwich.

Stimulated by the work of BONHOEFFER, TILLICH, and other modern theologians, *Honest to God* presented Robinson's highly personal concept of God. He rejected the image of God as 'an old man in the sky' in favour of God 'in depth' – within each of us. This untraditional interpretation was heralded by many as a refreshing and more accessible approach to Christian doctrines than the more conservative views of those who decried it. His later books include *The Human Face of God* (1973) and *Can We Trust the New Testament?* (1977). Robinson became fellow and dean of chapel at Trinity College, Cambridge, in 1969. He has also figured prominently in such issues as the abolition of capital punishment and the Campaign for Nuclear Disarmament.

Robinson, Sir Robert (1886–1975) *British organic chemist, who was awarded the 1947 Nobel Prize for Chemistry for his work on the structure of a number of biologically active plant extracts. He was knighted in 1937, awarded an OM in 1949, and served as president of the Royal Society from 1945 to 1950.*

The son of a manufacturer of surgical dressings, who was also one of the inventors of cotton wool, Robinson was educated at Manchester University, where he obtained his doctorate in 1910. He held chairs of organic chemistry at Sydney (1912–15), Liverpool (1915–20), St Andrews (1921–22), Manchester (1922–28), and London (1928–30), before being appointed in 1930 Waynflete Professor of Chemistry at Oxford, where he remained until his retirement in 1955.

Robinson's early work concerned the structure and synthesis of plant pigments. He moved on to study plant alkaloids and in 1925 worked out the chemical structure of morphine. He also worked on strychnine, cholesterol, vitamin D, and the steroids. Robinson is regarded as one of the founders of modern organic chemistry and was responsible for a number of new reactions and syntheses.

Robinson, Sugar Ray (Walker Smith; 1920–89) *US boxer who was world welterweight champion (1946–51) and five times the middleweight champion (1951 (twice), 1955, 1957, and 1958–60).*

Sugar Ray Robinson was born at Detroit and as an amateur boxer never lost in 125 fights. He became a professional in 1940 and won the welterweight title in 1946. In 1951 he beat Jake Lamotta to take the middleweight title for the first time. In the course of 202 professional bouts, Robinson lost only 19 fights. Few boxers in the twentieth century have boxed with greater skill; he was also renowned for his defence. Robinson retired in 1965.

Robson, Dame Flora (1902–84) *British actress. She was made a DBE in 1960.*

Flora Robson was born in South Shields and trained at what was to become the Royal Academy of Dramatic Art (1919–21), winning a bronze medal. She made her stage debut in Clement Dane's *Will Shakespeare* (1921) but left the stage after a couple of tours to become a factory welfare officer in Welwyn Garden City (1924). She was saved from obscurity four years later by Tyrone GUTHRIE, who invited her to join the Festival Theatre Company in Cambridge. There she played many successful roles and in 1931 received critical acclaim for her role in *The Anatomist* at the opening of the Westminster Theatre, London. She joined the Old Vic in 1933 and appeared in various Shakespeare productions opposite Charles LAUGHTON. Here she also gave memorable performances as Gwendolen in *The Importance of Being Earnest* and as Mrs Foresight in *Love for Love*, two of the rare opportunities she was given to display her talent for comedy. Paulina in *The Winter's Tale*, Miss Tina in *The Aspern Papers*, and Mrs Alving in *Ghosts* were among her many other notable performances.

Successful on stage in Britain and the USA, Robson also made films in both countries. Empress Elizabeth in *Catherine the Great* (1934) and Queen Elizabeth in *Fire Over England* (1937) were among her best early film roles, and for *Saratoga Trunk* (1946) she received an Academy Award nomination. Television appearances included the award-winning *The Shrimp and the Anemone* (1977).

In her honour a theatre in Newcastle-upon-Tyne was named after her and she appeared in its opening production, *The Corn is Green*, in 1962.

Rockefeller, Nelson Aldrich (1908–79) *US Republican politician; governor of New York (1958–73) and vice-president of the USA (1974–77).*

The grandson of John D. Rockefeller (1839–1937), founder of the celebrated dynasty, Nelson Rockefeller graduated from Dartmouth College in 1930 before joining the Rockefeller Center. He later served as its chairman (1945–53). Following a spell at the State Department during the early 1940s, Rockefeller served in the EISENHOWER administration as chairman of the president's advisory committee on government organization and was also involved in health and foreign affairs. 1958 saw his election as Republican governor

of New York State and in 1964 he just failed to secure the Republican presidential nomination from Barry Goldwater (1909–   ). He was again unsuccessful in 1968. Resigning his governorship in December 1973, Rockefeller was appointed vice-president by Gerald FORD but did not seek further office when his term expired in 1977.

Rockefeller was for many years a trustee of the Museum of Modern Art in New York and also served as its president and chairman. In 1954 he founded the Museum of Primitive Art, also in New York.

Rodgers, Richard Charles (1902–79) *US composer. With his collaborators Lorenz Hart (1895–1943) and Oscar HAMMERSTEIN II, he composed the music for many popular American shows.*

Born in New York, he began working with Hart when they were both students at Columbia University. Between 1926 and 1930 they contributed to fourteen shows, moving away from song-and-dance acts towards musical dramas. In 1930 they went to Hollywood to work on films, but returned to Broadway in 1934. *On Your Toes* (1936) contained the ballet sequence *Slaughter on Tenth Avenue*, one of Rodgers's first extended compositions, choreographed by BALANCHINE, as well as one of their best songs, 'There's A Small Hotel'. *Babes in Arms* contained 'My Funny Valentine', 'The Lady is a Tramp', 'I Wish I Were in Love Again', and 'Where or When'. *Pal Joey* (1940) was a failure because the story was considered scandalous, but it was successfully revived in 1952 and filmed in 1957 starring Frank SINATRA. The score includes 'Bewitched, Bothered and Bewildered'. In 1943 Hart died; a film biography of Rodgers and Hart, *Words and Music*, was made in 1948.

In the meantime Rodgers went into partnership with Hammerstein, a partnership that led to a series of phenomenal successes. *Oklahoma!* (1943), *Carousel* (1945), *South Pacific* (1949), *The King and I* (1951), and *The Sound of Music* (1959) were all smash hits and all were made into successful films. *Oklahoma!* and *South Pacific* both won Pulitzer Prizes for drama. Rodgers also wrote scores for the television documentary series *Victory at Sea* (1952) and *Winston Churchill: The Valiant Years* (1960).

Rodrigo, Joaquín (1901–   ) *Spanish composer, who is known best for his guitar concerto.*

Blind from the age of three, Rodrigo soon developed musical skills. After local tuition he went to Paris to study composition under DUKAS at the École Normale de Musique (1927–32). The success of his *Concierto de Aranjuez* for guitar and orchestra (1939) led to similar compositions in concerto style. In 1944 he was appointed music adviser to Spanish Radio and in 1946 became professor of music history at Madrid University. His *Concierto Pastorale* (1978) for flute and orchestra was commissioned by the flautist James GALWAY. Rodrigo married the Turkish pianist Victoria Kamki in 1933.

Rogers, Ginger (Virginia McMath; 1911–) *US actress and dancer, whose name is linked in cinema history with that of her dancing partner of the 1930s, Fred ASTAIRE.*

Born in Independence, Missouri, Rogers started out in vaudeville as a dancer. She appeared for a time with her first husband, Jack Pepper, as 'Ginger and Pepper', before securing solo roles in such Broadway musicals as *Top Speed* (1929) and *Girl Crazy* (1930–31). At about the same time she began to play leads in such film musicals as *42nd Street* and *Gold-diggers of 1933* (both 1933). It was, however, her partnership with Fred Astaire that has made her one of the screen's immortals. They first danced together in *Flying Down to Rio* (1933), the number 'Carioca' stealing the show. After that came the main part of the Astaire–Rogers canon – *Roberta* (1935), *Top Hat* (1935), *Follow the Fleet* (1936), *Swing Time* (1936), *Shall We Dance?* (1937), and *Carefree* (1938) – in which the polish, grace, and magic of their dancing brought delight to audiences all over the world. Later, they came together again for *The Barkleys of Broadway* (1949).

In the meantime, Rogers went on to make a name for herself as a straight actress, winning an Academy Award for *Kitty Foyle* (1940). As a comedy actress she also made *Roxie Hart* (1942), *The Major and the Minor* (1942), and *Monkey Business* (1952). It was, however, on stage that her major successes of the 1960s came, with *Hello Dolly!* (1965) on Broadway and the London production of *Mame* (1969).

Rogers, Sir Richard George (1933–) *British architect, whose 'high-tech' functionalist designs have aroused some controversy. He was knighted in 1991.*

Born to British parents living in Florence, Rogers was educated at the Architectural Association School in London and at Yale University. In 1963 he was a founder member of the Team 4 architectural practice with Norman Foster, Wendy Cheesman, and his then wife

Su; their buildings include the factory for Reliance Controls Ltd in Swindon (1967). Rogers achieved international fame with his design for the Centre National d'Art et de Culture Georges Pompidou in Paris (1971–77, with Renzo Piano), a glass-walled edifice with a striking exterior display of structural and service elements, the latter enclosed in brightly painted tubes. Although the siting of such an aggressively modern building in the historic centre of Paris was widely decried, the Centre has become a tourist attraction second only to the Eiffel Tower. Equally controversial is the Lloyd's Building in London (1986), where service elements, such as air-conditioning units, plumbing, etc., again appear on the outside of the building. While Rogers' admirers stress the boldness and originality of his designs, his detractors find them arrogant and over-theoretical.

Rolland, Romain (1866–1944) *French novelist, dramatist, and essayist. He was awarded the Nobel Prize for Literature in 1915.*

Born at Clamecy, Mièvre, into a well-established middle-class family, Rolland studied at the École Normale Supérieure in Paris, later lecturing there and at the Sorbonne on history, art, and the history of music. His early plays were inspired by heroic historical events; they include *Les Loups* (1898; translated as *The Wolves*, 1937) about the Dreyfus affair, *Danton* (1900), and *Le Quatorze juillet* (1902).

Rolland's passion for genius led to biographies of Beethoven (1903), Michelangelo (1905), and Tolstoy (1911) and ultimately to his major novel, *Jean-Christophe* (1904–12). The life story of a German composer who becomes a mouthpiece for Rolland's views and whose friendship with a young Frenchman symbolizes the author's desire for harmony between nations, the novel first appeared in instalments in the journal *Cahiers de la Quinzaine*. It initiated the literary form known as the 'roman-fleuve', in which the story develops like the course of a flowing river, and was widely acclaimed throughout Europe.

Living in retirement in Switzerland during World War I, Rolland published a controversial pamphlet, *Au-dessus de la mêlée* (1915), exhorting the intellectuals of France and Germany to cooperate in the quest for truth, humanity, and peace. The fantasy *Colas Breugnon* (1919) was followed by a series of studies of Indian philosophy, notably *Mahatma Gandhi* (1924) and *La Vie de Vivekananda* (1930). To this period also belong the play *Le Jeu de l'amour et de la mort*

(1925), a second novel-cycle in seven volumes, *L'Âme enchantée* (1922–33), and the early part of a lengthy musicological study of Beethoven (1928–49).

In 1923 Rolland founded the review *Europe* and began to develop an interest in left-wing activities and the politics of the Soviet Union. He also made known his opposition to MUSSO-LINI and HITLER, but never officially affiliated himself with any political party. The latter part of his life was devoted to his *Mémoires*, published posthumously in 1956. His correspondence with great contemporaries such as Richard STRAUSS, EINSTEIN, and Bertrand RUS-SELL is collected in the *Cahiers Romain Rolland* (1948).

Rolls, Charles Augustus (1877–1910) *See* ROYCE, SIR FREDERICK HENRY.

Romains, Jules (Louis-Henri-Jean Farigoule; 1885–1972) *French novelist, dramatist, and poet. He was elected to the Académie Française in 1946.*

The son of a schoolmaster, Romains was born in Saint-Julien-Chapteuil, Haute-Loire, and studied science and philosophy at the École Normale Supérieure in Paris. He became involved with a group of writers and artists known as the Abbaye, who published his verse collection *La Vie unanime* (1908). In this early work Romains expounded his theory of 'unanimisme', the concept of a collective consciousness in which the emotions and impressions of the group as a whole take precedence over the psychology of the individual. The concept had been introduced in Romains's first prose work, *Le Bourg Régénéré* (1906), set in a village community, and reappeared in the novel *Mort de quelqu'un* (1910; translated as *Death of a Nobody*, 1914), which examines the influence of an insignificant individual on the collective consciousness, and the farcical tale *Les Copains* (1913).

Romains began writing for the theatre in 1911 with the unanimiste drama *L'Armée dans la ville*. Subsequent plays include the highly successful satirical farce *Knock ou le Triomphe de la médecine* (1923; translated as *Dr Knock*, 1925), which brought Romains lasting fame as a dramatist, and the comedy *Monsieur le Trouhadec saisi par la débauche* (1923). His most ambitious work was the epic novel cycle in twenty-seven volumes, *Les Hommes de bonne volonté* (1932–46; translated as *Men of Goodwill*, 1933–46), which examines the evolution of French society in the period from 1908 to 1933. It was inevitable that certain parts of such a vast saga would be

more successful than others; *Prélude à Verdun* (1937) and *Verdun* (1938) are generally considered to be the finest volumes. Elsewhere, Romains's skilful depiction of collective action in crowd scenes provides an excellent illustration of the application of unanimisme to the art of novel writing.

Romains spent the major part of World War II in the USA and Mexico, where he published six volumes of *Les Hommes de bonne volonté*. He returned to France in 1946 and continued to write, producing novels, poems, essays, and short stories.

Rommel, Erwin (1891–1944) *German field-marshal who commanded Axis forces during the North African campaign in World War II.*

The son of a teacher, Rommel joined the 124th Württemberg Infantry Regiment in 1910 as an officer cadet and two years later was commissioned as a lieutenant. He fought with outstanding valour during World War I, earning the 'Pour le Mérite' medal for gallantry on the Italian front at the Battle of Caporetto in October 1917. Between the wars he served as a regimental commander, taught at military academies, and wrote a tactical manual, *Infanterie greift an* (1937; 'Infantry Attacks').

Rommel started World War II as a colonel in command of HITLER'S HQ guard but was transferred in February 1940 to head the 7th Panzer Division. He displayed his tactical skills during the invasion of France (May–June 1940), leading his troops from the front with characteristic bravado. The following February he was posted to North Africa to retrieve the situation following the collapse of Germany's Italian allies in their campaign against WAVELL'S British forces in Egypt. Rommel quickly reversed the balance, deploying his combined Panzer Divisions – Afrika Korps – to maximum advantage in a series of brilliant surprise manoeuvres. His forces captured Tobruk in June 1942 but, dependent upon an increasingly vulnerable supply line, were repulsed at Alam al Halfa in August by MONTGOMERY's reinforced 8th Army. In October, Montgomery began his counter-offensive, defeating Rommel at El Alamein and forcing the withdrawal of his forces into Tunisia. In March 1943, Rommel was ordered home by the Führer to prevent his capture. The following year, now Inspector of Coastal Defences, Rommel began reinforcing coastal defences in an attempt to stop any Allied invasion of northern Europe actually on the beaches. However, his enthusiasm was not shared by his commander-in-chief, VON RUNSTEDT, who favoured a more flexible defence mounted from the hinterland.

Increasingly disenchanted with Hitler, in early 1944 Rommel gave his tacit support to a group planning to overthrow the Führer. Rommel's popular support in Germany made him a strong contender as Hitler's successor. On 17 July, while directing his troops following the Allied invasion, his car was strafed and he sustained serious injuries. Three days later, the plotters, led by Graf von STAUFFENBERG, failed in their assassination attempt; Rommel was implicated in the conspiracy. Hitler, aware of Rommel's prestige, sent two generals to visit the convalescing field-marshal. They gave him a stark choice: take poison or face trial with the prospect of execution for his family and staff. He took poison. His death was announced as resulting from his injuries and he was given a state funeral with full military honours.

Roosevelt, Franklin D(elano) (1882–1945) *US Democratic statesman and thirty-second president of the USA (1933–45). His 'New Deal' policies helped US recovery from the economic depression of the 1930s.*

Born into a prominent New York State family, Roosevelt attended Harvard University and Columbia University Law School, qualifying as a lawyer in 1907. Having entered a New York law firm, he was elected state senator for the Hudson River district in 1910. His political talents and illustrious background secured him a job of assistant secretary to the navy (1913–20) in the WILSON administration and in 1920 he stood as running mate to the unsuccessful Democratic presidential contender, James Cox. An attack of poliomyelitis in 1921 left Roosevelt partially paralysed. Although he slowly recovered, a chronic weakness remained to dog him, especially in later years. He resumed public life and in 1928 was elected governor of New York. Re-elected in 1930, he received the Democratic presidential nomination in 1932.

Roosevelt campaigned and won on a 'New Deal' platform to revive the depressed economy and reduce record unemployment. The first 'Hundred Days' of his administration saw an unprecedented volume of legislation. The Federal Emergency Relief Administration distributed government aid to relieve the desperate poverty in many states. The new Public Works Administration provided work in the construction of dams, bridges, and other major projects, while nearly three million young men joined the Civilian Conservation Corps in its first six years. Many smaller projects, employing several million people, were initiated by the Works Progress Administration set up in 1935. Roosevelt gradually raised farm incomes by cutting over-production, provided federal loans for industry, and increased business confidence by guaranteeing bank deposits. Codes of practice were introduced to ensure better working conditions and trade unions were given greater scope to organize. Throughout, Roosevelt cultivated the image of a homely approachable man by means of frequent press conferences and his 'fireside chats' to millions of radio listeners. Re-elected in 1936, he faced a growing conservative backlash and a further crisis in the still staggering economy. But soon after World War II broke out in 1939 the 'war economy' finally put an end to massive unemployment.

During the 1930s, Roosevelt had maintained a 'flexible neutrality' in spite of growing Japanese threats to US interests in the Far East. With the declaration of war between Britain and Germany, Roosevelt faced the 1940 election with the electorate in an isolationist mood. Although promising Britain 'all aid short of war', Roosevelt reaffirmed that US policy was nonintervention. However, having won the election and extended the 'lend-lease' facility to Britain in March 1941, Roosevelt was pitched into the war when the Japanese attacked the US fleet in Pearl Harbor on 7 December 1941. He met the British prime minister, Winston CHURCHILL, several times during the war. Both agreed that the prime objective was to defeat HITLER but they differed about the time and place of an Allied invasion of Europe. The Allied landings were delayed until 1944 at Churchill's insistence but took place on the Normandy coast of France in accord with US wishes. Roosevelt, Churchill, and STALIN met at Tehran in 1943 and again at Yalta in February 1945. In constructing a political framework for the postwar world, Roosevelt was criticized in the USA for having conceded too much to the Soviets, particularly in the Far East. But he was by now a sick man and he died shortly after Yalta, on 12 April.

Eleanor Roosevelt (1884–1962), whom he married in 1905, was his distant cousin and a niece of President Theodore Roosevelt (1858–1919). She played an active role in her husband's affairs and after his death served as US delegate to the United Nations (1945–53, 1961) and chaired the UN Commission on Human Rights (1946–51).

Rosenberg, Julius (1918–53) *US army engineer who was executed, with his wife Ethel (1915–53), for passing military secrets to Soviet intelligence agents.*

Rosenberg, a committed communist since his youth, was disillusioned with American society and what he saw as its glaring injustices. He graduated in electrical engineering in 1939 and in 1940 joined the US army signal corps as an engineer. Motivated by their ideological principles, the Rosenbergs started to convey military secrets via their contact, Harry Gold, to the Soviet vice-consul in New York. Rosenberg's brother-in-law, David Greenglass, was working for the army at the Los Alamos atom bomb test site, and he supplied the Rosenbergs with valuable information concerning the Anglo-US weapons programme.

The Rosenbergs were arrested on 23 May 1950, ostensibly as a result of incriminating testimony provided by the spy Klaus Fuchs, arrested by the British in 1949. It is probable, however, that the Rosenbergs were already implicated by Soviet intelligence messages that US cipher experts had decoded in 1949. Gold, Greenglass, and another associate, Morton Sobell, were also arrested, tried, and sentenced to long jail terms. The Rosenbergs were sentenced to death and, in spite of international appeals for mercy on their behalf, were sent to the electric chair on 19 June 1953. With their trial held at the height of the anticommunist hysteria orchestrated by senator Joseph Mc-Carthy, the Rosenbergs became the first US civilians to be executed for espionage.

Rossellini, Roberto (1906–77) *Italian film director, known for his neorealist films and his controversial marriage to Ingrid Bergman.*

Rossellini, who was born in Rome, began as a writer, amateur film-maker, and a maker of fascist documentaries. One of his earliest features, *Prélude à l'après-midi d'un faune* (1938), was banned as indecent; however, in the years immediately after World War II he emerged in the vanguard of the Italian neorealist movement. His first successful international film was the universally acclaimed *Rome: Open City* (1945), shot in the streets of Rome in quasi-documentary style. Although it starred Anna Magnani (1909–73), its cast was mainly nonprofessional. In a similar vein came *Paisà* (1946) and *Germany, Year Zero* (1947). *Stromboli* (1949) brought him and his future wife, Ingrid Bergman, together but the public outcry that surrounded their affair and marriage (1950) led to their films being banned in several countries. Rossellini had by now departed from the neorealism of his earlier successes with *L'amore* (1948), again starring Magnani.

Rossellini and Bergman separated in 1956 and their marriage was annulled in 1958, the year that saw the release of his documentary *India*. His flagging international reputation was finally restored with *Il generale della Rovere* (1959), for which he won the Grand Prix at the Venice Festival. He directed several more films and television productions, often working from his own screenplay, before his death.

Rostropovich, Mstislav Leopoldovich (1927–) *Soviet cellist, pianist, and conductor. An outstanding cellist, he has performed all over the world and finally left the Soviet Union in 1975. He was awarded the Lenin Prize in 1964 and appointed a Commandeur de la Légion d'honneur in 1987. In 1987 he also received an honorary KBE.*

Rostropovich's first teachers were his pianist mother and his father, who was professor of the cello at the Gnesin Institute in Moscow and a former pupil of Casals. He later attended the Moscow Conservatory (1943), studying composition with Shostakovich and the cello with Kozolupov. In the late 1940s he won competitions in Moscow, Prague, and Budapest. After the improvement of East–West cultural relations in the 1950s, Rostropovich toured in Europe and the USA, making both his London and New York debuts in 1956. In 1957 he was appointed professor of the cello at the Moscow Conservatory.

In 1960 he became friendly with Benjamin Britten, who dedicated to him his *Sonata for Cello and Piano* (1961), *Symphony for Cello and Orchestra* (1963), and the three *Suites* for unaccompanied cello (1964; 1967; 1971). Rostropovich gave the first performances of all these works. Shostakovich, Khachaturian, Myaskovsky, and Prokofiev have also dedicated works to him. As a pianist, Rostropovich frequently accompanies his wife, the soprano Galina Vishnevskaya (1926–), whom he married in 1955. He made his conducting debut in 1968 at the Bolshoi Theatre and since then has conducted in many cities, being particularly known as a conductor of opera. In 1970, in an open letter to the leading Soviet papers, he supported the proscribed Nobel Prize winner, Alexander Solzhenitsyn; this caused a curtailment of his own freedom and in 1975 he left the Soviet Union. Two years later he became the conductor of the National Symphony Orchestra in Washington.

Roth, Philip Milton (1933–) *US novelist who has received, among other honours, a Guggenheim Fellowship (1959), the National Book Award (1960), and the O. Henry Award (1960).*

Roth was born in Newark, New Jersey, into a middle-class Jewish family. He was educated at Bucknell University, Pennsylvania, and the University of Chicago, where he returned to teach English after a brief spell in the US army. It was during this period of his life that he began publishing short stories in various magazines; these were later collected, together with the novella *Goodbye Columbus*, in a single volume, which appeared in 1959. Irony and irreverence motivate these early stories and Roth continued in a similar vein in *Letting Go* (1962). *When She Was Good* (1967) explored new territory and was altogether less exuberant. In *Portnoy's Complaint* (1969) Roth tackles the problems associated with the Jewish mother figure and Jewish guilt with outspoken humour. Its publication resulted in his meteoric rise to both success and notoriety, largely on the mistaken assumption that it was pornographic. Subsequent novels have demonstrated Roth's considerable versatility and skill as a satirist; they include *Our Gang (Starring Tricky and His Friends)* (1971), which uncannily anticipated the Watergate scandal, *The Breast* (1972), *The Great American Novel* (1973), *My Life as a Man* (1974), *The Professor of Desire* (1977), *The Ghost Writer* (1979), *Zuckerman Unbound* (1981), *Zuckerman Bound* (1985), and *Deception* (1990).

Rothko, Marc (Marcus Rothkovitch; 1903–70) *Russian-born US abstract painter.*

Rothko's family emigrated from Russia to Portland, Oregon, when he was ten years old. He studied at Yale University (1921–23) and began to paint in 1925. Apart from a brief period at the Art Students' League, Rothko was largely a self-taught artist. In 1935 he was co-founder of the expressionist group 'The Ten' but by the mid-1940s he had adopted an abstract surrealist manner. The mature style for which he is best known dates from about 1947, when he began to produce pictures consisting of a few large soft-edged rectangular areas of thin but intense colour. Variations on this theme made up the main body of his work from this point. An almost luminous effect was often created by the interaction of the colours, and the pictures were usually on a large scale with the intention of enveloping the spectator in the emotion expressed in the picture, ranging from tragedy to ecstasy. Rothko regarded

painting as a religious experience, which he tried to communicate in his works. During the last decade of his life the colours of his paintings became more sombre, possibly reflecting the depression that ended in his suicide in 1970.

Rothschild, Nathaniel Mayer Victor, 3rd Baron (1910–90) *British zoologist, businessman, and scientific administrator.*

Brother of the zoologist Miriam Rothschild (1908–), Rothschild studied at Trinity College, Cambridge, and embarked on an academic career, becoming a fellow of Trinity College (1935–39). He succeeded to the barony on the death of his uncle in 1937. Distinguished service in counter-sabotage activities during World War II earned him the George Medal. He served as chairman of the Agricultural Research Council (1948–58), stressing the need for academic research to be free of narrowly defined objectives. He was appointed assistant director of research in the zoology department at Cambridge (1950–70), where his own research focused on sperm locomotion and how fertilization of the egg is achieved. He wrote *Fertilization* (1956) and *A Classification of Living Animals* (1961). In 1959 he joined the Shell Oil Company as a part-time advisor, becoming chairman of Shell Research Ltd (1963–70). In 1971 he was appointed by the HEATH government to be the first director-general of the Central Policy Review Staff – the so-called 'think tank' – with the aim of examining government policies in a broader and longer-term context than hitherto. It was seen as a controversial appointment in view of Rothschild's former allegiance to the Labour Party in the House of Lords. He also served on other government committees, notably as chairman of the Royal Commission on Gambling (1976–78), and was also chairman of Rothschilds Continuation and a director of the family merchant bank, N. M. Rothschild and Sons.

Rouault, Georges (1871–1958) *French expressionist and religious painter, whose pictures are frequently reminiscent of stained-glass windows.*

Rouault was the son of a Paris cabinetmaker. At fourteen he became a stained-glass apprentice and then worked for a craftsman who restored glass in medieval church windows. At the same time Rouault did evening courses at the École des Arts Décoratifs. In 1891 he began to study at the École des Beaux Arts and through one of his teachers met MATISSE, with whom he shared an interest in the power of colour. By 1903 he had turned from religious subjects to vigorously expressive paintings of prostitutes, entertainers, and social rejects – a change of direction that was controversial and to many contemporaries inexplicable in such a devout Catholic. In these paintings his use of colour within rough heavy outlines created a luminous effect similar to that of stained glass windows. In 1908 he added judges to his repertoire of subjects. It was not until 1940 that he returned to religious themes and after this date he received numerous honours.

As well as canvases he painted ceramics, and from 1914 to 1920 he engraved illustrations for books, such as *Les Fleurs du Mal*. In addition he created designs for tapestry and for opera.

Rous, (Francis) Peyton (1879–1970) *US pathologist who pioneered cancer research and discovered that cancer can be caused by a virus, though his work was not recognized until fifty-six years later when he was awarded the 1966 Nobel Prize for Physiology or Medicine.*

Rous qualified at the Johns Hopkins Hospital, Baltimore (1900), and became an instructor in pathology at the University of Michigan. In 1909 he began work on a programme of cancer research at the Rockefeller Institute and, after only a few weeks, he managed to transplant a naturally occurring connective tissue tumour (Rous sarcoma) from one hen to another by grafting tumour cells. More significantly, he showed that an injection of a cell-free filtrate of the tumour could still cause cancer, suggesting a viral cause. This discovery was regarded with suspicion for many years although Rous and his co-workers were able to show the nature of the growth and the causal agent (called the Rous sarcoma virus). Disheartened, Rous abandoned cancer research and carried out important work on the functions of the gall bladder and the difference in acidity of animal tissues; during World War I he devised methods of preserving blood for several weeks.

In 1933, after R. E. Shope discovered a growth in rabbits that often progressed to cancer, Rous took up his studies again and demonstrated several ways in which carcinogenic chemicals and tumour-inducing viruses act together to speed up tumour production. He also discovered that carcinogenic action consists of two phases – initiation, which gives a cell malignant potential; and promotion, in which the potential is realized as an actively growing cancer. In his later years Rous was awarded many honours, including the Nobel Prize,

which he shared with Charles B. Huggins (1901–   ).

Royce, Sir Frederick Henry (1863–1933) *British engineer and designer of the Rolls-Royce car and aeroengines. He was made a baronet in 1930.*

Born in Alwalton, Huntingdonshire, the son of a miller, Royce began his working life as a newspaper boy and telegraph messenger. In 1877 he began a three-year apprenticeship with the Great Northern Railway Works, Peterborough. Royce's first business ventures were selling small electrical devices of his own design. In this field he was just beginning to experience some success, manufacturing a dynamo of his own design, when his market collapsed under a flood of cheaper US and German imports. Searching for other markets to explore, Royce turned his attention to the motor car. His first car was ready for testing in 1904. One of Royce's early cars was shown to the Honourable Charles Rolls (1877–1910), the son of Lord Llangattock and winner of the Automobile Club's Thousand Mile Trial in 1900. So impressed was Rolls with the car that shortly afterwards, in 1906, they entered into the famous Rolls-Royce partnership. Royce was the engineer, Rolls the promoter. Their first major success was the Silver Ghost. Introduced in 1907, it remained in production for nineteen years. With a special expansion chamber for each cylinder, it demonstrated Royce's concern with silence, comfort, and reliability. For its production a new factory was opened in Derby, still the home of Rolls-Royce. The founder, however, saw little of this factory: Rolls was killed in a flying accident in 1910, while in the same year Royce's health collapsed and he was given three months to live. Told he must live by the sea and take things much easier, he never returned to his Derby factory, setting up drawing offices with his chief designers to work on his new models at homes in the south of France and at West Wittering on the south coast of England.

In addition to further work on the Silver Ghost, Royce began on the outbreak of World War I in 1914 to design a 12-cylinder water-cooled aeroengine. The result, the Eagle engine, was ready for testing in 1915 and was followed by the Hawk, Falcon, Condor, and Kestrel engines. Nearly 75 per cent of the aero-engines used in British aircraft during World War I were designed by Royce. A few months before his death he had started work on the Merlin, the engine that would later power the Spitfire in World War II. The design of this engine was continued by a team led by A. G. Elliott (1889–1978), an engineer whom Royce had himself trained and who later became the company's chief engineer and managing director.

Rubbra, (Charles) Edmund (1901–86) *British composer, whose wide range of lyrical compositions is wholly in the English tradition and was largely inspired by his love of six-teenth- and seventeenth-century English poetry.*
Born into a family who encouraged his early love of music but who could ill-afford to further it, Rubbra left school at the age of fourteen to become a railway clerk. A passion for the music of DEBUSSY and Cyril Scott (1879–1970), based on the scores he found in an uncle's music shop, led him to give a concert in 1917 devoted to the works of Scott in his native Northampton. When Scott heard of this event, he approached Rubbra with an offer of tuition. In 1920 Rubbra won a composition scholarship to Reading University and in the following year, on an open scholarship, went to the Royal College of Music, where his teachers included Gustav HOLST and R. O. Morris (1886–1948). Thereafter Rubbra lived by teaching, accompanying, and journalism (for many years he reviewed music for the *Monthly Musical Record*).

Rubbra's eleven symphonies span the years 1935 to 1979; the first four (1937–41) were described by the composer as 'different facets of one thought'. The fifth symphony (1947–48) is the one most frequently heard, while the ninth, *Sinfonia sacra* (1971–72), is in the nature of a choral Passion. The tenth symphony is a chamber symphony, *Sinfonia da camera* (1974). Much of Rubbra's music has a religious background, including his two masses (the first Anglican and the second Catholic, after his conversion to Roman Catholicism in 1948).

Rubinstein, Artur (1888–1982) *Polish-born US pianist, who had one of the longest performing careers of any musician.*
Born in Warsaw, he became a child prodigy, studying under Joseph Joachim, Heinrich Barth, and Max BRUCH. He made his Berlin debut playing the Mozart A major concerto in 1900, at the age of twelve, causing a considerable sensation. Thereafter he studied briefly with PADEREWSKI and toured extensively in Europe, the USA (1906), and South America (1916–17). After his marriage in 1932, Rubinstein retired from the concert platform for a period of renewed study of his technique and

repertoire. He became a US citizen in 1946 and for the next thirty years carried out numerous concert tours, playing concertos and chamber music and making numerous recordings, including the complete works of Chopin and three cycles of the Beethoven concertos. He gave his final London concert in 1976, at the age of eighty-eight. A man of great charm and wit, he was frequently interviewed on television and he also wrote two volumes of autobiography, *My Young Years* (1973) and *My Many Years* (1980).

Runyon, (Alfred) Damon (1884–1946) *US short-story writer and journalist.*
Born in Manhattan, Kansas, the son of a printer, Runyon grew up in a small town in Colorado. Thanks to his father, Runyon could claim to have seen some of his writing printed, in local newspapers, while he was still in his early teens. At fourteen he added the necessary four years to his age in order to enlist in the Spanish–American War and served in the Philippines. After returning home in 1900, he worked for ten years as a reporter for newspapers in Colorado Springs, Denver, and San Francisco before getting a job as sports writer for the New York *American* in 1911. During World War I he was sent to Mexico and Europe as a war correspondent for the Hearst newspapers and after the war continued to work as a Hearst columnist.

Runyon's stories, which grew out of his knowledge of the amiable low-life of Broadway and the New York sporting scene, were published over a number of years and first collected in *Guys and Dolls* (1932). The chorus girls, gamblers, and small-time crooks who populate the stories are presented as types rather than individuals; they have hearts of gold and are not overendowed with brains. The stories, distinguished by their slang and vivacity of language, are in fact late examples of the traditional American comic tale told in vernacular speech. Runyon collaborated with the director and actor Howard Lindsay (1889–1968) on a comedy, *A Slight Case of Murder* (1935), and published other collections of his stories, including *Blue Plate Special* (1934) and *Take It Easy* (1938). The musical *Guys and Dolls* was based on his work.

Rushdie, Salman (1947–) *British writer, born in India, whose novel* The Satanic Verses *was denounced as blasphemous by many Muslims, leading to a demand for the author's death from Ayatollah KHOMEINI.*
Born to wealthy parents in Bombay, Rushdie was educated in England at Rugby School and

King's College, Cambridge. He became a British citizen in 1964. On graduating in 1968 he worked briefly as an actor before starting a career as an advertizing copywriter. His first novel, *Grimus*, appeared in 1974, but Rushdie owes his reputation to his second book, *Midnight's Children* (1981), a rich novel that explores the experience of growing up in post-independence India. The book won the 1981 Booker Prize and comparisons with the work of GARCÍA MÁRQUEZ and KUNDERA. In 1984 he produced *Shame*, a complex narrative combining satire, fantasy, and political allegory. The 1980s also saw Rushdie's emergence as a journalist, writing widely on political, cultural, and racial issues.

The Satanic Verses (1988) is a sophisticated and sometimes abstruse treatment of the experience of Muslim immigrants to Britain, dealing especially with the tension between secular and religious ways of thinking. Because a character identified by many Muslims as the prophet Muhammed was treated in a manner they regarded as disrespectful, the book was denied publication in some Muslim countries; early in 1989 Bradford Muslims burnt the book in public as part of a campaign to have it banned in Britain. This was followed in February by KHOMEINI's notorious *fatwa* demanding the author's death; a reward of one million pounds was subsequently offered to any Muslim who carried out the murder. While Rushdie went into hiding with a permanent police guard, the book became a bestseller and a focus for heated debate about the concept of blasphemy and the limits of free expression in a multi-cultural society. The affair had wide international repercussions, including violent demonstrations in the Muslim world, the breaking of diplomatic ties between Iran and the EC states, and the murder of moderate Islamic leaders in Belgium. Although relations between Britain and Iran improved following the death of Khomeini, the *fatwa* was not revoked. In late 1990 Rushdie held private talks with sympathetic Muslim leaders and made his first public appearances in almost two years; by the end of the year he had announced his acceptance of the basic tenets of Islam and agreed not to publish the novel in paperback – a demand he had previously resisted. While dismaying many of his supporters, these concessions did nothing to soften hardline Muslim opinion.

Since the start of his ordeal Rushdie has published *Haroun and the Sea of Stories* (1990), a children's fable, and *Imaginary Homelands* (1991), a collection of essays that includes reflections on the controversy and the principles involved.

Rusk, (David) Dean (1909–) *US statesman who served as secretary of state under presidents KENNEDY and JOHNSON.*

Rusk studied politics at Davidson College, North Carolina, and subsequently won a Rhodes scholarship to St John's College, Oxford, where he received an MA in 1934. Back in the USA, he joined Mills College, Oakland, as assistant professor in government and international relations, becoming dean of the faculty in 1938. After serving in Indo-China during World War II, he joined the State Department and in 1948 was appointed director of the Office of UN Affairs. Later, as assistant secretary of state for eastern affairs, Rusk helped implement TRUMAN's policies following the 1950 invasion of South Korea by the North. In 1952, Rusk became president of the Rockefeller Foundation, in charge of its considerable aid programme to underdeveloped countries, remaining there until president-elect Kennedy appointed him secretary of state in 1960.

Somewhat eclipsed by the young president's active participation in foreign policy, Rusk's influence was greater during the Johnson administration. He viewed the Vietnam War as a crucial campaign to stem the tide of international communism, although critics charged that the assumptions underlying such a policy were outdated. Certainly his efforts to justify the war met with an increasingly hostile reaction at home. Relinquishing office with the Republican victory in 1968, Rusk returned to academic life three years later by becoming professor of international law at the University of Georgia.

Russell, Bertrand Arthur William, 3rd Earl (1872–1970) *British philosopher, logician, mathematician, and political campaigner, probably the most famous British philosopher of the century. His many honours included the OM (1948) and the Nobel Prize for Literature (1949).*

The second son of Viscount Amberley, Russell was born in Trelleck, Gwent, and orphaned at the age of three. He was brought up by his grandmother, the widow of the former prime minister, Lord John Russell (created 1st Earl Russell), and educated privately until he went up to Cambridge (Trinity College). Here he was a pupil of A. N. WHITEHEAD, met G. E. MOORE, and won a prize fellowship at Trinity with a thesis on the foundations of geometry. He continued to work on the foundations of mathematics and in 1903 published his first major work, *The Principles of Mathematics*. In the course of this work Russell encountered a number of paradoxes, all dealing in some ways with very large classes, which he found impossible to resolve. Clearly a deeper analysis of the foundations of mathematics was called for and eventually led Russell to his central insight that 'mathematics and logic are identical'. To demonstrate the truth of such a proposition Russell joined with his former tutor, Whitehead, in a formal and rigorous derivation of the whole of mathematics from purely logical principles. It was an enormous achievement, demanding ten years of their lives and resulting in the three very large volumes of *Principia Mathematica* (1910–13).

In the course of this labour Russell had been forced to consider a number of problems more properly belonging to the philosophy of logic and language than pure mathematics. In particular, his theory of types and his theory of denoting were to exercise considerable influence on logical thought for the rest of his life.

After the publication of *Principia* Russell turned to more traditional philosophical problems, which he dealt with in his *Problems of Philosophy* (1912) and *Our Knowledge of the External World* (1914). Both belong to the empiricist tradition and were concerned to show how the external material world can best be derived from our basic sensory experience. Russell was to write many other books tackling this same problem throughout his life: his final word on the subject can be found in *Human Knowledge: its Scope and Limits* (1948).

It was during World War I that Russell first engaged in serious and prolonged political campaigning. The issue was the right of the individual to choose not to fight. Russell campaigned strenuously for the rights of the conscientious objectors and in 1916 he was brought before the courts for his behaviour and fined. Two years later he was charged for further political activity and jailed for six months, during which time he wrote one of his most elegant works: *Introduction to Mathematical Philosophy* (1919).

Russell's most original work during this period was his attempt to develop, largely under the influence of his pupil WITTGENSTEIN, the theory of logical atomism. Its success proved to be short-lived. After this effort, although Russell published many books, they tended to be the reworking of old material rather than attempts to break new ground. He remained politically active, with his greatest effort di-

rected in his later years to the problems of nuclear war. Again, in 1961, he was imprisoned: this time for civil disobedience during the Campaign for Nuclear Disarmament.

In his later years, due to the publication of his remarkably frank *Autobiography* (3 vols, 1967–69) and the appearance of the papers of his long-dead friends, a detailed picture of Russell's life began to appear. Accounts of his unhappy childhood and first marriage (he was married three times) and his relationship with Lady Ottoline Morrell (1873–1938) and several other mistresses were recorded as apparently objectively and as candidly as he recorded his mathematical work.

Russell, Henry Norris (1877–1959) *US astronomer best known for his independent discovery of the Hertzsprung–Russell diagram.*
The son of a Presbyterian minister, Russell was educated at Princeton, where he gained his PhD in 1899. After two years further study in Britain, Russell returned to Princeton, where he became professor of astronomy and director of the observatory from 1911 until his retirement in 1947.

In his classic paper *The Spectrum Luminosity Diagram* (1914) Russell plotted stellar brightness (absolute magnitude) against spectral type. The result, independently suggested by Ejnar HERTZSPRUNG, is now known as the Hertzsprung–Russell diagram. Russell also suggested an evolutionary scheme in which hot giant stars become small cold dwarfs. Although this turned out to be too simple, the diagram itself has remained as the basis for an account of stellar evolution. In 1929 Russell made the first reliable estimates of the composition of the sun. Using solar spectra he argued that 60 per cent of the sun's volume was composed of hydrogen. Although now recognized as an underestimate, it was then such a surprisingly high figure that it posed a major challenge to cosmologists.

Russell, Ken (1927–) *British film director, whose work in the cinema and on television has made controversial use of violence and the macabre.*
Russell, who was born in Southampton, had a varied career before becoming a director. After attending Pangbourne Nautical College he served in the Merchant Navy (1945) and the Royal Air Force (1946–49). He then turned to the theatre, becoming a dancer with the Ny Norsk Ballet (1950) and an actor with the Garrick Players (1951), after which he worked as a freelance photographer (1951–57), contributing to such journals as *Picture Post*. His first

venture into films, as a prize-winning amateur, was with *Amelia and the Angels* (1957) and *Lourdes* (1958). These successes led to his BBC television films, including *Elgar* (1962), *Isadora Duncan* (1966), and *A Song of Summer* (1968), a film biography of DELIUS. Russell made an equally successful impact on the large screen with his third feature, *Women in Love* (1969), adapted from D. H. LAWRENCE'S novel. *The Music Lovers* (1971), a rather eccentric presentation of Tchaikovsky's adult life, *The Devils* (1971), adapted from Aldous HUXLEY's *The Devils of Loudun*, *Mahler* (1974), *Valentino* (1977), *Gothic* (1986), and *The Rainbow* (1989), another Lawrence adaptation, were among the films that followed. His other productions included the operas *The Rake's Progress* (1982) in Florence, *Madame Butterfly* (1983) in Spoleto, and *Faust* (1985) in Vienna. He published an autobiography, *A British Picture*, in 1989.

Ruth, Babe (George Herman Ruth; 1895–1948) *Outstanding US baseball player of the 1920s and early 1930s. He did much to popularize the game and swell the baseball crowds.*

Born of poor parents in Baltimore, he had a tough 'street' childhood before going into a boys' home. His reputation as a pitcher reached the Baltimore team where, in 1914, he was nicknamed 'Babe' by a coach. Baltimore sold Ruth to the Boston Red Sox late in 1914 for $2900 and in the next four seasons his left-arm pitching helped this team to six major titles. In 1918 he delivered a record sequence of scoreless pitching but it was as an immensely powerful hitter that he achieved stardom. In 1919 the Red Sox sold Ruth, to the outrage of their fans, to the New York Yankees for $125,000. Between 1920 and 1935 Ruth and the Yankees won seven League pennants and five world series titles. Altogether he scored a total of 714 home runs, which was not surpassed until 1974; in his best year, 1927, he hit 60 home runs.

Ruth was a hard man to manage because of his gambling, drinking, and womanizing and tendency to brawl with players and officials. He retired in 1936 and in 1948, shortly before his death, saw the Hollywood version of his life, *The Babe Ruth Story*.

Rutherford, Ernest, 1st Baron (1871–1937) *New Zealand-born British physicist, who was awarded the 1908 Nobel Prize for Chemistry for his discovery that radioactive elements transform into other elements. Rutherford went on to discover the atomic nucleus*

and the artificial transformation of the elements. He was awarded the OM in 1925 and made a baron in 1931.

The son of a farmer, Rutherford was educated at Canterbury College, Christchurch. From there in 1895 he won a scholarship to Cambridge, where he worked with J. J. THOMSON. In 1898 Rutherford was appointed to the chair of physics at McGill University in Canada, where he first attempted to work out the nature of the atom in the light of Thomson's discovery of the electron. With the aid of the chemist Frederick SODDY, Rutherford established that radioactive elements were spontaneously transformed into other elements. Details of Rutherford's work were published in the successive editions of his *Radioactivity* (1903 and 1905).

In 1907 Rutherford returned to Britain, to the chair of physics at Manchester. While there, in 1911, Rutherford and his colleagues found that when alpha particles were directed at thin metal foils some passed through, while others were deflected by varying amounts. Describing the phenomenon as the 'most incredible event' of his life, Rutherford concluded that most of the atom's mass was concentrated in a small central area and that much of the atom was empty space. The small central mass became known as the atomic nucleus and Rutherford became the father of nuclear physics.

After service during World War I, spent working on methods to detect submarines, Rutherford left Manchester in 1919 to become Cavendish Professor of Physics and director of the Cavendish Laboratory at Cambridge in succession to J. J. Thomson. In 1920 Rutherford announced his discovery that when he exposed nitrogen to bombardment by alpha particles, the nitrogen atoms disintegrated leaving oxygen and hydrogen atoms in their place. Rutherford had thus discovered and correctly identified the artificial transmutation of matter.

Under Rutherford the Cavendish became one of the leading world centres for the study of nuclear physics. Rutherford himself continued with his work in the 1920s and 1930s until he died suddenly and unexpectedly in hospital after an apparently successful operation for a strangulated hernia.

Rutherford, Dame Margaret (1892–1972) *British actress. Her ability to blend eccentricity with plausibility made her one of Britain's leading comic character actresses.*

Born in London, Margaret Rutherford was a latecomer to the stage – she taught music and elocution before a legacy allowed her to pursue the career of an actress at the age of thirty-three. After a season at the Old Vic (1925), when she played Juliet's mother to Edith EVANS's Nurse, she returned to teaching before her career proper began at the Lyric, Hammersmith, two years later. At the Oxford Playhouse (1930) she met the actor Stringer Davis (1896–1973), whom she married in 1945. They appeared together in many productions.

The major break in her career came as Bijou Furze in *Spring Meeting* (1938). There followed some of her most famous portrayals: Miss Prism in *The Importance of Being Earnest* (1939), Madame Arcati in *Blithe Spirit* (1941), and Miss Whitchurch in *The Happiest Days of Your Life* (1948), all of which she transferred from stage to screen (1952, 1945, and 1950 respectively). Not confined to comedy, however, Rutherford's range extended to such roles as Mrs Danvers in *Rebecca* (1940). She toured with ENSA (1944) and made her Broadway debut as Lady Bracknell in *The Importance of Being Earnest* (1947). She made her first film in 1936 and during the sixties starred as Agatha CHRISTIE's Miss Marple in a series of successful films. *The VIPs* (1963) brought her an Oscar as best supporting actress.

Ryle, Gilbert (1900–76) *British philosopher, a leading figure in contemporary Oxford linguistic philosophy.*
Born in Brighton, the son of a doctor, Ryle was educated at Queen's College, Oxford. Apart from the war years (1939–45) spent with the Welsh Guards, Ryle remained at Oxford for his entire academic career, serving as Waynflete Professor of Metaphysics from 1945 until his retirement in 1968.

Ryle initially showed some interest in modern German philosophy, but in his paper 'Systematically misleading expressions' (1931) he announced his conversion to the new linguistic philosophy. He further developed his views in his most original early paper, 'Categories' (1937). But by 1945 such work no longer seemed adequate to Ryle. It was time the new philosophical techniques proved themselves by being deployed at length against a major longstanding problem of philosophy. Ryle chose to tackle the mind–body problem and in his classic work, *The Concept of Mind* (1949), argued against the Cartesian position of the mind as a mysterious spiritual ghost in a material machine. Mental concepts were thus seen

by Ryle not as descriptions of events taking place in some private inner theatre but as mainly dispositional and consequently analysable in terms of observable behaviour. Other works of Ryle include *Dilemmas* (1954), a study of certain apparently irreconcilable theories; and *Plato's Progress* (1966), a speculative, controversial, yet closely argued account of the nature and development of Plato's thought.

In his many years at Oxford Ryle exercised an enormous influence on the organization and development of Oxford philosophy. It was largely due to the imagination of Ryle that Oxford, after 1945, became one of the leading centres in the world for philosophical research. Ryle's own style as a thinker, writer, and teacher were unique – capable of being parodied but impossible to imitate: 'Le style, c'est Ryle', J. L. AUSTIN accurately noted.

Ryle, Sir Martin (1918–84) *British radio astronomer who carried out pioneering surveys of the radio sky in the 1950s and 1960s. Ryle was awarded the 1974 Nobel Prize for Physics (the first astronomer to be so honoured). He was knighted in 1966 and appointed Astronomer Royal (1972–82).*

The son of a physician, Ryle was educated at Oxford University and spent World War II at the Telecommunications Research Establishment in Malvern, working on the development of radar and related projects. In 1945 he joined the staff of the Cavendish Laboratory, Cambridge, as a lecturer in physics. With interest growing in radio astronomy, Ryle used mostly war-surplus equipment to build a radio telescope to investigate the radio sky. His first survey, begun in 1950, noted some fifty radio sources. However, his third survey, dating from 1959, lists the positions and strengths of nearly five hundred sources. Although it has been supplemented by more penetrating surveys it still remains the standard catalogue for the most easily detected sources in the northern hemisphere. The surveys were carried out on an 180-acre site outside Cambridge that, in 1957, became the Mullard Radio Observatory with Ryle as director. Two years later Ryle became Cambridge's first professor of radio astronomy.

The Cambridge surveys played an important role in the cosmological debates of the period. Ryle was one of the first to realize that the distribution of radio sources throughout the universe favoured the big-bang theory as opposed to steady-state cosmology.

S

Saarinen, Eero (1910–61) *Finnish-born US architect, whose premature death deprived America of one of its most creative 'second-generation' modern architects. Perhaps his most famous building is the birdlike TWA terminal (1956–62) at Kennedy Airport, New York.*

Born in Kirkkonummi, Finland, Saarinen emigrated with his father, Eliel Saarinen (1873–1950), the well-known Finnish architect, to America in 1923. After training as an architect at Yale, Eero Saarinen joined his father's practice in 1938 and became a US citizen in 1940. His first independent design was the General Motors Technical Center at Warren, Michigan (1948–56), an early example of the application of Mies van der Rohe's ideas; his subsequent churches, university buildings, and opera houses are evidence of his versatility. Among his buildings are the law school at Chicago University, the Auditorium Building (1953–55) at the Massachusetts Institute of Technology, with its shell-concrete roof springing from the ground at three points, and the Vivian Beaumont Theater, forming part of New York's Lincoln Center. Perhaps less auspicious is the US Embassy (1955–61) in London's Grosvenor Square, which, in attempting to be both traditional and innovative, in the view of its detractors fails to be either.

Sabin, Albert Bruce (1906–) *US microbiologist, born in Poland, who developed the oral vaccine against poliomyelitis that is still widely used and is named after him.*

Born in Bialystok (now in Poland), Sabin emigrated with his family to the USA in 1921. He attended New York University, receiving his BS degree in 1928 and his MD three years later. In 1935 he joined the Rockefeller Institute for Medical Research and in 1939 was appointed associate professor of paediatrics at the University of Cincinnati College of Medicine, later becoming professor (1946–60) and, in 1971, emeritus professor.

During World War II, Sabin joined the US army medical corps to develop vaccines against dengue and Japanese encephalitis that were afflicting combat troops in the Pacific. He also isolated the protozoan responsible for toxoplasmosis. After the war, Sabin identified the three types of virus that were causing poliomyelitis in the USA and developed a vaccine using live viruses that had been rendered harmless by laboratory treatment. (This contrasted with the killed-virus vaccine produced by Jonas Salk in 1953.) After initial trials with volunteer prisoners, the first large-scale trials were conducted in the Soviet Union. Not only was the Sabin vaccine proved safe, it was also easily administered orally and conferred long-lasting immunity against the disease. By the 1960s it had superseded the Salk vaccine throughout the world.

Sabin also worked on the relationship between viruses and cancer. He was consultant to the World Health Organization from 1969 until his retirement in 1986 and served as research professor of biomedicine at the University of South Carolina Medical College (1974–82) and thereafter as emeritus professor.

Sadat, Mohammed Anwar al (1918–81) *Egyptian statesman and president (1970–81). A charismatic leader, he was awarded the Nobel Peace Prize jointly with Menachem Begin in 1978 for his courage in trying to bring about peace in the Middle East.*

Born in a small village in the Nile delta, the son of a hospital clerk, Sadat attended the Cairo Military Academy, from which he graduated in 1938. Dedicated to the overthrow of the British-dominated monarchy, he was imprisoned by the British for pro-German activity during World War II. Sadat joined Colonel Nasser's Free Officers Movement in 1950. After the successful 1952 coup, in which King Farouk (1920–65) was deposed, he was appointed to various important posts in the new republic. He served as minister of state (1955–56), vice-chairman (1957–60) and chairman (1960–68) of the National Assembly, and vice-president of the republic (1964–66; 1969–70). Upon Nasser's death in 1970 Sadat became acting president and was formally elected president in October of that year.

Determined not to be overshadowed by his predecessor, he introduced measures to decentralize Egypt's political structure and diversify the economy. He also broke with Nasser's foreign policy, expelling Soviet technicians and advisers in 1972 and abrogating the 1971 treaty of friendship with the Soviet Union in 1976. At the same time he courted the support of the USA. After launching the costly Arab-Israeli Yom Kippur War of 1973, in which Arab, and particularly Egyptian, honour was satisfied, he began to work, under US pressure, for peace in the Middle East. In 1977 he created an international sensation by flying to Jerusalem to initiate peace negotiations in the face of condemnation from the rest of the Arab world. For his efforts (which culminated in the Israeli-Egyptian peace treaty of 1979) he shared the Nobel Peace Prize with Menachem Begin of Israel in 1978. He was assassinated by Muslim extremists while watching a military parade in 1981, shortly after taking steps to contain domestic opposition; he was succeeded by his vice-president, Hosni MUBARAK.

Sagan, Françoise (Françoise Quoirez; 1935–) *French novelist.*

Sagan was born in Cahors, Lot, and educated in Paris. Her first novel, *Bonjour Tristesse* (1954; translated in 1957), won instant fame and the Prix des Critiques for its young author. In this and subsequent novels Sagan examined the transitory nature of love as experienced in brief amoral liaisons, showing a precocious awareness of life and a premature cynicism that surprised and shocked a number of her readers. Later novels include *Un Certain Sourire* (1956; translated as *A Certain Smile*, 1956), *Aimez-vous Brahms?* (1959; translated in 1960) and *Les Merveilleux Nuages* (1961; translated as *Wonderful Clouds*, 1961). In *Des Bleus à l'âme* (1972; translated as *Scars on the Soul*, 1974) Sagan attempted to defend herself against the accusations of her critics. Later novels include *Le Chien couchant* (1980), *La Femme fardée* (1981; translated as *The Painted Lady*, 1982), *The Still Storm* (1984), and *Un Sang d'aquarelle* (1987). Sagan has also written for the theatre, notably *Château en Suède* (1960; translated as *Castle in Sweden*, 1962), *Les Violons parfois* (1962), and *Le Cheval évanoui* (1966). Other publications include collections of short stories and the biographical *Dear Sarah Bernhardt* (1989).

Saint-Exupéry, Antoine(-Marie-Roger) de (1900–44) *French novelist and aviator. He was posthumously created Commandeur de la Légion d'honneur.*

Saint-Exupéry was born in Lyons of impoverished aristocratic parents. He hoped to enter the École Navale in Paris, but failed the entrance examination and was conscripted in 1921 into the air force, where he trained as a pilot. His accomplishments in the field of civil aviation included the pioneering of air-mail routes in North Africa and South America. During World War II, despite his disablement from previous flying accidents, he served as a military reconnaissance pilot until the fall of France in 1940. Having escaped to New York, he rejoined his unit three years later in North Africa. He was killed in active service in 1944, shot down on a reconnaissance mission over Corsica.

Most of Saint-Exupéry's novels are based on his flying experiences, both physical and spiritual, and celebrate the heroism, comradeship, and solidarity displayed by airmen. *Vol de nuit* (1931; translated as *Night Flight*, 1932) and *Terre des hommes* (1939; translated as *Wind, Sand and Stars*, 1939) commemorate his years in civil aviation; *Pilote de guerre* (1942; translated as *Flight to Arras*, 1942) was inspired by a daring mission over Arras in 1940, for which Saint-Exupéry was mentioned in dispatches. Saint-Exupéry's best-known work outside his flying novels is his fable for children and adults, *Le Petit Prince* (1943; translated as *The Little Prince*, 1943). *Citadelle*, a philosophical work left uncompleted at his death, was published posthumously in 1948 (the English translation, *The Wisdom of the Sands*, appeared in 1952).

Sakharov, Andrei Dimitrievich (1921–89) *Soviet physicist and dissident, who was often referred to as the father of the Soviet hydrogen bomb. He was awarded the 1975 Nobel Peace Prize.*

The son of a physics teacher, Sakharov was educated at Moscow University. After graduating in 1942 he was assigned to research work instead of military service. At the end of World War II, Sakharov returned to pure science and the study of cosmic rays. Two years later, however, he left the Lebdev Physics Institute in Moscow to work with a secret research group on the development of the hydrogen bomb. Sakharov is believed to have been principally responsible for the Soviets' success in exploding their first thermonuclear bomb in 1954. As a by-product of this work Sakharov and I. E. Tamm (1895–1971) proposed ways in which controlled thermonuclear fusion could be achieved by confining an extremely hot ionized plasma in a torus-shaped magnetic bottle, known as a tokamak device. This approach has been followed with some theoretical success in the USA, Britain, Japan, and the Soviet Union.

Sakharov was elected to the Soviet Academy of Science in 1953, at the early age of thirty-two. There were, however, signs that he was beginning to find his official role restrictive. In 1961 he protested publicly against Soviet atmospheric tests of nuclear weapons and began his work that culminated in the test-ban treaty of 1963. In 1968 Sakharov went further in his *Progress, Coexistence and Intellectual Freedom*, which was published abroad. With this publication, he became both an internationally recognized name and an embarrassment to his own government. With great courage Sakharov continued to speak freely on the faults he saw in his own society and to campaign for disarmament. His reception of the Nobel Peace Prize was rewarded at home by the hostility and constant harassment of his government. Exiled to the closed town of Gorkii in 1980, he continued to speak fearlessly on the issues of human rights and disarmament and staged several hunger strikes. After intense international pressure, he was released in 1986 and in 1988 allowed to tour overseas, where he was warmly received. In 1989 he became a member of the newly created Congress of People's Deputies and continued to be a leading spokesman for liberal reformers until his death a few months later.

Saki (H(ector) H(ugh) Munro; 1870–1916) *British short-story writer.*

Munro was born in Burma, where his father was inspector-general of police. After his mother's death he was sent to England to be brought up by his grandmother and aunts in a straitlaced household, the funny side of which he only appreciated in later life. After leaving Bedford Grammar School, Munro went back to Burma to join the police (1893), but failing health forced his resignation and return to England.

Earning his living as a journalist, he wrote political satire for the *Westminster Gazette* and travelled on the Continent as foreign correspondent of the *Morning Post* (1902–08). In 1904 his first collection of stories was published under the title *Reginald*. Other volumes of stories followed: *Reginald in Russia* (1910), *The Chronicles of Clovis* (1911), and *Beasts and Super-Beasts* (1914). His novel *The Unbearable Bassington* appeared in 1912. In all these 'Saki' tales his elegant and effete young heroes, Reginald and Clovis, take an often heartless and sometimes cruel delight in the discomfort or downfall of their conventional and pretentious elders in circumstances that are occasionally macabre but always highly amusing.

In World War I Munro enlisted as an ordinary soldier, refusing to accept a commission. In November 1916 he was killed in action.

Salam, Abdus (1926–   ) *Pakistani physicist who, with Sheldon GLASHOW and Steven WEINBERG, was awarded the 1979 Nobel Prize for Physics for their unification of the weak and electromagnetic forces. He received an honorary KBE in 1989.*

Abdus Salam was educated at the Punjab University and St John's College, Cambridge, where he was awarded his PhD in 1952. He taught at the Punjab University (1951–54) and briefly at Cambridge before moving in 1957 to Imperial College, London, as professor of physics.

Salam is a particle physicist whose most successful work has been the unification of the weak and electromagnetic forces. Similar but independent results were published at the same time by Weinberg in the USA. Coming from the third world, Salam was acutely aware of the intellectual isolation of promising young physicists and mathematicians. To stimulate the growth of science in the third world, in 1964 Salam set up the International Centre for Theoretical Physics in Trieste, where many young scientists from the underdeveloped world have met to exchange ideas and learn new techniques. Salam has been director of this centre since its inception and shares his time between Imperial College and Trieste.

Salazar, Antonio de Oliveira (1889–1970) *Portuguese statesman and prime minister (1932–68), who ruled Portugal as a virtual dictator for thirty-six years.*

Born in Vimieiro in the province of Beira Alta, the son of a small landowner, Salazar was educated at a seminary, where he took minor orders, before entering the University of Coimbra in 1910. Graduating in economics in 1914, he joined the teaching staff of the university. In 1918 he was appointed professor of economics, gaining a reputation for his writings on finance and economic matters. Salazar was elected to the Portuguese parliament as a Catholic Centre deputy in 1921. He resigned after only one session and returned to academic life until 1926, when he was offered the position of finance minister in the military dictatorship that had overthrown the parliamentary regime. He accepted the post, but resigned five days later because of confusion within government ranks; however, he again became minister of finance in 1928 at the request of the provisional government. In 1932 he was appointed prime minister and given almost total control of the government, retaining the post of finance minister until 1940. He remained in power as virtual dictator for thirty-six years, until he was incapacitated by a stroke in 1968.

An austere leader, Salazar introduced a modified version of fascism to Portugal, enacting a new constitution that combined authoritarianism with the values of the *Rerum novarum*, a nineteenth-century papal encyclical. As leader of the only permitted party, the Portuguese National Union, he firmly suppressed opposition (which occurred at home and in several colonies) but he did so without ostentatious displays of power. He maintained Portugal's neutrality during World War II, although he granted facilities to the British in the Azores.

Salk, Jonas Edward (1914–   ) *US microbiologist who developed the first effective vaccine against poliomyelitis, using killed viruses.*
Born in New York, the son of a garment maker, Salk received his BS degree from the College of New York in 1934 and five years later his MD from New York University College of Medicine. In 1942 he joined the University of Michigan, becoming assistant professor of epidemiology in 1946. Here he worked on the development of an influenza vaccine. In 1947 Salk moved to the University of Pittsburg as director of the virus research laboratory. Capitalizing on recent advances, he and his team cultured viruses from all three strains responsible for poliomyelitis in the USA. From these, Salk produced a formaldehyde-killed mixture of the three strains in a mineral-oil emulsion. He announced his vaccine in 1953 and after animal tests and preliminary clinical trials had been successfully completed, the first large-scale vaccination programmes commenced in the USA in 1954. The following year, some children injected with faulty vaccine contracted polio. In spite of this tragedy, the mass-vaccination went ahead, albeit with more stringent controls to ensure that potent viruses were never present in the vaccine.

The Salk vaccine made a major contribution to reducing the incidence of polio but by the 1960s it had largely been replaced by the SABIN vaccine, which conferred greater and longer-lasting immunity. In 1963, Salk became director (from 1984 Distinguished Professor in International Health Sciences) of what is now the Salk Institute for Biological Studies in San

Diego, California. His books include *Man Unfolding* (1972), *World Population and Human Values* (1981), and *Anatomy of Reality* (1983).

Samuelson, Paul Anthony (1915–   ) *US economist who won the 1970 Nobel Prize for his work in economic analysis and methodology, particularly the use of mathematical tools and derivation of new theorems.*
Born in Indiana, Samuelson was educated at the universities of Chicago and Harvard and was appointed professor of economics at the Massachusetts Institute of Technology in 1940 (emeritus professor since 1985). He worked for the US Treasury for seven years after World War II and has acted as a consultant to many government bodies. Much of his work has appeared in journal articles, but his introductory text *Economics* (1948) is now in its thirteenth edition and has been translated into twenty-four languages. Other publications are *The Foundations of Economic Analysis* (1947) and *Linear Programming and Economic Analysis* (1958, written jointly with Dorfman and Solow).

Samuelson studied a wide range of economic topics, including dynamics and equilibrium theory, the trade cycle, capital (upholding the neoclassical position in the debate on the subject with Cambridge, England, in the 1960s), welfare economics (he devised the Samuelson Tests for judging whether or not some charge will increase welfare), 'revealed preference' as a substitute for utility theory in consumer demand, public finance, and international economics. His 'factor-price equalization theorem' states the conditions under which free international trade tends to equalize the prices of a good in the trading countries and thereby to equalize the prices of factors of production, so that free trade can be seen as a substitute for the free mobility of the factors of production. He was the first to construct a model of the trade cycle, based on the interaction between the multiplier and the accelerator in an economy.

Sanger, Frederick (1918–   ) *British biochemist, who has been awarded two Nobel Prizes for Chemistry (1958 and 1980) for his work on the structure of complex biological molecules. He was awarded the CH in 1981 and the OM in 1986.*
The son of a physician, Sanger was educated at Bryanston and St John's College, Cambridge, where he gained his PhD in 1943. He has remained at Cambridge ever since, working in the Medical Research Council's Molecular Biology Laboratory (1951–83).

Sanger began his investigation of the chemical structure of the insulin molecule in 1944. Inventing many of the processes of protein analysis that are now in common use, he took ten years to discover the complete sequence of amino acids in the two chains that constitute the insulin molecule. His first Nobel Prize was awarded for this achievement. Sanger next undertook a comparable study of DNA (deoxyribonucleic acid). He chose the virus Phi X174, which infects the common bacteria *Escherichia coli* and consists of 5375 nucleotides assembled in a specific order. Again new techniques were required before the full sequence could be worked out (and published in 1977). This won for Sanger his second Nobel Prize.

Santayana, George (1863–1952) *Spanish-born US philosopher and poet.*

Born in Madrid, Santayana was brought to Boston at the age of eight. He was educated at Harvard and remained there on the faculty until 1912, when, with the aid of a small inheritance, he left the USA for Europe. He spent some time in England but in 1924 settled in Rome, where he remained for the rest of his life.

A noted stylist, Santayana presented his earlier philosophical views in the four volumes of *The Life of Reason* (1905–06). He rejected the idealism current at the Harvard of his student days and presented instead a naturalistic account of reality, consciousness, and values. In later works, *Scepticism and Animal Faith* (1923) and the four volumes of *The Realms of Being* (1927–40), Santayana sought to develop his views. Adopting an initial position of doubt, he was led to conclude that while we could always doubt that anything exists we were still compelled to accept that we were aware of certain essences. On the basis of these essences it was possible to pass to the world of existents but, Santayana insisted, we could never be aware of this world in the same direct way in which we were aware of the world of essences. Santayana also wished to argue for the existence of the realm of spirit, although, unlike the realm of matter, he argued, it lacked all power.

Santayana was also known for his poetry and for his best-selling novel *The Last Puritan* (1935).

Sapir, Edward (1884–1939) *German-born US linguist and anthropologist.*

Born in Lauenberg, Germany, Sapir went to the USA in 1889, at the age of five, and graduated from Columbia University, where he studied German philology, in 1904. He was then persuaded by the prominent anthropologist Franz Boas to study the languages of the American Indians from an anthropological point of view. After brief periods at California and Pennsylvania universities, Sapir moved to Ottawa in 1910 and spent the next fifteen years studying Nootka and other Canadian Indian languages in his capacity as chief of anthropology at the Canadian National Museum. It was during this time that Sapir wrote his book *Language* (1921), in which he presented his thesis that language should be studied within its social context. He argued that language and culture were interdependent and that the study of language might explain the diverse behaviour of people from different cultural backgrounds. Through this book and his own teaching, Sapir became one of the principal developers of a US school of structural linguistics and a founder of ethnolinguistics.

From 1925 to 1931 Sapir worked at the University of Chicago; in 1929 he suggested that the numerous languages of the American Indians could be classified into six divisions. In 1931 he accepted a professorship at Yale University, where he continued to write essays and articles on American Indian languages and cultures and established a department of anthropology. He was also active as a poet, scholar, and composer. Some of his essays were published in 1949 in *Selected Writings of Edward Sapir in Language, Culture, and Personality.*

Sarnoff, David (1891–1972) *US broadcaster and businessman, who pioneered the development of radio and television in the USA.*

Born in southern Russia, Sarnoff travelled with his mother and younger brothers at the age of nine to join his father, who had emigrated to the USA. His first job was that of delivery boy, and his life continued to display a rags-to-riches element. He became a wireless operator and met Marconi in 1906. Foreseeing the multiple possibilities of radio, he became commercial manager of American Marconi in 1917, having already predicted that radio would become 'a household utility in the same sense as the piano or phonograph'. Sarnoff went on to join the new Radio Corporation of America, successor to the Marconi group, in 1919, and became its general manager in 1921 and its president in 1930. He remained chairman and chief executive until 1950, controlling his empire with vigour and enthusiasm and steering it into the world of television, first black and white, then colour. He also worked through the National Broadcasting Corporation, the first American broadcasting network, founded in 1926. Sarnoff was known as much for his 'wisdom', collected in book form, as for his willingness to take risks and his drive.

Sartre, Jean-Paul (1905–80) *French philosopher, novelist, dramatist, and critic, a leading exponent of existentialism. He was awarded and rejected the Nobel Prize for Literature in 1964.*

Sartre was born in Paris. After his father's death he was brought up in the home of Charles Schweitzer, uncle of Albert SCHWEITZER and maternal grandfather of Jean-Paul. In 1929 Sartre met Simone DE BEAUVOIR, who was to become his lifelong companion; in the same year he graduated from the École Normale Supérieure as a teacher of philosophy. In 1933 he took a break from his teaching career to spend a year in Berlin, where he studied the work of the German phenomenologists Edmund HUSSERL and Martin HEIDEGGER, on whose theories Sartre's version of existentialism is based. He made his name as a writer with his first novel, *La Nausée* (1938; translated as *Nausea*, 1949); this was followed by *Le Mur* (1939), a collection of short stories.

During the latter part of World War II Sartre worked with the Resistance, having been captured by the Germans during the fall of France and released in 1941. In 1943 he expounded his existentialist theories in *L'Être et le néant* (translated as *Being and Nothingness*, 1956), which established him as one of the leading philosophers of the period. Sartre's existentialism stresses the precedence of existence over essence: the individual is born into a void, from which point his life progresses as a series of free choices, decisions, and actions, for which he alone is responsible. Theories of predestination and abstract concepts of the universe have no place in the existentialist world. Later philosophical works include *L'Existentialisme est un humanisme* (1946), which dealt with the social responsibility of freedom, and *Critique de la raison dialectique* (1960), an attempt to synthesize existentialism with Marxist sociology.

In 1945 Sartre founded the left-wing journal *Les Temps modernes* and became increasingly involved in political activities. In the same year he published *L'Âge de raison* (translated as *The Age of Reason*, 1947) and *Le Sursis* (translated as *The Reprieve*, 1947), the first two volumes of what was to have been a tetralogy, now known as the trilogy *Les Chemins de la liberté*. It follows the 'paths to freedom' of

a number of characters, notably the philosophy teacher Mathieu Delarue, through the early years of World War II. The last complete volume, *La Mort dans l' âme* (translated as *Iron in the Soul*, 1950), appeared in 1949; the projected fourth volume would have dealt with the period of the Occupation.

Sartre began writing for the theatre in 1943; his plays emphasize the importance to society of political commitment and action. *Les Mouches* (1943; translated as *The Flies*, 1946) was followed by *Huis-clos* (1944; translated as *In Camera*, 1946). Among his best-known postwar plays are *Les Mains sales* (1948; translated as *Crime Passionel*, 1949), *Le Diable et le Bon Dieu* (1951; translated as *Lucifer and the Lord*, 1953), *Kean* (1954), and *Les Séquestrés d'Altona* (1959; translated as *Loser Wins*, 1960). He also wrote film scenarios, notably *Les Jeux sont faits* (1947) and *L'Engrenage* (1948).

In *Les Temps modernes* Sartre published a number of his own critical essays, which covered a wide range of subjects, beginning with *Qu'est-ce que la littérature?* (1947). Subsequent critical studies dealt with such literary figures as Baudelaire (1947); Jean GENET, in *Saint Genet: comédien et martyr* (1952); and Gustave Flaubert, in the extensive but unfinished biography *L'Idiot de la famille* (1971–72). *Situations* (1947–76) is a compilation of articles, essays, and other miscellaneous writings. In the autobiographical work *Les Mots* (1963; translated as *Words*, 1964), for which he refused the Nobel Prize, Sartre describes the formative years of his childhood and the fantasy world he created in his early attempts at writing. Further insight into the more intimate details of the author's life can be gained from the memoirs of Simone de Beauvoir, which cover the five decades from their student days until Sartre's death and form a lasting memorial to one of the greatest intellectuals of the twentieth century.

Sassoon, Siegfried Lorraine (1886–1967) *British poet and writer. He was awarded the CBE in 1951.*

His father, Alfred Sassoon, belonged to a Jewish family that had made its way to England from Baghdad and Bombay. His mother, Theresa Thornycroft, came from a family of English sculptors. After his father left his family and subsequently died (1895), Sassoon was brought up entirely by his mother. He went to Marlborough College and Clare College, Cambridge, but did not take a degree. Afterwards he led the life of a man of leisure in the country.

He wrote at this period some promising but derivative poetry, which he had privately printed. As an officer in the Royal Welch Fusiliers from 1915, Sassoon's courage was legendary; he was recommended for the VC, but ended up with a bar to his previously won MC. Recuperating from a wound received in spring 1917, he wrote an attack, *A Soldier's Declaration* (1917), on the British authorities, accusing them of deliberately prolonging the war. He expected to be court-martialled but the authorities responded by sending him to Craiglockhart War Hospital near Edinburgh as a 'shell shock' case. There he met Wilfred Owen, a key event in the poetic development of both men. *The Old Huntsman* (1917) and *Counter-Attack* (1918) showed how the war had matured his poetic vision. These collections established his reputation, which was consolidated with *Selected Poems* (1925), *Satirical Poems* (1926), and *The Heart's Journey* (1927).

In *Memoirs of a Fox-Hunting Man* (1928), Sassoon – lightly disguised as the narrator George Sherston – wrote a prose elegy for the world of hunting and cricket that had been swept away by the war. It was an instant success and established itself as a classic. Sassoon followed it with *Memoirs of an Infantry Officer* (1930) and *Sherston's Progress* (1936). From the early 1930s, when he married Hester Gatty, Sassoon spent the remainder of his life in his country house near Warminster, Wiltshire. There he wrote a three-volume autobiography: *The Old Century and Seven More Years* (1938), *The Weald of Youth* (1942), and *Siegfried's Journey* (1945). He continued writing poetry and to mark his conversion to Roman Catholicism in 1957 published the anthology *The Path to Peace* (1960). An enlarged edition of his *Collected Poems* (1947) was published in 1961.

Satie, Erik Alfred Leslie (1866–1925) *French composer. His relatively small output, his humour, and his inventiveness endeared him to Les Six, who regarded him as their guiding light, and later to the School of Arcueil, which included Darius Milhaud.*

The son of a Parisian music publisher and his English wife, Satie attended the Paris Conservatoire in 1879. However, he soon left and became a pianist in a café in Montmartre. During this period he composed some of his early piano music, including *Trois Sarabandes* (1887) and *Trois Gymnopédies* (1888). After a brief affair with the painter Suzanne Valadon (1867–1938) and a flirtation with mystical sects, he retired in 1898 to Arcueil (a suburb of

Paris), where he lived a reclusive and solitary life. In 1905 (when he was approaching forty) he attended the Schola Cantorum and became a pupil of Vincent d'Indy. By 1911, Debussy and Ravel had brought his music to a wider audience and from 1915 Cocteau took an interest in his work and wrote the scenarios for his ballet *Parade* (1917), which Diaghilev produced and Picasso designed. Although he was regarded by some as a charlatan – *Parade* was scored, inter alia, for typewriters and aeroplane propellers – his *Socrates* (1918) for four sopranos and a chamber orchestra is regarded as a very serious and moving work. Debussy, with whom he had a long friendship, was deeply influenced by his work.

Sato Eisaku (1901–75) *Japanese prime minister (1964–72), who presided over the postwar economic growth of Japan and worked towards improving relations with its Asian neighbours. He shared the Nobel Peace Prize in 1974 for his opposition to nuclear weapons.*

Born in Tabuse, the son of a sake distiller, Sato was educated at the Tokyo Imperial University, where he graduated in law in 1924. After working for the ministry of railways he took charge of the Automobile Bureau in the ministry of transport. By 1948 he had become vice-minister for transportation, with a reputation for skilful negotiation on behalf of the government in industrial disputes.

Sato joined the Liberal Party in the same year and was appointed chief secretary to the cabinet of Yoshida Shigeru. He was elected to the parliament (Diet) in 1949 and held several ministerial positions during the 1950s, surviving involvement in a shipbuilding scandal (1954). He emerged as one of the leading members of the coalition Liberal Democratic Party and served as minister of finance when his brother, Kishi Nobusuke, became prime minister (1957). In 1964 he succeeded Ikeda Hayato as president of the Liberal Democratic Party and prime minister. Over the next eight years Japan's economy grew rapidly but Sato's reputation for shady political dealings and his overreliance on the USA made him personally unpopular. To counter this, he sought new markets in southeast Asia and Europe and signed a treaty with the USA (1969), which returned the Ryukyu Islands to Japan and ensured the removal of all nuclear weapons from the area. However, the USA remained in control of Okinawa, which was not returned to Japan until 1972, months before Sato's defeat by Tanaka for the party leadership and premiership.

Saussure, Ferdinand de (1857–1913) *Swiss linguist, noted for his early structuralist views on language.*

Born in Geneva, Saussure was educated at the universities of Leipzig and Berlin. He worked initially on the comparative philology of the Indo-European languages, which he taught in Paris, before returning in 1891 to Switzerland to take up the post of professor of linguistics at the University of Geneva. After his death Saussure's theoretical views became widely known and extremely influential, through the posthumous publication of his lectures in *Cours de linguistique générale* (1915; translated as *Course in General Linguistics*, 1959).

It was Saussure more than any other linguist who broke away from the previous century's obsession with historical linguistics. In addition to the diachronic study of language, he insisted, we should also consider its synchronic features. We should also be prepared to pay as much attention to language as *langue* (roughly, grammatical rules) as to language as *parole* (the actual act of speaking). Saussure also spoke of other distinctions – form and matter, paradigm and syntagm, for example – which have since become an accepted part of the structuralist's vocabulary. In more general terms Saussure saw language as a system of signs in which there is established a conventional relationship between the *signifié* (the thing signified, i e the meaning) and the *significant* (the signifier, i e the word). Language thus became a system of mutually dependent and interacting signs. Consequently no item could be understood on its own but is in fact like a chess piece, part of a complex and integrated structure. It is this notion of structure that proved to be one of the most suggestive of Saussure's innovations, influencing the work of (among others) the linguist Roman Jakobson (1896–1982) and the anthropologist Lévi-Strauss.

Savage, Michael Joseph (1872–1940) *New Zealand statesman and the first Labour prime minister (1935–40).*

Born near Benalla, Victoria, of Irish parents, Savage left school at fourteen to take a position in a store before working on a farm in New South Wales and later down a mine in Victoria. Travelling to New Zealand in 1907, he took a job as a cellarman in Auckland, becoming active in the trade union movement as a representative of Auckland brewery workers. Joining the Labour Party on its formation in 1916, he was elected national secretary in 1919, winning the seat of Auckland West for the Labour

Party the same year. In 1923 he was elected deputy leader of the parliamentary party and, although relatively unknown, succeeded to the leadership in 1933 following the death of the party's leader Henry Holland. Winning a resounding electoral victory in 1935, he became the first Labour prime minister, a position he held until his death in 1940.

Although not regarded as an intellectual, Savage was very popular within the electorate because of his appeal to the average person and is remembered for his public benevolence and humanitarianism. He introduced many reforms, including some far-reaching social security legislation that he dubbed 'applied Christianity'.

Sayers, Dorothy L(eigh) (1893–1957) *British writer, known especially for her detective novels.*

Dorothy Sayers was born at Oxford, where her father was headmaster of Christ Church Choir School. She was educated at the Godolphin School, Salisbury, and won a scholarship to Somerville College, Oxford. After a year's teaching, she became an advertising copywriter, a job that she kept until 1931. In the meantime she began methodically to learn the skill of writing detective novels; *Whose Body?* (1923), her first attempt at the genre, introduced her aristocratic detective Lord Peter Wimsey. After that she kept up a steady stream of ingenious novels, ending with *Busman's Honeymoon* (1937), in which Lord Peter marries the lady writer of detective stories whom he had cleared of a murder charge in an earlier novel, *Strong Poison* (1930).

Two plays written for the Canterbury Festival – *The Zeal of Thy House* (1937) and *The Devil to Pay* (1939) – marked Dorothy Sayers's change from thriller writing to Christian apologetics. Her best-known production in this field was the radio drama *The Man Born to be King* (first broadcast 1941–42). She also wrote the essays entitled *The Mind of the Maker* (1941) and *Introductory Papers on Dante* (1954). The translation of Dante occupied much of the last period of her life; the *Inferno* appeared in 1949, the *Purgatorio* in 1955, and the *Paradiso* (1962) was incomplete at the time of her death.

Scheel, Walter (1919–) *West German statesman and president of the Federal Republic of Germany (1974–79). He earned a reputation for his economic expertise and his advocacy of the 'aid by trade' policy in relations with developing countries.*

Born in Solingen, the son of a wheelwright and carriage builder, Scheel was educated at the Reform-Gymnasium in Solingen. Graduating in 1938, he began an apprenticeship with a local mercantile bank before serving in the German air force during World War II.

Scheel joined the Free Democratic Party (FDP) in 1946 and was elected to the city council in Solingen in 1948. He won a seat in the North Rhine-Westphalia state legislature in 1950, becoming a member of the Bundestag (lower house of the federal legislature) in 1953. In 1955 he was chosen as the Federal Republic representative in the European parliament, a position he held until 1969. He was appointed minister for economic cooperation in the coalition government under ADENAUER in 1961 and was elected vice-president of the Bundestag in 1967. Becoming chairman of the FDP and leader of the opposition against Kiesinger in 1968, he assumed the positions of vice-chancellor and foreign minister in the BRANDT coalition government. He resigned these posts in 1974 to take up the presidency, which he held until 1979.

Schiaparelli, Elsa (1896–1973) *French couturière, born in Italy, noted for her innovative use of nylon and other synthetic fabrics.*
Schiaparelli was born in Rome into an illustrious academic family. She worked initially in the USA as a film scriptwriter, translator, and sculptress before moving to Paris in the mid-1920s. She began designing knitwear in her hotel bedroom and in 1927 moved to a studio in the Rue de la Paix. Her introduction in 1931–32 of the built-up shoulder was to influence women's fashions for the entire decade. Her designs were often inspired by contemporary artists, such as Salvador DALI, with touches of the surreal. A notable feature was her bold use of colour: a particularly strong pink, first employed by her, became known as shocking pink.

Schiaparelli moved into new premises in the Place Vendôme in 1934 and her business expanded into the whole spectrum of women's wear as well as cosmetics and perfumes. During World War II she toured the USA giving lectures and in 1949 opened a branch in New York. Her autobiography, *Shocking Life*, was published in 1954.

Schiele, Egon (1890–1918) *Austrian painter.*
Schiele joined the Vienna Academy of Art in 1906 but left in 1909 because his teacher disapproved of his paintings, which were influenced by the art nouveau artist Gustav Klimt (1862–1918). By 1910 he had evolved a ma-

ture style of his own, which was expressionist in character although Schiele never identified himself with the expressionist movement. During the next eight years he produced the oils, watercolours, and drawings – mostly portraits – upon which his reputation rests. Distorted, neurotic, and unappealing, the portrait figures nevertheless reveal the artist's great sensitivity and masterly draughtsmanship. The element of eroticism in some of them resulted in Schiele spending a month in prison in 1911. In 1913 he was drafted into the army, four days after his wedding; he spent part of World War I as a war artist. In 1918 at the exhibition of the Vienna Secession his pictures received international acclaim, but his death the same year of Spanish influenza deprived the world of a highly original talent.

Schlesinger, John Richard (1926–) *British film, television, and theatre director, who has also worked in the USA.*
Born in London and educated at Uppingham and Balliol College, Oxford (he was made an honorary fellow in 1981), Schlesinger began as an actor, first in university productions and then in the professional theatre and in such films as *The Battle of the River Plate* (1956) and *Brothers-in-Law* (1957). As a director he began with BBC television programmes (1958–60), working on *Tonight* and *Monitor*; he then won the Golden Lion Award at the Venice Film Festival with the British Transport film *Terminus* (1960).

His feature film *A Kind of Loving* (1962), adapted from Stan Barstow's novel, won the Golden Bear Award at the Berlin Film Festival. Subsequent films have included *Billy Liar* (1963), *Darling* (1965), *Far From the Madding Crowd* (1967), *Midnight Cowboy* (1969), which was made in New York and won the best film and best director Oscars, *Sunday Bloody Sunday* (1971), *Marathon Man* (1976), *Yanks* (1978), *Honky Tonk Freeway* (1980), *The Believers* (1987), and *Pacific Heights* (1990). His television plays, *Separate Tables* (1982) and *An Englishman Abroad* (1983), were also enthusiastically received. He was an associate director of the National Theatre from 1973 to 1988 and has directed numerous plays as well as *The Tales of Hoffmann* (1980) at Covent Garden.

Schlick, Moritz (1882–1936) *German philosopher and physicist, founder of the famous school of logical positivists known as the Vienna Circle.*
Schlick was born in Berlin and educated at the university there, where he studied physics

under Max PLANCK, gaining his PhD in 1904. He taught physics at the University of Rostock from 1911 to 1922, when he moved to Vienna as professor of philosophy. Influenced by Ernst Mach (1838–1916) and WITTGENSTEIN, Schlick sought to develop a positivist account of science.

In this he was fortunate to reside in Vienna, which at that time commanded a large number of young, talented, and industrious mathematicians, philosophers, and scientists who were as interested in the problem as Schlick. Consequently such figures as CARNAP, GÖDEL, Waismann, Neurath, and many others began to meet with Schlick in the late 1920s once a week to discuss their common problems. The meetings continued until 1936, when Schlick was murdered by one of his graduate students. In some accounts the student is described as mad, but others speak of jealousy, revenge for a failed doctoral thesis, and even political motives. In any case the great days of the Circle were already over. Many of its members were Jews or had left-wing sympathies and, with the rise of HITLER and the impending *Anschluss*, had begun to leave Vienna for saner and safer institutions in Britain and the USA.

Schlick's own philosophical output was relatively modest. He published a work on ethics, *Fragen der Ethik* (1930), one on epistemology, *Allgemeine Erkenntnislehre* (1919; translated as *General Theory of Knowledge*, 1974), and his collected papers were published posthumously in 1938 as *Gesammelte Aufsätze*.

Schmidt, Helmut (1918–   ) *German statesman and chancellor of the Federal Republic of Germany (1974–82).*

Born in Hamburg, the son of a teacher, Schmidt was educated at Lichtwark high school and at the University of Hamburg. During World War II he served on the Russian and western fronts but was captured by the British towards the end of the war. From 1949 to 1953 he managed the transport administration of Hamburg state.

Schmidt joined the Social Democratic Party (SPD) in 1946, becoming the first national chairman of the Socialist Students League the following year. Elected as a deputy to the Bundestag (lower house of the federal legislature) in 1953, he lost his seat in 1962 and assumed the post of director of the department of the interior in Hamburg state. He returned to the Bundestag in 1965, serving as chairman of the SPD parliamentary party from 1967 to 1969. Appointed defence minister in the BRANDT government in 1969, he became finance minis-

ter in 1972, a position he held until 1974, when he succeeded Brandt as chancellor. He lost the chancellorship in 1982, when he was defeated at the polls by Helmut KOHL, and finally left the Bundestag in 1986. Since 1983 he has been the publisher of the newspaper *Die Zeit*.

Schnabel, Artur (1882–1951) *Austrian-born US pianist, known particularly for his interpretations of Beethoven and Schubert.*

Schnabel studied the piano in Vienna with Theodor Leschetizky (1830–1915), who summarized his pupil's approach to music in the words: 'You will never be a pianist; you are a musician'. However, Schnabel made his debut as pianist in 1890 and thereafter embarked on a long and successful career as a concert pianist. In 1905 he married the contralto Therese Behr, and their partnership in Lieder recitals culminated in a historic series of Schubert programmes in Berlin in 1928. Schnabel also partnered many of the greatest string players of his day in chamber music; his pupils included Clifford CURZON and Claude Frank.

Until 1933 Schnabel spent much of his life in Berlin, touring Europe and the USA extensively; he was also much sought after as a teacher. When the Nazis came to power, he emigrated to the USA, where he became a naturalized citizen in 1944. He also composed three symphonies, five string quartets, a piano concerto, and many songs and piano pieces.

Schoenberg, Arnold Franz Walter (1874–1951) *Austrian-born US composer, who invented serial composition and became one of the most controversial figures in twentieth-century music.*

Born in Vienna, Schoenberg had little formal musical education, teaching himself to play any instruments at hand; he eventually sought advice about his works from the conductor and composer Alexander von Zemlinsky (1872–1942), whose sister he married in 1901. During this period Schoenberg was composing mainly large-scale works in a rich chromatic style, influenced by both Brahms and Wagner. Examples of his early work include the mono-drama *Transfigured Night* (1899) for string sextet (which he later orchestrated), *Gurrelieder* (1900–11) for solo voices, chorus, and large orchestra, and the tone poem *Pelleas und Melisande* (1902–03). Although he was forced to work for a period in a bank, he was soon able to earn a living orchestrating theatre music and teaching. In 1910 he became a teacher at the Vienna Academy, where Alban BERG and Anton WEBERN became his students. Both

Richard STRAUSS and Gustav Mahler had by then taken an interest in his work.

Schoenberg's second string quartet (1907–08), ostensibly in F sharp minor, is a landmark in the development of western music as the final movement introduced the concept of atonality. The *Three Piano Pieces* of 1909 is the first work in which Schoenberg abolished the distinction between consonance and dissonance, while in *Pierrot Lunaire* (1912) for voice and five instrumentalists he consolidated his use of these techniques and introduced the device of *Sprechstimme*, a pitch-notated form of speech. From these experiments Schoenberg gradually evolved a formal system of composition in which all twelve notes of the chromatic scale are of equal importance; a series of up to twelve of these notes could be used horizontally or vertically, starting on any note of the chromatic scale, the complete series of notes being used before any of them appear again. This technique was known as serialism or twelve-tone composition. The third and fourth movements of the *Serenade* (1923) for seven instruments and bass voice are the first clear examples of this technique; *Variations for Orchestra* (1927–28) is a mature serial work.

In 1933, like so many other Jewish Austrians, Schoenberg left Europe for the USA. He had been previously converted to Catholicism but now returned to his Jewish faith; his unfinished opera *Moses und Aron* (1932–51), using his own libretto, is an indication of his religious affiliations in this period. In 1936 he became professor of composition at California University, where he remained until his retirement at seventy; he became a naturalized US citizen in 1940. During these years in America Schoenberg returned to tonal composition from time to time, but also continued to develop his serial techniques, which gradually began to find acceptance by younger composers in the 1960s.

Schrödinger, Erwin (1887–1961) *Austrian physicist, who was awarded the 1933 Nobel Prize for Physics for the wave equation that he formulated.*

The son of a prosperous factory owner, Schrödinger was educated privately until he was eleven and then at the Vienna grammar school and university. After serving as an artillery officer in World War I, Schrödinger taught at the universities of Stuttgart and Kiel until 1927, when he succeeded Max PLANCK as professor of physics at Berlin. By this time Schrödinger had published his wave equation,

enabling physicists accurately to describe the motion of atomic and nuclear particles.

In 1933 the Nazis gained power in Germany and Schrödinger, long opposed to fascism, immediately left Germany for exile in Oxford. He returned to Austria, to the University of Graz, in 1936 but the *Anschluss* of 1938 forced him to flee first to Italy and eventually to the USA, where he taught briefly at Princeton. In 1939 he received an offer from DE VALERA, prime minister of Eire, to join the newly founded Institute of Advanced Studies in Dublin. Schrödinger remained in Dublin until his retirement in 1956, when he returned to his native Austria. During his period at Dublin, Schrödinger wrote a number of general works. One of these, *What is Life?* (1944), proved to be remarkably influential among scientists of all disciplines.

Schumacher, Ernst Friedrich (1911–77) *German-born economist and conservationist, known mainly for his book* Small is Beautiful *(1973), a critique of western methods of mass production and specialization.*

Schumacher studied economics at New College, Oxford, coming to the UK with a Rhodes scholarship. At the age of twenty-two he taught at Columbia University, New York, before turning to business, farming, and journalism. He was economic adviser to the British Control Commission in Germany (1946–50) and economic adviser to the National Coal Board (1950–70), but his principal interest was in developing the concept of intermediate technology for the developing countries, acting as an adviser on rural development. In 1966 he founded the Intermediate Technology Development Group Ltd. Concerned with the conservation of natural resources, he was president of the Soil Association, which is involved with organic farming methods; he was also a director of the Scott-Bader Company, a successful experiment in common ownership that is at the forefront of polymer chemistry.

Schumacher published *Export Policy and Full Employment* (1945), *Roots of Economic Growth* (1962), *Small is Beautiful: Economics as if People Mattered* (1973), and *Good Work* (1979), a collection of speeches published posthumously.

Schuman, Robert (1886–1963) *French statesman, prime minister (1947; 1948), and founder of the European Coal and Steel Community (1952).*

Born in Luxembourg into a wealthy Lorraine family, Schuman was educated at the universities of Bonn, Berlin, Munich, and Strasbourg, where he gained a doctorate in law. He started to practise law in Metz, but spent most of World War I in a German prison because he refused to serve in the German army.

Schuman was elected to the French Chamber of Deputies as the representative for Moselle in 1919. A member of the moderate conservative Parti Democratique Populaire (PDP), he served on the parliamentary finance commission for seventeen years, becoming undersecretary of state for refugees in the Reynaud government in 1940. Imprisoned by the Germans the same year, he escaped in 1942 and fought with the French resistance movement during the final days of World War II. In 1944 Schuman founded the Mouvement Républicain Populaire (MRP), a liberal Catholic party that replaced the PDP. Elected a deputy for the MRP in 1945, he was appointed finance minister in 1946–47, becoming prime minister in late 1947. He was forced to resign after only seven months in office but held the post briefly again in 1948, before serving as foreign minister in ten successive governments (1948–52).

An ardent supporter of European integration and Franco-German reconciliation, Schuman gained recognition for the so-called Schuman Plan of 1950, which proposed that French and German coal and steel production should be combined under one authority. This authority, the European Coal and Steel Community, which emerged in 1952, was the forerunner to the European Common Market.

Schumann, Elisabeth (1885–1952) *German-born US singer, who specialized in the soprano roles of Mozart and Richard STRAUSS.*
She studied in Dresden, Berlin, and with Alma Schadow in Hamburg, where she made her opera debut at the State Theatre in 1909 as the Shepherd in Wagner's *Tannhäuser.* In 1919, Richard Strauss persuaded her to join the Vienna State Opera, and in 1921 Schumann toured America with Strauss in recitals of his songs. She remained in Vienna until 1938, when the Nazis forced her to leave. From then on she lived in New York, becoming a US citizen in 1944.

Schuschnigg, Kurt von (1897–1979) *Austrian statesman and chancellor (1934–38), who as a fervent anti-Nazi did his utmost to maintain Austrian independence against the will of the people.*
Born in Riva on Lake Garda (now in Italy), the son of an army officer, Schuschnigg was educated at a Jesuit school in Feldkirch before serving in the Austro-Hungarian army during World War I. He was taken prisoner in 1918 and was later decorated for bravery; after the war he graduated in law from the University of Innsbruck (1922).

Schuschnigg was elected as a Christian Social deputy to the Austrian parliament in 1927. Appointed minister of justice (1932–34) and minister of education (1933–34), he became chancellor in 1934 following the assassination of DOLLFUSS. In the face of German encroachment he tried to maintain Austrian independence, which he managed to achieve until he lost MUSSOLINI's support in 1937. In 1938 he attempted to forestall HITLER's demands for Austrian Nazis to be included in the government but was forced to resign shortly before German troops invaded and annexed Austria. The *Anschluss* was later ratified by plebiscite. Schuschnigg was imprisoned by the Germans until 1945, when he emigrated to the USA, where he became a professor of political science at St Louis University, Missouri. He remained in the USA for twenty years before returning to Austria.

Schwarzkopf, (Olga Maria) Elisabeth (Friederike) (1915–) *German singer, who is known for her soprano roles in Mozart and Richard STRAUSS and for her Lieder singing.*
Schwarzkopf studied under Maria Ivogün and in 1938 joined the Berlin Municipal Opera. In 1942 she made her debut as a recital singer in Berlin, which considerably enhanced her reputation. In 1943, at Karl Böhm's invitation, she joined the Vienna State Opera, making her Covent Garden debut in 1947, when the Vienna Opera Company visited London. Thereafter she sang at Covent Garden for five seasons as coloratura soprano in both the Italian and German repertory. Schwarzkopf sang for the first time at the Salzburg Festival in 1949, as the Countess in Mozart's *The Marriage of Figaro*, and her debut at La Scala in Milan, in the same part, took place in 1949. Of the wide range of operatic roles she sang, Schwarzkopf was best known as the Marschallin in Richard Strauss's *Der Rosenkavalier*, the part she sang in her farewell performance in Brussels in 1972.

As a Lieder singer, Schwarzkopf was supreme. In 1953 she married Walter Legge, then artistic director of EMI records, and made many recordings under his direction. Her master classes in the USA and Britain are widely acclaimed.

Schwarzkopf, H Norman (1935–) *US general, who became internationally known as* *overall commander of the victorious US-led alliance against Iraq in the Gulf War of 1990.*

Son of the police chief Herbert Norman Schwarzkopf, who was in charge of the investigation into the kidnapping of the LINDBERGH baby in 1932, he was given the first name 'H' as his father disliked the family name 'Herbert'. The junior Schwarzkopf won a reputation at school and subsequently in the army for his intelligence (he was credited with an IQ of 170) and his unpredictable temper, which earned him the nicknames 'The Bear' and 'Stormin' Norman'. In contrast to his ebullient and sometimes controversial behaviour as a soldier, he also became known for his love of music, ballet, and amateur magic and for his concern to protect the lives of the men under his command. During the Vietnam War he won two Purple Hearts and three Silver Stars (for courage in battle). When the crisis in the Gulf erupted in 1990, following Iraq's invasion of Kuwait, Schwarzkopf proved a major success both as a military strategist and as a popular and forceful leader who could handle relations with his own and foreign troops, as well as with the world's press, with consummate ease. After the triumphant liberation of Kuwait in 1991, with minimal loss of life to allied troops, he returned home to a hero's welcome and speculation that he might be considered a presidential candidate. He also received an honorary knighthood from ELIZABETH II.

Schweitzer, Albert (1875–1965) *German theologian, musician, and medical missionary. He was awarded the Nobel Peace Prize in 1952 and became an honorary OM in 1955.*

Born in Alsace, the son of a pastor, Schweitzer studied theology and philosophy at the universities of Strasbourg, Paris, and Berlin and quickly made his reputation as an original and powerful theologian. In his best-known work, *Von Reimarus zu Wrede* (1906; translated as *The Quest for the Historical Jesus*, 1910), he emphasized the need to understand the thought of Jesus within the context of the Jewish thought of his day. Schweitzer, who had studied the organ under Charles Widor (1844–1937) in Paris, was at the same time working on his *Bach, le musicien-poète* (1905), an equally impressive study.

But Schweitzer's theological and musical careers were coming to an end: in 1896 he had decided that he would devote only his first thirty years to scholarship and music. Thereafter his life would be spent in service to humanity. He consequently began to study medicine,

and after qualifying in 1913 Schweitzer and his wife (who trained as a nurse) left for Lambaréné, a missionary station on the Ogowe River in what is now Gabon. Apart from periodic visits to Europe to raise funds and give organ recitals, Schweitzer remained at Lambaréné for the rest of his life, demonstrating by example, in the phrase he made famous, his 'reverence for life'. Schweitzer did, however, continue writing, and such works as *On the Edge of the Primeval Forest* (1922) brought his own position and views before a wide public. He became in fact one of the most famous figures in the world and Lambaréné the best-known hospital in Africa.

Scorsese, Martin (1942–) *US film director, who emerged as a leading figure in the cinema in the 1970s.*

Born on Long Island, New York, Scorsese studied film at New York University, where he became an instructor in 1963; he made his film debut as a director in 1968 with *Who's That Knocking at My Door?*. However, his reputation dates from 1973, when he made *Mean Streets*, a violent and realistic study of New York's Italian community that marked the beginning of Scorsese's long collaboration with the actor Robert DE NIRO; it also established his interest in the themes of brutality and conflict that characterize many of his subsequent films. In *Alice Doesn't Live Here Anymore* (1974) he tackled the subjects of bereavement and self-discovery, while in *Taxi Driver* (1976) he depicted the violent life of a psychopathic New York cab driver, played by De Niro. He continued to win public and critical acclaim working with De Niro in *New York, New York* (1977), *Raging Bull* (1980), and *The King of Comedy* (1983) was well as with Paul NEWMAN in the Oscar-winning *The Color of Money* (1980). Scorsese's career has not been without controversy however; *The Last Temptation of Christ* (1988) caused an international uproar, being considered blasphemous by many Christian pressure groups. His most recent films include *GoodFellas* (1990), in which he resumed his partnership with De Niro.

Scott, Paul Mark (1920–75) *British novelist.* Scott was born in London and educated at Winchmore Hill Collegiate School. From 1940 to 1946 he served in the army, mainly in India and Malaya, which form the background to his best-known novels. In 1941 *I, Gerontius*, a book of poems, was published. On leaving the army he worked first for a publishing firm (1946–50) and then became a literary agent. He resigned in 1960 in order to devote himself to his own writing. Meanwhile he had published *Johnnie Sahib* (1952) and written several plays for television and radio.

During the 1960s Scott's reputation as a novelist grew steadily with such books as *The Chinese Love Pavilion* (1960), a story set in Malaya that deals with a characteristic Scott theme – love and guilt in sexual relations across racial boundaries. *The Bender* (1963) was a new departure in that it was set in London. *The Jewel in the Crown* (1966) was the first volume of the epic *Raj Quartet*, four interconnected novels set in the last days of British rule in India. It was followed by *The Day of the Scorpion* (1968), *The Towers of Silence* (1971), and *A Division of the Spoils* (1975). In 1977 *Staying On*, a coda to the *Raj Quartet*, was awarded the Booker Prize. The *Raj Quartet* became immensely popular when it was televised as a serial in 1983.

Scott, Sir Peter Markham (1909–89) *British naturalist and painter who made a major contribution to nature conservation, in particular by founding the Wildfowl Trust (subsequently renamed the Wildfowl and Wetlands Trust) in Britain. He was knighted in 1973.*

The son of the polar explorer Robert Falcon SCOTT, Peter Scott first studied science at Trinity College, Cambridge (1927), but soon switched to art history and architecture. Later he studied at the State Academy School, Munich, and the Royal Academy, London, and in 1933 gave his first one-man exhibition of paintings, which were mainly of geese and other wildfowl. Wildlife remained his favourite subject although he also painted some portraits, including Queen ELIZABETH II and Princess Margaret. Inspiration for his work derived from his favourite hobby – wildfowling. In the late 1930s he made several trips abroad to observe geese on their migrations and he became increasingly concerned with conserving wildfowl rather than shooting them. He was also a keen yachtsman, winning a bronze medal in the 1936 Olympic Games and the Prince of Wales Cup in 1937, 1938, and 1946. He also captained the British entry in the 1964 America's Cup. During World War II he served in the Royal Naval Supplementary Reserve and became lieutenant-commander of a gunboat, which he renamed *The Grey Goose*.

After the war in 1946 Scott founded the Wildfowl Trust at Slimbridge on the River Severn in Gloucestershire and later opened another reserve at Peakirk, Lincolnshire, both sites becoming major centres for the breeding and study of wildfowl and other birds. Scott made many expeditions to study wildlife around the world. He was presenter of the TV series *Look* and wrote many books, including *Morning Flight* (1935), *Key To Wildfowl of the World* (1949), *The Eye of the Wind* (1961), and *Travel Diaries of a Naturalist* (1983). He was chairman of World Wildlife Fund International (1962–82) and vice-president of its British branch.

Scott, Robert Falcon (1868–1912) *British explorer who, on the second of his two Antarctic expeditions, reached the South Pole with several companions only to perish on the return journey.*

Born in Plymouth, Scott entered the Royal Navy, becoming a first lieutenant; in 1900 he was appointed leader of the National Antarctic Expedition and promoted to commander. His ship *Discovery* departed in August 1901 and, after studies of the Ross Ice Shelf, moored for the winter at McMurdo Sound. The expedition, which included E. H. SHACKLETON, spent two years making scientific observations and surveying South Victoria Land. On their return in 1904, Scott was promoted to captain.

After various naval commands and admiralty duties, Scott's plans for a further Antarctic expedition came to fruition on 1 June 1910 when his vessel *Terra Nova* departed for the Ross Sea. They established base at Cape Evans in January 1911 and, after setting up forward supply bases, set out for the Pole on 1 November 1911. The motor sledges were soon abandoned; several of the ponies succumbed and the rest were shot for food. At the Beardmore Glacier, only five members continued, pulling their sledges by hand: Scott, Dr E. A. Wilson, Captain Lawrence Oates, Lieutenant H. R. Bowers, and Petty Officer Edgar Evans. They reached the Pole on 17 January 1912 only to find the Norwegian flag erected by AMUNDSEN's team one month earlier. Exhausted and dejected, Scott's group started the return journey. On 17 February, Evans died. On 17 March, Oates, too badly frostbitten to continue, gallantly crawled out of their tent into a blizzard to his death in order not to slow down the rest of the party. Scott wrote the final entry in his diary on 29 March as they camped in a blizzard only eleven miles from their southernmost supply depot. Their bodies, together with Scott's diaries and letters and the geological specimens they had gathered, were found by a search party the following November.

Scriabin, Alexander Nikolayevitch (1872–1915) *Russian composer and pianist,* *whose reputation relies on his early piano music.*

Son of an aristocratic Moscow family, Scriabin studied at the Moscow Conservatory under Sergei Taneyev (1856–1915) at the same time as RACHMANINOV. After graduation he toured Europe playing his own piano preludes, impromptus, dances, studies, and early sonatas, which owe a great deal to Chopin and Liszt in their form and content. After a period as professor of the piano at the Moscow Conservatory (1898–1904), Scriabin concentrated almost entirely on composition, writing a number of tone poems and sonatas in a new, complex, and somewhat eccentric style that emerged from his preoccupation with mysticism and theosophy. In this category are *The Divine Poem* (1903, his third symphony), *The Poem of Ecstasy* (1908), and *Prometheus: The Poem of Fire* (1909–10), which he envisaged as using a keyboard to project colours as well as a large orchestra, piano, organ, and chorus. Scriabin regarded his later works as exercises in preparation for his ultimate 'Mystery', a union of the arts that would transform the world into a paradise and himself into a messiah. This work had not progressed beyond a few preliminary sketches before he died.

Scullin, James Henry (1876–1953) *Australian statesman and Labor prime minister (1929–31).*

Born near Ballarat, Victoria, the son of a railway worker, Scullin left school to work in a grocery store and attended night classes, where he won prizes as a debator and public speaker. Scullin joined the Labor Party in 1903 and was elected to the House of Representatives as the member for Corangamite in 1910. He lost this seat in 1913 but was elected as the member for Yarra in 1921, becoming leader of the party in 1928. He was elected prime minister in 1929 at the height of the Depression. Beset by difficulties of the Depression, ministerial inexperience, and an opposition-controlled senate, he lost the 1931 election but retained the Labor leadership until 1935, when he resigned because of illness.

An effective parliamentary speaker who was greatly respected for his integrity and sincerity, Scullin was a close advisor to his successor, CURTIN, during World War II.

Seaborg, Glenn Theodore (1912–   ) *US chemist, who was awarded the 1951 Nobel Prize for Chemistry for his discovery of a large number of transuranic elements.*

Seaborg's father, a machinist, emigrated to the USA from Sweden, Seaborg himself being educated at the University of California, Berkeley, where he gained his PhD in 1937. He remained at Berkeley where in 1940, with several colleagues, he isolated the new element 94, later known as plutonium. With the outbreak of World War II, Seaborg moved to the Metallurgical Laboratory at Chicago, where he worked on the production of plutonium. The techniques worked out by Seaborg in Chicago were later used at the plant in Hanford, Washington, to produce the plutonium for the first atomic bomb. Seaborg continued to work on the transuranic elements and between 1944 and 1958 his group at Berkeley was involved in the discovery of all the elements from 95 (americium), discovered in 1944, up to and including 102 (nobelium), in 1958.

Seaborg returned to Berkeley after the war as professor of chemistry at the Radiation Laboratory. He left Berkeley again in 1961 to serve for a ten-year period as chairman of the Atomic Energy Commission, during which time the US nuclear-power industry underwent considerable growth. In 1971 he returned to Berkeley to the Lawrence Radiation Laboratory, as it had been named. His publications include *Nuclear Milestones* (1972) and *Stemming the Tide* (1987), about nuclear arms control.

Seferis, George (Giorgios Seferiadis 1900–71) *Greek poet, critic, and diplomat. He was awarded the Nobel Prize for Literature in 1963.*

Seferis was born at Smyrna but at the outbreak of World War I left it, never to return (the Greek community did not survive there after 1922). A sense of exile and of the loss of his birthplace colours much of his work. He was educated at Athens and the Sorbonne, graduating in law in 1924. After two years spent studying English in London he joined the Greek diplomatic corps, working in London until 1934 and during World War II serving in various posts with the Greek government-in-exile. From 1957 until his retirement in 1962 he was ambassador to Britain. His denunciation of the dictatorship of the Papadopoulos regime in 1969 was an act of great courage and at his funeral in Athens he was honoured not only for his cultural but also for his political achievement.

While vice-consul in London in 1931 he read and was influenced by the works of Ezra POUND and T. S. ELIOT. His first volumes, *Strophe* (1931; 'Turning-Point') and *I Sterna*

(1932; 'The Cistern'), also owe something to Rimbaud, Laforgue, and Valéry. Seferis's ability to combine traditional elements, every-day diction, and demotic and folk-poetry materials had an immensely invigorating effect on modern Greek poetry. His own work had the added authority of his first-hand familiarity with the great historical events of his time. His collections include *Mythistorema* (1935) and a series of poetic 'logbooks', *Himeralogion katastromatos* (I, 1940; II, 1944; III, 1965). *Dokimes* (1962; translated as *On the Greek Style*, 1966) is generally held to be the best work of literary criticism on modern Greek. Seferis's *A Poet's Journal: days of 1945–1951* (1974) has been translated by A. Anagnostopoulos, and *Collected Poems, 1924–55* by E. Keeley and P. Sherrard.

Segovia, Andrés (1893–1987) *Spanish guitarist, who enjoyed an unrivalled reputation as the greatest classical guitarist of the century and was largely responsible for the revival of the instrument's popularity.*

Self-taught and against family opposition, Segovia gave his first public recital at the age of fifteen in Grenada. Its success provided the encouragement he needed to make the guitar his career. Recitals in Madrid and Barcelona were followed by a tour of Latin America (1916), and his Paris debut (1924) led to recital tours in Europe and the Soviet Union. In 1928 he visited the USA for the first time and subsequently he enjoyed a worldwide reputation.

His self-taught technique, using finger tips and nails to provide an exceptionally wide range of tone-colour, was extremely influential on younger players. Because the repertory of the guitar is relatively modest, Segovia made a large number of transcriptions of classical music, including Bach, to increase it. He also commissioned works for his instrument from such composers as VILLA-LOBOS, RODRIGO, FALLA, and Castelnuovo-Tedesco.

Sellers, Peter (1925–80) *British comedy actor, whose gift of impersonation made him one of the best known faces of the 1960s.*

Born in Southsea, Sellers appeared in his parents' theatre, with ENSA, at the famous London Windmill Theatre, and ultimately in the popular radio series of the 1950s, *The Goon Show*. He then went on to make such British comedy films as *The Lady Killers* (1955) and *The Smallest Show on Earth* (1957). With the Boulting Brothers' *I'm All Right Jack* (1959) he attained stardom, which was reinforced by *The Millionairess* (1960), in which he played an Indian doctor. In the 'Pink Panther' series of the 1960s and 1970s, he created the part of the French Inspector Clouseau. *Lolita* (1962), *What's New Pussycat?* (1965), and *There's a Girl in my Soup* (1970) were among the many other films that helped to make him an international superstar. One of his last films, *Being There* (1979), earned him an Oscar nomination as best actor.

An early heart attack left him vulnerable and he died, prematurely, in his prime. His children by his first wife, Anne Howe, whom he married in 1951, wrote a biography of him, *PS, I Love You* (1981). He also married the actresses Britt Ekland (1964) and Lynne Frederick (1977).

Senanayake, Don Stephen (1884–1952) *First prime minister of Ceylon (1947–52).*

Born in Ceylon (now Sri Lanka), the son of a prosperous landowner, Senanayake was educated at St Thomas College, Colombo, before joining the civil service as a clerk. He worked for a number of years on his father's rubber and coconut plantation.

Senanayake first became involved in politics in 1919, when he co-founded the Ceylon National Congress. In 1922 he was elected to the legislative council of the British Crown Colony of Ceylon and campaigned for legal and constitutional reforms. In 1931 he was elected to the council of state and was appointed minister of agriculture and lands. His book *Agriculture and Patriotism*, dealing with his projection of Ceylon's future, was published in 1935. He held both posts until 1947 and was leader of the council from 1942 to 1947. Becoming prime minister the same year, he presided over the achievement of full dominion status by Ceylon in 1948, and in 1950 he was appointed a member of the Commonwealth Privy Council. During his term in office he won the respect of all ethnic groups in Ceylon and was instrumental in the passage of legislation to improve the conditions of peasants and agricultural labourers, resettling 250 000 people in previously uninhabited areas. He remained in office until his death in a riding accident (1952), when he was succeeded by his son, Dudley Senanayake (1911–73).

Senghor, Léopold Sédar (1906–) *President of the Republic of Senegal (1960–80). An outstanding personality in French-speaking black Africa, Senghor has established himself as a world political figure and a poet of great power.*

Born in Joal, the son of a Sere peasant family, Senghor was educated in Catholic mission schools before being awarded a scholarship in 1928 to study in France. Receiving his Licence des Lettres in 1931, Senghor taught at schools in Tours and Paris between 1935 and 1948. He was a prisoner of war in Germany for two years in World War II. During his period in Paris he met Aimé Césaire (1913–), the West Indian poet, with whom he founded the concept of French-speaking black literature known as *négritude*.

Senghor entered politics in 1945 when he participated in the French constituent assemblies that shaped the Fourth Republic. He was elected as a deputy from Senegal to the French national assembly in 1946 and joined the parliamentary group of French socialists (SFIO) under Lamire Gueye, the mayor of Dakar, in 1947. Breaking away in 1948 to form the Indépendants d'Outremer (IOM) and the Bloc Démocratique Sénégalais in Senegal, Senghor gained decisive electoral victories in 1951 and 1952 against the SFIO. In 1957 he won office in Senegal, reuniting with Gueye in 1958 to form the Union Progressive Sénégalaise (UPS). A strong advocate of francophone West African federation, Senghor founded the Parti de Regroupement Africain (PRA) in opposition to the Rassemblement Démocratique Africain (RDA), which supported the creation of autonomous territories. Following DE GAULLE's 1958 referendum, which hastened the passage to independence of the West African territories and fragmented the PRA, Senghor attempted to establish the Mali Federation. Following its breakdown in 1960, Senghor became president of the now independent Senegal, a position he held until 1980 when he retired.

His poetry won attention with the publication of *Chants d'ombre* (1945) and *Hosties noires* (1948). In a 1964 anthology of the works of French-speaking black poets, SARTRE praised Senghor and his group, saying that the black francophone poetry was the most powerful revolutionary verse then being written. His most recent publications include *Poèmes* (1984).

Sennett, Mack (Michael Sinnott; 1880–1960) *Canadian-born US film director, producer, and actor still remembered as Hollywood's 'King of Comedy'.*

Born in Richmond, Quebec, Sennett drifted into films as an apprentice to D. W. GRIFFITH at Biograph. He also worked as an actor, playing opposite such leading ladies of the day as Mary PICKFORD, Mabel Normand (1894–1930), and Blanche Sweet (1895–1986). In 1912 he estab-

lished his own comedy studio, called Keystone, and went on to produce the 'Keystone Cops', whose celebrated incompetence and high-speed chases made Sennett famous. In addition to these hilarious gems of the silent screen, Keystone also produced Charlie CHAPLIN, Frank CAPRA, and Roscoe 'Fatty' Arbuckle (1887–1933). After Keystone Sennett set up Mack Sennett Comedies, where Harry Langdon (1884–1944) made his name. Many of Sennett's silent films also starred his close friend Mabel Normand.

By the time sound films had appeared, Sennett's career had passed its peak and he lost his entire fortune. In 1939 he joined Twentieth Century Fox as an associate producer. Two years earlier, in recognition of his contributions to the pioneering days of film comedy, he was presented with a Special Academy Award.

Sessions, Roger Huntington (1896–1985) *US composer, theorist, and teacher, whose early neoclassical works gave way to more chromatic music using serial techniques. He was an influential and highly regarded teacher.*

Precocious musically and intellectually, Sessions entered Harvard University at the age of fourteen. After graduating he went on to Yale, where he studied with Horatio Parker (1863–1919). He later continued his study of composition in New York with Ernest BLOCH, becoming Bloch's assistant at the Cleveland Institute of Music (1921–25). From 1925 to 1933 Sessions travelled and studied extensively in Europe on Guggenheim Fellowships, the Prix de Rome (US), and a Carnegie Fellowship. On his return to the USA he held a series of teaching appointments, including professorships at the universities of California (1944–52; 1966–67), Princeton (1953–65), and Harvard (1968–69), as well as at the Juilliard School.

Sessions's early work was in the tradition of STRAVINSKY's neoclassicism; his later work, however, incorporated elements of atonality and serialism and is remarkable for its highly personal style. He said of his own music: 'in its very aloofness from the concrete preoccupations of life, [it] strives…to contribute form, design, a vision of order and harmony'. Sessions's compositions include nine symphonies (1927–68), two concertos (1935; 1956), three string quartets (1936; 1951; 1957), two operas, including the large-scale *Montezuma* (1964), and a double concerto (1972), as well as chamber and piano music. His many books on contemporary music theory and composers include studies of HINDEMITH and Ernst Krenek (1900–). He also encouraged the performance of contemporary music; the Copland–Sessions Concerts of Contemporary Music in New York (1928–31) were of considerable importance.

Shackleton, Sir Ernest Henry (1874–1922) *British explorer who led two expeditions to Antarctica but failed in both attempts to reach the South Pole. His heroic exploits were rewarded by a knighthood (1909).*

Shackleton was born in Kilkee, County Kildare, of English and Irish ancestry. After attending Dulwich College he joined the merchant navy in 1890. As a lieutenant in the Royal Naval Reserve, he was chosen by Robert F. SCOTT to accompany his expedition to Antarctica in 1901. Much to his chagrin, Shackleton was invalided home in 1903. Four years later he announced plans for his own Antarctic expedition and in August 1907 his ship *Nimrod* set sail, reaching the Ross Ice Shelf in January 1908. They established base at Cape Royds in the shadow of Mount Erebus. Climbing to its summit (3743 m), a party led by Professor Edgeworth David found it to be volcanic. On 16 January 1909, a group led by Douglas Mawson reached the southern magnetic pole but Shackleton's simultaneous attempt to reach the South Pole failed 155 km short of its goal.

By 1914, Shackleton had raised the money for a further expedition with the aim of transantarctic crossing. In December 1914, his ship *Endurance* left South Georgia for the Weddell Sea but soon encountered ice. By January she was trapped and drifting helplessly. Slowly crushed by the ice, the ship was abandoned the following October. After camping on the ice, in April 1916 the members of the expedition made their way in the ship's boats to the uninhabited Elephant Island in the South Shetlands. Shackleton and five others set out in an 8-metre-long boat to seek help and on 10 May landed on South Georgia. Meanwhile, a party from the expedition's other ship, *Aurora*, was stranded on the Ross Ice Shelf and not rescued until the following year. The events are related in Shackleton's book, *South* (1919). Shackleton died of a heart attack in South Georgia at the start of another Antarctic expedition and was buried at the whaling station at Grytviken.

Shaffer, Peter Levin (1926–) *British playwright.*

Born in Liverpool, Shaffer was educated at St Paul's School, London, and Trinity College, Cambridge. He worked in the mines during the war as a 'Bevan Boy' (1944–47); among his earliest writings were the novels *How Doth the Little Crocodile?* (1951) and *Woman in the Wardrobe* (1952), on which he collaborated with his twin brother Anthony under the pseudonym of Peter Anthony. After a period as an employee of the New York Public Library (1951–54), he joined a firm of music publishers (1954–56) as a literary and music critic. *Five-Finger Exercise* (1958), a domestic drama, established him as a playwright; it was followed by the double bill *The Private Ear* and *The Public Eye* (1962). *The Royal Hunt of the Sun* (1965; filmed 1969), an epic treatment of the conquest of Peru, was his next major theatrical success; to confirm his versatility as a dramatist he followed this with the witty and resourceful *Black Comedy* (1967). *Equus* (1973; filmed 1977), a play about a boy who blinds horses, was another success, and *Amadeus* (1979; filmed 1984), on the Mozart and Salieri story, won acclaim on both sides of the Atlantic. The film version received eight Oscars in 1985, including one for Shaffer himself for the best screenplay adaptation. His plays since then include *Yonadab* (1985) and the comedy *Lettice and Lovage* (1987). Shaffer has also written screenplays for William GOLDING's *Lord of the Flies* (1963) and his own *The Public Eye* (1972), as well as many plays for radio and television.

Shamir, Yitzhak (Yitzhak Jazernicki; 1915–) *Israeli statesman; leader of the Likud party (1983–) and prime minister (1983–84, 1986–).*

Born in Poland, Shamir was educated at the Hebrew school in Białystock and at Warsaw University, where he studied law and was active in the Zionist youth movement. He emigrated to Palestine in 1935. While studying at the Hebrew University in Jerusalem he joined the Irgun Zvai Leumi, an underground zionist organization, becoming a member of the Stern Gang, a fanatical splinter group, from 1940. Involvement in terrorist activities led to his arrest by the British in 1941 and again in 1946, when he was exiled to British Eritrea. Following his escape from internment camp he made his way to France, where he was granted asylum.

Shamir returned to Israel following independence in 1948. From 1955 he spent ten years working as a secret service operative for Mossad, mainly in Europe. After several years devoted to his business interests he re-entered politics in 1970, becoming active in the Herut party led by his Stern Gang colleague, Menachem BEGIN. He was elected to the Knesset in 1973 and became chairman of the Herut party executive committee in 1975. When the 1977 elections resulted in a victory for the Likud grouping (formed under Begin's leadership in 1973) Shamir became speaker (1977–80). As minister of foreign affairs (1980–83) he gained a reputation as a hardliner on such questions as the future of the occupied territories.

On Begin's retirement in 1983 Shamir became prime minister, but his party was narrowly defeated in elections a year later. In the resulting coalition he served as deputy prime minister under the Labour leader Shimon PERES until 1986, when the two men changed places. Elections in 1988 produced a slim victory for Likud and the coalition was renewed with Shamir as prime minister. In 1990, however, Shamir's resistance to a US plan for an Israeli-Palestinian peace conference led him to sack Peres, who supported this plan, and the coalition crumbled. Shamir lost a vote of confidence in the Knesset but eventually succeeded in forming a new government with the support of extreme right-wing and religious parties. On taking office once more he restated his determination never to concede any land to a Palestinian state. Under his leadership Israel did not retaliate when attacked by Iraqi missiles during the Gulf War of 1991, thus lessening the risk of an Arab coalition being formed in support of Saddam HUSSEIN. Later the same year Israel entered into tentative negotiations with the Palestinians, notably in arranging exchanges of prisoners.

Shankar, Ravi (1920–) *Indian sitar player and composer. He has been largely responsible for the interest in Indian music in the western world during the last thirty years.*

A precocious dancer and musician, Shankar had little formal education until he was taken to Paris at the age of ten by his elder brother, Uday Shankar. In 1936 he studied with Ustad Allauddin Khan in Mai-her, a fellow-pupil being his future wife, Annapurna. By the mid-1940s, Shankar was giving well-attended recitals and in 1946 he joined the All-Indian Radio as director of its instrumental section. In 1956 he toured Europe and the USA (1956–57), appearing in concerts and recitals. He played at the UNESCO concert in Paris in 1958, the Edinburgh Festival in 1963, and the UN Human Rights Day concert in New York in 1967 with Yehudi MENUHIN, with whom he has frequently appeared on television. Shankar has

founded a school of Indian music (the Kinnara School) in Los Angeles. His compositions include two concertos for sitar and orchestra, music for films, and *Ghanashyam – A Broken Branch* (1989). In 1986 he became a member of parliament in India.

Shaw, Artie (Arthur Jacob Arkshawsky; 1910–) *US jazz clarinettist, bandleader, and writer. His bands were among the most popular in the late 1930s and 1940s.*

A freelance musician for many years, he formed a band of his own, featuring a strong string section, in 1936; this failed. A more conventional swing band, formed the following year, had a surprise hit with 'Begin the Beguine', which was intended to be the 'B' side of a record. In his autobiography, *The Trouble with Cinderella* (1952), he revealed that he soon became disenchanted with popular music, but his excellent taste and musical skill kept him in the business. His rendering of 'Frenesi' featured a string section, and was number one in the charts for twenty-three weeks in 1940; 'Summit Ridge Drive' was made by a small group using a harpsichord instead of a piano, and 'Concerto for Clarinet' was a serious attempt at a jazz concerto (both 1941). 'Gloomy Sunday', featuring his hauntingly sad clarinet, was said to have caused more suicides than any other record. Like the other clarinettist of the period, Benny GOODMAN, Shaw employed black talent (trumpeter Roy Eldridge; singer Billie HOLIDAY) when it was brave and inconvenient to do so.

He retired from music in the 1960s, first running a dairy farm, then trading in property. He also wrote adaptations for the theatre. His novel, *I Love You, I Hate You, Drop Dead*, was published in 1965.

Shaw, George Bernard (1856–1950) *Irish playwright, critic, and propagandist. He accepted the Nobel Prize for Literature (1925), but declined all other public honours.*

Shaw's parents were an ill-assorted impoverished Dublin couple with pretensions to gentility. The discovery of his father's secret drinking was a traumatic experience for the boy. At the Wesley School, which he entered in 1867, he failed to distinguish himself, but nonetheless his mother's interest in music broadened his horizons and he read extensively. In 1871 Shaw became a clerk in an estate agent's office; in 1875 his mother and two sisters left Dublin for good, leaving Shaw with his father. Despite promising progress in his career, Shaw too left Dublin in 1876 to join his mother in London. For some years he lived in straitened circumstances, using periods of unemployment to write unsuccessful novels and evolve his socialist philosophy. By determined practice he became an effective orator. In 1884 he became one of the earliest members of the Fabian Society. At the instigation of drama critic William Archer (1856–1924), an acquaintance from the British Museum Library, Shaw became book reviewer on the *Pall Mall Gazette* (1885–88) and art critic on the *World* (1886–89). In 1888 he became music critic on the *Star*, writing under the name 'Corno di Bassetto', and in 1890 moved to the same post on the *World*. In 1895 he became drama critic on the *Saturday Review*.

His first play, originally conceived in 1885 as a collaboration with Archer, was *Widowers' Houses* (1892), but many of the plays he wrote in the 1890s were not performed until much later; for instance, *Mrs Warren's Profession*, printed in *Plays Pleasant and Unpleasant* (1898), was banned from the stage until 1925. The seven plays in this collection each had a Preface and thereafter Shaw invariably included a long elegant Preface with each of his plays, using them as vehicles for his diverse opinions, which were not always connected with the plays they prefaced. Other plays met with little success, although productions in the USA of *Arms and the Man* (1894) and *The Devil's Disciple* (1897) were better received. In 1898 he married a wealthy young Irish girl, Charlotte Payne-Townshend, who had nursed him through an illness caused by overwork. The marriage, based more on economic convenience than passion, lasted until her death in 1943. In 1899 Shaw's friend Mrs Patrick CAMPBELL starred in his *Caesar and Cleopatra*. He did not, however, achieve true recognition in the London theatre until 1904, with the production of *John Bull's Other Island*, followed by *Man and Superman* (completed 1903). Other prewar successes included *Major Barbara* (1905; filmed with additional scenes by Shaw in 1940), *The Doctor's Dilemma* (1906), *Androcles and the Lion* (1913), and *Pygmalion* (1914; filmed 1938 and made into a musical, *My Fair Lady*, in 1964). Shaw's essay *Common Sense About the War* (1914) attracted much odium, but his popularity recovered after the war. In 1921 *Heartbreak House* was staged in London, a year after its New York premiere. *Saint Joan* (1924) had Sybil THORNDIKE in the title role, and *The Apple Cart* (1929), a political extravaganza, was first performed in Warsaw.

In his old age Shaw continued to pursue contemporary issues with the same vigour with which he had promoted Ibsen and Wagner in the 1890s. *The Intelligent Woman's Guide to Socialism* (1928) and *The Adventures of the Black Girl in her Search for God* (1932) were both widely read. He was also a gifted letter writer, his correspondence with Mrs Patrick Campbell and Ellen TERRY being particularly interesting.

Sherrington, Sir Charles Scott (1857–1952) *British neurophysiologist, whose work on the mechanisms of integration in the nervous system earned him the 1932 Nobel Prize for Physiology or Medicine. He was knighted in 1922.*

Born in London, Sherrington studied physiology at Gonville and Caius College, Cambridge, before going on to study medicine at St Thomas's Hospital, London. He spent some time in Europe, including a spell in Robert Koch's laboratory in Berlin, before his appointment in 1887 as lecturer in systematic physiology at St Thomas's. He served (1891–95) as physician-superintendent of the Brown Institution, a London veterinary hospital. He then joined Liverpool University as professor of physiology (1895–1912) and subsequently held a similar post at Oxford University until his retirement in 1935.

Sherrington's early work was in neuroanatomy, mapping the pathways of nerves and their connections. In the late 1890s he established the phenomenon of reciprocal innervation, which showed how, when the flexor muscle of a joint is stimulated to contract, the motor nerves supplying the extensor muscle are simultaneously inhibited – and vice versa. In 1897 he introduced the term 'synapse' for the point of contact between two nerve cells. Starting from his early studies of simple spinal reflex actions, such as the knee jerk, Sherrington proceeded to establish how these are integrated with control centres in the spinal cord and brain and how simple reflexes are compounded to produce more complex reflexes, such as the scratch reflex in dogs. His book *The Integrative Action of the Nervous System* (1906) was an important text in neurophysiology. Sherrington's work was interrupted by World War I, during which he worked for a time in a munitions factory to gain first-hand experience of the effects of industrial fatigue. In 1924 he published his findings on the stretch reflex in muscle. Stretching of muscles, he found, stimulates stretch receptors in the muscle to trigger motor nerves to the muscle, causing it to contract, thus providing a built-in mechanism for maintaining posture.

Shockley, William Bradford (1910–89) *US physicist, who developed the junction transistor from the point-contact transistor. He was awarded the 1956 Nobel Prize for Physics.*

The son of a mining engineer, Shockley was educated at the California Institute of Technology and the Massachusetts Institute of Technology, where he gained his PhD in 1936. He immediately joined the research staff of Bell Telephone Laboratories and in 1953 became director of the transistor physics department. Shockley also became connected with a number of private companies all concerned with the commercial exploitation of the transistor. In 1963 he was appointed to the Poniatoff Professorship of Electrical Engineering at Stanford, remaining a consultant with Bell until he retired from both positions in 1975.

In 1947 Shockley's colleagues at Bell, J. BARDEEN and W. J. BRATTAIN, invented the point-contact transistor. This, however, was a theoretical rather than a practical breakthrough. Shortly afterwards Shockley developed the more practical junction transistor, which transformed the electronics industry. Shockley shared his Nobel Prize with Bardeen and Brattain. Subsequently Shockley argued his minority views on genetics, gaining considerable publicity. Believing that blacks are less intelligent than whites, and that the current population explosion is spreading 'bad' genes at the expense of 'good', Shockley enthusiastically supported such schemes as a sperm bank produced by Nobel prizewinners, restrictions on mixed marriages, and voluntary sterilization.

Sholem Aleichem (Sholom Rabinowitz; 1859–1916) *Yiddish writer.*

Born in the Jewish village of Pereyaslav, Poltava province, in the Ukraine, Sholem Aleichem was educated at nearby Voronkov, whose inhabitants were the models for the characters of the fictional community Kasrilevke, the setting of many of his stories. His childhood was a troubled one: both his mother and stepmother died and his father had to struggle with debts. He served as a government rabbi in Lubin for several years, eventually marrying and moving to Kiev when he inherited money from an uncle. Here he briefly edited a literary annual, *Die Yiddishe Folksbibliothek* (1888–89), but he was generous to a fault in paying the contributors and always unlucky in commercial matters, a failing also of the disastrously impractical hero of his epistolary novel, *Menakhem Mendl* (1895). Bankrupt, he moved to Odessa in 1890. By this

time he was well known to readers of Yiddish but his financial affairs did not improve. He continued to write his popular humorous sketches and despite unremitting money problems he travelled widely. He moved to New York at the start of World War I and died there.

More than any other writer, Sholem Aleichem (his pseudonym, adopted in 1883, is the traditional Jewish greeting, 'Peace be with you') taught his fellow Jews the value of humour in adversity. When his own affairs were at their lowest point, he published a collection of sketches that embody the saving grace of humour, *Tovye der Milkhiger* (1894; translated as *Tovye Stories*, 1965), about the impoverished driver of a milkwagon who endures hardship and maintains his love of life. (The musical *Fiddler on the Roof* was based on these stories.) Among his other works are *Stempenyu* (1889), *Yosele Solovey* (1890), and *Mottel Peyse dem Khazns* (1907–16; translated as *Adventures of Mottel, the Cantor's Son*, 1953), about a spirited character who refuses to come to terms with the world of adults.

Sholokhov, Mikhail Aleksandrovich (1905–84) *Soviet novelist. He was awarded the Nobel Prize for Literature in 1965.*

Born on a farm near the village of Veshenskaya in the Don Military District, Sholokhov was the son of a merchant-class father (not of Cossack origin). His education was minimal, though he later attended seminars on writing in Moscow. At the outbreak of the October Revolution (1917), the Don region was controlled by Whites; Sholokhov supported the Reds (and served in the Red Army during the civil war following the revolution), but he had intimate knowledge of the Don Cossacks' way of life and understood their attempt to preserve it, though he also realized that it could not survive the industrial and urbanizing changes about to occur. His first stories were collected under the title *Donskiye rasskazy* (1924; translated as *Tales from the Don*, 1961) and were officially criticized for being sympathetic to the Cossacks. Sholokhov in fact was completely objective, neither portraying the Cossacks as utter villains or the Reds as blameless heroes.

His major work is the epic novel of the revolution and the Cossacks' slow and painful adjustment to change in the years 1912–22, *Tikhiy Don* (1928–40; translated as *And Quietly Flows the Don*, Moscow, four vols, 1960). *Podnyataya tselina* (1932–59; translated in two volumes as *Virgin Soil Upturned* and *Harvest on the Don*) concerns the collectivization of the Don region. A third novel, *Oni srazhalis za rodinu* (1943–59; translated as *They Fought for Their Country*, 1959) is a less successful account of World War II in southern Russia.

Sholokhov was awarded both the Lenin and Stalin Prizes, as well as the Nobel Prize; he remained steadfastly loyal to the Soviet state and critical of its critics. Both in the 1920s and more recently, however, Sholokhov was accused of plagiarism in *Tikhiy Don*. SOLZHENITSYN in the 1970s published an article by an anonymous author alleging that large sections of Sholokhov's classic were written by a Cossack writer, F. D. Kryukov (1870–1920). Sholokhov did not deign to reply to the charge and technical stylistic studies have not done much to support it, but the controversy remains.

Shostakovich, Dmitri Dmitriievich (1906–75) *Soviet composer. One of the most important figures in twentieth-century music, he was awarded many honours and prizes in the Soviet Union, including the Order of Lenin twice (1956 and 1966).*

Born in St Petersburg Shostakovich studied at the conservatory there under Maximillian Steinberg (1883–1946) and Alaxander Glazunov (1856–1936). His first symphony (1924–25) was written as a graduation piece and subsequently performed successfully in St Petersburg (1926), which had recently been renamed Leningrad, Berlin (1927), and Philadelphia (1928). As a pianist he won honourable mention at the 1927 International Chopin Contest in Warsaw but chose to make his career as a composer, frequently playing his own piano parts in performance.

Shostakovich was a prolific composer, writing fifteen symphonies, fifteen string quartets, two operas and an operetta, six concertos for various instruments, works for piano and strings, vocal and piano music (including *24 Preludes and Fugues*, 1951), cantatas, oratorios, three ballets, thirty-six film scores, and incidental music for eleven plays. As a committed Soviet citizen, Shostakovich accepted criticism whenever his music failed to conform to the imposed standards of Soviet realism. For example, his opera *Lady Macbeth of the Mtsensk District* (1930–32; revised as *Katerina Ismailova*, 1956) was first hailed as a masterpiece (1934) but later condemned for its undesirable modernism (1936). Shostakovich's fifth symphony (1937) was the 'creative reply of a Soviet artist to justified criticism'. In fact, the fifth symphony failed to be as popular and optimistic as Soviet music

was then expected to be, and Shostakovich expressed himself privately in his chamber music until he was able to produce the required patriotism in the seventh, his *Leningrad Symphony* (1941). After STALIN's death Shostakovich acquired a certain immunity, and in his later works used such techniques as atonality and twelve-tone themes to some extent, but always returned to a basic tonality. However, he again suffered official disapproval after his thirteenth symphony appeared in 1962, with a text by YEVTUSHENKO criticising Soviet antisemitism. His son Maxim Shostakovich (1938–), who is a conductor and an enthusiastic performer of his father's works, defected to the West in 1981.

Sibelius, Jean Johan Julius Christian (1865–1957) *Finnish composer, whose early programme music gave way to the more abstract style of his later symphonies. He was greatly revered in Scandinavia and high respected in England and America.*

Although his musical talents developed early, Sibelius studied law at Helsingfors University. Not until the age of twenty-one did he decide that music was to be his career, transferring to the conservatory and later studying in Berlin (1889) and Vienna (1890). He returned to Finland in 1893, a time of political unrest. Sibelius's own affinity for the Finnish countryside and legends, especially the epic *Kalevala*, expressed themselves in a series of tone poems and in seven symphonies spanning the years 1898 to 1924. Sibelius's music was well received by Finnish (and later English) audiences; as early as 1897 he was granted a state pension enabling him to live in the countryside north of Helsingfors with his wife and five daughters. His uneventful life there was punctuated only occasionally by trips abroad, to the USA (1913) and to England (1912; 1921). A prolific composer for about thirty years, from about 1925 until his death he wrote nothing, living quietly as the master of Finnish music.

In his highly original style, Sibelius used the classical forms of symphony, sonata, and suite, his great tunes evolving from a series of fragments that eventually shape into sweeping melodic curves. Sibelius's tone poems include the early *En Saga* (1892; revised 1901), the popular *Finlandia* (1899; revised 1900), and *Four Legends from the Kalevala – Lemminkäinen and the Maidens of Saari* (1895), *Lemminkäinen in Tuonela* (1895), *The Swan of Tuonela* (1893), and *Lemminkäinen's Homecoming* (1895). His later tone poems were *Pohjola's Daughter* (1906) and *Tapiola*

(1925), one of his last works. Of the symphonies, the fifth (1915) is the most expansive, while the seventh (1924) is in a single movement. A violin concerto dates from 1903 (revised 1905). Of the large catalogue of more intimate works, the string quartet *Voces intimae* (1909) and many of the unaccompanied songs are still popular.

Sickert, Walter Richard (1860–1942) *British painter, etcher, and influential teacher, who dominated British art in the first three decades of this century.*

Born in Munich, Sickert was the son of a Danish painter, Oswald Adalbert Sickert, who settled in England in 1868. His mother was half English, half Irish. After a classical English education and a brief period as an actor Sickert studied at the Slade School of Art under Legros in 1881 and worked as assistant to Whistler in 1882–83. In Paris in 1883 he met Degas, who had a major influence on his style. During the period 1887–89 he painted a number of canvases depicting music hall scenes, such as *Le Lion comique*, and in 1888 he joined the New English Art Club. He became one of the most important British proponents of French impressionism and from the mid-1890s was highly regarded as a teacher. From 1899 to 1905 he lived in Dieppe, occasionally travelling to Venice, and both places are depicted in rich sombre tones in his work of this time.

The years between 1905, when Sickert settled in London, and 1914 are known as his 'Camden Town period'. During this time he painted scenes of the streets, lodging houses, and rented rooms of the area, for example *Ennui* (1913). His paintings enlivened dull or sordid scenes, frequently focusing on intimate moments caught unaware. His studio became a focal point for the group of artists who formed the Camden Town Group in 1911 and the London Group in 1913. During the 1920s his reputation continued to increase. From 1938 he lived and worked in Bath. In 1947 a collection of his writings was published posthumously under the title *A Free House*.

Sikorski, Władysław (1881–1943) *Polish general and statesman. He was prime minister of Poland (1922–23) and leader of the government-in-exile (1939–43) in London.*

Born in Galicia, Sikorski was educated at the universities of Kraców and Lvov, where he studied engineering. In 1908 he founded a nationalist military organization, which fought with the Austrian army against Russia during World War I. He served in the Polish-Soviet war (1920–21) and in 1921 was promoted head of the Polish general staff.

Sikorski was appointed prime minister of Poland in 1922–23 and in 1924–25 became minister of military affairs with responsibility for updating the army. Following the rise to power of Jósef Piłsudski (1867–1935) in 1926, he was forced to resign from his military command. When, in 1939, Poland was occupied by the Germans, he headed the government-in-exile in London, where he earned recognition and support from Allied powers. Only the Soviet Union broke off diplomatic relations (in 1943), over allegations that several thousand Polish officers had been murdered by Russian soldiers at Katyn. Sikorski was killed in an air crash over Gibraltar in July 1943. The circumstances of this accident, and the allegation that it had been engineered by CHURCHILL, form the subject matter of the play *Soldiers* (1966), by Rolf HOCHHUTH.

Sikorsky, Igor Ivanovich (1889–1972) *Russian-born US aircraft designer responsible for the first commercially successful helicopter.*

Born in Kiev, the son of a psychology professor, Sikorsky trained as an engineer at the Naval Academy in St Petersburg. His interest in flying was stimulated by accounts of Leonardo da Vinci's designs for helicopters, and he built two rotating-wing machines in 1908; neither rose from the ground. At this point he turned to more conventional fixed-wing designs and accepted the job of chief designer of the aircraft division of the Baltic Railway Car Factory in Petrograd. Here in 1913 he constructed the world's first four-engined aeroplane to fly. Known as the Bolshoi, it served as the basis of Sikorsky's Ilya Mourametz, the first four-engined bomber, seventy-three of which were delivered to the military before 1917.

Strongly opposed to the Russian revolution of 1917, Sikorsky emigrated to the USA, becoming a naturalized citizen in 1928. After some years teaching, he set up his own company and began, initially, to design and produce seaplanes. In 1929 his company became a subsidiary of United Aircraft, for whom he built the S-42 flying boat. In the 1930s Sikorsky turned once more to the design of helicopters. On 14 September 1939 his VS-300 first flew on a tethered test rig. Sikorsky found that his helicopter, when released, could fly in all directions except forwards. Dismissing this as a 'minor engineering problem', he continued with the machine's development and, suitably modified, it became the first helicopter to be produced in any quantity. Large numbers were purchased by the US army and in 1950 the RAF bought their first helicopter, the Westland-Sikorsky Dragon Fly. Sikorsky continued to develop his original helicopter design creating, for example, the S-61, an amphibious version, and the S-64, a cargo model. The growing demand for helicopters, both military and civilian, made Sikorsky one of the richest engineers in aviation history.

Silone, Ignazio (Secondo Tranquilli; 1900–78) *Italian novelist and critic. The humanity of his work reflects his loyalty to ideals and his belief in the importance of action (rather than theorizing) in helping the oppressed.*

Born into the family of a smallholder at Pescina, Abruzzi, Silone was educated in Catholic schools. His lifelong concern with social justice, however, was early and decisively acquired through familiarity with the harsh conditions of the poor peasantry of the Abruzzi. After his mother and five of his brothers were killed in an earthquake in 1915, Silone and his sole surviving brother moved to Rome, where for a time he edited a socialist weekly paper. He became an active communist, making several journeys to the Soviet Union as a member of the Italian party delegation. He and his brother then edited a Trieste political daily, *Il lavoratore*, which was banned by the fascists. Fleeing arrest, Silone first went into hiding in the Abruzzi and in 1930 managed to cross into Germany, eventually settling in Switzerland, where he remained until after the war (1944), when it was safe to return to Italy. (His brother had been captured by the police and died from beatings received in prison.)

Silone left the Communist Party in 1931, unable to accept a theory that could be twisted to allow even a temporary truce with fascism. From 1956 to 1968 Silone and the critic Nicola Chiaromonte edited the international cultural monthly *Tempo presente*. While in exile in Switzerland, Silone wrote three novels, a play, and two nonfiction works concerned with fascism. His first novel, *Fontamara* (1930), the story of an Abruzzi village under fascism, was enthusiastically received and translated into seventeen languages in the year of publication. (This and Silone's other books remained unknown in Italy, of course.) *Pane e vino* (1937; translated as *Bread and Wine*, 1962) and *Il seme sotto la neve* (1941; translated as *The Seed Beneath the Snow*, 1965) depict an intellectual freeing himself from the deceptions of

theory to share the life of the poor in reality. Of Silone's other works, the essays and reminiscences in *Uscita di sicurezza* (1965; translated as *Emergency Exit*, 1968) form a coherent intellectual autobiography of the greatest interest.

Simenon,   Georges(-Joseph-Christian) (1903–89) *Belgian novelist, creator of the detective Maigret.*

Simenon was born in Liège, Belgium. At the age of sixteen he took a job as junior reporter on a local newspaper; his first novel was published under a pseudonym in 1920. Moving to Paris in 1922, he continued to write short stories and popular novels using a variety of pseudonyms; *Pietr-le-Letton* (1931) was the first novel to be published under his own name.

The Maigret stories, such as *Les Inconnus dans la maison* (1940) and *Maigret et le clochard* (1963), began to appear in the early 1930s. Commissaire Maigret of the Paris *police judiciaire* is a pipe-smoking detective whose method of solving crime relies not on scientific deduction, but rather on his intuitive grasp of the criminal's motives. This emphasis on the psychological, together with Simenon's skilful evocation of atmosphere, raise the Maigret novels above the level of ordinary detective fiction. They are read in translation worldwide and have been successfully adapted for television and the cinema.

Simenon was a prolific writer, producing at the peak of his career more than ten new works each year. At the same time, he maintained a consistent quality that caused him to be admired by established French novelists, such as André GIDE. His enormous output of over five hundred works also includes many psychological novels, such as *Chez Krull* (1939; translated as *A Sense of Guilt*, 1955), *La Neige était sale* (1948; translated as *The Stain on the Snow*, 1953), and the semiautobiographical *Pedigrée* (1948). *Trois Chambres à Manhattan* (1946) is set in the USA, where Simenon spent ten years after the end of World War II. He subsequently lived in France and Switzerland, officially retiring in 1973. He was as proud of his reputation as a sexual athlete as he was of his reputation as a writer.

Simon, John Allsebrook, 1st Viscount (1873–1954) *British lawyer and politician, who headed the Simon Commission on the future of India (1927). He was created a viscount in 1940.*

Born in Manchester, Simon was called to the bar and practised as a barrister before becoming a Liberal MP (1906–18; 1922–40). He was then solicitor-general (1910), attorney-general (1913–15), and home secretary (1915–16), before resigning in protest against the introduction in World War I of conscription.

The report of the Simon Commission on India was largely his work; although highly regarded on technical grounds, it was completed too late to have much effect. Self-government had already been promised to India by the time the report was delivered (1930) but the subsequent Round Table conferences between Britain and Indian leaders were held in response to Simon's recommendation. In the 1930s he led those Liberals who supported the coalition national government – the Liberal Nationals – and was successively foreign secretary (1931–35), home secretary (1935–37), chancellor of the exchequer (1937–40), and lord chancellor (1940–45). His conduct of foreign policy was criticized for its weakness, particularly towards Japanese aggression in Manchuria, and Simon subsequently supported Neville CHAMBERLAIN's appeasement of the fascist powers. LLOYD GEORGE said of him: 'Simon has sat on the fence so long that the iron has entered into his soul.'

Simpson, George Gaylord (1902–84) *US palaeontologist, whose work did much to shape the postwar development of palaeontology and its impact on evolutionary theory.*

A native of Chicago, Simpson obtained his first degree in geology from Yale University and stayed there to obtain his doctorate in 1926, with a thesis on fossil mammals of the Mesozoic era. The following year he joined the American Museum of Natural History, New York, later becoming curator of the palaeontology department (1944–59). Here he published many papers, mostly concerning mammalian palaeontology, especially the primitive mammals of the Cretaceous and early Tertiary periods. He wrote the standard text *American Mesozoic Mammalia* (1929). During World War II he served with the US army in North Africa. After the war he was appointed professor of vertebrate palaeontology at Columbia University (1945–59), during which time he contributed to the debate on the nature of evolution, particularly in his books *Tempo and Mode in Evolution* (1944) and *The Major Features of Evolution* (1953). He also had influential views on the philosophical implications of evolutionary theory, published in *The Meaning of Evolution* (1949). He led several fossil excavations, including that of a major collection of ancestral horses, known as *Eohippus*, in Colorado. Simpson showed that horse evolution

had undergone not a steady continuous development but considerable fluctuation in form in the face of changing environmental conditions.

In 1959 Simpson was appointed Alexander Agassiz Professor of Vertebrate Palaeontology at Harvard University, a post he held until 1970. He was also professor of geosciences at the University of Arizona (1967–82).

Simpson, Wallis Warfield (1896–1986) See WINDSOR, DUKE OF.

Sinatra, Frank (Francis Albert Sinatra; 1915–) *US singer, actor, and international superstar.*

Born to Italian parents in Hoboken, New Jersey, Sinatra made his radio debut in 1938 and sang with the bands of Harry James (1916–82) and Tommy Dorsey (1904–57); from 1943 he had his own radio shows. In the mid-1940s the hysteria he engendered among teenage 'bobby-soxers' was a precursor of the reception Elvis PRESLEY later received. Between 1943 and 1951 he had twenty-eight hit records and appeared in more than a dozen films, winning an Academy Award for his part in *From Here to Eternity* in 1952. His phrasing, which he claimed to have learned from Dorsey's trombone playing, his mastery of the microphone, and his use of top arrangers, such as Nelson Riddle (1921–85), established him as one of the finest interpreters of American show tunes.

In the 1960s he formed his own record company and became a skilled producer, making albums such as *I Remember Tommy* (1962) and others with the bands of Count BASIE and Duke ELLINGTON. He made another thirty films, both as a singer and a dramatic actor. Although he has announced his retirement on several occasions, he seems unable to stay away from the footlights and is now an elder statesman of international show business.

Sinclair, Upton Beall (1878–1968) *US novelist whose political fiction helped to bring about social reform in the USA. He was awarded the Pulitzer Prize in 1943.*

As a child Sinclair was a prolific reader and enthusiastic scholar, who supported himself at the New York City College from the age of fourteen by writing in his spare time. After a short attendance at Columbia University he moved to Quebec, Canada, where in conditions of extreme poverty he wrote his first novel, *Springtime and Harvest* (privately published in 1901 and later renamed *King Midas*). By the age of twenty he was a committed socialist.

Fame came suddenly with *The Jungle* (1906), which exposed the appalling conditions in a meat-packing house in Chicago. The affluence accompanying success was short-lived, however, for Sinclair used his money to found a cooperative community that was destroyed by fire in 1907. This fiasco established a recurring pattern; any money Sinclair made by writing he invariably sank into a variety of social schemes. He was also active in US politics but never achieved any real success; his main contribution to political life lay in the message he expounded in his novels. He was a prolific writer, producing more than a hundred books, among them *King Coal* (1917), which followed the same formula as *The Jungle* but dealt with the plight of the Colorado mine workers; perhaps best known of all his works were the eleven volumes of the Lanny Budd series, *World's End* (1940–53).

Singer, Isaac Bashevis (1904–91) *Polish-born US Yiddish writer and novelist. He was awarded the Nobel Prize for Literature in 1978.*

Born in Radzymin, Poland, the son and grandson of rabbis, Singer grew up in Warsaw. He was educated at the Warsaw Rabbinical Seminary but chose to become a writer rather than a rabbi. Because a Nazi invasion of Poland seemed imminent, Singer immigrated to the USA in 1935. He settled in New York and began publishing in *Der Forverts* (the *Jewish Daily Forward*), the New York Yiddish newspaper. His fictional contributions to the paper were signed 'Isaac Bashevis', while his non-fictional journalism appeared under the name 'Isaac Warhofsky'. He had originally written in Hebrew, but quite early changed to Yiddish and his early work was much indebted to the Yiddish writers SHOLEM ALEICHEM and Sholem ASCH. For years, also, the writings of his elder brother, Israel Joshua Singer (1873–1944) – especially *Di Brider Ashkenazi* (1936; translated as *The Brothers Ashkenazi*, 1936) – were much better known than his own, both in Yiddish and in English. Singer, however, was an extremely prolific writer, producing approximately one book a year, and he gradually established himself as the leading Yiddish writer.

The subject of his fiction is the virtually medieval society of Polish Jews, which was utterly destroyed in World War II. This remote world teems with fantastic incidents, magic, violence, and eroticism, but is seen from a highly sophisticated and enlightened point of view. Singer was a natural storyteller and his

genius for depicting this unfamiliar territory is perhaps more evident in his stories than in his novels. Among the latter are *Satan in Goray* (1955; an early version appeared in 1938 but attracted little attention), *The Magician of Lublin* (1960), *The Slave* (1962), *The Manor* (1967), *The Penitent* (1984; published in Yiddish, 1973), *The King of the Fields* (1988), and *Scum* (1991). Among the short-story collections are *Gimpel the Fool* (1957), the title story of which is his best-known work, *The Spinoza of Market Street* (1961), *Short Friday* (1964), *The Séance* (1968), *A Friend of Kafka* (1972), *The Image* (1986), and *The Death of Methuselah* (1988). His father, the rabbi Pinchos Mendel, is a prominent figure in two autobiographical volumes, *In My Father's Court* (1966) and *A Little Boy in Search of God* (1976); he published further memoirs in 1985 as *Love and Exile*.

Sitwell, Dame Edith Louisa (1887–1964) *British poet and critic. Created a DBE in 1954, this tall, exotically dressed, bejewelled, and invariably turbanned eccentric was a familiar feature of London literary life.*

The only daughter of a difficult antiquarian baronet, Sir George Sitwell, Edith Sitwell had an unhappy childhood as neither her looks nor her interests were compatible with the conventional models favoured by her parents. Encouraged by Helen Rootham, who became her governess in 1903, she began slowly to develop her talents. In 1914 she and Miss Rootham took a flat together in London and in 1915 her first volume of poetry, *The Mother*, was published. She collaborated with her brothers Sir Osbert (1892–1969) and Sacheverell (1897–1988) on various projects, including the annual anthology *Wheels* (1916–21). In 1923 she attracted some attention with the first public performance of *Façade*, a highly innovative entertainment in which her poems were set to the music of William WALTON. Her other poems of the early 1920s are in a quieter mood – *Bucolic Comedies* (1923), *Sleeping Beauty* (1924), and *Troy Park* (1925) – but *Gold Coast Customs* (1929) recaptures some of the verve of *Façade*.

In the 1930s Edith Sitwell lived for a while in Paris with Helen Rootham until the latter's death in 1938. She wrote little poetry in this period but among her prose writings was her popular book on *The English Eccentrics* (1933). During World War II she again published volumes of verse, *Street Songs* (1942) and *Green Song* (1944); she also compiled the anthology *A Poet's Notebook* (1943). Even her last collections, *Gardeners and Astronomers* (1953), *The Outcasts* (1962), and *Song of the Cold* (1964) exhibit continuing inventiveness. Among her other prose works were *Fanfare for Elizabeth* (1946), *The Queens and the Hive* (1962), and her posthumously published autobiography, *Taken Care Of* (1965).

Skinner, B(urrhus) F(rederic) (1904–90) *US psychologist and one of the leading members of a US-based school of experimental psychology that arose in the 1930s and was concerned with investigating the learning ability of animals by using carefully controlled laboratory experiments.*

Skinner received his MA (1930) and PhD (1931) from Harvard University and remained there as research fellow until 1933, when he joined the University of Minnesota as an instructor in psychology, later becoming assistant professor (1937–39) and associate professor (1939–45). During World War II Skinner investigated the feasibility of training pigeons to pilot missiles but this was never put into practice. After the war, while professor of psychology at Indiana University (1945–48), he devised his 'Air-Crib', a hermetically sealed chamber which, he claimed, provided the optimum environment for the growing infant. Moving to Harvard as professor of psychology (1948–75), Skinner designed laboratory apparatus to examine learning in experimental animals, such as rats and pigeons. For instance, rats placed in a 'Skinner box' learnt to press a lever to receive a pellet of food. Other experiments involved learning the correct route through mazes or accomplishing other tasks, usually employing food as a reward for success and electric shocks as punishment for failure. Ethologists, such as Konrad LORENZ, criticized such studies as being artificially restricted and often irrelevant to animals in their natural surroundings. However, Skinner used his behaviourist principles in programmed learning courses designed for schools and colleges, using a teaching machine to provide stepwise instruction in certain subjects. His books include *The Behavior of Organisms* (1938), *Science and Human Behavior* (1953), *Beyond Freedom and Dignity* (1971), and *Skinner for the Classroom* (1982). He also wrote a novel, *Walden Two* (1948), which describes life in a utopian community based on Skinner's own ideas about social engineering.

Slim, William Joseph, 1st Viscount (1891–1970) *British general whose troops fought a memorable campaign to recapture*

Burma from the Japanese in World War II. He was created a viscount in 1960.

The son of a Bristol iron merchant, Slim volunteered for service at the start of World War I and joined the Royal Warwickshire Regiment. He saw action at Gallipoli, on the western front, and in Mesopotamia, being wounded twice and receiving the MC. Between the wars, Slim spent several years in India with the 6th Gurkha Rifles and later as a staff officer. In 1934 he was transferred to Camberley Staff College, becoming a lieutenant-colonel in 1935.

On the outbreak of World War II, Slim was a brigade commander in the 5th Indian division stationed in Eritrea. By May 1941, now major-general, he was commanding the 10th Indian division. The following March, Slim was posted to Burma, where British forces faced the likelihood of capture by the Japanese invaders. With his Chinese allies, Slim strengthened defensive lines to enable a more orderly retreat up country. Rangoon and then Mandalay fell to the enemy and the remnants of Slim's forces eventually made their way across the border into India. Slim was appointed head of the 14th Army in October 1943 under commander-in-chief, Lord Louis MOUNTBATTEN. In early 1944, Slim's forces advanced into Arakan province and soon faced formidable opposition. Their commander ordered massive reinforcement by air and, in so doing, turned possible defeat into a morale-boosting victory. In March the Japanese 15th army threatened the Indian towns of Imphal and Kohima to the north. Again Slim made full use of air supply to reinforce the garrisons and fierce fighting raged until June when, with their supply lines cut and with the onset of the monsoon, the Japanese were finally defeated.

Building on this success, the 14th Army advanced deeper into Burma in late 1944 and 1945. Slim deceived the Japanese into thinking that the bulk of his forces were concentrated through central Burma in the direction of Mandalay; in fact, under radio silence, substantial forces moved around the southwestern flank, breaking the enemy rail supply link from Rangoon and capturing Meiktila and then Mandalay. On 2 May, Rangoon fell to the Allies. Slim was promoted to general two months later and appointed commander-in-chief of Allied forces for southeast Asia.

After the war he became army chief of staff (1948), was promoted to field-marshal, and served as governor-general of Australia (1953–60). His account of World War II, *Defeat Into Victory*, was published in 1956.

Smith, Bessie (1895–1937) *Black US blues singer. Known as the 'Empress of the Blues', this tall handsome woman created a style that was followed by many jazz singers, including Louis* ARMSTRONG *and Billie* HOLIDAY.

Born in Chattanooga, Tennessee, she came from a very poor family. As a child she began her singing career, in which she was supported by another black singer, Ma Rainey (1886–1939). She started touring with Ma Rainey's Rabbit Foot Minstrels in about 1912. Even though the troupe was successful, Bessie Smith was ignored by record companies as too primitive. However, in 1923 she was discovered by Clarence Williams and her 'Downhearted Blues' became a best-seller in the new market for 'race' music. In the late 1920s she was one of the most popular black entertainers in America but by the time she died as a result of injuries received in a car crash, the blues were out of fashion and her alcoholism had made her difficult to engage.

She made a short film, *St Louis Blues*, in 1929. About 160 of her recordings survive, all of which were reissued in the early 1970s. In 1960 a play by Edward ALBEE, *The Death of Bessie Smith*, was based on the suggestion made at the time that if she had been white, more prompt and efficient medical attention after her accident could have saved her life.

Smith, (Robert) Harvey (1938–) *British showjumper.*

Born in Bingley, Yorkshire, Harvey Smith is effectively a self-taught showjumper. After competing successfully in junior events, including an International Young Riders Competition in London, he represented Britain for the first time in 1958. Since then he has represented Britain in many successful Nations Cup teams and in the 1968 and 1972 Olympics teams; in 1983 he was a member of the team that won the silver medal at the European championships. He has also frequently been leading rider at the Royal International Horse Show and Horse of the Year Show. In 1984, taking part in the Dublin Horse Show for the twenty-sixth time, he was leading rider and won the Grand Prix.

Smith was the first British rider to turn professional (1972), not a popular move at the time. Since 1978 he has ridden for the Sanyo UK team. Aggressive and outspoken, Harvey Smith nevertheless holds an impressive record in individual and team events. He now competes regularly with his son Robert and has also worked as a commentator on showjumping broadcasts.

Smith, Ian Douglas (1919–) *Prime minister of Rhodesia (now Zimbabwe) (1964–79). Unable to accept black majority rule, Smith made a Unilateral Declaration of Independence that provoked a civil war and in the end failed to achieve its ends.*

Born in Selukwe, Southern Rhodesia, the son of a butcher, Smith was educated at the Chaplin School, Gwelo, and Rhodes University, Grahamstown, where he graduated with a bachelor of commerce degree. Joining the RAF as a pilot in 1941 during World War II, he was shot down over Italy but returned to Rhodesia in 1946. Smith's political career began in 1948 when he was elected as the Rhodesia Party member for Selukwe in the Southern Rhodesia legislative assembly. Entering the federal parliament in 1953 as the United Federal Party member for Midlands, he was appointed the government's chief whip in 1956.

Resigning from the United Federal Party in 1961, Smith founded the right-wing Rhodesian Front (renamed Republican Front in 1981) in 1962, becoming leader and prime minister of Rhodesia in 1964. Pledged to fight for Rhodesia's independence on the basis of white supremacy, he would not accept Britain's condition that rule by the black majority should be prepared for. He therefore made a Unilateral Declaration of Independence (UDI) in 1965, despite the imposition of sanctions by the UK and the UN. Rhodesian black nationalists responded with a guerrilla war (1972–74), which cost thousands of lives. In 1979 Smith resigned as prime minister to make way for a multiracial government, serving as minister without portfolio under Bishop Muzorewa (1979–80). His opposition to the imposition of sanctions against South Africa resulted in his suspension from the House of Assembly in 1987, the year in which he also resigned as president of the Republican Front.

Smith, Stevie (Florence Margaret Smith; 1903–71) *British poet, artist, and novelist, who won the Queen's Gold Medal for Poetry in 1969.*

Stevie Smith was born in Hull but her family moved to London when she was still an infant; she was educated at North London Collegiate School, after which she became a secretary to a publisher. She continued to live as a spinster in Palmers Green for the rest of her life. She first attracted notice with her *Novel on Yellow Paper* (1936), written with the blend of seriousness and farce that also characterized her poetry. Her first two volumes of verse were *A*

Good Time Was Had By All (1937) and *Tender Only to One* (1938). Another novel, *Over the Frontier*, also appeared in 1938. She also contributed book reviews to several literary journals. The third of her three novels, *The Holiday*, was published in 1949.

In 1953 Stevie Smith relinquished her publishing post in order to care for her now bedridden aunt, who until then had kept house for her. Her best volume of poetry, *Not Waving But Drowning* (1957) – also the title of perhaps her most famous poem – was followed by *Some Are More Human than Others* (1958), a sketch book of comic, but disquieting, drawings. Her *Selected Poems* were published in 1962. In the 1960s the rising interest in poetry readings brought Stevie Smith into the public eye as a performer of her own poetry, both live and on radio. The simple ballad or hymn metres of her verse were well adapted to her idiosyncratic vocal delivery and she achieved a considerable following.

Smuts, Jan Christian (1870–1950) *South African prime minister (1919–24; 1939–48), soldier, and philosopher. He was a South African nationalist and international statesman of considerable stature, but his reputation in South Africa itself was less exalted. Many Afrikaners resented his part in taking South Africa into World War II and most black South Africans regarded him as a segregationalist.*

Smuts was born near Riebeck West, Cape Colony, the son of a farmer and member of the colonial parliament. He was educated at Victoria College, Stellenbosch, before being sent to study law at Cambridge University in 1891. In 1896 he set up a legal practice in Johannesburg and was appointed state attorney of the Transvaal in 1898. Smuts first rose to prominence when he led the Boer forces during the first Boer War (1899). He was present at the Vereeniging peace negotiations in 1902, during which he supported reconciliation. As one of the founders of the Het Volk party, he was made a minister in the Transvaal government in 1907. In 1910 he became minister of defence, mines, and the interior in the first Union government under Louis BOTHA. At the outbreak of World War I he was appointed to lead the campaign in East Africa. Impressing the British government with his leadership skills, he was invited to attend the Imperial War Conference in 1917 and participated in the peace conference at Versailles in 1919.

Shortly after, he succeeded Botha as leader of the United Party and became prime minister, but he lost to HERTZOG's Nationalist Party

in 1924. Deputy prime minister to Hertzog from 1933 to 1939, during the fusion of the two parties, Smuts became prime minister again in 1939 following disagreement over entry into World War II. He was appointed a field-marshal in the British army in 1941 and served in North Africa; he became a close advisor to CHURCHILL throughout the remainder of the war. As a leading participant in discussions on the founding of the United Nations, he wrote the preamble to the United Nations Charter. In 1948 he lost office to MALAN's reconstituted Nationalist Party, but remained leader of the parliamentary opposition until his death in 1950.

Apart from his activities as a world statesman, Smuts was the founder of the philosophical doctrine of holism. His *Holism and Evolution* was published in 1926 and he was elected a fellow of the Royal Society in 1930. He was appointed rector of St Andrew's University in 1931.

Snow, C(harles) P(ercy), Baron (1905–80) *British writer, scientist, and administrator. He was knighted in 1957 and created a life peer in 1964.*

The son of a church organist in Leicester, C. P. Snow attended school and university there, obtaining an MSc in physics (1928). He then moved to Christ's College, Cambridge, where he became a fellow (1930–50). His first novel, *The Search* (1934), draws on his experience as a research scientist. The title novel of his eleven-volume series *Strangers and Brothers* appeared in 1940; centred on the character of Lewis Eliot, whose history has many parallels with Snow's own rise from provincial obscurity, the sequence includes *The Masters* (1951), *The New Men* (1954), and *The Affair* (1960), which were successfully adapted for the stage, and *Corridors of Power* (1964).

During World War II Snow worked in the Ministry of Labour, becoming director of technical personnel (1942), and afterwards became a civil service commissioner with responsibility for scientific recruitment (1945–60). In 1950 he married the novelist Pamela Hansford Johnson (1912–81) and the same year his play *View Over the Park* was produced. Besides the later volumes of *Strangers and Brothers*, Snow wrote several studies of science and society and in 1959 his Rede Lecture *The Two Cultures and the Scientific Revolution* sparked off a famous controversy with the critic F. R. LEAVIS. From 1964 to 1966 he was parliamentary secretary in the Ministry of Technology. In 1970 the final volume of *Strangers and*

Brothers, Last Things, was published; the novels provide a fascinating insight into the minds of men in power and made Snow an international best-seller.

Sobers, Gary (Sir Garfield St Aubrun Sobers; 1936–) *West Indian cricketer and one of the game's greatest all-rounders of all time. He played 93 times for his country (39 times as captain), scoring more than 8000 runs (including 26 centuries) and taking 235 wickets. He was knighted in 1975.*

Born in Barbados, he went to Bay Street School and first played for his island in 1952. He was only seventeen when he made his first test appearance (1953); four years later, playing against Pakistan, he hit a record test score of 365 not out. He played for his country from 1953 to 1974, captaining the team from 1965. In the early sixties he played for South Australia. He came to England in 1968, captaining the Nottinghamshire team until 1974.

A left-hand bat of exceptional skill and natural talent, in his first-class career he scored 28 315 runs, including 86 centuries. His bowling, whether slow or fast, was equally impressive. In all, he took 1043 wickets and held 407 catches to illustrate the quality of his fielding. Gary Sobers has been – and for many still is – the undisputed hero of every West Indian boy. He has published several books on cricket.

Soddy, Frederick (1877–1956) *British chemist who was awarded the 1921 Nobel Prize for Chemistry for his work on radioactivity and isotopes.*

The son of a corn merchant, Soddy was educated at University College, Aberystwyth, and Oxford University, where he graduated in 1900. He applied immediately for the chemistry chair in Toronto but, failing to be appointed, accepted a junior post in the chemistry department of McGill University. Coincidentally, Ernest RUTHERFORD had been appointed to the chair of physics there in 1898, which led to a fruitful collaboration in which they established many of the basic properties of radioactivity. It was, they argued, an atomic phenomenon that lay 'outside the sphere of known atomic forces', produced new types of matter, and was 'a manifestation of subatomic chemical change'.

In 1903 Soddy left Canada to become a lecturer in physical chemistry at Glasgow University. While at Glasgow, Soddy tackled the problem of separating such apparently distinct elements as lead and radium D, a decay product of radium. Soddy suggested that elements were 'mixtures of several homogeneous elements of similar but not completely identical atomic weights'. For these homogeneous elements Soddy invented the name 'isotopes'. Soddy went on to formulate his displacement law, showing that the emission of alpha and beta particles transformed one element into another.

Soddy was appointed in 1914 to the professorship of chemistry at Aberdeen. He moved to Oxford in 1919 as Dr Lee's Professor of Chemistry, but thereafter he abandoned scientific research completely. Instead he wrote extensively on social and economic questions, pursuing in such works as *Cartesian Economics* (1922) and *Wealth, Virtual Wealth and Debt* (1926) views that attracted little attention. Soddy became increasingly bitter, objecting to the 'Cambridge clique' at the Cavendish for falsifying history and claiming all the limelight for themselves. He had equally strong views on the Royal Society and British science in general. On the death of his wife in 1936, he resigned from his Oxford chair. Described by colleagues as arrogant, rigid, and unsocial, he clearly had a lighter side. A paper published in *Nature* in 1936 on the geometrical problem of constructing four circles in contact with each other was in verse and had the title 'Kiss Precise'. In *The Story of Atomic Energy* (1949) he described himself as the 'sole surviving participator' in the ideas that led to the explosion of the first atomic bomb in 1945.

Solti, Sir Georg (1912–) *Hungarian-born naturalized British conductor. He was appointed a KBE in 1971.*

Solti studied at the Liszt Academy of Music in Budapest, under Ernst von Dohnányi (1877–1960), BARTÓK (piano), and KODÁLY, and began his career as répétiteur at the Budapest Opera. However, being Jewish he left Hungary in 1939 for Switzerland, where he won the 1942 Geneva International Piano Contest. In 1946 Solti returned to conducting, as musical director of the Bavarian State Opera, and soon established a European reputation. From 1952 until 1961 he conducted the Frankfurt Opera, moving to London as musical director of Covent Garden from 1961 to 1971. He was director of the Chicago Symphony Orchestra from 1969 to 1991. He also conducted the Orchestre de Paris from 1972 to 1975, and the London Philharmonic Orchestra from 1979 to 1983. Solti is particularly renowned for his interpretations of the works of Richard STRAUSS, Verdi, and Wagner.

Solzhenitsyn, Aleksandr Isayevich (1918–) *Soviet writer, who was awarded the*

Nobel Prize for Literature in 1970. He was expelled from the Soviet Union in 1974 and went to live in the USA, but in 1991 said he would return to his homeland now liberal reforms had taken place.

Solzhenitsyn was born at Rostov-on-Don, the son of an office worker. He graduated from Rostov University in mathematics before being called up for the army, in which he served with distinction during World War II. In 1945 he was arrested for making derogatory comments about STALIN and spent the next eight years in labour camps. Although he was released on Stalin's death (1953), he had to remain in exile for a further three years. Officially rehabilitated in 1957, he settled near Ryazan and became a schoolteacher.

Solzhenitsyn's first book, *One Day in the Life of Ivan Denisovich*, was completed in 1960 and published, with KHRUSHCHEV's permission, in 1962 in the prestigious literary journal *Novy Mir* ('New World'). Drawn from his own experiences, it describes the conditions in a labour camp in northern Kazakhstan and caused a sensation in the Soviet Union and abroad. The liberalization was short-lived; in 1968 the semi-official *Literary Gazette* attacked Solzhenitsyn for aligning himself with the Soviet Union's ideological enemies, and neither of Solzhenitsyn's next two documentary novels – *The First Circle* (1968) and *Cancer Ward* (1968) – could be published in his native country. In 1970 he was expelled from the Soviet Writers' Union. *August 1914* was published in Paris in 1971.

The harassment of Solzhenitsyn by the authorities culminated in his arrest and deportation (1974) after the publication abroad of the first of three volumes of *The Gulag Archipelago* in 1973; the first Russian-language edition finally appeared in 1989. In exile in Switzerland Solzhenitsyn retained his standing as one of the foremost critics of the Soviet system and its denial of human rights, but his later writings and speeches in the USA were not uncritical of the decadence and materialism of the West. His most recent publications include *October 1916* (1985).

Sommerfeld, Arnold Johannes Wilhelm (1868–1951) *German physicist who modified the Bohr theory of the atom, suggesting that orbital electrons travelled in elliptical orbits.*

The son of a physician, Sommerfeld was educated at the University of Königsberg. After teaching briefly at the universities of Göttingen, Clausthal, and Aachen he was appointed professor of physics at the University of Munich in 1906. Sommerfeld should have retired in 1936 in favour of his pupil, Werner HEISENBERG. Opposition from the Nazi party to Heisenberg's appointment prolonged Sommerfeld's tenure and it was not in fact until late 1939 that he finally retired, to be succeeded not by Heisenberg but by Wilhelm Müller, a Nazi aerodynamicist without a single publication in physics to his credit. Although Sommerfeld and Heisenberg were not Jewish, they were regarded by the Nazis as Jewish sympathizers. Sommerfeld, however, survived the war and returned to his Munich chair in 1945, continuing to work at physics until he died in a car accident in 1951.

Sommerfeld's main contribution to physics concerned the model of the atom proposed by Niels BOHR in 1913. Sommerfeld argued in 1916 that the fine structure of the spectrum of the hydrogen atom could be explained if it was assumed that electrons adopt elliptical rather than the circular orbits proposed by Bohr. This involved adopting a second quantum number, referred to as the azimuthal quantum number. Sommerfeld is also remembered as the author of the influential textbook, *Atombau und Spektrallinien* (1919; translated as *Atomic Structure and Spectral Lines*, 1923).

Sondheim, Stephen Joshua (1930–) *US composer and lyricist. As a songwriter he is celebrated for his witty and cynical lyrics.*

He wrote several full-length musicals while still at college and studied with composer Milton BABBITT. In 1956 he wrote incidental music for *Girls of Summer*; this was followed by lyrics for *West Side Story* (1958) with Leonard BERNSTEIN, *Gypsy* (1959) with Jule Styne (1905–), and *Do I Hear a Waltz?* (1965) with Richard RODGERS. He wrote words and music for his own shows *A Funny Thing Happened on the Way to the Forum* (1962), *Anyone Can Whistle* (1964), *Company* (1970), *Follies* (1971), *A Little Night Music* (1973), *Sweeney Todd* (1979), and *Into the Woods* (1987). The last five won both New York Drama Critics' Circle awards and Tony awards for best musical score. *Sunday in the Park with George* (1983) won a Pulitzer Prize. He has also written film music (*Stavisky*, 1974; *The Seven Percent Solution*, 1977; *Reds*, 1981). *Side by Side by Sondheim*, a show using a collection of his songs, was staged in London in 1976; *Marry Me A Little* (1981) and *You're Gonna Love Tomorrow* (1983) were also based on anthologies of his songs. In 1990 he spent six months as Visiting Professor of Drama at Oxford University.

Sopwith, Sir Thomas Octave Murdoch
(1888–1989) *British aeronautical engineer,
founder of Sopwith Aviation. He was knighted
in 1953.*

The son of a wealthy civil engineer, Sopwith
had an introduction to aviation that has become
legendary. In 1910 he purchased for £630 a
Howard-Wright biplane, assembled it, flew it
without any training, and immediately crashed
it. A little later he purchased a second kit,
assembled it, flew it, and gained his pilot's
licence – all on the same day. In the same year
he won the Baron de Forest prize of £4000 for
a long-distance flight of 169 miles. In 1912
Sopwith founded Sopwith Aviation Company
at Kingston-on-Thames. During World War I
he produced several successful planes, including the Sopwith Scout, Pup, and Camel. The
Camel, with 2790 'kills' to its credit, was the
most destructive fighter of the war. After the
disappearance of his company in 1920,
Sopwith remained in aviation, becoming chairman of Hawker-Siddeley in 1935, an office he
held until 1963. During this period he was
responsible for the production of the Lancaster
bombers and Hurricane fighters of World War
II. He also built the Gloster, which was modified to test Frank WHITTLE's newly designed jet
engine in 1941.

Soustelle, Jacques (1912–90) *French anthropologist and politician, who opposed Algerian independence.*

Born in Montpellier into a working-class family, Soustelle was educated at the École
Normale Supérieure, where he graduated with
a masters degree in philosophy in 1932. He
gained a further doctorate from the Sorbonne
in 1937, becoming assistant curator at the
Musée de l'Homme the same year; in 1938 he
was the youngest professor to hold the chair of
American Antiquities at the Collège de France.
Between 1932 and 1939 he participated in several anthropological missions to Mexico.

During World War II Soustelle worked with
DE GAULLE's government-in-exile, heading
several missions to Latin American and Caribbean countries and joining de Gaulle's staff in
Algiers in 1943. After the war he was elected
(1945) to the first constituent assembly. He
became secretary-general of the Rassemblement du Peuple Français (RPF) in 1947 and
won the seat of Rhône in the national assembly
in 1951. After his appointment as governor-
general in Algeria (1955–56), he advocated
French retention of Algeria and in 1958 supported the army coup there. He served as minister of information (1958–59) and minister for

the Sahara and atomic questions (1959–60)
under de Gaulle, but was dismissed in 1960 for
his opposition to de Gaulle's Algerian policy.
His membership of the OAS (Organisation de
l'Armée Secrète), a terrorist group opposed to
Algerian independence, led to the issue of a
warrant for his arrest in 1962. He lived in exile
from 1961 until 1968, when he returned to
France after the case had been dismissed. He
wrote several books on a wide range of topics
from Algeria to the Aztecs.

Soutine, Chaim (1893–1943) *Russian-born
expressionist painter associated with the
School of Paris.*

The tenth child of a Jewish tailor, Soutine
spent his childhood in poverty in a ghetto in
Minsk, Lithuania. He left home to study at the
École des Beaux-Arts at Vilna and in 1913
travelled to Paris, where he lived a bohemian
existence among a group of painters and poets
that included CHAGALL, LIPCHITZ, and LÉGER.
He formed a close friendship with MODIGLIANI,
whose death in 1920 devastated him. Between
1919 and 1922 Soutine lived mainly in Céret,
where he painted over two hundred canvases.
Mainly landscapes, they were executed in
thick paint with apparently feverish intensity.
During the 1920s Soutine also produced pictures of grotesque figures with twisted faces
and deformed bodies, for example *Woman in
Red* (1924–25). The subjects, close-up and
usually full-faced studies, were most often
women or uniformed young boys, such as
choir boys and page boys.

After 1923, following a successful exhibition, Soutine acquired some recognition and
financial stability. From 1925 he increasingly
painted still lifes, including plucked fowl and
flayed carcasses, using a violence of colour
and brushstroke that reflected his restless unstable temperament. When France was invaded in 1940, he refused to emigrate to the
USA and went to live in Touraine, where he
continued to paint often tempestuous landscapes until his death.

Spaak, Paul-Henri (1899–1972) *Belgian
statesman and prime minister (1938–39;
1946; 1947–49), best remembered as a
staunch advocate of European unity*

Born in Schaerbeck, the son of a writer, Spaak
was interned by the Germans during World
War I before attending the University of Brussels, where he graduated in law in 1922. He
was elected as a socialist deputy to the Belgian
parliament in 1932. Appointed minister of foreign affairs in 1936, he was elected as
Belgium's first socialist prime minister in

1938, a position he held for eleven months. Following the invasion of Belgium in 1940, Spaak served in London as foreign minister for the government-in-exile, playing a leading role in the establishment of Benelux, a customs union between Belgium, the Netherlands, and Luxembourg, formally created in 1948. On his return to Brussels in 1944 he served in several postwar coalition governments as foreign minister (1945–47; 1954–57), prime minister (1947–49), and deputy prime minister (1961). He retired from politics in 1966.

Spaak developed a reputation for his commitment to international organizations and for the role he played in promoting the economic recovery of western Europe. In the postwar years he was elected first president of the United Nations General Assembly in 1946 and was later president of the Consultative Assembly of the Council of Europe (1949–51), president of the General Assembly of the European Steel and Coal Community (1952), and secretary-general of NATO (1957–61). Spaak was chairman of the six-nation team that drafted the Treaty of Rome, establishing the EEC, and it was largely through his efforts that the treaty was signed in 1957.

Spark, Muriel (1918–   ) *British novelist.*
Muriel Sarah Maud was born in Edinburgh, where she was educated at James Gillespie's School and the Heriot Watt College. She then spent some years in Africa and in 1938 married S. O. Spark (the marriage was later dissolved). During World War II she worked in the Foreign Office's political intelligence department (1944–45). Later she edited the prestigious *Poetry Review* (1947–49) and her earliest published works were mainly critical or biographical studies of English writers. Her career as a fiction writer began when she won the 1951 *Observer* short story competition. From then on her stories regularly appeared in British and US magazines, and a collected edition was published in 1967.

In 1954 Muriel Spark converted to Roman Catholicism and her awareness of the paradoxes and ironies of the faith informs some of her best work, notably her most successful novel, *The Mandelbaum Gate* (1965), which won the James Tait Black Prize the following year. Her first novel, *The Comforters*, appeared in 1957, followed by *Robinson* (1958), *Memento Mori* (1959), *The Ballad of Peckham Rye* (1960), and *The Prime of Miss Jean Brodie* (1961). The last, about the havoc caused by the ideas of an emancipated schoolmistress in her own and her pupils' lives, was

made into a successful play (1966) and also filmed (1969). Her play *Doctors of Philosophy* was produced in London in 1963. Her later publications include *The Girls of Slender Means* (1963), *The Hothouse by the East River* (1973), *The Abbess of Crewe* (1974), about the Watergate affair, *Territorial Rights* (1979), *Loitering with Intent* (1981), *The Only Problem* (1984), a wry novel about a rich man who has problems with the Book of Job, *A Far Cry from Kensington* (1988), and *Symposium* (1990).

Spassky, Boris Vasselievich (1937–   ) *Soviet chess player and world champion from 1969 to 1972.*
Born in Leningrad (now St Petersburg), Spassky learnt to play chess when he was five years old; by eleven he was regarded as a prodigy. He was eighteen when he won the world junior championship and finished equal eighth in the Gothenburg interzonal tournament. This qualified him for the Candidates Series, the competition to produce a challenger to the reigning world champion. He finished equal third, a fine performance considering his limited experience. In 1956 he came equal first in the Soviet championship.

Spassky won the 1966 Candidates Series but failed to beat Petrosian for the world championship. However, in 1969 he succeeded and became world champion. The 1972 world championship challenge was overshadowed by the temperamental antics of Bobby FISCHER. Under enormous pressure Spassky remained dignified throughout the contest: although he lost the title he won the approbation of the world. In the next few years Spassky's career was a remarkable mixture of good wins and crushing defeats. He lost to the eventual champion, KARPOV, in the semifinal of the 1974 Candidates Series.

Spence, Sir Basil (1907–76) *British architect, who made his name with the Sea and Ships Pavilion at the Festival of Britain (1951) and with his designs for Coventry Cathedral. He was knighted in 1960 and awarded the OM in 1962.*
Educated in his native Edinburgh at George Watson's College and at the universities of Edinburgh and London, Spence began his architectural career assisting Sir Edwin LUTYENS in his plans for the Viceroy's House in New Delhi. Thereafter his own prewar practice was concerned mostly with the design of large Scottish country houses. During the war he served in the army. After the war he set up in London, concentrating on exhibition work for

the British Industries Fair (1947–49) and the Festival of Britain. In 1951 he won the competition for the new Coventry Cathedral, creating a vast concrete hall church, adjacent to the ruins of the old perpendicular cathedral, that he embellished with the works of EPSTEIN, Graham SUTHERLAND, John PIPER, and others. Much in demand for the new universities in the 1950s and 1960s, he created the spectacular Knightsbridge Barracks for the Household Cavalry in 1970 and the British Embassy in Rome in 1971.

Spencer, Sir Stanley (1891–1959) *British painter, noted for his religious and visionary works in the modern setting of his native village of Cookham, in Berkshire. He was knighted in 1959.*

Spencer studied at the Slade School of Art, London, gaining a scholarship and other prizes. During World War I he served first in the medical corps and then with the infantry in Macedonia. These war experiences played an important part in his postwar paintings, such as the large *Resurrection, Cookham* (1926), which is set in Cookham churchyard.

Independent of contemporary art movements, Spencer's work represents an individual and intensely spiritual view of experience expressed in a style of slightly naive realism. He was most prominent in British art between the world wars and in 1932 completed his most ambitious work – the series of murals for the Oratory of All Souls in the village of Burghclere, Berkshire. After World War II, when he was commissioned to paint the Clyde shipyards, Spencer turned to large-scale religious subjects, particularly on the theme of the Resurrection.

Spender, Sir Stephen Harold (1909–) *British poet and critic. He was made CBE in 1962 and knighted in 1983.*

Spender was born in London, where he attended University College School. His mother died when he was twelve and his father a few years later, leaving his maternal grandmother to look after the then teenage Spender children. At University College, Oxford, Spender, who was already writing poetry, met several of his outstanding literary contemporaries, notably W. H. AUDEN and Christopher ISHERWOOD, with whose left-wing views he was associated throughout the 1930s. His own first book of poems appeared in 1930.

Between 1930 and 1933 Spender divided his time almost equally between Germany and London; his critical study *The Destructive Element* (1936) shows his awareness of the

threatening forces being unleashed in Europe during this period. In the meantime two more verse collections appeared: *Poems* (1933) and *Vienna* (1934). *Forward From Liberalism* (1937) charts Spender's increasing commitment to communism and 1937 found him in Spain on an assignment from the *Daily Worker*. *Poems for Spain* and *The Still Centre* were published in 1939, the same year that Spender co-founded *Horizon* magazine and his marriage to Inez Pearn, whom he had married in 1936, broke down.

Spender did his war service as a fireman in the National Fire Service (1941–44). In 1941 he married the pianist Natasha Litvin. He continued writing poetry and *Ruins and Visions* (1941) and *Poems of Dedication* (1946) contain some of his most sensitive verse. His autobiography *World Within World* (1951) describes his life up to this time, and in the immediate postwar period he wrote again about Germany in the documentary *European Witness* (1946). His portrait of Israel, *Learning Laughter*, appeared in 1952. In 1953 he published a major study *The Creative Element* and the same year accepted the Elliston Chair of Poetry at Cincinnati University, thus beginning a period when he spent much time teaching in various US institutions. From 1953 to 1967 he was co-editor of *Encounter* magazine. In 1970 he accepted a chair at University College, London (1970–77). Among his later writings were a tribute to W. H. Auden (1975), *China Diary* (1982) in which he collaborated with David HOCKNEY, a translation (1983) of the *Oedipus* trilogy of plays, a collection of his journals (1985), and the novel *The Temple* (1988).

Spengler, Oswald (1880–1936) *German philosopher of history.*

Spengler was educated at the universities of Munich, Berlin, and Halle, completing in 1904 a PhD thesis on Heraclitus. He worked as a grammar-school teacher until 1911, when he devoted himself full-time to his own writings.

His famous work *Der Untergang des Abendlandes* (2 vols, 1918–22; translated as *The Decline of the West*, 1926–28), appearing at the end of World War I, appeared highly relevant to Spengler's contemporaries. He spoke of the inevitable decline of all previous civilizations, from the Egyptian onwards. Clearly, Spengler implied, our own civilization was unlikely to be an exception to the course of history. Choosing to present his case in a series of striking images rather than basing it on argument and a careful analysis of the historical record, Spengler identified our present civilization as the successor of the Greco-Roman, or Apollonian. He termed it Faustian and characterized it in terms of its command of space, its distinctive and destructive weapons, and its industrial power. Deploying biological and meteorological analogies without constraint, Spengler argued that civilizations undergo a seasonal cycle of about a thousand years. Faustian culture, which had experienced its spring during the Renaissance, was in its autumnal stage and about to move to its final wintry end.

Spielberg, Steven (1947–) *US film director and producer, who emerged as the most successful commercial director of the 1970s and 1980s.*

Born in Cincinnati, Ohio, he made amateur films while still as school and subsequently worked on television productions with Universal Pictures. The most successful of these television movies was *Duel* (1971), in which a motorist is terrorized by an enormous heavy goods vehicle: it won several European awards and launched Spielberg's career in mainstream cinema. *Jaws* (1975), about a man-eating white shark, became one of the most successful films ever made and spawned several sequels. Spielberg continued to concentrate on the sensational and fantastic themes that appealed to mass audiences with *Close Encounters of the Third Kind* (1977), about UFOs, *Poltergeist* (1982), a supernatural thriller which he co-wrote and produced, and *E.T.* (1982), another highly successful space fantasy about a creature from space, which he directed and produced. There followed a series of adventure films beginning with *Raiders of the Lost Ark* (1981) and continuing with *Indiana Jones and the Temple of Doom* (1984) and *Indiana Jones and the Last Crusade* (1989). As well as such other box-office hits as *Gremlins* (1984), *Back to the Future* (1986), and *Who Framed Roger Rabbit?* (1988), all of which he produced, he expanded his scope as a director with more serious films, notably *The Color Purple* (1985) and *Empire of the Sun* (1988). All his films, whether populist or serious, have been admired for their technical brilliance, particularly their special effects, and for their generally high standards of production.

Spitz, Mark Andrew (1950–) *US swimmer, winner of seven gold medals in the 1972 Munich Olympics and holder of twenty-seven world records for freestyle and butterfly (1967–72).*

Spitz began swimming at the age of eight and was soon training for seventy-five minutes a day. Later his father took a job as general manager of a scrap metal company near Santa Clara, California, which enabled Spitz to train with coach Haines. The benefits of this arrangement were soon evident when Spitz reached the national championships at the age of fourteen and had broken five world records in international competitions by the age of seventeen. At the 1968 Olympic Games in Mexico City he won two gold medals for swimming. The following year Spitz went to the University of Indiana, where he came under swimming coach James Counsilman. By 1972, more mature both physically and mentally, Spitz realized his full potential by winning gold medals in the 100 m and 200 m freestyle and butterfly, the 4×100 m and 4×200 m freestyle relays, and the 4×200 m medley relay. Since leaving university Spitz has made a career in real estate.

Spock, Benjamin McLane (1903–) *US paediatrician whose ideas on child rearing influenced a generation of parents after World War II. His book* The Common Sense Book of Baby and Child Care *(1946) has sold more than 30 million copies (more than any other book except the Bible).*

Born in New Haven, Connecticut, Spock was educated at Yale and Columbia universities and practised paediatrics at Cornell Medical College, New York (1933–47). He served in the US navy during World War II and afterwards joined the staff of the Rochester Child Health Institute, Minnesota (1947–51). From 1951 to 1955 he worked at the University of Pittsburgh, organizing the teaching of child development and psychiatry, and in 1955 became professor of child development at the Western Reserve University, specializing in the application of psychoanalytic principles to paediatrics.

Spock's theories of baby and child upbringing, which contrasted with traditional ideas of discipline and rigid routine, were often controversial but have influenced middle-class parents, especially, for many years. His books written for general readership include *A Baby's First Year* (1953), *Problems of Parents* (1962), *A Teenager's Guide to Life and Love* (1970), and *Raising Children in a Difficult Time* (1974). With increasing age Spock has become somewhat less permissive in the advice he gives to parents and later editions of his best-seller advocate a limit to the freedom accorded to small children. Spock was a prominent opponent of US involvement in the Vietnam War (*Dr Spock on Vietnam*, 1968) and was convicted for encouraging draft evasion. He subsequently became a leading member of the National Committee for Sane Nuclear Policy (SANE) and was the People's Party candidate for the US presidency in 1972.

Springsteen, Bruce (1949–) *US rock singer and songwriter, noted for his songs about working-class life in the USA and for his energetic stage performances.*

Born in Freehold, New Jersey, Springsteen formed his first rock band at high school. From 1969 he played in New York and east-coast clubs, building up a reputation as an exciting live performer. Although the songs on Springsteen's early albums – *Greetings from Asbury Park* (1973) and *The Wild, the Innocent, and the E-Street Shuffle* (1974) – are chiefly exuberant romanticized tales of New Jersey street life, their lyricism earned comparisons with Bob DYLAN. An accomplished third album, *Born to Run* (1975), was greeted with extravagant publicity declaring Springsteen 'the future of rock 'n' roll'. He was unable to build on this new fame, however, owing to a legal dispute with a former manager, which prevented him from recording for several years; instead he concentrated on touring.

Springsteen returned to recording with *Darkness on the Edge of Town* (1978) and *The River* (1981), albums that take a much bleaker view of US working-class life than his earlier material. In 1984 the *Born in the USA* album enjoyed extraordinary success, making Springsteen a household name and a national hero, with politicians (including President REAGAN) quoting his songs. This sudden success owed much to the loyal following he had built up through outstanding live performances over the years; a five-album collection of concert material was released in 1986. *Tunnel of Love* (1988) was an album of sombre love songs.

Stalin, Josef Vissarionovich (Josef Vissarionovich Dzhugashvili; 1879–1953) *Soviet statesman, general secretary of the Soviet Communist Party (1922–53) and Soviet prime minister (1941–53). The absolute ruler of the Soviet Union during World War II and after, he instigated a foreign policy that has largely determined the configuration of postwar Europe. Since his death he has been discredited in communist countries.*

Born in Gori, Georgia, the son of a shoemaker, Stalin attended a theological seminary in Tiflis but was expelled in 1899 for expounding sub-versive views. As a member of the Social Democratic Party (he joined in 1898), he became actively involved in revolutionary politics, for which he was imprisoned (but escaped) several times between 1903 and 1912.

Stalin joined the Bolsheviks under LENIN in 1903. Founding the party's newspaper *Pravda* in 1911, he became the leader of the Bolsheviks in the Duma (1913) but was exiled to Siberia between 1913 and 1917. He was appointed commissar for nationalities after the October Revolution and distinguished himself by defending Tsaritsyn (later Stalingrad, now Volgograd) during the civil war. He was elected general secretary of the central committee of the Communist Party in 1922 and succeeded Lenin as chairman of the Politburo in 1924. He secured enough support within the party to eliminate TROTSKY and other rivals, who disagreed with his theory of building socialism in the Soviet Union as a base from which communism could spread. By 1927 he was the uncontested leader of the party and government, and the following year he initiated the first of his five-year plans for the industrialization and collectivization of agriculture. During the 1930s he instigated his infamous 'purge trials' to rid himself of his opponents in the government and the army; by the outbreak of World War II he was in complete control of the country. In 1941 he became chairman of the Council of People's Commissars (prime minister) and took command of the armed forces when HITLER violated the 1939 nonagression pact; he assumed the title of marshal in 1943 and generalissimo in 1945. He attended the Allied conferences at Tehran (1943), Yalta (1945), and Potsdam (1945), emerging from all of them as a dominant figure. After 1945 he maintained his firm grip on the Soviet political machine and attempted to exercise similar control over other socialist states; only Yugoslavia under TITO succeeded in diverging. He died in office in 1953.

Stalin was a ruthless and authoritarian leader, who governed the Soviet Union with a tyrannical hand. However, at the time of his death the Soviet Union had become the second most important industrial country in the world. After his death, his severe one-man rule, the 'cult of personality', was denounced among eastern bloc nations. In 1961 his embalmed body was removed from Lenin's mausoleum and reburied in a nondescript grave adjacent to the Kremlin.

Stanislavsky, Konstantin (Konstantin Sergeyevich Alekseyev; 1863–1938) *Russian actor, director, and theoretician. He co-founded the Moscow Art Theatre (1898) and his theories later formed a basis for the development of 'method' acting.*

Born in Moscow, the son of a manufacturer, Stanislavsky started acting at the age of fourteen in his family's amateur dramatic group. The theatre gradually became an absorbing interest and in 1888 Stanislavsky formed a permanent company of amateur actors who staged their own productions. These attracted the attention of writer and director Vladimir Nemirovich-Danchenko (1859–1943), and in 1898 the two men founded the Moscow Art Theatre and staged an outstanding production of Chekhov's *The Seagull*. Stanislavsky was determined to develop a more naturalistic mode of acting, breaking with the stylized artificiality of the theatre of his day, and to this end demanded of his actors a much more psychological approach to the development of the characters. He went on to direct the first production of Chekhov's other major plays, including *Uncle Vanya* (1899), *The Three Sisters* (1901), and *The Cherry Orchard* (1904). As well as directing he also appeared in several Chekhov plays, Ibsen's *An Enemy of the People*, and Turgenev's *A Month in the Country*, among others; he continued as the leading actor of his theatre until 1928. In all some three studios were attached to the Moscow Art Theatre and through world tours its reputation spread. Stanislavsky also became increasingly involved with the Opera Studio at the Bolshoi Theatre; his production of Tchaikovsky's *Eugene Onegin* (1922) was highly acclaimed.

Stanislavsky's theories and methods were to have far-reaching effects, particularly through his publications. These included *My Life in Art* (1924), *An Actor Prepares* (1936), and *Building a Character* (1950). 'The Method', a naturalistic style of acting evolved in the USA in the thirties, which blossomed at the Actors' Studio during the forties and fifties, was based on the teachings of Stanislavsky.

Stark, Dame Freya Madeline (1893–) *British traveller and writer. She was made a DBE in 1972.*

Freya Stark was born in Paris and brought up mainly in Italy by her unconventional mother, who lived apart from her sculptor husband. During World War I Freya Stark nursed for the Red Cross in Italy. In the 1920s, despite ill health, she determined to escape from her demanding mother and to travel in the East. She learnt Arabic and in 1927 set off for Beirut, as described at the close of her first volume of autobiography, *Traveller's Prelude* (1950).

She first attracted attention as a travel writer with *Baghdad Sketches* (1933) and *The Valleys of the Assassins* (1934), which she followed with accounts of journeys in Arabia (1934–38) – *The Southern Gates of Arabia* (1936), *Seen in the Hadhramaut* (1938), and *A Winter in Arabia* (1940). In 1939 she joined the ministry of information and spent most of World War II in the Middle East helping to counter German influence in Aden, the Yemen, Egypt, and Iraq. In 1944 she travelled to North America and then on to India. Her travels in the 1930s and the events of the war years are described in three further volumes of autobiography: *Beyond Euphrates* (1951), *The Coast of Incense* (1953), and *Dust in the Lion's Paw* (1961). In 1947 she married the orientalist Stewart Perowne (1901–89).

The archaeology and history of the East continued to dominate Freya Stark's interests. Among her later publications were books on Turkey – *Ionia: A Quest* (1954) and *Turkey* (1971) – and Afghanistan – *The Minaret of Djam* (1970) – and the eastern frontiers of the classical world – *Alexander's Path* (1958) and *Rome on the Euphrates* (1966). Always a witty and prolific letter writer, she published the first volume of her collected letters in 1974; a volume of selected letters, *Over the Rim of the World*, was published in 1988.

Starling, Ernest Henry (1866–1927) *British physiologist whose important discoveries include, with Sir William Bayliss (1860–1924), the first identification of a hormone.*

A Londoner, Starling received his medical qualification from Guy's Hospital in 1889 and the following year started working at University College, London. Here he studied the mechanism of lymph secretion and blood circulation through the fine capillary blood vessels of the body. In 1899 Starling was appointed professor of physiology at University College. In collaboration with Bayliss, he demonstrated the nervous control of peristalsis (the waves of muscle contraction that propel the contents of the gut) and in 1902 they discovered the substance secreted by the duodenum into the blood, which they called secretin. This stimulates the pancreas to secrete digestive juice into the duodenum. Starling's generic term for secretin, hormone, introduced in 1904, is now used to describe all such messenger substances released into the blood by endocrine glands. Equally important were

Starling's investigations of the factors affecting heartbeat. Using isolated heart–lung preparations in experimental animals, he measured blood flow and pressures with varying arterial resistance and venous inflow. He established the principle, known as Starling's law, that the strength of contraction depends on the extent to which the muscles of the heart chambers are stretched during filling. This provides a self-correcting mechanism for maintaining constant heart rate despite varying quantities of venous blood entering the heart.

Starling's work was interrupted by World War I, when he served in the Royal Army Medical Corps and contributed to developments in gas mask design. His *Principles of Human Physiology* (1912) was recognized as a major text of modern physiology. Starling was elected a fellow of the Royal Society in 1899 and in 1922 was appointed the first Foulerton Research Professor of the Royal Society.

Starr, Ringo (Richard Starkey; 1940–   ) *British rock drummer, singer, and actor. As drummer with the Beatles (1962–70), he was one of the best-known people in the world; since the break-up of the group his career has been less exciting.*

Born in Liverpool, he became an apprentice engineer in 1959 after an undistinguished school career. After receiving a set of drums as a Christmas present, he joined a skiffle group. This led to a tour in Germany in 1961 with Rory Storme's Hurricanes, where he met John LENNON, Paul MCCARTNEY, and George HARRISON. The following year he replaced Pete Best as the Beatles' drummer and his style became synonymous with the 'Mersey beat' that swept the world of pop music. As one of the Beatles he occasionally sang vocals, often in songs borrowed from country music, such as 'Honey Don't' and 'Act Naturally'. His first song to be performed was 'Don't Pass Me By' (1968); the same year he sang the title song in the feature-length cartoon film *Yellow Submarine*.

His solo career began in 1970 with the disappointing album *Sentimental Journey*, a collection of pre-rock standards. Of the albums that followed the most successful were *Ringo* (1973) and *Goodnight Vienna* (1974). In 1973 he directed a documentary film called *Born to Boogie*; he has also appeared in several other films, notably *That'll Be The Day* (1973).

Staudinger, Hermann (1881–1965) *German chemist, who was awarded the Nobel Prize for Chemistry in 1953 for his work on macromolecules in chemistry and biology.*

The son of a philosophy professor, Staudinger was educated at the universities of Darmstadt, Munich, and Halle, where he obtained his PhD in 1903. He taught chemistry at the University of Strasbourg, the Technische Hochschule in Karlsruhe, and the Eidgenossische Technische Hochschule in Zürich, before moving to the University of Freiburg in 1926. He remained there until his retirement in 1951.

Staudinger established his reputation in 1907 when he succeeded in synthesizing a constituent of natural rubber. According to the orthodox chemistry of the time, complex molecules, such as rubber and protein, were regarded as aggregates of molecules of low molecular weight held together by secondary bonds, rather than large molecules. Staudinger's work on rubber made him sceptical of this view and in 1922 he introduced the term 'macromolecules' to intense opposition from the majority of his colleagues. When Staudinger argued his case to the Zürich Chemical Society in 1925, an eye-witness described him shouting Luther's words in defiance of his critics: 'Hier stehe ich, ich kann nicht anders' ('Here I stand, I can do no other'). It was left to the inventor of the ultracentrifuge, Theodor Svedberg (1884–1971), to confirm the existence of macromolecules and Staudinger did not receive his Nobel Prize for nearly thirty years.

Stauffenberg, Claus, Graf Schenk von (1907–44) *German army officer whose plot to assassinate HITLER failed and led to his own execution.*

Born in Jettingen, of Swabian aristocratic descent, Stauffenberg attended the infantry school at Dresden and the cavalry school at Hanover. Commissioned as a second lieutenant in 1930, he entered the General Staff College in 1936, becoming a captain in 1938. He was promoted to lieutenant-colonel and appointed senior staff officer of a panzer division in North Africa in the early stages of World War II. He was seriously wounded in 1943, after which he was posted back to the General Staff Office in Berlin and became chief of staff to General Olbricht with the Home Army. As chief of staff he had access to a vast amount of political and military information and became associated with a circle of conspirators opposed to Hitler, who managed to enlist the support of ROMMEL in their plans to overthrow the Führer. In 1944, following two unsuccessful attempts to assassinate Hitler, he left a bomb in a briefcase in Hitler's headquarters at Rastenberg. The bomb failed to kill Hitler, and

Stauffenberg – with several of his fellow conspirators – was executed on the same day.

Steel, Sir David Martin Scott (1938–)
British politician, leader of the Liberal Party (1976–88), who became a co-founder of the new Social and Liberal Democrats in 1988. He was knighted in 1990.

Born in Edinburgh and educated in Nairobi, Kenya, and then in Edinburgh, Steel became an MP for Roxburgh, Selkirk, and Peebles in 1965 (and for the new constituency of Tweeddale, Ettrick, and Lauderdale in 1983). He was president of the Anti-Apartheid Movement in Britain from 1966 to 1969. He sponsored the private member's bill that became the Abortion Act 1967. From 1970 to 1975 he was Liberal chief whip and in 1976 became the first Liberal leader to be elected by the party outside parliament. In 1977 Steel negotiated the Lib–Lab Pact with the minority Labour government and in 1982 formed the Liberal/SDP Alliance with the new Social Democratic Party. He shared the leadership of the alliance with David OWEN until 1988, when the Social and Liberal Democrats were formed as a single merged party, with Paddy ASHDOWN as leader. In 1989 he campaigned unsuccessfully for a place in the European Parliament, representing the Central Region of Italy.

Stein, Gertrude (1874–1946) *US writer, who from 1903 lived in Paris, where she became a focus for the American expatriate literary community.*

Born in Pennsylvania, Gertrude Stein graduated from Radcliffe College in 1897 where, under the influence of William James (1842–1910), she developed an interest in psychology. She spent several years at Johns Hopkins medical school studying the anatomy of the brain, but eventually grew bored with her studies and, in 1902, decided to follow her brother Leo to Europe. By 1903 she had settled in Paris where she was shortly joined by Alice B. Toklas, who remained her companion for the rest of her life. Apart from lecture tours – to Britain in the 1920s and to America in 1934 – she lived in France until her death. During World War II she withdrew from Paris to the country, where she lived with blithe unconcern at the risk to herself and to others who protected her from the Gestapo. At her Paris flat she presided as a kind of cult figure for the young, especially such young American writers as Ernest HEMINGWAY, whose prose style she influenced. An undeniable talent for self-promotion assured that her image – cropped hair and baggy shapeless clothes – was im-

mediately recognizable everywhere; indeed, this has remained familiar, while her difficult and often tedious writings have never been widely read.

With her brother Leo, who had profited from an acquaintance with Bernard BERENSON, Stein very early started buying the paintings of MATISSE, BRAQUE, and PICASSO before they were generally known, and she liked to take credit for the later growth of cubism. This claim was vehemently contested by most of the artists involved, one of whom noted that her knowledge of French was so poor that she could not have had the slightest inkling of what their concerns were. In any case she amassed a good collection of their pictures and did much to promote modern art.

Most of Stein's writing is experimental, an attempt to create a continual present by means of repetitions with slight variations and by other techniques. Her books include *Three Lives* (1909), relatively straightforward stories, *Tender Buttons* (1914), poetry concerned with rendering objects, *The Autobiography of Alice B. Toklas* (1933), which is in fact her own autobiography, the libretto for *Four Saints in Three Acts* (1934), an opera by Virgil THOMSON, *Matisse, Picasso, and Gertrude Stein* (1938), and *Lectures in America* (1935), on her theory of composition and the influence on it of William James and BERGSON.

Steinbeck, John Ernst (1902–68) *US novelist whose work achieved both popularity and critical acclaim. He was awarded a Pulitzer Prize (1940) and the Nobel Prize for Literature (1962).*

Steinbeck was born in Salinas, California, which provided the setting for many of his books. After studying marine biology at Stanford University, he attempted to earn a living as a journalist in New York, before returning to his native state. His first novel was *Cup of Gold* (1929), a romanticized adventure story, and neither it nor the two following books, *The Pastures of Heaven* (1932) and *To a God Unknown* (1933), attracted much attention. However, *Tortilla Flat* (1935), which described with affection and whimsy the lives of the California 'paisanos', became a best-seller; it established a successful formula that Steinbeck was to repeat in later years, as in *Cannery Row* (1945) and *Sweet Thursday* (1954). A change of tone was apparent in *In Dubious Battle* (1936) and in his best-known novels – *Of Mice and Men* (1937; filmed 1939) and *The Grapes of Wrath* (1939; filmed 1940). These novels combined an ease of style, to ensure a popular market, with serious social comment. Subsequent works include *Sea of Cortez* (1941), the nonfiction *Bombs Away* (1942) and *Once There Was a War* (1958), the short story *The Red Pony* (1945), *The Pearl* (1947), and the highly successful *East of Eden* (1952; filmed 1955).

Steiner, Rudolf (1861–1925) *Austrian founder of anthroposophy.*

Born in Kraljeve, Croatia (now in Yugoslavia), the son of a Catholic stationmaster, Steiner was educated at the University of Vienna, where he studied natural science. From 1890 to 1897 he worked at Weimar, editing Goethe's writings on natural history. At this point Steiner's career as an academic scholar ended when he came under the influence of Annie Besant (1847–1933) and the theosophist movement. For some ten years Steiner served the movement but in 1912 he broke away to found his own school of anthroposophy. Steiner established the headquarters of his new movement at the so-called 'Goetheanum' at Dornach near Basel.

Like its theosophical ancestor, Steiner's anthroposophy is essentially eclectic, with elements taken from eastern religions, early Christian gnosticism, mystic literature, and classical German philosophy. It has consequently been quite self-contained. More influential have been his educational theories with their emphasis on play and creative activity in the learning process; Steiner schools are operating in many parts of the western world.

Stern, Isaac (1920–) *Soviet-born US violinist. He was appointed an Officier de la Légion d'honneur in 1975.*

Stern was born in Kreminiecz in the Soviet Union, but spent his childhood in San Francisco. He studied at the San Francisco Conservatory from 1928 to 1931 and later with Naoum Blinder, a violinist of the Russian School and his most influential teacher (1932–37). Stern made his recital debut in 1935 and his debut with the San Francisco Symphony Orchestra in 1936. After his first appearance in New York (1937) he retired for further study, making a second appearance there in 1939, with unqualified success. Stern toured Europe in 1948, quickly becoming a popular figure at the major festivals (Prades 1950–52; Edinburgh 1953). In 1960 he formed a trio with Eugene Istomin and Leonard Rose, which was highly acclaimed. As chairman of the American-Israel Cultural Foundation since 1964, he is instrumental in furthering opportunities for young musicians. His films include

Mao to Mozart: Isaac Stern in China (1981), which won an Academy Award, and *Carnegie Hall – The Grand Reopening* (1987), for television.

Stern, Otto (1888–1969) *German-born US physicist who was awarded the 1943 Nobel Prize for Physics for his use of molecular beams to establish the existence of atomic magnetic moments.*

The son of a prosperous grain merchant, Stern was educated at the University of Breslau, where he obtained his PhD in 1912. He worked with EINSTEIN as a postdoctoral student in Prague and Zürich and, after service with the German army during World War I, taught at the University of Rostock (1921–23). In 1923 he moved to the University of Hamburg as professor of physical chemistry, but with the rise of the Nazis in 1933, as a Jew, he decided to emigrate to the USA. He took up an appointment with the Carnegie Institute of Technology, Pittsburgh, which he held until his retirement in 1945. On his retirement Stern moved to Berkeley, California, where he died of a heart attack, aged eighty-one, on his daily visit to the cinema.

In 1920–21, in cooperation with Walter Gerlach, Stern attempted to check Arnold SOMMERFELD's suggestion that some atoms have magnetic moments, as required by his version of the quantum theory. Stern passed beams of silver atoms through a nonuniform magnetic field. As predicted, the beam split into two separate parts, thus establishing Sommerfeld's theory. Using similar techniques Stern went on to measure the magnetic moment of the proton (1933) and the deuteron (1934).

Stevens, Wallace (1879–1955) *US poet, whose influential work achieved a new poetic insight into the nature of reality.*

Stevens was born at Reading, Pennsylvania. He attended Harvard, and later New York Law School, being admitted to the US bar in 1904. He was employed as a lawyer by an insurance company for many years, finally becoming vice-president of the firm in 1934. His poetry was a private activity, of which many of his colleagues had no knowledge.

After the appearance of several poems in various anthologies and a couple of one-act free-verse plays, Stevens's first collection of poetry, *Harmonium*, was finally published in 1923, when he was forty-four. Subsequent collections include *Ideas of Order* (1936), *The Man With the Blue Guitar* (1937), *Parts of a World* (1942), *The Necessary Angel* (1951; a collection of essays), and *Collected Poems* (1954). Stevens was preoccupied with the interaction between external reality and man's perception of it. This theme permeated his poetry, which is characterized by a brilliantly original vocabulary. Critics, perhaps confused by the contrast between Stevens's conservative lifestyle and his innovative poetry, have had some difficulty in categorizing his work; generally he is placed within the imagist branch of modernism. Although his early poems have a pronounced European flavour (and he probably was influenced by Baudelaire and Mallarmé), he remained essentially a spokesman for his own environment; indeed, he never left the USA, even for a holiday abroad.

Stevenson, Adlai E(wing) (1900–65) *US politician who twice stood unsuccessfully as Democratic presidential candidate and who served as US ambassador to the United Nations under President KENNEDY.*

Grandson of the former vice-president of the same name, Stevenson read history at Princeton University, then law at Harvard and Northwestern University, qualifying in 1926. While practising law in Chicago, he became active in public service, with a special interest in civil rights and world affairs. After World War II, during which he acted as special assistant to the navy (1941–44), Stevenson was appointed in 1945 assistant to the secretary of state and became involved in the formation of the United Nations. He was senior adviser with the US delegation to the first general assembly held in London in 1946 and a delegate at subsequent assemblies in New York. In 1948 he was elected governor of Illinois and in 1952 received the Democratic presidential nomination. Although an eloquent and witty speaker, he lost in both 1952 and 1956 elections to Dwight D. EISENHOWER. He was appointed US ambassador to the UN in 1960 and held that office until his death.

Stewart, Jackie (John Young Stewart; 1939–) *British motor racing driver and three times World Champion (1969, 1971, and 1973).*

The son of a garage proprietor from the Dunbarton district, Stewart started racing locally in 1961 and in 1964 was racing Cooper-BMC cars for Ken Tyrrell as well as a Lotus-Cosworth in Formula Two events. His success led to a place in the BRM Formula One team in 1965, victory in the Italian Grand Prix, and third place in the Drivers' Championship. He also drove Cooper-BRMs and Matras in

Formula Two, again for Ken Tyrrell, and in 1968, Tyrrell and Stewart started their successful partnership in Formula One. Driving a Matra-Ford, Stewart achieved second place in the championship, in spite of a cracked wrist bone sustained after crashing his Formula Two car during practice for the Spanish Grand Prix. However, in 1969 he drove the Matra-Ford to six Grand Prix victories and his first World Championship. He repeated this success in 1971, driving a Tyrrell-Ford, and again in 1973, when he retired from racing with a record tally of twenty-seven Grand Prix wins.

Stewart was a champion even before his motor-racing career – as a champion clay pigeon marksman, he was a reserve for the 1960 British Olympic team. Following his retirement, he has pursued his business interests and acted as a commentator on motor sport and other motoring topics for television and the press.

Stewart, James (1908–) *US film star, who typically portrayed slow-speaking honest heroes.*

Born in Indiana, Pennsylvania, Stewart showed an early interest in entertainment as an amateur magician and actor in Boy Scout productions. With his distinctive drawl, Stewart must have seemed an unlikely candidate for stardom when in 1932, as a gangling young Princeton graduate in architecture, he joined the university's theatre group with his friends Henry FONDA and Margaret Sullavan (1911–60). He worked with only moderate success in the theatre until making his first film, *The Murder Man* (1935). Films that followed included *You Can't Take It With You* (1938), *It's a Wonderful World* (1939), *Mr Smith Goes to Washington* (1939), for which he received the New York Film Critics best actor award, and possibly his best film, *The Philadelphia Story* (1940), which earned him an Oscar. During World War II he saw active service as a bomber pilot and in 1968 retired from the US Air Force Reserve with the rank of brigadier-general. After the war he had memorable roles in *The Naked Spur* (1953), HITCHCOCK's *Rear Window* (1954) and *Vertigo* (1958), and such westerns as *The Man from Laramie* (1955), which demonstrated the extent of his range. Later films included *Airport* (1977) and *The Magic of Lassie* (1978). He received a special Academy Award in 1984 for his contribution to the cinema.

Stieglitz, Alfred (1864–1946) *US photographer who established photography as a fine art*

in the USA and, through his galleries and publications, introduced Americans to modern art.

Stieglitz was born in New Jersey, the son of a wool merchant, and initially studied engineering, first in New York and then (in 1881) in Berlin. Here he purchased his first camera and straightaway switched to courses in photochemistry relevant to photography. From the outset, Stieglitz was determined that photography should be recognized as a legitimate art form. He experimented with such innovations as night-time photography and working in rain and snow. In 1890 he returned to the USA and in the years following he edited *American Amateur Photographer*, won a host of prizes with his work, and gained an international reputation. In 1902, he and a group of fellow photographers, including Edward Steichen, formed Photo-Secession with the aim of establishing photography as art through exhibitions and the quarterly *Camera Work*. Stieglitz opened Little Galleries at 291 Fifth Avenue, New York, in 1905. Known as '291', the gallery exhibited not only photographs but also modern paintings and sculpture, with work by Rodin, Cézanne, MATISSE, and PICASSO, as well as contemporary US painters, such as Max WEBER and Georgia O'Keeffe (1887–1986), whom Stieglitz married in 1924. Thus it was Stieglitz who first exposed the American public to the 'shock of the new'. Following the closure of '291' in 1917, he set up The Intimate Gallery (1925–29) and in 1930 An American Place opened on Madison Avenue.

Stieglitz was an adherent of 'straight photography' with a minimum of darkroom trickery. Among his best-known earlier works are *The Terminal* (1893), an atmospheric study of a horse-drawn tram in the snow, and *The Steerage* (1907), a poignant view of passenger decks on a ship. In the 1920s and 1930s his work consisted largely of a series of portraits of Georgia O'Keeffe and his 'equivalents' – pictures of sun and clouds that transcend form to reflect the artist's own hopes and fears. Ill health forced Stieglitz to cease taking pictures in the late 1930s but he continued to attend his New York gallery up to his death.

St Laurent, Yves Mathieu (1936–) *French couturier noted for his youth-culture motifs in high fashion in the early 1960s.*

The son of a lawyer, St Laurent was born in Oran, Algeria, and after his secondary education moved to Paris, determined to break into the fashion industry. When he won first prize in a design competition, his talent was recognized by Christian DIOR, who in 1953 hired St

Laurent as an assistant. After Dior's death in 1957, St Laurent succeeded him as head designer; following his first collection in 1958, which introduced the 'little-girl look', he was acclaimed Dior's rightful heir. In 1960 St Laurent was called up for national service and suffered a nervous collapse. He was replaced at Dior by his assistant, Marc Bohan, a move that led to a suit for damages, which were awarded in favour of St Laurent. In 1961, in collaboration with the US businessman J. Mack Robinson, St Laurent opened his own fashion house and showed his first collection in February 1962. Four years later, the first of a worldwide chain of Rive Gauche boutiques was opened to sell ready-to-wear garments. He has also designed costumes for the stage and cinema.

Stockhausen, Karlheinz (1928–) *German composer and pioneer of electronic music.*

Orphaned during World War II, Stockhausen had to work to finance his own musical education. He studied first with the Swiss composer Frank Martin (1890–1974) at the Cologne Musikhochschule (1947–50) and later with Olivier MESSIAEN in Paris (1952–53). He also studied physics (especially acoustics) and phonetics at Bonn University (1953–56). His early works for conventional instruments include *Kreuzspiel* (1951), for oboe, bass clarinet, piano, and percussion, and *Kontra-Punkte* (1952; revised 1953), for ten solo instruments, which is the first of Stockhausen's pieces to use groups of notes as units within a total serialism. *Gruppen* (1955–57), for three orchestras, reflects Stockhausen's preoccupation with the spatial aspect of music.

Stockhausen's electronic works date from 1953, when he joined the newly founded electronic music studio of the West German Radio at Cologne. Here, working in conjunction with its director, Herbert Eimert, he used three signal generators to produce a form of music. In *Gesang der Jünglinge* (1956), the human voice is combined with electronic sound. In the 1960s he developed means of electronic transformation, as in *Kontakte* (1960) and *Momente* (1962). In *Stimmung* (1968) and *Mantra* (1970) much is left to the intuition of the performer or the player. At this time his music was also influenced by eastern mysticism. Stockhausen invented a method of music notation for his works, using graphs and geometrical figures, in which each note or group of notes is precisely notated, although it also allows for controlled aleatoric elements. With

Donnerstag (1980), he began a planned heptalogy for performance on each evening of a week; *Samstag* followed in 1984 and *Montag* in 1988.

Stockton, Earl of (1894–1986) See MACMILLAN, (MAURICE) HAROLD EARL OF STOCKTON.

Stockwood, (Arthur) Mervyn (1913–   ) *Anglican clergyman who, as Bishop of Southwark (1959–80), voiced the concern of progressive elements in the Church on a wide range of contemporary moral and political issues.*

Stockwood read history and theology at Cambridge and, after attending Westcott House theological college, Cambridge, was ordained in 1936. When he was appointed curate to the parish of St Matthew Moorfields, Bristol, the poverty of his area convinced Stockwood of his duty, as a Christian, to work for justice through social change. He became friendly with the prominent Labour politician and Bristol MP, Sir Stafford CRIPPS, and in 1938 Stockwood joined the Labour Party, although Cripps and his supporters were expelled from the party shortly afterwards. Installed as vicar of St Matthew Moorfields in 1941, Stockwood breathed new life into the parish through his imaginative blend of the sacred and secular, although his steps towards cooperation between different churches often ran into difficulties. His pastoral skills were further displayed as vicar (1955–59) of the University Church in Cambridge – Great St Mary's – and in 1959 he was appointed Bishop of Southwark.

Ministering to the vast south London diocese, Stockwood was inevitably drawn into the furore surrounding the publication of *Honest to God*, written by the Bishop of Woolwich, John ROBINSON. Less dramatic was Stockwood's introduction of the Southwark Ordination Course for auxiliary priests, which set a precedent for other dioceses. Stockwood was a familiar figure in the House of Lords, championing many causes, including the abolition of capital punishment and apartheid, and homosexual law reform. His dismay at the ignorance of journalists and others regarding Marxist philosophy and Christian theology resulted in his book *The Cross and the Sickle* (1978). His autobiography, *Chanctonbury Ring*, was published in 1982.

Stokowski, Leopold Anthony (Antoni Stanislaw Boleslawowich; 1882–1977) *British-born US conductor. He reached a wide* audience in Walt DISNEY's film *Fantasia and through his many records.*

Born in London to a Polish father and an Irish mother, Stokowski studied at the Royal College of Music from the age of thirteen. After appointments as organist and choirmaster at St James's, Piccadilly (1902), and St Bartholomew's, New York (1905), he made his conducting debut in Paris in 1908, standing in at short notice for an indisposed colleague. He was then appointed conductor of the Cincinnati Symphony Orchestra (1909–12) and subsequently of the Philadelphia Symphony Orchestra (1912–38). During this period he became a naturalized American (1915) and achieved world fame for himself and his orchestra, whose standard he raised to world-class.

In 1941 he cooperated with Walt Disney in making the film *Fantasia* (1940), arranging and conducting the score as well as appearing in the film. Subsequently he founded the Hollywood Bowl Symphony Orchestra (1945) and the American Symphony Orchestra (1961). In these years he toured America and Europe extensively. He also made many orchestral transcriptions, some of which were criticized for their vulgarity.

Stopes, Marie Charlotte Carmichael (1880–1958) *British botanist and social worker, who pioneered the establishment of birth control clinics in Britain and probably has done more than any other individual to control the population explosion in twentieth-century Europe.*

Born in Edinburgh, Stopes received a degree in botany and geology from University College, London, in 1902 and two years later her PhD from Munich University for a thesis on cycad ovules. In the same year she was appointed assistant lecturer and demonstrator in botany at Manchester University – the first woman appointee to the science faculty. In 1905 she received her DSc, her special interest being fossil plants. She embarked on a fossil-hunting tour of Japan in 1907 and subsequently was appointed lecturer in palaeobotany at Manchester. Following her marriage and move to London in 1911, she occupied a similar post at University College, London (1913–20). During this time she published *Catalogue of Cretaceous Flora* (2 vols, 1913–15) based on fossil collections held by the British Museum, and, with R. V. Wheeler, *The Composition of Coal* (1918).

Stopes's first marriage was annulled in 1916, an event that is said to have alerted her to sexual problems and contraception. Two years later she married H. V. Roe, an aircraft manufacturer. Increasingly concerned about the prevailing lack of candour about sex, she wrote *Married Love* (1918), a frank treatment of sexuality within marriage, which proved very popular. This was quickly supplemented by *Wise Parenthood* (1918), a short book on contraception. In 1921 she established the Mothers' Clinic for Birth Control in Holloway, London, which set a precedent in advising women about birth control. Two years later she published *Contraception: Its Theory, History and Practice*, the first comprehensive work on the subject. Her work drew fierce attacks from the establishment, in particular the Catholic Church and the medical profession. She wrote many other books about birth control and sex but in later years, when contraception had been accepted, she turned to writing romantic poetry and prose.

Stoppard, Tom (1937–   ) *Czech-born British playwright. He was awarded the CBE in 1978 and won the Shakespeare Prize the following year.*

Stoppard was born in Czechoslovakia and spent part of his childhood in Singapore and India until coming to England in 1946. He attended schools in Nottingham and Yorkshire and then became a journalist working in Bristol (1954). After a period as a freelance journalist (1960–63), he devoted himself to drama. His play *A Walk on the Water*, written in 1960 and televised three years later, was not produced in London until 1968 when, in the wake of the immense success of *Rosencrantz and Guildenstern are Dead*, the stage version was performed as *Enter a Free Man*. *Rosencrantz and Guildenstern are Dead* had its première at the 1966 Edinburgh Festival and was staged at the National Theatre in 1967; it was filmed in 1991. Its ingenious plot (based on the characters in *Hamlet*) and verbal pyrotechnics attracted sophisticated audiences all over the world and won its author several awards. Stoppard consolidated his reputation with the one-act comedy *The Real Inspector Hound* (1968), about two theatre critics drawn into a second-rate murder mystery they were reviewing. Later plays include *Jumpers* (1972), *Travesties* (1974), *Dirty Linen* (1976), *Night and Day* (1978), *On the Razzle* (1981), *The Real Thing* (1982), and *Hapgood* (1988). In addition to his plays he has also written radio and television drama, film scripts, short stories, and a novel, *Lord Malquist and Mr Moon* (1965).

Strachey, Christopher (1916–75) *British mathematician, best known for his work as a computer designer.*

Born in London, the son of a cryptographer (and nephew of the writer Lytton STRACHEY), Strachey was educated at King's College, Cambridge. He was a schoolteacher for several years before joining (1951) the staff of the National Research and Development Corporation, for whom he was responsible for the overall design of the Ferranti Pegasus Computer. He later worked in Canada on the computer simulation for the St Lawrence Seaway project, before returning to a research fellowship in 1962 at Churchill College, Cambridge. In 1965 Strachey moved to Oxford to set up the Programming Research Group, being appointed professor of computing in 1971.

Strachey's main research interest was in the field of programming. He developed the language CPL (Computer Programming Language) for Manchester University's ATLAS, which he later developed into the more adaptable BCPL (Basic-CPL). This work led Strachey to develop a general account of the languages used to program computers. It remained incomplete at his death.

Strachey, (Giles) Lytton (1880–1932) *British biographer.*

The son of the distinguished soldier and Indian administrator Sir Richard Strachey, Lytton Strachey took after his mother in his artistic leanings. He was educated at Abbotsholme School, Derbyshire, and Leamington College, before going to Liverpool University (1897–99). After this he moved to Trinity College, Cambridge, where he made many of the friends who later formed part of the Bloomsbury set. On leaving Cambridge, Strachey supported himself by journalism in London, working for the *Spectator* and other journals. While he found this uncongenial, it enabled him to remain in the intellectual circles that he admired. After his first book, *Landmarks in French Literature* (1912), appeared, his friends combined to free him from financial problems and provide him with accommodation in their country houses so that he could devote himself to writing. He was a conscientious objector during World War I.

Eminent Victorians (1918) caused a considerable stir because Strachey's incisive portraits of his four subjects seemed disrespectful in the opinion of those who believed that biography should be adulatory. His subsequent studies, *Queen Victoria* (1921), *Books and Characters, French and English* (1922), *Elizabeth and*

Essex (1928), and *Portraits in Miniature* (1931), enhanced his reputation and established a new school of biographical writing. He died of cancer after prolonged ill health. His devoted friend, the painter Dora Carrington (Mrs Ralph Partridge), who had nursed him in his last illness, committed suicide shortly afterwards.

Strauss, Richard (1864–1949) *German composer and conductor. He carried on the German operatic tradition of Mozart and Wagner, of whose works he was a greatly admired conductor.*

Born in Munich into a conventional middle-class family (his father was a horn player at the Munich opera house), Strauss had a precocious musical talent that was firmly guided along classical lines. He entered Munich University in 1882, having already published a *Festival March* (1876) and some chamber works. By the age of twenty-one he had been deeply influenced by Berlioz, Wagner, and Liszt and was himself beginning to compose a series of tone poems, which proved to be the ideal vehicle for his powerful fantasy and brilliant orchestration. They include *Aus Italien* (1886; *From Italy*), *Tod und Verklärung* (1889; *Death and Transfiguration*), *Till Eulenspiegels lustige Streiche* (1895; *Till Eulenspiegel's Merry Pranks*), *Also sprach Zarathustra* (1895–96; *Thus Spake Zarathustra*), *Don Quixote* (1897), and the autobiographical *Ein Heldenleben* (1898; *A Hero's Life*). His *Symphonia Domestica* (1902–03; *Domestic Symphony*), was also autobiographical.

Strauss was equally successful as a composer of opera. During the performance of the early *Guntram* (1892–93) he met his future wife, Pauline de Ahna, who in Weimar in 1894 sang the leading role. *Salome* (1904–05), a study in lust, prepared the way for *Elektra* (1906–08), a study in revenge. This was the first of Strauss's operas in collaboration with the librettist Hugo von Hofmannsthal (1874–1929), an association that also produced *Der Rosenkavalier* (1909–10), *Ariadne auf Naxos* (1911–12; *Ariadne on Naxos*), *Die Frau ohne Schatten* (1914–17; *The Woman Without a Shadow*), *Die Aegyptische Helena* (1924–27; *The Egyptian Helen*), and *Arabella* (1930–32). The association ended with Hofmannsthal's untimely death. Strauss also composed over a hundred songs in the tradition of the German Lieder. Examples include 'Allerseelen' (1885), 'Morgen' (1894–95), and 'Schlechtes Wetter' (1919). His wife was an indefatigable performer of these songs to

his own accompaniment. In 1921 Strauss and Elisabeth SCHUMANN made a recital tour of the USA in programmes of his songs.

As a conductor Strauss was pre-eminent, particularly in the works of Mozart, and held a number of appointments in the opera houses of Europe. In 1933 he accepted an official music post in HITLER's Germany and later conducted concerts that others would not. However, he resigned from his post as president of the Reichsmusikkammer when he was criticized by the Nazis for using a libretto by the Jewish writer Stefan ZWEIG. Although the rest of his life was contaminated by this contact with the Nazis, he was cleared by a postwar denazification trial and is said to have cooperated with the Nazis in order to protect his Jewish daughter-in-law. His old age was spent in a villa in the Bavarian Alps, where he produced *Metamorphosen* (1944–45; *Metamorphoses*), for twenty-three solo string instruments, and *Four Last Songs* (1948), for high voice and orchestra.

Stravinsky, Igor Feodorovich (1882–1971) *Russian composer who became a naturalized US citizen. He made his name as a composer of ballets for DIAGHILEV's Ballets Russes; in subsequent works he progressed from a neoclassical to a serialist style.*

The son of the leading bass singer at the St Petersburg Opera, Stravinsky was discouraged from a career in music and sent to the University of St Petersburg to read law. While he was there however, he met Rimsky-Korsakov and subsequently studied composition with him for three years. His music attracted the attention of Sergei Diaghilev, who commissioned a ballet for his Ballets Russes in Paris. The success of the result, *L'Oiseau de feu* (1910; *The Firebird*), encouraged Stravinsky to produce *Petrushka* (1911) and *Le Sacre du printemps* (1913; *The Rite of Spring*), the first performance of which created an uproar because of the unfamiliar modernity of the score. A year later, however, a successful concert performance of the work was given, again under the baton of Pierre MONTEUX.

At the outbreak of World War I, Stravinsky and his wife and children left Russia to settle in Switzerland. There he started composing scores of more modest proportions; for example, *L'Histoire de soldat* (1918; *The Soldier's Tale*) and *Pulcinella* (1919), which was his first composition in what came to be called the neoclassic style. Further examples of this style are seen in *The Fox* (1915–16), a dance scene, and *Les Noces* (1923; *The Wedding*), a ballet

including four solo singers, chorus, four pianos, and percussion. During this period Stravinsky also became interested in jazz, elements
of which occur in *Ragtime for Eleven Instruments* (1918) and *Piano-Rag Music* (1919).
His instrumental works of the 1920s mostly
follow the principle of contrasting tonal
groups, as in a concerto grosso. This period of
increasing austerity and abstration culminated
in the piano concerto (1924) and the opera-or-
atorio *Oedipus Rex* (1927), with a text by Jean
Cocteau translated into the ritualistic impersonality of Latin. With *The Symphony of
Psalms* (1930), commissioned by the Boston
Symphony Orchestra, Stravinsky moved into a
richer and more colourful musical idiom.

In 1920 Stravinsky left Switzerland for
France, where he remained until 1939, when
he crossed the Atlantic to the USA, becoming
a US citizen in 1945. To this period in America
belong his *Mass* (1948), meant for liturgical
use in the orthodox church, and the opera *The
Rake's Progress* (1951), based on Hogarth's
paintings. In the 1950s he began to explore the
use of serial techniques in a number of pieces,
including *In Memoriam Dylan Thomas* (1954)
for tenor, string quartet, and four trombones, in
memory of an operatic collaboration that never
took place.

Strawson, Sir Peter Frederick (1919–)
*British philosopher, a leading figure in con-
temporary Oxford linguistic philosophy. He
was knighted in 1977.*

Strawson was born in London and educated at
St John's College, Oxford; after service in the
army (1940–46), he returned to Oxford in 1947
as a fellow of University College. He was
Wayneflete Professor of Metaphysics from
1968 to 1987.

Strawson's first major publication, 'On
Referring', appeared in 1950; it was a powerful challenge to Bertrand Russell's theory of
descriptions in which Strawson argued that
Russell and others had ignored a vital distinction between sentences and statements. The
former, while meaningful, could not themselves be true or false; they could, however, be
used to make true or false statements. Strawson developed this and many other arguments
in his *Introduction to Logical Theory* (1952).
He argued that formal logicians, by limiting
themselves to context-free propositions, such
as 'All men are mortal', ignored or misrepre-
sented much of importance. The implication
relation analysed by logicians, for example,
ignored many of the complex uses bestowed
on the 'If ... then ...' construction in ordinary

language. It was time such complex cases were
studied further.

In *Individuals* (1958) Strawson once more
broke new ground with what he termed 'an
essay in metaphysics'. He attempted to chart
our conceptual scheme at its most general and
argued that we accept as basic particulars not
the atoms of scientists, or private Cartesian
experiences, but material objects. He further
insisted that the basic particulars to which we
ascribe consciousness were persons. Such
themes, both metaphysical and logical, were
developed further in Strawson's later works:
The Bounds of Sense (1966; a study of Kant)
and *Subject and Predicate in Logic and Grammar* (1974).

Streep, Meryl (Mary Louise Streep;
1949–) *US film actress, who became a leading star in the 1980s.*

Born in Summit, New Jersey and educated at
Vassar and Yale, she began as a stage actress,
appearing in the New York Shakespeare Festival in 1976 and subsequently on Broadway in
Tennessee Williams's *27 Wagons Full of Cotton*. In 1978 she won an Emmy for her role in
the television series *Holocaust*. Her debut on
the big screen came in *Julia* (1977), but her
reputation as a promising contemporary film
performer was made in *The Deerhunter*
(1978). A year later she appeared with Woody
Allen in *Manhattan*, although she is generally
considered less successful in comedy than in
demanding tragic roles, often based on actual
events. In *Kramer vs. Kramer* (1980) she won
an Academy Award for Best Supporting Actress in a story about a particularly painful
divorce, while in *The French Lieutenant's
Woman* (1981), she won a BAFTA award for
her performance as the enigmatic woman of
the title, opposite Jeremy Irons. Subsequent
roles, in which she was praised for her uncompromising and sincere performances, have included those of a survivor of the Nazi
concentration camps in *Sophie's Choice*
(1982), for which she won an Academy
Award, a contaminated nuclear worker in *Silk-
wood* (1983), for which she received an Oscar
nomination, the writer Isak Dinesen in *Out of
Africa* (1986), for which she received another
Oscar nomination, and a mother accused of
murdering her own baby in *A Cry in the Dark*
(1989).

Streisand, Barbra Joan (1942–) *US
singer and actress, whose wit and personality
earned her a special Broadway Tony Award in
1970 as 'actress of the decade'.*

Streisand was born in Brooklyn, New York, and her career began modestly when she won a Greenwich Village nightclub contest. With her Broadway debut in *I Can Get It For You Wholesale* (1962), however, came a New York Critics Award and instant acclaim. Her stardom was confirmed by the Broadway musical *Funny Girl*, in which she also made her film debut in 1968, winning an Academy Award for her performance. *Hello Dolly!* (1969) and *On A Clear Day You Can See Forever* (1970) are two other musicals that followed. However, she managed to break away from musicals by starring in such films as *The Owl and the Pussycat* (1970) and *What's Up Doc?* (1972). Further evidence of the diversity of her talent was provided when she starred in, wrote some of the music for, and was executive producer of *A Star Is Born* (1976), the song 'Evergreen' (music by Streisand, lyrics by Paul Williams) winning an Academy Award. More recent films include *All Night Long* (1981), *Yentl* (1983), which she also produced and directed, and *Nuts* (1987).

With Paul NEWMAN and others, Streisand was involved in establishing the production company First Artists.

Sukarno, Achmad (1901–70) *The first president of independent Indonesia (1945–67).* Sukarno helped form the Indonesian National Party in 1927 and was imprisoned by the Dutch for his nationalist activities (1929–31). He was Indonesian leader during the Japanese occupation in World War II (1942–45) and then led the struggle for independence from the Netherlands, which was formally granted only in 1949–50. Sukarno's outstanding achievement was to weld Indonesia's disparate populations scattered on islands extending over 5150 km in the Pacific and Indian Oceans into a single nation, and he was the object of devotion for the mass of the people. In 1957, however, he replaced the 1945 constitution with so-called 'guided democracy', and his credibility was further undermined with his adoption, in a less convincing form, of Chinese communist views, especially against American colonialism. In 1965 an unsuccessful communist coup was suppressed with great ferocity by anticommunist military forces led by General Suharto (1921–), followed by appalling massacres of communists. Sukarno's powers gradually dwindled and he finally surrendered all his powers to Suharto in 1967.

Sun Yat-sen (Sun Zhong Shan; 1866–1925) *Chinese nationalist leader. Sun Yat-sen has been called the father of modern China. His doctrines, encapsulated in his 'Three Principles of the People' (nationalism, democracy, and the people's livelihood) have been accepted by both the Kuomintang (Nationalist Party) and the Communist Party.*

Born in Kwantung province, the son of a peasant, Sun Yat-sen was brought up in Hawaii, where he received a Christian education and became familiar with western traditions. He became a Christian and went on to study medicine in Canton and Hong Kong, graduating in 1892.

Sun Yat-sen first became actively involved in politics in 1894, when he founded the Revive China Society, a forerunner of several revolutionary groups (including the Kuomintang) that he founded. When an attempted uprising in Canton failed in 1895, he went into exile for sixteen years. Visits to the USA and his brief imprisonment (1896) in the Chinese legation in Britain brought him worldwide publicity and support. He also founded the Alliance Society in Japan (1905) and issued an early version of his 'Three Principles of the People' (1907). He returned to China during the 1911 revolution that overthrew the Manchu dynasty and was elected president of the provisional government. He resigned from that position in 1912, after opposition from more conservative members of the government, and after leading an unsuccessful revolt the following year, left the country. With assistance from the Soviet Union he finally regained power in 1923 and was elected president of the republican government in Canton. During his presidency he reorganized the Kuomintang along the lines of the Soviet Communist Party, which achieved the cooperation of the Chinese communists, and expounded his revolutionary doctrine, before dying of cancer in 1925. His widow, Soong Ching-ling, was appointed vice-chairman of the People's Republic of China in 1950.

Surtees, John (1934–) *British motor cyclist and motor racing driver who won seven world championship titles in motor cycling besides the 1964 world motor racing drivers' championship.*

The son of a Croydon motor-cycle dealer, Surtees served a five-year apprenticeship with the Vincent motor cycle company and started his racing career in 1951. After riding for the Norton works team in 1955, he joined the dominant MV-Augusta team and gained his first 500 cc title. 1958, 1959, and 1960 saw Surtees win the championships in both the 500 cc and

300 cc classes. He also won Tourist Trophy (TT) events on the Isle of Man in 1956, 1958, 1959, and 1960. However, in 1959 he was also racing Cooper cars in Formula Junior and Formula Two, as well as a Lotus in certain Grands Prix. He retired from motor cycling after the 1960 season to concentrate on motor racing. After driving Lolas in 1962 Formula One events, he joined the Ferrari works team in 1963 and won the 1964 drivers' title with victories in the German and Italian Grands Prix.

In 1965, Surtees suffered serious injuries after crashing his Lola sports car in a Can-Am race but he recovered to win the Can-Am series the following year. In 1967 he joined the Honda works team making their debut in Formula One, winning the 1967 Italian Grand Prix. After an unsuccessful season with BRM in 1969, Surtees concentrated on developing his own cars and the Surtees TS7 made its first appearance at the 1970 British Grand Prix. Although not successful in Formula One, a Surtees driven by Mike Hailwood (1940–81) won the 1972 Formula Two championship; the cars were also prominent in Formula 5000. Their creator retired from driving in 1973 and by 1978 the Surtees team had pulled out of Formula One racing.

Sutcliffe, Herbert (1894–1978) *Yorkshire and England cricketer, who during his first-class career scored 149 centuries, 16 of them for his country.*
Born in Pudsey, Yorkshire, Sutcliffe did not play for his county until 1919, after war service during which he was commissioned in the Sherwood Foresters. In a first-class career that lasted until 1945 he scored a remarkable total of 50 138 runs at an average of 51.95. His test match batting average was 60.73; he played for England 54 times (1924–35). He is especially remembered for his sustained opening partnership with Jack HOBBS, with whom he developed an exceptional affinity. After World War II, he retained his interest in Yorkshire cricket by becoming a member of the committee. From 1959 to 1961 he was an England selector. He was also a successful businessman.

Sutherland, Graham Vivian (1903–80) *British painter. He was awarded the OM in 1960.*
Born in London, Sutherland studied engraving at the Goldsmiths' College School of Art (1921–26) after a year as an engineering apprentice in Derby. He became a Catholic in 1926 and the following year took up a teaching post at the Chelsea School of Art, which he kept until 1939. At a time when etchings were

popular, Sutherland enjoyed considerable success until the early 1930s, when he began painting in oils and watercolours. His highly individual landscapes, painted mainly in the wilder areas of Britain, were not of the scenic type but rather expressed the strangeness of nature. The depiction of forms was based on principles of organic growth: Sutherland was 'fascinated by the ... tensions produced by the power of growth'. This approach resulted in the 'portraits' of strangely formed trees and boulders of the late 1930s and an increasingly semiabstract style.

During World War II Sutherland worked as a war artist, producing powerful pictures of devastated buildings and twisted metal in bombed cities as well as pictures of mines and foundries. His postwar work was varied: in addition to semiabstract paintings on plant and insect themes, such as his *Thorn Trees*, he produced a number of *Crucifixions* to be hung in churches and, in a contrasting style reminiscent of Byzantine art, the huge tapestry for Coventry Cathedral, *Christ in Majesty*, designed in the late 1950s. Sutherland also painted some powerful portraits, for example *Somerset Maugham* (1949) and *Winston Churchill* (1954); the latter was destroyed on the instruction of Lady Churchill as her husband disliked it intensely. From 1956 Sutherland lived in the south of France.

Sutherland, Dame Joan (1926–) *Australian soprano. She was created a DBE in 1979.*
A pupil of John and Aida Dickens, she sang her first operatic role in Purcell's *Dido and Aeneas* in 1947. Her stage debut, in the title role of Eugene GOOSSENS's *Judith*, was in Sydney in 1951. Joan Sutherland then came to England to study in London at the Royal College of Music with Clive Carey. She made her first appearance at Covent Garden in Mozart's *The Magic Flute* in 1952. Her success was due largely to the brilliance of her vocal technique in the Italian bel canto style, which she developed under the guidance of her accompanist and conductor husband, Richard Bonynge (1930–). Of the many coloratura roles in which she attained international fame, perhaps the best known was that of Lucia in Donizetti's *Lucia di Lammermoor*, which she first sang at Covent Garden in 1959. She retired in 1990.

Sutton, Walter Stanborough (1877–1916) *US geneticist who gave the first clear formulation of the theory that chromosomes carry the physical units that determine inheritance.*
Sutton first studied engineering at the University of Kansas but switched to biology, gradu-

ating in 1900. He then moved to Columbia University to work with the geneticist Edmund Beecher Wilson (1856–1939). Sutton discovered that the chromosomes of the grasshopper *Brachystola magna* could be identified individually and their behaviour followed during the cell division (meiosis) prior to gamete formation. He confirmed previous suggestions that maternal and paternal chromosomes form hoTologous pairs in the dividing cell and he established that these pairs are quite randomly separated during cell division, leading to random assortment of maternal and paternal chromosomes in the gametes. Sutton saw that this random assortment of homologous chromosomes was consistent with Mendel's principle of independent assortment of characters and accordingly proposed that the chromosomes carried the physical determinants of heredity. Sutton also realized that this implied linkage of determinants carried by the same chromosome – a theory later modified by T. H. MORGAN's discovery of crossing over.

Sutton's famous work was published in three papers during the years 1900–03. Thereafter he turned to medicine, receiving his MD in 1907. After two years at the Roosevelt Hospital, New York, he entered private practice as a surgeon in Kansas City.

Suzuki, Daisetsu Teitaro (1870–1966) *Japanese Buddhist philosopher, who was responsible for familiarizing the West with the principles of Zen Buddhism.*

Suzuki was born in Kanazawa. After attending Ishikara College and Tokyo University, he spent the period 1892–97 as a novice at the Buddhist monastery at Kamakura. From 1897 to 1909 he lived in the USA, where he deployed his remarkable linguistic skills translating Pali, Sanskrit, Chinese, and Japanese texts into English for the Open Court Press. On his return to Japan in 1909 he taught English and philosophy, first at Tokyo University and from 1921 to 1940 at Otani University. Through his lectures and numerous books Suzuki became a familiar name in the West. It was in fact such books of his as *Essays in Zen Buddhism* (3 vols, 1927–34), *Manual of Zen Buddhism* (1935), and *Studies in Zen* (1955) that first introduced the basic principles of Zen Buddhism to a receptive western audience in the 1960s and 1970s.

Synge, J(ohn) M(illington) (1871–1909) *Irish playwright.*

Born near Dublin, Synge was educated privately before going to Trinity College, Dublin, in 1888. On leaving university (1893) he in-

tended to make music his profession, but after studying in Germany he changed his mind and moved to Paris early in 1895. Over the next few years he spent periods in France, Ireland, and Italy; in Paris in 1899 he met W. B. YEATS, who encouraged him to drop his critical studies of French and English literature and to find a subject of his own. Synge had already paid one visit to the Aran Islands (1898), and, acting on Yeats's suggestion, he returned there in 1899, 1900, 1901, and 1902. The lives of the people there, and in Wicklow and Kerry, provided Synge with the materials for books and articles but above all for his plays.

Synge's one-act plays *The Shadow of the Glen* (1903) and *Riders to the Sea* (1904) were published in one volume in 1905. Meanwhile he began writing *The Tinker's Wedding*, which he did not finish until 1906, and *The Well of the Saints*, which was first performed in 1905 in the newly opened Abbey Theatre, of which Synge was one of the directors. *The Playboy of the Western World* (1907), his best-known play, caused a riot at its first performances but eventually contributed to make both Synge's reputation and that of the Abbey Theatre. He died of cancer while putting the final touches to *Deirdre of the Sorrows*, which was posthumously published and produced in 1910. Synge also left poems, translations from Petrarch, and the books *The Aran Islands* (1907) and *In Wicklow and West Kerry* (1908).

Szilard, Leo (1898–1964) *Hungarian-born US physicist, who played an important role in the development of the atomic bomb.*

Born in Budapest, the son of an architect, Szilard studied electrical engineering in Budapest and physics at the University of Berlin, where he gained his PhD in 1922. With the rise of HITLER, Szilard decided to leave Germany and, after a brief period in England, emigrated to the USA in 1938. He became a naturalized US citizen in 1943.

As early as 1934 Szilard had grasped the essentials of a nuclear chain reaction, and had consequently applied for the patents rights, which he assigned to the British Admiralty. Mistakenly, however, Szilard thought the chain reaction was most likely to be sustained by the element beryllium. As a result of the work of Otto HAHN and Lise MEITNER, by 1939 he realized that it was uranium rather than beryllium that was the key element in a chain reaction. Determined that the USA rather than Nazi Germany would be the first to develop atomic weapons, Szilard approached Albert EINSTEIN and persuaded him to write a letter to

President ROOSEVELT urging him to provide the backing for research into uranium fission. The letter, dated 2 August 1939, initiated the programme that eventually produced the Hiroshima and Nagasaki atom bombs. Szilard worked on the project himself, mainly with Enrico FERMI in Chicago on the development of the uranium–graphite pile.

Although he was instrumental in its development, Szilard attempted to prevent the dropping of the bomb on Japanese cities by organizing the 'Szilard petition' signed by many leading scientists and addressed to politicians. After the failure of this petition, Szilard foresaw the nuclear arms race and campaigned vigorously, and again unsuccessfully, for a programme of disarmament. At this point Szilard, with a number of other disillusioned physicists, turned away from physics, to the emerging discipline of molecular biology. Attending classes, he learnt the basic laboratory techniques and soon developed sufficient competence to design new instruments and to contribute to the development of theory. Academic appointments meant little to Szilard. He was connected with the University of Chicago until his later years, when he moved to the newly formed Salk Institute at La Jolla, California.

T

Taft, William Howard (1857–1930) *US Republican statesman and twenty-seventh president of the USA (1909–13).*

The son of judge and former cabinet minister Alphonso Taft (1810–91), William Taft studied at the universities of Yale and Cincinnati, qualified as a lawyer, and in 1880 became a member of the Ohio bar. He held various public offices before his appointment as an Ohio superior court judge in 1887, and in 1890 he became US solicitor-general under President Benjamin Harrison. But he resigned after two years to resume life as a federal circuit judge. President McKinley appointed Taft to head a commission concerned with the restoration of civil administration in the Philippines and he subsequently became the first civil governor (1901–04), successfully administering the reallocation of land. Appointed secretary of war by Theodore Roosevelt, Taft became a close friend and adviser to the president, lending his calm judicial approach to a variety of crises. In 1908 he was persuaded to run for the presidency himself, and won.

Taking office in March 1909, Taft soon witnessed a widening gulf between conservative and progressive elements in the Republican Party. Lacking the political astuteness of Roosevelt, Taft pleased no-one with his feeble compromise legislation on tariffs – the Payne–Aldrich Act of 1909 – while his dismissal of the forest service chief, Gifford Pinchot, alienated liberal Republicans. However, he vigorously enforced antitrust legislation and established the Department of Labour. In an acrimonious 1912 election campaign, Theodore Roosevelt formed his breakaway Progressive Party and ran against Taft's official Republican ticket, thus splitting the vote and enabling victory for the Democrat, Woodrow WILSON.

In 1913, Taft was appointed professor of constitutional law at Yale University. Later, during World War I, he was joint chairman of the US War Labor Board. In 1921 came the climax of his legal career – his appointment as the US Supreme Court's chief justice, whereupon he embarked on improving the coordination and efficiency of the judicial system.

Tagore, Rabindranath (1861–1941) *Indian writer, who was awarded the 1913 Nobel Prize for Literature. In 1915 he was knighted but repudiated the honour in protest against the Amritsar Massacre (1919).*

Tagore was born into a distinguished Bengali family in Calcutta; his father was the Maharishi Debendranath Tagore, the Hindu reformer and mystic. Privately educated, he read Bengali and English poets and wrote poetry from an early age. In 1877 he went to England to study law but soon returned, married (1883), and undertook the management of his family's estates. By the early 1890s he was the chief contributor to leading Bengali journals. He also published his first poetic collections – *Manasi* (1890), *Chitra* (1895), and *Sonar Tari* (1895), in which he pioneered the use of colloquial Bengali instead of the archaic literary idiom then approved for verse, and wrote his first plays.

In 1901 he founded Shantiniketan at Bolpur near Calcutta: this famous educational establishment was a blend of traditional ashram and western schools. His best novel, *Gora*, appeared in 1908 and his most famous collection of lyrics, *Gitanjali*, followed in 1909. *The King of the Dark Chamber* (1910) is one of his most successful dramas, though generally his plays were too symbolic and literary to exert a lasting influence. On a visit to England (1912) Tagore showed his own English version of *Gitanjali* to William Rothenstein and W. B. YEATS, under whose auspices it was published in England. Lecture tours in the USA (1912–13) and Britain (1913) followed; Tagore was hailed as a sage and lionized in western intellectual circles.

Tagore used his Nobel Prize money to improve Shantiniketan, adding an agricultural school (1914). He continued writing, with the lyric collection *Balaka* appearing in 1914 and the novel *Home and the World* in 1916. Not naturally drawn to politics, he shunned active resistance to British rule, seeking instead ways of harmonizing eastern and western world views. To this end he added an international university to the Shantiniketan complex (1921) and divided the rest of his life mainly between its affairs and travelling abroad on lecture tours. He sought to interpret Indian philosophy to other cultures. His 1930 Hibbert Lectures at Oxford were published as *The Religion of Man* (1931).

Tange Kenzo (1913–   ) *Japanese architect, who was one of the first twentieth-century Japanese architects to establish an international reputation. Deeply influenced by LE CORBUSIER, he manages to adapt western glass and concrete trends to the indigenous Japanese idiom.*

Born in Imabari, Tange studied at Tokyo University before joining the office of Kunio Maekawa (1905–86), who had in turn worked in the office of Le Corbusier. Tange's notable buildings include the Peace Centre (1955), built on the site of the epicentre of the Hiroshima bomb, the Kawaga town hall (1958), the Atami Gardens Hotel (1961), St Mary's Roman Catholic Cathedral in Tokyo (1965), and the roof for the National Gymnasium for the 1964 Tokyo Olympic Games. Later in his career, Tange moved into town planning and in 1960 produced a plan (unexecuted) for relieving congestion in Tokyo by extending the city on piles into Tokyo Bay. Since the 1970s Tange has been active in Europe and the Middle East. His most recent designs include those for Yokohama's City Museum (1983), Singapore's Indoor Stadium (1985), and the New Tokyo City Hall Complex (1986).

Tanguy, Yves (1900–55) *French-born US surrealist painter.*

The son of a ship's captain, Tanguy embarked on a career as a merchant seaman, but after two years at sea followed by military service, during which he met the surrealist poet Jacques PRÉVERT, he began to look for an alternative vocation. He was inspired to take up painting in 1923, after seeing a picture by Giorgio de CHIRICO in the window of a gallery in Paris. In 1925 Tanguy joined the surrealist movement and painted some of the earliest surrealist works. His paintings portrayed bleak unearthly landscapes with stark inanimate objects or ghostly life forms, as in *He Did What He Wanted* (1927) and *Days of Slowness* (1937). They were an attempt to display on canvas the irrational products of the subconscious mind. Rock formations played an important part in the early pictures and even more so after his trip to Africa in 1930, where he was fascinated by the structure of the rock formations that he saw. In his later work the forms he depicted tended to be more abstract. In 1940 Tanguy married the US painter Kay Sage and in 1942 he settled with her in Woodbury, Connecticut, where he remained for the rest of his life, becoming a US citizen in 1948.

Tanizaki Jun'ichiro (1886–1965) *Japanese novelist and playwright.*

Born in Tokyo, Tanizaki went to university there but left in 1910 without a degree. The same year he published his first short novel, *Shisei*, which attracted attention by its affinities with contemporary European trends and the works of Poe and Baudelaire. The other early novels that followed, such as *Otsuya-goroshi* (1915), confirmed Tanizaki as a writer with the ability to shock and enthral the reader. In 1922 his plays *Ai sureba koso* and *Eienno gūzō* appeared, followed by *Mumyō to Aizen* (1924), all three of which were soon translated into French.

A move to Osaka after the great Tokyo earthquake of 1923 brought about a change of mood in Tanizaki's work. *Tade kuu mushi* (1929; translated as *Some Prefer Nettles*, 1955), one of his best novels, weighed the old and the new in Japanese values regarding personal relations, a theme continued in the long novel *Sasame-yuki* (1943–48; translated as *The Makioka Sisters*, 1957). In 1932 he began his highly acclaimed adaptation of the eleventh-century classic *Genji Monogatari* (*The Tale of Genji*) in modern Japanese. His critical views at this period were elegantly embodied in his manual of style, *Bunsho no Dokuhon* (1934).

Tanizaki's preoccupation with eroticism retained its ability to shock his Japanese readership in the postwar period. *Kagi* (1956; translated as *The Key*, 1961, and *La Confession impudique*, 1963) and *Fūten Rōjin Nikki* (1961–62; translated as *The Diary of a Mad Old Man*, 1965) are frank and intimate accounts of sexual relations.

Tansley, Sir Arthur George (1871–1955) *British botanist and one of the pioneers of plant ecology, who contributed much to the development of scientific methods for analysing plant communities. He received a knighthood in 1950.*

After reading natural sciences at Trinity College, Cambridge, Tansley was appointed demonstrator at University College, London, where he later became assistant professor of botany (1893–1906). He made an extensive tour of Ceylon, the Malay Archipelago, and Egypt (1900–01), studying vegetation types, and in 1902 founded the botanical journal *New Phytologist*, which he edited for twenty-one years. In 1904 Tansley assembled a team of botanists with the aim of describing plant communities in Britain. He was editor and co-author of the resulting *Types of British*

Vegetation (1911). This led to the formation in 1913 of the British Ecological Society, with Tansley as its first president (1913–15). He also edited the *Journal of Ecology* following its launch in 1917. Tansley moved to Cambridge in 1906 as lecturer in botany and in 1923 he resigned to pursue his interest in psychology. He wrote *The New Psychology and Its Relation to Life* (1920) and studied under FREUD in Vienna (1923–24). In 1927 he was appointed Sherardian Professor of Botany at Oxford University and established the importance of ecology in the teaching of plant sciences. *The British Islands and Their Vegetation* (1939) is recognized as a major work on plant communities and their ecology. He also wrote *Practical Plant Ecology* (1923) and was co-author of *Aims and Methods in the Study of Vegetation* (1926).

Although he retired in 1937, Tansley subsequently played a major role in planning postwar government policy on nature conservation and served as the first chairman of the Nature Conservancy Council (1949–53).

Tati, Jacques (Jacques Tatischeff; 1908–82) *French film actor and director. Despite making only five feature films in over twenty years, he became one of the world's leading film comedy actors.*

Tati, who was born in Le Pecq, became a professional entertainer after a short period as a professional Rugby footballer, when he often performed at rugger suppers. As a professional he made his name in the 1930s with his music-hall act doing comic mimes of leading sportsmen. Some of these were later made into film shorts. His first feature-length film was *Jour de fête* (1949), in which he played a village postman endeavouring to apply new high-speed postal methods with disastrously funny results. Co-scripted by Tati with Henri Marquet, it won the best script award at the Venice Film Festival. The film was basically an extended version of a number of shorts Tati had already done, including *L'École des facteurs* (1947). As the innocent and endearing Monsieur Hulot, in sharp contrast to the absurdities of the modern world, he won international acclaim in *Les Vacances de Monsieur Hulot* (1953; *Mr Hulot's Holiday*), *Mon Oncle* (1958), *Playtime* (1968), and *Traffic* (1971).

Tatum, Art(hur) (1910–56) *Black US jazz pianist. His prodigious technical virtuosity attracted the admiration of such classical pianists as Walter Geiseking and Vladimir HOROWITZ, who came to night clubs to hear him.*

Born with cataracts in both eyes, after surgery Tatum was blind in one eye and had diminished sight in the other. He attended a school for the blind in Columbus, Ohio, and studied music for two years in Toledo. He first played as a professional locally; subsequent performances on local radio (1929–31) led to engagements all over the country and by 1938 he was playing in Europe. He usually performed in a trio with bass and guitar. His last big concert was in the Hollywood Bowl in 1956.

In the mid-1950s he made more than 120 solo recordings and several trio and quartet records. The quartet set, with tenor saxophonist Ben Webster (1909–73), is particularly highly regarded. Perhaps his most revealing solo recordings were made after hours in Harlem night clubs on portable recording equipment in 1940–41; these were released in the album *God is in the House* (1973).

Tatum, Edward Lawrie (1909–75) *US biochemist who, in collaboration with George Wells Beadle (1903–89), produced convincing evidence that a single gene codes for a single specific enzyme. For this work, they were awarded the 1958 Nobel Prize for Physiology or Medicine.*

Tatum was educated at the University of Wisconsin, receiving his PhD in microbiology in 1934. In 1937 he was appointed research associate at Stanford University and in 1940 began his work with Beadle. They irradiated conidia (spores) of the bread moulds *Neurospora crassa* and *N. sitophila* and allowed them to germinate on various culture media. In this way they discovered three mutant strains that had lost the ability to synthesize specific vitamins, implying that in each case the necessary enzyme was missing or nonfunctional. In subsequent crosses with normal strains, the mutant characters were shown to differ from normal by only a single gene. This gave considerable weight to the theory, now widely accepted, that one gene codes for one enzyme. Tatum then applied his technique to the bacterium *Escherichia coli* and found similar nutritionally deficient mutants. This led to his discovery, with Joshua LEDERBERG, of the phenomenon of genetic recombination, in which part of the bacterial chromosome is transferred from 'male' cells to 'female' cells by a form of sexual reproduction called conjugation. This represented a major step in the development of microbial genetics and the use of *E. coli* in recombinant DNA technology.

Tatum moved to Yale in 1945, becoming professor of microbiology, but returned to

Stanford after three years. In 1957 he joined the Rockefeller Institute for Medical Research in New York.

Tavener, John Kenneth (1944–) *British composer. Most of his works are religious and deal with the theme of spiritual triumph over disaster.*

The son of a north London builder, he was still a student at the Royal Academy of Music (1961–65), a pupil of Lennox BERKELEY and David Lumsdaine, when his dramatic cantata *Cain and Abel* won the 1965 Prince Rainier of Monaco Prize. Another cantata, *The Whale* (1965–66), was performed by the London Sinfonietta in their inaugural concert in 1968. This work opens with the reading of an entry on whales from an encyclopedia. The reading is progressively engulfed by orchestral sound and the cantata goes on to retell the story of Jonah and the whale. The cantata's success led to a BBC commission in 1968, *In alium* for high soprano, orchestra, and prerecorded tape. Tavener's subsequent works were influenced to some extent by STRAVINSKY'S and MESSIAEN'S compositions. They include the *Celtic Requiem* (1969), *Ultimos ritos* (1969–72), based on poems of St John of the Cross, *Little Requiem for Father Malachy Lynch* (1972), which was followed by a full *Requiem for Father Malachy* (1973), the opera *Thérèse* (1973–76), based on the life of St Thérèse of Lisieux (produced at Covent Garden in 1979), *A Gentle Spirit* (1976), *The Last Prayer of Mary Queen of Scots* (1977), and *The Immurement of Antigone* (1979).

From 1960 Tavener was organist at the Presbyterian church of St John in Kensington and in 1969 he was appointed professor at Trinity College of Music. He was received into the Russian Orthodox Church in 1977 and some of his most recent works, such as *Akhmatova rekviem* (1979–80) and *The Protecting Veil* (1987), have been associated with the eastern rite. He completed *Dance Lament of the Repentant Thief*, for strings, clarinet, and percussion, in 1991.

Taylor, Elizabeth (1932–) *US actress, whose long list of husbands is only exceeded by the long list of her films.*

Born in London to American parents, Taylor was taken to Hollywood at the outbreak of World War II. Her first films, *There's One Born Every Minute* (1942), *Lassie Come Home* (1943), and *National Velvet* (1944), made her a child star.

As an adult she was nominated for Academy Awards for *Raintree County* (1957), *Cat on a* *Hot Tin Roof* (1958), and *Suddenly Last Summer* (1959), and received Oscars for *Butterfield 8* (1960) and *Who's Afraid of Virginia Woolf?* (1966). In the latter she co-starred with Richard BURTON, whom she had married two years earlier. The Burton–Taylor partnership, which began on the set of the extravagant *Cleopatra* (1963), became one of the world's most publicized relationships. Their other films together included *The VIPs* (1963), *The Taming of the Shrew* (1967), and *Under Milk Wood* (1971). Before Burton, Taylor had been married to hotelier Nick Hilton, actor Michael Wilding (1912–79), producer Mike Todd (1907–58), and singer Eddie Fisher (1928–). She and Burton were divorced briefly but subsequently remarried and redivorced. She has been married twice since. Taylor and Burton came together for the last time professionally on stage in *Private Lives* (1983). Her most recent films include *Winter Kills* (1985).

Tedder, Arthur, 1st Baron (1890–1967) *British air marshal who during World War II was deputy to EISENHOWER, both in the Mediterranean campaign and in the Allied invasion of France. He was created a baron in 1946.*

Tedder studied history at Magdalene College, Cambridge, taking his BA in 1912. In June 1916, he joined 25 Squadron of the Royal Flying Corps in France, performing reconnaissance and bombing missions, and the following year he took command of 70 Fighter Squadron. Later he served as a flight instructor and at the end of the war was commissioned in the fledgling RAF as a squadron-leader. Tedder's interest always lay more towards the theoretical aspects of air power. He was assistant commandant of the RAF Staff College (1929–31), besides working at the Air Ministry and in the Far East. Promoted to air vice-marshal in 1937, the following year he became director-general of research and development at the Air Ministry, later working under the minister for aircraft production, Lord BEAVER-BROOK.

In 1941 Tedder was appointed commander-in-chief of the RAF in the Middle East. He sought to overcome the chronic shortage of aircraft and supplies that constantly hampered his air support for the army and navy and his campaign against enemy planes and their bases. His pattern bombing of ROMMEL's tank formations – known as 'Tedder's carpet' – helped the advance of MONTGOMERY's forces into Tunisia. In February 1943, Tedder was appointed air C-in-C (Mediterranean) under

Eisenhower. Their harmonious working relationship contributed to the Allied victory in North Africa (May 1943) and the subsequent invasion of Sicily and the Italian mainland. Their partnership was maintained when, in late 1943, Tedder was recalled to Britain as Eisenhower's deputy to plan the forthcoming Allied invasion of France. Tedder coordinated Allied air forces in destroying enemy supply lines in advance of D-Day and used his immense diplomatic powers to overcome many instances of friction between Allied commanders. Tedder's was one of the signatures on the German surrender of 8 May 1945. He was promoted to marshal of the RAF and appointed air chief of staff in 1946. His memoirs, *With Prejudice*, were published in 1966.

Teilhard de Chardin, (Marie Joseph) Pierre (1881–1955) *French Jesuit theologian, philosopher, and scientist who developed a synthesis of evolutionary theory and Christianity, which although unacceptable to the Vatican has had a considerable influence.*

Teilhard's family were landowners and had associations with the Church. He studied at the Jesuit College at Mongré and in 1899 began his training for the priesthood, as well as studying philosophy and science. He spent some time in Cairo teaching physics and chemistry at a Jesuit college (1905–08) and in 1911 was ordained as a priest. But increasingly his studies were devoted to palaeontology. After serving as a stretcher-bearer during World War I and receiving the Légion d'honneur for gallantry, Teilhard returned to Paris and in 1922 received his PhD in palaeontology from the Sorbonne. He joined the Catholic Institute in Paris as assistant professor of geology but, because of his unorthodox views, was requested to leave in 1926. Teilhard travelled to China to study the fossil-bearing rocks and in 1929 discovered the fossil remains of a hominid, later known as Peking Man (*Homo erectus*). He returned to Paris in 1946. The final three years of his life were spent at the Wenner-Grenn Foundation for Anthropological Research in New York.

From his knowledge of palaeontology and evolution, Teilhard proposed three key steps in evolution: the evolution of matter into a 'geosphere', namely the earth; the evolution of living organisms ('biosphere'); and the advent of thinking man, thereby giving the world an intellectual dimension, or 'noosphere'. He visualized this progression of complexity culminating in an 'omega point' – when the natural and supernatural will achieve unity in God.

Although he was a devout Catholic, the Church refused him permission to publish his books during his lifetime. His major full-length works were published posthumously and drew wide attention and considerable acclaim. The best known are *Le Phénomène humain* (1955; translated as *The Phenomenon of Man*, 1959) and *Le Milieu divin* (1957; translated as *The Divine Milieu*, 1960).

Te Kanawa, Dame Kiri Janette (1944–) *New Zealand singer, who has made her name in Europe and the USA singing soprano roles, principally in Mozart and Italian opera. She was created a DBE in 1982.*

Kiri Te Kanawa came to London, after winning many prizes in New Zealand and Australia, to study with Vera Rosza at the London Opera Centre. Having sung with the Northern Opera and the Chelsea Opera Group, she made her Covent Garden debut in 1970 as a Flower Maiden in Wagner's *Parsifal*. Shortly afterwards she won praise in the role of Countess Almaviva in Mozart's *The Marriage of Figaro* there. She made her US debut as Desdemona in Verdi's *Otello* in New York (1974), since when she has appeared in the world's leading opera houses on both sides of the Atlantic in a wide repertoire of roles. She reached a wider audience in 1981, when she sang at the wedding of the Prince of Wales, since when she has made numerous television appearances. In 1989 she published *Land of the Long White Cloud: Maori myths and legends*.

Teller, Edward (1908–) *Hungarian-born US physicist, sometimes known as the 'father of the hydrogen bomb'.*

The son of a lawyer, Teller was educated at the Budapest Institute of Technology and the universities of Karlsruhe, Munich, Göttingen, and Leipzig, where he obtained his PhD in 1930. With the rise to power of HITLER, as a Jew, Teller emigrated to the USA in 1935. He became a naturalized US citizen in 1941. Teller has held chairs of physics at George Washington University (1935–41), Columbia (1941–42), Chicago (1946–52), and the University of California, Berkeley (1954–75). Since 1975 he has been a research fellow of the Hoover Institution.

During World War II Teller worked at Los Alamos on the development of nuclear weapons. Initially, he worked on the problem of designing a fission bomb, but his main interest lay in the development of the fusion bomb. Eventually, the Los Alamos director, J. R. OPPENHEIMER, allowed him to concentrate on the hydrogen bomb (known in the jargon of the day as the 'Super'). By the end of the war Teller had designed a hydrogen bomb, but it required a large refrigeration plant to make it work. In 1949, Oppenheimer and the General Advisory Committee advised the Atomic Energy Commission not to pursue the Super. To the violently anticommunist Teller this seemed like madness. However, on 29 August 1949 the Soviet Union exploded its first fission bomb; political attitudes in Washington changed overnight. On 30 January 1950 President TRU-MAN announced a crash programme to develop the Super, with Teller returning to Los Alamos as assistant director for weapons development. The breakthrough came in early 1951 when Teller and the Polish mathematician Stanislaw Ulam worked out the Teller–Ulam configuration, in which X-rays produced by an initial atomic explosion trigger a thermonuclear explosion. It was successfully tested in November 1952 on an island in the Eniwetok Atoll. By this time Teller had left Los Alamos to set up his own laboratory at Berkeley, known as the Livermore Laboratory.

In April 1954 the AEC heard Oppenheimer's appeal against the loss of his security clearance. Alone among the senior scientists from Los Alamos, Teller testified against Oppenheimer declaring that 'one would be wiser not to grant clearance.' When, shortly afterwards, he visited Los Alamos he was either ignored or greeted with a formal politeness. He did not return for another ten years. Thereafter Teller became increasingly political, arguing against the test ban treaty of 1963 and denying that fallout from nuclear testing was 'worth worrying about'. His publications include *The Structure of Matter* (1949), *Nuclear Energy in a Developing World* (1977), and *Better a Shield than a Sword* (1987).

Temple, Shirley (1928–) *US child film star and diplomat.*

Born in Santa Monica, California, she was appearing in one-reelers and small parts in feature films by the time she was four years old. After *Stand Up and Cheer* (1934) her rise to fame was meteoric and by the end of 1934 she was the recipient of a Special Academy Award. Her diminutive size, curly hair, dimpled smile, and ability to sustain a competent, if somewhat embarrassing, song and dance routine made her the idol of mothers and daughters throughout the world. Shirley Temple dresses, hair styles, and look-alike competitions became a feature of life in the 1930s wherever the silver screen had penetrated. She

starred in *Little Miss Marker* (1934) and from 1935 to 1939 was a top box-office draw with such films as *Curly Top* (1935), *Dimples* (1936), *Wee Willie Winkie* (1937), *Heidi* (1937), *Rebecca of Sunnybrook Farm* (1938), and *Little Princess* (1939). Temple continued to make films until 1949, including *Since You Went Away* (1944), *Fort Apache* (1948), and *A Kiss for Corliss* (1949), but as a young woman never achieved the charisma of her childhood. Two television series, *The Shirley Temple Story Book* (1958) and *The Shirley Temple Show* (1960), proved disappointing.

As Shirley Temple Black (she married her second husband, Charles Black, in 1950) she became active in politics. Major appointments included US representative to the United Nations (1969–70), US ambassador to Ghana (1974–76), chief of protocol at the White House (1976–77), and US ambassador to Czechoslovakia (1989–).

Temple, William (1881–1944) *British clergyman and theologian who served as Archbishop of York (1929–42) and of Canterbury (1942–44); he was one of the principal architects of Anglican theology in the twentieth century.*

The son of a former Archbishop of Canterbury, Frederick Temple (1821–1902), William Temple read classics at Balliol College, Oxford, receiving his degree in 1904. He then joined Queen's College as fellow and lecturer in philosophy. His doubts concerning certain orthodox Christian doctrines, particularly the Virgin Birth, threatened to impede his ordination but he finally entered the Anglican Church in 1909. The previous year, Temple had been elected first president of the Workers' Educational Association, an organization devoted to nonpartisan nonsectarian adult education, a post in which he remained until 1924. Throughout his career, Temple maintained his concern with contemporary social issues, particularly education and unemployment. He joined the Labour Party in 1918 and intervened, controversially, in the General Strike of 1926.

After serving as head of Repton School (1910–14), Temple was rector of St James's, Piccadilly, during World War I, then briefly canon of Westminster before his appointment in 1920 as Bishop of Manchester. During this time he wrote *Christus Veritas* (1924), which stressed the Incarnation of Christ as the key to an understanding of the Divine. In the 1930s, however, Temple increasingly turned from metaphysical issues to an evangelism quickened by the threat of war. Enthroned as Archbishop of York in 1929, he continued to preach and lecture widely, was an advocate of Church unity, and chaired the Church of England's long-running Commission on Doctrine. In 1942 he succeeded Cosmo Lang as Archbishop of Canterbury. Apart from providing spiritual leadership through the war, he played a part in drafting the 1944 Education Act.

Tenzing Norgay (c. 1914–86) See HILLARY, SIR EDMUND PERCIVAL.

Teresa, Mother (Agnes Gonxha Bojaxhiu; 1910–   ) *Albanian-born missionary, now an Indian citizen, who has dedicated her life to helping the poor and underprivileged in India and other countries. She was awarded the 1979 Nobel Peace Prize and in 1980 she received the Bharat Ratna (Star of India); in 1983 she received an honorary OM.*

Born in Skopje (now in Yugoslavia), the daughter of an Albanian grocer, Agnes Bojaxhiu decided at the age of twelve to become a missionary. In 1928 she joined the Sisters of Loretto, a community of Irish nuns in Dublin, and shortly afterwards sailed for India. While acting as principal of St Mary's High School in Calcutta, she was appalled by the living conditions of the city's many destitute people. After training in medicine, she founded her Order of Missionaries of Charity in 1948, which two years later received official recognition from the Catholic Church. The first Nirmal Hriday (Pure Heart) shelter for the dying opened in 1952 and later the leper colony of Shanti Nagar was built near Asanol, partly funded by a raffle for the pope's car donated to Mother Teresa by Pope PAUL VI on his visit to India in 1964. Her order has opened schools, orphanages, clinics, and rehabilitation centres besides the Shishu Bavan homes for retarded, crippled, and abandoned children. Mother Teresa's organization now operates in many other Indian cities and other deprived regions of the world. Tough, diminutive, and with a purpose born of an unshakeable Christian conviction, Mother Teresa has received numerous awards and prizes for her work, all of which she has used to finance her order.

Terry, Dame (Alice) Ellen (1847–1928) *British actress, known for her theatrical partnership with Henry Irving (1838–1905).*

Born in Coventry into a theatrical family, Ellen Terry made her debut at the age of nine in *The Winter's Tale*, produced by Charles Kean (1811–68) at the Princess's Theatre, London. She appeared in several more of Kean's productions before joining the repertory company playing at the Theatre Royal, Bristol. In 1864 she gave up the stage to marry the artist G. F. Watts, but within a year the marriage failed and she returned to acting, making her first appearance opposite Henry Irving in 1867, in *Katharine and Petruchio*. The following year Terry left the stage again to live with architect and theatrical designer Edward Godwin, by whom she had two children, one of whom – Gordon Craig (1872–1966) – became a distinguished theatre designer and producer. In 1875 the couple parted and Terry returned to the stage to support her children; she secured her place as one of Britain's leading actresses as Portia in *The Merchant of Venice* (1875) and Olivia in *The Vicar of Wakefield* (1876). In 1877 she married actor Charles Kelly but they separated in 1881.

In 1878 Ellen Terry began her distinguished association with Henry Irving, which lasted for twenty-four years. Based at the Lyceum Theatre in London, she played opposite Irving in all the major Shakespearean and other classical roles and the partnership became famous both at home and on tour abroad. During the 1890s Terry began her famous correspondence with George Bernard SHAW (first published in 1931) and after she left Irving in 1902, three years before his death, she appeared in several of Shaw's plays in parts written especially for her, notably Lady Cicely Waynflete in *Captain Brassbound's Conversion* (1905).

Terry's fifty years on stage (1906) were celebrated at Drury Lane, when over twenty members of her family and leading actors participated in a gigantic matinée. In 1907 she married the US actor James Carew, remaining with him until 1910. Acting less, she lectured on Shakespeare, touring the USA and Australia (1910–11). Her last performance came in *Crossings* (1925) at the Lyric, Hammersmith.

Teyte, Dame Maggie (Margaret Tate; 1888–1976) *British soprano, who introduced French impressionist vocal music to Britain and the USA. She was created a Chevalier de la Légion d'honneur (1957) and a DBE (1958).*

She studied at the Royal College of Music in London and in Paris (1903–07) with Jean de Reszke (1850–1925), making her debut there at the Opéra-Comique in the 1906 Mozart festival. In 1908 DEBUSSY chose her to succeed Mary Garden (1877–1967) as Mélisande in his opera *Pelléas et Mélisande*, coaching her in the part himself and accompanying her in recitals of his songs. In 1937, after a substantial career in opera and operetta, she was commissioned to record an album of Debussy's songs with Alfred CORTOT as pianist; an album *French Song from Berlioz to Debussy* followed in 1940. Thereafter she was much in demand as a recitalist until her final concert at the Festival Hall in 1955.

Thant, U (1909–74) *Burmese secretary-general of the United Nations (1961–72). In this capacity his equanimity and composure helped to defuse several international crises.*

Born in Pantanaw, the son of a wealthy landowner and rice miller, Thant was educated at the University College, Rangoon, before taking up the position of headmaster at the National High School in Pantanaw. During World War II, when Burma was under Japanese occupation, he worked for the independence movement and in 1947 joined the government service as a press director.

Thant was appointed director of broadcasting and in 1948 became a close adviser to the prime minister, U Nu, in the first Burmese republican government. Sent to the General Assembly of the United Nations (UN) as a Burmese delegate in 1952, he was appointed permanent representative to the UN in 1957. After the death of Dag HAMMARSKJÖLD in 1961, he was elected secretary-general of the UN, in which capacity he helped resolve several crises of international security, including the Cuban missile crisis (1962), the Congo crisis (1963), the civil war in Cyprus (1964), the Indian-Pakistan conflict in Kashmir (1965), and the Arab-Israeli Six Day War (1967). He retired as secretary-general of the UN in 1972. U Thant was the author of several books on such subjects as Burmese history, education, and the League of Nations.

Thatcher, Margaret Hilda (1925–) *British Conservative politician and the country's first woman prime minister (1979–90).*

Born in Grantham, Lincolnshire, Margaret Roberts became interested in politics as a child through the influence of her father, a prosperous grocer and Methodist lay-preacher, who was twice mayor of Grantham. At Somerville College, Oxford, she read chemistry and was president of the University Conservative Association. After graduating in 1947, she worked as a research chemist, and in 1949 she stood for parliament for the first time, but was not elected. She married Denis Thatcher in 1951, began to study law, and took her bar finals in 1953, the year her twin son and daughter were born. She practised as a taxation lawyer until entering parliament as MP for Finchley, London, in 1959. From 1961 to 1964 she was joint parliamentary secretary for the Ministry of Pensions and National Insurance, and then, in opposition (1964–70), spokesman on education. When the Conservatives returned to power she served as secretary of state for education and science (1970–74), being remembered for her toughness in, for example, ending the provision of free school milk. Following the Conservative defeat at the polls in 1974, she won the party leadership from Edward HEATH in 1975 and in 1979 became prime minister.

Standing on the right wing of the Conservative Party, Mrs Thatcher advocated a reduction in state control and encouragement to private enterprise. Her government adopted a monetarist policy of regulating the money supply by the tight control of public expenditure (except ing that on defence and the police) and high interest rates.

During her first term of office she was successful in reducing the rate of inflation, although the policy brought ever-increasing unemployment. Dubbed the 'Iron Lady' by the Soviets, Mrs Thatcher proved intransigent in her handling of foreign affairs, notably in dispatching a large task force in response to the Argentine invasion of the Falkland Islands (1982). Though the Falklands War was bitterly attacked in some quarters, the British victory was thought to have contributed to her re-election as prime minister, with a very large majority, in 1983.

Thatcher's second term was nearly brought to a premature end in 1984, when an IRA bomb planted in the Grand Hotel during the Conservative Party conference in Brighton narrowly failed to wipe out the entire cabinet. The next three years witnessed the consolidation of her policies of fiscal caution, privatization, and self-help, which came to be known as 'Thatcherism'. As well as pushing ahead with her ambitious privatization programme, with the floating of such major industries as British Gas as public companies, she also placed increased restraints upon trade-union power, defeated a year-long miners' strike (1984–85), reduced taxation, and continued to restrain public spending (overriding accusations that Thatcherite Britain favoured the better off). Although unemployment remained high, the economy was considered leaner and fitter after years of inefficiency and the City boomed (until the crash of 1987). In international relations, Thatcher acquired a reputation for her forthright opinions on the European Community. She also continued to enjoy close personal contact with the US president, Ronald REAGAN, although some critics attacked her readiness to endorse US policy on virtually all issues. By 1987 – the year of her third election victory – she had become one of the world's most respected leaders. In 1988 she became the longest serving UK prime minister of the century.

Legislation embarked on in Thatcher's third term included radical proposals in such fields as education, health care, and public-sector housing. However, disaffection within Conservative ranks and in the country at large grew more intense in reaction to Thatcher's increasing hostility towards the European Community, her autocratic attitude towards her cabinet colleagues, and perhaps most of all her determination to impose the highly unpopular community charge (or poll tax) to replace the rates in 1990 (introduced in Scotland in 1989). The resignation of her long-time chancellor of the exchequer Nigel Lawson (1989) was followed by the departure of her deputy prime minister Sir Geoffrey Howe (1990), who made a hostile resignation speech in the House of Commons. This led Michael Heseltine to challenge her for the leadership of the party. After failing to win an outright victory over Heseltine in a ballot for the leadership she succumbed to party pressure and resigned. John MAJOR, her chancellor of the exchequer, succeeded her as prime minister but was hampered by Thatcher's continuing publicly expressed opposition to closer integration (particularly monetary union) within the EC. In 1991 she announced that she would retire from the Commons.

Theodorakis, Mikis (1925–) *Greek composer and politician, who led a revival of Greek folk music in the 1960s and wrote a song that became the rallying call of opposition to the military government.*

Theodorakis studied at the Athens Conservatory and, after war service and deportation during the civil war (1947–52), in Paris with Olivier MESSIAEN. He wrote a symphony (1948–50) and a number of song settings, including *Five Cretan Songs* (1950) for chorus and orchestra, before turning to such chamber music as his two sonatinas for violin and piano (1957; 1958). He then composed incidental music for the theatre, notably for the ballet *Antigone* (1958), produced at Covent Garden in 1959, and for Euripides' *Phoenician Women*. He first became known outside Greece, however, when he wrote the score for the film *Zorba the Greek* (1965), with its theme tune 'Zorba's Dance', which became an

international best-seller. He also wrote pop oratorios, such as *Axion Esti* (1966).

Theodorakis returned to Greece in 1961 and was elected a member of parliament in 1964. In 1967 he was arrested for his left-wing political activities, but was released in response to worldwide appeals in 1970. While in detention he composed the music for *Z*, a film about political repression. He resigned from the Communist Party in 1972 and in 1981 was again elected member of parliament, this time representing the Moscow-orientated Communist Party of Greece (KKE). His most recent works include three more symphonies and the opera *Kostas Kariotakis* (1985).

Thomas, Dylan Marlais (1914–53) *Welsh poet, writer, and broadcaster.*

Dylan Thomas was born in Swansea and as a boy at Swansea Grammar School showed signs of an exceptional poetic talent. After a short spell as a reporter in Swansea (1931–32), he concentrated on writing poetry. In 1934 *Eighteen Poems* was published and at the end of that year Thomas moved to London. *Twenty-five Poems* appeared in 1936. The following year Thomas married Caitlin Macnamara and they settled at Laugharne. *Portrait of the Artist as a Young Dog* (1940) is an affectionate account of the poet's boyhood in Wales and contains some of his most delightful prose.

During World War II Thomas lived mainly in London, where he wrote for the BBC; these pieces were collected and published posthumously as *Quite Early One Morning* (1954). As a radio actor and reader of poetry he achieved great popularity, especially with his best-known radio play *Under Milk Wood* (first broadcast 1953). He also wrote film scripts and an unfinished novel, *Adventures in the Skin Trade* (1955). *Deaths and Entrances* (1946) contains some of his finest poetry. In 1950 Dylan Thomas went on a lecture tour of the USA, principally in order to make money. His health was already seriously undermined by years of heavy drinking. In 1952 his *Collected Poems* was published. Late in 1953 he went again to New York to take part in a performance of *Under Milk Wood*. It was hailed as a triumph, but a few days later Thomas died of alcoholic poisoning.

Thompson, Daley (Francis Morgan Thompson; 1958–) *British athlete, whose performances in the decathlon in the 1980s established him as one of the most successful all-round athletes ever.*

Born in Notting Hill, London, he competed in his first decathlon event in 1975, in the Welsh Open Championship. Two years later he won the European Junior Decathlon Championship and, in 1978, went on to gain the Commonwealth Decathlon title and a silver medal in the European competition. He repeated his Commonwealth victory in 1982 and 1986 and won the gold medal in the European event in the same years. His other successes included gold medals in the Olympic games in 1980 and 1984; until 1987 he had an unbeaten record. His performance in the decathlon at the Olympics in 1984, in which he scored 8847 points, set a new world record. He was also world decathlon champion in 1983.

Thomson, Sir George Paget (1892–1975) *British physicist who demonstrated the wave-particle nature of elementary particles. He received the 1937 Nobel Prize for Physics and was knighted in 1943.*

The son of J. J. THOMSON, Thomson was educated at Cambridge, where he later taught (1914–22). In 1922 he was appointed to the chair of physics at Aberdeen, and in 1930 accepted a similar post at Imperial College, London. In 1952 Thomson returned to Cambridge as master of Corpus Christi, a post he held until his retirement in 1962.

In 1927, in collaboration with Alexander Reid, Thomson passed a beam of electrons through a thin metal foil in a vacuum. A photographic plate behind the metal foil revealed a clear diffraction pattern. This could mean only that the electrons had passed through the metal foil as waves. Thomson's work thus confirmed the suggestion made by DE BROGLIE in 1923. Thomson worked on nuclear physics in the 1930s and after CHADWICK's discovery of the neutron in 1932, he began to explore some of its properties. Consequently, at the outbreak of World War II Thomson was ideally suited to head the Maud Committee set up to advise the British government on the feasibility of constructing an atomic bomb. The committee advised that a device using uranium-235 could be built. The project was transferred to the USA and Thomson spent the latter part of the war as adviser to the Air Ministry.

Thomson, Sir Joseph John (1856–1940) *British physicist, who was awarded the 1906 Nobel Prize for Physics for his discovery of the electron in 1897. He was knighted in 1908.*

The son of a bookseller and publisher, Thomson was educated at Owens College in his home town of Manchester and Trinity College, Cambridge, where he studied mathematics. He

remained in Cambridge working at the Cavendish Laboratory under Lord Rayleigh (1842–1919), whom he succeeded as Cavendish Professor of Experimental Philosophy in 1884. Thomson remained at the Cavendish until he was succeeded by his pupil Ernest RUTHERFORD in 1919, when he himself was appointed master of Trinity College.

By passing cathode rays through evacuated discharge tubes, Thomson identified a particle that could be deflected by both electric and magnetic fields. He went on to measure the ratio of the particle's charge to its mass and found it to be considerably higher than that worked out for light atoms. The new particle was given the name 'electron'. Thomson's work was reported in his *Conduction of Electricity Through Gases* (1903). Having identified the electron Thomson went on to explore the nature of 'positive rays' (protons), work he described in his *Rays of Positive Electricity*. Not the least of Thomson's achievements after 1900 was the Cavendish Laboratory itself. Seven of his pupils won Nobel Prizes, fifty-five became professors, and the Cavendish became the leading laboratory in the world for research in atomic physics. He was buried in Westminster Abbey close to his Trinity College predecessor, Sir Isaac Newton, and his Cavendish successor, Lord Rutherford.

Thomson, Roy Herbert, 1st Baron (1894–1976) *Canadian-born British newspaper publisher and businessman, who, as chairman of the Thomson Organization, was proprietor of* The Times *and* The Sunday Times. *He was created Baron Thomson of Fleet in 1964.*

Thomson was born in Toronto, the son of a barber of Scottish descent. He started work in 1908 as a clerk in a coal yard, then joined a cordage company, rising to branch manager. Poor eyesight precluded military service during World War I. After a brief and unsuccessful farming venture (1919–20), he started selling car parts and equipment. However, bad debts brought him close to bankruptcy and he was obliged to move to Ottawa and become a salesman for a radio distributor. So successful was he that he paid off his debts and, in 1931, opened a radio station in North Bay, Ontario.

He acquired his first newspaper three years later when he bought the weekly *Press* in the mining town of Timmins, northern Ontario. He opened more radio stations and bought further newspapers, mainly provincial titles in Ontario and Saskatchewan; in 1952 he bought his first US newspaper, the *Independent* of St

Petersburg, Florida. The following year, Thomson purchased the prestigious Edinburgh-based daily, *The Scotsman*, and as a result, moved his home to Edinburgh. His expansion into the British media continued with a controlling interest in Scottish Television, which started broadcasting in 1956, and with the acquisition in 1959 of the Kemsley newspaper group, which included *The Sunday Times* and a host of provincial titles. He added *The Times* to his collection in 1967, by which time the Thomson Organization owned numerous companies throughout the world, principally in publishing but also in printing, radio, television, and travel.

Thomson, Virgil (1896–1989) *US composer whose lyrical style shows the influence of both his American roots and his contact with the French composers of Les Six.*

Born and brought up in Kansas City, Thomson went to Harvard University in 1919, after army training, before studying for a year (1921) in Paris with Nadia BOULANGER. In 1925 he returned to Paris to compose and was inspired by Erik SATIE and Les Six to write in a simple, elegant, precise, and humorous style. His friendship with Gertrude STEIN resulted in her writing the libretti for his first opera, the successful *Four Saints in Three Acts* (1928), and *The Mother of Us All* (1947).

Thomson gradually evolved a style that combined the lyricism of the hymn tunes and folk music of his childhood with the influence of contemporary French artistic movements. He wrote several notable film scores, in particular for Pare Lorenz's *The Plow that Broke the Plains* (1936) and Robert FLAHERTY's *Louisiana Story* (1948). In 1940 Thomson returned to the USA and was appointed music critic of the New York *Herald Tribune*, establishing himself as one of the major critical writers of his time. Fourteen years later he resigned and devoted himself to composition. Thomson's other works include the opera *Lord Byron* (1961–68), several ballets, three symphonies (1928; 1931, revised 1941; 1972), and orchestral, chamber, and choral works.

Thorndike, Dame Sibyl (1882–1976) *British actress. She was created a DBE in 1931 and a CH in 1970.*

Daughter of an Anglican canon, Sibyl Thorndike was born in Gainsborough, Kent, and in 1903 began her long, uninterrupted, and distinguished career with a four-year tour of Britain and the USA with Ben Greet (1857–1936), appearing in numerous Shakespearean roles. In 1908 she married the actor Lewis Casson

(1875–1969), whom she met while working with Miss Horniman's Repertory Company in Manchester. In 1914 she became the leading actress at the Old Vic with Lilian BAYLIS, about whom she co-authored a memoir (1937) with her brother, actor Russell Thorndike (1885–1972). After World War I, during which her husband served in the army, the Cassons' careers were virtually inseparable. They performed in numerous plays together, went into management (1922–27), led the Old Vic Company tour of Wales (1940), and played in many parts of the world. Among Thorndike's many memorable performances were Hecuba in *The Trojan Women* (1919), the title role in the first London showing of SHAW's *Saint Joan* (1924), which she is said to have inspired, and Miss Moffat in Emlyn Williams's *The Corn is Green* (1938), playing opposite the author.

Between 1921 and 1963 she made eighteen films, including *Dawn* (1928), as Edith CAVELL, *Nicholas Nickleby* (1947), as Mrs Squeers, and OLIVIER's *The Prince and the Showgirl* (1957). In her eighties she was still appearing on stage, for example in the musical *Vanity Fair* (1962) and playing opposite former pop-star Adam Faith (1940–) in *Night Must Fall* (1968). At eighty-seven she gave her last performance, in *There Was an Old Woman* at the Thorndike Theatre, Leatherhead, named in her honour.

Thurber, James Grover (1894–1961) *US humorist and artist, who had a long association with the* New Yorker.

Born in Columbus, Ohio, Thurber was educated at Ohio State University, taking a degree in 1919. (Because of an accident in childhood, he was blind in one eye; his poor eyesight prolonged his studies and had an effect on his work.) He worked for a time as a code clerk in the Department of State in Washington. He was subsequently sent to the embassy in Paris and later found work on the Paris edition of the Chicago *Tribune*. He remained in Paris until 1926. On returning to America, he sought the help of the writer E. B. White in finding a job at the newly founded *New Yorker* magazine, where he was employed as managing editor for six months. Thereafter he found a niche for himself writing 'The Talk of the Town' section at the front of the magazine. His association with the *New Yorker* was permanent and he became the chronicler of its heyday under its most famous editor, Harold Ross, in *The Years with Ross* (1958).

Thurber's humour, often marked by a certain sadness and resignation, takes the form of stories, fables, and essays, frequently illustrated by seemingly rudimentary line drawings chiefly of timid men, domineering women, and animals cryptically appearing in unexpected places. (Thurber said that, while producing a drawing, he was never sure what the outcome would be.) Among the collections of his pieces, almost all first published in the *New Yorker*, are *Is Sex Necessary?* (1929), a collaboration with E. B. White satirizing sex manuals, *The Owl in the Attic and Other Perplexities* (1931), *The Seal in the Bedroom and Other Predicaments* (1932), *My Life and Hard Times* (1933), *The Middle Aged Man on the Flying Trapeze* (1935), *Fables for Our Times, and Famous Poems Illustrated* (1940), *My World – And Welcome to It* (1942), which contains his famous story 'The Secret Life of Walter Mitty', and *The Thurber Carnival* (1945). He wrote several books for children, including *The Thirteen Clocks* (1950), and collaborated with Elliott Nugent on a comedy, *The Male Animal* (1940).

Tillich, Paul (1886–1965) *US theologian and philosopher, born in Germany, whose work addressed the crucial implications for Christianity raised by existentialist philosophers of the twentieth century and who explored the relationship between theology and culture.*

The son of a village pastor from Starzeddel, Brandenburg province, Tillich moved with the family to Berlin in 1900. He studied theology at the universities of Berlin, Tübingen, and Halle, and received his doctorate from Breslau in 1910. After serving as a military chaplain during World War I, he became a lecturer in theology at Berlin University and in 1924 was appointed professor of theology at Marburg University, moving to Dresden the following year. He also taught at Leipzig and in 1929 he accepted the chair of philosophy at Frankfurt.

During the 1920s, Tillich published numerous articles and papers examining religion and culture and preparing the ground for his magnum opus, *Systematic Theology* (three vols; 1951–63). For him, the key role of theologians was to 'correlate' the answers provided by the revelation of Christianity to the problems of being and nonbeing raised by existential philosophical analysis. He regarded God as 'the ground of being' and our existence, in effect, as part of God. Throughout his work, Tillich repeatedly emphasized the importance of culture, particularly symbolism, myth, and the arts in general.

Tillich was dismissed from his post at Frankfurt for his opposition to the Nazis and, at the invitation of the distinguished theologian Reinhold Niebuhr, moved to the Union Theological Seminary, New York, as professor of philosophical theology. He later taught at Harvard (1955–62) and also at the University of Chicago.

Tinbergen, Jan (1903–) *Dutch economist, joint winner of the 1969 Nobel Prize for Economics with the Norwegian economist Ragnar Frisch (1895–1973) for their work in econometrics.*

The elder brother of ethologist Niko Tinbergen, he was educated at Leiden University and worked at the Central Bureau of Statistics from 1929 to 1945; he has been professor at the Netherlands School of Economics since 1933 (emeritus since 1973). Tinbergen has acted as adviser to many organizations and from 1936 to 1938 was on the staff of the League of Nations. As head of the Central Planning Bureau in The Hague he used econometric techniques for forecasting and planning for the Netherlands; he is also concerned with planning in developing countries, and has contributed to the concept of 'shadow prices' – imputed prices of goods whose prices cannot be decided by market forces.

Business Cycles in the USA, 1919–39 (1939) is an important study of cyclical fluctuations in the US economy, in which Tinbergen identified and quantified the importance of the different factors involved. His *Theory of Economic Policy* (1952) discussed the need for a wide range of policies to suit the increasing number of economic goals. Other major works include *Economic Policy: Principles and Design* (1956), *Shaping the World Economy* (1962), *Development Planning* (1968), *Income Distribution: Analysis and Policies* (1975), and *World Security and Equity* (1990).

Tinbergen, Niko(laas) (1907–88) *British zoologist, born in the Netherlands, whose studies of animal behaviour helped establish ethology as a distinct discipline and earned him the 1973 Nobel Prize for Physiology or Medicine.*
Tinbergen studied zoology at Leiden University and received his PhD in 1932 for a memorable thesis showing that digger wasps (*Philanthus*) locate the entrance to their burrows by reference to landmarks in the vicinity. With his wife, Tinbergen joined an expedition to Greenland, where he studied the significance of territorial disputes in the breeding success of snow buntings. In 1936 he joined Leiden University as a lecturer, becoming pro-

fessor of experimental zoology in 1947. Tinbergen not only studied animal behaviour in the field but also under experimental conditions in the laboratory. For instance, he used a range of model fish to investigate the precise nature of the visual stimuli that elicited the aggressive response or the courtship response in male sticklebacks. He moved to Oxford University in 1949 and pursued his longstanding interest in the behaviour of herring gulls and other seabirds. He showed how an abnormally large (supernormal) stimulus produces a stronger than normal response in certain behaviour; for instance, a giant egg produced a stronger brooding response than a normal egg. In 1952 he introduced the term 'displacement activity' to describe seemingly irrelevant actions that often intersperse behaviour, such as needless necktie adjustments by men in the company of strangers. Much of Tinbergen's early work was published in *The Study of Instinct* (1951) and *The Herring Gull's World* (1953).

Tinbergen stressed the relevance of ethology to human psychology and sociology, for instance in helping to distinguish basic drives, such as fear and sexual attraction, that determine our behaviour. In collaboration with his wife, he developed a revolutionary therapy for autistic children that assumes a largely emotional basis for the disease, published as *'Autistic' Children: new hope for a cure* (with E. A. Tinbergen; 1983). He also wrote *Animal Behaviour* (1965) and *The Animal in its World* (2 vols, 1972–73). Tinbergen was appointed professor of animal behaviour at Oxford in 1966 and emeritus professor in 1974. He became a fellow of the Royal Society in 1962.

Tippett, Sir Michael Kemp (1905–) *British composer, who has emerged as one of the leading figures in late twentieth-century music. He was knighted in 1966, created a CH in 1979, and received the OM in 1983.*

After studying at the Royal College of Music (1923–28), Tippett taught French (1929–34) while he developed his own style of composition. His sympathies with the unemployed of the 1930s led him to work at summer camps and he became conductor of the South London (Morley College) Orchestra for unemployed musicians (1933). In 1940, Tippett became music director at Morley College and also joined the Peace Pledge Union; he was subsequently imprisoned for his pacifist beliefs. In 1951 Tippett resigned from Morley College in order to devote more time to composition.

Drawing upon such diverse sources as jazz, madrigals, spirituals, and the music of Beethoven, Tippet experimented with the traditional forms of sonata and symphony in his early work, which included a concerto for double string orchestra (1939) and the oratorio *A Child of Our Time* (1944), his first public success. Tippett's sure sense of theatre has shown itself in five operas, *The Midsummer Marriage* (1955), *King Priam* (1962), *The Knot Garden* (1970), *The Ice Break* (1977), and *New Year* (1989), for all of which he also wrote the libretti. Expressing his own social and philosophical ideas, these operas have been highly acclaimed for their strong humanist message and structural sophistication. His other works include the oratorio *The Mask of Time* (1983), four symphonies, several song cycles (including *The Heart's Assurance*, 1950–51), chamber music, and four piano sonatas (1938, revised 1942; 1962; 1972–73; 1984). He is also the author of a collection of essays, *Moving into Aquarius* (1959), and *Music of Angels* (1980). He became president of the London College of Music in 1983.

Tito, Josip Broz (1892–1980) *Yugoslav marshal and statesman. As president (1953–80) for nearly thirty years, he maintained a socialist government that was independent of the Soviet Union and pursued a foreign policy of nonalignment.*

Born in Croatia into a peasant family, Tito was a metalworker before serving in the Austrian army during World War I. Taken prisoner by the Russians, he fought with the Red Army during the civil war but returned to Yugoslavia in 1923. There he joined the clandestine Communist Party, for which he was imprisoned (1929–34).

Tito visited the Soviet Union in 1935 and in 1937 was made secretary-general of the Yugoslav Communist Party. When, in 1941, HITLER invaded Yugoslavia, he organized a partisan resistance movement to conduct guerrilla warfare against the occupying forces. His success in resisting the Germans earned him the title of marshal (1943) and gave him enough support to lead the new federal government as prime minister and commander-in-chief after the war. In 1948, in a bid to preserve Yugoslavia's independence, he refused to bow to STALIN's pressure to conform to the Soviet model of socialism and was consequently expelled from the Cominform. As a result Tito was able to develop a decentralized system of workers' self-government, introduce a federal constitution that recognized the major nationalities within Yugoslavia as separate republics, and adopt a policy of nonalignment abroad. Under the new constitution (1953) he became president; after five subsequent re-elections he was made president for life (1974). He was succeeded by a collective leadership, which he had himself instigated.

Tizard, Sir Henry Thomas (1885–1959) *British chemist and scientific administrator. Chairman of the famous Tizard Committee, he was knighted in 1937.*

The son of a naval officer, Tizard studied chemistry at Oxford and Berlin where, in the laboratory of Walther NERNST, he first met Frederick Lindemann (see CHERWELL, VISCOUNT). He returned to Oxford in 1911 to teach chemistry but, after spending World War I in the Royal Flying Corps, abandoned scientific research for its administration. Recognizing that he would never be outstanding as a pure scientist, he took the post of assistant secretary to the Department of Scientific and Industrial Research (DSIR). In 1929 he moved to Imperial College, London, where he served until 1942 as rector.

In 1935 Tizard was appointed chairman of the Air Ministry's Committee for the Scientific Survey of Air Defence, known as the Tizard Committee. Through this committee, Tizard backed the development of radar, supported the jet engine of Frank WHITTLE, and in 1940 set up the Maud Committee to examine the feasibility of making nuclear weapons. In several of these decisions he was opposed by Lindemann, who had become friend and scientific adviser to the future prime minister, Winston CHURCHILL. Consequently, with the rise of Lindemann, a most unforgiving man, Tizard's influence declined and in 1942 he abandoned Whitehall, becoming master of Magdalen College, Oxford. With the fall of Churchill in 1945, Tizard returned to Whitehall to advise the newly elected Labour government on defence problems. He resigned in 1950 and died from a cerebral haemorrhage nine years later.

Todd, Alexander Robertus, Baron (1907–) *British biochemist who received the 1957 Nobel Prize for Chemistry in recognition of his work in determining the chemical structure of nucleotides – constituents of the nucleic acids DNA and RNA. He was knighted in 1954, created a life peer in 1962, and awarded the OM in 1977.*

Born in Glasgow, Todd received his first degree in 1928 from Glasgow University and his PhD from the University of Frankfurt am Main in 1931. Following further research at Oxford

and Edinburgh, he joined the Lister Institute of Preventive Medicine in London (1936–38), where he studied the structure of a range of plant compounds, notably thiamine (vitamin B_1). In 1938 he was appointed professor of chemistry at Manchester University. He elucidated the structure of vitamin E, which he also synthesized, and made a study of the pharmacologically important compounds derived from the cannabis plant, *Cannabis sativa*. Todd moved to Cambridge University as professor of organic chemistry (1944–71) and began his work on the structure and synthesis of the nucleotides – work that contributed to the eventual discovery of the molecular structure of the genetic carrier molecules, DNA and RNA. In 1949, Todd synthesized adenosine triphosphate (ATP), a nucleotide that acts as an energy carrier in the metabolic reactions of living organisms. This was followed by the synthesis of the coenzyme flavin adenine dinucleotide (FAD), also in 1949, and the nucleotide uridine triphosphate (UTD) in 1954. In the following year, Todd published the structure of cyanocobalamin (vitamin B_{12}).

Todd was elected a fellow of the Royal Society in 1942 and later served as its president (1975–80).

Togliatti, Palmiro (1893–1964) *Italian politician and secretary of the Italian Communist Party (1926–64), who was committed to gaining power by constitutional measures.*

Born in Genoa, the son of a bookkeeper, Togliatti won a scholarship to the University of Turin, where he studied law and philosophy. As a socialist he worked in the factories of Turin, before serving in the medical corps during World War I. In 1919 he co-founded the review *L'Ordine Nuovo*, which asserted that the experience of the Russian revolution was not necessarily limited to Soviet Russia. Two years later, along with GRAMSCI and several others, he broke away from the Socialist Party during the Livorno conference, to form the Communist Party.

Togliatti became secretary of the Communist Party in 1926 following the arrest and imprisonment of Gramsci. When the fascists came to power, he went into exile and lived abroad, mostly in Moscow, for the next eighteen years. During this time he became secretary of Comintern (1935) and was in charge of the communist units fighting in the Spanish civil war (1936–39). He returned to Italy in 1944, after the fall of MUSSOLINI, and joined the Badoglio government. Over the next twenty years he rebuilt the Communist Party, transforming it into the second largest party in Italy and the largest Communist Party in western Europe. He remained leader of the party until his death in 1964.

Togo Heihachiro, Marquis (1846–1934) *Japanese naval commander. A hero of the Russo-Japanese War, he was made a count in 1907 and a marquis in 1934.*

Born in Satsuma (now Kagoshima) of samurai descent, Togo joined the Satsuma provincial army at the age of eighteen, serving in the Anglo-Satsuma War (1863) and the Boshin civil war (1868). From 1870 to 1878 he studied in England at the Royal Naval College, Greenwich, returning to Japan as a lieutenant in 1878. During the Sino-Japanese War (1894–95) he commanded the warship *Naniwa* and won international attention for the sinking of the British steamer *Kwoching*, which was carrying Chinese troops.

In 1903, Togo was appointed commander-in-chief of the combined Japanese fleet. During the Russo-Japanese War (1904–05) he carried out a ten-month blockade of the Russian Far Eastern naval forces at Port Arthur (now Lüshun) and succeeded in defeating the Russians in a series of battles, the most decisive of which was the Battle of Tshushima (1905), in which the Russian Baltic fleet was destroyed and Japanese victory in the war was secured. Togo's strategy in the battle was later studied by the European navies and incorporated into their own tactics. After the war Togo was appointed head of the Naval General Staff (1905–09) and in 1913 he was promoted to the rank of admiral of the fleet. From 1914 to 1921 he served as president of Togu Gogakomonjo, a body responsible for providing education for the crown prince. During the 1930s he was acknowledged as an elder statesman. He remained a member of the high military council until his death in 1934, when he was given a state funeral and honoured with the erection of a Togo Shrine in Herajuku, Tokyo.

Tojo Hideki (1884–1948) *Japanese war minister (1940–44) and prime minister (1941–44). Known as 'Razor Tojo' within the Japanese army, he was an adroit commander in the field and a skilful administrator whose uncompromising attitude towards the USA helped to precipitate hostilities between the two countries.*

Born in Tokyo, of samurai descent, Tojo was educated at the Imperial Military Academy and the Military Staff College, where he graduated in 1915. Advancing rapidly through the ranks, he was appointed commander of the infantry brigade in 1934 and sent to Manchuria in 1935 as commander-in-chief of the Kwantung Army. By 1937 he had risen to chief of staff of the Kwantung Army.

Tojo was appointed vice-minister of war (1938–39) in the first cabinet of Prince KONOE FUMIMARO. Serving as commissioner of aviation (1940), he was given the post of minister of war in the second and third Konoe cabinets (1940–41). He succeeded Konoe as prime minister in 1941 and shortly afterwards initiated the attack on the US base at Pearl Harbor. This was followed by a number of major victories in southeast Asia and the Pacific before the Allied forces became organized. By 1944, as chief of the general staff, he had assumed virtual control of all political and military decision-making but was unable to prevent a succession of military losses, which culminated in the loss of the Mariana Islands. Opposition from within the army forced him to resign later that year. Following Japan's surrender in 1945, he attempted to commit suicide but failed and was arrested as a war criminal. He was brought to trial before the International Military Tribunal for the Far East and in 1948 was hanged.

Tolkien, J(ohn) R(onald) R(euel) (1892–1973) *British scholar and author, whose novels* The Hobbit *and* The Lord of the Rings *trilogy became international best-sellers.*

Tolkien was educated at King Edward VI School, Birmingham, and Exeter College, Oxford. From 1915 to 1918 he served with the Lancashire Fusiliers and in 1920 became reader in English Language at Leeds University, where he later held the chair (1924–25). In 1925 he returned to Oxford where he spent the rest of his life, first as Rawlinson and Bosworth Professor of English Language and Literature (1945–59). For many years he was one of the group of friends known as 'the Inklings' which gathered around C. S. LEWIS.

Tolkien made his reputation as a Middle English scholar in the 1920s with *A Middle-English Vocabulary* (1922) and the standard text of *Sir Gawain and the Green Knight* (1925), the latter edited with E. V. Gordon. Among his critical works was the 1936 lecture *Beowulf: the Monsters and the Critics*. In 1962 he published an important edition of the Middle English manual for nuns, the *Ancrene Wisse*.

As far back as 1917 Tolkien was engaged in creating the mythology, geography, inhabitants, and languages of an imaginary world that had its roots in his study of ancient Germanic cultures. The first published fruit of this extraordinary feat of the imagination was *The Hobbit* (1937), but it received its full expression in *The Lord of the Rings* trilogy: *The Fellowship of the Ring* (1954), *The Two Towers* (1954), and *The Return of the King* (1955). *The Silmarillion* (1977), edited by Tolkien's son Christopher, is a compilation of myths and legends making up the 'prehistory' of *The Hobbit* and *The Lord of the Rings* narratives. Tolkien also wrote stories and verse for children.

Toller, Ernst (1893–1939) *German playwright and poet.*

Born near Bromberg into a middle-class Jewish family, Toller was educated at the University of Grenoble. He was aware from an early age of the antisemitism that was increasing in Germany, but he welcomed the opportunity to volunteer in World War I. The experience of trench warfare, however, caused a breakdown and he was discharged from the army in 1916. This experience left him committed to pacifism and socialism. He moved to Munich, where he took part in literary life there and was also connected with the short-lived communist government (the Bavarian Socialist Republic) of 1918. In 1919 he was imprisoned for this for five years, during which period he wrote his most important work. In all Toller wrote thirteen plays, four books of poetry, and seven volumes of prose works, which include his autobiography *Eine Jugend in Deutschland* (1933; translated as *I Was a German*, 1934). Forced to flee the country in 1933, Toller emigrated to America, but he became increasingly depressed at the growth of Nazism. He committed suicide in New York soon after the invasion of Czechoslovakia.

Toller's first volume of verse, *Gedichte der Gefangenen* (1921), was followed by the expressionist dramas for which he is best known. The first to be successfully produced, and his masterpiece, was *Masse-Mensch* (published in 1921; translated as *Masses and Man*, 1923). Its angry message is typically presented in seven disconnected episodes, rather than in a logically developed plot, and deals with a revolution led by the one named character, Sonja, who is eventually submerged by violence and shot by the state as the revolution fails. The denunciation of violence, by whatever side in the political struggle, is also a theme in *Die Maschinenstürmer* (1922; translated as *The Machine-Wreckers*, 1923), a verse play about the Luddites (1812–15) inspired by Lord Byron's maiden speech in the House of Lords (1812). The workers' leader, Jimmy Cobbett,

is killed as a traitor by his own class after opposing the wrecking of the steam loom. Toller's best-known realistic play, *Der Deutsche Hinkemann* (1923; translated as *Brokenbrow*, 1926), concerns the tragic predicament, common enough at the time, of a German veteran wounded in the war. In *Das Schwalbenbuch* (1924; translated as *The Swallow Book*), his finest book of verse, the swallows that nest in his prison cell are seen as emblems of the renewal of life.

Tolstoy, Aleksey Nikolayevich (1883–1945) *Soviet novelist, playwright, poet, and essayist.*

Born into a noble family (his own title was Count) in Nikolayevsk, Samara, Tolstoy was distantly related to the author of *War and Peace*. He was educated at the Technological Institute of St Petersburg and in 1905 published his first work – symbolist poetry. He was a war correspondent with the White Army during World War I and after the October Revolution (1917) chose to live in exile in Paris and Berlin, where he wrote his first fiction. He returned to Russia in 1923, bringing with him several works that firmly established his reputation. The novel *Syestry* (1921; 'The Sisters') was the first part of his trilogy *Khozhdeniye po mukam* (1921–40; translated as *The Road to Calvary*, 1946), which was awarded the Stalin Prize in 1942. His other works written in exile were the novella *Detstvo Nikity* (1920; translated as *Nikita's Childhood*, 1945) and the utopian novel *Aelita* (1922). Over the next decade Tolstoy continued to publish adventure novels, science fiction, and other works that make him one of the most popular writers in the Soviet Union. He also made a great contribution to Soviet cultural life in encouraging the publication of works in the regional languages of Russia and in preserving and publishing folklore material. He was elected to the Academy of Sciences in 1939.

His masterpiece and by far his most important work was the historical novel *Pyotr Pervy* (1929–43; translated as *Peter the Great*). It was at first the centre of controversy since its opening parts were published when the dominant school of historiography (the Pokrovsky school) rejected the idea that great men shape history. Tolstoy took an unpolitical traditional approach in portraying the tsar as a shaper of history. Both within and outside the Soviet Union the book was soon generally acknowledged as the greatest historical novel ever written by a Soviet author.

Tortelier, Paul (1914–90) *French cellist, composer, conductor, and teacher.*

Tortelier studied at the Paris Conservatoire, where he won first prize at the age of sixteen, and made his debut a year later at the Concerts Lamoureux. After further study, he moved to the USA and was engaged as cellist with the Boston Symphony Orchestra (1937–40). After further appearances in Paris and Amsterdam he made his British debut in 1947 under BEE-CHAM as solo cellist in Richard STRAUSS'S *Don Quixote*; subsequently he became established as a world-class cellist. As well as being in worldwide demand as a soloist, he also gave recitals in concert with other members of his family, including his wife Maud Martin Tortelier (1926–), also a cellist, their son Jan Pascal Tortelier (1947–), a violinist, and his daughter Maria de la Pau (1950–), a pianist.

Tortelier's compositions include the *Israel Symphony*, written after a year spent on a kibbutz (1955–56). He was appointed professor at the Paris Conservatoire in 1957 and taught such distinguished cellists as Jacqueline DU PRÉ. He also presented a series of television masterclasses (1970) and wrote a number of books for cello and piano, including *How I Play, How I Teach* (1973).

Toscanini, Arturo (1867–1957) *Italian conductor, who was one of the finest musical interpreters of his day.*

At the age of nine Toscanini entered the Parma Conservatory, where he studied cello, piano, and composition, graduating in 1885 with maximum marks. He began his professional life as an orchestral cellist and played in the first performance of Verdi's *Otello* (1887). While touring in Brazil with an Italian opera company at the age of nineteen (1886), Toscanini substituted as conductor for a performance of Verdi's *Aida*, which he conducted entirely from memory, with great success. After conducting works, including the premieres of *I pagliacci* (1892) and *La Bohème* (1896), at various Italian opera houses, he became musical director of La Scala, Milan (1898–1903; 1906–08). Specializing in the music of Verdi and Puccini, he became conductor at the Metropolitan Opera, New York (1908–21), before returning to La Scala (1921–29), where he conducted the first performance of Puccini's *Turandot* (1926) and introduced the German operas of Wagner and Richard STRAUSS to the Italian public. He subsequently became conductor of the New York Philharmonic Orchestra (1929–36) and finally of the National Broadcasting Company Orchestra (1937–53).

An uncompromising perfectionist, Toscanini discouraged short applause when detrimental to the music and demanded total commitment from his performers. When conducting he relied on his phenomenal memory to overcome increasing shortsightedness.

Tournier, Michel (1924–) *French novelist. His novels blend traditional storytelling with elements of fantasy and eroticism.*

Born in Paris, Tournier enjoyed a successful and varied career as a broadcaster, translator, and publisher before turning to writing. His first novel, *Vendredi ou les Limbes du Pacifique* (1967; translated as *Friday and Robinson: Life on Speranza Island*, 1972), based on the classic story of Robinson Crusoe, was awarded the Grand Prix du Roman of the Académie Française. *Le Roi des Aulnes* (1970; translated as *The Erl King*, 1972), in which Tournier explores the myths of St Christopher and the Erl King, won for its author the Prix Goncourt. His other works include *Les Météores* (1975; translated as *Gemini*, 1981), *Gaspard, Melchior et Balthazar* (1980; translated as *The Four Wise Men*, 1982), an extended version of the story of the Magi, *Gilles et Jeanne* (1983); and *La Goutte d' Or* (1986).

Toynbee, Arnold Joseph (1889–1975) *British historian. He was appointed a CH in 1956.*

Toynbee was educated at Winchester and at Balliol College, Oxford, having won scholarships to both. After teaching at Balliol (1912–15), he entered government service for the duration of the war and in 1919 attended the Paris Peace Conference as a member of the British delegation. He resumed academic life as Koraes Professor of Byzantine and Modern Greek Language, Literature, and History at London University (1919–24) and from 1925 until his retirement (1955) held the posts of research professor of international history there and director of studies in the Royal Institute of International Affairs. During World War II he directed the Foreign Office Research Department (1943–46) and at its conclusion was a member of the British delegation at the peace conference in Paris.

Throughout his long and distinguished career Toynbee wrote with authority on a wide range of subjects that included politics, ancient history, travel, and religion. His main work, however, was *A Study of History*, published in twelve volumes between 1934 and 1961, in which he surveyed the rise and fall of many civilizations and evolved his own grand philosophy of the progress of human history. Early in his career he established himself as an expert on Middle Eastern affairs, producing volumes on Greece and Turkey in the 1920s, and he edited the *Survey of International Affairs* covering the years 1920 to 1946. An enthusiastic traveller, he wrote several travel books, including *Between Oxus and Jumna* (1961). He chose *The World and the West* (1953) as his subject for the 1953 Reith Lectures and also delivered the Gifford Lectures (1953–54) on *An Historian's Approach to Religion* (1956). *Comparing Notes: a Dialogue Across a Generation* (1963) was written with his son, the journalist and novelist Philip Toynbee (1916–81).

Tracy, Spencer (1900–67) *US film star who had the distinction of being awarded Oscars two years in succession.*

Born in Milwaukee, Wisconsin, Spencer Tracy turned to acting while at Ripon College, enrolling at the American Academy of Dramatic Arts, New York, in 1922. He made his debut as a robot in Karel ČAPEK's *RUR* in the same year and continued in the theatre until 1930, when his performance as a killer in *The Last Mile* led to a screen test. His rough features and stocky build caused him to be cast as a tough guy in such films as *Twenty Thousand Years in Sing Sing* (1933). Later roles, however, such as the priest in *San Francisco*, with Jeanette MacDonald (1902–65) and Clark GABLE, and the innocent victim in *Fury* (both 1936), exploited his talent more fully, as did *Northwest Passage* and *Edison the Man* (both 1940). Humour, too, became an important part of his repertoire, as in *Woman of the Year* (1942) and *Adam's Rib* (1949), two of the many films he made with his close friend Katherine HEPBURN. The role of the firm but understanding father figure became a feature in later life.

Tracy's two Oscars were awarded for *Captains Courageous* (1937) and *Boys' Town* (1938). He also received no fewer than seven other Oscar nominations: for *San Francisco* (1936), *Father of the Bride* (1950), *Bad Day at Black Rock* (1955), *The Old Man and the Sea* (1958), *Inherit the Wind* (1960), *Judgment at Nuremberg* (1961), and his last film, *Guess Who's Coming to Dinner* (1967).

Trenchard, Hugh Montague, 1st Viscount (1873–1956) *British air marshal who, as chief of staff between the wars, built up the Royal Air Force into the third major element in Britain's armed services. He was created a baronet in 1919, a baron in 1929, and a viscount in 1936. He was awarded the OM in 1951.*

A lawyer's son from the west of England, Trenchard passed his militia entrance examination at the third attempt in 1893, having failed to enter the navy and the Royal Military Academy. He was posted to India as a second lieutenant in the Royal Scots Fusiliers and later served as a captain in the Boer War, during which he was badly wounded. After a seven-year spell in the Southern Nigeria Regiment, he returned home in 1910 due to illness, later rejoining his old regiment in Ireland.

In 1912, on the prompting of an aviator friend, Trenchard took private flying lessons and quickly obtained his pilot's certificate. He was transferred to the newly formed Royal Flying Corps in the rank of assistant commandant and on the outbreak of World War I found himself in command of 1 Wing attached to the 1st Army Corps on the western front. Soon he was head of all RFC units in France. Trenchard built up the size of his units in a constant battle for air supremacy with the Germans. His pilots also gave support to ground operations and by the end of 1917 were flying bombing raids into Germany. With the creation of the Air Ministry, Trenchard was appointed its first chief of staff in January 1918 but resigned in April over differences with the minister, Lord Rothermere (1868–1940). He returned to France to command the Inter Allied Independent Air Force under the auspices of the Allied generalissimo, FOCH, and directed strategic bombing of Germany.

After the war, Trenchard was appointed air chief of staff of the Royal Air Force, formed in April 1918. He more or less built the service from scratch, fiercely defending its independence against the predatory designs of its larger sister services. With his large build and powerful voice, Trenchard proved a formidable champion of the RAF, doubtless justifying his nickname, 'Boom' Trenchard, in many a committee. He was appointed first marshal of the RAF in 1927. After a brief 'retirement', Trenchard became Metropolitan Police commissioner in 1931, presiding over the establishment of Hendon Police College and forensic laboratories.

Trevelyan, George Macaulay (1876–1962) *British historian. He was appointed CBE in 1920 in recognition of his war work and awarded the OM in 1930.*

Trevelyan was the third son of the statesman and historian Sir George Otto Trevelyan (1838–1928), whose biography he wrote in 1932. He was educated at Harrow School and Trinity College, Cambridge, where he obtained a first-class degree (1896). His dissertation on *England in the Age of Wycliffe* (1899) won him a fellowship at Trinity, where he taught until moving to London in 1903. His next book, *England Under the Stuarts* (1904), confirmed his ability to combine scholarship with popular appeal, but it was with his trilogy on Garibaldi (1907–11) that he really established his reputation as a historian. During World War I he commanded an ambulance unit of the British Red Cross on the Italian front. After the war Trevelyan consolidated his success as a popular historian with *British History in the Nineteenth Century* (1922) and *History of England* (1926). In 1927 he became Regius Professor of Modern History at Cambridge, and his major work in the next decade was a three-volume history of *England under Queen Anne* (1930–34). He also wrote a notable biography of *Grey of Fallodon* (1937). From 1940 to 1951 he was master of Trinity College, Cambridge, a post in which he gained much respect and affection. Written during the war, his *English Social History* (1944) became the most widely read of all his books.

Besides his vocation as a historian, Trevelyan had many other interests. His 1954 Clark Lectures, *A Layman's Love of Letters*, summed up his deep love and knowledge of English literature. He was an influential conservationist and supporter of the National Trust and in 1931 became president of the newly founded Youth Hostels Association. In 1949 he published his *Autobiography and Other Essays*.

Trotsky, Leon (Lev Bronstein; 1879–1940) *Russian revolutionary leader who, from the October Revolution to the death of LENIN, was the second most powerful man in the Soviet Union. He lost ground to STALIN after Lenin's death and was assassinated in exile.*

Born in Yelisavetgrad (now Kirovograd), the son of a farmer, Trotsky was educated at a local Jewish school before attending schools in Odessa and Nikolaev, from which he graduated in 1897. Active in the South Russian Workers Union, he was exiled to Siberia but escaped, joining Lenin in London, where he began to write for *Iskra*. He first supported the Menshevik faction of the Russian Social Democratic Workers' Party against the Bolsheviks led by Lenin; during the 1905 revolution he was elected as speaker for the Mensheviks in the St Petersburg Soviet. He was exiled to Siberia a second time in 1907 but escaped to Vienna, where he worked as the editor of the newspaper *Pravda*.

Trotsky abandoned the Mensheviks to rejoin Lenin and the Bolsheviks at the Sixth Party Congress in July 1917; with Lenin he helped to organize the October Revolution and the overthrow of the provisional government. When the Bolsheviks came to power he was named commissar for foreign affairs in the new Soviet government and led the negotiations with the Germans at the Brest-Litovsk Peace Conference. During the civil war, in the capacity of minister for war, he built up the Red Army and directed the campaign against the White Army. His rapid advancement and his views on the nature of the revolutionary process (he believed that socialism inside Russia was not possible until revolution had occurred in western Europe) created bitter enemies. When Lenin died in 1924, Trotsky attempted to form an alliance with Lev Kamenev (1883–1936) and ZINOVIEV but was defeated in the struggle for power by Stalin. He was expelled from the party in 1927 and, exiled to central Asia in 1928, was forced to leave the Soviet Union. He eventually settled in Mexico (1936), where he was assassinated in 1940, probably by Soviet agents.

In exile Trotsky continued to write and speak out against Stalin; his best-known publications include *My Life* (1930), *The History of the Russian Revolution* (3 vols, 1932–33), and *The Revolution Betrayed* (1937).

Trudeau, Pierre Elliott (1919–   ) *Canadian statesman and Liberal prime minister (1968–79; 1980–84). He became a CH in 1984.*

Born into a wealthy French-Canadian family, Trudeau studied law at Montreal University and was called to the bar in 1943. He then went to Harvard, the London School of Economics, and Paris University to broaden his knowledge of economics and political science. After travelling around the world, he became a lawyer in 1949, working for the Privy Council and later specializing in civil rights and labour disputes. In 1950 he helped found *Cité Libre* ('Free City'), a magazine for liberal protest against the repressive Duplessis administration in Quebec province. After serving as associate professor of law at Montreal University (1961–65), Trudeau joined the Canadian Liberal Party, becoming parliamentary secretary and (in 1967) justice minister to prime minister Lester PEARSON. Following Pearson's resignation in 1968, the youthful and charismatic Trudeau was appointed his successor on a wave of popular support and won a decisive election victory.

A committed federalist, Trudeau constantly had to face the strong separatist movement in French-speaking Quebec. In October 1970, members of the militant Quebec Liberation Front kidnapped a British trade official and later abducted and murdered Quebec's labour minister. Trudeau invoked the War Measures Act to suspend civil rights in Quebec and round up over 450 suspects. The consequent widespread resentment was a setback for Trudeau's aims. However, he did make both French and English official languages throughout the Canadian administration. His popularity was eroded by increasing economic difficulties and in 1979 the Liberals fell to a minority Conservative government. Returned to power in 1980, Trudeau held a provincial referendum in Quebec, which rejected independence, and presided over the transfer of remaining constitutional powers from Britain to Canada. He resigned as Liberal leader in 1984.

Trueman, Freddie (Frederick Sewards Trueman; 1931–) *Yorkshire and England cricketer who took 2304 first-class wickets; he was the first bowler in the history of the game to take 300 wickets in test matches. He played 67 times for his country and was one of the outstanding postwar 'characters' in the game. He received an OBE in 1989.*

Born in Stainton, Yorkshire, he made his debut for his county in 1949 and two years later won his cap; he first played for England in 1952. Known affectionately as 'Fiery Fred', because of his early bowling style and blunt manner, he took his 300th test wicket at the Oval in 1964. He could also be a useful, if unpredictable, batsman, scoring more than 9000 runs (including three centuries). His impetuous behaviour occasionally led to trouble with the authorities; after his retirement he became an establishment figure as a test-match commentator and sports writer.

Truffaut, François (1932–84) *French film director and producer, who was a leading force in the 'New Wave' of film directors.*

Born in Paris, Truffaut had an unhappy childhood and spent some time in a reformatory. Later his military service ended in desertion and prison. As a film critic with the journal *Cahiers du Cinéma*, however, he soon established a reputation for his attacks on conventional film-making. For a time he worked with Roberto ROSSELLINI, before setting up his own company. With his first feature-length film, the quasi-autobiographical *Les Quatre Cents Coups* (1959), he earned a prize at the Cannes

Festival. This was the first of a series of 'New Wave' films written around the character of Truffaut's alter ego, Antoine Doinel, played by Jean-Pierre Léaud (1944–). Among the many award-winning films that followed were *Jules et Jim* (1961), *Baisers volés* (1968), *L'Enfant sauvage* (1970), and *La Nuit américaine* (1973; *Day for Night*), which won an Academy Award for best foreign film.

Two directors particularly influenced him: Jean RENOIR, after whose *Le Carrosse d'or* he named his company Films du Carrosse, and Alfred HITCHCOCK, whose conversations with Truffaut were published in the latter's *Hitchcock* (1967).

Trujillo (Molina), Rafael Leónidas (1891–1961) *Dictator of the Dominican Republic (1930–61). Called Generalissimo, he ruled the country for thirty-one years with the aid of a strong and ruthless police force.*

Born in San Cristóbal into a poor family, Trujillo enlisted in the army in 1918. He was trained by US marines, who occupied the Republic (1916–24), and in 1921 he graduated from the Haira Military Academy. Rising rapidly through the ranks, he became a general in 1927 and chief of staff in 1928.

Trujillo became president in 1930. Although he was formally president for only two periods (1930–38 and 1942–52), he maintained absolute power from 1930 until his death, either directly or indirectly through puppets. As a dictator he improved the social services and increased material benefits for his people, but he crushed all opposition, allowed only one political party to exist, and extended his command of the army through nepotism. Towards the end of his regime, however, he faced increasing opposition from the middle classes, the Roman Catholic Church, and neighbouring states (particularly Cuba). He was killed in 1961 by an unknown assassin near Trujillo City.

Truman, Harry S. (1884–1972) *US Democratic statesman and thirty-third president of the USA (1945–52). He was responsible for the dropping of the atom bomb on Japan and for the involvement of the USA in the Korean war.*

Born into a farming family in Missouri, Truman was awarded a place at West Point military academy but was unable to take it up because of bad eyesight. After World War I, during which he served in France as an artillery captain, he started a men's clothing store in Kansas City, but the venture failed and Truman was made bankrupt. At the age of thirty-eight Truman entered the political world through the influence of Thomas J. Pendergast, the boss of Democratic politics in Kansas, who arranged for his election as a county judge. He took up the study of law and showed such honesty and ability in his office that in 1934 Pendergast chose him as candidate for the Senate, where he proved a reliable supporter of ROOSEVELT's New Deal. When re-elected in 1940 he headed a committee to investigate waste in military spending and succeeded in cutting a thousand million dollars from the US defence budget. The popularity of the Truman Committee helped secure his successful nomination for the vice-presidency in 1944, and on the death of Roosevelt in April 1945, Truman succeeded to the presidency.

Foreign affairs dominated the first few weeks of Truman's term of office. While attending the Potsdam conference (July 1945) he authorized the use of the atom bomb against Hiroshima and Nagasaki to end the war with Japan. After the war, deteriorating relations with the Soviet Union led to the start of the 'Cold War'; Truman's policy of containment towards the Soviets, characterized by the 'Truman Doctrine' to assist countries resisting communism and evidenced in aid to Greece and Turkey, was criticized by those who preferred liberation. However, the Berlin airlift and the establishment of the CIA in 1947 restored lost confidence. In 1947 Truman's administration introduced the MARSHALL Plan of emergency aid to war-shattered European countries and in 1949 helped to set up NATO. Truman pioneered the Point 4 programme to give technical aid to underdeveloped countries, and at the Bonn and Paris conventions advocated the inclusion of West Germany in the European Defence Community.

Although he did much to stabilize postwar Europe, Truman's policies failed in the Far East. He negotiated a peace settlement with Japan but was unable to end the Chinese civil war. In order to back up his professed commitment to collective security, he was forced to involve the US army in the Korean War. His courageous dismissal of General MACARTHUR from the Far East command in 1951, however, underlined the power of civil over military authority. Truman was nevertheless brought down by his inability to bring the same firm hand to domestic affairs. His victory over T. E. Dewey in the 1948 election was a personal triumph, but during both his terms Congress blocked much of the legislation to implement his 'Fair Deal' policy of social reform (based on and extending Roosevelt's New Deal). His second term was further overshadowed by the

communist victory in China, allegations of communist subversion at home culminating in Senator Joseph MCCARTHY's anticommunist purge, and corruption and incompetence in his administration. Truman did not run for re-election in 1953, and the Democratic presidential candidate, Adlai STEVENSON, was defeated by the Republican EISENHOWER.

Trumper, Victor Thomas (1877–1915) *Australian cricketer and one of the great batsmen of the game's golden age. He died at the age of thirty-seven, having played in 48 test matches (1899–1912).*

Born in Sydney, Trumper played for New South Wales until 1913–14, captaining the side from 1907–08. As a member of the Australian team touring England in 1902, he scored 2570 runs including 11 centuries – one, at Old Trafford, before lunch. His all too brief career – before his fatal illness – was punctuated by unprecedented feats of batsmanship and as late as 1913, in New Zealand, he made 293 in three hours as part of an eighth-wicket stand of 433. Trumper was also known for his unselfish attitude to the game – on several occasions he sacrificed his wicket to ensure that his team-mates were given 'some batting practice'.

Tshombe, Moise (1919–69) *Prime minister of the Belgian Congo (1964–65) and leader of the Katanga secession movement (1960–63).*

Born in Musumba, Katanga (now Shaba) province, the son of a wealthy businessman and related by blood to the royal Lunda family of Mwatiamvu, Tshombe was educated at an American mission school before entering the family business owning a chain of stores and a hotel. Popular with his fellow Lunda tribesmen, he became president of the Confederation of Mutual Associations of the Lunda Empire in 1956. Tshombe began his political career in 1951 when he was elected to the Katanga Provincial Council. Becoming president in 1959 of the Confédération des Associations du Katanga (Conakat), a political movement supported by the Lunda Association, allied tribal bodies, and the Belgian mining company Union Minière du Haut Katanga (which controlled the province's rich copper mines), he attended the independence conference in Brussels in 1960. Declaring Katanga an independent republic eleven days after the Congo achieved independence, Tshombe was elected head of state by the Katanga provincial assembly, a position he maintained for two and a half years despite lack of diplomatic recognition by any other country.

Forced to surrender in 1963 following military intervention by the UN, Tshombe went into exile in Spain. Recalled by President Kasavubu (*c.* 1917–69) the next year to form a government, Tshombe became prime minister of the Congo but was dismissed in October 1965 following political opposition. Obliged to stay out of Congolese politics after the coup led by Sese Seko Mobutu (1930–   ) in November 1965, Tshombe returned to Spain. In 1966 he was alleged to have been behind a mutiny of Katanganese units in Stanleyville and was charged with high treason. Kidnapped and taken to Algeria in 1967, he remained there under house arrest until his death from a heart attack in 1969.

Tsubouchi Shōyō (Tsubouchi Yūzō; 1859–1935) *Japanese writer, scholar, and translator.*
Tsubouchi was born at Ōt, near Nagoya, into a samurai family. He graduated from Tokyo Imperial University in 1883 and the same year began teaching at the institution that was later to become Waseda University, where he was to remain for nearly fifty years. He soon won fame both as a novelist and as a translator of the English classics into Japanese. *Tōsei shoseikatagi* (1885–86), about the follies of contemporary university students, is probably his best-known novel; in it he puts into practice the principles of realism that he urged in his influential critical work *Shōsetsu shinzui* (1885). In 1891 he became the founder-editor of the literary journal *Waseda bungaku*.
Influenced by modern European trends in the theatre, Tsubouchi began writing plays in the 1890s. The best of these are *Kiri hitoha* (1896), *Shinkyoku Urashima* (1904), and *En no gyōja* (1916). He was also the moving spirit behind the shingeki (new theatre) movement, which revitalized Japanese drama. From 1907 to 1928 he was engaged upon the mammoth task of a complete Japanese translation of the works of Shakespeare.

Tubman, William Vacanarat Shadrach (1895–1971) *President of Liberia (1944–71). Tubman presided over this tiny African country for longer than anyone else. During his presidency valuable deposits of high-grade iron ore were discovered and the country flourished.*
Born in Liberia's Maryland county, the son of a minister and former speaker of the Liberian house of representatives, Tubman was educated at a Methodist seminary. After reading law, he was appointed county attorney in Maryland in 1919. Tubman first entered politics when he was elected to the national legislature in 1923. After losing his seat in 1927, he became a Methodist lay preacher but returned to politics on his re-election to the senate in 1929. Following a scandal involving forced labour, he resigned in 1930 and failed to regain his seat in 1934. In 1937 he was appointed an associate justice of the Supreme Court of Liberia and was chosen by the outgoing President Barclay to stand as the Whig Party presidential candidate in 1943. Winning office in 1944, Tubman remained in power, despite a number of assassination attempts, until his death in 1971.
Tubman succeeded in reducing antagonism between the Americo-Liberian people and the indigenous Africans, introducing universal adult suffrage and representation of all people in the national legislature. A supporter of the African independence movements of the 1950s and 1960s, he was the sponsor and organizer of the Monrovia Conference of African Heads of State, held in 1961 to promote unity and cooperation in Africa. He also attended the first conference of the Organization for African Unity in Addis Ababa in 1963.

Tupolev, Andrei Nikolaievich (1888–1972) *Soviet aeronautical engineer, who was responsible for the world's first supersonic airliner. He was twice awarded the Order of Lenin and was elected an honorary fellow of the Royal Aeronautical Society.*
The son of a lawyer exiled for revolutionary activities, Tupolev was educated at the local gymnasium and Moscow Technical High School. Arrested for political activities in 1911, he was expelled from technical college, although he was allowed to return in 1914 to complete his studies. Tupolev spent the war working in Duks aircraft factory. After the 1917 revolution, he joined the Central Aerodynamical Institute in Moscow, being promoted in 1921 to the directorship of the aircraft design department.
Although he produced many safe and conventional piston-engined planes, Tupolev only became known outside the Soviet Union for his turbojets. He produced the first Soviet jet bomber, the TU-12, in 1946 and went on to design the TU-104, the first jet passenger plane for Aeroflot. The TU-114, a forerunner of the Jumbo jets, was introduced into service in 1961. His most spectacular triumph, however, was the TU-144, the supersonic plane designed by Tupolev and his son Alexei. It flew first on 31 December 1968, some two months

before the maiden flight of its rival, the Anglo-French Concorde.

Turing, Alan Mathison (1912–54) *British mathematician, who introduced the concept of the Turing machine.*

Born in London, he was educated at King's College, Cambridge, where he was elected to a fellowship in 1936. In the following year he published *On Computable Numbers*, a work that quickly gained him a worldwide reputation and introduced into mathematics the notion of a Turing machine. Turing's work derived from the 23rd problem posed (in 1900) by David HILBERT, i e how to decide whether the propositions of predicate logic are true or false – the *Entscheidungsproblem*. Turing described a universal machine capable of modelling the process of computation. It would consist of no more than a continuous tape divided into cells. The machine would be capable of moving the tape to the left and to the right, to halt it, to print the numbers 0 and 1, and to erase. Turing was able to show that there are noncomputable functions. He was further able to link this result with first-order logic and so demonstrate that predicate logic was essentially undecidable.

With the outbreak of World War II Turing found himself at Bletchley Park, where he worked on deciphering the German *Enigma* codes. Although much remains to be written of this period it appears that Turing's work on the project was vital and that few individuals, of whatever rank, made a greater contribution to the final Allied victory. It was also work that demonstrated the importance of computers in solving otherwise intractable problems. Turing, in no sense an abstract mathematician, was aware of both the theoretical potentialities and the engineering constraints of a computing machine and consequently sought ways to develop and promote his concept of the modern computer. After the war (1945) he therefore joined the staff of the National Physical Laboratory at Teddington to work on the development of ACE (Automatic Computing Engine). However, without the discipline imposed by war Turing found it impossible to submit to the bureaucratic procedures of the civil service and in 1948 moved to Manchester University to work on MADAM (Manchester Automatic Digital Machine). While there he published his widely read paper *Computing Machinery and Intelligence* (1950), in which he invited his readers, unsuccessfully he predicted, to propose ways to distinguish between computers and intelligent minds.

With sufficient backing Turing could well have helped establish a soundly based computer industry in Britain in the 1950s. As it turned out, he was charged with, by the standards of the day, a minor homosexual offence, which led to his suicide. He died by sucking on an orange he had previously injected with cyanide.

Tutu, Desmond Mpilo (1931–) *Black South African Anglican cleric, internationally known for his stand against apartheid. He received the Nobel Peace Prize in 1984.*

The son of a teacher, Tutu was educated at mission schools and training college before entering his father's profession in 1954. Following his ordination in 1960, he studied theology at King's College, London, and lectured on the subject in South Africa. In 1972 he became associate director of the theological education fund of the World Council of Churches. He was appointed bishop of Lesotho in 1977 and general secretary of the South African Council of Churches the following year.

It was in the late 1970s that Tutu, previously regarded as a somewhat conservative figure, emerged as a leading voice in the struggle against apartheid. His calls for economic sanctions against South Africa and his emphasis on nonviolent routes to change gained him an international reputation and ultimately the Nobel Peace Prize (1984). In 1985 he was appointed Anglican bishop of Johannesburg, becoming archbishop of Cape Town, head of the Anglican church in South Africa, a year later: in both cases he was the first black to hold the position. He has published several collections of sermons, lectures, and speeches, including *Crying in the Wilderness* (1982) and *Hope and Suffering* (1983).

Tutuola, Amos (1920–) *Nigerian writer.*

Tutuola was born at Abeokuta, western Nigeria, and received a primary education at mission schools. He then worked on his father's farm before training as a coppersmith. He later became a government messenger in Lagos but in 1945 took up the post of storekeeper with the Nigerian Broadcasting Corporation in Ibadan. His first and probably best book was *The Palm-Wine Drinkard* (1952), a fantastic blend of realism and Yoruba fairy-tale written in a highly personal idiom. His subsequent publications include *My Life in the Bush of Ghosts* (1954), *Simbi and the Satyr of the Jungle* (1955), *The Brave African Huntress* (1958), *The Feather Woman of the Jungle* (1962), *Ajaiyi and his Inherited Poverty*

(1967), *The Witch-Herbalist of the Remote Town* (1981), and *Pauper, Brawler and Slanderer* (1987).

Tyson, Mike (1966–   ) *US boxer and world heavyweight champion (1986–90).*
Born into a deprived area of New York City, Tyson became the youngest-ever world heavyweight champion in 1986, when he decisively defeated Trevor Berbick in a WBC fight at Las Vegas. Subsequently he employed his ferocious hitting power to take the rival WBA title from James 'Bonecrusher' Smith and the IBF title from Tony Tucker, in 1987. These victories over formidable opponents prompted boxing experts to claim that he was one of the most impressive heavyweight boxers of all time, with a long reign as champion ahead of him.

He continued to demolish all challengers to his titles with comparatively little trouble until 1990, when he was unexpectedly knocked out in a fight in Tokyo against the US boxer James 'Buster' Douglas. The result was highly controversial, as Douglas had – according to Tyson's camp – already been floored for 12 seconds in the eighth round. The decision went in Douglas's favour, however, and he became the new champion in one of the biggest upsets in boxing history. Nonetheless, Tyson – though troubled by press comment on his private life (including an allegation that he raped a contestant in a beauty competition) and stunned by his defeat – remained in most commentators' eyes the man to beat.

U

Ulanova, Galina Sergeyevna (1910–)
Soviet prima ballerina who has embodied the spirit of Soviet ballet in the twentieth century.

Ulanova was trained at the Leningrad State School of Choreography by her mother, the dancer Maria Romanova, and then by Agrippina Vaganova. After graduation she joined the Kirov Ballet in 1928 and danced the *Blue Bird* pas de deux, *The Sleeping Beauty*, and *Swan Lake*. However, Ulanova's first major breakthrough came as Maria in Zakharov's *The Fountain of Bakhchisarai* (1934), in which her performance was highly acclaimed. There followed a series of ballets in which she developed a style of profound purity, simplicity, and flowing movement that added great emotional effect to her interpretation of the classics. All three of Prokofiev's ballets – *Romeo and Juliet, Cinderella,* and *The Stone Flower* – were composed with Ulanova in mind and she danced them both for the Kirov Ballet and for the Bolshoi Ballet, to which she transferred in 1944.

Ulanova was first seen in the West in 1951 when she performed in Florence; she was subsequently greeted with immense enthusiasm in London and wherever else she toured with the company. She performed less after 1959 and officially retired in 1962, although she continues to coach younger dancers.

Ulbricht, Walter (1893–1973) *East German statesman, first general secretary of the East German Socialist Unity (Communist) Party (1946–71), and chairman of the East German Council of State (1960–71).*

Born in Leipzig, the son of a tailor, Ulbricht was apprenticed as a carpenter. At the age of thirteen he joined the socialist youth movement, becoming a member of the leftist Woodworkers Union in 1910. In 1912 he entered the Socialist Party, joining the Spartacus League, a radical group advocating revolution, in 1918. In 1919 he co-founded the German Communist Party. He spent the next few years in the Soviet Union learning about the organization of the Soviet Communist Party.

Ulbricht was elected to the Reichstag in 1928 as the Communist member for South Westphalia. When HITLER came to power in 1933 he was forced to leave the country and sought refuge in the Soviet Union. Between 1936 and 1938 he was in charge of a republican army unit in Spain, entrusted by Moscow with the elimination of all anti-Stalinists. He returned to Germany in 1945 and founded the Socialist Unity Party the following year. As general secretary of the party (a position he held for twenty-five years) he proclaimed the Democratic German Republic in 1949, adopting a policy of 'socialization' in eastern Germany. In 1960 he became chairman of the Council of State and head of the armed forces, thereby combining leadership of the government and the party. The next year, to stem the flood of refugees fleeing to the West, he erected the Berlin Wall, which divided the city into separate East and West entities. He retired from office in 1971.

Updike, John Hoyer (1932–) *US novelist and poet. His witty and elegant style camouflages but does not conceal the serious things he has to say.*

Updike was born in Shillington, Pennsylvania, and after graduating from Harvard spent a year at Oxford University. He became a reporter for the *New Yorker* in 1955, when he was provided with an outlet for his early writing – including nonfiction pieces and poetry as well as short stories. His first published book was a collection of verse, *The Carpentered Hen and Other Tame Creatures* (1958), followed a year later by his first novel, *The Poorhouse Fair*. Possibly his most famous book, however, is *Rabbit, Run* (1960), with its sequels *Rabbit Redux* (1971), *Rabbit is Rich* (1981), which won a Pulitzer Prize (1982), and *Rabbit at Rest* (1990), which won a second Pulitzer Prize (1991). His other novels, including *The Centaur* (1963), *Of the Farm* (1965), *Couples* (1969), *The Witches of Eastwick* (1984; filmed 1987), and *S.* (1988), reveal a deep insight and a technical skill that matches that of any of his contemporaries. Other recent publications include the poems of *Facing Nature* (1985), the short-story collection *Trust Me* (1987), and the autobiography *Self-Consciousness: Memoirs* (1989).

Urey, Harold Clayton (1893–1981) *US chemist whose discovery of deuterium earned him the 1934 Nobel Prize for Chemistry.*

The son of a teacher and lay minister, Urey was educated at the University of Montana, where he studied zoology. His interest in chemistry was aroused during World War I, when he was working on the manufacture of war materials. After the war he spent a year studying under Niels BOHR in Copenhagen before beginning his academic career as an associate in chemistry at Johns Hopkins University in 1924. Thereafter he held a series of appointments at Columbia (1929–45), Chicago (1945–52), and La Jolla, California (1958–70).

Urey established his reputation in 1931 with his isolation of deuterium by fractional distillation. He continued to work on isotope separation and in 1937 was able to isolate nitrogen-15. This work was of considerable practical importance when it became necessary to separate large quantities of the rare isotope uranium-235 for use in the fission bomb and fission reactor. A variety of separation methods were investigated for this purpose. Urey was made responsible in 1942 for the gaseous diffusion approach, which turned out to be by far the most successful.

In 1951, with Stanley Miller (1930–), he performed a classic series of experiments to demonstrate that an electric discharge in a sealed container of methane, ammonia, and water could yield a number of amino acids, the basic building blocks of proteins. The implications of their work on the origin of life are still being investigated.

Ustinov, Sir Peter Alexander (1921–) *British actor, playwright, and director. He was made a CBE in 1975 and received a knighthood in 1990.*

Of Russian and French extraction, Ustinov was born in London, educated at Westminster School, and trained under Michel Saint-Denis (1897–1971) at the London Theatre School. He made his first professional appearance as Waffles in *The Wood Demon* (1938) and his London debut at the Players' Theatre in his own sketches in *Late Joys* (1939). Although he has performed in a variety of plays, for much of his career Ustinov has appeared in or directed his own, beginning with *Fishing for Shadows* (1940), adapted from Sarment's *Le Pêcheur d'ombres*. His other plays in which he acted include *The Love of Four Colonels* (1951), playing Carabosse, and *Romanoff and Juliet* (1957), as the General (he also appeared in, directed, and produced the film version of

this in 1961). Later plays include *Photo Finish* (1962), *Who's Who in Hell* (1974), and *Beethoven's Tenth* (1983). Lear at the Stratford Festival, Ontario (1979), was his first Shakespearean role. Since his screen debut in 1940 Ustinov has made many films, including *Quo Vadis* (1951), *Beau Brummell* (1954), *Spartacus* (1960) and *Topkapi* (1964), both of which brought Academy Awards, *Death on the Nile* (1978), *Evil under the Sun* (1981), and *Appointment with Death* (1988).

As a man of many talents, Ustinov is well known as an impersonator, raconteur, and broadcaster and has written novels, short stories, and reflections upon life in the Soviet Union. *Dear Me*, his autobiography, was published in 1977. He is also respected for his work on behalf of UNICEF.

Utrillo, Maurice (1883–1955) *French painter of townscapes.*

The natural son of the painter Suzanne Valadon (1867–1938), he was adopted by the Spanish architect and writer Miguel Utrillo in 1891. While still a boy he underwent the first of numerous treatments for alcoholism and it was on his release from hospital that his mother encouraged him to take up painting as a form of therapy.

His early paintings in thick rough textures and sombre colours depict Parisian street scenes, particularly of the Montmartre and Montmagny districts, as in *Les Toits à Montmagny* (1906–07), and the years 1904–08 have been called his 'Montmagny period'. Between 1909 and 1914 Utrillo used white extensively in his work. His sensitive interpretations transform mundane settings, conveying an aura of strength and serenity as in *L'Impasse Cottin* (*c*. 1910). It is during this 'white period' that Utrillo is considered to have reached the peak of his achievement. There followed a period extending into the 1930s in which heightened colour and increased stress on line was combined with the use of figures in such paintings as *Boulevard in the Suburbs of Paris* (1924).

In 1935 Utrillo married Lucie Pauwels, who introduced a greater degree of sobriety and order into his life, and they went to live in Le Vesinet, where he remained until his death. His later works (1930–55) were inconsistent in quality, frequently with poor drawing and colouring, and they are generally regarded as inferior to his previous work.

V

Valentino, Rudolph (Rodolpho Gugliemi di Valentina d'Antonguolla; 1895–1926) *Italian-born US actor who, in the persona of the Latin lover, became a legend of the silent screen.*

Born in Castellaneta and arriving in the USA in 1913, Valentino failed selection as an army officer and worked as a gardener, waiter, and exhibition dancer before presenting himself at Hollywood. His first film was *Alimony* (1918), but it was *The Four Horsemen of the Apocalypse* (1921), some sixteen films later, that took him to stardom and into the sexual fantasies of many 1920s ladies, especially those with a penchant for tangoing Italian waiters. Following this success he secured his position with such films as *The Sheikh* (1921), *Blood and Sand* (1922), *Monsieur Beaucaire* (1924), and *The Son of the Sheikh* (1926), his last film before his sudden death from peritonitis.

Thousands of women flocked to his funeral and there were reports of mass hysteria and several suicides. Had he lived longer, he would have had to prove himself in the talkies, a test that he may well have failed. Since his death many books have been written about him and films made of his life, including Ken RUSSELL's *Valentino* (1977) with Rudolf NUREYEV in the title role.

Valéry, Paul(-Ambroise) (1871–1945) *French poet, essayist, and critic. He was elected to the Académie Française in 1925.*

Valéry was born in the small Mediterranean town of Sète, the son of a customs officer. He studied law at the University of Montpellier and in the early 1890s made the acquaintance of the writers André GIDE and Pierre Louÿs (1870–1925), who published Valéry's first poems in his symbolist review *La Conque* and introduced him to the poet Stéphane Mallarmé (1842–94). In 1892, however, Valéry suddenly decided to abandon poetry and devoted himself for some twenty years to more intellectual writings, notably *Introduction à la méthode de Léonard de Vinci* (1895); abstract speculation, embodied in *La Soirée avec Monsieur Teste* (1896; translated as *An Evening with Monsieur Teste*, 1925); and to his careers in the War Office (1897–1900) and the Havas News Agency (1900–22).

Valéry's return to the world of poetry was marked by the publication of *La Jeune Parque* (1917), a long symbolic poem that had taken five years to write and won him instant acclaim. It was soon followed by the collections *Album de vers anciens* (1920) and *Charmes* (1922); the latter contains one of Valéry's best-known poems, 'Le Cimetière marin' (translated by C. DAY LEWIS as 'The Graveyard by the Sea', 1946), a philosophical meditation on death in the cemetery of his native Sète. It is on these three works, in which lyrical sensuousness is skilfully blended with intellectual eloquence, that Valéry's fame as a poet rests. Apart from his lectures at the Collège de France, where he became professor of poetry in 1937, he dedicated the rest of his life to prose writings.

In the Socratic dialogue *Eupalinos ou l'Architecte* (1923; translated as *Eupalinos, or the Architect*, 1932) Valéry revived his early interest in architecture. His most notable literary essays appeared in the collections *Variété* (1924–44) and *Tel quel* (1941–43), and *Regards sur le monde actuel* (1931; translated as *Reflections on the World Today*, 1948) contains his views on politics and other aspects of modern civilization. His only play, *Mon Faust*, was published in unfinished form in 1946, after his death. Throughout his life Valéry rose early in the morning to record the workings of his mind and his meditations on a variety of subjects: mathematics, religion, politics, philosophy, language; these notes were published posthumously as *Cahiers* (1958–62).

A well-known and highly revered society figure, much in demand for his erudition and conversational skills, Valéry travelled all over Europe on lecture tours, mingling with writers, scientists, and political leaders with equal ease. Widely recognized as one of the greatest literary figures of the twentieth century, he was given a state funeral on his death in the summer of 1945. His grave can be found at Sète, in the cemetery of his famous poem.

Vallejo, César Abraham (1892–1938) *Peruvian poet. Although his distinctive and original writing was shamefully ignored in his lifetime, it has since had a considerable influence on Latin American poets.*

Born in Santiago de Chuco into a large mestizo family (he claimed that his parents were both illegitimate children of a priest and an Indian woman), Vallejo studied literature at the university of Trujillo (1913–15) and later law (1915–17). Published in Lima, his first book of poems, *Los heraldos negros* (1918; translated as *The Black Heralds*, 1963), was Christian in tone and influenced by the antinaturalistic symbolist-inspired *modernismo* associated with Rubén Darío (1867–1916); it was largely unnoticed. After returning home in 1920, Vallejo was apparently unjustly accused of involvement in a political disturbance and imprisoned for two and a half months. Much of his important second book of verse, *Trilce* (1922; translated as *Ten Versions from Trilce*, 1970), was composed in prison. It was a highly experimental work, with innovative typography and coinages (the title, for example, coalesces *triste*, sad, and *dulce*, tender).

After publishing an unremarkable novella (*Fabula salvaje*, 1923), Vallejo moved to Paris in 1923. Apart from two trips to Russia (1928, 1929) and visits to Spain, he lived there in poverty with his wife Georgette until his death. He joined the Communist Party in 1931. His novel, *El Tungsteno* (1931), published in Madrid, has the defects of a socialist-realist work in which characters are not fully realized. His final poetic works, however, contain some of the best things he wrote. Both *España, aparta de mí este cáliz* (1938), a poem on the Spanish civil war, and the collection *Poemas humanos* (1932; translated as *Human Poems*, 1968) were published posthumously.

Van Allen, James Alfred (1914–) *US physicist, who discovered the Van Allen belts surrounding the earth.*

Van Allen was educated at the Iowa Wesleyan College and the University of Iowa, where he gained his PhD in 1939. He worked at the Carnegie Institute, Washington, but with the outbreak of World War II served in the US navy, becoming a lieutenant-commander. On release from the navy in 1946, Van Allen joined the High Altitude Research Group at Johns Hopkins University, where he remained until 1951. He was then Carver Professor of Physics at the University of Iowa (1951–85; subsequently emeritus professor).

After initial work on cosmic rays using captured German rockets, Van Allen was, in 1955, put in charge of the instrumentation to be carried by the planned US satellites. To coincide with the International Geophysical Year, Explorer I was launched in 1958; on board were some of Van Allen's Geiger counters. These counters failed to record any radiation at a height of about 800 kilometres and similar readings were recorded in Explorers II and III. Van Allen, suspecting that the radiation was so intense that the counters jammed, prepared special lead-shielded Geiger counters for Explorer IV. This time he found two toroidal belts (the Van Allen belts) circling the earth's equator at heights of 1000–5000 km and 15 000–25 000 km respectively. They have been shown to consist of charged particles trapped in the earth's magnetic field. Van Allen has continued to explore their shape, dimensions, and properties.

Van der Post, Sir Laurens Jan (1906–) *South African explorer, conservationist, and writer. He was knighted in 1981.*

Van der Post was born of Afrikaner stock at Philippolis, Orange Free State, where he himself later farmed. The profound sympathy for Africa and its peoples that imbues his writings developed from his early youth on the veld. As a young man he also made a voyage aboard a Japanese whaling vessel, which marked the start of his enduring interest in Japan. In World War II he showed exceptional courage and resourcefulness on active service with the British army in Ethiopia, North Africa, and the Far East (1940–43). While commanding a guerrilla unit in Java he was captured by the Japanese; his experiences as a prisoner of war (1943–45) formed the basis for the film *Merry Christmas, Mr Lawrence* (1983). After liberation he remained with the British in Batavia until 1947.

In 1949 the British government sent him to explore some remote areas of Nyasaland (now Malawi). *Venture to the Interior* (1952), his book about his journey, was hailed as a masterpiece; in it, besides brilliant descriptions of African terrain and fauna, he first propounded his personal philosophy regarding the necessary balance between 'unconscious, feminine' Africa and 'conscious, masculine' Europe. This theme is developed in *The Lost World of the Kalahari* (filmed 1956; published 1958, revised edition 1988) and *The Heart of the Hunter* (1961) and *Testament to the Bushmen* (1984), both about the Bushmen of the Kalahari desert, to which he led an expedition in 1952. *A Mantis Carol* (1975) and *Jung and the Story of Our Time* (1976) further examine Bushman legends in Jungian terms. Bushmen are also sympathetically portrayed in the heavily autobiographical novels *A Story Like the Wind* (1972) and *A Far-Off Place* (1974). Van der Post's other writings include the adventure story *Flamingo Feather* (1955) and the travel books *Journey into Russia* (1964), *A Portrait of All the Russias* (1967), and *A Portrait of Japan* (1968).

Varah, (Edward) Chad (1911–) *Church of England clergyman who founded the Samaritans, the British telephone service for befriending the suicidal and despairing.*

Born in Barton-on-Humber, of which his father was vicar, he was educated at Worksop College and Keble College, Oxford. As a priest in south London he recognized the problem of isolation that human beings can suffer, even in the midst of a family. This was brought home to him when he had to bury a young girl who had taken her own life because she believed herself to be terminally ill when she began to menstruate. An article he wrote on sexual problems for the magazine *Picture Post* also brought a large mail from people who had no-one with whom to discuss their problems. In response to this apparent need he was eventually able to set himself up as a counsellor in 1953 in the crypt of the Lord Mayor's parish church, St Stephen Walbrook, of which he is still rector. Recognizing that many of the people who responded to his initial advertisements were helped by talking to his untrained, but carefully selected, helpers, he founded the non-religious movement later dubbed 'the Samaritans' by the media. This now has 187 branches in the UK, manned by 22 000 volunteers. Chad Varah was president of Befrienders International (Samaritans Worldwide), which seeks to spread the movement abroad, from 1983 to 1986. He has travelled widely abroad, including a visit to the USSR, to spread the Samaritan principles. He was appointed a prebendary of St Paul's Cathedral in 1975 and his autobiography was published in 1992.

Varèse, Edgard (1883–1965) *French-born US composer and conductor, who pioneered the use of electronic and taped sound in composing.*

Varèse initially planned to become an engineer, but at eighteen, having composed his first opera at the age of twelve, he entered the Schola Cantorum in Paris (1904), where he studied with D'INDY and Albert Roussel (1869–1937). He also attended Widor's composition classes at the Conservatoire (1905) before destroying all his own compositions and moving to Berlin (1909), where he founded a choir for the performance of ancient music and met Ferruccio BUSONI and Richard

STRAUSS. After being discharged from the French army on medical grounds in 1915, Varèse went to the USA and established himself as an advocate and conductor of contemporary music, co-founding the International Composers' Guild for the presentation of new works (1921).

With the encouragement of Leopold STOKOWSKI, who conducted several of his works against vociferous public opposition, Varèse began to combine electronic and taped sound with wind and percussion instruments to develop a unique style that he called 'organized sound'. Of his early works, *Hyperprism* (1923), written for a chamber ensemble of wind and percussion, *Octandre* (1924), and *Intégrales* (1925) attracted the greatest controversy for their rejection of traditional musical forms. *Ionisation* (1931), his most celebrated work, was scored for forty-one percussion instruments and two sirens; in contrast was *Density 21.5* (1935), which featured an unaccompanied flute. Notable works that made striking use of taped and electronic sound include *Déserts* (1954) and *Poème électronique* (1958), which was written to be relayed in a constantly recurring eight-minute cycle over four hundred loudspeakers at the Brussels Exposition.

Vasarély, Victor (1908–) *Hungarian-born French painter, sculptor, and graphic artist and the main originator of op art.*

Vasarély studied medicine in Budapest before changing to study art (1927–29), partly under the tuition of MOHOLY-NAGY. In 1930 Vasarély moved to Paris, where he concentrated on graphic work until this gave way to painting in 1943. The style of geometric abstraction for which he is best known dates from about 1947. Through these paintings and the manifestos on the use of optical phenomena in art, written after 1955, Vasarély was largely responsible for the growth of the op art movement and the founding in 1961 of the Groupe de Recherche d'Art Visuel. His painting method, which he described as 'cinétisme', created an impression of movement mainly by means of visual ambiguities that required the precisely calculated clear-cut geometric forms characteristic of his pictures. The aim of these pictures was not beauty but perceptual discomfort producing intense visual experience.

Vasarély also produced sculptures, experimented with kinetic art, and collaborated with architects on numerous projects, such as the French pavilion at Expo 67 in Montreal. He received many international art prizes in the

1960s and academic and national honours in the 1970s and 1980s.

Vaughan Williams, Ralph (1872–1958)

British composer, musician, and teacher. A prominent composer of the first half of the twentieth century, he founded a new nationalist movement in English music. He was admitted to the OM in 1935.

The son of a Gloucestershire clergyman, Vaughan Williams studied with Sir Hubert Parry (1841–1903) and Sir Charles Stanford (1852–1924) at the Royal College of Music, at Trinity College, Cambridge, with BRUCH in Berlin (1897), and with RAVEL in Paris (1908). Like his close friend Gustav HOLST, he became interested in English music of the Tudor period and began to collect folksongs (1903), from which he later derived much material. From 1904 to 1906 he was musical editor of the *English Hymnal*; in this period he also started on the first of his nine symphonies, *A Sea Symphony* (1903–09), which was based upon the poems of Walt Whitman. His distinctive modal style found its first full expression, however, in *Fantasia on a Theme by Tallis* (1910), in which he broke decisively with the prevalent German academic tradition.

After war service, he was appointed professor of composition at the Royal College of Music (1919) and continued to compose in a wide variety of moods and forms. Subsequent symphonies included the evocative *Pastoral Symphony* (1922) and the *Sinfonia Antartica* (1952), based on his score for the film *Scott of the Antarctic*. Among many other notable works were the operas *The Pilgrim's Progress* (1951) and *Sir John in Love* (1929), the ballets *Old King Cole* (1923) and *Job* (1931), several concertos, and many songs and smaller pieces. Large-scale choral works, such as the mass in G minor (1922), the oratorio *Sancta Civitas* (1926), and the cantata *Hodie* (1954), also formed a significant part of his achievement.

Vavilov, Nikolai Ivanovich (1887–1943)

Soviet plant geneticist who made a vast collection of plant species from around the world and who proposed the theory that this diversity arose from several centres of origin located throughout the world.

Born in Moscow, Vavilov studied under the geneticist William BATESON at Cambridge University and at the John Innes Horticultural Institution (1913–14). On his return to the Soviet Union he was appointed professor of botany at the University of Saratov (1917–20) and then became director of the Bureau of Applied Botany, Petrograd (now St Petersburg). Under his

leadership, over four hundred research establishments were set up throughout the Soviet Union. Vavilov undertook a series of plant-collecting expeditions worldwide and eventually amassed over 50 000 species to be tested for their potential as crop plants. In *Centres of Origin of Cultivated Plants* (1920), he proposed six centres of origin in the world – regions where species exhibited maximum adaptability, i e genetic diversity. He later increased his number of such 'genecentres' to twelve and distinguished between primary genecentres (sites where ancestral forms persist) and secondary genecentres (sites where more recent proliferation of species has occurred). His theories remain an important concept in studies of plant populations.

Although his efforts brought international recognition for Soviet genetics, Vavilov came under increasing criticism from his rival geneticist, T. D. LYSENKO, whose influence over Soviet science was growing. The International Congress of Genetics planned for 1937 was cancelled on Lysenko's orders and in 1939 Lysenko made a vitriolic public denunciation of Vavilov as a purveyor of 'bourgeois' biology. The following year, Lysenko ousted Vavilov as director of the USSR Academy of Sciences' Institute of Genetics and later that year Vavilov was arrested and imprisoned. He was moved in 1942 to a concentration camp at Magadan and an unknown fate. In the 1950s, his reputation as one of the foremost Soviet scientists was restored as Lysenko's influence waned.

Venturi, Robert (1925–)

US architect and writer on architecture, regarded as the leading exponent of the postmodernist style.

Venturi was born in Philadelphia and educated at Princeton University. In the 1950s he worked as a designer in the offices of the leading modernist architect Eero Saarinen (1910–61) and Louis I. Kahn (1901–74); he also became (1954–56) a fellow of the American Academy in Rome, the first of several academic appointments. Venturi's rejection of the dogmas of the international style first became apparent from the idiosyncratic house he built for his mother in 1962 (the Vanna Venturi House, Philadelphia). In 1966 he published his *Complexity and Contradiction in Architecture*, which decried the element of puritanism in modernist theory and championed eclecticism, ornamentation, and a sense of wit. A still more provocative book was *Learning from Las Vegas* (1972; with Denise Scott Brown and Steven Izenour) in which Venturi defended the garish environment of Las Vegas and other US cities against aesthetic strictures.

Since the mid-1960s Venturi has worked in partnership with John Rauch and Denise Scott Brown, his wife from 1967. His buildings of the 1970s, such as the Brant-Johnson House in Vail, Colorado (1976), and the Gordon Wu Hall, Butler College, Princeton (1980), are celebrated for their playful use of historical references, which sometimes involves a deliberate element of kitsch. Venturi's work was perhaps the leading influence on the design of offices and shopping malls in 1980s Britain.

When the intervention of Prince CHARLES led to the rejection of a proposed modernist design for the extension to the National Gallery in London, Venturi won the commission to produce an alternative. The new wing opened amid great publicity, mostly favourable, in 1991.

Verwoerd, Hendrik Frensch (1901–66)
South African prime minister (1958–66), remembered for his quiet advocacy of extreme and brutal policies.

Born in Amsterdam, the son of Dutch missionaries, Verwoerd was educated in Cape Province, Southern Rhodesia, Holland, and Germany. He held the chairs of applied psychology and sociology at Stellenbosch University from 1927 to 1937. Verwoerd's political career began in 1938, when he became founder and editor of the Nationalist paper *Die Transvaler*. Making no secret of his anti-British pro-German sentiments, he was active in Afrikaner groups and was a member of the Broederbond during World War II. Entering the senate as a government-nominated member in 1948, Verwoerd was appointed minister of Bantu affairs in 1950. In this role, he developed the segregation policy of apartheid, based on the belief that every race should have the opportunity to develop its own potentialities in its own environment. In practice this meant forcing urban Africans to return to homelands on tribal reserves. Elected prime minister in 1958, Verwoerd survived an assassination attempt in 1960, shortly after he had banned the African National Congress and the Pan-African Congress, in response to the Sharpeville demonstration. Having pledged to establish a republic when appointed prime minister, Verwoerd withdrew South Africa from the British Commonwealth in 1961. He was assassinated in 1966 by a parliamentary messenger in the house of assembly.

Vidal, (Eugene Luther) Gore (1925–   )
US novelist, playwright, and essayist.

Vidal was born in New York at West Point, the son of an instructor of aeronautics at the US Military Academy. He grew up in Washington, where his father became director of Air Commerce under ROOSEVELT and his grandfather, Senator T. P. Gore, represented Oklahoma in Congress. He was educated at Philips Exeter Academy, New Hampshire. In 1943 he enlisted in the army, serving as a warrant officer and then as first mate on an army supply vessel in the Aleutians, an experience drawn on in his first novel, *Williwaw* (1946), written when he was nineteen. His third book, *The City and the Pillar* (1948), dealt with homosexuality and was considered outspoken at the time. *The Judgment of Paris* (1952) made use of the myth in relating a story about Americans in Europe. In 1960 Vidal ran for Congress; although not elected, he made a respectable showing, receiving more votes, he noted, than had John KENNEDY in similar circumstances. His play, *The Best Man* (1960), drew on this experience. Vidal's intimate knowledge of American politics is matched by a talent for historical writing, as in the political novels *Washington D.C.* (1967), *Burr* (1973), *1876* (1976), *Lincoln* (1984), and in *Julian* (1964), perhaps his best historical novel, on Julian the Apostate. Hollywood and sexual ambivalence are the subjects of *Myra Breckinridge* (1968) and *Hollywood* (1989). His other fiction includes *Creation* (1981), a novel set in the fifth century BC involving the memoirs of Zoroaster's grandson, and *Duluth* (1983), a satire on Middle American political, cultural, and sexual peculiarities. Vidal's many essays cover a wide range of subjects and form a distinguished and witty commentary on cultural and political life in America. Among the collections are *Rocking the Boat* (1962), *Reflections upon a Sinking Ship* (1969), *The Second American Revolution and Other Essays (1976–1982)* (1982), and *At Home* (1988). *A Thirsty Evil* (1956) is a collection of short stories. He has also written a travel book, *Vidal in Venice* (1987), and several screenplays.

Villa-Lobos, Heitor (1887–1959) *Brazilian composer and educationalist, who became the most celebrated Latin-American composer of the twentieth century.*

Born in Rio de Janeiro, Villa-Lobos was taught to play the cello by his father, and as a youth made a living by playing in cafés and cinemas. At eighteen he journeyed into the Brazilian interior collecting folk music, returning to study at the National Institute of Music in Rio until 1915. A concert of his own pieces in that year led to a commission that became his third symphony (*A guerra*), performed in 1922 under his own baton in celebration of the centenary of Brazilian independence. The pianist Artur RUBINSTEIN was instrumental in his acquiring a government award to fund a visit to Europe, where he lived in Paris (1923–30) and had sensational success as a composer with such exotic pieces as *Rudepoema* (1926). On his return to Brazil in 1930, Villa-Lobos was appointed director of musical education at São Paulo, and in 1932 he took charge of the teaching and performance of music throughout the country. He established a chain of music schools in Brazilian cities, and in 1945 founded the Brazilian Academy of Music in Rio, of which he was president until his death.

A highly prolific writer, Villa-Lobos is credited with at least two thousand works, all of which he kept, regardless of quality. The best of his music is his series of fourteen *Chôros* (1920–29) and his nine *Bachianas Brasileiras* (1930–45), in which he attempted to combine Brazilian folk music with the style of European composers, particularly Bach and Puccini. As well as these attractive pieces, he wrote twelve symphonies (1916–57), seventeen string quartets (1915–57), concertos for various instruments, and dramatic, choral, vocal, instrumental, and educational works.

Visconti, Luchino (Luchino Visconti di Modrone; 1906–76) *Italian film and stage director.*

Visconti was born into an aristocratic family in Milan and his early interests lay more in art and horse-breeding than theatre and film. By the time he was thirty, however, he was a committed Marxist and had begun to work as an amateur designer in the theatre. In 1936 he went to France to work on the films *Les Basfonds* and *Une Partie de campagne* with Jean RENOIR. His first independent venture, the neorealist film *Ossessione* (1942), adapted from *The Postman Always Rings Twice*, ran into difficulties with the Italian authorities and it was only later that it was seen as the forerunner of cinema neorealism. Subsequently, however, he moved away from the naturalism of his first film. Notable among his later films were *Senso* (1954), *The Leopard* (1963), for which he received the Golden Palm at the Cannes Film Festival, and *Vaghe stelle dell' orsa* (1965), which won the Venice Golden Lion Award. *The Damned* (1969) and *Death in Venice* (1971), both starring Dirk BOGARDE, were also highly acclaimed. His last film was *L'innocente* (1976). In the theatre Visconti directed many successes, including plays, operas, and ballets, in a number of countries.

Vlaminck, Maurice de (1876–1958) *French painter, who exhibited with the fauves.*

Vlaminck's parents were both musicians and he himself was a racing cyclist as a young man. In 1900 he met DERAIN and began to paint seriously, earning his living by playing the violin. At this time he also contributed to anarchist magazines. His early work was influenced by the Van Gogh exhibition of 1901 with its violent use of pure strong colour. Like MATISSE and Derain he used colour as an emotive force rather than solely a descriptive medium, applying paint straight from the tube to the canvas and dispersing it with rough brushstrokes. His work was exhibited with that of the fauves at their historic exhibition of 1905. Vlaminck was largely self-taught as a painter and gloried in his undisciplined talent – he derided traditional culture and claimed never to have set foot in the Louvre.

Vlaminck's style began to change after he had seen the retrospective exhibition of Cézanne's paintings in 1907. Over the next five years both colour and brushwork became more subdued and the structure more defined, as in *Self Portrait* (1912). His later work is more directly representational. He moved from Paris to the country, where he indulged his passion for racing cars and painted predominantly landscapes and still lifes in a vigorously expressive realistic style. He also illustrated books and wrote several autobiographical works.

von Braun, Werner Magnus Maximilian (1912–77) *German-born US rocket engineer, responsible both for the V-rockets that bombed London in World War II and the Saturn rockets that put Americans on the moon in 1969.*

Born in Wirsitz in Germany (now Wyrzysk, Poland), the son of a senior government official, von Braun came from an affluent background. He was educated in Zürich and Berlin, where he gained his doctorate in 1934 with a thesis on liquid-fuelled rockets. An early and enthusiastic member of the Verein für Raumschiffahrt (Society for Space Travel) in Berlin, von Braun was recruited by the German Ordnance Department and began to develop for them a series of powerful liquid-fuelled rockets. Barred from developing more orthodox weapons by the Versailles treaty, the German High Command saw rockets as a way around these restrictions and ample funds were available for rocket research. In 1937 von Braun moved to

Peenemunde on the Baltic coast as technical director of the newly established experimental weapons research centre. Here he developed a number of powerful weapons including the V-1 (the flying bomb) and the V-2 (the first ballistic missile to carry a warhead). Successfully tested in 1942, some 1200 were launched against London in 1944, each carrying 2000 lbs of high explosive. Yet for von Braun there was more to the V-2 than its destructive power; it was also a powerful rocket that one day might be capable of launching a payload into space.

With the collapse of the Third Reich, von Braun and his team were taken to the US army's White Sands Proving Ground in New Mexico, where they began the long process of turning the V-2 into an intercontinental ballistic missile capable of delivering a nuclear warhead anywhere in the world. At White Sands, and from 1950 at Redstone Arsenal in Huntsville, Alabama, they worked on Redstone, a short-range nuclear missile. The launch of Sputnik I in October 1957 changed the direction of von Braun's work. Within a remarkably short time a Redstone rocket had been sufficiently modified to launch the first American satellite into orbit on 31 January 1958. Thereafter von Braun's main task was to provide the rocket powerful enough to satisfy the presidential call to place an American on the moon during the 1960s. The task was successfully accomplished with a Saturn V rocket in July 1969 at a cost of twenty-four billion dollars.

Shortly after the success of the moon landing, von Braun left Huntsville for Washington to plan new space programmes for NASA. It soon, however, became apparent that the political will for imaginative and costly space ventures no longer existed in NIXON's Washington. Von Braun consequently resigned in 1972 and joined Fairchild Industries. Four years later, after an unsuccessful operation for cancer, he resigned his post, a few months before his death.

von Laue, Max Theodor Felix (1879–1960) *German physicist who was awarded the 1914 Nobel Prize for Physics for his discovery of the diffraction of X-rays by the atoms or ions in crystals.*

Born in Koblenz, the son of a civil servant, von Laue studied at the universities of Strasbourg, Göttingen, and Berlin, where he was awarded his doctorate in 1903 for work done under the supervision of Max PLANCK. He remained in Berlin until 1909, when he moved to SOMMERFELD's Institute of Theoretical Physics

in Munich. Brief periods at the universities of Zürich and Frankfurt were followed by service during World War I at Wurzburg, working on improving military communications. By this time von Laue had already gained an international reputation and been awarded a Nobel Prize for his proposal in 1912 that, as crystals were composed of regular arrays of atoms or ions, X-rays when passed through a crystal would behave comparably to light falling on a diffraction grating. Experiments carried out by his students W. Friedrich and P. Knipping, in which copper sulphate crystals were irradiated with X-rays, revealed the presence of regularly spaced dots on the photographic plate placed behind the sample. It was soon realized that these dots could be used to explore the structure of complex molecules.

At the end of the war in 1919, von Laue was appointed to the chair of theoretical physics at Berlin University. In the 1930s he was one of the few senior German scientists to protest at the behaviour of the Nazi government. He strongly condemned the expulsion of EINSTEIN from the German Physical Society and refused to allow the Nazi physicist J. Stark membership of the Prussian Academy. Von Laue retired from his Berlin post in 1943. With the end of World War II, he spent most of his retirement attempting to revive German science. As a sign of his standing in the international community he was invited to England shortly after the war.

von Neumann, John (1903–57) *Hungarian-born US mathematician, creator of the theory of games and pioneer in the development of the modern computer.*

Born in Budapest, the son of a wealthy banker, von Neumann was educated at the universities of Berlin, Zürich, and Budapest, where he obtained his PhD in 1926. After teaching briefly at the universities of Berlin and Hamburg, von Neumann moved to the USA in 1930 to a chair in mathematical physics at Princeton. In 1933, he joined the newly formed Institute of Advanced Studies at Princeton as one of its youngest professors. By this time he had already established a formidable reputation as one of the most powerful and creative mathematicians of his day. In 1925 he had offered alternative foundations for set theory, while in his *Mathematischen Grundlagen der Quantenmechanik* (1931; translated as *Mathematical Foundations of Quantum Mechanics*, 1933) he removed many of the basic doubts that had been raised against the coherence and consistency of quantum theory.

In 1944, in collaboration with Oskar Morgenstern (1902–77), von Neumann published *The Theory of Games and Economic Behaviour*. A work of great originality, it is reputed to have had its origins at the poker tables of Princeton and Harvard. The basic problem was to show whether it was possible to speak of rational behaviour in situations of conflict and uncertainty as in, for example, a game of poker or wage negotiations. In 1927 von Neumann proved the important theorem that even in games that are not fully determined, safe and rational strategies exist.

With entry of the USA into World War II in 1941 von Neumann, who had become an American citizen in 1937, joined the Manhattan project (for the manufacture of the atom bomb) as a consultant. In 1943 he became involved at Los Alamos on the crucial problem of how to detonate an atom bomb. Because of the enormous quantity of computations involved, von Neumann was forced to seek mechanical aid. Although the computers he had in mind could not be made in 1945, von Neumann and his colleagues began to design Maniac I (*M*athematical *a*nalyser, *n*umerical *i*ntegrator, *a*nd *c*omputer). Von Neumann was one of the first to see the value of a flexible stored program: a program that could be changed quite easily without altering the computer's basic circuits. He went on to consider deeper problems in the theory of logical automata and finally managed to show that self-reproducing machines were theoretically possible. Such a machine would need 200 000 cells and 29 distinct states.

Having once been caught up in affairs of state von Neumann found it difficult to return to a purely academic life. Thereafter much of his time was therefore spent, to the regret of his colleagues, advising a large number of governmental and private institutions. In 1954 he was appointed to the Atomic Energy Commission. Shortly after this, cancer was diagnosed and he was forced to struggle to complete his last work, the posthumously published *The Computer and the Brain* (1958).

von Rundstedt, (Karl Rudolf) Gerd (1875–1953) *German field-marshal who commanded his country's forces against the Allied invasion of northwest Europe in 1944.*

From an aristocratic Prussian family, von Rundstedt rose through the ranks of the Imperial German army to become an eminent staff officer; during World War I he assisted in the reorganization of the Turkish general staff. Between the wars he participated in the clandestine rearmament programme until his retirement in 1938. Recalled by HITLER in 1939, von Rundstedt first headed the German southern wing in the invasion of Poland and then, in 1940, commanded an army group in the invasion of France. His decision to delay the advance of the Panzer divisions on Dunkirk in order to consolidate his positions enabled the evacuation of the trapped British Expeditionary Force across the English Channel.

Promoted to field-marshal in the spring of 1941, von Rundstedt commanded the southern army group in the invasion of the Soviet Union. By November the offensive was bogged down and he requested a strategic withdrawal to the Mius River. Hitler refused and von Rundstedt relinquished his command. However, in July 1942, Hitler appointed him commander-in-chief, west, with responsibility for the defence of the northern European coast and the French Mediterranean coast against possible Allied attack. Nominally under von Rundstedt's command was ROMMEL, who energetically strengthened the coastal defences; von Rundstedt, with his forces constantly depleted by the war in the east, was doubtful of the value of this. In July 1944 he was again dismissed by Hitler for urging a strategic withdrawal to the Seine following the Allied invasion in June. Two months later he was reappointed yet again but played little part in the planning or execution of the Ardennes offensive of December 1944, which initially threatened to break the Allied advance. Von Rundstedt was relieved of his command for the third time in March 1945 and in May was captured by US troops only to be released shortly afterwards on health grounds.

Voroshilov, Kliment Yefremovich (1881–1969) *Soviet statesman and marshal. One of the oldest of the Bolsheviks, he became president (1953–60) after STALIN's death.*

Born in Verkhne, Dnepropetrovsk, the son of a miner, Voroshilov worked in the mines as a child before attending school. He later worked in a locomotive factory, from which he was dismissed for organizing a strike (1899). In 1903, while employed as an electrical fitter, he became chairman of the Lugansk branch of the Social Democratic Party (1905), joining the Bolsheviks in 1906. He worked in munitions factories during World War I before fighting in the civil war, in which he distinguished himself in 1919 as a commander at Tsaritsyn (now Volgograd). Now settled in a military career, he served in the Polish-Russian war (1920), fought in the Far East (1921), and became

W

Waddington, Conrad Hall (1905–75) *British geneticist who proposed the theory of genetic assimilation to account for cases in which apparently acquired characteristics are brought under some degree of genetic control. He was awarded the CBE in 1958.*

Waddington studied geology at Sidney Sussex College, Cambridge, receiving his degree in 1926. He was later appointed lecturer in zoology and embryology at Strangeways Laboratory, Cambridge (1933–45), and in 1947 moved to Edinburgh University as professor of animal genetics. His early research in embryology led to his interest in the relationship between development and genetics. Waddington recognized how the development of organisms follows a well-defined course in spite of variations in conditions, i e development is canalized. In explaining his concept of genetic assimilation, introduced in 1942, he cited the example of calluses on the rumps of ostriches. These first developed not by mutation but as a result of abrasion by the ground. Waddington suggested that selection over a long period stabilized the shape and size of the calluses and eventually this canalized development came to be triggered by a gene mutation, thereby ensuring the spread of the mutation throughout the population. He also performed experiments with fruit flies (*Drosophila*), which he claimed demonstrated this phenomenon occurring, but these have been criticized as merely instances of artificial selection and genetic fixation of hitherto undisclosed degenerative characters.

Much concerned with the impact of science on society and the conservation of global resources, Waddington was a founder member of the Pugwash Conference and the Club of Rome. His books include *Introduction to Modern Genetics* (1939), *Principles of Embryology* (1956), *The Ethical Animal* (1960), and a discussion of the influence of science on modern art, *Behind Appearances* (1970). Waddington became a fellow of the Royal Society in 1947.

Wain, John Barrington (1925–) *British poet, novelist, and critic. He became a CBE in 1984.*

John Wain was educated at Newcastle-under-Lyme High School and St John's College, Oxford, where he held a fellowship from 1946 to 1949. While lecturing in English literature at Reading University (1947–55) he made a considerable impact with his first novel, *Hurry on Down* (1953), with its satirical portrait of a contemporary university graduate. His next novel, *Living in the Present* (1955), enabled him to resign his lectureship to become a freelance writer. He subsequently held a number of temporary academic positions, including professor of poetry at Oxford (1973–78). More recent novels include *The Pardoner's Tale* (1978), *Young Shoulders* (1982), which won the Whitbread Prize, *Where the Rivers Meet* (1988), and *Comedies* (1990).

In 1951 the first of several collections of Wain's poetry, *Mixed Feelings*, was published. It was followed by *A Word Carved on a Sill* (1956), *Weep Before God* (1961), *Wildtrack* (1965), *Letters to Five Artists* (1969), *Feng* (1975), *Poems 1949–79* (1981), and *Open Country* (1987). *Nuncle and Other Stories* (1960), *Death of the Hind Legs* (1966), and *The Life Guard* (1971) contain short stories. In addition to his fiction and poetry, Wain wrote or edited collections of critical essays and compiled anthologies of poetry and selections from various writers, including Samuel Johnson, of whom he wrote a prize-winning biography (1974). *Professing Poetry* (1977) was the result of his tenure of the Oxford chair. His autobiographical writings include *Sprightly Running* (1962) and *Dear Shadows* (1986).

Waksman, Selman Abraham (1888–1973) *Russian-born US biochemist, who won the 1952 Nobel Prize for Physiology or Medicine for his discovery of the antibiotic streptomycin.*

After emigrating to the USA in 1910, Waksman studied at Rutgers University and the University of California, from which he received his PhD in 1918. He returned to Rutgers as a lecturer in soil microbiology, becoming professor (1930–40) and then professor of microbiology (1940–58). He was made emeritus professor on his retirement in 1958.

Waksman's special field was soil microbiology, in particular the role of fungi and bacteria in the decomposition of organic matter and humus formation. He wrote *Principles of Soil Microbiology* (1927), one of the most comprehensive works on the subject at that time. The discovery of penicillin's therapeutic potential encouraged Waksman to investigate the soil microorganisms called actinomycetes in the hope of finding new antibiotics (a term introduced by Waksman in 1941). In 1944 he announced the discovery of streptomycin, which he had isolated from *Streptomyces griseus*. This was the first safe antibiotic found that was effective against Gram-negative bacteria, including the species responsible for tuberculosis, which are resistant to penicillin. He later discovered another antibiotic, neomycin, obtained from *Streptomyces fradiae*. This is used to treat bowel infections and local skin or eye infections. Waksman's autobiography, *My Life with the Microbes*, was published in 1954.

Waldheim, Kurt (1918–) *Austrian diplomat and secretary-general of the United Nations (1972–82). He became president of Austria in 1986 despite controversy over his war record.*

Born in Sankt Andrae-Woerden, the son of a civil servant, Waldheim served in the Austrian army (1937–38) before studying at the Vienna Consular Academy. He began his law studies at the University of Vienna but was drafted into the German army at the outbreak of World War II. Wounded and discharged in 1942, he resumed and completed his degree in 1944.

Waldheim entered the diplomatic service in 1945. From 1945 to 1947 he was a member of the Austrian delegation to the Austrian State Treaty negotiations. He served in Paris and Vienna (1948–55) before accepting the post of permanent observer to the United Nations (1955–56). After serving as envoy and ambassador to Canada (1956–60) and director-general of political affairs in Austria (1960–64), he returned to the United Nations as a permanent representative in 1964, chairing the outer space committee (1965–68). In 1968 Waldheim became minister of foreign affairs in Austria but following his dismissal from the post in 1970 after a change in leadership, again assumed the position of permanent representative to the United Nations. In 1971 he attempted unsuccessfully to re-enter Austrian politics by standing as the Conservative People's Party candidate for the presidency. He was appointed secretary-general of the United Nations in 1972, a position he held until 1982 when he resigned to become research professor of diplomacy at Georgetown University, Washington, DC.

In 1986 he was chosen as the People's Party candidate for the presidency of Austria, which he secured after a run-off election. His election was highly controversial due to the discovery of documents that seemed to show that as an interpreter and intelligence officer in the German army in Greece and Yugoslavia he probably knew of atrocities committed against Yugoslav partisans and of the mass deportation of Jews from Salonika to the concentration camps in 1943. Waldheim was obliged to admit that he had not been completely open about his war record but maintained that he knew nothing of these war crimes. He was subsequently cleared in court of charges relating to his war record, but his election remained controversial and an embarrassment in international relations.

Wałesa, Lech (1943–) *Polish trade-union leader who became president of Poland in 1990. He was awarded the Nobel Peace Prize in 1983.*

Born in Popowo, the son of a carpenter, Wałesa was educated in Lipino before moving to Gdansk in 1966, where he worked as an electrician in the Lenin Shipyards. In 1976 he was dismissed from his position for taking part in a protest over the erosion of economic concessions made to workers by the government after the food riots in 1970.

For the next four years Wałesa was unemployed, although he edited an underground paper and participated in meetings of the Workers' Self-Defence Committee. In 1980, when workers demanded higher wages as compensation for price increases and scattered strikes occurred across Poland, Wałesa scaled the fence at the Lenin Shipyards to join workers occupying the yard. Taking charge of the strike, which spread across the Baltic region, he negotiated an agreement with the government providing for the right of workers to form independent unions and to strike. After some ten million workers had registered, he was appointed chairman of the National Coordinating Committee of Independent Autonomous Trade Unions, known as Solidarity. In 1981 martial law was imposed by the government in response to worker militancy over the failure to implement the Gdansk agreement and to pressure from the Soviet Union; this led to the banning of Solidarity. Wałesa was detained for eleven months (1981–82) and was awarded the 1983 Nobel Peace Prize for his efforts to establish workers' rights.

In the face of the worsening economic situation in Poland, the government of General

Wojciech Jaruzelski (1923–) began negotiations with Wałesa and other Solidarity leaders in 1988 and in 1989 Solidarity was given legal status once more. A partially free election followed, with Solidarity's candidates winning all but one seat of those contested. Jaruzelski became president but a year later Wałesa won a landslide victory in the presidential elections. His triumph was welcomed throughout the West and he has since been received by most of the heads of state of the western democracies in his campaign to secure economic assistance for his country. His attempts to increase the president's influence over parliament were thwarted, however, in 1991. His autobiography, *A Path of Hope*, was published in the West in 1987.

Wallace, (Richard Horatio) Edgar (1875–1932) *British journalist, dramatist, and thriller writer.*

The son of actor parents, Wallace was brought up by a Billingsgate fish porter and at the age of twelve completed his formal education, leaving his Peckham elementary school to earn a living by a variety of menial jobs. He joined the army and was drafted (1896) to South Africa, where he wrote for local newspapers and became a correspondent first for Reuter and then for the *Daily Mail*. In 1902 he became first editor of the Johannesburg *Rand Daily Mail*.

Returning to England he continued to work for the *Daily Mail* and wrote the first of his best-selling novels, *The Four Just Men* (1905). *Sanders of the River* (1911) is perhaps the best known of his books with an African setting. In all he published over a hundred and seventy books; some of these, such as *The Crimson Circle* (1922) and *The Green Archer* (1923), were immensely successful and had huge worldwide sales.

In the 1920s, while continuing his journalism and novel writing, Wallace also achieved considerable success as a playwright, starting with *The Ringer* (1926). Always a rapid worker, he wrote one of his most powerful plays, *On the Spot* (1931), in a single weekend. In 1931 he stood unsuccessfully for parliament as a Liberal candidate, and shortly afterwards went to Hollywood to work on film scripts. Among these was *King Kong*, which was produced shortly after his unexpected death.

Wallenberg, Raoul (1912–c.47) *Swedish diplomat, who disappeared in 1945 after saving the lives of some 95,000 Jews threatened by Nazi persecution in Hungary during World War II.*

Wallenberg was born into a wealthy Swedish banking family and in 1939 joined the staff of a foodstuff firm run by a Jewish refugee from Budapest. He soon became a familiar figure in Budapest's business circles and subsequently undertook diplomatic work on behalf of the Swedish government, becoming an official at the Budapest legation in 1944, secretly charged with the protection of Hungarian Jews in the city.

In order to save the Jewish community from mass deportation to Auschwitz, Wallenberg issued them with thousands of Swedish passports and arranged shelters in Sweden for Jewish refugees. His heroic work on behalf of the Jews, especially those living in the Budapest ghetto, won him an international reputation. In 1945, however, when the Red Army arrived in Budapest, Wallenberg, who was seeking further relief on behalf of the Jewish inhabitants of the city, was arrested and taken to Moscow. No further definite information about his fate has ever been released, despite intense pressure from many countries and humanitarian organizations upon the Soviet Union. In 1989 the Soviet authorities conducted an investigation into Wallenberg's fate and concluded that he had been shot in the infamous Lubyanka prison in 1947. Many Wallenberg supporters, however, refused to accept this version of events and continued to maintain he was still alive in a Soviet prison. Wallenberg was made an honorary US citizen in 1981 and is honoured at Yad Vashem, the memorial to the Holocaust in Jerusalem, as the foremost of the 'Righteous Gentiles'.

Waller, Fats (Thomas Wright Waller; 1904–43) *Black US jazz pianist, composer, bandleader, and singer. An inspired clown and a popular entertainer, he was the foremost exponent of the New York stride school of piano playing.*

Born in New York, the son of a Baptist minister who regarded jazz as 'music from the devil's workshop', he won a talent contest in 1918 playing James P. Johnson's 'Carolina Shout'. He later took lessons from Johnson (1891–1955), dean of Harlem piano players and founder of the stride school. By the age of fifteen, in spite of his father's disapproval, he was playing in cabarets and theatres. He also learnt to accompany silent films on the organ. A long collaboration with lyricist Andy Razaf (1895–1973) began in the 1920s and eventually resulted in many hit songs, including 'Ain't Misbehavin'' and 'Honeysuckle Rose' (which began as a piano variation on 'Tea for

Two'). He contributed to several shows and revues, including *Keep Shufflin'* (1928) and *Hot Chocolates* (1929; recently revived). He played and recorded with the bands of Jack Teagarden (1905–64), Fletcher HENDERSON, and McKinney's Cotton Pickers, and in 1927–29 recorded steadily as Fats Waller and his Buddies.

Stride piano uses tenths in the left hand for a strong bouncy bass line; as a master of it, Waller had a great influence on many jazz pianists, especially Art TATUM. From 1934 onwards it was as leader of the small group Fats Waller and His Rhythm that he achieved popular success. Waller could make hits out of unlikely material; if it was a song he liked, he would do it straight ('I'm Gonna Sit Right Down and Write Myself a Letter'); if it was a novelty or second-rate, he would clown his way through it ('Your Feet's Too Big') paraphrasing the lyrics and improvising tag lines ('One never knows, do one?'). Despite the jokes, his records are a source of splendid jazz centred around his outstanding piano playing.

He appeared in several films, including *King of Burlesque* (1936) and *Stormy Weather* (1943). His huge appetite for food, drink, and hard work led to his early death from pneumonia, on a train between engagements.

Wallis, Sir Barnes Neville (1887–1979) *British aviation engineer, designer of the R 100 airship, the Wellington bomber, bouncing bombs, and swing-wing aircraft. He was knighted in 1968.*

Born in Ripley, Derbyshire, the son of a doctor, Wallis was apprenticed with the Thames Engineering Company (1904–08). In 1913 he joined the firm of Vickers to work on the design of airships and was responsible for the construction of the R 100, completed in 1930 as a rival to the government-backed R 101. Although the R 100 crossed the Atlantic without difficulty and was faster than the R 101, the crash of its rival on its test flight with the loss of forty-eight lives meant the end of the airship as a commercial proposition. Consequently, in 1930, Wallis, as chief designer of structures at Vickers, turned his attention to the design of military aircraft. His first plane was the revolutionary Wellesley, with its novel goedetic structure. Although this was turned down by the Air Ministry, the same lattice structure was incorporated in Wallis's Wellington bomber, one of the most successful British bombers of the war. Wallis is, however, best known for his wartime work, chronicled in the book (1951) and film (1955), *The Dam Busters*. Normal

bombing, he argued, would have little impact on Germany's dispersed industrial capacity. It would therefore be sensible to aim at an essential target, such as a dam, that could not be dispersed. This required more powerful and sophisticated bombs than were then available and Wallis set about designing bombs that would destroy such substantial structures as the Möhne and Eder dams. The result was the ten-ton bouncing bombs, which successfully destroyed the Ruhr dams in 1943.

Wallis next attracted attention with his idea of swing-wing variable-geometry aircraft, capable of flying supersonic and subsonic speeds efficiently at both high and low altitudes. Vickers backed the project and in the 1950s an experimental version of Wallis's design, the Swallow, was built. When the British government lost interest in the project, Wallis's plans were abandoned. Some of his ideas were, however, incorporated in several other planes, notably the US fighter, the F 111. When Wallis was in his nineties, he was working on plans for a 'square' plane capable of flying several times faster than the speed of sound with a range of 10 000 miles. A model was made by BAC and although wind-tunnel tests looked promising, Wallis's last design remained unexploited.

Walpole, Sir Hugh Seymour (1884–1941) *British novelist. He was knighted in 1937.*

Walpole was born in Auckland, New Zealand, and educated at King's School, Canterbury, Durham School, and Emmanuel College, Cambridge. On leaving Cambridge he discovered that he had no vocation for the priesthood, which he had orginally planned to enter. After some years travelling in Europe and teaching, he settled in London (1909) and in the same year published his first novel, *The Wooden Horse*. He also made many literary friends, in particular Henry James. Other novels quickly followed, most notable of which was *Mr Perrin and Mr Traill* (1911), about life at a boy's school.

During World War I Walpole served first with the Russian Red Cross in Galicia and then directed Anglo-Russian propaganda in Petrograd. From these experiences he distilled his highly esteemed Russian novels, *The Dark Forest* (1916) and *The Secret City* (1919). In 1919 he undertook the first of his successful lecture tours in the USA; during the next two decades he remained a prominent figure on the literary scene on both sides of the Atlantic, writing over forty works of fiction as well as critical studies of Conrad (1916) and Trollope

(1928) and three autobiographical volumes: *The Crystal Box* (1924), *The Apple Trees* (1932), and *Roman Fountain* (1940). Among his successes were the Jeremy trilogy comprising *Jeremy* (1919), *Jeremy and Hamlet* (1923), and *Jeremy at Crale* (1927) and the Herries quartet comprising *Rogue Herries* (1930), *Judith Paris* (1931), *The Fortress* (1932), and *Vanessa* (1933). In his capacity as chairman of the Society of Bookmen, he did much to encourage young or struggling authors.

Walter, Bruno (Bruno Walter Schlesinger; 1876–1962) *German-born US conductor, particularly renowned for his interpretations of Mozart, Mahler, Brahms, and Bruckner.*

Having made his debut as a pianist at the age of ten, he studied at the Stern Conservatory, Berlin, and decided to become a conductor after hearing a concert conducted by von Bülow in 1889. In 1894 he made his first appearance as a conductor, in Cologne, and worked at the Hamburg Opera under his close friend Mahler for the first time. In 1901 Mahler engaged him as director of the Vienna State Opera and during the next decade Walter consolidated his European reputation. After Mahler's death in 1911, Walter gave the first performance of *Das Lied von der Erde* (composed 1908) in Munich, and in 1912 the first performance of Mahler's ninth symphony (1909) in Vienna.

From 1914 to 1922 he was director of the Munich Opera, after which he toured Europe and the USA before returning to Berlin in 1925 as music director of the State Opera and beginning his long association with the Salzburg Festival. In 1929 he became conductor of the Leipzig Gewandhaus Orchestra but suffered antisemitic persecution from the Nazis and left Germany for Austria in 1933. Forced to leave Austria in 1938, following the *Anschluss*, he moved to France and finally settled in the USA (1939), where he conducted the Metropolitan Opera and the New York Philharmonic (1947–49). After World War II he again performed in Europe, continuing to champion the works of Mahler and also appearing as accompanist with Kathleen FERRIER.

Walton, Ernest Thomas Sinton (1903–   ) *Irish physicist who, with Sir John* COCKROFT, *was responsible for the first experimental splitting of the atomic nucleus. For this they received the 1951 Nobel Prize for Physics.*

The son of a clergyman, Walton studied at the Methodist College, Belfast, and (from 1922 to 1926) at Trinity College, Dublin, where he graduated in mathematics and experimental

science. In 1927 he went to Cambridge University on a research scholarship. It was here, working under Lord RUTHERFORD in the Cavendish Laboratory, that he collaborated with Cockroft in experiments with accelerated charged particles. In 1932 they demonstrated that lithium nuclei could be transformed into two helium nuclei by the impact of protons.

Walton became a fellow of Trinity College, Dublin, in 1934 and from 1947 to 1974 was Erasmus Smith's Professor of Natural and Experimental Philosophy. In 1974 he became a Fellow emeritus. His other research interests have included work on hydrodynamics and microwaves.

Walton, Sir William Turner (1902–83) *British composer, who first came to prominence as a protégé of the Sitwells. He was knighted in 1951 and admitted to the OM in 1967.*

The son of a music teacher, Walton became a chorister at Christ Church, Oxford, and later an undergraduate there; as a composer, he was largely self-taught. While still an undergraduate, he formed a close friendship with Osbert and Sacheverell Sitwell, with whom he travelled and lived during the decade after he left Oxford. During these formative years he wrote the humorous and highly original *Façade*, a setting of Edith SITWELL's verse for reciter and instrumental ensemble (1921–22, final revision 1942). Subsequent works were, however, more lyrical and romantic, while the overture *Portsmouth Point* (1925) established Walton's vein of witty neoclassicism. The music of Paul HINDEMITH and Igor STRAVINSKY strongly influenced Walton's first mature work, the viola concerto (1928–29; revised 1961), first performed with the composer conducting and Hindemith playing the solo viola (1929).

Other works of importance were the violin concerto (1938–39), the cello concerto (1956–57), two symphonies (1932–35; 1959–60), the operas *Troilus and Cressida* (1950–54; revised 1975–76) and *The Bear* (1965–67), and the oratorio *Belshazzar's Feast* (1930–31), with biblical text selected by Osbert Sitwell. Noted for his craftsmanship and superb orchestration, Walton also wrote the film scores for *Escape Me Never* (1934), *Major Barbara* (1941), *Henry V* (1943–44), *Hamlet* (1947), *Richard III* (1955), and *The Battle of Britain* (1969). In 1948 he married and lived in Italy until his death.

Wang Ching-wei (Wang Jing Wei; 1883–1944) *Chinese revolutionary. A hero of the 1911 Chinese revolution, he was branded* *a traitor for his collaboration with the Japanese during World War II and his political vacillation.*

Born in Canton, Wang Ching-wei was educated at the Tokyo College of Law in Japan, where he joined the revolutionary party of SUN YAT-SEN. He returned to China in 1910 and joined a plot to assassinate the prince regent, which failed. Faced with the death sentence, Wang Ching-wei so impressed the regent with his courage that the sentence was commuted to life imprisonment; he was released in 1911 following the establishment of the republic.

Wang Ching-wei then served as a principal assistant to Sun Yat-sen (1911–25) and, after his death, became the new chairman of the government. However, the right wing of the Kuomintang favoured CHIANG KAI-SHEK, who set up his own regime in Nanking. In response, the left-wing faction of the Kuomintang, in alliance with the communists, formed a rival regime headed by Wang Chei-wei in Wuhan. In 1932 Wang Chei-wei achieved a reconciliation with Chiang Kai-shek and the nationalists and became president of the administrative council. Nonetheless his political allegiances continued to fluctuate; in 1938 he left the Nationalist government over his support for a peace settlement with Japan, and fled to Hanoi. Two years later he was installed by the Japanese as head of a puppet regime in Nanking and made several fruitless attempts to persuade the Kuomintang to stop fighting the Japanese. He remained in this post until his death in 1944.

Wankel, Felix (1902–88) *German engineer, who invented the Wankel rotary engine.*

The son of a civil servant who was killed in World War I, Wankel received no more than a high-school education. He began his working life in 1921 with a Heidelberg bookseller, acquiring his engineering skills by private study and correspondence courses. In 1924, claiming that the reciprocating engine was aesthetically unpleasing, Wankel became interested in the idea of a rotary engine. In that year he opened a workshop in Heidelberg and in 1929 he took out his first patent. In 1928 Wankel met Wilhelm Keppler, one of HITLER's economic advisers. Although Wankel had been a member of the Hitler Youth, he left the party in 1932. In the following year he exposed the corrupt behaviour of a party official and found himself in prison. Keppler secured his release, and shortly afterwards Wankel was working for the Luftwaffe in his Wankel Versuchswerkstatten at Lindau on Lake Constance, producing rotary valves and improving his rotary engine. Some of these were actually tested by Junkers, who were sufficiently impressed to place a firm order before the war ended.

Wankel was imprisoned by the French in 1945; although released in 1946, he was forbidden to resume production. In 1951, after Keppler's release from prison, sufficient funds became available to set up a new establishment in Lindau and Wankel soon had a number of manufacturers interested in his new engine. Rights were sold to the Curtiss-Wright Co. in the USA, to NSU in Germany, and to Toyo Kogyo in Japan. By 1963 cars powered by Wankel engines were available, but problems with the design of the engine have never been satisfactorily resolved and Wankel was one of the few to make money from his invention.

Warburg, Otto Heinrich (1883–1970) *German biochemist whose investigations of the role of cytochromes in cell respiration earned him the 1931 Nobel Prize for Physiology or Medicine.*

Warburg came from an illustrious family: his father, Emile Warburg, was professor of physics at Berlin University. Otto Warburg received a doctorate in chemistry from Berlin in 1906 and his medical degree from Heidelberg University in 1911. During World War I he served in the Prussian Horse Guards and was awarded the Iron Cross for gallantry. Having joined the Kaiser Wilhelm Institute for Biology, Berlin-Dahlem, before the war, he became a professor in 1918. In 1931 he moved to head the new Kaiser Wilhelm Institute for Cell Physiology in Berlin (later renamed the Max Planck Institute).

Warburg was a prodigious researcher throughout his long career; one of his early innovations was a technique for measuring the oxygen uptake of thin slices of living tissue, using a special manometer – the Warburg manometer. In this way he detected that carbon monoxide blocked oxygen uptake by cells and in 1924 he proposed that cytochromes (proteins involved in cell respiration) contain haem groups similar to haemoglobin, the oxygen carrier in blood. In 1932–33, Warburg discovered that the so-called 'yellow enzymes' (flavoproteins) are complexes of the enzyme and a coenzyme, flavin adenine dinucleotide (FAD), which he isolated in 1938. He also discovered nicotinamide adenine dinucleotide (NAD) and showed that both it and FAD function as hydrogen acceptors in the metabolic reactions of living cells. He also studied photosynthesis, especially the thermodynamic efficiency of the conversion of light to chemical energy. In cancer research, Warburg was the first to demonstrate that the metabolism of cancer cells was increased in the absence of oxygen. This prompted his controversial proposal in 1966 that cancer was the result of impaired oxidative metabolism and its prevention and treatment should entail supplementation of the diet with iron, B-vitamins, and other components required to synthesize the respiratory enzymes and coenzymes.

Ward, Dame Barbara Mary, Baroness Jackson (1914–81) *British economist, conservationist, and journalist. She was created a DBE in 1974 and a life peer, Baroness Jackson of Lodsworth, in 1976.*

Born in York, Barbara Ward graduated from Somerville College, Oxford, in politics, philosophy, and economics. In 1939 she joined *The Economist*, becoming a foreign editor in 1940. She was subsequently a governor of the BBC (1946–50), Schweitzer Professor of International Economic Development at Columbia University (1968–73), and president of the Conservation Society (from 1973). In 1950 she married Commander Robert Jackson, an Australian who worked for the United Nations and other organizations.

Barbara Ward was an influential and prolific writer on economic, ecological, and political subjects, particularly the need for countering communism in the third world with western social policies of the type that had reduced poverty and promoted social justice in the West. She was an early enthusiast for European economic union. Deeply concerned with the preservation and conservation of the world's natural resources, she wrote *Spaceship Earth* (1966) and *Only One Planet* (1972, with René Dubos), which brought her views to a wider public. Her other major books include *The International Share-Out* (1938), *The Rich Nations and the Poor Nations* (1962), and *The Home of Man* (1976).

Warhol, Andy (Andrew Warhola; 1930–87) *US pop artist.*
The son of Czechoslovak immigrant parents, Warhol was born and brought up in Pennsylvania. After studying pictorial design at the Carnegie Institute of Technology, Pittsburgh (1945–49), he was a successful commercial artist in New York before becoming a painter.

Ten years after his first one-man exhibition of 1952 he was the most widely known and most controversial of the pop artists. For subjects Warhol took the most banal and familiar illustrations from magazines and reproduced them, frequently employing repetition with slight variations, as in his *100 Soup Cans* (1962). He aimed at impersonality in his work and gloried in mechanical mass-production; typical of his remarks were 'I want to be a machine' and 'I think it would be terrific if everybody was alike'. He made extensive use of silk-screen printing, most of which was done by assistants after the early sixties – he called his studio 'The Factory'. Warhol also made a cult of being boring and superficial. 'I like boring things' he said of his six-hour film *Sleep* (1964), in which there is virtually no discernible movement.

Having retired as an artist in 1965, he devoted himself to film: his *Chelsea Girls* was the first underground film to be seen in commercial cinemas. He continued to produce drawings and constructions and also managed rock groups, including The Velvet Underground. In 1975 Warhol wrote in his *The Philosophy of Andy Warhol*: 'Business Art is the step that comes after Art.'

Warner, Rex (1905–86) *British novelist and classical scholar.*
Warner was born in Warwickshire, the son of a Church of England clergyman. He was educated at St George's School, Harpenden, and Wadham College, Oxford, and subsequently taught in schools in England and Egypt. His first *Poems* were published in 1937, the same year that his novel *The Wild Goose Chase* won him recognition as a serious novelist. The Kafkaesque vein in his fiction was further exhibited in *The Professor* (1938), *The Aerodrome* (1941), and *Why Was I Killed?* (1943). During the war Warner remained a schoolmaster and afterwards became director of the British Institute in Athens (1945–47).

After the war Warner established a reputation as a translator of Greek classical literature with his translations of the plays of Euripides – *Medea* (1944), *Hippolytus* (1950), and *Helen* (1951) – and Aeschylus – *Prometheus Bound* (1947). He also translated Xenophon's *Anabasis* (1949), Thucydides (1954), Plutarch (1958), Caesar (1959), St Augustine's *Confessions* (1962), and Xenophon's *Hellenica* (1966). He published a translation of the poems of the modern Greek poet SEFERIS in 1960. In addition to his translations, Warner also wrote a number of studies of classical subjects, including *The Young Caesar* (1958) and *Pericles the Athenian* (1963). From 1964 to 1974 he was a professor at the University of Connecticut.

Wassermann, August von (1866–1925)
*German bacteriologist and immunologist who
discovered a practical diagnostic test for syphilis (the Wassermann reaction) that, together
with other tests, is still used today.*

Wassermann was born in Bamberg, Bavaria,
and studied at Erlangen, Vienna, Munich, and
Strasbourg. He began his career as a physician
in Strasbourg in 1888 but two years later became an assistant to Robert Koch at the Koch
Institute for Infectious Diseases in Berlin, later
becoming director of the department of experimental therapy and serum research
(1906–13).

In 1906, in the wake of a number of important discoveries in immunology by his contemporaries, Wassermann developed his test for
syphilis. It depends on the fact that the blood
of an infected person contains antibodies to the
causative agent, *Treponema pallidum*, that
form a complex with known antigens. This
complex can be detected by its ability to fix
complement, a component of the blood. In
1913 Wassermann became director of the department of experimental therapy at the Kaiser
Wilhelm Institute, where he developed a diagnostic test for tuberculosis and researched the
possibilities of diagnosing cancer by testing
reactions taking place in the blood serum.

Watson, James Dewey (1928–   ) *US biochemist who, with Francis CRICK, first proposed the molecular structure of DNA, the
essential material of chromosomes and genes.
He was awarded the 1962 Nobel Prize for
Physiology or Medicine.*

Watson was born in Chicago, Illinois, and educated at the universities of Chicago and Indiana, obtaining his PhD in 1950. Following
work on viruses at the University of Copenhagen (1950–51), he moved to the Cavendish
Laboratory, Cambridge, and turned his attention to the structure of the large molecules that
determine inheritance and control the functions of living cells – deoxyribonucleic acids
(DNA). Much information about DNA was
already available, including work by Erwin
Chargaff (1905–   ) on the relative proportions
of the constituent bases and X-ray diffraction
studies of its three-dimensional structure by
Maurice WILKINS and Rosalind FRANKLIN.
Using this, Watson and Crick devised a model
for DNA comprising two complementary helically coiled chains linked by hydrogen bonds.
This double helix fitted the data and fulfilled
the biological requirements, i e it could repli-
cate itself and carry the genetic code. They
published their findings in 1953. Watson

moved to the California Institute of Technology (1953–55) and then to Harvard, becoming
a professor in 1961 and continuing work on the
genetic code. He published *Molecular Biology
of the Gene* (1965) and *The Double Helix*
(1968). In 1968 he was appointed director of
the Cold Spring Harbor Laboratory, where his
major work was cancer research. In 1989 he
also became director of the National Center for
Human Genome Research. In 1981 he published *The DNA Story*, in 1983 *The Molecular
Biology of the Cell*, and in 1984 *Recombinant
DNA*.

Watson, John B(roadus) (1878–1958) *US
psychologist who founded the school of psychology known as behaviourism, which dominated American psychology in the 1920s and
1930s and has greatly influenced contemporary views.*

Watson qualified at the University of Chicago
(1903) and continued there as an instructor,
carrying out research on behaviour in a variety
of animals. In 1908 he became professor of
psychology at the Johns Hopkins University,
Baltimore, where he began to formulate his
behaviourist ideas. Watson's view was that all
theories of animal behaviour must be objective
and based on experiments and observations
made in the laboratory. In *Psychology as a
Behaviourist Views It* (1913) he applied his
ideas to humans, proposing that as psychology
is the science of human behaviour, it should be
studied under the same objective conditions.
His experiments explored the relationships between stimuli and responses (behaviour), regarding human behaviour in terms of
conditioned responses only and disregarding
any contribution of reasoning or original
thought.

Watson's first major work was *Behaviour:
An Introduction to Comparative Psychology*
(1914), in which he defined instinct as a series
of reflexes activated by heredity. He published
his definitive version of behaviourist theory in
*Psychology from the Viewpoint of a
Behaviourist* in 1919. A year later, following
adverse publicity concerning his divorce, he
abandoned his academic career and worked in
the advertising business until his retirement in
1946.

Watson, John Christian (1867–1941) *Australian statesman and the first Labor prime
minister (1904).*

Born in Valparaiso, Chile, Watson was educated in New Zealand and left school at thirteen to work as a printer's helper on the local
newspaper. He arrived in Australia in 1886 and

started his political career when he became the president of the Labor Party Conference in 1891. Entering federal parliament in 1901, he was elected as the first Labor leader in the House of Representatives and formed the first Labor government in 1904. Powerless in office because he lacked a majority, Watson was forced to resign four months later. He gave up the leadership of the party in 1907 and his membership in 1910, and was expelled from the party in 1916 for supporting conscription.

Known for his sound judgment and fairness to opponents, Watson became a director of several companies and was chairman of Ampol (Australian Motorists' Petroleum and Oil Limited) in the late 1930s.

Watson-Watt, Sir Robert Alexander (1892–1973) *British engineer, who was responsible for the development of radar. He was knighted in 1942.*
Watson-Watt was educated at the University College of Dundee. He first worked at the Meteorological Office in London on the radio-location of thunderstorms. After the end of World War I Watson-Watt worked for the DSIR and the National Physical Laboratory on research into the reception and transmission of radio waves. In 1935, however, he was approached by the Air Ministry and asked to advise on the practicality of developing any form of death ray. Replying that any such ray would require so much power that it would be impractical, Watson-Watt added that disturbances had been noticed in radio reception when aircraft flew by, and he wondered if this could be used to warn of the presence of enemy aircraft. Watson-Watt was invited to set up a research station for the Air Ministry at Bawdsey in Suffolk, where he and his team were able to develop airborne radar, well before the outbreak of World War II in 1939. They were also able to train RAF pilots in its use, which gave them an invaluable advantage over the much larger Luftwaffe in the Battle of Britain.

Watson-Watt was later appointed scientific adviser on telecommunications to the Air Ministry, a post he retained until his retirement in 1952. For his work on radar he was awarded an ex gratia payment of £50,000.

Waugh, Evelyn (Arthur St John) (1903–66) *British novelist.*
Born in London, the son of a publisher, Waugh was sent to Lancing College, from which he won a scholarship to Hertford College, Oxford. On leaving university he spent three unhappy years schoolmastering, a period of failure crowned by an unconvincing attempt at suicide by drowning. A privately printed essay on the Pre-Raphaelites (1926) brought Waugh a commission for a book on Rossetti (1928), but still desperately needing money in order to marry the Hon Evelyn Gardner, he wrote *Decline and Fall* (1928), the first of his novels. This was an instant success, and incidentally a revenge by caricature upon the Denbighshire school in which he had taught so unsuccessfully. Waugh married in 1928 and began working on his next novel, *Vile Bodies* (1930). His wife was unfaithful to him and in 1930 he obtained a divorce; however, his marital position was complicated by his entry into the Roman Catholic Church in the same year. The necessary annulment was not granted until 1936; in 1937 he married again, to Laura Herbert, by whom he had six children. In the meantime he had visited Ethiopia for the coronation of HAILE SELASSIE (1930); this journey and his subsequent travels in Africa were reported in *Remote People* (1931) and they also provided the background for the comic novel *Black Mischief* (1932). He returned to Ethiopia as a war correspondent in 1935 and 1936, visits that were the genesis of the novel *Scoop* (1938). In 1935 his biography of the Jesuit martyr Edmund Campion won the Hawthornden Prize.

In 1939 Waugh obtained a commission in the Royal Marines, with whom in 1941 he went to the Middle East, taking part in the battle for Crete. *Put Out More Flags* (1942) depicted the upper-class characters of Waugh's fictional world adapting to the rigours and opportunities of wartime. *Brideshead Revisited* (1945), perhaps Waugh's best-known book, centres on an aristocratic English Roman Catholic family. It was later made into an immensely successful television series.

After the war Waugh resumed his travels. *The Loved One* (1948) satirizes American attitudes to death and arose from a trip to California in 1947 to discuss the possibility of a film of *Brideshead Revisited*. *Helena* (1950), about the mother of Constantine the Great, was his one historical novel. He then turned to writing his war trilogy, eventually published under the title *The Sword of Honour* (1962), comprising *Men at Arms* (1952), *Officers and Gentlemen* (1955), and *Unconditional Surrender* (1961). It was hailed as one of the greatest works of fiction to emerge from World War II and was also made into a TV series. *The Ordeal of Gilbert Pinfold* (1957) is a thinly fictionalized account of the hallucinations Waugh himself suffered after taking excessive doses of sleeping tablets. In 1959 Waugh's biography of his friend Monsignor Ronald KNOX appeared. His last travel book was *A Tourist in Africa* (1960). He then turned to writing his autobiography but completed only the first volume, *A Little Learning* (1964), before his death.

Wavell, Archibald Percival, 1st Earl

(1883–1950) *British field-marshal who, as commander-in-chief of the Middle East forces, conducted a successful campaign against Italian forces in North Africa during the early stages of World War II. He was created Earl Wavell of Eritrea and Winchester in 1947.*

The son of an army major, Wavell attended the Royal Military College, Sandhurst, and then joined the illustrious Black Watch regiment. He saw action in the Boer War, served on the Northwest Frontier in India, and lost the sight of an eye at Ypres in World War I. In the 1930s he commanded a brigade, then a division, and in 1939 was appointed supreme commander of Allied forces in the Middle East. The Italians entered the war in June 1940, promptly invaded North Africa, and in September crossed the Libyan border into Egypt. On 9 December, Wavell's troops, far outnumbered by Graziani's Italians, broke through the Italian lines at Sidi Barrani and routed the enemy in a victory that greatly boosted British morale. Simultaneously, Wavell's troops were fighting the Italians in east Africa. In January 1941 the British invaded Ethiopia and defeated the Italians at Agordat, invaded Somaliland, and by 4 April had entered Addis Ababa. Wavell was also fighting on two other fronts, against Iraq and against Vichy-controlled Syria, when in March 1941 the German commander ROMMEL arrived with his Panzer divisions. Wavell's forces were further depleted by the transfer of 60 000 troops to the ill-fated defence of Greece. Rommel swiftly took advantage of the situation and was at the Egyptian border by 11 April. In July, CHURCHILL – never a great admirer of the sometimes taciturn Wavell – replaced him with Claude Auchinleck (1884–1981), commander-in-chief for India, whose post Wavell now filled. Unfortunately, Wavell faced an equally hopeless position following his appointment as supreme Allied commander in the southwest Pacific in December 1941. His impoverished forces were no match for the Japanese, who quickly overran Burma and Singapore. In June 1943, Wavell returned to India as viceroy to deal with serious famine in Bengal and political

conflict between Hindus and Muslims. He retired in 1947. Wavell was also a writer of some merit; in 1944 he edited a volume of poetry, *Other Men's Flowers*.

Wayne, John

(Marion Michael Morrison; 1907–79) *US film star, known all over the world as 'Duke' or 'Big John'. Wayne spent fifty years in films bringing a vicarious enjoyment to millions as the virile quick-drawing hero of the American West.*

Wayne was born in Winterset, Iowa, and educated at the University of Southern California. Vacation jobs at Fox led to bit parts by 1928 until John FORD secured him the lead in *The Big Trail* (1930). A host of minor westerns followed, but in 1939 came the major breakthrough to stardom with Ford's *Stagecoach*. *Red River* (1948), *She Wore a Yellow Ribbon* (1949), *Sands of Iwo Jima* (1949), for which he was nominated for an Oscar, *The Quiet Man* (1952), and *The Alamo* (1960), which Wayne also produced and directed, are a few of the more notable of his many films.

Wayne received an Academy Award for *True Grit* (1969), the film with which he became most closely associated. His last film was *The Shootist* (1976), in which, somewhat ironically, he played an old gunslinger dying of cancer, the disease that killed Wayne himself. At his death the US Congress authorized the striking of a special gold medal in his honour.

Webb, Sidney James, Baron Passfield

(1859–1947) *British social historian, reformer, and politician, who with his wife (Martha) Beatrice Webb (1858–1943) wrote widely on socialism and founded the London School of Economics (1895). He was raised to the peerage in 1929.*

Born in London and educated at the City of London College, Sidney Webb joined the civil service and graduated in law from London University in 1885, the year in which he joined the Fabian Society through an introduction by his friend George Bernard SHAW. He wrote *Fabian Essays in Socialism* (1889) and the first edition of the Fabian Society's *Facts for Socialists* (1887). He met Beatrice Potter while she was studying the cooperative movement (she published *The Cooperative Movement in Great Britain* in 1891). They married in 1898 and from then on their work was very much of a joint nature. They wrote together *The History of Trade Unionism* (1894), *Industrial Democracy* (1897), and the mammoth work *English Local Government* (1906–29). In 1913 they founded the socialist periodical *The New Statesman*.

Sidney Webb's major interest was education, and while serving on the London County Council (1892–1910) he was responsible for creating state secondary schools and scholarships for elementary school pupils. He also helped reorganize the University of London, having earlier with Beatrice founded the LSE. Beatrice had written widely on poverty and served on the Royal Commission on the Poor Laws. They both became influential figures in the Labour Party and Sidney drafted *Labour and the New Social Order* in 1918, a major statement of Labour policy. Elected Labour MP for Seaham Harbour in Durham in 1922, Sidney served in two Labour governments, becoming president of the Board of Trade in 1924 and colonial secretary in 1929, the year he accepted his peerage. In 1932 he and Beatrice visited the Soviet Union, publishing *Soviet Communism: a new civilization?* in 1935.

Weber, Max (1864–1920) *German social scientist and political economist who became a founding father of modern sociology.*

Weber studied legal and economic history at several German universities. After a brief period as a legal assistant and on completion of his doctoral dissertation, he was appointed professor first (1894) at the University of Freiburg and then (1897) at Heidelberg. Despite a severe nervous breakdown several years later, Weber produced a body of work that established him as the foremost figure in social thought of the twentieth century.

Weber's study was in three main areas. His study of the sociology of religion led to his celebrated book, *Die protestantische Ethik und der Geist des Kapitalismus* (1904; translated as *The Protestant Ethic and the Spirit of Capitalism*, 1930), in which he linked the psychological effects of Calvinism and Lutheranism with the development of European capitalism. Secondly, his interest in political sociology, presented in such works as the unfinished *Wirtschaft und Gesellschaft* (1922; translated as *Economy and Society*, 1968), led him to major discussions on types of economic activity and the relationship between social and economic organization. Finally, he laid down various systems of inquiry in authoritative essays published posthumously in translation as *Methodology of the Social Sciences* (1949).

In all his work Weber tried to trace links between different types of social activity and stressed that the bureaucratization of political and economic society was the most significant development in the modernization of western civilization. He also produced major analyses

of society in Israel, China, and India, in each of which he showed the mutual dependence of culture and society. Towards the end of his life, Weber became politically active and served on the committee that drafted the constitution of the Weimar Republic in 1918.

Webern, Anton Friedrich Wilhelm von (1883–1945) *Austrian composer and conductor who, with Alban BERG, was the leading exponent of SCHOENBERG's classical form of serial composition.*

Born in Vienna, Webern first became interested in music after hearing Wagner's music-dramas at the Bayreuth festival. As a student of musicology at the University of Vienna (1902–06), Webern studied the complex polyphonic style of the fifteenth-century composer Heinrich Isaac, which – together with the music of his teacher Schoenberg – strongly influenced his own style of composition. After leaving university he worked as a conductor in various provincial theatres and in Prague (1917). In 1918 he was closely associated with Schoenberg and Berg when the former founded the Society for Private Musical Performances.

Webern followed Schoenberg's evolution from tonality to atonality and serialism and became a master of compositional technique and brevity. The *Passacaglia for Orchestra* (1908) was his longest movement, lasting about ten minutes; such works as the *Five Movements for String Quartet* (1909), *Six Orchestral Pieces* (1910), *Six Bagatelles for String Quartet* (1913), and *Five Orchestral Pieces* (1913) lasted no more than a minute each. In these atonal works Webern developed the 'Klangfarbenmelodie' ('sound melody') technique, in which each note of a theme has a different instrumental timbre and often a separate dynamic marking; this was later to influence such composers as BOULEZ and STRAVINSKY. He also used Schoenberg's twelve-tone technique to great effect in such pieces as *Three Sacred Folk Songs* (1924) and *Variations for Orchestra* (1940). He died in Vienna in 1945, when he was mistakenly shot by a US soldier during the Allied occupation.

Weil, Simone (1909–43) *French philosopher, whose life and thought fit into no obvious categories.*

Born in Paris, the daughter of a doctor, she came from an agnostic intellectual Jewish background. Her brother André Weil (1906–) is a mathematician of considerable importance. Simone distinguished herself at the École Normale Supérieure as a student from 1928 to 1931 but totally rejected the attractions of an orthodox academic career. Instead, in a similar way to George ORWELL in England, although with more passion and commitment, she chose to identify herself with and live the life of the poor and the oppressed. Consequently she served in the Spanish civil war on the republican side, worked as a manual labourer on farms and in the Renault car factory, and later, during World War II, with the resistance movement in England. Finally, she resolved to restrict herself, while continuing to work, to the same diet served to the inmates of the Nazi labour camps. Far from robust when she undertook such a regime, she soon died from tuberculosis.

Simone Weil published little during her life. Her reputation rests therefore on the posthumous appearance of such works as *La Pesanteur et la grâce* (1949; translated as *Gravity and Grace*, 1952) and, above all, *Cahiers* (1952–55; translated as *Notebooks*, 1956). In no sense an orthodox Christian or socialist – she chose, in fact, to remain outside all religious and political organizations – Weil reveals in her writings that she had been strongly influenced by the Christian mystic and gnostic traditions, as well as by early socialist analyses of modern industrial society.

Weill, Kurt (1900–50) *German composer, who developed a new genre of satirical opera in collaboration with Bertolt BRECHT.*

Weill studied at the Berlin High School of Music and worked as an opera coach and conductor before studying with BUSONI (1921–24). Turning to composition, he developed a style that was forcefully direct, economical, and harmonically simple and lent itself well to political satire. Although his early operas, including *Der Protagonist* (1926; *The Protagonist*), were well received, it was his first collaboration with Bertolt Brecht, *Aufstieg und Fall der Stadt Mahagonny* (1927; *The Rise and Fall of the City of Mahagonny*), that established Weill as Germany's most promising young composer. This reputation was consolidated with his next work with Brecht, *Die Dreigroschenoper* (1928; *The Threepenny Opera*), based upon John Gay's *Beggar's Opera* (1728). Brecht was again the librettist for *Der Jasager* (1930), and *Die Sieben Todsünden der Kleinbürger* (1933).

Persecuted by the Nazis, on account of his artistic activities and because he was a Jew, Weill moved to Paris (1933), then to London, and finally to New York (1935). There he embarked upon a series of successful Broadway musicals, including *Johnny Johnson* (1935), *The Eternal Road* (1937), *Knickerbocker Holiday* (1938), *Lady in the Dark* (1941), *One Touch of Venus* (1943), *Street Scene* (1947), and *Lost in the Stars* (1949). His wife, the singer Lotte Lenya (1900–81), appeared in many of his works. Weill died at the height of his fame of a heart attack. Other works included two symphonies, a violin concerto, and *Lindberg's Flight* (1928) and *The Seven Deadly Sins* (1933), both to libretti by Brecht.

Weinberg, Steven (1933–) *US physicist, who shared the 1979 Nobel Prize for Physics with Abdus SALAM and Sheldon GLASHOW for his unification of the weak and the electromagnetic forces.*

The son of a New York court stenographer, Weinberg was educated at the Bronx High School of Science, where Sheldon Glashow was a classmate. He later attended Cornell and Princeton universities, gaining his PhD at Princeton in 1957. After holding appointments at Columbia (1957–59), Berkeley (1959–69), and MIT (1969–73), Weinberg moved to Harvard, where he was Higgins Professor of Physics (1973–83); since 1982 he has been Josey Regental Professor of Science at the University of Texas.

In 1967 Weinberg showed how two of the four basic forces operating on elementary particles, the weak and the electromagnetic, could be united to yield a single force, an achievement comparable to the unification of electrical and magnetic forces by James Clerk Maxwell. One consequence of the theory is a hitherto unobserved neutral current between elementary particles, which was first detected in 1973. A similar unification was independently proposed by Abdus Salam in 1968 and later extended by Glashow. Since his initial success Weinberg has been seeking a unified field theory covering all four forces of nature, but so far has not achieved an acceptable hypothesis.

Weinberg is also well known for his work in astrophysics and cosmology. He is the author of a standard monograph on the subject, *Gravitation and Cosmology* (1972), as well as a successful popular work, *The First Three Minutes* (1978); other publications include *Elementary Particles and the Law of Physics* (1988).

Weissmuller, Johnny (1904–84) *US record-breaking swimmer, who won three Olympic gold medals in 1924 and two in 1928. Later he* *achieved wider recognition as the star of the Tarzan films of the 1930s and 1940s.*

At the age of seventeen Weissmuller had already become a swimming star. The 1924 Olympics brought him gold medals in the 100 m and 400 m freestyle and the 4 × 200 m freestyle relay and a bronze medal in the water polo. He was the first man to swim 100 metres in less than one minute and 440 yards in less than five minutes, and his 100 yards record of fifty-one seconds stood for seventeen years. In all, he set twenty-eight world records in freestyle events. In 1929 he turned professional. Weissmuller's great success was due to the work he and his coach at Illinois Athletic Club, William Bachrach, devoted to improving the American crawl. Weissmuller rode higher in the water than his contemporaries and adopted an unusual breathing pattern by turning to either side and watching his opponents without disrupting the stroke.

Despite limited acting ability, his superb physique and agility made him an obvious choice for the part of Tarzan in the early films. After his last Tarzan film (1948) he starred in another film series playing the character of Jungle Jim.

Weizmann, Chaim Azriel (1874–1952) *First president of Israel (1949–52). Considered one of the founding fathers of Israel, he was widely respected not only for his political involvement in the cause of Zionism but also for his achievements as a scientist.*

Born near Pinsk in Russian Poland, Weizmann studied chemistry at the universities of Darmstadt and Berlin before gaining a doctorate in science from the University of Freiburg in 1900. Between 1900 and 1916 he taught chemistry and biochemistry at the universities of Geneva and Manchester, becoming a British subject in 1910. While working as the director of the Admiralty Laboratories (1916–19) he discovered a process for the manufacture of acetone, which proved invaluable to the British munitions industry.

Weizmann became an ardent supporter of the Zionist cause in the early 1900s. During World War I his connections with the British authorities and his active involvement in the Zionist movement enabled him to participate in the negotiations that resulted in the Balfour Declaration (1917), which outlined British support for a Jewish homeland in Palestine. He was president of the World Zionist Organization (1920–31; 1935–46) and then became president of the Jewish Agency for Palestine (1929), in which capacity he appeared before the United Nations Special Committee for Palestine in 1947. In 1949, after the declaration of independence, he became Israel's first president. He remained in office until his death in 1952.

The Sief Research Institute in Rehovoth, which he founded in 1932 and in which he served as director (1932–52), was renamed the Weizmann Institute of Science in his honour. Many pilgrims have visited his grave at Rehovoth. His autobiography, *Trial and Error*, was published in 1949.

Welensky, Sir Roy (1907–) *Prime minister of the Central African Federation (1956–63). Totally dedicated to the concept of the Federation, he regarded the British agreement to the independence of Malawi and Zambia as an abdication of responsibility. He was knighted in 1953.*

Born in Salisbury (now Harare), the son of a Lithuanian father and an Afrikaner mother, Welensky was educated in Salisbury before joining the railways as a fireman and engine driver in 1924. Moving to Broken Hill (now Kabwe) in 1933, he became chairman of the Railway Workers Union, a position he held until 1953. From 1926 to 1928 he was the heavyweight boxing champion of the Rhodesias. Welensky entered politics when he was elected to the Northern Rhodesian legislative council in 1938. He was promoted to the executive council in 1946 and became leader of the unofficial opposition in 1947. Cooperating with Huggins of Nyasaland to establish the Central African Federation in 1953, he was elected to the federal legislature the same year, becoming deputy prime minister under Huggins. On Huggins's retirement in 1956, Welensky became federal prime minister, heading the United Federal Party. He remained in his position until dissolution in 1963. Following the unsuccessful attempt to win the seat of Arundel in the Southern Rhodesian parliament in 1964, he retired from politics.

Welles, (George) Orson (1915–85) *US actor and director, whose screen career started with the blockbusting* Citizen Kane, *an achievement he never repeated.*

Born in Kenosha, Wisconsin, Welles first became interested in drama with school productions. After attending the Chicago Art Institute he appeared at the Gate Theatre, Dublin, in 1931 and went on to tour in Africa and America. He subsequently founded his own company, Mercury Theatre, producing plays and making Shakespeare recordings for schools. The Mercury Theatre's radio broadcast of H.

G. WELLS's *The War of the Worlds* (1938) caused widespread panic among American listeners, who believed a real Martian invasion was taking place. This episode brought Welles considerable public attention.

Welles produced, directed, wrote, and acted in his first film, the critically acclaimed *Citizen Kane* (1941), based on the life of William HEARST, the newspaper magnate. The impact of this film was so enormous that Welles had difficulty in living up to it. However, his second film, *The Magnificent Ambersons* (1942), was well, if not enthusiastically, received. Notable of the films that followed were *Jane Eyre* (1943), in which Welles played Rochester, and *The Lady from Shanghai* (1948), which also featured his second wife, Rita Hayworth (1918–87).

Welles then left Hollywood for a long stay in Europe. In Britain he appeared briefly in Carol REED's masterpiece *The Third Man* (1949), as the philosophical racketeer Harry Lime, and on a visit to the USA made *Touch of Evil* (1958), which won a Brussels World's Fair prize. Other films include *Macbeth* (1947), *Othello* (1956), and *Chimes at Midnight* (1966), for which he drew on Shakespeare's 'Falstaff' plays. Lear and Othello were among his stage roles, his London debut being made in the latter in 1951. He was awarded a Special Oscar (1971) and the Life Achievement Award of the American Film Institute (1975).

Wells, H(erbert) G(eorge) (1866–1946) *British novelist and social philosopher.*

Wells was born in Bromley, Kent, the son of an unsuccessful tradesman and a lady's maid. As the result of a boyhood accident he became an omnivorous reader, which provided an opportunity to supplement his meagre elementary-school education. When he was fourteen his father became insolvent and the Wells home broke up, with H. G. Wells, after several false starts, being apprenticed to a draper. Although he loathed the job it gave him time to study. In 1883 he managed to obtain a place as a student teacher at Midhurst Grammar School, from which he gained a free studentship to what later became the Imperial College of Science, South Kensington, where he was taught biology by T. H. Huxley. At this time he was also interested in socialism and became an early member of the Fabian Society.

After failing his third-year examinations, Wells became a teacher but his health intermittently broke down. In 1890 he finally took his BSc. In 1891 he married his cousin but abandoned her in 1893 for one of his students, whom he married in 1895. Meanwhile he had already begun his career as a writer and found a ready public for his stories. *The Time Machine* (1895) was the first in a long series of so-called 'Wellsian' scientific romances, embodying his fascination with technological change and the potentialities of a reorganized human society. His knack of being able to anticipate scientific developments, such as the splitting of the atom in *The World Set Free* (1914), won him the status of a popular prophet. In the 1890s he also started to write his realistic novels, the first of which was *The Wheels of Chance* (1896). He followed this with *Love and Mr Lewisham* (1900), *Kipps* (1905), *Tono-Bungay* (1909), *Ann Veronica* (1909), and *The History of Mr Polly* (1910); all these drew strongly upon his own youthful experiences and were immensely popular, although *Ann Veronica* caused something of a scandal by its advocacy of greater sexual freedom for women. Social change, especially as affecting the status of women, was the theme of several of Wells's books in the early 1900s, including *A Modern Utopia* (1905), *The New Machiavelli* (1911), and *Marriage* (1912). He resigned from the Fabian Society in 1908 but continued to pursue various radical crusades. He was a poor orator so he relied upon his books, of which he wrote over a hundred, to make his views known.

The futile tragedy of World War I caused Wells to focus his attention sharply on a strategy to enable mankind to avoid self-destruction. After a brief flirtation with theism, expressed in the novel *Mr Britling Sees It Through* (1916), he pursued the idea of a world state. In some of his most powerful later treatises – *The Outline of History* (1920), *The Science of Life* (1931), and *The Work, Wealth and Happiness of Mankind* (1932) – he laid the theoretical groundwork for such a state.

Wesker, Arnold (1932–) *British playwright and director. He became a Fellow of the Royal Society Literature in 1985.*

Wesker was born in the East End of London into a family of communist Jewish immigrants. He was educated at Upton House Central School, Hackney, which he left in 1948. For the next decade he pursued a variety of unskilled or semiskilled trades, including farm labourer and pastrycook, broken by a spell in the RAF (1950–52). His first play, *The Kitchen*, was produced at the Royal Court Theatre, London, in 1959 and filmed in 1961. The trilogy *Chicken Soup with Barley* (1957),

Roots (1959), and *I'm Talking about Jerusalem* (1960) were first performed at the Belgrade Theatre, Coventry, and subsequently at the Royal Court (1959–60). *Chips With Everything* (1962) was his first West End success and was produced on Broadway the following year. These five plays established Wesker as an important force in the British theatre.

Between 1961 and 1970 Wesker was director of Centre 42, an institution based on labour-movement support and aiming to enable the arts to reach beyond their traditional middle-class audience. In this period Wesker began to direct his own plays, with *The Four Seasons* in Cuba in 1968 and the world première of *The Friends* in Stockholm in 1970. Since that time many of his plays have been premièred abroad; *The Journalists*, for example, appeared first on Yugoslav TV. *Love Letters on Blue Paper*, published as a story in 1974, was staged in 1978 as a play at the National Theatre, London, with the author directing. *Caritas* (1981) was also first produced at the National Theatre. In 1980 his *Collected Plays* appeared in four volumes. More recent work includes *Four Portraits* (1982), *Annie Wobbler* (1983), *When God Wanted a Son* (1986), and *Beorhtel's Hill* (1989).

West, Mae (1892–1980) *US actress, who became a buxom burlesque of a sex symbol with a great sense of comedy.*

Born in Brooklyn, New York, West went into show business as a child and progressed to vaudeville. Early stage successes included *À la Broadway* (1911) and *Sometime* (1913). *Sex* (1926), the first play she wrote, was considered so sensational that the authorities closed it down and imprisoned her for ten days for obscenity. Her public following grew, however, with such plays as *Diamond Lil* (1928) and *The Constant Sinner* (1931). *Night After Night* (1932) was her first film, with *She Done Him Wrong* (1933), *Klondike Annie* (1936), *Go West Young Man* (1936), and *My Little Chickadee* (1940) among the many successes that quickly followed.

During the 1940s West returned to the stage and, later, appeared in cabaret. Her film career, however, was not over; at seventy-eight she made *Myra Breckinridge* (1970) and at eighty-five *Sextette* (1978), believing to the end that 'It's better to be looked over than overlooked'. Her autobiography, *Goodness Had Nothing to Do with It*, was published in 1959. Her name has entered the language as a type of lifejacket worn round the chest.

West, Nathanael (Nathan Wallenstein Weinstein; 1903–40) *US novelist whose satirical fiction is considerably more popular now than at the time of his premature death.*

West was born and educated in New York. He completed his first novel, *The Dream Life of Balso Snell* (1931), in Paris, where he lived for two years. *Miss Lonelyhearts* (1933), perhaps the most famous of his books, fulfilled the promise of the first novel. Its theme was the gradual disintegration, culminating in murder, of the hero, who is unable to cope with the misery he confronts in the letters he receives as an agony aunt. It was followed by *A Cool Million* (1934) and *The Day of the Locust* (1939). The former was a satire on political opportunism and the latter, also highly acclaimed, an uncompromising and objective account of the sordid side of Hollywood, where West spent the last five years of his life writing scripts. He was killed, together with his wife, in a car crash at the age of thirty-seven. Much of West's writing is now highly regarded – the black comedy and unremitting pessimism of his books were not as acceptable in the 1930s as they are today.

West, Dame Rebecca (Cicily Isabel Fairfield; 1892–1983) *British writer. She was made a DBE in 1959.*

Rebecca West was born in Kerry and educated at Watson's Ladies College in Edinburgh. She trained as an actress in London, but from 1911 onwards became increasingly attracted to journalism, particularly in the cause of women's suffrage. In 1913 she began her stormy liaison with H. G. WELLS and their son Anthony was born the following year. In 1916 her first significant work, a critical biography of Henry James, appeared.

In 1923 Rebecca West broke with Wells and went to the USA, where her journalistic talent was soon recognized. She had also begun writing novels, publishing *The Judge* (1922) and *Harriet Hume* (1929), among others. In 1930 she married Henry Maxwell Andrews, who died in 1968. She was sent to report on Yugoslavia in 1937 and the outcome of her observations there is perhaps her most famous book: *Black Lamb and Grey Falcon* (1942). After World War II the *New Yorker* magazine commissioned her to write about the trial of 'Lord Haw-Haw' (William JOYCE); this prompted the major study of the psychology of traitors, *The Meaning of Treason* (1949). *A Train of Powder* (1955) grew out of her observations on the Nuremberg trials of German war criminals. She pursued the theme of treachery in two later publications, *The Vassall Affair* (1963) and *The New Meaning of Treason* (1964).

The Thinking Reed (1936), *The Fountain Overflows* (1957), *The Birds Fall Down* (1966), and *This Real Night* (a sequel to *The Fountain Overflows* published posthumously in 1984) evinced Rebecca West's continuing interest in fiction, but her reputation and most later books lie in the field of nonfiction. In *The Court and the Castle* (1958) she turned to the study of religion and politics in imaginative literature. *1900* (1982) is a skilful evocation of the world of her childhood.

Wheeler, John Archibald (1911–) *US theoretical physicist, known for his work in nuclear physics and cosmology.*

Born in Jacksonville, Florida, he was educated at Johns Hopkins University, where he obtained his PhD in 1933. After two years in Copenhagen working with Niels BOHR he returned to America, teaching first at the University of North Carolina and then (1938) at Princeton, where he was professor of physics (1947–76). In 1976 he moved to the University of Texas as director of the Center for Theoretical Physics.

It was with Bohr that Wheeler made his first major contribution to physical theory. In 1939 they developed the liquid-drop model of the atomic nucleus and quickly saw that it would have implications for the recently discovered phenomenon of nuclear fission. Calculations showed that the rare isotope uranium-235 would be more fissionable than the common isotope uranium-238. It was this insight that made the atomic bomb a possibility, although it imposed on its builders the enormous task of uranium enrichment. Wheeler himself worked on the project at the Metallurgical Laboratory, Chicago, where he identified the problem of reactor poisoning and proposed design changes to deal with it.

After the war, Wheeler worked (1945–49) with FEYNMAN on the problem of action at a distance. Later, in the early 1950s, he returned to the development of nuclear weapons; this time the problem was to design a hydrogen bomb. With the success of the project Wheeler turned to cosmology and general relativity. In his *Geometrodynamics* (1962) he attempted to unify the gravitational and electromagnetic fields by introducing the concept of a geon. This proved no more successful than comparable attempts by others. More durable, perhaps, has been his work on black holes, a term he coined. In this field he formulated the so-called 'No Hair' theorem: black holes are bald, i e

their only known properties are their mass, charge, and angular momentum.

In 1979 Wheeler became the centre of a prolonged and continuing controversy, when at the annual meeting of the American Association for the Advancement of Science (AAAS) he called for the expulsion of the recently admitted parapsychologists. They had produced no hard results, he argued, and until such results had been produced parapsychology should not be granted the respectability of AAAS membership.

Wheeler, Sir (Robert Eric) Mortimer (1890–1976) *British archaeologist, noted for his work on Romano-British sites, who developed new methods of excavation, recording, and interpretation that greatly influenced archaeological field work. He received a knighthood in 1952 and was made a CH in 1967.*

After receiving his BA degree in 1910 from University College, London, and his MA two years later, Wheeler was appointed assistant to the Royal Commission on Historical Monuments. During World War I he served in the Royal Artillery and afterwards (1920–26) was appointed keeper of archaeology at the National Museum of Wales, Cardiff. He excavated several sites in the 1920s, including Sergontium, Brecon Gaer, Caerleon, and Lydney, and wrote *Prehistoric and Roman Wales* (1925). In 1926 he became curator of the London Museum, during which time he wrote several books about the capital's history. From 1930, Wheeler and his wife, Tessa Verney, spent three years excavating the site of Verulamium, near St Albans, and discovered evidence of a pre-Roman settlement controlled by the Belgic kings. Perhaps his most famous discoveries were made at Maiden Castle, Dorset, a prehistoric hilltop fort. Wheeler's excavations here (1934–37) revealed the existence of an Iron Age fortified village dating from the fourth century BC. His findings were published in *Maiden Castle* (1943). In 1938 and 1939 he worked on Roman sites in Brittany and Normandy.

During World War II Wheeler served with MONTGOMERY's forces in North Africa, becoming a brigadier in 1943. The following year he was appointed director-general of archaeology to the Indian government. His work on the citadel mound at Harappa in the Punjab is described in *The Indus Civilization* (1953). In 1948 he returned to Britain as professor of the Roman provinces at the London Institute of Archaeology and in 1965 he became professor of ancient history to the Royal Academy.

Above all, Wheeler wrought changes in methodology. *Archaeology from the Earth* (1954) covers his innovations in planning, photography, recording, and preserving finds, which have had a lasting impact.

White, Patrick Victor Martindale (1912– 90) *Australian novelist and playwright. He won the 1973 Nobel Prize for Literature.*

White was born in London during a visit to Europe by his parents but was brought up on his father's sheep station in Australia. He returned to England as a pupil at Cheltenham College and went on to King's College, Cambridge. Settling in London with the intention of becoming a novelist, he destroyed his first two novels. *Happy Valley* (1939), his next novel, became an instant success. White served in the RAF during World War II as an intelligence officer, based mainly in Greece and the Middle East. His second published novel, *The Living and the Dead*, appeared in 1941, and *The Ham Funeral*, his first play, followed in 1947. In 1948 he returned to settle in Australia on a farm near Sydney and in the same year published his novel *The Aunt's Story*.

The Tree of Man (1955) and *Voss* (1957) established him in the forefront of contemporary writers; *Voss*, the story of the doomed attempt made in 1845 by a German immigrant to cross the Australian continent, epitomizes the grandeur and the pathos White perceived in human enterprises. In the early 1960s he produced another major novel, *Riders in the Chariot* (1961), and three plays: *The Season at Sarsaparilla* (1961), *A Cheery Soul* (1962), and *Night on Bald Mountain* (1962), published in one volume in 1965. He also wrote the short stories *The Burnt Ones* (1964). Later novels include *The Solid Mandala* (1966), *A Fringe of Leaves* (1976), and *The Twyborn Affair* (1980). His work has been translated into many languages. Towards the end of his life he campaigned for complete Australian independence from the UK and the USA; he returned his Order of Australia in 1976 and spoke out against the 200th anniversary celebrations of white settlement in Australia in 1988.

White, T(erence) H(anbury) (1906–64) *British writer.*

Born in Bombay, White was educated at Cheltenham College and Queen's College, Cambridge. From 1930 to 1936 he taught at Stowe School, where his collection of unusual pets marked him out from the ordinary run of schoolmasters. His first book to be published was the verse collection *Loved Helen* (1926), but it was not until ten years later that he won recognition with the autobiographical *England Have My Bones* (1936). *The Sword and the Stone* (1938) was the first volume in White's highly original reworking of the Arthurian legend.

During World War II White lived in seclusion in Ireland and in 1945 moved to the Channel Island of Alderney. Besides the further volumes in the Arthurian series, he wrote *Mistress Masham's Repose* (1946), *The Elephant and the Kangaroo* (1947), and a social history entitled *The Age of Scandal* (1950). *The Goshawk* (1951) is an account of his experiences as a falconer. In 1958 White's Arthurian novels appeared in a one-volume revised version under the title *The Once and Future King*. This formed the basis of the musical *Camelot* (1960) and brought White fame and considerable wealth. He died while on a Mediterranean cruise.

Whitehead, A(lfred) N(orth) (1861–1947) *British mathematician, logician, and metaphysician.*

The son of a clergyman, Whitehead was born in Ramsgate, Kent, and educated at Trinity College, Cambridge, where he remained as student, lecturer, and fellow until 1910, when he took up an appointment at Imperial College, London, as professor of applied mathematics. In 1924, at an age when most people are contemplating retirement, Whitehead resigned his chair and moved to the USA to begin a new career as professor of philosophy at Harvard.

Whitehead's reputation was initially established by his work in mathematical logic. In 1898 he published his *Universal Algebra* and shortly afterwards began, in collaboration with his most famous pupil, Bertrand Russell, an ambitious project to demonstrate that the whole of mathematics was derivable from purely logical assumptions. A decade of intense labour finally saw the results of their collaboration in the publication of *Principia Mathematica* (3 vols, 1910–13), the most important logical work since the days of Aristotle. A fourth volume, on geometry, by Whitehead alone, was planned but never completed.

For the next decade Whitehead wrote on problems concerning the nature and development of science. In the 1920s, however, he turned his attention to the construction of a metaphysical system, the details of which he published in his *Process and Reality* (1929). Whitehead rejected traditional attempts to identify the ultimate constituents of reality with atoms or points of space, proposing instead that events are the ultimate stuff of nature. In developing his philosophy of organism Whitehead thus had to show how a world of apparently permanent objects could be represented by transitory events. He sought to describe such a world rather than to deduce its features or to argue for its properties. The result is a work of considerable difficulty, replete with neologisms and obscurities.

Whitlam, (Edward) Gough (1916–) *Australian statesman and Labor prime minister (1972–75). He became a Companion of the Order of Australia in 1978.*

Born in Melbourne, Victoria, Whitlam was educated in Canberra and Sydney, graduating from the University of Sydney with BA and LLB degrees. He served in the RAAF as a flight lieutenant (1941–45) and was admitted to the New South Wales bar in 1947.

Whitlam joined the Australian Labor Party in 1945. After standing unsuccessfully for a state seat in 1950, he won the Federal seat of Werriwa in a by-election (1952). He became deputy leader of the parliamentary Labor Party in 1960 and leader of the party (in opposition) in 1967. Although unsuccessful in the 1969 House and 1970 Senate elections, he won the House of Representatives election in 1972, becoming the first Labor prime minister for twenty-three years. In 1974 Whitlam called an election for both houses in response to a challenge from the Senate. He won a majority in the House of Representatives, but not in the Senate. When, in 1975, the opposition blocked finance bills in the Senate, he refused to call a general election. In circumstances of great controversy, he was dismissed as prime minister by the governor-general, Sir John Kerr. Although defeated in the subsequent election, he remained leader of the Labor Party until 1977: he retired from parliament in 1978.

A classical and historical scholar, Whitlam became Visiting Professor of Australian Studies at Harvard University in 1979 and First National Fellow of the Australian National University in 1980. Subsequent appointments included Australian president of the International Commission of Jurists (1982) and Australian representative to UNESCO in Paris (1983–86); he joined the executive board of UNESCO in 1985.

Whittle, Sir Frank (1907–) *British aeronautical engineer responsible for the design and development of the first successful jet engine. He was knighted in 1948 and admitted to the OM in 1986.*

The son of a mechanic, Whittle assisted his father in his work from an early age. Unlike his father, however, he received a first-class technical education as a Royal Air Force apprentice at the RAF College, Cranwell and later (1934–36) at Cambridge. While at Cranwell Whittle wrote a thesis entitled 'Future Developments in Aircraft Design', in which he predicted that for planes of the future to fly significantly faster they would need to ascend to great heights, where air resistance was negligible. He accepted, however, that at such great heights the traditional propellor would be useless. Whittle thought almost immediately of jet propulsion but could not visualize the way in which it should work. It was only when, in 1929, it occurred to him to couple a compressor to a power-driven turbine rather than a piston engine that the turbojet seemed a feasible proposition. Although Whittle's plans were dismissed by the Air Ministry as impractical, he took out a patent in 1930. As the Air Ministry had no interest in the patent application, it was not considered necessary to place it on the secret list. When the patent was due for renewal in 1935, Whittle felt unwilling to invest a further £5 in the renewal fee. However, the German aircraft industry, less conservative than the British Air Ministry, began to develop Whittle's engine.

With the encouragement and support of two former RAF colleagues, R. Dudley Williams and C. Tinling, further patents were taken out, £20,000 of private capital raised, and the company Power Jets formed (in 1936). Only in mid-1939, by which time jet engines had actually been built, did the Air Ministry recognize Power Jets and provide funds. With a Whittle jet engine fitted, a specially built Gloster E28/39 made its first flight on 15 May 1941. It entered service with the RAF in 1944. In the same year Power Jets was taken over by the government; unhappy with ministry interference, Whittle resigned in 1945. For his work on the development of the jet engine he was awarded a tax-free sum of £100,000. He resigned from the RAF, with the rank of air commodore, in 1948; over the next thirty years he served as a consultant with such companies as Bristol Siddeley Engines and British Overseas Airways Corporation. In 1977 he moved to the USA to take up the post of research professor at the US Naval Academy, Annapolis.

Wiener, Norbert (1894–1964) *US mathematician, the founder of cybernetics.*

The son of a Harvard professor of Slavonic languages and literature, Wiener was born in Columbia, Montana. A mathematical infant prodigy, he graduated from Tufts at the age of fourteen and went on to gain his Harvard PhD at eighteen. He completed his education in Europe, studying under Bertrand RUSSELL in Cambridge and David HILBERT in Göttingen. On his return to the USA, however, he experienced considerable difficulty in finding a job. It was not until 1919, with the wartime shortage of staff, that he obtained an appointment at the Massachusetts Institute of Technology, where he prudently remained for the rest of his career. Wiener worked initially on mathematical logic, later turning to the theory of random processes, but he is best known for his work on harmonic analysis and Fourier transforms.

This work, carried out in the 1920s and 1930s, came to be overshadowed in the 1940s by his growing interest in 'the science of control and communication in the animal and machine', a subject for which he coined the name 'cybernetics'. His book *Cybernetics* (1948) was somewhat ahead of its time and in some respects gave the impression of a writer seeking a subject rather than of one presenting a thoroughly worked out thesis. Nevertheless it managed to introduce to an eager public a whole range of new concepts. Such words as 'homeostasis', 'negative feedback', and 'cybernetics' itself rapidly became part of the language and prepared the way for the computer revolution, which in 1948 still lay a generation ahead. Wiener has left a fascinating account of his strange childhood in his first volume of autobiography, *Ex-Prodigy* (1953). The second volume, *I Am a Mathematician*, is of less interest.

Wiesenthal, Simon (1908–) *Austrian Jewish war crimes investigator, whose activities have brought to justice many perpetrators of atrocities against the Jewish people during the Nazi era.*

Born in Buczacz, Poland, he trained as an architect but subsequently spent three years (1942–45) in Nazi labour and concentration camps. Miraculously both he and his wife survived the war and Wiesenthal began his long campaign to bring Nazi war criminals to justice. He aided the US army in gathering evidence for the war trials and in 1947 opened the Documentation Centre on the fate of the Jews and their Persecutors in Linz, Austria. The centre, which also helped survivors of the camps as well as contributing to the war trials, closed in 1954, but Wiesenthal continued to

work alongside Israeli agents to seek out unprosecuted Nazis. In 1959 he played a major role in the discovery of Adolf EICHMANN in Argentina. In 1961 he opened the Jewish Documentation Centre – also called the Wiesenthal Centre – in Vienna and continued to track down Nazi criminals when other countries had ceased to pursue their cases. Although his refusal to consider any kind of amnesty for now ageing Nazis has caused heated moral debates over the years, with some critics suggesting that such a desire for retribution is as undesirable as the evil it aims to expose, Wiesenthal has received many international honours for his work, including the rank of Chevalier de la Légion d'honneur. His books about his experiences include *I Hunted Eichmann* (1961) and *Every Day Remembrance Day* (1986).

Wigner, Eugene Paul (1902–) *Hungarian-born US physicist, who was awarded the 1963 Nobel Prize for Physics for his introduction of the concept of parity into nuclear physics.*

The son of a Budapest businessman, Wigner was educated at the Berlin Technische Hochschule, where he obtained a doctorate in engineering. After a short period at Göttingen University, Wigner moved to the USA in 1930 and taught mathematical physics at Princeton before being appointed Thomas D. Jones Professor of Mathematical Physics, a post he held from 1938 until his retirement in 1971. Wigner became a naturalized US citizen in 1937.

In 1927 Wigner introduced the concept of parity as a property of nuclear reactions that is conserved. A basically mathematical idea relating to certain transformations of the wave function, it has the effect that nuclear processes look the same whether they are observed directly or seen in a mirror. The virtue of Wigner's proposal is that it enables the existence of some processes to be predicted. It was later found (1956) that parity is not conserved in weak interactions.

Wigner joined with his Hungarian colleagues, SZILARD and TELLER, to persuade Albert EINSTEIN in 1939 to write his famous letter to President ROOSEVELT advising him of the possibilities of nuclear weapons. During World War II Wigner himself worked on the bomb in Chicago and after the war briefly was director (1946–47) of the Atomic Energy Commission's laboratory at Oak Ridge, Tennessee.

Wilder, Thornton Niven (1897–1975) *US novelist and playwright.*

Born in Madison, Wisconsin, Wilder lived with his parents for a time in China, where his father was consul-general at Shanghai and Hong Kong. He graduated from Yale in 1920, after serving for a year in World War I and studying at the American Academy in Rome. He became housemaster at Lawrenceville School while doing postgraduate studies at Princeton, where he received an MA in 1925. He wrote a novel and a play before publishing the book that solidly established his reputation, *The Bridge of San Luis Rey* (1927), which won a Pulitzer Prize. The story, which traces the lives of unrelated characters up to the fatal and providential moment that unites them (the collapse of a Peruvian footbridge), contains a philosophical interest and range that was to characterize his later work. From 1930 to 1936 Wilder was professor of English at the University of Chicago. He had already written his first important play, *The Angel that Troubled the Waters* (1928), but his two major dramatic works were to follow. *Our Town* (1938) was an innovatory play, performed on an empty stage with the stage manager appearing as a character in various roles as required; it presented the day-to-day lives of two families in the microcosm of a New Hampshire town. The historical view took a much broader form in *The Skin of Our Teeth* (1942), which relates the near-miraculous survival of the archetypal Antrobus family through such catastrophes as the Ice Age, the Flood, and finally a devastating war. Wilder's comedy *The Matchmaker* (1954; originally entitled *The Merchant of Yonkers*, 1938) is an adaptation of a farce by the Austrian playwright Nestroy; more recently it provided the plot for the musical *Hello, Dolly!* Among his other prose works are *The Ides of March* (1948), consisting of fictional letters written by Julius Caesar, and the novels *The Eighth Day* (1967) and *Theophilus North* (1974).

Wilkins, Maurice Hugh Frederick (1916–) *New-Zealand-born British molecular biologist, who contributed to the elucidation of the structure of DNA. He shared the 1962 Nobel Prize for Physiology or Medicine with Francis CRICK and J. D. WATSON.*

Born in Pongaroa, New Zealand, Wilkins came to Cambridge University to read physics, where he was much influenced by the dominating intellect of J. D. BERNAL. After graduating in 1938 he moved to Birmingham and with John Randall worked on the development of radar. To his later regret Wilkins spent the later part of World War II in the USA at the Univer-

sity of California, working on the atomic bomb. Disillusioned with physics after the dropping of the bombs on Japan in 1945, and stimulated by Erwin SCHRÖDINGER's book *What is Life?* (1944), Wilkins decided to deploy his physicist's techniques on a new range of problems. He therefore joined his former colleague, John Randall, in a biophysics research unit at King's College, London, becoming professor of molecular biology (1963–70) and later professor of biophysics, from 1970 until his retirement in 1981.

In 1950 Wilkins and Rosalind FRANKLIN started their investigation into the structure of DNA. As a result of personal difficulties between Wilkins and Franklin, limited progress on the problem was achieved, but Wilkins communicated their results to Francis Crick and J. D. Watson in Cambridge, who were finally able to determine the structure of DNA. Once the double helix had been proposed by Watson and Crick, Wilkins and his colleagues published evidence confirming their hypothesis. Later work by Wilkins has been concerned with the structure of nerve membranes. He has also been a leading figure in the British Society for Social Responsibility in Science and in the Food and Disarmament International organization.

Williams, John Christopher (1941–) *Australian guitarist and composer. He was admitted to the Order of Australia in 1987.*

Born in Melbourne, Williams studied with his father, himself a distinguished guitarist, and in England at the Royal College of Music. His playing came to the notice of SEGOVIA in 1955 and he was invited to attend guitar courses at the Accademia Musicale Chigiana in Siena (1957–59). Becoming resident in the UK, he made his official debut at the Wigmore Hall, London, in 1958, since when his attractive and stylish playing has made him a popular recitalist. He was professor of guitar at the Royal College of Music (1960–73) and became visiting professor at the Royal Northern College of Music in 1973. He tours widely, playing in concert with various ensembles and soloists, including Julian BREAM. Williams is particularly interested in broadening the scope of the guitar to include both European and oriental forms of music-making; in 1979 he formed the pop group Sky (disbanded in 1984). A composer himself, he has also had works written for him, notably by Dodgson, Torroba, and Previn. He has been artistic director of several major music festivals.

Williams, J(ohn) P(eter) R(hys) (1949–   )
Welsh rugby player and doctor. He played for
Wales (fifty-two times) and the British Lions
and was the outstanding rugby union full-back
of the 1970s.

The son of doctors in Bridgend, Williams attended the local grammar school before going
on to Millfield School. He was a fine tennis
player and won the junior championship at
Wimbledon in 1966. After leaving school he
went to study at St Mary's Hospital Medical
School in London, where he joined the London
Welsh rugby club. His defence tactics and attacking flair soon attracted the international
selectors and he made his debut for Wales, as
a nineteen-year-old, against Scotland in 1969.
For ten years Williams was an integral part of
the Welsh teams. In 1979 he captained both
Wales to their fourth consecutive Triple
Crown and his club, Bridgend, to the Welsh
Rugby Football Union Cup.

He will also be remembered for his contributions to the victorious 1971 and 1974 British
Lions tours: Willie John McBRIDE, captain of
the undefeated 1974 Lions in South Africa,
said of Williams that his performances have
'never been surpassed'. He has now retired to
concentrate on orthopaedic medicine, particularly sporting injuries.

Williams, Tennessee (Thomas Lanier Williams; 1911–83) *US playwright. He made a*
lasting contribution to the theatre, not least in
introducing a highly effective kind of theatrical
poetry in his fine control of language and sense
of atmosphere.

Born in Columbus, Mississippi, where his
grandfather was the Episcopal rector, Williams was taken from these relatively idyllic
surroundings to St Louis at the age of twelve,
a change that first made him aware of being
poor. His education was interrupted by the
Depression (he later graduated from the University of Iowa in 1938 or 1940) and he was
forced to take a routine job. However, he continued to write throughout his youth and in
1940 won a grant from the Rockefeller Foundation to work on a play. This, entitled *Battle*
of Angels (1940; later successfully revised as
Orpheus Descending, 1958), was a failure, and
Williams again supported himself by various
menial jobs while he continued to write.
The Glass Menagerie (1945; filmed 1950)
won the New York Drama Critics' Award and
began Williams's career as one of the most
successful modern American dramatists. The
play, like many of his best works, portrayed a
sensitive vulnerable woman whose poetic fan-

tasy world is essential to her survival but is
also tragically fragile. This theme reaches full
development in *A Streetcar Named Desire*
(1947), which was to become something of a
modern American classic. The neurotic heroine, sympathetically portrayed, is shattered
(specifically, raped) by brute (but innocent and
healthy) reality in the form of her brother-in-
law (a role that made Marlon BRANDO famous
in the 1952 film). In *The Rose Tattoo* (1951;
filmed 1956) the atypically vigorous heroine is
a Sicilian widow about to remarry. *Camino*
Real (1953) introduced expressionist techniques but was not successful. In *Cat on a Hot*
Tin Roof (1955; filmed 1958), which won a
Pulitzer Prize, Williams returned to a distinctly
Southern setting and the themes of conflict of
his earlier successes. Progressively more sensational elements appear in other plays, for
example cannibalism in *Suddenly Last Sum-*
mer (1958; filmed 1959) and castration in
Sweet Bird of Youth (1959; filmed 1963).
Among his nondramatic works are the novel
The Roman Spring of Mrs Stone (1950; filmed
1961) and the story collections *Hard Candy*
(1954) and *The Knightly Quest* (1969). A volume of *Memoirs* was published in 1975.

Williams, William Carlos (1883–1963) *US*
poet, who received the Dial award (1926), the
National Book Award (1950), and the
Bollingen Prize in Poetry (1952).

The son of an English father and a Puerto
Rican mother, Williams was born in Rutherford, New Jersey, and educated in France and
Switzerland before returning to the USA to
study medicine at the University of Pennsylvania. After a short spell at Leipzig University
working as a paediatrician, he set up as a general practitioner in Rutherford, where he remained until his retirement in 1951.

Despite his connections with Europe,
Williams's best poetry focused on his native
land. His first published work was *Poems*
(1909); subsequent collections of poetry included *The Tempers* (1913), *Kora in Hell*
(1920), and *Sour Grapes* (1921). *Spring and*
All (1923) marked a turning point in
Williams's attempt to evolve a definitive
American poetic technique. Various influences from the world of art contributed to its
emergence, including expressionism, dadaism,
and cubism. In literature, Williams was briefly
associated with the objectivists, an offshoot of
the imagist movement.

The bulk of his poetry appeared in *Collected*
Later Poems (1950) and *Collected Earlier*
Poems (1951). Probably his most important

achievement in the latter half of his career was the epic four-volume *Paterson* (1946–51). Williams also produced two collections of essays, *The Great American Novel* (1923) and *In the American Grain* (1925), as well as short stories and several novels, including the trilogy comprising *White Mule* (1937), *In the Money* (1940), and *The Build-Up* (1952).

Williamson, Henry (1895–1977) *British novelist, best known for* Tarka the Otter *and his other animal stories.*

Williamson was born and brought up in Bedfordshire. While still very young he enlisted as a private in the army, serving throughout World War I. This experience left him unfit to pursue an active career, and a short and unhappy spell as a reporter in London was a failure. After living in desperate poverty for a while in London, he retired to a cottage in Devon, where he was happy and able to write. *The Flax of Dreams*, his first novel sequence, comprised *The Beautiful Years* (1921), *Dandelion Days* (1922), *The Dream of Fair Women* (1924), and *The Pathway* (1928) and won Williamson a small but discerning circle of admirers, among them T. E. LAWRENCE, whose biography Williamson later wrote (1941). He also wrote the short stories published under the title *The Old Stag* (1926). It was, however, with *Tarka the Otter* (1927) that he became famous as a nature writer. It won the Hawthornden Prize and provided an immediate solution to Williamson's financial difficulties.

He confirmed his reputation as a superb observer of animal life with *Salar the Salmon* (1935), and for some time combined the careers of farmer and writer. In 1937 he published selections from the works of the nineteenth-century naturalist Richard Jefferies, who had been a formative influence on his life. *A Chronicle of Ancient Sunlight*, his fifteen-volume novel sequence, occupied him for nearly two decades, beginning with *The Dark Lantern* (1951) and ending with *The Gale of the World* (1969); the narrator of this fictional autobiography, Phillip Maddison, leads a life very similar to Williamson's own.

Williamson, Malcolm Benjamin Graham Christopher (1931–) *Australian composer, pianist, and organist, resident in England. He has been Master of the Queen's Music since 1975.*

Williamson studied the piano, the violin, the horn, and composition at the Sydney Conservatory under Eugene GOOSSENS and later learnt to play the organ in order to perform his own

works. In 1950 he came to Europe and studied with Elisabeth Lutyens (1906–83), after which he worked as a church organist (1955–60). Strongly influenced by Roman Catholicism and the music of MESSIAEN, he has written music in a wide variety of forms, combining serial techniques with modal and tonal writing. Williamson's earliest opera, *Our Man in Havana* (1963), is the most musically convincing of his dramatic works, which include *The Violins of Saint-Jacques* (1966) and a number of children's pieces, such as *The Happy Prince* (1965). He has written seven symphonies, the third being a choral work entitled *The Icy Mirror* (with text by Ursula Vaughan Williams), various concertos, and chamber music. His *Mass of Christ the King* (1977) was written for Queen ELIZABETH II's silver jubilee. More recent works include the ballet *Heritage* (1985) and *Songs for a Royal Baby* (1985) for chorus and orchestra. He served as president of the Royal Philharmonic Orchestra from 1977 to 1982.

Wilson, Sir Angus (Frank Johnstone) (1913–91) *British writer. He was knighted in 1980.*

Angus Wilson was born in England but spent part of his childhood in South Africa, his mother's homeland, before completing his education at Westminster School and Merton College, Oxford. In 1936 he went to work in the department of printed books in the British Museum (now the British Library). During World War II he worked for the Foreign Office (1942–46) but then returned to the Museum to supervise the replacement of books lost in the war. His first volumes of short stories, *The Wrong Set* (1949) and *Such Darling Dodos* (1950), won immediate critical recognition. His emerging reputation as an accomplished satirist was confirmed by his first novel, *Hemlock and After* (1952). In 1955 he resigned from the museum to devote himself to writing.

Apart from his technically skilled and witty fiction – which includes *Anglo-Saxon Attitudes* (1956), *The Middle Age of Mrs Eliot* (1958), *The Old Men at the Zoo* (1961), *As If By Magic* (1973), and *Setting the World on Fire* (1980) – he also wrote several television dramas and the play *The Mulberry Bush*, which was produced at the Royal Court Theatre, London, in 1956. In the 1960s he taught and lectured at universities in Britain and the USA; *The Wild Garden* (1963), based on lectures given at Los Angeles in 1960, is an interesting examination of his own fiction. He also published a study of Émile Zola (1952), *The*

World of Charles Dickens (1970), and *Portable Dickens* (1983); *The Collected Stories of Angus Wilson* was published in 1987.

Wilson, Charles Thomson Rees

(1869–1959) *British physicist, who was awarded a share in the 1927 Nobel Prize for Physics for his invention of the cloud chamber.*
The son of a Scottish sheep farmer, Wilson was educated at Owens College, Manchester, and Sidney Sussex College, Cambridge. In 1900 he was elected to a fellowship at his college and spent the rest of his professional life at Cambridge, becoming Jacksonian Professor of Natural Philosophy (1925–34). On his retirement in 1934 Wilson returned to his native Scotland where he indulged his passion for mountain climbing until he was well into his eighties.

Before going up to Cambridge, Wilson spent a few weeks in 1894 as an observer at the Ben Nevis Observatory. It aroused in him a lifelong interest in meteorology, and one aspect of this work led Wilson to seek ways to form clouds in his laboratory by allowing moist air to expand in glass vessels. As a by-product of this work, Wilson realized that his cloud chamber could, under certain conditions, be used to show the tracks of charged atomic particles as they passed through the chamber. Wilson published his first photographs of the tracks of charged particles in 1911. The chamber thereafter became a vital tool of particle physics until it was replaced by the more efficient spark and bubble chambers of the 1950s.

Wilson, Colin Henry

(1931–) *British writer.*
Colin Wilson was born and educated in Leicester. After leaving school at sixteen he had a variety of menial jobs, served in the RAF (1949–50), spent time in Paris, Strasbourg, and London, and in 1954 began writing *The Outsider* (1956). Conceived as an investigation into the spiritual malaise of contemporary society, the book catapulted Wilson to fame as an 'Angry Young Man'. From then on he devoted himself to writing, producing a steady flow of philosophical and critical works, as well as several novels and plays. *Ritual in the Dark* (1960) was his first novel. His other books, which show the range of his interests, include *Origins of the Sexual Impulse* (1963), the musical essays *The Brandy of the Damned* (1964), *Rasputin and the Fall of the Romanovs* (1964), *Introduction to the New Existentialism* (1966), *Bernard Shaw: A Reassessment* (1969), *Poetry and Mysticism* (1970), *The Occult* (1971), *Mysteries* (1978), *Quest for Wilhelm Reich*

(1981), and *Jack the Ripper* (1987). *Voyage to a Beginning: A Preliminary Autobiography* was published in 1969.

Wilson, (James) Harold, Baron

(1916–) *British politician and Labour prime minister (1964–70; 1974–76), noted for his tactical skills in maintaining positive government with a very small majority. He was knighted in 1976 and created a life peer in 1983.*

Born in Huddersfield, the son of a works chemist, Wilson was educated at Jesus College, Oxford, where he read economics. During World War II he worked as a civil servant: he was economics assistant to the war cabinet secretariat (1940–41) and then to the mines department (1941–43), serving finally as director of economics and statistics at the Ministry of Fuel and Power. In 1945 he was elected to parliament and was president of the Board of Trade from 1947 to 1951, when he resigned in protest against the proposed introduction of social service cuts. He was spokesman for economic affairs (1955–59) and then for foreign affairs (1961–63) in the Labour shadow government, succeeding GAITSKELL as party leader in 1963 (having unsuccessfully challenged his leadership in 1960).

Wilson took Labour to victory, but with a very small majority, in the election of 1964. Almost immediately he was faced with the problem of Rhodesia (now Zimbabwe), and Ian SMITH's Unilateral Declaration of Independence (1965). The government's response was to impose ever harsher economic sanctions. Wilson achieved a personal triumph in obtaining a greatly increased majority in the 1966 election, but the country's economic difficulties demanded severe and unpopular measures. Wilson's statutory incomes policy was attacked both within the party and by the unions, and proposals for reforms in industrial relations had to be shelved. Wilson was also criticized for supporting US policy in Vietnam and for imposing further restrictions on immigration. His popularity was to some extent restored in the last year of the government, and its defeat in 1970 came as a surprise. Returning to power in 1974, with another small majority, Wilson renegotiated Britain's terms of entry into the European Economic Community, dealing skilfully with opposition within the party. The new terms were confirmed by a national referendum in 1975. In the following year, Wilson unexpectedly resigned.

Wilson is the author of *The Labour Government 1964–70: A Personal Record* (1971) and *The Governance of Britain* (1976), among other books; *Harold Wilson Memoirs 1916–64* appeared in 1986.

Wilson, (Thomas) Woodrow (1856–1924) *US Democratic statesman and twenty-eighth president of the USA (1913–21). He paid a heavy personal and political price for his commitment to the peace negotiations after World War I.*

The son of a Presbyterian pastor and professor, Wilson acquired his taste for politics at Princeton University, from which he graduated in 1879. After studying law, he was admitted to the bar in 1882. His book *Congressional Government* (1885) earned him a PhD from Johns Hopkins University in 1886. After teaching at Bryn Mawr College and Wesleyan University, Connecticut, he returned to Princeton as professor (1890) and produced a series of scholarly works on history and politics. In 1902 he was appointed president of the university and proceeded to reform the curriculum and teaching methods. But proposed changes to the social fabric of varsity life encountered fierce opposition. Wilson was elected governor of New Jersey in 1910 and two years later, after many ballots, received the Democratic presidential nomination. With the Republican vote split between the incumbent TAFT and Theodore Roosevelt, Wilson won the election.

The 'New Freedom' programme of the Wilson administration reduced tariffs, levied the first federal income tax, recognized the banking system and created the Federal Reserve Board, and increased protection for trade unions. Wilson exercised restraint in foreign policy, although Mexico was a persistent thorn in America's side and in 1917 General Pershing (1860–1948) led a punitive expedition into Mexico following an attack by Pancho Villa's forces on the New Mexican town of Columbus.

The USA maintained its neutral status following the outbreak of World War I in Europe in 1914. Wilson was re-elected in 1916. In spite of German assurances to respect US shipping, four US vessels were sunk by German submarines early in 1917. In April, the USA declared war on Germany and soon US troops were sent to Europe. While the fresh input of men and materials helped turn the course of the war, Wilson enumerated his own conditions for an eventual peace treaty in the 'Fourteen Points' speech of January 1918, emphasizing the principles of democracy and self-determination. Following the armistice and cessation of hostilities, Wilson went to the Paris Peace Conference. But his idealism fell victim to the *realpolitik* of other Allied leaders and he succumbed to frustration and exhaustion. However, his support for a League of Nations did bear fruit. Returning home, he fought to win Senate ratification of the Versailles Treaty, which was signed by the other Allies and Germany in 1919. Driving himself remorselessly, Wilson suffered a stroke in October 1919 and never fully recovered. Failure to ratify the treaty prevented US participation in Wilson's cherished League of Nations. The Republicans won the 1920 presidential election and Wilson spent his last years in continuing poor health.

Windsor, Duke of (1894–1972) *Eldest son of GEORGE V. He became King Edward VIII of the United Kingdom on the death of his father (1936) but abdicated eleven months later in order to marry Mrs Wallis Simpson (1896–1986).*

Born Prince Edward of York at White Lodge, Richmond Park (and known to his family and friends as David), he was tutored privately until he became a naval cadet at the age of thirteen. On his sixteenth birthday he was created Prince of Wales and a year later was invested at Caernarvon. In 1912 he went up to Magdalen College, Oxford, where he remained until the outbreak of World War I. During the war the prince served in the army and contrived to spend some time at the front, in spite of efforts to keep him away from danger. After the war he visited many towns and cities both in the UK and the Empire, acquiring an enormous popularity with his father's subjects, not only because of his personal charm but also as a consequence of his compassionate speeches during the Depression. During the latter part of this period he met, and fell in love with, Mrs Wallis Simpson, an American whose first marriage had ended in divorce and whose second marriage was still legally intact.

On 20 January 1936, George V died and the Prince of Wales was proclaimed king. However, by the autumn of that year the American and continental papers were carrying sensational articles about the king's relationship with Mrs Simpson, especially in view of her impending second divorce. Stanley BALDWIN (the prime minister) made it clear to the king that Mrs Simpson would not be acceptable as a queen consort either to the British people or to the established church. On 11 December Edward VIII abdicated in favour of his brother, the Duke of York, who immediately became King GEORGE VI. The same evening the ex-king broadcast to his former Empire: '...you must believe me when I tell you', he said in his address, 'that I have found it impossible to carry the heavy burden of responsibility and to discharge my duties as king as I would wish to do without the help and support of the woman I love.' Edward VIII was the only British monarch to have resigned the Crown of his own free will.

Immediately after the abdication George VI declared his brother Duke of Windsor, with the personal title of Royal Highness – a title that was not to apply to either his wife or his descendants. In May 1937 Mrs Simpson's divorce became absolute and in June the couple were married privately in France. They remained abroad until Winston CHURCHILL appointed the duke governor of the Bahamas in 1940, an office he held for four and a half years. After World War II the duke and duchess returned to Paris, where he remained for the rest of his life.

Wingate, Orde Charles (1903–44) *British major-general who pioneered techniques of guerrilla warfare used by his Chindit columns, which operated behind Japanese lines in Burma during World War II.*

Wingate, the son of an Indian army colonel, was born into a strict puritan family of Plymouth Brethren. After attending the Royal Military Academy, Woolwich, he joined the Royal Artillery. He was essentially a loner and something of an intellectual, who held profound religious convictions. A keen interest in oriental studies and Arabic prompted his trip to the Sudan, where he served in the Sudan Defence Force. He was attached to military units in England (1933–36) before he joined intelligence staff in Palestine. Wingate became an ardent Zionist and champion of the Jewish settlers in their struggle against the Arabs. Befriended by Jewish leaders, Wingate organized Jewish guerrilla groups, the so-called 'Night Squads'. His scant regard for orthodoxy and wilful nature were often seen as verging on insubordination by his military superiors. In spite of this, in 1940, after serving as brigade major with an anti-aircraft unit since the outbreak of World War II, he was summoned by WAVELL to the Middle East. Wingate created a guerrilla group – the Gideon Force – which, between January and May 1941, raided Italian garrisons inside Ethiopia with considerable success. However, Wingate returned to Cairo exhausted and depressed and even attempted suicide.

After convalescence in Britain, he was again called by Wavell, now C-in-C India. Wingate originated a bold proposal for long-range penetration units to operate behind enemy lines. Eventually the 3000-strong 77th Indian Infantry Brigade was formed and, after intensive training in India, entered Burma in February 1943. Supplied by air, they successfully harried enemy forces for several weeks until exhaustion and disease caught up with them. Named after Chinthe, a mythical Buddhist temple guardian, Wingate's Chindits caught the imagination of both the public and CHURCHILL, who authorized a much larger force to enter Burma in February 1944 as part of SLIM's offensive against the Japanese. Wingate was killed in a plane crash in Assam on 24 March in the same year.

Wittgenstein, Ludwig (1889–1951) *Austrian-born philosopher and logician, author of* Tractatus Logico-Philosophicus *(1922) and* Philosophical Investigations *(1953), two of the most influential philosophical works of the century. He became a naturalized British citizen in 1938.*

One of nine children, Wittgenstein was born in Vienna, the son of an immensely wealthy industrialist. Trained originally as an engineer in Berlin, he went to Manchester in 1908 to study aeronautical engineering. In the course of this work he became aware of some intractable problems in the foundations of mathematics and, seeking guidance, obtained an intruduction to Bertrand RUSSELL at Cambridge. Russell quickly recognized Wittgenstein's genius and gave him every encouragement to pursue problems in logic puzzling to both of them. Before this work could be completed, however, World War I broke out and Wittgenstein as an Austrian returned home to fight for his country against Britain. He served on the eastern front and late in 1918 was taken prisoner by the Italians.

By this time he had actually finished the *Tractatus*, which he was keen to publish. Though a prisoner of war, he managed to send a copy of the work to Russell, who met him in 1919 and eventually arranged to have it published in 1922 in both its original German together with an English translation. It is a remarkable work. Written as a series of numbered propositions rather than continuous prose, it claims to solve all philosophical problems. It did in fact present a picture theory of meaning combined with the view that logical truths are tautologies. From these two points many of the characteristic later doctrines of the Vienna Circle (see SCHLICK, MORITZ) were to emerge.

Wittgenstein, however, consistent with his claim to have solved all philosophical problems, abandoned the subject and from 1920 to 1926 worked as a schoolteacher in remote Austrian villages. By the late 1920s he had begun to doubt the success of the *Tractatus* and seemed willing to return to Cambridge to consider once more the problems of philosophy. Help was needed. Although he had inherited a great fortune, he had given it all away on the grounds that such wealth could only inhibit genuine philosophical thought. Funds were found and Wittgenstein returned to Cambridge in 1929, when he was awarded a PhD for the earlier published *Tractatus*. Cambridge remained Wittgenstein's base for much of the rest of his life. From 1939 to 1947 he served there as professor of philosophy, although for much of World War II he was absent working as a hospital porter, first in London and later in Newcastle.

Although Wittgenstein's thought had developed considerably since the *Tractatus*, none of it appeared in print before the posthumous publication of the *Investigations*. Much of this later philosophy had, however, been transmitted in lectures and circulated in the form of cyclostyled notes dictated to selected pupils. In place of the picture theory of meaning Wittgenstein proposed that we no longer ask what a term means but how it is used. He also made the notion of a rule, and how to follow one, central to his philosophy. From such simple beginnings Wittgenstein drew conclusions that seriously challenged some of the deepest assumptions of the earlier positivism he inspired. Among his papers Wittgenstein had left the manuscripts of many works in various states of completion. By now most of it has been published, making the Wittgensteinian canon one of the richest, most complex, and largest in contemporary philosophy. It is not just the philosophy of Wittgenstein that has attracted attention. The character and life of the man have proved to be as puzzling and complex as his thought. Consequently he has become the subject of many memoirs and anecdotes while much of his correspondence has also been published.

Wodehouse, Sir P(elham) G(renville) (1881–1975) *British-born humorous writer, who adopted US citizenship in 1955. He was knighted in 1975.*

Wodehouse was born in Guildford but his father's work took him and his wife to the Far East, leaving the boy to be brought up by relatives – the prototypes for the aunts and clergymen who populate his novels. After Dulwich College, he first went into banking but, recognizing this as an unsuitable occupation, he soon turned to journalism. His early successes with school stories for boys were written for the magazine *Captain*. After 1909 Wodehouse lived mainly abroad, eventually settling in the USA, where his talent blossomed. Known to his friends as 'Plum', he wrote or contributed lyrics to a number of successful musical comedies, but it was his humorous novels that made him into a cult figure. The tales of Bertie Wooster and his archetypical manservant Jeeves, of Psmith, of the members of the Drones Club, of the dotty aristocrat Lord Ickenham, and the whole mindless world of Blandings Castle made Wodehouse's books international best-sellers.

In 1940 Wodehouse was captured in Le Touquet and interned by the Nazis. He was released after a short time but not allowed to leave Germany; during this period he made a number of ill-judged and unexplained radio broadcasts from Berlin, which subsequently led to accusations that he was pro-Nazi. Feelings ran so high that he was unable to return to Britain after the war, so he went first to Paris and then back to the USA. Here he continued to write, eventually bringing the total of his books to over a hundred. Among these were the autobiographical *Performing Flea* (1953) and *Over Seventy* (1957). A few weeks before his death he was knighted, a sign that the establishment considered that he was not guilty of collusion with the Nazis.

Wolfe, Thomas Clayton (1900–38) *US novelist whose epic fiction is highly regarded.*

Wolfe was born in Asheville, North Carolina, where his father was a stone-cutter and his mother ran a boarding house. He was educated at the University of North Carolina and Harvard, where he studied the art of playwriting. After two years in Europe, he returned to the USA to teach English at New York University, a position he held from 1924 to 1930. He relinquished it after the success of his first novel, *Look Homeward, Angel* (1929). This novel is often regarded as Wolfe's finest achievement; its authenticity was confirmed by the storm of outrage it provoked in his home town, which provided the setting for the book. Much of the credit for its success was due to Maxwell Perkins, Wolfe's editor at Scribner's, his publisher, who pruned and shaped the book from the unwieldy bulk of its original draft. Perkins performed the same function with *Of Time and the River* (1935), but after this Wolfe left Scribner's. Edward C. Aswell of Harpers carried out the same procedure with the mass of material left after Wolfe's early death from a cerebral infection following pneumonia. The result was *The Web and the Rock* (1939) and *You Can't Go Home Again* (1940).

There is little doubt that Wolfe's writing suffered from a lack of self-discipline and a surfeit of unnecessary rhetoric. However, the power, vigour, and lyricism of his best work have encouraged several critics to regard him as a genius.

Wolfit, Sir Donald (1902–68) *British actor-manager. He was knighted in 1957.*

Born in Newark-on-Trent, Wolfit made his stage debut in 1920 and first appeared in London as Phirous in *The Wandering Jew* (1924). After joining the Old Vic in 1929 he had a number of engagements, including a tour of Canada as Robert Browning in *The Barretts of Wimpole Street* and a couple of seasons at Stratford-on-Avon (1936–37). A man of immense dedication to his profession, Wolfit founded his own company in 1937 and began the work for which he is perhaps best remembered. Performing mainly Shakespeare's plays, the company toured extensively as well as appearing in the West End. During the Battle of Britain (1940) he continued to stage performances when nearly all the other London theatres were closed. He also undertook tours with ENSA to entertain troops in Europe and the Middle East.

Among Wolfit's many notable roles were Hamlet, Lear, Shylock, Malvolio, Macbeth, Volpone, Richard III, and Tamburlaine, as well as Pastor Manders in *Ghosts*, which he also played on television. He gave numerous Shakespeare recitals throughout the world, either solo or in partnership with his third wife, actress Rosalind Iden. He made many broadcasts and some twenty films (1934–68), one of his most memorable portrayals being the title role of *Svengali* (1954).

Wolfson, Sir Isaac (1897–1991) *British businessman and philanthropist. He was created a baronet in 1962 and made a fellow of the Royal Society in 1963.*

Educated at Queen's Park School, Glasgow, Wolfson joined Great Universal Stores in 1932 and two years later was appointed managing director. He became chairman in 1946 and honorary life president in 1987. The Wolfson Foundation was established in 1955 with the aims of promoting and funding medical research, particularly into cancer and children's diseases, and of encouraging development in

education and religion. In 1966 Wolfson endowed the Oxford college that now bears his name and is devoted to graduate studies. In 1972 University College, Cambridge, also received an endowment from Wolfson and changed its name to Wolfson College.

Wood, Sir Henry Joseph (1869–1944) *British conductor, organist, and composer, best remembered as the founder of the annual London Promenade Concerts. He was knighted in 1911 and created a CH in 1944.*

Having worked as an organist in various London churches, Wood studied at the Royal Academy of Music (1886–88); he intended to become a composer, but turned to conducting when he realized his ability. He then toured as a conductor of opera (1889), supervised rehearsals of Sullivan's *Ivanhoe* (1890), and oversaw a season of Italian opera at the Olympic Theatre, London. In 1894 he organized a series of Wagner concerts at the newly built Queen's Hall, and the following year conducted the first series of Promenade Concerts there. Over the next fifty years that he conducted the 'Proms', Wood systematically broadened the popular repertory to include such contemporary composers as DEBUSSY, SCRIABIN, and BARTÓK. In 1927 the BBC took over the management of the concerts and began to broadcast them. When the Queen's Hall was bombed in World War II (1941), the 'Proms' made the Albert Hall their permanent home. Although he was criticized by some for his arrangements for symphony orchestra of pieces by Bach, Handel, Debussy, and others, Wood's achievement in popularizing orchestral music in Britain has never been questioned. VAUGHAN WILLIAMS wrote *A Serenade to Music* for sixteen solo voices and orchestra in honour of his jubilee concert at the Albert Hall (1938).

Woolf, (Adeline) Virginia (1882–1941) *British novelist and critic. A mentally sick woman, plagued by depression that finally led to her suicide, she rejected in her writing the superficial materialism of her contemporaries and broke new ground by writing with great poetic intensity about the feelings that constituted reality for her and her characters.*

Virginia Woolf was born Virginia Stephen, daughter of the eminent man of letters Sir Leslie Stephen (1832–1904) by his second wife. As a nervous and delicate child, she was educated at home, mainly by her father. Her mother's death in 1895 and her father's death in 1904 caused her first major breakdowns. Virginia, her brother Thoby, and her sister

Vanessa (1879–1961) then set up house in the Bloomsbury district of London, gathering around them the Bloomsbury group of writers and artists, many of whom had been at Cambridge with Thoby. The group included the writer Leonard Woolf (1880–1969) and Clive Bell (1881–1964), whom Vanessa later married. Other members of the group were Roger FRY, J. M. KEYNES, and Lytton STRACHEY. In 1906 the death of Thoby from typhoid caused Virginia to suffer another prolonged breakdown.

In 1912 Virginia married Leonard Woolf; it was his devotion and care throughout her bouts of mental illness that provided the stability she needed in order to write during her calm periods. Her first novels *The Voyage Out* (1915) and *Night and Day* (1919) were not strikingly innovative, but *Jacob's Room* (1922) launched her as a highly original writer. In the 1920s her husband was literary editor of the *Nation* (1923–30) and they both played an active part in running the Hogarth Press, the publishing house they had founded together in 1917. In this period Virginia Woolf produced two of her best novels, *Mrs Dalloway* (1925) and *To the Lighthouse* (1927). In both these novels the plot is very slight and of considerably less importance than the relationships and thoughts of her characters. In addition to these novels, her first series of essays appeared in this period, published as *The Common Reader* (1925; second series 1932). *A Room of One's Own* (1929) acknowledges her interest in feminism while her sixteenth-century fantasy *Orlando* (1928), perhaps her most imaginative work, caused a considerable stir. This strange evocation of her lesbian lover, the writer Vita Sackville-West (1892–1962) to whom the book is dedicated, is set in the ancestral home of the Sackvilles and extends over four centuries – with Orlando, the main protagonist, undergoing a sex change during the reign of Charles II.

Despite the nervous exhaustion that writing caused her, Virginia Woolf continued a steady output of books. *Flush* (1933) was a whimsical biography of Elizabeth Barrett Browning's spaniel. She also wrote three more major novels: *The Waves* (1931), *The Years* (1937), and the posthumously published *Between the Acts* (1941), as well as essays and short stories. Shortly after the outbreak of World War II the Woolfs left London for their Sussex home at Rodmell. However, Virginia's depressions became increasingly incapacitating and in the spring of 1941 she drowned herself. Her edited letters and diaries, as well as Leonard's five-volume autobiography (1960–69), reveal the extent of her friendships among the literary and artistic circles of her time as well as the seriousness of her illness.

Woolley, Sir Charles Leonard (1880–1960) *British archaeologist responsible for excavating the ancient Sumerian city of Ur (in modern Iraq), in particular its celebrated royal treasures. He received a knighthood in 1935.*

Woolley read classics and theology at New College, Oxford, but, advised by his warden, Dr William Spooner, he decided to take up archaeology and in 1905 became assistant keeper of the Ashmolean Museum, Oxford. Between 1907 and 1911, Woolley worked on sites in the Nubia region of southern Egypt and northern Sudan. In 1912 he started excavations at the site of Carchemish on the Upper Euphrates in eastern Turkey, in collaboration with T. E. LAWRENCE. They unearthed notable relics of the Hittite civilization but work was interrupted by World War I. Woolley served with British intelligence in Egypt and was captured and imprisoned in a Turkish camp (1916–18). He returned to Carchemish in 1919 but then moved to work for the Egypt Exploration Society at Tell el-Amarna. In 1922 he was appointed director of a joint British Museum and University of Pennsylvania expedition to Ur. Woolley found successive cities on the site dating from around 4000 BC to the fourth century BC. Most spectacular was his discovery of the royal cemetery and its occupants buried with the royal treasures and servants. The site also yielded thousands of clay tablets whose inscriptions provided a wealth of detail concerning everyday life in the Sumerian city. Woolley also uncovered a deposit of clay that is claimed as evidence of the great flood referred to in the Old Testament. His book *Ur of the Chaldees* (1929) is a popular account of the excavation.

Following this success, Woolley found evidence of a hitherto unknown mixed Semitic and Hurrian culture, which flourished between 1800 and 1500 BC at Tell Atchana in southeastern Turkey. His two excavations, 1937–39 and 1946–49, were recorded in *A Forgotten Kingdom* (1953). Although some of his chronology is no longer accepted, Woolley was undoubtedly one of the leading archaeologists of the twentieth century.

Wootton, Barbara Frances, Baroness (1897–1988) *British educationalist and economist. She was created a life peer in 1958.*

Barbara Wootton studied at Cambridge University, where she later lectured in economics.

In 1920 she became a research officer for the Trades Union Congress and Labour Party Joint Research Department and began to acquire experience of the workings of the welfare state. This knowledge led to her appointment to several royal commissions and parliamentary committees, most notably those tackling penal reform. In 1927 Barbara Wootton became director of studies for tutorial classes at the University of London, where from 1948 to 1952 she held the chair of social studies. She also served as justice of the peace in the Metropolitan courts and wrote extensively on issues of social concern. Her books *Testament for Social Science* (1950), which advocated the use of scientific methods in sociology, and *Social Science and Social Pathology* (1959) were particularly influential.

Wright, Billy (William Ambrose Wright; 1924–) *British Association footballer for Wolverhampton Wanderers and England. He was the first player to make more than one hundred appearances for his country in full internationals.*

Billy Wright joined Wolverhampton as a groundstaff boy and went on to captain both his club side and England. He helped Wolverhampton win the FA Cup in 1949 and was a key member of the team that won the Football League championship in 1954, 1958, and 1959. Wright played first as a wing half and later as a centre half. When he retired from playing, he managed the FA youth team for a time as well as Arsenal. Later he became a television executive.

Wright, Frank Lloyd (1869–1959) *US architect of outstanding originality, whose important buildings are in America. He designed nothing in Europe.*

Born in Richland Center, Wisconsin, Wright studied architecture in Chicago under Louis H. Sullivan (1856–1924), one of the first architects to build skyscrapers and a master whom Wright continued to admire throughout his long career. Wright's first independent buildings were his long low 'prairie' houses, constructed in the Chicago suburbs. At about this time he also designed and built his first office block, the highly original Larkin office building in Buffalo, New York (1903), and his first church, the Unity Temple of Oak Park, Illinois (1904), in which he used reinforced concrete for the first time. During and after World War I the size of his buildings expanded with his reputation; the Imperial Hotel in Tokyo (1916–20), no longer standing, was an example of his work in this period. Condemning the

advancing congestion of cities, Wright began his advocacy of a dispersed way of life, aimed at bringing people closer to nature. These naive ideas encouraged him to found a community of young architects on his own estate at Taliesin East, Wisconsin, with its winter annex, Taliesin West, in the Arizona desert. Here Wright held court with his disciples. Recognition as an architect of international status, however, was slow to emerge. In Racine, Wisconsin, he built the fine Johnson Wax office block (1936) and laboratory tower (1949); in New York City, he produced the Guggenheim Museum (1956), an entirely novel design based on a spiral ramp (not then as familiar a feature of urban carparks as it is now); and in San Raphael, California, he designed the highly original Marin Civic Center.

Frank Lloyd Wright also wrote some twenty books, including *An American Architecture* (1955) and *An Autobiography* (1932; revised 1943 and 1962).

Wright, Orville (1871–1948) and **Wilbur** (1867–1912) *US aviation pioneers, the first to achieve controlled powered flight.*

The sons of a bishop of the United Brethren Church, the Wright brothers received no more than a high-school education. Their engineering skills were consequently largely self-taught. They set up first as cycle manufacturers, founding their own Wright Cycle Company in 1892 in their home town of Dayton, Ohio. Their interest in flight was to some extent aroused by the death of Otto Lilienthal (1848–96) in a gliding accident. After a careful study of the literature, in 1900 they felt sufficiently confident to build their first full-sized glider. With this plane they soon discovered that much of the published data was quite unreliable. Consequently, in 1901, they built their own wind tunnel in which to test wings and airframes of different shapes and designs. Unable to use the excessively heavy car engines then available, they also designed and built their own water-cooled 12-hp engine. By the summer of 1903 they had built a flyable biplane, which they moved in pieces to Kill Devil Hills; on 14 December 1903 Orville succeeded in flying the plane a distance of 120 feet in a time of 12 seconds. On the final flight of the day, with Wilbur at the controls, the plane travelled 852 feet in 59 seconds. The success of Flyer I, as the Wrights called their plane, was due in part to their carefully designed and tested wings, propellers, and engine but also to their solution to the problem of control. For Flyer I was unlike any of its predecessors in having rudder, elevator, and ailerons. This layout has, in general, been followed by all subsequent fixed-wing aircraft. By 1905 the Wrights were able to sustain flight of almost an hour in their latest aircraft, Flyer III. It was also capable of executing a figure-of-eight.

There still remained the problem of interesting the outside world in their machine. The US War Department rejected the offer of the plane and all their patents in 1905. It was left to the Wrights, as skilful as publicists as they were as engineers, to demonstrate the value of their invention. Wilbur went to Europe and revealed to excited crowds in Paris and elsewhere the range and adaptability of powered flight. Orville remained in the USA and continued to demonstrate to the military the increasing predictability of the flight of their aeroplanes. In 1909 the Signal Corps bought their first plane.

At the height of their success Wilbur died suddenly and unexpectedly of typhoid. Thereafter Orville became something of a recluse in his large Dayton house, continuing his various researches and occasionally emerging into the outside world to defend his patents.

Wright, Sewall (1889–1988) *US geneticist and mathematician noted particularly for his mathematical analyses of population genetics, especially his introduction of the concept of genetic drift.*

Wright received his MS degree from the University of Illinois in 1912 and his ScD from Harvard University three years later. He then worked as senior animal husbandman with the US Department of Agriculture (1915–25) before joining the University of Chicago as associate professor of zoology (1926–29), later becoming a full professor (1930–54). His early work involved the experimental inbreeding and crossbreeding of guinea pigs, from which he derived the concept of inbreeding coefficient – an index expressing the relatedness of an individual's parents. His mathematical treatment extended to describing and predicting changes in the frequency of genes in populations, the selection pressures and other factors governing such changes, and the process of evolution itself. Wright also proposed how in relatively small isolated populations, genes with only slight or no adaptive value can undergo random changes in their frequency,

even leading to their extinction or predominance, with potentially profound evolutionary consequences. He called this phenomenon genetic drift (also called the Sewall Wright effect).

Wright was appointed professor of genetics at the University of Wisconsin in 1955 and made emeritus professor in 1960. His major work is *Evolution and the Genetics of Populations* (vols 1–4, 1968–78).

Wynne-Edwards, Vero Copner (1906–)
British ethologist who made the contentious proposal that natural selection may act not only on individuals within social groups but also on the fitness of each group as a whole in relation to its environment. He was made a CBE in 1973.

Wynne-Edwards read zoology at New College, Oxford, obtaining his degree in 1927. After working at Oxford and Bristol universities he became assistant professor of zoology at McGill University, Montreal (1939–44), and later associate professor (1944–46). Here he made a study of the dispersion of seabirds at sea and, following his move to Aberdeen University as professor of natural history (1946–74), he extended this work to other species. In 1962 he published his notable book, *Animal Dispersion in Relation to Social Behaviour*. In it he made the radical claim that certain social behaviour within animal groups makes members of the group aware of their population density and so regulates their breeding in accordance with the food and other resources available to them. As examples of such group-regulating behaviour he cited the establishment of breeding territories and communal roosting, which in some birds is accompanied by elaborate communal displays (which he called epideictic displays). Wynne-Edwards argued that natural selection acts on these behavioural mechanisms and the ability of the group to respond for the common good. Critics argue strongly that natural selection operates on individuals only.

Wynne-Edwards has served on various bodies concerned with nature conservation. He was chairman of the Natural Environment Research Council (1968–71) and was elected a fellow of the Royal Society in 1970.

X

Xenakis, Yannis (1922–   ) *Greek-born naturalized French composer, who originated the 'stochastic' style of composition, based on mathematical probabilities.*
Born in Romania, as a child Xenakis absorbed local folk music and the Byzantine music of the Greek Orthodox Church. In 1932 his family returned to Greece, and Xenakis later enrolled at the Athens Polytechnic to study engineering. For five years during World War II he was active in the Greek resistance, being badly wounded and losing the sight of an eye. He escaped to France and after the war took French citizenship. From 1947 to 1959 Xenakis worked for the architect LE CORBUS-IER, collaborating in many of his great designs, including the Philips pavilion at the 1958 Brussels Exposition. During this period he became seriously interested in musical composition and was encouraged by Olivier MESSIAEN and by Hermann Scherchen (1891–1966), who conducted several first performances of his works.

In composing his music, Xenakis evolved what is known as the 'stochastic' system, in which a random sequence of notes is produced according to certain probabilities, which at any one point depend on the occurrence of preceding notes. Using a computer to assist him, Xenakis has also applied this system to determine the duration, speed, and intensity of his works. Examples of his compositions are *Pithoprakta* (1955–56), based on the Maxwell–Boltzmann law of the movement of gas molecules, in which the pizzicati and glissandi of separate instruments become totally subordinated to the overall sound mass; *Duel* (1959), for fifty-four instruments and two conductors; and *Terretektorh* (1966), in which eighty-eight musicians (each with additional percussion instruments) are seated among the audience, thus breaking down the barrier between performer and audience. More recent works, in which he has made use of folk music, include *Cendrées* (1974), for chorus and orchestra, *Shar* (1983) and *Thallein* (1985), for orchestra, *Chant des soleils* (1983), a choral work, and the string quartet *Tetras* (1983).

Y

Yamamoto Gombe (1852–1933) *Japanese naval minister (1898–1906) and prime minister (1913–14; 1923–24). Principally remembered for his role in expanding the Japanese navy, Yamamoto also advocated the participation of the military in Japanese government.*

Born in Satsuma (now Kagoshima), Yamamoto joined the forces loyal to the crown during the Boshin civil war at the age of sixteen. In 1870 he was assigned to the naval training section and then to the naval training barracks, where he graduated in 1874. Three years later he joined a German warship to further his naval training.

Yamamoto was appointed chief secretary of the ministry of the navy in 1891. When the Sino-Japanese War (1894–95) broke out, he was sent to the Imperial headquarters as an adviser on behalf of the naval ministry. His appointment as a rear admiral (1895) and his skilled leadership gave him great influence within the Imperial headquarters over the navy's role in the war. In 1898 he was promoted to vice-admiral and chosen to become naval minister in the cabinet of Yamagata Aritomo, a position he retained until 1906, by which time he was a full admiral. During this period he initiated steps towards the Anglo-Japanese alliance of 1902, oversaw the Japanese victory in the Russo-Japanese War (1904–05), and initiated Japanese involvement in Chinese Manchuria. In 1913 he became prime minister, with the support of the Seiyukan Party, but resigned the following year over a scandal relating to naval construction. He became prime minister again in 1923, declaring martial law throughout the Tokyo region after the Great Kanto Earthquake. The attempted assassination of the crown prince, for which his cabinet assumed collective responsibility, led to his resignation the same year.

Yamamoto Isoroku (1884–1943) *Japanese naval commander. Considered a national hero after he was killed in action, he was a brilliant strategist who masterminded the attack on Pearl Harbor.*

Born in Nagaoka (now Niigata Prefecture), of samurai descent, Yamamoto attended the Naval Academy and the Naval Staff College.

He served in the navy during the Russo-Japanese War (1904–05) and in 1905 was commissioned as an ensign. From 1919 to 1921 he was stationed in the USA, becoming an instructor at the Naval Staff College on his return to Japan. After serving as an instructor and executive officer at the Kasumigaura Naval Air Station (1924), he returned to the USA, where he served as a naval attaché to the Japanese embassy (1925–28). In 1929 he was elevated to the rank of rear admiral and attended the London Disarmament Conference as part of a Japanese delegation.

Yamamoto was appointed commander of the air squadron in 1931 and became a vice-admiral five years later. In 1938 he was given the command of the First Fleet, becoming an admiral (1940), and ultimately commander of the Combined Fleet (1941). Although he initially opposed war with Britain and the USA, he drew up plans to destroy the US fleet at Pearl Harbor and establish Japanese domination of the Pacific. As commander of the Combined Fleet he directed the successful enactment of the plans, but was killed when the Allies shot down his plane near the Shortland Islands in 1943. He was given a state funeral and was posthumously awarded the title of admiral of the fleet.

Yang, Chen Ning (1922–) *Chinese-born US physicist who, with Tsun-Dao LEE, was awarded the 1957 Nobel Prize for Physics for their discovery that parity is not conserved in the weak interaction.*

The son of a mathematician, Yang was educated at the National Southwest Associated University, Yunnan, where he first met his future collaborator Lee. At the end of World War II, Yang moved to the USA and gained his PhD from the University of Chicago in 1948. After teaching for a year at Chicago, Yang joined the staff of the Institute of Advanced Studies at Princeton, remaining there until his appointment in 1966 as the Einstein Professor of Physics at the State University of New York.

In 1956 Yang, in collaboration with Lee, published a paper proposing that the conservation of parity does not apply in the weak interaction. Experiments they suggested were performed shortly afterwards at Columbia and confirmed their proposal. Despite the publication of thirty-five joint papers, the friendship between Yang and Lee did not long survive the Nobel award. The last joint paper appeared in 1962 and thereafter they parted company for good. Yang has claimed that the original idea of the violation of parity was his alone and that 'Lee at first resisted the idea.' This version of the events has been disputed by Lee.

Yeats, W(illiam) B(utler) (1865–1939) *Irish poet, critic, and playwright. He was awarded the Nobel Prize for Literature in 1923.*

Yeats was born near Dublin into a Protestant family, the son of a painter and brother of the painter Jack Yeats (1871–1957). He was brought up partly in London and partly in his mother's home county of Sligo, where he absorbed the beauty of the countryside and discovered the Irish legends that played a large part in his early writing. In 1881 Yeats was sent to Dublin High School but when his family moved back again to London he moved with them (1887). The reception of his first book of poems, *The Wanderings of Oisin* (1889), encouraged him to leave art school for literature. His first poetic play, *The Countess Cathleen* (1892), the stories in *The Celtic Twilight* (1893), as well as the poetry in *Poems* (1895; revised edition 1899) and *The Wind Among the Reeds* (1899) demonstrate both his love for traditional Irish themes and his growing accomplishment as a writer. In this period he also became interested in the Irish republican movement, largely influenced by the Fenian, John O'Leary, and Maud Gonne, who was also the inspiration of his love poetry. His own account of his early years is given in *Autobiographies* (1926), comprising *Reveries over Childhood and Youth* (1915) and *The Trembling of the Veil* (1922).

By 1897 Yeats, encouraged by Lady Gregory, was anxious to establish an Irish national theatre. The performance of *The Countess Cathleen* in Dublin in 1899 marked the inauguration of this enterprise, which developed into the Abbey Theatre. Yeats's prose play *Cathleen ni Houlihan* (1902), with Maud Gonne in the title role, was one of the Abbey's earliest successes. Yeats remained a director of the theatre until his death and wrote many plays for it, among them *Deirdre* (1907), *The Green Helmet* (1910), *The Cat and the Moon* (1924), which shows the influence of Japanese Nōh drama, and *The Words Upon the Window Pane* (1934), about Swift. Besides his plays

and poems, he also wrote critical and literary essays collected in, among other volumes, *The Cutting of an Agate* (1912), *Per Amica Silentia Lunae* (1918), and *Plays and Controversies* (1923). *A Vision* (1925) showed his growing absorption with the occult.

The nostalgia of Yeats's early poetry was replaced by more robust themes as the theatre and his friends involved him in the realities of Irish political life. *Responsibilities* (1914) and *The Wild Swans at Coole* (1917) show this change, consolidated in *Michael Robartes and the Dancer* (1921). From 1922 to 1928 Yeats was a senator in the Irish Free State, but philosophy and mysticism became more important than Irish politics and he devoted his last years to writing. *The Tower* (1928) and *The Winding Stair* (1933) contain some of his greatest, as well as his most difficult, verse. In 1933 his *Collected Poems* were published; his *Collected Plays 1892–1934* followed in 1934.

Yeltsin, Boris Nikolayevich (1931–) *Soviet statesman and president of the Russian SSR (1991–), who became a leader of reformists in the Soviet Union in the late 1980s.*

Born in Sverdlovsk in the Urals, he worked in the construction industry and joined the Soviet Communist Party in 1961, becoming a full-time party worker in 1968. He became first secretary of the Sverdlovsk District Central Committee in 1976 and subsequently a deputy to the Supreme Soviet. In 1985 he acquired a national reputation when he was appointed first secretary of the Moscow City Party Committee as one of the new generation of liberal reformers introduced after Gorbachov became general secretary. Flamboyant and radical, Yeltsin soon clashed with other senior figures and became a political embarrassment to Gorbachov, accusing him of hindering the progress of *perestroika* and preventing the constituent republics of the Soviet Union from pursuing their political destinies. He built up a big following in the Russian SSR and in 1990 won elections for the Russian presidency, a humiliating blow to Gorbachov, who had attempted to block his candidacy. Subsequently, Gorbachov attempted to establish a working relationship with Yeltsin, his most serious rival for power, but Yeltsin continued to demand freedom from interference by the Kremlin in Russia's affairs and threatened to declare unilateral independence. When reactionary elements staged a coup in 1991, Yeltsin rallied opposition to the new leaders and obliged them to stand down. Gorbachov was reinstated as president of the Soviet Union, but was forced to compromise with Yeltsin, who emerged with new stature. Yeltsin subsequently seized all Communist Party assets in Russia for the state and demanded a share in the control of the Soviet Union's nuclear weapons. In return he agreed to preserve the Soviet Union, but retained Russia's right to "review its borders".

Yevtushenko, Yevgeni Aleksandrovich
(1933–   ) *Soviet poet.*

Yevtushenko was born of Ukrainian stock in Zima, a small town on the Trans-Siberian Railway near Lake Baikal, and his childhood was divided between there and Moscow. After being expelled from school at fifteen he divided his attention between football and writing poetry. He first achieved publication in 1949. During his years as a student at Moscow's Gorki Literary institute (1951–57) Yevtushenko wrote prolifically: *Third Snow* (1955), *Zima Junction* (1956), *The Highway of Enthusiasts* (1956), and *The Promise* (1957) were among the works that encapsulated the feelings and aspirations of the younger generation then struggling out of the shadow of Stalinism. His poetry readings attracted audiences of many thousands, but some of his poetry incurred official hostility, notably *Babiy Yar* (1961) with its controversial statement about Soviet antisemitism. The poem was set to music by SHOSTAKOVICH. In the early 1960s Yevtushenko began writing on more international themes and travelled abroad to give poetry recitals that won him an enthusiastic audience in the West. His *Precocious Autobiography* appeared in 1963 in Paris and the freedom with which Yevtushenko expressed his thoughts in it about Soviet society caused him to be recalled to the Soviet Union in disgrace. He regained favour with *Bratsk Station* (1965). More recently he published a book of photographs, *Invisible Threads* (1981), and entered politics as a member of the new Congress of People's Deputies (1989).

Yonai, Mitsumasa (1880–1948) *Japanese naval minister (1937–40; 1944) and prime minister (1940). A liberal leader, he opposed Japanese rearmament and involvement in World War II.*

Born in Iwate Prefecture, the son of a samurai, Yonai was educated at a school in Monioka before attending the Naval Academy and the Naval Staff College. Commissioned as an ensign in 1903, he reached the rank of rear admiral by 1925, serving in Russia (1915–17) and Berlin (1920–22).

Yonai was promoted to vice-admiral in 1930. He was subsequently appointed commander of Saseto Naval Station (1933–34) and commander-in-chief of the second squadron (1934) and of the Imperial Japanese fleet (1936). In 1937 he became minister of the navy, a position he retained until 1940, when he succeeded Hiranuma as prime minister. Throughout this period he favoured closer relations with Great Britain and the USA and opposed the military alliance with Germany and Italy. He was forced out of office after seven months by army officers who favoured a stronger militaristic form of government and demanded rearmament and the implementation of a more aggressive foreign policy, which resulted in war. Yonai became minister of the navy again in 1944 and was alone in realizing that Japanese defeats in the Pacific had sealed the country's fate; from that time he had advocated surrender. He served in successive cabinets until his death in 1948.

Yoshida Shigeru (1878–1967) *Japanese Liberal prime minister (1946–47; 1948–54). The longest continuously serving prime minister in Japanese history, he was a virtual dictator who oversaw the ending of the occupation and Japan's re-emergence as an independent nation.*

Born in Tokyo, the adopted son of a merchant, Yoshida was educated at the Tokyo Imperial University, from which he graduated in law in 1906. He then entered the foreign service and held a number of minor diplomatic posts in the Far East, Europe, and the USA. He became vice-minister of foreign affairs (1928–30) and, when the army vetoed his appointment as foreign minister, served as ambassador to Britain (1936–39).

Retired by the military in 1939, he was imprisoned towards the end of World War II (1945) for advocating an early peace settlement with the Allies. After the war he was reappointed foreign minister; when the president of the Liberal Party, Hatoyama Ichiro, was purged from political activity by the occupation authorities, Yoshida reluctantly assumed the party leadership and became prime minister. In this office he assumed extraordinary powers in order to implement a new constitution (1947) and reorder the economy: he instituted considerable land reform, established a Police Reserve Force (later the Self-Defence Forces), and negotiated (1951) the Peace Treaty and the Security Treaty with the USA. He lost office in 1947 but returned to power in 1948, remaining as prime minister until 1954, when he resigned. In retirement, he continued to exercise considerable political influence and adopted the role of elder statesman within conservative circles.

Younghusband, Sir Francis Edward (1863–1942) *British explorer and diplomat who, in addition to contributing to the geography of the Himalayan region, negotiated the 1904 treaty between Britain and Tibet, for which he was knighted.*

Younghusband was born in Muree, India, the son of a British army major. He attended the Royal Military College, Sandhurst, and was commissioned in the Dragoon Guards based at Meerut, India. He undertook a military reconnaissance of the Afghan border and Kashmir and in 1886 embarked on the first of several expeditions to the Himalayan region, journeying overland from Peking to Rawalpindi via hitherto unmapped mountain passes. In 1889 he explored the passes of the Karakoram range and, in the early 1890s, those of the Pamirs. His travels are described in *The Heart of a Continent* (1896).

Younghusband acted as correspondent for the London *Times* with the Chitral relief force in 1895 and during his stay in southern Africa (1895–97). In 1903 he was requested by Lord Curzon (1859–1925), viceroy of India, to head a diplomatic mission to negotiate a treaty with Tibet in an attempt to counter growing Russian influence in the region. At first denied entry, an armed brigade was sent to facilitate, often bloodily, Younghusband's progress to Lhasa, where, in September 1904, a treaty was signed with the Dalai Lama. After serving as British Resident in Kashmir (1906–09), Young-

husband returned to Britain. He was president of the Royal Geographical Society (1919–22) and helped organize early expeditions to Mount Everest. His interest in eastern religions led to several works, including *Life in the Stars* (1927) and *The Living Universe* (1933). In 1930 he founded the World Congress of Faiths.

Yukawa Hideki (1907–81) *Japanese physicist, who was awarded the 1949 Nobel Prize for Physics for his prediction in 1935 of the existence of the pi-meson (pion).*

The son of a geologist, Yukawa was educated at Kyoto University and Osaka University, where in 1938 he was awarded his doctorate. He taught at both Osaka and Kyoto from 1932 to 1938, when he was appointed to the chair of physics at Kyoto.

In the 1930s Yukawa sought to elucidate the nature of the strong force that holds together the protons and neutrons in the atomic nucleus. Just as the electromagnetic interaction can be visualized as an exchange of photons, Yukawa suggested that the strong interaction within the nucleus could be visualized as an exchange of a particle as yet unidentified. Yukawa calculated that such a particle would have a mass some two hundred times greater than that of the electron. A particle fitting this description was detected in 1937; it turned out, however, to be what is now called the muon. Yukawa's prediction, made in 1935, had to wait a further ten years before C. F. POWELL discovered the pion in debris caused by cosmic rays.

Z

Zapata, Emiliano (1879–1919) *Mexican agrarian reformer, whose slogan was 'Land, liberty, and death to the landowners'.*
Born in Anencuilco, of Indian blood, Zapata had little formal education. He worked on the land until 1911, when he joined Francisco Madero's successful uprising against Porfirio Díaz (1830–1915). In 1912, when Madero refused to redistribute land, he initiated his own programme of agrarian reform, known as the Plan of Ayala, and attempted to implement it by means of guerrilla warfare.

Zapata continued the struggle following the overthrow of Madero by Victoriano Huerta in 1913, eventually forcing Huerta to flee the country. Over the next five years, together with Pancho Villa (1878–1923), he fought against the forces of General Gonzalez in an attempt to win control of the country. He was ambushed and assassinated in 1919 by a follower of Gonzalez, who was posing as one of Zapata's supporters. Following his death many of the reforms he demanded were implemented by the government of Alvaro Obregón.

Zátopek, Emil (1922–) *Czech long-distance runner.*
Zátopek's first Olympic gold medal came in London in 1948 in the 10 000 metres event. Four years later in Helsinki he completed an extraordinary Olympic treble, with gold medals in the 5000 metres, the 10 000 metres, and the marathon. During his career he set world records for nine different distances: in the 10 000 metres event alone he set a new world record five times. Despite a hernia operation, he took part in the Olympic marathon of 1956 and finished sixth. His wife, Zatopkova, was also an outstanding athlete.

Zeeman, Pieter (1865–1943) *Dutch physicist, who discovered the Zeeman effect. For this he was awarded the 1902 Nobel Prize for Physics.*
The son of a Lutheran minister, Zeeman was educated at the University of Leiden, where he obtained his PhD in 1893. He taught initially at Leiden before moving to the University of Amsterdam in 1897. Zeeman remained at Amsterdam until his retirement in 1935, having served as professor of physics since 1900.

Like many physicists at the turn of the century, Zeeman was interested in the interaction between light and magnetic fields. His doctoral thesis had been concerned with this problem. Zeeman continued with this work and in 1896 explored the effect on the light emitted by a sodium flame when it is placed in a strong magnetic field. He found what at first seemed to be a broadening of the spectral lines. On a more careful examination, however, using more powerful magnets, he found certain spectral lines split into groups of two or three lines. Now known as the Zeeman effect, it turned out to be a quantum effect, only fully explicable in terms of quantum theory.

Zeffirelli, G. Franco (1923–) *Italian film and theatre director and designer.*
After studying architecture in his native Florence, Zeffirelli became an actor and set designer, often working with Visconti. During the 1950s he emerged as one of Italy's leading opera and stage designers. As Visconti's assistant he began his film career working on such films as *Bellissima* (1951) and *Senso* (1954). He also worked with other directors, including De Sica and Rossellini, before making his first film as director in the late 1960s. Often admired for their visual richness, Zeffirelli's films include *The Taming of the Shrew* (1967), *Romeo and Juliet* (1968), which earned him an Academy Award nomination, *Brother Sun, Sister Moon* (1973), *Jesus of Nazareth* (1978), made for television, *La Traviata* (1983), *Otello* (1986), and *The Young Toscanini* (1988). He has also worked at La Scala, Milan, Covent Garden, Glyndebourne, and the New York Metropolitan Opera, as well as in other leading opera houses throughout the world. He also directed and designed the plays *Othello* at Stratford-on-Avon (1961) and *Hamlet* at the Old Vic (1964) and the ballet *Swan Lake* at La Scala (1985). His autobiography, *Zeffirelli*, was published in 1986.

Zhu De (Chu Teh; 1886–1976) *Founder and commander-in-chief of the Chinese People's Liberation Army (1930–54). He is principally remembered for his outstanding military achievements in support of the communist cause.*

Born in Szechwan province, the son of a peasant, Zhu was educated at a local high school before attending the sports school in Chengdu (1906–07), where he became a teacher. In 1909 he enrolled at the Military Academy of Yunnan Province, forming the Revolutionary Expedition Corps in 1911. Promoted to major in 1912, he became a member of the Kuomintang (Chinese National Party) and was given the command of forces on the Indo-Chinese border (1913–14). Over the next eight years he continued to advance through the military ranks.

Zhu De joined the Chinese Communist Party in 1922 while in Germany, where he was studying at the University of Göttingen. Expelled from Germany in 1926 for his political activities, he returned to China in 1927 and helped to organize the Nanchang communist uprising against the Kuomintang. After its failure he led the survivors to Fujian province, joining MAO TSE-TUNG in 1928 and forming the People's Liberation Army (PLA). In 1930 he became commander-in-chief of the communist military forces, leading them on the epic 'long march' to Shensi (1934–35). During the war against the Japanese (1937–45), when the communists and the Kuomintang formed an alliance, he continued as the senior communist commander; after World War II he led the PLA against the nationalists in the civil war (1946–49). Although he was never an ambitious political leader, he served on the Politburo for many years. When he relinquished command of the PLA in 1954 he became vice-chairman of the central people's government council and, in 1959, chairman of the standing committee of the people's congress, the nominal legislative assembly.

Zhukov, Georgi Konstantinovich (1895–1974) *Soviet marshal who commanded his country's army against the German invasion during World War II. He received the Order of Lenin six times.*

Born into a peasant family, Zhukov was conscripted into the tsar's army during World War I. Following the 1917 revolution, he joined the Red Army as a junior cavalry commander, becoming a Communist Party member in 1919. He studied military science at the Military Academy at Frunze (which reverted to its orginal name Bishkek in 1991) and by the late 1930s was commanding five armoured brigades along the Mongolian–Manchurian border in a successful campaign to expel the Japanese, for which he received his first Order of Lenin.

In January 1941, Zhukov became chief of general staff. The Germans invaded in June and STALIN appointed Zhukov deputy commissioner for defence. Zhukov personally supervised the defence of Leningrad (now St Petersburg) and Moscow and in October was made commander-in-chief of the entire western front. After the Soviet counter-offensive of January 1942, the Germans broke through in the south to threaten Stalingrad (now Volgograd) by August. With dreadful losses on both sides, the Battle of Stalingrad raged until the November offensive by Soviet forces encircled the German 6th Army. Its commander, Paulus, surrendered in January. Throughout 1943, Zhukov held the Germans at massive cost, repulsing the Kursk offensive, until finally his forces gained the upper hand.

In September 1944, during the Soviet advance through eastern Europe, Zhukov's army halted outside Warsaw while the Polish uprising in the city was savagely crushed by the Germans. The delay in entering the capital cost many Polish lives and is one of the more controversial incidents in Zhukov's career. The final Soviet push to Berlin culminated in a week-long battle for the city and the formal German surrender to Zhukov on 8 May.

Zhukov was now a national hero. Stalin, recognizing a potential rival, effectively banished Zhukov to regional commands. Only after Stalin's death in 1953 did he return to favour, becoming deputy defence minister and, under KHRUSHCHEV in 1955, defence minister as well as a member of the all-powerful Presidium of the Communist Party. But his defence policies led to conflict with Khrushchev and his dismissal in 1957. Zhukov was restored to favour only after Khrushchev's death in 1964.

Zia ul-Haq, Mohammad (1924–88) *President of Pakistan (1978–88). Condemned worldwide for the execution of his predecessor, Zulfikar BHUTTO, he instituted a repressive regime in Pakistan.*

Born in Jullurdan to a middle-class family, Zia was educated at Stephen's College, Delhi, before attending the Royal Indian Military Academy at Dehra Dun. During World War II he served in Burma, Malaya, and Indonesia, becoming a commissioned officer in 1945. Joining the Pakistani army in 1947 after partition, he was appointed to various instructional, staff, and command posts over the next nineteen years. In this period he attended the command and staff colleges at Quetta (1955) and Fort Leavenworth, USA (1959 and 1963).

Between 1966 and 1976 Zia rose from commander of the Cavalry Regiment to general and chief of the army staff. He served as an adviser to the Royal Jordanian Army (1969–71) and as a deputy division commander during the war with India (1971), which saw the loss of East Pakistan (now Bangladesh). In 1977 he led a bloodless coup, deposing prime minister Bhutto, who had appointed him to his senior positions. He retained those posts after the military takeover, adding to them chief martial law administrator, and in 1978 was sworn in as president. In 1979 he ignored worldwide appeals for clemency when he refused to commute Bhutto's death sentence on charges of conspiracy to murder. He also banned all political parties and embarked upon a programme for the total Islamization of Pakistan. Zia was re-elected as president in a referendum in December 1984. In 1985 he announced constitutional changes increasing the powers of the president and brought martial law to an end, though in practice he still maintained strict political control. He died in an air crash, possibly as the result of sabotage.

Ziaur Rahman (1936–81) *Bengali nationalist and president of Bangladesh (1977–81). A master of political intrigue, he rose to power on a wave of coups and countercoups, yet managed to win popular support despite his reputation as an authoritarian leader. He was assassinated during an unsuccessful military coup.*

Born in Sylhet (then in East Bengal), Ziaur enlisted in the Pakistan army in 1953, at the age of seventeen, shortly after finishing school. He was commissioned two years later and served as a company commander during the Indian-Pakistani War (1965). He then became a military instructor and was involved in the Bengali nationalist movement. In 1971, having risen to the rank of lieutenant-colonel, he led the mass uprising against Pakistan that resulted in the independence of Bangladesh.

Ziaur was appointed to a senior rank in the newly established Bangladesh army under Sheik MUJIBUR RAHMAN. Following Mujibur's assassination in 1975, he became chief of the army staff under President Ahmed's successor, Sayem. In 1976 he was appointed chief martial law administrator and president after a national referendum (1977). A year later he founded the Bangladesh National Party and organized the country's first general election, in which his rule was endorsed. He remained in this post until 1981, when he was assassi-

nated during a military coup led by General Ershad.

Ziegfeld, Florenz (1869–1932) *US theatre manager. Through the 'Ziegfeld Follies', which began in 1907 and were produced annually for over twenty-three years, Ziegfeld gave the USA its equivalent of the French Folies-Bergère.*

Born in Chicago, Ziegfeld began his career at the World's Columbian Exposition in Chicago (1893), where he acted as manager for the strongman Eugene Sandow. He subsequently became a theatrical manager and in 1907 produced the first of his famous reviews in New York City, *The Follies of 1907*. Elaborate sets, scantily clad beautiful girls, and comedy acts were the ingredients of these highly popular revues staged under such slogans as 'Glorifying the American Girl'. As well as the annual follies Ziegfeld also produced musical shows, such as *Show Boat* (1927), *Rio Rita* (1927), and *Bitter Sweet* (1929). His famous Ziegfeld Theatre on the corner of 6th Avenue and 54th Street, New York, was opened in 1927. (After his death it had a chequered career before being demolished in the mid-sixties.) Among the stars promoted by Ziegfeld were Fanny Brice (1891–1951), W. C. Fields (1879–1946), Eddie Cantor (1892–1964), Irene Dunne (1898–1990), Paulette Goddard (1911–90), Maurice CHEVALIER, and Fred ASTAIRE.

Zinoviev, Grigori Yevseevich (1883–1936) *Russian revolutionary and chairman of Comintern (1919–26).*

Born in Yelisavetgrad (now Kirovograd) into a middle-class family, Zinoviev joined the Russian Social Democratic Workers Party in 1901. He aligned himself with the Bolsheviks in 1903 and took part in the unsuccessful 1905 revolution, becoming closely associated with LENIN in the years that followed.

Zinoviev played a significant part in the October Revolution (1917), and as one of the principal Bolshevik leaders he was put in charge of the Petrograd (now St Petersburg) party organization and chairman of the Petrograd Soviet. He became chairman of the executive committee of the newly formed Communist International (Comintern) in 1919 and in 1921 was made a full member of the Politburo. In the years after Lenin's death, he formed an alliance with Lev Kamenev (1883–1936) and STALIN to oppose TROTSKY. However, when Trotsky had been expelled from the party, Stalin turned against Zinoviev and, defaulting on the alliance, forced him to resign from the Politburo and Comintern

(1926). He was expelled from the party in 1927 and although he was readmitted shortly after, was expelled again in 1932 and 1934. In 1935 he was arrested and convicted of treason and complicity in the assassination of Sergei Kirov (1888–1934); he was executed in 1936. As head of Comintern he achieved international notoriety for a letter (known as the 'Zinoviev letter'), allegedly written by him and published in the British press, which instructed British communists to engage in revolutionary activity. This letter subsequently contributed to the downfall of the Labour government in 1924.

Zuckerman, Solly, Baron (1904–84) *British biologist and civil servant, born in South Africa, who did much to influence the postwar development of science and technology in Britain. He received a knighthood in 1956, an OM in 1968, and a life peerage in 1971.*

Zuckerman received a degree in anatomy from the University of Cape Town in 1923 and remained there as demonstrator until moving to Britain in 1926. He studied medicine at University College Hospital, London, and qualified as a doctor in 1928. He served as research anatomist to the Zoological Society of London (1928–32) and was research associate at Yale University (1932–34) before moving to Oxford University as demonstrator and lecturer in human anatomy (1934–45). Zuckerman's studies of primates in captivity were among the first to relate their sexual and social behaviour to their reproductive physiology, especially hormone levels, and resulted in *The Social Life of Monkeys and Apes* (1932). During World War II he acted as scientific adviser, initially investigating the effects of bomb explosions and conducting a survey of air raid casualties. He was later concerned with planning air attacks, both in the Mediterranean and in northwestern Europe.

After the war he became professor of anatomy at Birmingham University (1945–68) and, in 1955, honorary secretary of the Zoological Society of London, which he sought to revitalize. In 1977 he was appointed its president. Zuckerman served on numerous government committees, notably as chief scientific advisor to the Ministry of Defence (1960–66) and to the government (1964–71). He was elected a fellow of the Royal Society in 1943. His books include *A New System of Anatomy* (1961), *Scientists and War* (1966), and *Nuclear Reality and Illusion* (1982).

Zweig, Stefan (1881–1942) *Austrian biographer, essayist, and playwright.*

Born in Vienna into a well-to-do Jewish family, Zweig studied at the universities of Berlin and Vienna, travelled extensively, and during his lifetime cultivated a wide friendship with fellow artists and intellectuals (including GORKI, RILKE, Romain ROLLAND, Rodin, TOSCANINI, FREUD, and Richard STRAUSS). He served in World War I and emerged from it a pacifist. Between the wars he lived mainly in Salzburg; in 1934 he went into exile, living briefly in England and New York before moving to Brazil. A humanist, Zweig lived long enough to see the world he knew utterly destroyed. In despair he and his wife committed suicide near Rio de Janeiro in 1942.

Zweig's writing covers a wide range of genres. His early work includes translations of Verlaine, Baudelaire, and Émile Verhaeren and collections of his own romantic poetry (1901, 1906), which was indebted to von Hofmannsthal. The influence of Freud's work can be seen in his short-story collections: *Erstes Erlebnis* (1911), *Amok* (1922), and *Verwirrung* (1927; translated as *Conflicts*, 1927). He wrote several plays, the earliest, an antiwar play entitled *Jeremias* (1917), while still in uniform. He translated Jonson's *Volpone* (1925) and *Epicoene* (*Die schweigsame Frau*, 1935, which provided the libretto of the opera by Richard STRAUSS). His main work of fiction and also his last creative work was *Schachnovelle* (1942; translated as *The Royal Game*), in which the chess game is a metaphor for the disintegration of an intellectual being interrogated by the Gestapo.

But Zweig's reputation finally rests on his biographical essays and full-length biographies, which are among the best literary examples of Freud's influence, especially in penetrating the workings of the creative process. The shorter essays, first published in groups, were collected as *Die Baumeister der Welt* (1934; translated as *Master Builders*, 1939) and include the lives of Balzac, Dickens, Dostoievsky, Hölderlin, Kleist, Nietzsche, Casanova, Stendhal, and Tolstoy. His longer biographies are *Romain Rolland* (1921), *Marie Antoinette* (1932), *Maria Stuart* (1935), and, perhaps his greatest, *Triumph und Tragik des Erasmus von Rotterdam* (1935; translated as *Right to Heresy*, 1951). A book of 'historical miniatures', *Sternstunden der Menschheit* (1927; translated as *The Tide of Fortune*, 1955), is concerned with moments that changed history. His autobiography, *Die Welt von Gestern* (1942; translated as *The World of Yesterday*, 1943), focuses on Europe before World War I.